T0177755

GRAVITATIONAL WAVES

Gravitational Waves

Volume 2
Astrophysics and Cosmology

Michele Maggiore

Département de Physique Théorique
Université de Genève

OXFORD

UNIVERSITY PRESS

OXFORD
UNIVERSITY PRESS

Great Clarendon Street, Oxford, OX2 6DP,
United Kingdom

Oxford University Press is a department of the University of Oxford.
It furthers the University's objective of excellence in research, scholarship,
and education by publishing worldwide. Oxford is a registered trade mark of
Oxford University Press in the UK and in certain other countries

First Edition published in 2018

Reprinted 2018, 2019

Impression: 8

Published in the United States of America by Oxford University Press
198 Madison Avenue, New York, NY 10016, United States of America

British Library Cataloguing in Publication Data
Data available

Library of Congress Control Number: 2017943894

Set ISBN   978–0–19–875528–9
Volume 1   978–0–19–857074–5
Volume 2   978–0–19–857089–9

only available as part of a set

DOI 10.1093/oso/9780198570899.001

Printed and bound by CPI Group (UK) Ltd, Croydon, CR0 4YY

Contents

Preface to Volume 2

In the preface to Vol. 1 we wrote: *"The physics of gravitational waves is in a very special period. [...] As a result of these experimental efforts, there are good reasons to hope that the next decade will witness the direct detection of gravitational waves and the opening of the field of gravitational-wave astronomy and, possibly, cosmology."* The writing of Vol. 2 took 10 more years, and during that time these hopes have indeed been fulfilled. The first direct detection of gravitational waves (GWs) took place in September 2015, when the two detectors of the LIGO Observatory, at the very beginning of the first science run of advanced LIGO, observed the signal from a coalescing black-hole binary. The result was announced by the LIGO and Virgo collaborations in February 2016. For this discovery Reiner Weiss, Barry Barish and Kip Thorne were awarded the 2017 Nobel Prize for Physics "for decisive contributions to the LIGO detector and the observation of gravitational waves". Several more binary black-hole coalescences have now been observed, including a triple coincidence between the two LIGO interferometers and the Virgo interferometer. Furthermore, in August 2017, the GWs from the coalescence of a neutron star binary were observed by LIGO and Virgo, in coincidence with a γ-ray burst detected by Fermi-GBM and by INTEGRAL. The electromagnetic counterpart was then identified and observed by several telescopes in all bands of the electromagnetic spectrum. Thus, the GW window has been opened, and we are starting to look through it. With the planned improvements in the sensitivities of ground-based interferometers, we expect that GWs from astrophysical sources will soon be observed routinely while, in the near and mid-term future, we expect that several other detectors, such as pulsar timing arrays, the planned LISA space interferometer, and possibly third-generation ground-based detectors such as the Einstein Telescope, will explore the Universe up to cosmological distances. Thus, we are indeed in a very exciting period for GW astrophysics and cosmology.

The first volume of this book dealt with the theory of GWs, in Part I, and with experiments, in Part II. This second volume is devoted to what we can learn from GWs, in astrophysics (Part III) and in cosmology (Part IV). Our main aim has been to systematize a large body of theoretical and observational developments that have taken place in GW physics over the last few decades. Even though we expect the field to evolve rapidly, thanks to new experimental discoveries as well as to theoretical advances, we believe that most of these methods will still form the backbone of our understanding for many years to come, and will be the basis for future developments. As with Vol. 1, we have typically tried to rederive afresh all theoretical results, trying to give a coherent and consistent picture of the field, and clarifying or streamlining the existing derivations whenever possible. An Errrata web page will be maintained at http://theory.physics.unige.ch/~maggiore/home.html

Various people have read and commented on the drafts of some chapters. I am particularly grateful to Alessandra Buonanno, Emanuele Berti, Thibault Damour, Ruth Durrer, Valeria Ferrari, Stefano Foffa, Kostas Kokkotas, Michael Kramer, Martin Kunz, Georges Meynet, Andrea Passamonti, Eric Poisson, Luciano Rezzolla, Alberto Sesana and Riccardo Sturani, for their careful reading of selected chapters, and to Camille Bonvin, Daniela Doneva, Alex Kehagias and Vittorio Tansella for useful comments.

I also wish to thank Sonke Adlung, Harriet Konishi and the staff of Oxford University Press for their competent and friendly help, and Mac Clarke for the careful and competent copyediting.

Geneva, December 2017

Part III

Astrophysical sources of gravitational waves

Stellar collapse

10

We begin our study of astrophysical sources of gravitational waves (GWs) with a chapter devoted to stellar collapse. From the point of view of GW physics, this is interesting for at least two different reasons. First, stellar collapse through supernova explosion is itself a potentially interesting source of GW production. Second, stellar collapse can leave as remnant a compact object, such as a white dwarf (WD), a neutron star (NS) or a stellar-mass black hole (BH). As we will see in detail in later chapters, these compact objects are among the most interesting sources from the point of view of GW astrophysics.

Supernovae (SNe) are among the most spectacular events in the Universe. As we will discuss in detail in this chapter, there are two very different mechanisms leading to SN explosions: (1) The gravitational collapse of the core of a star, once the nuclear fuel that feeds the thermonuclear reactions inside the core is exhausted. As we will see in Section 10.2.1, depending on the properties of the progenitor, this leads to events classified as type Ib, Ic or type II SNe, and leaves behind a compact remnant, usually a neutron star (which is sometimes observable as a pulsar) or possibly a black hole (BH).[1] (2) The thermonuclear explosion of a white dwarf that accretes mass from a companion, going beyond its Chandrasekhar limit. This gives rise to type Ia SNe. In this case the star that explodes is dispersed in space and its remnant is not a compact object. A typical core-collapse SN can release an energy $\sim 10^{53}$ erg. Of this energy, 99% is emitted in neutrinos, about 1% goes into kinetic energy of the ejected material, and less than 0.01%, i.e. about 10^{49} erg, is released in photons. The corresponding peak luminosity in photons can be of order a few times $10^9 \mathcal{L}_\odot$ or higher. Thus, a typical core-collapse SN at its peak has an optical luminosity that rivals the cumulative light emitted by all the stars in its host galaxy. As we will see in detail in this chapter, type Ia SNe have similar electromagnetic luminosities. Such events, when they take place in our Galaxy (barring obscuration from dust in the Galactic plane beyond a few kpc), can lead to truly impressive optical displays that have been observed by mankind since ancient times. A number of milestones in astronomy are associated with nearby SN explosions. Tycho's SN in 1572 and Kepler's in 1604 marked a new epoch in astronomy, while recent discoveries such as the pulsar in the remnant of the 1054 Crab SN and the detection of neutrinos from the 1987A SN in the nearby Large Magellanic Cloud rank among the milestones of modern astronomy. We therefore find it appropriate to begin this chapter with an introduction to historical SNe.

[1] There are also other mechanisms that can trigger the explosion, as in electron-capture SNe, as well as a more refined classification of SN type, as we will discuss in more detail later in the chapter. It is also in principle possible that the collapse produces no visible SN event and all the matter is swallowed by the final BH.

Gravitational Waves, Volume 2: Astrophysics and Cosmology. Michele Maggiore.
© Michele Maggiore 2018. Published in 2018 by Oxford University Press.
DOI 10.1093/oso/9780198570899.001.0001

Table 10.1 Summary of the historical SNe, and the source of their records, adapted from Green and Stephenson (2003). Remnants updated from Ferrand and Safi-Harb (2012). (The identifier of the remnants, as for example in G120.1+01.4, refers to its Galactic coordinates). We have classified SN 1604 as type Ia, following for example Blair *et al.* (2007), Reynolds *et al.* (2007) and references therein. The nature of the Crab SN (core collapse or electron capture) is debated; see the Further Reading.

Date	Length of visibility	Remnant	Classification	Historical records				
				Chinese	Japanese	Korean	Arabic	European
AD1604 (Kepler's)	12 months	G004.5+06.8	type Ia	few	–	many	–	many
AD1572 (Tycho's)	18 months	G120.1+01.4	type Ia	few	–	two	–	many
AD1181	6 months	G130.7+03.1 ?		–	few	few	–	–
AD1054 (Crab)	21 months	G184.6−05.8	debated	many	few	–	one	–
AD1006	3 years	G327.6+14.6	type Ia	many	many	–	few	two
AD393	8 months	G347.3−00.5	–	one	–	–	–	–
AD386?	3 months	G011.2−00.3	–	one	–	–	–	–
AD369?	5 months	–	–	one	–	–	–	–
AD185	8 or 20 months	G315.4−02.3	type Ia ?	one	–	–	–	–

10.1   Historical Supernovae

The reader eager to plunge into more technical issues can simply skip this section

As we will see in Section 10.2.1, in a galaxy like ours the estimated rate of SN explosions is of order of two per century. However, in the Galactic plane the extinction of visible light significantly limits our view; for instance, the extinction of visible light from the Galactic center is about 30 magnitudes. Thus, a Galactic SN visible to the naked eye is a very rare event. However, when a nearby star becomes a SN, the effects are quite spectacular. In the historical period, this has happened a handful of times.

Evidence for the oldest SNe observed by mankind is based on historical records. Since a Galactic SN remains visible for months, and sometimes up to a few years, one can focus on historical records of stars that suddenly appeared and remained visible for at least four months, in order to eliminate most novae[2] and the possibility of confusion with comets (comets being eliminated also because of their evident proper motion).

[2]Novae arise when a white dwarf (WD) in a binary system accretes mass from a companion at a rate of about $10^{-9} - 10^{-8} M_\odot$/yr. The material accreted is rich in hydrogen and accumulates on the surface of the WD, where it is compressed by further accretion and heated. When a layer of about $10^{-5} - 10^{-4} M_\odot$ of hydrogen has been accreted, a runaway thermonuclear reaction takes place, releasing an energy of order 10^{45} erg, which is therefore several orders of magnitude smaller than in SN explosions. The WD then gradually cools down and goes back to its quiescent state, where the accretion process can start again, leading to recurrent nova explosions. A clear distinction between classical novae and supernovae was first made by Baade and Zwicky in 1934.

Table 10.1 gives a summary of SNe that have been seen by the naked eye in historical times, and for which we have written records, in the last 2000 years.

SN 185

The oldest recorded SN is SN 185, which was first observed on December 7, 185 at the imperial observatory of Lo-Yang, in central China. The Chinese astronomers recorded its appearance and its gradual fading, providing the oldest plausible historical account of a SN. It finally disappeared after either 8 or 20 months (depending on the interpretation of a sentence in the record to mean "next year" or "year after next"). The identification of the recorded position of the event with the region between α and β Cen suggests that the remnant of SN 185 could be identified with the SN remnant RCW 86 (G315.4−02.3), at an estimated distance of about 2.8 kpc. Recent observations with Chandra and XMM-Newton have indeed strengthened the case for this identification. The study of its X-ray synchrotron emission has in fact suggested an estimate for the age of this remnant consistent with the explosion date, within the uncertainties of the determination. The regular shape of the remnant shell, together with the absence of a pulsar, would point toward a type Ia SN, which is also suggested by the Chandra X-ray observations.

Another possible remnant candidate has been proposed, G320.4−01.2 (RCW 89), which contains the pulsar PSR B1509-58, and therefore would corresponds to a core-collapse SN. The timing measurement of this pulsar are also consistent with the age of SN 185.

SN 1006

A number of other possible SNe were recorded before AD 1000; see Table 10.1. However, the first event for which we have extensive historical records is a SN that exploded in AD 1006 in the direction of the Lupis constellation, and was recorded in China, Japan, Europe, Egypt and Iraq. It disappeared a first time from view after 17 months, and remained occasionally visible at dawn for 3 years.

The identification of the likely remnant of this SN came in 1965, by searching radio catalogues covering part of the region of sky suggested by the historical records. The remnant is now identified with the radio source PKS 1450-51 (whose Galactic coordinates are G327.4+14.6). Its distance from us, as we now infer from the remnant, was about 2.2 kpc.

From the spectrum of the remnant it is believed that it was a type Ia SN. Using the standard luminosity of type Ia SNe together with the known distance to the remnant (and plausible assumptions about absorption) allows us to estimate that it reached an apparent visual magnitude $V \simeq -7.5$.[3] By comparison, the full Moon has a maximum apparent magnitude $V \simeq -12.94$, and the limiting brightness for an object to be visible with the naked eye when the Sun is high is $V \simeq -4$, dropping to $V \simeq -2.5$ when the Sun is less than 10 degrees above the horizon. Comparing with the visual magnitudes given in Table 10.5 on page 70, we see that, at its maximum brightness, SN 1006 was by far the brightest object in the night sky after the Moon, and easily visible during daytime. The historical records confirm that it was indeed visible during the day, and it was even bright enough to read at night by its light! The most detailed reports by far are from Chinese astronomers, who determined the position of this "guest star" to good accuracy. Two accounts of this SN appeared also in Europe, in chronicles of the monasteries of St. Gallen, in Switzerland, and of Benevento, in Italy.

[3]The standard astronomical notions of luminosity, magnitudes and color index of stars are recalled in the Complement Section 10.6.

SN 1054 (the Crab supernova)

Shortly after, in 1054, the Crab SN exploded and was recorded by Arab, Chinese and Japanese astronomers. It gave rise to the Crab nebula, whose remnant neutron star we observe today as the Crab pulsar, PSR B0531+21. Its distance from us is 2.0 ± 0.5 kpc.

The identification of the remnant of the 1054 SN with the Crab nebula, based on the expansion speed of the cloud measured spectroscopically, as well as on various concordant historical records, is quite firm. The first suggestion that the Crab nebula was the remnant of SN 1054 was due to Hubble. In 1928, when the nature of nebulae was still unknown, he observed that the Crab nebula was expanding radially and deduced that it was most probably the remnant of an explosion. From the expansion rate he could infer that the explosion took place nine centuries earlier, consistent with SN 1054. This analysis was refined and strengthened by Mayall and Oort in 1939 and 1942. From the analysis of the historical records, it is believed that it reached an apparent magnitude between -7 and -4.5 and again it was visible during daylight. It remained visible to the naked eye for about a year and a half. It is puzzling that, for such an impressive phenomenon, there is no unambiguous European record, particularly considering that, half a century earlier, SN 1006 was recorded in Europe, so European chroniclers of that epoch did not lack interest in such phenomena.[4]

Today, after almost 1000 years, the Crab nebula still has a luminosity of $8 \times 10^4 \mathcal{L}_\odot$, mostly in the form of polarized synchrotron radiation, indicating the presence of highly relativistic electrons spiraling around magnetic field lines. The source of this energy, so long after the explosion, remained a puzzle until the discovery of the pulsar. As we will discuss in Section 11.1.1, ultimately, this huge luminosity comes from the pulsar's spindown. The Crab nebula was first detected in radio waves in 1963 and in X-rays in 1964. The discovery of the pulsar inside the Crab nebula, in 1968, was a milestone in astronomy and made the Crab the most studied and famous SN remnant, together with the recent SN 1987A.

SN 1181

The next visible Galactic SN appeared in AD 1181, and was recorded in China and Japan. It remained visible for six months. The SN remnant 3C58, with Galactic coordinates G130.7+03.1, at a distance of about 3.2 kpc, has been proposed as the remnant of this supernova. Within this nebula has been discovered the radio and X-ray pulsar J0205+6449. Recent work, however, has questioned this identification of the remnant. If the identification were correct, its young age of less than 900 years would suggest a rapid expansion of the nebula, while the observed expansion seems considerably lower, and rather suggests that the SN remnant 3C58 could be at least 2700 years old.

[4]The subject is quite controversial among historians of science. Some European documents have been suggested as referring to SN 1054, but look quite imprecise, and overall the evidence that they actually refer to the SN is not very convincing. It has been suggested that the lack of written documents was due to censorship from the Church (given that the chronicles of the epoch were compiled in monasteries), possibly connected with the fact that the SN appeared just at the time of the East-West Schism that gave rise to the separation of the Church of the Roman Empire into what later became the Roman Catholic Church and the Eastern Orthodox Church. In fact, the date of the excommunication of the Patriarch of Constantinople Michael I Cerularius is July 16, 1054. The usually accepted date for the appearance of the SN is July 4, so by July 16 it was just near its maximum luminosity. There have also been suggestions that this SN is depicted, together with the Moon, in cave paintings from native americans in Arizona. It has been proposed that the paintings represents a conjunction between the Moon and the SN, made possible by the fact that soon after it appeared, on July 5, the SN as seen from Earth was indeed on the plane of the ecliptic and in a direction close to that of the Moon. However, the dating of the paintings is quite imprecise (between the 10th and 12th centuries), and only one of them shows the crescent moon with the correct orientation in relation to the SN, so it is difficult to make definite statements.

SN 1572 (Tycho's supernova)

The next SN visible to the naked eye appeared after about four centuries, in AD 1572, in the direction of the constellation of Cassiopea, and was observed in Asia and Europe. The SN was probably first detected in Europe on November 6 (or possibly a few days earlier) by the abbot of Messina and by other observers in Europe. Very detailed observations of it were made by the Danish astronomer Tycho Brahe, who realized that it was a new star and begun observing it on November 11.

He carefully followed it over the following year, establishing that its position was fixed, and determining it to within a few arcminutes. He was also the only astronomer to monitor carefully the decline of its brightness in the following months. This SN is thus often referred to as Tycho's SN.[5] It remained visible for 18 months. The fact that its position remained fixed, without a daily parallax with respect to the fixed stars, allowed Brahe to infer that the object was far beyond the Moon; furthermore, the absence of apparent proper motion over a period of months showed him that it was much further away than the planets, and was indeed a star. This was a highly non-trivial conclusion for the epoch, which contradicted the accepted Aristotelian doctrine that changes in the sky could only take place in the sublunar region, and ultimately led to the notion of the immutability of the heavens being abandoned. It also opened the way to modern astronomy, revealing the importance of performing accurate astrometric measurements.

Chinese observers recorded that it was visible in daylight, while Korean documents compared its brightness with that of Venus. This suggests that at its maximum the SN was brighter than the threshold $V \simeq -4$ for visibility during daylight, but not by much. The remnant is identified with the radio and X-ray source 3C10 (G120.1+01.4).[6] Today the remnant is a shell of angular diameter 8 arcmin (by comparison, the full Moon has an angular diameter between approximately 29 and 34 arcmin, depending on its exact distance from Earth along its orbit).

Historical records of the light curve suggest a type Ia SN. No compact remnant has been detected at its center, which also points toward a type Ia SN. Finally, the classification as type Ia has been beautifully confirmed thanks to the detection of a light echo from the SN, due to scattering and absorption/re-emission of the SN flash by interstellar dust near the remnant. In this way it has been possible to observe some of the light emitted by the SN near maximum brightness, which has arrived more than four centuries after the direct flash, and it has been possible to perform a spectroscopic analysis. The spectrum has confirmed that SN 1572 was a typical type Ia SN. Its distance from us has usually been quoted in the range $2.3 - 2.8$ kpc, although the absolute brightness inferred from the light echo, together with the historical information that it had $V \simeq -4$, rather gives a distance $3.8^{+1.5}_{-1.1}$ kpc.

A search for the companion star from which the white dwarf accreted mass, going beyond the Chandrasekhar limit, has revealed a G2 star in the direction of the center of the nebula, and at a distance between

[5]As Tycho reports, "On the 11th day of November in the evening after sunset, I was contemplating the stars in a clear sky. I noticed that a new and unusual star, surpassing the other stars in brilliancy, was shining almost directly above my head; and since I had, from boyhood, known all the stars of the heavens perfectly, it was quite evident to me that there had never been any star in that place of the sky, even the smallest, to say nothing of a star so conspicuous and bright as this. I was so astonished of this sight that I was not ashamed to doubt the trustworthiness of my own eyes. But when I observed that others, on having the place pointed out to them, could see that there was really a star there, I had no further doubts. A miracle indeed, one that has never been previously seen before our time, in any age since the beginning of the world." (From Burnham, 1978).

[6]Beautiful images of this and other SN remnants in X-rays have been taken by Chandra, see http://chandra.harvard.edu/photo/2011/tycho/

2.5 and 4.0 kpc, so compatible with the distance to the nebula. This star has a peculiar phase-space velocity, more than three times as large as that of any other star in this region, which could be interpreted as arising from the kick received by the companion in the explosion. Other interpretations, in terms of a star unrelated to the SN and belonging to the thick disk population, are, however, possible.

SN 1604 (Kepler's supernova)

The next SN visible to the naked eye appeared in 1604, and was first observed on October 9 in Italy, by astronomers in Verona and in Cosenza. It remained visible until October 1605, and was extensively followed by European, Chinese and Korean astronomers. Johannes Kepler was among those who observed it. Due to poor weather conditions in Prague, he could start observing it only on October 17. However, his treatise *De Stella Nova in Pede Serpentarii* ("On the new star in the foot of Ophiuchus", Serpentarius being a less familiar name for the constellation Ophiuchus) was the most important European description, in particular for the accurate positional measurements, so this SN is also called Kepler's SN. The most accurate positional measurements are in general from European astronomers, with an accuracy of about 1 arcmin (compared with a precision of about 1 degree from Chinese and Korean astronomers). Its luminosity curve can be inferred both from European observations and from Korean documents recording its luminosity on a day-by-day basis for the first six months, and matching the European observations. The SN was detected about 20 days before maximum luminosity, thanks also to the fortuitous circumstance that at that time Mars and Jupiter happened to be in conjunction. This conjunction was being carefully observed, and the SN appeared at just 3 degrees NE from the position of these two planets (and 6 degrees East of Saturn). We therefore have a detailed record of the luminosity curve from appearance, thorough maximum luminosity and fading; see Fig. 10.1. According to a reconstruction of the luminosity curve due to Baade, it reached a maximum visual magnitude $V \simeq -2.25$, so it was not as spectacular as SN 1006, with its inferred $V \simeq -7.5$, the Crab (maximum V between -7 and -4.5) or Tycho's SN (maximum $V \simeq -4$). Still, in the night sky it was brighter than any star, and among the planets only Venus was significantly brighter than the SN. When the SN appeared it was as bright as Mars, and then surpassed Jupiter in a few days. The fact that this SN was relatively high with respect to the Galactic plane reduced significantly the obscuration by dust, allowing the SN to be visible by the naked eye.

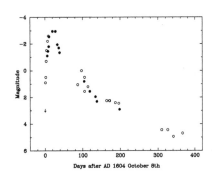

Fig. 10.1 The light curve of the SN of AD 1604 from European (◦) and Korean (•) observations, with a European upper limit on October 8. From Stephenson and Green (2002).

The remnant (whose Galactic coordinates are G004.5+06.8) was identified by Baade in 1943, and has been studied at infrared, optical, radio and X-ray wavelengths. Its distance from us is quite uncertain, with estimates ranging from about 3 kpc to more than 6 kpc. It is located in the direction of the Galactic center, so, taking 4 kpc as an estimate of its distance, it is roughly half-way from the Galactic center. It is relatively high above the Galactic plane, at a vertical height $z \simeq 590\,(d/5\,\mathrm{kpc})$ pc, where d is its distance from us.

The classification of this SN has been controversial for decades. Baade suggested that the historical light curve was consistent with a type Ia, but his claim was later questioned. The presence of dense circumstellar material implies significant mass loss from the precursor star, which is more suggestive of a core-collapse SN (although there are a few examples of circumstellar material around spectroscopically confirmed type Ia SNe). Since the 1990s, however, evidence accumulated in favor of a type Ia SN. Analyses of X-ray spectra with Exosat, ASCA, XMM-Newton and, more recently, Chandra indicate an overabundance of Fe and Si in the ejecta and relatively little oxygen emission. Such a composition is inconsistent with a core-collapse SN, and is instead typical of a type Ia SN. Furthermore, no neutron star has been detected near its center, either in radio or in X-rays. In X-rays, the limit on the flux from a hypothetical central compact object shows that, if it were present, its intrinsic X-ray luminosity would be more than 100 times smaller than that of the compact remnant of a typical core-collapse SN such as Cas A. Infrared studies with Spitzer point again toward the conclusion that this was a type Ia SN, although with significant circumstellar material. Recent hydrodynamical simulations show that the morphology of the remnant and its expansion characteristics can be reproduced by a model of a type Ia SN originating from a white dwarf that accreted mass by wind accretion from an asymptotic giant branch star with mass $(4-5)M_\odot$.

The population of Galactic SN remnants

Another interesting Galactic SN remnant is Cas A, which is observed as a bright source in radio and X-rays. At visible wavelengths this remnant displays a ring of expanding knots. From their expansion velocity, and assuming no deceleration, one can estimate that this SN exploded around AD 1671. Its distance from us is estimated at $3.4^{+0.3}_{-0.1}$ kpc. It is quite puzzling that for such a recent and relatively nearby SN there is no historical record.

Galactic SNe of course occur at a much higher rate than SNe visible to the naked eye, given that, for most lines of sight close to the Galactic plane, obscuration by dust limits our view in visible light to a few kpc. As of 2017, the catalog of Galactic SN remnants lists 381 objects.[7] The remnant age is often uncertain or undetermined, but $O(20-30)$ are estimated to be less than 2000 yr old. As we will see in Section 10.2.1, typical estimates for the rate of SN explosions in our galaxy are of order 2 SNe per century. According to this estimate, there should be about 40 remnants younger than 2000 yr (raising to about 60 if we take 3 SNe per century as an estimate of the rate). This is not inconsistent with the number of known young remnants, although it suggests that more young remnants have still to be discovered. Comparing with the

[7]See http://www.physics.umanitoba.ca/snr/SNRcat, which is regularly updated.

7 SNe in Table 10.1 for which convincing historical records exist, we can estimate that about $10-20\%$ of the galactic SNe exploding in the last 2000 years have been recorded in writing.

Among the known Galactic SN remnants, the record for the youngest is currently held by G001.9+00.3. This remnant is located in the direction of the Galactic center, at a distance from the center of about 300 pc in projection. From its expansion rate, assuming no deceleration, the explosion date is estimated as the year 1855 ± 11. Since some deceleration must have occurred, this is actually an upper bound on its age, and the explosion date was most likely around 1900.[8] The remnant was first observed by VLA (Very Large Array) radio observations, and in X-rays by Chandra.

10.2   Properties of Supernovae

Modern research on SNe began in 1885 when Hartwig discovered the first extragalactic SN, in the Andromeda galaxy. It was first interpreted as a nova. However, when in 1919 Lundmark estimated the distance to Andromeda, it became clear that Hartwig's "nova" had actually been $O(10^3)$ times brighter than a classical nova.[9] However, it was only with the pioneering paper by Baade and Zwicky in 1933 that the notion of SNe was put forward and a first clear distinction was made between novae and supernovae. In 1933, just one year after Chadwick's discovery of the neutron, Baade and Zwicky put forward the idea of a neutron star[10] and further speculated that SNe are an end-state of stellar evolution, with the source of energy being provided by the gravitational energy released by the collapse of the progenitor star to a neutron star.[11]

In the following years systematic searches, performed particularly by Zwicky at Palomar Mountain, led to the discovery of 54 SNe up to 1956, and 82 more from 1956 to 1963. In recent years, largely stimulated by the importance of type Ia SNe for determining the expansion rate of the Universe, intensive searches have dramatically increased the number of detected SNe. Browsing the Asiago Supernova Catalog[12] one finds that in the period 1980–1989 were observed $O(200)$ SNe, in the period 1990–1999 were found over 900 SNe, and in the period 2000–2009 were observed $O(3500)$ SNe, obviously all of them extragalactic. As of 2017 the total number of SNe in the catalog is about 9000, and is increasing at a rate of about two SNe per day. This large sample, combined with the improved quality of the observations, has allowed remarkable progress in our understanding of the properties of SNe. Future large synoptic surveys are expected to detect possibly $O(10^5)$ SNe per year, at redshifts $z \sim 0.5-1$.

Much progress has also been made in recent years in identifying the progenitor stars of local SN explosions, by comparisons with archival data from HST or ground-based telescopes. Currently $O(20)$ progenitors have been identified. To draw statistical conclusions from this sample, most of the information comes from progenitors at a distance below

[8]Needless to say, by this we mean that the signal from the SN reached Earth around 1900. The actual explosion took place about 25,000 yrs ago, given the time taken by light to reach us from the Galactic center. The same applies to all the other SN explosion dates mentioned above.

[9]Or, the other way around, Shapley in 1919 used this SN as an argument against the island universe hypothesis (i.e. the existence of other galaxies, separated from ours by huge distances), arguing that SN 1885A in Andromeda would have $M = -16$, which was "out of the question". The typical absolute magnitude of most novae is around -8.8.

[10]Actually, even before the discovery of the neutron, in 1931, neutron stars were somewhat anticipated by Lev Landau, who wrote about stars where "atomic nuclei come in close contact, forming one gigantic nucleus".

[11]"With all reserve we advance the view that supernovae represent the transition from ordinary stars into neutron stars, which in their final stages consist of closely packed neutrons", W. Baade and F. Zwicky, Meeting of the American Physical Society, December 1933, published in *Phys. Rev.* **45** (1934) 138.

[12]   See   http://graspa.oapd.inaf.it/asnc.html for an on-line version.

28 Mpc, although objects outside this distance, such as the massive progenitor of SN2005gl at 60 Mpc, have also been identified.

Nowadays we understand that there are two basically different mechanisms behind SN explosions: either the core collapse of a star, powered by the energy released in the gravitational collapse (just as proposed by Baade and Zwicky) or the thermonuclear explosion of a white dwarf in a binary system that accretes mass from a companion, going beyond its Chandrasekhar limit.[13] However, as we now discuss, a classification of SNe that is more direct from an observational point of view is based rather on the spectrum of the light emitted at maximum luminosity.

[13] As we will later discuss, another possibility that has been proposed in the literature is a "double-degenerate" scenario, in which the thermonuclear explosion results from the merger of two white dwarfs.

10.2.1 SN classification

The first distinction, proposed by R. Minkowski in 1941, was between type I and type II SNe. Type I are defined by the fact that their spectrum near maximum luminosity does not show evident H lines, while type II near their peak luminosity have prominent hydrogen lines. Subsequently, in the late 1980s it was realized that type I SNe did not constitute a homogeneous class, and that it was necessary to distinguish between two fundamentally different classes, which were denoted as type Ia and type Ib. Type Ia SNe are defined as type I SNe that, in their spectrum taken near maximum luminosity, show a deep absorption trough, typically around 6000–6150 Å, produced by blue-shifting of two Si II lines with rest-frame wavelengths $\lambda = 6347$ and 6371 Å (often collectively denoted as $\lambda 6355$). The significance of these Si lines, as well as lines of other intermediate-mass elements such as Ca, Mg, S and O that are present in the spectrum of type Ia SNe at maximum light, is that the part of the ejecta that is exposed shortly after the explosion has undergone thermonuclear processing, leading to the production of intermediate-mass elements. This thermonuclear processing has, however, been incomplete, since Si, O, etc. have not been burnt further to eventually produce elements of the iron group.

The mean velocities of the ejecta, obtained from the blueshift of the spectral lines, are of order 5000 km/s, with peak velocities as large as 20,000 km/s $\simeq c/15$. The typical spectrum of a type Ia SN is shown in the upper curve of the left panel in Fig. 10.2. It shows strong Si II absorption at about 6150 Å, while H lines are absent. At late times, say after more than 4 months, the typical spectrum of a type Ia SN changes, and rather than being dominated by lines of intermediate-mass elements, it is now dominated by a blend of dozens of Fe emission lines, mixed with some Co lines. This is a consequence of the fact that, with time, the initially very dense ejecta become more and more transparent because of their expansion or, in other words, the photosphere (i.e. the surface from where the photons can start to free stream toward us) recedes, exposing the products of thermonuclear burning in the inner core.

Type Ib SNe, in contrast, are defined as type I SNe whose spectrum near maximum brightness does not show the Si II absorption line that defines type Ia SNe. In fact, it is useful to distinguish further between

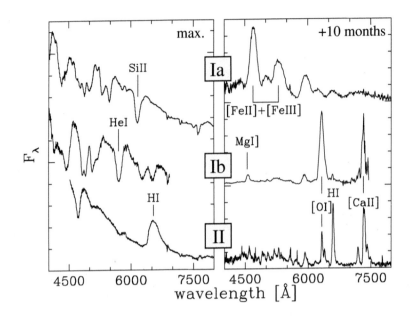

Fig. 10.2 Left panel: Spectra of basic SN types near maximum luminosity. Type II are defined by a clear signature of Hα. Type Ia is defined by a lack of H lines and a strong Si II absorption at about 6150 Å. Type Ib/c lack both H and Si lines, but type Ib has He lines, which are not present in type Ic. Ten months later (right panel) SN Ia show strong emissions of [FeII] and [FeIII], SN Ib/c are dominated by [CaII] and [OI]. These same lines plus strong Hα emission are typical of SN II at late times. From Cappellaro and Turatto (2001).

type Ib and type Ic SNe. Neither type shows H or Si lines near maximum luminosity. However, near maximum luminosity, type Ib SNe show He lines, while type Ic SNe have neither H nor He lines. At late time, the spectra of type Ib and type Ic become very similar, indicating that the distinction among Ib and Ic (sometimes collectively denoted as Ibc) is less fundamental than the distinction between type Ia and type Ibc.

Type II SNe are defined by the fact that they show prominent H lines in the spectrum taken near maximum luminosity. They are further divided into two classes, depending not on further spectral features, but rather on the decline with time of their light curves. Those that, after reaching maximum luminosity, show a plateau of almost constant luminosity for about 2–3 months are called II-P, while those that show a decline approximately linear in magnitude (and exponential in luminosity, recalling that the magnitude is obtained from the logarithm of the luminosity; see the Complement Section 10.6) are called II-L. A comparison between the spectra of type Ia, Ib and II SNe, both at maximum luminosity (left panel) and after 10 months (right panel), is shown in Fig. 10.2. A decision tree based on these criteria is shown in Fig. 10.3.

The separation into type I and type II, however, does not really reflect the underlying explosion mechanisms. Rather, type Ia SN are believed

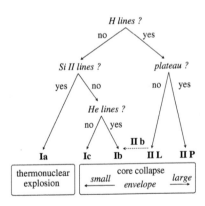

Fig. 10.3 A decision tree for the classification of SNe, based on spectral features near maximum light and, for type II SNe, on the temporal evolution of the light curve.

to originate from the thermonuclear explosion of a carbon–oxygen white dwarf (WD) in a binary system. In the simplest and commonly accepted scenario, called single-degenerate, a white dwarf (WD) accretes matter from a companion such as a red giant (or even a main sequence star for a sufficiently close binary system) until its mass goes beyond the Chandrasekhar limit $M_{\rm Ch} \simeq 1.39 M_\odot$. Then the star undergoes a disruptive thermonuclear explosion, and is dispersed in the external space in the form of a gaseous nebula. No compact remnant is formed in the process. An alternative scenario, called double-degenerate, rather involves the coalescence of two WDs, which again results in a thermonuclear explosion, and no compact remnant.

In contrast, types Ib, Ic, II L and II P all correspond to core collapse SNe, i.e. to the gravitational collapse of the inner part of the star. When a star has exhausted the nuclear fuel in its core, it is no longer able to balance its self-gravity by the pressure generated by the thermonuclear reactions, and a complicated dynamics begins. As we will discuss in detail in Section 10.3, this can lead to a SN explosion, with ejection of the external layers of the stars, while the core collapses to a NS or a BH.

The difference between the various types of core collapse can be traced to the mass and size of the progenitor star, and to the size and composition of its envelope. In the early phase of the explosion the ejecta are very dense and opaque to electromagnetic radiation. This implies that the light emitted in this early stage, including the spectrum near maximum luminosity, only probes the outermost layers and is therefore mostly sensitive to the size and composition of the envelope.

The mass of the H envelope of a star can vary, depending on many factors. For instance, stars can lose their envelope through strong stellar winds, as in the Wolf–Rayet stars.[14] Furthermore, when a star is a member of a close binary system, the interaction with the companion can have important effects on its envelope; the two stars can undergo a phase of a common envelope evolution, or material can be transferred from a star to the other through Roche-lobe overflow. It is generally accepted that type II SNe derive from the explosion of stars with mass in the range $10 - 30\, M_\odot$. In a typical SN of type II-P the progenitor is a red supergiant whose H envelope has a large mass, of order $5 - 10\, M_\odot$. The energy of the explosion completely ionizes the H envelope. The gas then enters a stage of prolonged recombination, which maintains an approximately constant temperature, typically between 5000 and 8000 K, and the photons emitted during recombination produce the plateau of type II-P SNe. The duration of this plateau depends on the mass of the H envelope. As the photosphere recedes, the dominant mechanism for photon emission becomes the decay of various radioactive nuclei newly synthesized in the explosion, which at first can extend the plateau for a brief time. Then, the SN enters in the phase where its luminosity is powered uniquely by radioactive decays. Since radioactive decays follow an exponential law $N(t) = N_0 e^{-t/\tau}$ (where $N(t)$ is the number of nuclei at time t, and τ is related to the half-life $\tau_{1/2}$ by $\tau_{1/2} = \tau \ln 2$), the corresponding luminosity $\mathcal{L}$ decreases exponentially, with a slope determined

[14]Wolf–Rayet (WR) stars originate from luminous OB stars with strong and broad emission lines and an atmosphere made mostly of He, while H is deficient or totally absent. It is believed that they are very massive evolved stars, whose H layer has been blown away by strong stellar winds, produced by radiation pressure. As a result, WR stars are continually ejecting gas at high speed, with line-of-sight velocities that can be over 2000 km/s. The resulting Doppler broadening is responsible for the observed broadness of their emission lines. These winds are also responsible for a surface composition that shows the presence of products of the nuclear burning in the inner regions.

by the half-life of the dominant process. Thus in this stage $\log \mathcal{L}$, and hence the negative of the apparent magnitude, decreases linearly.

Type II-L supernovae are due to the core collapse of stars with an H envelope of about $1-2\,M_\odot$. In this case, because of the smaller envelope mass, the velocity of expansion of the H-rich envelope is larger, and the SN enters directly into the phase dominated by radioactive decays, with little or no plateau, and a linear decrease of the magnitude. Intermediate cases exist between typical type II-P and typical type II-L supernovae, depending on the size of the H envelope.

In the phase powered by radioactive decays an important role is played by ^{56}Ni, which is the main product of burning at nuclear statistical equilibrium in the conditions of temperature and density typically present in SNe. In particular, ^{56}Ni produces ^{56}Co through electron capture

$$^{56}\text{Ni} + e^- \rightarrow\, ^{56}\text{Co} + \nu_e + \gamma\,, \tag{10.1}$$

with a half life of 6.1 days, emitting photons with energies of 750, 812 and 158 keV. The main source of luminosity later becomes the radioactive decay of ^{56}Co to ^{56}Fe, which goes mostly through electron capture (81%)

$$^{56}\text{Co} + e^- \rightarrow\, ^{56}\text{Fe} + \nu_e + \gamma\,, \tag{10.2}$$

and, to a lesser extent (19%), through β-decay $^{56}\text{Co} \rightarrow\, ^{56}\text{Fe}+e^+ +\nu_e+\gamma$. The process has a half-life of 77.7 days and the photons from electron capture are emitted at energies of 1.238 MeV and 847 keV. These γ-ray photons are down-scattered and thermalized in the ejecta until they emerge as optical or near-infrared photons.[15] The subsequent important radioactive decay is that of a different isotope of cobalt, ^{57}Co, which is initially less abundant but has a longer lifetime, $\tau_{1/2} \simeq 271$ days.

In general, because of the very high expansion velocity of the ejecta, the lines of type II SNe show significant Doppler broadening. However, approximately 10–15% of type II SNe show, on top of the broad lines, some narrow lines, and are then classified as type IIn, where "n" stands for narrow. They are interpreted as type II SNe with significant circumstellar material, which is most likely due to episodes of strong mass losses before the SN explosion. When the ejecta hits this circumstellar material, some of its kinetic energy is converted to radiation, and produces lines whose broadening rather reflects the much smaller velocity of the circumstellar material.

The close relation between type Ib and type II SNe is clearly seen by comparing the left and right panels of Fig. 10.2. On the left panel, which shows the spectra near maximum luminosity, the spectra of type Ib and type II SNe look very different. However, after about 10 months, we see that the spectra of type Ib SNe are similar to those of type II, with strong lines from neutral oxygen and singly-ionized calcium. In contrast, the spectrum of a type Ia SN remains very different, and after 10 months is rather dominated by iron lines. The meaning of these patterns can be understood by observing that the light emitted near maximum luminosity only carries informations about the external layers

[15] Hard X-rays and γ-rays were also directly detected from the type II SN1987A in the Large Magellanic Cloud. The ^{56}Ni decay chain also powers the light curve of type Ia SNe. For the type Ia SN2014J, which is at a distance of just 3.3 Mpc, it has been possible to detect with the INTEGRAL satellite the two main γ-ray lines due to the ^{56}Co decay; see Churazov *et al.* (2014) and Diehl *et al.* (2015).

of the stars, given that the optical depth of the ejecta at that time is high. After 10 months we are instead receiving light free-streaming from deeper regions of the ejecta, and we find that type Ib and type II SNe now look very similar. The natural interpretation of these facts is that type Ib SNe are the result of the core collapse of stars that have lost their H envelope, because of stellar winds or interaction with a companion. This explains the lack of H lines, which instead dominate the spectrum at maximum luminosity in type II SNe, and the similarity of the spectra at later times. This interpretation is further supported by the existence of cases intermediate between type Ib and type II; the missing link was first provided by SN 1993J, whose spectrum evolved from that of a type II to type Ib in just a few weeks. This is interpreted as the result of the core collapse of a star whose H envelope had a rather small mass, of about $0.2\,M_\odot$. There are other examples of this type, so SNe that quickly evolve from type II to type Ib are sometimes classified as members of a new class, called type IIb. In Fig. 10.3 we represent this class with an arrow connecting type II-L with type Ib.

Type Ic SNe also fit well in this scheme. Again their spectrum at late time is the same as that of type Ib and type II, showing that they are core-collapse SNe. The lack of both H and He lines can be understood on recalling that a massive star, during its lifetime, goes through different stages of nuclear burning, which results in a onion-like structure. In particular, the most external layers of the star consist of a H envelope, on top of a He envelope. Type Ic events are then naturally interpreted as the core collapse of stars that have lost both their H and their He envelopes.[16]

Type Ib, Ic and II SN are then collectively denoted as core-collapse SNe, and can be thought as a sequence Ic–Ib–IIb–IIL–IIP, from smaller to larger H and He envelopes. Type Ic, Ib and IIb SNe are also collectively called "stripped-envelope SNe". An interesting subclass of type Ic SNe, denoted type Ic-BL (for "broad-line") consists of type Ic SNe with unusually broad absorption lines, and photospheric velocities in excess of 20,000 km/s.

The difference between core-collapse and type Ia SNe is further highlighted by the fact that core-collapse SNe have only been observed in spiral galaxies, near sites of recent star formation. This again indicates that their progenitors are massive short-lived stars. Type Ia SNe, in contrast, have been observed both in spiral and in elliptical galaxies. The latter show little or no sign of recent star formation, showing that type Ia SNe are not related to massive stars, since these have a short life-time. Rather, their progenitors must be low-mass stars.

[16]Likely type Ic have some Helium too in their outer layer, but this helium produces no line due to the lack of favorable physical conditions, see Dessart, Hillier, Li and Woosley (2012).

10.2.2 Luminosities

Luminosity of Type Ia SNe

The thermonuclear explosion of type Ia SNe takes place when the mass of the WD goes beyond a fixed threshold, given by the Chandrasekhar mass. As a result, their intrinsic luminosity, to a first approximation, is uniform. Type Ia SNe rise to maximum luminosity in a period of about 20 days, reaching a peak value of the absolute blue magnitude $M_{B,\text{peak}}$ whose average over several type Ia events is given by[17]

$$\langle M_{B,\text{peak}} \rangle \simeq -19.31 \pm 0.03 + 5 \log_{10}(h_0/0.70) \,. \qquad (10.3)$$

Using eqs. (10.156) and (10.161) in the Complement Section 10.6, for the typical peak luminosity in the blue band of a type Ia SN we get

$$\mathcal{L}_{B,\text{peak}} \simeq 8 \times 10^9 \, \mathcal{L}_{B,\odot} \,. \qquad (10.4)$$

This huge luminosity allows us to see type Ia SNe at cosmological distances. Then, the fact that their intrinsic peak luminosity is, to a first approximation, quite uniform, makes type Ia SNe potential standard candles for cosmological purposes. Actually, the observed spread in peak magnitudes of type Ia SNe by itself would still be too large for them to be used as accurate standard candles in cosmology. However, the residual differences in the intrinsic luminosities of type Ia SNe can be corrected thanks to empirical relations between the peak luminosity and the shape of the light curve, with faster declining SNe being fainter. The original correlation, known as the Phillips relation (or the Pskovskii–Phillips relation), is expressed in terms of $\Delta m_{15}(B)$, the decline rate in B-band magnitude after 15 days, and has the form[18]

$$(M_B)_{\text{peak}} = -21.727 + 2.698 \, \Delta m_{15}(B) \,, \qquad (10.5)$$

with similar linear correlations in the V- and I-bands. Many refinements of this idea have been developed. For instance, the correlation between peak luminosity and the shape of the light curve can be expressed either using a "stretch parameter", defined introducing a stretch in the time axis relative to a standard luminosity template (Perlmutter *et al.* 1997), or using a multi-parameter fit in multiple colors (Riess *et al.* 1996), which also makes use of the fact that fainter type-Ia SNe appear redder than brighter objects. With these corrections the dispersion M_B can be reduced to about $\Delta M_B \simeq 0.1$ and, as a result, the distance to type Ia SNe can be deduced to about 10% accuracy. This has allowed the use of type Ia SNe at cosmological distances as accurate probes of the expansion of the Universe. In this way it is has been shown that in the recent cosmological epoch the expansion of the Universe is accelerating, providing clear indication in favor of the existence of a dark energy component, that dominates the total energy budget of the Universe in the recent cosmological epoch.[19]

Luminosity of core-collapse SNe

In contrast, the luminosities of typical core-collapse SNe span a broad range of absolute peak magnitudes, from about -15 to -20.5. We see, from eq. (10.156) in the Complement Section 10.6, that a difference in magnitude $\Delta M \simeq 5$ corresponds to a factor 100 in luminosity. Exceptionally, core-collapse SNe with magnitude up to about $M = -22$ have been observed.

---

[17]We quote the value from P. Astier *et al.* [The SNLS Collaboration] (2006). The dependence on h_0 comes from the fact that, for a given flux $\mathcal{F}$, the luminosity is $\mathcal{L} = 4\pi \mathcal{F} d_L^2$, where $d_L(z)$ is the luminosity distance. For SNe at cosmological distances $d_L \propto h_0^{-1}$, so $\mathcal{L} \propto h_0^{-2}$. Then, from $M = -(5/2) \log_{10} \mathcal{L} + \text{const.}$ (see eq. (10.155) in the Complement Section 10.6) we get $M = M(h_{0,\text{ref}}) + 5 \log_{10} h_0/h_{0,\text{ref}}$, where $h_{0,\text{ref}}$ is the reference value chosen for the reduced Hubble constant h_0.

[18]The exact numerical values of the coefficients in this relation depend on the sample of type Ia SNe used. Here we have used the numbers from the original paper by Phillips (1993), obtained however with a sample that is quite limited compared with more recent ones.

[19]The discovery of the accelerated expansion using type Ia SNe has been awarded with the 2011 Nobel Prize in Physics to Saul Perlmutter, Brian Schmidt and Adam Riess.

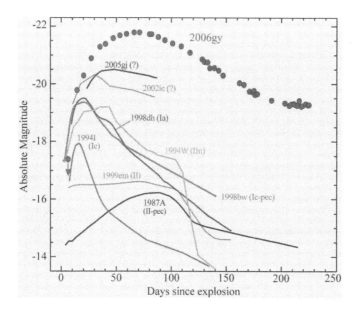

Fig. 10.4 Comparison of the absolute R-band light curve of various SNe. The curve labeled SN 1998dh is a typical SN Ia. All other curves represent core-collapse SNe, except possibly the most luminous SN 2006gy, which could be a pair-instability SN. From Smith *et al.* (2007).

Contrary to type Ia SNe, there does not seem to be a correlation, such as the Phillips relation, between the decline rate of the light curve and the peak absolute magnitude. From systematic studies of R-band photometry,[20] taking into account the host galaxy extinction, which is often significant, one finds that the populations of type Ib and type Ic SNe are statistically indistinguishable, with average values of the peak magnitude in the red band

$$\langle M_{R,\mathrm{peak}} \rangle = -18.2 \pm 0.9 \,. \tag{10.6}$$

Individual type Ib and Ic SNe, however, can have fluctuations in the peak absolute magnitude of approximately ± 2 magnitudes around these average values. Type Ic-BL SNe have a somewhat higher luminosity, with a red-band peak magnitude

$$\langle M_{R,\mathrm{peak}} \rangle = -19.0 \pm 1.1 \,, \tag{10.7}$$

and a distribution scattered by about ± 1 mag around the mean.

Figure 10.4 shows the R-band luminosity curves of various SNe, displaying a typical type Ia SN (SN 1998dh) together with a variety of core-collapse SNe. We see the typical plateau of a type II-P SN in the light curve of SN 1999em, while SN 1987A, in the Large Magellanic Cloud, the best observed supernova ever, is considered an extreme case of type II-P, with a steady increase in magnitude lasting about 3 months, and a low peak magnitude. As we will see in Section 10.3.1, the low luminosity of SN 1987A is due to the fact that, at the time of explosion, its

[20] See the Complement Section 10.6 for definitions of the different color bands.

progenitor was a blue supergiant, rather than a red supergiant. Among the other curves in the figure, SN 1994W is a type IIn SN that is powered by a strong interaction with its circumstellar material. Particularly noteworthy is the upper curve, which refers to SN 2006gy, a most remarkable SN explosion. This type IIn SN exploded in Sept. 2006 in the galaxy NGC 1260, at a distance of about 73 Mpc. It reached a peak visual magnitude of about -21.8, and stayed brighter than -21 for about 100 days. It was also characterized by a very slow rise, reaching its maximum luminosity in about 70 days, instead of the 20 days of a typical SN. Its total radiated electromagnetic energy is estimated at $\sim 10^{51}$ erg, two orders of magnitudes larger than typical core-collapse SNe. A yet more luminous supernova, SN 2005ap, has been recorded, about twice as bright as SN 2006gy at peak luminosity, although not as energetic overall, since its light curve declined in a few days. A few other ultra-luminous SNe with comparable radiated energy have been found in recent years. The mechanism that powers these ultra-luminous events is still debated. A possible explanation is that SN 2006gy is a pair-instability SN, discussion of which we defer to page 33.

10.2.3   Rates

For the purposes of direct observation with GW detectors we are particularly interested in the rate of SNe in our Galaxy (although 'third-generation' GW interferometers such as the Einstein Telescope could detect SNe at extragalactic distances, and SNe at cosmological distances are also potentially interesting sources of stochastic backgrounds of GWs). However, most information on SN rates comes from observations of SNe in other galaxies. To translate the information obtained from other galaxies into predictions for the SN rate in our Galaxy, the rates must be normalized to some quantity correlated to the stellar mass of the galaxy. The most commonly used quantity is the B-band luminosity of the galaxy in question, which is a measure of the stellar mass of the galaxy, at least for galaxies of the same morphological type. The classical units is the SNu (SN unit in B-band), defined as the number of SN events per century, per $10^{10} \mathcal{L}_{B,\odot}$, where $\mathcal{L}_{B,\odot}$ is the solar luminosity in the B-band. When one wishes to emphasize that the normalization has been performed to the B-band luminosity, this unit is also denoted by SNuB. Another useful normalization is to the far-infrared (FIR) band luminosity, and the corresponding unit is denoted by SNuIR. Its usefulness is that the FIR luminosity of a galaxy is generated by dust heated by massive stars and is therefore proportional to the star formation rate. To transform from SNu to rates in our Galaxy we need its B-band luminosity, which is

$$\mathcal{L}_{B,\text{Gal}} \simeq 2.3 \times 10^{10} \mathcal{L}_{B,\odot}, \tag{10.8}$$

so the rate of supernova explosion in our Galaxy, expressed in terms of the SNu unit, is about 2.3 SNu. Recent determination of the SN rates using the existing databases of extragalactic SNe, for the different types of SNe and for different classes of host galaxies, are shown in Table 10.2.

Table 10.2 The local SN rates, per galaxy type, in SNu. As usual, $h_0 = H_0/(100\,\mathrm{km\,s^{-1}\,Mpc^{-1}})$. From Cappellaro and Turatto (2001).

Galaxy type	SN type			
	Ia	Ib/c	II	All
E-S0	$0.32 \pm .11\ h_0^2$	$< 0.02\ h_0^2$	$< 0.04\ h_0^2$	$0.32 \pm .11\ h_0^2$
S0a-Sb	$0.32 \pm .12\ h_0^2$	$0.20 \pm .11\ h_0^2$	$0.75 \pm .34\ h_0^2$	$1.28 \pm .37\ h_0^2$
Sbc-Sd	$0.37 \pm .14\ h_0^2$	$0.25 \pm .12\ h_0^2$	$1.53 \pm .62\ h_0^2$	$2.15 \pm .66\ h_0^2$
All	$0.36 \pm .11\ h_0^2$	$0.14 \pm .07\ h_0^2$	$0.71 \pm .34\ h_0^2$	$1.21 \pm .36\ h_0^2$

We see that core-collapse SNe are not observed in elliptical galaxies, consistently with the fact that their progenitors are massive, short-lived stars. In contrast, the rate of type Ia SNe is constant, within the error, from elliptical galaxies to late spirals. Taking our Galaxy to be type Sb-Sbc (which is accounted for by performing an interpolation of the rates on a grid of galaxy types), using eq. (10.8) and setting $h_0 = 0.70$, one finds that the expected rates $\mathcal{R}_G$ in our Galaxy are

$$\mathcal{R}_G(\text{type Ib} + \text{Ic} + \text{II}) = (1.7 \pm 1.0)\ \text{events/century}\,, \qquad (10.9)$$

and

$$\mathcal{R}_G(\text{type Ia}) = (0.5 \pm 0.2)\ \text{events/century}\,. \qquad (10.10)$$

Thus, overall we get

$$\mathcal{R}_G(\text{all}) = (2.2 \pm 1.2)\ \text{events/century}\,, \qquad (10.11)$$

i.e. approximately between one and three Galactic SNe per century. This rate would imply that there should be about 44 ± 24 SN remnants in our Galaxy younger than 2000 years, which is consistent with the $O(20 - 30)$ that are presently known; see the discussion on page 9.

Somewhat higher rates are predicted using the five historical SNe observed in the last 1000 yr, and generating with a Monte Carlo simulation a number of SN events distributed in the Galaxy, to see how many events are necessary in order to have five events with apparent magnitude $m_V < 0$. Using a realistic model of the Galaxy, which includes the thin and thick disks as well as the stellar halo, plus a realistic distribution of dust to compute the extinction, one finds that about 39 events would be needed to produce the five visible historical SNe in the last 1000 yr, corresponding to a rate[21]

$$\mathcal{R}_G(\text{all}) = (3.9 \pm 1.7)\ \text{events/century}\,. \qquad (10.12)$$

This is higher than the value given in eq. (10.11), but consistent within the errors, especially in view of the uncertainties in the modelization used and of the low statistics of historical SNe.

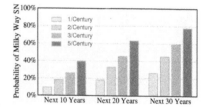

Fig. 10.5 Probabilities for one or more SNe in the Galaxy over different time spans, depending on the assumed SN rate. From Kistler, Yuksel, Ando, Beacom and Suzuki (2011).

[21] See Tammann, Loeffler and Schroder (1994).

[22]See Diehl *et al.* (2006).

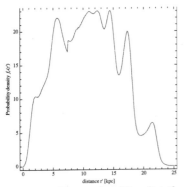

Fig. 10.6 The probability distribution for explosion of a Galactic SN as a function of the distance from Earth, in kpc. From Ahlers, Mertsch and Sarkar (2009).

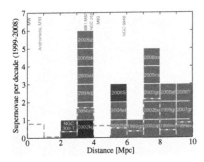

Fig. 10.7 Estimates of the core-collapse supernova rate in the nearby universe, based on that expected from 22 supernovae observed in 1999–2008 (bins), compared with the theoretical prediction using B-band luminosity (dashed line). From Kistler, Yuksel, Ando, Beacom and Suzuki (2011).

An upper bound on the Galactic SN rate can be obtained from the absence of neutrino bursts of Galactic origin at neutrino detectors such as Baksan, Mont Blanc, IMB and (Super)-Kamiokande. The detection of neutrinos from SN 1987A in the Large Magellanic Cloud shows that neutrinos from a Galactic core-collapse SNe would be clearly detected. The absence of any detection of neutrino bursts in the last 30 years puts an upper bound

$$\mathcal{R}_G(\text{type Ib} + \text{Ic} + \text{II}) < 7.7 \text{ events/century} \qquad (90\% \text{ c.l.}). \qquad (10.13)$$

A direct estimate of the rate of core-collapse SNe in our Galaxy is obtained from the γ-rays produced by the radioactive decay of ^{26}Al. It can be shown that a substantial fraction of ^{26}Al is of Galactic origin. The present-day equilibrium mass in ^{26}Al produced by ongoing nucleosynthesis throughout the Galaxy can be estimated from this γ-ray flux, making some assumptions on its three-dimensional spatial distribution. Since ^{26}Al is mostly produced by massive stars, from this one can infer that the rate of core-collapse SNe in the Galaxy is equal to[22]

$$\mathcal{R}_G(\text{type Ib} + \text{Ic} + \text{II}) = (1.9 \pm 1.1) \text{ events/century}, \qquad (10.14)$$

in excellent agreement with eq. (10.9). Given a rate, the corresponding prediction for the probability of observing at least one galactic event during a given time-span is given by the Poisson distribution for that rate. These probabilities are shown in Fig. 10.5 for different choices of the rate, ranging from 1 event per century up to a 5 per century, and for different time spans.

Beside the rate of Galactic SNe, another relevant issue, particularly for GW astronomy, is what is the distribution of distances from the Earth at which we can expect such events. This question can be addressed by performing a Monte Carlo calculation with a large number of sources drawn from a probability distribution that encodes our knowledge of the distribution of supernova remnants in the Galaxy. To obtain a realistic distribution it is also necessary to take into account the spiral arm structure of the Galaxy. The result, shown in Fig. 10.6, indicates that the distribution of events as a function of distance from the Earth is very broad, and distances between, say, 5 and 18 kpc can be considered as typical.

Both for GW astronomy and for neutrino astronomy it is important to know what distances a detector should reach, in order to expect at least one event per year. Figure 10.7 shows the number of core-collapse SNe actually detected, in the local universe, at distances up to 10 Mpc, during the period 1999–2008, while the dashed line gives the prediction obtained from the rates in Table 10.2, after assigning the appropriate blue luminosity and Hubble type to the approximately 40 major galaxies within 10 Mpc. First of all one observes that the theoretical prediction is lower by a factor of approximately 2–3, compared with the number of observed events. Furthermore, the sample of detected galaxies is likely incomplete, because SN surveys under-sample small galaxies and the Southern hemisphere, so the actual number of SN events that occurred in this period within 10 Mpc could be even larger. A possible explanation for this discrepancy is that the B-band luminosity is not an accurate indicator of the number of high-mass stars, and hence of the core-collapse SN rate. This is due to the fact that different galaxies can have different dust obscuration in the B-band, and that B-band light receives contribution from both high-mass and low-mass stars. In any case, using the actually observed SNe shown in Fig. 10.7, we see that the core-collapse SN rate reaches 1 event per year at a distance of about 6 Mpc, while at 10 Mpc we can expect 2 events per year. We also observe that all the 22 SNe observed in this period were core-collapse SNe. This sets a bound on the ratio of type Ia to core-collapse events in the local universe of order 0.2 (at 90% c.l.). The same information is shown, in terms of the cumulative core-collapse SN rate as a function of distance, in Fig. 10.8.

The rate of core-collapse SNe at even larger distances is shown in the top panel of Fig. 10.9 (denoted there by R_{CCSN}). On the left side of the figure the rate is plotted against the distance (in Mpc), while beyond $z = 0.1$ the rate is shown against the redshift. Since massive stars are short-lived, the rate of core-collapse SNe should follow the star formation rate. The central solid line gives the prediction obtained using a fiducial cosmic star formation history, with the upper and lower lines giving the uncertainty range of the theoretical model. The bottom panel shows the ratio of type Ia to core-collapse events. In the local region we have an upper bound (shown by the lines at 90% and 99% c.l.), coming from the fact that the 22 SNe observed in the local region in 1999–2008, and shown in Fig. 10.7, were all core-collapse SNe. At cosmological distances this ratio takes values between 0.2 and 0.4, consistent with the values reported in Table 10.2.

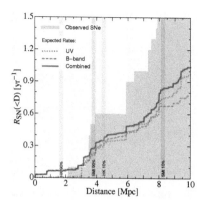

Fig. 10.8 The cumulative rate of core-collapse supernovae in the nearby universe, based on that expected from 22 supernovae observed in 1999–2008, compared with the theoretical prediction using B-band emission from 451 galaxies (dashed), UV emission from 315 galaxies (dotted), and the 589 galaxies of the combined catalog (solid). The vertical bands corresponds to the reach of different possible neutrino detectors. From Kistler, Haxton and Yüksel (2013).

10.3 The dynamics of core collapse

10.3.1 Pre-SN evolution

The evolution of a massive star is governed by the competition between the gravitational contraction under its own gravity, on the one hand, and the pressure exerted by the radiation, by the thermal gas and, in the late stages of its evolution, by the partially degenerate electron gas, on the other. Energy is lost mainly to radiation during the core H- and He-burning phases, and to neutrino emission afterwards, and this energy is provided by the thermonuclear reactions taking place in the inner parts of the star. These reactions are very sensitive to the temperature in the core. High temperatures are necessary to overcome the Coulomb repulsion of the positively charged nuclei and bring them close enough. At that point the attractive part of the short-range strong force between nucleons can take over and create a bound nuclear state. The higher the electric charge of the nuclei, the higher is the Coulomb barrier to be overcome. When a protostellar cloud collapses under its own self-

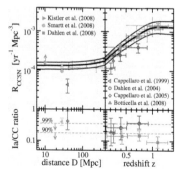

Fig. 10.9 Evolution of the core-collapse supernova rate (top panel) and the ratio of Type Ia to core-collapse SNe (bottom panel). From Horiuchi, Beacom and Dwek (2009).

gravity, the first thermonuclear reaction that can ignite is the fusion of deuteron and proton to give ^{3}He, $D + p \rightarrow {}^3\text{He} + \gamma$, and afterwards the fusion of protons into ^{2}He nuclei. If the mass of the protostellar cloud is smaller than about $0.08 M_\odot$, the central temperature is never large enough to allow for the ignition of stable thermonuclear reactions, and radiation is only emitted as the cloud contracts and cools down. The outcome is then a planet or a brown dwarf. If instead $M \gtrsim 0.08 M_\odot$ the star starts burning steadily H to He in its core. For a star of mass $M \simeq M_\odot$ this phase lasts about 10^{10} yr, and in this stage the star is on the main sequence of the Hertzsprung–Russell (HR) diagram. More generally, in a large range of masses, the evolutionary time-scale on the main sequence can be approximately reproduced by

$$t_{\text{evol}} \simeq 7.3 \times 10^9 \, (M/M_\odot)^{-5/2} \text{ yr} . \tag{10.15}$$

When the H fuel in the core is nearly exhausted the star enters a complicated evolutionary phase, which depends strongly on its mass. We focus here on massive stars, $M \gtrsim (6-9)M_\odot$, since a value in this range is believed to be the minimum progenitor mass necessary for a SN explosion.[23] When the H fuel is almost exhausted the He core can no longer support itself, and begins to contract. As a result, the temperature at the edge of the core increases sufficiently that the H in a shell around the core can again undergo fusion into He. At this stage the star consists of an inert He core surrounded by a burning H shell. The core no longer produces energy, so it has a vanishing luminosity. Since the luminosity produced at radius r is proportional to the temperature gradient $dT(t)/dr$, the vanishing of the luminosity implies that the He core is isothermal. However, the luminosity produced by the H shell is larger than the one that was produced during core-H burning. Part of this energy goes into a slow expansion of the envelope, and as a result the star becomes larger and redder.

The burning H shell continues to produce He, and these "nuclear ashes" increase the size of the He core. This phase of the evolution ends when the He core is no longer able to support the weight of the matter over it. Indeed, for a star of mass M with an isothermal core of mass qM and an envelope of mass $(1-q)M$, there is a maximum value of q beyond which the isothermal core can longer sustain the weight of the layers above it. This critical value, known as the Chandrasekhar–Schönberg limit, is given by $q_{\text{max}} \simeq 0.37(\mu_{\text{env}}/\mu_{\text{ic}})^2$, where μ_{env} and μ_{ic} are the mean molecular weights in the envelope and in the core, respectively.[24] For a star with solar-mass composition, $q_{\text{max}} \simeq 0.08$. For low-mass stars at this stage the electron gas starts to be degenerate, and electron degeneracy pressure increases the value of q_{max}. For instance, for a 1 $M_\odot$ star the resulting value is $q_{\text{max}} \simeq 0.13$. In contrast, for a massive star, say $M \gtrsim 5M_\odot$, the effect of electron degeneracy in the core at this stage is still negligible, so when the mass in the He core becomes about 8% of the total mass of the star the He core begins to contract, on a time-scale much shorter than the typical time-scale of the main-sequence evolution. For a star with $M = 5M_\odot$, this phase lasts

[23] A more detailed discussion of the value of this minimal mass, as well as its dependence on metallicity, is given on page 30. The notion of star metallicity is reviewed in the Complement Section 10.6.

[24] See e.g. Padmanabhan (2001), Section 3.4.3, or Carroll and Ostlie (2007), Section 13.1, for derivations and detailed discussions.

about 3×10^6 yr. The contraction raises the temperature in the core, and when the inner core temperature reaches a value $T \sim 10^8$ K, He burning ignites and stops the contraction of the core. At this stage the star has two sources of nuclear energy. An inner core, where He is burnt to carbon and oxygen, and a shell where H is burnt to He. The two are separated by a He-rich region, where the temperature is not large enough to burn He.

Carbon is produced through the triple-alpha process in which two α particles, i.e. nuclei of $^4_2$He, collide to produce beryllium,

$$^4\text{He} + {^4}\text{He} \leftrightarrow {^8}\text{Be} \, . \tag{10.16}$$

The ^{8}Be nucleus is unstable and will decay very quickly back to two α particles, unless it reacts immediately with a third α particle to produce C,

$$^4\text{He} + {^8}\text{Be} \rightarrow {^{12}}\text{C} + \gamma \, . \tag{10.17}$$

In this sense, the triple-alpha process is really a three-body collision. After sufficient C has been produced, in the He-burning core there are the conditions for producing oxygen, through

$$^{12}\text{C} + {^4}\text{He} \rightarrow {^{16}}\text{O} + \gamma \, . \tag{10.18}$$

When most of the He fuel has been burnt, thermonuclear reactions can no longer support the C–O core, which contracts. Then a He shell around it ignites.

For masses in the range $(6 - 9) M_\odot$ the subsequent evolutionary scenario is not so well known, with a final output that can be (from the lower to the upper mass range), a CO white dwarf, a NeMgO white dwarf, an electron-capture SN before the neon photo-disintegration phase, and finally nuclear burning all the way up to the production of an iron core.[25]

For a star with a mass higher than a critical value M_{up} in the range $(7 - 9) M_\odot$, after further contraction the burning of C in the core begins. The theoretical prediction for the critical mass depends on issues such as the treatment of convection. For $M > M_{\text{up}}$ the star now has a core where C is burnt, a burning He shell and an outer H-burning shell, separated by inert regions. C burns through the direct fusion of two ^{12}C nuclei into a highly excited state of ^{24}Mg, which then decays into various channels, including decay to ^{20}Ne. After C burning, the composition of the core is mostly dominated by the ^{16}O produced during He burning and by ^{20}Ne and ^{24}Mg produced by C burning. Even though O has a lower Coulomb barrier, neon burning ignites first through the process $^{20}\text{Ne} + \gamma \rightarrow {^{16}}\text{O} + {^4}\text{He}$, induced by high-energy photons from the tail of the Planckian distribution, and later O burning ignites.

Ne ignites for stars with a mass above a critical value that is somewhat uncertain, but is estimated to be in the range $(10 - 12) M_\odot$. After the ignition of Ne the star goes through the subsequent stages of O and silicon burning. Si burning eventually produces a core of elements of the iron group, mostly ^{54}Fe, ^{56}Fe and ^{56}Ni. These nuclei have the highest binding energy per nucleon, so no further energy can be extracted by

[25]See Woosley and Heger (2015), and the discussion on page 31.

fusion of these elements. Their relative abundance at a given temperature and density is determined not only by their binding energy per nucleon, but also by the ratio of neutrons to protons in the core. For temperatures $T \lesssim 10^{10}$ K, the equations of nuclear statistical equilibrium favor the most tightly bound nucleus for a given neutron excess (i.e. the relative excess of neutrons of protons). For a composition with approximately an equal number of neutrons and protons this is ^{56}Ni while for a neutron excess $\eta \simeq 0.07$ it is ^{56}Fe, and for even higher η it shifts toward heavier isotopes.

Neutrino losses also play an important role in the dynamics of the star in these stages. When the temperature exceeds 5×10^8 K, thus after He burning, neutrino losses due to pair annihilation become the dominant mechanism for the energy loss of the star, more important than radiative diffusion and convection. When the Si-burning phase is reached neutrino losses due to electron capture also become important. Neutrino production through pair annihilation is very sensitive to temperature, with a loss term roughly proportional to T^9. As we go through the process of exhausting a nuclear fuel, core contraction and ignition of the next nuclear fuel, the core temperature rises, and neutrino losses raise very rapidly. This leads to an acceleration of the burning process. As we see from Table 10.3, for a star with $M = 15 M_\odot$ the main-sequence lifetime of steady H burning lasts 11 Myr, He burning 2 Myr, and C burning 2000 yr, while Si burning is completed in just about 18 days! The corresponding values for a star with $M = 25 M_\odot$ are given in Table 10.4. Thus, the internal structure of the star evolves on a time-scale much shorter that that needed to propagate these changes to the surface. After C burning has started, the surface luminosity and temperature basically have no time to adjust to the changes in the interior and do not evolve significantly, until the star explodes as a SN.

During its supergiant phase, when the star has an extended envelope supported by He burning (either in a shell or in the core), there are two possible solutions for the equations of stellar structure, a red supergiant or a blue supergiant, separated by thermally unstable solutions. Red supergiants have a convective envelope, lower temperatures and a much larger radius than blue supergiants. Whether a star evolves to a blue or to a red supergiant depends on its mass, metallicity and mass loss rate. In general, for a star to spend most of its core He-burning phase as a blue supergiant directly after the main-sequence phase, it must have $M \lesssim 20 M_\odot$, low metallicity and low mass losses. Whether the progenitor of a SN was a red or a blue supergiant has important implications for the luminosity of the SN explosion. Since a blue supergiant is more compact, its envelope sits in a deeper gravitational well and a larger fraction of the energy of the explosion goes into ejecting the envelope of the star out of this potential well. As a result, the electromagnetic luminosity released in the SN explosion of a blue supergiant is lower.

A clear example of this is provided by the explosion of SN1987A in the Large Magellanic Cloud. As we see from Fig. 10.4 on page 17, SN1987A was a sub-luminous supernova. The sub-luminous nature of SN1987A is

Table 10.3 Core temperature, core density and duration of various core-burning stages of a star with $M = 15 M_\odot$. From Table I of Woosley, Heger and Weaver (2002).

Fuel	T (K)	ρ (g/cm^3)	τ
H	$3.5 \cdot 10^7$	5.8	11 Myr
He	$1.8 \cdot 10^8$	$1.4 \cdot 10^3$	2 Myr
C	$8.3 \cdot 10^8$	$2.4 \cdot 10^5$	2 kyr
Ne	$1.6 \cdot 10^9$	$7.2 \cdot 10^6$	0.7 yr
O	$1.9 \cdot 10^9$	$6.7 \cdot 10^6$	2.6 yr
Si	$3.3 \cdot 10^9$	$4.3 \cdot 10^7$	18 d

Table 10.4 As in Table 10.3, for $M = 25 M_\odot$. From Table I of Woosley, Heger and Weaver (2002).

Fuel	T (K)	ρ (g/cm^3)	τ
H	$3.8 \cdot 10^7$	3.8	6.7 Myr
He	$2.0 \cdot 10^8$	$0.8 \cdot 10^3$	0.8 Myr
C	$8.4 \cdot 10^8$	$1.3 \cdot 10^5$	0.5 kyr
Ne	$1.6 \cdot 10^9$	$3.9 \cdot 10^6$	0.9 yr
O	$2.1 \cdot 10^9$	$3.6 \cdot 10^6$	0.4 yr
Si	$3.6 \cdot 10^9$	$3.0 \cdot 10^7$	0.7 d

explained by the fact that its progenitor was a blue supergiant. In fact, in this case we know the progenitor, which was catalogued as Sk 202-69, and it is indeed known that at the time of the explosion it was a 12-th magnitude blue supergiant, with $M \simeq 20M_\odot$. Actually, observations of the circumstellar material show that the star was a red supergiant until about 30,000 years before the explosion. The transition from red to blue supergiant shortly before the explosion could take place in a single star by a combination of low metallicity and fast rotation, resulting in an envelope richer in He. Such a heavier envelope favors a blue solution. Another possibility is that the progenitor was originally in a binary system, with a red supergiant of about $(16 - 18)M_\odot$ and a companion with a mass of order $3M_\odot$, probably a main-sequence star. In this scenario the red supergiant expands and engulfs the companion star. In the following phase of common envelope evolution part of the common envelope is ejected while the main sequence star is tidally disrupted. The resulting larger mass of the envelope drives the transition from red to blue supergiant.

10.3.2 Core collapse and neutrino-driven delayed shock

The collapse of the core

After the burning of Si in the core is completed, the star has an onion-like structure with a Fe–Ni core (usually generically referred to simply as the iron core), surrounded by a shell of Si burning and further concentric shells of O burning, Ne burning, C burning, He burning and H burning, separated by inert regions. The pressure in the core at this stage is dominated the degeneracy pressure of the relativistic electrons, and stability is only possible as long as the mass of the core is below the Chandrasekhar mass M_{Ch}. However, the burning of Si in a shell outside the core keeps producing more Fe–Ni "ashes", which increase the mass of the core. When the mass of the partially degenerate Fe–Ni core exceeds the critical threshold, the core begins to collapse. The Chandrasekhar mass is given by

$$M_{\mathrm{Ch}} \simeq 5.828 \, Y_e^2 M_\odot \,, \qquad (10.19)$$

where Y_e is the number of electrons per baryon,

$$Y_e = \frac{n_e}{n_B} \,, \qquad (10.20)$$

and n_e and n_B are the number densities of electrons and baryons, respectively. If we take equal number densities for protons and neutrons (whether free or bounds in nuclei), we have $n_B = n_p + n_n = 2n_p$. Furthermore, by charge neutrality $n_p = n_e$, so we get $Y_e = 1/2$, which gives $M_{\mathrm{Ch}} \simeq 1.457M_\odot$. This value is reduced to $1.42M_\odot$ by general-relativistic corrections. Actually, because of an excess of neutrons over protons, Y_e is somewhat smaller than $1/2$, and across the iron core typically ranges from a value $Y_e \simeq 0.42$ at the center to a value $Y_e \simeq 0.48$ at the edge.

Taking an average value $Y_e \simeq 0.45$ gives $M_{\rm Ch} \simeq 1.18 M_\odot$, which is then reduced to $1.15 M_\odot$ by general-relativistic corrections. This is the critical mass for a completely degenerate electron gas, i.e. for a gas where the electron temperature is negligible compared with the Fermi energy. In the iron core, however, the electrons are only partially degenerate and a significant contribution to the pressure still comes from the electron temperature, raising the critical mass to a value

$$M_{\rm Ch}(T) \simeq M_{\rm Ch}(T=0) \left[1 + \left(\frac{\pi kT}{\epsilon_F} \right)^2 \right], \qquad (10.21)$$

where ϵ_F is the Fermi energy,

$$\epsilon_F \simeq 1.1 \, Y_e^{1/3} \left(\frac{\rho}{10^7 \, {\rm g/cm^3}} \right)^{1/3} {\rm MeV}. \qquad (10.22)$$

During Si-shell burning, the mass of the core increases because of the production of Fe–Ni ashes. At the same time the critical value (10.21) decreases because the large energy losses due to neutrino emission lower the value of the T-dependent term. This finally brings the iron core beyond the critical value, and the collapse begins. For a star with $M = 15 M_\odot$ the mass of the core at time of collapse is about $M_{\rm core} \simeq 1.34 \, M_\odot$, while heavier stars collapse with heavier iron cores. As a result, one can expect that more massive stars produce as remnants more massive neutron stars, although the mass of the remnant also depends on the subsequent collapse dynamics, as we will discuss in detail in Section 10.3.3.

The collapse is then accelerated by two instabilities. First, as the density in the collapsing core increases, electrons reach Fermi energies of the order of MeV, and the process of electron capture by heavy nuclei becomes energetically favorable. This produces neutron-rich nuclei via inverse β-decay and removes electrons. Therefore Y_e decreases, and consequently the electron pressure that was trying to oppose the collapse also decreases, or, in other words, the maximum mass that can be stabilized by electrons decreases. Indeed, we see from eq. (10.21) that the term $M_{\rm Ch}(T=0)$ is proportional to Y_e^2 while the T-dependent term is proportional to $M_{\rm Ch}(T=0)/\epsilon_F^2$ and therefore to $Y_e^{4/3}$, so both terms decrease. A second instability is due to photo-disintegration. At the temperatures now present in the core, which are of order 10^{10} K (so $k_B T \sim$ MeV), photons are sufficiently energetic to destroy even some of the strongly bounds nuclei of the iron group, with processes such as

$$^{56}{\rm Fe} + \gamma \rightarrow 13 \, ^4{\rm He} + 4n. \qquad (10.23)$$

This process is highly endothermic, since the ^{56}Fe nucleus has higher binding energy per nucleon than the α particles. It therefore reduces the thermal energy of the core, and decreases the T-dependent term in eq. (10.21). It should be noted, however, that due to the exponentially large number of excited nuclear states available at large temperatures, the partition function still favors the presence of bound heavy nuclei, and prevents their complete disintegration into α particles, or even further into free nucleons.

Thus, the collapse accelerates; in the inner core the collapse is homologous, i.e. the collapse velocity is proportional to the radial distance, until a radius where the velocity exceeds the local sound speeds. Beyond this radius the collapse can no longer continue to increase linearly with distance, but still this outer core is supersonic. In about one second, the core collapses from the typical size of a white dwarf, i.e. about the Earth's radius, down to a radius of about 50 km.

The next remarkable step takes place when the density reaches a value $\rho \simeq 10^{12}\,\mathrm{g/cm^3}$. Then the neutrinos, which are being copiously produced by electron capture and, to some extent, by the β-decays of some of the nuclei, become trapped. Neutrinos in fact scatter coherently over the nuclei, and their diffusion time becomes larger than the typical collapse time-scale.

Bounce and neutrino-driven delayed shock

Eventually the whole core reaches nuclear densities. At this point the whole collapsing core can be seen as a single gigantic nucleus. At such a density the strong interaction between nucleons becomes repulsive. When the density reaches the value $\rho \simeq 8 \times 10^{14}\,\mathrm{g/cm^3}$, approximately three times larger than the density of an atomic nucleus, the nuclear matter bounces back, sending a shock wave into the outer core, which is still falling inward at supersonic speed.

Originally it was thought that this bounce would have caused the explosion of the star. Nowadays, numerical simulations have conclusively shown that this "prompt" bounce-shock is not sufficient, by itself, to drive the SN explosion. The reason is that when the shock wave meets the infalling outer core most of its energy is lost, and goes into heating the outer core. The violence of the encounter is such that the iron of the outer core is completely disintegrated into protons and nucleons. This removes from the shock about 10^{51} ergs for each $0.1 M_\odot$ of the outer core. Furthermore, this change in the composition of the outer core induces further energy losses, because the capture rates of electrons on free protons, through inverse β-decay $e^- + p \to n + \nu_e$, is larger than on heavy nuclei, because of blocking from nuclear-shell effects. A fraction of the neutrinos produced in this way leaves the star quickly, producing a neutrino burst, which carries away further energy that could otherwise have contributed to the explosion. The shock then stalls, and after about 10 ms from the core bounce one is left with an almost stationary shock wave, at a radial distance of 100–200 km, accreting the infalling matter of the outer core at a rate of about $(1 - 10) M_\odot\,\mathrm{s^{-1}}$.

A second energy source is therefore needed at this stage to drive the explosion. While there is a consensus on the fact that the prompt explosion mechanism does not work, the details of the mechanism that does finally drive the explosion are not yet known. It is, however, clear that a crucial role in reviving the shock is played by the neutrinos. In the inner core region, because of the high density, neutrinos cannot stream freely. Therefore, just as a star has a photosphere that delimits the region where the photons can begin to stream freely to the exterior, in this case a "neutrinosphere" develops below the shock front. Neutrinos deposit energy in the matter between the nascent neutron star in the inner core and the photosphere, through scattering processes such as $\nu_e + n \to e^- + p$ and $\bar{\nu}_e + p \to e^+ + n$. Since most of the gravitational energy liberated in the collapse goes into neutrinos, these processes are quite significant, and heat up the matter below the shock front, generating a further pressure that can revive the shock, on a time-scale of 0.1 s. In fact, to revive the shock, it is sufficient that a few percent of the total neutrino energy is converted into thermal energies of the nucleons, leptons and photons. If this happens, the shock wave resumes and lifts the external layers of the star in a SN explosion.

Hydrodynamical instabilities

While the scenario just described is theoretically appealing and convincing, a proof that this is what actually happens must rely on numerical studies. Simulations of SN explosions are, however, of the outmost complexity. Present simulations unavoidably make important approximations of some sort, such as Newtonian rather than general-relativistic gravity and spherical symmetry instead of full three-dimensional simulations, but the outcome turns out to depend strongly on details. Spherically symmetric simulations, even including accurate hydrodynamic codes, neutrino transport and an extended set of neutrino reactions, fail to obtain SN explosions unless the progenitor is in the mass range $\sim (7-9)M_\odot$.[26] For larger masses, this points toward the importance of instabilities in a full three-dimensional setting. Rayleigh–Taylor fingering instabilities have been considered as a possible boost for neutrino luminosity. More generally, since the heating from neutrinos is strongly dependent on the distance from the center, convective instabilities are expected to play an important role. This is also confirmed by the detailed observations that have been made possible by the explosion of SN 1987A in the Large Magellanic Cloud, which have shown that radioactive nuclei produced deep in the core of the star have been mixed with the hydrogen envelope of the star. Three-dimensional numerical simulations have indeed shown that violent convective overturns appear. To assess whether this really leads to SN explosion, it is necessary to implement accurate energy-dependent neutrino transport codes, which in the full three-dimensional setting is beyond present computer power.

However, three-dimensional simulations have revealed the existence of non-radial instabilities that might be very important for the explosion mechanism. In particular it has been found that, under the effect of accretion from the infalling material, the shock front displays a non-radial instability termed SASI (standing accretion shock instability), which in a decomposition in spherical harmonics is especially evident in the $l = 1, 2$ mode, i.e. it produces a dipole and a quadrupole deformation

[26]This is basically related to the fact that, as we will discuss on page 31, stars in this mass range explode when they have a Ne–O core, and at the edge of such a core the density gradient is very steep. In contrast, heavier stars explode after burning their fuels all the way to iron, and therefore have an iron core. The profile of iron cores is much flatter, leading to larger infall and ram pressure from the external layers. Iron cores are therefore much harder to explode than Ne–O cores.

of the accretion shock front. This causes a back-and-forth sloshing of the shock front, which transfers energy outward more efficiently than convection, and ultimately can lead to an asymmetric explosion. At the same time, such a highly anisotropic explosion can result in a kick of the newborn neutron star, with velocities that in some numerical simulations have reached 1000 km/s, in agreement with the measured velocities of young pulsars.

Magnetohydrodynamic instabilities might also play an important role in driving the explosion. The rotational energy stored in the nascent proto-neutron star is huge,

$$E_{\rm rot} \sim 2.4 \times 10^{52}\,{\rm erg} \left(\frac{M_{\rm NS}}{1.5M_\odot}\right) \left(\frac{R_{\rm NS}}{10\,{\rm km}}\right)^2 \left(\frac{1\,{\rm ms}}{P_{\rm NS}}\right). \qquad (10.24)$$

Magnetohydrodynamic processes can in principle extract efficiently part of this rotation energy, and explode the star. The numerical investigation of these mechanisms is, however, very challenging, since it requires full three-dimensional simulations, accurate neutrino transport codes and high numerical resolution.

Asymmetric explosions and pulsar kicks

The proper motion of a pulsar can be measured using pulsar timing or interferometers. To translate the proper motion into a velocity one also needs to infer the pulsar's distance. Unless annual parallaxes are available, the distance is estimated using the dispersion measure, defined in eq. (6.37), together with a model of the Galactic electron density. From the study of 73 young pulsars (with ages less than 3 Myr) one finds that the average value of their inferred three-dimensional velocity is

$$\langle v_{\rm 3D} \rangle = (400 \pm 40)\,{\rm km/s}, \qquad (10.25)$$

while the average value of their two-dimensional velocity in the plane transverse to the line of sight is

$$\langle v_{\rm 2D} \rangle = (307 \pm 47)\,{\rm km/s}. \qquad (10.26)$$

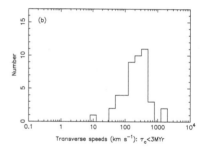

Fig. 10.10 The distribution of two-dimensional velocities for young pulsars. From Hobbs, Lorimer, Lyne and Kramer (2005).

The two-dimensional velocity distribution is shown in Fig. 10.10. Observe that some pulsars have velocities larger than 1000 km/s. In particular, PSR B2011+38 and PSR B2224+64 have two-dimensional speeds of about 1600 km/s. These proper-motion velocities are statistically much higher than those of their progenitor stars, the massive OB stars. This suggests that pulsars might get a significant kick at birth, during the SN explosion. Various mechanisms have been proposed to explain these kicks. In an asymmetric explosion the ejecta exert on the proto-neutron star a direct "contact" force through hydrodynamical interactions during the explosion, which, however, can hardly produce a kick of about 200 km/s. More importantly, numerical simulations of supernova explosions suggest that the ejecta also exert a long-lasting asymmetric gravitational acceleration on the proto-neutron star, due to the fact that the gravitational force exerted by ejecta that leave faster in one direction does not compensate that of the slower and typically denser ejecta leaving in the opposite direction. This mechanism can build velocities larger than 1000 km/s. As we will study in Section 14.3.5, even-higher recoil velocities can be imparted to the remnant in the coalescence of spinning compact binaries.

10.3.3   The remnant of the collapse

An important question is what remnant, if any, is left behind by a SN explosion. The answer depends on the mass of the collapsing core at the time of explosion and on the amount of material that falls back after the explosion. The mass of the collapsing core in turn depends on the initial mass $M_{\rm prog}$ of the progenitor star (more precisely, on the mass that the progenitor had at the time when it first reached the main sequence, known as the zero-age main sequence, or ZAMS) and on the history of its mass loss. For single stars the only mass loss mechanism is provided by stellar winds, which are functions of the star's metallicity. Stars with low metallicity have weaker stellar winds and retain most of their initial mass, while stars with solar-type metallicities have stronger winds and hence larger mass losses.[27] For stars in a binary system the situation is complicated by the various possible mass loss mechanisms due to the interaction with the companion.

The exact value of the minimal progenitor mass $M_{\rm prog}^{\rm min}$ for which the core collapses and a SN explosion takes place has been subject to much investigation. Earlier work suggested that it is around $9M_\odot$, independent of metallicity. More recent studies[28] suggest that the transition indeed takes place around $9M_\odot$ for solar metallicities $Z = Z_\odot \simeq 0.016$, but it drops to about $6.3M_\odot$ at $Z = 10^{-3}Z_\odot$, with a possible fit

$$M_{\rm prog}^{\rm min} = \begin{cases} 9.0 + 0.9\log_{10}\left(Z/Z_\odot\right) & \text{if } Z > 10^{-3}Z_\odot \\ 6.3M_\odot & \text{if } Z < 10^{-3}Z_\odot. \end{cases} \tag{10.27}$$

For masses below this critical value there is no SN explosion and the remnant is a white dwarf (WD), with a composition that depends on the elements that the progenitor star has been able to burn. For the remnant, we can distinguish the following possible regimes (taking into account that the values of the masses separating these regimes are still somewhat uncertain, and should only be taken as indicatives).

(a) $M_{\rm prog} \lesssim 7M_\odot$, WD remnant. On the lower end of the mass distribution one has WDs whose spectrum indicates that they are made of He, and have masses $M_{\rm remn} \lesssim 0.45M_\odot$. It is believed that their progenitors were members of binary systems that lost their envelope to a companion before reaching the tip of the giant branch phase, i.e. before He burning started. As a result the central pressure was relieved and these stars never burned He to C and O. The lightest known WD has a mass $M \simeq 0.17M_\odot$.[29] If the envelope of the progenitor is not stripped by binary interaction, the star can go through successive stages

of nuclear burning. As we saw in Section 10.3.1, for progenitor masses below a critical value M_{up}, which for definiteness we take here to be $\simeq 7 M_\odot$, the star is only able to burn He, producing C and O. Thus, typical WDs are made of carbon and oxygen, and are called CO WDs. The corresponding remnant mass M_{remn} increases with the mass of the progenitor, from typical values of order $0.6 M_\odot$ up to the Chandrasekhar mass $M_{\mathrm{Ch}} \simeq (1.3 - 1.4) M_\odot$.

(b) $7 M_\odot \lesssim M_{\mathrm{prog}} \lesssim 9 M_\odot$: *NeO WD or electron-capture SNe.* For progenitor masses in this range the star reaches the stage where C is burnt to O and Ne. Then a degenerate Ne–O core forms. The subsequent dynamics depends on the mass losses. If sufficient material is expelled, the star will end up as a NeO WD. Otherwise, the star undergoes core collapse. Electron degeneracy is reached before the ignition of Ne, and the collapse of the core is triggered by electron capture on Mg and Ne, which reduces the degeneracy pressure. Therefore we do not have the collapse of an iron core, as in more massive stars, but rather of a Ne–O core (or, more precisely, of a Ne–O–Mg core) and the resulting SN explosion is called an electron-capture SN. The remnant of the collapse in this case is a neutron star (NS). The mass of the collapsing core is close to the Chandrasekhar mass, which for a Ne–O core is $M_{\mathrm{Ch}} \simeq 1.38 M_\odot$, and the mass of the NS remnant will then be equal to the Chandrasekhar mass minus the mass of the ejected material. Simulations of the collapse predict ejected masses typically around $0.02 M_\odot$ (although values in the range $(0.01 - 0.20) M_\odot$ could be possible), leading to a typical mass of the NS remnant of about $1.36 M_\odot$.

In typical models[30] one finds that, for solar metallicities, electron-capture SNe takes place in a window of progenitor masses $\Delta M_{\mathrm{prog}} = 0.25 M_\odot$, near $M_{\mathrm{prog}} \simeq 9 M_\odot$. Convolving this result with the mass distribution of progenitor stars given by the Salpeter initial mass function, one finds that electron-capture SNe should be about 4% of all SNe in the local Universe. With more extreme choices of the mass loss rate the window of progenitor masses can widen up to $\Delta M_{\mathrm{prog}} = 1.4 M_\odot$, and the fraction of electron-capture SNe can rise to 20%. Furthermore, for lower metallicities, the window for electron-capture SN widens and shifts toward lower masses.

[30] See e.g. Poelarends, Herwig, Langer and Heger (2008).

(c) $9 M_\odot \lesssim M_{\mathrm{prog}} \lesssim 10.3 M_\odot$. For stars in this regime the dynamics can be quite complicated. Ne and O ignite off-center (while C ignites centrally). The off-center ignition results in so-called "convectively bounded flames". Si can also ignite off-center, possibly leading to a localized deflagration. These complications make it difficult to realistically model the evolution in this mass range. Unless an electron-capture SN takes place, the evolution proceeds to the formation of an iron core.

(d) $10.3 M_\odot \lesssim M_{\mathrm{prog}} \lesssim (30 - 40) M_\odot$: *core collapse to NS or BH.* Starting from a mass around $10.3 M_\odot$, all fuels ignite centrally. These stars burn

the nuclear fuels all the way up to iron and form an iron core that undergoes collapse, giving rise to the classic core-collapse SNe. As we discussed in Section 10.3.2, the SN explosion is most probably driven by instabilities, such as convective instabilities or the standing accretion shock instability (SASI), that enhance the conversion of the neutrino energy into kinetic energy of the nucleons, leptons and photons. The material accelerated in this way moves outward, pushing the layers above it and producing the explosion. In the process, because of the work done lifting the upper layers, some of the material belonging to the shock front loses enough energy to be decelerated below the local escape velocity. This material will then fall back onto the newborn proto-NS, adding to the mass of the remnant. For progenitor masses below $11 M_\odot$ the amount of fallback is believed to be negligible. In contrast, for progenitors heavier than about $11 M_\odot$ it can become very important.

Most of the fallback takes place in the first 10–20 s after the explosion. The amount of fallback depends on the explosion mechanism and on details of the stellar structure, so it gives rise to the main theoretical uncertainty in the computation of the remnant mass for stars in this mass range. In general, however, we can consider three possible scenarios: (1) A successful SN explosion is launched, with most of the star's material being ejected, although some material falls back onto the the proto-NS, increasing the mass of the NS remnant. (2) A successful SN explosion is launched, but the amount of material that falls back is sufficient to push the proto-NS beyond its Chandrasekhar limit. In this case, on a time-scale ranging from minutes to hours, we have black hole (BH) formation by fallback. (3) The explosion is not sufficiently energetic to eject the external layers. This gives rise to a failed SN, and all the material of the star will collapse, giving rise to the formation of a proto-NS followed by BH formation by rapid accretion. In this case BH formation takes place within a few seconds after core collapse. The mass of the resulting BH depends on whether the total mass of the star at collapse goes into the BH. The other possibility is that, even for failed SNe, a considerable fraction of the H envelope, and possibly all of it, is anyhow stripped because it is extremely loosely bound. In that case the mass of the resulting BH is rather given by the mass of the He core of the pre-collapse star. Cases (2) and (3) are referred to as the *collapsar* model for BH formation.

Numerical simulations indicate that, for stars of solar metallicity, progenitor masses below approximately $15 M_\odot$ always give rise to a NS, while for heavier progenitors the remnant can be either a NS or a BH, depending on details of the explosion. As we saw above, the collapse of stars with $M_{\rm prog} \lesssim (9-11) M_\odot$ is expected to produce NS remnants with masses narrowly peaked around $1.36 M_\odot$. In contrast, because of the role of fallback, the mass distribution of NS created in the collapse of stars with $M_{\rm prog} \gtrsim (9-11) M_\odot$ is broader, and can cover the range $[1.2 - 2.0] M_\odot$.

Observationally, a few NS are known with a mass around $2 M_\odot$. One is the millisecond pulsar PSR J1614-2230. This pulsar is a member of a binary system. This allows a precise determination of its mass, which turns out to be $M_{NS} = (1.97 \pm 0.04)\, M_\odot$.[31] The formation mechanism for systems such as this binary millisecond pulsar suggests that this NS can have accreted only $O(10^{-2})M_\odot$ from the companion. Thus, unless different mechanisms of mass transfer from the companion can be found, such a high value for a NS mass can only be explained by fallback at birth. A second example is given by PSR J0348+0432, a pulsar with a spin-down period of 39 ms and a mass $M_{NS} = (2.01 \pm 0.04)\, M_\odot$.[32] This pulsar is also a member of a binary system. Its companion is a very low-mass WD, with a mass $M = 0.172 \pm 0.003 M_\odot$, and the orbital period is only 2.46 hr. This system is therefore particularly interesting since it is very relativistic, and at the same time the large pulsar mass creates a strong gravitational field, so it can be used to test relativistic gravity in strong fields.

Numerical simulations suggest that, while for $M_{prog} \lesssim 15 M_\odot$ all core collapses produce NS, for heavier progenitors there is no obvious correlation between the mass of the progenitor and whether the star collapses to a NS, to a BH by fallback or to a BH by direct collapse. This is basically due to the fact that the stellar structure varies non-monotonically with the mass, and even small changes in the progenitor mass can give rise to important differences in the explosion. Failed explosions with BH formation have been obtained for progenitors below $20M_\odot$, while successful SNe with NS formation are found also between 20 and 40 $M_\odot$.[33] However, independently of whether the remnant is a BH or a NS, the remnant mass is an almost monotonic function of the mass enclosed below the O–Si burning shell.

(e) $(30 - 40)M_\odot \lesssim M_{prog} \lesssim 95 M_\odot$. In this regime the main theoretical uncertainty comes from mass losses of the progenitors due to stellar winds. Mass losses increase with the initial progenitor mass. For $M_{prog} \gtrsim (30 - 40)M_\odot$ the whole H envelope can be lost and the He core is uncovered before the star collapses, leading to a Wolf–Rayet star. In the WR stage the mass losses due to stellar winds are particularly strong, and their theoretical modelization is quite uncertain. For sufficiently large mass loss rate the star gives rise to a NS remnant, otherwise the star collapses to a BH.

(f) $95M_\odot \lesssim M_{prog} \lesssim 130 M_\odot$: *pulsational pair-instability SNe.* For these extremely massive stars, and for low metallicities, a mechanism different from iron core collapse can set in, and can lead to a "pair-instability supernova", which is an extremely luminous type of SN explosion. The instability arises because, after burning He and C, if the resulting O core is heavier than $(40 - 50)M_\odot$, one reaches a central temperature T_c of order 10^9 K at relatively low density, and in these conditions it is possible to efficiently create electron–positron pairs from high-energy photons before the ignition of O. Part of the thermal energy of the star is therefore drained and transformed into the rest mass of the $e^\pm$ pairs. As a result the pressure drops and rapid contraction takes place. This leads

[31] See Demorest, Pennucci, Ransom, Roberts and Hessels (2010). The measure is based on the observation of Shapiro time delay from the pulsar. As we see from eq. (6.95) of Vol. 1, the range r and shape s of the Shapiro time delay determine the mass of the companion and the inclination of the orbit, respectively. Furthermore, for a spectroscopic binary the observation of the Doppler shift of the lines of the companion (which in this case turns out to be a WD with $M = (0.500 \pm 0.006) M_\odot$) fixes the Keplerian mass function, i.e. a combination of the masses of the two stars. Having fixed the companion mass with Shapiro time delay, one can then obtain the pulsar mass.

[32] See Antoniadis *et al.* (2013).

[33] See Ugliano, Janka, Marek and Arcones (2012) and Janka (2012).

to a temperature increase and, since the rate of pair production increases as T_c^9, we get a runaway collapse followed by a thermonuclear explosion due to explosive ignition of O. This can eject several solar masses of material but, for stars that at birth had a mass in the approximate range $(95 - 130)M_\odot$, it does not unbind the star. The residual core contracts and, on a time-scale ranging from a few days to a few decades, it starts burning its nuclear fuel again. For progenitor masses in the range $(95 - 130)M_\odot$, at the time of death the He core has a mass in the range $(40 - 60)M_\odot$, if it has not lost much mass to stellar winds. If, after this first pulse, the He core mass remains above about $40M_\odot$, the star encounters the pair instability again. In this way the star can go through a series of pulses, and one has a "pulsational pair-instability SN". Later pulses are more energetic than the first one, since the first pulse lost part of its energy in expelling the H envelope (and possibly part of the He envelope). Then, the ejecta of the second and later pulses catch up with the first shell. In the resulting shock all the kinetic energy is dissipated in radiation, since the collision takes place at large radii, where adiabatic losses due to the expansion are negligible. This should be compared with the classic iron-core collapse where, as we saw, only about 1% of the kinetic energy is transformed into radiation. As a result, one can estimate that $\sim 10^{50}$ erg can be radiated in light, about an order of magnitude more than in a typical SN explosion. It is possible that this mechanism explains superluminous SN events.

In this way a star can go through a series of SN explosions, until the core starts burning Si to Fe, and in the end the core will undergo the standard iron-core collapse, and the remnant will typically be a BH. Observe that the pair-instability mechanism hinges on the possibility of a He core with mass over $40M_\odot$. Most probably this cannot be achieved for stars of solar metallicity, because of mass loss to stellar winds. Thus, pair-instability SNe are most probably restricted to low-metallicity stars.

(g) $130M_\odot \lesssim M_{\mathrm{prog}} \lesssim 260M_\odot$: *pair-instability SNe.* Stars with low metallicity, and particularly Pop III stars, can have masses in this range, since for low Z these large masses can be preserved against stellar winds. For progenitors masses approximately in the range $(130 - 260)M_\odot$ (corresponding to He cores between 60 and $137M_\odot$) the thermonuclear explosion that follows the pair-instability collapse is sufficiently energetic to completely disrupt the star, leaving no remnant behind. This gives rise to events with explosion energies in excess of 10^{53} erg, and the production of more than $50M_\odot$ of ^{56}Ni, possibly the biggest stellar explosions in the Universe. These events have sometimes been referred to as hypernovae, although nowadays they are more commonly called pair-instability (or pair-capture) SNe. These events also emit a huge energy in neutrinos, of order $(0.01 - 0.03)Mc^2 \sim 10^{55}$ erg. A number of ultra-bright SNe explosions (such as SN 2006gy, whose light curve is shown in Fig. 10.4) have been suggested as candidates for pair-instability SNe, although other explanations might be possible for their extreme luminosities, such as the interaction of ejected material with a dense circum-

stellar medium, or the extraction of extra energy from the spin-down of a highly magnetized proto-NS. In our Galaxy, the star η-Carinae is a potential candidate for a future pair-instability SN.

(h) $M_{\rm prog} \gtrsim 260 M_\odot$. Finally, for even heavier stars the nuclear burning of the core is unable to halt the collapse induced by the pair instability and reverse it with a thermonuclear explosion. Thus the star collapses directly to a BH, with a mass comparable to that of the progenitor star. If progenitors with these masses were an important component of Pop III stars, they will have produced a population of BHs with masses of several hundred solar masses.[34]

[34]Pair-instability SNe also have peculiar nucleosynthesis signatures, which should be reflected in the chemical composition of some very old low-mass stars. At the moment there are not many observations of very old low-mass stars bearing such signatures. See, however, Aoki, Tominaga, Beers, Honda and Lee (2014) for one possible case.

10.4   GW production by self-gravitating fluids

In this section we will first of all set up the relevant equations for the generation of GWs by a self-gravitating fluid, adapting to this case the general formalism for GW production, developed in particular in Chapters 3 and 5 of Vol. 1. Although we are developing it here in the context of GWs from stellar collapse, this formalism is of more general interest.

10.4.1   Energy–momentum tensor of a perfect fluid

In general relativity (GR) a perfect fluid is defined by the condition that, at every point in space-time, there is a locally inertial frame (i.e. a frame where, in a sufficiently small region around the space-time point P considered, the metric is flat), comoving with the fluid element (so that in this frame the fluid velocity vanishes at the point P), where the fluid is isotropic. In this frame the energy–momentum tensor of the fluid at the point P must therefore satisfy $T^{0i} = 0$ and $T^{ij} \propto \delta^{ij}$, i.e.

$$T^{00} = \rho c^2\,, \qquad T^{0i} = 0\,, \qquad T^{ij} = p\delta^{ij}\,, \tag{10.28}$$

so (in this frame) the energy density of the fluid is ρc^2 and p is the pressure. To find the form of $T^{\mu\nu}$ in a frame that is still locally inertial, but not necessarily comoving with the fluid, we introduce the four-velocity field u^μ at the space-time point P, defined by requiring that (1) in the comoving locally inertial frame of the fluid u^μ takes the value $u^0 = c, u^i = 0$; (2) it transforms as a four-vector under Lorentz transformations of the locally inertial frame constructed at P.

Observe that the normalization condition on u^μ can be written in a Lorentz-invariant form as

$$\eta_{\mu\nu} u^\mu u^\nu = -c^2\,. \tag{10.29}$$

In terms of this four-vector the rest-frame expression for the energy–momentum tensor (10.28) can be written compactly as

$$T^{\mu\nu} = \left(\rho + \frac{p}{c^2}\right) u^\mu u^\nu + p\eta^{\mu\nu}\,. \tag{10.30}$$

[35] Note, however, that the physical energy density is $-T^0_0$, i.e. it is the (00) component of a tensor with one upper and one lower index (this definition will carry over also to curved space; see Section 18.3). The Lorentz scalar $\rho(x)c^2$ coincides with the physical energy density $-T^0_0(x)$ at a point x only in a frame where $u^\mu(x) = (c,0,0,0)$. Otherwise, in a generic frame (but still in flat space),

$$-T^0_0 = \rho(u^0)^2 + p[(u^0)^2/c^2 - 1].$$
$$(10.31)$$

If we now define ρ and p as the *Lorentz-invariant* quantities defined by their values in the comoving locally inertial frame, the right-hand side of eq. (10.30) is a Lorentz tensor, and therefore eqs. (10.29) and (10.30) give the generalization of the energy–momentum tensor of a perfect fluid for a generic locally inertial frame (so in particular in flat space).[35]

Of course, u^μ is a function of the space-time point P used in the construction, so it is actually a four-vector field $u^\mu(t, \mathbf{x})$. At the level of the particles making up the fluid, u^μ corresponds to the average, over many particles in a small volume element, of the particle four-velocities, defined as the derivative of the particle trajectory $x^\mu_0(\tau)$ with respect to proper time τ, rather than with respect to coordinate time. In fact, for a particle on a trajectory $x^\mu_0(\tau)$, proper time τ is defined by

$$c^2 d\tau^2 = -\eta_{\mu\nu} dx^\mu_0(\tau) dx^\nu_0(\tau),$$
$$(10.32)$$

and therefore

$$\eta_{\mu\nu} \frac{dx^\mu_0}{d\tau} \frac{dx^\nu_0}{d\tau} = -c^2.$$
$$(10.33)$$

Comparison with eq. (10.29) shows that u^μ is the average of $dx^\mu_0/d\tau$. The coordinate velocity of the particles is rather $v^i \equiv dx^i_0/dt$. Writing $v^i = (dx^i_0/d\tau)(d\tau/dt)$ and observing that u^0 is the average of $cdt/d\tau$, for the corresponding velocity fields we have

$$v^i(t, \mathbf{x}) = \frac{cu^i(t, \mathbf{x})}{u^0(t, \mathbf{x})}.$$
$$(10.34)$$

Using

$$\frac{dt}{d\tau} = \left(1 - \frac{v^2}{c^2}\right)^{-1/2},$$
$$(10.35)$$

we have

$$u^0(t, \mathbf{x}) = c\left[1 - \frac{v^2(t, \mathbf{x})}{c^2}\right]^{-1/2},$$
$$(10.36)$$

and, from eq. (10.34),

$$u^i(t, \mathbf{x}) = \left[1 - \frac{v^2(t, \mathbf{x})}{c^2}\right]^{-1/2} v^i(t, \mathbf{x}).$$
$$(10.37)$$

Thus, in terms of v^i, we can rewrite eq. (10.30) as

$$T^{00} = \frac{1}{1 - (v^2/c^2)}\left(\rho c^2 + p\frac{v^2}{c^2}\right),$$
$$(10.38)$$

$$T^{0i} = \frac{\rho c^2 + p}{1 - (v^2/c^2)}\frac{v^i}{c},$$
$$(10.39)$$

$$T^{ij} = p\delta^{ij} + \frac{(\rho c^2 + p)}{1 - (v^2/c^2)}\frac{v^i v^j}{c^2}.$$
$$(10.40)$$

These expressions could also have been obtained directly from eq. (10.28) by performing a Lorentz boost with velocity v^i.

For non-relativistic matter, the pressure p is $O(\rho v^2)$ and is therefore suppressed with respect to ρc^2 by a factor $O(v^2/c^2)$, where v^2 is the average of the square velocity of the particles in the fluid at the point under consideration. Then, to leading order in the fluid velocity field v/c, eqs. (10.38)–(10.40) reduce to[36]

$$T^{00} = \rho c^2 \left[1 + O(v^2/c^2)\right] , \qquad (10.41)$$

$$T^{0i} = \rho c v^i \left[1 + O(v^2/c^2)\right] , \qquad (10.42)$$

$$T^{ij} = (p\delta^{ij} + \rho v^i v^j) \left[1 + O(v^2/c^2)\right] . \qquad (10.43)$$

In curved space the generalization of eqs. (10.29) and (10.30) to a generic frame, not necessarily locally inertial, is simply obtained by constructing an object that transforms as a tensor under general coordinate transformations and that, in a locally inertial frame where $g_{\mu\nu} = \eta_{\mu\nu}$, reduces to its flat-space counterpart. This is obtained by writing

$$\boxed{T^{\mu\nu} = \left(\rho + \frac{p}{c^2}\right) u^\mu u^\nu + p g^{\mu\nu} ,} \qquad (10.44)$$

with ρ and p now defined as the *diffeomorphism-invariant* quantities equal to their values in the locally inertial frame. Similarly, u^μ is now defined as the four-vector under general coordinate transformation that, in the comoving locally inertial frame, takes the value ($u^0 = c, u^i = 0$). Thus, it now satisfies

$$g_{\mu\nu} u^\mu u^\nu = -c^2 . \qquad (10.45)$$

Equations (10.44) and (10.45) provide the general expression for the energy–momentum tensor of an ideal fluid in curved space.

If the fluid is not ideal, in the comoving locally inertial frame it will still satisfy $T^{0i} = 0$, since this condition can be taken as the definition of a comoving frame, i.e. of being at rest with respect to the fluid motion, at point P. However, if the fluid is not isotropic, T^{ij} will no longer be proportional to δ^{ij}. Thus, in the comoving locally inertial frame we now have

$$T^{00} = \rho c^2 , \qquad T^{0i} = 0 , \qquad T^{ij} = p\delta^{ij} + \Sigma^{ij} , \qquad (10.46)$$

where Σ^{ij} is a symmetric and traceless tensor (since any trace part can be reabsorbed into the definition of p). The tensor Σ^{ij} is called the anisotropic stress tensor, and describes the effect of viscosity. In a frame that is still locally inertial but not necessarily comoving, eq. (10.30) generalizes to

$$T^{\mu\nu} = \left(\rho + \frac{p}{c^2}\right) u^\mu u^\nu + p\eta^{\mu\nu} + \Sigma^{\mu\nu} , \qquad (10.47)$$

where $\Sigma^{\mu\nu}$ is defined as the Lorentz tensor that, in the comoving frame, has $\Sigma^{00} = \Sigma^{0i} = 0$, while its space-space components have the values Σ^{ij} given in eq. (10.46). Similarly eq. (10.44) is generalized by adding to the right-hand side the quantity $\Sigma^{\mu\nu}$, now defined as the tensor under general coordinate transformations that reduces to Σ^{ij} in the locally inertial comoving frame, so that[37]

$$T^{\mu\nu} = \left(\rho + \frac{p}{c^2}\right) u^\mu u^\nu + p g^{\mu\nu} + \Sigma^{\mu\nu} . \qquad (10.48)$$

[36]Note that the condition $p = O(\rho v^2)$ valid for non-relativistic matter implies that, as far as the powers of v are concerned, the terms $p\delta^{ij}$ and $\rho v^i v^j$ in eq. (10.43) are of the same order. In the rest frame of the fluid at the point P, v^i vanishes, while p is proportional to the average of the square velocity, which is a scalar, so in the rest frame at P only $p\delta^{ij}$ survives, giving eq. (10.28), while at a generic point such that $\langle v_i \rangle^2 \sim \langle v^2 \rangle$ the two terms are of the same order. However, it is clear that the term $p\delta^{ij}$ cannot contribute to GW production, since it is isotropic, and its contribution to $h_{ij}^{\rm TT}$ vanishes when we perform the projection onto the transverse-traceless part, as we will indeed check explicitly in eq. (10.76). Therefore, the leading contribution of T^{ij} to GW production comes from the term $\rho v^i v^j$.

[37]The energy–momentum tensor for a fluid in curved space will be further discussed in Part IV. Let us anticipate, from Section 18.3, that in curved space the physical energy density will be given by $-T^0_0$ (which, in a generic metric, is of course not the same as T^{00}). Let us also mention that, in a cosmological setting, for a generic fluid in curved space we will rather use a different convention, and we will define ρ from $-T^0_0 = \rho c^2$ in any frame (and not just in the locally inertial frame comoving with the fluid element), so that ρ will be the component of a tensor with an upper and a lower index, rather than a scalar; cf. Note 10 on page 374. As we will see in Section 18.3.2, in cosmological perturbation theory the difference between these two definitions of ρ only shows up at second order.

10.4.2 GW production from gravitating Newtonian fluids

In principle, once we have $T^{00}(t,\mathbf{x}) = \rho(t,\mathbf{x})c^2$, the emission of GWs to lowest order in v/c could be computed simply plugging it into the quadrupole formula (3.59) of Vol. 1. This however requires the computation of the second time derivative of the quadrupole moment. When $\rho(t,\mathbf{x})$ is obtained from numerical simulations, as for SN explosions, this requires replacing the derivative by finite differences, and introduces significant numerical noise, particularly at high frequencies. To make things worse, in the computation of the quadrupole moment the density ρ is weighted with a factor r^4 (a factor r^2 coming from $d^3x = r^2 dr d\Omega$ and a factor r^2 coming from the term $x^i x^j$ in the definition of the second mass moment). Thus, a large contribution to the quadrupole moment comes from the low-density outer layers of the star, where the low values of ρ is compensated by the large value of r^4. However, even if the contribution of the outer layers to the quadrupole moment is large, their contribution to its second time derivative is suppressed because the motion in the outer low-density layers is slower and less coherent, and in fact the main contribution to the GW signal comes from the inner region. Thus, when taking the second time derivative, the contribution of the external regions should undergo partial cancellations due to the less coherent nature of the matter flow (compare with Section 4.3.2 of Vol. 1, where we discussed GW radiation produced by coherent and by incoherent motions). The numerical error in the evaluation of the second time derivative can, however, spoil these delicate cancellations (even more considering that in the outer region the numerical grid is coarser) and introduces significant numerical artifacts.

A better procedure is to use the space–space components of the energy–momentum tensor as a source for h_{ij}^{TT}. Recall indeed from eq. (3.26) that, in the flat-space treatment of Chapter 3, h_{ij}^{TT} was originally given by a spatial integral of the space–space components T^{ij}, evaluated at retarded time. Performing a multipole expansion of this expression, we got eq. (3.34). Only upon repeated use of energy–momentum conservation was this expression transformed into a second time derivative of T^{00}. In our case, to avoid numerical problems with the time derivatives, it is clearly convenient not to perform these manipulations, and start directly from the expression of h_{ij}^{TT} in terms of the space–space components of the energy–momentum tensor.

However, a complication inherent to the use of the space–space components is that to compute h_{ij}^{TT} for a gravitating system we cannot just naively use the flat-space formulas of Chapter 3. In fact, from eqs. (10.41) and (10.43) we see that T^{ij} is suppressed with respect to T^{00} by a factor $O(v^2/c^2)$. To get correctly all terms of the same order we must therefore compute the (i,j) component of the source term to order v^2. To this order, the Newtonian potential U also contributes since, by the virial theorem, $U \sim v^2$. We have studied these issues in full generality in Chapter 5. However, in order to get the correct Newtonian result it is not necessary to go through the full formalism of Chapter 5. We can simply recall, from Section 5.2, that the Einstein equations can be recast in the exact form

$$\Box h^{\alpha\beta} = \frac{16\pi G}{c^4} \tau^{\alpha\beta}, \qquad (10.49)$$

where $\Box$ is the *flat-space* d'Alembertian, $h^{\alpha\beta} \equiv (-g)^{1/2} g^{\alpha\beta} - \eta^{\alpha\beta}$, and we have fixed the gauge $\partial_\beta h^{\alpha\beta} = 0$. The tensor $\tau^{\alpha\beta}$ is given by

$$\tau^{\alpha\beta} = (-g)(T^{\alpha\beta} + t_{LL}^{\alpha\beta}) + \partial_\mu \partial_\nu \chi^{\alpha\beta\mu\nu}, \qquad (10.50)$$

where $t_{LL}^{\alpha\beta}$ is the Landau–Lifshitz energy–momentum pseudotensor, which is quadratic in the derivatives of $h^{\alpha\beta}$ and is given in eq. (5.75). In the last term, $\chi^{\alpha\beta\mu\nu} = (c^4/16\pi G)(h^{\alpha\mu} h^{\beta\nu} - h^{\mu\nu} h^{\alpha\beta})$. As a consequence of $\partial_\beta h^{\alpha\beta} = 0$, $\tau^{\alpha\beta}$ satisfies the exact conservation equation

$$\partial_\mu \tau^{\mu\nu} = 0, \qquad (10.51)$$

with an ordinary rather than covariant derivative; see eq. (5.82).

We now compute the lowest-order term in the post-Newtonian expansion of $\tau^{\alpha\beta}$ in the near-source region. This can be done using the Newtonian form of the metric perturbation,

$$h_{00} = -2\phi, \quad h_{ij} = -2\phi \delta_{ij}, \quad h_{0i} = 0; \qquad (10.52)$$

see eq. (5.11). Writing $g_{\mu\nu} = \eta_{\mu\nu} + h_{\mu\nu}$, in the limit of small $h_{\mu\nu}$ we found in Section 5.2 that $h^{\alpha\beta} \simeq -h^{\alpha\beta} + \frac{1}{2}\eta^{\alpha\beta} h$. Using the Newtonian expression for $h_{\mu\nu}$ we find that the Newtonian expression for $h_{\mu\nu}$ is

$$h_{00} = 4\phi, \quad h_{0i} = h_{ij} = 0. \qquad (10.53)$$

We can now plug this expression for $h^{\alpha\beta}$ into the Landau–Lifshitz energy–momentum pseudotensor given in eq. (5.75). To lowest order, the factors $g_{\mu\nu}$ in terms such as $g_{\lambda\mu} g^{\nu\rho} \partial_\nu h^{\alpha\lambda} \partial_\rho h^{\beta\mu}$ can simply be replaced by $\eta_{\mu\nu}$, and similarly the factors $-g$ can be replaced by 1. Furthermore, we recall that in the near region a time derivative brings one more power of v/c than spatial derivatives, $\partial_0 \sim (v/c)\partial_i$. Therefore, at leading order we can neglect all terms involving ∂_0 and keep only spatial derivatives. We express the result in terms of the Newtonian potential U (defined so that $U > 0$), related to ϕ by $U = -\phi c^2$. Observe that U satisfies the Poisson equation

$$\nabla^2 U = -4\pi G\rho; \qquad (10.54)$$

see eqs. (5.17) and (5.21), with $^{(0)}T^{00} = \rho c^2$. Then, using eq. (10.53), a simple calculation gives $\chi^{\alpha\beta\mu\nu} = 0$ and

$$t_{LL}^{ij} = \frac{1}{4\pi G}\left(\partial^i U \partial^j U - \frac{1}{2}\delta^{ij}\partial_k U \partial^k U\right). \qquad (10.55)$$

Note that t_{LL}^{ij} is of the same order as the space-space components T^{ij} of the fluid energy–momentum tensor.[38] In fact, if L is the typical length-

[38] Observe that $t_{LL}^{ij} = -t^{ij}$, where t^{ij} is the tensor defined in eq. (5.224). In terms of t_{LL}^{ij}, eq. (5.223) reads

$$m_a \frac{d^2 z_a^i}{dt^2} = -\int_{S_a} dS_j \, t_{LL}^{ij}.$$

scale of variation of U, we have

$$\frac{1}{G}\partial_i U\partial_j U \sim \frac{U}{GL^2}U\,. \tag{10.56}$$

Since $\boldsymbol{\nabla}^2 U = -4\pi G\rho$, we have $U/(GL^2) \sim \rho$. Furthermore, because of the virial theorem, $U \sim v^2$, and therefore

$$\frac{1}{G}\partial_i U\partial_j U \sim \rho v^2\,. \tag{10.57}$$

We can similarly verify that also t_{LL}^{00} and t_{LL}^{0i} are of order ρv^2. However, since T^{00} and T^{0i} are of order ρ and ρcv, respectively, the contribution of the Landau–Lifshitz pseudotensor to τ^{00} and τ^{0i} is subleading. Thus, to leading order,

$$\tau^{00} = \rho c^2\,, \tag{10.58}$$
$$\tau^{0i} = \rho cv^i\,, \tag{10.59}$$
$$\tau^{ij} = p\delta^{ij} + \rho v^i v^j + \frac{1}{4\pi G}\left(\partial^i U\partial^j U - \frac{1}{2}\delta^{ij}\partial_k U\partial^k U\right)\,. \tag{10.60}$$

The result agrees with what we already found in eqs. (5.111)–(5.113), where we also gave the result up to 1PN order for τ^{00} and τ^{0i}.[39]

[39] Observe that V in eqs. (5.111)–(5.113) is the potential given by the retarded integral (5.39). However, to lowest order in v/c it reduces to U, as we can see observing that to lowest order $\sigma = (1/c^2)(T^{00} + T^{ii}) \simeq T^{00}/c^2$, and using the expansion (5.40) in the retarded integral.

To understand the physical meaning of the U-dependent term in τ^{ij} we can apply the conservation equations (10.51) to the Newtonian expression for τ^{ij} given in eqs. (10.58)–(10.60). The equation $\partial_\mu \tau^{\mu\nu} = 0$ with $\nu = 0$ gives the continuity equation

$$\partial_t \rho + \partial_i(\rho v^i) = 0\,, \tag{10.61}$$

while the equation with $\nu = i$ gives the Euler equation in a gravitational potential,

$$\rho(\partial_t v^i + v^j \partial_j v^i) = -\partial^i p + \rho\partial^i U\,, \tag{10.62}$$

where we have used eq. (10.54). Thus, we see that the term in τ^{ij} that depends on U correctly gives the Newtonian force in the Euler equation. The quadrupole radiation can now be computed from

$$\Box \mathsf{h}^{ij} = \frac{16\pi G}{c^4}\tau^{ij}\,, \tag{10.63}$$

The problem is now formally identical to that studied in Section 3.1 of Vol. 1, with $\bar{h}_{\mu\nu}$ replaced by $\mathsf{h}_{\mu\nu}$ (which for $h_{\mu\nu} \ll 1$ is the same as replacing $h_{\mu\nu}$ by $-h_{\mu\nu}$, since in the weak-field limit $\mathsf{h}_{\mu\nu} \simeq -\bar{h}_{\mu\nu}$) and $T_{\mu\nu}$ replaced by $-\tau_{\mu\nu}$ [because of the different relative sign between the right-hand sides of eqs. (10.63) and (3.3), which indeed reflects the fact that $\mathsf{h}_{\mu\nu}$ reduces to $-\bar{h}_{\mu\nu}$ in the weak-field limit]. Therefore eq. (3.34) now becomes, for the term corresponding to quadrupole radiation,

$$h_{ij}^{\rm TT}(t,\mathbf{x}) = \frac{1}{r}\frac{4G}{c^4}\Lambda_{ij,kl}(\hat{\mathbf{n}})\int d^3x'\,\tau^{kl}(t - r/c, \mathbf{x}')\,, \tag{10.64}$$

where, as usual, $r = |\mathbf{x}|$, $\hat{\mathbf{n}} = \mathbf{x}/r$ is the unit vector in the direction (θ, ϕ), and the Λ tensor is the projector that extracts the TT part, introduced in Section 1.2 of Vol. 1. This, together with eq. (10.60), gives the GW amplitude as a function of the values of ρ, p, v^i and U on a fixed time-slice. The potential U, in turn, can be obtained from ρ by integrating the Poisson equation, which gives

$$U(t, \mathbf{x}) = G \int d^3x' \, \frac{\rho(t, \mathbf{x}')}{|\mathbf{x} - \mathbf{x}'|} \,. \tag{10.65}$$

While the problem of dealing with time derivatives has been eliminated, eq. (10.64) still has the drawback that, even if the source ρ has a spatially compact support, $\tau^{kl}(t_{\rm ret}, \mathbf{x})$ does not because $U(t, \mathbf{x})$ is non-vanishing even outside the source, as we see from eq. (10.65). This can be problematic in numerical work, since the evolution of the system is only studied, for computational feasibility, in a small region around the source, and then one should reconstruct $U(t, \mathbf{x})$ even outside the region covered by the numerical grid. While this can in principle be done using the known multipole moments of the source, in practice it is more convenient to use an alternative approach, that further reduces the computation of the waveform to an integral at fixed time over a compact spatial support. The strategy is to start from the usual quadrupole formula (3.55),

$$\left[h_{ij}^{\rm TT}(t, r, \theta, \phi) \right]_{\rm quad} = \frac{1}{r} \frac{2G}{c^4} \Lambda_{ij,kl}(\hat{\mathbf{n}}) \ddot{M}^{kl}(t - r/c) \,, \tag{10.66}$$

where

$$M^{kl}(t) = \frac{1}{c^2} \int d^3x \, T^{00}(t, \mathbf{x}) \, x^k x^l \,. \tag{10.67}$$

We then observe that in this expression, to lowest order in v/c, we are free to replace T^{00} by τ^{00}, since $\tau^{00} = T^{00}[1 + O(v^2/c^2)]$,

$$M^{kl}(t) = \frac{1}{c^2} \int d^3x \, \tau^{00}(t, \mathbf{x}) \, x^k x^l \,. \tag{10.68}$$

We can now use the fact that $\tau^{\mu\nu}$ satisfies the conservation equation (10.51). For $\nu = 0$ we therefore have

$$\partial_0 \tau^{00} = -\partial_i \tau^{0i} \,, \tag{10.69}$$

while for $\nu = k$

$$\partial_0 \tau^{0k} = -\partial_i \tau^{ik} \,. \tag{10.70}$$

Therefore (recalling that the $\dot{M} = dM/dt = c\partial_0 M$)

$$\begin{aligned} \dot{M}^{kl} &= \frac{1}{c} \int d^3x \, \partial_0 \tau^{00}(t, \mathbf{x}) \, x^k x^l \\ &= -\frac{1}{c} \int d^3x \, \partial_i \tau^{0i}(t, \mathbf{x}) \, x^k x^l \\ &= \frac{1}{c} \int d^3x \, [\tau^{0k}(t, \mathbf{x}) x^l + \tau^{0l}(t, \mathbf{x}) x^k] \,, \end{aligned} \tag{10.71}$$

where in the last line we have integrated by parts. Similarly

$$\begin{aligned} \ddot{M}^{kl} &= \int d^3x \, [(\partial_0 \tau^{0k}) x^l + (\partial_0 \tau^{0l}) x^k] \\ &= -\int d^3x \, [(\partial_i \tau^{ik}) x^l + (\partial_i \tau^{il}) x^k] \,. \end{aligned} \tag{10.72}$$

We now split $\tau^{ik} = T^{ik} + t^{ik}_{LL}$. In the term involving T^{ik} we integrate again by parts, obtaining

$$-\int d^3x\,[(\partial_i T^{ik})x^l + (\partial_i T^{il})x^k] = 2\int d^3x\,T^{kl}\,. \tag{10.73}$$

In contrast, in the term involving t^{ik}_{LL} we compute explicitly the spatial derivative using eq. (10.55),

$$\partial_i t^{ik}_{LL} = \frac{1}{4\pi G}\left[(\boldsymbol{\nabla}^2 U)\partial^k U - \frac{1}{2}\partial^k(\partial_i U \partial^i U)\right]$$

$$= -\rho\partial^k U - \frac{1}{8\pi G}\partial^k(\partial_i U \partial^i U)\,, \tag{10.74}$$

where in the second line we have used the Poisson equation for U. The first term in eq. (10.74) is proportional to ρ and therefore has compact support. The second, once inserted into eq. (10.72) and integrated by parts, gives a contribution proportional to δ^{kl}. Thus

$$\ddot{M}^{kl} = \int d^3x\left[2(\rho v^k v^l + p\delta^{kl}) + \rho\,(x^k\partial^l + x^l\partial^k)U - \frac{1}{4\pi G}\delta^{kl}\partial_i U \partial^i U\right]. \tag{10.75}$$

In this expression, only the last term does not have compact support. However, when we multiply by $\Lambda_{ij,kl}$ to extract the transverse-traceless part, the terms proportional to δ^{kl} give a vanishing contribution. The GW radiation in the quadrupole approximation is then given by

$$\boxed{h^{\mathrm{TT}}_{ij} = \frac{1}{r}\frac{4G}{c^4}\Lambda_{ij,kl}\int d^3x\,\rho\,(v^k v^l + x^k\partial^l U)\,,} \tag{10.76}$$

where the right-hand side is evaluated at the retarded time $t - r/c$. We have therefore succeeded in writing the result in terms of a integrand with compact support. One can similarly compute the third time derivative of M^{kl}, which determines the radiated energy.

10.4.3   Quadrupole radiation from axisymmetric sources

If the core is both collapsing and rotating, the quadrupole moment of its mass distribution has a non-vanishing second time derivative even under the assumption of axial symmetry, as will become clear below. To exploit the axial symmetry it is convenient to expand the amplitude in spherical components, as in Section 3.5.2. Recall from eq. (3.275) that, in general, the GW amplitude in the wave zone can be written as

$$h^{\mathrm{TT}}_{ij} = \frac{1}{r}\sum_{l=2}^{\infty}\sum_{m=-l}^{l}\left[A^{E2}_{lm}(t - r/c)(\mathbf{T}^{E2}_{lm})_{ij} + A^{B2}_{lm}(t - r/c)(\mathbf{T}^{B2}_{lm})_{ij}\right], \tag{10.77}$$

where the tensors $\mathbf{T}^{E2}_{lm}(\theta,\phi)$ and $\mathbf{T}^{B2}_{lm}(\theta,\phi)$ are the pure-spin tensor spherical harmonics discussed in Section 3.5.2.[40] As we saw in eqs. (3.291)

[40]To conform to standard notation in the literature on GWs from SNe, we denote here by A^{E2}_{lm} and A^{B2}_{lm} the amplitudes denoted in eq. (3.275) by $(G/c^4)u_{lm}$ and $(G/c^4)v_{lm}$, respectively.

and (3.293), A_{lm}^{E2} is given by the l-th time derivative of the mass multipoles while A_{lm}^{B2} is given by the l-th time derivative of the current multipoles. Thus, if we restrict to the contribution of the mass quadrupole, the only non-vanishing contribution comes from A_{lm}^{E2} with $l = 2$, while all $A_{lm}^{B2} = 0$. Furthermore, if we assume axial symmetry, there is no dependence on ϕ and only the $m = 0$ mode contributes. Therefore in these approximations eq. (10.77) becomes

$$h_{ij}^{\rm TT}(t, r, \theta) = \frac{1}{r} A_{20}^{E2}(t - r/c)(\mathbf{T}_{20}^{E2})_{ij}(\theta). \tag{10.78}$$

The tensor $(\mathbf{T}_{lm}^{E2})_{ij}$ was defined in eq. (3.271), so in particular

$$(\mathbf{T}_{20}^{E2})_{ij}(\theta) = \frac{1}{8}\sqrt{\frac{15}{\pi}} r^2 \Lambda_{ij,kl}(\hat{\mathbf{n}})\partial_k\partial_l(z^2/r^2). \tag{10.79}$$

Since $\mathbf{T}_{20}^{E2}$ is independent of ϕ, without loss of generality we can compute it choosing the axes in the (x, y) plane so that $\phi = 0$, i.e.

$$\hat{\mathbf{n}} = (0, \sin\theta, \cos\theta) = (1/r)(0, y, z). \tag{10.80}$$

Then, using the explicit expression of $\Lambda_{ij,kl}(\hat{\mathbf{n}})$ given in eq. (1.39) and evaluating $\partial_k\partial_l(z^2/r^2)$, we get[41]

$$(\mathbf{T}_{20}^{E2})_{ij}(\theta) = -\frac{1}{8}\sqrt{\frac{15}{\pi}}\sin^2\theta \begin{pmatrix} 1 & 0 & 0 \\ 0 & -\cos^2\theta & \sin\theta\cos\theta \\ 0 & \sin\theta\cos\theta & -\sin^2\theta \end{pmatrix}_{ij}, \tag{10.81}$$

which is symmetric and traceless, as it should be. This gives the tensor structure of h_{ij} in the frame where the z axis is the symmetry axis of the source. In order to identify the two polarizations of the GW we must rotate to the frame where the z axis coincides with the propagation direction. This can be performed using the rotation matrix

$$\mathcal{R} = \begin{pmatrix} 1 & 0 & 0 \\ 0 & \cos\theta & -\sin\theta \\ 0 & \sin\theta & \cos\theta \end{pmatrix}. \tag{10.82}$$

Applying this rotation to the vector $\hat{\mathbf{n}} = (0, \sin\theta, \cos\theta)$ we get $n_i' = \mathcal{R}_{ij}n_j = (0, 0, 1)$, so this rotation indeed brings the propagation direction onto the z axis. Under this rotation a generic tensor T_{ij} becomes $T_{ij}' = \mathcal{R}_{ik}\mathcal{R}_{jl}T_{kl} = (\mathcal{R}T\mathcal{R}^T)_{ij}$. We then find that in this new frame

$$(\mathbf{T}_{20}^{E2})_{ij} = -\frac{1}{8}\sqrt{\frac{15}{\pi}}\sin^2\theta \begin{pmatrix} 1 & 0 & 0 \\ 0 & -1 & 0 \\ 0 & 0 & 0 \end{pmatrix}_{ij}, \tag{10.83}$$

and therefore

$$h_{ij}^{\rm TT}(t, r, \theta) = -\frac{1}{8}\sqrt{\frac{15}{\pi}}\frac{1}{r}A_{20}^{E2}(t - r/c)\sin^2\theta \begin{pmatrix} 1 & 0 & 0 \\ 0 & -1 & 0 \\ 0 & 0 & 0 \end{pmatrix}_{ij}. \tag{10.84}$$

[41] The actual computation is much less cumbersome than what one might fear by looking at eq. (1.39). For instance,

$$r^2 n_k n_l \partial_k \partial_l (z^2/r^2)$$
$$= (y^2\partial_y^2 + 2yz\partial_y\partial_z + z^2\partial_z^2)(z^2/r^2)$$
$$= r^{-6}2x^2z^2(x^2 - 3y^2 - 3z^2),$$

and this vanishes because we have chosen $\hat{\mathbf{n}}$ so that $x = 0$. So the terms $n_k n_l \delta_{ij}$ and $n_i n_j n_k n_l$ in eq. (1.39) do not contribute. Defining $t_{ij} = \Lambda_{ij,kl}(\hat{\mathbf{n}})\partial_k\partial_l(z^2/r^2)$, we have for instance

$$t_{xx} = [\partial_x^2 - (1/2)\boldsymbol{\nabla}^2](z^2/r^2)|_{x=0}$$
$$= -y^2/r^4 = -(1/r^2)\sin^2\theta,$$

and similarly for the other components.

[42]Observe that eq. (10.84) gives h_{ij}^{TT}
with respect to the system of axes
$(\hat{u}, \hat{v}, \hat{n})$ already shown in Fig. 3.2, in
Vol. 1. From the point of view of the
observer, for which $\hat{n}$ is an incoming
rather than an outgoing vector, the polarizations are rather read with respect
to a system of axes where $(\hat{u}, \hat{v})$ are rotated clockwise by $\pi/2$, which results in
an overall minus sign in the polarization, and therefore reabsorbs the minus
sign in eq. (10.84). Compare also with
Note 27 on page 208 of Vol. 1.

This shows that the waveform generated by an axisymmetric source in
the quadrupole approximation has purely the + polarization, i.e.[42]

$$h_\times = 0, \tag{10.85}$$

while

$$h_+(t, r, \theta) = \frac{1}{8} \sqrt{\frac{15}{\pi}} \frac{A_{20}^{E2}(t - r/c)}{r} \sin^2 \theta. \tag{10.86}$$

Finally, we must compute $A_{20}^{E2}(t)$ in terms of the energy–momentum
tensor of the source. Using eqs. (3.291) and (3.296), as well as the explicit
form for the spherical harmonic $Y_{20}(\theta) = (5/16\pi)^{1/2}(3\cos^2\theta - 1)$, gives

$$A_{20}^{E2}(t) = \frac{G}{c^4} \frac{d^2 I_{20}}{dt^2}, \tag{10.87}$$

with

$$I_{20}(t) = \frac{8\pi^{3/2}}{\sqrt{15}} \int_{-1}^{1} d\cos\theta \, (3\cos^2\theta - 1) \int_0^\infty dr \, r^4 \, \rho(t, r, \theta), \tag{10.88}$$

and $\rho = (1/c^2)T^{00}$. Just as discussed in the previous section, this formula has the disadvantage that it involves a second time derivative,
which is not convenient for numerical work. We can, however, proceed
as in eqs. (10.66)–(10.76), replacing ρc^2 by T_{00} and computing the second time derivative by using $\partial_\mu T^{\mu\nu} = 0$ twice. Rewriting $I_{20}(t)$ in the
form

$$I_{20}(t) = \frac{4\sqrt{\pi}}{c^2\sqrt{15}} \int d^3x \, \tau^{00}(t, \mathbf{x})(3z^2 - r^2), \tag{10.89}$$

we get

$$\ddot{I}_{20}(t) = \frac{4\sqrt{\pi}}{\sqrt{15}} \int d^3x \, \left[T^{ij}\partial_i\partial_j(3z^2 - r^2) - (\partial_j t_{LL}^{ij})\partial_i(3z^2 - r^2) \right]. \tag{10.90}$$

Since $\nabla^2(3z^2 - r^2) = 0$, the term in T^{ij} proportional to $p\delta^{ij}$ drops, and
similarly, after an integration by parts, for the term $\partial^i(\partial_j U \partial^j U)$ from
$\partial_j t_{LL}^{ij}$, see eq. (10.74). In the computation with the Cartesian rather
than spherical components of the tensor that lead to eq. (10.76) this
corresponds to the fact that the contraction of δ^{kl} with $\Lambda_{ij,kl}$ gives zero.
We are then left with

$$\ddot{I}_{20}(t) = \frac{4\sqrt{\pi}}{\sqrt{15}} \int d^3x \, \rho(t, r, \theta) \left[v^i v^j \partial_i \partial_j (3z^2 - r^2) + \partial^i U \partial_i (3z^2 - r^2) \right]. \tag{10.91}$$

The contraction $v^i v^j \partial_i \partial_j (3z^2 - r^2)$ is equal to $2(2v_z^2 - v_x^2 - v_y^2)$. We
can express it terms of the polar components of the velocity, writing
$v_x = v_r \sin\theta + v_\theta \cos\theta$, $v_z = v_r \cos\theta - v_\theta \sin\theta$ and $v_\phi = -v_y$ (without loss
of generality, we have considered the velocity vector of a fluid element

located at a position with polar coordinates $(r, \theta, \phi = 0)$. Then we get

$$
\begin{aligned}
\ddot{I}_{20}(t) = \frac{16\pi\sqrt{\pi}}{\sqrt{15}} \int_{-1}^{1} &d\cos\theta \int_{0}^{\infty} r^2 dr\, \rho(t, r, \theta)\, \left[v_r^2 (3\cos^2\theta - 1) \right. \\
&+ v_\theta^2 (2 - 3\cos^2\theta) - 6 v_r v_\theta \sin\theta\cos\theta - v_\phi^2 \\
&\left. + (3\cos^2\theta - 1)\, r\partial_r U - 3\sin\theta\cos\theta\, \partial_\theta U \right] .
\end{aligned}
\tag{10.92}
$$

This provides an expression for the amplitude of the GW that does not involve the numerical evaluation of the second time derivative, has a compact support [since the integrand is proportional to $\rho(t, r, \theta)$ and therefore vanishes outside the source], and furthermore exploits the fact that in axial symmetry only the $m = 0$ mode contributes. Apart from its usefulness for numerical extraction of the waveform, eq. (10.92) also nicely shows how a collapsing and rotating configuration acquires a non-vanishing second derivative of the mass quadrupole. In the absence of rotation $v_\theta = v_\phi = 0$, so the terms in the second line vanish. Furthermore, if the collapse is spherical, ρ, v_r and U do not depend on θ and then the terms proportional to $(3\cos^2\theta - 1)$ integrate to zero, while the term $\partial_\theta U$ vanishes. So eq. (10.92) correctly reproduces the fact that in a spherical collapse there is no GW production. As soon as we switch on rotation, however, the terms in the second line proportional to v_θ and v_ϕ are non-vanishing. Furthermore, since the centrifugal force is zero on the rotation axis, a rotating fluid assumes the form of an oblate spheroid flattened on the poles. Therefore the radial velocity v_r and the potential U acquire a dependence on θ, and contribute to the integral in eq. (10.92).

An alternative derivation of eq. (10.86), which does not make use of the full apparatus of spin-2 tensor harmonics, can be obtained as follows. For an axisymmetric source, setting the symmetry axis in the z directions, the mass quadrupole Q_{ij} satisfies $Q_{11} = Q_{22}$ by axial symmetry and $2Q_{11} + Q_{33} = 0$ because of the zero-trace condition, while the off-diagonal elements Q_{ij} with $i \neq j$ vanish. This can also be seen from eq. (3.222), $Q_{ij} = \sum_{m=-2}^{2} Q_m \mathcal{Y}_{ij}^{2m}$, observing that in axial symmetry only the spherical component Q_m with $m = 0$ is non-vanishing, and using the expression (3.218) for the tensor $\mathcal{Y}_{ij}^{20}$.

We then use eq. (3.72), which gives the polarizations h_+ and $h_\times$ of the quadrupole radiation for a generic propagation direction.[43] This shows immediately that $h_\times = 0$, since $h_\times$ depends only on the combination $\ddot{Q}_{11} - \ddot{Q}_{22}$ and on the off-diagonal components $\ddot{Q}_{12}$, $\ddot{Q}_{13}$ and $\ddot{Q}_{23}$. For h_+, using $Q_{11} = Q_{22}$, (and reversing again the sign of h_+ to refer it to the axes of the observer, as in Note 42) eq. (3.72) gives

[43] Observe that eq. (3.72) can be written equivalently in terms of M_{ij} or of $Q_{ij} = M_{ij} - (1/3)\delta_{ij} M$; see the Note on page 110 of Vol. 1. This can also be explicitly checked replacing $M_{ij} = Q_{ij} + (1/3)\delta_{ij} M$ in eq. (3.72), and verifying that all occurrences of M cancel.

$$
\begin{aligned}
h_+ &= -\frac{1}{r}\frac{G}{c^4} \sin^2\theta\, (\ddot{Q}_{11} - \ddot{Q}_{33}) \\
&= -\frac{1}{2r}\frac{G}{c^4} \sin^2\theta\, \frac{d^2}{dt^2} (M_{11} + M_{22} - 2M_{33}),
\end{aligned}
\tag{10.93}
$$

where we have again used axial symmetry to write M_{11} as $(1/2)(M_{11} +$

M_{22}). Using the explicit expression for the mass multipole, this becomes

$$
\begin{aligned}
h_+ &= -\frac{1}{2r}\frac{G}{c^4}\sin^2\theta\,\frac{d^2}{dt^2}\int d^3x'\,\rho(t,\mathbf{x}')(x'^2+y'^2-2z'^2) \\
&= -\frac{1}{2r}\frac{G}{c^4}\sin^2\theta\,\frac{d^2}{dt^2}\int d^3x'\,\rho(t,\mathbf{x}')(r'^2-3z'^2) \qquad (10.94)\\
&= \frac{1}{2r}\frac{G}{c^4}\sin^2\theta\,\frac{d^2}{dt^2}2\pi\int_{-1}^{1}d\cos\theta'\,(3\cos^2\theta'-1)\int_0^\infty r'^4 dr'\,\rho(t,r',\theta'),
\end{aligned}
$$

which is indeed equivalent to eq. (10.86) with A_{20}^{E2} given by eqs. (10.87) and (10.88).

10.5   GWs from stellar collapse

In this section we examine several mechanisms for the production of GWs during stellar collapse. As we already saw in the previous sections, modeling the stellar collapse and the subsequent post-bounce evolution is extremely complicated. As a consequence, even just a qualitative assessment of the mechanisms that actually produce a significant amount of GWs requires detailed numerical modeling. These numerical simulations should be three-dimensional (rather than just axisymmetric), since we have seen that the dynamics of collapse, bounce and explosion crucially involves three-dimensional phenomena such as the SASI and other hydrodynamical instabilities. They should ideally cover length-scales from a few meters (where small-scale turbulence develops) up to $O(10^3)$ km, which is the size of the typical stellar core. They should have a sufficiently small time step in the computation to be able to resolve well the dynamics on the scale of a few milliseconds, which is the typical time-scale on which the waveform can show structures, corresponding to GW signals in the bandwidth of a few hundred Hz. This implies computational time steps $\lesssim 10^{-6}$ s. On the other hand, they must follow the evolution for at least $1-2$ s, which is the time necessary for the development of the explosion. The evolution should be studied using full general relativity rather than just Newtonian dynamics. Indeed, as we will see in this section, in certain cases general relativity does not merely provide numerical corrections to Newtonian dynamics, but can even change qualitatively the collapse scenario and the corresponding waveform. Similarly, the details of the microphysics and the treatment of neutrino transport play an essential role. In its full generality the problem is therefore formidable, and beyond the limit of existing computer power. Still, remarkable progress has been made in recent years, thanks to the effort of various groups and to the increase in computing power.

The details of the waveform can also depend strongly on various properties of the progenitor. An obvious parameter that enters is the progenitor mass (or, better, the mass of the core at bounce). Other parameters, however, play an important role, such as the rotation rate of the progenitor, and more generally the dependence of the angular velocity on the distance from the center of the star. The properties of the equation of state (EoS) at and above nuclear densities can also be quite relevant. These quantities are not very well constrained by observations nor by theory. Thus, extensive parameter studies are required, and rather than computing a single "template" for the waveform from SN explosions, the best one can do is to explore the possible waveforms under a variety of realistic conditions.

We will examine a number of different physical mechanisms for the production of GWs during supernova explosion, focusing in particular on GW emission during collapse and bounce of a rotating core, and GWs produced by bar-modes instabilities, by convective instabilities, and by anisotropic emission of neutrinos. We will also discuss the magneto-rotational scenario for collapse, where a strong magnetic field induces a strongly anisotropic explosion, and the associated GW production. Another potentially important mechanism is the GW emission during the pulsations of the new-born proto-NS However, we defer its treatment to Section 11.2.1, after the discussion of the normal modes of neutron stars.

10.5.1   GWs from collapse and bounce of rotating cores

As we will see in this and in the following subsections, there are various mechanisms that can produce sizable GWs during SN explosions. If the core rotates sufficiently fast, already during collapse and bounce there can be significant GW production, and in this subsection we discuss this production mechanism (which, historically, is also the most studied one for GW production from SNe). It should, however, be borne in mind that a strong GW emission is only possible during collapse and bounce if the core rotates fast enough. Stellar evolution computations suggest that the required conditions only take place in a small fraction of the population of massive stars, possibly of the order of 1%.[44] Thus, the rate of events where a GW signal from core collapse and bounce could be observed is correspondingly suppressed. In the following subsections we will then discuss production mechanisms that rather involve the post-bounce phase, and that are expected to take place more generally in SN explosions.

As discussed above, the lack of knowledge about details of the microphysics and of the structure of the progenitor star can only be dealt by extensive parameter study, in which one varies the mass, EoS and rotational structure of the differentially rotating progenitor. The EoS of the initial collapsing core configuration can be taken to be a polytropic one,

$$p = K\rho^{\gamma}, \qquad (10.95)$$

where p is the pressure, ρ is the energy density, and γ is called the adiabatic index. The adiabatic index γ is often written as

$$\gamma = 1 + \frac{1}{n}, \qquad (10.96)$$

[44]See Heger, Woosley and Spruit (2005).

where n is called the polytropic index. For an ideal Fermi gas at zero temperature the EoS is indeed polytropic, with $\gamma = 4/3$ in the high-density ultra-relativistic limit (which is the appropriate limit for core collapse) and $\gamma = 5/3$ in the low-density non-relativistic limit. In a first approximation one can use the EoS of a degenerate Fermi gas at zero temperature, because even the extremely high thermal temperatures $T \sim 10^{10}$ K found in the interior of the collapsing iron core are small compared with the typical Fermi temperature, $T_{\text{Fermi}} \sim 10^{12}$ K. For a more realistic treatment one must, however, use a non-zero-temperature EoS. As discussed in Section 10.3.2, at present numerical simulations of SNe successfully obtain an explosion only for progenitors in the mass range $(7-9)M_{\odot}$. In studies of GW production the collapse is then typically artificially triggered starting from a differentially rotating polytrope with $\gamma = 4/3$, and decreasing the adiabatic index below some critical value. Early studies, using Newtonian gravity and simplified microphysics, found (at least) three different typologies for collapse and the associated gravitational waveform. The first type, denoted in the literature as type I, is due to the stiffening of the EoS when the core reaches densities of the order of the nuclear density,

$$\rho_{\text{nucl}} \simeq 2 \times 10^{14} \, \text{g/cm}^3 \,. \tag{10.97}$$

Then the core overshoots and reaches inner densities above ρ_{nucl}. At these densities the adiabatic index γ of the nuclear EoS becomes larger than $4/3$ and the core bounces back. In this "pressure-dominated" scenario the dynamics of the collapse is not strongly affected by the core rotation. The bounce is followed by a "ringdown" phase in which the core dissipates away its extra energy and quickly settles down into its post-bounce equilibrium configuration. The GW signal correspondingly shows a single large spike at bounce, followed by a damped ring-down phase. Type II signals, in contrast, were seen in collapses with sufficiently large rotation, such that centrifugal forces stopped the collapse well below nuclear densities. This typically gave rise to multiple sequences of expansion and contraction, and the waveform consisted of repeated sequences of large spikes corresponding to the multiple bounces. Type III models appeared in pressure-dominated bounces characterized by very fast collapse, due to a soft EoS or very efficient electron capture, and small masses of the collapsing core.

With the increasing sophistication of numerical simulations it has been found that the waveforms are exclusively of type I, and the type II and III waveform do not appear. Even when the model undergoes a centrifugal bounce, the core does not re-expand to densities much smaller than the bounce density, but rather quickly settles into its equilibrium configuration after a short ring-down phase, so even the waveform has a single spike followed by ringdown.

During core collapse the general-relativistic corrections to the Newtonian potential do not exceed 30%, so one might think that they just give numerical corrections of this order on the waveform obtained from Newtonian simulation. However, the potential well that includes the relativistic corrections is deeper than the Newtonian one, and this counter-balances the stabilizing effect of centrifugal forces. As a result, most models that in a Newtonian treatment showed the multiple bounces associated with type II collapse, become of type I once GR corrections are included. At the level of microphysics, it also turns out to be important to include processes in which electrons are captured on free protons and heavy nuclei. This reduces the electron fraction Y_e, a process referred to as deleptonization. This decreases the mass of the inner core that undergoes homologous collapse, since from eqs. (10.19) and (10.21) we see that $M_{\rm Ch}$ is approximately proportional to Y_e^2. Furthermore, deleptonization also decreases the effective adiabatic index in the pre-collapse core to a value $\gamma_{\rm eff} \simeq 1.29$. Once GR and deleptonization were both included in the numerical simulations, none of the models investigated showed a type II waveform, and type III no longer occurred either. Thus, the waveform from core collapse and bounce is generically of type I.

Observe also that, because of their deeper gravitational potential, general-relativistic models of core collapse have higher central densities and, on average, larger infall and rotational velocities compared with a model that has the same microphysics and initial rotational structure but that is evolved using Newtonian gravity.[45] This, however, does not necessarily imply a stronger GW emission since, as already remarked in Section 10.4.2, the quadrupole emission depends on the second time derivative of the quadrupole moment, and the latter is given by the density ρ weighted by a factor r^4, due to a factor r^2 coming from $d^3x = r^2 dr d\Omega$ and a factor r^2 coming from the definition of the second mass moment. A more compact configuration implies higher velocities and therefore enhances the time derivatives, but also means that the quadrupole moment itself is smaller, so the two effects go in the opposite directions.

Figure 10.11 shows the evolution of the maximum density, which in these models corresponds to the central density (in principle the maximum could also be off-center, giving rise to a toroidal distribution of matter). The upper panel shows the result for a model with low pre-collapse rotation (pre-collapse angular velocity at the center $\Omega_{c,i} = 1.43\,\mathrm{rad/s}$) and two different EoS. For the Lattimer–Swesty EoS (higher curve) the density reaches a value at bounce of order $4 \times 10^{14}\,\mathrm{g/cm^3}$, which is about twice the nuclear density. Then the core slightly re-expands and stabilizes quickly at a value of the central density $\rho_c \sim 1.5\rho_{\rm nucl}$. The middle panel shows a model with moderate pre-collapse rotation, $\Omega_{c,i} = 4.56\,\mathrm{rad/s}$. The lower panel shows a model with very rapid pre-collapse rotation ($\Omega_{c,i} = 11.01\,\mathrm{rad/s}$), again for the Lattimer–Swesty EoS (curve with the largest value at maximum), and for the EoS of Shen et al. This model undergoes a centrifugal bounces, with a value of the central density at bounce somewhat below nuclear density, and a series of more pronounced oscillations.

Typical waveforms of models with slow and moderate rotation are shown in Fig. 10.12. The model with moderate rotation (right panel) shows a marked negative peak at bounce, followed by a rather regular ring-down. The model in the left panel, which has a slow and almost

[45]For a detailed comparison between GR and Newtonian models see Dimmelmeier, Font and Müller (2002).

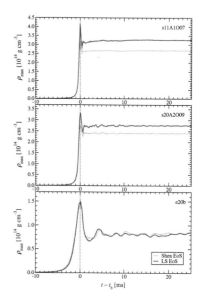

Fig. 10.11 Evolution of the maximum density in a model with slow pre-collapse rotation (upper panel), moderate pre-collapse rotation (middle panel) and rapid pre-collapse rotation (lower panel), for two different EoS. From Dimmelmeier, Ott, Marek and Janka (2008).

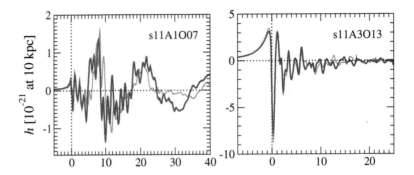

Fig. 10.12 GW amplitude in a model with low pre-collapse rotation displaying large post-bounce convection (left panel) and in a model with moderate pre-collapse rotation (right panel), for the EoS of Lattimer–Swesty and of Shen et al. The horizontal axis gives time in ms, with $t = 0$ chosen at the time of bounce. From Dimmelmeier, Ott, Marek and Janka (2008).

uniform pre-collapse rotation, develops significant post-bounce convection due to convective overturn of the shock-heated layers immediately after the shock-wave sent out by the bounce stalled (recall the discussion in Section 10.3.2). These convective motions are responsible for the low-frequency modulation of the waveform seen in the left panel. Models with fast rotation, undergoing a centrifugal bounce, still give a waveform qualitatively similar to that on the right panel in Fig. 10.12.

These plots display the GW amplitude h in the plus polarization in the equatorial plane, obtained from eq. (10.86) setting $\sin\theta = 1$, for a fiducial distance $r = 10$ kpc, i.e. for a typical Galactic source. The strongest peak values of h are obtained for moderate rotation velocities, with a value at $t = 0$ (defined as the value at bounce) in the range

$$h_{\text{max}} \simeq (0.5 - 10) \times 10^{-21} \left(\frac{10 \, \text{kpc}}{r} \right) , \qquad (10.98)$$

and a total radiated energy in the range

$$E_{\text{GW}} \simeq (0.1 - 5) \times 10^{-8} \, M_\odot c^2 . \qquad (10.99)$$

For lower rotation rates the signal due to collapse and bounce decreases because, as we saw in Section 10.4.3, for axisymmetric sources a combination of collapse and rotation is required in order to have a non-vanishing second time derivative of the quadrupole moment. For sources with pre-collapse central angular velocities $\Omega_{c,i} \lesssim (1-1.5)$ rad/s, the contribution from core collapse and bounce to the GW signal typically gives $|h_{\text{max}}| \lesssim 0.5 \times 10^{-21}$ at $r = 10$ kpc, as we see indeed from the value at bounce (corresponding to $t = 0$) on the left panel in Fig. 10.12. In this model the signal is rather dominated by a different physical mechanism, namely post-bounce convection. For rapidly rotating models, on the other hand, the bounce is dominated by centrifugal forces and takes place at a sub-nuclear central densities, resulting again in a decrease of the GW amplitude.

The Fourier transform of typical GW amplitudes is shown in Fig. 10.13, for slow rotation (upper panel), moderate rotation (middle panel) and fast rotation (lower panel). At low rotation the spectrum has a large low-frequency component due to post-bounce convection, while the collapse and bounce phase is responsible for a higher-frequency contribution. For this model, the latter peaks at 737 Hz for the EoS of Shen et al., and at 806 Hz for the Lattimer–Swesty EoS. With increasing rotation (middle panel) the amplitude of the low-frequency contribution due to convection decreases, while the peak due to collapse and bounce is at 710 and 744 Hz, respectively, for these two EoS. The lower panel shows an example of centrifugal bounce. In this case the peak due to collapse and bounce shifts toward lower frequencies, around 490 Hz, and there is a large low-frequency component, which in this case is not due to post-bounce convection but rather to the rotationally slowed dynamics of the model and to its stronger post-bounce oscillations.

Observe that the Fourier modes $\tilde{h}(f)$ of low frequency do not contribute much to the spectrum of radiated energy, since $dE/d\log f \propto f^3|\tilde{h}(f)|^2$; see eq. (1.160). As a result, for models with slow rotation the total energy radiated in GWs is

$$E_{\rm GW} \lesssim 1 \times 10^{-9} \, M_\odot c^2 \,, \qquad (10.100)$$

while for fast rotation leading to centrifugal bounce typically

$$7 \times 10^{-10} \, M_\odot c^2 \lesssim E_{\rm GW} \lesssim 5 \times 10^{-9} \, M_\odot c^2 \,, \qquad (10.101)$$

to be compared with the value for moderate rotation and typical pressure-dominate bounce given in eq. (10.99).[46]

10.5.2   GWs from bar-mode instabilities

The signal from rotating core collapse, that we discussed in the previous section, is present even in axisymmetric configurations. A further contribution to GW production can come if the rotating core develops non-axisymmetric instabilities, i.e. modes with an azimuthal dependence $e^{im\phi}$, with $m = \pm 1, \pm 2$, etc. In particular, the $m = 2$ mode correspond to a bar-like deformation.[47]

The development of these instabilities is controlled by the parameter $\beta = T_{\rm rot}/|W|$, where $T_{\rm rot}$ is the rotational kinetic energy and W the gravitational potential energy of the rotating fluid. We can distinguish between dynamical instabilities and secular instabilities. Dynamical instabilities are driven by hydrodynamics and gravity, and grow on time-scales of the order of the typical oscillation time-scale of the system. In contrast, secular instabilities are associated with dissipative processes related to viscosity or gravitational radiation, and grow on much longer time-scales.

From the classical theory of MacLaurin spheroids,[48] i.e. rigidly rotating axisymmetric configurations with uniform density, one knows that a rotating incompressible axisymmetric fluid body whose dynamics is

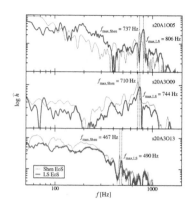

Fig. 10.13 The Fourier transform $\tilde{h}(f)$, on a logarithmic scale, in a model with slow pre-collapse rotation $\Omega_{c,i} = 1.01\,{\rm rad/s}$ (upper panel), a model with moderate pre-collapse rotation, $\Omega_{c,i} = 8.99\,{\rm rad/s}$, representative of typical pressure-dominate bounce (middle panel), and a model with rapid pre-collapse rotation $\Omega_{c,i} = 11.30\,{\rm rad/s}$, representative of a centrifugal bounce (lower panel). From Dimmelmeier, Ott, Marek and Janka (2008).

[46]These values are provided in Ott (2009), eliminating the convective contribution from the analysis, since the deleptonization scheme used might over-estimate them. However, for the total radiated energy this should not affect the result too much, again because of the f^3 factor in $dE/d\log f$, which suppresses the low-frequency contribution.

[47]In terms of the star normal modes (which, for neutron stars, will be discussed in Section 11.2.1), the $l = m = 2$ bar instability corresponds to an instability in the f-mode; see Passamonti and Andersson (2015).

[48]See e.g. Chandrasekhar (1969).

treated with Newtonian gravity has a dynamical instability leading to a bar-like deformations if

$$\beta \geqslant \beta_{\mathrm{dyn}} \simeq 0.27 \,. \tag{10.102}$$

Numerical simulations indicate that this value of β_{dyn} is approximately correct even for differentially rotating compressible fluids in GR, and depends only weakly on the stiffness of the EoS, with typical values of β_{dyn} in the range $0.24 - 0.25$. Once the bar has developed, at least for $\beta \gg \beta_{\mathrm{dyn}}$, the initial exponential growth of the unstable mode saturates within a time of order of the dynamical time-scale of the system, and is followed by the formation of spiral arms, which further leads to the shedding of mass and angular momentum. For such a short-lived bar, the corresponding GW signal consists of a relatively short bursts.

However, if β is only slightly larger than β_{dyn}, a long-lived bar structure can form, which emits quasi-periodic GWs and only damps on the time-scales determined by radiation reaction. Furthermore, on a longer time-scale, the $m = 2$ mode also has a secular instability driven by dissipation, which sets in at

$$\beta \geqslant \beta_{\mathrm{sec}} \simeq 0.14 \,, \tag{10.103}$$

and again gives rise to a long-lived triaxial ellipsoidal deformation. We already computed in eqs. (4.223) and (4.224) of Vol. 1 the amplitude of the GW radiated by an ellipsoid rotating around a principal axis. We use as reference values for the rotating core a mass $M = 0.7 M_\odot$, a radius $R = 12\,\mathrm{km}$, a rotational frequency $f = 500$ Hz (corresponding to emission of GWs at a frequency $f_{\mathrm{gw}} = 2f = 1$ kHz) and an ellipticity $\epsilon = 0.1$.[49] This gives

$$h_+ = h_0 \, \frac{1 + \cos^2 \iota}{2} \, \cos(2\pi f_{\mathrm{gw}} t) \,, \tag{10.104}$$

$$h_\times = h_0 \, \cos \iota \, \sin(2\pi f_{\mathrm{gw}} t) \,, \tag{10.105}$$

where ι is the inclination angle with respect to the line of sight, and

$$h_0 = \frac{4\pi^2 G}{c^4} \, \frac{I_3 f_{\mathrm{gw}}^2}{r} \, \epsilon \tag{10.106}$$

$$\simeq 8.47 \times 10^{-21} \left(\frac{\epsilon}{10^{-1}} \right) \left(\frac{M}{0.7 M_\odot} \right) \left(\frac{R}{12\,\mathrm{km}} \right)^2 \left(\frac{10\,\mathrm{kpc}}{r} \right) \left(\frac{f}{500\,\mathrm{Hz}} \right)^2 ,$$

where we have used $I_3 = (2/5) M R^2$ for the moment of inertia of the ellipsoid, which is valid to lowest order in ϵ. Comparing with eq. (10.98) we see that the development of bar-like instabilities could produce a GW signal comparable to or larger than that generated in axisymmetric collapse and bounce, with the further advantage of being a potentially long-lasting signal.

The issue is therefore whether values of β in excess of the thresholds for dynamical or secular instabilities can be reached in realistic models. First of all one can observe that, for a virialized configuration, we have

$\beta \leqslant 0.5$. A more stringent bound comes from the requirement that, if the core rotates too fast, it will shed mass. This happens when its equator rotates at the same frequency as the Kepler rotation frequency of a particle in circular orbit around it. This mass-shedding limit is reached at rotation frequency

$$\Omega_K \simeq \frac{2}{3} \sqrt{\pi G \rho_0} \,, \tag{10.107}$$

where ρ_0 is the average density of the corresponding non-rotating stellar core. For a uniformly rotating configuration with rotational frequency ω, whose equation of state is given by a Newtonian polytrope, one can estimate that

$$\frac{T_{\rm rot}}{|W|} \simeq \frac{1}{9} \left(\frac{\omega}{\Omega_K} \right)^2 \,, \tag{10.108}$$

so the mass-shedding limit corresponds to $\beta_K \simeq 1/9 \simeq 0.11$. Using more realistic EoS at supra-nuclear densities, but still assuming that the core is rigidly rotating, provides a maximum value of β in the range $0.09 - 0.13$. However, for a star that is differentially rotating, the upper bound on β coming from the mass-shedding limit can be considerably larger, since the star could rotate much faster in the interior, while still keeping a rotational velocity at the surface below the mass-shedding limit. It is indeed expected that differential rotation arises in the proto-NS at birth, and can be further enhanced by accretion from fallback material.

Of course, if β_K is smaller than the threshold $\beta_{\rm dyn}$ or $\beta_{\rm sec}$, the corresponding instabilities cannot develop. On the other hand, even if β_K is larger than $\beta_{\rm dyn}$ or $\beta_{\rm sec}$, this does not mean that the post-bounce core of a realistic SN explosion actually reaches a value of β that exceeds the thresholds for secular or dynamical instabilities. In particular, numerical simulations indicate that even the most extreme models do not reach values of β comparable to the value $\beta_{\rm dyn}$ given in eq. (10.102), so this specific instability is probably not relevant.

The situation becomes much more promising, however, in differentially rotating configurations. Indeed, dynamical instabilities in the $m = 2$ mode that sets in at values of β as low as $\beta = O(0.01)$ have been found in configurations with strong differential rotation.[50] Furthermore, in differentially rotating configurations numerical investigations have also revealed the existence of dynamical instabilities that are dominated by the $m = 1$ mode. The critical value of β depends on the profile of the differentially rotating configuration. This instability was first observed in a configuration that developed a dynamical instability at $\beta \simeq 0.14$,[51] and again requires a high degree of differential rotation.

The $m = 1$ instabilities give rise to a one-armed spiral structure and trigger a secondary bar-like $m = 2$ instability of smaller amplitude, which then emits GWs described by eqs. (10.104) and (10.105). This gravitational radiation therefore shows a strong correlation between the h_+ and $h_\times$ polarizations, which oscillate at the same frequency and are out of phase by $\pi/2$ [in contrast to the radiation from axisymmetric sources which, as shown in eqs. (10.85) and (10.86), only has the plus polarization]. Its angular distribution is such that the signal is stronger at

[50]See Shibata, Karino and Eriguchi (2002), and the Further Reading section.

[51]See Centrella, New, Lowe and Brown (2001).

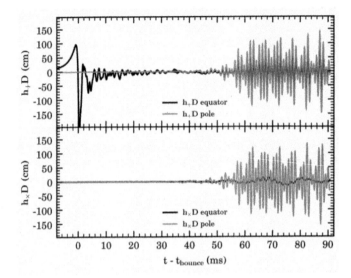

Fig. 10.14 The GW amplitudes h_+ (upper panel) and $h_\times$ (lower panel), multiplied by the distance r to the source (here denoted by D). In each panel are superimposed the signal in the equatorial plane and the signal in the z direction; see the text. From Ott (2009).

$\cos\iota = 0$, i.e. along the polar axis (again contrary to the radiation from axisymmetric sources, which is stronger in the equatorial plane and vanishes on the polar axis). The amplitude of the non-axisymmetric modes reaches a sizable amplitude after about 30–40 ms and starts producing significant GWs after about 50–60 ms. In contrast, the GW signal from collapse and bounce is concentrated in the first ~ 5 ms after bounce; see for example the right panel in Fig. 10.12.

Thus, in a model in which the bounce of a rotating core is followed by the development of a bar-like instability, the overall gravitational signal consists of a component concentrated in the first few ms, due to core bounce, present uniquely in the plus polarization, and stronger in the equatorial plane. The signal then damps out and is revived after about 50–60 ms by the contribution from the bar-like instability. The duration of this second phase can be from a few times 10 ms up to $O(100)$ ms or more, depending on the lifetime of the bar. Now h_+ and $h_\times$ have comparable amplitudes, and the signal is stronger along the z axis. An example from a differentially rotating model with a mass $20 M_\odot$ is shown in Fig. 10.14. In this model the dynamical instability develops at $\beta \simeq 0.13$, and is predominantly in the $m = 1$ mode, which then triggers the growth of the $m = 2, 3$ modes. In this figure, as is customary in the literature, the plot is of the quantity hr, measured in cm, for $h = h_+$ (upper panel) and $h = h_\times$ (lower panel). This can be converted to a value for h at a reference distance r such as 10 kpc, using $1\,\mathrm{cm}/(10\,\mathrm{kpc}) \simeq 3.24 \times 10^{-23}$, so

$$h \simeq 3.24 \times 10^{-23} \left(\frac{hr}{\text{cm}} \right) \left(\frac{10\,\text{kpc}}{r} \right) . \qquad (10.109)$$

Thus, the values of hr in excess of 100 shown in the figure correspond to values of h of several times 10^{-21} for a source at the reference distance $r = 10$ kpc, and we see from the figure that the peak amplitude due to the bar radiation is of the same order as the prompt signal due to core collapse and bounce. However, since the bar signal lasts much longer, it dominates the energy loss. Indeed, in typical models that undergo bar formation, the maximum GW amplitude during the bar phase is in the range

$$h_{\text{max}} \sim (1 - 5) \times 10^{-21} \left(\frac{10\,\text{kpc}}{r} \right) , \qquad (10.110)$$

and the energy radiated in GWs during this phase can be of order

$$E_{\text{GW}} \sim (5 - 15) \times 10^{-8}\, M_\odot c^2 \,, \qquad (10.111)$$

with most of the energy radiated in a bandwidth $\Delta f_{\text{peak}} \sim 50 - 200$ Hz around a peak frequency $f_{\text{peak}} \sim 900$ Hz. Comparing with eqs. (10.99), (10.100) or (10.101) we see that the energy radiated in GWs by the bar-mode instability can be one or two orders of magnitude larger than that radiated during core collapse and bounce. Of course, exact numbers depend strongly on the number of cycles for which a coherent rotating bar-like structure survives.

10.5.3   GWs from post-bounce convective instabilities

As we discussed in Section 10.3.2, convective instabilities such as the SASI are believed to play an important role in the dynamics of SNe, in particular in reviving the stalled shock and launching the explosion. These instabilities induce non-radial coherent motions of large amounts of matter, and are therefore potentially interesting sources of GWs.

One can distinguish between different convective mechanisms that operate on different temporal and spatial scales. "Prompt convection" takes place in the first 10–50 ms after the bounce, and is due to negative entropy gradients behind the stalled shock. "Post-shock convection" is driven by negative lepton gradients near the surface of the newly formed proto-neutron star (PNS), as well as by neutrino-driven convection in the post-shock heating region. PNS convection sets in 20–50 ms after the bounce and can last up to about $O(1)$ s after the bounce, and it is limited to the inner 10–30 km. Neutrino-driven convection develops on a similar time-scale, 30–50 ms, but its spatial extents is of the order of the gain region (i.e. the region in which neutrino heating dominates over neutrino cooling), i.e. up to a region $O(100)$ km, and is due to the fact that neutrino heating below the stalled shock establishes a negative radial entropy gradient. A further convective mechanism arises when the SASI instability discussed in Section 10.3.2 becomes non-linear, producing a violent back-and-forth sloshing of the material in the post-shock region, as well as enhancing funnels of material that plunge at high rates onto the PNS. This takes place approximately 200–800 ms after the bounce, depending also on the rotation rate of the core, and is spatially superimposed on neutrino-driven convection.

Figure 10.15 shows various features in the waveform due to post-bounce convection, for a model with a progenitor mass $15M_\odot$ and a ν_e luminosity $L_{\nu_e} = 3.7 \times 10^{52}$ erg/s. In this model collapse has been assumed to proceed spherically, so there is no gravitational signal associated with core collapse and bounce, and the whole contribution to the waveform comes from the post-bounce dynamics. To compare with Fig. 10.14 observe that in Fig. 10.15 the time axis covers more than 1 s after bounces, while in Fig. 10.14 it only covers the first 90 ms after bounce. Observe also that the scale on the vertical axis is quite different. The signal from core bounce and from bar instabilities reaches values of $hr \sim O(100 - 150)$ cm, while the signal from convection only reaches values $hr \sim O(5 - 10)$ cm. In Fig. 10.15 we can distinguish an initial phase, of duration $\lesssim 50$ ms, in which the GW signal is due to prompt convection, and $hr \sim O(5)$ cm. During the subsequent post-shock convection phase, neutrino-driven convection and convection near the PNS give a weaker signal, $hr \sim O(1)$ cm. The signal rises again, to values of hr of order 10 cm and higher, when the SASI enters in the non-linear regime. There, one can also identify SASI "plumes", i.e. spikes in the GW amplitude coincident in time with narrow plumes that strike the PNS surface, at about 50 km from the center. The final rise of the GW signal, which takes place approximately 800 ms after bounce, is associated with the explosion. During explosion, large asymmetric mass ejection produces a steadily rising GW signal. In models where the explosions is approximately spherical, this part of the signal is absent.

In general, depending on the progenitor model, the maximum GW amplitudes obtained from convection are in the range

$$h_{\max} \sim \text{a few} \times (0.1 - 1) \times 10^{-21} \left(\frac{10\,\text{kpc}}{r} \right). \qquad (10.112)$$

Values of $h_{\max}$ due to convection of order 10^{-21} can for instance be seen in the left panel of Fig. 10.12. The energy radiated in GWs from convection is of order

$$E_{\text{GW}} \sim \text{a few} \times (10^{-11} - 10^{-10})\, M_\odot c^2. \qquad (10.113)$$

These values can be compared with eqs. (10.98) and (10.99) for core collapse and bounce with moderate rotation, and with eqs. (10.110) and (10.111) for bar instabilities.

In frequency space the overall signal from post-bounce convection has a broad spectrum, with significant power between a few Hz and a few kHz, and a broad maximum typically around 300–400 Hz (with higher peak frequencies corresponding to higher progenitor masses).

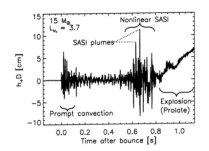

Fig. 10.15 The contribution to the GW amplitude from various phases of post-bounce convection. From Murphy, Ott and Burrows (2009).

10.5.4   GWs from anisotropic neutrino emission

We have seen in the previous sections that SNe emit about 99% of their energy in the form of neutrinos, while only about 1% of the energy of the explosion goes into the kinetic energy of the ejecta. Until now we have only discussed the GWs generated by bulk motions of matter during collapse. It is clear, however, that also the energy carried by the neutrinos could be a potentially interesting sources of GWs. Just as the motion of matter only generates GWs if the collapse is not spherical, neutrinos can generate GWs only if their emission during collapse is anisotropic. Anisotropies in the neutrino emission can, however, arise from a number of physical mechanisms. For instance, we have seen that realistic collapses are driven by strongly anisotropic processes, such as SASI. Anisotropic neutrino emission could also arise from rotationally deformed proto-neutron stars, and of course from any asymmetry in the pre-collapse matter distribution.

To compute the corresponding GW production we need the energy–momentum tensor associated with the neutrino radiation field. To compute it we idealize the situation, assuming that a point-like source in $\mathbf{x} = 0$ radially emits neutrinos with a time-dependent luminosity $L_\nu(t) = dE_\nu/dt$, and we further assume that the neutrinos stream out freely at the speed of light. These assumptions are meaningful at distances larger than the neutrinosphere. For orientation, at the neutrinosphere the matter density is of order $\rho \sim 10^{12}\,\mathrm{g/cm^3}$, to be compared with the typical nuclear densities $\rho \sim 10^{14}\,\mathrm{g/cm^3}$ reached at bounce. The typical dynamical time-scale of a gravitating system with density ρ is of order $\tau \sim 1/\sqrt{G\rho}$, so the typical frequency of the source is $f \sim \sqrt{G\rho}$. Since the density at the neutrinosphere is two orders of magnitudes lower than the central density, we infer that, in order of magnitude, the frequency of the GWs radiated by neutrinos will be lower by a factor ~ 10 compared with the typical frequencies of GWs produced during core collapse, i.e. from neutrinos we expect GWs with typical frequencies from a few Hz up to $O(100)$ Hz.

Recall from eq. (3.121) that the energy–momentum tensor of a set of particles, labeled by the index A and following the trajectories $\mathbf{x}_A(t)$, is given by

$$T^{\mu\nu}(t, \mathbf{x}) = \sum_A \frac{E_A}{c^2} \frac{dx_A^\mu}{dt} \frac{dx_A^\nu}{dt} \delta^{(3)}(\mathbf{x} - \mathbf{x}_A(t)) \qquad (10.114)$$

(where we write E_A/c^2 instead of $\gamma_A m_A$, so the result is valid even for massless particles). Thus

$$T^{00}(t, \mathbf{x}) = \sum_A E_A \delta^{(3)}(\mathbf{x} - \mathbf{x}_A(t)) \qquad (10.115)$$

and

$$T^{ij}(t, \mathbf{x}) = \sum_A \frac{E_A}{c^2} v^i v^j \delta^{(3)}(\mathbf{x} - \mathbf{x}_A(t)). \qquad (10.116)$$

For a set of particles moving radially at the speed of light we have $v^i = c\,n^i$, where $n^i = \mathbf{x}/|\mathbf{x}|$ is the unit vector in the radial direction at the point $\mathbf{x}$. Thus

$$T^{ij}(t,\mathbf{x}) = n^i n^j T^{00}(t,\mathbf{x}). \qquad (10.117)$$

To compute the energy density $T^{00}(t,\mathbf{x})$ in neutrinos at $(t,\mathbf{x})$ (where $\mathbf{x} = r\hat{\mathbf{n}}$) we consider the neutrinos emitted by the source between times t_0 and $t_0 + dt$, into a cone with infinitesimal opening angle $d\Omega$ centered around the direction $\hat{\mathbf{n}}$. After a time r/c these neutrinos have reached a distance r from the source,[52] and occupy a volume $dV = r^2 d\Omega \times cdt$. Thus, if the total energy of the neutrinos contained in this volume is dE_ν, the energy density in this volume at time $(t_0 + r/c)$ is

$$T^{00}(t_0 + r/c, \mathbf{x}) = \frac{1}{c\,r^2}\frac{dE_\nu}{d\Omega dt}. \qquad (10.118)$$

We now observe that dE_ν/dt is the luminosity of the source at the time t_0 at which the neutrinos were emitted. Therefore $T^{00}(t_0 + r/c, \mathbf{x}) = (1/cr^2)dL_\nu(t_0, \hat{\mathbf{n}})/d\Omega$ or, equivalently,

$$T^{00}(t,\mathbf{x}) = \frac{1}{c\,r^2}\frac{dL_\nu}{d\Omega}(t - r/c, \hat{\mathbf{n}}). \qquad (10.119)$$

Then eq. (10.117) gives

$$T^{ij}(t,\mathbf{x}) = \frac{n^i n^j}{c\,r^2}\frac{dL_\nu}{d\Omega}(t - r/c, \hat{\mathbf{n}}). \qquad (10.120)$$

We now use the equations of linearized theory to compute the production of GWs, using this energy–momentum tensor as a source. It should be observed that this source is non-compact, since $T^{ij}(t,\mathbf{x})$ does not vanish beyond some distance, in contrast to what happens for localized matter sources. Rather, it goes to zero as $1/r^2$ at large distances. Using eq. (3.9) of Vol. 1 we have

$$h_{ij}^{\rm TT}(t,\mathbf{x}) = \frac{4G}{c^4}\Lambda_{ij,kl}(\hat{\mathbf{n}})\int d^3x'\,\frac{1}{|\mathbf{x}-\mathbf{x}'|}\,T_{kl}\left(t - \frac{|\mathbf{x}-\mathbf{x}'|}{c}, \mathbf{x}'\right)$$

$$= \frac{4G}{c^5}\Lambda_{ij,kl}(\hat{\mathbf{n}})\int_0^\infty dr'\int d\Omega'\,\frac{n'_k n'_l}{|\mathbf{x}-\mathbf{x}'|}\frac{dL_\nu}{d\Omega'}\left(t - \frac{r'}{c} - \frac{|\mathbf{x}-\mathbf{x}'|}{c}, \hat{\mathbf{n}}'\right),$$
$$(10.121)$$

where as usual $\Lambda_{ij,kl}(\hat{\mathbf{n}})$ is the projector onto the TT gauge defined in eq. (1.36), $\Lambda_{ij,kl}(\hat{\mathbf{n}}) = P_{ik}P_{jl} - \frac{1}{2}P_{ij}P_{kl}$, with $P_{ij}(\hat{\mathbf{n}}) = \delta_{ij} - n_i n_j$.[53]

Observe that, in the argument of $dL_\nu/d\Omega'$ in eq. (10.121), there are two different retardation effect. The term $-|\mathbf{x}-\mathbf{x}'|/c$ is the usual retardation effect describing the delay by which the GW produced by the energy–momentum tensor at $\mathbf{x}'$ reaches the observer at $\mathbf{x}$. The term $-r'/c$, in contrast, is the time delay that reflects the fact that the energy–momentum tensor at time t and at the position $\mathbf{x}'$ originates from neutrinos that were emitted by the source, located at the origin, at time $t - r'/c$. The two effects are shown in Fig. 10.16. We now introduce the variable

$$t' \equiv t - \frac{r'}{c} - \frac{|\mathbf{x}-\mathbf{x}'|}{c}, \qquad (10.122)$$

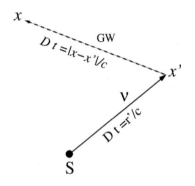

Fig. 10.16 Neutrinos propagate from the source S in $\mathbf{x} = 0$ to the point $\mathbf{x}'$. The energy–momentum tensor of the neutrinos at $\mathbf{x}'$ generates GWs that propagate to the observer at $\mathbf{x}$. The respective time delays are r'/c and $|\mathbf{x}-\mathbf{x}'|/c$.

[52]Of course this is only valid if r is much larger than the radius of the neutrinosphere.

[53]A subtle point here is that the TT gauge can actually only be reached *outside* the source, i.e. where $T^{ij}(t,\mathbf{x}) = 0$, see the discussion in Section 1.2. In the case at hand the source is non-compact and strictly speaking this is never possible. However, we are of course interested in the GW at the observer position where, because of its $1/r^2$ behavior, the neutrino energy density has by now been diluted beyond any meaningful value, and we can simply neglect it.

and replace the integral over r' by an integral over t'. Observe that $r' = 0$ corresponds to $t' = t - r/c$ while $r' = \infty$ corresponds to $t' = -\infty$. Differentiating eq. (10.122) and making use of the identity

$$|\mathbf{x} - \mathbf{x}'| = (r^2 + r'^2 - 2rr'\cos\alpha)^{1/2},\qquad (10.123)$$

where α is the angle between the directions of $\mathbf{x}$ and $\mathbf{x}'$ (so that $\hat{\mathbf{n}} \cdot \hat{\mathbf{n}}' = \cos\alpha$), we get

$$\frac{dr'}{|\mathbf{x} - \mathbf{x}'|} = -\frac{cdt'}{c(t - t') - r\cos\alpha}.\qquad (10.124)$$

Therefore eq. (10.121) can be rewritten as

$$h_{ij}^{\mathrm{TT}}(t, \mathbf{x}) = \frac{4G}{c^4}\int_{-\infty}^{t - r/c} dt' \int d\Omega' \frac{(n_i' n_j')^{\mathrm{TT}}}{c(t - t') - r\cos\alpha}\frac{dL_\nu}{d\Omega'}(t', \hat{\mathbf{n}}'),$$

$$(10.125)$$

where $(n_i' n_j')^{\mathrm{TT}} \equiv \Lambda_{ij,kl}(\hat{\mathbf{n}})n_k' n_l'$. Equation (10.125) is the general formula for the amplitude of GWs generated by an anisotropic flux of radiation (in this case of neutrinos, but of course the formalism is in principle valid for any other form of radiation).

A peculiar aspect of this result is that it is given by an integral over t' from $-\infty$ to $t - r/c$, and is therefore sensitive to the whole past behavior of the luminosity of the source. To understand this physically it can be useful to use a quantum language, in which neutrinos radiate gravitons by bremsstrahlung. Then, as it is clear from Fig. 10.16, the shortest possible time delay between emission of the neutrino at the origin and observation of a GW signal at $\mathbf{x}$, which is r/c, can be obtained in two possible ways: either the gravitons are produced by the neutrinos already very close to $\mathbf{x} = 0$ (or, more physically, as soon as they leave the neutrinosphere) and then travel straight to the observer, or more generally, they are produced by neutrinos that were flying in the direction of the observer and which then emitted gravitons along their direction of motion. These gravitons therefore proceeded in straight line to the observer, always at the speed of light (we neglect here the very small correction due to the fact that neutrinos are not exactly massless). Any other kinematical configuration, in which the neutrino starts at an angle compared with the direction to the observer and then emits gravitons toward the observer, as in the path shown in Fig. 10.16, results in a longer time delay between the emission of the neutrino at the origin and the arrival of the gravitational radiation at the point $\mathbf{x}$. The radiation observed at time t at the point $\mathbf{x}$ is the superposition of all these possible paths, and includes paths that started in the far past and made a long detour before reaching the point $\mathbf{x}$. Equation (10.125) therefore has the form of a "hereditary term", similar to those discussed in the context of the post-Newtonian expansion [see eq. (5.171) and the discussion of memory and tail terms on pages 272–274 of Vol. 1]. In that case, however, the delay was due to the non-linearities of the gravitational field itself or, in a quantum language, to gravitons that only reached the observer at $\mathbf{x}$ after scattering one or more times against the background geometry.

Observe also that, in contrast to the usual treatment valid for localized sources discussed in Section 3.1, we never assumed that $r' \ll r$, i.e. we never used the approximation (3.10), $|\mathbf{x} - \mathbf{x}'| \simeq r - \mathbf{x}' \cdot \hat{\mathbf{n}}$. Indeed, the GWs observed at the point $\mathbf{x}$ in Fig. 10.16 can in principle receive significant contributions from the integration over values of $\mathbf{x}'$ such that r' is comparable or even larger than r. In the literature, however, a different approximation is usually performed in eq. (10.125), assuming that, in the denominator of eq. (10.125), we can replace the integration variable t' by $t - r/c$, i.e. by the upper end of the integration domain. As we will see, this approximation breaks down when t is parametrically larger than r/c.[54] However, for $t \sim r/c$, the approximation gives the correct order of magnitude, and has the advantage of neatly factorizing a $1/r$ term from the GW amplitude, which now reads

$$h_{ij}^{\mathrm{TT}}(t,\mathbf{x}) \simeq \frac{4G}{c^4 r} \int_{-\infty}^{t-r/c} dt' \int d\Omega' \frac{(n_i' n_j')^{\mathrm{TT}}}{1 - \cos\alpha} \frac{dL_\nu}{d\Omega'}(t',\hat{\mathbf{n}}') . \quad (10.126)$$

To perform the angular integral it is convenient to use a reference frame related to the source properties, chosen so that for an axisymmetric source the z axis coincide with the symmetry axis of the source. In this frame the propagation direction is parametrized by the angles θ and ϕ,

$$\hat{\mathbf{n}} = (\sin\theta\cos\phi, \sin\theta\sin\phi, \cos\theta). \quad (10.127)$$

Similarly $\hat{\mathbf{n}}' = (\sin\theta'\cos\phi', \sin\theta'\sin\phi', \cos\theta')$, and $\cos\alpha$ can be expressed in terms of $\theta, \phi, \theta', \phi'$ using $\cos\alpha = \hat{\mathbf{n}} \cdot \hat{\mathbf{n}}'$. Using eq. (1.39),

$$\Lambda_{ij,kl}(\hat{\mathbf{n}}) n_k' n_l' = -\frac{1}{2}\delta_{ij}\sin^2\alpha + \frac{1}{2}n_i n_j(1 + \cos^2\alpha) + n_i' n_j'$$
$$- (n_i' n_j + n_j' n_i)\cos\alpha, \quad (10.128)$$

so in general we can write

$$h_{ij}^{\mathrm{TT}}(t,\mathbf{x}) \simeq \frac{4G}{c^4 r} \int_{-\infty}^{t-r/c} dt' \int d\cos\theta' d\phi' \, \Psi_{ij}(\theta,\phi;\theta',\phi')\frac{dL_\nu}{d\Omega'}(t';\theta',\phi') , \quad (10.129)$$

for some function $\Psi_{ij}(\theta,\phi;\theta',\phi')$. The plus and cross polarizations can be extracted by rotating the frame so that the new z axis coincides with the propagation direction, proceeding similarly to what we did in the derivation of eq. (3.72), so the result takes the form

$$h_{+,\times}(t,\mathbf{x}) \simeq \frac{2G}{c^4 r} \int_{-\infty}^{t-r/c} dt' \int d\cos\theta' d\phi' \, \Psi_{+,\times}(\theta,\phi;\theta',\phi')\frac{dL_\nu}{d\Omega'}(t';\theta',\phi') , \quad (10.130)$$

where $\Psi_{+,\times}$ are some functions of the angles.[55] It is also convenient to define the angular averages of $dL_\nu/d\Omega'$, weighted with the functions

[54]Observe also that this approximation does not mean that we are disregarding memory effects. The resulting expression for $h_{ij}^{\mathrm{TT}}(t,\mathbf{x})$, eq. (10.126), is still given by an integral over the past history of the source, which is the definition of a memory-type contribution.

[55]See Müller and Janka (1997) for the explicit expressions of $\Psi_{+,\times}(\theta,\phi;\theta',\phi')$ in the equatorial plane $\theta = \pi/2$ and along the polar axis $\theta = 0$.

$\Psi_{+,\times}$,

$$\alpha_{+,\times}(t;\theta,\phi) = \frac{1}{L_\nu(t)} \int d\cos\theta' d\phi' \, \Psi_{+,\times}(\theta,\phi;\theta',\phi')\frac{dL_\nu}{d\Omega'}(t;\theta',\phi') \,, \tag{10.131}$$

where $L_\nu(t)$ is the total neutrino luminosity,

$$L_\nu(t) = \int d\cos\theta' d\phi' \, \frac{dL_\nu}{d\Omega'}(t;\theta',\phi') \,. \tag{10.132}$$

The "anisotropy parameters" $\alpha_{+,\times}$ are a measure of the anisotropy of the neutrino radiation, and vanish for isotropic emission. Then

$$h_{+,\times}(t,r,\theta,\phi) \simeq \frac{2G}{c^4 r} \int_{-\infty}^{t-r/c} dt' \, \alpha_{+,\times}(t';\theta,\phi)L_\nu(t') \,. \tag{10.133}$$

The fact that this result has the structure of a time integral over the past history of the source means that, as long as the source emits neutrinos, the signal typically grows steadily [even if the functions $\alpha_{+,\times}(t)$ are not necessarily positive and could oscillate and change sign with time, so the growth of the integral could in principle be closer to a random walk rather than linear in t]. Even if at some time $t_{\rm end}$ the source shuts off and neutrino emission stops, $h_{+,\times}(t)$ does not vanish but remains at the value $h_{+,\times}(t_{\rm end})$ for all times $t > t_{\rm end}$, and in this sense it retains the "memory" of the burst that passed.

At the conceptual level one might be puzzled by the fact that this "memory" persists for an infinite time after the passage of the neutrino bursts. Closer examination shows that this is actually an artifact of the approximation leading from eq. (10.125) to eq. (10.126). To see this, consider the case where the neutrino luminosity can be modeled as a Dirac delta in time,

$$L_\nu(t;\theta,\phi) = \tau\,\delta(t-t_0)L_\nu^* f(\theta,\phi) \,, \tag{10.134}$$

where t_0 is the time at which the burst takes place, the constant L_ν^* is the peak luminosity of the burst, $f(\theta,\phi)$ is a function of the angles (normalized so that it gives 1 when integrated over $d\Omega$) that provides the anisotropy of the emission and τ is a time-scale [which represents the effective duration of the burst; compare with eq. (7.99)]. Then eq. (10.125) gives

$$h_{ij}^{\rm TT}(t,\mathbf{x}) = \frac{4G}{c^4}\,\tau L_\nu^*\theta(t-r/c-t_0)\int d\Omega'\,\frac{(n_i'n_j')^{\rm TT}f(\theta,\phi)}{c(t-t_0)-r\cos\alpha} \,, \tag{10.135}$$

where θ is the step function. We see from this expression that, for values of t such that $c(t-t_0) \gg r$, the amplitude is proportional to $1/[c(t-t_0)]$ rather than to $1/r$. In the regime $c(t-t_0) \gg r$, we have

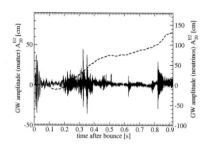

Fig. 10.17 The GW signal due to matter (solid line) and to neutrinos (dashed) in a model with a progenitor with a mass $M = 11.2M_\odot$. Observe that the scale for the GW produced by matter (left vertical axis) is different from that of the GWs produced by neutrinos (right vertical axis). From Mueller, Janka and Marek (2013)

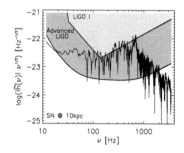

Fig. 10.18 Spectral energy distribution of the quadrupole radiation due to convective mass flow and anisotropic neutrino emission for a rotating model with $M = 15M_\odot$, as a function of frequency. From Müller, Rampp, Buras, Janka and Shoemaker (2004).

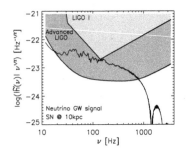

Fig. 10.19 The contribution to Fig. 10.18 due to anisotropic neutrino emission. From Müller, Rampp, Buras, Janka and Shoemaker (2004).

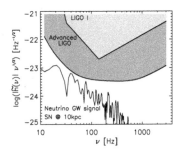

Fig. 10.20 The contribution due to anisotropic neutrino emission in a non-rotating progenitor model with $M = 11.2 M_\odot$. From Müller, Rampp, Buras, Janka and Shoemaker (2004).

$1/[c(t - t_0)] \ll 1/r$, so the amplitude is much smaller than the value suggested by the approximation (10.126), and vanishes in the limit $t - t_0 \to \infty$. Thus, strictly speaking the gravitational radiation induced by anisotropic neutrino emission has the form of a tail term (of the type discussed on pages 272–274 of Vol. 1), rather than of a memory term. Of course, in practice in this tail regime the signal is even smaller, so the observer only has a chance of observing the signal when $t - t_0$ is indeed of order r/c. Thus, from a practical point of view the approximation (10.126) is appropriate.

We can now use eq. (10.133) to get some order-of-magnitude estimate of the expected signal. Inserting typical values for the neutrino luminosity and for the duration Δt of the neutrino burst, for the asymptotic value reached by the GW amplitude we get

$$h_{+,\times} \sim \frac{2G}{c^4 r} \Delta t \, \alpha_{+,\times} L_\nu \tag{10.136}$$

$$\simeq 5 \times 10^{-21} \left(\frac{10\,\mathrm{kpc}}{r} \right) \left(\frac{\Delta t}{1\,\mathrm{s}} \right) \left(\frac{\alpha_{+,\times}}{0.1} \right) \left(\frac{L_\nu}{10^{52}\,\mathrm{erg\,s^{-1}}} \right) .$$

Comparing with eq. (10.98) or with eq. (10.110), we see that the GW amplitude generated by neutrinos is competitive with the other main GW generation mechanisms if values of $\alpha_{+,\times}$ of order 0.1 can be obtained. Such value are indeed obtained in numerical simulations. Figure 10.17 compares the GW amplitude from post-bounce convection with that from neutrinos, in a model with a progenitor with a mass $M = 11.2 M_\odot$ that undergoes several violent shock oscillations before the explosion. The amplitude of the GWs generated by neutrinos (dashed) grows steadily with time, as a consequence of its integral form (10.133), and reaches values much higher that that produced by convective motions of matter.

Figures 10.18 shows the frequency spectrum of the GW signals due to the sum of the contribution from convective matter motion and from neutrinos, and compares it with the sensitivity of the initial and advanced LIGO detectors, for a source at 10 kpc and a rotating progenitor with $M = 15 M_\odot$. Figure 10.19 shows separately the contribution due to anisotropic neutrino emission in this progenitor model. It should be observed that the strength of the neutrino signal depends, of course, on the progenitor model, and for different progenitors the signal for a SN at 10 kpc can be below the advanced LIGO sensitivity curve, as for instance in the model shown in Fig. 10.20. Observe also that, as expected from the argument discussed at the beginning of this section, the signal from anisotropic neutrino emission is stronger at frequencies $f \lesssim O(100)$ Hz.

10.5.5 GWs from magneto-rotational core collapse

The observation of very asymmetric SN explosions, as well as the interpretation of magnetars (highly magnetized NS, which will be discussed further in Section 11.1.5) as the outcome of the collapse of massive stars with very strong magnetic fields, suggests that magnetic fields can play an important role in SN explosions. In particular, magneto-hydrodynamic forces can play an important role in launching the explosion, and can drive tightly collimated relativistic outflows along the rotation axis. Magnetic fields can also have an important effect on the spin of the remnant NS, since they provide an efficient means of transporting angular momentum toward the exterior of the collapsing material, thereby slowing down the rotation rate of the nascent proto-neutron star.

Because of the conservation of magnetic flux, an initial seed magnetic field in the stellar core will be significantly amplified during core collapse. For rapidly rotating cores the initial magnetic field can be further amplified by the winding of the magnetic field lines due to differential rotation of the core, which amplifies the toroidal component of the field. There is also a magneto-rotational instability that can provide yet another mechanism for magnetic field amplification. In order of magnitude, if B reaches values of order 10^{16} G the pressure due to the magnetic field becomes comparable to the matter pressure in the collapsing core. Actually, already at somewhat smaller values of B the explosion is driven by the magnetic field. MHD simulations indicate that the combination of high rotation,

$$\beta = \frac{T_{\rm rot}}{|W|} \gtrsim 10^{-3}\,, \qquad (10.137)$$

(where, as in Section 10.5.2, $T_{\rm rot}$ is the rotational kinetic energy and W the gravitational potential energy), and large magnetic field energy,

$$\frac{E_{\rm m}}{|W|} \gtrsim 10^{-3} \qquad (10.138)$$

(where $E_{\rm m}$ is the energy stored into the magnetic field), leads to a jet-like explosion, in which a fast MHD shock is launched in the direction of the rotational axis. Even if the magnetic pressure is much smaller than the matter pressure in the inner core, the shock itself is driven by magnetic pressure since in the shock region, far away from the center, the matter density drops quickly while the magnetic field is almost constant along the rotational axis. This scenario is referred to as "magneto-rotational core collapse".

Typical pre-collapse magnetic fields are in the range $B_{\rm in} \sim 10^7 - 10^{10}$ G, and these pre-collapse values cannot be amplified to the values required for magneto-rotational collapse. In general, initial magnetic fields $B_{\rm in} \lesssim 10^{11}$ G have no significant effect on the dynamics of the core and on the GW signal. One rather needs extreme values of the pre-collapse magnetic fields $B_{\rm in} \gtrsim 10^{12}$ G, which from stellar evolution calculations might be attained in about 1% of the massive star population. Such initial fields can be amplified to values $B_{\rm fin} \gtrsim 10^{15}$ G at the surface of the core. Pre-collapse magnetic fields in the range $B_{\rm in} \sim 10^{12} - 10^{13}$ G can power explosions that release an energy of order 10^{51} erg or higher, of the order of the most luminous SN events observed.

To compute the GW production in the presence of the magnetic field one observes that the space-space components of the energy–momentum

[56]See e.g. eq. (3.160) of Maggiore (2005). We are using here unrationalized units for the electric charge, so that $\alpha = e^2/(\hbar c) \simeq 1/137$, as is usually done in classical electrodynamics. In quantum field theory it is more common to use rationalized units where the electric charge is rather defined from $\alpha = e^2/(4\pi\hbar c) \simeq 1/137$, and then there is no factor $1/(4\pi)$ in eq. (10.139).

tensor of the electromagnetic field are[56]

$$T^{ij} = \frac{1}{4\pi}\left[-E^i E^j - B^i B^j + \frac{1}{2}\delta^{ij}(E^2 + B^2)\right]. \tag{10.139}$$

Thus, in the presence of a magnetic field, the energy–momentum tensor (10.60) becomes

$$\tau^{ij} = \left(\rho v^i v^j - \frac{1}{4\pi}B^i B^j\right) + \delta^{ij}\left(p + \frac{B^2}{8\pi}\right)$$
$$+ \frac{1}{4\pi G}\left(\partial_i U \partial_j U - \frac{1}{2}\delta_{ij}\partial_k U \partial_k U\right). \tag{10.140}$$

Repeating the steps leading from eq. (10.90) to eq. (10.91), when computing $\ddot{I}_{20}$ the part proportional to δ_{ij} drops out, so in the computation of the GW amplitude the effect of the magnetic field is simply to replace $\rho v^i v^j$ by the combination

$$f_{ij} \equiv \rho v^i v^j - \frac{1}{4\pi}B_i B_j. \tag{10.141}$$

Thus, eqs. (10.87) and (10.92) become

$$A_{20}^{E2}(t) = \frac{16G\pi\sqrt{\pi}}{c^4\sqrt{15}}\int_{-1}^{1}d\cos\theta\int_0^\infty r^2 dr\,\left[f_{rr}(3\cos^2\theta - 1)\right.$$
$$+ f_{\theta\theta}(2 - 3\cos^2\theta) - 6f_{r\theta}\sin\theta\cos\theta - f_{\phi\phi}$$
$$\left. + \rho(3\cos^2\theta - 1)\,r\partial_r U - 3\rho\sin\theta\cos\theta\,\partial_\theta U\right]. \tag{10.142}$$

This amplitude can be naturally separated into its hydrodynamical, gravitational and magnetic parts,

$$A_{20}^{E2} = A_{20\,\text{(hyd)}}^{E2} + A_{20\,\text{(grav)}}^{E2} + A_{20\,\text{(mag)}}^{E2}, \tag{10.143}$$

where

$$A_{20\,\text{(hyd)}}^{E2} = \frac{16G\pi\sqrt{\pi}}{c^4\sqrt{15}}\int_{-1}^{1}d\cos\theta\int_0^\infty r^2 dr\,\rho(t,\mathbf{x}) \tag{10.144}$$
$$\times\left[v_r^2(3\cos^2\theta - 1) + v_\theta^2(2 - 3\cos^2\theta) - 6v_r v_\theta\sin\theta\cos\theta - v_\phi^2\right],$$

$$A_{20\,\text{(grav)}}^{E2} = \frac{16G\pi\sqrt{\pi}}{c^4\sqrt{15}}\int_{-1}^{1}d\cos\theta\int_0^\infty r^2 dr\,\rho(t,\mathbf{x}) \tag{10.145}$$
$$\times\left[(3\cos^2\theta - 1)\,r\partial_r U - 3\sin\theta\cos\theta\,\partial_\theta U\right],$$

$$A_{20\,\text{(mag)}}^{E2} = -\frac{4G\sqrt{\pi}}{c^4\sqrt{15}}\int_{-1}^{1}d\cos\theta\int_0^\infty r^2 dr \tag{10.146}$$
$$\times\left[B_r^2(3\cos^2\theta - 1) + B_\theta^2(2 - 3\cos^2\theta) - 6B_r B_\theta\sin\theta\cos\theta - B_\phi^2\right].$$

Observe that these equations are only valid to lowest order in v/c, since we have used the non-relativistic limit (10.43) of T^{ij}.[57] Correspondingly, the total GW amplitude can be written as

$$h = h_{\text{(hyd)}} + h_{\text{(grav)}} + h_{\text{(mag)}}. \tag{10.147}$$

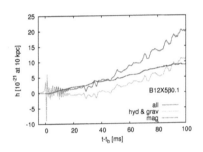

Fig. 10.21 The GW amplitudes $h_{\text{(hyd)}} + h_{\text{(grav)}}$ (lower curve), $h_{\text{(mag)}}$ (middle curve, monotonically rising) and their sum, for $B = 10^{12}$ G. From Takiwaki and Kotake (2011).

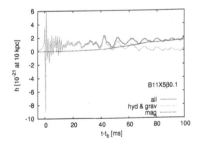

Fig. 10.22 The same as Fig. 10.21 for $B = 10^{11}$ G. From Takiwaki and Kotake (2011).

[57]The special-relativistic form of the MHD equations is discussed in Takiwaki and Kotake (2011).

Figure 10.21 shows the GW amplitude obtained from numerical simulation in a model with initial magnetic field $B = 10^{12}$ G, $\beta = T_{\rm rot}/|W| = 0.1\%$ and moderate differential rotation, for a SN at 10 kpc. The lowest curve is the sum of the hydrodynamic and gravitational contribution, $h_{\rm (hyd)} + h_{\rm (grav)}$. After the usual initial spikes due to core collapse and bounce, it starts rising, with some oscillations. The middle smooth curve, which rises monotonically, is $h_{\rm (mag)}$, while the upper curve is the total sum $h = h_{\rm (hyd)} + h_{\rm (grav)} + h_{\rm (mag)}$. Fig. 10.22 shows the same quantities in a model with the same parameters, except that the initial magnetic field is $B = 10^{11}$ G.

These plots, in agreement with similar numerical simulations, show that a pre-collapse magnetic field of order 10^{11} G or smaller has little effect on the GW emission. In contrast, for extreme values of the pre-collapse magnetic fields, of order $B = 10^{12}$ G, the GW signal acquires a rising part, both in $h_{\rm (hyd)} + h_{\rm (grav)}$ and in $h_{\rm (mag)}$. Comparison with the velocity field of the matter in the simulation shows that this rising part comes from bipolar outflows driven by the MHD explosion, which indeed only appear when $B \sim 10^{12}$ G. With such extreme values of the pre-collapse magnetic field and fast rotation, the corresponding magneto-rotational SN explosion releases more than 10^{51} erg. Therefore this rising part in the GW amplitude only appears in the most powerful SN explosions.

Figs. 10.23 and 10.24 show the characteristic strain

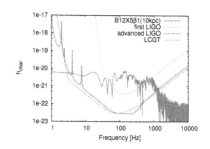

Fig. 10.23 The characteristic strain $h_c(f)$ in a model with $B = 10^{12}$ G, for a SN at 10 kpc. From Takiwaki and Kotake (2011).

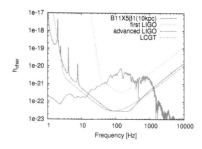

Fig. 10.24 The characteristic strain $h_c(f)$ in a model with $B = 10^{11}$ G, for a SN at 10 kpc. From Takiwaki and Kotake (2011).

$$h_c(f) \equiv \frac{1}{r}\sqrt{\frac{2}{\pi^2}\frac{G}{c^3}\frac{dE}{df}}, \qquad (10.148)$$

and compare it with the sensitivity of first- and second-generation interferometers, for a SN at 10 kpc. Fig. 10.23 shows the signal for a model with $B = 10^{12}$ G, $\beta = 1\%$ and again moderate differential rotation while, in Fig. 10.23, $B = 10^{11}$ G. The two figures show a significant difference only in the low-frequency region, since the rising part that appears for $B = 10^{12}$ G is slowly varying in time and therefore only contributes at low frequencies, $f \lesssim O(20)$ Hz.

10.5.6   GWs from fragmentation during collapse

In Section 10.3.3 we have discussed the collapsar model for BH formation. In the case of delayed BH formation the SN explosion is successfully launched, but it is not sufficiently energetic to eject all the matter in the layers over the initial proto-neutron star. Between 0.1 and $5M_\odot$ of material fall back onto the initial remnant, with typical accretion rates estimated to be in the range $(10^{-3} - 10^{-2})M_\odot$/s, which is also of the right order of magnitude required in MHD models of long gamma-ray bursts (GRBs). If the amount of fallback material is sufficiently large, the proto-neutron star is turned into a BH. Furthermore, an accretion disk is established around the BH.

In contrast, prompt BH formation takes place in a failed SN explosion, when all the material (execpt possibly the loosely bound H envelope) falls back onto the remnant. In this case, the accretion rate can be about two orders of magnitude larger than in the delayed formation scenario, although the process has a shorter duration, and the total accreted mass might be comparable. The BH forms within a few seconds from collapse, and an accretion disk is also formed, on about the same time-scale, and with a mass of order of a few solar masses. After that, viscosity drives the material in the disk toward the BH, at an accretion rate larger than $0.1 M_\odot/\text{s}$. In the collapsar model, this flow onto the BH is identified as the energy source that powers a relativistic jet, eventually responsible for a long GRB.

Actually, GRB observations show that X-ray flares often appear $10^2 - 10^4$ s after the initial GRB. A possible explanation is that the accretion disk could be gravitationally unstable and fragments at large radii, where the typical time-scale due to viscosity is comparable to the observed delay of the X-ray flares with respect to the initial GRB. The study of the dynamics of the accretion disk under the conditions prevailing in collapsars suggests that fragments will collapse until they are stabilized by neutron degeneracy pressure, forming NS with masses $(0.1 - 1)$ $M_\odot$.[58] These fragments will be driven toward the central BH by viscosity and GW losses, until tidal disruption. This would give a strong GW signal, similar to the merger phase in the coalescence of a compact binary system. As we will discuss in Chapter 14, the energy radiated in GWs in the coalescence of two spinless compact objects with masses m_1, m_2 can be estimated using eq. (14.164). Thus, for instance, a $1 M_\odot$ NS fragment falling into a $5 M_\odot$ BH would radiate in GWs an energy

$$E_{\text{GW}} \simeq 4 \times 10^{-2} M_\odot c^2 \,, \tag{10.149}$$

while, for a fragment with mass $0.1 M_\odot$ falling into a $5 M_\odot$ BH,

$$E_{\text{GW}} \simeq 6 \times 10^{-3} M_\odot c^2 \,. \tag{10.150}$$

These estimates exceed by several orders of magnitudes even the largest values obtained from the other mechanisms that we have discussed; compare for example with eq. (10.111). Thus, even if the fragmentation scenario is for the moment somewhat speculative, and does not have the same degree of support from numerical simulations, still it is potentially very interesting.

10.6   Complements: luminosity, color and metallicity of stars

In this Complement Section we recall some basic notion of astronomy such as stellar luminosity, magnitude, color and metallicity, that we have often used in this chapter.

Luminosity

The bolometric luminosity $\mathcal{L}$ is defined at the total electromagnetic energy radiated per unit time, integrated over all wavelengths. For ex-

[58] See Piro and Pfahl (2007).

ample, the nominal bolometric luminosity of the Sun is $\mathcal{L}_\odot = 3.828 \times 10^{33}$ erg/s. The solar mass is[59]

$$M_\odot = 1.98848(9) \times 10^{33}\,\text{gr}\,, \tag{10.151}$$

so

$$M_\odot c^2 = 1.79647(8) \times 10^{54}\,\text{erg}\,. \tag{10.152}$$

The bolometric solar luminosity can then be written as

$$\mathcal{L}_\odot \simeq 6.7 \times 10^{-14}\,M_\odot c^2/\text{yr}\,. \tag{10.153}$$

Explosive phenomena involve much higher luminosities. In chapters 14 and 15 we will see that the GW emission in some of the most energetic phenomena in the Universe, such as the coalescence of BH–BH binaries, can result in a instantaneous GW luminosity of the order of $200 M_\odot$ per second!

The life-time of the Sun as a main-sequence star is about 9×10^9 yr. This figure is obtained from eq. (10.153) taking into account that only about 9% of the mass of the Sun belongs to the inner core that will be involved in thermonuclear reactions, and that the efficiency for conversion of mass into radiation of thermonuclear reactions in the Sun is about 7×10^{-3}, resulting in a total efficiency of conversion of mass into radiation of about 6×10^{-4}. We will see in chapter 14 that, in the coalescence of compact binaries, the efficiency of conversion of mass into gravitational radiation is much higher.

The total bolometric luminosity of a star, together with its radius R, can be used to define the star's effective temperature. Recalling that a perfect black body at temperature T emits a radiation per unit area given by σT^4, where $\sigma = 5.670367(13) \times 10^{-5}$ erg/(s cm² K⁴) is the Stefan–Boltzmann constant, we can define the effective temperature T_{eff} of a star with bolometric luminosity $\mathcal{L}$ and radius R by

$$\mathcal{L} = 4\pi R^2 \sigma T_{\text{eff}}^4\,. \tag{10.154}$$

The Sun has a nominal equatorial radius $R_\odot = 6.957 \times 10^{10}$ cm so, using the solar luminosity, for the effective temperature of the Sun we get $T_{\text{eff},\odot} \simeq 5778$ K. If stars were perfect black bodies, this would correspond exactly to their surface temperature. In practice, stars are not perfect black bodies and T_{eff} can differ from the actual surface temperature (determined by the analysis of their spectrum) by possibly a few hundred degrees.

Given the great difference in luminosities between stars, it is convenient to use a logarithmic scale, defining the *absolute (bolometric) magnitude* M_{bol} (usually denoted simply by M, not to be confused with the mass of the star) from

$$M = -\frac{5}{2}\log_{10}\mathcal{L} + \text{constant}\,, \tag{10.155}$$

where, for historical reasons, the constant is chosen so that, for the Sun, the absolute bolometric magnitude is $M_{\text{bol},\odot} = 4.76$.[60] From eq. (10.155)

[59]We use the values from Patrignani *et al.* [Particle Data Group Collaboration] (2016).

[60]The factor 2.5 in eq. (10.155) is a convention rooted in a history that goes back to the ancient Greek astronomers. Indeed, the great Greek astronomer Hipparchus divided the apparent magnitude m of stars, whose modern definition is given in eq. (10.157), into six classes, such that the brightest stars were of first magnitudes, and the faintest seen by the naked eye were of sixth magnitude. The difference between first and sixth magnitudes (which corresponds to $\Delta m = 5$) is roughly equivalent to a factor of 100 in intensity, which motivates the factor 5/2 in front of the logarithm in eqs. (10.155) and (10.157).

it follows that the luminosity $\mathcal{L}$ of an object with magnitude M can be expressed in terms of the solar luminosity and the solar magnitude as

$$\mathcal{L} = 10^{\frac{2}{5}(M_\odot - M)}\, \mathcal{L}_\odot\,. \tag{10.156}$$

The luminosity $\mathcal{L}$ gives a measure of the intrinsic energy output of the star. What we actually detect is the flux $\mathcal{F} = \mathcal{L}/(4\pi d^2)$, where d is the distance to the star.[61] The convention is to define a fictitious luminosity, $\mathcal{L}_{10pc}$, as the luminosity that would produce the observed flux $\mathcal{F}$, if the star were located at a distance of 10 pc, rather than at its actual distance d, i.e. $\mathcal{L}_{10pc} \equiv \mathcal{L} \times (10\,\mathrm{pc}/d)^2$. The *apparent (bolometric) magnitude* m is then defined as

$$m = -\frac{5}{2}\log_{10}\mathcal{L}_{\mathrm{bol,10pc}} + \mathrm{constant}$$

$$= M + 5\log_{10}\left(\frac{d}{10\,\mathrm{pc}}\right), \tag{10.157}$$

[where the constant in the first line is the same as in eq. (10.155)], so the absolute magnitude is equal to the apparent magnitude that the star would have if it were at a distance of 10 pc. The *distance modulus* is defined as

$$\mu = m - M\,, \tag{10.158}$$

and, from eq. (10.157), it is related to the distance to the object by

$$d = 10^{(5+\mu)/5}\,\mathrm{pc}\,. \tag{10.159}$$

In practice, before extracting the distance d from the distance modulus, one must apply a correction due to interstellar extinction. Because of absorption and scattering of photons in the interstellar space the stars appear dimmer than they actually are, i.e. their apparent magnitude is increased. This correction depends strongly on the wavelength. Furthermore, for objects moving rapidly away from us, as for instance objects at a cosmological distance, one must also take into account the redshift of the emitted photons. So, in general,

$$m - M = 5\log_{10}\left(\frac{d}{10\,\mathrm{pc}}\right) + A + K\,, \tag{10.160}$$

where A takes into account interstellar extinction, and the term K, called the K-correction, takes into account the relative velocity between us and the object. Optical and UV wavelengths are strongly affected by interstellar extinction. As a rough order of magnitude, in the disk of the Galaxy the extinction at visible light is of order one magnitude per kpc. This value, however, increases significantly if the line-of-sight includes for example giant molecular clouds. For instance, the distance of the Sun from the Galactic center is $R_0 = 8.0 \pm 0.5$ kpc, and the extinction at visible light from sources near the Galactic center is more than 30 magnitudes. Therefore, the study of the Galactic plane in the optical and UV wavelengths is limited to relatively nearby stars. Of course, the

[61] Actually, for objects at a cosmological distance, the quantity d that enters in this expression defines the luminosity distance; see Section 4.1.4 of Vol. 1. Within the Galaxy, however, we can use the Newtonian definition of distance.

extinction is dramatically lower in the directions away from the Galactic plane, where most dust resides. Even in the Galactic plane, however, there exist some line of sight where extinction is minimal. In particular the Baade's window, which is 3.9° below the Galactic center and whose line of sight passes just 550 pc from the center, allows us to see stars located beyond the Galactic center.

Color index

Another important characterization of the energy output of a star is given by its relative luminosity in different wavebands, which furthermore corresponds better to what a single detector actually measures. The visual (or V) band, the blue (or B) band and the ultraviolet (or U) band are defined as the bands centered at the wavelengths $\lambda = \{551, 445, 365\}$ nm for V, B and U, respectively, and effective bandwidth $\Delta\lambda/\lambda \simeq 0.2$ (more precisely, $\Delta\lambda = \{88, 94, 66\}$ nm, respectively). The corresponding luminosities are denoted by L_V, L_B and L_U, the absolute magnitudes by M_V, M_B, and M_U, and the apparent magnitudes by m_V, m_B, and m_U. It is also common to use the notation $m_V = V$, $m_B = B$ and $m_U = U$. With the increased importance of infrared astronomy, this UVB system has been extended toward the red and the infrared, with filters named R, I, J, H, K, L, M.[62] In particular the K band is centered around 2.2 μm (in the near infrared) with $\Delta\lambda/\lambda \simeq 0.2$, and in this window the atmosphere is relatively transparent. The arbitrary constant in eq. (10.155), is chosen in a different way in each band, with the convention that the star Vega has $U = B = V = R = \ldots = M = 0$ (except that, for historical reasons, the apparent magnitude of Vega in the visible band actually is $V = +0.03$).

With this convention, the absolute magnitudes of the Sun in the U, V and B bands are

$$M_U \simeq 5.61\,, \qquad M_B \simeq 5.48\,, \qquad M_V \simeq 4.83\,. \qquad (10.161)$$

The corresponding apparent magnitudes are obtained from eq. (10.157), with

$$d = 1\,\mathrm{AU} \simeq 4.848 \times 10^{-6}\,\mathrm{pc}\,, \qquad (10.162)$$

and are

$$U \simeq -25.96\,, \qquad B \simeq -26.09\,, \qquad V \simeq -26.74\,. \qquad (10.163)$$

From the definition, higher intrinsic luminosities corresponds to more negative values of the apparent magnitude, while faint stars have large positive apparent magnitudes. For orientation, Sirius, which is at a distance of just 2.7 pc and is the brightest star visible in the night sky, has $V = -1.45$, while the faintest stars visible to the naked eye have $V \simeq +6$. Typical type Ia SNe at cosmological distances, which have played a crucial role in establishing the fact that the expansion of the Universe at the present epoch is accelerating, have visual magnitudes ranging from, say, $V \simeq 15$ to $V \simeq 22$. Of course this definition of

[62] For the precise definition, and comparison with different filter schemes, such as those used on the Hubble Space Telescope, see Binney and Merrifield (1998), Section 2.3.1. In particular, the Sloan Digital Sky Survey (SDSS) used a photometric scheme with five filters, basically non-overlapping, covering the wavelengths from 300 nm (corresponding to the atmospheric UV cutoff) to the sensitivity limit of silicon CCD, near 1100 nm, which has become quite standard. These five color bands are denoted by u', g', r', i' and z'. The u' filter peaks at 350 nm, in the near UV, and has a full width at half maximum of 60 nm; g' is a blue-green band centered at 480 nm with a width of 140 nm; r' is a red band centered at 625 nm, again with a width of 140 nm; i' is a far red filter centered at 770 nm with a width of 150 nm, and z' is a near-infrared filter centered at 910 nm with a width of 120 nm; see Fukugita *et al.* (1996). This photometric system evolved from the Thuan-Gunn photometric system, with bands denoted u, g, r, i and z.

Object	$V_{\min}$	$V_{\max}$
Sun		-26.74
Moon		-12.94
Venus	-3.82	-4.89
Jupiter	-1.61	-2.94
Mars	+1.84	-2.91
Sirius		-1.47

[63]In this context, astronomers denote collectively as "metals" all elements different from hydrogen and helium, even when these elements are not metals in the usual chemical sense.

absolute and apparent magnitudes can also be applied to galaxies. The faintest objects seen in visible light with the Hubble Space Telescope have $V \simeq 31.5$. The maximum and minimum apparent visual magnitude of the Sun, Moon, Sirius and some planets are reported in Table 10.5.

Comparison of the luminosities in different bandwidths can be used to define the color of a star. Since stars emit approximately as black bodies, the color of a star gives a measure of its temperature and can be characterized by the ratio of luminosities in two different wavebands, for example by L_B/L_V. Equivalently, since magnitudes are logarithms of luminosities, the color can be expressed by the difference in the apparent magnitudes $B - V$, which is called the color index, or simply the color, of the star. Observe that, increasing the temperature, the peak of a black-body spectrum moves from the visible band to the blue band, so L_B/L_V increases and

$$B - V = -2.5 \log(L_B/L_V) + \text{constant} \tag{10.164}$$

decreases.

Metallicity and Population I–II–III

Another revealing property of a star is its chemical composition. A first approximate classification of stars according to their chemical properties is based on their *metallicity Z*, defined as the proportion of elements other than hydrogen and helium in the chemical composition of the star.[63] More generally, one denotes by X the mass fraction of H, by Y the mass fraction of He, and by Z the mass fraction of all heavier elements. The physics behind this distinction is that primordial nucleosynthesis basically results in H and He, plus extremely tiny traces of other elements such as ^{7}Li. Thus, the first generation of stars created after the big bang should have had essentially zero metallicity. Elements heavier than He are synthesized in stars that, after exploding as SNe, eject these heavier elements into the interstellar medium. Subsequent generations of stars form from this metal-enriched gas, and therefore have higher metallicities. A first rough distinction can then be drawn among *Population I stars*, which are metal-rich (with typical values $X \simeq 0.70, Y \simeq 0.28, Z \simeq 0.02$), and therefore are the most recent stellar population, and *Population II stars*, which are older and therefore metal-poor (typically $X \simeq 0.76, Y \simeq 0.24, Z \simeq 0.001$), although a wide range of intermediate metallicities exist in stars. *Population III stars*, with essentially zero metallicity (except for a small quantity of elements produced during big bang nucleosynthesis, such as ^{7}Li), representing the very first generation of stars after the big bang, have not been found yet, and are actively being searched for. The Sun is a Pop I star, and its initial composition was $X_\odot \simeq 0.71, Y_\odot \simeq 0.27, Z_\odot \simeq 0.016$.

A more accurate definition of metallicity is obtained from the iron-to-hydrogen ratio in the atmosphere of the star, normalized to the Sun's value,

$$\left[\frac{\text{Fe}}{\text{H}}\right] \equiv \log_{10}\left(\frac{N_{\text{Fe}}}{N_H}\right) - \log_{10}\left(\frac{N_{\text{Fe}}}{N_H}\right)_\odot, \tag{10.165}$$

so the Sun by definition has [Fe/H] = 0, and older stars have negative values of [Fe/H]. As we have seen in this chapter, iron is produced in the most massive stars, and typically ejected in type Ia SN explosions. Observed metallicities range from values of [Fe/H] smaller than -5 for extremely metal-poor and old stars, to values [Fe/H] = $+1$ for young and very metal-rich stars. The current low-metallicity record is held by HE 1327-2327, with [Fe/H] = -5.4. Such a value of [Fe/H] means that the metallicity is $10^{-5.4} = 4 \times 10^{-6}$ times smaller than the solar value $Z \simeq 0.02$, and hence in this star the level of pollution by SN ejects is less than a part in 10^7.

Various other metallicity indicators are also used, such as [O/Fe], [Mg/Fe], [Si/Fe], [Ca/Fe], [Ti/Fe], [Cr/Fe], [Co/Fe] and [Al/Fe]. The equation of state, as well as the opacity of stars, depends on the chemical composition. As a result, even on the main sequence there are differences in the evolution of Pop I and Pop II stars. In particular, Pop II stars are comparatively hotter and therefore evolve faster than Pop I stars of the same mass.

Further reading

- Historical SNe and their remnants are discussed in the textbook by Stephenson and Green (2002) and in Green and Stephenson (2003). An interesting source of historical references is Burnham (1978).

- The identification of the remnant of SN 185 with RCW 86 was proposed by Clark and Stephenson (1977), and recently supported by the results of Vink *et al.* (2006). The alternative identification with G320.4−01.2 (RCW 89), which contains the pulsar PSR150958, was proposed in Thorsett (1992).

  The interpretation of the Crab SN in terms of an electron-capture SN event is discussed in Smith (2013), Tominaga, Blinnikov and Nomoto (2013) and Moriya *et al.* (2014). Arguments against this interpretation are discussed in Bietenholz and Nugent (2015).

  The association of SN 1181 with G130.7+03.1 (3C58) was proposed in Green and Stephenson (2003). It has, however, been questioned in Bietenholz (2006) and in Gotthelf, Helfand and Newburgh (2006) where it is suggested that, because of its expansion velocity, this remnant is several thousand years old.

  The detection of scattered light echoes from Cas A and Tycho's SNe are reported in Krause *et al.*

(2005) and Rest *et al.* (2008). The classifications of Cas A as type IIb and of Tycho's SN as type Ia from the light echos are discussed in Krause *et al.* (2008a) and (2008b). Ruiz-Lapuente *et al.* (2004) report the observation of a star, dubbed Tycho G, that could have been the companion of the white dwarf that exploded. The possibility that the kinematics of Tycho G could rather be explained assuming that it is a member of the thick disk population is discussed in Fuhrmann (2005) and further commented upon in González Hernández *et al.* (2009).

Evidence suggesting a type Ia classification of Kepler's SN can be found in Blair *et al.* (2007), Reynolds *et al.* (2007) and Chiotellis, Schure, and Vink (2012).

- Catalogs of Galactic SN remnants are maintained at http://www.mrao.cam.ac.uk/surveys/snrs/, described in Green (2014), and at http://www.physics.umanitoba.ca/snr/SNRcat, described in Ferrand and Safi-Harb (2012). The youngest known Galactic SN remnant, G001.9+00.3, is discussed in Reynolds *et al.* (2008) and Carlton *et al.* (2011).

  Progenitor stars of recent core-collapse SN explosions can be identified from comparison with archival data; see Smartt (2015) for a review.

- A discussion of SN types can be found in Cappellaro and Turatto (2001). For a discussion of optical light curves of SNe see Filippenko (1997), Leibundgut and Suntzeff (2003) and references therein. A comparative study of the luminosity of core-collapse SNe can be found in Richardson *et al.* (2002) and Richardson, Branch and Baron (2006). A systematic study of the luminosities in the *V*- and *R*-bands, taking into account host galaxy extinction, is presented in Drout *et al.* (2011). The Asiago Supernova Catalog is described in Barbon, Buondi, Cappellaro and Turatto (1999), and a regularly updated on-line version can be found at http://graspa.oapd.inaf.it/asnc.html.

- The remarkably bright SN 2006gy is studied in Smith *et al.* (2007). Further ultra-luminous events are discussed in Chomiuk *et al.* (2011) and references therein.

- The original correlation between peak luminosity and the shape of the luminosity curve of type Ia SNe was discussed in Phillips (1993). Accurate correlations have been obtained in terms of a multicolor fit in Riess, Press and Kirshner (1996), and introducing a stretch parameter in Perlmutter *et al.* (1997). Peak magnitudes of type Ia SNe are discussed in Guy, Astier, Nobili, Regnault and Pain (2005) and in Astier *et al.* (2006), using the SALT template. Type Ia SNe are reviewed in Hillebrandt and Niemeyer (2000).

- SN rates inferred from external galaxies are discussed in Cappellaro, Evans and Turatto (1999) and Cappellaro and Turatto (2001). For earlier work, including discussion of the methodology for estimating the rates from external galaxies, as well as estimates based on historical SNe, see van den Bergh and Tammann (1991) and Tammann, Loeffler and Schroder (1994). The estimate of the core-collapse SN rate from the radioactive decay of ^{26}Al is discussed in Diehl *et al.* (2006). Rates in the local universe, up to 10 Mpc, are discussed in Kistler, Yuksel, Ando, Beacom and Suzuki (2011) and Kistler, Haxton and Yüksel (2013). The redshift dependence of the SN rate is discussed in Horiuchi, Beacom and Dwek (2009). A discussion of SN rates at advanced GW detectors is given in Gossan *et al.* (2016).

- For a general discussion of the mechanism of SN explosions see e.g. Chapter 4 of Padmanabhan (2001) and Chapters 15 of Carroll and Ostlie (2007). Evolutionary computations of rotating massive stars are presented in Maeder and Meynet (2000) and Heger, Woosley and Spruit (2005). Numerical simulations of SN explosions are reviewed in Janka, Langanke, Marek, Martinez-Pinedo and Müller (2007). The effect of the SASI instability on the neutrino-delayed explosion mechanism is investigated in detail in Marek and Janka (2009). A detailed review on SN explosions is Janka (2012).

- Fryer and Kalogera (2001) used collapse calculations to estimate the mass of the remnant. In particular they found a distribution of NS masses, broader than that known at the time, but confirmed by subsequent observations. A review on the evolution and explosion of massive stars is Woosley, Heger and Weaver (2002). The evolution of progenitor stars in the range $(9 - 11) M_\odot$ is discussed in Woosley and Heger (2015). The remnants of SN explosions are discussed in Heger, Fryer, Woosley, Langer and Hartmann (2003). A discussion using updated SN explosion mechanisms is given in Fryer *et al.* (2012). Simulations investigating the relation between the progenitor star and the remnant are presented in Ugliano, Janka, Marek and Arcones (2012). Lovegrove and Woosley (2013) and Sukhbold, Ertl, Woosley, Brown and Janka (2016) study the scenario in which, even for failed SNe, the H envelope is stripped and the BH masses are reduced to the He-core masses of the pre-collapse stars.

- The proper motion of pulsars is studied in Hobbs, Lorimer, Lyne and Kramer (2005), using a catalogue of 233 pulsars. The pulsar kick due to the gravitational pull in asymmetric SN explosions is discussed in Scheck, Kifonidis, Janka and Müller (2006).

- Pair-instability SNe are discussed in Woosley, Heger and Weaver (2002). The interpretation of SN 2006gy as a pair-instability SN is discussed in Woosley, Blinnikov and Heger (2007). The observational evidence for pair-instability SNe is reviewed in Gal-Yam (2012).

- The detection of He WDs with masses below $0.2 M_\odot$ is reported in van Kerkwijk, Bergeron and Kulkarni (1996) and Kilic, Prieto, Brown and Koester (2007). See Tremblay, Bergeron and Gianninas (2011) for a systematic discussion of low-mass WDs using catalogs from the SDSS survey. The detection of NS with mass of about $2 M_\odot$ is reported in Demorest, Pennucci, Ransom, Roberts and Hessels (2010) for PSR J1614-2230 and in Antoniadis *et al.* (2013) for PSR J0348+0432.

- Expressions for quadrupole radiation obtained using directly the spatial components τ^{ij} are presented in Finn and Evans (1990), Blanchet, Damour and Schäfer (1990) and Mönchmeyer, Schäfer, Müller and Kates (1991).

- The first numerical study of the GWs emitted in the relativistic collapse of a rotating stars is due to Stark and Piran (1985), who found bounds $E_{\rm GW} < 7 \times 10^{-4} M_\odot c^2$. Zwerger and Müller (1997) performed systematic studies of GWs emitted by axisymmetric collapse of rotating polytropes using Newtonian gravity. The first extensive simulations of the collapse of rotating cores in axisymmetric configurations using GR were presented in Dimmelmeier, Font and Müller (2002). Ott *et al.* (2007) and Dimmelmeier, Ott, Janka, Marek and Müller (2007) presented the first treatment employing both axisymmetric and full three-dimensional simulations, using GR and including relevant microphysics such as finite-temperature EoS and an approximate treatment of deleptonization and the neutrino radiation effect, and found that the waveforms are exclusively of type I. A large model parameter study including both GR effects and all the known necessary microphysics is presented in Dimmelmeier, Ott, Marek and Janka (2008). A review on GW signals from core-collapse SNe is Ott (2009).

- For a review of GWs generated by instabilities in compact stars see Anderson (2003). Dynamical instabilities that set in already at $\beta \simeq 0.14$, dominated by the $m = 1$ mode, were found in numerical simulation by Centrella, New, Lowe and Brown (2001). In stars with an extreme degree of differential rotation, dynamical instabilities in the $m = 2$ mode were observed at values of β as low as $\beta \simeq 0.03$; see Shibata, Karino and Eriguchi (2002). Further studies of these low-β dynamical instabilities include Saijo, Baumgarte and Shapiro (2002), Ott, Ou, Tohline and Burrows (2005) (who found a one-armed spiral-shaped dynamical instability at $\beta \simeq 0.08$ using a pre-collapse iron core evolved from the evolution of a realistic $20 M_\odot$ star) and Cerdá-Durán, Quilis and Font (2007). In Ott *et al.* (2007) the low-β instability, which was previously studied within Newtonian dynamics, is confirmed using GR simulations. This low-β instability can be interpreted as a dynamical

shear instability, as discussed in Watts, Andersson and Jones (2005) and confirmed numerically in GR simulations of differentially-rotating stars in Corvino, Rezzolla, Bernuzzi, De Pietri and Giacomazzo (2010). The effect of the magnetic field on the bar-mode instability is studied in Franci, De Pietri, Dionysopoulou and Rezzolla (2013), and in Muhlberger *et al.* (2014).

- Detailed numerical studies of GWs from post-bounce convective instabilities are performed in Müller, Rampp, Buras, Janka and Shoemaker (2004), Marek, Janka and Müller (2009), Murphy, Ott and Burrows (2009) and Mueller, Janka and Marek (2013).

- The formalism for the computation of GWs generated by anisotropic neutrino emission was developed in Epstein (1978) and Turner (1978). Numerical studies include Burrows and Hayes (1996), Müller and Janka (1997), Müller, Rampp, Buras, Janka and Shoemaker (2004) and Müller, Marek and Janka (2013). See also the reviews by Kotake, Sato and Takahashi (2006) and Ott (2009).

- The formalism for the computation of GWs in the magneto-rotational collapse is developed in Kotake *et al.* (2004), Obergaulinger, Aloy and Müller (2006), and Takiwaki and Kotake (2011). Numerical studies of collapse dynamics and GW production in the magneto-rotational collapse include Kotake *et al.* (2004), Yamada and Sawai (2004), Obergaulinger, Aloy and Müller (2006), Shibata, Liu, Shapiro and Stephens (2006), Scheidegger, Kaeppeli, Whitehouse, Fischer and Liebendoerfer (2010) and Takiwaki and Kotake (2011). The effect of magnetic fields on SN collapse is reviewed in Kotake, Sato and Takahashi (2006) and Kotake (2013).

- The collapsar model is discussed in Woosley (1993), Woosley and Weaver (1995), Paczynski (1998), MacFadyen and Woosley (1999), Fryer (1999), Popham, Woosley and Fryer (1999), Fryer, Woosley and Hartmann (1999), MacFadyen, Woosley and Heger (2001). The possibility of disk fragmentation in the collapsar model is discussed in Perna, Armitage and Zhang (2005). GW production through fragmentation is discussed in Fryer, Holz and Hughes (2002) and in Piro and Pfahl (2007).

11

Neutron stars

In this chapter we move to a discussion of another fascinating source of GWs, namely neutron stars (NS). We will begin by discussing first their discovery as pulsars, their characteristics, as summarized in the so-called $P - \dot{P}$ diagram, and their properties as a population. We will also discuss magnetars, i.e. highly magnetized NS. We will then turn to a discussion of the various mechanism of GW emission in isolated NS. The GW emission from binaries involving NS will be discussed in Section 14.4 from the theoretical point of view, and in Section 15.3 from the observational point of view.

11.1   Observations of neutron stars

11.1.1   The discovery of pulsars

As we mentioned in Section 10.2, the existence of NS was already suggested in the early 1930s by Landau, and by Baade and Zwicky. These theoretical ideas were, however, much ahead of their time, and until the beginning of the 1960s the idea of a NS (basically a single gigantic nucleus of stellar mass) was considered very speculative. The situation changed in the 1960s, thanks to the spectacular advances in radio astronomy and in X-ray astronomy.

The first pulsar, now known as PSR B1919+21, was discovered as a byproduct of an investigation, lead by Antony Hewish, whose purpose was the study of the interplanetary scintillation of radio sources (i.e. the radio "twinkling" of radio sources due to the interaction of their emission with the interplanetary plasma). The actual discovery of the first pulsar was made by Jocelyn Bell (at the time his PhD student) in August 1967, within a month of the start of regular data taking. The distinctive property of the signal was the appearance of regular pulses with a period of about 1.337 s. Fluctuations of a radio signal on such a fast time-scale were routinely observed due to interference related to human activity, such as power-line discharges, while at that time they were not expected from astronomical sources. Thus, previously signals of this sort had simply been rejected.[1] The crucial new elements that made possible its identification as an astronomical source were the fast time resolution and a repetitive observing routine. This showed that the signal was appearing four minutes earlier each day, as expected for a celestial object observed with a transit telescope. After this first discovery, pulsars started to be actively searched for by various radio

[1] Indeed, it later turned out that signals from other pulsars had already been recorded but not recognized in other radio telescopes. See Chapter 1 of Lyne and Graham-Smith (2006) for a nice historical review.

Gravitational Waves, Volume 2: Astrophysics and Cosmology. Michele Maggiore.
© Michele Maggiore 2018. Published in 2018 by Oxford University Press.
DOI 10.1093/oso/9780198570899.001.0001

telescopes, and within a few months a number of other pulsars were discovered.

The identification of pulsars with NS came thanks to the work of Pacini and of Gold. Already in 1967, just a few months before the discovery of the first pulsar, Pacini had proposed that the magnetic field associated with a rapidly rotating NS was at the origin of the energy source of the Crab nebula. If the pulsar emits a beam of radiation at an angle θ with the rotation axis (see Fig. 6.2 in Vol. 1), an observer in a direction swept by the beam will see periodic pulses, similarly to a lighthouse, with a period given by the rotational period of the NS. The relevant theory was developed in the months after the discovery, both by Gold and by Pacini. For a few months after the discovery the main competing theory was based on the idea that the signal could be due to oscillations of white dwarfs (WDs) in their normal modes. At the time WDs were already a familiar concept to astrophysicists, while NS were rather exotic objects. However, WD normal modes cannot have a period much shorter than about 1 s. This explanation was then ruled out in 1968 with the discoveries of the Crab pulsar, which has a period of 33 ms, and of the Vela pulsar, with a period of about 89 ms. At the same time, as we will see in more detail later in this chapter, the normal-mode oscillations of NS have typical periods in the range 1–10 ms, so in this case too small. Thus, the signal could not be related to oscillations, either of neutron stars or of white dwarfs. On the other hand, a WD is not compact enough to rotate with such a short period, which left rotating neutron stars as the most natural explanation. Further confirmation came from the observation of the slowdown in the Crab pulsar. If the source of the radiated energy ultimately comes from the rotational energy of the pulsar, the latter must slow down in time, and the period P of the pulses will accordingly increase (whereas for oscillations no significant change in the normal-mode frequency is induced by the emission). This slowdown was indeed confirmed in 1968–69, from a few months of monitoring of the Crab pulsar, whose period increased uniformly at a rate of about 34 ns/day, i.e. $\dot{P} \simeq 4.2 \times 10^{-13}$.

11.1.2 Pulsar spindown and the $P - \dot{P}$ plane

More quantitatively, if we assume a simple model of a pulsar as a magnetic dipole $\boldsymbol{\mu}$ rotating at frequency Ω, we get a radiated electromagnetic power

$$\mathcal{L}_{\rm em} = \frac{2}{3c^3}\mu_\perp^2\Omega^4 \,, \tag{11.1}$$

where $\mu_\perp$ is the component of the dipole moment transverse to the rotation axis. Ultimately this energy is taken from the rotational energy of the pulsar $E_{\rm rot} = (1/2)I\Omega^2$, where I is the moment of inertia. The quantity $-dE_{\rm rot}/dt$ is called the *spindown luminosity* of the pulsar, and can be inferred from the measurement of P and $\dot{P}$ using

$$-\frac{dE_{\rm rot}}{dt} = -\frac{d}{dt}\left(\frac{1}{2}I\Omega^2\right)$$

$$= -I\Omega\dot{\Omega}$$

$$= 4\pi^2 I \frac{\dot{P}}{P^3} \,, \tag{11.2}$$

where we have used $\Omega = 2\pi/P$. As we will see, the masses and radii of NS are constrained in a short range. Furthermore, the larger the mass the smaller is the radius, so the moment of inertia of different NS does not change much. Taking a characteristic value $I = 10^{45} \, \text{g cm}^2$, we get

$$-\frac{dE_{\text{rot}}}{dt} \simeq 3.95 \times 10^{31} \, \text{erg s}^{-1} \left(\frac{I}{10^{45} \, \text{g cm}^2}\right) \left(\frac{\dot{P}}{10^{-15}}\right) \left(\frac{1\,\text{s}}{P}\right)^3 . \tag{11.3}$$

Writing energy conservation as $dE_{\text{rot}}/dt = -\mathcal{L}_{\text{em}}$, we have

$$\frac{d}{dt}\left(\frac{1}{2}I\Omega^2\right) = -\frac{2}{3c^3}\mu_\perp^2 \Omega^4 \,, \tag{11.4}$$

and therefore

$$\dot{\Omega} = -\kappa\Omega^n \,, \tag{11.5}$$

with $\kappa = 2\mu_\perp^2/(3Ic^3)$ and $n = 3$. Actually, this model of a pulsar as a magnetic dipole is over-simplified, and the radiation mechanisms in the pulsar magnetosphere are more complicated and not completely understood. Thus, one rather uses eq. (11.5) as a simple phenomenological description of pulsar slowdown, with κ and n taken as free parameters. The parameter n is called the braking index, and we have seen that magnetic dipole radiation corresponds to $n = 3$. To get the braking index we must be able to measure the second derivative $\ddot{\Omega}$, and then $n = \Omega\ddot{\Omega}/\dot{\Omega}^2$. However, this is in general difficult to measure, except in a few pulsars. One finds that n changes significantly from pulsar to pulsar, and in all cases where it has been reliably measured it is smaller than 3. For instance, for the Crab $n = 2.515 \pm 0.005$ and for Vela $n = 1.4 \pm 0.2$, while for the pulsar PSR J1734-3333, which has $P = 1.17$ s and $\dot{P} = 2.28 \times 10^{-12}$, one finds an even lower value $n = 0.9 \pm 0.2$. For many pulsars a phenomenological equation such as (11.5), with κ and n time-independent, might simply be inappropriate, because with time the pulsar magnetic field can decay or align itself with the pulsar spin, or because of frequent timing irregularities such as glitches or micro-glitches that we will discuss later.

Integrating eq. (11.5) from an initial time $t = 0$ at which the pulsar is born to time t we get

$$\kappa t = \frac{1}{n-1}\left[\frac{1}{\Omega^{n-1}(t)} - \frac{1}{\Omega_0^{n-1}}\right] \,, \tag{11.6}$$

where $\Omega_0 = \Omega(t_0)$ is the angular velocity at birth. If $\Omega(t_0)$ is large, then, because of slowdown, after a sufficiently long time $\Omega(t) \ll \Omega(t_0)$. Thus, the first term in the square brackets dominates and $t \simeq 1/[(n-1)\kappa\Omega^{n-1}(t)]$. Using again eq. (11.5) to write $\kappa\Omega^n = -\dot{\Omega}$, we can estimate the age τ of a pulsar as

$$\tau \simeq -\frac{1}{n-1}\frac{\Omega}{\dot{\Omega}}$$
$$= \frac{1}{n-1}\frac{P}{\dot{P}}, \tag{11.7}$$

where P is the pulsar period. From a measurement of P and $\dot{P}$ we can therefore get an estimate for the age of the pulsar. Clearly, the age estimated through eq. (11.7) is only approximate because it assumes, among other things, that n remained constant during the whole pulsar lifetime. Furthermore, for most pulsars n is not reliably known, so one typically defines a characteristic age τ_c by setting $n = 3$, as for magnetic dipole braking,

$$\tau_c \equiv \frac{P}{2\dot{P}}. \tag{11.8}$$

For instance, for the Crab pulsar $P \simeq 33$ ms and $\dot{P} \simeq 4.2 \times 10^{-13}$, so the characteristic age is $\tau_c \simeq 1240$ yr. In contrast, using eq. (11.7) with the measured value $n \simeq 2.515$, we get $\tau \simeq 1600$ yr. These estimates give the correct order of magnitude for the age of this pulsar, which, as discussed in Section 10.1, is the remnant of the SN that exploded in AD 1054. For the Vela pulsar $P \simeq 89$ ms, $\dot{P} \simeq 1.25 \times 10^{-13}$ and $\tau_c \simeq 1.1 \times 10^4$ yr.

Alternatively, if the age is known (for example by association with a historical SN) and n can be measured, then we can use eq. (11.6) to infer the value of Ω_0 and therefore the rotational period at birth.

Another important information that we can get from a measurement of P and $\dot{P}$ concerns the surface magnetic field B, assuming that spindown is dominated by magnetic dipole radiation. The dipole magnetic field B is related to the magnetic dipole moment μ by $B \simeq \mu/r^3$. Using eq. (11.4) and writing $\mu_\perp = \mu \sin \alpha \simeq B_S R^3 \sin \alpha$, where $B_S = B(r = R)$ is the magnetic field at the surface $r = R$ of the pulsar, we get

$$B_S \sin \alpha \simeq \left[\frac{3c^3}{8\pi^2} \frac{IP\dot{P}}{R^6} \right]^{1/2} \tag{11.9}$$

$$\simeq 1.0 \times 10^{12}\, \text{G} \left(\frac{I}{10^{45}\, \text{g cm}^2} \right)^{1/2} \left(\frac{10\, \text{km}}{R} \right)^3 \left(\frac{\dot{P}}{10^{-15}} \right)^{1/2} \left(\frac{P}{1\, \text{s}} \right)^{1/2}.$$

Thus $B \propto (\dot{P}P)^{1/2}$. To characterize the properties of the population of pulsars it is convenient to use the so-called $P - \dot{P}$ diagram shown in Fig. 11.1. From eq. (11.3) we see that lines of constant luminosity correspond to $\dot{P} \propto P^3$. From eq. (11.8), curves of constant characteristic age are given by $\dot{P} \propto P$, while eq. (11.9) shows that lines of constant magnetic field correspond to $\dot{P} \propto 1/P$. These lines are shown in the plot.

In this diagram most pulsars fall into the shaded region labeled "normal pulsars", with typical periods between 0.1 and a few seconds (the pulsar with the longest know period so far is PSR J2144-3933, with $P \simeq 8.5$ s, a value so large that it challenges models of emissions). The

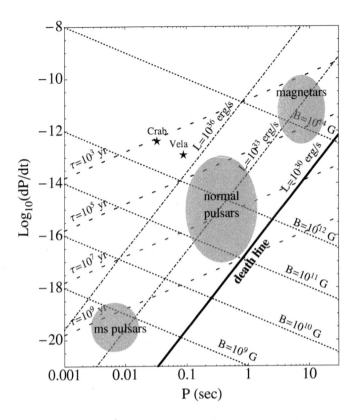

Fig. 11.1 The $P - \dot{P}$ diagram. We display some lines of constant magnetic field B (dotted), constant characteristic age τ_c (dashed) and constant luminosity L (dot-dashed). The line with $L = 10^{30}$ erg/s corresponds approximately to the "death line" (solid line), beyond which pulsars are no longer active. The positions of the Crab and Vela pulsars are also shown. The size and shape of the regions labeled "normal pulsars", "ms pulsars" and "magnetars" provide only a schematic representation. In particular, pulsars are found all along the evolutionary track leading from normal to ms pulsars; see the text.

typical spindown $\dot{P}$ is between 10^{-17} and 10^{-13}, typical magnetic fields are $10^{11} - 10^{13}$ G, and a characteristic spindown luminosity is 10^{33} erg/s. Young pulsars such as the Crab and Vela are shifted toward the upper-left corner, with respect to the majority of normal pulsars, reflecting their young age and correspondingly large luminosity. In particular the Crab pulsar has a spindown luminosity,

$$\mathcal{L} \simeq 4.5 \times 10^{38} \, \mathrm{erg \, s}^{-1} \simeq 1.2 \times 10^5 \, \mathcal{L}_\odot \qquad (11.10)$$

(assuming the conventional value $I = 10^{45} \, \mathrm{g \, cm^2}$). Out of this energy output, at least 1.5×10^{38} erg/s must be absorbed by the surrounding Crab nebula to account for the radiation that it emits, while the rest goes into the kinetic energy of the expanding nebula itself and as a wind of energetic particles. The radio emission that allows us to detect the

NS as a pulsar, in contrast, contributes to less than 1% to this energy budget. Several pulsars with a value of $\dot{P}$ even larger than the Crab's value are now known.

The evolutionary track of normal pulsars can be easily visualized on the $P - \dot{P}$ diagram. Pulsars are born in SN explosions and at birth have a short period, so, according to eq. (11.2), their spindown luminosity is large, and at birth they are on the upper-left corner of the $P-\dot{P}$ diagram. The emission of radiation slows them down, so in the $P - \dot{P}$ diagram they move rightward, toward the "normal-pulsar" region. Assuming a constant magnetic field, the pulsars move rightward along a line of constant B and the "normal-pulsar" region is reached over a time-scale of $10^5 - 10^6$ yr. Young pulsars are often associated with visible SN remnants. However, after they have reached the normal-pulsar region, the association with the original SN remnant is no longer recognizable, either because the remnant has faded away or because, over a time-scale of 10^5-10^6 yr, the "kick" imparted to the pulsar during the SN explosion has taken it far away from the original SN location. Presently, about 60 young pulsars in our Galaxy are associated with SN remnants, although not all these associations are firmly established.

Finally, over a time-scale of order $10^7 - 10^9$ yr, the pulsar moves across the "death line", where its radio emission is predicted to shut off, and the NS is no longer detectable as a pulsar. This line is estimated to correspond to

$$B \simeq 0.17 \times 10^{12} \text{G} \left(\frac{P}{1\,\text{s}} \right)^2 . \qquad (11.11)$$

Combining this with eq. (11.9) (with $\sin \alpha = 1$) and with eq. (11.3), this corresponds to a line of constant spindown luminosity 1.1×10^{30} erg/s. It should be remarked, however, that a few pulsars have been observed beyond this theoretical line. Two more very interesting regions, marked as millisecond pulsars and as magnetars, respectively, are shown in the $P - \dot{P}$ diagram. We discuss them in Sections 11.1.3 and 11.1.5, respectively.

Beside the slow spindown over astrophysical time-scales, some pulsars occasionally suffer sudden discontinuous changes of the period, known as glitches. These are believed to correspond to readjustment of the internal structure of the NS. The simplest mechanism of this sort takes into account the fact that a NS has an equilibrium value of its ellipticity, which depends on the rotation rate. As the NS slows down, the equilibrium value changes, but the rigid crust cannot follow continuously, and only readjusts itself in steps. Actually, this effect is not large enough to explain the observed glitches, and more elaborate mechanisms are believed to be at work. In particular, NS have a liquid, in fact superfluid, core plus a solid crust. These two components are only loosely coupled, and the slowdown of the crust due to the emission of electromagnetic radiation does not propagate immediately to the fluid core. The two components therefore rotate differentially, and beyond some critical value of the differential rotation some angular momentum is suddenly transferred between the two components, speeding up the crust and slowing down the core. Glitches are more frequently observed in young pulsars, such as the Vela and Crab pulsars. In particular, the glitches of the Vela pulsar are quite remarkable. We have seen that the Vela pulsar slows down at a uniform rate $\dot{P} \simeq 1.25 \times 10^{-13} \simeq 10.8\,\text{ns/day}$; on top of this it suffers large glitches, say about once every three years, where the period suddenly decreases by up to 200 ns, corresponding to a relative change in rotation frequency of order $\Delta f/f \sim 10^{-6}$.

11.1.3   Millisecond pulsars

Millisecond pulsars are a remarkable subclass of pulsars. As we see from Fig. 11.1, they occupy the lower-left corner of the $P - \dot{P}$ diagram, with periods of order a few ms to a few tens of ms, extremely small slowdown rates $\dot{P} \sim 10^{-21} - 10^{-19}$, very long ages of order 10^9 yr and weak magnetic fields, $B \sim 10^9$ G or lower. Furthermore, most of them are in binary systems. These distinctive properties are naturally explained in an evolutionary scenario in which the progenitor was a binary system made of two main-sequence stars. The more massive star evolves faster and undergoes the SN explosion first, leaving behind a NS remnant. If after the explosion the system remains bound, we have a binary made of a pulsar and a main-sequence star. The young pulsar evolves at first normally in the $P - \dot{P}$ diagram. Depending on the companion mass, by the time that the companion star ends its life on the main sequence and enters the red-giant phase, the pulsar has either reached the "normal-pulsar" region, or has already crossed the "death line" in Fig. 11.1. At this stage, matter from the evolved companion star is accreted by the NS. The matter falling onto the NS can generate a strong X-ray flux, making the system visible as an X-ray binary. The accreted matter forms an accretion disk, with an angular momentum that is inherited from the orbital angular momentum of the binary system. Upon accretion, this angular momentum is transferred to the NS, which therefore spins up and evolves from the "normal pulsar" region (or from the dead pulsar region) toward the region labeled "ms pulsar" in the $P - \dot{P}$ diagram. These pulsars, which escaped the fate of fading away beyond the death line, or came back from the dead-pulsar region, are also often called *recycled pulsars*. In particular, we see from the constant-luminosity lines of the $P - \dot{P}$ diagram that their luminosity can be comparable to that of young pulsars. However, all other parameters are very different. Their periods are much shorter than the periods of recently born pulsars, while their magnetic fields have significantly decayed with age, and therefore also their $\dot{P}$ values are much smaller. Most of the known millisecond pulsars are still bound in a binary system to their companion. Depending on the fate of the secondary stars, they give rise to pulsar–WD binaries or to double NS systems (as the Hulse–Taylor and the double pulsar that we discussed in detail in Chapter 6 of Vol. 1). A clear confirmation of this evolutionary scenario is indeed provided by the double pulsar system PSR J0737-3039. The component PSR J0737-3039A is a millisecond pulsar, with a period 22.7 ms, while its companion PSR J0737-3039B is a normal pulsar, with a period 2.7 s; see Table 6.2. Clearly the millisecond pulsar is the remnant of the star that exploded first, which was born as a normal pulsar and was later recycled by accretion from its companion. When eventually the companion also underwent a SN explosion, it gave rise to a normal pulsar.

A few isolated millisecond pulsars are also known, and are believed to result from the disruption of the binary system in the second SN explosion or of the ablation of the companion by the strong relativistic particle wind of the pulsar during the X-ray phase.

Historically, the search for millisecond pulsars was made possible by an increase in the sampling rate. Furthermore, high sampling rates in many different frequency channels become necessary in order to correct the signal for dispersion (see the discussion on pag. 312 in Vol. 1). Another complication is that, since most millisecond pulsars are in binary systems with a short orbital period, the observed pulsar period varies rapidly as the pulsar moves along its orbit, which requires a search in a wider parameter space. Indeed, many of the data analysis techniques necessary for GW detection of periodic signals, which we discussed in detail in Chapter 7, originated from pulsar astronomy. The first millisecond pulsar, PSR B1937+21, with a period of 1.557 ms, was discovered in 1982, and its signal was so strong that it was possible to use a very narrow-band receiver so that no de-dispersion was needed for its detection. It also remained the pulsar with the shortest known period for over 20 years. Currently, the pulsar with the shortest known period is J1748-2446ad, discovered in 2004 in the globular cluster Terzan 5, with a period $P \simeq 1.397$ ms. The special environment of globular clusters is indeed particularly favorable to the formation of millisecond pulsars. For instance, a total of 22 millisecond pulsars have been discovered in the globular cluster Tuc 47.

11.1.4   Pulsar demography

At present, over 2500 pulsars are known in the Galaxy, including about 300 millisecond pulsars, and new pulsars are discovered at a rate of about $50 - 100$ per year. This, however, only represents the tip of the iceberg of the actual population of Galactic pulsars. In fact, beyond a few kpc the flux density of most pulsars falls below current sensitivities, and only the brightest objects can be detected. One must also include the beaming correction, due to the fact that only a fraction f of active pulsars have beams that intercept our line of sight. The beaming fraction is estimated to be $f \sim 0.2$ for normal pulsars. However, short-period pulsars seem to have wider beams and therefore larger values of the beaming fraction, $f > 0.4$. After correcting for these sensitivity limitations and for beaming fraction, as well as for other selection effects, one estimates that in the Galaxy there should be about 1.6×10^5 active normal pulsars and 4×10^4 millisecond pulsars.[2] From the distribution of pulsars in the $P - \dot{P}$ plane one sees that most of the normal pulsars have a lifetime $\tau \sim 10^7$ yr, so one can estimate a Galactic birth rate of normal pulsars

[2] See Lyne and Graham-Smith (2006) and Lorimer (2008) for a detailed discussion of selection effects and pulsar demography.

of about $1.6 \times 10^{-2}\,\mathrm{yr}^{-1}$, i.e. about two per century, which is indeed consistent with the SN rates discussed in Section 10.2.3. Millisecond pulsars have a much longer characteristic age $\tau \sim 10^9 - 10^{10}\,\mathrm{yr}$, so the corresponding Galactic birth rate should rather be about 4×10^{-5} to $4 \times 10^{-6}\mathrm{yr}^{-1}$.

Most pulsars are concentrated within $\pm 600\,\mathrm{pc}$ from the Galactic plane, with a distribution in the vertical scale that is approximately an exponential, and are typically moving away from it. This is consistent with the fact that they were born close the Galactic plane, where their progenitors, which are massive stars, reside. Their mean velocity at birth is estimated to be in the range 400–500 km/s, more than an order of magnitude larger than the mean velocity of massive stars (which is typically in the range 10–50 km/s, as deduced from proper motion measurements), and some pulsars have velocities in excess of 1000 km/s. This is consistent with the fact that they received a large kick at birth, during the SN explosion. These large velocities are responsible for the fact that the scale height ~ 600 pc of the vertical distribution of pulsars is much higher than that of most Pop I stars. Indeed, the so-called thin-disk component of the Galaxy, where star formation is currently ongoing, rather has a vertical scale height of about $300 - 350$ pc.

Younger pulsars are found closer to the Galactic plane, again in agreement with their birth in SN explosions. The measure of the transverse velocity v_z of a pulsar can be used to give a different estimate of the pulsar age, obtained dividing its height h over the Galactic plane by v_z. This defines the "kinetic age" of the pulsar. For young pulsars it turns out to agree, within a factor $O(1)$, with the characteristic age τ_c. However, for kinetic ages larger than a few Myr, the kinetic age can be smaller than the characteristic age by a factor $O(10)$. This was initially interpreted as evidence for decay of pulsar magnetic fields on a time-scale of 10 Myr, which would explain why the characteristic age overestimates the actual age. However, it was later realized that the discrepancy might rather be due to selection effects in the computation of the kinetic ages, for example due to pulsars that move mostly parallel to the Galactic plane, i.e. with a transverse velocity v_z small compared with the modulus v of the velocity. Removing these pulsars from the analysis leaves us with no evidence for magnetic field decay over a time-scale $O(100)$ Myr. Thus, at present, it is not clear whether there is a significant decay of the pulsar's magnetic field on such time-scales, which are of the order of the lifetime of normal pulsars (while, for recycled pulsars, the low value of the magnetic field can be explained by magnetic field decay during the accretion of material from the companion). In contrast, there is strong evidence that the angle between the magnetic field and the pulsar rotation axis evolves in time, and the magnetic field aligns with the spin axis on a time-scale of order 10^7 yr.

Although pulsars were first observed at radio frequencies, they are now known to emit radiation over a huge bandwidth. For instance, the spectrum of the Crab pulsar has been observed over 18 orders of magnitudes, from radio waves up to X- and γ-rays. In contrast to accretion-powered

X-ray sources, which are binary systems where the X-ray emission is due to accretion onto a NS of matter from a companion star, normal X-ray pulsars are powered by rotation of the NS, and exhibit the usual periodic behavior of pulsars. On top of this, a fainter unpulsed thermal X-ray radiation can be detected in some isolated pulsars, and is most probably a thermal relic of the energy released during the gravitational collapse.

About 20 pulsars have also been detected in the Magellanic Clouds. Combined with the relatively shallow potential well of the Magellanic Clouds and with the large proper motion velocities of pulsars, this indicates that a large population, $O(10^5)$ pulsars, must have escaped the Magellanic Clouds and are wandering in the local intergalactic medium.

11.1.5   SGRs and magnetars

Soft gamma repeaters (SGRs) are astrophysical objects that emit bursts of soft γ-rays and hard X-rays. While γ-ray bursts are one-shot events, SGRs exhibit recurrent activity, with flares separated by months or years of more-or-less quiescent activity, and occasional giant flares. "Normal" bursts from SGRs have a typical peak luminosity, in X- and γ-rays, in the range $10^{38} - 10^{42}$ erg/s (assuming isotropic emission) and duration in the range $0.01 - 1$ s, with a log-normal distribution peaked at $0.1 - 0.2$ s, displaying a fast rise time and a slower decay. Recalling that the bolometric luminosity of the Sun (i.e. the luminosity integrated over all wavelengths) is of order $\mathcal{L}_\odot \simeq 3.8 \times 10^{33}$ erg/s, we see that a SGR burst can radiate in less than a second an energy comparable to that radiated by the Sun in one year. Their spectrum is relatively soft, with typical energies of order 10 keV, and can be well fitted by a two-black-body model with two different temperatures, implying emitting areas with radii of about 14 km and 2 km, respectively, corresponding to a size slightly larger than the surface of a NS, and to the size of its polar caps, respectively.

Giant flares are much rarer, and even more impressive, and to date only three of them have been observed. They can last for a few minutes, and their total radiated energy can exceed 10^{46} erg, with a peak luminosity in X- and γ-rays of order $10^{44} - 10^{46}$ erg/s. The energy of these events is therefore huge compared with most other astrophysical phenomena, although still much smaller than that of SN explosions. Indeed, we saw in Chapter 10 that a typical core-collapse SN radiates about 10^{49} erg in the electromagnetic spectrum, and the strongest SNe, such as SN 2006gy, reach $\sim 10^{51}$ erg. The energy spectrum of giant flares is harder than that of normal flares, extending up to the MeV range and beyond. A distinctive feature of giant flares, which as we will see below is crucial to understanding the nature of SGRs, is the fact that their fading tails display periodic oscillations, with a period of a few seconds. From this (rather limited) sample of three giant flares one can tentatively estimate that each SGR could emit a giant flare about 1–3 times per century.

[3]This giant flare was detected by a number of spacecraft that were cruising in the Solar System, such as the Russian Venera 11 and 12, the NASA Helios 2 probe, the Pioneer Venus Orbiter's detector, and several others. Its total radiated energy was about 5.2×10^{44} erg, which is about 10^4 times larger than a typical thermonuclear flash. Its bright spike lasted about 0.25 s, reached a peak luminosity of 3.6×10^{44} erg/s, and emitted an isotropic energy of 1.6×10^{44} erg. Its flux on Earth was about 100 times stronger than any burst of γ-ray previously detected. It was followed by a fading tail that lasted about 3 minutes, where the signal oscillated with a period of 8.00 ± 0.05 s, and which radiated a further 3.6×10^{44} erg. Fourteen hours later another weaker burst came from the same direction, and in the course of the next four years 16 more bursts were detected from this source. Its last burst was observed in 1983, and since then the source, identified as SGR 0526-66, remained quiescent.

[4]See the SGR/AXP Online Catalog at http://www.physics.mcgill.ca/~pulsar/magnetar/main.html for updated information, and Olausen and Kaspi (2014) for a description of the catalog.

The first "normal" SGR burst was detected on January 7, 1979, from a source located in the direction of Sagittarius, now identified as SGR 1806-20, although at that time the distinction between SGRs and the more common γ-ray bursts was not yet clearly recognized. The first giant flare was detected shortly afterwards, on March 5, 1979, from a different source.[3] Observations by various spacecraft allowed astronomers to triangulate the source direction, which was found to be consistent with that of a young SN remnant, SNR N49 in the Large Magellanic Cloud (although the association is not totally secure, with a probability of chance alignment of about 0.5%). Assuming the correctness of this identification, the NS was somewhat displaced from the center of the SN remnant, indicating a kick velocity at birth of order 1000 km/s. The high luminosity, the association with a SN remnant and the periodic pulsations in the fading tail clearly suggest that the source was a young NS with a rotation period of about 8 s. The $P - \dot{P}$ diagram of Fig. 11.1 shows that a young NS that has already been slowed down to such a value of P must have an ultra-strong magnetic field, of order 10^{15} G. At about the same time, a period of intense reactivation of SGR 1806-20 in 1983 made it clear that these bursts, which had initially been classified as a subclass of γ-ray bursts characterized by short duration and soft spectrum, really formed a different class. Indeed, no source of "classical" γ-ray bursts has ever been shown to emit multiple bursts over the years, and in all current theoretical models the source of a classical γ-ray burst, whether a coalescing binary or an exploding star, is destroyed in the explosion, leaving a remnant such as a final BH, and therefore cannot burst again.

A few more SGRs have been identified since; see Table 11.1.[4] In particular, the sources SGR 1900+14 and SGR 1806-20 were both discovered in 1979 and have remained active since, with hundreds of bursts detected, while SGR 1801-23 was discovered in 1997 and SGR 1627-41 in 1998, when it emitted over 100 bursts. Again, they are associated with young SN remnants.

After the giant flare of March 1979, another giant flare was observed on August 27, 1998, coming this time from SGR 1900+14. Again, the tail of the signal displayed a periodic structure with a period $P \simeq 5.15$ s. Its energy release in the initial spike was larger than 1.4×10^{44} erg, and a further 1.2×10^{44} erg were released in the subsequent fading tail. Compared with the March 1979 giant flare from SGR 0526-66, the source was closer, at an estimated distance of 12.5 kpc (while, as already mentioned, SGR 0526-66, barring chance alignment, was located in the Large Magellanic Cloud, at a distance of about 50–55 kpc), and the corresponding γ-ray flux was the most intense ever recorded from an astrophysical source. The incoming flux hit the Earth on its night side, and was so strong that it significantly ionized the upper atmosphere. Indeed, for a period of about five minutes the inner edge of the ionosphere plunged from 80 km down to 60 km, which is about its day-time level. This was quite a remarkable effect for a source located at about 12 kpc. The record for the most energetic burst was then broken on December 27, 2004, when a

Name	P (s)	$\dot{P}$ (10^{-11})	B $(10^{14}$ G)	τ_c (kyr)	Distance (kpc)	Association	Band	Notes
SGR 0418+5729	9.07838822(5)	0.0004(1)	0.061	36000	~ 2	—	X	Low-B magnetar
SGR 0501+4516	5.76209653(3)	0.582(3)	1.9	16	0.8(4)?	SNR HB9?	X,O	
SGR 0526-66	8.0544(2)	3.8(1)	5.6	3.4	50	SNR N49, LMC?	X	Giant flare, 1979
SGR 1627-41	2.594578(6)	1.9(4)	2.2	2.2	11.0(3)	SNR G337.0-0.1	X	
SGR 1745-2900	3.7635603(68)	0.65(14)	1.6	9.2	~ 8.5	Galactic Center, Sgr A*?	X,R	
SGR 1806-20	7.6022(7)	75(4)	24	0.16	$8.7^{+1.8}_{-1.5}$	Massive star cluster	X,H,O	Giant flare, 2004
SGR 1822-1606	8.43771977(4)	0.0254(22)	0.46	540	1.6(3)	M17 (HII region)?	X	Low-B magnetar
SGR 1833-0832	7.5654084(4)	0.35(3)	1.6	34	—	—	X	
SGR 1834-0846	2.4823018(1)	0.796(12)	1.4	4.9	4.2(3)	SNR W41	X	
SGR 1900+14	5.19987(7)	9.2(4)	7.0	0.90	12.5(1.7)	Massive star cluster	X,H,(O?)	Giant flare, 1998

Table 11.1 Data on 10 confirmed SGRs, from the SGR/AXP Online Catalog at http://www.physics.mcgill.ca/~pulsar/magnetar/main.html. The column "band" gives the wavebands in which persistent emission has been detected: H = soft γ-rays/hard X-rays (> 10 keV); X = X-rays ($1 - 10$ keV); O = optical/near-infrared; R = radio. B is the estimated magnetic field at the surface of the NS.

giant flare came from SGR 1806-20. This source is located at a distance of about 8.7 kpc, in the direction of the Galactic center.[5] The initial spike of this giant flare released a total energy of a few 10^{46} erg in about 0.1 s, corresponding to a luminosity of about 10^{47} erg/s $\simeq 3 \times 10^{13} \mathcal{L}_\odot$, and about 10^{44} erg were released in the subsequent pulsating tail, which oscillated with a period of about 7.6 s. Furthermore, soon after the end of the pulsating tail, a hard X-ray bump was observed, whose intensity peaked at about 700 s after the start of the giant flare, and which lasted about one hour. It is interpreted as an afterglow, analogous to what happens in classical γ-ray bursts. The energy emitted in the fading tail was of order 10^{44} erg for all three giant flares, while the energy in the initial spike of SGR 1806-20 was at least two orders of magnitude larger than in the other two giant flares. Finally, quasi-periodic oscillations (QPOs) have been observed in the X-ray tails of the giant flares, on top of the periodic oscillation at the NS period, but with larger frequencies; see the discussion on page 96.

A class of objects that are now believed to be closely related to SGRs are the anomalous X-ray pulsars (AXPs). These differ from normal X-ray pulsars in many respects. They have a narrow range of spin periods ($5-12$ s), and a luminosity that exceeds the spindown luminosity, so they cannot be powered by rotation. On the other hand, the lack of evidence for an orbital Doppler shift is not consistent with a binary interpretation, and also the fact that some are associated with SNR remnants, and their small scale height above the Galactic plane, indicate that they are too young to be low-mass binaries. They have a persistent X-ray luminosity of order $10^{35} - 10^{36}$ erg/s, a steady spindown, and a relatively soft X-ray spectrum. The first AXP was discovered in 1981, and to date about a dozen are known.[6]

[5] The initial estimate of about 12-15 kpc was later revised downward, see Bibby, Crowther, Furness and Clark (2008), which reduced correspondingly the estimate for the radiated energy.

[6] See the SGR/AXP Online Catalog at http://www.physics.mcgill.ca/~pulsar/magnetar/main.html

In the magnetar model discussed in the next subsection, AXPs are explained as NS powered by huge magnetic fields, just like SGRs. In fact, the distinction between AXPs and SGRs seems to be rather a matter of how these objects were first detected, whether from their bursts or from their persistent luminosity. SGRs were first detected from their bursting activity but are now known to also have a persistent X-ray emission. On the other hand, AXPs were first identified through their persistent emission, but in fact the discovery by the Rossi X-ray Timing Explorer (RXTE) that they also emit short bursts, similar to those of SGRs, has confirmed the close relation between these two classes of objects. Bursts have now been observed from several AXPs. It has also been observed that both SGRs and AXPs can enter a state with low persistent luminosity, $10^{33} - 10^{34}$ erg/s. This has important consequences for the estimates of their populations, since it implies that many more SGRs and AXPs might exist, but are presently undetected because they are in their low-luminosity state.

The magnetar model

It is by now widely accepted that SGRs and AXPs are highly magnetized NS, with a surface magnetic field of order $10^{14} - 10^{15}$ G (see Table 11.1), and possibly an even larger internal magnetic field, and that the energy of the bursts is powered by these huge magnetic fields. Such highly magnetized NS are collectively denoted as magnetars. Actually, there are examples of SGRs with relatively low surface magnetic field (see again Table 11.1), but whose outburst properties clearly show that they belong to the SGR class. This indicates that the crucial features that give rise to the SGR activity are the strength and geometry of the magnetic field in the NS interior, rather than those at the surface.

The first observational evidence for such strong magnetic fields came from observations with RXTE, an X-ray satellite designed to be sensitive to temporal variations in the X-ray flux. The observation of the persistent emission of SGRs/AXPs revealed a periodic structure in the (folded) pulse profile of these objects, which is naturally interpreted as due to the rotational period P of the NS. Long-term monitoring over a few years then allows the observers to extract not only P but also its derivative $\dot{P}$. The values of P observed in the pulsating tail of the three SGRs that produced giant flares also agree very well with those extracted from the persistent emission. From the measured values of P and $\dot{P}$ one can reconstruct the magnetic field at the surface, B, through eq. (11.9), which leads to the values shown in Table 11.1.

By itself, this estimate of B is subject to some uncertainty, since it is based on simplified assumptions. In particular, one is assuming that (1) the NS braking is due mainly to the magnetic field and (2) the magnetic field can be modeled as a magnetic dipole rotating in vacuum. Actually, we saw in Section 11.1.2 that the braking index n in general differs substantially from the value $n = 3$ predicted by magnetic braking, so the first assumption in general is certainly not completely correct. An important contribution to the pulsar spindown can come from other processes such as the ejection of a wind of relativistic particles. The assumption that the magnetic field can be modeled as a dipole can be tested against more realistic models of the NS magnetosphere, which, however, give estimates in agreement with the rotating dipole model, within a factor of about 2.

However, evidence for such ultra-strong magnetic fields also comes from a number of other arguments. The first is based on the energies of the bursts and of the giant flares. Comparing with eq. (11.3) we see that the rotational energy is way too small to account even for the persistent X-ray luminosities $10^{35} - 10^{36}$ erg/s of most SGRs and AXPs (not to mention the huge energy release in giant flares in SGRs). This is indeed the reason why AXPs were originally recognized as a separate class from normal rotationally powered pulsars. In contrast, the energy density associated with a magnetic field is

$$\rho_{\text{mag}} = \frac{1}{8\pi} \left(\frac{B}{1\,\text{G}} \right)^2 \text{erg/cm}^3 \, . \qquad (11.12)$$

Thus the total magnetic energy due to an approximately constant magnetic field B in a volume $(4/3)\pi R^3$ is of order

$$E_{\text{mag}} \sim 2 \times 10^{47} \left(\frac{B}{10^{15}\,\text{G}} \right)^2 \left(\frac{R}{10\,\text{km}} \right)^3 \text{erg} \, . \qquad (11.13)$$

Actually, the magnetic field in the NS interior will be even larger than at the surface, possibly by a factor of order 10. Thus, values of the surface magnetic field in the range $10^{14} - 10^{15}$ G might correspond to average values of B in the NS interior of order $10^{15} - 10^{16}$ G and total magnetic energies of order $10^{47} - 10^{49}$ erg. A theoretical upper limit to the magnetic field inside the NS is about 3×10^{17} G. Beyond this value, the fluid inside the star would mix and the field would dissipate.

The idea that SGRs and AXPs have surface magnetic fields in the range $10^{14} - 10^{15}$ G provides a quite consistent pictures from the energetic point of view. With such a large magnetic field, after birth these NS move very fast in the $P - \dot{P}$ plane, slowing down to periods of order of a few seconds in just 10^4 yr. This explains why SGRs and AXPs with shorter periods are not observed, and is also consistent with the young age of these objects suggested by the observation of some associations with SN remnants or with clusters of massive stars, as well as by their small scale height distribution over the Galactic plane. This huge reservoir of magnetic energy, of order $10^{47} - 10^{49}$ erg, is potentially able to sustain the persistent emission of SGRs and AXPs, at the level $10^{35} - 10^{36}$ erg/s, for 10^4 yr $\simeq 3 \times 10^{11}$ s, or more. The energy of giant flares is more challenging to explain, but still consistent with these numbers. Consider a giant flare such as that of SGR 1806-20, which released a few times 10^{46} erg. Assuming that it takes place of order once per century in a given SGR, as suggested by the existing statistics, over a lifetime of 10^4 yr we need an energy reservoir of $10^{48} - 10^{49}$ erg, compatible with the above values.

Another line of evidence for such strong magnetic fields comes from the existence of the pulsating tail in giant flares. This long-lasting emission is understood as due to the fact that part of the energy released by the burst is stored in a hot plasma, magnetically confined in the magnetosphere. The energy that can be confined in this way depends on the surface magnetic field, and to reproduce the observed values of the energy radiated by the pulsating tail (which, as we have seen, was of order 10^{44} erg in all three giant flares) we need a surface magnetic field of a few times 10^{14} G in all three SGRs, in agreement with the values inferred from P and $\dot{P}$ given in Table 11.1.[7] Two possible mechanisms have been proposed to explain the existence of such ultra-strong magnetic fields. In the original magnetar model, proposed by Duncan and Thompson, such a magnetic field is generated during the collapse of the stellar core that gives rise to the NS. This takes place by dynamo amplification due to turbulent fluid motion, either in the convective region of the progenitor star or in the differentially rotating proto-NS. However, a coherent large-scale magnetic field can only be generated if the rotation rate of the proto-NS is comparable to or faster than the convection rate. For shorter rotation rates the dynamo mechanism only acts locally, rather than globally, and cannot generate a strong large-scale magnetic field. This requires that the proto-NS rotates very rapidly, with a period smaller than the convective overturn time, which is in the range $3 - 10$ ms. Population studies of radio pulsars suggest that such fast-spinning new-born NS might be rare. If this happens, however, significant amplification is possible. The dynamo mechanism acts only for about 10 s or less, but can generate a magnetic field

$$B \sim 3 \times 10^{17} \left(\frac{1 \, \mathrm{ms}}{P_0} \right) \mathrm{G} \,, \qquad (11.15)$$

where P_0 is the rotational period of the proto-NS. Another option is that these strong magnetic fields originate from progenitor stars that were themselves highly magnetized, with $B \sim 10^3$ G, and are then amplified through the usual mechanism of flux conservation during the star's collapse. Strong magnetic fields are expected on average for the more massive progenitors. Thus, in this "fossil field" scenario, magnetars are the remnants of stars with masses above about $40 M_\odot$.

The mechanism invoked to explain the burst activity is that the magnetic field lines in magnetars drift through the liquid interior of the NS, stressing the crust from below and generating strong shear strains. For magnetic fields stronger than about 10^{14} G, these stresses are so large that they cause the breaking of the 1 km thick NS crust, whose elastic energy is suddenly released in a large starquake, which generates a burst of soft γ-rays. The giant flares, in contrast, might be generated by large-scale rearrangements of the magnetic field lines.

Further evidence for the role of crustal stress in SGRs comes from the observation that the energy distribution of SGR events is the same as that of earthquakes, i.e. the number of events $N(E)$ that release an

[7]It is interesting to compare the strength of magnetic fields in magnetars with the critical value for which the energy between adjacent Landau levels becomes equal to the electron rest energy. Recalling that the Landau energy levels for an electron of mass m_e in a magnetic field B are given by $E_n = (e\hbar B/m_e c)(n + 1/2)$, the energy of a transition with $\Delta n = 1$ is $\Delta E = e\hbar B/m c$, and the condition $e\hbar B_c/m_e c = m_e c^2$ gives a critical magnetic field

$$B_c = \frac{m_e^2 c^3}{e\hbar} \simeq 4.4 \times 10^{13} \, \mathrm{G} \,. \quad (11.14)$$

In most magnetars the magnetic field exceeds this quantum critical value.

energy E has a power-law distribution

$$dN \sim E^{-\gamma}dE \qquad (11.16)$$

(up to a maximum cutoff, which reflects the finite energy contained in
any energy reservoir). For earthquakes, this is the well-known Gutenberg–
Richter law. Remarkably, the value of the index γ from different seismi-
cally active regions is approximately the same, $\gamma \simeq 1.6$, with variation
± 0.2 for different active regions. Even more striking is the fact that the
distribution of energies of 111 events from the SGR 1806−20 follows the
same law, with about the same value of the index, $\gamma \simeq 1.66$. This result
has been confirmed for SGR 1627−41 and with much larger statistics
from observations, by BATSE and RXTE, of SGR 1900+14, which suddenly became extremely active between May 1998 and January 1999,
after a long period of sporadic activity. With this large statistics (about
10^3 events) one finds $\gamma = 1.66 \pm 0.05$ over four orders of magnitudes in
energy. The underlying physical reason behind this universal behavior is
self-organized criticality, which is characterized by the fact that an agent
drives the system toward a critical state, until the energy of the system is
suddenly, and often catastrophically, released. Typical members of this
class are avalanches, earthquakes, and solar flares. Phenomena showing self-organized criticality have a certain degree of universality, in the
sense that their statistical properties are largely independent of the details of the physical mechanisms involved. Indeed, a power-law with the
same value of γ has also been obtained in computer simulations of frac-
tures in a stressed elastic medium. A similar analogy between SGR and
earthquakes has been observed in the statistics of waiting times, i.e. the
times between an event and the next.[8] These observations, together with
the fact that the huge energies released can only be explained in terms
of a compact object such as a neutron star, provide a strong evidence
that the activity of SGRs is related to starquakes on neutron stars.

[8] For both earthquakes and SGRs this statistics is very different from that of uncorrelated events. Earthquakes, SGRs and other phenomena related to self-organized criticality have periods of intense bursting activity, during which the events arrive in bunches or there is a large event followed by showers of smaller events; these intense periods are then followed by long, and some-time extremely long, periods of quies-cence. The statistics of waiting times of SGR (and of earthquakes) is well repro-duced by a log-normal function, with a mean that depends on the detector sen-sitivity, since for a more sensitive de-tector, more events go above the detec-tion threshold and the average waiting time is therefore shorter. No correla-tion is observed between the burst in-tensity and the waiting time.

11.2   GW emission from neutron stars

11.2.1   NS normal modes

Just like any elastic body, a NS excited by a perturbation oscillates into
a set of normal modes. These normal modes can be excited in a number
of astrophysical processes, such as the formation of a proto-NS in a SN
explosion, or the last stages of the collapse of a compact binary system
containing a NS. Furthermore, in the coalescence of a NS–NS binary, the
initial remnant of the coalescence can be a massive, hot and differentially
rotating proto-NS, which will initially show strong oscillations, excited
by the merger process. Other situations where normal modes can be
excited include quakes in the interior of the NS, such as those leading
to glitches in the timing residual of a pulsar, or major rearrangements
in the internal structure, such as those taking place during the giant
flares of magnetars. In turn, these coherent vibrations can in principle

be a significant source of gravitational radiation. For instance, during core collapse and bounce, the density of the core undergoes a series of compressions and re-expansions; see for example Fig. 10.11 on page 49. This induces oscillations in the new-born proto-NS mostly in the $l = 0$ mode, with variations in its central density of several percent. By itself the spherically symmetric $l = 0$ mode does not radiate GWs, but the non-linear coupling between the normal modes of the proto-NS also excites non-radial modes, such as the quadrupole ($l = 2$) mode, which in turn produce GWs.

The computation of the normal modes of a NS is quite complicated since it involves both perturbations of the metric around and inside the NS and perturbations of the fluid variables that describe the NS matter itself. The perturbations of space-time will be discussed in great detail when we study black hole (BH) quasi-normal modes in Chapter 12. In the case of NS the spectrum of normal modes is even richer because we must further include the matter perturbations.

Equilibrium configuration and Tolman–Oppenheimer–Volkov equations

To study the perturbations around an equilibrium configuration we must of course begin with the determination of the equilibrium configuration itself. For a spherically-symmetric and non-rotating star[9] this is obtained as follows. The metric, both inside and outside the star, can be written in the form

$$ds^2 = -e^{2\Phi(r)}c^2 dt^2 + e^{2\Lambda(r)} dr^2 + r^2(d\theta^2 + \sin^2\theta d\phi^2). \qquad (11.17)$$

The NS matter is modeled as a perfect fluid, whose energy–momentum tensor $T_{\mu\nu}$ is given in eq. (10.44). In this metric, the radial component of energy–momentum conservation gives[10]

$$\frac{dp}{dr} = -\left(\rho c^2 + p\right)\frac{d\Phi}{dr}. \qquad (11.18)$$

In the non-relativistic limit $p \ll \rho c^2$, this reduces to $dp/dr = \rho dU/dr$, where $U = -\Phi c^2$ is the sign-reversed gravitational potential [compare with eqs. (5.11) and (5.12) in Vol. 1]. We therefore recover the Newtonian equation of hydrostatic equilibrium in the gravitational potential. The (00) component of the Einstein equations gives

$$e^{-2\Lambda(r)} = 1 - \frac{2Gm(r)}{rc^2}, \qquad (11.19)$$

where

$$m(r) = 4\pi \int_0^r dr' \, r'^2 \rho(r') \qquad (11.20)$$

is the mass (or, more precisely, the energy divided by c^2) enclosed within radius r. The (rr) component of the Einstein equations gives

$$\frac{d\Phi}{dr} = \frac{G}{c^2}\left[m(r) + \frac{4\pi r^3 p(r)}{c^2}\right]\frac{1}{r[r - 2Gm(r)/c^2]}, \qquad (11.21)$$

[9]The approximation of spherical symmetry is excellent for the equilibrium configuration of NS. We will see in Section 11.2.4 that realistic values for the typical ellipticity of a NS can be of order 10^{-7}. In contrast, rotation is in general important and we will discuss its effect later.

[10]See Section 23.5 of Misner, Thorne and Wheeler (1973) for detailed derivations.

which in the Newtonian limit reduces to $dU/dr = -Gm(r)/r^2$. Combining eqs. (11.18) and (11.21) we get the relativistic equation for hydrodynamic equilibrium,

$$\frac{dp}{dr} = -\frac{G}{c^2}\left(\rho c^2 + p\right)\left[m(r) + \frac{4\pi r^3 p(r)}{c^2}\right]\frac{1}{r[r - 2Gm(r)/c^2]}. \quad (11.22)$$

Equations (11.18), (11.20) and (11.22) are the Tolman–Oppenheimer–Volkov (TOV) equations and, together with an equation of state (EoS) $p = p(\rho)$, determine the equilibrium configuration of a non-rotating star. Using the EoS to eliminate $p(r)$, eq. (11.22) becomes a first-order equation for $\rho(r)$, which can be integrated outward once we give as initial condition the central value of the density $\rho(0)$. Therefore the equilibrium configurations of a spherically symmetric star, modeled as a perfect fluid, form a one-parameter family.

Perturbation from equilibrium

We next compute perturbations around the equilibrium configuration. The metric perturbations are written as $g_{\mu\nu} = \bar{g}_{\mu\nu} + h_{\mu\nu}$, where $\bar{g}_{\mu\nu}$ is the background metric (11.17) (or its generalization for a rotating star). The metric perturbation $h_{\mu\nu}$ is conveniently written in a basis of tensor spherical harmonics, as we will see in great detail in Section 12.2.1. The perturbations of the energy–momentum tensor (10.44) of the perfect fluid are given by the density perturbation $\delta\rho$, the pressure perturbation δp and the Lagrangian change in the fluid velocity δu^μ.[11] Since ρ is a scalar under rotations, $\delta\rho$ can be expanded in the usual (scalar) spherical harmonics. For a perfect fluid, given an EoS $p = p(\rho)$, the perturbation δp is determined from $\delta\rho$ by

$$\delta p = \frac{dp}{d\rho}\delta\rho \equiv c_s^2\delta\rho, \quad (11.23)$$

where c_s is the speed of sound in the fluid. Finally, the velocity perturbations can be expanded in vector spherical harmonics, as in Section 3.5.2. We then obtain a set of coupled wave equations for the matter and the space-time variables, from which the normal modes of the system can be determined (the exact definition of the normal modes, or more precisely quasi-normal modes of the system, and the techniques used for computing the corresponding frequencies, will be described in detail in Section 12.3).[12]

In this way we find that NS have a rich spectrum of normal modes. One can distinguish between fluid modes, where mostly the matter variables $\delta\rho, \delta p$ or δu^μ are excited, and space-time modes, where mostly the metric perturbations are excited. Fluid modes are often computed in the Cowling approximation, which amounts to neglecting the space-time perturbations. Both types of modes can be further classified as polar or axial, depending on the intrinsic parity. We have already encountered this classification in Section 8.4, when we studied the vibration of an elastic sphere using Newtonian dynamics, and we saw that the corresponding normal modes can be classified into spheroidal modes and

[11]In fluid mechanics, the Lagrangian point of view corresponds to defining the variables as function of the position of a given fluid element. The Eulerian point of view, in contrast, corresponds to looking at the motion of the fluid at a fixed point in space. We will further discuss the Lagrangian and Eulerian points of view in the context of cosmological perturbation theory in Section 17.7.

[12]It is also interesting to observe that, for non-rotating stars, the metric perturbations can be decoupled from the matter perturbations, so the frequencies of the quasi-normal modes can be found just solving the equations for the space-time perturbations, as first found by Chandrasekhar and Ferrari (1991a).

toroidal modes. Spheroidal modes are proportional to a combination of $Y_{lm}(\theta,\phi)\hat{\mathbf{r}}$ (which describes a radial displacement) and of $\boldsymbol{\nabla} Y_{lm}(\theta,\phi)$ (which describes a transverse displacement); see eq. (8.242). These quantities are true vectors under parity, i.e. they describe polar modes. For generic l, polar modes have both radial and transverse components [except for $l=0$, in which case they are purely radial —see eq. (8.249)— since in this case $\boldsymbol{\nabla} Y_{lm}(\theta,\phi)$ vanishes]. Toroidal modes, in contrast, are proportional to $\mathbf{r} \times \boldsymbol{\nabla} Y_{lm}$, which is a pseudovector under parity, and therefore describe axial modes. Observe that $\mathbf{r} \times \boldsymbol{\nabla} Y_{lm}$ is purely transverse. For non-rotating stars, polar and axial modes are decoupled (which is no longer true for rotating stars). In the simplest treatment the NS can be modeled by a polytropic EoS as in (10.95), with no rotation nor magnetic field. In this approximation, one finds three main classes of fluid modes: the f-modes, the p-modes and the g-modes, which are all spheroidal, and which we discuss first. We will then discuss normal modes associated with rotating NS, such as the r-modes, and their instabilities, and we will finally discuss the w-modes, which are associated with space-time perturbations.

f-modes

For each $l \geqslant 2$, there is a single mode called the fundamental mode (or f-mode), with no overtones and a frequency that, depending on the EoS, is typically in the range 1–3 kHz.[13] For each l, the eigenfunction of this mode has no node inside the star and increases toward the surface of the star. It describes purely non-radial oscillations and can emit GWs efficiently. This is the only fluid mode that exists even if we model the star as a sphere of uniform density (in which case the pressure and gravity modes discussed later disappear), and is basically due to the mere existence of an interface separating the star from the external space.[14] It is instructive to first understand it in the simple setting of a constant-density self-gravitating Newtonian fluid. Its existence can then be readily understood from the continuity equation, whose linearization gives

$$\partial_t(\delta\rho) + \boldsymbol{\nabla}\cdot(\rho\,\delta\mathbf{v}) = 0\,. \tag{11.24}$$

For constant ρ this reduces to $\boldsymbol{\nabla}\cdot\delta\mathbf{v} = 0$, which is solved by $\delta\mathbf{v} = \boldsymbol{\nabla}\chi$ with $\boldsymbol{\nabla}^2\chi = 0$. Requiring a non-singular behavior for $r \to 0$ we obtain the solution (valid for $r < R$)

$$\chi(t,\mathbf{x}) = \sum_l A_l(t) r^l P_l(\cos\theta)\,, \tag{11.25}$$

where we can restrict attention to $l > 0$ since the $l = 0$ term is spatially constant and therefore does not contribute to $\delta\mathbf{v}$. This solution shows that, for any l, these modes have no node inside the star, i.e. for $0 < r < R$, and that their eigenfunctions increase toward the surface. For a star of uniform density, using again a Newtonian treatment of gravity, its frequency can be computed exactly for each l (by linearizing the Euler

---

[13]There is no $l = 0$ f-mode. The $l = 1$ mode also does not exist when one considers the full set of equations, with fluid and space-time variables, since it just corresponds to a displacement of the center of mass. However, in the Cowling approximation, where the space-time perturbations are neglected, an f-mode with $l = 1$ appears, since in this approximation momentum conservation is no longer satisfied, and the fixed external metric acts as a restoring force.

[14]This normal mode of a self-gravitating Newtonian fluid was already studied by Lord Kelvin in 1863.

equation in the presence of a gravitational potential). The result of this classic computation is[15]

[15]See Chandrasekhar (1961).

$$f(^lf) = \frac{1}{2\pi}\left[\frac{2l(l-1)}{2l+1}\frac{GM}{R^3}\right]^{1/2} \qquad (l \geqslant 2).\qquad(11.26)$$

For $l = 2$ this gives $f \simeq 2.0\,(\bar{M}/\bar{R}^3)^{1/2}$ kHz, where, here and in the following, we use the reference values

$$\bar{M} \equiv \frac{M}{1.4M_\odot}, \qquad \bar{R} \equiv \frac{R}{10\,\text{km}}.\qquad(11.27)$$

The oscillations of the f-modes with $l \geqslant 2$ are damped by GW emission. A first order-of-magnitude estimate of the damping time of the $l = 2$ mode is obtained observing that, in an oscillation of amplitude δR, the quadrupole moment of the star, $Q \sim MR^2$, has a variation $\delta Q \propto MR\delta R$ and therefore $\dddot{Q} \propto MR\omega^3\delta R$. Inserting this into the quadrupole formula (3.75), the radiated power is

$$P_{\text{quad}} \propto \frac{G}{c^5}(MR\omega^3\delta R)^2.\qquad(11.28)$$

The elastic energy contained in the oscillation is of order $E \sim M(\omega\delta R)^2$ [compare for example with eq. (8.15)]. The time τ in which this energy is dissipated via GW emission is therefore given by

$$\tau(^2f) \simeq \frac{E}{P_{\text{quad}}} \propto \frac{c^5}{G^3}\frac{R^4}{M^3},\qquad(11.29)$$

where we have used the estimate $\omega^2 \sim GM/R^3$ from eq. (11.26). In terms of $R_S = 2GM/c^2$, this order-of magnitude estimate becomes

$$\tau(^2f) \propto \frac{R}{c}\left(\frac{R}{R_S}\right)^3.\qquad(11.30)$$

The factor R/c is of order of the travel time of light across the star, and we see that, compared with this time-scale, the damping time is larger by a factor $(R/R_S)^3$. For realistic applications to NS eqs. (11.26) and (11.30) must be corrected by including a general-relativistic treatment of gravity and replacing the approximation of a constant-density sphere by a realistic EoS. This changes the numerical values, although in order of magnitude eqs. (11.26) and (11.30) still remain correct. For instance, for the $l = 2$ f-mode, performing a general-relativistic computation for non-rotating stars and fitting the results obtained from various EoS gives[16]

$$f(^2f) \simeq (0.79 \pm 0.09) + (1.50 \pm 0.09)\left(\frac{\bar{M}}{\bar{R}^3}\right)^{1/2}\text{ kHz}.\qquad(11.31)$$

This provides an estimate of the frequency of about 2.3 kHz for $\bar{M} = \bar{R} = 1$, and of about 1.7 kHz for $\bar{M} = 1, \bar{R} = 1.4$. Similarly, for the damping time one finds the fitting function

$$\tau(^2f) \simeq \frac{\bar{R}^4}{\bar{M}^3}\left[(2.31 \pm 0.05) - (1.49 \pm 0.05)\left(\frac{\bar{M}}{\bar{R}}\right)\right]^{-1}0.1\,\text{s},\qquad(11.32)$$

[16]We write the result in the form originally given in Andersson and Kokkotas (1998), using, however, the numerical values provided by Benhar, Ferrari and Gualtieri (2004) with updated EoS and including also the error on the parameters of the fit, which reflects the spread across the various EoS considered. Note also that Andersson and Kokkotas (1998) denote by ω the quantity that, according to standard notation, we rather denote by f. It should also be stressed that this fit is only valid for non-rotating stars.

giving $\tau(^2\mathrm{f}) \simeq 0.12$ s for $\bar{M} = \bar{R} = 1$, and $\tau(^2\mathrm{f}) \simeq 0.31$ s for $\bar{M} = 1, \bar{R} = 1.4$. In summary, the $l = 2$ f-mode has a typical frequency in the range

$$f(^2\mathrm{f}) \simeq 1.5 - 3\,\mathrm{kHz}\,, \qquad (11.33)$$

and a typical damping time

$$\tau(^2\mathrm{f}) \simeq 0.1 - 0.3\,\mathrm{s}\,, \qquad (11.34)$$

depending on $\bar{M}$, $\bar{R}$ and the EoS.[17] It is interesting to compare these values with the fundamental $l = 2$ quasi-normal mode of a Schwarzschild BH. As we will see in Chapter 12, in this case $f \simeq 12\,\mathrm{kHz}\,(M_\odot/M)$ and $\tau \simeq 5.5 \times 10^{-5}\,\mathrm{s}\,(M/M_\odot)$; see eqs. (12.258) and (12.259). Thus a BH with mass $(5 - 10)M_\odot$ has $f \sim 1 - 2$ kHz, which is of the same order as the typical f-mode frequency of a NS. However, the two signals are clearly distinguished by their damping times, which for such a BH is of order of a few tenths of a millisecond, while for the f-mode of a NS it is a few tenths of a second. The origin of this difference can be understood writing eq. (11.30) as $\tau(^2\mathrm{f}) \propto (R_S/c)\,(R/R_S)^4$. Compared with the time-scale R_S/c that characterizes a BH with the same mass, the decay time of the ^{2}f mode of a NS is longer by a factor $(R/R_S)^4$.

It should also be kept in mind that eqs. (11.26) and (11.30) are obtained by first computing accurately the frequencies for a given EoS, comparing them with the mass and radius of the NS obtained from the same EoS, and trying to find fitting functions $f(M, R)$ and $\tau(M, R)$ that work across several EoS. The spread across EoS is significant, and if one is interested in a specific EoS then one should of course use the value derived from the direct computation. The fitting formulas can, however, be used to obtain the mass and radius of a NS from a detection of some of its quasi-normal oscillations, without assuming a specific EoS.

p-modes

A second family of modes is given by the pressure modes (or p-modes). In this case pressure is the restoring force, and these modes basically describe acoustic waves inside the star. For these modes $\delta\rho$ and δp are significant, and the radial component of the velocity perturbation is usually quite a bit larger than the tangential component. The frequencies of these acoustic waves are determined by the travel time inside the star. These modes start from $l = 0$ (although purely radial oscillations by themselves do not generate GWs) and for each l there is now a full tower of overtones, denotes by $^l\mathrm{p}_n$, where $n = 1, 2, \ldots$ is the overtone number. The lowest $l = 2$ p-mode ($^2\mathrm{p}_1$) is at rather high frequencies, $5 - 7$ kHz, depending on the EoS.[18] The typical damping times are of order a few seconds. The frequency and the damping time both increase with the overtone number. It is more difficult to summarize the value of the frequency obtained from different EoS in a single fitting formula depending only on M and R, since EoS that produce similar values for M and R can show substantial scatter in the frequency of the $^2\mathrm{p}_1$ mode,

[17] See Table III of Benhar, Ferrari and Gualtieri (2004) for the precise values for several EoS.

[18] See again Table III of Benhar, Ferrari and Gualtieri (2004) for the precise values for several EoS. The range of values $5 - 7$ kHz refers to conventional scenarios with hadronic degrees of freedom in the NS interior. For the more exotic scenario of a quark star the frequency of the $^2\mathrm{p}_1$ mode becomes of order $10 - 11$ kHz.

and even more in its damping time. A formula that catches the result of a number of EoS, for the $l = 2$ p-mode frequency, is given by

$$f(^2\mathrm{p_1}) \simeq \frac{1}{\bar{M}}\left[(-0.7 \pm 0.4) + (7.9 \pm 0.4)\frac{\bar{M}}{\bar{R}}\right] \mathrm{kHz}\,. \qquad (11.35)$$

Again, eq. (11.35) is only valid for non-rotating stars. Observe that the frequency of the p-mode increases linearly with $\bar{M}/\bar{R}$, while we saw in eq. (11.31) that the frequency of the fundamental modes scales with $(\bar{M}/\bar{R}^3)^{1/2}$. Similarly to the f-mode, the p-modes with $l \geqslant 2$ are damped mainly by GW emission.

g-modes

The gravity modes (or simply g-modes) are modes for which the restoring force is gravity, which acts to restore thermally-induced buoyancy or discontinuities in the density profile of the matter distribution. Therefore, they can be seen only when one accounts for the non-zero temperature of the star, or in the presence of a composition stratification. The perturbations associated with g-modes have small values of $\delta\rho$ and δp, and are rather characterized by fluid motion δu^i in the tangential direction, corresponding to the fact that gravity tends to smooth out inhomogeneities along equipotential surfaces. Their frequencies are more difficult to estimate, since they depend sensitively on the thermal structure of the equilibrium model, with the frequency increasing approximately linearly with the central temperature. However, they are at lower frequencies compared with the f- and p-modes. For typical models of NS, the lowest $l = 2$ g-mode in the non-rotating limit has a frequency $f = O(10 - 500)$ Hz, depending on the EoS and central temperature.

The fact that for these modes $\delta\rho$ is small implies that they induce a very small time-varying quadrupole moment, and therefore their damping times due to GW emission are very long, of order days to years, so other dissipation mechanisms dominate. For instance, they can involve displacements near the surface of the NS, in which case they induce a "shaking" of the NS magnetosphere, which generates electromagnetic radiation. In general, damping can also result from neutrino emission and from internal friction and viscosity. For each l, they come with a full tower of overtones, with a frequency that decreases with overtone number. Observe that the f-mode has a frequency intermediate between the g-modes and the p-modes.

Normal modes in a liquid-core–crust–ocean model of NS

In a more realistic treatment the NS can be modeled as having a liquid core, a solid crust, and a surface fluid ocean of molten material. In this case the spectrum of fluid modes becomes richer. The frequencies of the f- and p-modes are not much affected by the existence of a crust, since their sound speed in the fluid is much larger than the speed of shear waves in the crust, so the existence of a crust has little effect on them. In contrast, g-modes are sensitively affected, because the solid crust does not sustain much thermally induced buoyancy. As a result, the g-modes now split into two groups: core g-modes, which are mostly confined to the fluid core below the solid crust, and surface g-modes, which only affect the thin superficial layer of molten material above the solid crust. The core g-modes are at considerably lower frequencies, for example $f \lesssim 0.1$ Hz for $l = 2$, while the lowest $l = 2$ surface g-mode has $f \lesssim 20$ Hz (with exact values depending again on the central temperature of the star). In the solid crust, in contrast, there propagate shear waves, termed s-modes, with frequency above 1 kHz, depending on the crust thickness. Furthermore, waves propagate on the solid–fluid interfaces, and the corresponding normal modes are denoted as interfacial, or i-modes. Their frequency depends strongly on the density and temperature at the interface. In a NS with a solid crust there also appear toroidal modes, describing torsional oscillations of the crust, with a frequency of about 50 Hz. Toroidal oscillations have $\delta\rho = 0$ and no mass-varying quadrupole moment, so are damped by viscosity and other mechanisms rather than by GW emission.

Quasiperiodic oscillations (QPOs)

Quasiperiodic oscillations (QPOs) have been observed in the X-ray tails of the giant flares. For instance, in the giant flare of SGR 1806-20, QPOs were detected at 18, 92.5, 150, 625 and 1480 Hz, while in the giant flare of SGR 1900+14 QPOs were detected at 28, 54, 84 and 155 Hz. They are typically divided into low-frequency QPOs ($f \lesssim 150$ Hz) and high-frequency QPOs ($f > 500$ Hz). The latter have only been observed in SGR 1806-20.

These oscillations appear to provide the first observational evidence of NS oscillations. The identification of the normal modes that are responsible for them is, however, non-trivial. QPOs were initially identified with crustal torsional modes. However, further studies have revealed a more complex picture. Indeed, because of the strong magnetic field of the magnetar, the crustal modes decay on a time-scale of at most a second due to the emission of Alfvén waves into the NS interior. There is also an important magnetic coupling between the elastic crust and the fluid core, so that pure torsional crust modes are not appropriate for the description of the coupled system. Rather, this strong coupling generates a new family of so-called "magneto-elastic" modes. Superfluidity in the core also plays an important role in the quantitative understanding of these modes, in particular for explaining the high-frequency QPOs. Indeed, the full spectrum of QPOs, including the low-frequency and the high-frequency QPOs, can be reproduced by numerical simulations of axisymmetric torsional magneto-elastic oscillations of magnetars with a superfluid core; see the Further Reading.

Normal modes of hot proto-neutron stars

The properties of f-, p- and g-modes discussed above hold for a normal, cold, NS. However, the proto-neutron star (PNS) that initially forms in the gravitational collapse has physical properties that are very different from the final cold NS. During the SN explosion, after the initial core bounce and while the shock is stalled (see the discussion in Section 10.3) the initial PNS has a radius of 20–30 km and is hot, lepton-rich and optically thick to neutrinos, which are initially trapped inside it. On a scale of 10–15 s the diffusion of very energetic neutrinos, with energies of order 200 MeV, deleptonizes and later cools the PNS. After about 50 s, the PNS has cooled down to temperatures of order $1-5$ MeV, and becomes a normal NS.

During the first few seconds of its evolution, the quasi-normal mode structure of the hot PNS is quite different from that of an old and cold NS.[19] The $l = 2$ f-mode frequency no longer scales with $(M/R^3)^{1/2}$. Indeed, during the first second the radius of the PNS decreases while its mass remains approximately constant. Thus, if the frequency of the $l = 2$ f-mode scaled as $(M/R^3)^{1/2}$, it should increase with time, whereas, on the contrary, in the first second it decreases. During the first second of the life of the PNS the frequencies of the $l = 2$ f-mode, as well as of the fundamental p_1 mode (and of the fundamental w_1 mode belonging to the family of w-modes that we will discuss below) all cluster in the region 900–1500 Hz, and are therefore much smaller than in the final cold NS.

At the same time, the frequency of the g-mode in the PNS is much higher than in a cold NS. The g-modes become particularly interesting in hot new-born PNS, since the entropy gradients that can source them are more important. Furthermore, we have seen that g-modes can also be excited by composition gradients. In a cold NS the g-modes due to composition gradients in the inner core have low frequencies, $f < 200$ Hz. In a hot PNS the composition gradients can be larger, and generate g-modes with frequencies that, depending on the EoS, can reach 800-900 Hz in the first second. As a result, for a hot PNS, the frequencies of the fundamental mode and of the p- w- and g-modes can be similar, and they can all make comparable contributions to GW emission.

Rotation and r-modes

A full general-relativistic study of normal modes in rapidly rotating stars is technically very challenging, and most studies are performed for slow rotation, with Newtonian or post-Newtonian dynamics, and in the Cowling approximation. In the non-rotating case the modes can be classified by the indices (l, m) of the corresponding spherical harmonics, and are degenerate in m. In the presence of rotation, however, the perturbation equation cannot be separated into a radial and an angular part. Of course, at a given time, one can always decompose a given perturbation in spherical harmonics. However, if at the initial time the perturbation has a definite value of l, the time evolution will eventually generate

[19] See Ferrari, Miniutti and Pons (2003) for an extended discussion.

modes with all values of l. To lowest order in the angular velocity Ω of the star, a polar mode with angular momentum l mixes with axial modes with angular momentum $(l \pm 1)$, and an axial l-mode mixes with the $(l \pm 1)$ polar modes. So the modes themselves no longer carry a definite value of l. However, modes can still be classified by the value of l of the mode to which they reduce in the limit of zero rotation. Another important change is that the degeneracy with respect to the index m is removed, and the mode frequency now depends on m, as illustrated in Fig. 11.2.

Furthermore, when we include rotation a new interesting class of modes appears, the r-modes. For these modes the restoring force is the Coriolis force, and they are analogous to the Rossby waves in the Earth's ocean. For Newtonian stars they describe convective currents in which the Eulerian velocity perturbation has the form

$$\delta \mathbf{v}(t, r, \theta, \phi) = \alpha R \Omega \left(\frac{r}{R}\right)^l \mathbf{Y}^B_{lm}(\theta, \phi) e^{i\omega t}, \qquad (11.36)$$

where $\mathbf{Y}^B_{lm}$ is the "magnetic-type" vector spherical harmonic introduced in eq. (3.255), R is the radius of the rotating star, Ω is its angular velocity and α is a constant. In contrast, to first order in Ω the perturbations of the density and pressure vanish, and $\delta p = \delta \rho = O(R^2\Omega^2/c^2)$. Thus, to first order in Ω these modes do not involve variation in the mass density, and they do not generate time-varying mass multipoles. Their lowest-order coupling to GWs is rather through the current quadrupole.

In the limit of zero rotation they are purely axial, and have zero frequency. For a Newtonian star rotating at angular velocity Ω, in a frame co-rotating with the star, to first order in Ω their frequency is given by

$$f(^{lm}\mathrm{r}) = \frac{1}{2\pi} \frac{2m\Omega}{l(l+1)}, \qquad (11.37)$$

where the factor m in the numerator (not to be confused with a mass!) is the index of the corresponding spherical harmonics Y_{lm}.

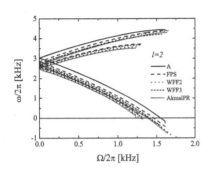

Fig. 11.2 The frequencies of the ^{2}f-mode in the inertial frame for $m = +2$ (lower branch) and $m = -2$ (upper branch), as a function of the star angular velocity, for different EoS. For each EoS, the curve stops when the angular velocity corresponding to the mass-shedding limit is reached. From Doneva, Gaertig, Kokkotas and Krüger (2013).

w-modes

The modes discussed so far are all fluid modes, and exist even in the Cowling approximation, i.e. when we freeze the space-time fluctuations. However, there are also NS normal modes that are basically perturbations of the space-time variables outside the NS and do not induce significant fluid motion. In this sense they are similar (although not identical) to the BH quasi-normal modes that will be discussed in detail in Chapter 12. The most important space-time modes are the w-modes, which exist for generic EoS. For a non-rotating star, a fit to their frequency and damping times for different EoS gives, for $l = 2$ and overtone number $n = 1$,

$$f(^2\mathrm{w}_1) \simeq \frac{1}{R} \left[(21.55 \pm 0.13) - (9.80 \pm 0.14) \left(\frac{\bar{M}}{\bar{R}}\right) \right] \text{kHz}, \quad (11.38)$$

$$\tau(^2w_1) \simeq \bar{M}\left[(17.4 \pm 9.2) + (72 \pm 20)\left(\frac{\bar{M}}{\bar{R}}\right)\right.$$

$$\left. + (-47.5 \pm 0.3)\left(\frac{\bar{M}}{\bar{R}}\right)^2\right]^{-1} \text{ms}. \tag{11.39}$$

Thus, the typical frequency is between 10 and 12 kHz, depending on the EoS, and the typical damping time is about $0.02 - 0.03$ ms, quite comparable to the values that we will find in Chapter 12 for the quasi-normal modes of a BH with the same mass. There are both polar and axial w-modes. Polar w-modes induce negligible fluid motion, while for axial w-modes the fluid motion is completely absent.

Other families of space-time modes are known. Trapped modes exist only for very compact stars, with $R \leqslant 3GM/c^2$, and are due to the existence of a barrier in the effective potential that we will study in Chapter 12. However, no known realistic EoS can produce such a compact NS, so these modes are probably not astrophysically relevant. A second branch of w-modes, known as interface modes or w_{II} modes, also exists. For each l there is a finite number of them, with frequencies from 2 to 15 kHz and damping times similar to that of w-modes. Their existence can be associated to the discontinuity at the surface of the star.

11.2.2 The CFS instability

Beside affecting the structure of the modes, rotation can also induce instabilities. To understand this point consider a mode that, in the frame co-rotating with the star, has the form

$$\delta f(t, r, \theta, \varphi_c) = A(r, \theta)e^{im\varphi_c}e^{i\omega_c t}, \tag{11.40}$$

where φ_c and ω_c are the polar angle and angular velocity in the co-rotating frame, and δf denotes generically the perturbation of any of the fluid variables. According to eq. (11.40), lines of constant phase are given by $m\varphi_c + \omega_c t = \text{const.}$, i.e. $\varphi_c = \varphi_c(0) - (1/m)\omega_c t$. Therefore [with the sign conventions used in eq. (11.40)] modes with $m > 0$ are counter-rotating, with respect to the observer that co-rotates with the star. An observer at infinity will rather use polar coordinates (r, θ, φ_i), with φ_i related to φ_c by

$$\varphi_i = \varphi_c + \Omega t, \tag{11.41}$$

where again Ω is the angular velocity of the star. Therefore, for such an observer

$$\delta f(t, r, \theta, \varphi_i) = A(r, \theta)e^{im(\varphi_i - \Omega t)}e^{i\omega_c t}$$
$$= A(r, \theta)e^{im\varphi_i}e^{i(\omega_c - m\Omega)t}. \tag{11.42}$$

Therefore the angular velocity ω_i measured by the inertial observer at infinity is given by

$$\omega_i = \omega_c - m\Omega. \tag{11.43}$$

In eq. (11.40) we have followed the sign convention more commonly used in the literature on NS normal modes. If we rather define the mode writing the time dependence as $e^{-i\omega_c t}$, as elsewhere in this book, the counter-rotating modes are those with $m < 0$. At the same time eq. (11.43) becomes $\omega_i = \omega_c + m\Omega$, so now ω_i can go through zero and switch sign for $m < 0$. Thus, of course, the physical statement that for counter-rotating modes ω_i can switch sign remains true, independently of the sign convention used.

For counter-rotating modes ($m > 0$), as we increase Ω, ω_i therefore eventually goes through zero and then changes sign. This means that a mode that is counter-rotating with respect to the star can be seen as co-rotating by an observer at infinity, since it is dragged along by the rotation of the star.[20]

This fact has the following very interesting consequence. A source rotating around the z axis radiates GWs that carry away angular momentum. If the angular momentum J_z of the source, as seen from an observer at infinity, is positive, the emitted radiation carries a positive J_z, too. Therefore the value of J_z of the source decreases toward zero. Vice versa, if the source has J_z negative, the GWs emitted carry away a negative J_z, so that again the value of $|J_z|$ of the source decreases. A mode that with respect to the star is counter-rotating, but with respect to the inertial observer at infinity is co-rotating, emits GWs that carry away a positive J_z at infinity. Therefore GW emission removes a positive angular momentum from the perturbation. However, in the frame co-rotating with the star, the angular momentum of the mode was negative, so, because of GW emission, it becomes even more negative, and larger in absolute value. The mode will therefore emit even more GWs, and a secular instability can develop. This is known as the Chandrasekhar–Friedman–Schutz (CFS) instability.

A more quantitative way of understanding the emergence of this instability, and computing the characteristic time-scale over which it develops, is as follows. We model the star as a self-gravitating Newtonian fluid with mass density ρ and fluid velocity v^i, and we add to its dynamics the back-reaction due the GWs. Let us at first work only to lowest order in the post-Newtonian expansion. In this case we can take into account the back-reaction due to mass quadrupole radiation using the Burke–Thorne potential Φ_{GW}: as we saw in eqs. (3.114) and (3.115), the GW back-reaction is then described by a force $\mathbf{F} = -m\boldsymbol{\nabla}\Phi_{GW}$ with

$$\Phi_{GW}(t, r, \theta, \phi) = \frac{G}{5c^5} x_i x_j \frac{d^5 Q_{ij}}{dt^5}$$
$$= \frac{G}{5c^5} r^2 \sum_{m=-2}^{2} Y_{2m}(\theta, \phi) \frac{d^5 Q_m}{dt^5}, \qquad (11.44)$$

where we wrote $x_i = r n_i$ and we used eq. (3.223). To compute Q_m as an integral over the source we use eqs. (3.222) and (3.219) to get the relation $Q_m = (8\pi/15) Q_{ij}(\mathcal{Y}_{ij}^{2m})^*$, which, combined with eqs. (3.296) and (3.302), gives

$$Q_m(t) = \frac{8\pi}{15} \int d^3 x' \, r'^2 \rho(t, \mathbf{x}') Y_{2m}^*(\theta', \phi'). \qquad (11.45)$$

Therefore

$$\Phi_{GW}(t, r, \theta, \phi) = \frac{8\pi G r^2}{75 c^5} \sum_{m=-2}^{2} Y_{2m}(\theta, \phi) \frac{d^5}{dt^5} \int d^3 x' \, r'^2 \rho(t, \mathbf{x}') Y_{2m}^*(\theta', \phi').$$
$$(11.46)$$

We now write down the equations governing the fluid motion, including the back-reaction. These are the continuity equation,

$$\partial_t \rho + \partial_i(\rho v^i) = 0 \,, \tag{11.47}$$

plus the Euler equation in the presence of the gravitational potential. The latter includes both the static Newtonian potential Φ and the back-reaction potential $\Phi_{\rm GW}$. Neglecting for the moment dissipation, the Euler equation reads

$$\partial_t v^i + v^j \partial_j v^i = -\frac{\partial^i p}{\rho} - \partial^i(\Phi + \Phi_{\rm GW})\,. \tag{11.48}$$

Finally, the static potential satisfies

$$\nabla^2 \Phi = +4\pi G \rho\,. \tag{11.49}$$

It is useful, as usual, to introduce here the enthalpy per unit mass, $h(p) = \int^p dp'/\rho(p')$, so that $\partial_i h = \partial_i p/\rho$ and the Euler equation becomes

$$\partial_t v^i + v^j \partial_j v^i = -\partial^i(h + \Phi + \Phi_{\rm GW})\,. \tag{11.50}$$

We can now linearize eqs. (11.47), (11.49) and (11.50) around the equilibrium configuration. On the equilibrium configuration $v_i = \epsilon_{ijk}\Omega_j x_k$, where $\boldsymbol{\Omega}$ is the constant angular velocity vector, so $\partial_i v^i = 0$. Thus, we get

$$\partial_t(\delta\rho) + \partial_i(\rho\delta v^i) + v^i\partial_i(\delta\rho) = 0\,, \tag{11.51}$$
$$\nabla^2\delta\Phi = 4\pi G\delta\rho \tag{11.52}$$
$$\partial_t(\delta v^i) + v^j\partial_j(\delta v^i) + \delta v^j\partial_j v^i = -\partial^i(\delta h + \delta\Phi + \delta\Phi_{\rm GW})\,, \tag{11.53}$$

where $\delta h = \delta p/\rho$. We now expand in modes with respect to the inertial observer, writing each perturbed quantity δf in the form

$$\delta f(t, r, \theta, \varphi_i) = e^{i\omega_i t + im\varphi_i}\delta f(r, \theta)\,. \tag{11.54}$$

Observe that we use a complex notation for the modes of the various perturbed quantities. The energy density associated with a given mode of the fluid perturbations is[21]

$$E_{\rm fluid} = \frac{1}{2}\int d^3x\left[\rho\delta v^i \delta v_i^* + \frac{1}{2}(\delta\rho\delta U^* + \delta\rho^*\delta U)\right]\,, \tag{11.55}$$

where $\delta U = \delta p/\rho + \delta\Phi$. Using eqs. (11.51)–(11.53) one can verify that, in the absence of $\delta\Phi_{\rm GW}$, this energy is conserved. Adding the GW back-reaction, however, one finds

$$\frac{dE_{\rm fluid}}{dt} = -\omega_i(\omega_i + m\Omega)N_2\omega_i^4|\delta D_{2m}|^2\,, \tag{11.56}$$

where $N_2 = 8\pi G/(75c^5)$ and

$$\delta D_{2m} = \int d^3x'\, r'^2\delta\rho(t, \mathbf{x}')Y_{2m}^*(\theta', \phi')\,. \tag{11.57}$$

[21]This result is derived observing that the linearized Euler equation (11.53) can be obtained as the equation of motion of a Lagrangian in which the field variable is the Lagrangian displacement vector field ξ^i, which relates a fluid element in the equilibrium configuration to the corresponding fluid element in the perturbed configuration. The energy of a field configuration $\xi^i(t, \mathbf{x})$ is then obtained with standard field-theoretical methods, i.e. from Noether's theorem. For the particular background configuration corresponding to a rotating fluid, and for perturbations with a dependence on t and φ_i of the form $\delta f = \delta f(r, \theta)e^{i\omega_i t + im\varphi_i}$, the relation between the velocity perturbation δv^i and the displacement field ξ^i can be inverted and used to eliminate ξ^i in favor of δv^i from the field-theoretical expression for energy, leading to eq. (11.55). See Friedman and Schutz (1978a, 1978b) and Ipser and Lindblom (1991).

In the context of linearized theory eq. (11.56) can be formally generalized to include all mass and current multipoles, using a formal expression for the back-reaction potential valid for all multipoles; see Thorne (1969b). In this case one gets

$$\frac{dE_{\text{fluid}}}{dt} = -\omega_i(\omega_i + m\Omega)\sum_{l\geqslant 2} N_l \omega_i^{2l}\left(|\delta D_{lm}|^2 + |\delta J_{lm}|^2\right), \qquad (11.58)$$

where

$$N_l = \frac{8\pi G}{c^{2l+1}}\frac{(l+1)(l+2)}{l(l-1)[(2l+1)!!]^2}, \qquad (11.59)$$

$$\delta D_{lm} = \int d^3x'\, r'^l \delta\rho(t,\mathbf{x}')Y_{lm}^*(\theta',\phi'), \qquad (11.60)$$

$$\delta J_{lm} = 2\sqrt{\frac{l}{l+1}}\int d^3x'\, r'^l(\rho\delta v^i + v^i\delta\rho)(\mathbf{Y}_{lm}^B)_i^*, \qquad (11.61)$$

and $\mathbf{Y}_{lm}^B$ are the vector spherical harmonics defined in eq. (3.255). Observe that (apart from a different normalization, included in N_l), δD_{lm} and δJ_{lm} are the linearizations of the quantities I_{lm} and S_{lm} defined in eqs. (3.296) and (3.299).[22]

The crucial aspect of eqs. (11.56) and (11.58) is the sign of the term $\omega_i(\omega_i + m\Omega) = \omega_i\omega_c$ (since the remaining factors ω_i^{2l} and $|\delta D_{lm}|^2 + |\delta J_{lm}|^2$ are of course always positive). For a given positive ω_c, when the inertial frequency $\omega_i = \omega_c - m\Omega$ becomes negative, dE_{fluid}/dt switches sign and becomes positive. Then, as a consequence of the GW back-reaction, the energy in the perturbation increases, rather than decreasing! Of course, overall energy is conserved because the total energy balance involves the rotational energy of the unperturbed star, the energy E_{fluid} of the perturbation, and the energy carried away by the GWs. When $\omega_i < 0$, it is the rotational energy of the unperturbed star that pays for both the GWs radiated at infinity and for the increase of the energy stored in the perturbation.

Equation (11.43) suggests that, with increasing Ω, every mode with $m > 0$ eventually hits the value $\omega_i = 0$ and undergoes the CFS instability. However, first of all Ω cannot be increased beyond the mass-shedding limit [compare with eq. (10.107)]. Furthermore, the instability can actually set in only if its growth time is shorter than the typical damping time due to viscosity. Let us define

$$\frac{1}{\tau_{\text{GR}}} = -\frac{1}{2E_{\text{fluid}}}\frac{dE_{\text{fluid}}}{dt}. \qquad (11.62)$$

In the usual situation the fluid loses energy because of GW emission, so $dE_{\text{fluid}}/dt < 0$ and $\tau_{\text{GR}} > 0$. The perturbations evolve as $e^{i\omega t - t/\tau_{\text{GR}}}$ and are therefore damped. When $dE_{\text{fluid}}/dt > 0$, in contrast, $\tau_{\text{GR}} < 0$ and the perturbation grows. Adding the standard viscosity terms to the Euler equation (11.63) we have

$$\rho(\partial_t v^i + v^j\partial_j v^i) = -\partial^i p - \rho\partial^i(\Phi + \Phi_{\text{GW}}) + 2\partial_j(\eta\sigma^{ij}) + \partial^i(\zeta\sigma), \qquad (11.63)$$

[22]One should, however, be aware that, beyond the lowest multipole (i.e. the mass quadrupole and the current quadrupole) the correct tool is the systematic PN expansion described in Chapter 5. In this case the back-reaction is no longer fully described by a modification $\Delta h_{00} = -2\Phi_{\text{GW}}$. Rather, now also h_{0i} and h_{ij} get corrections from the back-reaction [see eqs. (5.187)–(5.189), where the result is written in terms of the quantities $h_1^{\mu\nu}$ defined in eq. (5.84)]. The explicit expressions up to leading and next-to-leading order are given in eqs. (5.190) and (5.191), and we see that they also include non-local tail terms.

where σ^{ij} and σ are the shear and expansion, respectively,

$$\sigma^{ij} = \frac{1}{2}\left(\partial^i v^j + \partial^j v^i - \frac{2}{3}\delta^{ij}\partial_k v^k\right),\qquad (11.64)$$

$$\sigma = \partial_i v^i,\qquad (11.65)$$

while η and ζ are called the shear and bulk viscosity coefficients, respectively.[23] Then eq. (11.58) becomes

$$\frac{dE_{\text{fluid}}}{dt} = -\omega_i(\omega_i + m\Omega)\sum_{l\geqslant 2} N_l\omega_i^{2l}\left(|\delta D_{lm}|^2 + |\delta J_{lm}|^2\right)$$

$$-\int d^3x\left(2\eta\delta\sigma^{ij}\delta\sigma_{ij}^* + \zeta\delta\sigma\delta\sigma^*\right).\qquad (11.66)$$

Thus,

$$\frac{1}{2E_{\text{fluid}}}\frac{dE_{\text{fluid}}}{dt} = -\left(\frac{1}{\tau_{\text{GR}}} + \frac{1}{\tau_\zeta} + \frac{1}{\tau_\eta}\right),\qquad (11.67)$$

where

$$\frac{1}{\tau_{\text{GR}}} = \frac{\omega_i(\omega_i + m\Omega)}{2E_{\text{fluid}}}\sum_{l\geqslant 2} N_l\omega_i^{2l}\left(|\delta D_{lm}|^2 + |\delta J_{lm}|^2\right),\qquad (11.68)$$

$$\frac{1}{\tau_\zeta} = \frac{1}{2E_{\text{fluid}}}\int d^3x\,\zeta\delta\sigma\delta\sigma^*,\qquad (11.69)$$

$$\frac{1}{\tau_\eta} = \frac{1}{E_{\text{fluid}}}\int d^3x\,\eta\,\delta\sigma^{ij}\delta\sigma_{ij}^*.\qquad (11.70)$$

Thus, defining τ_V from

$$\frac{1}{\tau_V} = \frac{1}{\tau_\zeta} + \frac{1}{\tau_\eta},\qquad (11.71)$$

the CFS instability only takes place if $|\tau_{\text{GR}}|$ is sufficiently small that $\tau_{\text{GR}}^{-1} + \tau_V^{-1} < 0$, i.e. the time-scale $|\tau_{\text{GR}}|$ over which the instability grows must be smaller than the time-scale τ_V over which viscosity damps out the perturbation.

We finally need the values of η and ζ. For a NS made basically of neutrons, protons and electrons there are two main mechanisms leading to dissipation. Bulk viscosity dissipates energy via neutrino emission, through the process $n + n \to p + n + e^- + \bar{\nu}_e$, which is no longer compensated by the inverse reaction in the presence of the perturbation. Shear viscosity instead dissipates energy through nn scattering (or ee scattering, in a superfluid core). Detailed computations give[24]

$$\zeta = 6 \times 10^{25}\left(\frac{1\,\text{Hz}}{\omega_c}\right)^2\left(\frac{\rho}{10^{15}\,\text{g cm}^{-3}}\right)^2\left(\frac{T}{10^9\,\text{K}}\right)^6\,\text{g cm}^{-1}\text{s}^{-1},\ (11.72)$$

$$\eta = 2 \times 10^{18}\left(\frac{\rho}{10^{15}\,\text{g cm}^{-3}}\right)^{9/4}\left(\frac{T}{10^9\,\text{K}}\right)^{-2}\,\text{g cm}^{-1}\,\text{s}^{-1}.\qquad (11.73)$$

We see that these coefficients depend strongly on temperature. The shear viscosity coefficient is proportional to T^{-2} and dominates at low

[23] In this section we have written all the equations with respect to Cartesian coordinates, for notational simplicity. In the end, for a rotating star, spherical or cylindrical coordinates will typically be more useful. Then, all the partial derivatives ∂_i become covariant derivative ∇_i with respect to the spherical or cylindrical coordinates, and δ^{ij} in eq. (11.64) is replaced by the corresponding metric g^{ij}.

[24] See Cutler and Lindblom (1987) and Sawyer (1989).

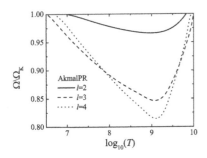

Fig. 11.3 The instability window for the ^{l}f-modes with $l = 2, 3, 4$ and $m = l$, for the AkmalPR EoS. From Doneva, Gaertig, Kokkotas and Krüger (2013).

[25]We will, however, see in Section 11.2.3 that rapidly rotating NS, formed after a NS–NS merger, and which do not have stable non-rotating counterparts, can have a much smaller value of $|\tau_{GR}|$, and can therefore be unstable for smaller values of Ω/Ω_K.

T, while bulk viscosity grows as T^6 and dominates at high temperature. There is a window, for temperatures around 10^9 K, where the viscosity is low and the instability can take place. This is indeed illustrated in Fig. 11.3, which shows the minimum value of the rotational frequency Ω (normalized to the Kepler angular velocity Ω_K that corresponds to the mass-shedding limit) for which $|\tau_{GR}| \leqslant \tau_V$, for $l = 2, 3, 4$ and $m = l$ (for most EoS, modes with $m \geqslant 5$ are not very relevant for the CFS instability).

We can see from fig. 11.3 that there is a region of temperatures where the instability window reaches about 80–85% of the mass-shedding limit for the modes with $l = 3, 4$, while for the $l = 2$ mode it is very close to the Keplerian limit.[25] In general, NS are born at temperatures above 10^{10} K, and rapidly cool down. In the standard scenario the main cool-down mechanism is the modified URCA process ($n+n \to n+p+e^-+\bar{\nu}_e$ followed by $n+p+e^- \to n+n+\nu_e$, resulting in the creation of a $\nu_e\bar{\nu}_e$ pair), and in this case one can estimate that the central temperature T_c of the NS evolves with time as

$$\frac{T}{10^9\,\text{K}} \simeq 19.156 \left(\frac{1\,\text{s}}{t}\right)^{0.159}, \qquad (11.74)$$

so that the NS cools down to 10^9 K in about one year. If the proton fraction in the NS is sufficiently large, it is also possible that the star cools down by the direct URCA process ($n \to p + e^- + \bar{\nu}_e$ followed by $p+e^- \to n+\nu_e$, resulting again in the creation of a $\nu_e\bar{\nu}_e$ pair), in which case

$$\frac{T}{10^9\,\text{K}} \simeq 2.115 \left(\frac{1\,\text{s}}{t}\right)^{0.25}, \qquad (11.75)$$

and the central temperature cools down to 10^9 K in just 20 s. However, even if the star core cools rapidly, the external layers of the star, which are more relevant for the onset of the instability, will stay above 10^9 K for about one year.

As the NS cools, its rotation slows down. The issue is whether the NS can pass through this instability window when it is still sufficiently young and rapidly spinning. However, for the f-modes the rotational frequency that is required to excite the CFS instability is a sizable fraction of the Keplerian frequency. The situation is more favorable for the r-modes, as we discuss next.

The r-mode instability

For r-modes, using eqs. (11.37) and (11.43), the frequency in the inertial frame is

$$\omega_i = -m\Omega \left(1 - \frac{2}{l(l+1)}\right), \qquad (11.76)$$

which is always negative, for all $l \geqslant 2$. In particular, for the $l = 2$ mode, $\omega_i = -(2/3)m\Omega$. Thus, already for an infinitesimal value of Ω these modes are co-rotating with respect to the star but counter-rotating for the inertial observer at infinity, and therefore in the absence of viscosity all these modes are CFS-unstable. To see whether the instability persists when viscosity is included, one must compute the time-scale τ_{GR} associated with the CFS instability. The r-modes emit GWs by current quadrupole radiation. Repeating the computation discussed above one finds, for the ^{l}r-mode,

$$\frac{1}{\tau_{\mathrm{GR}}(\Omega)} = -\frac{32\pi G \Omega^{2l+2}}{c^{2l+3}} \frac{(l-1)^{2l}}{[(2l+1)!!]^2} \left(\frac{l+2}{l+1}\right)^{2l+2} \int_0^R dr\, r^{2l+2} \rho(r) \;.$$

$$(11.77)$$

We see that $\tau_{\mathrm{GR}}(\Omega)$ is negative for all Ω, and $|\tau_{\mathrm{GR}}(\Omega)| \propto \Omega^{-(2l+2)}$. Thus, if the star's rotation rate is too small, the time needed to develop the instability diverges, and once we include viscosity the CFS instability will not develop. A minimal value of Ω is therefore required. Comparing the above result for τ_{GR} with the viscosity time-scale obtained by evaluating eqs. (11.69) and (11.70) for the r-mode perturbation, one finds the result shown in Fig. 11.4. The instability window is now much larger in temperature, and for the $l=2$ r-mode the CFS instability can set in already for values of Ω/Ω_K of order $0.1-0.2$.

The above results suggest that the r-mode instability could be an interesting source of gravitational radiation. However, even if the instability develops, this does not yet mean that there will be sizable GW production. To compute the GW emission we also need to know how the r-mode amplitude α, defined in eq. (11.36), evolves with time and in particular to what value it saturates, and we also need the time evolution of the rotation frequency Ω (the same of course holds for the f-mode, if an instability develops in this mode). The evolution of these two parameters can in principle be obtained combining eq. (11.67) for the energy loss with the corresponding equation for the loss of angular momentum due to the GWs. Precise estimates of the GW signal and of its detectability are not easy, however, since they require knowledge of the saturation amplitude of the instability, which is determined by complicated non-linear effects. Detailed estimates of the saturation amplitude based on the non-linear coupling between the r-mode and a "sea" of other modes suggest that the saturation amplitude is several orders of magnitude smaller than that assumed in early work. Still, the GW signal from the r-mode instability in a young NS might be detectable at advanced interferometers up to distances of order 100–200 kpc.[26] A further complication is given by the interaction of the r-modes with the magnetic field of the NS. Indeed, the r-modes generate differential rotation in the star, since they induce angular momentum losses whose rate depends on the mode shape and varies across the star. In turn, the differential rotation amplifies the magnetic field, and the energy lost to the magnetic field damps the r-mode instability.

Apart from the possibility of being directly detectable, the GW emission by r-modes is likely to have other important astrophysical consequences. In particular, it leads to a rapid spindown of the young NS, increasing the NS period to about 20 ms within a year from birth. This would explain why NS never seem to be born with periods of just a few

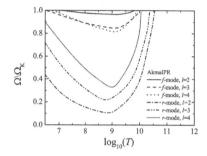

Fig. 11.4 The instability window for the ^{l}r-modes with $l = 2, 3, 4$ and $m = l$, compared with that for the ^{l}f-modes with $l = 2, 3, 4$, for the AkmalPR EoS. From Doneva, Gaertig, Kokkotas and Krüger (2013).

[26]See Arras *et al.* (2003).

ms, despite the fact that such values are consistent with the maximum theoretical value allowed by typical EoS, and that simple considerations based on angular momentum conservation would suggest that it is plausible to find newborn NS spinning near such a theoretical maximum.

11.2.3 GWs from post-merger NS remnants

As we will discuss in more detail in Section 14.4, in the coalescence of a NS–NS binary, for a broad range of masses and EoS of the initial stars, the initial remnant will be a massive, hot and differentially rotating NS. A differentially rotating proto-NS is also the general outcome of stellar collapse, and can give rise to bar instabilities, as we saw in Section 10.5.2. A differentially rotating NS can sustain an angular momentum much larger than its uniformly rotating counterpart, because it can have rapid rotation in the interior without exceeding the mass-shedding limit on the surface. Because of its large angular momentum, this NS remnant can oppose gravitational collapse even if its total mass is significantly larger than the maximum mass allowed for a non-rotating NS. Such a configuration, stabilized by differential rotation, is referred to as a hypermassive neutron star (HMNS). Eventually, differential rotation will be smoothed out by matter redistribution in the remnant, and furthermore the NS will loose a significant fraction of its angular momentum because of GW emission. If, depending on the EoS, this NS remnant has a stable non-rotating counterpart, it will eventually stabilize to a normal NS. Otherwise, the process will eventually lead to a delayed collapse to a BH. The time-scale for delayed collapse is difficult to compute, and current estimates can range from about 10 ms to a few hours, depending on the masses and the EoS, and on the intrinsic numerical uncertainties in the simulations. Indeed, the lifetime of a metastable object such as a HMNS is very sensitive to numerical artifacts. Only for a sufficiently large total mass of the initial stars does a prompt gravitational collapse to BH take place.

When a NS remnant is initially formed, it is highly excited by the merger process, and emits GWs. The GW emission during the coalescence of a NS–NS binary will be discussed in detail in Section 14.4, while here we focus on the GW emission from the NS remnant. Figure 11.5 shows the GW spectrum, on a scale emphasizing the post-merger phase, obtained from numerical simulations of the coalescence of a NS–NS binary, for two different EoS, and a source at 50 Mpc. In this example we see a main peak at a frequency f_2 in the $2 - 3$ kHz region, depending on the EoS. For a given total mass of the inspiralling binary (whose value can be deduced from the GW signal in the inspiral phase), there is a rather tight empirical relation between the frequency of this peak and the radius of the HMNS remnant, shown in Fig. 11.6. A measurement of f_2 (denoted by $f_{\rm peak}$ in Fig. 11.6) can therefore give important information on the EoS of the post-merger NS. Secondary peaks also appear in the spectrum shown in Fig. 11.5, and they also carry important information. In particular the position of the second largest peak, denoted by

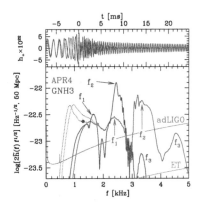

Fig. 11.5 The GW amplitude of a NS–NS coalescence as a function of time (upper panel) and the corresponding frequency spectrum $2\tilde{h}(f)f^{1/2}$ (lower panel), on a scale that emphasizes the post-merger phase. The results for two different equations of state are shown. The source is taken to be at a distance of 50 Mpc. The sensitivity of advanced LIGO at its final design, and the expected sensitivity of the Einstein Telescope are also shown. Courtesy of Luciano Rezzolla, based on work presented in Takami, Rezzolla and Baiotti (2014).

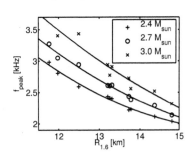

Fig. 11.6 The relation between $f_{\rm peak}$ and the radius of the remnant (computed neglecting the remnant rotation). From Bauswein, Stergioulas and Janka (2016).

f_1 in Fig. 11.5, turns out to be correlated to a parameter describing the tidal deformability of the NS. This correlation turns out to be largely independent of the EoS (in contrast to the correlation between the first peak and the radius, which, as we see from Fig. 11.6, depends on the NS mass and hence on the EoS); see the Further Reading.

Differential rotation in the remnant is also important because in a differentially rotating proto-NS there are instabilities controlled by the parameter $\beta = T_{\rm rot}/|W|$, where $T_{\rm rot}$ is the rotational kinetic energy and W the gravitational potential energy. We have discussed them in Section 10.5.2, where we have also seen that, in uniformly rotating stars, the threshold for the development of $m = 2$ bar-like instabilities is probably too high to be realized in typical astrophysical scenarios. However, differentially rotating proto-NS can develop a shear instability in the $m = 2$ modes already at low values of β (as well as instabilities in the $m = 1$ mode), which can be an interesting source of GWs. See the Further Reading for studies of the possibility of developing such instabilities in the merger of a NS–NS binary.

The fact that the NS post-merger remnant can have a mass significantly higher than that of a non-rotating NS also opens up another potentially interesting possibility for GW production, through the f-mode CFS instability discussed in Section 11.2.2. Indeed, Fig. 11.7 shows the growth time $|\tau_{\rm GR}|$ of the CFS instability, for a range of masses and of the ratio $T_{\rm rot}/|W|$ expected for the post-merger NS remnant. In this figure the upper black solid line is the Kepler limit beyond which, for a given mass, we cannot raise $T_{\rm rot}/|W|$, otherwise we enter in the mass-shedding regime. The lower black solid line is the limit beyond which, for a given $T_{\rm rot}/|W|$, we cannot increase the mass, otherwise we get prompt BH formation. With sufficient rotation, for the EoS considered, remnant NS masses up to about $2.5 M_\odot$ are possible. We see from the figure that, for such large masses, one can obtain values of $|\tau_{\rm GR}|$ as low as 10^2 s. This should be compared with the typical values of the viscosity time-scales τ_V, which, for instance, at $T = 10^9$ K are of order 10^7 s. Therefore the f-mode CFS instability arises on a time-scale much shorter than the time-scale of viscosity and in eq. (11.67) the negative term $1/\tau_{\rm GR}$ dominates over the positive viscosity term $1/\tau_V$ for quite some time, producing a large instability window. The f-mode amplitude will then initially increase exponentially. Then, non-linear coupling mechanisms stabilize it to a saturation amplitude, until eventually the star exits the instability window or collapses to a BH.

Once again, the crucial issue is the value of the saturation amplitude, which involves complicated non-linear physics and is therefore highly uncertain, even in orders of magnitude. Figure 11.8 shows the signal-to-noise ratio at advanced LIGO (left vertical scale), and also at a possible future third-generation ground-based interferometer such as the Einstein Telescope (ET, right vertical scale), for a source at 20 Mpc, assuming a value for the saturation amplitude $\alpha^f = 10^{-6} M c^2$, suggested by hydrodynamical simulations. For such a saturation amplitude, the corresponding GW emission can be quite significant. More detailed investigations

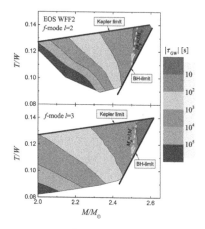

Fig. 11.7 Contour plots of the growth time $|\tau_{\rm GR}|$ for CFS unstable f-modes with $l = m = 2$ and $l = m = 3$. From Doneva, Kokkotas and Pnigouras (2015).

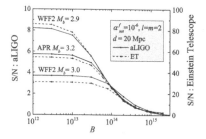

Fig. 11.8 The signal-to-noise ratio for advanced LIGO (solid line, left vertical scale) and ET (dashed line, right vertical scale), for three different EoS, for the $l = m = 2$ f-mode of a remnant NS star with baryon mass around $3 M_\odot$, assuming a saturation amplitude $\alpha^f = 10^{-6} M c^2$. From Doneva, Kokkotas and Pnigouras (2015).

[27]See also Pnigouras and Kokkotas (2016), where quadratic mode coupling is included, and it is found that the saturation amplitude changes as the NS evolves in the instability window.

involving also mode–mode couplings, are, however, necessary to allow robust predictions on the saturation amplitude to be made.[27] The result also depends on the surface dipole magnetic field B of the remnant NS, since magnetic dipole radiation drains rotational energy from the star and slows it down. Therefore, for larger values of B, less energy is emitted in GWs. The magnetic field of the initial merging NS is expected to be amplified during the collapse by two to four orders of magnitudes, so one might expect that, for the remnant post-merger NS, values of B in the range $10^{13} - 10^{14}$ G could be quite typical. We see that, for this saturation amplitude and $B < 10^{14}$ G, and a source at a distance $d = 20$ Mpc, the signal could be seen with a signal-to-noise ratio S/N of order $4 - 8$ (depending on the EoS) at advanced LIGO, and of order $40 - 80$ at ET. This means that at ET a signal of this type could be above a detection threshold of, say, $S/N = 5$, up to distances of order 200 Mpc.

11.2.4   GWs from deformed rotating NS

As we have seen in the previous subsections, the emission of GWs through quasi-normal modes is already present when we treat the NS as a self-gravitating fluid. A rotating self-gravitating fluid also develops a characteristic flattening of the poles, leading to an ellipticity proportional to Ω^2, where Ω is the rotational angular velocity. However, for a uniform density distribution such a deformation is axisymmetric, with respect to the rotation axis. The resulting quadrupole moment is time-independent, and does not contribute to GW production. GWs are produced if there are non-axisymmetric deformations, and the corresponding GW emission has been computed in Section 4.2 of Vol. 1. Non-axisymmetric deformations in a NS can be produced either because of deformations frozen in the solid crust or by a magnetic field that is not aligned with the rotation axis.

The existence of a solid crust means that, to some extent, a NS can sustain 'mountains' on its surface. In the simplest scenario leading to the formation of asymmetries, the NS crust solidifies early in the NS lifetime, as the star begins to cool, and takes the equilibrium shape corresponding to the relatively high angular velocity that the NS has at that time. As the NS evolves it spins down, and the corresponding equilibrium configuration changes. However, the crust is frozen in the shape in which it first solidified, until the energy difference between the new equilibrium shape and the actual crust shape is sufficiently large, and the crust can no longer sustain the stress. At this point the crust breaks down and adapts to its new equilibrium position. This scenario was originally developed to explain NS glitches. However, it was soon realized that this mechanism does not provide the energy release required for explaining the strongest glitches. As we saw on page 79, glitches are rather explained in terms of the interaction between the crust and the superfluid core. Still, this scenario provides a mechanism that can generate a non-axisymmetric ellipticity in the NS. Modeling of the NS crust suggests that the maximum ellipticity that the crust can sustain is[28]

$$\epsilon < 2 \times 10^{-5} \left(\frac{u_{\text{break}}}{0.1} \right) \tag{11.78}$$

[28]See the discussion in Andersson *et al.* (2011) and references therein.

where u_{break} is the crustal breaking strain (defined as the maximum stress that the crust can stand, normalized to the shear modulus). Molecular dynamics simulations of the Coulomb solid representing the NS crust show that in NS u_{break} is of order 0.1. This is a huge value compared with "hard" terrestrial materials. For comparison, for steel $u_{\text{break}} \simeq 0.005$. It is also interesting to observe that, while normal metals begin to yield continuously at low strain through the formation of dislocations, in the NS dislocations do not form, basically because of the huge pressure and density, and the crust breaks abruptly and collectively. This of course makes the breaking process more interesting from the point of view of the possible associated GW production. The values of the ellipticity that is actually generated in a NS, however, can be smaller than the maximum value that the crust can sustain before breaking.

Beside crustal stresses, other mechanisms can generate asymmetries. As first understood by Chandrasekhar and Fermi in 1953, in the presence of a strong internal magnetic field a star will not remain spherical and, if the magnetic field is not aligned with the rotation axis, it will generate a non-axisymmetric deformation. Using polar coordinates, the magnetic field can be split into a poloidal component $\mathbf{B}_p = (B_r, B_\theta, 0)$ and a toroidal component $\mathbf{B}_t = (0, 0, B_\phi)$. We denote by z the direction of the magnetic axis (which we take to differ from the rotation axis, to generate deformations that are non-axysimmetric with respect to the rotation axis). Then $\nabla \times \mathbf{B}_t$ points along the z axis. We are interested in the ellipticity

$$\epsilon = \frac{I_{xx} - I_{zz}}{I_0}, \tag{11.79}$$

where I_0 is the moment of inertia of the spherical star and I_{jk} is the inertia tensor defined in eq. (4.197). For a star with a $n = 1$ polytropic EoS and a poloidal magnetic field $\mathbf{B}_p$, assuming a normal fluid core, the magnetic field generates an ellipticity[29]

[29]See Haskell, Samuelsson, Glampedakis and Andersson (2008), with the correction in their Erratum.

$$\epsilon \simeq 3 \times 10^{-11} \left(\frac{R}{10\,\text{km}} \right)^4 \left(\frac{1.4 M_\odot}{M} \right)^2 \left(\frac{\bar{B}_p}{10^{12}\,\text{G}} \right)^2, \tag{11.80}$$

where $\bar{B}_p$ is the volume-averaged poloidal magnetic field. This is too small in normal pulsars, but can become significant for magnetars. The plus sign in eq. (11.80) corresponds to an oblate configuration, i.e. a configuration flattened at the poles (with respect to the z axis defined by the magnetic field).

In contrast, a toroidal magnetic field B_t gives

$$\epsilon \simeq -1.5 \times 10^{-12} \left(\frac{R}{10\,\text{km}} \right)^4 \left(\frac{1.4 M_\odot}{M} \right)^2 \left(\frac{\bar{B}_t}{10^{12}\,\text{G}} \right)^2, \tag{11.81}$$

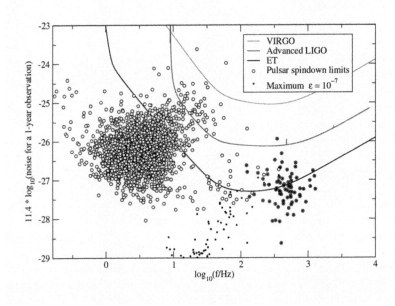

Fig. 11.9 Spindown upper limits on known radio pulsars. The open circles assume 100% coversion of spindown energy into GWs, regardless of how large an ellipticity is required to power such emission. In contrast, the black circles limit the allowed ellipticity to be no greater than 10^{-7}. An integration time of one year is assumed. From Andersson *et al.* (2011).

[30]To study the overall shape of the NS one must also include the axisymmetric deformation induced by the angular velocity Ω, for instance generated when the proto-NS was hot, before the crust solidified. Rotation always flattens the poles, i.e. gives an oblate contribution with respect to its rotation axis. So, when the magnetic field and the rotation axis are aligned, depending on the values of the angular velocity and magnetic field, even for a purely toroidal magnetic field we can have an oblate deformation (i.e. $\epsilon > 0$), as well as intermediate cases with an oblate exterior and a prolate interior. However, what matters for GW generation are only the non-axisymmetric contributions.

[31]See Table 5 of Frieben and Rezzolla (2012) for the dependence on the EoS in the case of toroidal fields.

which (for the same value of B) is smaller than the value generated by a poloidal field and has the opposite sign, i.e. induces a prolate deformation.[30] The precise value of the coefficients in eqs. (11.80) and (11.81) depend on the EoS used, and can vary by almost an order of magnitude over different typical equations of state.[31]

The situation further changes in the case of a superconducting core, where one can reach values

$$\epsilon \sim 10^{-9} \left(\frac{B}{10^{12}\,\mathrm{G}} \right) \left(\frac{H_c}{10^{15}\,\mathrm{G}} \right) , \qquad (11.82)$$

where H_c is a critical field value. Higher values of ϵ can be obtained with more exotic EoS that allow plastic deformations of the core (for instance for a strange quark core).

The above results suggest that a realistic value for the typical ellipticity of a NS could be for example of order 10^{-7}. For normal pulsars such a value is several orders of magnitude below the spindown limit, while for millisecond pulsar a value of order 10^{-7} is quite close to the spindown limit. The corresponding perspectives for GW detection from the known pulsars are summarized in Fig. 11.9. The open circles represent the values computed assuming, for each pulsars, a value of ϵ at its spindown limit. The black circles corresponds to setting an upper limit $\epsilon \leqslant 10^{-7}$. The three continuous lines represent the sensitivities of the first-generation VIRGO interferometer, of advanced LIGO, and of a possible third-generation interferometer such as the Einstein Telescope (ET). We see that, once a realistic ellipticity is assumed, the young pulsars disappear from the diagram and go below the sensitivity level even of the ET. In contrast millisecond pulsars can still be detectable, particularly at the ET sensitivity.[32]

Of course, many more NS exist than those currently observed as radio pulsars. As we saw in Section 11.1.4, in the Galaxy it is estimated that there are about 1.6×10^5 active normal pulsars and 4×10^4 millisecond pulsars, while we have so far only detected about 2500 pulsars, out of which $O(100)$ are millisecond pulsars. Thus we expect many rotating NS at a distance much closer than the known closest pulsars, which would give rise to a much stronger GW amplitude on Earth. However, their search can only performed with the blind search techniques discussed in Section 7.6.3, with the corresponding large increase in computer time for data analysis.

[32] It is quite impressive to realize that detecting GWs due to an eccentricity $\epsilon = 10^{-7}$ in a NS means that we would detect the effect due to a "mountain" on a NS, with a height of about $10^{-7} \times 10\,\mathrm{km} = 1\,\mathrm{mm}$!

Further reading

- The discovery of the first pulsar, now known as PSR B1919+21, was announced in Hewish, Bell, Pilkington, Scott and Collins (1968). The idea that a rotating NS could power the Crab nebula was advanced by Pacini (1967) a few months before the discovery of the first pulsar. Further classic papers that led to the identification of PSR B1919+21 with a NS are Gold (1968, 1969), Pacini (1968) and Goldreich and Julian (1969). Within a few months of the discovery, several more pulsars were found by various observatories. The Crab pulsar was discovered by Staeilin and Reifenstein (1968) and the Vela pulsar by Large, Vaughan and Mills (1968). The slowdown in the period of the Crab pulsar was found by Richards and Comella (1969).

- The first millisecond pulsar, PSR B1937+21, was discovered by Backer, Kulkarni, Heiles, Davis and Goss (1982). The observation of the currently shortest-period pulsar J1748-2446ad, was reported in Hessels *et al.* (2006). The discovery of 22 millisecond pulsars within the globular cluster Tuc 47 was reported in Freire *et al.* (2001a), and from their dynamics and dispersion measure it was possible to obtain the first detection of ionized gas within the cluster; see Freire *et al.* (2001b). On the

long-period side, the discovery of the radio pulsar PSR J2144-3933, with a period 8.5 s was reported in Young, Manchester and Johnston (1999). Such a long period challenges existing emission models. The braking index of the Crab pulsar was measured in Lyne, Pritchard and Smith (1988), that of Vela in Lyne, Pritchard, Graham-Smith and Camilo (1996) and that of PSR J1734-3333 in Espinoza, Lyne, Kramer, Manchester and Kaspi (2011).

- Pulsars are discussed in great detail in the books by Lorimer and Kramer (2005) and Lyne and Graham-Smith (2006), which also provide a nice historical discussion. Reviews of pulsars are also given in Lorimer (2008, 2011). The formation and evolution of binary and millisecond pulsars is reviewed in Bhattacharya and van den Heuvel (1991). Pulsar catalogs are maintained at http://www.atnf.csiro.au/research/pulsar/psrcat/ (The ATNF Pulsar Catalogue), see Manchester, Hobbs, Teoh and Hobbs (2005).

- Two different pulsar spindown models are compared in Ridley and Lorimer (2010). For discussions of the decay of magnetic fields in pulsars see e.g. Lorimer, Bailes and Harrison (1997), Kaplan and van Kerkwiik (2009), Popov, Pons, Miralles, Boldin and Posselt (2010) and references therein.

- Pulsars in globular clusters are discussed in Camilo and Rasio (2005) and Bogdanov *et al.* (2006). The internal dynamics of globular clusters is reviewed in Meylan and Heggie (1997). A review of globular clusters, stressing their importance for the formation of close binary systems and their role in GW physics, is Benacquista (2006).

- The magnetar model was proposed in Duncan and Thompson (1992) and Thompson and Duncan (1995, 1996). For reviews on magnetars, see Woods and Thompson (2006), Mereghetti (2008, 2013), Hurley (2010), and Mereghetti, Pons and Melatos (2015). An interesting popular discussion of magnetars is provided in Kouveliotou, Duncan and Thompson (2003). A catalog of observed magnetars is maintained at http://www.physics.mcgill.ca/~pulsar/magnetar/main.html, and is described in Olausen and Kaspi (2014). The statistics of SGR bursts and its relation with self-organized criticality is discussed in Cheng, Epstein, Guyer and Young (1995), Woods *et al.* (1999) and Gögüs *et al.* (2000). The "fossil field" scenario is discussed in Ferrario and Wickramasinghe (2006).

- The Tolman–Oppenheimer–Volkov equations are reviewed in Misner, Thorne and Wheeler (1973). The theory of oscillations of relativistic stars has a long history, going back to the pioneering paper by Thorne and Campolattaro (1967), Price and Thorne (1969) and Thorne (1969a). Quadrupole f-modes were studied, with different EoS, by Lindblom and Detweiler (1983). Torsional oscillations of NS were studied by Schumaker and Thorne (1983), g-modes in NS were studied by McDermott, van Horn, Hansen and Buland (1985) and Finn (1987), while g-modes due to stratification in NS were discussed by Reisenegger and Goldreich (1992). A study of fluid modes in the Newtonian approximation, for NS with a fluid core, a solid crust and a thin surface of molten fluid was performed by McDermott, van Horn and Hansen (1988) and generalized to slowly rotating relativistic stars in Vavoulidis, Kokkotas and Stavridis (2008). Detailed studies of polar and axial non-radial normal modes were performed by Chandrasekhar and Ferrari (1991a, 1991b). In particular, for non-rotating stars, Chandrasekhar and Ferrari (1991a) found that the metric perturbations can be decoupled from the matter perturbations, so the frequencies of the quasi-normal modes can be found just by solv-

ing the equations for the space-time perturbations. Allen, Andersson, Kokkotas and Schutz (1998) formulated the equations in a form suitable for evolution in the time domain (rather than in frequency space).

- A survey of normal modes for uniform-density stars is given in Andersson, Kojima and Kokkotas (1996). Fits of the frequency of NS quasi-normal modes as a function of the mass and radius of the NS are given, for different EoS, in Andersson and Kokkotas (1998) and Benhar, Ferrari and Gualtieri (2004). The existence of w-modes was first argued in a simple toy model in Kokkotas and Schutz (1986), and then found in realistic stellar models in Kojima (1988), Kokkotas and Schutz (1992) and Leins, Nollert and Soffel (1993). The axial w-modes were discovered in Chandrasekhar and Ferrari (1992). Trapped modes were found by Chandrasekhar and Ferrari (1992) and further studied by Kokkotas (1994), and interface modes were found by Leins, Nollert and Soffel (1993). Normal modes of NS are reviewed in Kokkotas and Schmidt (1999) and Nollert (1999).

- Early work on normal modes of rotating relativistic stars include Hartle, Thorne and Chitre (1972), Chandrasekhar and Friedman (1972), and Hartle and Friedman (1975). The CFS instability was first found for uniformly rotating Maclaurin spheroids in Chandrasekhar (1970), while Friedman and Schutz (1978a, 1978b) showed that it also takes place for generic self-gravitating perfect fluids. The computation of the time-scale τ_{GR} of the CFS instability and the comparison with the viscosity time-scale are performed in Ipser and Lindblom (1991); see also the reviews by Friedman and Ipser (1992) and Andersson and Kokkotas (2001). The computation uses the expression for the back-reaction potential including all mass multipoles in linearized theory, given in Thorne (1969b). The viscosity coefficients for NS are computed in Cutler and Lindblom (1987) and Sawyer (1989).

The effect of rotation on NS normal modes and the associated coupling of axial and polar perturbations were first studied by Chandrasekhar and Ferrari (1991b). Kojima (1992, 1993) studied them to first order in the rotation, and Cutler and Lindblom (1992) studied them using the post-Newtonian expansion. A study of the onset of the CFS instability for rapidly rotating NS in full GR is performed in Stergioulas and Friedman (1998), who show that the critical rotational frequency changes by about 15% when one uses GR instead of Newto-

nian gravity. Progress in numerical relativistic hydrodynamics has also made possible the direct numerical simulation of NS oscillations. Results in the
Cowling approximation are presented in Font, Dimmelmeier, Gupta and Stergioulas (2001), and simulations of bar-mode and non-axisymmetric instabilities for NS in full GR are presented in Baiotti,
De Pietri, Manca and Rezzolla (2007) and Manca,
Baiotti, De Pietri and Rezzolla (2007). Fits to the
frequency and damping time of the f-modes of rotating stars, for different EoS, are given in Gaertig and Kokkotas (2011). A study of the CFS instability window for f- and r-modes with modern
EoS is presented in Doneva, Gaertig, Kokkotas and
Krüger (2013).

For reviews on instabilities in relativistic stars see
Stergioulas (2003), Andersson (2003) and Friedman
and Stergioulas (2014).

- The frequency of the r-modes was computed, in
the Newtonian limit, by Papaloizou and Pringle
(1978). For a relativistic star the existence of the
CFS instability for r-modes was found numerically
by Andersson, (1998) and confirmed analytically by
Friedman and Morsink (1998). The instability window for r-modes was first discussed in Lindblom,
Owen and Morsink (1998) and Andersson, Kokkotas and Schutz (1999). The GWs produced by the
r-mode instability in a young rotating NS were first
computed in Owen *et al.* (1998). These initial computations of the GW emission were revised toward
lower values with more detailed estimates of the
saturation amplitude of the r-mode instability; see
Bildsten and Ushomirsky (2000) and Arras *et al.*
(2003). The role of differential rotation and the interaction with the magnetic field in damping the
r-modes is studied in Rezzolla, Lamb and Shapiro
(2000), Sá and Tome (2005) and Friedman, Lindblom and Lockitch (2016).

- The evolution of the normal-mode frequencies of
a newborn hot proto-NS during the period that
follows core bounce is studied in Ferrari, Miniutti
and Pons (2003) and in Burgio, Ferrari, Gualtieri
and Schulze (2011). The excitation of the normal
modes of the proto-NS and the associated GW production is studied in Passamonti, Stergioulas and
Nagar (2007). A Newtonian study of g-modes in
stratified NS is presented in Passamonti, Haskell,
Andersson, Jones and Hawke (2009), while relativistic g-modes are studied in Gaertig and Kokkotas (2009) and Krüger, Ho and Andersson (2015).
A detailed study of g-modes in young NS is given
in Passamonti, Andersson and Ho (2016).

- The relation between QPOs in magnetars and
crustal shear modes, and the role of Alfvén waves
and the magnetic coupling between the crust
and the core, are studied in Levin (2006) and
Glampedakis, Samuelsson and Andersson (2006).
In particular it is found that, for magnetized models with a crust, the interaction between magnetic
and elastic waves gives rise to a single family of
magneto-elastic modes; see also Gabler, Cerdá-
Durán, Stergioulas, Font and Müller (2012). The
roles of core superfluidity and of composition gradients are discussed in Passamonti and Lander (2013,
2014). The numerical simulations by Cerdá-Durán,
Stergioulas, Font and Müller (2013) show that the
full spectrum of QPOs can be reproduced by axisymmetric, torsional, magneto-elastic oscillations
of magnetars with a superfluid core. See Passamonti and Lander (2014) for a detailed discussion
of the literature.

- Post-merger oscillations of NS and the prospects
for using them to extract information on the
EoS of NS are studied in Shibata (2005), Stergioulas, Bauswein, Zagkouris and Janka (2011),
Bauswein and Janka (2012), Bauswein, Baumgarte
and Janka (2013), Takami, Rezzolla and Baiotti
(2014, 2015), Bauswein, Stergioulas and Janka
(2016) and Bauswein, Clark, Stergioulas and Janka
(2016). Simulations of NS–NS mergers leading to
prompt or delayed BH formation are studied in
Baiotti, Giacomazzo and Rezzolla (2008). The formation of an accretion torus in the coalescence
of unequal-mass binaries is studied in Rezzolla,
Baiotti, Giacomazzo, Link and Font (2010). The
universality of the relation between the second
largest peak and the tidal deformability of the NS
is discussed in Read *et al.* (2013), Bauswein and
Stergioulas (2015), Takami, Rezzolla and Baiotti
(2014, 2015), Bernuzzi, Dietrich and Nagar (2015),
Rezzolla and Takami (2016) and Hotokezaka, Kyutoku, Sekiguchi, and Shibata (2016). The f-mode
CFS instability in the post-merger NS is studied
in Doneva, Kokkotas and Pnigouras (2015) and
Pnigouras and Kokkotas (2016). The possibility
of exciting the low-β, $m = 1$, instability in NS–NS
mergers is studied in East, Paschalidis, Pretorius
and Shapiro (2016) and Radice, Bernuzzi and Ott
(2016).

See also the Further Reading in Chapter 10 for references on the instabilities controlled by the param-
eter $\beta = T_{\rm rot}/|W|$, and in particular on the low-β
instability. A detailed review on binary NS–NS coalescence is Baiotti and Rezzolla (2017).

- GW production during giant flares from magnetars was studied by Ioka (2001), who suggested, on the basis of energetic considerations, that a large GW signal could be produced; see also Corsi and Owen (2011). However, the dynamical analytic studies in Levin and van Hoven (2011), as well as the general-relativistic hydrodynamical simulations presented in Zink, Lasky and Kokkotas (2012) and in Ciolfi and Rezzolla (2012) suggest that (at least for the EoS and magnetic field configurations considered) the actual GW production is much smaller, since only a small fraction of the energy of the flare is converted into f-mode oscillations. The signal is then expected to be below the sensitivity of advanced detectors. GW production from pulsar glitches are studied in Sidery, Passamonti and Andersson (2010) and in Melatos, Douglass and Simula (2015).

- A review on the physics of the NS crust is Chamel and Haensel (2008). Estimates of the maximum values for NS ellipticities are discussed in Ushomirsky, Cutler and Bildsten (2000) and Haskell, Jones and Andersson (2006). Molecular-dynamics simulations of crustal strain in NS are performed in Horowitz and Kadau (2009). GW emission from the known accreting NS is studied in Watts, Krishnan, Bildsten and Schutz (2008). The emission of GWs from rotating magnetized NS is studied in Cutler (2002), Haskell, Samuelsson, Glampedakis and Andersson (2008) and Akgun and Wasserman (2008). Equilibrium configurations of rotating magnetized NS are studied in Frieben and Rezzolla (2012). GW emission from deformed NS is reviewed in Jones (2002) and Andersson *et al.* (2011).

Black-hole perturbation theory

12

In this chapter we study how a black hole reacts to external disturbances. We will discover in particular that BHs have a characteristic set of normal modes that can be excited, for instance by infalling matter, and then decay by emission of gravitational radiation (for this reason, they are rather called quasi-normal modes, to stress that they are damped).

The fact that an elastic body oscillates, and that its oscillations are characterized by a set of normal modes, is something to which we are well accustomed. For example, we studied in Chapter 8 how a resonant bar detector is set into oscillation by an external force, and we saw that these oscillations can be decomposed into normal modes. However, the same result is not at all obvious for BHs. BHs are purely space-time configurations, and have no matter that can sustain the oscillations. Nevertheless we will see in this chapter that, when perturbed, they do behave much like elastic bodies, and have (quasi)-normal modes, whose properties depend only on the mass and spin of the BH. The quasi-normal modes characterize the response of a BH to small perturbations, and are responsible for the emission of gravitational radiation by a perturbed BH. These oscillations, which have no matter to sustain them, are really oscillations of space-time itself (more precisely, of the space-time *outside* the horizon), and are a particularly fascinating topic, both by itself, and for its relevance to GW production in processes involving BHs. They are also crucial to get some analytical understanding of the waveform produced in the inspiral and merge phase of a BH–BH coalescence, and in the successive ringdown phase, as we will see in Chapter 14. Finally, even if these aspects will not be developed in this book (except for some consideration in Section 12.3.5), they also provide a possible bridge toward quantum gravity and even play an important role in the AdS/CFT correspondence. So, in this chapter we discuss BH perturbation theory in great detail.

12.1 Scalar perturbations

Before moving to our main subject, which is the study of gravitational perturbations of BHs, it is useful to consider the evolution of a scalar field in a BH background. Beside being interesting in its own right, from a technical point of view the study of scalar perturbations is also

Gravitational Waves, Volume 2: Astrophysics and Cosmology. Michele Maggiore.
© Michele Maggiore 2018. Published in 2018 by Oxford University Press.
DOI 10.1093/oso/9780198570899.001.0001

a warm-up exercise, where we can understand the basic features of the result, manipulating equations that are much simpler than the Einstein equations, and without the complication of the full machinery of tensor spherical harmonics, that we will use in the gravitational case.

Consider a massless scalar field ϕ. In a generic curved background with background metric $\bar{g}_{\mu\nu}$, its evolution is determined by the massless Klein–Gordon equation[1]

$$\Box\phi \equiv (-\bar{g})^{-1/2}\partial_\mu\left[(-\bar{g})^{1/2}\bar{g}^{\mu\nu}\partial_\nu\right]\phi = 0, \tag{12.1}$$

where, as usual, $\bar{g}$ denotes the determinant of $\bar{g}^{\mu\nu}$. For a Schwarzschild BH the metric is given by

$$\bar{g}_{\alpha\beta}dx^\alpha dx^\beta = -A(r)(cdt)^2 + B(r)dr^2 + r^2\left(d\theta^2 + \sin^2\theta d\phi^2\right), \tag{12.2}$$

where we have used the notation

$$A(r) = 1 - \frac{R_S}{r}, \quad B(r) = \frac{1}{A(r)}, \tag{12.3}$$

and $R_S = 2GM/c^2$ is the Schwarzschild radius. Given the spherical symmetry of the background metric, it is convenient to expand $\phi(x)$ in spherical harmonics,

$$\phi(t,r,\theta,\phi) = \frac{1}{r}\sum_{l=0}^\infty\sum_{m=-l}^l u_{lm}(t,r)Y_{lm}(\theta,\phi). \tag{12.4}$$

Inserting this expansion into eq. (12.1) we get

$$A\partial_r(A\partial_r u_{lm}) - \partial_0^2 u_{lm} - V_l(r)u_{lm} = 0, \tag{12.5}$$

with

$$V_l(r) = A(r)\left[\frac{l(l+1)}{r^2} + \frac{R_S}{r^3}\right]. \tag{12.6}$$

Of course, because of the spherical symmetry of the Schwarzschild metric, $V_l(r)$ does not depend on the azimuthal number m. It is convenient to introduce the so-called "tortoise coordinate" r_*,

$$r_* = r + R_S\log\frac{r-R_S}{R_S}. \tag{12.7}$$

Observe that r_* ranges from $-\infty$ to $+\infty$ as r goes from the horizon $r = R_S$ to $r = +\infty$. We also introduce the notation

$$\partial_* \equiv \frac{\partial}{\partial r_*}. \tag{12.8}$$

From the definition of r_* it follows that $\partial_* r = A$ and $A\partial_r = \partial_*$. Equation (12.5) then becomes

$$[\partial_*^2 - \partial_0^2 - V_l(r)]u_{lm}(t,r) = 0. \tag{12.9}$$

It is convenient to perform a Fourier transform with respect to the time variable, writing

$$u_{lm}(t,r) = \int_{-\infty}^{\infty} \frac{d\omega}{2\pi}\, \tilde{u}_{lm}(\omega,r)e^{-i\omega t}\,, \tag{12.10}$$

Then $\tilde{u}_{lm}(\omega,r)$ satisfies the equation

$$\left[-\frac{d^2}{dr_*^2} + V_l(r) \right] \tilde{u}_{lm} = \frac{\omega^2}{c^2}\, \tilde{u}_{lm}\,. \tag{12.11}$$

This is formally equivalent to a Schrödinger equation in one dimension, defined on the line $-\infty < r_* < \infty$, for a particle of mass m, written in units $\hbar^2/(2m) = 1$, with $V_l(r)$ playing the role of the potential and ω^2/c^2 that of the energy. In the following we will first see how very similar equations are obtained for metric perturbations, and we will then study their physical consequences.

12.2   Gravitational perturbations

To study the gravitational perturbations of a Schwarzschild BH we write

$$g_{\alpha\beta}(x) = \bar{g}_{\alpha\beta}(x) + h_{\alpha\beta}(x)\,, \tag{12.12}$$

where $\bar{g}_{\alpha\beta}$ is the Schwarzschild metric in Schwarzschild coordinates, given in eq. (12.2), and $h_{\alpha\beta}$ is the perturbation of the metric. The background Schwarzschild metric is a solution of the vacuum Einstein equations, subject to the conditions of spherical symmetry. We consider now the case in which some external matter, described by an energy–momentum tensor $T_{\alpha\beta}$, perturbs the black hole. Later we will consider in detail the case of a mass m falling radially into a black hole, but for the moment we do not specify $T_{\alpha\beta}$. We assume, however, that the perturbation is sufficiently small that linear perturbation theory is adequate. We will later check the validity of this approximation. Thus, we start from the Einstein equations

$$G_{\alpha\beta} = \frac{8\pi G}{c^4}\, T_{\alpha\beta}\,, \tag{12.13}$$

we write the metric as in eq. (12.12), and we expand the Einstein tensor $G_{\alpha\beta}$ to linear order in $h_{\alpha\beta}$,

$$G_{\alpha\beta} = \bar{G}_{\alpha\beta} + \Delta G_{\alpha\beta}\,. \tag{12.14}$$

Since the Schwarzschild metric is a solution of the vacuum Einstein equations, $\bar{G}_{\alpha\beta} = 0$, the equations governing the perturbations are

$$\Delta G_{\alpha\beta} = \frac{8\pi G}{c^4}\, T_{\alpha\beta}\,. \tag{12.15}$$

We saw in Chapter 1 of Vol. 1 that the description of GWs is simpler in the TT gauge. However, we will find in this chapter that the equations governing the perturbation $h_{\alpha\beta}$ are simpler in a different gauge. Thus, for the moment we do not fix the gauge, and we work with all 10 components of $h_{\alpha\beta}$. Since the background metric is spherically symmetric, we can separate the radial and angular dependence of $h_{\alpha\beta}$, and for this we need a basis of 10 independent tensor spherical harmonics, corresponding to the 10 components of $h_{\alpha\beta}$. In the next subsection we introduce such a basis.

12.2.1 Zerilli tensor harmonics

When we studied the separation of the angular dependence in the spatial components h_{ij}, in Section 3.5.2, we saw that a basis for the six components of a symmetric (but not necessarily traceless) matrix h_{ij} is given by the six tensor harmonics,

$$(\mathbf{T}_{lm}^{L0})_{ij}\,, \quad (\mathbf{T}_{lm}^{T0})_{ij}\,, \quad (\mathbf{T}_{lm}^{E1})_{ij}\,, \quad (\mathbf{T}_{lm}^{B1})_{ij}\,, \quad (\mathbf{T}_{lm}^{E2})_{ij}\,, \quad (\mathbf{T}_{lm}^{B2})_{ij}\,, \quad (12.16)$$

given in eqs. (3.268)–(3.272) and in eq. (3.276). We can extend them trivially to 4×4 matrices $(\mathbf{T}_{lm}^{L0})_{\mu\nu}$, etc. by stating that, in the rest frame of the BH, $(\mathbf{T}_{lm}^{L0})_{\mu\nu}$ is equal to $(\mathbf{T}_{lm}^{L0})_{ij}$ when both μ and ν are spatial indices, and is zero otherwise, and similarly for $\mathbf{T}_{lm}^{T0}, \ldots, \mathbf{T}_{lm}^{B2}$.

We next need four independent tensor harmonics that allow us to express the components $h_{0\mu}$. Since h_{00} is a scalar under rotations, it can be expanded in terms of the ordinary scalar spherical harmonics, so we define the tensor $(\mathbf{T}_{lm}^{tt})_{\mu\nu}$ by stating that, in the BH rest frame,

$$(\mathbf{T}_{lm}^{tt})_{\mu\nu} = \delta_\mu^0 \delta_\nu^0 Y_{lm}\,. \quad (12.17)$$

From the point of view of rotations, h_{0i} is a vector, and can be expanded in terms of the vector spherical harmonics $(\mathbf{Y}_{lm}^R)_i$, $(\mathbf{Y}_{lm}^E)_i$ and $(\mathbf{Y}_{lm}^B)_i$ defined in eqs. (3.254)–(3.256). We then define the tensor harmonics $(\mathbf{T}_{lm}^{Rt})_{\mu\nu}$, $(\mathbf{T}_{lm}^{Et})_{\mu\nu}$ and $(\mathbf{T}_{lm}^{Bt})_{\mu\nu}$ by stating that, in the BH rest frame, they are non-vanishing only when $\mu = 0$ and $\nu = i$ (or when $\mu = i$ and $\nu = 0$, with $(\mathbf{T}_{lm}^{Rt})_{0i} = (\mathbf{T}_{lm}^{Rt})_{i0}$, etc.), and

$$(\mathbf{T}_{lm}^{Rt})_{0i} = \frac{1}{\sqrt{2}}\,(\mathbf{Y}_{lm}^R)_i\,, \quad (12.18)$$

$$(\mathbf{T}_{lm}^{Et})_{0i} = \frac{1}{\sqrt{2}}\,(\mathbf{Y}_{lm}^E)_i\,, \quad (12.19)$$

$$(\mathbf{T}_{lm}^{Bt})_{0i} = \frac{1}{\sqrt{2}}\,(\mathbf{Y}_{lm}^B)_i\,. \quad (12.20)$$

Explicitly,

$$(\mathbf{T}_{lm}^{Rt})_{0i} = \frac{1}{\sqrt{2}}\,n_i Y_{lm}\,, \quad (12.21)$$

$$(\mathbf{T}_{lm}^{Et})_{0i} = [2l(l+1)]^{-1/2}\,r\partial_i Y_{lm}\,, \quad (12.22)$$

$$(\mathbf{T}_{lm}^{Bt})_{0i} = i[2l(l+1)]^{-1/2}\,L_i Y_{lm}\,. \quad (12.23)$$

As in Section 3.5.2, $\mathbf{L} = -i\mathbf{r} \times \boldsymbol{\nabla}$, and $\hat{\mathbf{n}}$ is the unit vector in the radial direction, normalized with respect to the flat-space metric, $\delta_{ij}n^i n^j = 1$.

The factors i in the definition of $\mathbf{T}_{lm}^{Bt}$, $\mathbf{T}_{lm}^{B1}$ and $\mathbf{T}_{lm}^{B2}$ [see eqs. (3.270), (3.272) and (12.23)] have been inserted to ensure that these tensors also satisfy the relation

$$(\mathbf{T}_{lm}^{a})_{\mu\nu}^{*} = (-1)^{m}(\mathbf{T}_{l\,\bar{m}}^{a})_{\mu\nu} \,, \qquad (12.24)$$

which is derived observing that $i\mathbf{L} = \mathbf{r} \times \boldsymbol{\nabla}$ is real.

As in eqs. (3.254)–(3.256), $\mathbf{T}_{lm}^{Rt}$ is defined for $l \geqslant 0$, while $\mathbf{T}_{lm}^{Et}$ and $\mathbf{T}_{lm}^{Bt}$ have $l \geqslant 1$. Similarly, we saw in Section 3.5.2 that a complete basis for the expansion of h_{ij} is provided by the six spherical harmonics (12.16), where T_{lm}^{L0} and T_{lm}^{T0} have $l \geqslant 0$, T_{lm}^{E1} and T_{lm}^{B1} have $l \geqslant 1$, while T_{lm}^{E2} and T_{lm}^{B2} have $l \geqslant 2$.[2]

From eqs. (3.258) and (3.277) we see that these 10 tensor harmonics are normalized so that

$$\int d\Omega \, \eta^{\mu\rho}\eta^{\nu\sigma}(\mathbf{T}_{l'm'}^{a})_{\mu\nu}^{*}(\mathbf{T}_{lm}^{b})_{\rho\sigma} = \epsilon_{a}\delta^{ab}\delta_{ll'}\delta_{mm'} \,, \qquad (12.25)$$

where the indices a, b run over the set

$$\{L0, T0, E1, B1, E2, B2, tt, Rt, Et, Bt\}\,. \qquad (12.26)$$

The coefficient ϵ_{a} is equal to -1 for $a = Rt, Et, Bt$, and $+1$ for all the other values of a, and of course there is no sum over a in $\epsilon_{a}\delta^{ab}$ on the right-hand side, since the index a also appears on the left-hand side.[3] Observe also that in eq. (12.25) the contractions of the Lorentz indices are performed with the flat Minkowski metric $\eta_{\mu\nu}$.

We can now separate $h_{\mu\nu}$ into a part that depends on the angular variables (θ, ϕ) and a part that depends on the variable (t, r),

$$h_{\mu\nu}(t, \mathbf{x}) = \sum_{a}\sum_{lm} H_{lm}^{a}(t, r)(\mathbf{T}_{lm}^{a})_{\mu\nu}(\theta, \phi)\,, \qquad (12.27)$$

where the label a runs over the 10 values (12.26), and it is understood that the sum over l runs over $l \geqslant 0$, $l \geqslant 1$ or $l \geqslant 2$, depending on the tensor spherical harmonic concerned. The $(\mathbf{T}_{lm}^{a})_{\mu\nu}$ are known as the Zerilli tensor spherical harmonics.

Observe that the tensors $(\mathbf{T}_{lm}^{a})_{\mu\nu}$ describe the expansion of the metric $h_{\mu\nu}$ in Cartesian coordinates, $x^{\mu} = (ct, x, y, z)$, i.e. they are a basis for the expansion of $h_{xx}, h_{xy}, \dots$. Since the background Schwarzschild metric is much simpler in polar coordinates (ct, r, θ, ϕ), we need the perturbation with respect to these coordinates, i.e. we need $h_{rr}, h_{r\theta}, h_{\theta\theta}, \dots$. These are related to the components in a Cartesian reference frame, $h_{xx}, h_{xy}, \dots$ as follows. We use the notation $x^{\mu} = (ct, x, y, z)$ and $x^{\alpha} = (ct, r, \theta, \phi)$, and we denote by $h_{\mu\nu}$ the Cartesian components of the metric perturbation, i.e. h_{xx}, h_{xy}, etc., and by $h_{\alpha\beta}$ the polar components, i.e. $h_{rr}, h_{r\theta}$, etc. By definition, the polar components $h_{\alpha\beta}$ are related to the Cartesian components $h_{\mu\nu}$ by

$$h_{\alpha\beta}dx^{\alpha}dx^{\beta} = h_{\mu\nu}dx^{\mu}dx^{\nu}\,, \qquad (12.28)$$

[2] In the literature, following the original papers of Regge and Wheeler (1957) and Zerilli (1970), for these 10 tensor harmonics the following notation is often used:

$$\begin{aligned}
a_{lm} &= \mathbf{T}_{lm}^{L0}\,, & b_{lm} &= \mathbf{T}_{lm}^{E1}\,, \\
c_{lm} &= -i\mathbf{T}_{lm}^{B1}\,, & d_{lm} &= -i\mathbf{T}_{lm}^{B2}\,, \\
f_{lm} &= \mathbf{T}_{lm}^{E2}\,, & g_{lm} &= \mathbf{T}_{lm}^{T0}\,, \\
a_{lm}^{(0)} &= \mathbf{T}_{lm}^{tt}\,, & a_{lm}^{(1)} &= i\mathbf{T}_{lm}^{Rt}\,, \\
b_{lm}^{(0)} &= i\mathbf{T}_{lm}^{Et}\,, & c_{lm}^{(0)} &= -\mathbf{T}_{lm}^{Bt}\,.
\end{aligned}$$

The notation that we adopt rather stresses the properties of the various spherical harmonics under rotation and parity. We saw in Section 3.5.2 that, for massless fields, $\mathbf{T}^{L0}$ describes a longitudinal scalar mode, $\mathbf{T}^{T0}$ a transverse scalar mode, $\mathbf{T}^{E1}$ a vector mode with electric-type parity, $\mathbf{T}^{B1}$ a vector mode with magnetic-type parity, $\mathbf{T}^{E2}$ a spin-2 mode with electric-type parity, and $\mathbf{T}^{B2}$ a spin-2 mode with magnetic-type parity.

Similarly, $\mathbf{T}^{tt}$ describes a scalar mode corresponding to the time-time component h_{00}, $\mathbf{T}^{Rt}$ describes the radial part of the vector h_{0i}, while $\mathbf{T}^{Et}$ and $\mathbf{T}^{Bt}$ describe the two transverse components of h_{0i}, the former with electric-type parity and the latter with magnetic-type parity.

[3] The sign given by ϵ_{a} (which was forgotten in the original Zerilli paper) comes from the fact that, when we raise the indices $(0i)$ with the Minkowski metric, we pick a minus sign. This does not influence T^{tt}, which only has the (00) component, or the six tensor harmonics in eq. (12.16), which only have spatial components (ij), but, for $a = Rt, Et$, or Bt, where we have only the $(0i)$ components, we have a factor $\eta^{\mu\rho}\eta^{\nu\sigma}$ with $\mu = \rho = 0$ and $\nu = \sigma = i$ and, for each fixed i, $\eta^{00}\eta^{ii} = -1$.

and similarly we can define the polar components of the Zerilli tensors $(\mathbf{T}_{lm}^a)_{\alpha\beta}$ from

$$(\mathbf{T}_{lm}^a)_{\alpha\beta}dx^\alpha dx^\beta = (\mathbf{T}_{lm}^a)_{\mu\nu}dx^\mu dx^\nu \,, \tag{12.29}$$

so that

$$h_{\alpha\beta}(t,r,\theta,\phi) = \sum_a \sum_{lm} H_{lm}^a(t,r)(\mathbf{T}_{lm}^a)_{\alpha\beta}\,. \tag{12.30}$$

Since the time variable is the same in Cartesian and in polar coordinates, the transformation is only on the spatial variables, and it is useful to define the matrix

$$A_a{}^i = \frac{\partial x^i}{\partial x^a} = \begin{pmatrix} \sin\theta\cos\phi & \sin\theta\sin\phi & \cos\theta \\ r\cos\theta\cos\phi & r\cos\theta\sin\phi & -r\sin\theta \\ -r\sin\theta\sin\phi & r\sin\theta\cos\phi & 0 \end{pmatrix}, \tag{12.31}$$

where $x^a = (r,\theta,\phi)$ and $x^i = (x,y,z)$. Then, for a generic tensor, $T_{ab} = A_a{}^i A_b{}^j T_{ij}$, while $T_{0a} = A_a{}^i T_{0i}$. When we compute these expressions for the spatial parts of the Zerilli tensor harmonics we find terms containing $A_a{}^i n_i$, $A_a{}^i \partial_i$, and $A_a{}^i L_i$. Starting from the expression for n_i in Cartesian coordinates, $n_i = (\sin\theta\cos\phi, \sin\theta\sin\phi, \cos\theta)$, and, using the explicit form of $A_a{}^i$, we get the obvious result $n_a \equiv A_a{}^i n_i = (1,0,0)$, i.e. $n_r = 1$, $n_\theta = n_\phi = 0$, and similarly $\partial_a \equiv A_a{}^i \partial_i = (\partial_r, \partial_\theta, \partial_\phi)$. For the angular momentum we get[4]

$$L_a \equiv A_a{}^k L_k = ir\left(0, \frac{1}{\sin\theta}\partial_\phi, -\sin\theta\,\partial_\theta\right)_a\,. \tag{12.36}$$

Using these results, the computation of the explicit expressions for all tensor harmonics in polar coordinates becomes straightforward. The result can be conveniently written as

$$(\mathbf{T}_{lm}^a)_{\alpha\beta} = c^a(r)(\mathbf{t}_{lm}^a)_{\alpha\beta} \tag{12.37}$$

(no sum over a), where $c^a(r)$ are given by

$$c^{L0} = c^{tt} = 1\,, \quad c^{T0} = \frac{r^2}{\sqrt{2}}\,, \quad c^{Rt} = \frac{1}{\sqrt{2}}\,, \tag{12.38}$$

$$c^{Et} = c^{E1} = -c^{Bt} = -c^{B1} = \frac{r}{[2l(l+1)]^{1/2}}\,, \tag{12.39}$$

$$c^{E2} = c^{B2} = r^2\left[\frac{1}{2}\frac{(l-2)!}{(l+2)!}\right]^{1/2}\,, \tag{12.40}$$

and the tensor harmonics $(\mathbf{t}_{lm}^a)_{\alpha\beta}$ are given, as matrices in the (α,β) indices, by

$$\mathbf{t}_{lm}^{tt} = \begin{pmatrix} 1 & 0 & 0 & 0 \\ 0 & 0 & 0 & 0 \\ 0 & 0 & 0 & 0 \\ 0 & 0 & 0 & 0 \end{pmatrix}Y_{lm}\,, \quad \mathbf{t}_{lm}^{Rt} = \begin{pmatrix} 0 & 1 & 0 & 0 \\ 1 & 0 & 0 & 0 \\ 0 & 0 & 0 & 0 \\ 0 & 0 & 0 & 0 \end{pmatrix}Y_{lm}\,,$$

$$\mathbf{t}_{lm}^{L0} = \begin{pmatrix} 0 & 0 & 0 & 0 \\ 0 & 1 & 0 & 0 \\ 0 & 0 & 0 & 0 \\ 0 & 0 & 0 & 0 \end{pmatrix}Y_{lm}\,, \quad \mathbf{t}_{lm}^{T0} = \begin{pmatrix} 0 & 0 & 0 & 0 \\ 0 & 0 & 0 & 0 \\ 0 & 0 & 1 & 0 \\ 0 & 0 & 0 & \sin^2\theta \end{pmatrix}Y_{lm}\,,$$

[4]To understand the meaning of this expression, and its relation to the more usual form of the angular momentum vector in polar coordinates, observe that eq. (12.36) gives the components L_a of the angular momentum with respect to the basis of 1-forms $dx^a = (dr, d\theta, d\phi)$. These are related to the Cartesian components L_i by $dx^i L_i = dx^a L_a$, i.e. $dx L_x + dy L_y + dz L_z = dr L_r + d\theta L_\theta + d\phi L_\phi$. By raising the index a with the metric g^{ab} we get $L^a = g^{ab} L_b$, whose explicit form is given by

$$L^a f = \frac{i}{r\sin\theta}(0, \partial_\phi f, -\partial_\theta f)^a\,, \tag{12.32}$$

where f is any differentiable function. These are the components of the vector $(L^a f)$, expressed in the basis of tangent space which is dual to the basis $(dr, d\theta, d\phi)$ in the space of 1-forms, i.e. in the basis $(\partial_r, \partial_\theta, \partial_\phi)$, so, as a tangent space vector,

$$\mathbf{L}f = \frac{i}{r\sin\theta}(\partial_\phi f)\partial_\theta - \frac{i}{r\sin\theta}(\partial_\theta f)\partial_\phi\,. \tag{12.33}$$

In polar coordinates, it is more convenient to use as a basis for tangent space

$$\mathbf{e}_a = \left(\partial_r, \frac{1}{r}\partial_\theta, \frac{1}{r\sin\theta}\partial_\phi\right)_a\,, \tag{12.34}$$

which is equivalent to the basis $(\hat{\mathbf{r}}, \hat{\boldsymbol{\theta}}, \hat{\boldsymbol{\phi}})$, in the sense that the components of a directional derivative $\partial_{\mathbf{u}}$ with respect to the basis $\mathbf{e}_a$ are the same as the components of the vector $\mathbf{u}$ in the basis $(\hat{\mathbf{r}}, \hat{\boldsymbol{\theta}}, \hat{\boldsymbol{\phi}})$. In the basis $\mathbf{e}_a$, the components of the vector $\mathbf{L}f$ become

$$L^a f = i\left(0, \frac{1}{\sin\theta}\partial_\phi f, -\partial_\theta f\right)^a\,. \tag{12.35}$$

This is the expression that can be obtained directly from $\mathbf{L}f = -i\mathbf{r} \times \boldsymbol{\nabla}f$, using the usual form of the gradient in polar coordinates.

$$\mathbf{t}_{lm}^{Et} = \begin{pmatrix} 0 & 0 & \partial_\theta & \partial_\phi \\ 0 & 0 & 0 & 0 \\ \star & 0 & 0 & 0 \\ \star & 0 & 0 & 0 \end{pmatrix} Y_{lm}, \qquad \mathbf{t}_{lm}^{E1} = \begin{pmatrix} 0 & 0 & 0 & 0 \\ 0 & 0 & \partial_\theta & \partial_\phi \\ 0 & \star & 0 & 0 \\ 0 & \star & 0 & 0 \end{pmatrix} Y_{lm},$$

$$\mathbf{t}_{lm}^{Bt} = \begin{pmatrix} 0 & 0 & (1/\sin\theta)\partial_\phi & -\sin\theta\,\partial_\theta \\ 0 & 0 & 0 & 0 \\ \star & 0 & 0 & 0 \\ \star & 0 & 0 & 0 \end{pmatrix} Y_{lm},$$

$$\mathbf{t}_{lm}^{B1} = \begin{pmatrix} 0 & 0 & 0 & 0 \\ 0 & 0 & (1/\sin\theta)\partial_\phi & -\sin\theta\,\partial_\theta \\ 0 & \star & 0 & 0 \\ 0 & \star & 0 & 0 \end{pmatrix} Y_{lm},$$

$$\mathbf{t}_{lm}^{E2} = \begin{pmatrix} 0 & 0 & 0 & 0 \\ 0 & 0 & 0 & 0 \\ 0 & 0 & W & X \\ 0 & 0 & \star & -\sin^2\theta\,W \end{pmatrix} Y_{lm},$$

$$\mathbf{t}_{lm}^{B2} = \begin{pmatrix} 0 & 0 & 0 & 0 \\ 0 & 0 & 0 & 0 \\ 0 & 0 & -(1/\sin\theta)X & \sin\theta\,W \\ 0 & 0 & \star & \sin\theta\,X \end{pmatrix} Y_{lm}, \qquad (12.41)$$

where rows and columns corresponds to t, r, θ and ϕ, in this order, the stars denote components obtained from the symmetry of the matrix, and we have defined the operators

$$X = 2\partial_\theta\partial_\phi - 2\operatorname{ctg}\theta\,\partial_\phi\,, \qquad (12.42)$$

$$W = \partial_\theta^2 - \operatorname{ctg}\theta\,\partial_\theta - \frac{1}{\sin^2\theta}\,\partial_\phi^2\,. \qquad (12.43)$$

We can now write the expansion of the metric perturbation as

$$h_{\alpha\beta}(x) = \sum_a \sum_{lm} h_{lm}^a(t,r)(\mathbf{t}_{lm}^a)_{\alpha\beta}(\theta,\phi)\,, \qquad (12.44)$$

where we have reabsorbed the factors $c^a(r)$, defining[5,6]

$$h_{lm}^a(t,r) = c^a(r)H_{lm}^a(t,r)\,. \qquad (12.46)$$

Polar and axial perturbations

Under a parity transformation, $\theta \to \pi - \theta$ and $\phi \to \phi + \pi$, the tensor harmonics with $a = L0, T0, E1, E2, tt, Rt$ and Et pick a factor $(-1)^l$, and are said to be polar, or even. In contrast, those with $a = B1, B2$ and Bt pick a factor $(-1)^{l+1}$, and are said to be axial, or odd. Since the Schwarzschild metric is invariant under parity, when we linearize around it, polar and axial perturbations do not mix. It is therefore useful to separate

$$h_{\alpha\beta}(x) = h_{\alpha\beta}^{\mathrm{pol}}(x) + h_{\alpha\beta}^{\mathrm{ax}}(x)\,, \qquad (12.47)$$

[5] Observe that the Cartesian components of the tensor (and vector) spherical harmonics, e.g. $(T_{lm}^{Et})_{0x}$, etc. are functions of θ and ϕ only, so in Cartesian coordinates it is indeed correct that in eq. (12.27) we have separated the dependence on (θ, ϕ) from the dependence on (t, r). However, the trasnformation from Cartesian to polar coordinates, e.g. to $(T_{lm}^{Et})_{0r}$, etc. introduces a further dependence on r, which is at the origin of the fact that the factors $c^a(r)$ depend on r. So a separation between a part that depends only on (θ, ϕ) and a part that depends only on (t, r) is obtained by reabsorbing the $c^a(r)$ into $h_a(t, r)$.

[6] A (trivial) comment on dimensional analysis. Denoting by l a quantity with dimensions of length, $h_{\alpha\beta}dx^\alpha dx^\beta$ has dimensions l^2. In polar coordinates $dx^\alpha = (cdt, dr, d\theta, d\phi)$ has dimension $(l, l, 1, 1)$ so $h_{tt}, h_{rr}, h_{tr} \sim 1$, while $h_{\theta\phi}, h_{\theta\theta}$ and $h_{\phi\phi}$ have dimensions $\sim l^2$, and the other elements of $h_{\alpha\beta}$ have dimensions $\sim l$. From this, and observing what are the non-vanishing matrix elements of each $(t_{lm}^a)_{\alpha\beta}$ in eq. (12.41), it follows that, dimensionally,

$$h_{lm}^{tt}, h_{lm}^{Rt}, h_{lm}^{L0} \sim 1\,,$$
$$h_{lm}^{Et}, h_{lm}^{E1}, h_{lm}^{Bt}, h_{lm}^{B1} \sim l\,, \qquad (12.45)$$
$$h_{lm}^{T0}, h_{lm}^{E2}, h_{lm}^{B2} \sim l^2\,,$$

as can also be checked from eq. (12.46) and the definitions (12.38)–(12.40) of the coefficients c^a.

where, explicitly,

$$
h_{\alpha\beta}^{\mathrm{pol}}(t,\mathbf{x}) = \sum_{l=0}^{\infty}\sum_{m=-l}^{l}\left[h_{lm}^{tt}(\mathbf{t}_{lm}^{tt})_{\alpha\beta} + h_{lm}^{Rt}(\mathbf{t}_{lm}^{Rt})_{\alpha\beta}\right]
$$
$$
+\sum_{l=0}^{\infty}\sum_{m=-l}^{l}\left[h_{lm}^{L0}(\mathbf{t}_{lm}^{L0})_{\alpha\beta} + h_{lm}^{T0}(\mathbf{t}_{lm}^{T0})_{\alpha\beta}\right]
$$
$$
+\sum_{l=1}^{\infty}\sum_{m=-l}^{l}\left[h_{lm}^{Et}(\mathbf{t}_{lm}^{Et})_{\alpha\beta} + h_{lm}^{E1}(\mathbf{t}_{lm}^{E1})_{\alpha\beta}\right]
$$
$$
+\sum_{l=2}^{\infty}\sum_{m=-l}^{l} h_{lm}^{E2}(\mathbf{t}_{lm}^{E2})_{\alpha\beta}\,,
\tag{12.48}
$$
$$
h_{\alpha\beta}^{\mathrm{ax}}(t,\mathbf{x}) = \sum_{l=1}^{\infty}\sum_{m=-l}^{l}\left[h_{lm}^{Bt}(\mathbf{t}_{lm}^{Bt})_{\alpha\beta} + h_{lm}^{B1}(\mathbf{t}_{lm}^{B1})_{\alpha\beta}\right]
$$
$$
+\sum_{l=2}^{\infty}\sum_{m=-l}^{l} h_{lm}^{B2}(\mathbf{t}_{lm}^{B2})_{\alpha\beta}\,.
\tag{12.49}
$$

The energy–momentum tensor

Since the perturbation $h_{\alpha\beta}$ is expanded in terms of the tensor harmonics $(\mathbf{t}_{lm}^{a})_{\alpha\beta}$, it is convenient to perform the same decomposition for the matter energy–momentum tensor $T_{\mu\nu}$. So we write

$$
T_{\alpha\beta}(x) = T_{\alpha\beta}^{\mathrm{pol}}(x) + T_{\alpha\beta}^{\mathrm{ax}}(x)\,,
\tag{12.50}
$$

where, similarly to eqs. (12.48) and (12.49),

$$
T_{\alpha\beta}^{\mathrm{pol}}(t,\mathbf{x}) = \sum_{l=0}^{\infty}\sum_{m=-l}^{l}\left[s_{lm}^{tt}(\mathbf{t}_{lm}^{tt})_{\alpha\beta} + s_{lm}^{Rt}(\mathbf{t}_{lm}^{Rt})_{\alpha\beta}\right]
$$
$$
+\sum_{l=0}^{\infty}\sum_{m=-l}^{l}\left[s_{lm}^{L0}(\mathbf{t}_{lm}^{L0})_{\alpha\beta} + s_{lm}^{T0}(\mathbf{t}_{lm}^{T0})_{\alpha\beta}\right]
$$
$$
+\sum_{l=1}^{\infty}\sum_{m=-l}^{l}\left[s_{lm}^{Et}(\mathbf{t}_{lm}^{Et})_{\alpha\beta} + s_{lm}^{E1}(\mathbf{t}_{lm}^{E1})_{\alpha\beta}\right]
$$
$$
+\sum_{l=2}^{\infty}\sum_{m=-l}^{l} s_{lm}^{E2}(\mathbf{t}_{lm}^{E2})_{\alpha\beta}\,,
\tag{12.51}
$$
$$
T_{\alpha\beta}^{\mathrm{ax}}(t,\mathbf{x}) = \sum_{l=1}^{\infty}\sum_{m=-l}^{l}\left[s_{lm}^{Bt}(\mathbf{t}_{lm}^{Bt})_{\alpha\beta} + s_{lm}^{B1}(\mathbf{t}_{lm}^{B1})_{\alpha\beta}\right]
$$
$$
+\sum_{l=2}^{\infty}\sum_{m=-l}^{l} s_{lm}^{B2}(\mathbf{t}_{lm}^{B2})_{\alpha\beta}\,.
\tag{12.52}
$$

We have denoted by $s_{lm}^{a}(t,r)$ the coefficients of the expansion of the energy–momentum tensor in tensor harmonics.[7]

[7]The relation to the functions $A_{lm}^{(0)}$, $A_{lm}^{(1)}$, etc. defined by Zerilli (1970) in his eqs. (A1) and (A2) is (omitting for notational simplicity the indices l, m),

$$s^{tt} = A^{(0)}\,,$$
$$s^{Rt} = (i/\sqrt{2})A^{(1)}\,,$$
$$s^{T0} = (r^2/\sqrt{2})G_{lm}\,,$$
$$s^{L0} = A\,,$$
$$s^{Et} = ir[2l(l+1)]^{-1/2}B^{(0)}\,,$$
$$s^{E1} = r[2l(l+1)]^{-1/2}B\,,$$
$$s^{Bt} = r[2l(l+1)]^{-1/2}Q^{(0)}\,,$$
$$s^{B1} = ir[2l(l+1)]^{-1/2}Q\,,$$
$$s^{B2} = -ir^2[2(l-1)l(l+1)(l+2)]^{-1/2}D\,,$$
$$s^{E2} = r^2[2(l-1)l(l+1)(l+2)]^{-1/2}F\,.$$

Observe that the dimensions of s^a are not all the same; compare with Note 6. In particular, for $a = tt, Rt, L0$, we have $Gt^a/c^4 \sim 1/l^2$, while for $a = Et, E1, Bt, B1$, we have $Gt^a/c^4 \sim 1/l$ and for $a = T0, E2, B2$ we have $Gt^a/c^4 \sim 1$.

12.2.2 The Regge–Wheeler gauge

We consider an infinitesimal coordinate transformation

$$x^\mu \to x'^\mu = x^\mu + \xi^\mu(x). \qquad (12.53)$$

We found in eq. (1.215) that, choosing ξ_μ so that $\bar{D}_\mu \xi_\nu$ is infinitesimal of the same order as $h_{\mu\nu}$, under this transformation the background metric $\bar{g}_{\mu\nu}$ is invariant while, to linear order in $\bar{D}_\mu \xi_\nu$, $h_{\mu\nu}$ transforms as $h_{\mu\nu}(x) \to h'_{\mu\nu}(x')$, with $h'_{\mu\nu}(x')$ such that

$$h'_{\mu\nu}(x) = h_{\mu\nu}(x) - (\bar{D}_\mu \xi_\nu + \bar{D}_\nu \xi_\mu). \qquad (12.54)$$

This generalizes to curved space the flat-space gauge transformation (1.8). The defining property of $\bar{D}_\mu$ is that it transforms covariantly under coordinate transformations, so, when we transform from Cartesian coordinates, which we label with indices $\mu, \nu = (t, x, y, z)$, to polar coordinates, labeled by $\alpha, \beta = (t, r, \theta, \phi)$, eq. (12.54) obviously remains invariant in form,

$$h'_{\alpha\beta}(x) = h_{\alpha\beta}(x) - (\bar{D}_\alpha \xi_\beta + \bar{D}_\beta \xi_\alpha). \qquad (12.55)$$

We can exploit this gauge freedom to simplify the expansions (12.48) and (12.49). First of all, observe that a generic four-vector field $\xi_\alpha(x) = (\xi_0(x), \xi_a(x))$ can be expanded in scalar and vector spherical harmonics as follows:

$$\xi_0(x) = \sum_{l=0}^{\infty} \sum_{m=-l}^{l} \xi_{lm}^{(t)}(t,r) Y_{lm}(\theta,\phi), \qquad (12.56)$$

$$\xi_a(x) = \sum_{l=0}^{\infty} \sum_{m=-l}^{l} \xi_{lm}^{(R)}(t,r) Y_{lm}(\theta,\phi) n_a \qquad (12.57)$$

$$+ \sum_{l=1}^{\infty} \sum_{m=-l}^{l} \left[\xi_{lm}^{(E)}(t,r) \partial_a Y_{lm} + \xi_{lm}^{(B)}(t,r) \frac{i}{r} L_a Y_{lm} \right],$$

compare with eq. (3.259) and eqs. (3.254)–(3.256). Clearly, $\xi_{lm}^{(B)}$ generates an axial term in the metric, while $\xi_{lm}^{(t)}, \xi_{lm}^{(R)}$ and $\xi_{lm}^{(E)}$ generate polar terms.[8]

We consider first the axial gauge transformation, i.e. the transformation with $\xi_{lm}^{(t)} = \xi_{lm}^{(R)} = \xi_{lm}^{(E)} = 0$, and we denote $\xi_{lm}^{(B)}$ simply as Λ_{lm}, so

$$\xi_0^{\rm ax}(x) = 0, \qquad \xi_a^{\rm ax}(x) = \sum_{l=1}^{\infty} \sum_{m=-l}^{l} \Lambda_{lm}(t,r) \frac{i}{r} L_a Y_{lm}. \qquad (12.58)$$

Using eq. (12.36) we have

$$\xi_\alpha^{\rm ax} = \sum_{l=1}^{\infty} \sum_{m=-l}^{l} \Lambda_{lm}(t,r) \left(0, 0, -\frac{1}{\sin\theta} \partial_\phi Y_{lm}, \sin\theta\, \partial_\theta Y_{lm} \right). \qquad (12.59)$$

We can now compute the covariant derivative

$$\bar{D}_\alpha \xi_\beta^{\text{ax}} = \partial_\alpha \xi_\beta^{\text{ax}} - \bar{\Gamma}_{\alpha\beta}^\gamma \xi_\gamma^{\text{ax}}, \qquad (12.60)$$

where $\bar{\Gamma}_{\alpha\beta}^\gamma$ are the Christoffel symbols of the Schwarzschild metric.[9] Expressing $\bar{D}_\alpha \xi_\beta^{\text{ax}} + \bar{D}_\beta \xi_\alpha^{\text{ax}}$ in the basis $(\mathbf{t}_{lm}^a)_{\alpha\beta}$ given in eq. (12.41), we get

$$\bar{D}_\alpha \xi_\beta^{\text{ax}} + \bar{D}_\beta \xi_\alpha^{\text{ax}} = -\sum_{l=1}^\infty \sum_{m=-l}^{l} (\partial_0 \Lambda_{lm})(\mathbf{t}_{lm}^{Bt})_{\alpha\beta}$$

$$-\sum_{l=1}^\infty \sum_{m=-l}^{l} \left(\partial_r \Lambda_{lm} - \frac{2}{r} \Lambda_{lm} \right) (\mathbf{t}_{lm}^{B1})_{\alpha\beta}$$

$$+\sum_{l=2}^\infty \sum_{m=-l}^{l} \Lambda_{lm}(\mathbf{t}_{lm}^{B2})_{\alpha\beta}, \qquad (12.62)$$

where in the last line we could replace the sum over $l \geq 1$ with a sum over $l \geq 2$ since $(\mathbf{t}_{lm}^{B2})$ vanishes for $l = 1$, and we use the notation $\partial_0 \equiv (1/c)\partial_t$. We observe first of all that, as expected, ξ^{ax} induces a transformation in $h_{\alpha\beta}$ that can be expressed in terms of the axial tensor harmonics $\mathbf{t}_{lm}^{Bt}, \mathbf{t}_{lm}^{B1}, \mathbf{t}_{lm}^{B2}$, so it leaves $h_{\alpha\beta}^{\text{pol}}$ invariant and only affects $h_{\alpha\beta}^{\text{ax}}$. Comparing with eq. (12.49) we see that

$$h_{lm}^{Bt} \to h_{lm}^{Bt} + \partial_0 \Lambda_{lm}, \qquad (l \geq 1), \qquad (12.63)$$

$$h_{lm}^{B1} \to h_{lm}^{B1} + \left(\partial_r - \frac{2}{r} \right) \Lambda_{lm}, \qquad (l \geq 1), \qquad (12.64)$$

$$h_{lm}^{B2} \to h_{lm}^{B2} - \Lambda_{lm}, \qquad (l \geq 2). \qquad (12.65)$$

The Regge–Wheeler (RW) gauge choice, for the axial perturbations, consists in choosing Λ_{lm} so that $h_{lm}^{B2} = 0$, for all $l \geq 2$. Since h_{lm}^{B2} is defined only for $l \geq 2$, while in eq. (12.58) there enter Λ_{lm} with $l \geq 1$, this still leaves us the freedom of choosing Λ_{lm} for $l = 1$, which we use to set h_{1m}^{Bt} to zero.[10] In conclusion, after gauge fixing, in the RW gauge, for the axial perturbation $h_{\alpha\beta}^{\text{ax}}$ we have no degree of freedom with $l = 0$, one degree of freedom for $l = 1$, given by h_{1m}^{B1}, and two degrees of freedom for each $l \geq 2$, h_{lm}^{Bt} and h_{lm}^{B1},

$$h_{\alpha\beta}^{\text{ax}}(t, \mathbf{x}) = \sum_{l=2}^\infty \sum_{m=-l}^{l} h_{lm}^{Bt}(\mathbf{t}_{lm}^{Bt})_{\alpha\beta} + \sum_{l=1}^\infty \sum_{m=-l}^{l} h_{lm}^{B1}(\mathbf{t}_{lm}^{B1})_{\alpha\beta}.$$

$$(12.66)$$

Observe that the quantity

$$(k_1)_{lm} \equiv -h_{lm}^{B1} - \left(\partial_r - \frac{2}{r} \right) h_{lm}^{B2}, \qquad (12.67)$$

[9]In terms of $A(r)$ defined in eq. (12.3), the non-zero components of $\bar{\Gamma}_{\alpha\beta}^\gamma$ in the Schwarzschild metric are

$$\bar{\Gamma}_{rr}^r = -R_S/[2r^2 A(r)],$$
$$\bar{\Gamma}_{\theta\theta}^r = -rA(r),$$
$$\bar{\Gamma}_{\phi\phi}^r = -rA(r)\sin^2\theta,$$
$$\bar{\Gamma}_{tt}^r = A(r)R_S/(2r^2),$$
$$\bar{\Gamma}_{r\theta}^\theta = 1/r, \qquad (12.61)$$
$$\bar{\Gamma}_{\phi\phi}^\theta = -\sin\theta\cos\theta,$$
$$\bar{\Gamma}_{r\phi}^\phi = 1/r,$$
$$\bar{\Gamma}_{\theta\phi}^\phi = \text{ctg}\,\theta,$$
$$\bar{\Gamma}_{tr}^t = R_S/[2r^2 A(r)],$$

together with those obtained by $\bar{\Gamma}_{\alpha\beta}^\gamma = \bar{\Gamma}_{\beta\alpha}^\gamma$.

[10]There still remains a residual gauge freedom, since a further gauge transformation with a function $\Lambda_{1m}(r)$, independent of t, does not spoil the condition $h_{1m}^{Bt} = 0$; see eq. (12.63). We therefore have the residual freedom of adding a time-independent term to h_{1m}^{B1}. We will make use of this freedom later; see Note 18 on page 128.

(defined for $l \geqslant 2$) is invariant under the transformations (12.63)–(12.65), i.e. under linearized gauge transformations. In the RW gauge

$$h_{lm}^{B1} = -(k_1)_{lm} \, . \tag{12.68}$$

For the polar gauge transformation we can proceed in the same way. Using eq. (12.57) we have

$$\xi_\alpha^{\rm pol} = \sum_{l=0}^\infty \sum_{m=-l}^l (\xi_{lm}^{(t)}(t,r)Y_{lm}, \xi_{lm}^{(R)}(t,r)Y_{lm}, 0, 0)$$

$$+ \sum_{l=1}^\infty \sum_{m=-l}^l \xi_{lm}^{(E)}(t,r)(0, 0, \partial_\theta Y_{lm}, \partial_\phi Y_{lm}) \, . \tag{12.69}$$

This gives[11]

$$h_{lm}^{tt} \rightarrow h_{lm}^{tt} - \left[2\partial_0 \xi_{lm}^{(t)} - \frac{A(r)R_S}{r^2} \xi_{lm}^{(R)} \right] , \tag{12.70}$$

$$h_{lm}^{Rt} \rightarrow h_{lm}^{Rt} - \left[\partial_0 \xi_{lm}^{(R)} + \partial_r \xi_{lm}^{(t)} - \frac{R_S}{r^2 A(r)} \xi_{lm}^{(t)} \right] , \tag{12.71}$$

$$h_{lm}^{L0} \rightarrow h_{lm}^{L0} - \left[2\partial_0 \xi_{lm}^{(R)} + \frac{R_S}{r^2 A(r)} \xi_{lm}^{(R)} \right] , \tag{12.72}$$

$$h_{lm}^{T0} \rightarrow h_{lm}^{T0} - \left[2r A(r) \xi_{lm}^{(R)} - l(l+1) \xi_{lm}^{(E)} \right] , \tag{12.73}$$

$$h_{lm}^{Et} \rightarrow h_{lm}^{Et} - \left[\xi_{lm}^{(t)} + \partial_0 \xi_{lm}^{(E)} \right] , \tag{12.74}$$

$$h_{lm}^{E1} \rightarrow h_{lm}^{E1} - \left[\left(\partial_r - \frac{2}{r} \right) \xi_{lm}^{(E)} + \xi_{lm}^{(R)} \right] , \tag{12.75}$$

$$h_{lm}^{E2} \rightarrow h_{lm}^{E2} - \xi_{lm}^{(E)} \, , \tag{12.76}$$

For the polar perturbation, the RW gauge consists in choosing, first of all, $\xi_{lm}^{(E)}$, with $l \geqslant 2$, so that $h_{lm}^{E2} = 0$. We can next choose $\xi_{lm}^{(R)}$, with $l \geqslant 1$, so that $h_{lm}^{E1} = 0$, and $\xi_{lm}^{(t)}$, again with $l \geqslant 1$, so that $h_{lm}^{Et} = 0$. We still have the freedom to choose $\xi_{00}^{(t)}$, $\xi_{00}^{(R)}$ and $\xi_{1m}^{(E)}$. The gauge function $\xi_{1m}^{(E)}$ can be used to set $h_{1m}^{T0} = 0$, $\xi_{00}^{(R)}$ can be chosen so that $h_{00}^{T0} = 0$, while $\xi_{00}^{(t)}$ allows us to set $h_{00}^{Rt} = 0$.[12] This completely specifies the RW gauge. Thus, in the RW gauge the polar perturbations have the form

$$h_{\alpha\beta}^{\rm pol}(t, \mathbf{x}) = \sum_{l=0}^\infty \sum_{m=-l}^l \left[h_{lm}^{tt}(\mathbf{t}_{lm}^{tt})_{\alpha\beta} + h_{lm}^{L0}(\mathbf{t}_{lm}^{L0})_{\alpha\beta} \right]$$

$$+ \sum_{l=1}^\infty \sum_{m=-l}^l h_{lm}^{Rt}(\mathbf{t}_{lm}^{Rt})_{\alpha\beta} + \sum_{l=2}^\infty \sum_{m=-l}^l h_{lm}^{T0}(\mathbf{t}_{lm}^{T0})_{\alpha\beta} \, . \tag{12.77}$$

[11]As in eq. (12.48), the equations for h_{lm}^{L0}, h_{lm}^{T0}, h_{lm}^{tt} and h_{lm}^{Rt} hold for $l \geqslant 0$, those for h_{lm}^{Et} and h_{lm}^{E1} for $l \geqslant 1$, and that for h_{lm}^{E2} for $l \geqslant 2$.

[12]Observe however, from eq. (12.71), that this still leaves us the freedom of performing a further gauge transformation with $\xi_{00}^{(t)}$ such that

$$\partial_r \xi_{00}^{(t)} - \frac{R_S}{r^2 A(r)} \xi_{00}^{(t)} = 0 \, ,$$

which has the solution

$$\xi_{00}^{(t)}(r, t) = A(r) F(t)$$

with $F(t)$ an arbitrary function of time. From eq. (12.70) we see that the function $F(t)$ can be used to subtract an arbitrary function of time from $h_{00}^{tt}/A(r)$. We will make use of this freedom later.

Using the explicit form (12.41) of the tensors $\mathbf{t}_{lm}^a$, the polar perturbations in the RW gauge, at a given level (l, m) with $l \geqslant 2$, can be written as

$$h_{\alpha\beta}^{\text{pol}} = \begin{pmatrix} h_{lm}^{tt} & h_{lm}^{Rt} & 0 & 0 \\ h_{lm}^{Rt} & h_{lm}^{L0} & 0 & 0 \\ 0 & 0 & h_{lm}^{T0} & 0 \\ 0 & 0 & 0 & h_{lm}^{T0} \sin^2\theta \end{pmatrix} Y_{lm} \,. \tag{12.78}$$

Since the perturbation (12.78) must be compared with the background Schwarzschild metric (12.2), after choosing the RW gauge it is convenient to define new functions $H_{lm}^{(0)}$, $H_{lm}^{(1)}$, $H_{lm}^{(2)}$ and K_{lm} from

$$h_{lm}^{tt}(t, r) = A(r) \, H_{lm}^{(0)}(t, r) \,, \tag{12.79}$$

$$h_{lm}^{L0}(t, r) = B(r) \, H_{lm}^{(2)}(t, r) \,, \tag{12.80}$$

$$h_{lm}^{T0}(t, r) = r^2 K_{lm}(t, r) \,. \tag{12.81}$$

We also rename[13]

$$h_{lm}^{Rt}(t, r) = H_{lm}^{(1)}(t, r) \,, \tag{12.83}$$

$$h_{lm}^{Bt}(t, r) = -h_{lm}^{(0)}(t, r) \,, \tag{12.84}$$

$$h_{lm}^{B1}(t, r) = -h_{lm}^{(1)}(t, r) \,. \tag{12.85}$$

In conclusion, in the RW gauge the most general metric of a perturbed Schwarzschild BH, $g_{\alpha\beta} = \bar{g}_{\alpha\beta} + h_{\alpha\beta}$, can be written as

$$g_{\alpha\beta} dx^\alpha dx^\beta = -A(r) \left[1 - \sum_{l=0}^{\infty} \sum_{m=-l}^{l} H_{lm}^{(0)} Y_{lm} \right] (cdt)^2$$

$$+ 2cdt dr \left[\sum_{l=1}^{\infty} \sum_{m=-l}^{l} H_{lm}^{(1)} Y_{lm} \right]$$

$$+ B(r) dr^2 \left[1 + \sum_{l=0}^{\infty} \sum_{m=-l}^{l} H_{lm}^{(2)} Y_{lm} \right]$$

$$+ r^2 (d\theta^2 + \sin^2\theta d\phi^2) \left[1 + \sum_{l=2}^{\infty} \sum_{m=-l}^{l} K_{lm} Y_{lm} \right]$$

$$- 2cdt d\theta \frac{1}{\sin\theta} \left[\sum_{l=2}^{\infty} \sum_{m=-l}^{l} h_{lm}^{(0)} \partial_\phi Y_{lm} \right]$$

$$+ 2cdt d\phi \sin\theta \left[\sum_{l=2}^{\infty} \sum_{m=-l}^{l} h_{lm}^{(0)} \partial_\theta Y_{lm} \right]$$

$$- 2dr d\theta \frac{1}{\sin\theta} \left[\sum_{l=1}^{\infty} \sum_{m=-l}^{l} h_{lm}^{(1)} \partial_\phi Y_{lm} \right]$$

$$+ 2dr d\phi \sin\theta \left[\sum_{l=1}^{\infty} \sum_{m=-l}^{l} h_{lm}^{(1)} \partial_\theta Y_{lm} \right] \,, \tag{12.86}$$

[13] These redefinitions are chosen so as to make contact with the original notation used by Regge and Wheeler and by Zerilli, and in much of the subsequent literature. Before imposing the RW gauge, the relation between our functions h_{lm}^a and the functions h_0^{ax}, h_1^{ax}, h_2^{ax}, H_0, H_1, H_2, K, G, h_0^{pol} and h_1^{pol} defined originally by Regge and Wheeler is (omitting for notational simplicity the indices l, m)

$$h^{Bt} = -h_0^{\text{ax}} \,,$$
$$h^{B1} = -h_1^{\text{ax}} \,,$$
$$h^{B2} = -(1/2)h_2^{\text{ax}} \,,$$
$$h^{tt} = A(r)H_0 \,,$$
$$h^{Rt} = H_1 \,,$$
$$h^{L0} = B(r)H_2 \,,$$
$$h^{T0} = r^2[K - (1/2)l(l+1)G] \,,$$
$$h^{Et} = h_0^{\text{pol}} \,,$$
$$h^{E1} = h_1^{\text{pol}} \,,$$
$$h^{E2} = (1/2)r^2 G \,. \tag{12.82}$$

In the RW gauge $h^{E2} = G = 0$, and the relation $h^{T0} = r^2[K - (1/2)l(l+1)G]$ reduces to $h^{T0} = r^2 K$.

where $A(r)$ and $B(r)$ are given in eq. (12.3). The functions $H^{(0)}, H^{(1)}, H^{(2)}$ and K describe polar perturbations, while $h^{(0)}$ and $h^{(1)}$ describe axial perturbations.[14]

From these explicit expression we see that the perturbation $h_{\alpha\beta}$ in the RW gauge satisfies

$$h_{\theta\phi} = 0 , \tag{12.87}$$

$$h_{\phi\phi} = h_{\theta\theta} \sin^2 \theta , \tag{12.88}$$

$$\partial_\phi h_{t\phi} = -\sin\theta\, \partial_\theta(h_{t\theta} \sin\theta) , \tag{12.89}$$

$$\partial_\phi h_{r\phi} = -\sin\theta\, \partial_\theta(h_{r\theta} \sin\theta) . \tag{12.90}$$

In other words, in the RW gauge the four gauge functions $\xi_{lm}^{(t)}, \xi_{lm}^{(R)}, \xi_{lm}^{(E)}$ and $\xi_{lm}^{(B)}$ are chosen so that these four conditions are satisfied. Using eqs. (12.87)–(12.90), one can immediately check whether a given perturbation is in the RW gauge, without the need to decompose it first in tensor harmonics.

Observe that the imposition of the RW gauge has eliminated the terms with the highest derivatives in the angles θ, ϕ, in particular the terms associated with $\mathbf{t}_{lm}^{E2}$ and $\mathbf{t}_{lm}^{B2}$. However, these are just the tensor harmonics that, in the TT gauge, describe GWs at infinity; compare with eq. (3.275). Then, after solving the linearized equations in the RW gauge, where the equations for the perturbations are simpler, we must transform the solution (or at least its expression in the far zone) back to the TT gauge, to read the waveform of the GW radiated at infinity by a perturbed black hole. We will perform this gauge transformation explicitly in Section 12.2.6.

12.2.3 Axial perturbations: Regge–Wheeler equation

We can now write down the perturbation equations (12.15) in the RW gauge. The computation of $\Delta G_{\alpha\beta}$ using the form of $h_{\alpha\beta}$ in the RW gauge is algebraically long, but in principle straightforward (nowadays, it is easily performed with the help of any symbolic manipulation program). The three equations involving the axial perturbations are obtained from the components $(\alpha\beta) = (t\phi)$, $(r\phi)$ and $(\theta\phi)$ of eq. (12.15) and, in the RW gauge, read[15]

$$\partial_r^2 h_{lm}^{(0)} - \left(\partial_r + \frac{2}{r}\right)\partial_0 h_{lm}^{(1)} + \frac{1}{A}\left[\frac{2}{r}\frac{dA}{dr} - \frac{l(l+1)}{r^2}\right]h_{lm}^{(0)} = +\frac{16\pi G}{c^4 A} s_{lm}^{Bt} , \tag{12.91}$$

$$\partial_0^2 h_{lm}^{(1)} - \left(\partial_r - \frac{2}{r}\right)\partial_0 h_{lm}^{(0)} + A\frac{(l-1)(l+2)}{r^2}h_{lm}^{(1)} = -\frac{16\pi G}{c^4} A s_{lm}^{B1} , \tag{12.92}$$

$$\frac{1}{A}\partial_0 h_{lm}^{(0)} - \partial_r(A h_{lm}^{(1)}) = -\frac{16\pi G}{c^4} s_{lm}^{B2} , \tag{12.93}$$

where eqs. (12.91) and (12.92) hold for $l \geqslant 1$ while eq. (12.93) holds for $l \geqslant 2$.[16] For each (l, m) with $l \geqslant 2$ we therefore have three differential

[14]Furthermore, according to Notes 10 and 12 we still have a residual gauge freedom that allows us to add an arbitrary function of r to $h_{1m}^{(1)}(r,t)$, and an arbitrary function of t to $H_{00}^{(0)}(r,t)$.

[15]The linearized equation in the absence of source were first computed by Regge and Wheeler (1957). The contribution from the external energy–momentum tensor for axial perturbations was first computed in Zerilli (1970), which, however, contains several typos and errors. The correct equations have been given by Sago, Nakano and Sasaki (2003), and we agree with their results. Observe that Sago, Nakano and Sasaki (2003) define the tensor harmonic $\mathbf{d}_{lm}$ (and therefore the coefficient D_{lm}) with the opposite sign compared with Zerilli.

[16]Recall, however, that $h_{1m}^{(0)} = 0$, so the equations with $l = 1$ only involve $h_{1m}^{(1)}$.

equations for two unknown functions $h_{lm}^{(0)}$ and $h_{lm}^{(1)}$ [or, equivalently, h_{lm}^{Bt} and h_{lm}^{B1}; see eqs. (12.84) and (12.85)], since gauge invariance allowed us to set to zero the third axial function h_{lm}^{B2}. Thus, gauge invariance must also imply that these three equations are not independent. For the vacuum equations (i.e. when the right-hand sides are set to zero) consistency is indeed ensured by the Bianchi identities, $\bar{D}_\alpha \bar{G}^{\alpha\beta} = 0$, where $\bar{D}_\alpha$ is the covariant derivative with respect to the Schwarzschild metric and $\bar{G}^{\alpha\beta}$ is the Einstein tensor in the Schwarzschild metric. Indeed, when we set to zero s_{lm}^{Bt}, s_{lm}^{B1} and s_{lm}^{B2}, the time derivative of eq. (12.91) is automatically satisfied, upon use of eqs. (12.92) and (12.93).[17] So, in the vacuum case, eqs. (12.92) and (12.93) imply

[17] This is easily obtained by using eq. (12.93) to eliminate $\partial_0 h_{lm}^{(0)}$ from eq. (12.92) and from the time derivative of eq. (12.91), and then using eq. (12.92) to eliminate $\partial_0^2 h_{lm}^{(1)}$.

$$\partial_r^2 h_{lm}^{(0)} - \left(\partial_r + \frac{2}{r}\right)\partial_0 h_{lm}^{(1)} + \frac{1}{A}\left[\frac{2}{r}\frac{dA}{dr} - \frac{l(l+1)}{r^2}\right]h_{lm}^{(0)} = f_{lm}(r), \quad (12.94)$$

where $f_{lm}(r)$ is an arbitrary function of r, independent of t. This arbitrary function can be reabsorbed into a shift of $h_{lm}^{(0)}$ of the form

$$h_{lm}^{(0)}(t,r) \to h_{lm}^{(0)}(t,r) + F_{lm}(r). \quad (12.95)$$

Therefore, eqs. (12.92) and (12.93) imply eq. (12.91), except that they leave the freedom of adding a time-independent part to $h_{lm}^{(0)}$. This is also clear from the fact that $\partial_0 h_{lm}^{(0)}$ appears in eqs. (12.92) and (12.93), but $h_{lm}^{(0)}$ itself does not. This time-independent part can, however, be fixed by imposing the boundary condition that the BH is unperturbed at $t \to -\infty$. So, eq. (12.91) is redundant.[18] In the presence of a non-zero source term, the consistency between the three equations is ensured by the covariant conservation of the matter energy–momentum tensor, $\bar{D}_\alpha T^{\alpha\beta} = 0$, which of course makes the Einstein equations (12.13) consistent with the Bianchi identity $\bar{D}_\alpha G^{\alpha\beta} = 0$. This means that the matter source must move along the geodesics of the Schwarzschild metric.

[18] With more general boundary conditions, one would simply solve eqs. (12.92) and (12.93), and then use eq. (12.91) to fix this time-independent part. Furthermore, as we discussed in Note 10, the possibility of adding a time-independent term to $h_{1m}^{(0)} = -h_{1m}^{Bt}$ corresponds to the freedom of performing a residual gauge transformation, so even in this case eq. (12.91) is only needed for $l \geqslant 2$.

Using eq. (12.93) we can now eliminate $\partial_0 h_{lm}^{(0)}$ from eq. (12.92), obtaining an equation involving $h_{lm}^{(1)}$ only. We introduce the Regge–Wheeler function

$$Q_{lm}(t,r) = \frac{1}{r}A(r)h_{lm}^{(1)}(t,r), \quad (12.96)$$

which, like $h_{lm}^{(1)}$, is defined for $l \geqslant 1$. It is convenient to use the tortoise coordinate r_* defined in eq. (12.7), and $\partial_* = \partial/\partial r_*$. Then eq. (12.92) becomes

$$(\partial_*^2 - \partial_0^2)Q_{lm} - V_l^{\mathrm{RW}}(r)Q_{lm} = S_{lm}^{\mathrm{ax}}, \quad (12.97)$$

where

$$V_l^{\mathrm{RW}}(r) = A(r)\left[\frac{l(l+1)}{r^2} - \frac{3R_S}{r^3}\right] \quad (12.98)$$

is the Regge–Wheeler potential (actually, dimensionally it is rather an inverse length squared), and the source term is

$$S_{lm}^{\text{ax}}(t,r) = \frac{16\pi G}{c^4} \frac{A(r)}{r} \left\{ A(r)s_{lm}^{B1}(t,r) + \left(\partial_r - \frac{2}{r}\right)\left[A(r)s_{lm}^{B2}(t,r)\right] \right\}.$$

(12.99)

Comparing with the result (12.6) that we found for the scalar case, we see that the equation for scalar perturbations and that for axial gravitational perturbations have the same form, with a slightly different effective potential. The latter can be written in a unified manner as

$$V_l(r) = A(r)\left[\frac{l(l+1)}{r^2} + \frac{(1-\sigma^2)R_S}{r^3}\right],$$

(12.100)

with $\sigma = 0$ for scalar perturbations and $\sigma = 2$ for axial gravitational perturbations. Quite remarkably, the equation for electromagnetic perturbations can also be put in the same form, with $\sigma = 1$.

Equation (12.97) is the *Regge–Wheeler equation*.[19] Just as in the scalar case, we can perform a Fourier transform with respect to the time variable, writing

$$Q_{lm}(t,r) = \int_{-\infty}^{\infty} \frac{d\omega}{2\pi} \tilde{Q}_{lm}(\omega,r)e^{-i\omega t},$$

(12.101)

and similarly for $S_{lm}(t,r)$. Then $\tilde{Q}_{lm}(\omega,r)$ satisfies an equation of Schrödinger type,

$$\frac{d^2}{dr_*^2}\tilde{Q}_{lm} + \left[\frac{\omega^2}{c^2} - V_l^{\text{RW}}(r)\right]\tilde{Q}_{lm} = \tilde{S}_{lm}^{\text{ax}},$$

(12.102)

where $r = r(r_*)$ is obtained by inverting eq. (12.7). In Fig. 12.1 we show $V_l^{\text{RW}}(r)$, as a function of r, for $l = 2, 3, 4$, and in Fig. 12.2 we plot it as a function of r_*.

Observe that, in the RW gauge in which we are working,

$$h_{lm}^{(1)} \equiv -h_{lm}^{B1} = +(k_1)_{lm}$$

(12.103)

[see eq. (12.68)] and $(k_1)_{lm}$ is gauge-invariant under linearized gauge transformations. Therefore, the definition (12.96) of $Q_{lm}(t,r)$, which was made in the RW gauge, can be replaced by the more general definition

$$Q_{lm}(t,r) = \frac{1}{r}A(r)(k_1)_{lm}(t,r),$$

(12.104)

showing that $Q_{lm}(t,r)$ is gauge-invariant, to linear order in the parameter ξ_μ of the gauge transformation. Thus, $Q_{lm}(t,r)$ has a direct, gauge-invariant, meaning.

Before studying the RW equation in detail, we turn our attention to the equation governing the polar modes.

[19]As we have already mentioned, Regge and Wheeler (1957) actually derived the homogeneous equation, while the source term was first derived by Zerilli (1970). We have corrected some sign errors (both the overall sign and a relative sign of the term $\partial_r[A(r)s_{lm}^{B2}]$ with respect to the other terms in S_{lm}) in eq. (11) of Zerilli (1970). We agree with Sago, Nakano and Sasaki (2002).

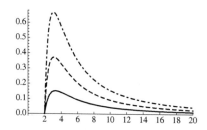

Fig. 12.1 The Regge–Wheeler potential $V_l^{\text{RW}}(r)$, as a function of r, for $l = 2$ (solid line), $l = 3$ (dashed) and $l = 4$ (dot-dashed). The radial coordinate r is measured in units of GM/c^2 (so the horizon is at $r = 2$) and the potential is measured in units of $(GM/c^2)^{-2}$.

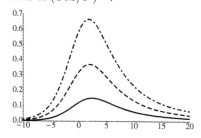

Fig. 12.2 The same as Fig. 12.1, with V_l^{RW} plotted as a function of the tortoise coordinates r_* (again measured in units of GM/c^2). The horizon is at $r_* = -\infty$.

12.2.4   Polar perturbations: Zerilli equation

Linearizing the Einstein equations around the Schwarzschild metric, one also finds the equations involving the four polar perturbations $H_{lm}^{(0)}$, $H_{lm}^{(1)}$, $H_{lm}^{(2)}$ and K_{lm}. The derivation is somewhat more involved, since we now have seven equations, some of which are dynamical, while others are constraints on the initial data. We discuss them in great detail in Problem 12.1, and here we give only the main results.

For simplicity, we set $G = c = 1$. First of all, it is convenient to perform the Fourier transform with respect to time, defining

$$\tilde{K}_{lm}(\omega, r) = \int_{-\infty}^{\infty} dt\, K_{lm}(t, r) e^{i\omega t}, \qquad (12.105)$$

and similarly for $H_{lm}^{(0)}$, $H_{lm}^{(1)}$, $H_{lm}^{(2)}$ and for the coefficients $s_{lm}^a(t, r)$ of the energy–momentum tensor,

$$\tilde{s}_{lm}^a(\omega, r) = \int_{-\infty}^{\infty} dt\, s_{lm}^a(t, r) e^{i\omega t}. \qquad (12.106)$$

We also define

$$\lambda = \frac{(l-1)(l+2)}{2}. \qquad (12.107)$$

We introduce the *Zerilli function* $\tilde{Z}_{lm}(\omega, r)$,

$$\tilde{Z}_{lm}(\omega, r) = \frac{r^2}{\lambda r + 3M} \tilde{K}_{lm}(\omega, r) + \frac{rA(r)}{i\omega(\lambda r + 3M)} \tilde{H}_{lm}^{(1)}(\omega, r). \qquad (12.108)$$

Then all dynamical equations for the polar perturbations collapse to a single equation for $\tilde{Z}_{lm}(\omega, r)$,

$$\boxed{\frac{d^2}{dr_*^2}\tilde{Z}_{lm} + \left[\omega^2 - V_l^Z(r)\right]\tilde{Z}_{lm} = \tilde{S}_{lm}^{\text{pol}},} \qquad (12.109)$$

where

$$\boxed{V_l^Z(r) = A(r)\frac{2\lambda^2(\lambda+1)r^3 + 6\lambda^2 Mr^2 + 18\lambda M^2 r + 18M^3}{r^3(\lambda r + 3M)^2},}$$

$$(12.110)$$

is called the Zerilli potential, and the source term is

$$\tilde{S}_{lm}^{\text{pol}} = A\partial_r(AJ_{lm}) - \frac{16\pi\tilde{s}_{lm}^{Et}}{i\omega}\frac{A[\lambda(\lambda+1)r^2 + 3Mr\lambda + 6M^2]}{r(\lambda r + 3M)^2}$$
$$- \frac{8\pi\tilde{s}_{lm}^{Rt}}{i\omega}\frac{\lambda r^2 A^2}{(\lambda r + 3M)^2} + 8\pi(r\tilde{s}_{lm}^{L0} + 2\tilde{s}_{lm}^{E1})\frac{rA^2}{\lambda r + 3M} - \frac{32\pi A}{r}\tilde{s}_{lm}^{E2},$$

$$(12.111)$$

where

$$J_{lm} = \frac{r}{i\omega(\lambda r + 3M)} 8\pi(r\tilde{s}_{lm}^{Rt} + 2\tilde{s}_{lm}^{Et}).$$ (12.112)

Equation (12.109) is the *Zerilli equation*. To reinstate the correct powers of G and c in the above equations, one must replace $M \to GM/c^2 = R_S/2$, $\omega \to \omega/c$ and $\tilde{s}_{lm}^a \to (G/c^4)\tilde{s}_{lm}^a$.

It is remarkable that, after the very long computation performed in Problem 12.1, we end up with just a single Schrödinger-type equation, just as for axial perturbations. Furthermore, despite its more complicated analytic form, the Zerilli potential $V_l^Z(r)$ is numerically almost indistinguishable from the RW potential, especially at large l, as we show in Fig. 12.3. They also have the same asymptotic behavior at large r,

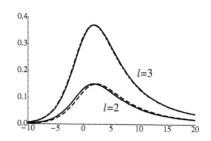

$$V_l^Z(r) \simeq V_l^{RW}(r) \simeq \frac{l(l+1)}{r^2}, \qquad (r \to \infty),$$ (12.113)

while, as $r \to R_S$, both V_l^Z and V_l^{RW} go to zero as $A(r)$ (although with a different proportionality constant). We will also see in Section 12.3.4 that the two potentials even have exactly the same spectrum of quasi-normal modes.

Fig. 12.3 The Zerilli potential V_l^Z (solid lines) and the RW potential V_l^{RW} (dashed lines) as functions of r_*, for $l = 2$ (lower curves) and $l = 3$ (upper curves). Observe that V_l^Z and V_l^{RW} are almost indistinguishable. As in Fig. 12.1, the potential is measured in units of $(GM/c^2)^{-2}$.

12.2.5 Boundary conditions

The problem of BH perturbations has therefore been reduced to the solution of two differential equations. In order to specify the problem completely, however, we must assign boundary conditions to these equations. In general, there are various possible ways of choosing the boundary conditions of a second-order differential equation defined on the line $-\infty < r_* < \infty$, and the appropriate choice depends on the physics of the problem.

First of all, we look at the possible asymptotic behaviors of the solutions of the RW and Zerilli equations as $r_* \to \pm\infty$. Consider first $r_* \to +\infty$. We restrict ourselves to energy–momentum tensors that, for $r \to \infty$, go to zero faster than $1/r^2$, so that the total energy of the perturbation, obtained by integrating over $d^3x = r^2 dr d\Omega$, is finite. Both the RW and Zerilli potentials vanish as $1/r^2$ for large r [see eq. (12.113)], so to leading order in $1/r$ the RW and Zerilli equations become free equations. The general solution $Z_{lm}(t,r)$ of the Zerilli equation for $r_* \to +\infty$ will therefore be a superposition of plane waves,

$$Z_{lm}(t,r) \to \int_{-\infty}^{\infty} d\omega \left[A_{lm}^{out}(\omega)e^{-i\omega(t-r_*/c)} + C_{lm}^{in}(\omega)e^{-i\omega(t+r_*/c)} \right],$$ (12.114)

where

$$A_{lm}^{out}(\omega) = [A_{lm}^{out}(-\omega)]^*,$$ (12.115)

and $C_{lm}^{in}(\omega) = [C_{lm}^{in}(-\omega)]^*$, since $Z_{lm}(t,r)$ is real. The solution proportional to $\exp\{-i\omega(t - r_*/c)\}$ is an outgoing radial wave, i.e. radiation escaping to infinity, while $\exp\{-i\omega(t + r_*/c)\}$ is an incoming wave, describing radiation coming from infinity toward the black hole.

The boundary conditions to be imposed depend on the physics of the problem. The situation in which we are interested is a Schwarzschild space-time that has been perturbed by some external agent, so at some time t_0, which we take as an initial time for the subsequent evolution, $Z_{lm}(t_0, r)$ and $Q_{lm}(t_0, r)$ are non-vanishing, and we want to know how this perturbation will evolve in time. In general, part of the perturbation will propagate toward infinity, and part will propagate toward the BH horizon. Thus, at $r_* \to +\infty$, we will have a purely outgoing wave. We are not interested in the case where gravitational radiation is impinging on the BH from infinity, and we therefore set $C_{lm}^{\mathrm{in}}(\omega) = 0$ in eq. (12.114), so the boundary condition at $r = +\infty$ is

$$Z_{lm}(t, r) \to \int_{-\infty}^{\infty} d\omega\, A_{lm}^{\mathrm{out}}(\omega) e^{-i\omega(t - r_*/c)}, \qquad (r_* \to +\infty). \quad (12.116)$$

Similarly, for the axial perturbations we require

$$Q_{lm}(t, r) \to \int_{-\infty}^{\infty} d\omega\, \frac{\omega}{c} B_{lm}^{\mathrm{out}}(\omega) e^{-i\omega(t - r_*/c)}, \qquad (r_* \to +\infty). \quad (12.117)$$

The factor ω/c in eq. (12.117) is part of our definition of $B_{lm}^{\mathrm{out}}(\omega)$, and is inserted so that $B_{lm}^{\mathrm{out}}(\omega)$ has the same dimensions as $A_{lm}^{\mathrm{out}}(\omega)$. To determine the possible solutions at $r_* \to -\infty$, we observe that the potentials $V_l^{\mathrm{RW}}(r)$ and $V_l^Z(r)$ near the horizon vanish as $A(r)$. Close to the horizon, $r \simeq R_S$, the relation between r_* and r becomes

$$r_* \simeq R_S + R_S \log \frac{r - R_S}{R_S}, \quad (12.118)$$

so

$$A(r) \simeq e^{(r_* - R_S)/R_S}, \quad (12.119)$$

which vanishes exponentially in r_* as $r_* \to -\infty$. Therefore (assuming that also the energy–momentum tensor of the perturbation vanishes near the horizon, as will be the case in all the physically interesting cases that we will meet), as $r_* \to -\infty$ we again have a free wave equation, and therefore we again find incoming and outgoing radial waves. Incoming waves describe the fact that part of the initial perturbations $Z_{lm}(t_0, r)$ and $Q_{lm}(t_0, r)$ propagates toward the BH. However, we require that nothing comes out from the BH horizon. Thus, near the horizon we select purely incoming-wave boundary conditions

$$Z_{lm}(t, r) \to \int_{-\infty}^{\infty} d\omega\, A_{lm}^{\mathrm{in}}(\omega) e^{-i\omega(t + r_*/c)}, \qquad (r_* \to -\infty), \quad (12.120)$$

and

$$Q_{lm}(t, r) = \int_{-\infty}^{\infty} d\omega\, \frac{\omega}{c} B_{lm}^{\mathrm{in}}(\omega) e^{-i\omega(t + r_*/c)}, \qquad (r_* \to -\infty). \quad (12.121)$$

Observe that the values of $A_{lm}^{\mathrm{in,out}}(\omega)$ and $B_{lm}^{\mathrm{in,out}}(\omega)$ are left unspecified, and will be determined by the differential equation itself, given the source term (or by the initial values $Z_{lm}(t_0, r)$ and $Q_{lm}(t_0, r)$, if we study the evolution of a given initial perturbation). In particular, we cannot impose $A_{lm}^{in}(\omega) = A_{lm}^{out}(\omega) = 0$ [and similarly for $B_{lm}^{in,out}(\omega)$]. In fact, this would imply that the function $\tilde{Z}(\omega, r)$ vanishes at $r_* \to \pm\infty$. In the language of the equivalent Schrödinger equation, defined as in the discussion following eq. (12.11), this amounts to looking for bound-state solutions. However, the Zerilli and RW potentials are everywhere positive, and clearly admit no bound state, whatever the value of ω.

Observe also that the behavior of each outgoing Fourier mode at $r \to \infty$ is proportional to $\exp\{-i\omega(t - r_*/c)\}$, rather than to $\exp\{-i\omega(t - r/c)\}$ as expected for a plane wave. For large values of r, r_* differs from r by a logarithmic term, $r_* \simeq r + R_S \log(r/R_S)$, and therefore there is a logarithmic phase shift in the outgoing wave. This is completely analogous to what happens in the scattering problem in a Coulomb potential, and simply reflects the long-range behavior of the electromagnetic or gravitational interaction, whose potential, at large r, decreases only as $1/r$.

12.2.6   The radiation field in the far zone

We now discuss how to transform the solution from the RW gauge to the TT gauge, at least in the far region where $\omega r \gg 1$, in order to obtain the GWs radiated at infinity. Recall from eq. (3.275) that in the TT gauge, in the far region, the gravitational field is given by

$$h_{\mu\nu}^{TT} = \frac{1}{r} \frac{G}{c^4} \sum_{l=2}^{\infty} \sum_{m=-l}^{l} \left[u_{lm}(t - r/c)(\mathbf{T}_{lm}^{E2})_{\mu\nu}(\theta, \phi) \right.$$
$$\left. + v_{lm}(t - r/c)(\mathbf{T}_{lm}^{B2})_{\mu\nu}(\theta, \phi) \right] + O\left(\frac{1}{r^2}\right). \quad (12.122)$$

For each (l, m), the two functions of retarded time u_{lm} and v_{lm} describe the two degrees of freedom of gravitational waves. We have written the metric perturbation in the covariant form $h_{\mu\nu}$, rather than writing h_{ij} as in eq. (3.275), but recall that in the rest frame of the BH $(\mathbf{T}_{lm}^{E2})_{\mu\nu}$ and $(\mathbf{T}_{lm}^{B2})_{\mu\nu}$ are non-zero only when both μ and ν are spatial indices; see the discussion following eq. (12.16).

The power radiated in GWs was computed, in terms of u_{lm} and v_{lm}, in eq. (3.301),[20]

$$\frac{dE}{dt} = \frac{G}{32\pi c^5} \sum_{l=2}^{\infty} \sum_{m=-l}^{l} \langle \dot{u}_{lm}^2 + \dot{v}_{lm}^2 \rangle. \quad (12.123)$$

Let us make the (obvious) remark that the fact that a radiation field in the far zone has a $1/r$ behavior holds only when we consider the Cartesian components of the metric, i.e. h_{xx}, h_{xy}, etc. If we instead use polar components such as h_{rr}, $h_{r\theta}$, etc., there can be extra factors of r or r^2, depending on the particular component considered. For instance, a term $h_{r\theta}$ contributes to the interval ds^2 as $2h_{r\theta}dr d\theta$. When we go back to

[20] Observe that, in the definition (3.271) and (3.272) of $\mathbf{T}_{lm}^{E2}$ and $\mathbf{T}_{lm}^{B2}$, we have chosen the factors of i so that they are real, and therefore also u_{lm} and v_{lm} are real, and in eq. (3.301) we can replace $|u_{lm}|^2$ and $|v_{lm}|^2$ with u_{lm}^2 and v_{lm}^2, respectively. The equations of Section 3.5.2 were, however, written so that they are valid for a generic choice of the phases in the tensor spherical harmonics.

Cartesian coordinates x^i, for dimensional reasons $drd\theta$ becomes proportional to $1/r$ times factors $dx^i dx^j$. Therefore, the contribution of $h_{r\theta}$ to the Cartesian components h_{ij}, is actually proportional to $(1/r)h_{r\theta}$. So, $h_{r\theta}$ contributes to the radiation field only if it becomes $O(1)$ as $r \to \infty$. Similarly, $h_{\theta\theta}$ contributes to the radiation field only if it becomes $O(r)$ as $r \to \infty$, and so on. In order to avoid this nuisance, when studying the radiation field at infinity it is convenient to go back to Cartesian components, i.e. to the expansion (12.27) in terms of the amplitudes $H^a_{lm} = h^a_{lm}/c^a(r)$; see eq. (12.46). For all values of the index a, H^a_{lm} contributes to the radiation field if it goes like $1/r$ at large r. We find it convenient to work at first with h^a, but we will then express our final results in terms of the amplitudes H^a. We now search for a gauge transformation that transforms the solution from the RW gauge to the TT gauge, at least for large r.

Axial perturbations

We first consider axial perturbations. Their behavior under gauge transformation has been given in eqs. (12.63)–(12.65). It is convenient to perform the Fourier transform with respect to the time variable, working in terms of $\tilde{h}^a(\omega, r)$ rather than $h^a(t, r)$. Starting from the perturbation in the RW gauge, $(\tilde{h}^a)^{\mathrm{RW}}$, and performing a gauge transformation with a gauge function Λ, we arrive at a new, gauge equivalent perturbation $(\tilde{h}^a)^{\mathrm{new}}$, given by

$$
\begin{aligned}
(\tilde{h}^{Bt}_{lm})^{\mathrm{new}} &= (\tilde{h}^{Bt}_{lm})^{\mathrm{RW}} - (i\omega/c)\tilde{\Lambda}_{lm} \\
&= -\tilde{h}^{(0)}_{lm} - (i\omega/c)\tilde{\Lambda}_{lm}\,,
\end{aligned} \tag{12.124}
$$

$$
\begin{aligned}
(\tilde{h}^{B1}_{lm})^{\mathrm{new}} &= (\tilde{h}^{B1}_{lm})^{\mathrm{RW}} + \left(\partial_r - \frac{2}{r}\right)\tilde{\Lambda}_{lm} \\
&= -\tilde{h}^{(1)}_{lm} + \left(\partial_r - \frac{2}{r}\right)\tilde{\Lambda}_{lm}\,,
\end{aligned} \tag{12.125}
$$

$$
\begin{aligned}
(\tilde{h}^{B2}_{lm})^{\mathrm{new}} &= (\tilde{h}^{B2}_{lm})^{\mathrm{RW}} - \tilde{\Lambda}_{lm} \\
&= -\tilde{\Lambda}_{lm}\,,
\end{aligned} \tag{12.126}
$$

where we have used the notation (12.84) and (12.85) for the perturbations in the RW gauge, and the fact that $(\tilde{h}^{B2}_{lm})^{\mathrm{RW}} = 0$. We now choose

$$
\tilde{\Lambda}_{lm} = -\frac{c}{i\omega}\,\tilde{h}^{(0)}_{lm}\,, \tag{12.127}
$$

so that $(\tilde{h}^{Bt}_{lm})^{\mathrm{new}} = 0$. Equation (12.125) then becomes

$$
(\tilde{h}^{B1}_{lm})^{\mathrm{new}} = -\left[\tilde{h}^{(1)} + \frac{c}{i\omega}\left(\partial_r - \frac{2}{r}\right)\tilde{h}^{(0)}_{lm}\right]\,. \tag{12.128}
$$

Using the equation of motion (12.92) written in Fourier space,

$$
(i\omega/c)^2\tilde{h}^{(1)}_{lm} + (i\omega/c)\left(\partial_r - \frac{2}{r}\right)\tilde{h}^{(0)}_{lm} + A\frac{(l-1)(l+2)}{r^2}\tilde{h}^{(1)}_{lm} = \frac{16\pi G}{c^4}\,A\tilde{s}^{B1}_{lm}\,, \tag{12.129}
$$

and the fact that the components s^a_{lm} of the energy–momentum tensor vanish faster than $1/r^2$ for large r, we see that $(\tilde{h}^{B1}_{lm})^{\text{new}}$ vanishes at large distances as $\tilde{h}^{(1)}_{lm}/r^2$. From eq. (12.96) we see that, at large r, $\tilde{h}^{(1)}_{lm} \sim r\tilde{Q}$, while the boundary condition (12.117) tells us that for large r,

$$\tilde{Q}(\omega,r) \to \frac{\omega}{c} B^{\text{out}}_{lm}(\omega) e^{i\omega r_*/c}, \tag{12.130}$$

which is $O(1)$ with respect to r. Thus, at large r we have $\tilde{h}^{(1)}_{lm} \sim r$, $(\tilde{h}^{B1}_{lm})^{\text{new}} \sim 1/r$ and

$$(\tilde{H}^{B1}_{lm})^{\text{new}} = \frac{1}{c^{B1}(r)}(\tilde{h}^{B1}_{lm})^{\text{new}} \sim \frac{1}{r^2}. \tag{12.131}$$

Thus, $(\tilde{H}^{B1}_{lm})^{\text{new}}$ does not contribute to the radiation field in the far zone.

We consider next $\tilde{H}^{B2}_{lm}$. While in the RW gauge it is absent, the transformation (12.126) generates it, giving

$$(\tilde{H}^{B2}_{lm})^{\text{new}} = \frac{c}{i\omega c^{B2}(r)}\tilde{h}^{(0)}_{lm}. \tag{12.132}$$

Using eq. (12.93) in Fourier space we find, to leading order in the large-r limit,

$$\frac{-i\omega}{c}\tilde{h}^{(0)}_{lm} = \partial_r(r\tilde{Q}_{lm}). \tag{12.133}$$

Using the asymptotic behavior of $\tilde{Q}_{lm}$ given in eq. (12.130) we get, for large r,

$$\tilde{h}^{(0)}_{lm}(\omega,r) = -\frac{\omega r}{c} B^{\text{out}}_{lm}(\omega) e^{i\omega r_*/c} + O(1), \tag{12.134}$$

and therefore, using the explicit expression (12.40) for $c^{B2}(r)$,

$$(\tilde{H}^{B2}_{lm})^{\text{new}}(\omega,r) = \frac{i}{r}\frac{1}{\sqrt{2}}\left[\frac{(l-2)!}{(l+2)!}\right]^{1/2} B^{\text{out}}_{lm}(\omega) e^{i\omega r_*/c} + O\left(\frac{1}{r^2}\right). \tag{12.135}$$

Thus, $(\tilde{H}^{B2}_{lm})^{\text{new}}$ decreases as $1/r$ in the far zone, and contributes to the radiation field.

Polar perturbations

The gauge transformation of polar perturbations was given in eqs. (12.70)–(12.76). Consider first eq. (12.72). In Fourier space, to leading order in $1/r$, it gives

$$(\tilde{h}^{L0}_{lm})^{\text{new}} = \tilde{H}^{(2)}_{lm} + \frac{2i\omega}{c}\tilde{\xi}^{(R)}_{lm}, \tag{12.136}$$

where, as in eq. (12.80), the value of h^{L0}_{lm} in the RW gauge has been denoted by $B(r)H^{(2)}_{lm}$, and we have used the fact that, to leading order in $1/r$, $B(r) = 1$. We now choose $\tilde{\xi}^{(R)}_{lm} = -(c/2i\omega)\tilde{H}^{(2)}_{lm}$, so $(\tilde{h}^{L0}_{lm})^{\text{new}} = 0$. We next consider eq. (12.70),

$$(\tilde{h}^{tt}_{lm})^{\text{new}} = A(r)\tilde{H}^{(0)}_{lm} + \frac{2i\omega}{c}\tilde{\xi}^{(t)}_{lm} + \frac{A(r)R_S}{r^2}\tilde{\xi}^{(R)}_{lm}. \tag{12.137}$$

Thus, choosing

$$
\begin{aligned}
\tilde{\xi}_{lm}^{(t)} &= -\frac{c}{2i\omega} A(r) \left[\tilde{H}_{lm}^{(0)} + \frac{R_S}{r^2} \tilde{\xi}_{lm}^{(R)} \right] \\
&= -\frac{c}{2i\omega} A(r) \left[\tilde{H}_{lm}^{(0)} - \frac{R_S}{r^2} \frac{c}{2i\omega} \tilde{H}_{lm}^{(2)} \right],
\end{aligned}
\tag{12.138}
$$

we set to zero also $(\tilde{h}_{lm}^{tt})^{\mathrm{new}}$. We next consider eq. (12.74). Recalling that $\tilde{h}_{lm}^{Et}$ vanishes in the RW gauge, we get

$$
(\tilde{h}_{lm}^{Et})^{\mathrm{new}} = -\tilde{\xi}_{lm}^{(t)} + \frac{i\omega}{c} \tilde{\xi}_{lm}^{(E)}.
\tag{12.139}
$$

We then choose

$$
\tilde{\xi}_{lm}^{(E)} = \frac{c}{i\omega} \tilde{\xi}_{lm}^{(t)},
\tag{12.140}
$$

so that $(\tilde{h}_{lm}^{Et})^{\mathrm{new}} = 0$. The three gauge functions $\tilde{\xi}_{lm}^{(t)}$, $\tilde{\xi}_{lm}^{(R)}$ and $\tilde{\xi}_{lm}^{(E)}$ have now been fixed, and we can compute the resulting expressions for $(\tilde{h}_{lm}^{T0})^{\mathrm{new}}$ and $(\tilde{h}_{lm}^{E1})^{\mathrm{new}}$ using eqs. (12.73) and (12.75). Just as we did earlier for axial perturbations, we can check that, upon use of the equations of motion, they do not contribute to the radiation field, i.e. $(\tilde{H}_{lm}^{T0})^{\mathrm{new}}$ and $(\tilde{H}_{lm}^{E1})^{\mathrm{new}}$ are $O(1/r^2)$. We are left with $(\tilde{h}_{lm}^{E2})^{\mathrm{new}}$. Using eq. (12.76) and the results found earlier for $\tilde{\xi}_{lm}^{(E)}$ and $\tilde{\xi}_{lm}^{(t)}$, to leading order in $1/r$ we find

$$
\begin{aligned}
(\tilde{h}_{lm}^{E2})^{\mathrm{new}} &= -\frac{c}{i\omega} \tilde{\xi}_{lm}^{(t)} \\
&= -\frac{c^2}{2\omega^2} \tilde{H}_{lm}^{(2)} \left[1 + O\left(\frac{1}{r^2} \right) \right].
\end{aligned}
\tag{12.141}
$$

[Recall that $\tilde{H}^{(0)} = \tilde{H}^{(2)} + O(1/r^2)$; see eq. (12.371).] Similarly to what we did for axial perturbations, the asymptotic behavior of $\tilde{H}_{lm}^{(2)}(\omega, r)$ can be computed in terms of the asymptotic behavior of the function $\tilde{Z}(\omega, r)$. Using the fact that all source terms decrease faster than $1/r^2$, and therefore do not contribute to these asymptotic relations, from eqs. (12.371) and (12.372) we have

$$
\tilde{H}_{lm}^{(2)}(\omega, r) \rightarrow (1 + r\partial_r)\tilde{K}_{lm} + \frac{l(l+1)}{2i\omega r} \tilde{H}_{lm}^{(1)},
\tag{12.142}
$$

where the arrow means that we have neglected terms that are subleading for large r. From eqs. (12.388) and (12.389)

$$
\tilde{K}_{lm} \rightarrow \frac{\lambda+1}{r} \tilde{Z}_{lm} + \partial_r \tilde{Z}_{lm},
\tag{12.143}
$$

$$
\tilde{H}_{lm}^{(1)} \rightarrow -i\omega(1 + r\partial_r)\tilde{Z}_{lm}.
\tag{12.144}
$$

Using the asymptotic behavior (12.116), i.e.

$$
\tilde{Z}_{lm}(\omega, r) \rightarrow A_{lm}^{\mathrm{out}}(\omega)\, e^{i\omega r_*/c},
\tag{12.145}
$$

we see that in eq. (12.143) the leading term is actually given by $\tilde{K}_{lm} \rightarrow r\partial_r \tilde{Z}_{lm}$ and in eq. (12.142) it is sufficient to retain $\tilde{H}_{lm}^{(2)}(\omega, r) \rightarrow r\partial_r \tilde{K}_{lm}$, so we get

$$\tilde{H}_{lm}^{(2)}(\omega, r) \rightarrow -\frac{\omega^2}{c^2} r A_{lm}^{\text{out}}(\omega) \, e^{i\omega r_*/c} \,. \tag{12.146}$$

As a result

$$(\tilde{h}_{lm}^{E2})^{\text{new}} \rightarrow \frac{1}{2} r A_{lm}^{\text{out}}(\omega) \, e^{i\omega r_*/c} \,. \tag{12.147}$$

Using eq. (12.40),

$$(\tilde{H}_{lm}^{E2})^{\text{new}}(\omega, r) = \frac{1}{r}\frac{1}{\sqrt{2}} \left[\frac{(l+2)!}{(l-2)!}\right]^{1/2} A_{lm}^{\text{out}}(\omega) \, e^{i\omega r_*/c} + O\left(\frac{1}{r^2}\right) \,. \tag{12.148}$$

Therefore, this term decreases as $1/r$ and contributes to the radiation field at infinity.

In conclusion, in this new gauge $\tilde{H}_{lm}^{E2}$ and $\tilde{H}_{lm}^{B2}$ decrease as $1/r$, while all other components are $O(1/r^2)$. We see that this new gauge is nothing but the TT gauge. Comparing with eq. (12.122), and approximating r_* with r at large distances, we get[21]

$$\frac{G}{c^4} u_{lm}(t - r/c) = \frac{1}{\sqrt{2}} \left[\frac{(l+2)!}{(l-2)!}\right]^{1/2} \int_{-\infty}^{\infty} \frac{d\omega}{2\pi} A_{lm}^{\text{out}}(\omega) \, e^{-i\omega(t-r/c)} \,. \tag{12.149}$$

Writing

$$A_{lm}^{\text{out}}(\omega) = |A_{lm}^{\text{out}}(\omega)| \, e^{i\delta_{lm}(\omega)} \,, \tag{12.150}$$

and using the reality condition (12.115), which implies $|A_{lm}^{\text{out}}(-\omega)| = |A_{lm}^{\text{out}}(\omega)|$ and $\delta_{lm}(-\omega) = -\delta_{lm}(\omega)$, we can rewrite eq. (12.149) as an integral over the positive frequencies,

$$u_{lm}(t - r/c) = \frac{c^4}{G} \sqrt{2} \left[\frac{(l+2)!}{(l-2)!}\right]^{1/2} \tag{12.151}$$

$$\times \int_0^\infty \frac{d\omega}{2\pi} |A_{lm}^{\text{out}}(\omega)| \, \cos\left[\omega(t-r/c) - \delta_{lm}(\omega)\right] \,.$$

Similarly, using eq. (12.135), and writing

$$B_{lm}^{\text{out}}(\omega) = |B_{lm}^{\text{out}}(\omega)| \, e^{i\gamma_{lm}(\omega)} \,, \tag{12.152}$$

we get[22]

$$v_{lm}(t - r/c) = \frac{c^4}{G} \sqrt{2} \left[\frac{(l+2)!}{(l-2)!}\right]^{1/2} \tag{12.153}$$

$$\times \int_0^\infty \frac{d\omega}{2\pi} |B_{lm}^{\text{out}}(\omega)| \, \cos\left[\omega(t-r/c) - \gamma_{lm}(\omega)\right] \,.$$

We now plug these expressions into eq. (12.123) and integrate over time from $-\infty$ to $+\infty$, using the fact that, for $\omega > 0$ and $\omega' > 0$,

$$\int_{-\infty}^{\infty} dt \, \cos\omega t \cos\omega' t = \frac{1}{2} \int_{-\infty}^{\infty} dt \, e^{i(\omega-\omega')t}$$

$$= \frac{1}{2} \, 2\pi\delta(\omega - \omega') \tag{12.154}$$

[21] Observe that u_{lm} is a function of retarded time $t - r/c$ only to leading order in the large r limit. We see from this computation that, to next-to-leading order, u_{lm} becomes a function of $t - r_*/c$. We already met such logarithmic terms, in a different guise, in the coalescence of compact binaries [see eq. (5.265)], where the total accumulated phase $\phi(t) = \int dt' \omega_s(t')$ is shifted by a term proportional to $\omega_s(t) \log \omega_s(t)$ and hence, using Keplers law (5.252), proportional to $\omega_s(t) \log r(t)$.

[22] From the point of view of dimensional analysis $Z_{lm}(t, r)$ has dimensions of length while $Q_{lm}(t, r)$ is dimensionless. Correspondingly, $A_{lm}^{\text{out}}(\omega)$ has dimensions length/frequency, while $B_{lm}^{\text{out}}(\omega)$ has dimensions 1/frequency.

[23] Recall that our definition of Fourier transform is

$$\tilde{F}^{\rm our}(\omega) = \int_{-\infty}^{\infty} dt\, F(t) e^{i\omega t},$$

(see the Notation section in Vol. 1), while Zerilli (1970) uses

$$\tilde{F}^{\rm Zer}(\omega) = \frac{1}{(2\pi)^{1/2}} \int_{-\infty}^{\infty} dt\, F(t) e^{i\omega t}.$$

So $\tilde{F}^{\rm our}(\omega) = (2\pi)^{1/2}\tilde{F}^{\rm Zer}(\omega)$ and $(A_{lm}^{\rm out})^{\rm our} = (2\pi)^{1/2}(A_{lm}^{\rm out})^{\rm Zer}$. Therefore, with the definitions of Zerilli, in eq. (12.155) there appears

$$d\omega\, \omega^2 |A_{lm}^{\rm out}(\omega)|^2$$

instead of

$$\frac{d\omega}{2\pi} \omega^2 |A_{lm}^{\rm out}(\omega)|^2,$$

and similarly for the term $|B_{lm}^{\rm out}(\omega)|^2$.

(as is easily shown by writing the cosine in term of exponentials). We thus find[23]

$$E = \frac{c^3}{32\pi G} \sum_{l=2}^{\infty} \sum_{m=-l}^{l} \frac{(l+2)!}{(l-2)!}$$

$$\times \int_0^{\infty} \frac{d\omega}{2\pi} \omega^2 \left[|A_{lm}^{\rm out}(\omega)|^2 + |B_{lm}^{\rm out}(\omega)|^2 \right]. \quad (12.155)$$

The energy spectrum is therefore

$$\frac{dE}{d\omega} = \frac{c^3}{64\pi^2 G} \sum_{l=2}^{\infty} \sum_{m=-l}^{l} \frac{(l+2)!}{(l-2)!} \omega^2 \left[|A_{lm}^{\rm out}(\omega)|^2 + |B_{lm}^{\rm out}(\omega)|^2 \right].$$

$$(12.156)$$

12.2.7 Summary

After this long and technical section, it is appropriate to summarize the main results that we have obtained. Our aim is to find the gravitational radiation emitted when a BH is perturbed. We have therefore expanded the metric around the unperturbed Schwarzschild solution. Rather than using directly the TT gauge (or, more precisely, any gauge that reduces to the TT gauge in the far region), where the GWs can be read off immediately, we have used the Regge–Wheeler gauge, which has the advantage of considerably simplifying the equations. In this gauge the metric is expressed in terms of six functions: four with polar-type parity [see eq. (12.77)], and two with axial-type parity [see eq. (12.66)].

Linearizing the Einstein equations around the Schwarzschild background, we have then found the equations that govern these metric perturbation. These equations are apparently quite complicated. However, after a series of suitable redefinitions, we have been able to re-express them in terms of just two "master equations": the Regge–Wheeler equation (12.102) for axial perturbation, and the Zerilli equation (12.109) for polar perturbations. These equations are written in terms of two "master functions", the Regge–Wheeler function $\tilde{Q}_{lm}(\omega,r)$ and the Zerilli function $\tilde{Z}_{lm}(\omega,r)$. We have worked out in detail the relation between the original metric perturbations and these two functions, so in principle, once one has found $\tilde{Q}_{lm}(\omega,r)$ and $\tilde{Z}_{lm}(\omega,r)$, one could go back and reconstruct the full metric in the RW gauge.

A noteworthy point is that the relation between the Zerilli function and the polar metric perturbation is a simple linear relation only if we perform the Fourier transform with respect to time, and work with functions of ω and r. If we work instead with functions of t and r, these relations are non-local; this is due to factors $1/\omega$ such as that on the right-hand side of eq. (12.385). By multiplying this equation by ω and going back to functions of (t,r), we see that eq. (12.385) gives a linear relation between $\partial_t Z(t,r)$, $\partial_t K(t,r)$ and $H^{(1)}(t,r)$, so $Z(t,r)$ is determined by $K(t,r)$ and by the integral over time of $H^{(1)}(t,r)$. Similarly, the presence of factors $1/\omega$ in the source term (12.111) does not allow us to write the Zerilli equation, with a generic source term, as a simple second-order differential equation in time, and we must instead work in frequency space.

Even if the relation between these "master functions" and the original metric perturbations is quite complicated, if one is just interested in the radiation field at infinity, it is not necessary to go back through the sequence of redefinitions and reconstruct the full metric in the RW gauge. Indeed, we have seen in Section 12.2.6 that the radiation field in the far zone can be determined directly by the asymptotic behavior at infinity of the Zerilli and RW functions. In terms of the functions $A_{lm}^{\text{out}}(\omega)$ and $B_{lm}^{\text{out}}(\omega)$, defined in eqs. (12.116) and (12.117), the radiation field at infinity in the TT gauge, defined as in eq. (12.122), is obtained by taking the Fourier transforms of $A_{lm}^{\text{out}}(\omega)$ and $B_{lm}^{\text{out}}(\omega)$ with respect to retarded time [see eqs. (12.149)–(12.153)], while the energy spectrum is given directly in terms of $A_{lm}^{\text{out}}(\omega)$ and $B_{lm}^{\text{out}}(\omega)$ in eq. (12.156).

Finally, it is interesting to compare the formalism that we have developed here with the formalism of linearized theory, discussed in Chapter 3, and with the post-Newtonian formalism of Chapter 5. Conceptually, the logic and the limitations of the expansion that we have develped here are very much the same as those of Chapter 3. In both cases, we linearize the Einstein equations around a fixed background (flat space in Chapter 3 and the Schwarzschild metric here). The fact that the background is fixed implies that these results apply only to test masses moving in the given metric: the formalism of Chapter 3 allowed us to compute the GWs emitted by a test mass on a given geodesic of Minkowski space, while the formalism of this chapter allows us to compute the GWs emitted by a test mass on a given geodesic of the Schwarzschild metric [Just as in Chapter 3 (see in particular Section 3.3.5) the restriction to geodesic motion comes from the requirement of covariant conservation of the energy–momentum tensor]. On the other hand, no small-velocity approximation has been made, so the radiation emitted by such test masses can be computed exactly in v/c. In contrast, the post-Newtonian (PN) expansion is a small-velocity expansion, in which higher-order terms in v/c are more and more difficult to compute, but it is valid not just for test masses: in particular, the problem of the coalescence of a binary system of two BHs with masses m_1 and m_2 can be studied order by order in v/c using the PN expansion, as we saw in detail in Chapter 5. However, in the test-mass limit $m_1/m_2 \to 0$, the Zerilli and RW equations provide the exact result for the radiation emitted. We will indeed discuss an explicit computation of this type in Section 12.4.

Thus, the PN expansion and the linearization around the BH background give complementary information. In Chapter 14 we will discuss attempts to combine them.

12.3   Black-hole quasi-normal modes

We now study in some detail the RW and Zerilli equations, and we will see that they describe oscillations of space-time that can be interpreted as a characteristic "ringing" of black holes.

12.3.1   General discussion

We consider the situation in which the source term, in the RW or Zerilli equation, is zero, and we ask how an initial perturbation evolves. We therefore wish to study an equation of the form

$$\phi''(\omega, x) + [\omega^2 - V(x)]\phi(\omega, x) = 0 \,, \qquad (12.157)$$

where $x \equiv r_*/c$ ranges from $-\infty$ to $+\infty$, the prime denotes differentiation with respect to x, $\phi(\omega, x)$ is the Fourier transform with respect to time of either the RW or the Zerilli function, $V(x)$ is the RW or Zerilli potential, respectively, and we have reabsorbed a factor c^2 into $V(x)$. Since the RW and Zerilli potentials have the same qualitative form, we will treat the two cases in parallel. All that we will use, for the moment, is that $V(x)$ has the qualitative form shown in Fig. 12.3, and its asymptotic behavior for $x \to \pm\infty$.

Setting the source term to zero and looking for the evolution of an initial perturbation is the standard method for understanding the intrinsic properties of the system under study. We already used this method in Chapter 8 when we studied the response of a resonant bar (or, equivalently, a one-dimensional string), and we saw that in this case the behavior of the system is entirely characterized by a set of normal modes, i.e. by a set of solutions of the form

$$\phi_n(t, x) = e^{-i\omega_n t}\psi_n(x) \,. \qquad (12.158)$$

We also saw that the imposition of the appropriate boundary conditions [for example $\phi(t, \pm L/2) = 0$ for a string with fixed endpoint in $x = \pm L/2$, or $\partial \phi/\partial x(t, \pm L/2) = 0$ for a resonant bar] selects a discrete set of frequencies ω_n. These frequencies are called the normal-mode frequencies, and the corresponding solutions (12.158) are the normal modes of the system. For a one-dimensional string, the normal modes form a complete set, so the initial perturbation can be expanded in normal modes. Since each normal mode evolves with a simple harmonic dependence, the general evolution of a perturbation has the form

$$\phi(t, x) = \sum_n a_n e^{-i\omega_n t}\psi_n(x) \,, \qquad (12.159)$$

with the coefficients a_n fixed by the form of the perturbation at the initial time. In principle, one might expect something similar to emerge from the study of eq. (12.157). However, we will see in this section that there are important differences. From the physical point of view, we can understand these differences by observing that the normal modes of a finite string with fixed endpoints describe stationary states, where kinetic energy bounces back and forth and, in the absence of internal dissipation, the total kinetic plus potential energy stays constant. In our case, in contrast, we have an open system in which the energy associated with an initial perturbation will finally disappear, either escaping to infinity or approaching the horizon, at $x = -\infty$. Mathematically, this is reflected in the boundary conditions appropriate to the physics of the problem. For a string with endpoints fixed at $z = \pm L/2$, the boundary conditions are $\phi(t, \pm L/2) = 0$. For BH perturbations, we cannot impose that $\phi(t, x) = 0$ as $x \to \pm\infty$. As we already saw on page 133, the resulting problem would formally be equivalent to that of searching for bound-state solutions of an equivalent Schrödinger equation, with a potential given by the Zerilli or the RW potentials, and it is clear from Fig. 12.3 that these potentials do not admit bound states. Rather, from eqs. (12.116)–(12.121) it follows that the appropriate boundary conditions for a BH perturbation $\phi(\omega, x)$ are

$$\phi(\omega, x) \propto e^{+i\omega x} \qquad (x \to +\infty)\,, \qquad (12.160)$$

$$\phi(\omega, x) \propto e^{-i\omega x} \qquad (x \to -\infty)\,, \qquad (12.161)$$

or, more compactly,[24]

$$\phi(\omega, x) \propto e^{+i\omega|x|} \qquad (x \to \pm\infty)\,. \qquad (12.162)$$

Similarly to what happens to a string, the imposition of these boundary conditions selects some discrete values of ω, and the corresponding solutions are the normal modes of the system. A crucial difference, however, is that, as we will show, for BHs these normal mode frequencies have both a real and an imaginary part. To stress this point the corresponding solutions are called *quasi-normal modes* (QNMs), and the corresponding frequencies are called the quasi-normal mode frequencies.

To show that the boundary conditions (12.160) and (12.161) pick out some discrete values of ω, it is useful to first write eq. (12.157) as an equivalent equation in the time domain, as[25]

$$\left[-\frac{\partial}{\partial t^2} + \frac{d^2}{dx^2} - V(x) \right] \phi(t, x) = 0\,. \qquad (12.163)$$

Since $V(x)$ vanishes for $x \to \pm\infty$, at infinity the solutions of eq. (12.163) reduce to right-moving or left-moving plane waves, $\exp\{-i\omega(t \pm x)\}$. In this equivalent one-dimensional scattering problem we can consider the situation in which, at $x = -\infty$, we prepare a right-moving wavepacket

$$\phi_0(t, x) = \int_{-\infty}^{\infty} \frac{d\omega}{2\pi} A_0(\omega) \exp\{-i\omega(t - x)\}\,, \qquad (x \to -\infty). \quad (12.164)$$

This wavepacket will be partly reflected and partly transmitted by the potential $V(x)$. So, at $x = -\infty$ there will also be a reflected, left-moving, wavepacket

$$\phi_{\rm r}(t, x) = \int_{-\infty}^{\infty} \frac{d\omega}{2\pi} A_{\rm r}(\omega) \exp\{-i\omega(t + x)\}\,, \qquad (x \to -\infty), \quad (12.165)$$

[24]When comparing with the literature, one must take into account the sign convention for the Fourier transform. For us the Fourier transform is defined as

$$F(t) = \int_{-\infty}^{\infty} \frac{d\omega}{2\pi} \tilde{F}(\omega) e^{-i\omega t}\,;$$

see the Notation in Vol. 1. If one rather used $e^{+i\omega t}$ in the Fourier transform, for an outgoing radial wave the coefficients $A_{lm}^{\rm out}$ and $B_{lm}^{\rm out}$ in eqs. (12.116) and (12.117) would be rather multiplied by $e^{+i\omega(t-x)}$, and therefore at $x \to +\infty$ we would rather have $\phi(\omega, x) \propto e^{-i\omega x}$, and similarly at $x \to -\infty$ we would have $\phi(\omega, x) \propto e^{+i\omega x}$.

[25]Recall, from the discussion on page 139, that it is possible to rewrite the Zerilli equation as a differential equation local in the time only in the absence of the source term, which, however, is the situation in which we are interested here.

while at $x = +\infty$ there will be a right-moving wavepacket,

$$\phi_{\rm t}(t,x) = \int_{-\infty}^{\infty} \frac{d\omega}{2\pi} A_{\rm t}(\omega) \exp\{-i\omega(t-x)\}, \qquad (x \to +\infty). \quad (12.166)$$

Thus, in terms of the Fourier modes $\phi(\omega, x)$, the asymptotic solution at $x = -\infty$ will be

$$\phi(\omega, x) \simeq A_0(\omega)e^{+i\omega x} + A_{\rm r}(\omega)e^{-i\omega x}, \quad (12.167)$$

while at $x = +\infty$ we will have

$$\phi(\omega, x) \simeq A_{\rm t}(\omega)e^{+i\omega x}. \quad (12.168)$$

Conservation of probability requires $|A_0(\omega)|^2 = |A_{\rm r}(\omega)|^2 + |A_{\rm t}(\omega)|^2$. In this equivalent one-dimensional scattering problem, the amplitude for reflection is

$$S(\omega) = \frac{A_{\rm r}(\omega)}{A_0(\omega)}. \quad (12.169)$$

The boundary conditions (12.160) and (12.161) correspond to $A_0(\omega) = 0$ with $A_{\rm r}(\omega) \neq 0$, and therefore to poles of the scattering amplitude $S(\omega)$. So, the imposition of the boundary conditions (12.160) and (12.161) selects some discrete values of ω. In the language of scattering theory, these special frequencies are the resonances of the system.

These special values of ω, which we denote by $\omega_{\rm QNM}$, are complex, and we write them as

$$\omega_{\rm QNM} \equiv \omega_R + i\omega_I$$
$$\equiv \omega_R - i\frac{\gamma}{2}. \quad (12.170)$$

We will see later how this emerges from explicit computations. However, this result is easily understood, since it is physically clear that in this problem there are no stationary perturbations.[26] Rather, an initial BH perturbation decays and disappears either into gravitational radiation at $x = +\infty$ or approaching the horizon at $x = -\infty$. This means that at each fixed x the perturbation amplitude must eventually go to zero, so any solution of the form $e^{-i\omega_{\rm QNM}t}\psi_n(x)$ must have $\omega_{\rm QNM}$ of the form given in eq. (12.170), with $\omega_I < 0$, or $\gamma > 0$.[27,28]

Of course, in any realistic macroscopic system the normal-mode frequencies always have an imaginary part, because of dissipation, so in this sense one might think that the difference between BH QNMs and the usual normal modes of elastic bodies is not so much a difference of principle. However, in normal macroscopic bodies the mechanisms responsible for dissipation, and therefore for ω_I, are partly independent of the mechanisms giving rise to rigidity, i.e. to ω_R, and we can tune the parameters of the system, or external parameters such as the temperature, so that $|\omega_I| \ll \omega_R$. We saw for instance in Vol. 1 that the quality factor Q of resonant bars, which is related to ω_R and ω_I by $Q = \omega_R/(2|\omega_I|) = \omega_R/\gamma$, can be made as large as $O(10^7)$, choosing appropriate materials and working at cryogenic temperatures. In contrast,

---

[26] A formal proof can be obtained by looking for static solutions of the RW or Zerilli equation, and showing that there are none; see Vishveshwara (1970).

[27] A rigorous proof that linear perturbations of the Schwarzschild metric remain uniformly bounded in space, for all time, has been given in Wald (1979).

[28] Observe that the sign of ω_I depends on the convention used for the Fourier transform. For consistency with the rest of the book, we use as always

$$F(t) = \int_{-\infty}^{\infty} \frac{d\omega}{2\pi} \tilde{F}(\omega)e^{-i\omega t}.$$

Then a pole in $\tilde{F}(\omega)$ at $\omega = \omega_R + i\omega_I$ gives rise to a contribution to $F(t)$ oscillating as $e^{-i\omega_R t}e^{+\omega_I t}$, so a decaying mode has $\omega_I < 0$. In the literature on BH QNMs, the opposite convention on the Fourier transform is typically used,

$$F(t) = \int_{-\infty}^{\infty} \frac{d\omega}{2\pi} \tilde{F}(\omega)e^{+i\omega t},$$

Then $\omega_I > 0$ corresponds to a decaying mode. Because the gravitational perturbation is real, the values of ω_R come in pairs with opposite signs, $\omega_R = \pm|\omega_R|$, so for ω_R the convention on the Fourier transform is irrelevant. Compare also with Note 24 for related sign differences.

in BHs ω_R and ω_I are determined simultaneously by the same equation (the Zerilli or RW equation), and there is no parameter that we can tune to achieve $|\omega_I| \ll \omega_R$. We will indeed see that $|\omega_I|$ is always at least of the same order, or much larger, than ω_R.

Mathematically, the fact that the QNM frequencies are complex can be traced back to the fact that the boundary conditions (12.160) and (12.161) are complex. Drawing again on the formal analogy between the equation (12.157) governing BH perturbations and an equivalent time-independent Schrödinger equation, we can make use of standard results from quantum mechanics and observe that, in general, in quantum mechanics a Hamiltonian has real eigenvalues only if the boundary conditions imposed on the wavefunctions are also real. For instance, if we rather impose, as a boundary condition at infinity, that the wavefunction be an outgoing spherical wave (i.e. $\psi \sim (1/r)e^{ikr}$ in three dimensions), then the eigenvalues of the Hamiltonian are complex, $E = E_R - i\Gamma/2$, and the wavefunction oscillates as $e^{-iEt/\hbar} = e^{-iE_Rt/\hbar}e^{-\Gamma t/2}$, where $\Gamma > 0$ is the decay width of the initial state.[29]

While expected physically, the form (12.170) of the QNM frequencies has a peculiar consequence for the spatial dependence of the function describing a QNM (its "wavefunction", in the equivalent Schrödinger problem). By definition, at $x \to \pm\infty$, the QNMs satisfy the boundary conditions (12.160) and (12.161), i.e.

$$\phi(\omega, x) \propto e^{i\omega|x|} = e^{i\omega_R|x|}e^{+\gamma|x|/2}, \qquad (x \to \pm\infty). \qquad (12.171)$$

Since $\gamma > 0$, the QNMs diverge exponentially, both at spatial infinity and at the horizon! This is a clear indication of the fact that QNMs behave quite differently from the usual normal modes. First of all, a QNM cannot represent a physical state of the system at a given time, over all of space, since at any given time it carries an infinite energy. Rather, it can at most describe the behavior of $\phi(t, x)$ at sufficiently large values of t, at a *fixed* value of x. The larger the value of $|x|$, the larger is also the value of time at which this asymptotic behavior sets in, so the exponentially growing factor $e^{+\gamma|x|/2}$ is always compensated by the time dependence $e^{-\gamma t/2}$.[30]

Another consequence of this spatial dependence is that QNMs do not in general form a complete set. We can understand this intuitively since it seems unlikely that a physical finite-energy perturbation that goes to zero at $x \to \pm\infty$ could be expressed as a superposition of functions that diverge in these limits, and that carry infinity energy. We will examine this issue more formally in Section 12.3.3.

We conclude this introductory section by observing that there is a mathematical subtlety in the definition of QNMs that we have presented here, due to the fact that the imposition of the boundary conditions (12.162) is not sufficient to single out a unique solution, when ω is complex. In fact, consider a solution $\phi_1(\omega, x)$ that, for $x \to +\infty$, has the asymptotic behavior

$$\phi_1(\omega, x) = e^{i\omega x}\left[1 + O\left(\frac{1}{x}\right)\right]$$

[29]See, e.g. Landau and Lifshitz, Vol. III (1977), Section 134.

[30]In Section 12.3.3 we will discover that, beside the exponentially damped signal due to QNM, the response of a BH to an external disturbance also has a power-law tail, which dominates the asymptotic behavior as $t \to \infty$. Then, we will understand that the correct statement is that QNMs describe the response of the BH at a fixed location x for intermediate times, sufficiently large that the "prompt" signal caused by the incoming disturbance has passed, but not so large that the asymptotic power-like behavior has already set in.

$$= e^{i\omega_R x} e^{\gamma x/2} \left[1 + O\left(\frac{1}{x}\right) \right]. \qquad (12.172)$$

Since eq. (12.157) is invariant under $\omega \to -\omega$, there exists also a second solution whose asymptotic behavior is

$$\phi_2(\omega, x) = e^{-i\omega x} \left[1 + O\left(\frac{1}{x}\right) \right]$$

$$= e^{-i\omega_R x} e^{-\gamma x/2} \left[1 + O\left(\frac{1}{x}\right) \right]. \qquad (12.173)$$

Of course, any linear combination of the form $\phi_1(\omega, x) + \alpha\phi_2(\omega, x)$, with α a constant, still has the same asymptotic behavior as $\phi_1(\omega, x)$ when $x \to +\infty$, since $\phi_2(\omega, x) \propto e^{-\gamma x/2}$ is anyhow subleading with respect to $\phi_1(\omega, x)$, which is proportional to $e^{+\gamma x/2}$,

$$\phi_1(\omega, x) + \alpha\phi_2(\omega, x) = e^{i\omega_R x} e^{\gamma x/2} \left[1 + O\left(\frac{1}{x}\right) + \alpha e^{-2i\omega_R x} e^{-\gamma x} \right]$$

$$= e^{i\omega_R x} e^{\gamma x/2} \left[1 + O\left(\frac{1}{x}\right) \right]. \qquad (12.174)$$

Therefore the imposition of the boundary condition $\phi(\omega, x) \propto e^{i\omega x}$ at $x \to +\infty$ is not sufficient to single out a unique solution, and the same happens for the condition at $x \to -\infty$. For this reason, the definition of QNMs discussed in this section, based on the Fourier-transformed equation (12.157), is only heuristic. A rigorous treatment, which also illuminates other aspects of QNMs, can, however, be obtained by using the Laplace transform, as we now discuss.

12.3.2 QNMs from Laplace transform

To get a different perspective on QNMs, it is useful to go back to the homogeneous Zerilli or RW equations written in the time domain, eq. (12.163), which we recall here,

$$[\partial_x^2 - \partial_t^2 - V(x)]\phi(t, x) = 0, \qquad (12.175)$$

where again $x \equiv r_*/c$ and we will often use the prime to denote the derivative with respect to x.[31] We then study this equation by making use of the Laplace transform with respect to time, rather than by using the Fourier transform. We first show how eq. (12.175), together with the initial conditions $\phi(t, x)\,|_{t=0}$ and $[\partial_t \phi(t, x)]\,|_{t=0}$, can be formally solved using the Laplace transform. We will then see how the QNMs emerge in this picture.

[31]Observe also that $V(x)$ should be written as $V_l(x)$ and correspondingly all the solutions $\phi(x), \hat{\phi}_\pm(x)$, etc. discussed below should carry an index l, which we suppress for notational simplicity.

Solution via Laplace transform

Recall that the Laplace transform $\hat{\phi}(s, x)$ of a function $\phi(t, x)$ is defined, for s real and positive, by

$$\hat{\phi}(s, x) = \int_0^\infty dt\, e^{-st} \phi(t, x). \qquad (12.176)$$

If the function $\phi(t, x)$ is bounded, $\hat{\phi}(s, x)$ has an analytic continuation into the complex half-plane $\text{Re}(s) > 0$. Equation (12.176) can be inverted to give $\phi(t, x)$ with $t \geqslant 0$, by performing an integral over a contour parallel to the imaginary axis in the complex s-plane, and displaced into the right half-plane by an infinitesimal quantity $\epsilon > 0$,[32]

$$\phi(t, x) = \int_{\epsilon - i\infty}^{\epsilon + i\infty} \frac{ds}{2\pi i} \, e^{st} \hat{\phi}(s, x) \,. \qquad (12.177)$$

From eq. (12.175) it obviously follows that

$$\int_0^\infty dt \, e^{-st} [\partial_x^2 - \partial_t^2 - V(x)] \phi(t, x) = 0 \,, \qquad (12.178)$$

i.e.

$$\partial_x^2 \hat{\phi}(s, x) - V(x) \hat{\phi}(s, x) - \int_0^\infty dt \, e^{-st} \partial_t^2 \phi(t, x) = 0 \,. \qquad (12.179)$$

In the last term we integrate twice by parts. The boundary terms at $t = \infty$ give zero because of the convergence factor e^{-st}, while the boundary terms at $t = 0$ contribute. We then obtain

$$[\partial_x^2 - s^2 - V(x)] \hat{\phi}(s, x) = \mathcal{J}(s, x) \,, \qquad (12.180)$$

where

$$\mathcal{J}(s, x) = -\{s\phi(t, x) + \partial_t \phi(t, x)\} \,|_{t=0} \,. \qquad (12.181)$$

The source term $\mathcal{J}(s, x)$ therefore depends on the initial conditions at $t = 0$. This should be contrasted with the method of the Fourier transform, in which the equation in Fourier space is homogeneous and has no dependence on the initial conditions.

Let $\hat{G}(s, x, x')$ be any Green's function of eq. (12.180), i.e. any solution of

$$[\partial_x^2 - s^2 - V(x)] \hat{G}(s, x, x') = \delta(x - x') \,. \qquad (12.182)$$

Then the corresponding solution of eq. (12.180) is given by

$$\hat{\phi}(s, x) = \int_{-\infty}^\infty dx' \hat{G}(s, x, x') \mathcal{J}(s, x') \,. \qquad (12.183)$$

Of course, there are many possible Green's functions, while the solution of eq. (12.175), once we assign the initial data $\phi(t, x)|_{t=0}$ and $[\partial_t \phi(t, x)]|_{t=0}$, is uniquely determined. The issue is therefore to select the Green's function appropriate to our problem.

This can be done observing that, if $\hat{\phi}_+(s, x)$ and $\hat{\phi}_-(s, x)$ are any two linearly independent solutions of the *homogeneous* equation

$$[\partial_x^2 - s^2 - V(x)] \hat{\phi}(s, x) = 0 \,, \qquad (12.184)$$

a Green's function can be constructed by writing

$$\hat{G}(s, x, x') = \frac{1}{W(s)} \left[\theta(x - x') \hat{\phi}_-(s, x') \hat{\phi}_+(s, x) \right.$$

$$\left. + \theta(x' - x) \hat{\phi}_-(s, x) \hat{\phi}_+(s, x') \right] \,, \qquad (12.185)$$

[32] More generally, ϵ must be chosen so that it is larger than the real part of all the singularities in the complex s-plane of the function $\hat{\phi}(s, x)$. The condition that the function $\phi(t, x)$ is bounded ensures that there are no singularities of $\hat{\phi}(s, x)$ with $\text{Re}(s) > 0$ and therefore one can take $\epsilon > 0$ infinitesimal. If all singularities of $\hat{\phi}(s, x)$ are in the left half-plane (or if there are no singularities at all) we can set $\epsilon = 0$ and, upon the change of variables $s = i\omega$ (or $s = -i\omega$, which we will rather use below) the inversion formula becomes equivalent to a Fourier transform.

where $\theta(x)$ is the step function and $W(s)$ is the Wronskian of $\hat{\phi}_-$ and $\hat{\phi}_+$,

$$W(s) \equiv \hat{\phi}_-(s,x)\partial_x\hat{\phi}_+(s,x) - \hat{\phi}_+(s,x)\partial_x\hat{\phi}_-(s,x). \qquad (12.186)$$

From this definition, and making use of the fact that $\hat{\phi}_-$ and $\hat{\phi}_+$ are solutions of eq. (12.184), it follows that

$$\begin{aligned}
\partial_x W &= \hat{\phi}_-\partial_x^2\hat{\phi}_+ - \hat{\phi}_+\partial_x^2\hat{\phi}_- \\
&= \hat{\phi}_-(s^2+V)\hat{\phi}_+ - \hat{\phi}_+(s^2+V)\hat{\phi}_- \\
&= 0. \qquad (12.187)
\end{aligned}$$

Thus, the Wronskian depends on s but not on x. The fact that the function $\hat{G}(s,x,x')$ defined by eq. (12.185) gives indeed a Green's function can be verified straightforwardly, by applying the operator $[\partial_x^2 - s^2 - V(x)]$ to it and using the fact that, for any differentiable function $f(x)$,

$$\begin{aligned}
\partial_x^2[\theta(x-x')f(x)] &= \partial_x[\delta(x-x')f(x) + \theta(x-x')f'(x)] \\
&= \delta'(x-x')f(x) + 2\delta(x-x')f'(x) + \theta(x-x')f''(x) \\
&= \delta(x-x')f'(x) + \theta(x-x')f''(x), \qquad (12.188)
\end{aligned}$$

where in the first line we have used $\partial_x\theta(x-x') = \delta(x-x')$ and in the last line we have used one of the defining properties of the Dirac delta, $\delta'(x-x')f(x) = -\delta(x-x')f'(x)$.[33] Similarly

$$\partial_x^2[\theta(x'-x)f(x)] = -\delta(x-x')f'(x) + \theta(x-x')f''(x). \qquad (12.189)$$

Then, applying $[\partial_x^2 - s^2 - V(x)]$ to the function $\hat{G}(s,x,x')$ defined by eq. (12.185), we see that the terms of the generic form $\theta(x-x')f''(x)$, i.e. the terms where ∂_x^2 passed throught the step function, cancel using the fact that $\hat{\phi}_-$ and $\hat{\phi}_+$ are solutions of the homogeneous equation, while the remaining terms reconstruct $\delta(x-x')W(s)$ in the numerator, so that eq. (12.182) is indeed satisfied.

Combining eqs. (12.183) and (12.185), we can write the solution for $\hat{\phi}(s,x)$ as

$$\boxed{\begin{aligned}
\hat{\phi}(s,x) = \;& \frac{1}{W(s)}\hat{\phi}_+(s,x)\int_{-\infty}^{x} dx'\,\hat{\phi}_-(s,x')\mathcal{J}(s,x') \\
&+ \frac{1}{W(s)}\hat{\phi}_-(s,x)\int_{x}^{\infty} dx'\,\hat{\phi}_+(s,x')\mathcal{J}(s,x').
\end{aligned}}$$

$$(12.190)$$

A theorem in the theory of partial differential equations states that any Green's functions can be written in the form (12.185). The issue of choosing the appropriate Green's function for our problem therefore amounts to choosing the appropriate solutions $\hat{\phi}_-$ and $\hat{\phi}_+$ of the homogenoeous equation. To understand the properties of the solutions

[33]Of course, all the identities involving the Dirac delta are really valid in the sense of distributions, i.e. integrated over x. This relation should therefore be written as

$$\int dx\,\delta'(x-x')f(x)$$
$$= -\int dx\,\delta(x-x')f'(x),$$

on any integration interval.

$\hat{\phi}_\pm$, let us study their asymptotics behavior at $x \to \pm\infty$. In this limit the Zerilli and RW potentials both go to zero and, as long as Re(s) is non-zero, to leading order in eq. (12.184) we can neglect $V(x)$ with respect to s^2. So, for $x \to \pm\infty$ eq. (12.184) becomes

$$[\partial_x^2 - s^2]\hat{\phi}(s, x) \simeq 0 \,, \qquad (12.191)$$

which has the solutions $e^{\pm sx}$. In other words, we can choose the two linearly independent solutions of the second-order differential equation (12.184) so that, at $x \to +\infty$, one is proportional to e^{+sx} and the other to e^{-sx}. The most general solution of the homogeneous equation therefore has the asymptotic form

$$\hat{\phi}_{\mathrm{hom}}(s, x) \to a_1(s)e^{sx} + a_2(s)e^{-sx} \qquad (x \to +\infty) \,. \qquad (12.192)$$

Similarly, at $x \to -\infty$, the most general solution is of the form

$$\hat{\phi}_{\mathrm{hom}}(s, x) \to b_1(s)e^{sx} + b_2(s)e^{-sx} \qquad (x \to -\infty) \,. \qquad (12.193)$$

A generic solution, with $a_1(s) \neq 0$ and $b_2(s) \neq 0$, therefore diverges exponentially both at the horizon ($x = -\infty$) and at spatial infinity, since we are considering Re$(s) > 0$. The special solution with $b_2(s) = 0$ is finite at the horizon but, for generic values of s, when we evolve it to large positive x, it will match a solution of the form (12.192) with $a_1(s)$ and $a_2(s)$ both non-vanishing.

Since linear perturbations of the Schwarzschild space-time are uniformly bounded in x (see Note 27 on page 142), and since $\phi(t, x)$ is obtained, through eq. (12.177), from $\hat{\phi}(s, x)$ with Re $(s) > 0$, $\hat{G}(s, x, x')$ must be such that the solution obtained from eq. (12.183) is bounded everywhere and therefore also for $x \to \pm\infty$. Consider for mathematical simplicity an initial perturbation with compact support, so $\mathcal{J}(s, x')$ in eq. (12.183) is non-vanishing only for $x_L < x' < x_R$. Thus, if $x > x_R$ the second integral in eq. (12.190) vanishes, and

$$\hat{\phi}(s, x) = \frac{c_-(s)}{W(s)} \hat{\phi}_+(s, x) \qquad (x > x_R) \,, \qquad (12.194)$$

where

$$c_-(s) = \int_{x_L}^{x_R} dx' \, \hat{\phi}_-(s, x')\mathcal{J}(s, x') \,. \qquad (12.195)$$

Therefore the solution $\hat{\phi}(s, x)$ [with Re$(s) > 0$] is bounded as $x \to +\infty$ if, and only if, $\hat{\phi}_+(s, x)$ is bounded as $x \to +\infty$. Similarly, at $x < x_L$ we get

$$\hat{\phi}(s, x) = \frac{c_+(s)}{W(s)} \hat{\phi}_-(s, x) \qquad (x < x_L) \,, \qquad (12.196)$$

where

$$c_+(s) = \int_{x_L}^{x_R} dx' \, \hat{\phi}_+(s, x')\mathcal{J}(s, x') \,. \qquad (12.197)$$

So, in order to have $\hat{\phi}(s, x)$ bounded at $x \to -\infty$, we must require that $\hat{\phi}_-(s, x)$ stays bounded at $x \to -\infty$. Comparing with the asymptotic behaviors (12.192) and (12.193), we see that $\hat{\phi}_+(s,x)$ is fixed uniquely as that particular solution of the homogeneous equation that has $a_1(s) = 0$, and $\hat{\phi}_-(s,x)$ as that with $b_2(s) = 0$. Their asymptotic behaviors are therefore

$$\hat{\phi}_-(s,x) \simeq \begin{cases} e^{+sx} & (x \to -\infty), \\ a_1(s)e^{sx} + a_2(s)e^{-sx} & (x \to +\infty) \end{cases} \qquad (12.198)$$

and

$$\hat{\phi}_+(s,x) \simeq \begin{cases} b_1(s)e^{sx} + b_2(s)e^{-sx} & (x \to -\infty), \\ e^{-sx} & (x \to +\infty). \end{cases} \qquad (12.199)$$

We have chosen the normalization of $\hat{\phi}_-(s,x)$ so that, for $x \to -\infty$, it approaches e^{sx} with a proportionality coefficient of unity, and similarly for $\hat{\phi}_+(s,x)$ at $x \to +\infty$. Observe that the solution (12.190) is independent of the normalization of $\hat{\phi}_\pm(s,x)$, since a rescaling of $\hat{\phi}_\pm$ in the numerator is compensated by the same rescaling in the Wronskian in the denominator. Using the fact that $W(s)$ is independent of x, we can compute it at $x \to -\infty$, inserting the asymptotic forms for $\hat{\phi}_\pm$ from eqs. (12.198) and (12.199) into eq. (12.186), and we get

$$W(s) = -2sb_2(s). \qquad (12.200)$$

The same computation performed at $x \to +\infty$ gives

$$W(s) = -2sa_1(s), \qquad (12.201)$$

and, comparing the two results, we infer that $a_1(s) = b_2(s)$.

QNMs as poles in the complex s-plane

The foregoing analysis has shown that the Green's function is in principle uniquely determined and therefore, using eq. (12.183), the solution corresponding to given initial data is also in principle fixed. Below we will give the explicit form of the solutions $\hat{\phi}_\pm(s,x)$ for all x. However, it is possible to understand the emergence of QNMs without making reference to the explicit (and rather involved) solutions. To this purpose, we examine $\phi(t,x)$, which is obtained from $\hat{\phi}(s,x)$ using eq. (12.177). Again, we consider for mathematical simplicity an initial perturbation with compact support, i.e. $\mathcal{J}(s,x')$ is non-vanishing only for $x_L < x' < x_R$, and for definiteness we study the solution $\phi(t,x)$ at $x > x_R$, where it is given by eq. (12.194). Equation (12.177) then gives

$$\phi(t,x) = \frac{1}{2\pi i} \int_{\epsilon-i\infty}^{\epsilon+i\infty} ds\, e^{st} \frac{c_-(s)}{W(s)} \hat{\phi}_+(s,x). \qquad (12.202)$$

We now attempt to close the contour in the complex s-plane, in order to identify significant contributions to the integral. Since t is positive (we are interested in the response to an initial perturbation specified at $t = 0$, so we want the solution for $t > 0$), the factor e^{st} diverges on a circle at infinity in the right half of the complex plane, where $\text{Re}(s) > 0$, so we can only hope to be able to close the contour in the left half of the complex plane, where $\text{Re}(s) < 0$. We then consider the contour shown in Fig. 12.4. For the moment we assume that the function $\hat{\phi}_+(s,x)$, as well as $\hat{\phi}_-(s,x)$ and the source term $\mathcal{J}(s,x)$ that enter in $c_-(s)$, are analytic in s, so there are no essential singularities or branch cuts inside the contour shown in Fig. 12.4. We hasten to add that these assumptions turn out to be correct if the potential $V(x)$ has compact support (such as a square barrier) or goes to zero exponentially at both boundaries, but are wrong for the Zerilli and RW potentials, which have a power-like behavior at infinity. We will therefore come back to these assumptions later. First, however, it is instructive to see what would the result be if these assumptions were fulfilled. We also neglect for the moment the contribution of the semicircle at infinity.

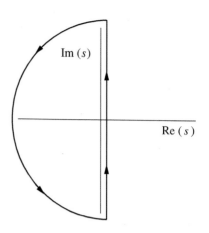

Fig. 12.4 The integration contour discussed in the text.

Under these assumptions, the only contributions to the integral can come from the poles of the integrand due to zeros of $W(s)$. Suppose that there is a set of simple zeros in the left half of the complex plane, at complex values $s = s_n$. Near these zeros we can write

$$W(s) = W'(s_n)(s - s_n) + O(s - s_n)^2 \qquad (12.203)$$

where

$$W'(s_n) \equiv \frac{dW}{ds}\bigg|_{s=s_n}. \qquad (12.204)$$

Picking the residues of the poles (and observing that the poles are encircled counterclockwise) we get

$$\phi(t,x) = \sum_n e^{s_n t} \frac{c_-(s_n)}{W'(s_n)} \hat{\phi}_+(s_n,x)$$

$$\equiv \sum_n c_n u_n(x) e^{s_n t}, \qquad (12.205)$$

where

$$c_n \equiv \frac{c_-(s_n)}{W'(s_n)}$$

$$= \frac{1}{W'(s_n)} \int_{x_L}^{x_R} dx' \, \hat{\phi}_-(s_n,x') \mathcal{J}(s_n,x'), \qquad (12.206)$$

and

$$u_n(x) \equiv \hat{\phi}_+(s_n,x). \qquad (12.207)$$

We see that, under the assumptions that we have made, $\phi(t,x)$ is expressed as a sum over a discrete set of terms.[34] To make contact between these discrete contributions and the QNMs introduced in the previous section, observe from eq. (12.186), that the condition $W(s_n) = 0$ implies that

$$\frac{\hat{\phi}'_+(s_n,x)}{\hat{\phi}_+(s_n,x)} = \frac{\hat{\phi}'_-(s_n,x)}{\hat{\phi}_-(s_n,x)} \qquad (12.208)$$

[34]To simplify the mathematics, we are assuming that $\mathcal{J}(s_n,x)$ has compact support. However, we see from eq. (12.198), that, when $\text{Re}(s_n) < 0$, $\hat{\phi}_-(s_n,x)$ diverges exponentially at both $x \to \pm\infty$, so removing the assumption of compact support is non-trivial; see Sun and Price (1988).

which integrates to

$$\hat{\phi}_+(s_n, x) = c(s_n)\hat{\phi}_-(s_n, x)\,. \tag{12.209}$$

This means that, for $s = s_n$ the two solutions of the homogeneous equation, $\hat{\phi}_+(s, x)$ and $\hat{\phi}_-(s, x)$, are no longer independent, and we have a single solution $\hat{\phi}_+(s_n, x) \propto \hat{\phi}_-(s_n, x) \equiv \hat{\phi}_*(s_n, x)$. As we know from eqs. (12.198) and (12.199), $\hat{\phi}_-(s, x)$ approaches e^{+sx} for $x \to -\infty$ and $\hat{\phi}_+(s, x)$ goes to e^{-sx} for $x \to +\infty$. Therefore

$$\hat{\phi}_*(s_n, x) \propto e^{-s_n|x|}\,, \qquad (x \to \pm\infty)\,. \tag{12.210}$$

Recall from the discussion following eq. (12.193) that, if we start at $x \to -\infty$ with a solution proportional to e^{sx}, when we evolve it to large positive x, for generic values of s it will match a solution proportional to a mixture of e^{sx} and e^{-sx}. We see that the values s_n are just those special values of s for which the solution with $b_2(s) = 0$ at $x \to -\infty$ matches a solution with $a_1(s) = 0$ at $x \to +\infty$.

To make clear the connection with the QNMs, discussed in the previous section in the language of the Fourier transform, and also to make contact with a large body of literature on BHs, it is convenient to write

$$s = -i\omega\,, \tag{12.211}$$

where

$$\omega = \omega_R + i\omega_I \tag{12.212}$$

is still a complex variable, and to introduce the notation

$$\hat{u}^{\rm in}(\omega, x) \equiv \hat{\phi}_-(-i\omega, x)\,, \tag{12.213}$$
$$\hat{u}^{\rm up}(\omega, x) \equiv \hat{\phi}_+(-i\omega, x)\,. \tag{12.214}$$

In this notation, eqs. (12.198) and (12.199) become

$$\hat{u}^{\rm in}(\omega, x) \simeq \begin{cases} e^{-i\omega x} & (x \to -\infty)\,, \\ A_1(\omega)e^{-i\omega x} + A_2(\omega)e^{i\omega x} & (x \to +\infty)\,, \end{cases} \tag{12.215}$$

where $A_{1,2}(\omega) \equiv a_{1,2}(-i\omega)$, and

$$\hat{u}^{\rm up}(\omega, x) \simeq \begin{cases} B_1(\omega)e^{-i\omega x} + B_2(\omega)e^{i\omega x} & (x \to -\infty)\,, \\ e^{i\omega x} & (x \to +\infty)\,, \end{cases} \tag{12.216}$$

where $B_{1,2}(\omega) \equiv b_{1,2}(-i\omega)$. Recalling that the time dependence is given by $\exp\{-i\omega t\}$, and that $x = r_*/c$, we see that $\hat{u}^{\rm in}$ is purely ingoing at the horizon, i.e. it describes a wave that, near the horizon, is falling toward the BH, while $\hat{u}^{\rm up}$ is purely outgoing at future null infinity.[35] In terms of $A_1(\omega)$, eq. (12.201) reads

$$W = 2i\omega A_1(\omega)\,, \tag{12.217}$$

and $A_1(\omega) = B_2(\omega)$. Similarly, computing the Jacobian between the "up" and "out" solutions (see Note 35) at $x \to +\infty$ and comparing it with the value at $x \to -\infty$, we get $B_1(\omega) = -A_2^*(\omega)$.

[35] In the literature there can also be found the function $\hat{u}^{\rm down} \equiv (\hat{u}^{\rm up})^*$ which, combined again with a time-dependence $\exp\{-i\omega t\}$, describes a purely incoming wave at future null infinity, and the function $\hat{u}^{\rm out} \equiv (\hat{u}^{\rm in})^*$, which is purely outgoing at the horizon. Away from the special values of ω where the Wronskian vanishes, any two of the four functions $\hat{u}^{\rm in}$, $\hat{u}^{\rm out}$, $\hat{u}^{\rm up}$ and $\hat{u}^{\rm down}$ can be chosen as the two independent solutions of the homogeneous equation.

Writing $s_n = -i\omega_n$ and $u_*(\omega_n, t) \equiv \phi_*(s_n = -i\omega_n, x)$, eq. (12.210) becomes

$$u_*(\omega_n, t) \simeq e^{+i\omega_n|x|} \qquad (x \to \pm\infty). \qquad (12.218)$$

These are just the boundary conditions satisfied by the solutions of the Fourier-transformed Zerilli or RW equation that we called the QNMs, see eq. (12.162). On the other hand, under the replacement $s = -i\omega$, the homogeneous Laplace-transformed equation (12.184) becomes identical to the Fourier-transformed equation (12.157), so the solutions of the former must turn into the solutions of the latter. Then we see that the solutions $\hat{\phi}_*(s_n, x)$ are just the QNMs, expressed in the language of the Laplace transform. Writing $\omega_{I,n} = -\gamma_n/2$, we have

$$s_n = \omega_{I,n} - i\omega_{R,n}$$

$$= -\frac{\gamma_n}{2} - i\omega_{R,n}. \qquad (12.219)$$

so eq. (12.205) can be rewritten as

$$\phi(t, x) = \sum_n c_n u_n(x) e^{-i\omega_n t}$$

$$= \sum_n c_n u_n(x) e^{i\omega_{R,n} t} e^{-\gamma_n t/2}. \qquad (12.220)$$

Since the poles s_n inside the integration contour have $\mathrm{Re}(s_n) < 0$, we have $\gamma_n > 0$, so this solution is a superposition of damped oscillations.

With regards to the spatial behavior, however, the fact that $\mathrm{Re}(s_n) < 0$ means that the QNM wavefunction $u_n(x)$ diverges exponentially both at spatial infinity and at the horizon as $\exp\{\gamma_n|x|/2\}$, see eq. (12.210), as indeed we already found in eq. (12.171).[36]

In Section 12.3.4 we will see how to compute the QNM frequencies. Then, the analytic result (12.220) can be compared with the direct numerical integration of the differential equation. Figure 12.5 shows the comparison for the scalar wave equation given by eqs. (12.9) and (12.6), for $l = 2$. Observe that, by including more QNMs, the agreement can be pushed toward earlier times.

12.3.3 Power-law tails

The results that we have just obtained assumed that $\hat{\phi}_\pm(s, x)$ are analytic functions of s in the left half of the complex s-plane. However, this turns out to be incorrect. Indeed, the RW equation (12.102) in the absence of source can be recast in a form known as a generalized spheroidal wave equation, whose solutions are known exactly. They turn out to have the form (in units $c = G = 2M = 1$)

$$\hat{\phi}_+(s, r) = (2is)^s e^{i\phi_+} (1 - 1/r)^s \qquad (12.221)$$

$$\times \sum_{L=-\infty}^{\infty} b_L [G_{L+\nu}(-is, isr) + iF_{L+\nu}(-is, isr)],$$

$$\hat{\phi}_-(s, r) = r^{-2s}(r - 1)^s e^{-s(r-2)} \sum_{n=0}^{\infty} a_n (1 - 1/r)^n, \qquad (12.222)$$

[36] Concerning the sign in front of $\omega_{R,n}$, for a real potential the solutions are real, so for each QNM with $s_n = \omega_{I,n} - i\omega_{R,n}$ there is necessarily another one with $s_n = \omega_{I,n} + i\omega_{R,n}$, so the factors $e^{i\omega_{R,n} t}$ and $e^{-i\omega_{R,n} t}$ give rise to the real combinations $\sin(\omega_{R,n} t)$ and $\cos(\omega_{R,n} t)$.

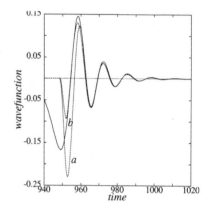

Fig. 12.5 Comparison of the numerical integration of the scalar wave equation for $l = 2$ (solid line) with the approximate contribution from the first $l = 2$ QNMs. The two dashed lines are for (a) the slowest damped QNM and (b) the sum of the first six modes. From Andersson (1997).

[37]See Leaver (1986b), eqs. (17) and (18), and Appendix A. The constant phase ϕ_+ is given in his eq. (19).

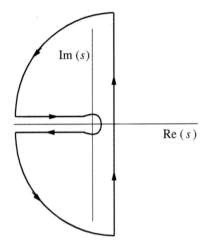

Fig. 12.6 The integration contour of Fig. 12.4, modified so that it goes around the branch cut at $\mathrm{Re}(s) < 0$.

[38]See e.g. Gradshteyn and Ryzhik (1980), 17.13.26. This result is obtained observing that the left-hand side of eq. (12.226) is the inverse Laplace transform of the identity function. Since the identity function obviously has no pole in the complex s-plane, we can set $\epsilon = 0$ (cf. Note 32). Then, rotating the contour in the complex plane, we find the standard integral representation of the Dirac delta, $\int (d\omega/2\pi) e^{-i\omega t} = \delta(t)$. In this and in similar rotations, as in eq. (12.232), if we perform the rotation writing formally $s = -i\omega$ (so that the factor e^{st} gives directly the factor $e^{-i\omega t}$, with the sign that corresponds to our convention for the Fourier transform), we must obviously not forget that, as s runs along the imaginary axis from $-i\infty + \epsilon$ to $+i\infty + \epsilon$, $\omega = is$ runs along the real axis in the "wrong" direction, i.e. from $+\infty + i\epsilon$ to $-\infty + i\epsilon$. Therefore

$$\int_{\epsilon - i\infty}^{\epsilon + i\infty} ds = \int_{+\infty + i\epsilon}^{-\infty + i\epsilon} (-i)d\omega$$

$$= +i \int_{-\infty + i\epsilon}^{+\infty + i\epsilon} d\omega. \qquad (12.225)$$

where $G_{L+\nu}$ and $F_{L+\nu}$ are Coulomb wavefunctions, and the coefficients a_n and b_L, as well as the parameter ν, are defined in terms of a three-term recurrence relation.[37] These explicit solutions are quite cumbersome, but the essential point for our purpose is that $\hat{\phi}_-(s, r)$ is analytic in s, while $\hat{\phi}_+(s, r)$, because of the factor $s^s = \exp\{s \log s\}$, has a branch cut in the complex s-plane, along the negative real axis. To close the contour in the complex s-plane we must therefore go around this branch cut, and the contour of Fig. 12.4 must be replaced by that shown in Fig. 12.6.

As a result, there are three distinct contribution to $\phi(t, x)$: (i) the contribution from the poles inside the contour; (ii) the contribution from the two quarter-circles at infinity; (iii) the contribution from the branch cut. The sum of these three contributions determines the response of the BH to an initial perturbation. The contribution from the poles inside the contour is just what we have computed in the previous section, and we have seen that it is due to the QNM. Now, we want to understand the remaining contributions.

In order to single out the properties intrinsic to the BH space-time, rather than using $\phi(t, x)$, which depends on the details of the initial conditions encoded in $\mathcal{J}(s, x)$, it is more convenient to work with the Green's function $G(t, x, x')$ in the time domain, which we now introduce.

The time-domain Green's function $G(t, x, x')$

In the discussion of the previous section, a key ingredient was the Green's function in s-space, $\hat{G}(s, x, x')$. For $t \geqslant 0$, we can go back to a function of time t using the inverse Laplace transform, as in eq. (12.177), i.e. we can define $G(t, x, x')$, when $t \geqslant 0$, by

$$G(t, x, x') = \int_{\epsilon - i\infty}^{\epsilon + i\infty} \frac{ds}{2\pi i} e^{st} \hat{G}(s, x, x'). \qquad (12.223)$$

We complete the definition of $G(t, x, x')$ by setting $G(t, x, x') = 0$ for $t < 0$. This defines the retarded Green's function. From eq. (12.182) it follows that

$$[\partial_x^2 - \partial_t^2 - V(x)]G(t, x, x') = \int_{\epsilon - i\infty}^{\epsilon + i\infty} \frac{ds}{2\pi i} e^{st} [\partial_x^2 - s^2 - V(x)]\hat{G}(s, x, x')$$

$$= \delta(x - x') \int_{\epsilon - i\infty}^{\epsilon + i\infty} \frac{ds}{2\pi i} e^{st}. \qquad (12.224)$$

From the properties of the Laplace transform it follows that[38]

$$\int_{\epsilon - i\infty}^{\epsilon + i\infty} \frac{ds}{2\pi i} e^{st} = \delta(t), \qquad (12.226)$$

and therefore

$$[\partial_x^2 - \partial_t^2 - V(x)]G(t, x, x') = \delta(x - x')\delta(t). \qquad (12.227)$$

So $G(t, x, x')$ is the (retarded) Green's function of the operator $[\partial_x^2 - \partial_t^2 - V(x)]$. Performing the inverse Laplace transform of eq. (12.183) and using the definition of $\mathcal{J}$ given in eq. (12.181), we find

$$\phi(t,x) = -\int_{-\infty}^{\infty} dx' \left[\phi(t=0,x) \int_{\epsilon-i\infty}^{\epsilon+i\infty} \frac{ds}{2\pi i} se^{st}\hat{G}(s,x,x') \right.$$
$$\left. + \partial_t\phi(t=0,x) \int_{\epsilon-i\infty}^{\epsilon+i\infty} \frac{ds}{2\pi i} e^{st}\hat{G}(s,x,x') \right]. \quad (12.228)$$

In the first integral we write $se^{st} = \partial_t e^{st}$, and we get

$$\phi(t,x) = -\int_{-\infty}^{\infty} dx' \left[\partial_t G(t,x,x')\phi(t=0,x) + G(t,x,x')\partial_t\phi(t=0,x) \right]. $$
$$(12.229)$$

This shows how the Green's function $G(t,x,x')$ allows us to reconstruct $\phi(t,x)$, given the initial data $\phi(t=0,x)$ and $\partial_t\phi(t=0,x)$.

In order to extract the characteristic properties of a BH, getting rid of mathematical complications, it is convenient to consider the situation in which the initial data have compact support in a region far away from the BH, i.e. $\mathcal{J}(x')$ is non-vanishing only for $x_L < x' < x_R$, with $R_S/c \ll x_L$ (recall that $x = r_*/c$). We then study the response seen by an observer at $x > x_R$ (so also $x \gg R_S/c$). In this setting we always have $x > x'$, so in eq. (12.185) only the term $\theta(x - x')$ contributes, and

$$\hat{G}(s,x,x') = \frac{1}{W(s)} \hat{\phi}_-(s,x')\hat{\phi}_+(s,x). \quad (12.230)$$

Furthermore, since both x and x' are in the asymptotic region, we can use the asymptotic behaviour at spatial infinity of $\hat{\phi}_\pm$ given in eqs. (12.215) and (12.216). Using also the expression (12.217) for the Wronskian, we get

$$\hat{G}(s=-i\omega,x,x') \simeq \frac{1}{2i\omega} \left[e^{i\omega(x-x')} + \frac{A_2(\omega)}{A_1(\omega)} e^{i\omega(x+x')} \right]. \quad (12.231)$$

This is called the "asymptotic approximation" to the Green's function. Inserting this into eq. (12.223), and rotating the contour from the imaginary to the real axis (see Note 38) gives

$$G(t,x,x') \simeq \int_{-\infty+i\epsilon}^{\infty+i\epsilon} \frac{d\omega}{2\pi} \frac{1}{2i\omega} \left[e^{i\omega(x-x'-t)} + \frac{A_2(\omega)}{A_1(\omega)} e^{i\omega(x+x'-t)} \right]. $$
$$(12.232)$$

We can now close the contour. In the complex ω-plane the contour of Fig. 12.6 becomes the one shown in Fig. 12.7, and $G(t,x,x')$ separates into three contributions,

$$G(t,x,x') = G_{\text{QNM}}(t,x,x') + G_B(t,x,x') + G_F(t,x,x'), \quad (12.233)$$

where $G_{\text{QNM}}(t,x,x')$ is the contribution from the poles inside the contour, $G_B(t,x,x')$ that from the branch cut, and $G_F(t,x,x')$ that from the two quarter-circles at infinity.

We examine the three contributions in turn. The one associated with QNMs has already been computed in the previous section directly for

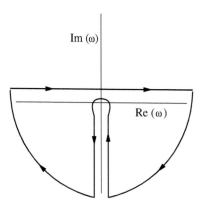

Fig. 12.7 The integration contour of Fig. 12.6, in the ω-plane.

$\phi(t,x)$, but it is useful to write it again in terms of the Green's function. The QNMs correspond to the poles of the Wronskian, i.e. of $A_1(\omega)$. Near a QNM with complex frequency ω_n we can write

$$A_1(\omega) = A_1'(\omega_n)(\omega - \omega_n) + O(\omega - \omega_n)^2. \qquad (12.234)$$

Because of reality, the QNM frequencies appear in pairs with $\omega_n = (\pm\omega_{R,n}) + i\omega_{I,n}$ and $\omega_{I,n} = -\gamma_n < 0$ (see Note 36). Then, in the asymptotic approximation, we get

$$G_{\mathrm{QNM}}(t,x,x') \simeq \mathrm{Re}\left[\sum_n B_n e^{-i\omega_n(t-x-x')}\right] \qquad (12.235)$$

$$= \mathrm{Re}\left[\sum_n B_n e^{-i\omega_{R,n}(t-x-x')}\, e^{-(\gamma_n/2)(t-x-x')}\right],$$

where (taking into account that the poles are circled clockwise)

$$B_n = -\frac{A_2(\omega_n)}{\omega_n A_1'(\omega_n)}, \qquad (12.236)$$

and the sum runs over all normal modes, with one representative for each pair $(\pm\omega_{R,n}) + i\omega_{I,n}$ (for example over the modes with $\omega_{R,n} > 0$).[39] The coefficients B_n are called the *quasi-normal excitation factors*. Observe that they depend only on the BH properties, and not on the initial perturbation.

We next examine the contribution from the branch cut. This can be computed using the exact solutions (12.221) and (12.222) and plugging them into eq. (12.230), as well as into the Wronskian.[40] The result for $G_B(t,x,x')$, at generic values of t, x and x', is quite complicated. However, it simplifies considerably in the limit $t \gg x+x'$. From eq. (12.235), we see that, in this limit, the QNM signal is already exponentially small. Physically, this is easily understood observing that, since the source is localized at a distance $\sim cx'$ (with $x_L < x' < x_R$; recall that $x \equiv r_*/c$) much larger than R_S, then x' is the typical time needed for a disturbance traveling at the speed of light to go from the source to the near-horizon region, where the potential barrier in the RW or Zerilli potential is large. So x' is the time-scale needed for the initial disturbance to propagate to the near-horizon region and excite the quasi-normal modes. Similarly, x is the time that the signal generated by the QNMs takes to go back from the near-horizon region to the observer. Thus, the signal due to QNMs ringing reaches the observer at time $\sim (x+x')$, and the condition $t \gg x+x'$ means that, at the fixed location x of the observer, the part of the BH signal due to the ringing of the QNMs has already passed through the observer position. We are therefore looking for any possible residual signal that is left, at the fixed position x, after the QNM signal has already faded away, decreasing exponentially as in eq. (12.235).

In the limit $t \gg x+x'$, the integral along the branch cut is dominated by small values of $|\omega|$. While taking $t \gg x+x'$, we can further distinguish two different limits:

[39] As we will see in Section 12.3.4, there are, however, "exceptional" QNMs with $\omega_R = 0$. Then, they enter in eq. (12.235) with a coefficient B_n reduced by a factor of 2.

[40] This computation is performed explicitly in Leaver (1986b).

(1) We can consider the limit $t \to \infty$ with x fixed. In this case one finds[41]

$$G_{\rm B}(t, x, x') \simeq (-1)^{l+1} \frac{2(2l+2)!}{[(2l+1)!!]^2} \frac{R_S}{c} \frac{(xx')^{l+1}}{t^{2l+3}} . \qquad (12.237)$$

Therefore there is a power-law tail at spatial infinity, i.e. a non-radiative tail. Actually, this is just the leading term, and there are higher-order tails. The full result is

$$G_{\rm B}(t, x, x') \simeq (-1)^{l+1} \frac{R_S}{c}$$
$$\times \sum_{m=0}^{\infty} \sum_{n=0}^{m} \frac{2^{1-m}(2l+2m+2)!}{n!(m-n)!(2l+2n+1)!!(2l+2m-2n+1)!!}$$
$$\times \frac{x^{l+2m-2n+1} x'^{l+1+2n}}{t^{2l+3+2m}} . \qquad (12.238)$$

As it is clear from Fig. 12.8, the inclusion of higher-order tails improves considerably the agreement with the numerical results.

(2) We can consider the limit $t \to \infty$ with t/x fixed. This means that we are looking for radiative terms at future null infinity. In this case, one finds[42]

$$G_{\rm B}(t, x, x') \simeq (-1)^{l+1} \frac{(l+1)!}{(2l+1)!!} \frac{R_S}{c} \frac{(x')^{l+1}}{u^{l+2}} , \qquad (12.239)$$

where $u \equiv t - x$. We therefore have again a power-law tail. Since we are at future null infinity, this is a radiative contribution, i.e. a radiative tail. Observe that both the overall amplitude of the tail term, and the exponent $l+2$ of the power-law decay, are independent of the spin of the perturbation, i.e. they are the same for scalar perturbations, governed by eq. (12.9) with the potential (12.6), and for gravitational perturbations, governed by the RW or Zerilli equation. The same result also holds for electromagnetic, and for higher-spin perturbations.

How these behaviors of the Green's function reflect on the solution $\phi(t, x)$ depends on the nature of the initial data. In particular, we see from eq. (12.229) that, if the initial data are such that $\phi(t = 0, x) = 0$, the time behavior of $\phi(t, x)$ is the same as that of $G(t, x, x')$, i.e. $\sim u^{-l-2}$ at future null infinity and $\sim t^{-2l-3}$ at spatial infinity.[43]

These results have been shown to agree very well with those found from direct numerical integration of eq. (12.175); see Fig. 12.8. Of course, an equation such as (12.175) comes from linearization of Einstein equations, and one might ask whether non-linearities are important. However, in the case of a spherically symmetric scalar field, direct numerical integration of the full Einstein equations shows that linearized theory works remarkably well, giving power-law tails and QNM ringing very close to that of the full theory.[44]

In the case of QNM ringing, the physical interpretation is relatively clear. This represents the oscillations of space-time in the near-horizon

[41] This result was first found by Price (1972a), from a direct study of the differential equation (12.175), rather than with the Green's function method. The higher-order correction given in eq. (12.238) is due to Andersson (1997).

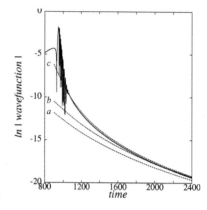

Fig. 12.8 The numerical evolution of the $l = 2$ scalar wave equation (solid line) compared with the power-law tail (dashed lines): (a) the leading-order power-law tail, (b) the tail approximation including the first two terms and (c) using the first 11 terms in the tail-sum. From Andersson (1997).

[42] See Leaver (1986b), eq. (45). Note that he uses units $c = G = 2M = 1$.

[43] A different situation arises when the initial data are specified on a *null* surface, rather than on the spacelike surface $t = 0$. This case is studied in Price (1972a) and, for the radiative tail, in Gundlach, Price and Pullin (1994a). The result is that $\phi(t, x)$ decays at future null infinity as u^{-l-1} (static initial field) or as u^{-l-2} (no static initial field), while at spatial infinity it is found that $\phi(t, x) \sim 1/t^{2l+2}$ (static initial field) or $\phi(t, x) \sim 1/t^{2l+3}$ (no static initial field).

[44] See Gundlach, Price and Pullin (1994b).

region, basically in the region inside the peak of the RW or Zerilli equation. We would like to develop a physical explanation also for the power-law tails. To understand where the effect comes from, one can study eq. (12.175) for different forms of the potential $V(x)$. It turns out that the presence of tails is related to the form of $V(x)$ as $x \to +\infty$. For instance, if the potential behaves asymptotically as $l(l+1)/r^2$ (corresponding to a free propagation in the original three-dimensional problem, since this is just the centrifugal potential that comes from writing the Laplacian in polar coordinates), with no subleading powers, there are no power-law tails. In contrast, the addition of higher powers such as $1/x^3$, $1/x^4$, etc. to the large-x limit of the potential generates tails. Systematic studies have shown that tails are absent if the potential has compact support, such as a potential barrier, or if it decreases exponentially at infinity. Only the asymptotic form of the potential for large x matters and, in particular, the existence of tails is independent of whether or not the metric has a horizon, so the same tails would be generated by a neutron star and by a black hole of the same mass.

These results indicate that the tails are generated by back-scattering off the space-time curvature at large distances. They correspond to a part of the signal that was emitted outward from the source at a radial distance cx', reached a large distance, and was then scattered back toward the observer at radial distance cx; see Fig. 12.9. Observe that the propagation direction of the tail is not the same as that of the QNM part of the signal, which rather propagates directly from the central object to the observer. The reader who has gone through Chapter 5 of Vol. 1, in particular pages 272–274, will recognize the close similarity with the tails found in the post-Newtonian expansion, which again reflected the effect of backscattering on the non-trivial background geometry or, equivalently, the fact that perturbations in a non-flat space-time travel not only on the light cone, but also inside it.

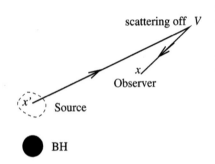

Fig. 12.9 A graphical illustration of the scattering off the potential, giving raise to the tail term.

Finally, we have the contribution from the two quarter-circles at infinity. These contributions are evaluated in the limit $|\omega| R_S/c \to \infty$. In this limit, $|\omega|$ is much larger than the height of the peak of the potential, so effectively they can be computed setting $V = 0$. This means that they reflect the free propagation of the high-frequency components of the initial perturbation. In particular, they describe the free propagation of the high-frequency components of that part of the initial perturbation which moves outward, and travels directly from the source to the observer.

To summarize, if at $t = 0$ we have an initial disturbance localized around a radial distance cx' from the BH, the signal seen by an observer located at a distance $cx > cx'$ (taken for simplicity in the same angular direction) can be separated into the following main contributions:

(i) For $t < x - x'$, for causality reasons, no signal has yet reached the observer.

(ii) At $t \simeq x - x'$ there arrives that part of the initial disturbance that was initially moving outward. This is propagated by $G_F(t, x, x')$. Because of dispersion effects, part of this signal will arrive at a later time, so this will give the main contribution until the next epoch, which, as we will see, is at $t \simeq x + x'$. This part of the signal is known as the *precursor*.

(iii) At $t \simeq x + x'$ there arrives the signal from the QNMs of the BH. In fact, the part of the signal that traveled toward the BH arrives in a time $\sim x'$ in the near-horizon region, where it excites the QNMs. The resulting excitation takes a time $\sim x$ to propagate outward to the observer. This *ringdown* phase lasts for a time given by the lifetime of the longest-lived QNM. As we will see in the next section, this is typically of the order of R_S/c, times numerical factors. The relevant part of the Green's function in this phase is $G_{\rm QNM}$.

(iv) When the signal from the QNMs has vanished exponentially, we can begin to see the late-time power-law tail (which in principle was present even before, but was masked by the stronger contribution from the QNMs). The relevant part of the Green's function in this phase is $G_{\rm B}$.

As a final comment, we can now understand that QNMs do not form a complete set, since an expansion of the form (12.220) does not represent the full signal. In the language of the Green's function in the compex ω-plane, we saw that, beside the contribution of the poles inside the integration contour, which corresponds to QNMs, we also have a contribution from the branch cut and from the quarter-circles at infinity. This confirms the physical arguments discussed on page 143, following eq. (12.171).

12.3.4   Frequency spectrum of QNMs

We now want to compute the frequency spectrum of a Schwarzschild BH. First of all, we show that the RW and Zerilli equations have exactly the same spectrum, for all QNMs. This can be proved as follows. Suppose that two operators L_1 and L_2 are related by

$$DL_1 = L_2 D\,, \qquad (12.240)$$

where D is another operator. Let ψ by an eigenfunction of L_1, $L_1\psi = \lambda\psi$. Then eq. (12.240) gives $L_2(D\psi) = \lambda(D\psi)$, so for each eigenfunction ψ of L_1 there is an eigenfunction $(D\psi)$ of L_2 with the same eigenvalue.[45]

We now take as L_1 the Zerilli operator for the l-th multipole

$$(L_1)_l = \frac{d^2}{dx^2} - V_l^Z\,, \qquad (12.241)$$

and as L_2 the RW operator,

$$(L_2)_l = \frac{d^2}{dx^2} - V_l^{\rm RW}\,, \qquad (12.242)$$

where, as usual, $x = r_*/c$ and we set for the moment $c = 1$. We seek a solution for the operator D in the form

$$D_l = \frac{d}{dx} - g_l(r)\,. \qquad (12.243)$$

[45] Provided that, when ψ satisfies the boundary conditions that we wish to imposes on the eigenvectors of L_1, $(D\psi)$ satisfies the boundary conditions that we impose on the eigenvectors of L_2. This will be trivially the case for the Zerilli and RW operators.

Then eq. (12.240) becomes

$$-\frac{dV_l^Z}{dx} - V_l^Z\frac{d}{dx} + g_l V_l^Z = -2\frac{dg_l}{dx}\frac{d}{dx} - \frac{d^2g_l}{dx^2} - V_l^{RW}\frac{d}{dx} + V_l^{RW}g_l\,. \quad (12.244)$$

Requiring cancellation of the terms proportional to d/dx gives

$$2\frac{dg_l}{dx} = V_l^Z - V_l^{RW}\,, \quad (12.245)$$

which integrates to

$$g_l(r) = -\frac{r - R_S}{r^2[1 + (2\lambda r)/(3R_S)]} + a_l\,, \quad (12.246)$$

with a_l an integration constant. Requiring that the remaining terms cancel gives

$$\frac{d^2g_l}{dx^2} + g_l(V_l^Z - V_l^{RW}) - \frac{dV_l^Z}{dx} = 0\,. \quad (12.247)$$

Using eq. (12.245), this becomes

$$\frac{d^2g_l}{dx^2} + 2g_l\frac{dg_l}{dx} - \frac{dV_l^Z}{dx} = 0\,, \quad (12.248)$$

which integrates to

$$\frac{dg_l}{dx} + g_l^2 - V_l^Z = C_l\,, \quad (12.249)$$

with C_l some constant. Substituting eq. (12.246), we can check that the left-hand side of eq. (12.249) is indeed a constant if we choose $a_l = -(2/3)\lambda(\lambda + 1)/R_S$. Therefore, for each l there exists an operator D_l which relates the RW and Zerilli operators as in eq. (12.240), and therefore the axial and polar modes of a Schwarzschild BH are isospectral. This is an important result that is specific to Schwarzschild BHs and does not hold, for instance, for neutron stars.

We now want to compute these QNM frequencies. Many techniques have been developed in the literature. A direct numerical approach to the problem does not work well. Indeed, one might try to integrate numerically the differential equation in the frequency domain, eq. (12.157), starting from a large negative x_1, imposing the boundary condition (12.161) (or, better yet, using an asymptotic series approximation to the solution valid for x_1 large and negative) and run the integration for many complex values of ω until one finds the values for which the asymptotic behavior at $+\infty$ is of the form (12.160), i.e. purely proportional to $\exp\{i\omega x\}$, with no contamination from the solution proportional to $\exp\{-i\omega x\}$. The problem with such an approach is that, as we have seen, the QNM frequencies have the form $\omega_n = \omega_{R,n} - i\gamma_n/2$ with $\gamma_n > 0$, so such a solution blows up exponentially both at $x = -\infty$ and at $x = +\infty$, and our task is to check that, in both limits, there is no contamination from an exponentially small solution. Of course, numerically this is a very difficult task, even more so for highly excited modes.

Various techniques have been devised to circumvent this problem. In the limit of $|\omega R_S/c|$ large, the RW or Zerilli equation can be approximated by a confluent hypergeometric equation.[46] From the known

---

[46]See Liu and Mashhoon (1996).

asymptotic behavior of the confluent hypergeometric functions, one can then extract the leading large-ω behavior of the coefficients $A_1(\omega)$ and $A_2(\omega)$ in eq. (12.215),

$$A_1(\omega) \simeq i \frac{\Gamma(1 - 2i\omega R_S/c)(2i\omega R_S/c)^{-1/2}}{\Gamma(1/2 - 2i\omega R_S/c)} \qquad (12.250)$$

$$A_2(\omega) \simeq \frac{\Gamma(1 - 2i\omega R_S/c)(2i\omega R_S/c)^{-1/2 + 2i\omega R_S/c}}{\sqrt{\pi}} e^{-2i\omega R_S/c}. \quad (12.251)$$

We have seen that the QNM frequencies are those for which the Wronskian $W(\omega)$ vanishes. From eq. (12.217), this means that they are defined by the condition $A_1(\omega) = 0$. Then we can read from eq. (12.250) the leading term of the QNM frequencies for large ω: the zeros of $A_1(\omega)$ correspond to the poles of $\Gamma(1/2 - 2i\omega R_S/c)$ and are given by

$$\frac{\omega_n R_S}{c} \simeq -\frac{i}{2}\left(n + \frac{1}{2}\right), \qquad (n \gg 1). \qquad (12.252)$$

This shows that for each l there is an infinity of QNMs,[47] and their leading asymptotic behavior is independent of l. Actually, at this level of approximation, in eq. (12.252) we do not have the right to keep the factor $1/2$ compared with n, since we are working only to leading order in the large-n limit, and we have also missed any $O(1)$ term in $\mathrm{Re}(\omega)$. An elegant analysis using analytic continuation in the complex r-plane[48] (where r is the radial coordinate) gives the leading and next-to-leading result, for a generic spin-j perturbation (so $j = 0$ corresponds to the scalar perturbation governed by eqs. (12.6) and (12.9), while $j = 2$ corresponds to the RW or Zerilli equation, and $j = 1$ to electromagnetic perturbations). In the limit of $\mathrm{Im}(\omega)$ large (and negative, with our sign conventions), and for $\mathrm{Re}(\omega) \geqslant 0$, the result is

$$e^{4\pi R_S \omega_n/c} \simeq -[1 + 2\cos(\pi j)]. \qquad (12.253)$$

For $j = 0$ and $j = 2$ we have $\cos(\pi j) = 1$, so eq. (12.253) gives

$$\frac{\omega_n R_S}{c} \simeq \frac{\log 3}{4\pi} - \frac{i}{2}\left(n + \frac{1}{2}\right) \qquad (n \gg 1). \qquad (12.254)$$

The same analysis for $\mathrm{Re}(\omega) < 0$ gives of course $\omega_n R_S/c = -\log 3/4\pi - (i/2)(n + 1/2)$, since we know that QNM frequencies appear in pairs related by $\omega_R \to -\omega_R$. Observe that even the next-to-leading-order terms, i.e. the factor $\log 3/(4\pi)$ and the factor $1/2$ in $(n + 1/2)$, are independent of l. The first l-dependence enters at the next order, which is given by[49]

$$\frac{\omega_{nl} R_S}{c} \simeq \frac{\log 3}{4\pi} - \frac{i}{2}\left(n + \frac{1}{2}\right) + (1 + i)\frac{\sqrt{2}\,\Gamma^4(1/4)}{144\pi^{5/2}}\frac{l(l+1) - 1}{\sqrt{n}}$$
$$+ O\left(\frac{1}{n}\right). \qquad (12.255)$$

[47]This fact was proved rigorously in Bachelot and Motet-Bachelot (1992).

[48]See Motl and Neitzke (2003).

[49]See Maassen van den Brink (2004).

For vector perturbations, $j = 1$, eq. (12.253) rather gives,

$$\frac{\omega_n R_S}{c} \simeq -\frac{i}{2} n + O\left(\frac{1}{\sqrt{n}}\right), \qquad (n \gg 1). \qquad (12.256)$$

so in this case the real part vanishes asymptotically. Observe that the RW and Zerilli equations have the same asymptotic frequencies, in agreement with the fact that the two operators are isospectral, as we have already discussed.

It is also possible to study the limit of large l, given an upper limit on n. In this case

$$\frac{\omega_{n,l} R_S}{c} \simeq (2l+1) - i(2n+1). \qquad (12.257)$$

The highly damped QNMs are especially interesting in the context of semiclassical quantum gravity and string theory, and might provide a bridge between classical and quantum gravity; see the Further Reading. From the point of view of GW experiments, however, the least-damped modes are more relevant, since they will dominate the signal. Various techniques have been developed to evaluate the QNM frequencies for the low-lying modes. In particular, the reduction of the RW equation to a generalized spheroidal wave equation, which gives rise to the exact solutions (12.221) and (12.222), allows one to recast the problem of the computation of QNM frequencies as a recursion relations in terms of continued fractions. The latter can be solved numerically with good accuracy. The results for some of the modes up to $n = 60$ are given in Table 12.1. Observe that for $l = 2$ the mode with $n = 9$ has $\omega_R = 0$ (within the numerical accuracy, which however degrades near the imaginary axis). In fact, for each l there is a special value $\bar{n}_l$ of n such that ω_R is numerically close to zero; for $l = 3$, $\bar{n}_l = 41$, and $\bar{n}_l$ rapidly increases with l. This special value separates two branches of the spectrum; one, for $n < \bar{n}_l$, in which ω_R decreases with n and one, for $n > \bar{n}_l$, in which ω_R increases with n, until it reaches the constant asymptotic value given by eq. (12.254).

From Table 12.1 we see that the least-damped mode emits GWs at a frequency $f = \omega_R/(2\pi)$ given by

$$f \simeq 0.747 \frac{c}{2\pi R_S} \simeq 12 \, \text{kHz} \left(\frac{M_\odot}{M}\right). \qquad (12.258)$$

Thus, for a $10 M_\odot$ BH we get $f \sim 1$ kHz, while a supermassive BH with $M = 10^6 M_\odot$ rings at $f \sim 10$ mHz. The ringdown signal vanishes exponentially with a characteristic time $\tau = 1/|\omega_I|$, which, for the least-damped mode, is

$$\tau \simeq \frac{R_S}{0.178c} \simeq 5.5 \times 10^{-5} \, \text{s} \left(\frac{M}{M_\odot}\right). \qquad (12.259)$$

Thus, the ringdown signal from a $10 M_\odot$ BH vanishes exponentially in a time of order a millisecond, while a supermassive BH with M of order a few $10^6 M_\odot$ rings for a few minutes.

Table 12.1 The frequencies of the QNMs of a Schwarzschild BH, for $l = 2$ and for $l = 3$, in units $R_S/c = 1$ [data from Leaver (1985)].

	$l = 2$	$l = 3$
n	(ω_R, ω_I)	(ω_R, ω_I)
1	$(0.747343, -0.177925)$	$(1.198887, -0.185406)$
2	$(0.693422, -0.547830)$	$(1.165288, -0.562596)$
3	$(0.602107, -0.956554)$	$(1.103370, -0.958186)$
4	$(0.503010, -1.410296)$	$(1.023924, -1.380674)$
5	$(0.415029, -1.893690)$	$(0.940348, -1.831299)$
6	$(0.338599, -2.391216)$	$(0.862773, -2.304303)$
7	$(0.266505, -2.895822)$	$(0.795319, -2.791824)$
8	$(0.185617, -3.407676)$	$(0.737985, -3.287689)$
9	$(0.000000, -3.998000)$	$(0.689237, -3.788066)$
10	$(0.126527, -4.605289)$	$(0.647366, -4.290798)$
20	$(0.175608, -9.660879)$	$(0.404157, -9.333121)$
30	$(0.165814, -14.677118)$	$(0.257431, -14.363580)$
40	$(0.156368, -19.684873)$	$(0.075298, -19.415545)$
50	$(0.151216, -24.693716)$	$(0.134153, -24.119329)$
60	$(0.148484, -29.696417)$	$(0.163614, -29.135345)$

It is instructive to compare the fundamental BH frequency, $\omega_{n=1,l=2}$, with the corresponding frequency of an elastic sphere of radius R, which we studied in Section 8.4. From eq. (8.241) and Table 8.1 [recalling that $(\mu/\rho)^{1/2} = v_\perp$ is the speed of sound of transverse elastic waves; see the text following eq. (8.229)], the $l = 2$, $n = 1$ spheroidal mode of an elastic sphere of radius R has a frequency $\omega_{n=1,l=2} \simeq 2.651\, v_\perp/R$. So, as far as the fundamental BH frequency is concerned, we can picture the space-time outside the BH as a cavity of radius $R \simeq (2.651/0.747) R_S \simeq 3.5 R_S$, that sustains oscillations with $v_\perp = c$. Of course, the precise number 3.5 has little significance, but still this indicates that QNMs correspond to oscillations of the region of the space-time within a few Schwarzschild radii from the horizon, which is the region where the Zerilli or RW potential is important.

Beside their frequency, the other important property of QNMs is their excitation factors B_n, defined in eq. (12.236). Indeed, we see from eqs. (12.229), (12.233) and (12.235) that the amplitude of the oscillation of the signal at a given QNM frequency, in response to an initial perturbation, is given by the coefficient B_n (which is independent of the initial perturbation) times an integral involving the initial perturbation. The coefficients B_n have been computed in the literature; see the Further Reading.

12.3.5 The physical interpretation of the QNM spectrum

We have seen that perturbations of BHs vanish in time as a superposition of damped oscillations, of the form

$$e^{\omega_I t}\left[a\sin(\omega_R t) + b\cos(\omega_R t)\right],\qquad(12.260)$$

with a spectrum of complex frequencies $\omega = \omega_R + i\omega_I$, where $\omega_I < 0$ (recall that the negative sign of ω_I is due to our convention on the Fourier transform; see Note 28 on page 142). The spectrum of QNMs encodes the response of the BH to a disturbance, due for example to infalling matter. Beside their importance in gravitational-wave astrophysics, QNMs are also important because BHs are often used as a testing ground for ideas in quantum gravity, and their QNMs are obvious candidates for an interpretation in terms of quantum levels.

To conclude this section on QNM frequencies, it is interesting to realize that the spectrum of QNMs has a quite peculiar structure. In Fig. 12.10 we show the frequencies of the $l = 2$ modes of a Schwarzschild BH, given in Table 12.1. We see that $|\mathrm{Im}\,\omega_n|$ grows monotonically with n, so the least damped mode corresponds to $n = 1$, and has $(R_S/c)|\mathrm{Im}\,\omega| \simeq 0.1779$. The next least-damped mode is $n = 2$, with $(R_S/c)|\mathrm{Im}\,\omega| \simeq 0.5478$, and so on. In contrast, the real part of ω is not monotonic with n. It rather decreases at first, until it becomes zero (within numerical error) for $n = 9$, and then starts growing again, reaches a maximum, and then decreases, eventually settling to the constant asymptotic value $(c/4\pi R_S)\ln 3$ given by eq. (12.254), approaching it from above with the $1/\sqrt{n}$ behavior given by eq. (12.255). The pattern shown in Fig. 12.10 repeats for higher l. There is always a value $\bar{n}_l$ of n such that, for $n < \bar{n}_l$, $\mathrm{Re}(\omega_{n,l})$ decreases with n until it becomes close to zero at $n = \bar{n}_l$, while above this critical value it first increases and then approaches from above the asymptotic value $(c/4\pi R_S)\ln 3$.

If we compare with the normal mode structure of familiar classical systems, such as a vibrating rod, we immediately realize that the structure displayed in Fig. 12.10, and particularly the "inverted branch" formed by the modes with $n \leqslant \bar{n}_l$, is quite peculiar. In classical systems, the least-damped mode is in general also the one with the *lowest* value of $\mathrm{Re}\,\omega$, and typically $\mathrm{Re}\,\omega$ and $|\mathrm{Im}\,\omega|$ both increase with n. In contrast, we see from Fig. 12.10 the least-damped mode is the one with the *highest* possible value of $\mathrm{Re}\,\omega$ and, for $n < \bar{n}_l$, $\mathrm{Re}\,\omega$ is a decreasing function of $|\mathrm{Im}\,\omega|$! Even the "normal" branch $n > \bar{n}_l$ is somewhat puzzling. In particular, the fact that it saturates to a finite value is surprising. In a normal macroscopic system the underlying reason why, for large n, $|\mathrm{Im}\,\omega_n|$ goes to infinity (and therefore these modes decay very fast) is that also $\mathrm{Re}\,\omega_n$ diverges, so, with increasing n, the wavelength $\lambda_n = 2\pi/\mathrm{Re}\,\omega_n$ gets smaller and smaller, and finally becomes of the same order as the lattice spacing of the underlying atomic structure. At this point the perturbation can no longer be sustained as a wave by the medium, and quickly disappears in the thermal agitation of the lattice nuclei.

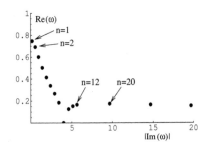

Fig. 12.10 $\mathrm{Re}\,\omega_{nl}$ against $|\mathrm{Im}\,\omega_{nl}|$ [both in units of (R_S/c)] for $l = 2$. The points displayed correspond to overtone numbers $n = 1, 2, \ldots 12$, and $n = 20, 30, 40$. From Maggiore (2008).

The QNM structure of Fig. 12.10 is no less puzzling if we attempt a semiclassical description and interpret it as the structure of excited levels of a quantum BH. In normal quantum systems the levels with high excitation energy, $E_n = \hbar \mathrm{Re}\,\omega_n$, are those that decay fast, first of all because, in a multipole expansion, the decay width Γ grows with $\mathrm{Re}\,\omega$ [for example $\Gamma \sim (\mathrm{Re}\,\omega)^3$ for a dipole transition and $\Gamma \sim (\mathrm{Re}\,\omega)^5$ for a quadrupole transition] and, second, because they can decay into many different channels, i.e. into all the levels with lower excitation energy not forbidden by selection rules. So, again it is very surprising that, for $n < \bar{n}_l$, we have an inverted structure, where the lifetime of the state *increases* with its excitation energy.

In order to get some understanding of this structure, it is useful to consider a damped harmonic oscillator $\xi(t)$, governed by the equation

$$\ddot{\xi} + \gamma_0 \dot{\xi} + \omega_0^2 \xi = f(t), \tag{12.261}$$

where γ_0 is the damping constant, ω_0 the proper frequency of the harmonic oscillator, and $f(t)$ an external force per unit mass. We examined this system in Vol. 1, when we studied resonant-bar detectors; see eq. (8.20). Solving this equation in Fourier transform we get

$$\xi(t) = -\int_{-\infty}^{\infty} \frac{d\omega}{2\pi} \frac{\tilde{f}(\omega)}{(\omega - \omega_+)(\omega - \omega_-)} e^{-i\omega t}, \tag{12.262}$$

where $\omega_\pm$ are the two roots of the equation $\omega^2 + i\gamma_0\omega - \omega_0^2 = 0$, i.e.

$$\omega_\pm = \pm\sqrt{\omega_0^2 - (\gamma_0/2)^2} - i\frac{\gamma_0}{2}. \tag{12.263}$$

Consider the response to a Dirac delta perturbation, $f(t) \propto \delta(t)$, so $\tilde{f}(\omega) \propto 1$. For $t < 0$ we can close the integration contour in eq. (12.262) in the upper half-plane and, since $\omega_\pm$ both lie below the real axis, we get zero, as required by causality. For $t > 0$ we close the contour in the lower half-plane and we pick the residues of the two poles. So the result for $\xi(t)$ is a superposition of a term oscillating as $e^{-i\omega_+ t}$ and of a term oscillating as $e^{-i\omega_- t}$. Therefore, (12.260) is reproduced by a damped harmonic oscillator, with the identifications

$$\omega_R = \sqrt{\omega_0^2 - (\gamma_0/2)^2}, \tag{12.264}$$

$$\omega_I = -\frac{\gamma_0}{2}, \tag{12.265}$$

which can be inverted to give

$$\omega_0 = \sqrt{\omega_R^2 + \omega_I^2}. \tag{12.266}$$

We see that the seemingly obvious identification $\omega_0 = \omega_R$ only holds when $\gamma_0/2 \ll \omega_0$, i.e. for very long-lived modes. For most BH QNMs we are in the opposite limit; in particular, for highly excited modes, we have $|\omega_I| \gg \omega_R$ (see Fig. 12.10), so $\omega_0 \simeq |\omega_I|$ rather than $\omega_0 \simeq \omega_R$.

If we model the BH perturbations in terms of a collection of damped harmonic degrees of freedom (which can be useful both classically, to have an intuitive physical picture of a BH as a whole, and in semi-classical quantum gravity, to get hints about the quantum structure of space-time) the correct identification for the frequency of the equivalent harmonic oscillator is given by eq. (12.266), together with $\gamma_0/2 = -\omega_I$.

In terms of ω_0 the energy level structure of a BH becomes physically very reasonable, and for $l = 2$ it is shown in Fig. 12.11 (a similar result holds for higher l). We see that the frequency $(\omega_0)_{nl}$ increases monotonically with the overtone number n. Recall that the damping coefficient $\gamma_0/2$ is equal to $-\omega_I$, so also $\gamma_0/2$ increases monotonically with n. Thus, in terms of the equivalent harmonic oscillators, the least-damped mode, which is still the $n = 1$ mode, is also the one with the *lowest* value of ω_0, and the larger the value of $(\omega_0)_{nl}$, the shorter is the lifetime, as we expected from physical intuition.

This result can be rewritten in a very suggestive form by introducing the Hawking temperature T_H from $k_B T_H = \hbar c/(4\pi R_S)$, where k_B is the Boltzmann constant, and defining $E_n \equiv \hbar(\omega_0)_n$. Setting for simplicity $\hbar = c = k_B = 1$ and using eq. (12.254), in the large-n limit, eq. (12.266) gives

$$E_n \simeq \sqrt{m_0^2 + p_n^2}\,, \tag{12.267}$$

where

$$m_0 = T_H \ln 3\,, \tag{12.268}$$

$$p_n = 2\pi T_H \left(n + \frac{1}{2}\right). \tag{12.269}$$

This expression for p_n is especially intriguing, since it corresponds to a particle quantized with antiperiodic boundary conditions on a circle of length $L = 1/T_H = 8\pi M$.

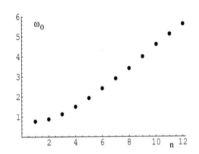

Fig. 12.11 $(\omega_0)_{nl}$ (in units of R_S/c) for $l = 2$, against n. From Maggiore (2008).

12.4   Radial infall into a black hole

We now want to understand how the QNMs can be excited by external disturbances. In this section we consider a test mass falling radially into a Schwarzschild BH, and we compute the GWs emitted, including the contribution from the QNMs. Actually, we already studied this problem in Section 4.3, in the simpler context of a particle moving on a radial Newtonian trajectory, and using the quadrupole formula for computing the gravitational radiation emitted. As we discussed, this was sufficient to compute the low-frequency part of the energy spectrum of the radiation, since this is emitted when the infalling particle is at large distances from the BH, where these assumptions are adequate. The quadrupole formula, however, is the first term of an expansion obtained working in a flat background space, and therefore does not know anything about QNMs. To get the full spectrum of the radiation emitted, and in particlar the QNM contribution, we must rather use the formalism that we have developed in this chapter.

This is a very instructive problem, and we discuss it in detail. We begin by computing the source term in the Zerilli and RW equations.

12.4.1   The source term

The energy–momentum tensor of a point-like particle of mass m was given in eq. (5.47),

$$T^{\mu\nu} = \frac{1}{\sqrt{-g}} m\gamma \frac{dx^\mu}{dt} \frac{dx^\nu}{dt} \delta^{(3)}(\mathbf{x} - \mathbf{x}_0(t)). \qquad (12.270)$$

where, in the general-relativistic case,

$$\gamma = \frac{dt}{d\tau}, \qquad (12.271)$$

and τ is proper time. For the Schwarzschild metric, in polar coordinates (t, r, θ, ϕ), we have $\sqrt{-g} = r^2|\sin\theta|$, so for a particle following a generic trajectory $x_0^\mu(t) = (t, r_0(t), \theta_0(t), \phi_0(t))$, we have

$$T^{\mu\nu} = \frac{1}{r^2|\sin\theta|} m\gamma \frac{dx_0^\mu}{dt} \frac{dx_0^\nu}{dt} \delta[r - r_0(t)]\delta[\theta - \theta_0(t)]\delta[\phi - \phi_0(t)]$$

$$= m\gamma \frac{dx_0^\mu}{dt} \frac{dx_0^\nu}{dt} \frac{\delta[r - r_0(t)]}{r^2} \delta[\cos(\theta) - \cos(\theta_0(t))]\delta[\phi - \phi_0(t)]$$

$$\equiv m\gamma \frac{dx_0^\mu}{dt} \frac{dx_0^\nu}{dt} \frac{\delta[r - r_0(t)]}{r^2} \delta^{(2)}[\Omega - \Omega_0(t)]. \qquad (12.272)$$

The coefficients s_{lm}^a in the expansion of the energy–momentum tensor in spherical harmonics, defined in eqs. (12.51) and (12.52), can now be obtained using the orthogonality property (12.25) together with the redefinition (12.37). This gives[50]

$$s_{lm}^a = \epsilon_a[c^a(r)]^2 \int d\Omega\, \eta^{\mu\rho}\eta^{\nu\sigma} [t_{lm}^a(\Omega)]_{\mu\nu}^* T_{\rho\sigma}(\Omega) \qquad (12.273)$$

$$= \epsilon_a[c^a(r)]^2\, m\gamma \frac{\delta[r - r_0(t)]}{r^2} [t_{lm}^a(\Omega_0(t))]_{\mu\nu}^* \eta^{\mu\rho}\eta^{\nu\sigma} g_{\rho\rho'}g_{\sigma\sigma'} \frac{dx_0^{\rho'}}{dt} \frac{dx_0^{\sigma'}}{dt}.$$

The computation of the various $s_{lm}^a(t, r)$ is now in principle straightforward. We must then perform the Fourier transform to get $\tilde{s}_{lm}^a(\omega, r)$. The computation is performed in detail in Problem 12.2. The result is that, as a consequence of the cylindrical symmetry of the problem, the quantities s_{lm}^{B1} and s_{lm}^{B2} that give the source term in the RW equation both vanish, so the axial modes are not excited. We are therefore concerned only with the polar modes, described by the Zerilli equation. Again because of the cylindrical symmetry of the problem, for the polar modes only the quantities $\tilde{s}_{l,m=0}^a(\omega, r)$ are non-vanishing (where the z axis has been chosen as the direction from which the particle is coming). Then for the source $\tilde{S}_{l,m=0}^{\rm pol} \equiv \tilde{S}_l$ of the Zerilli equation, defined in terms of the $\tilde{s}_{lm}^a(\omega, r)$ in eqs. (12.111) and (12.112), we find (setting for simplicity $c = 1$)

$$\tilde{S}_l(\omega, r) = -4m\sqrt{2\pi} \left(l + \frac{1}{2}\right)^{1/2} \frac{A(r)}{\lambda r + 3M} \qquad (12.274)$$

[50]Of course the factor m appearing as a prefactor in the combination $m\gamma$ is the mass of the particle, not to be confused with the index m of the spherical harmonics!

$$\times \left[\left(\frac{r}{R_S} \right)^{1/2} - \frac{2i\lambda}{\omega(\lambda r + 3M)} \right] e^{i\omega T(r)},$$

where the function $T(r)$ is given in eq. (12.400), in Problem 12.2.

12.4.2 Numerical integration of the Zerilli equation

We finally arrive at the equation

$$\frac{d^2}{dx^2}\tilde{Z}_l + \left[\omega^2 - V_l^Z(r)\right]\tilde{Z}_l(\omega, x) = \tilde{S}_l(\omega, r), \qquad (12.275)$$

where $\tilde{Z}_l(\omega, x) = \tilde{Z}_{l,m=0}(\omega, x)$, the Zerilli potential $V_l^Z(r)$ is given by eq. (12.110), the source term $\tilde{S}_l(\omega, r)$ is given by eq. (12.274), and we use units $c = 1$, so $x = r_*/c = r_*$. This equation can be integrated with the Green's function technique that we have already discussed in Section 12.3.2. Namely, we consider the solutions $\hat{u}_l^{\text{in}}(\omega, x)$ and $\hat{u}_l^{\text{up}}(\omega, x)$ of the *homogeneous* equation

$$\frac{d^2}{dx^2}\hat{u}_l + \left[\omega^2 - V_l^Z(r)\right]\hat{u}_l = 0, \qquad (12.276)$$

that satisfy the boundary conditions (12.215) and (12.216), respectively. With them we construct the Wronskian

$$W_l(\omega) = \hat{u}_l^{\text{in}}(\omega, x)\partial_x \hat{u}_l^{\text{up}}(\omega, x) - \hat{u}_l^{\text{up}}(\omega, x)\partial_x \hat{u}_l^{\text{in}}(\omega, x) \qquad (12.277)$$

and the Green's function

$$G_l(\omega, x, x') = \frac{1}{W_l(\omega)}\left[\theta(x - x')\hat{u}_l^{\text{in}}(\omega, x')\hat{u}_l^{\text{up}}(\omega, x) \right.$$
$$\left. + \theta(x' - x)\hat{u}_l^{\text{in}}(\omega, x)\hat{u}_l^{\text{up}}(\omega, x')\right]. \qquad (12.278)$$

The solution of eq. (12.275) is then

$$\tilde{Z}_l(\omega, x) = \frac{1}{W_l(\omega)}\hat{u}_l^{\text{up}}(\omega, x)\int_{-\infty}^{x} dx'\, \hat{u}_l^{\text{in}}(\omega, x')\tilde{S}_l(\omega, x')$$
$$+\frac{1}{W_l(\omega)}\hat{u}_l^{\text{in}}(\omega, x)\int_{x}^{\infty} dx'\, \hat{u}_l^{\text{up}}(\omega, x')\tilde{S}_l(\omega, x'); \qquad (12.279)$$

compare with eq. (12.190). In the limit $x \to \infty$ the second integral vanishes, while $\hat{u}_l^{\text{up}}(\omega, x) \simeq \exp\{i\omega x\}$; see eq. (12.216). Thus, for $x \to \infty$,

$$\tilde{Z}_l(\omega, x) \to e^{i\omega x}\frac{1}{W_l(\omega)}\int_{-\infty}^{\infty} dx'\, \hat{u}_l^{\text{in}}(\omega, x')\tilde{S}_l(\omega, x'). \qquad (12.280)$$

Comparing with eq. (12.116), we see that

$$A_l^{\text{out}}(\omega) = \frac{1}{W_l(\omega)}\int_{-\infty}^{\infty} dx'\, \hat{u}_l^{\text{in}}(\omega, x')\tilde{S}_l(\omega, x'), \qquad (12.281)$$

where we have written $A_{l,m=0}^{\text{out}}(\omega) = A_l^{\text{out}}(\omega)$. As in eq. (12.217), $W = 2i\omega A_1(\omega)$, where $A_1(\omega)$ is determined by the behavior of $\hat{u}_l^{\text{in}}(\omega, x)$ at $x \to +\infty$; see eq. (12.215).

A strategy for solving eq. (12.275) is therefore the following. For a given (real) ω we solve numerically the homogeneous equation (12.276), imposing the boundary condition[51]

$$\hat{u}_l(\omega, x) \to e^{-i\omega x} \qquad (12.282)$$

at $x \to -\infty$. This determines the solution $\hat{u}_l^{\text{in}}(\omega, x)$. The Wronskian is then obtained from the behavior of $\hat{u}_l^{\text{in}}(\omega, x)$ at $x \to +\infty$. We can now perform numerically the integral involving the source term, in eq. (12.281). We see that the Fourier modes of the Zerilli function, $A_l^{\text{out}}(\omega)$, can be determined directly in terms of the solution $\hat{u}_l^{\text{in}}(\omega, x)$ of the associated homogeneous equation.

Repeating the procedure for many values of ω, we have the function $A_l^{\text{out}}(\omega)$, on a sufficiently fine mesh over the real ω-axis. The energy spectrum $dE_l/d\omega$ of the radiation emitted at infinity in the multipole l is then given by eq. (12.156) (with $B_{lm}^{\text{out}}(\omega) = 0$, since the axial modes are not excited).

12.4.3   Waveform and energy spectrum

The result of such a numerical analysis for the energy spectrum is shown in Fig. 12.12, for the multipoles $l = 2$ and $l = 3$. Since the source $\tilde{S}_l(\omega, r)$ in the Zerilli equation is proportional to the mass m of the test particle, see eq. (12.274), the amplitude $A_l^{\text{out}}(\omega)$ obtained from eq. (12.281) is also proportional to m, and the radiated energy is proportional to m^2. In Fig. 12.12 we therefore plot the dimensionless quantity $(c/Gm^2)dE_l/d\omega$, which is independent of the mass m, for $l = 2$ and $l = 3$. Observe that the energy spectrum for $l = 2$ is much larger than for $l = 3$.

As we discussed in Section 4.3, in the limit $\omega R_S/c \ll 1$ the result for the mode $l = 2$ can be computed simply by using the quadrupole formula, with the trajectory determined by the Newtonian equation of motion. In this limit, we found that

$$\frac{dE_{l=2}}{d\omega} \simeq 0.177 \, \frac{Gm^2}{c} \left(\frac{\omega R_S}{c}\right)^{4/3} . \qquad (12.283)$$

In Fig. 12.12 we also show the curve computed in this approximation. Observe that it gives a good approximation to the actual $l = 2$ energy spectrum only for very low values of ω, say $\omega R_S/c \lesssim 0.1$. This is not surprising, since the Newtonian radial trajectory describing a particle falling into a point-like mass M is a reasonable approximation to the actual geodesic in the Schwarzschild metric only at distance significantly larger that R_S, say $r/R_S \gtrsim O(10)$.

The spectrum is then cut off exponentially beyond a critical value $\omega R_S/c = O(1)$. From the figure, we see that for $l = 2$ the maximum is around $\omega R_S/c \simeq 0.64$. The $l = 3$ mode (the dashed line in Fig. 12.12)

[51] In practice, in a numerical integration we start from a large but finite negative value, so it is better to include corrections in powers of $1/|x|$ to the asymptotic behavior. These can be computed analytically in terms of a recursion relation, see Chandrasekhar and Detweiler (1975).

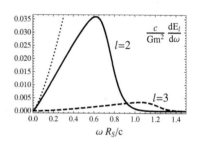

Fig. 12.12 The energy spectrum $(c/Gm^2)dE_l/d\omega$ as a function of $\omega R_S/c$, for $l = 2$ (solid line) and $l = 3$ (dashed). The lighter dotted line is the Newtonian approximation to the $l = 2$ spectrum. (Figures 12.12–12.16: numerical data courtesy of Ermis Mitsou.)

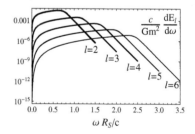

Fig. 12.13 The energy spectrum $(c/Gm^2)dE_l/d\omega$ as a function of $\omega R_S/c$, for $l = 2, 3, 4, 5, 6$, on a logarithmic vertical scale.

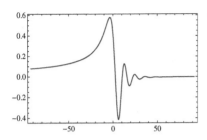

Fig. 12.14 The gravitational waveform $u_l(t)$ with $l = 2$ (in units $G = c = 1$).

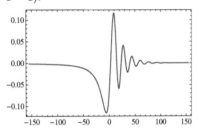

Fig. 12.15 The gravitational waveform $u_l(t)$ with $l = 3$ (in units $G = c = 1$).

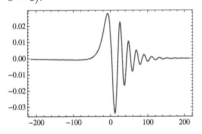

Fig. 12.16 The gravitational waveform $u_l(t)$ with $l = 4$ (in units $G = c = 1$).

gives a contribution that is much smaller, and peaks at higher frequencies, with a maximum around $\omega R_S/c \simeq 1.1$. Comparing with Table 12.1 we see that the peak in the spectrum is at a frequency of the order of the value of ω_R for the least-damped ($n = 1$) QNM, for both $l = 2$ and $l = 3$. The contribution from higher multipoles would not be visible on the linear scale of Fig. 12.12. In Fig. 12.13 we plot the contribution to the spectrum of the modes from $l = 2$ to $l = 6$ on a logarithmic scale. Observe that with increasing l the position of the maximum of the spectrum moves toward higher ω, and the value of $dE_l/d\omega$ at the maximum decreases monotonically with l. Integrating over the frequency, one finds that the total radiated energy is

$$\Delta E_{\rm rad} \simeq 0.0104 \left(\frac{m}{M}\right) mc^2\,. \tag{12.284}$$

We next examine the waveform. As discussed in Section 12.2.6, in the radiation zone the waveform in the TT gauge is given by the functions $u_{lm}(t-r/c)$ and $v_{lm}(t-r/c)$; see eq. (12.122). For the radial infall into a Schwarzschild BH only the polar modes with $m = 0$ are non-vanishing, i.e. the functions $u_l(t - r/c) \equiv u_{l,m=0}(t - r/c)$. These are obtained by performing the Fourier transform of $A_l^{\rm out}(\omega)$; see eq. (12.149). In Figs. (12.14)–(12.16) we show $u_l(t)$ as a function of time, for r fixed at a large value $r \gg R_S$, for $l = 2, 3$ and 4, respectively. We observe that with increasing l the waveform shows more and more oscillations. Observe also the change in vertical scale among the figures. We see that the last part of the waveform shows damped oscillations. This part can be very well fitted using the least-damped QNMs.

12.5   Perturbations of rotating black holes

Until now we have considered Schwarzschild black holes, which are spherically symmetric. Spherical symmetry considerably simplifies the theoretical description. In realistic astrophysical situations, however, one expects that all BHs will be rapidly rotating. For BHs created by the collapse of a massive star, this is an unavoidable consequence of the conservation of angular momentum. During collapse a star passes from its initial radius, which for a massive star will be several time the solar radius $R_\odot \simeq 7 \times 10^8$ m, down to its Schwarzschild radius (which, for the Sun, is $R_S \simeq 3 \times 10^3$ m). Angular momentum conservation implies that ωr^2 stays constant. This means that even a very small initial rotation, which is present in all stars, is amplified by a huge factor during the collapse, giving rise to a rapidly rotating remnant, whether a BH or a neutron star. Indeed, we already mentioned in Chapter 6 that this mechanism can spin up neutron stars up to rotation periods as short as 10 ms (while further spin-up can arise because of accretion from a companion). For the same reason, we expect that rotation will be important for BHs created by a collapsing star. We will see in Chapter 14 that the final BH created in the coalescence of a compact binary is also rapidly rotating. In this case the relevant metric is not the Schwarzschild metric, but the Kerr metric, which we now introduce.[52] We will then discuss the perturbations of the Kerr metric and the quasi-normal modes of rotating BHs.

12.5.1 The Kerr metric

A classic result of GR is that the most general stationary solution of the vacuum Einstein equation describing a BH with mass M and angular momentum J is given by the Kerr metric,

$$ds^2 = -\left(1 - \frac{R_S r}{\rho^2(r,\theta)}\right)c^2 dt^2 - \frac{2aR_S r \sin^2\theta}{\rho^2(r,\theta)}\, cdt\, d\phi + \frac{\rho^2(r,\theta)}{\Delta(r)}dr^2$$
$$+\rho^2(r,\theta)d\theta^2 + \left(r^2 + a^2 + \frac{a^2 R_S r \sin^2\theta}{\rho^2(r,\theta)}\right)\sin^2\theta\, d\phi^2\,, \quad (12.285)$$

where we still use the notation $R_S = 2GM/c^2$, and we introduced the *Kerr parameter*

$$a = \frac{J}{Mc}\,, \quad (12.286)$$

which has dimension of length, and the functions

$$\rho^2(r,\theta) = r^2 + a^2 \cos^2\theta\,, \quad (12.287)$$
$$\Delta(r) = r^2 - R_S r + a^2\,. \quad (12.288)$$

The coordinates (t, r, θ, ϕ) are called *Boyer–Lindquist coordinates*. When $a = 0$ the metric (12.285) reduces to the Schwarzschild metric, so the Boyer–Lindquist coordinates are a generalization of the Schwarzschild coordinates.

We briefly recall some basic features of this metric.[53] The Kerr metric is asymptotically flat. According to the rules of GR, the total mass and angular momentum are then determined by the orbits of test masses and by the precession of gyroscopes at large distances, and in this why one can show that M is the mass and J is the angular momentum associated to this metric.

From eq. (12.285) we see that the Kerr metric becomes singular when either $\rho = 0$ or $\Delta = 0$. The condition $\rho = 0$ means that $r = 0$ and $\theta = \pi/2$. Computing the Riemann tensor one finds that this is a real physical singularity, where the curvature diverges. It is a generalization of the singularity that the Schwarzschild metric has at $r = 0$. The condition $\Delta = 0$ is satisfied at

$$r_{\pm} = \frac{R_S \pm \sqrt{R_S^2 - 4a^2}}{2}\,. \quad (12.289)$$

The singularities at the surfaces $r = r_+$ and $r = r_-$ are coordinate singularities, just like the singularity of the Schwarzschild metric at $r = R_S$. Just as in the Schwarzschild case, one can find new coordinates in which the metric is no longer singular at these radii. Observe that, when $a = 0$, we get $r_+ = R_S$, so this surface reduces to the horizon of the Schwarzschild metric.

The structure of the metric at $r < r_+$ is quite interesting mathematically. However, for astrophysical purposes, we are only interested in what happens outside the horizon, defined as the boundary of the space-time region from which a light ray can escape to infinity from any point. Such a boundary surface must be a null three-surface, i.e. it must have a null tangent vector representing those light rays that "remain stuck" on this surface, and separates the light rays that escape to infinity from those that collapse into the interior. The surface $r = r_+$ is indeed the (outer) horizon of the Kerr metric, and will therefore play the role of the boundary of space-time for an observer at $r > r_+$. Light rays generated at $r < r_+$ cannot escape outside the horizon, and therefore the Kerr geometry describes a black hole.

Observe that the intrinsic geometry of the horizon is not spherically symmetric. Setting $r = r_+$ and $t = \text{const}$ in eq. (12.285) we get a two-dimensional surface with intrinsic metric

$$d\Sigma^2 = \rho_+^2(\theta)d\theta^2 + \left(\frac{R_S r_+}{\rho_+(\theta)}\right)^2 \sin^2\theta \, d\phi^2 \,, \qquad (12.290)$$

where $\rho_+^2(\theta) = r_+^2 + a^2\cos^2\theta$. This is not the geometry of a 2-sphere. Rather, a two-dimensional surface with the intrinsic metric (12.290), when embedded in flat three-dimensional space, looks like a "squashed sphere", flattened at the poles, just as one would expect for a rotating ball of fluid. For instance, when $\theta = \text{const.} = \pi/2$ the line element obtained from (12.290) is $d\Sigma = R_S d\phi$, so the equatorial circumference is $2\pi R_S$, while along a meridian $\phi = \phi_0$ we have $d\Sigma = \rho_+(\theta)d\theta$, so this circumference has a length

$$\Sigma = 2\int_0^\pi d\theta \, \sqrt{r_+^2 + a^2\cos^2\theta}$$
$$= 2\int_0^\pi d\theta \, \sqrt{R_S r_+ - a^2\sin^2\theta} \,, \qquad (12.291)$$

where in the last line we have used $r_+^2 + a^2 = R_S r_+$, which follows on setting $\Delta = 0$ in eq. (12.288). For $a \neq 0$, $r_+ < R_S$ and Σ in eq. (12.291) is always smaller than $2\pi R_S$.

The Kerr metric describes a BH only as long as

$$a \leqslant \frac{R_S}{2} = \frac{GM}{c^2} \,, \qquad (12.292)$$

(so, in units $G = c = 1$, one has the condition $a \leqslant M$), otherwise $r_\pm$ become complex and the horizons disappear, exposing a naked singularity at $\rho = 0$. This means that, for a given mass M, the angular momentum of a BH is limited,

$$J \leqslant \frac{GM^2}{c} \,. \qquad (12.293)$$

Black holes that saturate this bound are called *extremal*, and are on the verge of developing a naked singularity. Most astrophysical BHs are believed to be spun up by their accretion disk, possibly up to a value of J close to the extremal one (see Note 47 on page 253 for discussion of the observational evidence). It is believe that it is impossible to raise a to or above the critical value by throwing matter into a Kerr BH; see the Further Reading. This is an example that supports the *cosmic censorship* conjecture, which states that the singularities formed in gravitational collapse are always hidden behind the horizon of a BH.[54]

Motion of test masses in the Kerr geometry

Some aspects of the motion of test masses in the Kerr geometry are qualitatively very different from the motion in the Schwarzschild geometry. For instance, consider a particle that is falling into a BH from infinity and that initially moves radially. In the Schwarzschild geometry a geodesic that at infinity is radial will remain radial. In the Kerr geometry this is not true. One can write the geodesic equation (1.66) in the Kerr metric, using Boyer–Lindquist coordinates $x^\mu(\tau) = (t(\tau), r(\tau), \theta(\tau), \phi(\tau))$. For symmetry reasons it is clear that an orbit initially in the equatorial plane $\theta = \pi/2$, with initial velocity along the equatorial plane, will always remain in the equatorial plane. So, restricting for simplicity to orbits in the equatorial plane $\theta = \pi/2$, on integrating the geodesic equation we can get $dr/d\tau$ and $d\phi/d\tau$, and from these the equation of the orbit $d\phi/dr$.[55] The result is

$$\frac{d\phi}{dr} = -\frac{R_S a}{r\Delta}\left[\frac{R_S}{r}\left(1-\frac{a^2}{r^2}\right)\right]^{-1/2}. \qquad (12.296)$$

This shows that, for $a \neq 0$ and r finite, $d\phi/dr$ does not vanish, and an initially radial geodesic is bent in the direction in which the BH rotates. This is an example of the dragging of inertial frames in the Kerr geometry. When one gets sufficiently close to a Kerr BH this effect becomes extreme. From the Kerr metric (12.285) we see in fact that the metric component g_{tt} vanishes when $\rho^2 = R_S r$, i.e. when

$$r = r_e(\theta) \equiv \frac{R_S + \sqrt{R_S^2 - 4a^2\cos^2\theta}}{2}. \qquad (12.297)$$

Comparing with the definition of r_+ in eq. (12.289) we see that $r_e(\theta) > r_+$ except at the poles $\theta = 0, \pi$ where the two surfaces touch each other; see Fig. 12.17. The surface $r = r_e(\theta)$ is called the *static limit* and the region $r_+ < r < r_e(\theta)$ is called the *ergosphere* of a Kerr BH. In the ergosphere, the metric component g_{tt} becomes negative, so a four-velocity of the form

$$u^\mu = (u^t, 0, 0, 0) \qquad (12.298)$$

becomes space-like, and it cannot be identified with the four-velocity of any physical observer. This means that, for $r_+ < r < r_e(\theta)$, it is impossible to put an observer, or a test particle, in a trajectory with $dr/d\tau = 0$, $dr/d\theta = 0$ and $dr/d\phi = 0$, i.e. there are no physical trajectories in which the observer stays at a fixed value of the Boyer–Lindquist coordinates.

[54]Some counterexamples are know to this conjecture, although they involve unphysical equations of state, or very specific initial data, or higher-dimensional space-times; see e.g. Barausse, Cardoso and Khanna (2011) for references.

[55]Actually, the computation is much simpler if one use the fact that in the Kerr metric there are two Killing vectors ξ and η, which, in Boyer–Lindquist coordinates, have components $\xi^\mu = (1,0,0,0)$ and $\eta^\mu = (0,0,0,1)$, reflecting the fact that the Kerr metric is stationary and axisymmetric. As a consequence, there are two first integrals of motion,

$$e \equiv -g_{\mu\nu}\xi^\mu u^\nu \qquad (12.294)$$

and

$$l \equiv g_{\mu\nu}\eta^\mu u^\nu, \qquad (12.295)$$

where $u^\mu = dx^\mu/d\tau$ is the four-velocity. Physically, e is the energy per unit mass and l is the angular momentum per unit mass. For a test mass falling from infinity and moving initially radially with zero kinetic energy, $e = 1$ and $l = 0$. Equations (12.294) and (12.295), together with the normalization condition for a time-like trajectory ($g_{\mu\nu}u^\mu u^\nu = -1$), and the condition that the motion is in an orbital plane, i.e. $u^\theta = 0$, give three equations that can be immediately solved for the three unknown $u^t = dt/d\tau$, $u^r = dr/d\tau$ and $u^\phi = d\phi/d\tau$. See Hartle (2003), Section 15.4.

This is very different from what happens in the Schwarzschild metric. Of course, a trajectory with r fixed is not a geodesic of the Schwarzschild metric. However a rocket, with sufficient power, would be able to keep an observer at fixed r, resisting the gravitational attraction of the Schwarzschild BH. In the Kerr geometry, in contrast, inside the ergosphere no amount of rocket power can keep an observer at a fixed position. It is, however, possible to keep it at fixed θ and r, at the price of rotating in the ϕ direction. In fact, consider a four-vector of the form

$$u^\mu = u^t(1,0,0,\Omega)\,. \tag{12.299}$$

Observe that

$$\Omega = \frac{u^\phi}{u^t} = \frac{d\phi/d\tau}{dt/d\tau}$$
$$= \frac{d\phi}{dt}\,, \tag{12.300}$$

so Ω is the angular velocity with respect to an asymptotic observer, since the proper time of a faraway observer coincide with the coordinate time t. Computing the norm of u^μ in the Kerr metric, in the equatorial plane $\theta = \pi/2$, and requiring that $u^\mu u_\mu < 0$, we get the condition

$$[r(r^2 + a^2) + a^2 R_S]\Omega^2 - 2acR_S\Omega + (R_S - r)c^2 < 0\,. \tag{12.301}$$

In the plane $\theta = \pi/2$ the ergosphere starts at $r = R_S$ [see eq. (12.297)], so the coefficient $(R_S - r)$ is positive on the portion of the equatorial plane that lies inside the ergosphere. Clearly, the second-order polynomial in Ω on the left-hand side of eq. (12.301) can be negative only when it has two real zeros $\Omega_{\mathrm{min,max}}$, in which case it is negative for $\Omega_{\mathrm{min}} < \Omega < \Omega_{\mathrm{max}}$. The values of $\Omega_{\mathrm{min,max}}$ are

$$\frac{1}{c}\Omega_{\mathrm{min,max}} = \frac{aR_S \pm \sqrt{(aR_S)^2 - (R_S - r)[r(r^2 + a^2) + a^2 R_S]}}{r(r^2 + a^2) + a^2 R_S}\,. \tag{12.302}$$

When one enters the ergosphere on the equatorial plane one has $r = R_S$, so $\Omega_{\mathrm{min}} = 0$. This is the smallest value of r where an observer, on a rocket with arbitrarily large power, can stay at a fixed value of ϕ. As long as r becomes smaller than R_S, Ω_{min} becomes strictly positive, and the observer is necessarily dragged along by the rotation of the BH. At the same time, there is a maximum angular velocity admitted. When r approaches r_+, the range $[\Omega_{\mathrm{min}}, \Omega_{\mathrm{max}}]$ shrinks, until it becomes a point when $r = r_+$. At this point the quantity inside the square root vanishes (as it is easily shown using repeatedly the relation $r_+^2 + a^2 = R_S r_+$), and $\Omega_{\mathrm{min}} = \Omega_{\mathrm{max}} \equiv \Omega_H$, where

$$\Omega_H = \frac{ac}{R_S r_+}\,. \tag{12.303}$$

When we are exactly at $r = r_+$, $u^\mu u_\mu$ becomes zero and the trajectory whose four-velocity is u^μ becomes light-like. So Ω_H is the angular velocity, as seen by an observer at infinity, of the light rays that are trapped on the horizon and, in this sense, Ω_H is the angular velocity of the horizon itself.

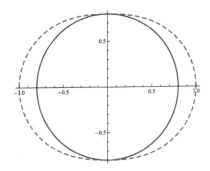

Fig. 12.17 The horizon of a Kerr BH, in Boyer–Lindquist coordinates and in units of R_S, for $a = 0.4R_S$ (solid line), and the static limit (dotter line). The region between these two surfaces is the ergosphere.

12.5.2 Null tetrads and the Newman–Penrose formalism

In order to determine the equations that govern the perturbations of the Kerr metric, it would seem natural to proceed in analogy to what we have done for Schwarzschild BHs. Namely, we could write the metric as the Kerr metric plus perturbation, try to find a clever gauge, and hope that the different metric components decouple, so that we would end up with a single master equation for a metric perturbation, analogous to the RW or Zerilli equation. Unfortunately, the Kerr metric is much more complicated than the Schwarzschild metric and, in this form, such a program has never been carried out. Furthermore, the fact that in the Kerr geometry we do not have spherical symmetry, but only axial symmetry, means that a separation of variables based on spherical harmonics is no longer possible, and in principle we should expect to end up with partial differential equations in both r and θ, rather than with an ordinary differential equation in r.

Luckily there exists an alternative route to BH perturbations, where, rather than considering perturbations of the metric, we directly consider perturbations of the curvature. This formalism makes use of the components of the Weyl tensor, projected along a null tetrad, a formalism first developed by Newman and Penrose. So, we next introduce the notion of null tetrad, we recall the definition and properties of the Weyl tensor and we then define the Newman–Penrose variables.

Null tetrads

Given a space-time with metric $g_{\mu\nu}$, a Newman–Penrose null tetrad is a set of four linearly independent four-vectors[56]

$$z_a^\mu \equiv (l^\mu, q^\mu, m^\mu, \bar{m}^\mu)\,, \qquad (12.304)$$

where the index a labels the four-vector. The four-vectors l^μ and q^μ are real, while m^μ is complex, and $\bar{m}^\mu$ is the complex conjugate of m^μ. These four-vectors are chosen to be null with respect to the metric $g_{\mu\nu}$,

$$g_{\mu\nu}l^\mu l^\nu = g_{\mu\nu}q^\mu q^\nu = g_{\mu\nu}m^\mu m^\nu = g_{\mu\nu}\bar{m}^\mu \bar{m}^\nu = 0\,. \qquad (12.305)$$

Furthermore,

$$g_{\mu\nu}m^\mu \bar{m}^\nu = 1\,, \qquad g_{\mu\nu}l^\mu q^\nu = -1\,, \qquad (12.306)$$

while all other scalar products are zero. From these relations, it follows that the metric can be written as

$$g^{\mu\nu} = m^\mu \bar{m}^\nu + m^\nu \bar{m}^\mu - l^\mu q^\nu - l^\nu q^\mu\,. \qquad (12.307)$$

In fact, since the four four-vectors z_a^μ are linearly independent, they form a basis, and we can prove the validity of eq. (12.307) by projecting it onto the basis z_a^μ. For instance, multiplying both sides of eq. (12.307) by l_μ and using eqs. (12.305) and (12.306) we get the identity $g^{\mu\nu}l_\nu = l^\mu$, and similarly for all other projected equations.

[56]In the original Newman–Penrose paper, and in most of the subsequent literature on QNMs, the four-vector that we denote by q^μ is denoted n^μ. In Section 14.3 we will, however, need both the null tetrad, and the unit normal to a space-like hypersurface, and we reserve the symbol n^μ for the latter.

Any four-vector or tensor can be projected onto the null tetrad. For instance, the Riemann tensor in this basis has components

$$R_{abcd} = z_a^\mu z_b^\nu z_c^\rho z_d^\sigma R_{\mu\nu\rho\sigma} \,. \tag{12.308}$$

To get the simplest example of a null tetrad, consider first flat space. Using Cartesian coordinates (t, x^i), we can define

$$l^\mu = (1, 0, 0, 1) \,, \tag{12.309}$$

$$q^\mu = \frac{1}{2}(1, 0, 0, -1) \,, \tag{12.310}$$

$$m^\mu = \frac{1}{\sqrt{2}}(0, 1, i, 0) \,, \tag{12.311}$$

$$\bar{m}^\mu = \frac{1}{\sqrt{2}}(0, 1, -i, 0) \,. \tag{12.312}$$

It can immediately be checked that these satisfy the orthogonality relations (12.305) and (12.306), and that the flat-space metric $\eta_{\mu\nu}$ can be expressed in terms of the null tetrad as

$$\eta^{\mu\nu} = m^\mu \bar{m}^\nu + m^\nu \bar{m}^\mu - l^\mu q^\nu - l^\nu q^\mu \,. \tag{12.313}$$

In the Kerr metric a possible choice of null tetrad is given, in Boyer–Lindquist coordinates, by

$$l^\mu = \frac{1}{\Delta}(r^2 + a^2, \Delta, 0, a) \,, \tag{12.314}$$

$$q^\mu = \frac{1}{2\rho^2}(r^2 + a^2, -\Delta, 0, a) \,, \tag{12.315}$$

$$m^\mu = \frac{1}{\sqrt{2}} \frac{1}{r + ia\cos\theta} \left(ia\sin\theta, 0, 1, \frac{i}{\sin\theta} \right) \,, \tag{12.316}$$

where the index μ runs over the values (t, r, θ, ϕ). This is known as the Kinnersley tetrad. Again, these four-vectors satisfy the orthogonality conditions (12.305) and (12.306), and the Kerr metric can be written in terms of them using eq. (12.307).

The basic quantities of the Newman–Penrose formalism are some projections of the Weyl tensor onto the null tetrad. Let us therefore first recall the definition and the main properties of the Weyl tensor.

The Weyl tensor

In a four-dimensional space-time endowed with a metric, the Weyl tensor $C_{\mu\nu\rho\sigma}$ is defined as

$$C_{\mu\nu\rho\sigma} = R_{\mu\nu\rho\sigma} - \frac{1}{2}(g_{\mu\rho}R_{\nu\sigma} - g_{\mu\sigma}R_{\nu\rho} - g_{\nu\rho}R_{\mu\sigma} + g_{\nu\sigma}R_{\mu\rho})$$
$$+ \frac{1}{6}R(g_{\mu\rho}g_{\nu\sigma} - g_{\mu\sigma}g_{\nu\rho}) \,. \tag{12.317}$$

Observe that $C_{\mu\nu\rho\sigma}$ vanishes upon contraction of any pair of its indices, so it is the trace-free part of the Riemann tensor. The Weyl tensor has the same symmetries as the Riemann tensor, i.e. antisymmetry with respect to the exchange of the first or of the second pair of indices,

$$C_{\mu\nu\rho\sigma} = -C_{\nu\mu\rho\sigma}\,, \tag{12.318}$$

$$C_{\mu\nu\rho\sigma} = -C_{\mu\nu\sigma\rho}\,, \tag{12.319}$$

symmetry with respect to the exchange of the first pair with the second,

$$C_{\mu\nu\rho\sigma} = +C_{\rho\sigma\mu\nu}\,, \tag{12.320}$$

and cyclicity with respect to the last three indices,

$$C_{\mu\nu\rho\sigma} + C_{\mu\sigma\nu\rho} + C_{\mu\rho\sigma\nu} = 0\,. \tag{12.321}$$

Just as for the Riemann tensor, these symmetries bring down the number of independent components to 20 (in four space-time dimensions). Furthermore, using the condition that all its traces are zero, one finds that in four dimensions the Weyl tensor has 10 independent components.

A conformal transformation of the metric (or, more precisely, a Weyl transformation) is a transformation in which we do not change the co-ordinates, while the metric is transformed as

$$g_{\mu\nu}(x) \to g'_{\mu\nu}(x) = e^{2\phi(x)} g_{\mu\nu}(x)\,, \tag{12.322}$$

with $\phi(x)$ an arbitrary function. Under this transformation,

$$C_{\mu\nu\rho\sigma}(x) \to C'_{\mu\nu\rho\sigma}(x) = e^{2\phi(x)} C_{\mu\nu\rho\sigma}(x)\,. \tag{12.323}$$

For this reason, the Weyl tensor it is also called the *conformal tensor*. From this result it is apparent that, if in a region of space-time there exists a coordinate system such that the metric can be written in the form

$$g_{\mu\nu}(x) = \exp\{2\phi(x)\}\eta_{\mu\nu}\,, \tag{12.324}$$

then the Weyl tensor vanishes, since the Weyl tensor computed with the metric $g_{\mu\nu}$ must be equal to $\exp\{2\phi(x)\}$ times the Weyl tensor computed with $\eta_{\mu\nu}$, and the latter is obviously zero.

The converse is also true: in a (simply connected) region of space-time it is possible to find a coordinate system such that the metric can be written in the form (12.324) if and only if the Weyl tensor vanishes.

The Newman–Penrose quantities

Projecting the Weyl tensor onto a null tetrad we can define the following Newman–Penrose quantities:[57]

$$\Psi_0 = C_{\mu\nu\rho\sigma} l^\mu m^\nu l^\rho m^\sigma\,, \tag{12.325}$$

$$\Psi_1 = C_{\mu\nu\rho\sigma} l^\mu q^\nu l^\rho m^\sigma\,, \tag{12.326}$$

$$\Psi_2 = \frac{1}{2} C_{\mu\nu\rho\sigma} l^\mu q^\nu (l^\rho q^\sigma + m^\rho \bar{m}^\sigma)\,, \tag{12.327}$$

$$\Psi_3 = C_{\mu\nu\rho\sigma} q^\mu l^\nu q^\rho \bar{m}^\sigma\,, \tag{12.328}$$

$$\Psi_4 = C_{\mu\nu\rho\sigma} q^\mu \bar{m}^\nu q^\rho \bar{m}^\sigma\,. \tag{12.329}$$

[57]In the original paper (Newman and Penrose, 1962), the signs in the definition of $\Psi_0, \ldots, \Psi_4$ are apparently the opposite to ours, e.g. $\Psi_0 = -C_{\mu\nu\rho\sigma} l^\mu m^\nu l^\rho m^\sigma$; see their eq. (4.3a). However, Newman and Penrose use a metric signature $(+, -, -, -)$, while we use $(-, +, +, +)$. Under a change in the sign of the metric, the Riemann tensor with all lower indices, $R_{\mu\nu\rho\sigma}$, changes sign, and so does the Weyl tensor. Thus our definition of $\Psi_0, \ldots, \Psi_4$ is finally the same as that of Newman and Penrose. In contrast, the Ricci tensor $R_{\mu\nu}$ is invariant if we flip the sign of the metric, which is why in eqs. (12.336)–(12.341) we have the same minus sign as in Newman and Penrose.

These are also known as the Weyl scalars. To understand the physical meaning of these quantities, consider a GW on a flat space-time, far from the source, propagating along the z axis. In this case we know from eq. (1.94) that

$$R_{0i0j} = -\frac{1}{2c^2} \ddot{h}_{ij}^{\mathrm{TT}}, \qquad (12.330)$$

where $h_{11} = -h_{22} = h_+$ and $h_{12} = h_{21} = h_\times$, while the other components of the Riemann tensor vanish. In vacuum $R_{\mu\nu} = 0$, so the Weyl tensor becomes the same as the Riemann tensor. Then, using the null tetrad (12.309)–(12.312), we find

$$\Psi_0 = -\frac{1}{2c^2}\left(\ddot{h}_+ + i\ddot{h}_\times\right), \qquad (12.331)$$

$$\Psi_4 = -\frac{1}{8c^2}\left(\ddot{h}_+ - i\ddot{h}_\times\right), \qquad (12.332)$$

[58] The difference in the numerical factor between Ψ_0 and Ψ_4 stems from the different normalizations of l^μ and q^μ. Observe that we could as well define these vectors by

$$l^\mu = \frac{1}{\sqrt{2}}(1,0,0,1) \qquad (12.333)$$

and

$$q^\mu = \frac{1}{\sqrt{2}}(1,0,0,-1) \qquad (12.334)$$

without spoiling the normalization condition $g_{\mu\nu}l^\mu q^\nu = -1$ or the other orthonormality conditions. With this more symmetric definition, eqs. (12.331) and (12.332) are replaced by

$$\Psi_4 = -\frac{1}{4c^2}\left(\ddot{h}_+ - i\ddot{h}_\times\right) \qquad (12.335)$$

and $\Psi_0 = \Psi_4^*$. Note also that a different tetrad, typically used in numerical relativity at asymptotically far distances from the source, will be introduced in eqs. (13.78)–(13.80). In that case, the proportionality constant in the relation between Ψ_4 and $\ddot{h}_+ - i\ddot{h}_\times$ becomes $+1$; see eq. (13.81).

while $\Psi_1 = \Psi_2 = \Psi_3 = 0$.[58] Similarly, from the Ricci tensor $R_{\mu\nu}$, one defines

$$\Phi_{00} = -\frac{1}{2}R_{\mu\nu}l^\mu l^\nu, \qquad (12.336)$$

$$\Phi_{01} = -\frac{1}{2}R_{\mu\nu}l^\mu m^\nu, \qquad (12.337)$$

$$\Phi_{02} = -\frac{1}{2}R_{\mu\nu}m^\mu m^\nu, \qquad (12.338)$$

$$\Phi_{11} = -\frac{1}{4}R_{\mu\nu}(l^\mu q^\nu + m^\mu \bar{m}^\nu), \qquad (12.339)$$

$$\Phi_{12} = -\frac{1}{2}R_{\mu\nu}q^\mu m^\nu, \qquad (12.340)$$

$$\Phi_{22} = -\frac{1}{2}R_{\mu\nu}q^\mu q^\nu. \qquad (12.341)$$

Observe that because of the orthonormality conditions (12.305) and (12.306), in the definitions (12.336)–(12.341) we could replace the Ricci tensor with its traceless part

$$Q_{\mu\nu} = R_{\mu\nu} - \frac{1}{4}g_{\mu\nu}R. \qquad (12.342)$$

Since, in vacuum, the Ricci tensor vanishes, all the Φ_{ab} are zero for a GW at large distance from the source. So, in GR the only two radiative degrees of freedom are described by the complex quantity Ψ_4 (or equivalently by Ψ_0, which, in the radiation zone, is proportional to Ψ_4^*).

Recall, from Problem 2.1 in Vol. 1, that a quantity has helicity s if, under rotation by an angle ψ around the propagation direction, it takes a multiplicative factor $\exp\{is\psi\}$. From eqs. (12.332) and (2.194) we see that Ψ_4 has helicity $s = -2$ while Ψ_0, which in the radiation zone, apart from normalizations, is its complex conjugate, has $s = +2$.

In extensions of GR where one also has scalar or vector radiative degrees of freedom, some other Newman–Penrose quantities can be non-vanishing in the radiation zone. Considering the most general form of the Riemann tensor for a weak GW and using the Bianchi identities for the linearized Riemann tensor, one can show that there are six functionally independent real components of the Riemann tensor. These can be conveniently chosen as the two real quantities Ψ_2 and Φ_{22}, and the complex quantities Ψ_4 and Ψ_3.[59] The real quantities Ψ_2 and Φ_{22} have zero helicity and describe radiative degrees of freedom that can appear in scalar–tensor theories of gravitation (for example in Brans–Dicke theory $\Phi_{22} \neq 0$ and $\Psi_2 = 0$), while Ψ_3 is complex and has $s = 1$, so its complex conjugate has $s = -1$. It describes the two degrees of freedom of a massless vector radiative degrees of freedom that can appear in extensions of GR. Finally, as we have seen, Ψ_4 and its complex conjugate describe the standard $s = \pm 2$ radiative degree of freedom of GR.[60]

[59]See Eardley, Lee, Lightman, Wagoner and Will (1973).

[60]An alternative description of the degrees of freedom with $s = 0, 1$ is provided by the scalar and vector Bardeen's variables that we will introduce in Chapter 18.

12.5.3 Teukolsky equation and QNMs of rotating BHs

The Einstein equations can be written in terms of the Newman–Penrose quantities. The explicit expression looks very cumbersome.[61] However, perturbation theory in this formalism is quite similar for the Kerr and the Schwarzschild metrics. To do perturbation theory one first specifies the background geometry by choosing a tetrad l_A^μ, q_A^μ, etc., where, following the notation of Teukolsky (1973), the subscript A denotes the background value. Perturbing the geometry means perturbing the tetrad to $l_A^\mu + l_B^\mu, q_A^\mu + q_B^\mu$, etc. where the subscript B denotes the perturbed quantities. All the Newman–Penrose quantities can then be written as $\Psi_a = \Psi_a^A + \Psi_a^B$ and $\Phi_{ab} = \Phi_{ab}^A + \Phi_{ab}^B$.

[61]See the original paper by Newman and Penrose (1962), or eqs. (2.2)–(2.8) of Teukolsky (1973).

As unperturbed tetrad one can chose the Kinnersley tetrad (12.316). This is such that the

$$\Psi_0^A = \Psi_1^A = \Psi_3^A = \Psi_4^A = 0 \,, \tag{12.343}$$

This means that these Ψ are invariant under linearized coordinate transformations. In fact, since the Ψ are scalars, under the diffeomorfism $x^\mu \to x'^\mu = x^\mu + \xi^\mu$ they transform as $\Psi(x) \to \Psi'(x')$ with

$$\Psi'(x') = \Psi(x) \,, \tag{12.344}$$

i.e., to linear order in ξ,

$$\Psi'(x) = \Psi(x) - \xi^\mu \partial_\mu \Psi \,. \tag{12.345}$$

Writing $\Psi = \Psi^A + \Psi^B$ and keeping only linear terms we get

$$\Psi'_B(x) = \Psi_B(x) - \xi^\mu \partial_\mu \Psi_A \tag{12.346}$$

and, if $\Psi^A = 0$, the perturbed quantity $\Psi_B(x)$ is invariant under gauge transformations, at linearized level.

Using this formalism, Teukolsky found that the perturbation equations for Ψ_0 and Ψ_4 decouple, and can be written in a unified form

$$\left[\frac{(r^2 + a^2)^2}{\Delta} - a^2 \sin^2 \theta \right] \partial_0^2 \psi + \frac{2aR_S r}{\Delta} \partial_0 \partial_\phi \psi + \left[\frac{a^2}{\Delta} - \frac{1}{\sin^2 \theta} \right] \partial_\phi^2 \psi$$

$$-\Delta^{-s}\partial_r\left(\Delta^{s+1}\partial_r\psi\right) - \frac{1}{\sin\theta}\partial_\theta\left(\sin\theta\partial_\theta\psi\right)$$

$$-2s\left[\frac{a(2r-R_S)}{2\Delta} + i\frac{\cos\theta}{\sin^2\theta}\right]\partial_\phi\psi$$

$$-2s\left[\frac{R_S(r^2-a^2)}{2\Delta} - r - ia\cos\theta\right]\partial_0\psi$$

$$+(s^2\mathrm{ctg}^2\theta - s)\psi = 4\pi G\rho^2 T\,, \tag{12.347}$$

[62]In the limit of a Schwarzschild BH, this reduces to a perturbation equation in terms of Newman–Penrose quantities that was first found by Bardeen and Press (1973).

where, as usual, $R_S = 2GM/c^2$. This is the *Teukolsky equation*.[62] The parameter s, called the spin weight, unifies different equations: for $s = +2$, the variable ψ in eq. (12.347) is equal to Ψ_0 (recall that since the background value of Ψ_0 vanishes, the perturbation of Ψ_0, denoted Ψ_0^B above, is the same as Ψ_0, to linear order), while setting $s = -2$ gives the equation for

$$\psi = (r - ia\cos\theta)^4\,\Psi_4\,. \tag{12.348}$$

The definition of the quantity denoted by T on the right-hand side of eq. (12.347) depends on s, too. To study the QNMs of Kerr BHs it is sufficient to consider the vacuum equations, in which case $T = 0$ on the right-hand side of eq. (12.347).[63]

[63]The explicit expression of T for $s = 2, -2$, as well as for vector and spinor perturbations is rather long, and can be found in Teukolsky (1973), Table I.

Quite remarkably, eq. (12.347) turns out to hold even for electromagnetic perturbations of Kerr BHs, as well as for massless spin-1/2 perturbations. Electromagnetic perturbations are described by eq. (12.347) with $s = \pm 1$. When $s = +1$, ψ is identified with the Newman–Penrose quantity

$$\phi_2 \equiv F_{\mu\nu}\bar{m}^\mu q^\nu\,, \tag{12.349}$$

while for $s = -1$, $\psi = (r - ia\cos\theta)^2\phi_2$. Massless spin-1/2 perturbations are described by eq. (12.347) with $s = \pm 1/2$.

In eq. (12.347), a separation of variables in terms of spherical harmonics is not possible, since spherical symmetry is replaced by axial symmetry. However, Teukolsky found that the equation can still be separated, writing

$$\psi^{(s)}(t,r,\theta,\phi) = \int_{-\infty}^{\infty} d\omega \sum_{l,m} R_{lm}^{(s)}(r,\omega)\,S_{lm}^{(s)}(\theta,\omega)e^{im\phi}e^{-i\omega t}\,, \tag{12.350}$$

[64]In the literature it is common to use the (rather unwieldy) notation $_s\psi$, $_sR_{lm}$ and $_sS_{lm}$ instead of $\psi^{(s)}$, $R_{lm}^{(s)}$ and $S_{lm}^{(s)}$. Similarly, the spin-weighted spherical harmonics $Y_{lm}^{(s)}(\theta,\phi)$ that we introduce in eq. (12.352) are usually denoted as $_sY_{lm}(\theta,\phi)$.

where we have added a superscript (s) to ψ to remind us that its definition depends on the spin weight s.[64] The angular problem reduces to solving the equation

$$\frac{1}{\sin\theta}\partial_\theta\left(\sin\theta\,\partial_\theta S_{lm}^{(s)}\right) + \left[\frac{a^2\omega^2}{c^2}\cos^2\theta - \frac{m^2}{\sin^2\theta} - \frac{2sa\omega}{c}\cos\theta\right.$$

$$\left. -2sm\frac{\cos\theta}{\sin^2\theta} - (s^2\mathrm{ctg}^2\theta - s) + A_{lm}^{(s)}(\omega)\right]S_{lm}^{(s)} = 0\,, \tag{12.351}$$

where $A_{lm}^{(s)}(\omega)$ is the separation constant. The functions $S_{lm}^{(s)}(\theta,\omega)$, defined by this equation together with the condition that they must be regular on the interval $\theta \in [0,\pi]$, are called the *spin-weighted spheroidal harmonics.* For $s = 0$, they reduce to the so-called spheroidal functions $S_{lm}(-a^2\omega^2/c^2, \cos\theta)$, while for $\omega = 0$ the functions

$$Y_{lm}^{(s)}(\theta, \phi) = S_{lm}^{(s)}(\theta)e^{im\phi} \tag{12.352}$$

are known as spin-weighted spherical harmonics.[65] The function $S_{lm}^{(s)}(\theta) \equiv S_{lm}^{(s)}(\theta, \omega = 0)$ is given by

$$S_{lm}^{(s)}(\theta) = (-1)^s \sqrt{\frac{2l+1}{4\pi}}\, d_{m,-s}^l(\theta)\,, \tag{12.354}$$

where $d_{m,m'}^j(\theta)$ is the Wigner (small) d-matrix,

$$d_{m,m'}^j(\theta) = \sqrt{(j+m)!(j-m)!(j+m')!(j-m')!}$$
$$\times \sum_{k=k_1}^{k_2} \frac{(-1)^k}{(j+m-k)!k!(m'-m+k)!(j-m'-k)!}$$
$$\times [\cos(\theta/2)]^{2j+m-m'-2k}\, [\sin(\theta/2)]^{2k+m'-m}\,, \tag{12.355}$$

and the sum runs over all values of k for which the arguments of the factorials are non-negative, i.e. $k_1 = \max(0, m - m')$ and $k_2 = \min(j + m, j - m')$.

For any given ω, the quantities $A_{lm}^{(s)}(\omega)$ can in principle be computed numerically as the eigenvalues of eq. (12.351). Observe also that, since $A_{lm}^{(s)}(\omega)$ depends on ω, the separation of the r and θ variables can be performed only in the frequency domain, and not in the time domain.

Inserting eq. (12.350) into eq. (12.347) we finally end up with an equation for the radial function $R_{lm}^{(s)}(r, \omega)$,

$$\Delta^{-s}\partial_r \left(\Delta^{s+1}\partial_r\right) R_{lm}^{(s)}(r, \omega) \tag{12.356}$$
$$+ \left(\frac{K^2(\omega) - is(2r - R_S)K(\omega)}{\Delta} + \frac{4is\omega r}{c} - \lambda_{lm}^{(s)}(\omega)\right) R_{lm}^{(s)}(r, \omega) = 0\,,$$

where

$$K(\omega) = (r^2 + a^2)\frac{\omega}{c} - am\,, \tag{12.357}$$

and

$$\lambda_{lm}^{(s)}(\omega) = A_{lm}^{(s)}(\omega) + \frac{a^2\omega^2}{c^2} - 2m\frac{a\omega}{c}\,. \tag{12.358}$$

Using this radial equation, the QNMs of Kerr BHs can in principle be computed in analogy to what we have done for the RW and Zerilli equations. However, the actual computation is more involved, due to the higher complexity of the equation. For each (l, m) there is an infinity of normal modes parametrized by $n = 1, 2, \ldots$, so we write the QNM frequencies as ω_{nlm}.[66]

The effect of rotation is to remove the degeneracy of the frequency with respect to the "quantum" number m. For GW emission the most relevant mode is the mode with $n = 1, l = m = 2$. Since the Kerr

---

[65] The spheroidal functions $S(\gamma^2, z)$ are solutions of the differential equation

$$\frac{d}{dz}\left[(1 - z^2)\frac{d}{dz}\right]S \tag{12.353}$$
$$+ \left[\lambda + \gamma^2(1 - z^2) - \frac{m^2}{1 - z^2}\right]S = 0\,.$$

For $s = 0$, eq. (12.351) reduces to this equation, with $z = \cos\theta$, $\gamma^2 = -a^2\omega^2/c^2$ and $\lambda = A_{lm} + a^2\omega^2/c^2$.

[66] More precisely, for each nlm there is a pair of complex solutions ω_{nlm} and ω'_{nlm} related by $\omega_{nlm} = -(\omega'_{nl-m})^*$, i.e. $(\omega_R)_{nlm} = -(\omega'_R)_{nl-m}$ and $(\omega_I)_{nlm} = +(\omega'_I)_{nl-m}$, while at the same time $(A_{lm}^{(s)})' = (A_{l,-m}^{(s)})^*$.

parameter a has dimensions of length, it is convenient to introduce the dimensionless Kerr parameter

$$\hat{a} \equiv \frac{a}{GM/c^2}$$

$$= \frac{cJ}{GM^2} \, .$$

(12.359)

The dependence of $(\omega_R)_{n=1,l=2,m=2}$ on $\hat{a}$ can be fitted to better than 5% over a range $\hat{a} \in [0, 0.99]$ by the formula

$$\frac{GM}{c^3}(\omega_R)_{n=1,l=2,m=2} \simeq 1.5251 - 1.1568(1 - \hat{a})^{0.1292} \, ,$$

(12.360)

while for the imaginary part, defining the quality factor

$$Q_{nlm} \equiv \frac{(\omega_R)_{nlm}}{2|(\omega_I)_{nlm}|} \, ,$$

(12.361)

one finds

$$Q_{n=1,l=2,m=2} \simeq 0.700 + 1.4187(1 - \hat{a})^{0.4990} \, .$$

(12.362)

Similarly, for the mode $n = 1, l = 2, m = 0$ an accurate fitting formula is[67]

$$\frac{GM}{c^3}(\omega_R)_{n=1,l=2,m=0} \simeq 0.4437 - 0.0739(1 - \hat{a})^{0.3350} \, ,$$

(12.363)

$$Q_{n=1,l=2,m=0} \simeq 4.000 - 1.9550(1 - \hat{a})^{0.1420} \, .$$

(12.364)

12.6   Solved problems

Problem 12.1. Derivation of the Zerilli equation

In this problem we give the explicit derivation of the Zerilli equation. For polar perturbations, in principle, there are seven equations. However, when expanded in tensor harmonics, four of them involve polar perturbations (and coefficients s_{lm}^a of the energy–momentum tensor) that start from $l = 0$, while two equations involve quantities with $l \geqslant 1$ and one involve quantities whose expansion starts from $l = 2$. To simplify the notation we momentarily set $G = c = 1$ and we drop all indices (l, m). Then, the four equations that hold for $l \geqslant 0$ are

$$A^2 \partial_r^2 K + A \left(3 - \frac{5M}{r} \right) \frac{1}{r} \partial_r K - A^2 \frac{1}{r} \partial_r H^{(2)} + \frac{A}{r^2}(K - H^{(2)})$$

$$- A \frac{l(l+1)}{2r^2}(K + H^{(2)}) = -8\pi s^{tt} \, ,$$

(12.365)

$$\partial_t \left(\partial_r K + \frac{1}{r}(K - H^{(2)}) - \frac{M}{r^2 A} K \right) - \frac{l(l+1)}{2r^2} H^{(1)} = -8\pi s^{Rt} \, ,$$

(12.366)

[67] See Berti, Cardoso and Will (2006b). These fitting formulas improve over the previous fitting formulas

$$(GM/c^3)(\omega_R)_{122} \simeq f(\hat{a}) \, ,$$

$$\tau_{122} \simeq \frac{4GM/c^3}{(1 - \hat{a})^{9/10}} f(\hat{a}) \, ,$$

where

$$f(\hat{a}) = 1 - 0.63(1 - \hat{a})^{3/10} \, ,$$

which are correct to about 10%, and were proposed in Echeverria (1989) and Fryer, Holz and Hughes (2002).

$$\frac{1}{A^2}\partial_t^2 K - \frac{1-M/r}{rA}\partial_r K - \frac{2}{rA}\partial_t H^{(1)} + \frac{1}{r}\partial_r H^{(0)} - \frac{1}{r^2 A}(K - H^{(2)})$$
$$+\frac{l(l+1)}{2r^2 A}(K - H^{(0)}) = -8\pi s^{L0}, \tag{12.367}$$

and

$$-\frac{1}{A}\partial_t^2(K + H^{(2)}) + A\partial_r^2(K - H^{(0)}) + \left(1 - \frac{M}{r}\right)\frac{2}{r}\partial_r K$$
$$+2\partial_r\partial_t H^{(1)} + \frac{2}{rA}\left(1 - \frac{M}{r}\right)\partial_t H^{(1)} - \frac{1}{r}\left(1 - \frac{M}{r}\right)\partial_r H^{(2)}$$
$$-\frac{1}{r}\left(1 + \frac{M}{r}\right)\partial_r H^{(0)} - \frac{l(l+1)}{2r^2}(H^{(2)} - H^{(0)}) = \frac{16\pi s^{T0}}{r^2}. \tag{12.368}$$

(Recall however that $H_{lm}^{(1)}$ vanishes for $l = 0$, and K_{lm} vanishes for $l = 0, 1$.) The two equations valid only for $l \geqslant 1$ are

$$\partial_r(AH^{(1)}) - \partial_t(K + H^{(2)}) = 16\pi s^{Et}, \tag{12.369}$$

$$\frac{1}{A}\partial_t H^{(1)} + \partial_r(K - H^{(0)}) - \frac{2M}{Ar^2}H^{(0)} - \frac{1-M/r}{Ar}(H^{(2)} - H^{(0)}) = -16\pi s^{E1}, \tag{12.370}$$

while the one valid only for $l \geqslant 2$ is

$$H^{(0)} - H^{(2)} = 32\pi s^{E2}. \tag{12.371}$$

We consider the case $l \geqslant 2$, which corresponds to the radiative degrees of freedom.[68] For each (l, m) with $l \geqslant 2$ we have seven equations and only four unknown functions $H^{(0)}$, $H^{(1)}$, $H^{(2)}$ and K. However, this is due to the fact that gauge invariance allowed us to reduce the polar perturbations from seven to four functions so, just as in the axial case, gauge invariance must also imply that the equations are not independent and a solution exists.

Equation (12.371) allows us to eliminate $H^{(0)}$ immediately, so we remain with six equations and three functions. At first sight an algebraic relation such as (12.371) might be surprising. In fact, let t_0 be the initial time at which we assign the initial conditions for our system of differential equations. At $t = t_0$ we can assign the value of the energy–momentum tensor, and therefore of s^{E2}, and we might expect that we are also entitled to choose the initial conditions on $H^{(2)}$ and $H^{(0)}$ ourselves. Instead, eq. (12.371) states that, given the external perturbation, the initial condition on $H^{(0)}$ fixes that on $H^{(2)}$. This result is in fact a consequence of the general structure of the Einstein equations. As it is well known, and as we will review in Section 13.1 when we discuss the ADM formalism, out of the 10 components of the Einstein equations, some are dynamical equations, while others are constraints on the initial data. Equation (12.371) indeed implies a constraints on the initial data. For instance, if at the initial time s^{E2} vanishes, we must have $H^{(0)} = H^{(2)}$ at the initial time. If a non-vanishing $s^{E2}(t, r)$ is specified, $H^{(0)}(t, r)$ is fixed at all subsequent times in terms of $H^{(2)}(t, r)$. In other words, eq. (12.371) is a constraint on the initial data, which is preserved by the equations of motion.

To deal with the remaining six equations it is convenient to perform first the Fourier transfom with respect to the time variable, as in eq. (12.101). We denote by $\tilde{K}(\omega, r), \dots, \tilde{H}^{(2)}(\omega, r)$ the Fourier transforms of $K(t, r), \dots, H^{(2)}(t, r)$, respectively, and by $\tilde{s}^a(\omega, r)$ the Fourier transforms of $s^a(t, r)$. Then, in eq. (12.366) the time derivative just brings a factor $(-i\omega)$, so we get

$$\partial_r\tilde{K} + \frac{1-3M/r}{rA}\tilde{K} - \frac{1}{r}\tilde{H}^{(2)} + \frac{l(l+1)}{2i\omega r^2}\tilde{H}^{(1)} = \frac{8\pi}{i\omega}\tilde{s}^{Rt}. \tag{12.372}$$

[68] The equations with $l = 0$ and $l = 1$ describe the evolution of the non-radiative degrees of freedom, i.e. of the mass, momentum and angular momentum of the system. For instance, in the infall of a particle of mass m into a BH of mass M, the $l = 0$ perturbation describes the change in the Schwarzschild mass of the system from M to $M + m$. The $l = 1$ axial perturbation, for a test particle on a trajectory $R(t)$, vanishes for $r < R(t)$ while, for $r > R(t)$, it has an angular momentum that is equal to the angular momentum of the orbiting particle. The $l = 1$ polar perturbations can be removed by a gauge transformation, which can be interpreted as a translation of the center of mass of the system. See Appendix G of Zerilli (1970).

Equation (12.370) can be solved for $\partial_r \tilde{H}^{(2)}$ [after eliminating $\partial_r \tilde{K}$ with the use of eq. (12.372), and eliminating $H^{(0)}$ using eq. (12.371)], and gives

$$\partial_r \tilde{H}^{(2)} + \frac{1 - 3M/r}{rA} \tilde{K} - \frac{1 - 4M/r}{rA} \tilde{H}^{(2)} + \left[\frac{i\omega}{A} + \frac{l(l+1)}{2i\omega r^2} \right] \tilde{H}^{(1)}$$
$$= \frac{8\pi}{i\omega} \tilde{s}^{Rt} + 32\pi \frac{1 - 3M/r}{rA} \tilde{s}^{E2} - 32\pi \partial_r \tilde{s}^{E2} + 16\pi \tilde{s}^{E1} . \tag{12.373}$$

The derivative $\partial_r \tilde{H}^{(1)}$ can be obtained from eq. (12.369), and is given by

$$\partial_r \tilde{H}^{(1)} + \frac{i\omega}{A} (\tilde{K} + \tilde{H}^{(2)}) + \frac{2M}{r^2 A} \tilde{H}^{(1)} = \frac{16\pi}{A} \tilde{s}^{Et} . \tag{12.374}$$

It is clear that, once we have given the external source and assigned the initial conditions on $\tilde{K}, \tilde{H}^{(1)}$ and $\tilde{H}^{(2)}$, eqs. (12.372), (12.373) and (12.374) fix the solution uniquely. Thus, the remaining equations of motion must either be redundant or be constraints on the initial values. Indeed, using eqs. (12.372)–(12.374), as well as the covariant conservation of the energy–momentum tensor, it is straightforward to check that eqs. (12.365) and (12.368), which are the only equations of second order in ∂_r, are automatically satisfied. The remaining equation is (12.367), which is of first order in ∂_r. Performing again the Fourier transform, so that $\partial_t \to -i\omega$, and inserting $\partial_r \tilde{K}$ from eq. (12.372) and $\partial_r \tilde{H}^{(0)}$ from eqs. (12.371) and (12.373), we discover that eq. (12.367) gives an algebraic relation between $\tilde{K}, \tilde{H}^{(1)}, \tilde{H}^{(2)}$, and the external source,

$$\frac{\lambda r^2 A + Mr - 3M^2 - \omega^2 r^4}{rA} \tilde{K} + \frac{(\lambda + 1)M - \omega^2 r^3}{i\omega r} \tilde{H}^{(1)}$$
$$-(\lambda r + 3M) \tilde{H}^{(2)} \tag{12.375}$$
$$= \frac{8\pi Mr}{i\omega} \tilde{s}^{Rt} + (\lambda r + 3M) 32\pi \tilde{s}^{E2} - 8\pi r^2 A \left(r\tilde{s}^{L0} + 2\tilde{s}^{E1} \right) .$$

where we have defined
$$\lambda = \frac{(l-1)(l+2)}{2} \tag{12.376}$$

(so that $l(l+1)/2 = \lambda + 1$). Just like eq. (12.371), this is a constraint on the initial data. Equivalently, we can see it as a first integral of the equations of motion (12.372)–(12.374). The consistency with eqs. (12.372) and (12.374) can be checked using eq. (12.375) to eliminate for example $\tilde{H}^{(2)}$ from two of the first-order equations (12.372)–(12.374), and verifying that the third one is automatically satisfied (for the right-hand side, as usual, one has to use the covariant conservation of $T_{\mu\nu}$).

So, using eq. (12.375), we eliminate one of the functions $\tilde{K}, \tilde{H}^{(1)}$ or $\tilde{H}^{(2)}$ from eqs. (12.372)–(12.374), and we remain with two first-order equations for two functions. Eliminating for instance $\tilde{H}^{(2)}$, eqs. (12.372) and (12.374) become

$$\partial_r \tilde{K} + a_1(r)\tilde{K} + a_2(r)\tilde{H}^{(1)} = j_1 , \tag{12.377}$$
$$\partial_r \tilde{H}^{(1)} + b_1(r)\tilde{K} + b_2(r)\tilde{H}^{(1)} = j_2 , \tag{12.378}$$

where

$$a_1(r) = \frac{2Mr - \lambda Mr - 6M^2 + \omega^2 r^4}{r(r - 2M)(\lambda r + 3M)} , \tag{12.379}$$

$$a_2(r) = \frac{\lambda(\lambda + 1)r + 2(\lambda + 1)M + \omega^2 r^3}{i\omega r^2 (\lambda r + 3M)} , \tag{12.380}$$

$$b_1(r) = i\omega r \, \frac{2\lambda r^2 + 4Mr - 4\lambda Mr - 9M^2 - \omega^2 r^4}{(r-2M)^2(\lambda r + 3M)} \,, \qquad (12.381)$$

$$b_2(r) = \frac{3\lambda Mr + Mr + 6M^2 - \omega^2 r^4}{r(r-2M)(\lambda r + 3M)} \,. \qquad (12.382)$$

The source terms are

$$j_1 = \frac{1}{i\omega} \frac{\lambda r + 2M}{\lambda r + 3M} 8\pi \tilde{s}^{Rt} + \frac{rA}{\lambda r + 3M} 8\pi (r\tilde{s}^{L0} + 2\tilde{s}^{E1}) - \frac{32\pi}{r} \tilde{s}^{E2} \,, \qquad (12.383)$$

$$j_2 = \frac{16\pi}{A} \tilde{s}^{Et} + \frac{Mr}{A(\lambda r + 3M)} 8\pi \tilde{s}^{Rt} + i\omega \frac{32\pi}{A} \tilde{s}^{E2}$$

$$- i\omega \frac{r^2}{\lambda r + 3M} 8\pi (r\tilde{s}^{L0} + 2\tilde{s}^{E1}) \,. \qquad (12.384)$$

We now transform this system of two first-order equations into a single second-order equation. We could for instance solve eq. (12.377) for $\tilde{H}^{(1)}$ and plug it into eq. (12.378). However, to get a simpler result, it is convenient to introduce a new function $\tilde{Z}(\omega, r)$ [actually, $\tilde{Z}_{lm}(\omega, r)$, but recall that we are not writing explicitly the indices (l, m)] defined by

$$\tilde{Z} = \frac{r^2}{\lambda r + 3M} \tilde{K} + \frac{rA}{i\omega(\lambda r + 3M)} \tilde{H}^{(1)} \,. \qquad (12.385)$$

The reason for this specific choice of coefficients will become clear later. Taking the derivative with respect to r of this equation, and using eqs. (12.377) and (12.378), we can express $\partial_r \tilde{Z}$ in terms of $\tilde{K}$, $\tilde{H}^{(1)}$ and source terms. The result is

$$\partial_r \tilde{Z} = \frac{-\lambda r^2 + 3\lambda Mr + 3M^2}{A(\lambda r + 3M)^2} \tilde{K} - \frac{\lambda(\lambda+1)r^2 + 3\lambda Mr + 6M^2}{i\omega r(\lambda r + 3M)^2} \tilde{H}^{(1)} + J, \qquad (12.386)$$

where

$$J = \frac{r^2}{\lambda r + 3M} j_1 + \frac{r - 2M}{i\omega(\lambda r + 3M)} j_2$$

$$= \frac{r}{i\omega(\lambda r + 3M)} 8\pi (r\tilde{s}^{Rt} + 2\tilde{s}^{Et}) \,. \qquad (12.387)$$

We can invert eqs. (12.385) and (12.386), and express $\tilde{K}$ and $\tilde{H}^{(1)}$ in terms of $\tilde{Z}$ and of its derivative, and of the source term J. We get

$$\tilde{K} = \frac{\lambda(\lambda+1)r^2 + 3\lambda Mr + 6M^2}{r^2(\lambda r + 3M)} \tilde{Z} + A\partial_r \tilde{Z} - AJ \,, \qquad (12.388)$$

$$\tilde{H}^{(1)} = -i\omega \frac{\lambda r^2 - 3\lambda Mr - 3M^2}{rA(\lambda r + 3M)} \tilde{Z} - i\omega r \partial_r \tilde{Z} + i\omega r J \,. \qquad (12.389)$$

Thus, given the external source, $\tilde{K}$ and $\tilde{H}^{(1)}$ are determined, once $\tilde{Z}$ is known. The remaining polar perturbations $\tilde{H}^{(0)}$ and $\tilde{H}^{(2)}$ are then obtained using eqs. (12.371) and (12.375). To obtain a differential equation that involves only $\tilde{Z}$, we take one more derivative of eq. (12.386). We express the resulting derivatives $\partial_r \tilde{K}$ and $\partial_r \tilde{H}^{(1)}$ in terms of $\tilde{K}$, $\tilde{H}^{(1)}$ using eqs. (12.377) and (12.378), and then the resulting terms $\tilde{K}$ and $\tilde{H}^{(1)}$ are written in terms of $\tilde{Z}$ and $\partial_r \tilde{Z}$ using eqs. (12.388) and (12.389). The result is given in the main text, eqs. (12.109), (12.110) and (12.111).

Observe that in eq. (12.109) we have rewritten $\partial_r^2 \tilde{Z}$ in terms of $\partial_*^2 \tilde{Z}$ using

$$\frac{1}{A^2}\partial_*^2\tilde{Z} = \partial_r^2\tilde{Z} + \frac{2M}{Ar^2}\partial_r\tilde{Z} \,. \tag{12.390}$$

Combining the term proportional to $\partial_r\tilde{Z}$ that comes from eq. (12.390) with those from the replacement of $\tilde{K}$ and $\tilde{H}^{(1)}$ in terms of $\tilde{Z}$ and $\partial_r\tilde{Z}$, the overall coefficient of $\partial_r\tilde{Z}$ in eq. (12.109) cancels. Indeed, the choice of coefficients in the definition of $\tilde{Z}$, given in eq. (12.385), was made just in order to obtain this cancellation. If, instead, we just solved eq. (12.377) for $\tilde{H}^{(1)}$ and plugged it into eq. (12.378), we would get an equation for $\tilde{K}$ that involves not only $\partial_*^2\tilde{K}$ but also $\partial_*\tilde{K}$.

Problem 12.2. The source term for radial infall

In this problem we perform the explicit computation of the source term of the Zerilli equation, for the radial infall of a test particle. We consider first s_{lm}^{L0}. From eq. (12.41) we see that $[t_{lm}^{L0}]_{\mu\nu}$ only has the component $\mu = \nu = r$, which is equal to Y_{lm}, and for $a = (L0)$ we have $\epsilon_a = +1$ and $c^a(r) = 1$; see eq. (12.38). In polar coordinates, $\eta_{\mu\nu} = (-1, 1, r^2, r^2\sin^2\theta)$, so $\eta^{rr} = 1$, while $g_{rr} = 1/A(r) = (1 - R_S/r)^{-1}$ Thus, eq. (12.273) gives

$$s_{lm}^{L0} = m\gamma \frac{\delta[r - r_0(t)]}{r^2} Y_{lm}^*[\Omega_0(t)](\eta^{rr})^2 g_{rr}^2 \left(\frac{dr_0}{dt}\right)^2$$

$$= m\gamma \left(\frac{dr_0}{dt}\right)^2 \frac{\delta[r - r_0(t)]}{r^2 A^2(r)} Y_{lm}^*[\Omega_0(t)] \,. \tag{12.391}$$

Similarly, observing that $[t_{lm}^{tt}]_{\mu\nu}$ only has the component $\mu = \nu = 0$, we find

$$s_{lm}^{tt} = m\gamma \left(\frac{dr_0}{dt}\right)^2 A^2(r) \frac{\delta[r - r_0(t)]}{r^2} Y_{lm}^*[\Omega_0(t)] \,. \tag{12.392}$$

To compute s_{lm}^{Rt} we observe that $[t_{lm}^{Rt}]_{\mu\nu}$ only has the components $(0r)$ (together with the component $(r0)$, which gives an overall factor of 2) and, using $\epsilon_{Rt} = -1$ and $c_{Rt} = 1/\sqrt{2}$,

$$s_{lm}^{Rt} = -mc\gamma \frac{\delta[r - r_0(t)]}{r^2} Y_{lm}^*[\Omega_0(t)]\eta^{00}\eta^{rr} g_{00}g_{rr}\frac{dr_0}{dt}$$

$$= -mc\gamma \frac{dr_0}{dt}\frac{\delta[r - r_0(t)]}{r^2} Y_{lm}^*[\Omega_0(t)] \,, \tag{12.393}$$

and similarly one can compute s_{lm}^{T0}. All other components are proportional to time derivatives of $\theta_0(t)$ and $\phi_0(t)$, as is clear from the form of the tensors $\mathbf{t}_{lm}^a$ in eq. (12.41). For example,

$$s_{lm}^{Et} = -\frac{r^2}{l(l+1)} mc\gamma \frac{\delta[r - r_0(t)]}{r^2}$$

$$\times \left[(\partial_\theta Y_{lm}^*)\eta^{00}\eta^{\theta\theta} g_{00}g_{\theta\theta}\frac{d\theta_0}{dt} + (\partial_\phi Y_{lm}^*)\eta^{00}\eta^{\phi\phi} g_{00}g_{\phi\phi}\frac{d\phi_0}{dt} \right]$$

$$= -\frac{mc\gamma}{l(l+1)} A(r)\delta[r - r_0(t)] \left[(\partial_\theta Y_{lm}^*)\frac{d\theta_0}{dt} + (\partial_\phi Y_{lm}^*)\frac{d\phi_0}{dt} \right]$$

$$= -\frac{mc\gamma}{l(l+1)} A(r)\delta[r - r_0(t)]\frac{dY_{lm}^*[\Omega_0(t)]}{dt} \,. \tag{12.394}$$

We now consider a test particle on a radial geodesic of the Schwarzschild metric, moving along the z axis, which as $t \to -\infty$ comes from $z = +\infty$, with zero initial velocity. For such a trajectory $\theta_0(t) = 0$ is constant. Furthermore, since $\theta = 0$, $Y_{lm}(\theta, \phi)$ is zero unless the index $m = 0$, so it is independent of ϕ. Thus, only s_{lm}^{L0}, s_{lm}^{tt}, s_{lm}^{Rt} and s_{lm}^{T0} are non-vanishing. Since the source term in the RW equation depends only on s_{lm}^{B1} and s_{lm}^{B2} [see eq. (12.99)], for the radial infall of a test mass the source in the RW equation vanishes, so the axial modes are not excited. We are therefore concerned only with the polar modes, described by the Zerilli equation.

Out of the remaining four quantities s_{lm}^{L0}, s_{lm}^{tt}, s_{lm}^{Rt} and s_{lm}^{T0}, only the two Fourier transform $\tilde{s}_{lm}^{L0}(\omega, r)$ and $\tilde{s}_{lm}^{Rt}(\omega, r)$ enter in the source term of the Zerilli equation; see eq. (12.111). To compute these Fourier transforms, we need the explicit form of $r_0(t)$. Let us then recall some basic results on radial orbits in the Schwarzschild metric.[69] The fact that the metric is independent of t leads to a first integral,

[69]See e.g. Hartle (2003), Section 9.3.

$$A[r_0(t)] \frac{dt}{d\tau} = \frac{E}{m} . \tag{12.395}$$

For a particle starting from infinity with zero initial velocity, the energy per unit mass $E/m = 1$, so

$$A[r_0(t)] \frac{dt}{d\tau} = 1 \tag{12.396}$$

or, equivalently, $\gamma = 1/A$. In terms of proper time τ, the equation for a radial geodesic in the Schwarzschild metric has the same form as the Newtonian equation (4.291),

$$\frac{dr_0}{d\tau} = -c \left(\frac{R_S}{r_0(\tau)} \right)^{1/2} . \tag{12.397}$$

Combining this with eq. (12.396),

$$\frac{dr_0}{dt} = -cA[r_0(t)] \left(\frac{R_S}{r_0(t)} \right)^{1/2} . \tag{12.398}$$

This can be integrated, to give

$$t = T[r_0(t)] , \tag{12.399}$$

where

$$T(r_0) = \frac{R_S}{c} \left[-\frac{2}{3} \left(\frac{r_0}{R_S} \right)^{3/2} - 2 \left(\frac{r_0}{R_S} \right)^{1/2} + \log \frac{(r_0/R_S)^{1/2} + 1}{(r_0/R_S)^{1/2} - 1} \right] . \tag{12.400}$$

For large r_0, this reduces to the Newtonian trajectory (4.299) (with the change of notation $\tau = -T$ and $r_0 = z$), as it should. If $r_0 \to R_S$, instead, the logarithm in eq. (12.400) diverges, so it takes an infinite coordinate time t to reach the horizon.

We can now compute the Fourier transform. Recalling that

$$Y_{l0}(\theta = 0) = \left(\frac{2l + 1}{4\pi} \right)^{1/2} , \tag{12.401}$$

and denoting $\tilde{s}_{l,m=0}^a \equiv \tilde{s}_l^a$, we have

$$\tilde{s}_l^{L0}(\omega, r) = \int_{-\infty}^{\infty} dt \, s_l^{L0}(t, r) \, e^{i\omega t} \tag{12.402}$$

$$= \frac{m(2l+1)^{1/2}}{\sqrt{4\pi}} \frac{1}{r^2 A^2(r)} \int_{-\infty}^{\infty} dt \, \gamma(t) \left(\frac{dr_0}{dt} \right)^2 \delta[r - r_0(t)] e^{i\omega t} .$$

Since $r_0(t)$ is a monotonic function of t, we can change integration variable from dt to dr_0. We also write $\gamma = dt/d\tau$, and we use eq. (12.397), obtaining

$$\tilde{s}_l^{L0}(\omega, r) = \frac{m(2l+1)^{1/2}}{\sqrt{4\pi}} \frac{1}{r^2 A^2(r)} \int_{R_S}^{\infty} dr_0 \left| \frac{dr_0}{d\tau} \right| \delta(r - r_0) e^{i\omega T(r_0)}$$

$$= \frac{mc(2l+1)^{1/2}}{\sqrt{4\pi}} \frac{1}{r^2 A^2(r)} \left(\frac{R_S}{r} \right)^{1/2} e^{i\omega T(r)}, \qquad (12.403)$$

valid for $r > R_S$. Similarly, using $\gamma(t) = 1/A(r_0)$, eq. (12.393) gives

$$\tilde{s}_l^{Rt}(\omega, r) = \frac{mc(2l+1)^{1/2}}{\sqrt{4\pi}} \frac{1}{r^2 A(r)} e^{i\omega T(r)}. \qquad (12.404)$$

Inserting these values in eqs. (12.111) and (12.112), and performing straightforward algebra, we get the final result given in the text, eq. (12.274).[70]

[70]To compare with eq. (4) of Davis, Ruffini, Press and Price (1971), observe that we have a different convention on the Fourier transform, so that $[S_l(\omega, r)]^{\text{ours}} = \sqrt{2\pi}[S_l(\omega, r)]^{\text{theirs}}$. We have, however, an overall minus sign compared with their result, and furthermore the overall factor is of course the particle mass m, as in our result, rather than the BH mass M.

Further reading

- Classic papers on BH perturbation theory are Regge and Wheeler (1957) and Zerilli (1970). In the former paper the problem of the stability of Schwarzschild BHs under small perturbations is posed for the first time and the equation for axial perturbations (in the absence of external matter) is derived. In the latter the equation for polar perturbations is derived, and this paper also includes the effect of the energy–momentum tensor of external matter, both for polar and for axial perturbations. A classic textbook on BH perturbation theory is Chandrasekhar (1983).

- BH QNMs were introduced by Vishveshwara (1970) and Press (1971). A gauge-invariant formalism was introduced by Moncrief (1974).

- The analytic expression for the QNM wave function is given in Leaver (1986a). The analysis of QNMs in terms of singularities in the complex frequency plane is performed in Leaver (1986b).

- Power-law tails at spatial infinity for scalar and gravitational perturbations were first discussed in Price (1972a) and Cunningham, Price and Moncrief (1978). Tails for higher-spin perturbations are discussed in Price (1972b). Radiative tails at future null infinity are computed in Leaver (1986b) and Gundlach, Price and Pullin (1994a). Analytic results have been compared with a full numerical integration of the Einstein equations in Gundlach,

Price and Pullin (1994b), where it was found that the prediction of linearized theory for QNM ringing and power-law tails is remarkably accurate. A simpler derivation of the late-time tail is given in Andersson (1997), using directly the low-frequency approximation to the exact solution to evaluate the contribution of the branch cut. This paper also discusses the convergence of the sum over QNMs.

- The excitation of QNMs is discussed in Leaver (1986b), Sun and Price (1988), Andersson (1995,1997), Nollert and Price (1999), Berti and Cardoso (2006), Berti, Cardoso and Will (2006a) and Zhang, Berti and Cardoso (2013). The detectability of QNM ringing from supermassive BHs is discussed in Berti, Cardoso and Will (2006b)

- By replacing the RW or the Zerilli potential by a more general potential $V(x)$, one can investigate the dependence of the tail terms on the shape of the potential. Ching, Leung, Suen and Young (1995a) and (1995b) find that the time dependences of the tails depend on the asymptotic behavior of the potential, and not on its local feature. For potentials with compact support, or decreasing exponentially both at the horizon and at infinity, there is no power-law tail.

- The fact that the RW and Zerilli equations have the same spectrum is shown in Chandrasekhar (1979) [see also Chandrasekhar (1983), Section 28].

Our treatment follows that of Anderson and Price (1991).

- The problem of computing the QNM frequencies was first addressed by Chandrasekhar and Detweiler (1975), trying to match the numerical integration starting from x large and negative, with the numerical integration from x large and positive, and seeking the values of ω for which the Wronskian vanishes. The method is, however, not very accurate, and only works for the lowest-lying modes. A very powerful method for computing QNM frequencies, using continued fractions, was developed by Leaver (1985). He used it to compute the first 60 QNM frequencies, and the method can be extended to higher modes. Phase integral computations of QNM frequencies are discussed in Schutz and Will (1985), Iyer and Will (1987), Iyer (1987) and Andersson and Linnaeus (1992). Calculations using the Laplace transform are presented in Nollert and Schmidt (1992). Fröman, Fröman, Andersson and Hökback (1992) suggested working in the complex r-plane, on a line such that ωr is purely real. In this way the exponentially growing and decreasing behaviors are turned into a purely oscillatory behavior, which avoids the problem of identifying an exponentially small solution in the presence of an exponentially large one. Andersson (1992) used this method to compute numerically the first 11 QNM frequencies for $l = 2, 3$.

- The asymptotic frequencies of the QNM of a Schwarzschild BH are discussed in Leaver (1985), Bachelot and Motet-Bachelot (1992), Nollert (1993), Andersson (1993), Liu and Mashhoon (1996), Motl (2003), Motl and Neitzke (2003) and Maassen van den Brink (2004). A simple calculation of the leading term in the QNM frequencies in the large damping limit, using the Born approximation, is discussed in Medved, Martin and Visser (2004), Padmanabhan (2004), and Choudhury and Padmanabhan (2004). The large l-limit given in eq. (12.257) was found in Ferrari and Mashhoon (1984).

- Bekenstein (1974) suggested that in quantum gravity the area of a BH is quantized, in units of $8\pi l_{Pl}^2$, where l_{Pl} is the Planck length. The connection between highly damped QNMs and the quantization of the area of the BH horizon in semiclassical quantum gravity is discussed in Hod (1998). A number of puzzles that emerged in this connections can be clarified using the interpretation of the spectrum of QNMs given in Maggiore (2008), and presented on pages 162–164.

- A clear discussion of the Laplace transform approach is given in Nollert (1999) and in Nollert and Schmidt (1992), where (in Appendix A) it is also shown that imposing asymptotic boundary conditions on the Fourier-transformed Zerilli or RW equation does not single out a unique solution.

- Reviews on BH quasi-normal modes are Nollert (1999), Kokkotas and Schmidt (1999), Chapter 4 of the textbook by Frolov and Novikov (1998) (this chapter was written jointly with N. Andersson), and Berti, Cardoso and Starinets (2009).

- The gravitational radiation emitted during the infall of a test mass into a BH is studied in Davis, Ruffini, Press and Price (1971), Davis, Ruffini and Tiomno (1972) and Wagoner (1979), and the subject is reviewed in Nakamura, Oohara and Kojima (1987). Improved numerical studies include Lousto and Price (1997) and Mitsou (2011). The effect of angular momentum of the test mass is studied in Detweiler and Szedenits (1979). Test-mass infall with generic energy and impact parameter is studied in Berti *et al.* (2010), where the gravitational radiation emitted in ultrarelativistic BH encounters is also studied semi-analytically. The generalization of the RW and Zerilli equations to higher-dimensional BHs is discussed in Kodama and Ishibashi (2003), and has been used in Berti, Cavaglia and Gualtieri (2004) to compute the radiation from the infall in a multi-dimensional BH.

- The cosmic censorship conjecture was proposed by Penrose (1969). Wald (1974) considered the possibility of pushing an extremal BH beyond the extremal limit and forming a naked singularity, by throwing into it particles with large angular momentum. However, he showed that, in the test-particle limit, particles that would have sufficient angular momentum to create a naked singularity are not captured, but simply scattered. Jacobson and Sotiriou (2009) found that *near-extremal* BHs can be over-spun beyond the extremal limit by the capture of non-spinning test bodies carrying sufficient orbital angular momentum, if one neglects conservative self-force effects, as well as radiation reaction (i.e. dissipative self-force effects). The inclusion of self-force appears, however, to prevent over-spinning; see Barausse, Cardoso and Khanna (2010, 2011) and Colleoni, Barack, Shah and van de Meent (2015). As discussed in the latter paper, to definitely settle the issue would require (currently unavailable) information about higher-order self-force corrections.

- The RW and Zerilli equations can be expanded in powers of v/c, and in this way they give back the post-Newtonian (PN) expansion, in the test-mass limit $\nu = 0$. The expansion is, however, non-trivial, and must be performed differently in the region near the horizon and in the far region, with the two expansions then being matched. For a review of the method, see Sasaki and Tagoshi (2003). In this way, results have been obtained up to 22PN order for Schwarzschild BHs, and up to 11PN for Kerr BHs; see Fujita (2012, 2015).

- The head-on collision of two BHs can be studied using the Zerilli equation, when the two BHs are sufficiently close that a common horizon surrounds them (close limit). One uses as initial configuration an analytic solution provided by Misner (1960), which describes the two throats of a momentarily static wormhole. This can be written as a perturbation of the Schwarzschild metric, which is then evolved using the Zerilli equation; see Price and Pullin (1994), Abrahams and Cook (1994) and Gleiser, Nicasio, Price and Pullin (1996). The gravitational radiation computed in this way is in agreement with the results of full numerical relativity simulations.

- The Newman–Penrose formalism was proposed in Newman and Penrose (1962). Using this formalism, Bardeen and Press (1973) gave a unified description of scalar, electromagnetic and gravitational perturbations of Schwarzschild BHs. The generalization to Kerr BHs is done in Teukolsky (1973). The relation between the Bardeen–Press equation and the Zerilli and RW equations is clarified in Chandrasekhar (1975). In the presence of sources extending to infinity, such as in the infall of a test mass, the Teukolsky equations needs to be regularized, see Poisson (1997). The numerical integration of the Teukolsky equation is discussed in Pazos-Avalos and Lousto (2005). A convenient formalism for studying numerically the gravitational radiation generated by a test particle falling into a Kerr BH has been developed by Sasaki and Nakamura (1982a, 1982b). In contrast to the case of the Teukolsky equation, in this formalism the equations have no divergence associated with the source term. See also Appendix A of Berti *et al.* (2010) for a review of the Sasaki–Nakamura formalism.

- QNMs of Kerr BHs are discussed in Leaver (1985), Seidel and Iyer (1990), Kokkotas (1991), Onozawa (1997) and Berti (2004). Fits to the dependence of the fundamental $l = m = 2$ mode as a function of a, computed numerically by Leaver (1985), are given in Echeverria (1989), Fryer, Holz and Hughes (2002) and Berti, Cardoso and Will (2006b); see also the review by Berti, Cardoso and Starinets (2009). Accurate numerical computations of the QNM frequencies for Kerr BHs are presented in Cook and Zalutskiy (2014).

- Cardoso, Franzin and Pani (2016) compare the radiation generated by the QNMs of a Schwarzschild BH with that emitted by a compact object without horizon, such as a traversable wormhole, in response to the radial infall of a test particle. They find that, despite the completely different structure of the QNMs, the initial ringdown signal is basically the same. This indicates that the ringdown signal is really a probe of the existence of a light ring, rather than of a horizon.

Properties of dynamical space-times

In Chapter 14 we will discuss in detail the current theoretical understanding of the inspiral and merger of compact binary systems, and the corresponding GW emission. In particular, we will see how the coalescence can be studied using numerical relativity. The output of a numerical integration of the Einstein equations is a set of gauge-dependent quantities such as the components of the metric. From these, we want to extract physical, gauge-invariant observables. There are two types of quantity in which we are especially interested. First, we want to know the global properties of the final space-time, such as the mass and angular momentum of the final BH. Second, we want to compute the gravitational radiation at infinity. The problem that we have to face is that in general relativity (GR) there is no local notion of energy density, and the definition of energy associated with a generic dynamical field configuration is a non-trivial issue. We have seen for instance in Section 1.3 of Vol. 1 that the definition of the energy associated with GWs requires an appropriate averaging procedure. In this chapter we will discuss how to associate mass, momentum and angular momentum with dynamical space-times, such as those generated by coalescing binary systems. In this way, we will prepare some theoretical tools that are needed for understanding the results obtained by numerical relativity, discussed in Section 14.3. Beside their direct application to the problem of GW emission by a coalescing binary, the issues discussed in this chapter have an intrinsic conceptual interest, and their study provides a deeper understanding of the structure of GR.

13.1   The 3+1 decomposition of space-time

The starting point is the Arnowitt–Deser–Misner (ADM) formalism, which is based on a $3 + 1$ decomposition of the metric. We use coordinates $x^\mu = (ct, x^i)$ and we write

$$ds^2 = -\alpha^2 (cdt)^2 + h_{ij}(dx^i + \beta^i cdt)(dx^j + \beta^j cdt) \,. \qquad (13.1)$$

Therefore the space-time metric $g_{\mu\nu}$ is written as

$$g_{\mu\nu} = \begin{pmatrix} -\alpha^2 + \beta_k \beta^k & \beta_j \\ \beta_j & h_{ij} \end{pmatrix} \,, \qquad (13.2)$$

Gravitational Waves, Volume 2: Astrophysics and Cosmology. Michele Maggiore.
© Michele Maggiore 2018. Published in 2018 by Oxford University Press.
DOI 10.1093/oso/9780198570899.001.0001

where spatial indices are lowered with the spatial metric h_{ij}, i.e. $\beta_i \equiv h_{ij}\beta^j$. The inverse metric is

$$g^{\mu\nu} = \begin{pmatrix} -\alpha^{-2} & \alpha^{-2}\beta^j \\ \alpha^{-2}\beta^j & h^{ij} - \alpha^{-2}\beta^i\beta^j \end{pmatrix},$$ (13.3)

where h^{ij} is the inverse of h_{ij}. Observe that $g_{ij} = h_{ij}$ but, if β^i is non-vanishing, $g^{ij} \neq h^{ij}$. In the following we use h_{ij} to lower spatial indices and h^{ij} to raise them.

The function α is called the lapse function and β^i is called the shift vector. The geometric meaning of eq. (13.1) is that we have foliated space-time with a set of space-like three-dimensional hypersurfaces Σ_t, parametrized by a parameter t. The lapse function and the shift vector are pure gauge degrees of freedom; the time-parameter t is only a label of the hypersurfaces, with no special physical meaning. If, for the purpose of numerical integration, we discretize t, the lapse function α tells us the spacing in proper time on which we choose to record the time evolution of the geometry. The choice of the lapse function is therefore entirely in our hands.

Observe that the lapse function depends both on t and on $\mathbf{x}$, so at each point in the hypersurface Σ_t we can choose to record the evolution with a different spacing in proper time. Since α is a pure gauge degrees of freedom, the physics does not depend on it. However, particularly in the presence of singularities (as we have in a BH space-time) the performance of a numerical code depends strongly on the gauge choice. In a numerical integration it is convenient to slow down the evolution in regions close to a numerical singularity in order to avoid, or at least to delay as much as possible, a situation in which the divergences encountered by the code at the singularity (which results in non-assigned numbers at some points of the grid), propagates to the rest of the grid, thereby spoiling the whole computation. This can be done by taking the lapse function locally very small.

Similarly, the shift vector β^i tells us how the coordinates x^i on a slice Σ_t are related to the coordinates x^i on the slice Σ_{t+dt}, so again β^i is a pure gauge degrees of freedom and is completely in our hands.

The spatial metric h_{ij} is instead regarded as the fundamental dynamical variable. The geometric meaning of h_{ij} is the following. Consider a generic three-dimensional surface Σ (such as one of the surfaces Σ_t, for given t), parametrized by generic coordinates y^i, $i = 1, 2, 3$. The embedding of the three-dimensional surface Σ into four-dimensional space-time is determined by a mapping $x^\mu = x^\mu(y^i)$, which specifies where a given point of the surface Σ, identified by its coordinates y^i, is located in four-dimensional space-time. We can then define three independent four-vector fields $e_i^\mu(y)$, where $i = 1, 2, 3$ labels the four-vector, by

$$e_i^\mu(y) = \frac{\partial x^\mu(y)}{\partial y^i}.$$ (13.4)

By definition, these three four-vectors are tangent to the curves contained in Σ. We also denote by $n^\mu(y)$ the normal to the surface Σ at the point y. The hypersurface Σ is called space-like, time-like or null if its normal n^μ is everywhere time-like, space-like, or null, respectively. If the surface is not null, as we will henceforth assume, its normal can always be normalized so that $n_\mu n^\mu = +1$ (if Σ is time-like) or $n_\mu n^\mu = -1$ (if Σ is space-like, as in the case of the surfaces Σ_t introduced above). In both cases n^μ, normalized so that $|n_\mu n^\mu| = 1$, is called the unit normal. Observe that the four-vectors e_i^μ are orthogonal to n^μ, $e_i^\mu(y)n_\mu(y) = 0$.

If we perform an infinitesimal displacement dx^μ along the hypersurface Σ, the length ds_Σ of the displacement is given by

$$ds_\Sigma^2 = g_{\mu\nu}dx^\mu dx^\nu$$
$$= g_{\mu\nu}\left(\frac{\partial x^\mu}{\partial y^i}dy^i\right)\left(\frac{\partial x^\nu}{\partial y^j}dy^j\right)$$
$$= h_{ij}dy^i dy^j\,, \tag{13.5}$$

where

$$h_{ij} = g_{\mu\nu}e_i^\mu e_j^\mu \tag{13.6}$$

is called the *induced metric* (or the first fundamental form) of the hypersurface.

The definitions (13.4) and (13.6) are completely general, and hold for an arbitrary three-surface Σ embedded in a four-dimensional space-time, and for a generic choice of the coordinates x^μ of space-time and of the coordinates y^i of Σ. Consider now the situation in which space-time is foliated by a set of (non-intersecting) three-dimensional surfaces Σ_t, where t is some parameter that labels the hypersurface. On each surface Σ_t we set up a coordinate system y^i. A particularly convenient choice for the space-time coordinates x^μ is $x^0 = ct$ and $x^i = y^i$. In this coordinate system we have $e_i^\mu = \delta_i^\mu$ and therefore $h_{ij} = g_{ij}$. Thus, the geometric meaning of the tensor h_{ij} that appears in eq. (13.2) is that it is the induced metric of the surfaces Σ_t.

Another important geometric quantity is the extrinsic curvature (also called the second fundamental form) K_{ij} of a hypersurface Σ. It is defined by[1]

$$K_{ij} \equiv -e_i^\mu e_j^\nu D_\mu n_\nu\,, \tag{13.7}$$

where, as usual, D_μ is the covariant derivative with respect to the metric $g_{\mu\nu}$. It therefore describes how the unit normal n^μ changes along the hypersurface. Observe that the induced metric h_{ij} is the metric of the hypersurface Σ, and as such it quantifies properties intrinsic to Σ, independently of whether this surface is embedded in a higher-dimensional space. In contrast, K_{ij} measures how the unit normal to the surface changes as we move along the surface itself. Of course the notion of unit normal only makes sense for a surface embedded in a higher-dimensional space, since (for a non-null surface) a displacement along n^μ moves us away from the hypersurface. Thus, K_{ij} is really a quantity that characterizes the embedding, and for this reason it is called "extrinsic" curvature. Together, h_{ij} and K_{ij} fully characterize the geometrical properties of the hypersurface. The explicit expression of K_{ij} in terms of α, β^i and h_{ij} is[2]

[1] The sign convention in the definition (13.7) of K_{ij} is quite common, and is used for instance in Misner, Thorne and Wheeler (1973). In this chapter we will often refer to results derived in the textbook by Poisson (2004a), where K_{ij} is instead defined as $+e_i^\mu e_j^\nu D_\mu n_\nu$, and we will therefore change the appropriate signs compared with the corresponding formulas in Poisson (2004a).

[2] See for example Poisson (2004a), Section 4.2.6, taking into account our different sign convention.

$$K_{ij} = \frac{1}{2\alpha}\left(\partial_i\beta_j + \partial_j\beta_i - 2\Gamma^k_{ij}\beta_k - \frac{1}{c}\partial_t h_{ij}\right)$$

$$= -\frac{1}{2\alpha}\left(\frac{1}{c}\partial_t h_{ij} - \bar{D}_i\beta_j - \bar{D}_j\beta_i\right), \qquad (13.8)$$

where $\bar{D}$ is the covariant derivative with respect to the three-dimensional metric h_{ij} (we put an overbar on it to distinguish it from the spatial component of the covariant derivative D_μ computed from the four-dimensional metric $g_{\mu\nu}$). Finally, we define the trace of the extrinsic curvature,

$$K = h^{ij}K_{ij}. \qquad (13.9)$$

The quantities α, β^i, h_{ij} and K_{ij} are the fundamental variables in a Hamiltonian treatment of GR. In the next section we will see that in a finite four-dimensional volume the correct definition of the gravitational action, and therefore of the Hamiltonian, includes a crucial boundary term. We will then see in Section 13.3 how the gravitational action and the Hamiltonian can be written in terms of α, β^i, h_{ij} and K_{ij}.

13.2 Boundary terms in the gravitational action

Our aim is to obtain a definition of the mass or energy of a dynamical space-time, and we will find that it is given by an integral over a boundary surface at infinity. In order to compute it properly, it is therefore necessary to work at first on a finite four-dimensional volume $\mathcal{V}$, whose boundary we denote by $\partial\mathcal{V}$. In an infinite four-dimensional volume the action of the gravitational field is given by the Einstein (or Einstein–Hilbert) action

$$S_E = \frac{c^3}{16\pi G}\int d^4x\,\sqrt{-g}\,R, \qquad (13.10)$$

that we already introduced in eq. (1.1). Naively, one might think that in a finite four-dimensional volume $\mathcal{V}$, the gravitational action is simply obtained by restricting the integration in eq. (13.10) to the volume $\mathcal{V}$. However, this is not correct. In fact, the action (13.10) depends not just on the metric and on its first derivatives, but also on its second derivatives, since R contains derivatives of the Christoffel symbol, which itself contains first derivatives of the metric.

[3]Recall that in Section 2.2 we used units $c = 1$, while here we are keeping c explicitly.

In fact, as we already saw in eq. (2.114),[3] the Einstein action can be written as

$$S_E = \frac{c^3}{16\pi G}\int d^4x\,\sqrt{-g}\,R$$

$$= \frac{c^3}{16\pi G}\int d^4x\,\left[\sqrt{-g}\,\mathcal{L}_2 - \partial_\mu K^\mu\right], \qquad (13.11)$$

where

$$\mathcal{L}_2 = g^{\mu\nu}\left(\Gamma^\alpha_{\beta\mu}\Gamma^\beta_{\alpha\nu} - \Gamma^\alpha_{\mu\nu}\Gamma^\beta_{\alpha\beta}\right) \qquad (13.12)$$

and

$$K^\mu = \sqrt{-g}\,\left(g^{\mu\nu}\Gamma^\alpha_{\alpha\nu} - g^{\alpha\beta}\Gamma^\mu_{\alpha\beta}\right). \tag{13.13}$$

The Lagrangian $\mathcal{L}_2$, known as the "$\Gamma\Gamma$" Lagrangian, is quadratic in the first derivatives of the metric, but does not contain second derivatives. In contrast, since K^μ already contains first derivatives (through the Christoffel symbol) the term $\partial_\mu K^\mu$ contains second derivatives of the metric.

In order to derive the equations of motion from a variational principle, we must require that the action be such that its variation with respect to the metric gives back the Einstein equations when on the boundary we hold $g_{\mu\nu}$ fixed, but without imposing constraints on its first derivative. This is just as in classical mechanics, when the variational principle for a classical degree of freedom $q(t)$ on the interval $t_i \leqslant t \leqslant t_f$ is obtained by requiring that $q(t)$ be held fixed at the boundary, i.e. at the initial and final times, $q(t_i) = q_i$ and $q(t_f) = q_f$. Once we have imposed these boundary conditions, for a system governed by a second-order equation of motion we no longer have the freedom of specifying the time derivative $\dot{q}(t)$ at the boundaries $t = t_i$ and $t = t_f$, otherwise the system becomes over-constrained, since the pair $(q(t_i), \dot{q}(t_i))$ already fixes $q(t_f)$ and $\dot{q}(t_f)$.

If one uses the Einstein action restricted to $\mathcal{V}$, the terms that depend on the second derivatives of the metric give an integral over the boundary $\partial\mathcal{V}$ that depends on the first derivatives of the metric, so, when written in terms of the metric and its first derivative only, the Einstein action consists of a volume term, plus a boundary term that depends on the first derivative. The variation of the volume term gives just the Einstein equations. However, since we are not allowed to hold $\partial_\rho g_{\mu\nu}$ fixed on the boundary, the variation of this boundary term is non-zero, so it spoils the condition $\delta S = 0$.

The solution to this problem is to add to the Einstein action a further boundary term whose variation cancels the unwanted term in the variation of the Einstein action, and therefore gives a well-defined variational principle. Thus, in a finite volume,

$$S_{\text{grav}} = S_{\text{E}} + S_{\text{boundary}}, \tag{13.14}$$

The simplest possibility would be to choose S_{boundary} equal to minus the boundary term in eq. (13.11), in which case we remain with the $\Gamma\Gamma$ action. The drawback of this procedure is that the $\Gamma\Gamma$ action by itself is invariant under diffeomorphisms only modulo boundary terms. There is, however, another form of the action, the so-called "Trace-K" action, which is explicitly invariant under coordinate transformations and has a well-defined variational principle. This is obtained by taking

$$S_{\text{boundary}} = -\frac{c^3}{8\pi G} \int_{\partial\mathcal{V}} d^3y\,\sqrt{|h|}\,\epsilon K, \tag{13.15}$$

where y^i are coordinates on $\partial\mathcal{V}$,[4] h is the determinant of the metric induced by $g_{\mu\nu}$ on $\partial\mathcal{V}$, K is the trace of the extrinsic curvature of $\partial\mathcal{V}$, and $\epsilon = +1$ on the regions of the boundary where $\partial\mathcal{V}$ is time-like and

[4]By definition, we take the coordinates y^i to have dimensions of length, so for a boundary in which one of the coordinate is time-like, we take it to be ct rather than t.

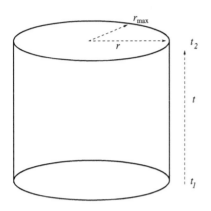

Fig. 13.1 The boundary ∂V discussed in the text. The spatial surfaces at fixed t are represented as two-dimensional, suppressing one angular variable, but are of course three-dimensional.

$\epsilon = -1$ where ∂V is space-like. We assume that ∂V is nowhere light-like. For example, for an asymptotically flat space-time we can choose V as the region of space-time such that $r \leqslant r_{\max}$ (where r is a radial coordinate that reduces asymptotically to the radial coordinate of flat space) and $t_1 \leqslant t \leqslant t_2$. In this case ∂V is a three-dimensional cylinder made of the two three-dimensional time-like hypersurfaces $\{t = t_1, r \leqslant r_{\max}\}$ and $\{t = t_2, r \leqslant r_{\max}\}$ (the "faces" of the three-dimensional cylinder) together with the space-like hypersurface $\{r = r_{\max}, t_1 \leqslant t \leqslant t_2\}$; see Fig. 13.1.

Thus, in a bounded four-dimensional space-time, we can take as gravitational action

$$S_{\text{grav}} = \frac{c^3}{16\pi G} \int_V d^4x \sqrt{-g}\, R - \frac{c^3}{8\pi G} \int_{\partial V} d^3y \sqrt{|h|}\, \epsilon K \,. \qquad (13.16)$$

The variation of this action gives the Einstein equation, through a well-defined variational principle. In the following, however, we will also be interested in the numerical value of the action on a given classical solution. Without spoiling the fact that the variation of the action reproduces the Einstein equations, one can still add to the action a constant or even an arbitrary boundary term that depends only on the induced metric h_{ij} (but not on its derivatives), since the fact that $g_{\mu\nu}$ is kept fixed on the boundary in the variational principle implies that h_{ij} is also kept fixed on the boundary. However, such a term would be relevant when we are interested in the numerical value of the action on a classical solution. To fix this term we require that the gravitational action vanish on flat space-times (which, as we will see, amounts to requiring that the gravitational mass of flat-space-time be zero). If $g_{\mu\nu} = \eta_{\mu\nu}$, we have $R = 0$ and the volume term in eq. (13.16) vanishes. The trace K of the extrinsic curvature vanishes on the hypersurfaces $\{t = \text{const}, r \leqslant r_{\max}\}$, but not on the hypersurface $\{r = r_{\max}, t_1 \leqslant t \leqslant t_2\}$, where $K = -2/r_{\max}$ (and $\epsilon = +1$). On this part of the boundary $d^3y \sqrt{|h|} = r_{\max}^2 d\Omega c dt$, so

$$\int_{\partial V} d^3y \sqrt{|h|}\, \epsilon K = -8\pi r_{\max} c(t_2 - t_1) \,, \qquad (13.17)$$

which is non-vanishing, and in fact even diverges when we eventually send $r_{\max} \to \infty$, and $t_2 - t_1 \to \infty$. The same happens for a generic asymptotically flat space-time. This problem can be cured by simply removing from the action the boundary term corresponding to flat space-time, i.e. by defining

$$\boxed{\frac{16\pi G}{c^3} S_{\text{grav}} = \int_V d^4x \sqrt{-g}\, R - 2 \int_{\partial V} d^3y \sqrt{|h|}\, \epsilon(K - K_0) \,,} \qquad (13.18)$$

where K_0 is the trace of the extrinsic curvature of ∂V when $g_{\mu\nu} = \eta_{\mu\nu}$. The term that we have added to the action is just a boundary term that depends only on h (because of $\sqrt{h}$ in the integration measure), but not on $\dot{h}_{ij}$, so it does not affect the variational principle. However, the action now has the property that it vanishes for flat space-time.

13.3 Hamiltonian formulation of GR

We can now plug the $3+1$ decomposition of the metric into the gravitational action (13.18). The Einstein–Hilbert term [i.e. the first term on the right-hand side of eq. (13.18)] then produces both a volume term and a surface term that adds to the explicit surface term given by the second term on the right-hand side of eq. (13.18). The volume term in the $3+1$ decomposition takes the form

$$S_{\rm Vol} = \frac{c^3}{16\pi G} \int_{\mathcal{V}} d^4x \, \alpha\sqrt{h} \left({}^3R + K_{ij}K^{ij} - K^2\right), \qquad (13.19)$$

where ${}^3R$ is the three-dimensional Ricci scalar computed from h_{ij}. To compute the boundary term in eq. (13.18) we must specify the geometry of the boundary $\partial\mathcal{V}$. We consider the same boundary as in Fig. 13.1, i.e. $\partial\mathcal{V}$ is a three-dimensional cylinder made of the two three-dimensional time-like hypersurfaces $\{t = t_1, \, r \leqslant r_{\rm max}\}$ and $\{t = t_2, \, r \leqslant r_{\rm max}\}$ together with the space-like hypersurface $\{r = r_{\rm max}, \, t_1 \leqslant t \leqslant t_2\}$. The latter space-like hypersurface, which is $(2+1)$-dimensional, can be foliated in terms of two-dimensional closed spatial surfaces S_t, defined as the regions of space-time with $r = r_{\rm max}$ at the given value of t, which are just two-spheres. Denoting the coordinates of S_t by θ^A (with $A = 1, 2$), the embedding of S_t into the three-dimensional surface Σ_t, whose coordinates are denoted by y^i, is determined by the function $y^i = y^i(\theta)$. Similarly to eq. (13.4), we can define the two three-vectors

$$e_A^i = \frac{\partial y^i}{\partial\theta^A}. \qquad (13.20)$$

The three-dimensional metric h_{ij} of Σ_t induces a two-dimensional metric σ_{AB} on S_t,

$$\sigma_{AB} = h_{ij}e_A^i e_B^i \qquad (13.21)$$

and the extrinsic curvature tensor k_{AB} of S_t (which has Euclidean signature) is defined by

$$k_{AB} = e_A^i e_B^j \bar{D}_i \hat{r}_j. \qquad (13.22)$$

Here $\hat{r}^i$ is the unit normal of S_t, and $\bar{D}_i$ is the covariant derivative computed with respect to the metric h_{ij}. The trace k of k_{AB} is given by

$$k = \sigma^{AB}k_{AB}. \qquad (13.23)$$

In terms of these quantities, combining the surface term coming from the Einstein–Hilbert action with the explicit surface term on the right-hand side of eq. (13.18) gives[5]

$$S_{\rm boundary} = -\frac{c^4}{8\pi G} \int_{t_1}^{t_2} dt \int_{S_t} d^2\theta \, \alpha\sqrt{\sigma}\,(k - k_0), \qquad (13.24)$$

[5]See Section 4.2.5 of Poisson (2004a) for a detailed computation.

where k_0 is the value of k for a flat space-time. So, in the end,

$$
\frac{16\pi G}{c^4} S_{\text{grav}} = \int_{t_1}^{t_2} dt \int_{\Sigma_t} d^3y \, \alpha\sqrt{h} \left({}^3R + K_{ij}K^{ij} - K^2\right)
$$
$$
-2 \int_{t_1}^{t_2} dt \int_{S_t} d^2\theta \, \alpha\sqrt{\sigma} \, (k - k_0) \, .
$$

(13.25)

Observe that this action does not depend on the time derivatives of the lapse and shift functions α and β^i, which therefore are not dynamical variables, in agreement with the discussion following eq. (13.1). The only dynamical variable is h_{ij}. The boundary action (13.24) does not depend on $\dot{h}_{ij}$, so the momentum conjugate to h_{ij}, denoted by π^{ij}, is given by[6]

$$
\pi^{ij} \equiv \frac{\delta}{\delta\dot{h}_{ij}} \left(\sqrt{-g}\,\mathcal{L}\right)
$$
$$
= -\frac{c^4}{16\pi G} \sqrt{h} \left(K^{ij} - K\,h^{ij}\right),
$$

(13.26)

where $\mathcal{L}$ is the "volume part" of the Lagrangian density,

$$
\frac{16\pi G}{c^4} \mathcal{L} = {}^3R + K_{ij}K^{ij} - K^2 \, .
$$

(13.27)

The "volume part" of the Hamiltonian density $\mathcal{H}$ is then given as usual by the Legendre transform

$$
\mathcal{H} = \pi^{ij}\dot{h}_{ij} - \sqrt{-g}\,\mathcal{L} \, .
$$

(13.28)

The full Hamiltonian H is obtained by integrating $\mathcal{H}$ in d^3x over Σ_t and adding a boundary term that is minus the boundary term of the Lagrangian. The explicit computation[7] gives

[7]See for example Section 4.2.6 of Poisson (2004a). Observe that the total boundary term in eq. (13.29) is given by the sum of the term proportional to $\alpha(k - k_0)$ (which comes from minus the boundary term in the Lagrangian) and the term proportional to $\beta_i\hat{r}_j(K^{ij} - Kh^{ij})$, which comes directly from $\pi^{ij}\dot{h}_{ij} - \sqrt{-g}\mathcal{L}$.

$$
\frac{16\pi G}{c^4} H = \int_{\Sigma_t} d^3x\sqrt{h} \left(-\alpha C_0 + 2\beta_i C^i\right)
$$
$$
+2\int_{S_t} d^2\theta\sqrt{\sigma} \left[\alpha(k - k_0) - \beta_i\hat{r}_j(K^{ij} - Kh^{ij})\right] ,
$$

(13.29)

where $\hat{r}^j$ is the unit normal to S_t, and we have defined

$$
C_0 = {}^3R + K^2 - K_{ij}K^{ij} \, ,
$$

(13.30)

$$
C^i = \bar{D}_j(K^{ij} - Kh^{ij}) \, ,
$$

(13.31)

where, as usual, $\bar{D}_j$ is the covariant derivative with respect to the metric h_{ij}.

The fact that this formulation provides a well-defined variational principle can also be checked directly from eq. (13.29). In fact, according to Hamilton's principle, Hamilton's equations of motion are

$$\dot{h}_{ij}(x) = \frac{\delta H}{\delta \pi^{ij}(x)} \, , \tag{13.32}$$

$$\dot{\pi}^{ij}(x) = -\frac{\delta H}{\delta h_{ij}(x)} \, , \tag{13.33}$$

where the functional derivatives are, by definition, the coefficients of δh_{ij} and of $\delta \pi^{ij}$ in a generic variation of the Hamiltonian. That is, if a generic variation of the Hamiltonian has the form

$$\delta H = \int d^3x \left[A^{ij}(x)\delta h_{ij}(x) + B_{ij}(x)\delta\pi^{ij}(x) \right] \, , \tag{13.34}$$

then Hamilton's equations of motion are $\dot{h}_{ij} = B_{ij}$ and $\dot{\pi}^{ij} = -A^{ij}$. In order to have a well-defined Hamilton principle, it is therefore necessary that the variation of the Hamiltonian take the form (13.34). However, if one only uses as Hamiltonian the volume term given in the first line of eq. (13.29), one finds that its variation also contains a boundary term, which is non-zero, and therefore spoils the variational principle. The boundary term in the second line of eq. (13.29) has just the property that its variation cancels the unwanted term, leaving us with a variation of the Hamiltonian of the form (13.34).[8]

[8]This is indeed the route that was followed by Regge and Teitelboim (1974), where the boundary term was first obtained.

13.4 Conserved quantities for isolated systems

In this section we define the energy (or mass), momentum and angular momentum of an isolated system, i.e. of a system such that, far away from it, space-time becomes flat. In order to give an accurate mathematical definition it is necessary to specify how rapidly space-time approaches a flat space-time. We will first discuss the definition of the Arnowitt–Deser–Misner (ADM) mass, which is a characterization of the mass (or energy) content of an asymptotically flat space-time. We begin by recalling the definition of asymptotic flatness.

Asymptotic flatness

We consider a (globally hyperbolic) space-time foliated by a family Σ_t of space-like three-dimensional hypersurfaces. As before, we denote by h_{ij} the metric induced by $g_{\mu\nu}$ on Σ_t [compare with eq. (13.1)], and by K_{ij} the extrinsic curvature of Σ_t. A space-time is asymptotically flat if there exists, on each space-like slice Σ_t, a metric f_{ij} such that:

(1) The Riemann tensor computed from f_{ij} vanishes (i.e. f_{ij} is a flat metric) except possibly on a compact domain B of Σ_t

(2) There exists a Cartesian-type coordinate system $x^i = (x, y, z)$ on Σ_t such that, outside B, $f_{ij} = \delta_{ij}$, and $r \equiv \sqrt{x^2 + y^2 + z^2}$ can take arbitrarily large values on Σ_t.

(3) In the limit $r \to \infty$ we have

$$h_{ij} = f_{ij} + O\left(r^{-1}\right), \tag{13.35}$$

$$\frac{\partial h_{ij}}{\partial x^k} = O\left(r^{-2}\right), \tag{13.36}$$

$$K_{ij} = O\left(r^{-2}\right), \tag{13.37}$$

$$\frac{\partial K_{ij}}{\partial x^k} = O\left(r^{-3}\right). \tag{13.38}$$

The region $r \to \infty$ is called "spatial infinity". Note that, strictly speaking, eq. (13.36) excludes the presence of GWs at spatial infinity, since for GWs propagating in the radial direction we have

$$h_{ij} = f_{ij} + \frac{F_{ij}(t - r/c)}{r}; \tag{13.39}$$

see for example eq. (3.14) or eq. (3.59). In this case, for large r,

$$\frac{\partial h_{ij}}{\partial x^k} = \frac{1}{r}\frac{\partial F_{ij}}{\partial x^k} + O\left(r^{-2}\right), \tag{13.40}$$

which, for a generic function F_{ij}, is $O(r^{-1})$ rather than $O(r^{-2})$. However, in practice, for a physical system that has existed (and has emitted GWs) only since a finite time in the past, this restriction is not relevant. Indeed, using the variable $u = t - r/c$, together with $(\partial u/\partial x^k)_t = -(1/c)\partial_k r = -(1/c)\hat{r}_k$, (where $\hat{r}_k = x_k/r$), we have

$$\frac{\partial F_{ij}(u)}{\partial x^k} = -\frac{1}{c}\hat{r}_k F'_{ij}(u), \tag{13.41}$$

where $F'_{ij}(u) = dF_{ij}/du$. As $r \to \infty$ at constant t, u goes to $-\infty$. In this limit the function $F'_{ij}(u)$ therefore has the same value that it has in the limit $t \to -\infty$ at fixed r. Any astrophysical source came into existence only after some time t_{birth}, so $F'_{ij}(u) = 0$ for any u that corresponds to a given r and $t < t_{\text{birth}}$, and therefore also in the limit $u \to -\infty$. An equivalent way of expressing the same conclusion is to observe that, for any source that only started to radiate after some initial time t_{birth}, the gravitational radiation has not yet reached spatial infinity, and therefore the radiation that it emits does not spoil the condition (13.36).

ADM mass

In order to define the ADM mass, we now evaluate the Hamiltonian (13.29) on a classical solution of the equations of motion. When we perform the variation of the Lagrangian (13.19) with respect to α to obtain the corresponding equation of motion, we clearly find $C_0 = 0$. Similarly, the variation with respect to β^i gives $C_i = 0$. Therefore $C_0 = C_i = 0$ are the constraint equations of the ADM formulation of

GR. So, when we consider the value of the Hamiltonian on a classical solution, the volume term in eq. (13.29) vanishes, and the contribution comes entirely from the boundary term.

The ADM mass of an asymptotically flat space-time (solution of the classical equations of motion) is defined as $1/c^2$ times the value of this Hamiltonian evaluated on the classical solution, when we choose $\alpha = 1$ and $\beta^i = 0$. This gauge choice makes the connection between energy and asymptotic translations with respect to the time coordinate t. Thus, from eq. (13.29),

$$M_{\text{ADM}} \equiv \frac{c^2}{8\pi G} \lim_{S_t \to \infty} \int_{S_t} d^2\theta \sqrt{\sigma}\,(k - k_0)\,. \tag{13.42}$$

Performing explicitly the computation of the extrinsic curvature k using the Cartesian-type coordinates involved in the definition of asymptotic flatness, one finds[9]

$$M_{\text{ADM}} = \frac{c^2}{16\pi G} \lim_{S_t \to \infty} \int_{S_t} d^2\theta \sqrt{\sigma}\,\hat{r}^i \left[\partial^j h_{ij} - \partial_i(\delta^{kl} h_{kl})\right]\,, \tag{13.43}$$

[9]See Hawking and Horowitz (1996) or Problem 4.5.7 of Poisson (2004a).

where the indices are raised and lowered with the flat metric. Observe that the integral converges thanks to the requirement of asymptotic flatness, eq. (13.36). Of course, the Cartesian coordinates at infinity are not always the most natural choice. For instance, for a spherically symmetric space-time, one might rather wish to use polar coordinates. If we use a generic coordinate system in the asymptotic region, then in eq. (13.43) the metric δ^{kl} must be replaced by the flat metric expressed in the coordinates that we are using, f^{kl}, and the partial derivatives ∂_i must be replaced by the covariant derivative $\mathcal{D}_i$ associated with the metric f_{ij}. Thus, if one uses a generic coordinate system in the asymptotic region, the ADM mass reads

$$M_{\text{ADM}} = \frac{c^2}{16\pi G} \lim_{S_t \to \infty} \int_{S_t} d^2\theta \sqrt{\sigma}\,\hat{r}^i \left[\mathcal{D}^j h_{ij} - \mathcal{D}_i(f^{kl} h_{kl})\right]\,. \tag{13.44}$$

Another instructive way of deriving the expression for the ADM mass is the following.[10] Consider first a source whose energy–momentum tensor $T_{\mu\nu}$ is everywhere sufficiently weak, so that a linearization of the Einstein equations is justified. Writing $g_{\mu\nu} = \eta_{\mu\nu} + h_{\mu\nu}$, the linearization of Einstein equations gives

$$\Box h_{\mu\nu} + \partial_\mu\partial_\nu h - \partial^\rho\partial_\nu h_{\mu\rho} - \partial^\rho\partial_\mu h_{\nu\rho} = -\frac{16\pi G}{c^4}\left(T_{\mu\nu} - \frac{1}{2}\eta_{\mu\nu}T\right)\,, \tag{13.45}$$

where $h = \eta^{\mu\nu}h_{\mu\nu}$ and $T = \eta^{\mu\nu}T_{\mu\nu}$.[11] Taking the trace of eq. (13.45) we

[10]We follow the derivation given in Townsend (1997).

[11]This is equivalent to the linearized equation (1.17) used in Vol. 1, as can be shown by subtracting from eq. (1.17) $(1/2)\eta_{\mu\nu}$ times the trace of eq. (1.17) itself, and using the definition of $\bar{h}_{\mu\nu}$ in terms of $h_{\mu\nu}$ given in eq. (1.15).

get

$$\Box h - \partial^\mu \partial^\nu h_{\mu\nu} = +\frac{8\pi G}{c^4} T. \tag{13.46}$$

Let us further assume, for now, that $T_{\mu\nu} = \mathrm{diag}(T_{00}, 0, 0, 0)$ (so $T = \eta^{\mu\nu} T_{\mu\nu} = -T_{00}$) and that the source is static, i.e. $dT_{00}/dt = 0$. Then, even $h_{\mu\nu}$ is time-independent, and the (00) component of eq. (13.45) becomes

$$\nabla^2 h_{00} = -\frac{8\pi G}{c^4} T_{00}, \tag{13.47}$$

while, writing $h = \eta^{\mu\nu} h_{\mu\nu} = -h_{00} + \delta^{kl} h_{kl}$ and using Cartesian coordinates, eq. (13.46) becomes

$$-\nabla^2 h_{00} + \nabla^2(\delta^{kl} h_{kl}) - \partial^i \partial^j h_{ij} = -\frac{8\pi G}{c^4} T_{00}. \tag{13.48}$$

Adding eqs. (13.48) and (13.47) we get

$$T_{00} = \frac{c^4}{16\pi G} \partial^i \left[\partial^j h_{ij} - \partial_i (\delta^{kl} h_{kl}) \right]. \tag{13.49}$$

For a static source, energy is locally conserved, and we can use the standard field-theoretical definition of the energy of a field configuration,

$$E = \int_{\Sigma_t} d^3x \, T_{00}, \tag{13.50}$$

(where the factor $\sqrt{-g}$ in the integral could be replaced by 1, since we are working to linear order in $h_{\mu\nu}$ and T_{00} is already linear in $h_{\mu\nu}$). The integral is performed over the hypersurface Σ_t at given t, but Noether's theorem (see for example Section 2.1.1 of Vol. 1) ensures that the result is actually time-independent. The crucial point now is that T_{00} is given in eq. (13.49) as a total derivative, and we can therefore transform the volume integral in eq. (13.50) into a surface integral at infinity, i.e. over the boundary S_t of Σ_t with $S_t \to \infty$. Thus

$$E = \frac{c^4}{16\pi G} \lim_{S_t \to \infty} \int_{S_t} d^2\theta \sqrt{\sigma} \, \hat{r}^i \left[\partial^j h_{ij} - \partial_i (\delta^{kl} h_{kl}) \right]. \tag{13.51}$$

Since this is a boundary term, the result depends only on the field h_{ij} at spatial infinity. So, we can change the source on the interior without changing the energy, as long as the asymptotic metric is unchanged. This means that we can relax our hypothesis of a weak-field and static source, and the result is valid in general, as long as the source is still such that the metric satisfies the conditions of asymptotic flatness, which also ensures that the integral in eq. (13.51) converges. Thus (after dividing by a factor c^2 to go from energy to mass) we have recovered eq. (13.43).

As an example, we evaluate the ADM mass for Schwarzschild spacetime. Using Schwarzschild coordinates the metric is given by eq. (12.2), so the spatial metric is

$$h_{ij} = \mathrm{diag} \left[\left(1 - \frac{2GM}{c^2 r} \right)^{-1}, r^2, r^2 \sin^2 \theta \right], \tag{13.52}$$

while the flat metric in polar coordinates is

$$f_{ij} = \text{diag}(1, r^2, r^2 \sin^2 \theta). \tag{13.53}$$

We take S_t to be the sphere at constant radius r on the hypersurface Σ_t, so $\theta^A = (\theta, \phi)$, $\sqrt{\sigma} d^2\theta = r^2 d \cos \theta d\phi \equiv r^2 d\Omega$ and $\hat{r}^i = (1, 0, 0)$ in the basis (r, θ, ϕ). Evaluating the Christoffel symbols with respect to f_{ij} and performing straightforward algebra gives

$$\hat{r}^i [\mathcal{D}^j h_{ij} - \mathcal{D}_i (f^{kl} h_{kl})] = \left(1 - \frac{2GM}{c^2 r}\right)^{-1} \frac{4GM}{c^2 r^2}$$

$$= \frac{4GM}{c^2 r^2} + O\left(\frac{1}{r^4}\right). \tag{13.54}$$

We insert this into eq. (13.44) and take the limit $S_t \to \infty$, i.e. $r \to \infty$. Then the term $O(1/r^4)$ in eq. (13.54) does not contribute, and we get

$$M_{\text{ADM}} = \frac{c^2}{16\pi G} \lim_{r \to \infty} \int r^2 d\Omega \frac{4GM}{c^2 r^2}$$

$$= M, \tag{13.55}$$

showing that the ADM mass is the same as the parameter M that appears in the Schwarzschild metric.

It is also easy to check that the ADM mass reduces to the Newtonian mass for a weak and nearly static Newtonian source. In this case, in fact, we know from eqs. (5.11) and (5.12) of Vol. 1 that $h_{00} = 2\phi$, $h_{0i} = 0$ and $h_{ij} = -2\phi \delta_{ij}$, where ϕ satisfies the Poisson equation $\nabla^2 \phi = (4\pi G/c^2)\rho$ and $\rho = T_{00}/c^2$ in the non-relativistic limit is the mass density of the source. Then, using Cartesian coordinates, eq. (13.43) gives

$$M_{\text{ADM}} = \frac{c^2}{4\pi G} \lim_{r \to \infty} \int_{S_t} d^2\theta \sqrt{\sigma} \, \hat{r}^i \, \partial_i \phi$$

$$= \frac{c^2}{4\pi G} \int d^3 x \, \nabla^2 \phi$$

$$= \int d^3 x \, \rho, \tag{13.56}$$

where in the second line we have transformed the surface integral of $\partial_i \phi$ into a volume integral of $\nabla^2 \phi$ using Stokes' theorem. We see that the ADM mass has the correct Newtonian limit.

We also expect that a properly defined mass will be positive-definite, under suitable positivity conditions on the matter energy–momentum tensor, corresponding to "normal" matter [of course, as we see from eq. (13.56), if we took a negative ρ, we would get a negative ADM mass]. In fact, the proof of the positivity of the ADM mass is quite non-trivial, and requires that the energy–momentum tensor of the matter that generates the space-time under consideration satisfy the dominant energy condition.[12] In that case, one can prove that $M_{\text{ADM}} \geqslant 0$, and $M_{\text{ADM}} = 0$ if and only if Σ_t is an hypersurface of Minkowski space-time.

[12]The dominant energy condition requires that, for any time-like and future-directed vector v^μ, the vector $-T^\mu{}_\nu v^\nu$ must be a future-directed time-like or null vector. This condition also implies the weak energy condition, namely that for any time-like and future directed vector v^μ, $T_{\mu\nu} v^\mu v^\nu > 0$. If v^μ is the four-velocity of an observer, $T_{\mu\nu} v^\mu v^\nu$ is the energy density measured by that observer.

Observe that the definition (13.42) makes explicit reference to time t, since the integral is computed on the surface S_t (which is then sent to spatial infinity, at fixed t), but the ADM mass is actually time-independent. This is due to the fact that the Hamiltonian (13.29) depends on the generalized coordinates α, β^i and h_{ij}, and on their conjugate momenta, but does not depend explicitly on time. It is therefore constant on any classical solution of the equations of motion, so

$$\frac{dM_{\text{ADM}}}{dt} = 0 . \tag{13.57}$$

For an isolated static system, such as the Schwarzschild BH, this is indeed what we expect from a proper definition of its mass, or of its energy $M_{\text{ADM}}c^2$. One might at first be puzzled to see that even the ADM energy of a dynamical space-time, which radiates GWs, is conserved. However, the ADM energy is only defined for asymptotically flat space-times. Recall from the discussion following eq. (13.38) that a system that radiates GWs only obeys the asymptotic flatness condition if it began to radiate at a finite time in the past (which is of course the case for any realistic astrophysical system) so that, at any finite value of t, the radiation that it emitted has not yet reached spatial infinity, where the ADM energy is evaluated. Therefore, for a dynamical space-time $M_{\text{ADM}}c^2$ contains both the energy of the source and the energy of the radiation field, and as such it is conserved. In numerical relativity, the integral defining the ADM mass is unavoidably evaluated at a large but finite radius from the source. In this case, M_{ADM} is not strictly conserved, and its change in time can be interpreted as due to the flux of matter and radiation that passes across the surface.

A different notion of mass can be defined through a different way of going to infinity. This is the Bondi–Sachs mass, and is still given by eq. (13.42), except that now the surface S_t is not taken to be at spatial infinity (i.e. $r \to \infty$ with t constant), but rather at null infinity. Defining as usual the null coordinates $u = t - r/c$ and $v = t + r/c$, future null infinity is defined by taking the limit $r \to \infty$ at fixed u rather than at fixed t. Using r and u as independent variables instead of r and t, and denoting by $S(u, r)$ the location in space-time of the boundary surface, the Bondi–Sachs mass is given by

$$M_{\text{BS}} \equiv \frac{c^2}{8\pi G} \lim_{r \to \infty} \int_{S(u,r)} d^2\theta \sqrt{\sigma}\,(k - k_0) . \tag{13.58}$$

For stationary space-times the ADM and Bondi–Sachs masses are identical. However, for a system radiating GWs (or any other type of radiation) it can be shown that the rate of change of the Bondi–Sachs mass is related to the flux F of the radiation emitted by[13]

[13]See Bondi, van der Burg and Metzner (1962) and Sachs (1962) for the rather involved proof.

$$\frac{d(M_{\text{BS}}c^2)}{du} = -\lim_{r \to \infty} \int_{S(u,r)} d^2\theta \sqrt{\sigma}\, F . \tag{13.59}$$

ADM momentum and angular momentum

We have seen that, for an asymptotically flat space-time, the ADM mass is the conserved quantity associated with time translations at spatial infinity. This connection is expressed by the fact that, to obtain the ADM mass, we set $\alpha = 1$ and $\beta^i = 0$ in eq. (13.29). Similarly, the Cartesian component P^i_{ADM} of the ADM momentum $\mathbf{P}_{\text{ADM}}$ is the conserved quantity associated with spatial translations at infinity along the i-th axis (defined with respect to a Cartesian-like frame at infinity) and is defined by setting $\alpha = 0$ and $\beta^i = 1$ in eq. (13.29), and adjusting the powers of c so to have a quantity with dimensions of momentum. Thus

$$
P^i_{\text{ADM}} = -\frac{c^3}{8\pi G} \lim_{S_t \to \infty} \int_{S_t} d^2\theta \sqrt{\sigma}\, \hat{r}_j (K^{ij} - K h^{ij}) \,.
\tag{13.60}
$$

The asymptotic flatness condition (13.38) ensures that the integral converges. Under a change of Cartesian coordinates $x^i \to x'^i$ that asymptotically corresponds to a rotation, the three quantities P^i_{ADM}, $i = 1, 2, 3$, transform as a vector. If one applies this definition to a Schwarzschild BH one finds that, with the foliation of space-time associated with the Schwarzschild coordinates (14.16), the extrinsic curvature K_{ij} of the $t = \text{const}$ hypersurfaces Σ_t vanishes, and therefore $P^i_{\text{ADM}} = 0$.[14]

The relation between the definition of ADM momentum and spatial translations at infinity can be better understood observing that, under an infinitesimal coordinate transformation $x^i \to x^i + \xi^i(x)$, the variation of the gravitational action (including the boundary term) is[15]

$$
\delta S_{\text{grav}} = -\frac{c^4}{8\pi G} \int_{t_1}^{t_2} dt\, \frac{d}{dt} \left[\lim_{S_t \to \infty} \int_{S_t} d^2\theta \sqrt{\sigma}\, \xi_i(x) \hat{r}_j (K^{ij} - K h^{ij}) \right]
$$
$$
+ \text{ (terms vanishing on the equations of motions)} \,.
\tag{13.61}
$$

If we take $\xi_i \to \epsilon_i$ constant as $r \to \infty$, we are dealing with an asymptotic translation, and under such a transformation the action is invariant, $\delta S_{\text{grav}} = 0$. Then, we see from eq. (13.61) that the three-quantities

$$
-\frac{c^4}{8\pi G} \lim_{S_t \to \infty} \int_{S_t} d^2\theta \sqrt{\sigma}\, \hat{r}_j (K^{ij} - K h^{ij}) \,,
\tag{13.62}
$$

with $i = 1, 2, 3$, are the conserved Noether charges associated with spatial translations at infinity. After adjusting the powers of c so as to obtain a quantity with dimensions of momentum, we recover the ADM momentum (13.60).[16]

This derivation of the ADM momentum can be generalized to define angular momentum. We consider a transformation $x^i \to x^i + \xi^i(x)$, where now asymptotically $\xi_i \to \epsilon_{ijk}(\delta\phi)^j x^k$, with $\delta\phi^j$ an x-independent infinitesimal rotation around the j-th axis. Plugging it into eq. (13.61) we see that the conserved charges associated with spatial rotations at infinity are given by

$$
J_i = \frac{c^3}{8\pi G} \lim_{S_t \to \infty} \int_{S_t} d^2\theta \sqrt{\sigma}\, \hat{r}_l \epsilon_{ijk} (K^{lj} - K h^{lj}) x^k \,.
\tag{13.63}
$$

[14] A non-zero value is, however, obtained for the "boosted" Schwarzschild solution; see York (1980).

[15] See Regge and Teitelboim (1974).

[16] As always with Noether's theorem, the overall normalization of the charges is arbitrary, since if a quantity is conserved, any multiple of it is conserved too. In our case the normalization of the ADM mass is fixed by the fact that it is the value of the Hamiltonian on the classical solution (and with this normalization it reproduces the Schwarzschild mass for a Schwarzschild space-time), and then the normalization of P^i_{ADM} is fixed by requiring that $P^\mu_{\text{ADM}} \equiv (M_{\text{ADM}} c^2, P^i_{\text{ADM}})$ transform as a four-vector under coordinate transformations that reduce to Lorentz transformations at infinity. This property was shown in the original paper by Arnowitt, Deser and Misner (1962).

We can now observe that, if S_t is a two-sphere, $\hat{r}_l \epsilon_{ijk} h^{lj} x^k = r \epsilon_{ijk} \hat{r}^j \hat{r}^k = 0$ (where we used the fact that indices are raised and lowered with h^{ij}, so $\hat{r}_l h^{lj} = \hat{r}^j$), so the second term in the parentheses does not contribute.[17] The result can also be written in a more compact form by defining

$$\phi_{(i),j} = \epsilon_{ijk} x^k \,, \tag{13.64}$$

[17] Taking S_t to be a two-sphere is the obvious choice for most practical applications. If, however, for some reason, one wishes to choose S_t as, say, a deformed two-sphere, then the second term in the parentheses contributes and must be kept.

where the label (i) identifies the axis around which we perform the rotation (this implies that we have chosen a Cartesian coordinate system at infinity), and writing the integration over the sphere in terms of $dS_l = d^2\theta \sqrt{\sigma}\, \hat{r}_l$. Then

$$J_i = \frac{c^3}{8\pi G} \lim_{S_t \to \infty} \int_{S_t} dS^k \, \phi^j_{(i)} K_{kj} \,. \tag{13.65}$$

Observe that, compared with the ADM mass or momentum, the integrand in eq. (13.65) has an extra factor $x^k = O(r)$, so it is not obvious that the asymptotic flatness conditions (13.35)–(13.38) are sufficient to guarantee the convergence of the integral. However, in practice for the space-times of interest, such as Kerr space-time, the integral indeed converges thanks to the fact that the expression $\epsilon_{ijk}(K^{lj} - Kh^{lj})x^k$ is projected onto the radial direction $\hat{r}_l$.[18]

[18] Actually, there are subtleties in this ADM-type definition of angular momentum. These are related to the fact that the asymptotic group of transformations at infinity that preserve the condition of asymptotic flatness is not simply given by the Poincaré transformations $x^\mu \to \Lambda^\mu{}_\nu x^\nu + a^\mu$, with a^μ constant, but more generally by $x^\mu \to \Lambda^\mu{}_\nu x^\nu + a^\mu(\theta, \phi)$. A transformation $x^\mu \to x^\mu + a^\mu$ where a^μ is not a constant but rather depends on the polar angles (θ, ϕ) is called a super-translation. It turns out that, because of the fact that super-translations are asymptotic symmetries, the three components of the angular momentum vector defined in eq. (13.65) do not transform as a three-vector. This problem can be solved using a coordinate system where the metric decays faster than in eqs. (13.35)–(13.38); see York (1979, 1980) and the review by Gourgoulhon (2007).

Just like the ADM mass, the ADM momentum and the angular momentum (13.65) are time-independent, since they are the conserved charges associated with translations and rotations. Therefore, in principle, they are not the appropriate quantities for a system that loses momentum and angular momentum through emission of gravitational radiation. However, as already discussed for the ADM mass, in practice for systems such as for instance a coalescing binary, they are useful observables either at early times, when the stars are still widely separated and therefore gravitational emission is small, or long after the coalescence, when the final BH has settled into its fundamental state, and again gravitational emission has become negligible.

Equations (13.42), (13.60) and (13.65) allow us to extract the total mass, momentum and angular momentum of space-time from the ADM metric. Observe that, even if in numerical relativity the simulations are performed in coordinate systems that are different from ADM coordinates, such as those discussed in Section 14.3, it is in principle straightforward to transform the results to ADM coordinates, and compute the ADM mass, momentum and angular momentum.

Mass and angular momentum of isolated BH horizons

The ADM mass, momentum and angular momentum are conserved global quantities that characterize the whole space-time. In a dynamical space-time including for instance two orbiting BHs, the ADM mass provides the total energy of the system, but gives no information on how much mass should be attributed to each BH individually, and how much energy is associated with the gravitational radiation. The same is true for the ADM momentum and angular momentum.

However, if the two compact objects are BHs, it is possible to associate with each of them a coordinate-invariant notion of mass and angular momentum by making use of the concept of apparent horizon, thereby providing a definition of mass and angular momentum that is quasi-local, rather than global. As in the Hamiltonian treatment of GR discussed in Section 13.3, it is convenient to start from a foliation of space-time, with spatial hypersurfaces Σ_t. Within such a three-dimensional hypersurface, the apparent horizon is a two-dimensional surface characterized by the fact that it is the outermost marginally trapped surface, i.e. the outermost surface on which the expansion of the outgoing light rays vanishes.[19] Within the whole four-dimensional space-time, the horizon is therefore a $(2+1)$-dimensional submanifold, which we denote by Δ.

One can first define the angular momentum associated with the apparent horizon. The procedure is basically an extension of the ADM formalism to the case where the space-time has an inner boundary. As we saw in Section 13.4, in the ADM formalism angular momentum is defined in terms of the Hamiltonian that generates diffeomorphisms that, at infinity, reduce to spatial rotations. Here one can proceed in a parallel way, using diffeomorphisms that reduce to spatial rotations on the apparent horizon. To define angular momentum, we must assume that the horizon surface S_H is axisymmetric, i.e. that on the horizon there is invariance under rotations around a symmetry axis, generated by a vector field φ^j tangent to S_H. Then one can show[20] that the component of the angular momentum along the symmetry axis associated with the horizon Δ, at time t, is given by

[20]See Dreyer, Krishnan, Shoemaker and Schnetter (2003).

$$J_\Delta = \frac{c^3}{8\pi G} \int_{S_H} dS^k\, \varphi^j K_{kj}\,, \qquad (13.66)$$

where S_H is the two-dimensional surface of the apparent horizon within the hypersurface Σ_t. This is the same formula that we found for angular momentum in the ADM formalism, eq. (13.65), except that now the integral is performed on the apparent horizon S_H, rather than on the two-sphere at infinity, and that the vector field $\phi^j_{(i)}$ of eq. (13.65), which describes rotations at infinity around the i-th axis, is replaced by the vector φ^j that describes rotations on the horizon around the symmetry axis.

Of course, what is interesting about this definition is the fact that it can be applied even to BHs that are not isolated. In this case the metric on the apparent horizon of a BH will be distorted, with respect to the Kerr metric, by the presence of the other bodies, and therefore, while in the ADM case $\phi^j_{(i)}$ is trivially given by eq. (13.64), determining φ^j can be more complicated. One must locate, numerically, the position of the distorted apparent horizon, and determine φ^j. Then, eq. (13.66) provides a coordinate-independent way of associating an intrinsic spin with each BH in a binary system.[21]

[22]See Ashtekar, Fairhurst, and Krishnan (2000) and Ashtekar, Beetle, and Lewandowski (2001).

Having defined the angular momentum associated with a isolated horizon, we can now define a mass M_Δ associated with an isolated horizon. A procedure analogous to the definition of the ADM mass, applied to the inner boundary Δ, gives[22]

$$\frac{GM_\Delta}{c^2} = \frac{1}{2R_\Delta}\left[R_\Delta^4 + \left(\frac{2GJ_\Delta}{c^3}\right)^2\right]^{1/2}, \qquad (13.67)$$

where we have defined R_Δ from $A_\Delta = 4\pi R_\Delta^2$, where A_Δ is the area of the apparent horizon. Observe that A_Δ is a coordinate-invariant quantity. In the space-time of a BH binary, the notion of horizon mass gives the possibility of associating physically meaningful masses, $M_{\Delta,1}$ and $M_{\Delta,2}$, with each of the two BHs, while the ADM mass provides the total mass of the system. The total energy of the system, given by the ADM mass, also contains the binding energy E_{binding} of the binary, plus the energy E_{GW} carried by the GWs produced by the system so, at least heuristically, we expect that

$$M_{\text{ADM}}c^2 = M_{\Delta,1}c^2 + M_{\Delta,2}c^2 - E_{\text{binding}} + E_{\text{GW}}. \qquad (13.68)$$

For a stationary space-time with a single BH, M_Δ and J_Δ become the same as the ADM mass M and angular momentum J. In particular, for a Kerr space-time,

$$A_\Delta = \int_{r=r_+} d\theta\, d\phi\, \sqrt{g_{\theta\theta}g_{\phi\phi}}. \qquad (13.69)$$

where r_+ is the outer horizon of the Kerr metric; see eq. (12.289). Inserting the Kerr metric (12.285) into eq. (13.69) we get

$$\begin{aligned}A_\Delta &= \int_{r=r_+} d\theta\, d\phi\, \sin\theta(r_+^2 + a^2)\\ &= 4\pi(r_+^2 + a^2),\end{aligned} \qquad (13.70)$$

and therefore, for an isolated Kerr BH,

$$R_\Delta^2 = r_+^2 + a^2, \qquad (13.71)$$

while $M_\Delta = M$ and $J_\Delta = J$, and eq. (13.67) reduces to the corresponding relation for Kerr BHs,

$$\frac{GM}{c^2} = \frac{1}{2R_\Delta}\sqrt{R_\Delta^4 + \left(\frac{2GJ}{c^3}\right)^2}, \qquad (13.72)$$

which can be derived using the explicit expression for r_+ in terms of M and J, given in eq. (12.289).

Irreducible BH mass

For a single isolated BH another useful quantity is the irreducible BH mass, which is defined by

$$\frac{GM_{\text{irr}}}{c^2} \equiv \sqrt{\frac{A_\Delta}{16\pi}} \, . \tag{13.73}$$

The irreducible mass of an isolated BH gives the energy that cannot be extracted from the BH with classical processes. For a Schwarzschild BH the ADM mass is the same as the irreducible mass, i.e. no energy can be extracted classically. From rotating (or charged) BHs one can extract some energy classically with processes such as the Penrose process or super-radiance, but it is still not possible to decrease the BH mass below the value of the irreducible mass. Writing again $A_\Delta = 4\pi R_\Delta^2$ in eq. (13.73), we see that

$$R_\Delta = \frac{2GM_{\text{irr}}}{c^2} \, , \tag{13.74}$$

i.e. R_Δ is the Schwarzschild radius associated with a mass M_{irr}. Using the explicit expression (13.71), together with eq. (12.289), we get

$$M_{\text{irr}}^2 = (r_+^2 + a^2)/4$$
$$= \frac{M^2}{2} + \frac{M}{2}\left[M^2 - \left(\frac{c^2 a}{G}\right)^2\right]^{1/2} . \tag{13.75}$$

Expressing R_Δ in eq. (13.72) in terms of M_{irr}, we get Christodoulou's formula

$$M^2 = M_{\text{irr}}^2 + \left(\frac{cJ}{2GM_{\text{irr}}}\right)^2 , \tag{13.76}$$

which can be inverted to give

$$2M_{\text{irr}}^2 = M^2 + \left[M^4 - (cJ/G)^2\right]^{1/2} . \tag{13.77}$$

We see that, for $J = 0$, $M_{\text{irr}} = M$, while, for $J \neq 0$, $M_{\text{irr}} < M$. In the extremal case cJ/G is equal to M^2 [see eq. (12.293)] and $M_{\text{irr}} = M/\sqrt{2}$.

13.5   GWs and Newman–Penrose scalar

Finally, we wish to be able to extract from a numerical simulation the gravitational radiation emitted in the far region. GWs take a simple form in the TT gauge, while, as we will see in Section 14.3, numerical simulations are performed in very different gauges. It is therefore convenient to have a characterization of GWs in terms of quantities that can be directly read from the Riemann tensor. These are the Newman–Penrose scalars, introduced in Section 12.5.2. In particular, to characterize the gravitational radiation emitted at large distances it is convenient to use the Newman–Penrose scalar Ψ_4, defined in eq. (12.329). Actually, to complete the definition of Ψ_4 we need to specify the null tetrad

$(l^\mu, q^\mu, m^\mu, \bar{m}^\mu)$, as discussed in Section 12.5.2. In numerical relativity, the most common choice is to construct the tetrad from the unit normal to the constant-time hypersurfaces, n^μ, and the radial unit vector $r^\mu = (0, \hat{r})$, writing

$$l^\mu = \frac{1}{\sqrt{2}}(n^\mu + r^\mu), \tag{13.78}$$

$$q^\mu = \frac{1}{\sqrt{2}}(n^\mu - r^\mu), \tag{13.79}$$

while m^μ is purely spatial and is given in terms of the unit vectors in the $\hat{\theta}$ and $\hat{\phi}$ directions by

$$m^\mu = \frac{1}{\sqrt{2}}(0, \hat{\phi} - i\hat{\theta}). \tag{13.80}$$

As usual $\bar{m}^\mu$ is the complex conjugate of m^μ, $\bar{m}^\mu = (1/\sqrt{2})(0, \hat{\phi} + i\hat{\theta})$. When using this tetrad, the relation between Ψ_4 and $\ddot{h}_+ - i\ddot{h}_\times$ [which was derived in eq. (12.332) using the tetrad given by eqs. (12.309)–(12.312)] becomes

$$\Psi_4 = \ddot{h}_+ - i\ddot{h}_\times. \tag{13.81}$$

Recall that this expression only holds at asymptotically far distance, since it was derived using the expression (12.330) for the Riemann tensor, which only holds far from the sources.[23]

If we perform a rotation by an angle ψ around the propagation direction, Ψ_4 transforms as

$$\Psi_4 \to e^{-2i\psi}\Psi_4, \tag{13.82}$$

as we can see from eq. (2.195). This is the transformation property of the spin-weighted spherical harmonics $Y_{lm}^{(s)}(\theta, \phi)$, defined in eq. (12.352), with $s = -2$. It is therefore convenient to expand Ψ_4 in terms of $Y_{lm}^{(-2)}(\theta, \phi)$,

$$\Psi_4(t, \mathbf{x}) = \sum_{l,m} \Psi_4^{lm}(t, r) Y_{lm}^{(-2)}(\theta, \phi), \tag{13.83}$$

where the sum starts from $l = 2$ since the spin-weighted spherical harmonics $Y_{lm}^{(s)}(\theta, \phi)$ vanish for $l < |s|$. From eq. (13.81) we see that Ψ_4 has dimensions of $(\text{time})^{-2}$ and that, in the wave zone, it is proportional to $1/r$. It is therefore convenient to consider the expansion of the dimensionless quantity $r(GM/c^4)\Psi_4$, where M is the total mass of the system, and define

$$r\left(\frac{GM}{c^4}\right)\Psi_4^{lm}(t, r) \equiv C_{lm}(t, r), \tag{13.84}$$

so $C_{lm}(t, r)$ is dimensionless and, in the large-r limit, becomes independent of r. Then eq. (13.83) becomes

$$\left(\frac{GM}{c^2}\right)\Psi_4(t, \mathbf{x}) = \frac{1}{r}\sum_{l,m} C_{lm}(t, r) Y_{lm}^{(-2)}(\theta, \phi). \tag{13.85}$$

[23] In numerical simulations Ψ_4 is measured at a finite extraction distance, typically of order $100(GM_{\text{ADM}}/c^2)$. One must therefore estimate the uncertainties arising from the finite extraction radius. This can be done by comparing the results at different extraction radii. An alternative could be to use a tetrad such that Ψ_4 describes the radiative degrees of freedom even at finite r. For a static Kerr BH this is the Kinnersley tetrad given in eq. (12.316). For the dynamical space-time representing the coalescence of two initial BHs into a final Kerr BH the best choice of tetrad is not obvious.

The waveform can be expanded in the same way, defining the quantities h_{lm} from

$$h_+ - ih_\times = \sum_{l,m} h_{lm}(t,r)\, Y_{lm}^{(-2)}(\theta,\phi)\,, \qquad (13.86)$$

and we see from eq. (13.81) that, in the radiation zone, $h_{lm}(t,r)$ can be obtained by integrating twice $\Psi_4^{lm}(t,r)$ with respect to time.

Further reading

- The 3+1 formulation of GR has a long history, which has its roots in the study of the Cauchy or initial value problem in GR. The first studies of the Cauchy problem in GR were made by Darmois in 1927, Lanczos in 1932 and Stellmacher in 1938. For historical reviews see Stachel (1988) and Choquet-Bruhat (2015). Classic work pre-dating the ADM formulation was done by Lichnerowicz in 1939 and 1944 and Choquet-Bruhat in 1948 and 1956; see for example the review by Choquet-Bruhat and York (1980). The Hamiltonian formulation of GR was first developed by Dirac (1958, 1959) [a review of Dirac's work in a historical perspective is given in Deser (2004)]. A book treating in detail the initial value problem in GR is Choquet-Bruhat (2009).

- The classic paper that developed the ADM formalism is Arnowitt, Deser and Misner (1962), published as a chapter of a now long out-of-print book. The original article has more recently been posted on the arXiv, as arXiv:gr-qc/0405109. Another classic paper on the subject is York (1979). In our treatment of the ADM formalism, asymptotic flatness, etc. we have mostly followed the very clear textbook by Poisson (2004a), which covers many geometrical aspects of GR, as well as the review by Gourgoulhon (2007). A recent book covering in detail many conceptual aspects of GR is by Poisson and Will (2014).

- Since the ADM mass gives the total energy of a gravitational system, it should be positive-definite (under appropriate conditions on the energy–momentum tensor). This was very hard to show, but it was eventually proved rigorously (the "positive-energy theorem"). The first complete proof was given by Schoen and Yau (1981). A simpler proof, which makes use of spinor fields, was later given by Witten (1981). A detailed discussion of the theorem can be found in the textbook by Straumann (2004).

- The role of the surface term in the gravitational Hamiltonian is discussed in Regge and Teitelboim (1974) and Hawking and Horowitz (1996). The latter paper also discusses the generalization of the ADM mass to space-times that are not asymptotically flat.

- The "ΓΓ" form of the action was used early on by Einstein (1916). The "trace-K" form of the boundary action was given by York (1972, 1986) and Gibbons and Hawking (1977); see also Brown and York (1993), and Brown, Lau and York (1997, 2000). The latter paper also contains a historical discussion and detailed references on the various forms proposed for the boundary term in the gravitational action.

- The Bondi–Sachs mass was introduced in Bondi, van der Burg and Metzner (1962) (in particular in part D of the paper, by H. Bondi) and in Sachs (1962). For a comparison of the ADM and Bondi–Sachs masses in dynamical space-times, see Section 4.3.5 of Poisson (2004a).

- Properties of the apparent horizons for isolated BHs and in dynamical space-times are discussed in Ashtekar, Fairhurst, and Krishnan (2000) and Ashtekar, Beetle, and Lewandowski (2001), and reviewed in Ashtekar and Krishnan (2004). For the definition of the angular momentum associated with the apparent horizon and its implementation in numerical relativity, see Dreyer, Krishnan, Shoemaker and Schnetter (2003). Christodoulou's formula is given in Christodoulou (1970).

14

GWs from compact binaries. Theory

In this chapter we present the theoretical advances that have been made in recent years, allowing us to obtain a detailed quantitative understanding of the coalescence of compact binaries and of the associated GW production. These advances have been made possible by new analytic techniques and by breakthroughs in numerical relativity. This theoretical understanding plays a crucial role in the detection of coalescing binaries at interferometers, by providing accurate templates. Furthermore, once a detection has been made, accurate predictions for the waveform are crucial for extracting physical information from the observed event, in particular for reconstructing the parameters of the source. The experimental observation of GWs from coalescing binaries will be the subject of Chapter 15, where we will see in detail how the waveforms computed with the techniques discussed in this chapter have been used for the detection and the follow-up study of the first observed GW event, called GW150914, which was produced by the coalescence of a BH–BH binary, and for the subsequent observed events.

The coalescence of a compact binary system can be roughly divided into three stages: the inspiral, the merger and the ringdown. First, we have a long inspiral phase where the emission of gravitational radiation drives the two bodies closer and closer. As we discussed in Chapter 6, the effect on the orbital motion of the loss of energy to gravitational radiation has been observed in binary pulsars, in particular in the Hulse–Taylor pulsar and in the double pulsar, providing a test of the Einstein quadrupole formula at a level of 0.05%. This inspiral phase can last for hundreds of millions of years after the formation of the compact binary system. For the double pulsar, for instance, the time to coalescence is 86 Myr; see Table 6.2. As the system evolves through the inspiral phase its orbital frequency increases, so the frequency of the radiation that it emits increases, as does the amplitude, leading to the characteristic "chirp" waveform computed in Section 4.1, and to a faster and faster evolution. Ground-based GW interferometers can in principle detect the very last part of the inspiral phase, as well as the subsequent merger and ringdown. For instance, as we saw in Section 4.1.1, for a binary system with masses $m_1 = m_2 \simeq 1.4 M_\odot$, typical of NS, the GW frequency reaches 10 Hz approximately 17 min before coalescence, while at 100 Hz we get the radiation emitted about 2 s before merger. For a binary BH system with about 30+30 solar masses, such as the GW150914 event that

Gravitational Waves, Volume 2: Astrophysics and Cosmology. Michele Maggiore.
© Michele Maggiore 2018. Published in 2018 by Oxford University Press.
DOI 10.1093/oso/9780198570899.001.0001

provided the first direct detection of GWs (and which will be studied in great detail in Chapter 15), the signal enters the bandwidth of present ground-based interferometers only about 0.2 s before coalescence.

The second stage is the merger of the two objects. Here v/c approaches a value close to 1 (for instance, for GW150914 the relative speed at merger was about $0.55c$), until the two compact bodies lose their individual identities and merge to form a final black hole. Finally, after the merger is complete, the final BH settles into its fundamental state by radiating away its excess energy through its quasi-normal modes. This is the ringdown phase.

The long inspiral phase can be studied analytically using the post-Newtonian and the post-Minkowskian expansions discussed in Chapter 5, while the ringdown phase can be studied analytically using the methods of BH perturbation theory developed in Chapter 12. In contrast, the merger is a regime where one might think that a theoretical understanding is very hard to obtain. However, in recent years there have been remarkable advances both in the analytic understanding of this regime and in numerical simulations of BH–BH coalescences in full GR, so nowadays we have a rather complete picture of the complex process of the coalescence of compact binary systems, and of the GWs that they emit.

The transition from the inspiral to the merger phase marks the breakdown of the perturbative post-Newtonian (PN) expansion. Analytic techniques address the problem by recasting the perturbative expansion in a form that is suitable for a resummation, motivated by some physical input. We will discuss in Section 14.1 the simplest example of this strategy, which makes use of information on the pole structure of some functions in the test-mass limit. This example is instructive, also because of its simplicity. However, the analytic approach that has been most successful is rather based on a reformulation of the two-body dynamics in terms of an effective one-body action, which will be discussed in Section 14.2. We will then present in Section 14.3 the breakthroughs in numerical relativity that allowed the simulation of compact binary coalescence and the extraction of the gravitational waveform, while in Section 14.4 we will discuss the extra effects that arise when the compact objects are neutron stars rather than BHs.

Analytic and numerical approaches to compact binary coalescence have complementary features. Numerical simulations give highly accurate waveforms covering inspiral, merger and ringdown, but are very time-consuming. A single simulation, depending on the values chosen for the masses and spins of the bodies, can take between one and six months (or more) on supercomputers. Therefore, they can be repeated only for a limited set of parameters of the binary systems, such as its masses and spins. Furthermore, numerical simulations become more and more difficult as the mass ratio becomes much larger than 1 (because of the need of accurately resolving the different length-scales corresponding to the Schwarzschild radii of the two BHs), as well as for large values of the dimensionless spin parameters $\hat{a}_i$ of the two BHs.[1]

[1] Currently, simulations that last for a large number of inspiral cycles have only been performed with mass ratios up to about 4–5, and spin parameter $\hat{a}$ below about 0.8 (while shorter simulation, covering only $O(15)$ inspiral cycles, have been performed for mass ratios up to 8, or for $\hat{a} \simeq 0.88$).

As we saw in Chapter 7, the search for a GW signal in the output data stream of an interferometer is done by performing matched filtering, in which the detector output is tested against a space of waveforms, spanned by the parameters of the binary system. In Chapter 15 we will analyze in detail how these techniques have been applied to the discovery of the first GW event, GW150914, and of the subsequent events, and we will see that in order to efficiently scan the parameter space given by the masses of the initial bodies and their spin, even restricting to an approximate treatment of the spins, still requires the use of about 250'000 templates, which must be efficiently generated almost on-line. Furthermore, once a GW event is discovered in this way, its follow-up and the estimation of the parameters of the coalescing binary system is performed on a finer mesh of a larger parameter space, and typically requires the computation of millions of likelihoods. Therefore, accurate waveforms must be generated very efficiently. Such accurate expressions for the waveform are obtained with a combination of analytic methods, tuned to the numerical relativity results, as we will see in this chapter.

14.1   Non-perturbative resummations. A simple example

In this section we illustrate the general idea of non-perturbative resummations by discussing how a physically-motivated resummation can be performed on the orbital energy function, which is one of the elements (together with the flux function) that enters the computation of the waveform. As we saw in Section 5.6, in the restricted PN approximation the waveform is proportional to $\cos 2\psi(t)$ and to $\sin 2\psi(t)$ for the plus and cross polarizations, respectively, where $\psi(t)$, apart from a logarithmic correction, is equal to the phase $\phi(t)$ describing the orbital motion of an inspiraling binary; see eq. (5.265). More generally, we saw that the amplitude also contains terms proportional to $\cos \psi$ and $\cos 3\psi$ for one polarization (and to $\sin \psi$ and $\sin 3\psi$ for the other polarization) corresponding to mass octupole and current quadrupole radiation, etc. In any case, the fundamental problem is the accurate computation of the phase $\phi(t)$. This phase can be obtained by integrating the energy-balance equation

$$\frac{dE}{dt} = -P_{\text{gw}} , \qquad (14.1)$$

where E is the orbital energy and P_{gw} is the power radiated in GWs. Both quantities are computed in a PN expansion, which is conveniently written in terms of the variable

$$x = \left(\frac{Gm\omega_s}{c^3} \right)^{2/3} , \qquad (14.2)$$

where $m = m_1 + m_2$ is the total mass of the system and ω_s is the orbital angular frequency of the source; see eq. (5.238). Actually, since the PN expansion involves both integer and half-integer powers of x, it can be convenient to use a velocity variable v defined by $v/c = x^{1/2}$, so we write $E = E(v)$ and $P_{\rm gw} = P_{\rm gw}(v)$ and, to conform to the notation commonly used in this context in the literature, we write $P_{\rm gw}(v) \equiv F(v)$. The energy balance equation (14.1) can be written as

$$dt = -\frac{dE}{F},\qquad(14.3)$$

which integrates to

$$t(v) = t_0 + \int_v^{v_0} dv\,\frac{E'(v)}{F(v)},\qquad(14.4)$$

where the prime denotes the derivative with respect to v, and t_0 is a reference point, where the velocity has the value v_0. A convenient choice for t_0 can be the value of time corresponding to the innermost stable circular orbit (ISCO), so $v_0 = v_{\rm ISCO}$. Using the definition $d\phi/dt = \omega_s$ and $v^3 = Gm\omega_s$, which follows from eq. (14.2) with $x = v^2/c^2$, we get

$$d\phi = \frac{v^3}{Gm}dt.\qquad(14.5)$$

Using eq. (14.3) and integrating, we then find

$$\phi(v) = \phi_0 + \frac{1}{Gm}\int_v^{v_0} dv\, v^3\,\frac{E'(v)}{F(v)}.\qquad(14.6)$$

Equations (14.4) and (14.6) together give a parametric representation of the phase $\phi(t)$. The results for the energy function $E(v)$ and the flux function $F(v)$ up to 3.5PN order have been given in eqs. (5.256) and (5.257).[2] For instance to lowest order, i.e. in the Newtonian approximation,

$$E(v) = -\frac{1}{2}\mu v^2,\qquad(14.7)$$

where as usual μ is the reduced mass, while

$$F(v) = \frac{32}{5Gc^5}\nu^2 v^{10},\qquad(14.8)$$

where $\nu = \mu/m$ is the symmetric mass ratio. Then eqs. (14.4) and (14.6) give (setting $v_0 = \infty$, since in the Newtonian approximation there is no least stable orbit, so we rather choose t_0 equal to the time $t_{\rm coal}$ at which the two objects coalesce, and the velocity in the Newtonian approximation formally diverges)

$$t(v) = t_0 - \frac{5}{256}\frac{Gmc^5}{\nu}\frac{1}{v^8},\qquad(14.9)$$

$$\phi(v) = \phi_0 - \frac{c^5}{32\nu}\frac{1}{v^5}.\qquad(14.10)$$

[2] The energy function has also been computed up to 4PN order in Damour, Jaranowski and Schäfer (2014).

Eliminating v we get

$$\phi(t) = \phi_0 - \left(\frac{5GM_c}{c^3}\right)^{-5/8} \tau^{5/8}, \qquad (14.11)$$

where $\tau = t_{\rm coal} - t$ is the time to coalescence and $M_c = m\nu^{3/5}$ is the chirp mass. The phase of the GW in the restricted PN approximation is $\Phi(t) = 2\phi(t)$, and we therefore recover eq. (4.30). Using the full 3.5PN results for $E(v)$ and $F(v)$ given in eqs. (5.256) and (5.257) one rather gets back the 3.5PN result for the phase, shown in eq. (5.260).

The problem is now how to go beyond the perturbative expansion of $E(v)$ and $F(v)$, in order to push the validity of the PN expansion closer to the merger phase. In this section we illustrate the general idea of non-perturbative resummations using a relatively simple technique. The idea is to perform a non-perturbative resummation of the energy function using as physical input the fact that, in the test-mass limit $\nu = 0$, we know the pole structure of some function related to the energy function $E(v)$. One can then enforce it at finite ν, using the technique of Padé approximants. By itself, this method is not used to provide accurate templates for compact binary coalescence. It is, however, an instructive and simple example of a resummation technique, and can also be used as an ingredient of the more elaborate EOB approach that we will discuss in Section 14.2, which is a method actually used in the search for and follow-up of GW events at interferometers.

Given a function $f(z)$ of a real or complex variable z, of which we know the Taylor expansion up to a given order n,

$$f(z) = f_0 + f_1 z + \ldots + f_n z^n + O(z^{n+1}), \qquad (14.12)$$

its Padé approximant of order (m, k), $P_k^m(z)$ (also denoted by $[f(z)]_k^m$), is given by

$$P_k^m(z) = \frac{N_m(z)}{D_k(z)}, \qquad (14.13)$$

where $k + m = n$, $N_m(z)$ is a polynomial of order m in z,

$$N_m(z) = n_0 + n_1 z + \ldots + n_m z^m, \qquad (14.14)$$

and $D_k(z)$ is a polynomial of order k, normalized so that $D(0) = 1$,

$$D_k(z) = 1 + d_1 z + \ldots + d_k z^k. \qquad (14.15)$$

Since $k + m = n$, the $k + (m + 1)$ coefficients d_i $(i = 1, \ldots, k)$ and n_j $(j = 0, \ldots, m)$ can be uniquely determined in terms of the $n + 1$ coefficients $f_0, \ldots, f_n$. The Padé approximants provide functions whose Taylor expansions coincide, by construction, with that of $f(z)$ up to order n. However, they have a specific structure of poles and zeros in the complex plane, determined by the zeros of the functions $D_k(z)$ and $N_m(z)$, respectively. Therefore, they are particularly useful resummations of a Taylor series when, beside knowing the Taylor expansion of $f(z)$ up to a given order, we also have some additional information on its pole structure in the complex plane, at least as far as the poles closest to the origin are concerned, or on its zeros in the complex plane. Actually, Padé approximants can even be employed to resum series that are not convergent, thereby providing a non-perturbative definition of $f(z)$.

In our problem the use of Padé approximants to "improve" the Taylor series is useful if we have some understanding of the pole structure of the functions $E(v)$ and $F(v)$ that enter in eq. (14.6). As an example of the use of the method, we can make use of the fact that in the test-mass limit, i.e. when the symmetric mass ratio $\nu = 0$, we can solve the problem exactly, so at least in this limit the pole structure of these functions is known. One can then assume that the same pole structure is valid at finite ν, and that the only modification is that the position of the poles depends on ν.

To implement this idea, we must first recall some elementary facts about circular orbits of test masses in the Schwarzschild metric.[3] We consider a test particle of mass m_2 moving in the Schwarzschild metric generated by a mass m_1 with $m_2/m_1 \to 0$, so $m = m_1+m_2 \to m_1$. Using Schwarzschild coordinates (t, r, θ, ϕ), the Schwarzschild metric has the usual form

[3]See e.g. Hartle (2003), Chapter 9.

$$ds^2 = \left(1 - \frac{2Gm}{c^2r}\right)(cdt)^2 + \left(1 - \frac{2Gm}{c^2r}\right)^{-1} dr^2 + r^2(d\theta^2 + \sin^2\theta d\phi^2).$$

$$(14.16)$$

The energy E_2 of the particle 2 is given by

$$E_2 = g_{\mu\nu}\xi^\mu p_2^\nu, \tag{14.17}$$

where ξ^μ is the Killing vector of the Schwarzschild metric corresponding to invariance under time translations, which in Schwarzschild coordinates is simply $\xi^\mu = (1, 0, 0, 0)$. This gives $E_2 = g_{00}p_2^0 = g_{00}m_2u^0$, where u^μ is the four-velocity. For circular orbits in the Schwarzschild metric,

$$u^0 = \left(1 - \frac{3Gm}{c^2r}\right)^{-1/2}, \tag{14.18}$$

so

$$\frac{E_2}{m_2c^2} = \left(1 - \frac{2Gm}{c^2r}\right)\left(1 - \frac{3Gm}{c^2r}\right)^{-1/2}. \tag{14.19}$$

Observe that, in the limit $r \to \infty$, we have $E_2 \to m_2c^2$, i.e. E_2 is the relativistic energy including the rest-mass contribution. For stable circular orbits in the Schwarzschild metric the orbital angular frequency $\omega_s = d\phi/dt$ of the test mass is given by

$$\omega_s^2 = \frac{Gm}{r^3} \tag{14.20}$$

(which has exactly the same form as the non-relativistic Kepler's law). Equation (14.2) therefore gives

$$x = \frac{Gm}{c^2r}, \tag{14.21}$$

and eq. (14.19) reads

$$\frac{E_2(x)}{m_2 c^2} = \frac{1 - 2x}{\sqrt{1 - 3x}}. \tag{14.22}$$

This shows that the energy per unit mass of a test particle in the Schwarzschild metric, as a function of x, has a square-root singularity in the complex x-plane at $x = 1/3$, i.e. at $r = 3Gm/c$. Observe that this is just the position of the so-called *light-ring*, i.e. the value of r for which circular orbits of light rays are possible in the Schwarzschild metric.

In the problem in which we are actually interested, m_2/m_1 is finite. In this case it is convenient to write the total relativistic energy E_{tot}, separating explicitly the rest-mass contribution,

$$E_{\text{tot}} = mc^2 + E, \tag{14.23}$$

where E is the orbital energy, whose PN expansion is given in eq. (5.256). We also introduce the new energy function

$$\epsilon(x) = \frac{E_{\text{tot}}^2(x) - (m_1^2 + m_2^2)c^4}{2m_1 m_2 c^4}, \tag{14.24}$$

which is symmetric under exchange of the two particles. In the test mass limit $\epsilon(x)$ reduces to $E_2(x)/m_2 c^2$, so it is given by eq. (14.22). To obtain a quantity that in the test-mass limit has a pole, rather than a square-root singularity, one can square this expression, defining

$$e(x) = \epsilon^2(x) - 1. \tag{14.25}$$

In the test-mass limit this function has the exact expression

$$e(x; \nu = 0) = -x \frac{1 - 4x}{1 - 3x}, \tag{14.26}$$

whose Taylor expansion is

$$e(x; \nu = 0) = -x + \sum_{k=1}^{\infty} a_k(\nu = 0) x^{k+1}, \tag{14.27}$$

where $a_k(\nu = 0) = 3^{k-1}$. To appreciate the power of the Padé resummation technique when the singularity structure of the function is known, suppose that we only knew the expansion of $e(x; \nu = 0)$ up to 2PN order, $e_{2\text{PN}}(x; \nu = 0) = -x[1 - x - 3x^2 + O(x^3)]$, and that we knew that the exact function had a simple pole. Then, it would be natural to consider a Padé resummation of the form

$$P_1^1(x) = -x \frac{n_0 + n_1 x}{1 + d_1 x}, \tag{14.28}$$

whose Taylor expansion is

$$P_1^1(x) = -x[n_0 + (n_1 - d_1 n_0)x + (d_1^2 n_0 - d_1 n_1)x^2 + O(x^3)]. \tag{14.29}$$

Requiring that this agrees with $-x(1-x-3x^2)$ up to $O(x^3)$ fixes uniquely $n_0 = 1$, $n_1 = -4$ and $d_1 = -3$. In this case, we have even reconstructed the exact result (14.26) from its 2PN expansion!

So, for $\nu = 0$ we know the function $e(x;\nu)$ exactly, and we know in particular its singularity structure in the complex plane: the function has a simple pole at $x = 1/3$. Our aim, of course, is to compute this function for $\nu \neq 0$. In this case, using eq. (5.256), its Taylor expansion up to 3PN order is[4]

$$
e_{3\mathrm{PN}}(x,\nu) = -x + \left(1 + \frac{\nu}{3}\right)x^2 + \left(3 - \frac{35\nu}{12}\right)x^3 \tag{14.30}
$$
$$
+ \left(9 - \frac{4309}{72}\nu + \frac{205\pi^2}{96}\nu + \frac{103}{36}\nu^2 - \frac{1}{81}\nu^3\right)x^4 ,
$$

which for $\nu = 0$ gives back the first few terms of eq. (14.27), as it should. One can then attempt a Padé resummation, assuming that even for $\nu \neq 0$ the structure of the singularities of $e(x,\nu)$ in the complex plane will be the same as the one of $e(x,\nu = 0)$, i.e. that the closest singularity will still be a simple pole. Of course the location of the pole will in general be shifted from the value that it has for $\nu = 0$, but at least for small enough ν we expect that it will be close to $1/3$. If this assumption is correct, then a Padé resummation of the PN series for $e(x,\nu)$ provides a result that is closer to the true value, and in fact even catches the main non-perturbative feature of the problem, which is the existence of a pole-like singularity or, in physical terms, of a light ring.[5]

In the next section we will present a different and more elaborate non-perturbative technique, the effective one-body method, (which can also be combined with Padé resummations), which is actually used in the development of analytic templates. We will see how it can be extended to cover the merger and ringdown phase, and that it can be tuned so that it compares very well with the results obtained from numerical relativity.

14.2 Effective one-body action

In this section we introduce the effective one-body (EOB) action, due to Buonanno and Damour, in which the two-body problem in GR is mapped into a one-body problem in an effective metric.

14.2.1 Equivalence to a one-body problem

To understand the method, we consider first the equations of motion of the two-body system up to 2PN order. We are therefore neglecting radiation-reaction effects, which, as discussed in Section 5.3.5, start at 2.5PN.[6] Then the equations of motion for the two bodies, labeled by the index $a = 1, 2$, have the generic form

$$
\frac{d\mathbf{v}_a}{dt} = \mathbf{A}_a^{2\mathrm{PN}}(\mathbf{x}_1, \mathbf{x}_2, \mathbf{v}_1, \mathbf{v}_2) , \tag{14.31}
$$

[4]The expansion to 4PN order can be obtained using eq. (5.5) of Damour, Jaranowski and Schäfer (2014). Observe, however, that to 4PN order $E(x)$ and $e(x)$ also contain a term proportional to $\nu x^5 \log x$.

[5]Similar arguments based on the test-mass limits can motivate a Padé resummation of the flux function $F(x)$ that enters in eqs. (14.4) and (14.6). See the original paper by Damour, Iyer and Sathyaprakash (1998) for details.

[6]As we will mention below, the EOB approach has been extended up to full 4PN order in the conservative dynamics. In the text, for the conservative part, we will use the 2PN results to illustrate the idea, and we will just mention the modifications that are required when extending the conservative dynamics to 3PN and to 4PN orders.

where $\mathbf{A}_a^{2\mathrm{PN}}$ has a Newtonian term, a 1PN term and a 2PN term and so can be written as

$$\mathbf{A}_a^{2\mathrm{PN}} = \mathbf{A}_{a,0} + \frac{1}{c^2}\mathbf{A}_{a,2} + \frac{1}{c^4}\mathbf{A}_{a,4}\,. \qquad (14.32)$$

In the center-of-mass frame, the explicit form of $\mathbf{A}_a^{2\mathrm{PN}}$ for circular orbits can be read from eqs. (5.251) and (5.252), while for general orbits the expression is very long and can be found in Blanchet (2006), eq. (182).

Recall from Section 5.1.4 that the computation leading to eqs. (14.31) and (14.32) is performed using the harmonic gauge, so these equations of motion are written in harmonic coordinates. One can now ask what is the Lagrangian from which these equations of motion can be derived. It turns out that at 2PN order the corresponding Lagrangian depends not only on the positions and velocities of the two bodies, but also on their accelerations, i.e. it has the form[7]

$$L(\mathbf{x}_a, \mathbf{v}_a, \mathbf{a}_a) = L_{0\mathrm{PN}}(\mathbf{x}_1 - \mathbf{x}_2, \mathbf{v}_1, \mathbf{v}_2) + L_{1\mathrm{PN}}(\mathbf{x}_1 - \mathbf{x}_2, \mathbf{v}_1, \mathbf{v}_2)$$
$$+ L_{2\mathrm{PN}}(\mathbf{x}_1 - \mathbf{x}_2, \mathbf{v}_1, \mathbf{v}_2, \mathbf{a}_1, \mathbf{a}_2)\,. \qquad (14.33)$$

It is, however, possible to eliminate the dependence on the accelerations in the 2PN Lagrangian by transforming from harmonic coordinates to the ADM coordinates introduced by Arnowitt, Deser and Misner in their Hamiltonian approach to GR; see Section 13.1. The transformation from harmonic to ADM coordinates has been carried out explicitly[8] and, denoting by $\boldsymbol{q}_a$, with $a = 1, 2$, the ADM coordinates of the two bodies, it gives rise to a 2PN Lagrangian $L(\boldsymbol{q}_a, \dot{\boldsymbol{q}}_a)$ that depends on positions and velocities, but not on accelerations. From this, performing the Legendre transform, one can find the corresponding Hamiltonian $H(\boldsymbol{q}_a, \boldsymbol{p}_a)$, where $\boldsymbol{p}_a$ are the momenta of the two particles. One can now work in the center-of-mass frame, introducing the relative position $\boldsymbol{q}' = \boldsymbol{q}_1 - \boldsymbol{q}_2$ and its conjugate momentum $\boldsymbol{p}' = \partial L / \partial \boldsymbol{q}'$. To write the Hamiltonian explicitly it is convenient to introduce the rescaled relative ADM coordinate $\boldsymbol{q}$ and the rescaled ADM momentum $\boldsymbol{p}$ defined by[9]

$$\boldsymbol{q} = \frac{\boldsymbol{q}'}{Gm}\,, \qquad (14.34)$$

$$\boldsymbol{p} = \frac{\boldsymbol{p}'}{\mu}\,, \qquad (14.35)$$

where as usual $m = m_1 + m_2$ is the total mass of the system and μ is the reduced mass. One further introduces the rescaled quantities

$$\widehat{t} = \frac{t}{Gm}\,, \qquad (14.36)$$

$$\widehat{H} = \frac{H^{\mathrm{NR}}}{\mu} = \frac{H^{\mathrm{R}} - mc^2}{\mu}\,, \qquad (14.37)$$

where H^{NR} and H^{R} denote the non-relativistic and relativistic Hamiltonians, respectively. Then the explicit form of the 2PN Hamiltonian in rescaled ADM coordinates is

$$\widehat{H}(\boldsymbol{q}, \boldsymbol{p}) = \widehat{H}_0(\boldsymbol{q}, \boldsymbol{p}) + \frac{1}{c^2}\,\widehat{H}_2(\boldsymbol{q}, \boldsymbol{p}) + \frac{1}{c^4}\,\widehat{H}_4(\boldsymbol{q}, \boldsymbol{p})\,, \qquad (14.38)$$

---

[7]See Damour and Deruelle (1981) for the explicit expression.

[8]See Damour and Schäfer (1985).

[9]In this section and in the next it will be unavoidable to introduce various different type of coordinates. In order not to get lost with the various definitions, let us summarize here the notation used. The harmonic coordinates of the two bodies are denoted by $\mathbf{x}_a$, where $a = 1, 2$ labels the body. ADM coordinates are denoted by $\boldsymbol{q}_a$, $a = 1, 2$. The relative coordinate is $\boldsymbol{q}' = \boldsymbol{q}_1 - \boldsymbol{q}_2$ and its conjugate momentum is denoted by $\boldsymbol{p}'$. The corresponding angular momentum is $\boldsymbol{L}' = \boldsymbol{q}' \times \boldsymbol{p}'$, and $r' = |\boldsymbol{q}'|$. The rescaled relative ADM coordinate, defined in eq. (14.34), is denoted by $\boldsymbol{q}$. Rescaled ADM polar coordinates in the orbital plane are denoted by (ρ, φ); see eq. (14.44). The coordinates of the effective one-body problem are $x_{\mathrm{eff}}^\mu = (t_0, R, \theta, \phi)$; see eq. (14.52). We will denote by r the dimensionless radial variable $r = Rc^2/(Gm)$; see eq. (14.88). Finally, the polar coordinates of the effective problem are denoted collectively by $\mathbf{Q} = (R, \theta, \varphi)$, and the corresponding conjugate momentum by $\mathbf{P}$; see eq. (14.89).

where

$$\widehat{H}_0(\boldsymbol{q},\boldsymbol{p}) = \frac{1}{2}\,\boldsymbol{p}^2 - \frac{1}{q}\,, \tag{14.39}$$

$$\widehat{H}_2(\boldsymbol{q},\boldsymbol{p}) = -\frac{1}{8}\,(1-3\nu)\,\boldsymbol{p}^4 - \frac{1}{2q}\,[(3+\nu)\,\boldsymbol{p}^2 + \nu(\boldsymbol{n}\cdot\boldsymbol{p})^2] + \frac{1}{2q^2}\,, \tag{14.40}$$

$$\widehat{H}_4(\boldsymbol{q},\boldsymbol{p}) = \frac{1}{16}\,(1-5\nu+5\nu^2)\,\boldsymbol{p}^6$$
$$+ \frac{1}{8q}\,[(5-20\nu-3\nu^2)\,\boldsymbol{p}^4 - 2\nu^2\,\boldsymbol{p}^2(\boldsymbol{n}\cdot\boldsymbol{p})^2 - 3\nu^2(\boldsymbol{n}\cdot\boldsymbol{p})^4]$$
$$+ \frac{1}{2q^2}\,[(5+8\nu)\,\boldsymbol{p}^2 + 3\nu\,(\boldsymbol{n}\cdot\boldsymbol{p})^2] - \frac{1}{4q^3}\,(1+3\nu)\,, \tag{14.41}$$

and we have also used the notation $q \equiv |\boldsymbol{q}|$ and $\boldsymbol{n} \equiv \boldsymbol{q}/q$. Having a well-defined Hamiltonian allows us to study the classical dynamics of the system using Hamilton–Jacobi theory. We introduce the action S, and the rescaled action

$$\widehat{S} = \frac{S}{G\mu m}\,. \tag{14.42}$$

The action S satisfies the Hamilton–Jacobi equation[10]

$$\frac{\partial S}{\partial t} + H\left(q_i, \frac{\partial S}{\partial q_i}\right) = 0\,. \tag{14.43}$$

[10]See any book on advanced classical mechanics, e.g. Goldstein (1980).

We consider the motion in the orbital plane, and we denote by (r,φ) the rescaled polar ADM coordinates in this plane, so

$$\boldsymbol{q} = (\rho\cos\varphi, \rho\sin\varphi, 0)\,. \tag{14.44}$$

Since the Hamiltonian (14.38) is invariant under time translations and rotations, there are two conserved quantities, the rescaled non-relativistic energy $\widehat{E}^{\mathrm{NR}}$ and the rescaled angular momentum $\boldsymbol{j}$,

$$\widehat{H}(\boldsymbol{q},\boldsymbol{p}) = \widehat{E}^{\mathrm{NR}}\,, \tag{14.45}$$

$$\boldsymbol{q}\times\boldsymbol{p} = \boldsymbol{j} \equiv \frac{\mathcal{J}}{G\mu m}\,. \tag{14.46}$$

The invariance of the Hamiltonian under rotations and time translations allows us to look for solutions of eq. (14.43) in the form of a separable function $S(t,\varphi,\rho)$. In terms of the rescaled action $\widehat{S}$ and of the constants of motion $\widehat{E}^{\mathrm{NR}}$ and j, we write the solution in the form

$$\widehat{S}(t,\varphi,\rho) = -\widehat{E}^{\mathrm{NR}}t + j\varphi + \widehat{S}_\rho(\rho; \widehat{E}^{\mathrm{NR}}, j)\,. \tag{14.47}$$

The Hamilton–Jacobi equation then becomes an algebraic equation for $dS_\rho/d\rho$, which can be solved iteratively up to 2PN order, inserting the explicit form (14.39)–(14.41) into eq. (14.43). One can then use the result to compute the (non-rescaled) radial action variable

$$I_R(E^{\mathrm{NR}}, \mathcal{J}) \equiv 2\int_{\rho_{\min}}^{\rho_{\max}} \frac{d\rho}{2\pi}\,\frac{dS_\rho(\rho; E^{\mathrm{NR}}, \mathcal{J})}{d\rho}\,, \tag{14.48}$$

where $\rho_{\min}$ and $\rho_{\max}$ are the minimum and maximum values of ρ for a motion with the given value of the constants of motion E^{NR} and $\mathcal{J}$. The factor of two in the numerator comes from the fact that the definition of the action variable involves an integral over a cycle, $\oint$, which in our one-dimensional problem is twice the integral from $\rho_{\min}$ to $\rho_{\max}$. Equation (14.48) can finally be inverted to get E^{NR} as a function of I_R and $\mathcal{J}$. It is convenient to define

$$\alpha = G\mu m\,, \tag{14.49}$$

and to use $\mathcal{J}$ and $\mathcal{N} = I_R + \mathcal{J}$ instead of $\mathcal{J}$ and I_R. The result of this long computation, working consistently up to 2PN order, and writing the result in terms of $E^{\mathrm{R}} = E^{\mathrm{NR}} + mc^2$, is

$$E^{\mathrm{R}}(\mathcal{N}, \mathcal{J}) = mc^2 - \frac{1}{2}\frac{\mu\alpha^2}{\mathcal{N}^2}\left[1 + \frac{\alpha^2}{c^2}\left(\frac{6}{\mathcal{N}\mathcal{J}} - \frac{1}{4}\frac{15-\nu}{\mathcal{N}^2}\right)\right. \tag{14.50}$$

$$\left. + \frac{\alpha^4}{c^4}\left(\frac{5}{2}\frac{7-2\nu}{\mathcal{N}\mathcal{J}^3} + \frac{27}{\mathcal{N}^2\mathcal{J}^2} - \frac{3}{2}\frac{35-4\nu}{\mathcal{N}^3\mathcal{J}} + \frac{1}{8}\frac{145-15\nu+\nu^2}{\mathcal{N}^4}\right)\right]\,.$$

Observe that, in the "old quantum theory", classical relations of this type, obtained by studying the two-body problem in a Coulomb potential, were the starting point for the Bohr–Sommerfeld quantization of the hydrogen atom: the action variables I_R and $\mathcal{J}$, or equivalently $\mathcal{N}$ and $\mathcal{J}$, are adiabatic invariants, and the passage to quantum theory was performed by quantizing them in units of $\hbar$, so $\mathcal{N}/\hbar$ becomes the principal quantum number, and $\mathcal{J}/\hbar$ becomes the total angular momentum quantum number. In eq. (14.50) the coupling constant $\alpha = G\mu m = Gm_1 m_2$ plays the role that in the Coulomb problem is played by the fine structure constant. As with any $1/r$ potential, the term $O(\alpha^2)$ in eq. (14.50) depends only on $\mathcal{N}$ and not on $\mathcal{J}$, while relativistic corrections, given by the PN expansion in the gravitational two-body problem, or by a Klein–Gordon or Dirac equation in the Coulomb problem (for spinless and spin-1/2 particles, respectively), generate corrections that start from $O(\alpha^4)$, and break the degeneracy in $\mathcal{J}$. Even if we are interested only in the classical gravitational two-body problem, this analogy with quantum theory can be quite instructive.

The crucial point is that, even if the computation has been performed in a specific gauge, eq. (14.50) is invariant under coordinate transformations, since it is a relation among the observable quantities E^{R}, $\mathcal{N}$ and $\mathcal{J}$. In other words, eq. (14.50) is a way to summarize, in a coordinate-invariant manner, the two-body dynamics at 2PN order.

The basic idea of the EOB method is to find a one-body problem in an external space-time that reproduces eq. (14.50). One considers a test particle of mass $\mu = m_1 m_2/(m_1 + m_2)$ governed by the action

$$S_{\mathrm{EOB}} = -\mu c \int ds_{\mathrm{eff}}\,, \tag{14.51}$$

where

$$ds_{\mathrm{eff}}^2 = g_{\mu\nu}^{\mathrm{eff}} dx_{\mathrm{eff}}^\mu dx_{\mathrm{eff}}^\nu$$

$$= -A(R)c^2 dt_0^2 + B(R)dR^2 + R^2(d\theta^2 + \sin^2\theta d\phi^2). \quad (14.52)$$

We therefore have a test mass μ moving along the geodesic of this metric. In eq. (14.52) $x_{\rm eff}^\mu = (t_0, R, \theta, \phi)$ are some effective coordinates, whose relation to the ADM coordinates of the real two-body problem will be discussed later,[11] and $A(R)$ and $B(R)$ are two functions to be determined, of the generic form

$$A(R) = 1 + a_1\frac{Gm}{c^2R} + a_2\left(\frac{Gm}{c^2R}\right)^2 + a_3\left(\frac{Gm}{c^2R}\right)^3 + O\left(\frac{1}{R^4}\right), \quad (14.53)$$

$$B(R) = 1 + b_1\frac{Gm}{c^2R} + b_2\left(\frac{Gm}{c^2R}\right)^2 + O\left(\frac{1}{R^3}\right). \quad (14.54)$$

Recall from Section 5.1 that a_1 determines the dynamics of a test mass at 0PN (i.e. Newtonian) order, a_2 and b_1 enter at 1PN order, and a_3 and b_2 enter at 2PN order, while higher-order coefficients a_4, b_3, etc. are not needed if we work at 2PN order.

First of all we know that, at the Newtonian level, the two-body problem with masses m_1 and m_2 is equivalent to the motion of a mass $\mu = m_1m_2/(m_1+m_2)$ in the gravitational potential of a mass $m = m_1+m_2$. This immediately fixes $a_1 = -2$, corresponding to the Schwarzschild metric generated by a mass m. To require that the equivalence between the real two-body problem and the effective one-body problem extends up to 2PN order, Buonanno and Damour require that the effective one-body problem reproduces eq. (14.50), since the latter formula is a coordinate-invariant way of summarizing the 2PN dynamics. To implement this requirement the first step is to derive the relation between energy and the action variable, for the effective one-body problem. This can be obtained by using the fact that the Hamilton–Jacobi equation for a test mass μ described by coordinates $x_{\rm eff}^\mu$ and moving in a metric $g_{\mu\nu}^{\rm eff}$ is[12]

$$g_{\rm eff}^{\mu\nu}\frac{\partial S}{\partial x_{\rm eff}^\mu}\frac{\partial S}{\partial x_{\rm eff}^\nu} + \mu^2 c^2 = 0. \quad (14.55)$$

In the static and spherically symmetric metric $g_{\mu\nu}^{\rm eff}$ given by eq. (14.52), restricting for simplicity to motion in the equatorial plane $\theta = \pi/2$, we can write the solution S of the Hamilton–Jacobi equation as

$$S(t_0, R, \phi) = -\mathcal{E}_0 t_0 + \mathcal{J}_0\phi + S_R^0(R; \mathcal{E}_0, \mathcal{J}_0). \quad (14.56)$$

The quantities $\mathcal{E}_0$ and $\mathcal{J}_0$ are the separation constants of the effective problem, and their relation to the energy $\mathcal{E}$ and angular momentum $\mathcal{J}$ of the real two-body problem will be determined later. Inserting eq. (14.56) into eq. (14.55) gives dS_R^0/dR in terms of $\mathcal{E}_0$ and $\mathcal{J}_0$,

$$-\frac{\mathcal{E}_0^2}{A(R)c^2} + \frac{1}{B(R)}\left(\frac{dS_R^0}{dR}\right)^2 + \frac{\mathcal{J}_0^2}{R^2} + \mu^2 c^2 = 0. \quad (14.57)$$

This allows us to compute the radial action variable of the effective problem,

$$I_R^0(\mathcal{E}_0, \mathcal{J}_0) \equiv 2\int_{R_{\rm min}}^{R_{\rm max}}\frac{dR}{2\pi}\frac{dS_R^0}{dR}, \quad (14.58)$$

[11]Note in particular that R is not the same as the relative radial distance in ADM coordinates.

[12]See e.g. Landau–Lifshitz, Vol. II (1979), eq. (9.19).

as a function of $\mathcal{E}_0$ and $\mathcal{J}_0$. After computing the integral,[13] we can invert this relation to express $\mathcal{E}_0$ in terms of $\mathcal{J}_0$ and I_R^0 or, better yet, in terms of $\mathcal{J}_0$ and $\mathcal{N}_0 = I_R^0 + \mathcal{J}_0$. The result of this computation, keeping terms up to 2PN order, i.e. up to $O(1/c^4)$, is

$$\mathcal{E}_0(\mathcal{N}_0, \mathcal{J}_0) = \mu c^2 - \frac{1}{2}\frac{\mu \alpha^2}{\mathcal{N}_0^2}\left[1 + \frac{\alpha^2}{c^2}\left(\frac{C_{3,1}}{\mathcal{N}_0 \mathcal{J}_0} + \frac{C_{4,0}}{\mathcal{N}_0^2}\right)\right.$$
$$\left. + \frac{\alpha^4}{c^4}\left(\frac{C_{3,3}}{\mathcal{N}_0 \mathcal{J}_0^3} + \frac{C_{4,2}}{\mathcal{N}_0^2 \mathcal{J}_0^2} + \frac{C_{5,1}}{\mathcal{N}_0^3 \mathcal{J}_0} + \frac{C_{6,0}}{\mathcal{N}_0^4}\right)\right], \quad (14.59)$$

where, as before, $\alpha = G\mu m$, while $C_{3,1}, \ldots, C_{6,0}$ are combinations of a_2, a_3, b_1 and b_2, whose explicit expression can be found in Buonanno and Damour (1999).

To compare this result with eq. (14.50) we still have to define the rules that relate the real two-body problem to the effective one-body problem, i.e. we must fix the relation between $\mathcal{E}, \mathcal{N}$ and $\mathcal{J}$ on one side, and $\mathcal{E}_0, \mathcal{N}_0$ and $\mathcal{J}_0$ on the other side. The choice of these rules is in our hands, since any choice that allows us to recover eq. (14.50) gives in principle a legitimate reformulation of the two-body problem in terms of an effective one-body dynamics, which, to 2PN order, is equivalent to the original problem. However, if we want to use the effective one-body reformulation as a starting point for a resummation of the PN expansion of the original two-body problem, as we will do in what follows, not all choices will be equally successful, and we must try to find some rather natural prescription for this mapping.

Even if the EOB method is purely classical, it is conceptually useful to look at eqs. (14.50) and (14.59) as semiclassical expressions for the quantum energy levels of the system. In this case the Bohr–Sommerfeld quantization conditions tells us that $\mathcal{N}$ and $\mathcal{N}_0$ are quantized in units of $\hbar$, i.e. $\mathcal{N} = n\hbar$, with n interpreted as the principal quantum numbers, while $\mathcal{J}$ and $\mathcal{J}_0$ are again quantized in units of $\hbar$ and interpreted as quantized angular momenta. So, it is very natural to require the identifications

$$\mathcal{N} = \mathcal{N}_0 \qquad (14.60)$$

and

$$\mathcal{J} = \mathcal{J}_0. \qquad (14.61)$$

Concerning the relation between $\mathcal{E}$ and $\mathcal{E}_0$, the issue is more subtle. The identification $\mathcal{E} = \mathcal{E}_0$ is not viable, because in the limit $\mathcal{N} \to \infty$ eq. (14.50) gives $\mathcal{E} \to mc^2$ while (given that $\mathcal{N} = \mathcal{N}_0$) in the same limit eq. (14.59) gives $\mathcal{E}_0 \to \mu c^2$. So, in general, the mapping between the energy of the real two-body problem and the energy of the effective one-body problem must be non-trivial,

$$\mathcal{E}_0 = f(\mathcal{E}). \qquad (14.62)$$

The issue is therefore to find a "natural" expression for the mapping (14.62), so that the corresponding one-body dynamics might be a useful starting point for a resummation of the PN expansion.

The simplest possibility is to write $\mathcal{E}_0 = \mathcal{E} - c_0$ with $c_0 = mc^2 - \mu c^2$, i.e. to identify the non-relativistic parts of $\mathcal{E}$ and $\mathcal{E}_0$. However, identifying the non-relativistic parts of eqs. (14.50) and (14.59) one finds that there are six equations (obtained by equating the coefficients of $\alpha^2/\mathcal{N}\mathcal{J},\ldots,\alpha^4/\mathcal{N}^4$) and only four unknowns a_2, a_3, b_1, b_2, and no solution exists. One might try to give up the requirement that the mass of the effective one-body theory is equal to the reduced mass μ of the two-body system, and introduce new free parameters that relate the mass of the effective one-body theory to μ. However, a much more natural solution emerges if one keeps the mass of the effective particle equal to the reduced mass μ, and rather introduces new parameters in the mapping (14.62), writing it as

$$\mathcal{E}_0^{\mathrm{NR}} = \mathcal{E}^{\mathrm{NR}}\left[1 + \alpha_1\frac{\mathcal{E}^{\mathrm{NR}}}{\mu c^2} + \alpha_2\left(\frac{\mathcal{E}^{\mathrm{NR}}}{\mu c^2}\right)^2 + \ldots\right], \qquad (14.63)$$

where $\mathcal{E}_0^{\mathrm{NR}} = \mathcal{E}_0 - \mu c^2$, $\mathcal{E}^{\mathrm{NR}} = \mathcal{E} - mc^2$, and α_1 and α_2 are two new free parameters. The terms denoted by dots are beyond the 2PN order, as it is clear from the explicit powers of c, and we can neglect them. The identification of the two energy level formulas now provide six equations for the six unknown $a_2, a_3, b_1, b_2, \alpha_1, \alpha_2$. Actually, it turns out that the equations are not all independent and we can choose the value of one variable. The natural choice is to set $b_1 = 2$, so the linearized metric coincides with the Schwarzschild metric. The remaining coefficients are then uniquely fixed, to the values $a_2 = 0$, $a_3 = 2\nu$, $b_2 = 4 - 6\nu$, $\alpha_1 = \nu/2$ and $\alpha_2 = 0$.

In conclusion, at 2PN order the real two-body problem with masses m_1, m_2, with $m_1 + m_2 = m$, can be mapped into an effective one-body problem for a mass μ equal to the reduced mass $m_1 m_2/m$, moving along the geodesics of the metric (14.52), with

$$A(R) = 1 - \frac{2Gm}{c^2 R} + 2\nu\left(\frac{Gm}{c^2 R}\right)^3,$$

$$B(R) = 1 + \frac{2Gm}{c^2 R} + (4 - 6\nu)\left(\frac{Gm}{c^2 R}\right)^2, \qquad (14.64)$$

supplemented by a mapping between the energy $\mathcal{E}_0$ of the effective problem and the energy $\mathcal{E}$ of the real problem, given by

$$\mathcal{E}_0^{\mathrm{NR}} = \mathcal{E}^{\mathrm{NR}}\left[1 + \frac{\nu}{2}\frac{\mathcal{E}^{\mathrm{NR}}}{\mu c^2}\right]. \qquad (14.65)$$

Inserting the definitions $\mathcal{E}_0^{\mathrm{NR}} = \mathcal{E}_0 - \mu c^2$ and $\mathcal{E}^{\mathrm{NR}} = \mathcal{E} - mc^2$, writing eq. (14.65) in terms of the energies $\mathcal{E}_0$ and $\mathcal{E}$ that include the rest-mass contribution, and recalling that $\nu = \mu/m$, we get

$$\mathcal{E}_0 = \frac{\mathcal{E}^2 - m_1^2 c^4 - m_2^2 c^4}{2mc^2}, \qquad (14.66)$$

whose inversion gives

$$\mathcal{E} = mc^2 \left[1 + 2\nu \left(\frac{\mathcal{E}_0 - \mu c^2}{\mu c^2} \right) \right]^{1/2}. \tag{14.67}$$

Equations (14.64) and (14.66), together with eqs. (14.51) and (14.52), define the equivalence of the 2PN two-body problem with an effective one-body problem.

Even though the derivation of this result has been quite technical, the final result is striking in its simplicity. We find that the full complicated two-body dynamics at 2PN order is equivalent to the geodesic motion of a particle of mass μ equal to the reduced mass of the system, moving along the geodesics of a metric that is a deformation of the Schwarzschild metric generated by a mass $m = m_1 + m_2$, with deformation parameter ν, and supplemented by a mapping between the energy $\mathcal{E}_0$ of the effective one-body problem and the energy $\mathcal{E}$ of the real problem, given by eq. (14.66).

Of course, the non-trivial part of this result resides in its ν dependence, since in the test-mass limit $\nu = 0$ we know everything exactly. In the EOB formulation, the ν-dependent corrections to the dynamics comes from two sources: the explicit ν dependence in the effective metric (14.64), and the dependence on m_1 and m_2, and therefore on ν, introduced by the mapping (14.66) or (14.67).

It is interesting to note that this mapping is the same (apart from an overall μ factor) as the one that was previously introduced in the context of Padé approximants; see eq. (14.24).[14] Observe that the mapping (14.66) can be rewritten in an explicitly Lorentz-invariant way by introducing the Mandelstam variable $s \equiv -(p_1 + p_2)^2 = \mathcal{E}^2$, where p_1 and p_2 are the four-momenta of the two bodies. The mapping is then given by the function

$$\varphi(s) \equiv \frac{s - m_1^2 c^4 - m_2^2 c^4}{2mc^2}. \tag{14.68}$$

Since $-(p_1 + p_2)^2 = m_1^2 c^4 + m_2^2 c^4 - 2(p_1 \cdot p_2)$, we see that $\varphi(s)$ is just equal to $-(p_1 \cdot p_2)/mc^2$. In the rest frame of m_1 we have $-(p_1 \cdot p_2) = m_1 c^2 \mathcal{E}_2$, where $\mathcal{E}_2$ is the energy of m_2 in the rest frame of m_1. In the test-mass limit $m_2 \ll m_1$ we have $m_1 \simeq m$ and therefore $m_1 \mathcal{E}_2 / m \simeq \mathcal{E}_2$. The function $\varphi(s)$ is therefore the simplest function that is at the same time Lorentz-invariant, symmetric under the exchange of the two bodies, and, in the test-mass limit $m_2 \ll m_1$, equal to the energy of the body with mass m_2 in the rest frame of the body with mass m_1. Because of its naturalness and its repeated appearance in different contexts involving two-body problems, it is quite tempting to assume that the mapping (14.66) is in fact exact, rather than restricted to 2PN order, so, as far as the ν-dependence in the mapping is concerned, one can hope that eq. (14.66) catches the full result. This hope is reinforced by the study of the equivalence between the real two-body problem and the effective one-body problem at 3PN and at 4PN order, as we will now see.[15]

[14]The same mapping was also introduced by Brézin, Itzykson and Zinn-Justin (1970) in quantum electrodynamics, to map the one-body relativistic Balmer formula onto the two-body one.

[15]Furthermore, to first order in the post-Minkowskian expansion (i.e. to first order in G) this mapping has been shown to be exact to all orders in v/c; see Damour (2016).

Extension to 3PN and 4PN order, and resummations

At 3PN order eqs. (14.53) and (14.54) must be replaced by[16]

$$A(R) = 1 + a_1 \frac{Gm}{c^2 R} + a_2 \left(\frac{Gm}{c^2 R}\right)^2 + a_3 \left(\frac{Gm}{c^2 R}\right)^3 + a_4 \left(\frac{Gm}{c^2 R}\right)^4 + \dots,$$

(14.69)

$$B(R) = 1 + b_1 \frac{Gm}{c^2 R} + b_2 \left(\frac{Gm}{c^2 R}\right)^2 + b_3 \left(\frac{Gm}{c^2 R}\right)^3 + \dots,$$

(14.70)

while eq. (14.63) becomes

$$\mathcal{E}_0^{\mathrm{NR}} = \mathcal{E}^{\mathrm{NR}} \left[1 + \alpha_1 \frac{\mathcal{E}^{\mathrm{NR}}}{\mu c^2} + \alpha_2 \left(\frac{\mathcal{E}^{\mathrm{NR}}}{\mu c^2}\right)^2 + \alpha_3 \left(\frac{\mathcal{E}^{\mathrm{NR}}}{\mu c^2}\right)^3 + \dots\right].$$

(14.71)

However, the addition of the three new parameters a_4, b_3, α_3 is not sufficient to fulfill the new equations that arise when one requires the equivalence of eqs. (14.50) and (14.59) including the new $O(\alpha^6/c^6)$ terms that arise at 3PN order. To obtain the equivalence it is necessary to add higher-order terms to the action (14.51), which then becomes of the generic form

$$S_{\mathrm{EOB}} = -\mu c \int ds_{\mathrm{eff}} \left[1 + Q_{\mu\nu\rho\sigma}^{(4)} u^\mu u^\nu u^\rho u^\sigma\right],$$

(14.72)

where $u^\mu = dx^\mu/ds_{\mathrm{eff}}$. Thus, the motion is no longer along geodesics of the effective metric. Still, remarkably, requiring the equivalence of eqs. (14.50) and (14.59) including the new $O(\alpha^6/c^6)$ terms that arise at 3PN order, fixes $\alpha_3 = 0$ in eq. (14.71) (under natural assumptions), independently of the form of the non-geodesic terms, so eq. (14.66) remains exact at 3PN order.[17] The same happens at 4PN order. In this case one must add to the action (14.72) a further term proportional to $Q_{\mu\nu\rho\sigma\alpha\beta}^{(6)} u^\mu u^\nu u^\rho u^\sigma u^\alpha u^\beta$. However, performing the matching, one finds again $\alpha_4 = 0$ in the energy mapping. Thus, it is natural to assume that the ν dependence in the energy mapping is indeed exact (at least in the non-spinning case that we are considering here).

The remaining ν dependence appears in the functions $A(R)$ and $B(R)$, as well as in the non-geodesic terms. Up to 5PN order, using the variable $u = Gm/(c^2 R)$ instead of R, the expansion of the function $A(u)$ has the form

$$A(u) = 1 - 2u + 2\nu u^3 + a_4(\nu)u^4$$

(14.73)

$$+[a_5^0(\nu) + a_5^1(\nu) \log u]u^5 + [a_6^0(\nu) + a_6^1(\nu) \log u]u^6.$$

The 3PN coefficient is given by

$$a_4(\nu) = \left(\frac{94}{3} - \frac{41\pi^2}{32}\right)\nu.$$

(14.74)

Recently the 4PN computation has also been completed. Observe that the 4PN result for $A(u)$ has both a part proportional to u^5 and a part proportional to $u^5 \log u$. The corresponding coefficients are[18]

[16] See Damour, Jaranowski and Schäfer (2000) for the 3PN result and Damour, Jaranowski and Schäfer (2015) for the 4PN EOB formulation.

[17] More precisely, at 3PN order one can set to zero the non-geodesic term and still find a solution, at the price of relaxing the condition $b_1 = 2$ (which, however, is quite unnatural). The resulting solution is, however, very complicated, and in this case the coefficient α_3 is no longer zero, so also the mapping becomes more complicated; see Appendix A of Damour, Jaranowski and Schäfer (2000). In any case the presence of non-geodesic terms become unavoidable at 4PN order, so the solution with $b_1 \neq 2$ is not used.

[18] See eq. (8.1) of Damour, Jaranowski and Schäfer (2015), where the non-geodesic terms are also given. A new feature of the 4PN result is the appearance of terms non-local in time, at the level of the two-body problem. In the effective one-body problem, these can be taken into account by adding further non-geodesic terms.

$$a_5^0(\nu) = \left(\frac{2275\pi^2}{512} - \frac{4237}{60} + \frac{128}{5}\gamma_E + \frac{256}{5}\ln 2 \right)\nu + \left(\frac{41\pi^2}{32} - \frac{221}{6} \right)\nu^2 ,$$

$$(14.75)$$

$$a_5^1(\nu) = \frac{64}{5}\nu .$$

$$(14.76)$$

Furthermore, even the logarithmic term at 5PN order is known,

$$a_6^1(\nu) = -\frac{7004}{105}\nu - \frac{144}{5}\nu^2 .$$

$$(14.77)$$

Similarly the 5PN result for the function $\bar{D}(u) = [A(u)B(u)]^{-1}$ has the form[19]

$$\bar{D}(u) = 1 + 6\nu u^2 + (52\nu - 6\nu^2)u^3 + [\bar{d}_4^0(\nu) + \bar{d}_4^1(\nu)\log u]u^4$$
$$+ [\bar{d}_5^0(\nu) + \bar{d}_5^1(\nu)\log u]u^5 ,$$

$$(14.78)$$

with

$$\bar{d}_4^0(\nu) = \left(-\frac{533}{45} - \frac{23761\pi^2}{1536} + \frac{1184}{15}\gamma_E - \frac{6496}{15}\log 2 + \frac{2916}{5}\log 3 \right)\nu$$
$$+ \left(\frac{123\pi^2}{16} - 260 \right)\nu^2 ,$$

$$(14.79)$$

$$\bar{d}_4^1(\nu) = \frac{592}{15}\nu ,$$

$$(14.80)$$

and also the logarithmic term at 5PN is known,

$$\bar{d}_5^1(\nu) = -\frac{1420}{7}\nu - \frac{3392}{15}\nu^2 .$$

$$(14.81)$$

[19]Observe that in the literature both of the functions $D(u) = A(u)B(u)$ and $\bar{D}(u) = 1/D(u)$ are used.

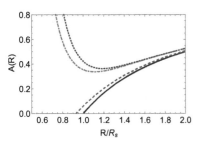

Fig. 14.1 The function $A(R) = 1 - R_S/R$ in the undeformed Schwarzschild case (solid line) compared with the EOB result for $A(R)$ for $\nu = 1/4$ at 2PN (dashed), 3PN (dot-dashed) and 4PN (dotted) order, before Padé resummation.

The truncations of $A(R)$ to 2PN, 3PN and 4PN order are shown, for $\nu = 1/4$, in Fig. 14.1 (as a function of R rather than of u), and compared with the undeformed Schwarzschild potential $A(R) = 1 - R_S/R$. We see that, at 3PN and at 4PN order, for $\nu = 1/4$, the function $A(R)$ no longer has a zero, which would correspond to the presence of a horizon. Note also that the potential becomes less and less attractive on increasing the PN order (at 3PN and 4PN order it even becomes repulsive for R/R_S below a critical value, but this will no longer be true after Padé resummation). One can try enforce the physical condition that the EOB effective metric should have a horizon, by making use of the Padé resummation technique discussed in Section 14.1, replacing the functions $A_{3PN}(u)$, $A_{4PN}(u)$ or $A_{5PN}(u)$ by their Padé approximants, constructed so that the numerator in eq. (14.13) is a polynomial of first degree. Thus, for instance, at 5PN order the function $A(u)$ can be replaced by its Padé approximant

[20]The 5PN coefficient $a_6^0(\nu)$, which currently is not known analytically, is fitted to the result of the numerical relativity simulations, as we will see later. Note that the coefficients $n_1, d_1, \ldots, d_5$ will in general depend also on $\log u$, starting from 4PN order, because of the $\log u$ terms in eq. (14.73). The fact that n_1 depends on $\log u$ also implies that the function $[A(u)]_5^1$ in general does not have just a simple zero, despite the apparently simple numerator. Similar Padé resummations have been employed also for the flux function; see the Further Reading.

$$[A(u)]_5^1 = \frac{1 + n_1 u}{1 + d_1 u + d_2 u^2 + d_3 u^3 + d_4 u^4 + d_5 u^5} ,$$

$$(14.82)$$

where $n_1, d_1, \ldots, d_5$ are fixed by requiring that the Taylor expansion of $[A(u)]_5^1$ up to $O(u^6)$ reproduces eq. (14.73).[20] Of course, since $u = Gm/c^2R$ is by definition positive, a zero of the function $A(u)$ only describes a horizon if it has a value $u_0 > 0$. Then, if the Padé-resummed function $A(u)$ has one or more zeros at a positive value of u, the largest of these zeros gives the position of the horizon. At 3PN order this turns out to be indeed the case for all ν in the range $0 \leqslant \nu \leqslant 1/4$. At 4PN order, in contrast, the function $[A(u)]_5^1$ has positive zeros only for $0 \leqslant \nu \leqslant \nu_0$, with $\nu_0 \simeq 0.14$, so for $\nu > \nu_0$ there is no horizon even in the Padé-resummed function. Observe that, for large u (i.e. small R), the function $[A(u)]_5^1$ in eq. (14.82) goes to zero as $1/u^4$, i..e as R^4, so in this case the Padé resummation of $A(R)$ has its first zero as $R \to 0^+$. Even in such cases, the Padé resummation is in general useful for helping the tuning to numerical relativity results, as we will discuss. One should, however, check that the Padé resummation does not introduce unwanted poles in the physical region $u > 0$; see also Section 14.2.5. Alternatively, one can use a logarithmic resummation, discussed in eq. (14.136) in the context of the spinning case, which enforces the presence of a horizon, without introducing spurious poles. One proceeds similarly for the function $D(u) = A(u)B(u)$, requiring that it stays positive.

We next study the physical consequences of the EOB formulation, focusing first on the conservative dynamics, and then adding the effects of radiation reaction.

14.2.2   Conservative dynamics

We now explore the conservative dynamics. If we limit ourselves to 2PN order, the function $A(R)$ in eq. (14.64) has a simple zero at positive R for all ν in the range $0 \leqslant \nu \leqslant 1/4$ (even before Padé resummation) and therefore, just as the Schwarzschild metric, the EOB metric has a horizon. The position of the horizon, $R_H(\nu)$, decreases monotonically from the value $R_H(\nu = 0) = 2Gm/c^2$ to the value

$$R_H(\nu = 1/4) \simeq 0.93 \left(\frac{2Gm}{c^2} \right) \tag{14.83}$$

in the equal-mass case. Similarly, the horizon area (which is an invariant quantity) moves toward smaller values as ν increases. At 2PN order, we see that the change in the position of the horizon is numerically quite small.

The analysis of the circular orbits in the EOB metric parallels exactly the analysis for the Schwarzschild metric. One finds in particular that there is an innermost stable circular orbit (ISCO, also called LSO, for "least stable orbit"). In the Schwarzschild case we have $R_{\text{ISCO}} = 6Gm/c^2$, i.e. $u_{\text{ISCO}} \equiv Gm/(c^2 R_{\text{ISCO}}) = 1/6 \simeq 0.167$. From the function $A(u)$ given in eq. (14.73) we see that this result is not modified at 1PN, since the term proportional to u^2 is absent. The dependence on ν first appears at 2PN order, and one then finds that the new ISCO, in the equal-mass case $\nu = 1/4$, moves at a slightly lower value than in the Schwarzschild case (consistent with the fact that the position of the horizon also decreases with ν), given by

$$R_{\text{ISCO,2PN}}(\nu = 1/4) \simeq 5.72 \left(\frac{Gm}{c^2} \right) , \tag{14.84}$$

i.e. $u_{\text{ISCO}} \simeq 0.175$. At 3PN order, after a Padé resummation of $A(R)$ that enforces the presence of a horizon, one finds

$$R_{\text{ISCO,3PN}}(\nu = 1/4) \simeq 4.95 \left(\frac{Gm}{c^2}\right), \qquad (14.85)$$

i.e. $u_{\text{ISCO}} \simeq 0.202$.

Another significant quantity in the Schwarzschild metric is the light-ring. For a massive test particle this is the radius of the smallest possible (unstable) circular orbit in the Schwarzschild metric, in contrast to the ISCO, which is rather the innermost *stable* circular orbit. If one rather studies the motion of a massless particle in the Schwarzschild metric, the radius corresponding to the smallest possible unstable circular orbit becomes the values of R where it is possible to trap a massless particle, such as a photon, on a circular orbit, and it then called the *light ring*.[21] In the Schwarzschild metric, the light ring is located at

$$R = \frac{3Gm}{c^2} \qquad (14.86)$$

From study of the orbits in the 2PN EOB metric one finds that, for $\nu \neq 0$, the position of the light ring is given by the real and positive solution of

$$r^3 - 3r^2 + 5\nu = 0, \qquad (14.87)$$

where we have introduced the dimensionless variable

$$r \equiv \frac{R}{(Gm/c^2)}. \qquad (14.88)$$

For $\nu = 1/4$, $r_{\text{light ring}} \simeq 2.85$, to be compared with $r = 3$ in the $\nu = 0$ case.

While the numerical value of R_{ISCO}, at a given PN order, does not change too much with ν, the physical role of the ISCO is very different in the two limiting cases $\nu \to 0$ and $\nu \to 1/4$. Consider first the case of very small ν. If ν is exactly zero, we have a test particle and, if we set it initially in a circular orbit, it remains in such an orbit forever. If we turn on ν, still keeping it very small, the geometry of the orbits does not change much compared with the Schwarzschild case, since the effect of the small mass μ on the background geometry is negligible compared with the effect of the large mass m. However, the mass will now emit gravitational radiation. The radiated power is proportional to the second derivative of its quadrupole moment, and therefore to μ, so it is small in the limit $\mu/m \ll 1$. However, on a time-scale proportional to $1/\nu$, this radiation reaction causes the particle to move through a sequence of quasi-circular orbits, spiraling inward toward the central BH. Once the ISCO is reached, there are no longer circular orbits on which the small mass can move, so it will be forced to plunge toward the central BH. Thus, for ν very small (but still not exactly zero), the dynamics can be quite clearly separated into a long inspiral phase where the light body moves on quasi-circular orbits, followed by a sudden plunge phase

[21] See e.g. Hartle (2003).

that is driven not by radiation reaction, but rather by the feature of the Schwarzschild metric that circular orbits are not allowed for $R < R_{\rm ISCO}$; in other words, for very small ν the plunge is really an effect due to GR in a strong static field. In the small-ν case, $R_{\rm ISCO}$ therefore has a rather clear physical meaning, separating the long inspiral phase driven by radiation reaction from the sudden plunge phase driven by static strong-field gravity.

As ν grows larger, however, this picture is modified. This is not much due to the change in the geometry felt by the moving masses, since we have seen that, in the equivalent one-body problem, features such as the positions of the horizon and of the ISCO do not change too much with ν, but rather by the fact that for larger ν the GW emission is stronger and therefore the back-reaction acts on a shorter time-scale. We can then expect that there is no longer a sharp transition between inspiral and plunge. To investigate this issue quantitatively, however, we must first introduce back-reaction in the EOB framework, as we will do in the next subsection.

In order to study the dynamics of generic orbits that are affected by back-reaction, and that therefore are not necessarily very close to circular, it is first of all convenient to rewrite the conservative dynamics in a more general form. We start from the Hamilton–Jacobi equation (14.57) and we recall that, in the Hamilton–Jacobi formalism, the radial momentum P_R is given by $\partial S/\partial R$, which is the same as dS_R^0/dR; see eq. (14.56). Furthermore $\mathcal{J}_0$ denotes the eigenvalue of angular momentum, i.e. of the momentum P_φ conjugate to the angular coordinate φ. Thus, denoting by H_0 the Hamiltonian of the effective one-body problem, eq. (14.57) shows that

$$H_0(\mathbf{Q},\mathbf{P}) = \mu c^2 \sqrt{A(R)\left[1 + \frac{P_R^2}{\mu^2 c^2 B(R)} + \frac{P_\varphi^2}{\mu^2 c^2 R^2}\right]}, \qquad (14.89)$$

where we denote collectively by $\mathbf{Q}$ the polar coordinates of the effective problem, so $\mathbf{Q} = (R,\theta,\varphi)$, and by $\mathbf{P}$ their conjugate momenta $(P_R, P_\theta, P_\varphi)$. From eq. (14.67), the Hamiltonian H of the real two-body problem is given by

$$H(\mathbf{Q},\mathbf{P}) = mc^2\left[1 + 2\nu\left(\frac{H_0(\mathbf{Q},\mathbf{P}) - \mu c^2}{\mu c^2}\right)\right]^{1/2}. \qquad (14.90)$$

Equations (14.89) and (14.90) give the EOB "improvement" of the original 2PN Hamiltonian of the real two-body problem. Observe that the Hamiltonian (14.90) of the real two-body problem is still expressed in terms of the coordinates and momenta $(\mathbf{Q},\mathbf{P})$ of the effective one-body problem. In principle, one can compute the transformation between these coordinates and the relative ADM coordinates of the real two-body problem.[22] However, it will be more convenient to work directly in effective coordinates. The Hamiltonian (14.90) determines, through Hamilton's equations, the evolution of any pair of canonical variables, and therefore also of the variables $(\mathbf{Q},\mathbf{P})$ of the effective problem, with

[22]To compute this mapping one observes that the action variables in the real and effective problems are identified [see eqs. (14.60) and (14.61)], and this ensures that the two problems are mapped by a canonical transformation. One can then find the generating function that maps the Hamiltonian of the real problem into the Hamiltonian of the effective problem. From the generating function one obtains the canonical transformation between the real and effective variables. See Section VI of Buonanno and Damour (1999).

respect to the time variable t of the *real* two-body problem,

$$\frac{dQ_i}{dt} = \frac{\partial H(\mathbf{Q}, \mathbf{P})}{\partial P_i}, \tag{14.91}$$

$$\frac{dP_i}{dt} = -\frac{\partial H(\mathbf{Q}, \mathbf{P})}{\partial Q_i}. \tag{14.92}$$

14.2.3 Inclusion of radiation reaction

Equations (14.91) and (14.92) take into account only the conservative part of the dynamics. As we saw in Chapter 5, radiation reaction enters for the first time at 2.5PN order and, at this order, it can be expressed in terms of a radiation-reaction force. The energy loss of the system on a generic orbit can be written as

$$\frac{d\mathcal{E}}{dt} = \dot{R}\mathcal{F}_R + \dot{\varphi}\mathcal{F}_\varphi, \tag{14.93}$$

in terms of the two flux functions $\mathcal{F}_R$ and $\mathcal{F}_\varphi$. On a quasi-circular orbit

$$\frac{d\mathcal{E}}{dt} \simeq \dot{\varphi}\mathcal{F}_\varphi + O(\dot{R}^2). \tag{14.94}$$

The flux $\mathcal{F}_\varphi$ is obtained from the PN expansion. We now introduce the dimensionless variables $r = Rc^2/(GM)$, $p_r = P_R/(\mu c)$, $p_\varphi = P_\varphi c/(\mu GM)$, $\widehat{H} = H/(\mu c^2)$ and $\hat{\mathcal{F}}_\varphi(r, p_r, p_\varphi) = \mathcal{F}_\varphi/\mu$. Using eqs. (14.89) and (14.90)

$$\widehat{H}(r, p_r, p_\varphi) = \frac{1}{\nu}\sqrt{1 + 2\nu\left[\sqrt{A(R)\left(1 + \frac{p_r^2}{B(r)} + \frac{p_\varphi^2}{r^2}\right)} - 1\right]}. \tag{14.95}$$

Radiation reaction on quasi-circular orbits can then be taken into account by writing the equations of motion in the form

$$\frac{dr}{d\hat{t}} = \frac{\partial \widehat{H}}{\partial p_r}, \tag{14.96}$$

$$\frac{d\varphi}{d\hat{t}} = \frac{\partial \widehat{H}}{\partial p_\varphi}, \tag{14.97}$$

$$\frac{dp_r}{d\hat{t}} = -\frac{\partial \widehat{H}}{\partial r}, \tag{14.98}$$

$$\frac{dp_\varphi}{d\hat{t}} = \hat{\mathcal{F}}_\varphi, \tag{14.99}$$

where $\hat{\mathcal{F}}_\varphi = \hat{\mathcal{F}}_\varphi(r, p_r, p_\varphi)$. The resulting equations of motion can be integrated numerically, and Fig. 14.2 shows the result for $\nu = 0.1$. As expected, for this relatively small value of ν, until the ISCO is reached the motion proceeds through a series of quasi-circular trajectories, and then suddenly changes into a plunge when the ISCO is reached. Figure 14.3 shows the result for $\nu = 1/4$. We see that in this case the transition from inspiral to plunge is less sharp, in agreement with the

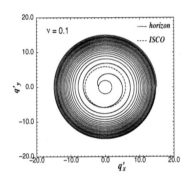

Fig. 14.2 Inspiraling circular orbits from 2PN EOB equations of motion with 2.5PN radiation reaction effects, with $\nu = 0.1$. The coordinates q'^i are defined by $q'^i = Q^i/Gm$, where $\mathbf{Q} = (R, \theta, \varphi)$ are the coordinate of the effective one-body problem. From Buonanno and Damour (1999).

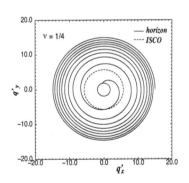

Fig. 14.3 The same as Fig. 14.2, with $\nu = 1/4$. From Buonanno and Damour (1999).

qualitative expectations discussed above. The treatment of the radiation reaction force can be further improved using Padé techniques for the radiative Hamiltonian, rather than using simply the 2.5PN result. This more refined treatment confirms that the transition between inspiral and plunge, and therefore the notion of the ISCO, becomes blurred when ν approaches $1/4$.[23]

Comparing the values of $p_r^2/B(r)$ and p_φ^2/r^2 on the numerical solution of the equations of motion, one finds that, for $\nu = 1/4$, the radial contribution is always very small down to $r = 3$, i.e. down to $R = 3GM/c^2$; see Fig. 14.4. This shows that, in the equal-mass case, the two bodies approach one another on a quasi-circular inspiral trajectory, rather than plunging toward each other radially.

14.2.4 The EOB waveform

Having obtained a reformulation of the dynamics of the two-body system that is a good candidate for an improvement on the perturbative PN expansion from which it was derived, we can now proceed to the computation of the GWs radiated at infinity. In the simplest approximation, the gravitational radiation emitted in this inspiral and plunge phase can be computed just by evaluating the quadrupole formula on the (numerical) solution of the equations of motion. This gives the waveform in the restricted PN approximation.[24] For definiteness we set the angle ι between the normal to the orbit and the line of sight to the value $\iota = \pi/2$, i.e. we observe the binary system edge-on. Using the expression given in eq. (5.262) for the waveform, together with the definition (5.238), we see that the polarization $h_\times$ vanishes, while the plus polarization of the GW is

$$h_+(t) = \mathcal{C}\omega_s^{2/3}(t)\cos[2\phi_s(t)]\,, \tag{14.100}$$

where $\phi_s(t)$ is the orbital phase of the source, computed numerically through the EOB equations of motion, $\omega_s = d\phi_s/dt$, and

$$\mathcal{C} = -2G\mu\frac{(Gm)^{2/3}}{c^4 R} \tag{14.101}$$

depends on the masses μ and m and on the distance R.[25] It is straightforward to go beyond the restricted PN approximation. In particular, the mass octupole and current quadrupole contributions, which both give terms oscillating as $\cos[\phi_s(t)]$ and as $\cos[3\phi_s(t)]$, are obtained by using eq. (5.266), and similarly for the PN corrections to the amplitude.

The issue is up to what value of time, or equivalently of distance between the two bodies, can we legitimately use eq. (14.100) (or its improvement beyond the restricted PN approximation). Below some distance it is clear that a change of description is necessary, related to the merger of the two bodies and the formation of the final Kerr BH. The basic observation is that the EOB resummation provides a smooth extension of the perturbative PN results down to values of R close to the light ring, $R \simeq 3Gm/c^2$. From the discussion in Chapter 12, and in particular from Fig. 12.1, we know that for a BH of mass M_{BH} the

[23] See Buonanno and Damour (2000) and the review by Damour and Nagar (2009a).

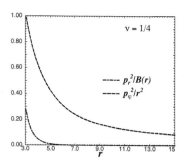

Fig. 14.4 A comparison of the two kinetic contributions to the Hamiltonian (14.95), the radial term $p_r^2/B(r)$ and the "azimuthal" term p_φ^2/r^2, evaluated on the solution of the EOB equations of motion with back-reaction, for $\nu = 1/4$. The smallness of the radial term shows that the motion is quasicircular. From Buonanno and Damour (2000).

[24] Present EOB results go beyond the restricted PN approximation, and include both the PN correction to the amplitude and higher multipoles, as discussed in Section 5.6.4.

[25] The overall sign depends on the convention on the axes chosen for defining the plus and cross polarizations, since under a rotation by π of these axes in the plane perpendicular to the propagation direction the plus and cross amplitude both change sign, see eq. (2.194).

peak of the barrier in the Regge–Wheeler and Zerilli equations is at a radial distance of order $3GM_{\rm BH}/c^2$, and the region at $R \lesssim 3GM_{\rm BH}/c^2$ is the one that is mostly responsible for the generation of the BH quasi-normal modes (QNMs). Therefore, at $R \simeq 3GM_{\rm BH}/c^2$ it makes sense to switch to a new description, in which we have a final Kerr BH with a mass $M_{\rm BH} \equiv M_f$ (equal to the total initial mass m of the binary minus the energy radiated away in GWs in the inspiral and plunge phase) and Kerr parameter $a_f = J_f/M_f c$ [see eq. (12.286)] determined by angular momentum conservation. It is therefore natural to postulate, within the EOB approach, that eq. (14.100) is valid up to a value of time close to the value $t = t_{\rm end}$ where the light ring is crossed.

The waveform at $t \geqslant t_{\rm end}$ is therefore the "ringdown" waveform, given by the oscillation of the BH in its QNMs, and the signal is a superposition of damped sinusoids. As discussed in Section 12.5, for a Kerr BH the fundamental QNM has $n = 1, l = 2, m = 2$, and its frequency $(\omega_R)_{122}$ and damping time τ_{122} are given in eqs. (12.360) and (12.361). Keeping for simplicity only the fundamental mode, we have, for $t \geqslant t_{\rm end}$,

$$h_+(t) = \mathcal{A} e^{-(t-t_{\rm end})/\tau_{122}} \cos[(\omega_R)_{122}(t - t_{\rm end}) + \mathcal{B}]. \qquad (14.102)$$

The constants $\mathcal{A}$ and $\mathcal{B}$ are fixed in terms of the constant $\mathcal{C}$ in eq. (14.100) requiring the continuity of $h_+(t)$ and of dh_+/dt at $t = t_{\rm end}$. It is in principle straightforward to improve the result by including higher QNMs.

The values of M_f and a_f are determined by integrating numerically the EOB equations of motion down to $t_{\rm end}$, computing the energy and angular momentum radiated at infinity, and applying the conservation of energy and angular momentum. For $\nu = 1/4$ this gives

$$M_f \simeq 0.9761m \qquad \text{(untuned)} \qquad (14.103)$$

and

$$a_f \simeq 0.7952 \frac{GM_f}{c^2}$$
$$\simeq 0.7762 \frac{Gm}{c^2} \qquad \text{(untuned)}, \qquad (14.104)$$

where the mention "untuned" stresses that, at this stage, no tuning to numerical relativity has yet been done, see below. Given the mass and spin of the final BH, the corresponding values of $(\omega_R)_{122}$ and τ_{122} are fixed by eqs. (12.360) and (12.361).

We then end up with a waveform covering inspiral, merger and ringdown, which can then be compared with the results of numerical relativity. Figures 14.5 and 14.6 show the waveform and the evolution of $\omega = (1/2)\omega_{\rm GW}$ computed from the EOB method, compared with early numerical relativity simulations (see also Fig. 14.14). It is remarkable that the EOB method gives a qualitatively correct waveform, covering inspiral, merger and ringdown, and, without any tuning, already provides predictions in agreement with numerical relativity at the level of about 10%.

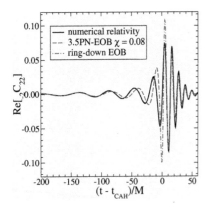

Fig. 14.5 The EOB waveform, without any tuning, compared with the result from numerical relativity. From Buonanno, Cook and Pretorius (2007). The total mass of the binary system is denoted by M.

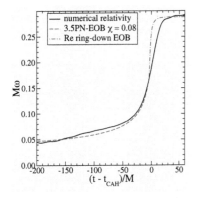

Fig. 14.6 The evolution of the frequency ω obtained from $\mathrm{Re}\,_{-2}C_{22}$ (which is equal to one-half of the GW frequency $\omega_{\rm GW}$) determined from the numerical simulation, compared with the EOB prediction with no tuning. From Buonanno, Cook and Pretorius (2007).

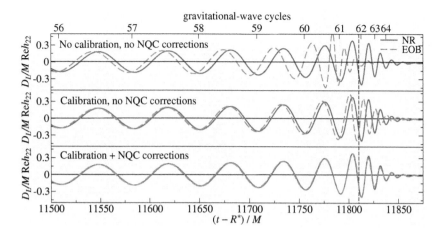

Fig. 14.7 The numerical-relativity (2,2) waveform for an equal-mass, non-spinning BH binary, compared with EOB inspiral-merger-ringdown waveforms: (i) with neither inspiral calibration nor non-quasi-circular (NQC) corrections (upper panel); (ii) with inspiral calibration, but without NQC corrections (middle panel); (iii) with both inspiral calibration and NQC corrections (lower panel). D_L is the distance to the source. Here $G = c = 1$. The waveforms are aligned in phase at low frequency. The vertical dashed line marks the position of the amplitude peak of the numerical relativity waveform. The calibration is performed here using the logarithmic resummation (14.136). The plot shows only the last cycles of a long numerical simulations, with about 60 inspiral cycles. From Taracchini (2014).

Still, we see from Fig. 14.5 that, at the quantitative level, the agreement with numerical relativity is not perfect. As we already discussed at the beginning of this chapter, despite the successes of numerical relativity, it is very important to have an accurate analytic representation of the waveform. Indeed, accurate simulations of BH–BH coalescences may require months of CPU time, and it is impossible to obtain in this way a bank of templates scanning with sufficient resolution the parameter space of BH–BH binaries, which includes at least the masses and spins of the two BHs. The best strategy is therefore to combine the flexibility of a successful analytic method such as the EOB method with the accuracy of numerical relativity, tuning the EOB waveforms to the numerical relativity results.

The agreement of the EOB waveforms with numerical relativity simulations can be improved as follows. First, one "calibrates" the EOB waveform, using the 5PN expression (14.73) for $A(R)$ and treating the yet unknown 5PN coefficient $a_6^0(\nu)$ as a free parameter, to be fixed by comparison with numerical relativity simulations at different values of ν.[26] As a second improvement, one can take into account the effect of deviations from a quasi-circular motion, by multiplying directly the $l = m = 2$ inspiral and plunge waveform by an additional "next-to-quasi-circular" (NQC) correction factor

$$f_{22}^{\mathrm{NQC}} = 1 + a_1(\nu)\frac{P_{R_*}}{(R\omega_0)^2} + a_2(\nu)\frac{\ddot{R}}{R\omega_0^2}, \qquad (14.105)$$

[26] Alternatively, we will see in eq. (14.136) that one can use a logarithmic resummation, and calibrate a function $K(\nu)$.

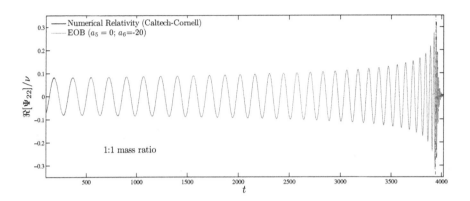

Fig. 14.8 The EOB waveform for equal-mass binaries, with calibration and NQC corrections, compared to numerical relativity, from Damour and Nagar (2009a). This EOB waveform was computed when the a_5 4PN coefficient was not yet known. Since in the calibration procedure a_5 and a_6 are largely degenerate, the calibration was performed setting $a_5 = 0$ and fitting a_6.

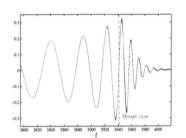

Fig. 14.9 A close-up of Fig. 14.8 near the time of merger. From Damour and Nagar (2009a).

where P_{R_*} is the momentum canonically conjugate to the EOB tortoise coordinate R_*, ω_0 the EOB orbital frequency, and the functions $a_1(\nu), a_2(\nu)$ are fitted to the numerical simulations. Figure 14.7 shows the effect of these corrections. We see that, after these tunings, the agreement with numerical relativity is excellent, thereby providing accurate analytic templates that can be efficiently used in the data analysis.

Another example is given in Fig. 14.8, again for equal-mass binaries (but using Padé resummation for the calibration), from a simulation with 35 inspiral cycles, shown on a scale that emphasizes the inspiral phase. The EOB waveform and the result obtained from numerical relativity agree so well that they are almost indistinguishable on the scale of this figure. The phase difference is $\Delta\phi \leqslant 0.01$ rad (which is at the level of the numerical error of the numerical relativity result) during the entire inspiral and plunge. Figure 14.9 shows a close-up around the merger time. It is clear that the agreement is striking, even around the time of merger.

It is interesting to remark that, before the advent of the EOB approach, and of the successes of numerical relativity that will be the subject of Section 14.3, the picture that was heuristically suggested for the waveform of BH–BH coalescence was quite different. It was expected that the waveform produced during the plunge and merger phase would have been completely different from that produced in the previous inspiral waveform, and very difficult to compute. This complicated merger waveform would then have gradually settled into the ringdown waveform.

The EOB method suggests a much simpler picture, in which the complicated intermediate part of the waveform is absent, and the radiation from the inspiral phase is just matched directly to the radiation from the ringdown phase. Numerical relativity confirms the correctness of this simpler picture. With hindsight, we can trace the simplicity of this result to the existence of a barrier in the Regge–Wheeler and Zerilli equations, peaked around $R \simeq 3GM/c^2$. This barrier affects the outward propagation of GWs, "filtering out" much of the low-frequency signal produced inside the barrier.[27] Thus, even if the final stage of the merger of the two bodies is a complicated process, the GWs that are produced during the merger phase are not very sensitive to the details of the motion of the bodies at such short separations, but rather are well described in terms of a final Kerr BH that settles down to its ground state, radiating away its excess energy in its QNMs[28]

It is interesting to analyze the behavior of the waveform, from inspiral to ringdown, in frequency space. Recall that in Section 4.1.1 we used, as a criterion for estimating the end of the inspiral phase, the condition that the orbital separation of the binary system reaches the innermost stable circular orbit (ISCO), which we computed in the test-mass approximation $\nu = 0$. In this way, we found in eq. (4.40) that, for a binary system with total mass m, the inspiral phase ends when the frequency f_s of the orbital motion reaches a value $f_s \simeq 2.2\,\mathrm{kHz}(M_\odot/m)$. The corresponding GW frequency, in the quadrupole approximation, is twice this value, so during the inspiral phase the GW frequency sweeps up to a value of order[29]

$$f_{\mathrm{ISCO}} \simeq 4.4\,\mathrm{kHz}\left(\frac{M_\odot}{m}\right) \qquad (\nu = 0). \tag{14.106}$$

The EOB method allowed us to compute the finite-ν corrections to the position of the ISCO; see eqs. (14.84) and (14.85). Then, from eq. (14.97), we can compute the value of the corresponding frequency of the real problem; using for simplicity the 2PN result, setting $\nu = 1/4$, corresponding to equal-mass binaries, and multiplying again by a factor of two to get the corresponding GW frequency, one finds[30]

$$\frac{Gm}{c^3}\omega_{\mathrm{ISCO}} \simeq 0.15 \qquad (\nu = 1/4). \tag{14.107}$$

In terms of the GW frequency $f = \omega/(2\pi)$, for equal-mass binaries the GW frequency at which the inspiral phase ends can then be estimated, in order of magnitude, as

$$f_{\mathrm{ISCO}} \simeq 4.7\,\mathrm{kHz}\left(\frac{M_\odot}{m}\right). \qquad (\nu = 1/4). \tag{14.108}$$

We next compute the characteristic frequency of the GWs emitted during ringdown. Using for this estimate the untuned EOB predictions (14.103) and (14.104) and plugging these values into eq. (12.360), we find that the fundamental QNM of the final Kerr BH emits GWs at the frequency

$$\frac{Gm}{c^3}\omega_{\mathrm{qnm}} \simeq 0.62, \qquad (\nu = 1/4), \tag{14.109}$$

corresponding to

$$f_{\mathrm{qnm}} \simeq 19.4\,\mathrm{kHz}\left(\frac{M_\odot}{m}\right). \qquad (\nu = 1/4). \tag{14.110}$$

[27]This point was already realized long ago in Davis, Ruffini, Press and Price (1971) and Davis, Ruffini and Tiomno (1972).

[28]It should be observed that this intuitive explanation actually works because the instantaneous orbital frequency during the plunge phase can be computed in the EOB from eq. (14.97), is approximately proportional to $A(r)/r^2$ [see eq. (12) of Buonanno *et al.* (2007)]. Therefore, after reaching a maximum near the light ring, it goes to zero as r approaches the horizon. Thus, once we switch to a description in terms of QNM ringing of the final BH, at distances smaller than the peak of the Regge–Wheeler and Zerilli potentials the characteristic frequencies of the source term in the Regge–Wheeler and Zerilli equations (12.102) and (12.109) are small compared with the height of the potential barrier. Then, the GW signal induced by such a perturbation, and seen at large distances, is indeed filtered by the potential barrier. In principle one can construct toy examples of source terms that, inside the barrier, have typical Fourier modes with frequencies large compared with the height of the barrier. In such examples, the GW signal seen outside will not be filtered, and will be sensitive to the details of the trajectory; see Price, Nampalliwar and Khanna (2016).

[29]We assume here that the binary is sufficiently close to neglect its cosmological redshift. Otherwise, the frequency in the observer frame is reduced by a factor $1/(1+z)$; see Section 4.1.4. See eqs. (16.24) and (16.25) for the expressions for f_{ISCO} and f_{qnm} normalized to the values of masses and redshifts typical of supermassive BH binaries.

[30]We stress that this result holds for spinless BHs. For co-rotating spinning BHs at 3PN order one rather has $(Gm/c^3)\omega_{\mathrm{ISCO}} \simeq 0.20$ [see Table I in Damour, Gourgoulhon and Grandclement (2002), after multiplying again for a factor of 2 to get the GW frequency]. Recall also that, in the equal-mass case, the notion of ISCO becomes somewhat blurred, so this estimate can only be taken as indicative.

Thus the GW frequency, which at the end of the inspiral phase has reached the value (14.108), during the plunge and merger phase keeps on increasing until it reaches the value given by eq. (14.110). We indeed see this behavior in Fig. 14.6 (where the orbital frequency must be multiplied by a factor of two to get the GW frequency, and the time at which the ISCO is crossed can be recognized from the change of slope of ω).

It is worth observing that an accurate determination of the waveform in the frequency range corresponding to plunge and merger is important for at least two reasons. First, as we discussed in Chapter 7, an accurate determination of the waveform is necessary in order to extract the signal from the noise, using the matched filtering technique. Consider a ground-based detector operating between $O(10)$ Hz and a few kHz. For binaries with a total mass m smaller than $10M_\odot$, the inspiral signal extends up to ~ 500 Hz, so it is well inside the detector bandwidth, and gives most of the contribution to the signal-to-noise ratio. For such binaries, an accurate computation of the merger waveform is not crucial for extracting the signal from the noise. As the total mass m of the binary increases, however, the portion of the inspiral signal accessible to the detector shrinks, and for $m \sim 50M_\odot$ the contribution to the signal-to-noise ratio from the inspiral phase is comparable to that from the merger and ringdown phase.[31]

Another reason why an accurate computation of the waveform emitted in the merger and ringdown phase is important is that the GWs emitted in this regime probe GR in an extreme situation, of very strong fields and highly non-linear gravity. They therefore carry some potentially very interesting information.

14.2.5   Spinning binaries

The formulation of the EOB method that we have presented so far is valid for Schwarzschild BHs. Realistic BHs are, however, expected to be spinning, and possibly with large spin. In order to obtain sufficiently accurate waveforms it is therefore important to include the spin of the initial BHs. In the PN expansion the spin enters at first with a leading-order spin–orbit coupling at 1.5PN order, while the spin–spin coupling enters at 2PN. Next, at 2.5PN, enters the next-to-leading-order correction to the spin–orbit coupling.

We work in ADM coordinates and we use the notation $q' = q_1 - q_2$ for the relative ADM coordinate and p' for its conjugate momentum. We also introduce $r' = |q'|$ and $L' = q' \times p'$ (see the summary of notations in Note 9 on page 218), and we denote the spins of the two BHs by S_1 and S_2. Then the spin-dependent part of the Hamiltonian, up to 2.5PN order, has the general form

$$H_{\rm spin} = H_{\rm SO}^{\rm 1.5PN} + H_{\rm SS}^{\rm 2PN} + H_{\rm SO}^{\rm 2.5PN}. \tag{14.111}$$

The 1.5PN spin–orbit Hamiltonian is

$$H_{\rm SO}^{\rm 1.5PN} = \frac{2G}{c^2}\frac{L'}{r'^3}\cdot\left[\left(1+\frac{3}{4}\frac{m_2}{m_1}\right)S_1 + \left(1+\frac{3}{4}\frac{m_1}{m_2}\right)S_2\right]$$

[31]For instance for the event GW150914, which we will discuss in Chapter 15, $m \simeq 65M_\odot$. The signal entered the detector bandwidth at $f \simeq 20$ Hz and the inspiral phase lasted up to $f \simeq 132$ Hz. The SNR obtained restricting the analysis to the frequency range $20 < f < 132$ Hz was ~ 19.5, while the SNR for $f > 132$ Hz was ~ 16.

$$= \frac{G}{c^3}\frac{\boldsymbol{L'}}{r'^3}\cdot\left(2\boldsymbol{\sigma}+\frac{3}{2}\boldsymbol{\sigma}^*\right),\tag{14.112}$$

where we have defined

$$\boldsymbol{\sigma}=c\left(\boldsymbol{S}_1+\boldsymbol{S}_2\right),\tag{14.113}$$

and[32]

$$\boldsymbol{\sigma}^*=c\left(\frac{m_2}{m_1}\boldsymbol{S}_1+\frac{m_1}{m_2}\boldsymbol{S}_2\right).\tag{14.114}$$

The 2PN spin–spin Hamiltonian is

$$H_{\rm SS}^{\rm 2PN}=\frac{G}{c^4}\frac{\nu}{2r'^3}\left[3(\boldsymbol{n'}\cdot\boldsymbol{\sigma}_0)^2-(\boldsymbol{\sigma}_0)^2\right],\tag{14.115}$$

where $\boldsymbol{\sigma}_0=\boldsymbol{\sigma}+\boldsymbol{\sigma}^*$, and $\boldsymbol{n'}=\hat{\boldsymbol{r}}'/r'$. Finally, the 2.5PN spin–orbit contribution is

$$
\begin{aligned}
H_{\rm SO}^{\rm 2.5PN}=\frac{G}{c^5}\frac{\boldsymbol{L'}}{r'^3}\cdot\Bigg[&\left(\frac{19}{8}\nu\,\hat{\boldsymbol{p}}'^2+\frac{3}{2}\nu(\boldsymbol{n'}\cdot\hat{\boldsymbol{p}}')^2-(6+2\nu)\frac{Gm}{c^2r'}\right)\boldsymbol{\sigma}\\
&+\left(\frac{16\nu-5}{8}\hat{\boldsymbol{p}}'^2+\frac{3}{4}\nu(\boldsymbol{n'}\cdot\hat{\boldsymbol{p}}')^2-(5+2\nu)\frac{Gm}{c^2r'}\right)\boldsymbol{\sigma}^*\Bigg],
\end{aligned}
\tag{14.116}
$$

where, as usual, $m=m_1+m_2$, and we have also defined $\hat{\boldsymbol{p}}'=\boldsymbol{p}'/\mu$.[33]

In the EOB framework this PN expansion, including the spin terms, is mapped into an effective one-body problem of a spinning test particle with mass $\mu=m_1m_2/(m_1+m_2)$ and suitable spin $\boldsymbol{S}$, moving in a deformation of a Kerr space-time with a suitable spin $\boldsymbol{S}_{\rm Kerr}$. As in the spinless case, the deformation parameter is the symmetric mass ratio ν. To perform the mapping explicitly, rather than going again through the comparison of the "energy level" formulas (14.50) and (14.59), we can proceed as follows. We start from a general axisymmetric metric in quasi-isotropic coordinates, i.e. a metric of the form

$$
\begin{aligned}
ds^2=&-e^{2\nu(R,\theta)}c^2dt^2+R^2\sin^2\theta B^2(R,\theta)e^{-2\nu(R,\theta)}\left[d\phi-\omega(R,\theta)cdt\right]^2\\
&+e^{2\mu(R,\theta)}(dR^2+R^2d\theta^2),
\end{aligned}
\tag{14.118}
$$

where B,ω,μ,ν are for the moment generic functions of R and θ. To recover the Kerr metric (12.285) in Boyer–Lindquist coordinates, for a Kerr BH of mass M, as a first step we introduce the Boyer–Lindquist coordinate r from[34]

$$\frac{dR}{dr}=\frac{R}{\sqrt{\Delta}}.\tag{14.119}$$

Explicitly this gives

$$r=R+\frac{GM}{c^2}+\frac{R_H^2}{R},\tag{14.120}$$

where $R_H=(1/2)\sqrt{(GM/c^2)^2-a^2}$ is the horizon radius in quasi-isotropic coordinates. Next, we set

$$B(r)=\frac{\sqrt{\Delta(r)}}{R(r)},\qquad \omega(r,\theta)=\frac{ar}{\Lambda(r,\theta)}\frac{2GM}{c^2},\tag{14.121}$$

$$e^{2\nu(r,\theta)}=\frac{\rho^2(r,\theta)\Delta(r)}{\Lambda(r,\theta)},\qquad e^{2\mu(r,\theta)}=\frac{\rho^2(r,\theta)}{R^2(r)},\tag{14.122}$$

[32]The factor c in the definition of $\boldsymbol{\sigma}$ and $\boldsymbol{\sigma}^*$ is convenient for BHs and for compact relativistically rotating stars. Indeed, for a BH we found in eq. (12.359) that the spin is given by $S=\hat{a}GM^2/c$, with $\hat{a}$ a dimensionless parameter between 0 and 1. Similarly, for a compact star rotating relativistically, $S\sim Mv_{\rm rot}r\sim McR_S\sim GM^2/c$. Therefore, for BHs or relativistically rotating compact stars, from the point of view of counting the powers of $1/c$ to understand the PN order of a given term, we have $\boldsymbol{\sigma},\boldsymbol{\sigma}^*=O(c^0)$.

[33]These expressions have been given in ADM coordinates because, as we saw in the discussion following eq. (14.33), starting from 2PN order in harmonic coordinates the orbital part of the Lagrangian contains both position, velocities and accelerations, while in ADM coordinates the dependence on the acceleration disappears and one can define a canonical Hamiltonian. Nevertheless, the 3PN equations of motion in harmonic coordinates are manifestly invariant under the Poincaré group and therefore one can define the corresponding conserved quantities. In particular, the expression for the energy corresponding to the 1.5PN spin–orbit term, in harmonic coordinates, is

$$E_{\rm SO}^{\rm 1.5PN}=\frac{2G}{c^2}\frac{\boldsymbol{L}}{r^3}\cdot\left(\frac{m_2}{m_1}\boldsymbol{S}_1+\frac{m_1}{m_2}\boldsymbol{S}_2\right).\tag{14.117}$$

See Kidder (1995) for the 2PN energy contribution from the spin–spin term in harmonic coordinates, and Faye, Blanchet and Buonanno (2006) for the contribution from the 2.5PN spin–orbit term. The transformation from harmonic to ADM coordinates is discussed, to 3PN order, in de Andrade, Blanchet and Faye (2001).

[34]Observe that in this section, following Barausse and Buonanno (2010), we are denoting by r the radial coordinate of the EOB problem in Boyer–Lindquist coordinates. In the limit $a=0$ this corresponds to the coordinate that was denoted by R in the spinless case, while here R denotes the radial coordinate in quasi-isotropic coordinates.

where

$$\Lambda(r,\theta) = (r^2 + a^2)^2 - a^2 \Delta(r) \sin^2\theta, \qquad (14.123)$$

while $\rho^2(r,\theta)$ and $\Delta(r)$ have been defined in eqs. (12.287) and (12.288). In particular, we can rewrite

$$\Delta(r) = r^2 \left[1 - 2u + \frac{a^2}{(GM/c^2)^2} u^2 \right], \qquad (14.124)$$

where inside the square brackets we have used the variable $u = GM/(c^2 r)$ instead of r.

In the EOB approach we are, however, interested in a deformation of the Kerr metric. One can then compute the Hamiltonian corresponding to a spinning particle in the metric (14.118), leaving for the moment the functions B, ω, μ and ν as generic functions of r and θ, to be fixed later by comparison with the PN expansion. This defines the Hamiltonian H_{eff} of the effective one-body problem. This Hamiltonian has a purely orbital part, a spin–orbit term and a spin–spin term. One then defines the Hamiltonian of the real problem, improved with respect to the PN expansion by the EOB resummation, using again the mapping (14.67), i.e.

$$H_{\text{real}} = mc^2 \left[1 + 2\nu \left(\frac{H_{\text{eff}} - \mu c^2}{\mu c^2} \right) \right]^{1/2}. \qquad (14.125)$$

One can now compare the PN expansion of this effective one-body Hamiltonian with the Hamiltonian computed in the PN expansion, with its orbital part computed to the desired order (for example 3PN or 4PN) plus the spin terms given in eqs. (14.111)–(14.116). Just as in the spinless case, one must take into account that the canonical coordinates and momenta of the EOB formulation will in general turn out to be different from the ADM coordinates in terms of which we have written the PN expansion of the Hamiltonian. The two Hamiltonians will therefore become equal (with a suitable choice of the EOB metric) only after a canonical transformation of the coordinates and conjugate momenta. In this way one finally finds[35] that the dynamics of two spinning BHs with masses m_1, m_2 and spins $\mathbf{S}_1, \mathbf{S}_2$ can be mapped onto the dynamics of a spinning particle of mass $\mu = m_1 m_2/(m_1 + m_2)$, moving in a deformation of the Kerr metric for a BH with a mass M equal to the total mass $m = m_1 + m_2$ of the BH binary in the real problem. Performing the matching, the deformed Kerr metric turns out to be obtained by modifying eqs. (14.119), (14.121), (14.122) and (14.123) to

$$\frac{dR}{dr} = \frac{R}{\sqrt{\Delta_r}}, \qquad (14.126)$$

$$B(r) = \frac{\sqrt{\Delta_t(r)}}{R(r)}, \qquad \omega(r,\theta) = \frac{ar}{\Lambda_t(r,\theta)} \frac{2GM}{c^2}, \qquad (14.127)$$

$$e^{2\nu(r,\theta)} = \frac{\rho^2(r,\theta)\Delta_t(r)}{\Lambda_t(r,\theta)}, \qquad e^{2\mu(r,\theta)} = \frac{\rho^2(r,\theta)}{R^2(r)}. \qquad (14.128)$$

[35]See Barausse, Racine and Buonanno (2009) and Barausse and Buonanno (2010) for the (rather long) explicit computations.

The new functions Δ_r, Δ_t and Λ_t are given by

$$\Delta_t(r) = r^2 \left[A(u) + \frac{a^2}{(GM/c^2)^2} u^2 \right], \qquad (14.129)$$

$$\Delta_r(r) = \bar{D}(r)\Delta_t(r), \qquad (14.130)$$

$$\Lambda_t(r,\theta) = (r^2 + a^2)^2 - a^2\Delta_t(r)\sin^2\theta, \qquad (14.131)$$

where $A(u)$ is the EOB-improved function given explicitly to 4PN order in eqs. (14.73)–(14.76), and $\bar{D}(u) = D^{-1}(u) = [A(u)B(u)]^{-1}$ was given to 4PN order in eqs. (14.78)–(14.80). The spin of the effective particle moving in this metric turns out to be

$$\boldsymbol{S} = \frac{\nu}{c} \left(\boldsymbol{\sigma}^* + \frac{1}{c^2}\boldsymbol{\Delta}_{\sigma^*} \right), \qquad (14.132)$$

where

$$\boldsymbol{\Delta}_{\sigma^*} = \frac{\nu}{12} \left[\frac{2GM}{R}(7\boldsymbol{\sigma}^* - 4\boldsymbol{\sigma}) + \hat{\boldsymbol{p}}^2(3\boldsymbol{\sigma} + 4\boldsymbol{\sigma}^*) - 6(\boldsymbol{n}\cdot\hat{\boldsymbol{p}})^2(6\boldsymbol{\sigma} + 5\boldsymbol{\sigma}^*) \right]. \qquad (14.133)$$

We have used the unit vector $\boldsymbol{n} = \hat{\boldsymbol{r}}/r$, where r is the EOB radial coordinate in Boyer–Lindquist coordinates (while r' in eqs. (14.112)–(14.116) was the radial ADM coordinate), and we have defined $\hat{\boldsymbol{p}} = \boldsymbol{p}/\mu$.[36] The spin of the Kerr BH in the effective description is

$$\boldsymbol{S}_{\mathrm{Kerr}} = \boldsymbol{S}_1 + \boldsymbol{S}_2, \qquad (14.134)$$

and its spin parameter is $a = |\boldsymbol{S}_{\mathrm{Kerr}}|/Mc$. As in the spinless case, one can further improve the EOB result by performing the appropriate resummations of the functions that appear in the metric of the effective problem, to enforce the presence of a horizon. However, a direct Padé resummation of the function

$$\Delta_u(u) \equiv A(u) + \frac{a^2}{(GM/c^2)^2} u^2 \qquad (14.135)$$

produces spurious poles. Therefore, in the literature, the Padé resummation is rather applied directly to the function $A(u)$ so that, for instance, at 5PN order one would replace $\Delta(u)$ by $[A(u)]_5^1 + [a^2/(GM/c^2)^2]u^2$. Alternatively, instead of the Padé resummation, one can use a logarithmic resummation of the full function $\Delta_u(u)$, of the form

$$\Delta_u(u) = \bar{\Delta}_u(u) \left[1 + \nu\,\Delta_0 + \nu\,\log\left(1 + \Delta_1\,u + \Delta_2\,u^2 + \Delta_3\,u^3 + \Delta_4\,u^4\right) \right], \qquad (14.136)$$

where

$$\begin{aligned}
\bar{\Delta}_u(u) &= \frac{a^2}{M^2} \left(u - \frac{M}{r_{\mathrm{H},+}^{\mathrm{EOB}}} \right) \left(u - \frac{M}{r_{\mathrm{H},-}^{\mathrm{EOB}}} \right) \\
&= \frac{a^2 u^2}{M^2} + \frac{2u}{\nu K(\nu) - 1} + \frac{1}{[\nu K(\nu) - 1]^2} \qquad (14.137)
\end{aligned}$$

[36] The terms $\hat{\boldsymbol{p}}^2$ and $(\boldsymbol{n}\cdot\hat{\boldsymbol{p}})^2$ can be further covariantized, writing the scalar products in terms of the deformed Kerr metric. This gives an expression equivalent to this PN order, which, however, slightly improves the EOB dynamics; see Barausse and Buonanno (2010). Observe also that the mapping can be performed in different ways, by adding a further term to the particle spin $\boldsymbol{S}$ and subtracting a corresponding term to the BH spin $\boldsymbol{S}_{\mathrm{Kerr}}$. The choice made in eqs. (14.133) and (14.134) gives a particularly simple result for the BH spin in the EOB description, which becomes equal to the sum of the two spins in the real problem.

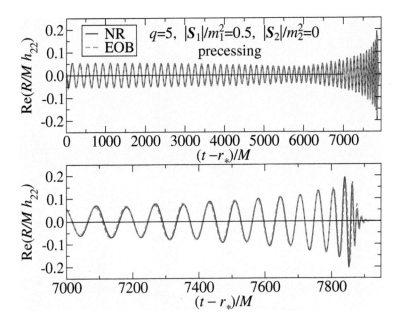

Fig. 14.10 Numerical relativity (NR) and EOB (2,2) precessing waveforms of a BH binary with a mass ratio $1 : 5$ and initial spins $|\mathbf{S}_1|/m_1^2 = 0.5$ in the orbital plane, and $|\mathbf{S}_2|/m_2^2 = 0$. The two waveforms are aligned at low frequency. R is the distance to the source. The lower panel shows a close-up near the time of merger. From Taracchini *et al.* (2014).

and

$$r_{\mathrm{H},\pm}^{\mathrm{EOB}} = \left(M \pm \sqrt{M^2 - a^2} \right) \left[1 - K(\nu)\,\nu \right]. \qquad (14.138)$$

Here, $r_{\mathrm{H},\pm}^{\mathrm{EOB}}$ are the EOB-modified Kerr horizons. The coefficients Δ_i are fixed by the result up to 4PN order, and $K(\nu)$ is an adjustable function that modifies the position of the Kerr horizons with respect to the $\nu = 0$ case. Its introduction is an equivalent way of parameterizing the effect of the unknown coefficients at higher PN order, and the function $K(\nu)$ is then fixed by calibration to numerical relativity simulations.

   One can now study the dynamics of the inspiral and plunge phase for spinning binaries, using the EOB formulation. In the real two-body problem with generic spin configurations, because of the spin–orbit and spin–spin coupling, the orbital angular momentum and the spins are not separately conserved, and they all precess. As a consequence, in general the orbits are no longer planar. However, they remain planar when the spins are aligned or anti-aligned with the orbital angular momentum. Just as in the non-spinning case, the radius of the equatorial circular light ring is a natural point for matching the inspiral and plunge phase to the phase dominated by the ringdown of the final BH. The resulting waveforms, covering inspiral, merger and ringdown, can then be tuned to numerical relativity simulations, as in Fig. 14.7. In this way, even in the spinning case one finds a remarkable agreement between the EOB and the numerical relativity waveforms; see Fig. 14.10. Observe that, in the inspiral phase, the waveform for spinning binaries does not have the simple chirp waveform, with a monotonically increasing amplitude, but rather displays modulations due to the spin.

14.3   Numerical relativity

The first attempts to solve the Einstein equations numerically for a binary BH system were carried out as long as the 1960s and 1970s. In the mid-1970s the head-on collision of two BHs was studied numerically. The fact that this problem has axial symmetry simplifies it considerably, but even in this simplified setting the numerical simulations were limited by the computing resources available at the time. The full three-dimensional problem of binary BH coalescence turned out to be extremely difficult and, for a long time, all numerical simulations suffered from stability problems that caused the numerical evolution to terminate after times relatively short compared with the typical orbital period. Despite intense effort this problem resisted for decades, until 2005 when, thanks to some technical breakthroughs, several independent groups were able to simulate the coalescence of a BH–BH system, from the last few orbits of the inspiral through merger and ringdown, and sufficiently long afterwards to extract the GWs emitted in the radiation zone.

14.3.1   Numerical integration of Einstein equations

From a mathematical point of view the Einstein equations, before fixing a gauge, are a set of 10 coupled second-order differential equations for the 10 components of the metric $g_{\mu\nu}$. These equations, however, have no definite mathematical character (hyperbolic, parabolic or elliptic).[37] The first step in their solution is to choose a coordinate system, thereby reducing the gauge invariance of the problem. The most common procedure is to choose the coordinates so that one, called t, is time-like, and the remaining three, called x^i, are space-like. This is not the only possible choice. For instance, one could rather introduce a null coordinate u (which in the flat-space case corresponds to retarded time $t - r/c$) or double-null coordinates u, v (corresponding to advanced and retarded times $t \pm r/c$ in flat space). Null coordinates are natural in a radiation problem, where ideally one wishes to measure gravitational radiation at future null infinity. However, in non-trivial dynamical space-times it turns out that caustics can form along null coordinates, so in practice $3 + 1$ schemes with three space-like coordinates and one timelike coordinate appear numerically more suitable for studying binary BH coalescences. The best known 3+1 split of space-time is given by the traditional Arnowitt–Deser–Misner (ADM) approach, where the metric is written as in eq. (13.1).

In this coordinate system, once α and β^i in eq. (13.1) have been assigned as explicit functions of time and space, the 10 Einstein equations split into six dynamical equations and four constraint equations. The

[37]Recall that a differential equation of the form $A\partial_x^2 f + B\partial_x\partial_y f + C\partial_y^2 f =$ (source terms) is called hyperbolic, parabolic or elliptic, if the quadratic form $Ap_x^2 + Bp_xp_y + Cp_y^2 = 0$ describes a hyperbola, parabola or ellipse, respectively (and similarly in higher dimensions). The prototype of a hyperbolic equation is the wave equation $(\partial_t^2 - \partial_x^2)f(t, x) = 0$, while a typical elliptic equation is $(\partial_x^2 + \partial_y^2)f(x, y) = 0$. A classic example of parabolic equation is the heat equation, $\partial_t f = k\partial_x^2 f$. A crucial features of hyperbolic equations is that their solutions are wave-like and propagate with a finite speed. In particular, a perturbation of the initial data (or of the boundary conditions) is felt at a generic point only after a finite time. In contrast, perturbations of the initial data in elliptic or parabolic equations are felt at once in the whole domain.

six dynamical equations are second-order partial differential equations for the six remaining functions h_{ij}. In order to properly specify the initial value problem, we must therefore assign both h_{ij} and its first time derivative on the initial hypersurface. Actually, instead of the first derivative, it is convenient to assign the extrinsic curvature, defined in eq. (13.7). From eq. (13.8) we see indeed that, given α, β_i and h_{ij}, the time derivative $\partial h_{ij}/\partial t$ can be traded for K_{ij}. The remaining four Einstein equations do not contain second time derivatives of h_{ij} and therefore are just relations between h_{ij} and its first derivative, or equivalently K_{ij}. If these equations are satisfied at the initial time, then the time evolution preserves them, so we only need to impose them on the initial conditions; in other words, these four equations are constraints on the initial data, rather than dynamical equations.

In this formulation, it appears that we must simply provide initial conditions that satisfy the constraint equations (and that describe the physical problem that we want to simulate, i.e. a BH–BH pair orbiting at some large separation), and evolve the system using only the six dynamical equations. In practice, due to truncation errors in the numerical integration, the evolution generates "constraint-violating modes", i.e. solutions of the six dynamical equations that do not satisfy the constraint equations. A significant role in the recent breakthroughs in numerical relativity has been played by the development of formulations of the field equations, different from the standard ADM formulation, that keep under control the growth rate of these constraint-violating modes.[38]

Two main schemes have been developed. One is the Baumgarte–Shapiro–Shibata–Nakamura (BSSN) formulation, which is a variant of the original ADM scheme, while the other departs completely from the ADM formulation, and is rather based on a generalization of harmonic coordinates.[39] Here, just to give a flavor of these approaches, we discuss the generalized harmonic coordinates used in Pretorius (2005), which is where the first successful numerical evolution of the entire process of BH–BH coalescence, including waveform extraction, were obtained. There are now several dedicated textbooks on numerical relativity, mentioned in the Further Reading section, and we refer the reader to them for more details, as well as for discussions of the BSSN formulation.

Recall from Section 5.1.4 that harmonic coordinates are defined by the condition

$$\Box x^{\mu} = 0\,, \tag{14.139}$$

where[40]

$$\Box \equiv \partial_{\mu}(\sqrt{-g}\,g^{\mu\nu}\partial_{\nu})\,. \tag{14.140}$$

The idea is to define generalized harmonic coordinate from the condition

$$\Box x^{\mu} = H^{\mu}(x)\,, \tag{14.141}$$

where $H^{\mu}(x)$ are functions to be specified (just as, in the usual ADM formulation, α and β^{i} are functions to be specified). The advantage of

---

[38] In more formal language, the original ADM formulation of the Einstein equations turns out to be only weakly hyperbolic, and replacing derivatives with finite differences in a weakly hyperbolic system results in a ill-posed problem.

[39] The relative advantages of the two approaches can be combined in the so-called CCZ4 formulation; see the Further Reading.

[40] The curved-space d'Alembertian on a scalar function ϕ is actually defined by $\Box\phi = (1/\sqrt{-g})\partial_{\mu}(\sqrt{-g}\,g^{\mu\nu}\partial_{\nu})\phi$, so the equation $\Box\phi = 0$ is equivalent to $\partial_{\mu}(\sqrt{-g}\,g^{\mu\nu}\partial_{\nu})\phi = 0$. In this section we just define $\Box$ as a notation for $\partial_{\mu}(\sqrt{-g}\,g^{\mu\nu}\partial_{\nu})$.

the usual harmonic coordinates is that, when substituted in the (exact) Einstein equation, they give equations of the form

$$g^{\rho\sigma}\partial_\rho\partial_\sigma g_{\mu\nu} + (\ldots) = -\frac{16\pi G}{c^4}T_{\mu\nu}\,, \tag{14.142}$$

where the dots denotes terms that are non-linear in the first derivative of the metric but do not contain second derivatives; compare with eq. (5.72).

The linear part of this equation is a hyperbolic wave equation,[41] with all complications relegated to the non-linear terms. The same happens with generalized harmonic coordinates. In this case the (trace-reversed) Einstein equations can be written as

$$g^{\rho\sigma}\partial_\rho\partial_\sigma g_{\mu\nu} + (\partial_\mu g^{\rho\sigma}\partial_\rho g_{\nu\sigma} + \partial_\nu g^{\rho\sigma}\partial_\rho g_{\mu\sigma}) + 2\Gamma^\rho_{\nu\sigma}\Gamma^\sigma_{\mu\rho}$$
$$+(\partial_\mu H_\nu + \partial_\nu H_\mu) - 2H_\rho\Gamma^\rho_{\mu\nu} = -\frac{16\pi G}{c^4}\left(T_{\mu\nu} - \frac{1}{2}T g_{\mu\nu}\right). \tag{14.143}$$

Defining

$$C^\mu \equiv H^\mu - \Box x^\mu\,, \tag{14.144}$$

the gauge condition fixes $C^\mu = 0$. In this formulation the four equations $C^\mu = 0$ are just the constraint equations. However, to deal with constraint-violating modes, one keeps C^μ generic for now, and modifies eq. (14.143) to

$$g^{\rho\sigma}\partial_\rho\partial_\sigma g_{\mu\nu} + (\partial_\mu g^{\rho\sigma}\partial_\rho g_{\nu\sigma} + \partial_\nu g^{\rho\sigma}\partial_\rho g_{\mu\sigma}) + 2\Gamma^\rho_{\nu\sigma}\Gamma^\sigma_{\mu\rho}$$
$$+(\partial_\mu H_\nu + \partial_\nu H_\mu) - 2H_\rho\Gamma^\rho_{\mu\nu} + \kappa[\partial_\mu C_\nu + \partial_\nu C_\mu - g_{\mu\nu}n^\rho C_\rho]$$
$$= -\frac{16\pi G}{c^4}\left(T_{\mu\nu} - \frac{1}{2}T g_{\mu\nu}\right), \tag{14.145}$$

where n^μ is the unit normal to the $t = \text{const.}$ hypersurfaces, and κ is an arbitrary parameter. Since in the end $C^\mu = 0$, this equation is equivalent to eq. (14.143). However, for C^μ generic one can show that, if the metric obeys eq. (14.145), the constraint C^μ defined by eq. (14.144) obeys

$$\Box C^\mu = -R^\mu{}_\nu C^\nu + \kappa\nabla_\nu(n^\nu C^\mu + n^\mu C^\nu)\,. \tag{14.146}$$

Among the terms on the right-hand side, the term $n^\nu\nabla_\nu C^\mu$ is a wave-equation damping term, and its effect is to moderate the growth of any spurious constraint-violating modes generated by numerical errors. Finally, one has to choose the function H^μ. A convenient choice is $H^i = 0$, while H_0 is taken to satisfy

$$\Box H_0 = -\xi_1\frac{\alpha - 1}{\alpha^\eta} + \xi_2 n^\nu\partial_\nu H_0\,, \tag{14.147}$$

where ξ_1, ξ_2 and η are parameters to be chosen appropriately. This wave equation helps to keep the lapse function α close to unity, alleviating an instability that was otherwise present.

Another important issue is how to deal with the BH singularity. Here one relies on the fact that no information can flow outside the BH horizon, which encloses the singularity. One way to deal with the singularity

[41] Actually, the hyperbolic character is determined by the light-cones of $g_{\mu\nu}$, which is itself the unknown quantity to be determined. Thus, more accurately, eq. (14.142) will become a hyperbolic wave equation once one finds a solution for $g_{\mu\nu}$ with the appropriate $(-, +, +, +)$ signature.

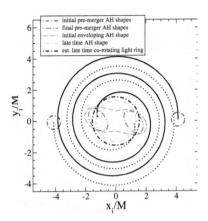

Fig. 14.11 The last few orbits of a BH–BH coalescence for an equal-mass binary, obtained from numerical relativity. The shapes of the apparent horizons at several key moments are also shown. From Buonanno, Cook and Pretorius (2007).

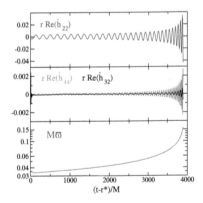

Fig. 14.12 Upper panel: the leading mode $\dot{h}_{22}$. Middle panel: the two next largest modes, $\dot{h}_{44}$ and $\dot{h}_{32}$ (smallest). Lower panel: the frequency ϖ obtained from $\dot{h}_{22}$. (In units $G = c = 1$.) From Boyle *et al.* (2008).

is therefore simply by excision, which means that a two-sphere inside the BH horizon and enclosing the singularity is considered the boundary of the computational domain. No boundary conditions are specified on the excision surface. Rather, here one simply solves the finite difference equations, replacing symmetric discretization of derivatives with "sideways" discretizations. Other successful methods for dealing with the BH singularity are the "moving puncture" method, or the imposition of singularity-avoiding slicing conditions; see the Further Reading.

14.3.2   Equal-mass non-spinning BH binaries

Figure 14.11 shows the last few orbits of a BH–BH coalescence for an equal-mass binary without spin, obtained from numerical relativity. The shapes and positions of the apparent horizons are also shown at some key moments. We see that, when the BHs are still far away, their apparent horizons are spherical. As they get closer and closer, the shapes of their apparent horizons are tidally deformed, until a new apparent horizon appears, which engulfs both BHs. We can take this as the definition of the time at which the two initial BHs merge into a final BH. Observe that, initially, the intersection of the apparent horizon enveloping the two BHs with the orbital plane is far from having a circular shape. However, the final BH eventually settles down into its stationary configuration, with a horizon whose intersection with the orbital plane is circular, as expected for a Kerr BH. The initial "potato-like" shape of the horizon evolves into its symmetric final shape by radiating away the excess energy into the QNMs of the final BH.

Once we have obtained a stable numerical evolution of the system, leading from an initial state with two separate BHs orbiting each other to a final state of a single BH, we can extract several pieces of information from it. We are interested both in quantities that characterize the final space-time itself (such as the ADM mass and angular momentum of the final BH), and in the gravitational radiation at infinity.

Figure 14.12 shows the separate contributions $\dot{h}_{lm}$ [where h_{lm} was defined in eq. (13.86)] for $l = 2$, $l = 3$ and $l = 4$, from a numerical simulation of an equal-mass binary on an initial quasi-circular orbit, in which the initial BHs have no spin, on a scale that emphasizes the long inspiral phase. We see that the signal is dominated by the quadrupole component $l = 2$, $|m| = 2$. The next leading contribution is $l = 4, |m| = 4$, and has an amplitude less that 10% of the quadrupole during inspiral, rising up to 20% at merging. Since energy is quadratic in the amplitude, this means that $h_{l=2,|m|=2}$ is responsible for about 99% of the energy radiated during inspiral, and over 95% of the energy radiated during merger.

The lower panel of Fig. 14.12 shows the characteristic frequency ϖ, defined as

$$\varpi = \frac{d\varphi}{dt}, \qquad (14.148)$$

where

$$\varphi = -\arg(\dot{h}_{22}). \qquad (14.149)$$

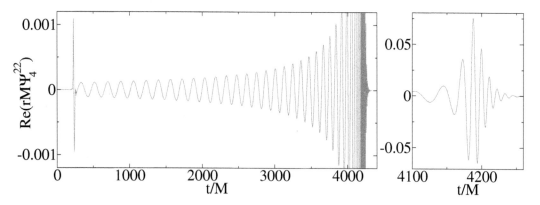

Fig. 14.13 The component $l = 2, m = 2$ of Ψ_4^{lm} for equal-mass non-spinning binaries. The left panel shows the inspiral waveform and the right panels is a zoom of the merger and ringdown phase (units $G = c = 1$). The initial spike at $t \simeq 200M$ (where M is the total mass of the binary) is an artifact of the initial conditions. From Scheel *et al.* (2009).

The increase in frequency is characteristic of the chirp amplitude. Figure 14.13 shows the waveform in greater detail, using as a variable the component $l = m = 2$ of the variable Ψ_4^{lm}, defined in eq. (13.83) by expanding the Newman–Penrose scalar Ψ_4 in the basis of the $s = 2$ spin-weighted spherical harmonics. The left panel displays the long inspiral waveform, while the right panel is a zoom of of the merger and ringdown phase. Note the difference in vertical scale between these two panels.

Figure 14.14 shows the evolution of the characteristic frequency from inspiral to ringdown. The quantity denoted here by ω_λ is conceptually the same as $\varpi/2$, except that it is extracted numerically from the Newman–Penrose scalar Ψ_4 rather than from $\dot{h}_{22}$. The factor $1/2$ transforms it from the characteristic GW frequency to the characteristic frequency of the orbital motion. The quantity ω_c is the orbital frequency obtained by tracking the coordinate locations of the centers of the apparent horizons of each individual BH. Clearly, this second definition goes astray when the two BHs merge, while ω_λ retains its physical meaning all through merger and ringdown. We can see from the figure that the two definitions indeed give consistent results until a common apparent horizon is detected. Then ω_c diverges and loses physical meaning, while ω_λ reaches a plateau at a value $\simeq 0.3$ (in units of Gm/c^3). The corresponding GW frequency $2\omega_\lambda$ then reaches a value $\simeq 0.6$, which is just the frequency of the fundamental QNM of the final Kerr BH; compare with eq. (14.109).

Higher overtones with $l = 2, m = 2, n > 0$, have lower frequencies and shorter decay times (in the case of Schwarzschild BHs we can see this from Table 12.1), so their contributions disappear more quickly. Thus, while the fundamental QNM with $l = 2, m = 2, n = 0$ is responsible for the plateau, higher overtones have the role of raising the frequency from the peak of the radiation to the plateau.

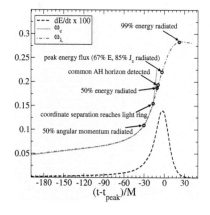

Fig. 14.14 Two upper curves: the orbital frequency ω_c evaluated from the motion of the apparent horizons of the individual BHs (solid line) and the orbital frequency ω_λ evaluated from the wave (dot-dashed). Lower (dashed) curve: the GW energy flux dE/dt. Also shown, marked with circles, are some key moments in the evolution. From Buonanno, Cook and Pretorius (2007).

Also shown in Fig. 14.14 are several key moments in the evolution, namely, the time when the common apparent horizon (AH) of the final BH first appears, when the binary separation reaches the light ring of the final BH, the peak of the radiation flux (which occurs around a time t given by $t \simeq t_{\rm peak} - \alpha GM/c^3$, where α is in the range 3–4, and $t_{\rm peak}$ marks the peak in the amplitude of the waveform), the time when 50% of the energy and angular momentum have been radiated, and the time when 99% of the energy has been radiated. The lower curve represents the radiated energy dE/dt (multiplied by a factor 100), with E measured in units of mc^2, where m is the total mass of the binary, and t measured in units of Gm/c^3. We see that at the peak the luminosity in GWs of an equal-mass BH–BH coalescence is

$$\left(\frac{dE}{dt}\right)_{\rm peak} \simeq 0.1 \times 10^{-2} \frac{mc^2}{Gm/c^3} = 10^{-3}\frac{c^5}{G}. \tag{14.150}$$

The quantity

$$\mathcal{L}_E \equiv \frac{c^5}{G} \simeq 3.629 \times 10^{59}\,{\rm erg/s} \tag{14.151}$$

is called the Einstein luminosity (or the Planck luminosity). Its physical meaning is that it provides an upper bound on the luminosity emitted by any astrophysical object. Indeed, consider an astrophysical body of mass M that converts a fraction x of its mass in radiation (of any sort, for instance gravitational, electromagnetic, or neutrino), with $x \leqslant 1$ by definition. An intense burst of radiation can only be obtained from coherent emission over a region comparable to the size of the body. For causality reasons, such coherent radiation cannot be emitted in a time shorter than the time $2R/c$ that light takes to go across an object of radius R, so even in this extreme case the luminosity cannot be larger than

$$\mathcal{L} = \frac{xMc^2}{2R/c}. \tag{14.152}$$

For a body of mass M we have $R \geqslant 2GM/c^2$, with the equality saturated by BHs, and therefore

$$\mathcal{L} \leqslant \frac{xc^5}{4G} = \frac{x}{4}\mathcal{L}_E. \tag{14.153}$$

We see that, even setting $x = 1$, one cannot exceed the value $0.25\mathcal{L}_E$. Another way to get a grasp of this huge luminosity is to express it in solar masses per millisecond, in which case eq. (14.151) reads

$$\mathcal{L}_E \simeq 203\,M_\odot/{\rm ms}. \tag{14.154}$$

From eq. (14.150) we see that the GW luminosity of an equal-mass BH–BH coalescence has a peak value that is about 10^{-3} of the Einstein luminosity, resulting in the huge value

$$\left(\frac{dE}{dt}\right)_{\rm peak} \simeq 3.6 \times 10^{56}\,{\rm erg/s}, \tag{14.155}$$

or, equivalently, a GW luminosity of about 200 solar masses per second. This makes BH–BH coalescences the astrophysical events with by far the highest luminosity in the Universe.

The final BH is a Kerr BH, and its mass M_f and Kerr parameter $a = J_f/(M_f c)$ can be determined from the numerical relativity results, using the properties of the apparent horizon as discussed in Section 13.4. Denoting by m the sum of the irreducible masses of the initial BHs, the numerical simulations show that, for an equal-mass non-spinning BH binary, the final BH mass is[42]

[42]See Scheel *et al.* (2009).

$$M_f = (0.95162 \pm 0.00002)\, m\,, \tag{14.156}$$

so about 4.8% of the total mass of the binary system is radiated away in GW. It should be remarked that this is a huge conversion efficiency. For comparison, in nuclear reactions the highest efficiency for conversion of nuclear binding energy is obtained in the conversion of hydrogen into helium in the $p-p$ chain, and releases about 0.7% of the initial mass.

For the spin of the final BH, again for equal-mass non-spinning binaries, the numerical simulations give

$$a_f = (0.68646 \pm 0.00004)\, \frac{GM_f}{c^2}\,. \tag{14.157}$$

The waveform in the ringdown regime can be compared with the results of BH perturbation theory, discussed in Chapter 12, as follows. First of all, having determined the mass and Kerr parameter of the final BH, the frequencies and decay times of the QNMs of the final BH are fixed. For instance, the frequency ω_{nlm} of the mode with $n=1, l=2, m=2$ is given in eq. (12.360), and the corresponding lifetime is obtained from eq. (12.361). BH perturbation theory in the Kerr background then allows us to compute the components of the metric as an expansion in spin-weighted spheroidal harmonics $S_{lm}(\theta,\omega)$, as in eq. (12.350). Therefore, in the wave zone, the Newman–Penrose scalar Ψ_4, which is related to the metric by eq. (13.81), can also be computed in an expansion in spin-weighted spheroidal harmonics. The ringdown signal is obtained from the closed contour in the complex ω-plane shown in Fig. 12.7, retaining only the contribution of the poles, as explained in Section 12.3.2. Thus, one finds

$$\left(\frac{GM_f}{c^4}\right)\Psi_4(t,r,\theta,\phi) = \frac{1}{r}\sum_{nlm}\left[\mathcal{C}_{nlm}\,S_{nlm}(\theta)e^{im\phi}e^{-i(\omega_{nlm}t+\varphi_{nlm})}\right.$$
$$\left. +\mathcal{C}'_{nlm}\,S^*_{nlm}(\theta)e^{-im\phi}e^{i(\omega^*_{nlm}t+\varphi'_{nlm})}\right]\,, \tag{14.158}$$

where

$$\omega_{nlm} = (\omega_R + i\omega_I)_{nlm} \tag{14.159}$$

are the complex QNM frequencies, $S_{nlm}(\theta)$ are the spheroidal harmonics $S_{lm}^{(s)}(\theta,\omega)$ with spin weight $s=-2$, evaluated at $\omega = \omega_{nlm}$,

$$S_{nlm}(\theta) \equiv S_{lm}^{(-2)}(\theta,\omega_{nlm})\,, \tag{14.160}$$

and $\mathcal{C}_{nlm}, \mathcal{C}'_{nlm}, \varphi_{nlm}$ and φ'_{nlm} are real constants that describe the amplitude and phase of each QNM contribution. The terms proportional to $\mathcal{C}'_{nlm}$ take into account the fact that for each nlm there are actually two QNM frequencies; see Note 66 on page 179. The spheroidal harmonics are rather complicated functions. However, we have seen in eq. (13.85) that Ψ_4, having spin-weight $s = -2$, admits an expansion in terms of the much simpler $s = -2$ spin-weighted *spherical* harmonics $Y_{lm}^{(-2)}(\theta, \phi)$ defined in eq. (12.352),

$$\left(\frac{GM_f}{c^2}\right) \Psi_4(t, r, \theta, \phi) = \frac{1}{r} \sum_{l,m} C_{lm}(t)\, Y_{lm}^{(-2)}(\theta, \phi), \qquad (14.161)$$

where we made use of the fact that, in the radiation zone, $C_{lm}(t, r)$ becomes independent of r. The two expansions (14.158) and (14.161) can be related by writing the spin-weighted spheroidal harmonics in terms of spin-weighted spherical harmonics. The expansion has the generic form[43]

$$S_{nlm}(\theta)e^{im\phi} = \sum_{l'} \mathcal{A}_{ll'mn} Y_{l'm}^{(-2)}(\theta, \phi). \qquad (14.162)$$

Inserting this into eq. (14.158) we see that, in principle, each C_{lm} depends on an infinite series of QNM oscillations labeled by l' and by n; for example C_{22} is a superposition of the terms in eq. (14.158), proportional to $\mathcal{C}_{n22}, \mathcal{C}_{n32}, \mathcal{C}_{n42}, \ldots$, as well as terms proportional to $\mathcal{C}'_{n2-2}, \mathcal{C}'_{n3-2}, \mathcal{C}'_{n4-2}$, etc. In practice, however, comparison with the numerical relativity results shows that the contribution to C_{22} of the terms $\mathcal{C}_{nl2}$ or $\mathcal{C}'_{nl-2}$ with $l > 2$ is negligible, and of the order of the numerical precision. Furthermore, even the terms with $m = -2$, i.e. those proportional to $\mathcal{C}'_{n2-2}$, are numerically negligible, so in practice one can simply fit the ringdown part of the signal to the first N QNMs with $l = 2$ and $m = 2$, writing

$$C_{22}(t) \simeq \sum_{n=1}^{N} \mathcal{C}_{n22} e^{-i(\omega_{n22}t + \varphi_{n22})}, \qquad (14.163)$$

with the complex frequencies $\omega_{n22}(M_f, a_f)$ fixed by knowledge of the mass and angular momentum of the final BH, and taking $\mathcal{C}_{n22}$ and φ_{n22} as fitting parameters. Better yet, for a given numerical relativity simulation, one can take also M_f and a_f as free fitting parameters, and compare their determination from the ringdown signal with the determination obtained from the apparent horizon of the final BH. The results of the two methods, using $N = 4$, i.e. including the fundamental QNM and its first three overtones, are in excellent agreement.

14.3.3 Unequal-mass non-spinning BH binaries

The case of equal-mass, non-spinning binaries on an initial orbit with zero eccentricity, discussed in the previous section, is the simplest configuration of a compact binary, and is therefore a convenient starting

[43] See Press and Teukolsky (1973).

point for studying the merger process, both analytically and numerically. However, in realistic coalescences of compact binaries, we must eventually remove these simplifying assumptions. For instance, while for NS the mass distribution is quite narrowly peaked around $1.4 M_\odot$, the masses of stellar BHs can range from a few solar masses to tens of solar masses, so in a BH–BH coalescence we have no reason to expect that the two BH masses will be approximately equal. Furthermore, realistic BHs are expected to be spinning. Further complications may come from the eccentricity of the orbit. We saw in Section 4.1.3 that radiation reaction circularizes the orbit efficiently, so during the last part of the inspiral the orbit can be taken to be circular. However, particularly in dense stellar environments, such as globular clusters, there is the possibility that the interaction of the binary system with a third body induces an eccentricity in the orbit of the binary system. The inclusion of these complications generates novel and interesting effects, which we explore in this subsection and in the next.

We consider here the case of unequal-mass binaries. As we have already mentioned, numerical relativity faces further technical challenges when the mass ratio m_2/m_1 (where we take $m_2 \leqslant m_1$) becomes very small. The reason is that, to achieve sufficient numerical resolution near the smaller BH, it is necessary to make an additional adaptive mesh refinement. Furthermore, the smaller BH also moves faster than the larger one, since it is farther away from their common center of mass and still completes one orbit around the center of mass in the same time as the largest BH, so a smaller time step is also required to follow its time evolution.

Radiated energy

In the equal-mass case we found in eq. (14.156) that the energy radiated in GWs is about 4.8% of the total mass of the binary system. For unequal-mass binaries, a smaller fraction of the total mass is converted into GWs. Combining analytic information in the test-mass limit with fits to numerical simulations, for non-spinning binaries the radiated energy is well fitted (to better than 1%) by the formula[44]

$$\frac{E_{\text{rad}}}{mc^2} \simeq \nu(1 - 4\nu)[1 - \tilde{E}_{\text{ISCO}}] + p_0(4\nu)^2 \,, \qquad (14.164)$$

where, as usual, $\nu = m_1 m_2/(m_1 + m_2)^2$ is the symmetric mass ratio. For spinless binaries, $\tilde{E}_{\text{ISCO}} = 2\sqrt{2}/3$, while the coefficient p_0 is obtained by fitting to numerical relativity, and one finds $p_0 \simeq 0.048$. Observe that, for ν close to the value $1/4$ that corresponds to equal-mass binaries, the term proportional to $(1 - 4\nu)$ is small and the radiated energy scales basically as ν^2, as in the radial infall into a BH; see eq. (4.296) or eq. (12.284). However, as $\nu \to 0$, the first term dominates and the radiated energy scales as ν.

An important difference from the equal-mass case is that a larger percentage of the energy is radiated in the multipoles with $l > 2$. Numerical

[44]See Barausse, Morozova and Rezzolla (2012), where a formula for spinning binaries is also given.

[45]See Berti *et al.* (2007).

[46]See González, Sperhake, Brügmann, Hannam and Husa (2007).

simulations show that, while in the equal-mass case about 98% of the total energy is radiated in the $l = 2$ mode, this percentage goes down to about 85% for a mass ratio $m_1/m_2 = 4$, while at the same time the percentage of energy radiated in the mode $l = 3$ raises from about 0.4% for equal masses up to about 11.6% for $m_1/m_2 = 4$.[45] This sensitivity to higher multipoles can be important for data analysis, since the dependence of the waveform on parameters of the binary such as the masses or the initial spins is different for different multipoles, and therefore the detection of the contribution from multipoles with $l > 2$ can help to remove degeneracies in the parameter estimation which are present in the $l = 2$ waveform, resulting in a more accurate estimation of the binary parameters.

Concerning the spin of the final BH, for non-spinning binaries numerical simulations are well reproduced by a linear dependence on ν,[46]

$$a_f \simeq (0.089 + 2.4\nu)\frac{Gm}{c^2} , \tag{14.165}$$

which is consistent with eq. (14.157) when $\nu = 1/4$.

14.3.4   Final BH recoil

A novel and interesting effect appears in the unequal-mass case, namely the final BH can receive a significant recoil. As we saw in Section 1.4.3, GWs carry linear momentum, and the flux of linear momentum taken away by GWs is given by eq. (1.164). For a circular orbit this linear momentum loss, averaged over one orbit, is zero for symmetry reasons. However, for an inspiraling orbit that eventually suddenly terminates because of merger, any asymmetry between the two constituents, such as that due to unequal masses or to different spins of the two bodies, results in a net loss of linear momentum to GWs, and therefore in a recoil of the final BH.

As long as the binary is in the non-relativistic regime we can compute the instantaneous recoil at any point of the orbit using the multipole expansion. As we discussed in Vol. 1, page 131, in the quadrupole approximation the loss of linear momentum vanishes. In the multipole expansion non-vanishing contributions come, however, from the interference between multipoles of different parity, with the leading contribution coming from the interference between the mass quadrupole and the sum of the mass octupole and the current quadrupole; see eq. (3.167). Since radiation reaction tends to circularize the orbit, we will limit ourselves to a binary in a circular orbit. Thus, as long as the binary is non-relativistic, the leading term of the momentum flux is obtained by computing the mass quadrupole, mass octupole and current quadrupole, taking the two bodies to be on a Newtonian circular orbit and using eqs. (1.164) and (3.167). The computation is similar to the one that we already performed in Problems 3.2 and 3.3. Denoting by R the radius of the circular orbit and by $\varphi(t)$ the angular position along the orbit, the result is that the momentum flux $d\mathbf{P}_{\rm GW}/dt$ carried by GWs, integrated over the solid angle, has the form

$$\frac{d\mathbf{P}_{\rm GW}}{dt} = \left|\frac{d\mathbf{P}_{\rm GW}}{dt}\right| (-\sin\varphi, \cos\varphi, 0)\,, \qquad (14.166)$$

so the vector $d\mathbf{P}_{\rm GW}/dt$ is in the plane of the orbit and at right angles to the radius vector. Its modulus is given by

$$\left|\frac{d\mathbf{P}_{\rm GW}}{dt}\right| = \frac{464c^4}{105G}\left(\frac{Gm}{c^2R}\right)^{11/2} f(\nu)\,, \qquad (14.167)$$

where, as usual, $m = m_1 + m_2$ is the total mass of the system and

$$f(\nu) = \nu^2(1-4\nu)^{1/2}\,. \qquad (14.168)$$

The evolution of the center-of-mass vector $\mathbf{x}_{\rm CM}$ is then determined by momentum conservation,

$$\begin{aligned}
m\frac{d^2\mathbf{x}_{\rm CM}}{dt^2} &= -\frac{d\mathbf{P}_{\rm GW}}{dt} \qquad\qquad\qquad (14.169)\\
&= \frac{464c^4}{105G}\left(\frac{Gm}{c^2R}\right)^{11/2} f(\nu)\,(\sin\Omega t, -\cos\Omega t, 0)\,,
\end{aligned}$$

where we have used the fact that, on a circular orbit, $\varphi(t) = \Omega t$, where $\Omega^2 = Gm/R^3$ is the Keplerian angular velocity. Integrating once we get

$$\frac{d\mathbf{x}_{\rm CM}}{dt} = -\frac{464}{105}c\left(\frac{Gm}{c^2R}\right)^4 f(\nu)\,(\cos\Omega t, \sin\Omega t, 0)\,. \qquad (14.170)$$

The center of mass therefore moves on a circular orbit, and its instantaneous velocity has a modulus

$$\frac{v_{\rm CM}}{c} = \frac{464}{105}\left(\frac{Gm}{c^2R}\right)^4 f(\nu)\,. \qquad (14.171)$$

This is known as the *Fitchett formula*. As anticipated, eq. (14.170) gives zero when averaged over one orbit. In order to get a first crude order-of-magnitude estimate of the recoil of a binary due to coalescence and merger, one can observe that eq. (14.171) holds approximately even for a slowly inspiraling orbit with radius $R(t)$, as long as $\dot{R}$ is much smaller than the tangential velocity. One could then try to extrapolate it down to a minimum value of the separation $R_{\rm min}$, assuming that after that point the system merges, linear momentum radiation suddenly shuts off, and the system recoils with the instantaneous center-of-mass velocity that it has at that moment. Of course such a procedure is quite crude because the precise value of $R_{\rm min}$ is somewhat arbitrary, and the numerical value obtained in this way is quite sensitive to the choice made, since $v_{\rm CM}$ in eq. (14.171) has a strong dependence on R, $v_{\rm CM} \sim R^{-4}$. Recall, however, that the EOB description suggests that a change of regime takes place around the light ring, $R \simeq 3Gm/c^2$. Setting

$R = 3Gm/c^2$ in eq. (14.171) gives a first estimate for the recoil velocity $v_{\rm rec}$,

$$v_{\rm rec} \simeq A\nu^2\sqrt{1-4\nu}\,, \qquad (14.172)$$

with $A \simeq 1.6 \times 10^4$ km/s. The function $f(\nu) = \nu^2\sqrt{1-4\nu}$ vanishes both in the test-mass limit $\nu = 0$ and in the equal-mass limit $\nu = 1/4$, and is maximum at $\nu = 0.2$, with $f(0.2) \simeq 0.0179$. This provides a first estimate of the maximum recoil velocity of order 290 km/s.

In the end, an accurate numerical prediction can only come from numerical relativity. The results of simulations for different values of ν turn out to be well fitted by

$$v_{\rm rec} \simeq A\nu^2\sqrt{1-4\nu}\,(1+B\nu)\,, \qquad (14.173)$$

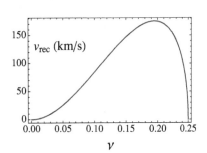

with $A \simeq 1.20 \times 10^4$ km/s and $B \simeq -0.93$. This recoil velocity is plotted as a function of the symmetric mass ratio ν in Fig. 14.15. The highest value is reached for $\nu \simeq 0.195 \pm 0.005$, corresponding to a mass ratio $m_2/m_1 \simeq 0.36$, and is

$$(v_{\rm rec})_{\rm max} \simeq 175 \pm 11\,{\rm km/s}\,, \qquad (14.174)$$

where the uncertainty is the estimate of the error of the numerical simulations. It is also common to write this result in terms of the mass ratio

$$q = \frac{m_2}{m_1}\,, \qquad (14.175)$$

Fig. 14.15 The recoil velocity $v_{\rm rec}$ as a function of the symmetric mass ratio ν.

where by definition m_2 is the lighter mass, so $0 < q \leqslant 1$. Then

$$\nu = \frac{q}{(1+q)^2}\,, \qquad (14.176)$$

and eq. (14.173) reads

$$v_{\rm rec} \simeq A\frac{q^2(1-q)}{(1+q)^5}\left[1+B\frac{q}{(1+q)^2}\right]\,. \qquad (14.177)$$

To get a more accurate analytic understanding of this result one can go beyond the Newtonian approximation used to derive eq. (14.172), and integrate the linear momentum losses during the inspiraling orbit using the post-Newtonian equation of motion of the orbit, and furthermore one can resum them non-perturbatively using the EOB approach. A further effect that has been studied analytically is an "anti-kick" due to the radiation emitted by the final BH in its QNMs. Indeed, the recoil increases monotonically during the inspiral and plunge phase up to a peak value, and then decreases quickly to a final asymptotic value, which can be smaller by about 30%, due to this "anti-kick". With these improvements, one obtains analytic predictions for the recoil velocity that agree with numerical relativity to about 10–15%, over a wide range of mass ratios.

14.3.5 Spinning BHs and superkicks

Realistic BHs are expected to be spinning. Indeed, for stellar mass BHs, the experimental observations that we will discuss in Chapter 15 show that the initial BHs in a coalescing binary in general are spinning. In particular for the event GW151226, which was the second confirmed discovery of a BH–BH coalescence, we know at 99% c.l. that at least one of the initial BHs had a dimensionless spin parameter greater than 0.2; the parameter estimation of the two dimensionless spin parameters $\hat{a}_i$ for GW150914 and for GW151226 gives best-fit values of order $0.2 - 0.5$; see e.g. Table 15.1 on page 289 or Table 15.2 on page 293.

For supermassive BHs, which will be studied in Chapter 16, X-ray observations of line broadening in active galactic nuclei (AGN) suggest that their spins are large, in fact quite close to the extremal Kerr limit, with a dimensionless spin parameter $\hat{a} > 0.9$.[47]

The inclusion of spin induces novel and interesting effects in the coalescence of a compact binary. The parameter space of a spinning BH binary is much larger than in the non-spinning case since, beside the mass ratio $q = m_2/m_1$, we now also have the three components of the spin of the first BH and the three components of the spin of the second. Numerical relativity simulations with spinning BHs are exploring this large parameter space, and have found a number of very interesting results.

Effect of the initial BH spins on the GW emission

The first simulations focused on configurations with the BH spins aligned or anti-aligned with the orbital angular momentum. Beside presenting the advantage of simplicity, such configurations are also believed to be generated in nature by a number of dynamical mechanisms. In particular, in a PN analysis, the spin–orbit coupling has a tendency to align the spins, starting from generic initial conditions. Dissipative effects due to gas within the binary can also induce alignment.

The result of the numerical simulations is that, in the configuration with both spins aligned with the orbital angular momentum, the merger is delayed with respect to the spinless case, while when both spins point in the direction opposite to the orbital angular momentum (i.e. anti-aligned spins) the merger takes place faster than in the spinless case. This can be understood already from the PN expansion, considering the leading spin–orbit Hamiltonian at 1.5PN in harmonic coordinates given in eq. (14.117). When the spins are aligned with the orbital angular momentum this Hamiltonian corresponds to an extra potential C/r^3, with $C > 0$, and is therefore a repulsive term. Conversely, when both spins are in the direction opposite to the orbital angular momentum, the spin–orbit term induces an attractive potential. As a result, in the aligned configuration this repulsion delays the merger and forces the BHs to orbit each other longer, resulting in a higher radiated energy. Conversely, in the anti-aligned case the attractive spin–orbit coupling speeds up the merger process. The radiated energy and angular momentum are then

[47]The idea behind these measurements is that the position of the ISCO is a monotonically decreasing function of the dimensionless Kerr parameter $\hat{a}$, which was defined in eq. (12.359) as $\hat{a} \equiv a/(GM/c^2)$, where M is the BH mass. For instance, for $\hat{a} = -1$ (i.e for an orbit that counter-rotates with respect to an extremal BH), $R_{\mathrm{ISCO}} = 9GM/c^2$, while for a Schwarzschild BH (i.e. $\hat{a} = 0$) we have already seen that $R_{\mathrm{ISCO}} = 6GM/c^2$, and for $\hat{a} = 1$, $R_{\mathrm{ISCO}} = GM/c^2$. In a typical accretion disk model, such as the standard Novikov–Thorne model, the inner part of the accretion disk is located at the ISCO and its dynamics is therefore influenced by the BH spin. A particularly revealing feature is the broadening of a $K\alpha$ iron line. A small ISCO radius means that, before plunging into the BH, the gas in the accretion disk reaches higher velocities, and therefore undergoes larger Doppler broadening. Furthermore, the line is also broadened by GR effects such as gravitational redshift and light bending and beaming, which result in a greatly elongated and skewed line profile, with a long low-energy tail. Again, these effects are enhanced for an accretion disk co-rotating with a near-extremal BH, because of its small ISCO value. The main uncertainty in the reconstruction of the spin comes from the possibility that a significant contribution to the broadening might also come from the matter inside the ISCO and plunging toward the horizon. Clearly, arbitrarily large redshifts could be obtained even around a non-rotating BH, if one assumes that significant contribution to the Fe $K\alpha$ line comes from emission that takes place from any radius arbitrarily close to the horizon. However, this does not seem physically plausible, since the region inside the ISCO is very tenuous and extremely highly photoionized. Assuming that no significant contribution comes from inside the ISCO, one can determine quite accurately the spin of some BHs in AGN. In particular, in this way Brenneman and Reynolds (2006), using *XMM-Newton* observations, determined the spin of the central BH in the galaxy MCG-06-30-15, obtaining $\hat{a} = 0.989^{+0.009}_{-0.002}$ at 90% c.l.; see also the Further Reading.

lower than in the spinless case.

Accurate quantitative results can be obtained from numerical relativity simulations. For two equal-mass BHs, in the spinless case we saw in eq. (14.156) that the energy radiated in GWs is about 4.8% of the total initial mass. If we rather consider two BHs, each with a dimensionless Kerr parameter [defined in eq. (12.359)] $\hat{a} = 0.757$ in the aligned configuration, numerical relativity shows that the total radiated energy rises to $(6.7 \pm 0.2)\%$ of the total initial mass while in the anti-aligned configuration it is reduced to $(2.2 \pm 0.1)\%$.[48]

The spin of the final BH of course depends on the value of the initial spins. In all cases, the result of the coalescence is a BH whose dimensionless Kerr parameter satisfies the condition $\hat{a} \leqslant 1$, necessary to have indeed a Kerr BH rather than a naked singularity. What allows one to maintain the condition $\hat{a} \leqslant 1$ is the fact that angular momentum is radiated away in GWs, and the radiated angular momentum is larger for the the aligned spin configuration, for the same reason for which the radiated energy is larger, thereby compensating for the fact that the initial total angular momentum is larger in this configuration. Taking for instance again the case of equal-mass BHs, in the spinless case the radiated angular momentum is $(26.9 \pm 0.1)\%$ of the total initial angular momentum, resulting in a final Kerr BH with $\hat{a} \simeq 0.686$; see eq. (14.157). For spinning BHs, each with a dimensionless Kerr parameter $\hat{a} = 0.757$, in the aligned configuration, the total radiated angular momentum rises to $(34 \pm 1)\%$ of the initial total angular momentum, resulting in a final BH with $\hat{a} \simeq 0.89$, while in the anti-aligned configuration the radiated angular momentum is $(26 \pm 2)\%$ of the initial total angular momentum, and the final BH has $\hat{a} \simeq 0.44$.

More general spin configurations, in which the spins of the two bodies are neither aligned nor anti-aligned with the orbital angular momentum, give rise to further effects. In particular, because of the spin–orbit and spin–spin couplings, both the spins and the orbital angular momentum will precess around the total angular momentum. Using the 1.5PN spin–orbit interaction and the 2PN spin–spin interaction (in harmonic coordinates; see Note 33 on page 237) and labeling by a and b the two BHs, one finds that the equation of motion of the spin $\mathbf{S}_a$ of the BH labeled by a is

$$\frac{d\mathbf{S}_a}{dt} = \mathbf{\Omega}_a \times \mathbf{S}_a , \qquad (14.178)$$

with

$$\mathbf{\Omega}_a = \frac{G}{c^2 r^3} \left[\left(2 + \frac{3m_b}{2m_a} \right) \mathbf{L} - \mathbf{S}_b + 3(\hat{\mathbf{x}} \cdot \mathbf{S}_b)\hat{\mathbf{x}} \right] , \qquad (14.179)$$

where $\mathbf{L}$ is the Newtonian orbital angular momentum, $\mathbf{x}$ is the relative separation of the two BHs, $r = |\mathbf{x}|$, and $\hat{\mathbf{x}} = \mathbf{x}/r$. Similarly, the total angular momentum will precess. If we neglect the angular momentum radiated away, the total angular momentum $\mathbf{J} = \mathbf{L} + \mathbf{S}_1 + \mathbf{S}_2$ is conserved and $d\mathbf{L}/dt = -d(\mathbf{S}_1 + \mathbf{S}_2)/dt$, so in this approximation the orbital plane precesses with the same angular velocity as the total spin $(\mathbf{S}_1 + \mathbf{S}_2)$. The

[48] See Campanelli, Lousto and Zlochower (2006).

direction of the spin of the final BH is determined by angular momentum conservation, including also the angular momentum radiated in GWs, and of course, in a generic configuration in which the initial spins are not aligned with the orbital angular momentum, it will in general be different from the spin of any of the two initial BHs. This rather sudden jump in the direction of the spin of the final BH, with respect to the initial BH spins, is referred to as a "spin flip". It can have significant astrophysical consequences since the direction of the spin of a BH determines the orientation of its inner accretion disk and the direction in which a jet can be launched. The spin flip can therefore be a possible explanation for the sudden change in the orientation of the jet that is observed in some AGN.

Spin of the final BH

Given that numerical relativity simulations are quite time-consuming, it is useful to have analytic formulas, tuned to numerical relativity, that reproduce the spin $\mathbf{S}_f$ of the final BH given the masses m_1, m_2 and the spins $\mathbf{S}_{1,2}$ of the initial BHs. To this end we write

$$\mathbf{S}_f = \mathbf{S}_1 + \mathbf{S}_2 + \Delta\mathbf{L} \,, \qquad (14.180)$$

where $\mathbf{S}_{1,2}$ are the spins of the two BHs at an initial (large) separation, while

$$\Delta\mathbf{L} = \mathbf{L} - \mathbf{J}_{\rm rad} \,, \qquad (14.181)$$

where $\mathbf{L}$ is the orbital angular momentum when the binary is at this large separation, while $\mathbf{J}_{\rm rad}$ is the angular momentum radiated in GWs during the whole inspiral and merger process. It is useful to introduce the dimensionless vectors

$$\boldsymbol{\alpha}_i = \frac{c}{G} \frac{\mathbf{S}_i}{m_i^2} \,, \qquad (14.182)$$

whose modulus is equal to the dimensionless Kerr parameter $\hat{a}_i$ of the i-th BH; see eqs. (12.286) and (12.359). Similarly, for the final spin we introduce $\boldsymbol{\alpha}_f = c\mathbf{S}_f/(GM_f^2)$, where M_f is the mass of the final BH. We also write $M_f = m(1 - \varepsilon_{\rm rad})$, where $\varepsilon_{\rm rad} = E_{\rm rad}/mc^2$ is the fractional energy radiated away in GWs. Then, multiplying eq. (14.180) by c/GM_f^2, we get

$$\boldsymbol{\alpha}_f = \frac{c(\mathbf{S}_1 + \mathbf{S}_2)}{Gm^2(1 - \varepsilon_{\rm rad})^2} + \frac{c\Delta\mathbf{L}}{Gm^2(1 - \varepsilon_{\rm rad})^2} \,. \qquad (14.183)$$

Since $\varepsilon_{\rm rad} \lesssim 0.1$ even in spinning configurations, in the term proportional to $\mathbf{S}_1 + \mathbf{S}_2$ one can expand in powers of $\varepsilon_{\rm rad}$. This gives

$$\boldsymbol{\alpha}_f = \frac{1}{(1 + q)^2} \left(\boldsymbol{\alpha}_1 + q^2 \boldsymbol{\alpha}_2 \right) + \nu \boldsymbol{\ell} \,, \qquad (14.184)$$

where

$$\boldsymbol{\ell} = \frac{c\Delta\mathbf{L}}{Gm_1 m_2(1 - \varepsilon_{\rm rad})^2} + [2\varepsilon_{\rm rad} + 3\varepsilon_{\rm rad}^2 + O(\varepsilon_{\rm rad}^3)] \frac{c\mathbf{S}}{Gm_1 m_2} \,, \qquad (14.185)$$

and $\mathbf{S} = \mathbf{S}_1 + \mathbf{S}_2$. Thus, the final dimensionless spin $\boldsymbol{\alpha}_f$ is given by the term $(1+q)^{-2}(\boldsymbol{\alpha}_1 + q^2\boldsymbol{\alpha}_2)$, which is just the sum of the initial spins written in terms of the variables $\boldsymbol{\alpha}_{1,2}$, plus a correction term $\nu\boldsymbol{\ell}$ that has two contributions: one coming from $\Delta\mathbf{L}$, and the other proportional to the radiated energy (and its higher powers) and to the total spin $\mathbf{S}$.

A relatively simple approximation to the final spin (for systems in quasi-circular orbits) is obtained by neglecting the term proportional to $\mathbf{S}$ in eq. (14.185), as well as the components of $\Delta\mathbf{L}$ orthogonal to $\mathbf{L}$, so that $\boldsymbol{\ell}$ points in the $\mathbf{L}$ direction. Its modulus is then written as

$$|\boldsymbol{\ell}| = \frac{s_4}{(1+q^2)^2}\left(\hat{a}_1^2 + \hat{a}_2^2 q^4 + 2\hat{a}_1\hat{a}_2 q^2 \cos\alpha\right) + \qquad (14.186)$$

$$\left(\frac{s_5\nu + t_0 + 2}{1+q^2}\right)\left(\hat{a}_1\cos\beta + \hat{a}_2 q^2 \cos\gamma\right) + 2\sqrt{3} + t_2\nu + t_3\nu^2\,,$$

where $\hat{a}_i = |\boldsymbol{\alpha}_i|$, and the angles α, β, γ are defined by

$$\cos\alpha = \hat{\mathbf{S}}_1{\cdot}\hat{\mathbf{S}}_2\,, \qquad \cos\beta = \hat{\mathbf{S}}_1{\cdot}\hat{\mathbf{L}}\,, \qquad \cos\gamma = \hat{\mathbf{S}}_2{\cdot}\hat{\mathbf{L}}\,. \qquad (14.187)$$

The term $2\sqrt{3}$ is obtained from a comparison with the point-particle limit, while the coefficients s_4, s_5, t_0, t_2, t_3 are obtained by fitting a large number of numerical-relativity simulations, and are given by $s_4 \simeq -0.129$, $s_5 \simeq -0.384$, $t_0 \simeq -2.686$, $t_2 \simeq -3.454$ and $t_3 \simeq 2.353$. This expression for the final spin is known as the *AEI formula*. For initial spins $\hat{a}_i$ smaller than about $0.6 - 0.7$ it reproduces the numerical results to better than 1% accuracy. For higher initial spins the accuracy becomes a few percent, due to the fact that most simulations used in the fit were at moderate spin.[49]

[49]For a more elaborate and more accurate formula for the final spin, see eqs. (2)–(6) and (13) of Hofmann, Barausse and Rezzolla (2016).

Effect of the initial BH spins on the final BH recoil

Any asymmetry in the binary system results in the fact that, at each point in the orbit, there is a non-vanishing linear momentum radiated away in GWs. As we discussed in the previous section, for a circular orbit such a loss of linear momentum averages to zero over an orbital period; however, because the orbit is shrinking and finally the process suddenly stops owing to the merger, a net momentum is imparted to the final BH. In the spinless case there is an asymmetry between the two bodies only if their masses are different. However, for general spinning bodies with spins of different magnitude or orientations, the system is asymmetric even in the equal-mass case, and therefore the final BH receives a further kick, which depends on the initial spins. A remarkable result obtained by numerical simulations is that the recoil of the final BH due to the initial spins can be much higher than that due to mass asymmetry. To understand the structure of this result, and its dependence on the mass ratio and on the spins of the two BHs, it is useful to proceed as in the previous section, and compute first the loss of linear momentum in the PN approximation. The spin contribution to the linear momentum loss of the system due to GW emission appears at 2PN order and is given by[50]

[50]See Kidder (1995).

$$\frac{d\mathbf{P}}{dt} = -\frac{8G^3}{15c^7}\frac{\mu^2 m}{r^5}\Big\{ 4(\hat{\mathbf{n}}\cdot\mathbf{v})(\mathbf{v}\times\boldsymbol{\Delta}) - 2v^2(\hat{\mathbf{n}}\times\boldsymbol{\Delta}) \tag{14.188}$$

$$-(\hat{\mathbf{n}}\times\mathbf{v})\left[3(\hat{\mathbf{n}}\cdot\mathbf{v})(\hat{\mathbf{n}}\cdot\boldsymbol{\Delta}) + 2(\mathbf{v}\cdot\boldsymbol{\Delta})\right]\Big\}.$$

Here $\hat{\mathbf{n}} = \mathbf{x}/|\mathbf{x}|$, where $\mathbf{x} = \mathbf{x}_2 - \mathbf{x}_1$ is the relative distance between the two bodies in harmonic coordinates, $\mathbf{v} = d\mathbf{x}/dt$, μ is the reduced mass, $m = m_1 + m_2$ is the total mass, and

$$\boldsymbol{\Delta} = m\left(\frac{\mathbf{S}_1}{m_1} - \frac{\mathbf{S}_2}{m_2}\right). \tag{14.189}$$

On a circular orbit of fixed radius $r(t) = R$ we have $\hat{\mathbf{n}}\cdot\mathbf{v} = 0$ and eq. (14.188) simplifies to

$$\frac{d\mathbf{P}}{dt} = \frac{16G^3}{15c^7}\frac{\mu^2 m v^2}{R^5}\left[(\hat{\mathbf{n}}\times\boldsymbol{\Delta}) + (\hat{\mathbf{n}}\times\hat{\mathbf{v}})(\hat{\mathbf{v}}\cdot\boldsymbol{\Delta})\right]. \tag{14.190}$$

In terms of the vectors $\boldsymbol{\alpha}_i$ defined in eq. (14.182) we have[51]

$$\boldsymbol{\Delta} = \frac{Gmm_1}{c}(\boldsymbol{\alpha}_1 - q\boldsymbol{\alpha}_2). \tag{14.191}$$

We can now proceed as in eqs. (14.166)–(14.171). We examine first the case in which the spins are parallel or antiparallel to the orbital angular momentum, i.e. $\boldsymbol{\alpha}_1 = \alpha_1\mathbf{z}$ and $\boldsymbol{\alpha}_2 = \alpha_2\hat{\mathbf{z}}$, where we take the orbit in the (x,y) plane, so $\hat{\mathbf{z}}$ is the direction of the orbital angular momentum, and $-1 \leqslant \alpha_i \leqslant 1$. Note that $|\alpha_i| = \hat{a}_i$ is the dimensionless Kerr parameter of the i-th BH. In this case the second term in the square brackets in eq. (14.190) vanishes, since $\mathbf{v}$ is in the plane of the orbit while $\boldsymbol{\Delta}$ is perpendicular to this plane, so $\hat{\mathbf{v}}\cdot\boldsymbol{\Delta} = 0$. For a circular orbit of radius $r = R$ we have $\hat{\mathbf{n}} = (x\hat{\mathbf{x}} + y\hat{\mathbf{y}})/R \equiv \hat{\mathbf{x}}\cos\varphi + \hat{\mathbf{y}}\sin\varphi$, so we get

$$\frac{d\mathbf{P}}{dt} = \pm\frac{16G^3}{15c^7}\frac{\mu^2 m v^2 \Delta}{R^5}(\sin\varphi, -\cos\varphi, 0), \tag{14.192}$$

where the $\pm$ sign depends on whether $\boldsymbol{\Delta}$ points in the $\hat{\mathbf{z}}$ or $-\hat{\mathbf{z}}$ direction, i.e. on the sign of $\alpha_1 - q\alpha_2$. Equation (14.192) shows that the radiated linear momentum is in the plane of the orbit, in the direction tangent to the orbit itself. Repeating the same derivation as in eqs. (14.169)–(14.171) we find that, for this spin configuration, the center of mass performs a circular motion in the (x,y) plane, with a velocity whose modulus is given by

$$\frac{v_{\rm CM}}{c} = \frac{1}{15}\left(\frac{Gm}{c^2 R}\right)^{9/2} f_{\rm SO}(q,\alpha_1,\alpha_2), \tag{14.193}$$

where

$$f_{\rm SO}(q,\alpha_1,\alpha_2) = \frac{16q^2}{(1+q)^5}|\alpha_1 - q\alpha_2|, \tag{14.194}$$

and the label "SO" in $f_{\rm SO}(q,\alpha_1,\alpha_2)$ reminds us that this term is due to the spin–orbit coupling. Observe that $f_{\rm SO}(q,\alpha_1,\alpha_2)$ reaches its maximum value when $q = 1$ and the spins are antiparallel to each other and

[51]To compare with Campanelli, Lousto, Zlochower and Merritt (2007b), where the results discussed below have been first presented, we have exchanged the labels 1 and 2 so that for us, for consistency with the notation used elsewhere in this book, the label 2 refers to the lighter particle.

maximal, i.e. for $\alpha_1 = -\alpha_2 = 1$ or $\alpha_1 = -\alpha_2 = 1$, and this maximum value is $f_{\rm SO,max} = 1$.

So, similarly to eq. (14.172), this PN computation suggests that, for spin parallel or antiparallel to the angular momentum, the final BH receives a recoil velocity perpendicular to the orbital angular momentum, and of the form

$$v_\perp = C\, f_{\rm SO}(q, \alpha_1, \alpha_2)\,, \qquad (14.195)$$

with C a constant. This argument also suggests the estimate

$$C \simeq \frac{c}{15}\left(\frac{Gm}{c^2 R_{\rm min}}\right)^{9/2}\,, \qquad (14.196)$$

where $R_{\rm min}$ is the minimum value for R for which this PN derivation can still be trusted. By taking $R_{\rm min} = 3Gm/c^2$, corresponding to the position of the light ring, we get $C \simeq 141$ km/s. Of course, a precise value can only be obtained from numerical relativity simulations. The simulations confirm the dependence on q, α_1 and α_2 given by eqs. (14.194) and (14.195) and, depending on the numerical simulations, give a value of C in the range $440 - 475$ km/s (formally corresponding to eq. (14.196) with $R_{\rm min} \simeq 2.3Gm/c^2$, where clearly the PN expansion is no longer valid). So, in particular, for an equal-mass binary ($q = 1$) with $\alpha_2 \equiv \hat{a}$ and $\alpha_1 = -\alpha_2$, one has

$$v_\perp \simeq (440 - 475)\,\hat{a}\,{\rm km/s}\,. \qquad (14.197)$$

For a near-extremal BH, $\hat{a} \simeq 1$, this kick is much larger than the maximum kick due to mass asymmetry, which, as we saw in eq. (14.174), is rather of order 175 km/s.

In eq. (14.192) we saw that, for these spin configurations, where the initial BH spins are parallel or antiparallel to the orbital angular momentum, the kick is in the plane of the orbit, just like the kick from mass asymmetry, which is present even in the spinless case. Comparison of eqs. (14.166) and (14.192) even suggests that these two contributions are collinear or anti-collinear. However, these PN expressions are only valid as long as the orbit is quasi-circular, and no longer hold during the plunge and merger phase. Thus, in general we can expect that these two contributions to the kick will have different directions within the orbital plane, as is indeed confirmed by numerical relativity simulations. We choose a system of orthogonal axes $(\mathbf{e}_1, \mathbf{e}_2)$ in the plane of the orbit, so that $\mathbf{e}_1$ points in the direction of the kick due to mass asymmetry. The contribution to the kick velocity from the spin will in general be in the direction $(\mathbf{e}_1 \cos\xi + \mathbf{e}_2 \sin\xi)$ for some angle ξ, so the total recoil velocity, when the initial BH spins are parallel or antiparallel to the orbital angular momentum, can be modeled as

$$\mathbf{v}_{\rm rec} = v_m \mathbf{e}_1 + v_\perp (\mathbf{e}_1 \cos\xi + \mathbf{e}_2 \sin\xi)\,, \qquad (14.198)$$

where v_m is the recoil velocity given in eq. (14.177) with $A \simeq 1.20 \times 10^4$ km/s and $B \simeq -0.93$ and, from eqs. (14.194) and (14.195),

$$v_\perp \simeq C\,\frac{16q^2}{(1+q)^5}\,|\alpha_1 - q\alpha_2|\,, \qquad (14.199)$$

with C in the range $440 - 475$ km/s.

In the most general configuration, the BH spins have components parallel to the orbital angular momentum, as well as components in the orbital plane. From eq. (14.190) we see that in this case both terms in the square brackets can contribute, and there is a component of the kick in the $\hat{\mathbf{z}}$ direction, i.e. orthogonal to the plane of the orbit. The configuration that maximizes the kick in the $\hat{\mathbf{z}}$ direction is obtained by taking $\boldsymbol{\Delta}$ to lie completely in the orbital plane and maximizing its modulus Δ, which, from eq. (14.189), means taking the spins to be anti-aligned. The spin configuration that maximizes the kick outside the orbital plane is therefore the one shown in Fig. 14.16, at two different phases of the orbit. Orienting the axes as in the figure, $\boldsymbol{\Delta}$ points in the positive x direction. At the phase of the orbit shown in Fig. 14.16(a), the relative position vector $\mathbf{x} = \mathbf{x}_2 - \mathbf{x}_1$, and therefore its unit vector $\hat{\mathbf{n}}$, point in the negative y direction, while the relative velocity $\mathbf{v}$ points in the negative x direction. Therefore the two terms in eq. (14.190) produce contributions to $d\mathbf{P}/dt$ that add up and point both in the positive z direction. At the phase of the orbit shown in Fig. 14.16(b), $\hat{\mathbf{n}}$ is antiparallel to $\boldsymbol{\Delta}$ while $\mathbf{v}$ is orthogonal to it, and both terms in eq. (14.190) vanish. Similarly, when the BHs have performed half an orbit with respect to the position (a), both terms point in the negative z direction, and vanish again when the BHs have performed half an orbit with respect to the position (b). So, the center of mass performs an oscillatory motion along the z direction. More precisely, writing $\hat{\mathbf{n}} = (\cos\varphi, \sin\varphi, 0)$, $\hat{\mathbf{v}} = (\sin\varphi, -\cos\varphi, 0)$ and $\boldsymbol{\Delta} = \Delta(1,0,0)$, the square bracket in eq. (14.190) becomes $-2\Delta\sin\varphi\,\hat{\mathbf{z}}$, so in this spin configuration the P_z component of the center-of-mass evolves accordingly to

$$\frac{dP_z}{dt} = -\frac{32G^3}{15c^7}\frac{\mu^2 m v^2 \Delta}{R^5}\sin\varphi. \tag{14.200}$$

Repeating the same analysis already performed for eq. (14.192), we see that the component of the kick velocity parallel to the orbital angular momentum, $v_{\parallel}$, has the form

$$v_{\parallel} = D' f_{\rm SO}(q, \alpha_1, \alpha_2), \tag{14.201}$$

where $f_{\rm SO}(q, \alpha_1, \alpha_2)$ is the same function as defined in eq. (14.194) and D' is a constant. In this PN approximation we would get $D' = 2C$, where C is defined in eq. (14.195). However, for a quasi-circular orbit describing the inspiral, there is a further contribution from the terms $(\hat{\mathbf{n}}\cdot\mathbf{v})$ in eq. (14.188), which adds up coherently. Furthermore, during the plunge and merger the quasi-circular orbit approximation is no longer adequate. Thus, once again, only numerical relativity can provide an accurate answer for the numerical value of the constant. Simulations show that D' has the form $D' = D\cos(\Theta - \Theta_0)$, where Θ is the angle between the projection of $\boldsymbol{\Delta}$ on the orbital plane and the infall direction at merger, Θ_0 is a constant, and $D \simeq (3750 \pm 60)$ km/s. This value of D is remarkably high, and it means that this spin configuration can

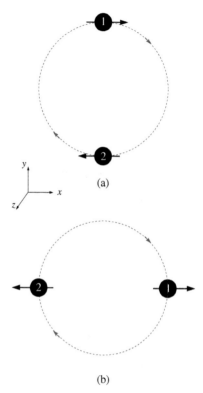

Fig. 14.16 The spin configuration that maximizes the kick in direction perpendicular to the plane of the orbit, seen at two different phases along the orbit. The black arrow denotes the direction of the spin of the BHs.

result in enormous kicks, of thousands of km/s, more than an order of magnitude larger than those due to the mass asymmetry.

We can now put together the results for the kicks from mass asymmetry and from spins in a generic configurations. If $\boldsymbol{\Delta}$ has an arbitrary direction, its component $\Delta_\parallel$ parallel to the orbital angular momentum will give rise to the kick $v_{\rm perp}$ in the plane of the orbit (i.e. perpendicular to the orbital angular momentum) computed in eq. (14.195), while its component $\Delta_\perp$ perpendicular to the orbital angular momentum will give rise to the kick $v_{\rm parallel}$ given in eq. (14.201).

To summarize, for a generic spin configuration, eqs. (14.198) and (14.199) generalize to

$$\mathbf{v}_{\rm rec} = v_m \mathbf{e}_1 + v_\perp (\mathbf{e}_1 \cos \xi + \mathbf{e}_2 \sin \xi) + v_\parallel \mathbf{e}_z \,, \qquad (14.202)$$

where

$$v_m \simeq A \frac{q^2(1-q)}{(1+q)^5} \left[1 + B \frac{q}{(1+q)^2} \right] \,, \qquad (14.203)$$

$$v_\perp \simeq C \frac{16q^2}{(1+q)^5} \left| \alpha_{\parallel,1} - q\alpha_{\parallel,2} \right| \,, \qquad (14.204)$$

$$v_\parallel \simeq D \cos(\Theta - \Theta_0) \frac{16q^2}{(1+q)^5} \left| \alpha_{\perp,1} - q\alpha_{\perp,2} \right| \,, \qquad (14.205)$$

and, according to the present numerical simulations, $A \simeq (1.20 \pm 0.08) \times 10^4$ km/s, $B \simeq -0.93$, $C \simeq (457 \pm 17)$ km/s and $D \simeq (3750 \pm 60)$ km/s; Θ is the angle between the projection of $\boldsymbol{\Delta}$ on the orbital plane and the infall direction at merger, and Θ_0 is a constant phase. Observe also that, while C and D are defined so that they multiply functions whose maximum is equal to 1, A multiplies a function whose maximum value is much smaller than 1, resulting in a maximum value for v_m given by eq. (14.174), $(v_m)_{\rm max} \simeq 175 \pm 11$ km/s.

14.3.6   Astrophysical consequences of BH recoil

Observations of the centers of nearby elliptical, lenticular or spiral-bulge galaxies have shown that most or possibly all of them contain a central supermassive BH. Such supermassive BHs are believed to have grown by successive coalescence and merger of central BHs in smaller galaxies, in a hierarchical mechanism of galaxy formation. We will discuss them in some detail in Chapter 16.

The presence of a supermassive BH has a significant effect on the dynamics of the central part of the galaxy in which it resides. In particular, a remarkable result that has been obtained from observations of the central regions of nearby galaxies is that there is a tight correlation between the mass M of the central BH and the velocity dispersion σ of the stars in the inner region of the galaxy. Using a reference value $\sigma_0 = 200 \mathrm{km\,s}^{-1}$, this "$M - \sigma$ relation" reads

$$\log_{10} \frac{M_{\rm BH}}{M_\odot} = \alpha + \beta \log_{10} \frac{\sigma}{\sigma_0} \,. \qquad (14.206)$$

Fits to the data give $\alpha = 8.32 \pm 0.05$ and $\beta = 5.64 \pm 0.32$ for all galaxies and $\alpha = 8.39 \pm 0.06$ and $\beta = 5.20 \pm 0.36$ for elliptical galaxies.[52]

The recoil of the final BH during the coalescence can have several significant astrophysical consequences, which we discuss here.

Ejection of the central BH

Figure 14.17 shows escape velocities in four types of stellar systems that could contain merging BHs: giant elliptical galaxies (E), dwarf ellipticals (dE), dwarf spheroidals (dSph) and globular clusters (GC). Elliptical galaxies are further separated into galaxies with a core in the density profile and power-law galaxies, which instead have a steeply rising featureless profile and lack cores. The solid line in Figure 14.17 shows the escape velocities from the dark matter (DM) halos associated with the luminous stellar systems.

We see from the figure that escape velocities from globular clusters and dwarf spheroidals (even including the DM contribution) are smaller than about 80 km/s. Escape velocities from the luminous matter in dwarf elliptical galaxies are typically in the range 20–100 km/s, and including their DM halos increases the average escape velocity to about 200 km/s. So, for most mass ratios, the recoil of the final BH in the coalescence of unequal-mass binaries can have a significant effect in globular clusters and in dwarf spheroidal and (to a lesser extent) dwarf elliptical galaxies, and might explain the lack of observation of massive BHs in these systems.

In contrast, the escape velocities from giant elliptical galaxies are much higher. In the sample shown,[53] the escape velocity is always higher than 450 km/s, and typically of order 1000 km/s. In this case, the recoil due to mass asymmetry cannot eject the final BH, while the recoil from spin asymmetry, for the spin configurations that maximize the recoil, could eject them.

Smoothing of galactic cores

Even when the recoil is not sufficient to eject the final BH, it can still displace it significantly from the center of the galaxy. In this case, the BH will fall back toward the center of the galaxy, dissipating energy via dynamical friction with the stars and gas, with most of the energy being dissipated during passages through the galactic center. This transfers energy to the nucleus of the galaxy and heats it up, thereby lowering its density. Assuming a central density profile $\rho \sim 1/r$, this has the effect of smoothing the cusp at small r into a core of roughly constant density, within a radius $\sim 2r_h$, where r_h is the radius of the BH's sphere of influence (defined as the radius of a sphere containing a mass in stars equal to twice that of the BH). In many nearby galaxies, a central region of order $2r_h$ can be resolved, so the recoil can have observable consequences on the structure of galactic nuclei.

Observations show that the density profiles of early-type galaxies fall into two classes: power-law galaxies, whose profile rises steeply toward

[52]These are the values given in McConnell and Ma (2013), which use new kinematical data and revised dynamical measurements of 72 BH masses to improve on earlier compilations (see in particular their Table II for fits to different subsets of galaxies). The precise values change from study to study. For instance, Gultekin *et al.* (2009) give $\alpha = 8.12 \pm 0.08$ and $\beta = 4.24 \pm 0.41$ for all galaxies and $\alpha = 8.23 \pm 0.08$ and $\beta = 3.92 \pm 0.42$ for elliptical galaxies. See also Table 16.1 on page 331 for more general relations between the central BH mass and properties of the host galaxy.

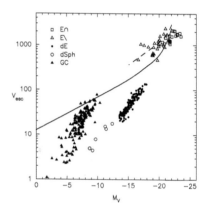

Fig. 14.17 Central escape velocities in km/s, in four types of stellar system that could harbor merging BHs, as a function of the magnitude M_V. Elliptic galaxies are shown with a separate symbol for core galaxies ($\square$) and power-law ($\triangle$) galaxies. The solid line is the mean escape velocity from the DM halos associated with the luminous matter. The dashed line is the escape velocity from the combined luminous + mean DM potentials for E galaxies. From Merritt, Milosavljević, Favata, Hughes and Holz (2004).

[53]This sample is taken from Hubble Space Telescope observations of the surface-brightness profiles of 61 elliptical galaxies and spiral bulges; see Faber *et al.* (1997).

the center as a featureless power-law, and core galaxies, which are usually brighter, and have a discernible break at small radii in their surface brightness profile. The recoil due to GW emission can therefore be a possible mechanism explaining the presence of such a core. Other mechanisms, however, can also result in the formation of a core. For instance, the formation of a BH binary can affect the stellar density in the inner region through the so-called "gravitational slingshot" mechanism, during which a star that gets close to the binary is ejected from the center, gaining energy at the expense of the orbital energy of the binary, as we will discuss in more detail in Section 16.2.1.

Off-nuclear AGN

After receiving the kick, the typical time that the displaced BH takes to fall back and settle into a stationary state depends of course on the precise value of the kick velocity and also on the velocity dispersion σ of the inner region of the galaxy, since the latter determines the dynamical friction with the stars.

The dependence on σ can be translated into a dependence on the BH mass using the $M - \sigma$ relation (14.206). Figure 14.18 shows the motion of a BH with mass $3 \times 10^6 M_\odot$ (assuming values for the coefficients in the $M - \sigma$ relation that give a velocity dispersion $\sigma \simeq 75$ km/s), assumed to lie on a radial orbit, for different values of the recoil velocity. We see that typical return times can be of order $10^6 - 10^9$ yr. Actually, the curves depend primarily on the dimensionless number $v_{\rm CM}/\sigma$ and so can be easily scaled to other values of BH masses and kick velocities. After making a number of damped oscillations, of the type shown in Fig. 14.18, the amplitude of the BH motion will eventually decay to the size of the galactic core. In this regime, numerical simulations show that the energy of the BH orbit continues to decay, but on a much longer time-scale, and the BH and the core oscillates around their common center of mass. Finally, when the BH energy reaches a value $(1/2)M_{\rm BH}V^2$ of the order of the typical kinetic energy of the other stars, the BH enters into a regime of gravitational Brownian motion where gravitational perturbations from stars can accelerate it as often as they can decelerate it.

This motion of the BH can then result in an active quasar displaced with respect to the galactic nucleus. In fact, a BH that acquires a recoil velocity $v_{\rm rec}$ retains gas that is orbiting around it up to a distance

$$r \sim \frac{GM_{\rm BH}}{v_{\rm rec}^2}$$

$$\sim 0.5 \,{\rm pc} \left(\frac{M_{\rm BH}}{10^8 M_\odot} \right) \left(\frac{10^3 \,{\rm km/s}}{v_{\rm rec}} \right)^2 . \tag{14.207}$$

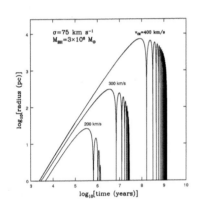

Fig. 14.18 The response of a $3 \times 10^6 M_\odot$ BH to a recoil velocity $v_{\rm CM}$ in an isothermal potential with dispersion $\sigma = 75$ km/s; radial orbits are assumed. The three curves refer to $v_{\rm CM} = 200, 300$ and 400 km/s. From Madau and Quataert (2004).

This means that an accretion disk, if present, would be retained by the BH after the kick. Furthermore, for $v_{\rm rec} \lesssim 10^3$ km/s, r is large enough to encompass the region where the broad emission lines of AGN are generated (in contrast, the narrow emission lines are generated in the gas moving in the gravitational potential of the host galaxy, and would not follow a recoiling BH). Galaxy merger is a leading mechanism for fueling quasar activity, and the typical lifetime of merger-driven activity is of the order of the Salpeter time-scale

$$t_S \simeq 4.5 \times 10^7 \, \mathrm{yr} \,. \tag{14.208}$$

Supermassive BHs with masses $M_{\mathrm{BH}} \simeq 10^8 M_\odot$ are hosted by galaxies with velocity dispersion $\sigma \simeq 200$ km/s or higher. Then we see from Fig. 14.18 (taking into account that the results basically depend only on v_{CM}/σ) that, for typical kick velocities of order of a few hundreds km/s, these BHs will reach a maximum displacement from the center of the galaxy of the order of only a few pc. Furthermore, their decay time due to dynamical friction will be smaller than 10^6 yr, and therefore they will spend most of their active quasar lifetime at distances much closer to the galactic center. Thus, the recoil effect due to the emission of gravitational waves is not sufficient to produce off-nuclear quasar activity in such massive BHs. The situation is, however, different for $M_{\mathrm{BH}} \sim (10^6 - 10^7) M_\odot$, for which a kick velocity of order 200–300 km/s can results in a decay time-scale comparable to the Salpeter time-scale, and in displacements from the nucleus of order 100 pc. This suggests that a small fraction of low-mass AGN could be displaced from the galactic nucleus, and their signature would be that their broad emission lines would have a Doppler shift, determined by the kick velocity, significantly different from that of the narrow lines, which are instead generated in a region that has not been dragged along by the recoiling BH.

Consequences for the formation and growth of massive BHs

Another potentially important effect of the final BH recoil concerns its impact on the history and growth of massive BHs. It is believed that the first massive black holes formed at high redshifts at the centers of DM halos of relatively low mass. These DM halos experienced successive episodes of mergers with other halos. As we will discuss in more detail in Section 16.2.1, when two DM halos coalesce, the BHs at the centers of the two halos fall into each other's potential well, losing energy because of dynamical friction and eventually forming a binary system. More precisely, when halos of masses M_1 and M_2 merge, with $M_1 < M_2$, it can be shown that the smaller halo (the "satellite") will merge with the central galaxy of the larger halo on a time-scale shorter than the Hubble time only if $M_1/M_2 \gtrsim 0.3 - 0.5$. If instead the satellite is much lighter than the heavier object, tidal stripping may leave the satellite's massive BH wandering in the halo, too far from the center of the remnant for the formation of a BH binary. If the interaction with the stellar or gaseous environment drives the binary to a distance closer than about 0.01 pc, then the emission of gravitational radiation takes over and becomes the dominant source of loss of energy and angular momentum, until the binary system coalesces, forming a heavier BH. This evolutionary history eventually leads to the population of bright quasars that we observe at $z \lesssim 6$ and to the supermassive BHs that nowadays are commonly observed

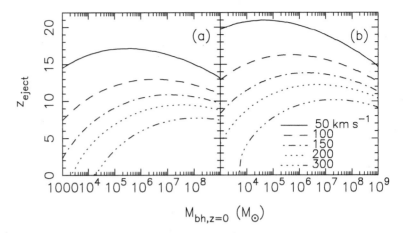

Fig. 14.19 The maximum redshift z_{eject} at which (a) DM halos only, and (b) DM halos and the central galaxies combined, can confine BHs as a function of the $z = 0$ BH mass, for five values of the kick velocity. From Merritt, Milosavljević, Favata, Hughes and Holz (2004).

at the centers of nearby galaxies.

The inclusion of BH recoil can have significant effects on this scenario. At high redshift the DM halos that are the progenitors of today's halos were less massive, and therefore their confining effect against recoils of the final BH was smaller. Using the known empirical relation between the present-day BH mass and the mass of the corresponding halo today, and evolving backward in time the history of the halo formation using numerical simulations of structure formation, it is possible to compute, for a given value of the BH mass today $M_{\text{BH}}(z = 0)$, the maximum redshift for which the progenitor of the present-day halo was still able to confine the progenitor of the present-day BH. The result is shown in Fig. 14.19, and shows that, in order to avoid the situation in which a single merger with recoil velocity $v_{\text{kick}} \sim 150$ km/s ejects the progenitor BH from the halo, forbidding further growth, the assembly of a BH that todays has a mass $M_{\text{BH}} \simeq 10^8 M_\odot$ must have started at $z < 11$. Therefore, models that grow supermassive BHs from seeds at $z \gtrsim 10$ are disfavored.

Wandering BHs

Another consequence of the gravitational recoil is that the mergers that result in the ejection of the final BH will produce a population of "wandering" BHs. Both from semi-analytic models and from inclusion of the recoil effect in numerical N-body simulations of structure formations, one finds that there is a population of BHs that have been unbound from their host galaxy and are wandering in the DM halo, and even a population of BHs whose recoil velocity was large enough to unbind them not only from their host galaxy but even from their halo. The typical masses of these wandering BHs are estimated to lie in the range $M_{\text{BH}} \sim 10^2 - 10^5 M_\odot$.

14.4   GWs from NS–NS binaries

In this section we discuss theoretical issues specific to the emission of GWs in the coalescence of NS–NS binaries. In Section 15.3 we will illustrate the observational results obtained from GW170817, the first detected NS–NS coalescence, which was observed by the advanced LIGO and advanced Virgo interferometers on August 17, 2017.

14.4.1   Inspiral phase and tidal effects

Just like BH–BH coalescences, the inspiral phase of NS–NS (or NS–BH) binaries can be studied with the PN expansion, eventually improved with the EOB technique. The new aspect, compared with BH–BH binaries, is that a NS is deformed by the tidal gravitational field of the companion (whether another NS or a BH). This tidal deformability depends on the equation of state (EoS) in the NS interior. This is a very interesting effect, which affects the phase of the GW signal in a way potentially detectable at advanced or at third-generation interferometers, opening the possibility of obtaining information on the NS EoS.

Newtonian theory of tidal deformations

The deformability of the NS can be characterized in terms of its Love numbers, as follows. Consider first Newtonian gravity. In this case, an external gravitational potential $U_{\rm ext}$ (such as that due to the companion in a binary system) generates a quadrupolar tidal field

$$\mathcal{E}_{ij} = -\partial_i \partial_j U_{\rm ext} \,. \tag{14.209}$$

Observe that, outside the source that generates it, $U_{\rm ext}$ satisfies $\boldsymbol{\nabla}^2 U_{\rm ext} = 0$, and therefore $\mathcal{E}_{ij}$ is both symmetric and traceless. In the absence of this external field, a non-rotating NS would be spherical (apart from effects related to crustal deformations, discussed in Section 11.2.4, that we neglect in the following discussion). The tidal field, however, generates a perturbation $\delta\rho$ in the equilibrium position of the self-gravitating fluid that forms the NS. From eq. (5.12) of Vol. 1 we see that the gravitational potential generated by the deformed NS, in the Newtonian approximation, is

$$U(t,\mathbf{x}) = G \int d^3x' \frac{\bar{\rho}(r') + \delta\rho(t,\mathbf{x}')}{|\mathbf{x} - \mathbf{x}'|} \,, \tag{14.210}$$

where we have written $\rho = \bar{\rho} + \delta\rho$, where $\bar{\rho}$ is the equilibrium configuration, and $\delta\rho$ the perturbation induced by the external gravitational field. For a non-rotating star we take the equilibrium configuration to be spherically symmetric, so $\bar{\rho}$ depends only on $r' \equiv |\mathbf{x}'|$. Outside the NS we can perform the usual multipole expansion, taking as origin of the reference frame the center of mass of the NS, and using

$$\frac{1}{|\mathbf{x} - \mathbf{x}'|} = \frac{1}{r} + \frac{x_i}{r^3} x_i' + \frac{3\hat{x}_i \hat{x}_j - \delta_{ij}}{2r^3} x_i' x_j' + \dots \,, \tag{14.211}$$

where $r \equiv |\mathbf{x}|$, $\hat{x}_i = x_i/r$, and the expansion is valid for $r' < r$ [and therefore in the NS exterior, since in eq. (14.210) $\mathbf{x}'$ is an integration variable ranging only over the NS interior]. Inserting this expansion into eq. (14.210), the first term gives the unperturbed monopole contribution Gm/r, where m is the NS mass; the second term is a dipole contribution, proportional to $\int d^3x' \, \rho(t, \mathbf{x}')x_i'$, which vanishes by definition of center of mass. The first term generated by the perturbation is therefore proportional to the second moment of the mass density,

$$M_{ij}(t) = \int d^3x' \rho(t, \mathbf{x}')x_i'x_j' \,. \tag{14.212}$$

Since, in U, M_{ij} is multiplied by the traceless tensor $3\hat{x}_i\hat{x}_j - \delta_{ij}$ that appears in eq. (14.211), we can replace M_{ij} by the traceless quadrupole moment (3.58),

$$Q_{ij}(t) = \int d^3x' \, \rho(t, \mathbf{x}') \left(x_i'x_j' - \frac{1}{3}r'^2\delta_{ij} \right) \,. \tag{14.213}$$

Note that actually only $\delta\rho(t, \mathbf{x}')$ contributes to Q_{ij}, because $\bar{\rho}$ has been assumed to be spherically symmetric. Then, outside the NS, adding the potential generated by the deformed NS to the external potential U_{ext}, written as an expansion around the center of mass of the NS as $U_{\text{ext}} = -(1/2)\mathcal{E}_{ij}x_ix_j + O(x^3)$, and using $Q_{ij}\delta_{ij} = 0$, we get

$$U(t, \mathbf{x}) = \frac{Gm}{r} + \frac{3G}{2r^3}\hat{x}_i\hat{x}_jQ_{ij}(t) + O\left(\frac{1}{r^4}\right) - \frac{1}{2}\mathcal{E}_{ij}(t)x_ix_j + O(x^3) \,. \tag{14.214}$$

If the external perturbation $\mathcal{E}_{ij}$ can be considered as static, then also the induced quadrupole Q_{ij} will be time-independent, and U will be a function only of $\mathbf{x}$. In the following we will restrict to static or quasi-static perturbations, so that t plays the role of an adiabatic parameter. In this case, to linear order the induced quadrupole moment $Q_{ij}(t)$ is proportional to the $\mathcal{E}_{ij}(t)$,

$$Q_{ij}(t) = -\lambda\mathcal{E}_{ij}(t) \,, \tag{14.215}$$

for some constant λ. The $l = 2$ tidal Love number k_2 is then defined as

$$k_2 = \frac{3}{2}\frac{G\lambda}{R^5} \,, \tag{14.216}$$

where R is the NS radius. Note that k_2 is dimensionless. Then

$$Q_{ij} = -\frac{2}{3G}k_2R^5\mathcal{E}_{ij} \,. \tag{14.217}$$

The factor $3/2$ in the definition eq. (14.216) is a conventional normalization of k_2, chosen so to cancel the factor $3/2$ in eq. (14.214), which then reads

$$U(t, \mathbf{x}) = \frac{Gm}{r} - \frac{1}{2}\mathcal{E}_{ij}(t)x_ix_j \left[1 + 2k_2 \left(\frac{R}{r} \right)^5 \right] + O\left(\frac{1}{r^4}\right) + O(x^3) \,. \tag{14.218}$$

The corresponding component g_{00} of the metric, in the Newtonian approximation, is given by $g_{00} = -1 + 2U/c^2$ [see eq. (5.11)], and therefore

$$g_{00} = -1 + \frac{2Gm}{c^2 r} - \frac{1}{c^2} \mathcal{E}_{ij}(t) x_i x_j \left[1 + 2k_2 \left(\frac{R}{r} \right)^5 \right] + O\left(\frac{1}{r^4} \right) + O(x^3).$$
(14.219)

The procedure can be extended to generic l, by considering higher-multipole contributions of the external field. These are described by the tensor

$$\mathcal{E}_{i_1 i_2 \dots i_l} \equiv -\frac{1}{(l-2)!} \partial_{i_1} \dots \partial_{i_l} U_{\text{ext}},$$
(14.220)

which is symmetric and trace-free in all its indices, since the contraction of any two indices produces a Laplacian acting on U_{ext}. We can then define a Love number k_l for each Newtonian multipole, writing[54]

$$g_{00} = -1 + \frac{2Gm}{c^2 r}$$
(14.221)
$$- \frac{1}{c^2} \sum_{l=2}^{\infty} \frac{2}{l(l-1)} \mathcal{E}_{i_1 i_2 \dots i_l} x^i \dots x^{i^l} \left[1 + 2k_l \left(\frac{R}{r} \right)^{2l+1} \right].$$

Observe that the term $x^i \dots x^{i^l}/r^{2l+1} = \hat{x}^i \dots \hat{x}^{i^l}/r^{l+1}$ is of order $1/r^{l+1}$. We can write these results more compactly using the multi-index notation introduced in Section 3.5.1 of Vol. 1,

$$\mathcal{E}_L = -\frac{1}{(l-2)!} \partial_L U_{\text{ext}}$$
(14.222)

and

$$g_{00} = -1 + \frac{2Gm}{c^2 r} - \frac{1}{c^2} \sum_{l=2}^{\infty} \frac{2}{l(l-1)} \mathcal{E}_L x^L \left[1 + 2k_l \left(\frac{R}{r} \right)^{2l+1} \right],$$
(14.223)

General-relativistic theory of tidal deformations

Of course, the Newtonian approximation discussed so far has only limited accuracy for NS. A full relativistic theory of tidal deformations can be developed as follows.

The appropriate framework for the definition and computation of tidal effects in the general-relativistic case is perturbation theory over the equilibrium configuration of the star, using the tools that we developed in Chapter 12 for BH perturbation theory. The problem naturally separates into the solution of the relevant perturbed equations outside the NS (the exterior problem) and inside the NS (the interior problem). In the exterior region we search for the general solution of the perturbed Einstein equations in vacuum. This is exactly the same problem that we have already discussed in Chapter 12 for BHs. In particular, we have seen that the metric perturbations separate into polar and axial modes, for each multipole. The coefficients of these modes (which, as long as we only consider the exterior problem, are just free parameters) define two

[54]There are different possible conventions on the normalization of the Love numbers. Here we are following the notation in Binnington and Poisson (2009).

[55]Compare also with the discussion in Section 5.3.1, where it was found that that the most general exterior vacuum solution is written as an expansion parametrized by two sets of multipoles: "mass-type", or, equivalently, "electric-type", multipoles M_L and "current-type", or, equivalently, "magnetic-type" multipoles S_L.

families of Love numbers, corresponding to polar and axial modes, and called "electric-type" and "magnetic-type" Love numbers, $k_l^{(e)}$ and $k_l^{(m)}$, respectively.[55] These multipole moments parametrize the most general vacuum solution for the perturbations of the components g_{00}, g_{0i} and g_{ij} of the metric. Recall, from the discussion following eq. (5.10) of Vol. 1, that the Newtonian limit of GR is obtained by setting $g_{ij} = \delta_{ij}$, $g_{0i} = 0$ and $g_{00} = -1 + 2U/c^2$, where U is the (sign-reversed) Newtonian potential. In this limit the magnetic-type multipoles disappear, and we remain only with a single family of Newtonian multipoles, parametrized by the Newtonian Love numbers introduced in eq. (14.223). In full GR, however, we have two sets of "electric-type" and "magnetic-type" Love numbers.

The information that the source of the perturbation is an external tidal field is inserted by requiring the appropriate behavior at large distances of the components of the metric. For instance, for the quadrupole, the tidal field is described in GR by

$$\mathcal{E}_{ij} = c^2 R_{0i0j} \,, \qquad (14.224)$$

[56]The factor c^2 ensures that $\mathcal{E}_{ij}$ has the same dimensions as its Newtonian counterpart (14.224).

where $R_{\mu\nu\rho\sigma}$ is the Riemann tensor.[56] In a reference frame with origin centered on the NS and asymptotically Cartesian coordinates x^i, such an external tidal field has indeed the effect of adding a term $-R_{0i0j}x^i x^j$ to g_{00}; compare with eq. (1.87) of Vol. 1. More generally, we can expand any external gravitational field in tensor spherical harmonics, and compute the effect of the polar and axial modes of the external field on the metric.[57]

[57]See Damour and Nagar (2009b) and Binnington and Poisson (2009).

In order to actually compute the Love numbers we must then solve the interior problem and match the interior and exterior solutions on the (perturbed) boundary of the NS. In the interior problem we start from a Tolman–Oppenheimer–Volkov (TOV) equilibrium configuration (that depends on the assumed EoS), and we consider the coupled system of the perturbed Einstein equations and the perturbed hydrodynamical equations. As we saw in Section 11.2.1, once the EoS and the value of the central density have been assigned, the TOV equations can be integrated outward, starting from the center of the star. The Love numbers are then computed by matching the external and internal solutions. In this way, for a range of realistic EoS one finds (see the Further Reading)

$$k_2 \simeq 0.05 - 0.15 \,. \qquad (14.225)$$

For BHs there are subtleties in the computation of the Love numbers. For non-rotating BHs these subtleties have been addressed and it appears that, in four space-time dimensions, the Love numbers vanish. The rotating case is still not settled; see the Further Reading.

Given the induced quadrupole moment, one can compute its effect in the equations of motion of the binary system. As we already estimated in eq. (5.237), this gives a correction $O[(v/c)^{10}]$ to the leading Newtonian contribution, i.e. is a 5PN correction. As we discussed in Section 5.6.4, the inspiral waveform in frequency space can be written, in the restricted

Newtonian approximation, as

$$\tilde{h}(f) = \mathcal{A} f^{-7/6} e^{i\Psi(f)}, \qquad (14.226)$$

where the explicit expression for the amplitude $\mathcal{A}$ is given in eq. (5.274), and the explicit expression for the phase $\Psi(f)$ up to 2PN is given in eq. (5.275). The 5PN correction to the phase due to tidal effects has been computed. It is convenient to introduce, for each NS, the quantity

$$\Lambda_i = \frac{2}{3} k_2 \left(\frac{R_i}{Gm_i/c^2} \right)^5, \qquad (14.227)$$

where R_i and m_i are the radius and mass of the i-th NS. Note that Λ_i, just like k_2, is dimensionless, but it is proportional to the fifth power of the radius R_i, in units of Gm_i/c^2.[58] Numerically,

$$\left(\frac{R_i}{Gm_i/c^2} \right)^5 \simeq 6.6 \times 10^3 \left(\frac{R_i}{12\,\mathrm{km}} \right)^5 \left(\frac{1.4\,M_\odot}{m_i} \right)^5. \qquad (14.229)$$

Thus, for a NS with a realistic EoS this factor can be of order $10^3 - 10^4$. Combined with eq. (14.225), we see that realistic EoS give values of Λ_i of order $10^2 - 10^3$, with the larger values in this range being obtained from EoS leading to less compact NS, and the smaller values being obtained for more compact NS. Given the values Λ_1 and Λ_2 of the two NS, we define

$$\tilde{\Lambda} = \frac{16}{13} \frac{(m_1 + 12m_2)m_1^4 \Lambda_1 + (m_2 + 12m_1)m_2^4 \Lambda_2}{(m_1 + m_2)^5}. \qquad (14.230)$$

The overall normalization $16/13$ is chosen so that, when $m_1 = m_2$, $\tilde{\Lambda} = (\Lambda_1 + \Lambda_2)/2$. Then one finds[59]

$$\Delta\Psi_{\mathrm{5PN}}^{\mathrm{tidal}} = -\frac{117}{256} \frac{m^2}{m_1 m_2} \tilde{\Lambda} \left(\frac{v}{c} \right)^5. \qquad (14.231)$$

Recalling that the leading (Newtonian) contribution to the phase is of order $(v/c)^{-5}$ [see eqs. (5.239) and (5.246)], we see that the tidal effect is a correction of order $(v/c)^{10}$ to the leading term, i.e. it is a 5PN contribution, as indeed we already found in eq. (5.237). We have seen, however, that Λ_1 and Λ_2 are numerically large, $O(10^2 - 10^3)$, so, especially in the late inspiral phase, the tidal effect can actually overcome lower-order terms in the PN expansion. This could make the tidal effect detectable at advanced or third-generation interferometers. As we will see in Section 15.3.1, the first observed event of GWs from a NS–NS coalescence, GW170817, provides an upper limit $\tilde{\Lambda} < 800$ (at 90% c.l., and assuming low NS spins).

NS spins

Another possible difference between NS binaries and BH binaries concerns the expected spin of the component stars. For stellar-mass BHs

[58]The inverse of this quantity,

$$C \equiv \frac{Gm_i}{c^2 R_i}, \qquad (14.228)$$

is called the compactness of the NS.

[59]See Flanagan and Hinderer (2008). The 6PN tidal contribution to the phase has also been computed; see Vines, Flanagan and Hinderer (2011).

there is no reason to expect low spin, and in fact we will see in Chapter 15 that, for the observed BH–BH coalescences, the best-fit values of the dimensionless Kerr parameters $\hat{a}_i$ of the initial BHs are typically in the range $0.2 - 0.5$ (although with a large error); see e.g. Table 15.1 on page 289 or Table 15.2 on page 293. In particular, for the event GW151226, whose parameters are shown in the latter table, one can conclude that, at 99% c.l., at least one of the initial BHs had a dimensionless spin parameter greater than 0.2. Supermassive BHs are rather expected to be nearly extremal, $\hat{a} > 0.9$, as we discussed in Note 47 on page 253.

For NS the situation is expected to be different. From the definition of $\hat{a}$ given in eq. (12.359), writing $J = I\omega$ and $\omega = 2\pi/P$, where I is the moment of inertia and P the rotational period of a NS with mass M and radius R, we have

$$\hat{a} = \frac{cI}{GM^2}\frac{2\pi}{P}\,. \tag{14.232}$$

The moment of inertia of a spherical NS can be written as

$$I = \frac{2}{5}\kappa M R^2\,, \tag{14.233}$$

where κ, depending on the EoS, is typically in the range $0.7 - 1$. Using typical reference values of M and R for NS, we get

$$\hat{a} \simeq 0.057\kappa \left(\frac{R}{12\,\text{km}}\right)^2 \left(\frac{1.4 M_\odot}{M}\right)\left(\frac{10\,\text{ms}}{P}\right)\,. \tag{14.234}$$

This shows that only millisecond pulsars can have a non-negligible value of $\hat{a}$. As we discussed on page 81, at present the pulsar with the shortest observed rotational period has $P \simeq 1.4$ ms, corresponding (for the above reference values of M and R, and $\kappa \simeq 0.7-1$) to $\hat{a} \simeq 0.3-0.4$. However, as long as the rotational period of the NS is above 10 ms, the spin parameter is small, and for a normal pulsar with a spin period of order a few seconds, $\hat{a}$ is of order 10^{-4}. From Table 6.2 on page 327 of Vol. 1 we see that, among the NS in the known confirmed NS–NS binaries, the fastest-spinning is J0737−3039A, in the double pulsar, with $P = 22.7$ ms and $M \simeq 1.34 M_\odot$. Even in this case, we have only $\hat{a} \simeq 0.02 - 0.03$.[60]

14.4.2 Merger phase and numerical relativity

We next turn to the merger phase. Here one must resort to numerical-relativity simulations. The coalescence of binary NS systems is even more complicated to investigate numerically than that of BH–BH binaries. Indeed, BHs are solutions of the Einstein equations in vacuum. In contrast, to simulate numerically a NS–NS binary coalescence, we also need to take into account the matter variables. In particular, this requires modeling of the NS through an EoS, as we already saw in Section 11.2 when we studied NS normal modes. In principle, the numerical study of the problem requires solving the full Einstein equations,

[60]The pulsar PSR J1807−2500B has a rotational period of 4.19 ms, and is in a binary system in a highly eccentric orbit with a companion that should be a massive white dwarf or a second NS; see Lynch, Freire, Ransom and Jacoby (2012). This millisecond pulsar has $\hat{a} \simeq 0.1$. However, for this binary system the time-scale for merger due to GW emission is longer than the Hubble time.

coupled to the equations of relativistic hydrodynamics. Furthermore, magnetic fields can also play an important role, in which case we need the equations of magnetohydrodynamics. We also need to take into account several microphysics aspects such as neutrino transport and nuclear reactions in the ejected matter. In recent decades there have been significant advances in these directions, and several groups have developed codes that provide realistic simulations of the complex process of the inspiral, merger and post-merger phase of NS–NS coalescence. The overall picture that has emerged from these simulations is as follows.[61]

If the total mass of the binary is larger than about twice the maximum mass $M_{\rm TOV}$ (for Tolman–Oppenheimer–Volkov, see the discussion on page 90) of an isolated non-rotating NS, we will have a prompt collapse to a BH. The dimensionless spin parameter of the final BH is found to be in the range $\hat{a} \simeq 0.7 - 0.8$, and the final BH is surrounded by a hot accretion disk with a mass of order $M_{\rm torus} \sim (0.01 - 0.1)M_\odot$.

However, one expects that, statistically, the most frequent case corresponds to an initial mass M of the binary that does not exceed this critical value of about $2M_{\rm TOV}$, and is rather of order $1.5M_{\rm TOV}$. In this case, the numerical simulations indicate that the final remnant is a hyper-massive neutron star (HMNS). As we already mentioned in Section 11.2.3, this is a NS that is stabilized by differential rotation, and therefore is able to exceed the maximum mass allowed for a uniformly rotating NS. Eventually, differential rotation is smoothed out by various viscosity mechanisms as well as by the emission of gravitational radiation, so in the end we have a delayed collapse to a rotating BH.[62] This is again associated with the formation of an accretion torus, with a relatively high density $\rho \sim 10^{12} - 10^{13}\,{\rm g/cm}^3$, extended radially for $O(10)$ km. The formation of an accretion torus is important, since it could be the source of the energy observed is short γ-ray bursts. The time-scale for delayed collapse to a BH is the main uncertainty in the process, since the lifetime of a metastable object such as a HMNS is very sensitive to numerical artifacts. Current estimates range between $O(10)$ ms and a few hours. Finally, in the opposite case (again expected to be statistically disfavored) of very low total initial mass, the final remnant could be a stable, non-rotating NS.

Figure 14.20 shows an example of a waveform obtained from the numerical simulation of a NS–NS coalescence. The part of the signal at $t < 0$ is the inspiral waveform. This part is well described analytically by the PN expansion supplemented, for the last part, by the EOB method. However, tidal effects, that, as we have seen, during the inspiral phase give only a small correction to the phase, become more and more important as the two initial NS get close to each other. At this stage, the numerical simulations show that tidal waves appear on the surfaces of the stars, and matter is stripped from these surfaces. The portion of the signal that in the figure is at $0 \lesssim t \lesssim 3$ ms is the merger phase, and it displays a rather complex shape. The subsequent epoch, corresponding approximately to $3\ {\rm ms} \lesssim t \lesssim 24$ ms, represents the quasi-periodic emission of the resulting bar-deformed HMNS. Finally, around $t = 24$ ms, we

[61] We follow the review by Baiotti and Rezzolla (2017), to which we refer the reader for more details.

[62] More precisely, the HMNS may directly collapse to BH once differential rotation has smoothed out, or else it can first go through the stage of supramassive NS, i.e. a uniformly rotating NS with mass higher than the limit for non-rotating NS. Eventually, angular momentum is removed by electromagnetic or neutrino emission, and the supramassive NS finally collapses to a BH.

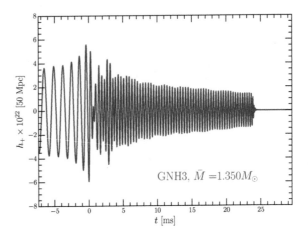

Fig. 14.20 The waveform from the coalescence of a NS–NS binary. Courtesy of Luciano Rezzolla, based on work presented in Takami, Rezzolla and Baiotti (2014).

have the collapse to a BH and the subsequent ringdown signal, which is much weaker. In comparison the BH–BH waveforms shown in the previous subsections are much simpler, with the EOB-improved inspiral waveform matching smoothly the ringdown phase. Here, in contrast, we have an intermediate phase that is more complicated, and correspondingly richer in information.

Figure 14.21 shows a similar waveform in frequency space, compared with the sensitivity of initial LIGO, advanced LIGO, and a future third-generation interferometer such as the Einstein Telescope. We see that strong tidal effects appear at a relatively high frequency, but still, even for a source at 100 Mpc, they could be within the sensitivity of advanced LIGO at its design sensitivity. In contrast the post-merger phase is not accessible to advanced LIGO for a source at this distance, and would require a third-generation instrument such as the Einstein Telescope. We have already shown a similar plot for the waveform generated in the post-merger phase of NS–NS coalescence in Fig. 11.5 on page 106, in that case for a source at 50 Mpc. For that distance, we saw in Fig. 11.5 that the post-merger signal can be detectable at advanced LIGO. The post-merger phase is particularly interesting, since it carries a strong imprint of the NS EoS and of the tidal polarizability of the NS. Indeed, we saw in Fig. 11.5 that the signal generated in this phase produces a series of peaks, which give information on the NS structure; in particular, Fig. 11.6 shows that, for a given mass, there is a tight correlation between the radius of the HMNS and the frequency of the main peak. In contrast, the frequency of the second-strongest peak is correlated to the tidal polarizability of the NS, in a way independent from the EoS (see the Further Reading in Chapter 11).

Several interesting questions concern the evolution of the magnetic fields in a NS–NS binary coalescence. Just before merger the two neutron

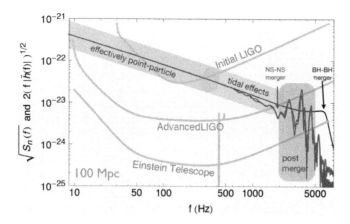

Fig. 14.21 The NS–NS waveform in frequency space, for a source at 100 Mpc, compared to the sensitivity of Initial LIGO, Advanced LIGO, and the Einstein Telescope. The curve labeled 'BH–BH merger' is the signal of BH–BH system with the same masses. Courtesy of Jocelyn Read, based on results presented in Read *et al.* (2013).

stars are very old. From Table 6.2 on page 327 of Vol. 1 we see that the typical time to coalescence of known NS–NS binaries is in the range $O(10^2 - 10^3)$ Myr. Then, from the $P - \dot{P}$ diagram on page 78 we see that typical magnetic fields before merger will be of order $10^8 - 10^{10}$ G. Different instabilities during the merger can amplify these fields. It is, however, difficult to quantify the degree of amplification, with simulations given values ranging from two or three orders of magnitudes up to five or six. The strength of the magnetic field is important, in order to understand the electromagnetic counterpart of NS–NS mergers.

Further reading

- The use of Padé approximants for improving the PN expansion was proposed by Damour, Iyer and Sathyaprakash (1998, 2000). The determination of the least-stable orbit using Padé resummations is discussed in Damour, Jaranowski and Schäfer (2000). The accuracy of the PN expansion is compared with that of Padé resummations in Section 9.6 of Blanchet (2006).

- The EOB approach was introduced by Buonanno and Damour (1999, 2000). The generalized Lagrangian that gives the 2PN equations of motion in harmonic coordinates was found in Damour and Deruelle (1981) and Damour (1982). The transformation to ADM coordinates that gives an ordi-

nary Lagrangian independent of accelerations is explicitly constructed in Damour and Schäfer (1985, 1988). Logarithmic terms in the EOB expansion, which appear starting from 4PN order, have been determined, using gravitational self-force computations, in Damour (2010), Blanchet, Detweiler, Le Tiec and Whiting (2010), Barausse, Buonanno and Le Tiec (2012), and Bini and Damour (2013).

The full 4PN result for the EOB conservative dynamics is obtained in Damour, Jaranowski and Schäfer (2015), using the results for the 4PN two-body problem found in Damour, Jaranowski and Schäfer (2014). A review of the EOB method is given in Damour and Nagar (2011).

- The first successful analytic prediction for the spin of the final BH, for non-spinning initial BHs, was given in Buonanno and Damour (2000), and without any tuning agrees with numerical relativity to an accuracy of $\sim 10\%$. By combining the EOB approach with test-mass-limit predictions for the energy and angular momentum released during the merger and ringdown phases, Damour and Nagar (2007a) have refined this prediction for non-spinning BHs, obtaining an accuracy of $\sim 2\%$.

- The leading (1.5PN) spin–orbit coupling and the 2PN spin–spin coupling were first given in Barker and O'Connell (1975, 1979) and applied to gravitational radiation from coalescing binaries in Kidder, Will and Wiseman (1993) and Kidder (1995). The 2.5PN spin–orbit terms are given in Faye, Blanchet and Buonanno (2006) and Blanchet, Buonanno and Faye (2006), using harmonic coordinates. A Hamiltonian formulation of the spin–orbit terms to 2.5PN order, using ADM coordinates, is provided in Damour, Jaranowski and Schäfer (2008a).

  Using these results, the EOB approach is extended to spinning initial bodies in Damour (2001), Buonanno, Chen and Damour (2006), Damour, Jaranowski and Schäfer (2008b), Barausse, Racine and Buonanno (2009), Barausse and Buonanno (2010), Taracchini *et al.* (2012, 2014), Pan *et al.* (2014) and Damour and Nagar (2014).

- Comparison of EOB waveforms with numerical relativity has been the subject of many investigations. For non-spinning equal-mass binaries see in particular Buonanno, Cook and Pretorius (2007), Pan *et al.* (2008), Damour, Nagar, Dorband, Pollney and Rezzolla (2008), Damour and Nagar (2009a) and Buonanno *et al.* (2009). For non-spinning unequal-mass binaries see Buonanno *et al.* (2007), Berti *et al.* (2007), Damour, Nagar, Hannam, Husa and Bruegmann (2008), Damour and Nagar (2008) and Damour, Nagar and Bernuzzi (2013). The extreme-mass-ratio limit is studied in Nagar, Damour and Tartaglia (2007) and Damour and Nagar (2007b); see also the review by Damour and Nagar (2011). Comparison of numerical relativity waveforms with PN waveforms is performed in Baker *et al.* (2007b), Boyle *et al.* (2007, 2008) and Hannam, Husa, Sperhake, Brügmann and Gonzalez (2008).

- The "flexibility parameters" that effectively describe higher-order PN effects in the EOB waveform are discussed in Damour (2001), Damour, Gourgoulhon and Grandclement (2002), and Damour, Iyer, Jaranowski and Sathyaprakash

(2003). The position of the ISCO at 3PN is computed, using Padé resummations, in Damour, Gourgoulhon and Grandclement (2002). Improved Padé resummations (which, for the flux function, involve no free parameter) are described in Damour, Iyer and Nagar (2009) and in Damour and Nagar (2009a), where 'next-to-quasi-circular' (NQC) corrections are also introduced. The calibration of EOB waveforms to numerical relativity is discussed in Buonanno *et al.* (2007) and in Damour and Nagar (2009a). The logarithmic resummation of $A(u)$ given in eq. (14.136) is introduced in Barausse and Buonanno (2010), to avoid spurious poles of the Padé resummation in the spinning case.

- Numerical relativity studies involving BHs have a long history, which started with the numerical simulation of the head-on collision of two BHs by Smarr (1979). The first simulation of a complete orbit of a BH–BH binary was presented in Brügmann, Tichy and Jansen (2004). The first successful numerical evolutions of the entire process of BH–BH coalescence, including waveform extraction, was presented in Pretorius (2005), using generalized harmonic coordinates. A few months later, two groups at Brownsville (Texas) and at NASA-Goddard independently succeded in successfully simulating the whole coalescence process and extracting the waveform using two completely different methods (equations in the BSSN form or moving punctures); see Campanelli, Lousto, Marronetti and Zlochower (2006) and Baker, Centrella, Choi, Koppitz and van Meter (2006). The accuracy of the simulations is increasing at a rapid pace. Accurate simulations are presented in Boyle *et al.* (2007, 2008), Husa, González, Hannam, Brügmann and Sperhake (2008) and Scheel *et al.* (2009). A numerical simulation spanning 350 GW cycles is presented in Szilágyi *et al.* (2015).

- For reviews of the numerical relativity approach to binary BH coalescence, including technical details, such as the choice of coordinate systems and the treatment of the BH singularity, that were crucial for successful simulations, see Pretorius (2009), Sperhake (2009) and Centrella, Baker, Kelly and van Meter (2010). Useful information can also be found in the reviews by Lehner (2001) and Baumgarte and Shapiro (2003), which were written before the 2005 breakthroughs. Initial data for numerical relativity are reviewed in Cook (2000). Constraint damping, in the form that was applied to generalized harmonic coordinates, was proposed in Gundlach, Martin-Garcia, Calabrese and Hinder (2005). There are several comprehensive textbooks dedicated to numerical relativity; see Alcubierre (2008), Bona, Palenzuela-Luque and Bona-Casas (2009), Baumgarte and Shapiro (2010), Gourgoulhon (2012), and Shibata (2015). Presently, many numerical relativity codes use the Z4c formulation developed in Bernuzzi and Hilditch (2010) or the CCZ4 formulation developed in Alic, Bona-Casas, Bona, Rezzolla and Palenzuela (2012).

- The analytic computation of the recoil of the final BH after the merger of an unequal-mass non-spinning BH–BH binary, at leading order in the PN expansion, is given in Fitchett (1983) and Fitchett and Detweiler (1984), while earlier classic papers on the loss of linear momentum are Peres (1962) and Bekenstein (1973). Improved analytic estimates are given in Favata, Hughes and Holz (2004), Blanchet, Qusailah and Will (2005), Damour and Gopakumar (2006) and Sopuerta, Yunes and Laguna (2006). The "anti-kick" due to the ringdown of the final BH is studied in Damour and Gopakumar (2006) and in Le Tiec, Blanchet and Will (2010). The first accurate numerical relativity studies of the recoil were presented in Baker *et al.* (2006), and a systematic investigation for different mass ratios, and the determination of the maximum kick, were performed in González, Sperhake, Brügmann, Hannam and Husa (2007). A similar anti-kick in the head-on collision of two BHs is discussed in Lovelace *et al.* (2010) and in Rezzolla, Macedo and Jaramillo (2010).

- Analitic estimates of the recoil for spinning BHs are given in Favata, Hughes and Holz (2004). The first full numerical simulations of the coalescence of spinning BH binaries were performed in Campanelli, Lousto and Zlochower (2006). The spin flip and precession in the merger of spinning binaries is studied in Campanelli, Lousto, Zlochower, Krishnan and Merritt (2007). Numerical relativity simulations of the BH recoil when the initial BHs have spins perpendicular to the orbital plane are performed in Herrmann, Hinder, Shoemaker, Laguna and Matzner (2007), Koppitz *et al.* (2007), Pollney *et al.* (2007) and Baker *et al.* (2007a).

  Expressions for the radiated energy are given in Barausse, Morozova and Rezzolla (2012). Analytic formulas for estimating the spin of the final BH have been proposed in Buonanno, Kidder and Lehner (2008), Tichy and Marronetti (2008), Rezzolla *et al.* (2008a, 2008b), Rezzolla (2009), Barausse and Rezzolla (2009) and Hofmann, Barausse

and Rezzolla (2016), combining information from the PN expansion, the test-mass limit, and fits to a large catalog of numerical relativity simulations.

- The possibility of "superkicks" when the initial BH spins are in the orbital plane was first suggested in Campanelli, Lousto, Zlochower and Merritt (2007a) and then confirmed by the numerical relativity simulations in González, Hannam, Sperhake, Brügmann and Husa (2007), Campanelli, Lousto, Zlochower and Merritt (2007b) and (2007c). A study of the contribution to the recoil from different multipoles is performed in Brügmann, Gonzalez, Hannam, Husa and Sperhake (2008) and in Schnittman *et al.* (2008).

- Techniques for measuring the spin of astrophysical BHs are reviewed in Narayan (2005). The measurement of the spin of the central BH in the galaxy MCG-06-30-15 is discussed in Brenneman and Reynolds (2006), where one can also find a discussion of the history of the measurements of this broad iron line and a discussion of the physical assumptions of the methods. A summary of measurements of BH spins with different techniques is given in Bambi, Jiang and Steiner (2016).

- Astrophysical consequences of the recoil of the final BH are discussed in Redmount and Rees (1989), Merritt, Milosavljević, Favata, Hughes and Holz (2004), Madau and Quataert (2004), Boylan-Kolchin, Ma and Quataert (2004), Volonteri and Rees (2006) and Gualandris and Merritt (2008). Wandering BHs are further discussed in Volonteri and Perna (2005) and, using N-body simulations that include the effect of the recoil on the final BH, in Libeskind, Cole, Frenk and Helly (2006).

- The $M - \sigma$ relation was proposed in Ferrarese and Merritt (2000) and Gebhardt *et al.* (2000). The values of the coefficients α and β that we have quoted are from McConnell and Ma (2013). Further compilations leading to fits to the coefficients in the $M - \sigma$ relation, as well as in the relation of the central BH mass with the mass or with the luminosity of the bulge, include Gultekin *et al.* (2009), Graham, Onken, Athanassoula and Combes (2011) and Sani, Marconi, Hunt and Risaliti (2011). A discussion of systematics in these relations is given in Tremaine *et al.* (2002).

- A possible candidate for a recoiling BH, the quasar SDSSJ0927+2943, is discussed in Komossa, Zhou and Lu (2008). This quasar shows an optical emission-line spectrum with two sets of emission lines: one set of very narrow lines, and a second set of broad lines that are blueshifted by 2650 km/s relative to the set of narrow lines. This observation is consistent with the hypothesis that the supermassive BH was ejected from the core of the galaxy because of a large kick due to gravitational radiation, carrying with it the broad-line gas while leaving behind the bulk of the narrow-line gas. A few more candidates have since been discovered; see Komossa (2012) for review.

- The fact that the tidal deformability of NS in a binary system gives effects at 5PN order was first discussed in Damour (1983, 1984). The equations of motion of weakly self-gravitating deformable bodies are studied in Damour, Soffel and Xu (1991, 1992), where the Newtonian Love numbers are generalized, in post-Newtonian theory, to "electric-type" and "magnetic-type" multipolar tidal coefficients. The phase shift at 5PN order due to the tidal deformation is computed in Flanagan and Hinderer (2008), who pointed out that the effect could be observable at ground-based interferometers; the computation of the tidal effect is extended to 6PN order in Vines, Flanagan and Hinderer (2011). The Love number of relativistic NS is computed, using polytropic EoS, in Flanagan and Hinderer (2008) and Hinderer (2008) and, using more realistic EoS, in Hinderer, Lackey, Lang and Read (2010). Accurate discussions of relativistic tidal properties of NS are given in Damour and Nagar (2009b) and in Binnington and Poisson (2009). The latter paper provides a definition of electric-type and magnetic-type Love numbers valid in arbitrarily strong gravitational fields. Favata (2014) discusses the systematic effects in the PN terms of lower order that must be kept under control to extract the tidal contribution.

- The Love numbers of BHs are discussed in Damour and Nagar (2009b) and in Binnington and Poisson (2009). In both computations the formal extrapolation of the result obtained for NS to the compactness of BHs gave $k_l = 0$, but different opinions were held on the correctness of this formal extrapolation, due to the possible appearance of divergences at 5PN, related to additional non-minimal couplings in the corresponding world-line action. Kol and Smolkin (2012), using effective field theory methods and working in general space-time dimensions, find that in $d = 4$, for non-rotating BHs, the coefficients k_l vanish, but they are indeed non-vanishing in higher dimensions.

- The EOB method can be extended to include tidal effects in NS–NS binaries; see Damour and Nagar (2010), Bernuzzi, Nagar, Dietrich and Damour (2015), Hinderer *et al.* (2016) and Steinhoff, Hinderer, Buonanno and Taracchini (2016).

- The moment of inertia for NS with typical EoS is computed in Cook, Shapiro and Teukolsky (1994). The corresponding maximum value of the Kerr parameter is given in Nissanke, Holz, Hughes, Dalal and Sievers (2010).

- The first GR simulations of NS–NS binaries were developed in Shibata (1999), Shibata and Uryu (2000, 2002), and Shibata, Taniguchi and Uryu (2003). The first simulations going stably beyond BH formation were presented in Baiotti, Giacomazzo and Rezzolla (2008). These results were obtained using simulations without excision of the singularity, but rather with the imposition of a suitable gauge condition, as proposed in Baiotti and Rezzolla (2006). Systematic studies of the effect of the NS EoS on the inspiral part of the signal are performed in Read *et al.* (2013) and Hotokezaka, Kyutoku, Sekiguchi, and Shibata (2016). Study of the post-merger phase is much more difficult numerically, because of the strong non-linearities involved, and has been tackled among others in Shibata and Uryu (2000, 2002), Yamamoto, Shibata, and Taniguchi (2008), Baiotti, Giacomazzo and Rezzolla (2008), Anderson *et al.* (2008), Giacomazzo, Rezzolla and Baiotti (2011), Bauswein and Janka (2012), Bauswein, Baumgarte and Janka (2013), Bauswein, Stergioulas and Janka (2016), Bauswein, Clark, Stergioulas and Janka (2016), Takami, Rezzolla and Baiotti (2014, 2015) and Rezzolla and Takami (2016). A very detailed review is given in Baiotti and Rezzolla (2017). See also the Further Reading of Chapter 11 for references on the study of the post-merger oscillations of NS.

GWs from compact binaries. Observations

15

As we saw in Vol. 1, the quest for the direct detection of GWs began as long as the 1960s. Finally, on September 14, 2015, the two detectors of the LIGO Observatory, at the very beginning of the first observing run in the advanced LIGO configuration, detected the GWs emitted by the coalescence of a BH–BH binary. After the necessary studies and verifications, the result was announced by the LIGO and Virgo collaborations on February 11, 2016.[1] The binary system was composed of two initial BHs with masses of about $36 M_\odot$ and $29 M_\odot$, which coalesced into a final BH with a mass of $62 M_\odot$, liberating in a few milliseconds an energy of about $3 M_\odot c^2$ in GWs. The source was at a cosmological distance, with a luminosity distance $d_L = O(400)$ Mpc and a redshift $z \simeq 0.09$ (see Table 15.1 for details of the parameter values and error estimates). The detection of this spectacular event was a historic moment, which took place just 100 years after the final formulation of general relativity (GR) by Einstein.[2]

The first observing run of the advanced LIGO detectors lasted until January 2016, and a second confirmed BH–BH coalescence was detected on December 26, 2015. Both events had a statistical significance higher than 5σ and are therefore confirmed GW detections. After further improvements in the sensitivity, the two advanced LIGO interferometers performed a second observing run, which lasted from November 30, 2016 to August 25, 2017. Two more BH–BH coalescences were detected, in January and in June 2017. Advanced Virgo started taking science-quality data in concert with the two LIGO detectors on August 1, 2017. On August 14 a BH–BH binary coalescence was detected in all three interferometers for the first time. Soon after, on August 17, the coalescence of two NS was detected in the two LIGO interferometers and in the Virgo interferometer. The gravitational signal was associated with a short γ-ray burst, which was detected independently by the Fermi Gamma-ray Bursts Monitor (Fermi-GBM) and confirmed by the INTEGRAL γ-ray satellite. The electromagnetic transient counterpart was then observed by more than 50 telescopes all over the electromagnetic spectrum. This event therefore also marked the opening of multi-messenger astronomy.

In this chapter we will discuss these detections in great detail. We will pay special attention to the analysis performed in order to assess the statistical significance of GW150914, both because of its historical importance as the first direct detection, and because a similar analysis

[1]LIGO and Virgo, while maintaining their separate identities as collaborations, exchange their expertise on the experimental side, analyze the data together, and publish the results together.

[2]For this discovery Reiner Weiss, Barry Barish and Kip Thorne were awarded the 2017 Nobel Prize for Physics "for decisive contributions to the LIGO detector and the observation of gravitational waves".

Gravitational Waves, Volume 2: Astrophysics and Cosmology. Michele Maggiore.
© Michele Maggiore 2018. Published in 2018 by Oxford University Press.
DOI 10.1093/oso/9780198570899.001.0001

has been applied to the subsequent detections. We will then examine the properties of the currently observed BH–BH coalescences and of the NS–NS binary coalescence, and discuss the theoretical and astrophysical implications of these discoveries.

15.1   GW150914. The first direct detection

The first event, labeled as GW150914 from the year/month/day of discovery, was detected in coincidence between the two LIGO detectors in Hanford, Washington (H1), and in Livingstone, Louisiana (L1). After a series of engineering and calibration runs, the first observing run of advanced LIGO, denoted by O1, was initially scheduled to begin on September 14 2015, at 15:00 UTC. On September 11, the start of O1 was delayed to September 18. However, calibration was completed by September 12 and when the event arrived, on September 14, at 09:50:45 UTC, the two LIGO detectors were in observation mode. The Virgo detector was still being upgraded to the advanced Virgo configuration, while GEO 600 was operating, but not in observational mode, and in any case was not sufficiently sensitive to detect this event. The wavefront arrived first at L1 and $6.9^{+0.5}_{-0.4}$ ms later at H1, and so was within the window of ± 10 ms given by the light travel-time between the two detectors.

The signal was sufficiently strong that it can be seen visually, even before performing any detailed data analysis, just by using a 35–350 Hz bandpass filter to select the bandwidth where the signal is concentrated; see the top row of Fig. 15.1. The second row in Fig. 15.1 shows the signal, reconstructed using the templates or wavelets (gray curves), and the numerical relativity result with the best-fit values. The last row gives a representation of the signal in the time–frequency plane, showing the characteristic behavior of the signal from a coalescing binary, with the frequency increasing in time, up to a maximum value, until the signal disappears. The event was first detected by a low-latency search for generic GW transients, and was reported within three minutes of data acquisition. It was then identified as a BH–BH binary and subsequently studied in more detail, by performing matched filtering to coalescing binary waveforms.

Even if the visual agreement with the predicted coalescing binary waveform is striking, in order to assess the statistical significance of the event we obviously have to go through a detailed statistical analysis. We developed the necessary tools in Chapter 7 of Vol. 1, in particular in Sections 7.3 and 7.7. In the next subsections we will see how these tools are applied to GW150914, and we will discuss the analysis of the LIGO and Virgo collaborations that led to the conclusion that the statistical significance of this event is higher that 5σ, allowing us to claim that it is indeed a GW event.[3]

---

[3]We will focus on the analysis that searched specifically for binary coalescences. GW150914 was also detected in a search for generic transients. For discussions of the latter search see Section V.A of the discovery paper [LIGO Scientific Collaboration and Virgo Collaboration], Abbott *et al.* (2016a), as well as the paper by Abbott *et al.* (2016g), where the event is studied with minimal assumptions.

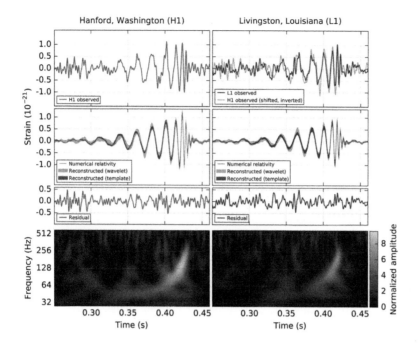

Fig. 15.1 The event GW150914 observed by the LIGO Hanford (H1, left column panels) and Livingston (L1, right column panels) detectors. For visualization, all time series are filtered with a 35–350 Hz bandpass filter to suppress large fluctuations outside the detectors' most sensitive frequency band. *Top row, left:* H1 strain. *Top row, right:* L1 strain. GW150914 arrived first at L1 and $6.9^{+0.5}_{-0.4}$ ms later at H1; for a visual comparison, in the right panel the H1 data are also shown, shifted in time by this amount and inverted (to account for the detectors' relative orientations). *Second row.* GW strain projected onto each detector in the 35–350 Hz band. Solid lines show a numerical relativity waveform for a system with parameters consistent with those recovered from GW150914. Shaded areas show 90% credible regions for two independent waveform reconstructions. *Third row:* Residuals after subtracting the filtered numerical relativity waveform from the filtered detector time series. *Bottom row:* A time–frequency representation of the strain data, showing the signal frequency increasing over time. From the discovery paper, [LIGO Scientific Collaboration and Virgo Collaboration], Abbott *et al.* (2016a). [Creative Commons Attribution 3.0 License.]

15.1.1   Evaluation of the statistical significance

The first analysis was performed on the data collected by the two LIGO detectors from September 12 to October 20, 2015. During these 38.6 days, the two detectors were in coincident operation for a total of 18.4 days. After discarding time removed by data-quality vetoes, the total coincident observation time remaining was $T = 16$ days. This time span was chosen because it was the minimum required for having a statistical significance over 5σ, as we will see. We restrict attention for the moment to the analysis of the data in this time span. The O1 run lasted until January 2016, and the results from the full run were released later, as we will discuss.

[4] Actually, $h(t)$ must also be corrected for frequency-dependent calibration errors, see [LIGO Scientific Collaboration and Virgo Collaboration], Abbott *et al.* (2016i).

As we saw in Chapter 7, a crucial step for data analysis is matched filtering. In a single detector the output $s(t)$ is a time series, of the form $s(t) = n(t) + h(t)$, where $n(t)$ is the noise and, hopefully, there might be a GW signal $h(t)$.[4] From the output, we form the quantity $\hat{s}$ defined in eq. (7.41),

$$\hat{s} = \int_{-\infty}^{\infty} dt \, s(t) K(t) \,, \tag{15.1}$$

where $K(t)$ is a filter function, chosen so to optimize the signal-to-noise ratio (SNR), for the specific waveform $h(t)$ that we are searching. We defined S as the expectation value of $\hat{s}$ over several realizations of the noise,

$$S = \langle \hat{s} \rangle \,, \tag{15.2}$$

and N as the variance of $\hat{s}$ when the GW is absent,

$$N = \left[\langle \hat{s}^2 \rangle - \langle \hat{s} \rangle^2 \right]_{h=0}^{1/2} \,. \tag{15.3}$$

We then showed in eq. (7.49) that it is possible to compute analytically the filter function $K(t)$ that maximizes S/N, for a given GW signal $h(t)$ that we are looking for. The result, in Fourier space, is given by

$$\tilde{K}(f) = \text{const} \times \frac{\tilde{h}(f)}{S_n(f)} \,, \tag{15.4}$$

where $S_n(f)$ is the spectral density of the noise. The corresponding value of S/N is given by eq. (7.51), which can be rewritten as

$$\left(\frac{S}{N} \right) = (h|h)^{1/2} \,, \tag{15.5}$$

by introducing the scalar product (7.46),

$$(A|B) = 4 \, \text{Re} \int_0^{\infty} df \, \frac{\tilde{A}^*(f)\tilde{B}(f)}{S_n(f)} \,. \tag{15.6}$$

Because of the factor $1/S_n(f)$, this scalar product gives a greater weight to the frequency region where the detector is more sensitive.

To perform a statistical analysis, we are interested not only in the average value S of the quantity $\hat{s}$, but also in its full statistical distribution, or better yet in the statistical distribution of the quantity ρ defined in eq. (7.76),

$$\rho = \frac{\hat{s}}{N} \,. \tag{15.7}$$

This quantity depends on the waveform $h(t)$ (the "template"), whose presence in the data we are testing, since the filter function in eq. (15.1) is chosen as in eq. (15.4), in order to maximize $\langle \rho \rangle = S/N$ for a given choice of $h(t)$. We must then compute ρ for "all" possible templates $h(t)$. This is a space of functions that depends on many parameters. One of them is the time of coalescence t_c, which can be defined for example as the time at which $h(t)$ reaches its maximum amplitude (equivalently, in Section 7.7.1, instead of t_c, we used as a parameter the time t_* at which the signal enters in the detector bandwidth). Furthermore, we have in principle the masses of the two initial stars, their spins (three components for each spin vector), the distance to the source, its position in the sky (two angles), the orientation of the orbit (two more angles, giving the orientation of the unit vector normal to the orbit), and the orbital phase at a reference time, so we have in principle a 15-dimensional parameter space.[5] Fortunately, several simplifications are possible. First of all, the time of arrival is taken into account by observing that different times of arrival simply induce a temporal shift in the waveform. As we discussed in Section 7.7.1, this allows one to take into account all possible arrival times, i.e. all possible values of t_c, with a single Fourier transform, which can be implemented efficiently with a fast Fourier transform.

Furthermore, recall from Section 7.7.1 that the maximization over the overall amplitude and the phase φ at coalescence can be performed analytically. The waveform has the general structure

$$h(\tau; \theta) = h_0(\tau; \theta) \cos \varphi + h_{\pi/2}(\tau; \theta) \sin \varphi \,, \qquad (15.8)$$

where $\tau = t - t_c$ and θ denotes the whole set of remaining parameters (excluding t_c) characterizing the signal. We can separate the overall amplitude a from the other parameters θ, writing $h(\tau; \theta) = a h_a(\tau; \xi)$, where ξ denotes collectively all the θ parameters except a. Then, the maximization over a and φ can be done analytically[6] and leads to

$$\rho^2 = \frac{(h_0|s)^2 + (h_{\pi/2}|s)^2}{(h_0|h_0)} \,, \qquad (15.9)$$

compare with eq. (7.176). The distance to the source, the arrival direction and the inclination of the orbit only enter in the overall amplitude [see eq. (7.168)], so we do not need to scan over these parameters. Thus, in the end in the search we only have to scan over the masses and spins of the initial stars. However, once an event is detected and we want to reconstruct its properties, we need in principle to construct the likelihood function over the full 15-dimensional parameter space.[7]

A technically complex issue is the inclusion of the full spin dynamics in the waveforms. As we discussed in Section 14.2.5, the inclusion of the spin components perpendicular to **L** has the effect of inducing a precession of the spins and of the plane of the orbit. The waveforms used in the search and in the subsequent follow-up studies will be discussed in more detail in Section 15.1.2. While the full follow-up study of GW150914, after the discovery, has been performed with precessing waveforms, the initial search that led to the discovery only used spins aligned with the orbital angular momentum. Even in this simpler setting, in order to scan the remaining parameter space and recover effectively a signal, the use of about 250,000 templates at each value of time (or, effectively, in steps Δt of order of the duration of the template) was necessary. It can, however, be shown that such templates effectively recover also systems with misaligned spins, in the parameter-space region of GW150914.

[5] We are neglecting the eccentricity of the orbit, which was set to zero in the search that we are discussing, since radiation reaction circularizes the orbit efficiently (see the computation on page 187 of Vol. 1). Future analyses might also include this parameter, since for instance three-body interactions in dense environments such as globular clusters might re-generate a non-vanishing eccentricity.

[6] More precisely, the maximization can be done analytically only for the $(2, 2)$ mode. The search indeed uses only the $(2, 2)$ mode, but includes the amplitude corrections discussed in Section 5.6.

[7] Among the parameters that enter in the amplitude, the arrival direction, with only two detectors, can be estimated from the difference in arrival time between the two detectors (as well as requiring consistency in the amplitude and phase of the signal at the two detectors). With more than two detectors a much more accurate triangulation is possible, as we will see for the triple LIGO-LIGO-Virgo detections of the BH–BH binary GW170814 and for the NS–NS binary GW170817, discussed in Sections 15.2.4 and 15.3.1, respectively.

[8] Actually, even if the trigger exceeds the threshold, it is only kept if there is no trigger with a higher value of ρ within a pre-defined time window, set to the characteristic length of the template, since otherwise we would get many triggers corresponding to the same event. This procedure is called "time clustering". Similarly, one performs a clustering in the space of templates, since, when a template produces a value of ρ over the threshold, it is likely that many similar templates will also produce a trigger, and only the one with the highest value of ρ is kept. Some care must be taken to avoid the possibility that a large noise glitch might mask a true signal that appears close in time. This can be done by regrouping the triggers both in time and in recovered masses.

[9] Indeed, this threshold is a compromise between the fact that, with a lower threshold, one would be overflooded by triggers, while on the other hand the higher the threshold, the higher is the probability of missing a real GW event.

[10] In the last searches of initial LIGO and initial Virgo, the value $p = 16$ was chosen. For the O1 run, it was found that the optimal choice for the number of bins depends on the template, and it was empirically found that a better suppression of noise transients is obtained with a value of p that depends on the peak frequency f_{peak}, according to $p = 0.4[(f_{\text{peak}}/\text{Hz})]^{2/3}$.

Thus, for each value of the arrival time t_c, one evaluates $\rho(t_c)$ over the whole bank of templates. One then monitors the time evolution of $\rho(t_c)$, looking for its maxima with respect to the arrival time t_c. Whenever a maximum of $\rho(t_c)$ exceeds a pre-assigned threshold ρ_t, this event is recorded as a "trigger".[8] A typical choice for the threshold (that was used in the O1 and O2 searches) is $\rho_t = 5.5$. As discussed in Sections 7.4.3 and 7.7.1, for purely Gaussian noise and in the absence of a signal, the distribution of ρ is given by a chi-square distribution with two degrees of freedom; see eq. (7.87). The two degrees of freedom correspond to the two polarizations of the GW, or, equivalently, to the two functions h_0 and $h_{\pi/2}$ in eq. (15.8). The corresponding probability that Gaussian random noise exceeds this threshold, using the chi-square distribution with two degrees of freedom (7.87), is $P(\rho > 5.5) \simeq 2.7 \times 10^{-7}$. Naively, one might think that this is such a small probability that Gaussian noise is eliminated. But, in fact, we must take into account that we are trying 250,000 templates at each value of time. More precisely, according to Note 8, we are effectively trying 250,000 templates in each time interval Δt corresponding to the typical duration of the template, which, for an event such as the coalescence of a $(30 + 30)M_\odot$ binary, is of the order 10 ms. Therefore, already after a few seconds, we have effectively tried of order 10^7 templates, and even noise fluctuations with a probability $O(10^{-7})$ will unavoidably show up. Thus, even Gaussian random noise produces triggers above the threshold, at a rate of order one every few seconds.[9] Furthermore, the actual noise in a detector is far from Gaussian (and also non-stationary), and produces a distribution for ρ with long tails, due for example to glitches in the detectors, which can extend up to values of ρ of order 1000 or more, and which therefore cannot be simply eliminated by setting a somewhat higher threshold.

As we discussed in Section 7.5.3, to substantially reduce the noise the next step is to perform coincidences between at least two detectors. For the two LIGO detectors, the light travel time is 10 ms. According to eq. (7.122), we must actually add the temporal resolution of the detector, particularly for signals with small SNR. In the search of the O1 run, a coincidence window of 15 ms was required. Requiring that the event shows up simultaneously in the two detectors within this coincidence window dramatically reduces the number of triggers that are kept. However, there will still be accidental coincidences due to noise. Then, one further requires consistency between the templates that generate the trigger in the two detectors. A further way of discriminating accidental coincidences from a true signal is provided by the fact that the signal from a binary coalescence has a very specific shape, in time and frequency. Thanks to the large bandwidth of interferometers, it is possible to split the data into p different frequency bands.[10] Defining ρ_i as the SNR, after optimal matched filtering, in the i-th frequency band, one computes the reduced chi-square

$$\chi_r^2 = \frac{p}{2p-2} \sum_{i=1}^{p} \left(\rho_i - \frac{\rho}{p} \right)^2 . \tag{15.10}$$

If χ_r^2 is near unity, this means that the SNR accumulates in the various frequency bands as expected for a binary coalescence. In contrast, a spurious signal might give a large value of ρ because its amplitude is large, and happens to match fairly well the template in one frequency band, even if it would match poorly the template in other frequency bands. This would, however, show up as a larger value of χ_r^2. The SNR is then re-weighted using the value of χ_r^2, so as to suppress the SNR from triggers with a large χ_r^2. In the O1 analysis this has been performed defining, empirically, a re-weighted SNR $\hat{\rho}$ as

$$\hat{\rho} = \frac{\rho}{[(1 + (\chi_r^2)^3)/2]^{1/6}} , \qquad (15.11)$$

if $\chi_r^2 > 1$, and $\hat{\rho} = \rho$ if $\chi_r^2 \leqslant 1$. Thus, we end up with a list of events that have passed the threshold in each detector, have passed the coincidence test both in time and in waveform (i.e. the triggers in the two detectors are obtained from consistent values of the parameters of the binary), and are characterized by a value $\hat{\rho}_1$ in one detector and a value $\hat{\rho}_2$ in the other. One now defines a global SNR indicator for the coincidences as $\hat{\rho}_c = (\hat{\rho}_1^2 + \hat{\rho}_2^2)^{1/2}$, and ranks the coincidences according to the value of $\hat{\rho}_c$.[11]

The next step is to determine the actual distribution of $\hat{\rho}_c$ due to noise only. This is done using the shifting technique already discussed on page 371 of Vol. 1. Namely, we take the list of triggers from one detector and the list from the other, and we artificially shift in time the output of one detector relative to the other, by an amount Δt much larger than the time of flight between the two detectors. We then go again through the above procedure for defining the coincidences and assigning them a value of $\hat{\rho}_c$. All the coincidences obtained in this way are just fake coincidences due to noise, because of the temporal shift. We can then repeat the procedure with a time shift $2\Delta t$, then $3\Delta t$, etc., accumulating in this way a huge amount of effective noise data. With a total observing period T, we can perform $T/\Delta t$ independent shifts, and therefore the equivalent time of pure noise data is $T^2/\Delta t$. The time of flight between the two detectors is 10 ms, so one can safely take $\Delta t = 0.1$ s. With the total coincident observing time $T = 16$ days of the first analysis of the O1 run, this gives an equivalent noise background time of 608,000 years! Actually, the analysis searched for three classes of signals, with different template lengths. To account for this trial factor of 3, the result in each search can be compared with about 203,000 years of equivalent noise.

The result of this analysis is shown in Fig. 15.2. GW150914 has a value of $\hat{\rho}_c \simeq 23.6$. This is larger than for any fake coincidence found in 203,000 years of noise data. The corresponding false-alarm probability, i.e. the probability P that, just because of noise, an event that is not seen in 203,000 years of noise will appear in the 16 days of actual coincident data, is therefore bounded as $P < 16\,\text{days}/203{,}000\,\text{yr} \simeq 2 \times 10^{-7}$. The black histogram with a long tail in Fig. 15.2 gives the expected average number of events per bin due to noise, rescaling the events found in

[11] Actually, two independent data analysis pipelines have been used in the analysis of the O1 data. For definiteness, our discussion refers to the PyCBC analysis. Consistent results are obtained from the GstLAL analysis. Both analyses employ matched filtering but differ in details, e.g. in the definition of the statistic ($\hat{\rho}_c$ for PyCBC, and a likelihood ratio for the signal and noise model in GstLAL).

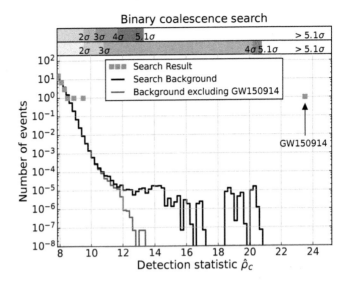

Fig. 15.2 Search results from the binary coalescence search, using only the first 16 days of coincident data of the O1 run. The histograms show the number of candidate events and the mean number of background events in the search class where GW150914 was found, as a function of the search detection statistic $\hat{\rho}_c$, and with a bin width of 0.2. The scales on the top give the significance of an event in Gaussian standard deviations based on the corresponding noise background. The tail in the background is due to random coincidences of GW150914 in one detector with noise in the other detector. The lower histogram is the background excluding those coincidences, which is used to assess the significance of the second-strongest event. From the discovery paper, [LIGO Scientific Collaboration and Virgo Collaboration], Abbott *et al.* (2016a). [Creative Commons Attribution 3.0 License.]

203,000 years of noise to the 16 days of observation. Of course, the distribution probability of $\hat{\rho}_c$ due to noise is far from Gaussian, as we see in particular from its long tail. However, since we are much used to thinking in term of standard deviations, and the standard criterion for discovery is "5σ", it is common to translate a probability into equivalent standard deviations by observing that, in a Gaussian distribution, a probability $P \simeq 2 \times 10^{-7}$ would correspond to 5.1σ. In this sense, the statistical significance of GW150914 is higher than 5.1σ, as shown in the lower of the two rows on the top of Fig. 15.2.

This analysis has then been repeated using the full data span of the O1 run. Using the final instrumental calibration had the effect of slightly decreasing $\hat{\rho}_c$, from the value 23.6 to the value 22.7. However, the coincident observation time, after data-quality cuts, now rises to $T = 46.1$ days for the PyCBC analysis (and 48.3 days for the GstLAL analysis; see Note 11). The false-alarm probability is given by the observation time T divided by the effective noise time $T^2/\Delta t$, where $\Delta t = 0.1$ s is the minimum shift used in the shifting technique. So the false-alarm probability is overall proportional to $1/T$ and for a true GW event it will therefore further decrease, when increasing the observation time.[12]

[12]However, the noises in the detectors are non-stationary, so in general one can apply the shifting procedure only as long as the detector properties at different times are comparable. Of course, an upper limit to T is given by the duration of a given science run, before the detector is upgraded.

Even in the longer time span of equivalent noise data corresponding to $T = 46.1$ days there is no fake coincidence with a value of $\hat{\rho}_c$ as large as the value $\hat{\rho}_c = 22.7$ of GW150914. Then, combining the lower value of $\hat{\rho}_c$ from the final instrumental calibration with the longer equivalent noise data, the statistical significance of GW150914 increases further, and becomes higher than 5.3σ.

It is also interesting to observe that all the events in the long tail of the noise distribution in Fig. 15.2 are due to coincidences between GW150914 in one detector, and noise in the shifted data stream of the other detector. If one eliminates GW150914 when performing the shifted coincidences, the background curve becomes the lower histogram in Fig. 15.2, which already for $\hat{\rho}_c \gtrsim 13$ has no events in 203,000 years of noise. To estimate the statistical significance of GW150914 we must include also GW150914 in the computation of the shifted coincidences, as indeed was done in the LIGO/Virgo analysis; otherwise the argument would be circular: eliminating GW150914 from the computation of the background amounts to assuming that it is a true GW event, which is rather what we want to prove. However, once an event has been firmly identified as a true GW event (by comparison with a background that was computed by including it), for the second-loudest event we can use for the background the distribution computed without including the loudest event, as we will do in Section 15.2.1 in the study of the other events of the full O1 run.

15.1.2   Properties of GW150914

After the search that resulted in its detection, the event GW150914 has been studied in greater detail, with a coherent search among the two detectors [see eq. (15.14)], a finer template sampling, and waveforms including a more complete treatment of the spins of the initial BHs. The signal entered the detector bandwidth at a frequency $f \simeq 30$ Hz, and sweept up in frequency, as expected for a coalescing binary, up to a maximum GW frequency $f_{\text{peak}} \simeq 144$ Hz, with about eight cycles of the inspiral phase in the detector bandwidth, followed by merger and ringdown. The instantaneous orbital frequency at merger, which is one-half of the GW frequency, was therefore about 72 Hz. When the signal swept through the LIGO bandwidth, the orbital velocity of the system evolved from a value $v/c \simeq 0.2$ to $v/c \simeq 0.55$ at merger. These figures already give us a feeling for how awesome is such an event. We have two BHs of about 30 solar masses each that, in the last stage of the coalescence, are rotating around each other with an instantaneous or-bital frequency corresponding to about 70 orbits per second, eventually merging at a fully relativistic speed. From the reconstruction of the full waveform, which we discuss in this section, one finds that an energy equal to $(3.0 \pm 0.5)M_\odot c^2$ was released in GWs in just a few tens of millisec-onds. The peak luminosity of the event was $3.6^{+0.5}_{-0.4} \times 10^{56}$ erg/s [compare with eqs. (14.150) and (14.151)], or equivalently $200^{+30}_{-20} M_\odot c^2$/s. This is about a factor of 10 larger than the estimated combined electromagnetic

[13]As an order of magnitude estimate, one can consider that the observable Universe contains of order 10^{11} galaxies, with on average 10^{11} stars, radiating with an average luminosity of a few times 10^{33} erg/s. This leads to an estimate for the total luminosity of order of a few times 10^{55} erg/s, see e.g. Wijers (2005).

[14]The hypothesis that the noise near the event is stationary and Gaussian can be checked on the actual data by using a time span close, but excluding, the time of the event, and also checking that the detector passes a number of environmental vetos.

luminosity of all stars and galaxies in the observable Universe.[13]

Having identified GW150914 as a true GW event, Bayesian parameter estimation can now be performed along the lines that we already discussed in Section 7.4.2. Namely, even if over the whole duration of the run the noise is highly non-Gaussian and non-stationary, when we restrict to a sufficiently short time period near the event GW150914, if the detector is running properly, the noise will be approximately stationary and Gaussian, with zero mean and a spectral density $S_n(f)$.[14] In the analysis of GW150914, $S_n(f)$ was determined using up to 1024 s of data adjacent to, but not containing, the GW signal. If the noise, at least over such a limited time interval, can be taken as Gaussian, then the Fourier components $\tilde{n}_0(f)$ of the noise have a probability distribution $\propto \exp\{-|\tilde{n}_0(f)|^2/[(1/2)S_n(f)]\}$. Therefore, as we already saw in the discussion around eq. (7.62), in a single detector the probability that the noise $n(t)$ has a value $n_0(t)$ is given by

$$p(n_0) = \mathcal{N} \exp\left\{ -\frac{1}{2} \int_{-\infty}^{\infty} df\, \frac{|\tilde{n}_0(f)|^2}{(1/2)S_n(f)} \right\}$$
$$= \mathcal{N} \exp\{-(n_0|n_0)/2\}, \tag{15.12}$$

where the scalar product $(n_0|n_0)$ has been defined in eq. (15.6). Since we have already established the presence of a GW signal, the total output of the detector is of the form $s(t) = h(t;\theta_t) + n_0(t)$, where $n_0(t)$ is the specific realization of the noise corresponding to this event, and θ_t are the (unknown) true values of the parameters θ [apart from the need to correct $h(t;\theta_t)$ for frequency-dependent calibration errors; see Note 4]. The likelihood function for the observed output $s(t)$, given the hypothesis that there is a GW signal corresponding to the parameters θ_t, with Gaussian and stationary noise at least in a short time span around the event, is obtained by plugging $n_0 = s - h(\theta_t)$ into eq. (15.12), leading to the likelihood function (7.63), which we recall here,

$$\Lambda(s|\theta_t) = \mathcal{N}\, e^{-\frac{1}{2}(s-h(\theta_t)|s-h(\theta_t))}. \tag{15.13}$$

Performing a coherent search between the two detectors amounts to defining a combined likelihood function

$$\Lambda(s|\theta_t) = \mathcal{N}\, e^{-\frac{1}{2}\sum_{k=1,2}(s_k-h_k(\theta_t)|s_k-h_k(\theta_t))}, \tag{15.14}$$

[15]Recall from eq. (7.22) that $h(t) = h_+(t)F_+(\theta,\phi) + h_\times(t)F_\times(\theta,\phi)$, where $F_{+,\times}(\theta,\phi)$ are the pattern functions of the detector. Therefore in general $h(t)$ depends on the detector in which it is observed, both through the pattern functions, and because the GW arrives at different times at different detectors.

where $k = 1, 2$ labels the detector, $s_k(t)$ is the output of the k-th detector, $h_k(t;\theta_t)$ is the GW signal in the k-th detector,[15] and we have assumed that the noises in the two detectors are uncorrelated. In the Bayesian approach, as described in Section 7.4, the posterior probability density function (PDF) is obtained by multiplying the likelihood by the priors. Parameter estimation is then performed by maximizing the posterior PDF, and the error in a parameter is obtained marginalizing the PDF over the other parameters to obtain the corresponding one-dimensional PDF (or two-dimensional PDF, if we want to study the correlation between two parameters), and looking for the 90% confidence limits. The corresponding multi-dimensional integrals were performed with numerical techniques tailored to the problem, and checked through a Markov-chain Monte Carlo, as well as nested sampling techniques. In the analysis of this event, the priors have been taken as quite uninformative, for example sources uniformly distributed in volume, and uniform priors in the masses and in the dimensionless spin parameters $\hat{a}_i \in [0, 1]$.

The final crucial element is the choice of a model for the BH–BH waveform. For a total mass larger than $4M_\odot$ two different models have been used in the analysis.[16] One is the EOB model tuned to numerical relativity simulation ("EOBNR"), discussed in Sections 14.2 and 14.3. EOB waveforms have been developed both for non-precessing spins and for double-precessing spins. However, the spin evolution is determined by a set of differential equations, which should in principle be solved for each of the millions of likelihood evaluations that are required, so fast numerical methods are necessary. At the time of the discovery, efficient numerical techniques for non-precessing spins were already available, while, for fully precessing spins, the development of computationally efficient algorithms was still under development. Thus, the analysis in the discovery paper used non-precessing EOBNR waveforms, in which for each spin only the component parallel to the angular momentum **L** is retained, leading to an 11-dimensional parameter space. Efficient numerical techniques for fully-precessing EOBNR waveforms were finalized shortly afterwards, and the analysis of the properties of GW150914 was then improved using spin-precessing EOBNR waveforms, thereby spanning the full 15-dimensional parameter space that covers all parameters of the binary system (except for the eccentricity; see Note 5 on page 281).

A second type of waveform that has been used in the analysis, called IMR-Phenom, models phenomenologically the inspiral, merger and ringdown phases so to capture the dominant spectral characteristics of the true waveform, as revealed by the PN expansion, EOB and numerical relativity. In a first approximation, the effect of the spin components parallel to the orbital angular momentum is parametrized using the "effective spin parameter"

$$\chi_{\text{eff}} \equiv \frac{c}{G}\left(\frac{\mathbf{S}_1}{m_1} + \frac{\mathbf{S}_2}{m_2}\right) \cdot \frac{\hat{\mathbf{L}}}{M}$$
$$= (1+\delta)\frac{\chi_1}{2} + (1-\delta)\frac{\chi_2}{2}, \qquad (15.15)$$

where $\delta = (m_1 - m_2)/M$, $M = m_1 + m_2$ and

$$\chi_i = \frac{c}{G}\frac{\mathbf{S}_i \cdot \hat{\mathbf{L}}}{m_i^2} \qquad (15.16)$$

is a commonly used notation for the projection of the dimensionless Kerr parameter $\hat{a}_i$ of the i-th body [defined in eq. (12.359)] onto the direction of the orbital angular momentum, $\chi_i = \hat{a}_i \cdot \hat{\mathbf{L}}$. Observe that χ_{eff} ranges from -1, when both BHs have maximal spin anti-aligned with $\hat{\mathbf{L}}$, and $+1$, for maximal spins aligned with $\hat{\mathbf{L}}$. The waveform is

[16]For $m_1 + m_2 > 4M_\odot$ the contribution to the signal in the LIGO bandwidth from the merger phase is important, while for smaller values of the total mass the signal in the bandwidth is dominated by the long inspiral phase, and is well described by the PN expansion.

particularly sensitive to χ_{eff} since this combination of spins affects the time to merger: if χ_{eff} is positive, the binary system completes more cycles before merger, starting from a given orbital distance, compared with a system with $\chi_{\mathrm{eff}} = 0$ that starts from the same distance, and, vice versa, a system with $\chi_{\mathrm{eff}} < 0$ merges within a smaller number of cycles. So, the IMR-Phenom waveforms are constructed to catch this main spin effects.

Precession has also been later included in IMR-Phenom waveforms, through further effective parameters, leading to a 13-dimensional parameter space. Comparing the results of parameter estimations from the EOBNR and the IMR-Phenom waveforms allows one to obtain an estimate of the error introduced by the waveform modeling. We will see that, for GW150914 (and also for the subsequent BH–BH detections), this error is below the statistical uncertainty (which is determined by the SNR).

Recall also from Section 4.1.4 that, for binaries at cosmological distances, with a redshift z, the actual chirp mass M_c of the system is not the quantity that is directly measured. Indeed, in the phase of the waveform, the chirp mass M_c appears together with the GW frequency in the source frame, $f_{\mathrm{gw}}^{(s)}$, in the dimensionless combinations $GM_c f_{\mathrm{gw}}^{(s)}/c^3$. What we measure is rather the GW frequency in the observer frame, $f_{\mathrm{gw}}^{(\mathrm{obs})}$, which, because of the cosmological redshift, is related to $f_{\mathrm{gw}}^{(s)}$ by

$$f_{\mathrm{gw}}^{(s)} = (1+z)f_{\mathrm{gw}}^{(\mathrm{obs})}\,, \qquad (15.17)$$

as in eq. (4.186). Then the factor $(1+z)$ can be reabsorbed into the chirp mass by introducing

$$\mathcal{M}_c \equiv (1+z)M_c\,. \qquad (15.18)$$

To higher order in the PN expansion the masses of the two initial bodies m_1 and m_2 also appear in the phase of the waveform through the symmetric mass ratio $\nu = m_1 m_2/(m_1 + m_2)^2$. Reabsorbing the factor $(1+z)$ into the chirp mass as in eq. (15.18), while keeping ν fixed is equivalent to introducing

$$m_i^{(\mathrm{obs})} = (1+z)m_i\,. \qquad (15.19)$$

The parameters $m_{1,2}^{(\mathrm{obs})}$ are the quantities that directly enter the waveforms, and which therefore are directly measured, and are called the "detector-frame masses", while $m_{1,2}$ (the "source-frame masses") are the actual masses of the stars. As we saw in Section 4.1.4, the same redefinitions can be performed in the amplitude of the waveform. In that case, after expressing $f_{\mathrm{gw}}^{(s)}$ in terms of $f_{\mathrm{gw}}^{(\mathrm{obs})}$, we can reabsorb the factors $(1+z)$ into the redefinition (15.18) of the chirp mass, while at the same time the factor $1/r$, where r is the comoving distance, becomes $1/d_L$, where d_L is the luminosity distance.

The actual masses can be reconstructed from the measured detector-frame masses if we know the redshift. From the likelihood method that

Table 15.1 Parameters of GW150914 (median values with symmetric 90% credible interval), for the two different models of the waveform, including precessing spins as discussed in the text, and using the full dataset of the O1 run. All masses given here are the actual ("source frame") masses. The dimensionless spin parameters are defined as $\hat{a}_i = cS_i/(Gm_i^2)$, where S_i is the spin of the i-th body, as in eq. (12.359). The redshift is obtained from the luminosity distance assuming a ΛCDM cosmological model with the cosmological parameters of the *Planck* 2015 data release. From [LIGO Scientific Collaboration and Virgo Collaboration], Abbott *et al.* (2016o), which improves on the analysis of the original discovery paper by using fully spin-precessing waveforms. In this and in the following tables, only a representative set of parameters is shown. See the original papers for estimates of other combination of these parameters, such as the detector-frame masses and the mass ratio, as well as the radiated energy and the peak luminosity.

Parameter	Precessing EOBNR	Precessing IMR-Phenom
Primary mass $m_1/M_\odot$	$35.6^{+4.8}_{-3.4}$	$35.3^{+5.2}_{-3.4}$
Secondary mass $m_2/M_\odot$	$30.0^{+3.3}_{-4.4}$	$29.6^{+3.3}_{-4.3}$
Chirp mass $M_c/M_\odot$	$28.3^{+1.8}_{-1.7}$	$28.1^{+1.7}_{-1.6}$
Total initial mass $M/M_\odot$	$65.6^{+4.1}_{-3.8}$	$65.0^{+4.0}_{-3.6}$
Final BH mass $M_f/M_\odot$	$62.5^{+3.7}_{-3.4}$	$62.0^{+3.7}_{-3.3}$
Primary spin $\hat{a}_1$	$0.22^{+0.43}_{-0.20}$	$0.32^{+0.53}_{-0.29}$
Secondary spin $\hat{a}_2$	$0.29^{+0.52}_{-0.27}$	$0.34^{+0.54}_{-0.31}$
Effective spin parameter $\chi_{\rm eff}$	$-0.02^{+0.14}_{-0.16}$	$-0.05^{+0.13}_{-0.15}$
Final spin $\hat{a}_f$	$0.68^{+0.05}_{-0.05}$	$0.68^{+0.06}_{-0.06}$
Luminosity distance d_L (Mpc)	440^{+160}_{-180}	440^{+150}_{-180}
Redshift z	$0.094^{+0.032}_{-0.037}$	$0.093^{+0.029}_{-0.036}$

we have discussed we can extract the luminosity distance, from which the redshift follows, once we assume a cosmological model. This has been taken to be ΛCDM, with the cosmological parameters of the *Planck* 2015 data release.

The results of the parameter estimation for GW150914 are shown in Table 15.1, for some of the parameters of the system. Observe that the chirp mass is the combination of masses determined with the highest accuracy, as expected from the fact that it is the combination that enters to the lowest order in the PN expansion. Among the dimensionless spin parameters, the initial spin parameters $\hat{a}_1$ and $\hat{a}_2$ are not accurately determined, since they only enter in the waveform at relatively high PN order. In contrast, the final spin is determined by angular momentum conservation, as the initial total angular momentum minus the angular momentum carried away by the GWs. The initial total angular momentum is dominated by the orbital part, so the uncertainty in the initial spins is not very important, and the angular momentum carried away by the GWs is automatically determined by the template. Thus, the spin of the final BH is known much more accurately that the initial spins. Observe also that the error on the luminosity distance of the source is quite large, since the distance d_L has an important degeneracy with the inclination angle ι of the orbit, compare with eqs. (4.189), (4.191) and (4.192). However, with more detectors which are not co-aligned one would in principle be able to measure separately h_+ and $h_\times$ and one could then disentangle the luminosity distance from the inclination of the orbit; see the discussion on page 199 of Vol. 1.[17]

Comparing the difference between the values in the two columns labeled EOBNR and IMR-Phenom in Table 15.1 we see that, for all the parameters, the differences due to the theoretical modeling of the waveform is smaller than the statistical uncertainty. For GW150914, the main source of statistical uncertainty is given by the finite value of the SNR and the limited number of inspiral cycles in the bandwidth (while the calibration error degrades primarily the accuracy in the source localization). It should also be remarked that the follow-up analysis discussed in this section yields an optimal coherent SNR $\hat{\rho}_c = 25.1$, higher than the value $\hat{\rho}_c = 23.6$ found in the search analysis discussed in Section 15.1.1, because of the finer sampling of a larger parameter space.

15.2   Further BH–BH detections

15.2.1   GW151226

A second event with a statistical significance higher than 5σ was observed by the two LIGO interferometers, still during the O1 run, on December 26, 2015, and was announced by the LIGO/Virgo collaboration on June 15, 2016. This event is quite different from GW150914. Its combined SNR is $\hat{\rho}_c \simeq 13.0$. Even if this is quite a bit smaller than the value $\hat{\rho}_c = 22.7$ of GW150914, the event still goes above the 5σ threshold. Indeed, having identified GW150914 as a GW event, we can now remove it from the estimate of the background obtained by performing the shifted coincidences. As we saw in Fig. 15.2, this drastically reduces the estimate of the background at $\hat{\rho}_c \gtrsim 12$, since all the shifted coincidences at higher values of $\hat{\rho}_c$ are due to random coincidences between GW150914 in one detector and noise in the other detector. Thus, the statistical significance of GW151226 has to be assessed against a much smaller background of accidental coincidences. The result is shown in Fig. 15.3. In the distribution obtained by eliminating GW150914 there is no random coincidence with a value of $\hat{\rho}_c$ equal to or higher than GW151226, so again we can only put a upper limit on the false-alarm probability, which in terms of equivalent Gaussian standard deviations turns out to correspond to a significance higher than 5.3σ.

Its reconstructed waveform is shown in Fig. 15.4. The inferred initial BH masses are of order $m_1 \simeq 14 M_\odot$ and $m_2 \simeq 7.5 M_\odot$, while its luminosity distance is $d_L = O(440)$ Mpc (see Table 15.2 for precise numerical values and error estimates). Thus, the initial BH masses are significantly

[17]We will see in Sections 15.2.4 and 15.3.1 how the localization and the estimate of d_L improve with three-detector observations performed by the two LIGO interferometer and the Virgo interferometer.

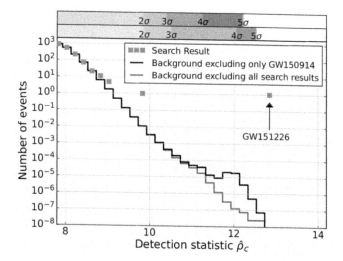

Fig. 15.3 Search result using the full O1 dataset. The event GW150914 is removed since it had already been confirmed as a real GW signal, and is not included in the computation of the background, which is given by the upper black solid line. The lower solid line is the background computed excluding not only GW150914, but also all candidate events that are shown as square markers. From [LIGO Scientific Collaboration and Virgo Collaboration], Abbott *et al.* (2016m).

smaller than in GW150914, while the luminosity distance is quite similar. As a result, the amplitude of the signal is smaller, and the event could only be extracted from the noise using matched filtering. In contrast, as we saw in Fig. 15.1, GW150914 was visible by eye over the noise, just by using a 35–350 Hz bandpass filter. Indeed, GW151226 was not detected in the generic search for transient events that initially detected GW150914, and was identified within 70 s by an online matched-filter search. The initial identification was then confirmed by the two independent off-line matched-filter analysis, the PyCBC analysis that we have discussed in detail in Section 15.1.1 and the GstLAL analysis mentioned in Note 11. Again, for definiteness, we will give here the results of the PyCBC analysis. The GstLAL analysis gives consistent results.

As we see from eq. (14.106), or from its finite-ν EOB improvement such as eq. (14.108), the frequency at which the inspiral phase ends is inversely proportional to the total mass of the system. Therefore, for GW151226 the inspiral phase lasts up to a much higher frequency compared with GW150914, and there are many more inspiral cycles in the detector bandwidth. Indeed, for the total mass corresponding to GW151226, eq. (14.108) (or rather its generalization for spinning binaries) shows that the GW frequency at the ISCO is $f \simeq 100$ Hz. The signal stayed in the detectors' bandwidth for approximately 55 cycles, which were swept in about 1 s. The contribution to the SNR from the early inspiral phase (about 45 cycles from 35 to 100 Hz) is about the same as the contribution from the late inspiral, plunge and merger (about 10

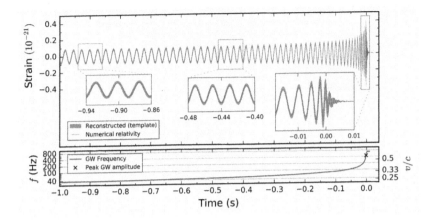

Fig. 15.4 The reconstructed waveform of GW151226. From [LIGO Scientific Collaboration and Virgo Collaboration], Abbott *et al.* (2016m).

cycles from 100 Hz to the frequency 450 Hz where the amplitude reaches its maximum). The signal then swept the detector bandwidth up to almost 800 Hz. This should be compared with GW150914, where the SNR was dominated by the last 10 cycles of inspiral and merger. Thus, the two events are quite complementary. The long inspiral phase observed for GW151226 allows us to test more accurately the predictions of the post-Newtonian expansion, and also provides a more accurate determination of the chirp mass, with an error of about 3.5% (see Table 15.2) compared with the error over 6% in GW150914. On the other hand, because of its smaller amplitude, GW151226 gives very little information on the merger and ringdown phases, compared with GW150914. For the same reason the individual masses of the initial BHs, which are determined by combining the chirp mass with the measurement of the symmetric mass ratio ν obtained from the late inspiral and merger phases, are determined less accurately, as we see again from Table 15.2. The posterior distribution is not consistent, at 99% c.l., with initial masses smaller than $4.5M_\odot$. This is well above the maximum possible value of the mass of a NS, even for the most extreme equations of state, and leaves BHs as the only known possibility. Quite interestingly, the posterior distribution for waveforms with both precessing and non-precessing spins shows that, again at 99% c.l., at least one of the initial BHs had a dimensionless spin parameter greater than 0.2. Equivalently, we see that the effective spin parameter $\chi_{\rm eff}$ is non-zero, with high significance.

The total energy radiated in GWs scales with the total mass of the system, so for GW151226 it is smaller than for GW150914, and has the value $E_{\rm rad} = 1.0^{+0.1}_{-0.2} M_\odot c^2$. However, the typical time-scale over which most of the energy is radiated, during the final merger, is determined by the Schwarzschild radius of the system, so again it scales with the mass, which means that the peak luminosity is basically independent

Table 15.2 Some selected parameters of GW151226 (median values with symmetric 90% credible interval), for spin-aligned EOBNR waveforms and precessing IMR-Phenom waveforms. From Table IV of [LIGO Scientific Collaboration and Virgo Collaboration], Abbott *et al.* (2016n).

Parameter	EOBNR	Precessing IMR-Phenom
Primary mass $m_1/M_\odot$	$14.0^{+10.0}_{-3.5}$	$14.5^{+6.6}_{-3.7}$
Secondary mass $m_2/M_\odot$	$7.5^{+2.3}_{-2.6}$	$7.4^{+2.3}_{-2.0}$
Chirp mass $M_c/M_\odot$	$8.87^{+0.35}_{-0.28}$	$8.90^{+0.31}_{-0.27}$
Total initial mass $M/M_\odot$	$21.6^{+7.4}_{-1.6}$	$21.9^{+4.7}_{-1.7}$
Final BH mass $M_f/M_\odot$	$20.6^{+7.6}_{-1.6}$	$20.9^{+4.8}_{-1.8}$
Primary spin $\hat{a}_1$	$0.42^{+0.35}_{-0.37}$	$0.55^{+0.35}_{-0.42}$
Secondary spin $\hat{a}_2$	$0.51^{+0.44}_{-0.46}$	$0.52^{+0.42}_{-0.47}$
Effective spin parameter $\chi_{\rm eff}$	$0.21^{+0.24}_{-0.11}$	$0.22^{+0.15}_{-0.08}$
Final spin $\hat{a}_f$	$0.73^{+0.05}_{-0.06}$	$0.75^{+0.07}_{-0.05}$
Luminosity distance d_L (Mpc)	450^{+180}_{-210}	440^{+170}_{-180}
Redshift z	$0.096^{+0.035}_{-0.042}$	$0.092^{+0.033}_{-0.037}$

of the mass of the system. Indeed, for GW151226 it is found to be $3.3^{+0.8}_{-1.6} \times 10^{56}$ erg/s, about the same as for GW150914, in agreement with eqs. (14.150) and (14.151).

We see from Fig. 15.3 that there is also a third interesting event (which was already present in the analysis of the first 16 days of effective coincident time, shown in Fig. 15.2). This candidate event, referred to as LVT151012,[18] has a value of $\hat{\rho}_c \simeq 9.6$. Its false-alarm rate can be assessed by comparing it against the background obtained by excluding, in the shifted coincidence, the two loudest events, GW150914 and GW151226, which are confirmed detections.[19] The resulting false-alarm rate is of one per 2.7 years, corresponding, in terms of equivalent Gaussian standard deviations, to a significance of 1.7σ in the PyCBC analysis and 2.0σ in the GstLAL analysis. This is not enough to claim a detection, although the event is sufficiently interesting to deserve a follow-up. The matched filtering analysis shows that this candidate event, if it is of astrophysical origin, is again a BH–BH binary, with masses 23^{+18}_{-6} and $13^{+4}_{-5}M_\odot$, at a distance 1000 ± 500 Mpc.

[18]The acronym LVT stands for "LIGO/Virgo trigger".

[19]Actually, for the value of $\hat{\rho}_c = 9.6$ corresponding to LVT151012, in the PyCBC analysis the estimate of the significance is essentially unaffected by the removal of the backgrounds associated to GW150914 and GW151226, while in the GstLAL analysis the removal of GW150914 changes somewhat the significance but the removal of GW151226 has little effect.

15.2.2   GW170104

After further commissioning resulting in improvements in the sensitivity, as well as in the duty factor and data quality, advanced LIGO resumed data-taking on November 30, 2016, with the O2 run. On January 4,

Table 15.3 Some selected parameters of GW170104, averaging over the results of two waveform models. From Table I of [LIGO Scientific Collaboration and Virgo Collaboration], Abbott *et al.* (2017b). Notation as in Table 15.1.

m_1	$31.2^{+8.4}_{-6.0}\ M_\odot$
m_2	$19.4^{+5.3}_{-5.9}\ M_\odot$
M_c	$21.1^{+2.4}_{-2.7}\ M_\odot$
M	$50.7^{+5.9}_{-5.0}\ M_\odot$
M_f	$48.7^{+5.7}_{-4.6}\ M_\odot$
$\chi_{\rm eff}$	$-0.12^{+0.21}_{-0.30}$
$\hat{a}_f$	$0.64^{+0.09}_{-0.20}$
d_L	880^{+450}_{-390} Mpc
z	$0.18^{+0.08}_{-0.07}$

2017, the signal from another BH–BH binary coalescence, GW170104, was detected in the two LIGO interferometers. The signal arrived at Hanford about 3 ms before Livingston, and so, well within the coincidence window of 15 ms, which takes into account the time of flight of 10 ms between the two detectors and the uncertainty in identifying the arrival time for weak signals. GW170104 was detected with a combined SNR $\hat{\rho}_c \simeq 13$. Using 5.5 days of coincident data collected after the event, the false-alarm rate is calculated as less than 1 event in 70,000 years of coincident observing time. GW170104 was also identified in a search that does not rely on detailed modeling of the waveform, and only looks for transient signals where the frequency rises over time. Of course, the increased generality of this search comes at a cost of a lower statistical significance, but still for GW170104 this analysis gives a false-alarm rate of about 1 in 20,000 years.

The source parameters, estimated from a coherent Bayesian analysis of the data from both detectors, are given in Table 15.3. The quoted results are obtained by averaging over two waveform models that differ in the treatment of spins: a model where inspiral spin precession is treated using an effective spin-precession parameter and a model with fully precessing spins. The two models give consistent results. We see that the system had a total initial mass slightly above $50M_\odot$, and the inferred median value of the luminosity distance is 880 Mpc. As we already discussed for GW150914, the error in the reconstruction of the luminosity distance is quite large. However, within this uncertainty, GW170104 is most likely at a greater distance than GW150914.

15.2.3 GW170608

A fourth BH–BH coalescence was detected on June 8, 2017 (and was announced in November 2017, actually after the announcements of the subsequent discoveries of the triple LIGO–LIGO–Virgo detection GW170814 and of the NS–NS detection GW170817, which we will examine below). The signal was first identified as an event with SNR $\simeq 9$ in the LIGO-Livingston detector from a low-latency matched-filter analysis. At the time of arrival of the event the LIGO-Hanford detector was undergoing a standard routine of noise minimization. In general, such times are not included in the search; however, it was found that the strain sensitivity above 30 Hz was not affected by this procedure, so the LIGO-Hanford data above this frequency could be used to check for a coincidence, as well as in the subsequent data analysis. The advanced Virgo interferometer was progressing toward the sensitivity that allowed it to later join the O2 run, on August 1; at the time of this event it was in observation mode, but its horizon for a signal such as GW170608 was 60–70 Mpc. This was not sufficient to detect this event, which has a luminosity distance $d_L = 340 \pm 140$ Mpc; see Table 15.4.

The combined analysis of the two LIGO detectors gave a network SNR of about 13. The background was estimated with the standard shift technique discussed in Section 15.1.1, using 2 days of data around the event. After removing periods of enhanced noise, 1.2 days of coincident data were retained, and used in two different pipelines. One gave a false-alarm rate of 1 event in 160,000 yr and the other provided an upper limit of less than 1 event in 3,000 yr. The event was also detected in the search for generic transients, which does not make use of the explicit waveforms. In this case the significance is necessarily lower, given the generic form of the signals searched, and in this search the false-alarm rate is 1 in ~ 30 yr.

The result of Bayesian parameter estimation using the two LIGO detectors is shown in Table 15.4. The values of the initial masses are particularly interesting. They are way too high to be consistent with NS, so this event is again interpreted as a BH–BH coalescence. However, they are significantly smaller than the masses of all other detected BH–BH events (except for GW151226, where, as we have seen, the estimated initial masses are 14.0 and $7.5 M_\odot$). The initial masses in GW170608 are rather of the order of the masses of the BHs observed in X-ray binaries, as we will discuss in Section 15.2.5. Thus this event, together with GW151226, provides a bridge between the population of BHs observed in the other BH–BH coalescences and the population of lighter BHs observed in X-ray binaries. The event was localized in an area of about $520 \deg^2$.

Table 15.4 Some selected parameters of GW170608. From [LIGO Scientific Collaboration and Virgo Collaboration], Abbott *et al.* (2017e). Notation as in Table 15.1.

m_1	$12^{+7}_{-2}\ M_\odot$
m_2	$7^{+2}_{-2}\ M_\odot$
M_c	$7.9^{+0.2}_{-0.2}\ M_\odot$
M	$19^{+5}_{-1}\ M_\odot$
M_f	$18.0^{+4.8}_{-0.9}\ M_\odot$
χ_{eff}	$0.07^{+0.23}_{-0.09}$
$\hat{a}_f$	$0.69^{+0.04}_{-0.05}$
d_L	340^{+140}_{-140} Mpc
z	$0.07^{+0.03}_{-0.03}$

15.2.4 GW170814: the first three-detector observation

Advanced Virgo joined the two advanced LIGO detectors in the search for GWs on August 1, 2017, and already on August 14 a BH–BH coalescence was detected in all three interferometers. The signal hit first the LIGO-Livingston interferometer and then swept through the LIGO-Hanford and Virgo detectors with delays of about 8 and 14 ms, respectively. The strain sensitivities of the detectors around the time of observation are shown in Fig. 15.5. Figure 15.6 shows the signal detected in the three interferometers. The single-detector SNRs in Hanford, Livingston and Virgo are 7.3, 13.7 and 4.4, respectively.

The SNR in the Virgo detector is smaller, given its higher strain sensitivity, so one should investigate the extent to which the detection in Virgo is statistically significant. Using only the two LIGO detectors, the event has a network SNR of 15. The corresponding false-alarm rate is 1 in 140,000 yr in a pipeline and 1 in 27,000 yr in a different pipeline; in both cases, the event is clearly identified as a GW signal. Using the best-fit waveforms obtained from the LIGO detectors only, and the noise realization in 5000 s of data around the event, one finds that the probability of a peak in the Virgo data as large as that observed, and within a time window determined by the maximum time of flight, is 0.3%. Furthermore, comparing the matched-filter marginalized likelihood for a model with a coherent BH–BH signal in all three detectors with that for a model that assumes pure Gaussian noise in Virgo and a BH–BH signal in LIGO, one finds that the model with a three-detector signal

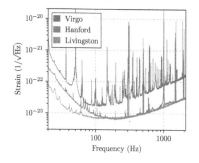

Fig. 15.5 Strain sensitivity of the LIGO-Livingston (lower curve), LIGO-Hanford (middle curve) and Virgo (upper curve) advanced interferometers, estimated using 4096 s of data around the time of GW170814. From [LIGO Scientific Collaboration and Virgo Collaboration], Abbott *et al.* (2017c). [Creative Commons Attribution 4.0 International license].

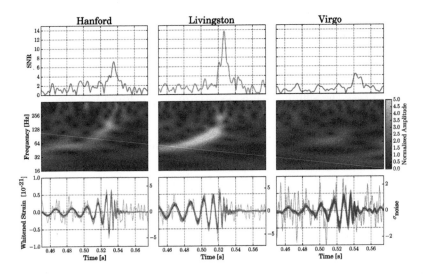

Fig. 15.6 The GW event GW170814 observed by LIGO-Hanford, LIGO-Livingston and Virgo. *Top row*: SNR time series produced in low-latency pipelines on August 14, 2017. The time series were produced by time-shifting the best-match template from the online analysis and computing the integrated SNR at each point in time. *Second row*: Time–frequency representation of the strain data around the time of GW170814. *Bottom row*: Time-domain detector data, and 90% confidence intervals for reconstructed waveforms, whitened by each instrument's noise amplitude spectral density between 20 Hz and 1024 Hz. The whitening emphasizes different frequency bands for each detector, which is why the reconstructed waveform amplitude evolution looks different in each column. From [LIGO Scientific Collaboration and Virgo Collaboration], Abbott *et al.* (2017c). [Creative Commons Attribution 4.0 International license].

Table 15.5 Some selected parameters of GW170814, averaging over the results of two waveform models. From Table I of [LIGO Scientific Collaboration and Virgo Collaboration], Abbott *et al.* (2017c). Notation as in Table 15.1.

m_1	$30.5^{+5.7}_{-3.0} \ M_\odot$
m_2	$25.3^{+2.8}_{-4.2} \ M_\odot$
M_c	$24.1^{+1.4}_{-1.1} \ M_\odot$
M	$55.9^{+3.4}_{-2.7} \ M_\odot$
M_f	$53.2^{+3.2}_{-2.5} \ M_\odot$
$\chi_{\rm eff}$	$0.06^{+0.12}_{-0.12}$
$\hat{a}_f$	$0.70^{+0.07}_{-0.05}$
d_L	540^{+130}_{-210} Mpc
z	$0.11^{+0.03}_{-0.04}$

is preferred with a Bayes factor of more than 1600. The Bayes factor B_{ij} is conventionally translated into a statement about the evidence of model i with respect to model j by using the Jeffreys' scale. Model i is favored with respect to model j if $B_{ij} > 1$. For $1 < B_{ij} < 3$ the evidence is deemed "weak", for $3 < B_{ij} < 20$ it is "definite", for $20 < B_{ij} < 150$ it is "strong" and for $B_{ij} > 150$ it is "very strong". Thus, one can convincingly claim detection also in the Virgo interferometer. The event was also identified in a coherent analysis searching for generic transients whose frequency increases with time. In this case, the combined analysis of the two LIGO detectors gives a false-alarm rate of about 1 event in 300 yr, while by adding Virgo one obtains a bound on the false-alarm rate of less than 1 in 5900 yr.

The parameters of the source are inferred through a Bayesian analysis performed coherently for the three-detector network [cf. eq. (15.14), where now $k = 1, 2, 3$], and are shown in Table 15.5.

The addition of a third detector to the network brings several advantages. First of all, the statistical significance of the event is enhanced. In particular, despite its lower sensitivity, adding Virgo improves the false-alarm probability by about an order of magnitude, compared with the

LIGO–LIGO network. A second important aspect is the sky localization of the event. For instance, with the two LIGO detectors the position of the first BH–BH detection, GW150914, could only be localized to an accuracy of about $600\,\mathrm{deg}^2$ (see the Further Reading), using time-of-flight differences as well as requiring consistency in the amplitude and phase of the signal at the two detectors. With three detectors a much more accurate triangulation is possible, employing the various time differences and amplitude ratios among the different detectors. This can reduce the uncertainty in the localization by an order of magnitude or more. Indeed, for GW170814 a first rapid localization, used to give the information to the various telescopes searching for an electromagnetic counterpart, gave a region of the sky of about $1160\,\mathrm{deg}^2$ (at 90% c.l.) when only the two LIGO detectors were included, which shrunk to about $100\,\mathrm{deg}^2$ on inclusion of the Virgo data. The final localization from the full parameter estimation reduced this further to a region of about $60\,\mathrm{deg}^2$. The detection in a three-detector network also allows us to perform tests of the absence of extra polarizations, compared with the two polarizations predicted by GR, as we will discuss in Section 15.4.

15.2.5   The population of BH–BH binaries

The GW events GW150914, GW151226, GW170104, GW170608 and GW170814 provide the first observations of BH–BH binaries. As we saw in Section 6.3, five confirmed NS–NS binaries are known from electromagnetic observations (plus at least three more systems whose identification with a NS–NS binary is quite likely, and the NS–NS coalescence detected through GWs, GW170817, that we will discuss in Section 15.3). In contrast, before the detection of GW150914, no BH–BH system was known. Indeed a BH–BH binary system, unless it has significant surrounding material in the form of accretion disks (which in the final stage of the coalescence is not expected, at least according to the standard lore), does not have significant electromagnetic emission. The detection of such systems through their GW emission is therefore a striking example of the fact that GW astronomy is opening a new window on the Universe, quite different from the electromagnetic window.

The large value of the masses of the two initial BHs in these events, especially in GW150914, GW170104 and GW170814, is also very interesting, and came somewhat as a surprise. Indeed, two populations of BHs are observationally known: stellar-mass BHs, which form as the remnants of stellar explosions, and supermassive BHs at the centers of galaxies, with masses of the order of millions to billions of solar masses, and that will be discussed in Chapter 16. Before the detection of GW150914, confirmed stellar-mass BHs had only been observed as X-ray binaries, in which the BH is in a binary system with a normal star and accretes matter from its companion, resulting in the emission of X rays. At present, 22 BHs in X-ray binaries have a reliable mass determination. All of them have masses smaller than $20 M_\odot$, with most being in the range $(5-10) M_\odot$. In contrast, in GW150914 we have observed for the first time two BHs with masses of about $30M_\odot$, and we have witnessed the birth of a BH with a mass over $60M_\odot$. The subsequent detections of GW170104 and GW170814 have further confirmed the existence of a population of BHs in binary systems with masses in the range $(20 - 30)M_\odot$, whose coalescence gives rise to BHs with masses $O(50 - 60)M_\odot$.

The initial BHs must have resulted from the collapse of their progenitor stars (unless they were primordial BHs, created by fluctuations in the early Universe). In that case, we see that stellar collapse can produce BHs with a mass of about $30M_\odot$. As we discussed in Section 10.3.3, the relation between the mass of the progenitor star and the type of remnant of the collapse (whether a BH or a NS, or no remnant at all), as well as the relation to the remnant mass, is highly non-trivial, and is influenced by several factors. In any case, the progenitor stars of GW150914, GW170104 and GW170814 must have been very massive to produce such massive BH remnants.

As we discussed in Section 10.3.3, stars lose mass by stellar winds, whose strength increases with the star's metallicity Z. This is due to the fact that a higher content of "metals" (in the sense of elements heavier than hydrogen and helium in the chemical composition of the star; see the Complement Section 10.6) increases the scattering cross-section of photons. This results in a larger momentum transfer from the radiation field to matter, and hence in larger stellar winds. In particular, as we discussed on page 33, for progenitor stars in the range $(30 - 40)M_\odot \lesssim M_{\text{prog}} \lesssim 95M_\odot$ the mass losses due to stellar winds provide the main theoretical uncertainty for determining the type of remnant and its mass, and indeed stellar winds can be so strong that, despite a very high progenitor mass, the remnant can be a NS. For GW150914, GW170104 and GW170814, the fact that the progenitor stars collapsed to such massive BHs, implies that they had low metallicities. Calculations of the remnant mass as a function of the metallicity of the progenitors are difficult, but no calculation finds BHs heavier than about $30M_\odot$ unless the metallicity is lower than the solar metallicity $Z_\odot$. These computations (including also the effect of stellar rotation) suggest that the progenitor system of these systems formed in an environment with $Z \lesssim (1/2)Z_\odot \simeq 0.008$, and possibly $Z \lesssim (1/4)Z_\odot$.

The existing BH–BH detections also provide the first evidence that binary BH systems form and can merge within the time-scale of the Universe at a detectable rate. This is non-trivial information. In general (again, if the BHs are of astrophysical rather than cosmological origin, as we assume in the following discussion), two types of formation mechanism for BH–BH binaries have been identified. The first, and most obvious, is that the progenitor was an isolated binary system, in which both progenitor stars evolved to BHs. The second possibility is a dynamical formation channel, in which a binary BH–BH system forms in a dense stellar environment, such as a cluster of massive stars, because of dynamical interaction between pre-existing isolated BHs. In particular, in a dense star cluster a massive BH would sink through dynamical friction to the center of the cluster. In such a dense environment it can interact with another BH and form a binary system, which then is often dynamically ejected by the cluster by further three-body interactions. For both formation channels, several mechanisms have been identified in the literature that could have strongly suppressed the formation of BH–BH binaries that merge at a detectable rate. For instance, in an isolated binary system, after the first star explodes and becomes a BH (or, depending on its mass, directly implodes to a BH), if the system is not disrupted then we remain with a BH in a binary system with a normal star. When, at the end of its main-sequence life, the second star begins to evolve, it has an extended envelope and the binary system can enter a phase of common-envelope evolution, in which the BH moves through the envelope of its companion. Some studies found that the majority of BH–star systems would merge during this common-envelope phase, because of friction in the envelope, thereby aborting the formation of a BH–BH binary. If the BHs in the observed coalescences are of astrophysical origin, rather than primordial BHs, models in which the binary system in general does not survive the common-envelope evolution, leading to very reduced predictions for the BH–BH coalescence rate, are ruled out.

Large natal kicks also pose a threat to the formation of close binary BH systems. In the scenario of isolated binary progenitors large natal kicks can disrupt the binary system, or at least widen it so much that it can no longer merge by GW emission within a Hubble time. Similarly, in the dynamical formation scenario in dense clusters, large natal kicks can eject the BH from the cluster before it has had time to capture a BH companion to form a BH–BH binary. Thus, the observation of these BH–BH events shows that natal kicks cannot always be larger than about 100 km/s.

From the current BH–BH observations one can estimate that the rate of BH–BH coalescences is $O(10 - 200)\,\mathrm{Gpc}^{-3}\,\mathrm{yr}^{-1}$. By comparison, before these detections, typical predictions for the rate, based on population synthesis models and on models of binary star evolution, were in the range $O(0.1 - 300)\,\mathrm{Gpc}^{-3}\,\mathrm{yr}^{-1}$. Thus, the actual rate turns out to be possibly quite close to the maximum value that was predicted by the theoretical models.

15.3   GW170817: the first NS–NS binary

15.3.1   GW observation

As we have discussed, the advanced Virgo interferometer joined the two advanced LIGO detectors near the end of the O2 run, on August 1, 2017, and the first three-detector observation of a BH–BH coalescence took place already on August 14. A few days later, on August 17, another major milestone in the history of GW research took place, when the three interferometers observed the signal from the coalescence of a NS–NS binary. Furthermore, the signal was associated with a short γ-ray burst (GRB) that was observed independently by Fermi-GBM and then confirmed by INTEGRAL. The subsequent identification of the transient electromagnetic counterpart by a large number of telescopes and its follow-up across the whole electromagnetic spectrum strengthened the association with a NS–NS merger. This remarkable event then marked the opening of the era of multi-messenger astrophysics.

The gravitational signal was first identified in the LIGO-Hanford detector, by a low-latency search using matched-filtering against post-Newtonian waveforms. The event did not generate an on-line trigger in the LIGO-Livingston detector, because of the temporal proximity of an instrumental disturbance, which could be safely removed in the off-line analysis. At the time of the event the horizon distance (i.e. the maximum distance at which a signal could be detected with SNR = 8, assuming optimal orientation of the source with respect to the detector) for a signal of this type was 218 Mpc for LIGO-Livingston, 107 Mpc for LIGO-Hanford and 58 Mpc for Virgo. The source turned out to be at about 40 Mpc (see Table 15.6), and so within the horizon distance of the three detectors, and the separate SNRs in the LIGO-Livingston, LIGO-Hanford and Virgo interferometers were 26.4, 18.8 and 2.0, respectively. The combined SNR across the three detectors was 32.4, making it the loudest signal yet observed. Apart from the higher sensitivity of LIGO, the lower value of the SNR in the Virgo detector is due to the fact that the signal arrived from close to a blind direction of Virgo; recall that GW interferometers in fact have an angular sensitivity, described by the pattern functions computed in Section 7.2. These pattern functions are quite broad, making GW interferometers almost 4π detectors, but still with blind directions. Thus the low value of the SNR in Virgo, with respect to its horizon distance for a signal of this type, actually contained important information that allowed a significantly better localization of the event. On the other hand, this also meant that the Virgo data did not contribute significantly to the estimation of the other parameters of the source.

Using 5.9 days of data near the event to estimate the background of accidental coincidences, the estimated false-alarm rate was found to be 1 in 1.1×10^6 yr in a pipeline, while another pipeline gave only an upper bound, of less than 1 in 8.0×10^4 yr, consistent with the former false-alarm estimate. This unambiguously identifies the event as a detection. Only the LIGO data were used to assess the statistical significance.

The signal swept through the interferometers' bandwidth for about 100 s, entering the detectors' bandwidth at a frequency of about 30 Hz, and is clearly visible by eye up to about 400 Hz in a time–frequency representation obtained by combining coherently the data from the two LIGO interferometers; see the bottom panel of Fig. 15.7 on page 303. The combination of LIGO and Virgo data allowed a precise localization of the source, within a region of 28 deg^2 (90% credibility region), while a first rapid localization using only the two LIGO detectors gave an area of 190 deg^2. Some reconstructed parameters of the source, using only the GW observation, are shown in Table 15.6. We see that the initial masses are in the typical range expected for NS. These are significantly smaller than the masses of known stellar-mass BHs, pointing to a NS–NS interpretation, in contrast to a NS–BH or a BH–BH binary. The presence of an electromagnetic counterpart, which we will discuss later, and which is not expected for a BH–BH binary, further favors the NS–NS interpretation, although these data cannot definitely exclude exotic objects more compact than a NS, or a NS–BH binary. We will see in Section 15.3.4 that the presence of lanthanide-poor ejecta with $v/c \simeq 0.3$ also favors a NS–NS interpretation.

The large number of inspiral cycles allows a very accurate determination of the chirp mass, even compared with GW151226, which previously was the event with the longest observed inspiral phase; compare with the relative error on M_c given in Table 15.2 on page 293. We also see that the source was at a relatively close distance, of order 40 Mpc. The energy radiated in GWs during the inspiral is found to be $0.025 M_\odot c^2$.

As we discussed in Section 14.4.1, in the inspiral phase of a NS–NS binary there are also tidal effects, which give information of the NS EoS. Adding to the waveform the tidal 5PN contribution (14.231), one can express the result in terms of the tidal deformability parameter $\tilde{\Lambda}$ defined in eq. (14.230). From Bayesian parameter estimation of GW170817 one finds the upper limit $\tilde{\Lambda} < 800$ (at 90% c.l.), assuming a low-spin prior $|\chi_z| \leqslant 0.05$ (where $\chi_z \equiv \hat{a}_z$) for each of the initial NS, as suggested by the discussion on page 269. This bound is already significant, and excludes EoS that give less compact NS.

In Table 15.6 we also show the values obtained for the effective spin parameter χ_{eff} defined in eq. (15.15). The low interval of values obtained, $\chi_{\text{eff}} \in (-0.01, 0.02)$, is consistent with the imposition of the low-spin prior. Assuming a prior $|\chi_z| \leqslant 0.89$ on the initial spins one obtains $\chi_{\text{eff}} \in (-0.01, 0.17)$, which is still quite stringent, and again consistent with a NS–NS interpretation, according to the discussion of NS spin on page 269.

The above results are based only on the post-Newtonian modelization of the signal. With the sensitivities of advanced LIGO/Virgo at the time of detection (which were still above the final design sensitivity), the post-merger signal of this event was not detectable.

15.3.2   The prompt γ-ray burst

Long and short GRBs

Gamma-ray bursts are classified into short GRBs, with typical duration below 2 s, and long GRBs, with a distribution peaked at about 30 s. On average, short bursts are made of photons of higher energy (i.e., harder) than long bursts. The separation of the observed GRBs into two populations, short-hard GRBs and soft-long GRBs, underlines a difference in their physical origin. It is now well established that long GRBs are produced in the core collapse of massive stars. A major breakthrough in the understanding of the origin of long GRBs came from the discovery by the BeppoSAX satellite, in 1997, that long GRBs are followed by

Table 15.6 Some selected parameters of GW170817, using only GW observations, from Table I of [LIGO Scientific Collaboration and Virgo Collaboration], Abbott *et al.* (2017d). Notation as in Table 15.1. These values assume a low-spin prior $|\chi_z| \leqslant 0.05$ (where $\chi_z \equiv \hat{a}_z \equiv \hat{\mathbf{a}} \cdot \mathbf{L}$) on the dimensionless spin parameters of the two component bodies; see the original paper for more parameters, and for the results with a high-spin prior $|\chi_z| \leqslant 0.89$ (a limit imposed by the availability of rapid waveform models). In particular, with the high-spin prior, the interval for the component masses is enlarged to $m_1 = (1.36 - 2.26) M_\odot$ and $m_2 = (0.86 - 1.36) M_\odot$. As we will see in Section 15.3.3, the redshift was later determined more accurately with the observation of the electromagnetic counterpart, leading to $z = 0.009680 \pm 0.00079$ and (assuming a local Hubble constant $H_0 = 73.24 \pm 1.74 \, \text{km s}^{-1} \, \text{Mpc}^{-1}$), $d_L = 40.4 \pm 3.4$ Mpc.

m_1	$(1.36 - 1.60) \, M_\odot$
m_2	$(1.17 - 1.36) \, M_\odot$
M_c	$1.188^{+0.004}_{-0.002} \, M_\odot$
M	$2.74^{+0.04}_{-0.01} \, M_\odot$
χ_{eff}	$(-0.01, 0.02)$
d_L	40^{+8}_{-14} Mpc
z	$0.008^{+0.002}_{-0.003}$

X-ray emission that can last for hours or days. This emission at longer wavelength, which lasts for a much longer period compared with the prompt GRB, is called the "afterglow". It allowed the detection of optical and radio counterparts and accurate localization of the source with sub-arcsecond accuracy. It has then been possible to identify the host galaxies of GRBs, proving that long GRBs are at cosmological distances, and even to identify the spectrum of Type Ic SNe in the afterglow. The association of long GRBs with Type Ic SNe, the fact that they are exclusively observed in star-forming galaxies, and their strong spatial correlation with bright star-forming regions within their host galaxies, convincingly show that long GRBs are generated in the core collapse of massive stars.

By contrast, understanding the origin of short GRBs remained more challenging. Again a breakthrough came from the detection of afterglows, first observed in 2005 by the *Swift* and HETE-2 satellites. This allowed it to be proved that short GRBs are also at cosmological distances. Concerning their origin, a first important piece of evidence is that in the existing sample of short GRBs there are several events at sufficiently low redshift that, if an associated SN existed, it would have been clearly detected, but none has been found. Another important clue is that about 20% of the observed short GRBs have been localized in early-type galaxies, where massive star formation ceased a long time ago. This allowed the collapse of massive stars to be excluded as the origin of short GRBs. The merger of NS–NS or NS–BH binaries has long been considered a very plausible candidate, because the duration of short GRBs requires a compact engine, with a characteristic time-scale of the order of tens of milliseconds, which indeed corresponds to the characteristic time-scale of a compact binary merger. General-relativistic magneto-hydrodynamical simulations of NS mergers, of the type discussed in Section 14.4.2, have indeed shown that these mergers can give rise to energy scales and jet opening angles consistent with the observations of short GRBs. Thus, at the time of the discovery of GW170817, NS merger was the favorite hypothesis for the origin of short GRBs.

The detection of the short GRB associated with GW170817 proves that indeed at least some short GRBs are produced in the merger of NS binaries. We will see, however, in Section 15.3.5 that the characteristics of the GRB associated with GW170817 appear to be different from those of the classic population of cosmological short GRBs, so this GRB could indeed be a new type of transient. At the time of writing the discovery is very recent, and further work will be necessary to fully elucidate its significance.

The observation of GRB 170817A

The merger time of the gravitational event GW170817 was August 17, 12:41:04 UTC. The gravitational signal was first identified approximately 6 minutes later in a low-latency search as a single-detector trigger in the

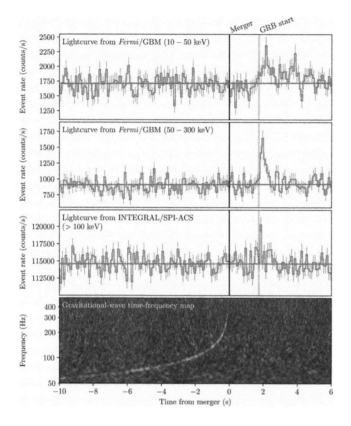

Fig. 15.7 Joint, multi-messenger detection of GW170817 and GRB 170817A. *Top panel*: Fermi-GBM light curve between 10 and 50 keV. The horizontal line is the background estimate. *Second panel*: Same as the top panel but in the 50–300 keV energy range. *Third panel*: Light curve from the SPectrometer on board INTEGRAL Anti-Coincidence Shield (SPI-ACS), with the energy range starting approximately at 100 keV and with a high energy limit of at least 80 MeV. *Bottom panel*: Time–frequency map of GW170817 obtained by coherently combining LIGO-Hanford and LIGO-Livingston data. From [LIGO Scientific Collaboration and Virgo Collaboration and Fermi-GBM and INTEGRAL Collaborations], Abbott *et al.* (2017). [Creative Commons Attribution 3.0 License.]

LIGO-Hanford detector, while the presence of temporally close transient noise prevented recording of the trigger on-line in the LIGO-Livingston detector.

Meanwhile, at 12:41:06 UTC, the Fermi-GBM onboard flight software had detected, classified and localized a short γ-ray burst, GRB 170817A, which was announced with an automatically generated Gamma-ray Co-ordinates Network (GCN) Notice at 12:41:20 UTC.[20] The temporal

[20] GCN Notices announce the detection of a signal. GCN Circulars distribute informations about follow-up observations.

proximity of the GW and GRB events was immediately identified by automatic procedures, and a GCN Notice based on the single low-latency Hanford trigger was issued by the LIGO/Virgo collaboration at 13:08:16 UTC, followed by a GCN Circular at 13:21:42 UTC. A rapid off-line re-analysis involving the two LIGO detectors and the Virgo detector confirmed the high significance of the coincident GW signal. The GRB was then detected also by INTEGRAL, through an off-line search triggered by the LIGO/Virgo and Fermi-GBM circulars, and was announced with a GCN Circular at 13:57:47 UTC.

Figure 15.7 shows the GRB observations of Fermi-GBM and of IN-TEGRAL, together with a time–frequency map of GW170817 obtained by combining coherently the LIGO-Hanford and LIGO-Livingston data. The signal in Fermi-GBM exceeds 5σ in three of the twelve NaI detectors on board. The difference between the merger time and the GRB time is

$$\Delta t = 1.734 \pm 0.054\,\text{s}\,. \tag{15.20}$$

This is consistent with theoretical expectations, since typical scenarios find that the mechanism responsible for launching a jet and producing the prompt γ-radiation acts on a time-scale of at most a few seconds after merger (this is also true for the cocoon scenario that we will discuss in Section 15.3.5).

The final Fermi-GBM analysis localizes the source within a broad region, covering about $1100\,\text{deg}^2$, consistent with the localization provided by the GW signal. Further refinement of the localization is provided from the difference in the arrival times of the GRB at Fermi-GBM and at INTEGRAL.

The statistical significance of the temporal coincidence between the Fermi-GBM and INTEGRAL observations is 4.2σ. This, together with the consistency between the observed event fluences (energy per unit area) and temporal structures, proves that both instruments are observing the same GRB. From the rate of observed short GRBs, one can evaluate that the probability of a chance temporal coincidence of the GW and GRB signals within a window $\Delta t = 1.7$ s is 5.0×10^{-6}, corresponding to 4.4σ in Gaussian statistics. The probability of a chance agreement in the spatial localization of the event, within the localization errors, turns out to be 0.01. Since these temporal and spatial probabilities are independent, the overall probability that the coincidence between the GW event and the GRB is due to chance is only 5.0×10^{-8}, corresponding to 5.3σ in Gaussian statistics. This unambiguously establishes the association of the GW event with the GRB.

The temporal profile of GRB 170817A depends on the energy range in which the signal is observed. The top panel in Fig. 15.7 shows that around the lower range of energies observed, $10 - 50$ keV, the signal has two distinct components. There is the initial triggering pulse, which lasts about half a second and is spectrally harder, which fits well with the typical duration and hardness of a short GRB. This first pulse is then followed by a subsequent weaker and spectrally softer emission that lasts a few seconds. The duration of the signal is much shorter if one restricts to energies above 100 keV, which is close to the minimum energy of ~ 75 keV accessible to INTEGRAL. The signal observed by INTEGRAL indeed consists of only two time bins of width 50 ms each, corresponding to a duration of at most 100 ms, including the rise time and the decay time. This firmly assigns GRB 170817A to the class of short GRBs. However, the two-component temporal structure in the lower-energy band is quite peculiar compared with standard GRBs, and could already hint at a different production mechanism; we will come back to this issue in Section 15.3.5.

Using only the signal in the $10 - 50$ keV range provides an independent source localization within a region of about $\pi\,(28\,\mathrm{deg})^2 \simeq 2460\,\mathrm{deg}^2$ (at 90% c.l.), whose location is consistent with the localization obtained from the data in the energy range $50 - 300$ keV, as well as from the GW signal. This indicates that this subsequent softer emission still belongs to the GRB and is not due to background fluctuations. The energy spectrum of the initial pulse is well fitted by a standard "Comptonized" function, given by a power law with a high-energy exponential cutoff, typical of short GRBs. In contrast, the energy spectrum of the subsequent soft emission is well fitted by a black-body spectrum with $k_B T = 10.3 \pm 1.5$ keV, although this emission is too weak to completely rule out a non-thermal spectrum. If its spectrum is indeed thermal, then this delayed, softer signal can be understood in terms of photospheric emission from a "cocoon". In this picture, a jet deposits a significant amount of energy in the surrounding dense material. This results in a hot "cocoon" that expands with mildly relativistic speeds and emits softer radiation, which does not undergo strong relativistic beaming. We will come back to this scenario in Section 15.3.5.

Some GRBs also show a precursor emission; this emission is less beamed, and for short GRBs the longest observed time lag between the precursor and the main peak is about 140 s. For GRB 170817A, no precursor emission has been observed. After the detection, INTEGRAL observed the same sky region without interruption for the next 10 hours (while Fermi-GBM was in an unfavorable position), but no other burst or continuous emission was detected in γ-rays over that period.

GRB 170817A is the closest short GRB with a measured redshift. Knowledge of its distance (from the GW observation and the subsequent identification of the electromagnetic counterpart discussed in Section 15.3.3) allows us to infer its isotropic energy release E_{iso}. The isotropic energy release of a GRB is a conventional quantity obtained by assuming that the source radiated isotropically, i.e. that the energy per unit area measured by the GRB detectors was the same in all directions. This will be a large overestimate of the actual total energy release if the emission is strongly beamed. Still, E_{iso} can be a useful quantity for comparing different GRBs, in the absence of more detailed information on the angular distribution of the emission. The isotropic energy release for GRB 170817A is $E_{\mathrm{iso,comp}} = (4.0 \pm 1.0) \times 10^{46}$ erg for the initial pulse with a Comptonized energy spectrum, and $E_{\mathrm{iso,BB}} = (1.3 \pm 0.3) \times 10^{46}$ erg for the subsequent softer signal, assumed to have a black-body spectrum, for a total of $E_{\mathrm{iso}} = (5.3 \pm 1.0) \times 10^{46}$ erg. This is a surprising result, since it is 4–6 orders of magnitude smaller than the isotropic energy of most short GRBs with known redshift. Previously, the weakest short GRB with a secure redshift association was GRB 150101B, with $E_{\mathrm{iso}} \simeq 2.3 \times 10^{49}$ erg, which still is about 400 times more energetic than GRB 170817A. We will discuss scenarios for understanding the energetics of GRB 170817A in Section 15.3.5.

15.3.3   The electromagnetic counterpart

Discovery of the electromagnetic counterpart

The announcement, through internal GCN circulars for the collaborations involved, of the coincident detection of the GW signal and of the GRB, started a massive campaign by a large number of ground- and space-based telescopes across all bands of the electromagnetic spectrum to search for the electromagnetic counterpart. The first LIGO/Virgo GCN circular, at 13:21:42 UTC, provided a broad localization within an area of $\pi(17.45\,\mathrm{deg})^2 \simeq 950\,\mathrm{deg}^2$. By 17:54:51 UTC the first three-detector skymap was available, which reduced the region to about $31\,\mathrm{deg}^2$ (at 90% credibility), while the luminosity distance was estimated to be 40 ± 8 Mpc. These values were already quite close to those of the final analysis. This restricted significantly the volume to be searched.

Of course, rapid detection of the electromagnetic counterpart was crucial, since fast evolution of the source could be expected. GW170817 was localized in the southern sky, making it inaccessible to the majority of telescopes located in the northern hemisphere. In the southern hemisphere the Atacama Desert, in northern Chile, is one of the best places for ground-based observations, and many of the best telescopes are located there. At the time of trigger, local time in Chile was 9.41 am, so observations could only start after about 10 hours. Furthermore, most of the 90% c.l. localization region was accessible only for the first two hours after twilight that evening. The electromagnetic counterpart was first detected and announced by the One-Meter, Two-Hemisphere (1M2H) team using the 1 m Swope Telescope at the Las Campanas Observatory in the Atacama Desert, observing in the near-infrared i-band.[21] The collaboration employed an effective search strategy, using a catalog of galaxies in the localization volume of the gravitational signal, prioritizing the galaxies on the basis of galactic stellar mass and star formation rate. The positions of the 100 highest-priority galaxies were examined to see if multiple galaxies could fit in a single Swope image. This allowed combination of 46 galaxies in 12 images. The observing schedule was then chosen so that 12 positions covering multiple galaxies were observed first, followed by the remaining individual galaxies in their order of priority. The observations started at 23:13 UTC, and at 23:33 UTC, in the ninth image, a source was detected, with i-band magnitude $i = 17.476 \pm 0.018$ mag, that was not present in archival images. The source was designated as Swope Supernova Survey 2017a

---

[21]See Note 62 on page 69 for discussion of photometric systems.

(SSS17a), and later received the IAU designation AT 2017gfo. It was located at right ascension $\alpha(\text{J2000.0}) = 13^{\text{h}}09^{\text{m}}48^{\text{s}}.085 \pm 0.018$ and declination $\delta(\text{J2000.0}) = -23°22'53''.343 \pm 0.218$, and had an offset of $10.6''$ (corresponding to 2.0 kpc at a distance of 40 Mpc) from the center of the galaxy NGC 4993, an elliptical S0 galaxy in the constellation Hydra, at a redshift

$$z = 0.009680 \pm 0.00079. \tag{15.21}$$

Assuming a local Hubble constant of $H_0 = 73.24 \pm 1.74 \,\text{km s}^{-1}\,\text{Mpc}^{-1}$, this corresponds to a luminosity distance $d_L = 40.4 \pm 3.4$ Mpc, consistent with the purely GW estimate. No other transient was found in the other images. After verification that the transient was not a previously known asteroid or supernova, the discovery was announced with a GCN Circular at 01:05 UTC. Five others teams discovered the transient independently before the announcement of SSS17a, and reported the discovery in a series of GCN Circulars.[22]

Follow-up electromagnetic observations

The discovery and precise localization of the electromagnetic counterpart opened the way for detailed follow-up observations in all bands of the electromagnetic spectrum, by ground- and space-based telescopes. In the following weeks, over 60 collaborations followed the source, in radio waves, IR, optical, UV, X-rays and γ-rays. These broadband follow-up observations were crucial for ruling out the possibility of a chance coincidence with an unrelated transient such as a supernova, and for a detailed understanding of the properties of the source.

The subsequent evolution of the UV, optical and IR light curves of the source were quite interesting. The curves showed a rapid decline in the luminosity, particularly in the bluest bands (u, g), where the decline was about 2 mag/day. Redder bands had a more complex behavior, with a decline of about 0.3 mag/day in the first 1.5 days, a shoulder at about 4 days, and subsequent decline at about 8 days. Correspondingly, the peak of the emission moved from the initial UV–blue emission toward the red and IR bands. This evolution is different from that of any previously observed transient source.

Spectroscopic observations started immediately after the discovery of SSS17a. The spectra obtained just 30 minutes after the first image show a blue and featureless continuum between 400 and 1000 nm. This could have been consistent with a young supernova, in which case the blue luminosity should have subsequently increased. In contrast, the spectra taken in the next 24 hr by several groups reported a rapid decline of the blue spectrum, without any feature identifiable with line absorption. This rules out the possibility of a young supernova, since supernova remnants display several lines due to the ionized elements in the ejecta. After about 2 weeks, the spectra taken in the near-IR showed the emergence of broad spectral features, related to radioactive decay of rapid neutron capture (r-process) nucleosynthesis elements. The rapid dimming of the luminosity, the shift of the spectra from the UV to the

[22]The teams involved, and the relative times of announcements, were the Dark Energy Camera (01:15 UTC), the Distance Less Than 40 Mpc Survey (DLT40; 01:41 UTC), Las Cumbres Observatory (LCO; 04:07 UTC), the Visible and Infrared Survey Telescope for Astronomy (VISTA; 05:05 UTC) and MASTER (05:38 UTC); see the Further Reading.

[23]See Abbott *et al.* [LIGO Scientific Collaboration and Virgo Collaboration and Fermi GBM and INTEGRAL and IceCube and other collaborations] (2017) for detailed listings of the observations performed across the electromagnetic spectrum.

IR and the signature of r-processes are in broad agreement with the kilonova scenario that we will discuss in Section 15.3.4.

Follow-up γ-ray observations of the source by many telescopes provided an accurate temporal coverage of the source, across 10 orders of magnitude in energy, from a few hundred seconds before merger, with observations continuing for the next month.[23] No further burst was observed, and limits on persistent emission were set.

Pointed X-ray observations of the source started quite soon, at $T_0 + 0.62$ days (where we denote by T_0 the merger time) with the X-ray telescope on the *Swift* satellite and at $T_0 + 0.70$ days with NuSTAR, but no X-ray emission from the counterpart was observed. The source was monitored by *Swift* in the following days, but again no X-ray counterpart was detected. Observations with the more sensitive *Chandra* X-ray observatory at about $T_0 + 2.3$ days also detected no significant emission, and set a bound on the (isotropically equivalent) X-ray luminosity $L_X \lesssim 3.2 \times 10^{38}$ erg/s. Finally, nine days after merger, an X-ray signal from the counterpart was detected with *Chandra*, corresponding to a isotropic luminosity $L_X \simeq 9 \times 10^{38}$ erg/s. Further observations were made by *Chandra* 15 days after merger, confirming the continued emission of X-rays. The X-ray source brightened with time, reaching $L_X \simeq 1.1 \times 10^{39}$ erg/s at 15.1 days after merger.

Finally, radio emission was the last to show up. Initial observation of radio emission, at first attributed to the counterpart, turned out to be due to an AGN in NGC 4993. Coordinated observations among the Karl G. Jansky Very Large Array (VLA) radio telescope, the VLA Low Band Ionosphere and Transient Experiment (VLITE), the Australia Telescope Compact Array (ATCA) and the Giant Metrewave Radio Telescope (GMRT), as well as with the Atacama Large Millimeter/submillimeter Array (ALMA), enabled monitoring of the target on almost a daily basis, but no signal was found until, at last, radio emission from the counterpart was detected with the VLA on September 2 and 3, at two different frequencies, with two independent observations. The signal was detected again in the following days, and was regularly monitored for the following months.

The search for a neutrino counterpart was carried out with IceCube, ANTARES and the Pierre Auger Observatory, but no candidate directionally coincident with the final localization of GW170817 was found, neither in a time span $T_0 \pm 500$ s nor in the 14-day period after the merger.

After these first observations, the source became unobservable to optical and X-ray telescopes, because of Sun blocking, for the next ~ 100 days, while it was still monitored in radio waves. The radio signal increased steadily over the next ~ 100 days, which provided crucial clues for the understanding of this event. Resumed X-ray observations after 110 days, just after Sun blocking was lifted, also showed a brightening of the X-ray emission, at a rate consistent with that from the radio light curve. In Sections 15.3.4 and 15.3.5 we will discuss how to interpret these observations.

15.3.4 Kilonovae and r-process nucleosynthesis

We have seen in Section 10.3 that nuclear fusion in stellar cores can synthesize elements up to the iron group (while big-bang nucleosynthesis only produces deuterium, ^{3}He, ^{4}He and ^{7}Li; see Section 22.7.1). An important question is how even heavier elements are synthesized. In general, this can happen through a sequence of neutron captures in a neutron-rich environment followed by neutron β-decay. Two main processes can be identified. One is the s-process, in which the rate of neutron capture is sufficiently slow; in this case, if the nucleus resulting from the capture of a neutron is radioactive, neutron capture is followed by β-decay (or, possibly, α-decay) of the radioactive nucleus, otherwise further neutron capture will take place. A second possibility is the r-process, in which the rate of neutron capture is so rapid that, even when a radioactive nucleus is formed, another neutron is absorbed before the radioactive nucleus has had a chance to decay, until, eventually, after further neutron captures, an unstable nucleus is reached that undergoes α- or β-decays fast enough before further neutron capture can occur. These two processes produce different sequences of elements and isotopes; about half of the isotopes heavier than iron can only be reached through the s-process, and half only through the r-process (although there is some overlap for certain isotopes). For instance, uranium, platinum and gold are produced by the r-process. In stars, under usual conditions, the rate of neutron capture is slow, and only s-process isotopes are synthesized. In particular, the s-process occurs in red giant stars, seeded by neutron capture on Fe nuclei produced by earlier generations of stars and that were already part of the material from which the star formed. The r-process, in contrast, requires more extreme conditions, with a high flux of free neutrons. These conditions could be met either in the ejecta of a core-collapse supernova or in a NS merger.

The detection of the NS–NS merger GW170817 and its follow-up in all bands of the electromagnetic spectrum, and particularly in the optical and near-IR wavelengths, has provided a unique opportunity for testing the idea that r-process nucleosynthesis takes place in binary NS mergers. The observations of the electromagnetic counterpart of GW170817 turn out to be in very good agreement with the predictions of the *kilonova* model. A kilonova is a thermal transient object, resulting from a NS–NS or NS–BH merger, that emits approximately isotropic radiation, mostly in the optical and near-IR. This radiation is powered by the decay of radioactive isotopes synthesized by the r-process in the ejecta of the merger. Several observational aspects, mentioned in Section 15.3.3, point toward a kilonova interpretation of the electromagnetic counterpart of GW170817: the initial featureless blue spectrum, the rapid evolution of the luminosity, and the subsequent emergence of spectral features in the near-IR consistent with the signature of an r-process.

Detailed computations of kilonova light curves have been presented in the literature, allowing for detailed quantitative comparison with the observation of the electromagnetic counterpart of GW170817. The key parameters are the mass and velocity of the ejected material, and its opacity. If the ejecta are dominated by elements with atomic mass number $A \lesssim 140$, the opacity is similar to that of iron-peak elements. However, even a small fraction of lanthanides (or actinides) affects the opacity significantly. This is due to the fact that these elements have their outer valence electrons in the f-shell. This configuration permits many transitions with excitation of just a few eV, and therefore greatly enhances the opacity to optical photons. The presence of lanthanides therefore has the effect of shifting the emission of radioactive decays produced in the r-process, which would peak in the blue band, toward the red or the IR.

The spectra observed in the electromagnetic counterpart of GW170817 and the transition from the initial blue spectrum toward a redder one cannot be reproduced by a simple supernova model, powered by the radioactive decay of ^{56}Ni and assuming typical iron-peak opacities. A kilonova model with just one component, either rich or poor in lanthanides, again does not provide a good fit to the data, failing to capture simultaneously in a quantitative manner the early blue regime and the late red regime. All papers that performed a detailed comparison of theoretical kilonova light curves with the observations, and that appeared simultaneously after the announcement of GW170817 (see the Further Reading), find that the data are well fitted by a kilonova model with (at least) two components: one rich in lanthanides, with a fractional lanthanide abundance $X_{\mathrm{lan}} \sim 10^{-2}$, a total ejecta mass $M_{\mathrm{ej}} \simeq 0.04 M_\odot$ and an expansion velocity $v_{\mathrm{ej}} \simeq 0.1c$, responsible for the red component; and a "blue" component, lanthanide-poor ($X_{\mathrm{lan}} \sim 10^{-4} - 10^{-5}$), with $M_{\mathrm{ej}} \simeq 0.01 M_\odot$ and $v_{\mathrm{ej}} \simeq 0.3c$.

Lanthanide-poor ejecta can form either through shock-heating of material at the contact point in the collision of two NS, or later via an accretion wind disk. However, only the former mechanism can produce outflow velocities larger than $0.1c$. Thus, the blue component is expected to be due to material ejected at the contact interface in the final collision of the two NS. In turn this implies the existence of a solid surface where the collision between the two compact stars took place, pointing to a NS–NS, rather than a NS–BH binary. The material from such a collision is mostly ejected within an angle $\theta \lesssim 45°$ from the normal to the orbital plane of the binary.

By contrast, lanthanide-rich ejecta are likely due to tidal forces, spiral mode instabilities or winds in the equatorial plane. According to this modelization, whether one sees a blue or a red kilonova therefore depends also on the viewing angle; at angles $\theta \gtrsim 45°$ (defined with respect to the normal of the orbital plane) the larger opacity of the red component would obscure the blue component, while at $\theta \lesssim 45°$ one would see first the blue component, which would, however, fade more rapidly, leaving the red component to dominate in the next few days.[24]

The mass of the ejecta estimated through comparison with the optical and near-IR light curves can be compared with that obtained from the parameter estimation of the binary system through the GW observation

[24] Indeed, the best kilonova candidate before GW170817 was GRB 130603B, which appeared to be a more isotropic red kilonova; see Berger (2014) for discussion and references.

only. Indeed, there are phenomenological fits, based on numerical simulations of (non-spinning) NS–NS mergers, that give the mass of the ejecta as a function of the initial gravitational masses of the two NS, their baryonic masses[25] and their radii. Assuming an EoS, the baryonic masses and radii can be obtained from the measured gravitational masses. In this way, it can be estimated that the mass of the dynamical ejecta (neglecting the contribution of winds, which is not taken into account in the simulations, but can be important) is in the range $10^{-3} - 10^{-2} M_\odot$, consistent, within the large systematic uncertainty of this estimate, with the estimate from the comparison with the kilonova light curves.

Comparing these ejecta masses with the local rate of binary NS mergers inferred by this first detection, which is $R = 1540^{+3200}_{-1200} \, \mathrm{Gpc}^{-3} \, \mathrm{yr}^{-1}$, one finds that, if about 10% of the matter dynamically ejected is converted to r-process elements, NS mergers can fully account for the amount of r-process elements observed in the Galaxy.

15.3.5 The cocoon scenario

We now turn to the mechanism responsible for the delayed radio and X-ray emission of GW170817, mentioned in Section 15.3.3, and we will see how this, in combination with the near-IR, optical and UV observations used for establishing the kilonova model, can be used to gather an overall picture of the event, and in particular to provide an explanation for the subluminous nature of GRB 170817A.

The prompt γ-ray emission in a short GRB is believed to be generated by internal processes, such as shocks or magnetic reconnection, in a ultra-relativistic and highly collimated jet. The standard explanation for the X-ray afterglow is that, after the prompt emission fades, on a time-scale of less than 2 s, the relativistic jet is decelerated by the ambient medium, and in the process broadband synchrotron radiation is emitted. This leads to a X-ray emission that decays with time as $1/t^2$, on a time-scale of $10^5 - 10^6$ s. This radiation is collimated in an angle $\theta \sim 1/\Gamma$, where Γ is the Lorentz factor of the jet. A typical ultra-relativistic jet has $\Gamma \gtrsim 100$. As the jet slows down, its Lorentz factor decreases and its opening angle widens. For a broad range of values of the energy of the jet and of the typical densities in the post-merger environment, the jet decelerates to $\Gamma \sim 10$ in about 1 day. Thus, an observer that is just on-axis should see an afterglow that decays with time. An observer that is off-axis would, however, miss the initial part of the afterglow, when it is too collimated to intercept the line of sight, but would start to see it after a delay, when the opening angle becomes broad enough. After the beam enters the observer's line of sight, the observer will see the intensity decrease with time, just as an on-axis observer would do.

The interaction of the jet with the post-merger environment also produces radio emission, which again is collimated in the beaming angle of the jet. Furthermore, in a binary NS merger radio emission is also produced due to the tidal ejection of $0.01 - 0.05 M_\odot$ of matter at speeds

[25] The baryonic mass of a NS is the sum of the masses of the baryons of which it is composed. The actual gravitational mass of the NS is smaller, because of the gravitational binding energy. For typical EoS, in NS the binding energy per neutron is about 100 MeV.

$v \sim 0.1c$. We have seen in Section 15.3.4 that the decay of the radioactive isotopes of this material is responsible for the red component of the kilonova, on a scale of days. On top of this, this material forms a blast wave that plows through the interstellar medium (ISM) surrounding the merger, producing synchrotron radiation that peaks at radio frequencies and that can last for months to years. Therefore this radio emission provides information on the energetics of this blast wave, and on the environment of the merger. In contrast to the radio emission from a relativistic jet, this radio emission is not strongly collimated.

We can now analyze different scenarios for understanding the electromagnetic observations discussed in Section 15.3.3. The first hypothesis that naturally comes to mind is that the GRB detected by Fermi-GBM and by INTEGRAL was observed on-axis, or just slightly off-axis but still close enough to our line of sight, so that the observed γ-rays were a component of the standard prompt GRB emission. This scenario, however, is ruled out. First, we have seen that GRB 170817A was about four orders of magnitude less energetic than typical cosmological short GRBs, and more than two orders of magnitude less energetic than the weakest GRB previously known. The assumption that it was just an extremely weak GRB does not work, since such a weak jet would not penetrate successfully through the ejecta and break out, unless the density of the mass of the ejecta was unacceptably low. Numerical simulations of a jet of this energy launched into expanding ejecta show that, even if the jet was well collimated, it would break out only for $M_{\rm ej} < 3 \times 10^{-6} M_\odot$, and this upper limit becomes an order of magnitude lower for a jet with opening angle $\theta_j \simeq 30°$.[26] However, we have seen that the UV/optical/IR signal requires $M_{\rm ej} \simeq 0.04 M_\odot$ of lanthanide-rich ejecta, launched within 45° of the polar axis (along which the jet will be launched). Furthermore, all numerical simulations find that the mass ejected at high latitude is larger than $10^{-3} M_\odot$. Thus, already from these observations, an on-axis GRB can be ruled out. On top of this, an on-axis GRB is not consistent with the lack of early detection of X-rays and radio waves. As we have seen, for on-axis emission the X-ray flux decreases as $1/t^2$. From the flux measured by *Chandra* 9 and 15 days after merger it follows that, for an on-axis jet, the signal at 2 days after merger should have been more than 10^3 higher than the upper limit from the non-detection. Similar constraints come from the radio observations. Thus an on-axis GRB is firmly ruled out.

A second hypothesis is that we are simply observing the jet off-axis. Since the profile of the GRB drops quite sharply with the angle, the energetics of the GRB could be understood with a somewhat fine-tuned angle. However, as we have mentioned, with the typical density of the ISM ($n \gtrsim 10^{-3}\,{\rm cm}^{-3}$), the Lorentz factor of the jet should decrease to $\Gamma \sim 10$ in about a day; at this point the jet is no longer well collimated and the X-ray and radio-wave afterglow should enter into the line of sight of the observer. The lack of detection of X-ray radiation for nine days would imply an implausibly low value $n \simeq 10^{-6}\,{\rm cm}^{-3}$, more typical of the intergalactic medium than of the interstellar medium within a

[26]See Kasliwal *et al.* (2017).

galaxy. Thus, already with the data available shortly after the event, this hypothesis was unlikely.

A third possibility is the formation of a "cocoon". As the jet drills through the ejecta, it deposits energy there, and the surrounding material inflates to form a hot cocoon that expands outward. Numerical simulations show that the cocoon has a wide opening angle $\theta_c \sim 40°$ and mildly relativistic velocities, with Lorentz factors $\Gamma \simeq 2 - 3$. If the jet is well collimated, say $\theta_j \simeq 10°$, and is sufficiently energetic, it will penetrate through the ejecta and break out, to produce a classical GRB for an observer located on axis. Otherwise, if the jet has a wide angle or is not sufficiently energetic, it will become choked, and will deposit all its energy in the cocoon. Given a dissipation mechanism (which could be provided by the interaction of the cocoon with the ejecta), the cocoon will generate γ-rays at wide angle. Relativistic hydrodynamical simulations[27] show that, if just a small fraction of the ejecta, corresponding to $3 \times 10^{-9} M_\odot$, moves at $v \simeq 0.8c$ (which gives a Lorentz factor $\Gamma \simeq 1.7$), for a wide range of ejecta and jet properties, it is possible to reproduce the observed γ-ray flux, including both the initial harder component and the softer tail. This weaker emission would in general be undetectable from GRBs at typical cosmological distances, and was detected in GRB 170817A only thanks to its much closer distance.

As the cocoon expands it also produce radio emission, through the forward shock propagating in the ISM. The signal rises with time, reflecting the fact that the cocoon must sweep enough material to produce a sizable signal. This can explain the long delay between the merger and the first detection of the radio emission. The result depends on the Lorentz factor of the cocoon, its energy, and the density of the material in the post-merger environment. Detailed computations[28] show that, for $\Gamma = 2$, the predictions are consistent with the observed radio emission for $n \sim 3 \times 10^{-3} \, \mathrm{cm}^{-3}$, while $\Gamma = 3$ requires $n \sim 3 \times 10^{-4} \, \mathrm{cm}^{-3}$. The X-ray data are also similarly explained by this scenario. We have also mentioned in Section 15.3.2 that the soft tail of the GRB observed in the energy range 10–50 keV (see the top panel in Fig. 15.7) can also be naturally explained in terms of cocoon emission.

The off-axis model and the cocoon model make very different prediction for the subsequent evolution of the radio and X-ray flux. In the off-axis model, the signal rises sharply and becomes visible as soon as the Lorentz factor has decreased sufficiently, so that the beaming angle intercepts the line of sight of the observer. After that, both the radio and X-ray signals decay with time, just as for an on-axis jet, since their intrinsic intensities are decreasing. In contrast, in the cocoon model, the radio and X-ray signals rise steadily in time as long as the jet continues to deposit energy in the cocoon, i.e. until it breaks out or it is choked. Unless this happens just when the signal has risen above the detectability threshold, the observer will see a steady increase of the signal for a long period of time, which can be of the order of months.

With the data available shortly after the discovery, both the off-axis

[27] See Kasliwal *et al.* (2017) and Gottlieb, Nakar, Piran and Hotokezaka (2017).

[28] See Hallinan *et al.* (2017).

and cocoon scenarios were consistent with the data, although, as we have seen, the off-axis model was less plausible. As we have already mentioned in Section 15.3.3, after the first observations the source became unobservable to optical and X-ray telescopes until early December because of Sun blocking. This is not a problem for radio telescopes, which continued monitoring the source. The data taken over the next 93 days, up to November 18, 2017, show that the radio signal continued to increase steadily, in full agreement with the predictions of the cocoon model. The lack of signatures of an off-axis jet suggests that the jet was likely choked. Soon after Sun blocking, *Chandra* resumed observations of the source, at about 110 days after merger, and found that the X-ray emission had also undergone significant brightening, at a similar rate to the radio light curve, confirming that the X-ray and radio emissions have a common origin.

This understanding of the dynamics leading to the observed features of the electromagnetic counterpart has several important implications. First, the emission mechanism for this GRB was clearly different from that of a classic short GRB at cosmological distances seen on-axis. Thus, while the observation of GW170817 and the associated GRB 170817A shows that GRBs are indeed emitted in NS–NS mergers, the relation between the mechanism identified in this event and the mechanism powering the classic short GRBs is still to be fully understood. The peculiarity of GRB 170817A is that it is the closest short GRB with measured redshift ever observed, and in this event we have been able to see a signature that would be undetectable for a GRB at much larger distances. So, in this sense, GRB 170817A appears to belong to a new class of transients. Another important consequence is that, for NS–NS mergers within the reach of advanced LIGO/Virgo, the electromagnetic counterparts could be much more common than predicted on the basis of classic short GRBs, where the beaming would limit the possibility of observing a GRB counterpart to only about 10% of the events. In contrast, the cocoon emission is at a wide angle, and provides a signal potentially accessible at all wavelengths.

15.4   Tests of fundamental physics

Several tests of GR and of fundamental physics are already possible with these first detections. With the increased sensitivity of the detectors planned for the near future we expect that many more events will be detected, including some event with a much higher SNR, potentially leading to much more stringent tests.

15.4.1   BH quasi-normal modes

A particularly interesting test of GR can be performed by studying whether the ringdown phase of the waveform is consistent with that expected from the quasi-normal modes (QNMs) of the final BH, given in eq. (14.102). This test can be performed with GW150914, given its

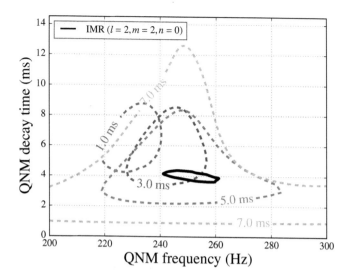

Fig. 15.8 The posterior 90% confidence regions from Bayesian parameter estimation for a damped-sinusoid model, assuming different start-times $t_0 = t_M + 1, 3, 5, 7$ ms (dashed lines). The black solid line shows contours of 90% confidence region for the frequency f_0 and decay time of the $l = 2$, $m = 2$ and $n = 0$ (i.e. the least-damped) QNM obtained from the inspiral-merger-ringdown waveform for the entire detectors bandwidth. From [LIGO Scientific Collaboration and Virgo Collaboration], Abbott *et al.* (2016e).

relatively large amplitude and the fact the peak frequency of its GW amplitude falls in the frequency range where the LIGO detectors have their best sensitivity. Having determined the mass and spin of the final BH from the reconstruction of the full waveform, GR predicts the frequency and decay time of the BH QNMs and in particular of the dominant, least-damped, QNM; see eqs. (12.360) and (12.362). We can then compare this prediction with the damped oscillating behavior of the final part of the waveform. There is, however, a certain ambiguity as to the exact definition of the "final part" of the waveform, which should be compared with the prediction (14.102). If we denote by t_M the time of merger, we can compare the predicted ringdown waveform with the data starting from time $t_0 = t_M + \Delta t$. If we take Δt too small, the signal is, however, still dominated by the merger phase, rather than by ringdown. If we take Δt too large, we are indeed in the region dominated by ringdown, but the signal vanishes exponentially, so the accuracy of the test quickly degrades. Figure 15.8 shows the results for $\Delta t = 1, 3, 5$ and 7 ms, comparing the frequency and decay time of the dominant QNM extracted by fitting the final part of the waveform to a damped sinusoid (dashed lines) with the values obtained by determining the final mass and spin from the full waveform, and then using the predictions (12.360) and (12.362) of BH perturbation theory. We see that, starting from $\Delta t = 3$ ms, the 90% confidence region obtained by fitting the waveform to a damped sinusoid overlaps with the 90% confidence region obtained from BH perturbation theory. Of course, increasing Δt further, the dashed contour becomes larger and larger, since the accuracy of the reconstruction of the frequency and decay time from the final part of the waveform degrades. We see, however, that, within the available accuracy, the GR prediction for the frequency and decay time of the least-damped BH QNM is consistent with the data. A more stringent test would be obtained if, with higher SNR, one could extract from the waveform the frequency and decay time of two QNMs, since from this information one could obtain an independent reconstruction of the mass and spin of the final BH, and compare them with the values obtained from the full waveform.

15.4.2   Tests of post-Newtonian gravity

Another interesting test of GR is obtained by considering the PN expansion for the phase of the waveform during the inspiral. As a function of the frequency (which is itself a function of time), it has the general form

$$\Psi(f) = 2\pi f t_c - \Phi_0 - \frac{\pi}{4} + \sum_{j=0}^{7} \left[\phi_j + \phi_j^{(l)} \log f \right] f^{(j-5)/3} . \qquad (15.22)$$

In eq. (5.275) we already wrote this result up to 2PN order, i.e. retaining terms up to $j = 4$. The result up to 3.5PN, as well as the numerical value of the coefficients ϕ_j and $\phi_j^{(l)}$, can be obtained from eq. (5.261). Observe that (up to $j = 7$), the coefficients $\phi_j^{(l)}$ of the logarithmic terms are non-zero only for $j = 5, 6$, and come from the terms $x^{5/2} \log x$ and $x^3 \log x$ in the term in braces in eq. (5.261). In order to test GR, we can replace phenomenologically the coefficients ϕ_j and $\phi_j^{(l)}$ by $\phi_j(1 + \delta\phi_j)$ and $\phi_j^{(l)}(1 + \delta\phi_j^{(l)})$, respectively. For the 0.5PN term, corresponding to $j = 1$, the GR value is $\phi_1 = 0$, so in this case $\delta\phi_1$ is taken as an absolute, rather than a relative, shift. The parameters $\delta\phi_j$ and $\delta\phi_j^{(l)}$ are then taken as free fitting parameters, determined by the data, and GR corresponds to $\delta\phi_j = \delta\phi_j^{(l)} = 0$. In practice one cannot put stringent bounds on these parameters if they are allowed to vary freely simultaneously, since the changes in one parameter are strongly degenerate with changes in the others. The best one can do, with just a few events, is to vary these parameters one at the time, leaving the others fixed at their GR values $\delta\phi_j = \delta\phi_j^{(l)} = 0$. The result of this analysis, performed using GW150914 and GW151226, is shown in Fig. 15.9. We see that the strongest limits comes from GW151226, because of its large number of inspiral cycles in the detector bandwidth.

We can compare this result with a similar analysis performed for the double pulsar PSR J0737–3039, whose properties we discussed in Section 6.3. In the double pulsar the orbital period changes basically at a constant rate, so the double pulsar is only sensitive to the 0PN term. However, because of the very long observation time (of order 10 yr, to be compared with a few tens of milliseconds for GW150914 or 1 s for

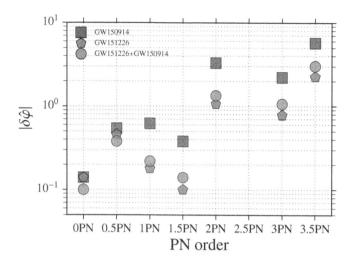

Fig. 15.9 90% upper bounds on the fractional variations for the PN coefficients compared with their known values in GR, from GW150914, from GW151226 and from a joint analysis of the two events (the 0.5PN term vanishes in GR and is then taken to be an absolute, rather than a relative, shift). From [LIGO Scientific Collaboration and Virgo Collaboration], Abbott *et al.* (2016n).

GW151226), the limit obtained from the double pulsar on the deviations from GR of the 0PN coefficients is stronger than that obtained from GW150914 and GW151226, and is $\delta\varphi_{0PN} \lesssim 10^{-2}$. On the other hand, beyond 0PN the limits obtained from the double pulsar are not very informative,[29] $\delta\varphi_{0.5PN} \lesssim 10$ and $\delta\varphi_{1PN} \lesssim 10^3$, while from GW150914 and GW151226 we get limits at the level of $(10-40)\%$ on several other PN coefficients.

[29]See Fig. 6 of [LIGO Scientific Collaboration and Virgo Collaboration], Abbott *et al.* (2016e).

15.4.3 Propagation and degrees of freedom of GWs

Speed of gravitational waves

In GR, GWs propagate at the speed of light, and their propagation is affected by the background metric in the same way as the propagation of electromagnetic waves.[30] The detection of the GW signal in the NS–NS coalescence GW170817, and of the associated GRB 170817A, provides a very stringent test of this fundamental property. Indeed, the GRB arrived with a delay Δt of about 1.7 s with respect to the GW signal; see eq. (15.20). As we have seen, this delay is entirely consistent with typical models of GRB emission. Furthermore, the fact that we have such a small time delay over a propagation distance of about 40 Mpc, i.e. of more than 100 million light-years, puts a very stringent limit on any difference in the propagation velocities of GWs and of light. Suppose that GWs propagate at a speed c_{gw} different from the speed of electromagnetic waves $c_{em} \equiv c$, and define $\Delta c = c_{gw} - c$. Then, if the electromagnetic signal and the GRB were emitted simultaneously at

[30]The evolution of tensor perturbations over a FRW background will be studied in Section 19.5, where we will also discuss the propagation of tensor modes in modified gravity.

a time $t = 0$, and we denote by t_{gw} and t_{em} the corresponding arrival times, we have $c_{gw} t_{gw} = d = c\, t_{em}$, where d is the distance to the source [ignoring the cosmological expansion, which is a good approximation in this case, given the small value (15.21) of the redshift]. Then, to first order in $\Delta c/c$,

$$\frac{\Delta c}{c} = \frac{c\Delta t}{d}, \qquad (15.23)$$

where $\Delta t = t_{em} - t_{gw}$. The distance to GW170817 has been fixed from the observation of the electromagnetic counterpart to $d_L = 40.4 \pm 3.4$ Mpc [assuming $H_0 = 73.24 \pm 1.74\,\mathrm{km\,s^{-1}\,Mpc^{-1}}$; see the discussion following eq. (15.21)]. Taking $\Delta t = 1$ s and $d_L = 40$ Mpc as reference values, the relation between Δc and the delay Δt induced by the different propagation speeds is therefore

$$\frac{\Delta c}{c} \simeq 2.43 \times 10^{-16} \left(\frac{\Delta t}{1\,\mathrm{s}}\right) \left(\frac{40\,\mathrm{Mpc}}{d}\right). \qquad (15.24)$$

Of course, to make any statement on $\Delta c/c$ from the measured delay Δt given in eq. (15.20), we need to know the intrinsic delay between the GW and GRB emission times. As we have mentioned, the observed delay $\Delta t \simeq 1.7$ s is fully consistent with typical mechanisms of GRB emission, so $|\Delta c|/c$ is consistent with zero, and only an upper bound on it can be inferred. The exact value of this upper bound depends on the assumptions that we make on the emission mechanism. In general, all models predict that the GRB is emitted after the merger time that marks the peak in the GW signal. Thus, a positive vale of Δc, which (with our sign conventions $\Delta c = c_{gw} - c$ and $\Delta t = t_{em} - t_{gw}$) would induce a positive delay Δt, cannot be compensated by the emission mechanism. In this case, an upper bound on $\Delta c/c$ is obtained by attributing the whole observed delay $\Delta t \simeq 1.7$ s to the different propagation speeds. By contrast, a negative value of $\Delta c = c_{gw} - c$ induces a negative Δt, which could be partially compensated by a delay in the emission mechanism, thereby masking the effect of Δc. In this case the precise value of the bound depends on the amount of fine tuning that we are willing to accept between these two completely unrelated effects. We can assume for instance that the GRB signal was emitted 10 s after the GW signal, and that the observed delay of 1.7 s resulted from a partial cancellation with the faster propagation of the GRB. Then,[31]

$$-2 \times 10^{-15} \lesssim \frac{\Delta c}{c} \lesssim 4 \times 10^{-16}. \qquad (15.25)$$

The lower limit can be compared with that obtained from the absence of gravitational Cherenkov radiation in cosmic rays. Indeed, if $c_{gw} < c$, particles with speed v such that $c_{gw} < v < c$ would emit gravitational Cherenkov radiation. The highest-energy cosmic rays, propagating over astronomical distances, would then lose most of their energy through this radiation, and would not be observed. If the highest-energy cosmic rays originate within our own galaxy, this implies a bound[32]

$$-2 \times 10^{-15} \lesssim \frac{\Delta c}{c}, \qquad (15.26)$$

[31] Our bounds are slightly more stringent that those given in [LIGO Scientific Collaboration and Virgo Collaboration and Fermi-GBM and INTEGRAL Collaborations], Abbott *et al.* (2017), because we have used the (90% c.l.) bound $d_L \gtrsim 37$ Mpc on the distance d_L, as later determined by the observation of the electromagnetic counterpart, rather than the 90% c.l. bound $d_L \gtrsim 26$ Mpc determined using only GW observations.

[32] See Moore and Nelson (2001).

which numerically is about the same as the lower bound in eq. (15.25), while, if they are extragalactic, the bound tightens to

$$-2 \times 10^{-19} \lesssim \frac{\Delta c}{c} \,. \tag{15.27}$$

In contrast, no significant upper limit on $\Delta c/c$ existed before the detection of GW170817/GRB 170817A. The fact that the GW in GW150914 (as well as in the subsequent observed BH–BH coalescences) indeed swept the distance between the two LIGO detectors within the expected temporal window of 10 ms gives an upper limit that, from eq. (15.23) with $d/c \simeq 10$ ms equal to the travel time between the LIGO detectors, and Δt equal to the actually observed delay, is only $\Delta c/c \leqslant O(1)$. The bound (15.25) rules out several scalar–tensor or vector–tensor modifications of GR; see the Further Reading.

A related test concerns the fact that, in GR, in a generic curved background, both electromagnetic waves and GWs follow the same geodesics of the metric. Therefore perturbations of the background metric, such as those due to the gravitational potentials generated by matter between the source and the observer, will affect their propagation in the same manner. We can see this property as a manifestation of the equivalence principle. A possible way to test this prediction is through the Shapiro time delay, that we discussed in Section 6.2.2 of Vol. 1. There we found that a Newtonian gravitational potential $\phi(\mathbf{x})$ induces a time delay, in the propagation from the position $\mathbf{r}_s$ of the source to the position $\mathbf{r}_{\rm obs}$ of the observer, given by

$$\Delta t = -\frac{2}{c} \int_{\mathbf{r}_{\rm obs}}^{\mathbf{r}_s} |d\mathbf{x}| \, \phi(\mathbf{x}) \,. \tag{15.28}$$

Deviations from this prediction can be expressed in terms of a parameter γ, writing[33]

$$\Delta t = -\frac{1+\gamma}{c} \int_{\mathbf{r}_{\rm obs}}^{\mathbf{r}_s} |d\mathbf{x}| \, \phi(\mathbf{x}) \,, \tag{15.29}$$

so GR predicts $\gamma = 1$, both for electromagnetic waves and for GWs. We can further study possible differences between the propagation of electromagnetic waves and that of GWs by introducing two separate parameters $\gamma_{\rm em}$ and $\gamma_{\rm gw}$. The best limit on the deviation of $\gamma_{\rm em}$ from the GR prediction comes from the measurements of the *Cassini* spacecraft at radio wavelengths,

$$\gamma_{\rm em} - 1 = (2.1 \pm 2.3) \times 10^{-5} \,. \tag{15.30}$$

From GW170817/GRB 170817A we can rather put a bound on $\gamma_{\rm gw} - \gamma_{\rm em}$. The strongest bound would be obtained by modeling the potential $\phi(\mathbf{x})$ along the whole line of sight. A more conservative bound is obtained by taking into account only the effect of the Galaxy, outside a sphere of 100 kpc. This gives[34]

$$-2.6 \times 10^{-7} \leqslant \gamma_{\rm gw} - \gamma_{\rm em} \leqslant 1.2 \times 10^{-6} \,, \tag{15.31}$$

[33] Actually, γ is defined more generally in terms of the parametrized post-Newtonian (PPN) formalism; see Will (2014).

[34] See [LIGO Scientific Collaboration and Virgo Collaboration and Fermi-GBM and INTEGRAL Collaborations], Abbott *et al.* (2017).

where, as in eq. (15.25), the asymmetry of the bound is due to the fact that a delay in the GRB arrival time compared with that of the GW could be partially compensated by a delay in the emission, taken again to be at most 10 s.

Massive gravitons

Another test of fundamental physics concerns the issue of whether the graviton can have a small mass. We discussed massive gravitons, at the linearized level, in Section 2.3 of Vol. 1. At the non-linear level, significant theoretical developments have taken place in massive gravity since the writing of Vol. 1; see the Further Reading. Independently of the intricacies of the construction of a consistent fully non-linear theory with a massive graviton, at a purely phenomenological level one can test whether gravitons are massive by observing that the dispersion relation of a massive graviton is $E^2 = p^2c^2 + m_g^2c^4$. This leads to a dependence of the velocity v on the frequency ω,

$$\frac{v^2}{c^2} = 1 - \left(\frac{m_g c^2}{\hbar \omega}\right)^2 ; \qquad (15.32)$$

see Note 23 on page 83 of Vol. 1. This means that different Fourier modes of the GW travel at different speeds and therefore reach the Earth at different times. The effect can be incorporated into a correction to the phase of the waveform,

$$\Delta\Psi(f) = -\frac{\pi d_L c}{\lambda_g^2(1+z)f} , \qquad (15.33)$$

where $\lambda_g = 2\pi\hbar/(m_g c)$ is the Compton wavelength of the graviton. Formally this has the same frequency dependence as a 1PN effect; compare with eq. (5.275). Observe that $\Delta\Psi$ is proportional to the luminosity distance d_L, since the effect accumulates over the distance traveled by the GW, from the source to the Earth. Given the cosmological distances of the observed events, this effect allows us to be sensitive to very small graviton masses. In general, in a consistent theory of massive gravity one can expect both a modification to the production mechanism, compared with GR, and in the subsequent GW propagation. Equation (15.33) only takes the latter into account. However, this provides an effect that is proportional to the propagation distance d_L, which would not be the case for modifications in the production mechanism, and therefore by itself already provides a meaningful test of deviations from GR.

The strongest constraint to date comes from GW170104, since this is currently the source at the greatest distance. Using eq. (15.33) and comparing with the signal from GW170104 gives a bound on the graviton mass,

$$m_g \lesssim 7.7 \times 10^{-23}\,\text{eV}/c^2 , \qquad (15.34)$$

at 90% c.l. By comparison, from GW150914 we get the slightly less stringent bound $m_g \lesssim 1.2 \times 10^{-22}\,\text{eV}/c^2$. Equation (15.34) corresponds to a bound $\lambda_g \gtrsim 0.5$ pc on the Compton wavelength of the graviton. By itself, this bound is not competitive with those discussed in Section 2.3 of Vol. 1 from the absence of a Yukawa screening of the gravitational force at the level of interaction between galaxies or within galaxy clusters, which give $\lambda_g \gtrsim O(1)$ Mpc. However, conceptually, these two bounds have a somewhat different meaning. The bound from BH–BH coalescences is directly a bound on the mass of the propagating tensor modes of the gravitational field. The bound from the limit on a Yukawa suppression of the gravitational force is rather a bound on the static interaction mediated by the component h_{00} of the gravitational field. Whether these masses are the same depends on the details of the modified gravity theory that we are considering. For instance, in a massive gravity theory that violates Lorentz invariance there is a priori no relation between the mass of h_{00} and the mass of the tensor modes. The only other direct bound on the mass of the tensor modes comes from the binary pulsars, and is about three orders of magnitude weaker than eq. (15.34).

This analysis can be generalized by studying the possibility of more general modifications of the dispersion relation for gravitons, for instance of the form $E^2 = p^2 c^2 + A p^\alpha c^\alpha$, with $\alpha \geqslant 0$ (so the inclusion of a graviton mass term corresponds to $\alpha = 0$); see the Further Reading.

Extra polarizations

A further test of GR can be performed by checking whether the gravitational signal has only the two polarization states predicted by GR, or if extra polarizations are present. For instance, a massive graviton would have five polarization states, corresponding to the two states with helicities ± 2 predicted by GW, two states with helicities ± 1, and one state with helicity 0. Scalar–tensor modifications of GR have in general extra longitudinal or transverse scalar modes. With only the two LIGO detectors, a test of the polarization content of the signal is not really possible, because the two detectors have the same orientation. Indeed, for GW150914, such a test was inconclusive, with the Bayes factor for two different hypotheses (purely GR polarization or purely scalar modes) being statistically equivalent. However, the addition of the Virgo detector, which is oriented differently, makes possible a significant test of the polarization content. The pattern functions of interferometers for the tensor polarizations were computed in Section 7.2 of Vol. 1 and are given by

$$F_+ = \frac{1}{2}(1 + \cos^2 \theta)\cos 2\phi\,, \qquad F_\times = \cos\theta \sin 2\phi\,. \tag{15.35}$$

In contrast, the pattern function for a transverse scalar polarization is

$$F_s(\theta, \phi) = -\sin^2\theta \cos 2\phi\,. \tag{15.36}$$

These tensor and scalar pattern functions are shown in Fig. 15.10. For GW170814, which has been detected in the two LIGO detectors and in the Virgo detector, one can perform a Bayesian analysis, coherent

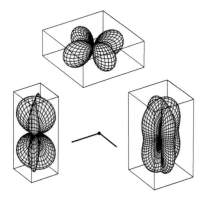

Fig. 15.10 The pattern function of interferometers for a transverse scalar mode (upper plot) compared with the standard pattern function for the two tensor modes (lower plots), relative to the orientation of the axes of the interferometers shown in the middle. From Maggiore and Nicolis (2000).

across the three detectors, using either the pattern function for tensor polarizations or that for scalar (or vector) polarizations. In this case, one finds a Bayes factor larger than 200 in favor of purely tensor polarizations against purely vector polarizations, and larger than 1000 for purely tensor against purely scalar polarization.

Further reading

- The first observation of the GWs from a binary BH coalescence, GW150914, was officially announced on February 11, 2016. The discovery paper is [LIGO Scientific Collaboration and Virgo Collaboration], Abbott *et al.* (2016a), and was accompanied by 11 more papers, which discuss several theoretical and experimental aspects of the discovery, as we summarize in the following.

- The main features of the LIGO detectors in the O1 run that led to the discovery of GW150914 are discussed in [LIGO Scientific Collaboration and Virgo Collaboration], Abbott *et al.* (2016b). The characterization of transient noise is discussed in [LIGO Scientific Collaboration and Virgo Collaboration], Abbott *et al.* (2016h). The detector calibration [which is important also for correctly reconstructing the GW signal $h(t)$, which can be distorted by frequency-dependent calibration errors] is discussed in [LIGO Scientific Collaboration and Virgo Collaboration], Abbott *et al.* (2016i).

- The result of the search leading to the discovery of GW150914 is discussed in greater detail in [LIGO Scientific Collaboration and Virgo Collaboration], Abbott *et al.* (2016c). The general search strategy for the matched-filtering analysis is discussed in particular in Allen, Anderson, Brady, Brown and Creighton (2012) and in Babak *et al.* (2013). See also Cannon *et al.* (2010), Dal Canton *et al.* (2014) and Usman *et al.* (2016) for various details and improvements of the search pipeline.

- Beside the search performed by matching the signal against coalescing binaries waveforms, GW150914 was also identified in a search for unmodeled bursts. The corresponding analysis is presented in [LIGO Scientific Collaboration and Virgo Collaboration], Abbott *et al.* (2016g). The Coherent WaveBurst (cWB) algorithm used in this search is described in Klimenko, Yakushin, Mercer and Mitselmakher (2008). Recent improvements in the algorithm

are presented in Klimenko *et al.* (2016). The low-latency search algorithm used in the analysis (oLIB) is described in Lynch, Vitale, Essick, Katsavounidis and Robinet (2017).

- After having identified GW150914 as a GW event in the search analysis, a follow-up study and Bayesian parameter estimation of this event have been performed, using a coherent analysis between the two detectors, more refined waveforms and a finer sampling of parameter space. The results are reported in [LIGO Scientific Collaboration and Virgo Collaboration], Abbott *et al.* (2016d).

This study used two kind of waveforms, called "EOBNR" and "IMR-Phenom", respectively. The EOBNR waveforms are constructed from the EOB method tuned to numerical relativity (see the Further Reading of Chapter 14 for general references on the EOB method and on numerical relativity). In particular, the EOBNR waveforms used in the discovery paper are those developed in Taracchini *et al.* (2014), with non-precessing spins, tuned to the catalog of numerical relativity simulations of Mroué *et al.* (2013). This model (which scans over 11 parameters, out of the 15 in a full treatment of precessing spins) is formulated as a set of differential equations that are computationally too expensive to solve for the millions of likelihood evaluations required, and it is then combined with a frequency-domain reduced-order model developed in Pürrer (2014, 2016). EOB waveforms with fully precessing spins, and therefore spanning the whole 15-dimensional parameter space of the binary (except for the eccentricity, which is still set to zero), have been developed in Pan *et al.* (2014) and used in an improved study of GW150914 in [LIGO Scientific Collaboration and Virgo Collaboration], Abbott *et al.* (2016o).

The other model used in the analysis, denoted IMR-Phenom, has been developed in Ajith *et al.*

(2007, 2008) and Pan *et al.* (2008). The most recent version of the model, used in the analysis, has been developed in Husa *et al.* (2016) and Khan *et al.* (2016). In the simplest model the spins are treated as non-precessing, using the parameter $\chi_{\rm eff}$ defined in eq. (15.15), which was introduced in Ajith *et al.* (2011) and Santamaria *et al.* (2010). Precession effects are then modeled through further effective precession spin parameters [see Hannam *et al.* (2014) and Schmidt, Ohme and Hannam (2015)], resulting in a model with 13 parameters.

- Classic papers on the methodologies for extracting the physical parameter of a coalescing binary are Finn and Chernoff (1993), Cutler and Flanagan (1994) and Poisson and Will (1995). A more recent study for spinning binaries is that in van der Sluys *et al.* (2008), while the spin determination is discussed in Vitale, Lynch, Veitch, Raymond and Sturani (2014).

- Astrophysical implications of GW150914 are discussed in [LIGO Scientific Collaboration and Virgo Collaboration], Abbott *et al.* (2016j), while the inferred rate of binary BH coalescences using the first three detections is discussed in [LIGO Scientific Collaboration and Virgo Collaboration], Abbott *et al.* (2017b). Scenarios for the formation of GW150914 by gravitational interaction in the core of a dense cluster are explored in Rodriguez, Haster, Chatterjee, Kalogera and Rasio (2016). The possibility that the BHs observed in the coalescing binaries are of primordial origin is discussed in Clesse and García-Bellido (2017). The implication for stochastic GW backgrounds from binary black holes are presented in [LIGO Scientific Collaboration and Virgo Collaboration], Abbott *et al.* (2016k).

- The localization of the signal and the search for an electromagnetic counterpart to the BH binary GW150914 is discussed in [LIGO Scientific Collaboration and Virgo Collaboration], Abbott *et al.* (2016l). Following previous agreements, when the GW signal was found an alert was sent to a network of 25 teams of observers in different bands of the electromagnetic spectrum, with both satellite and ground-based telescopes. The search for an EM counterpart started two days after the GW event was recorded, and used both archival data and targeted observation in the broad region of sky, of about $600\,{\rm deg}^2$, where the event was localized. Stellar-mass BH binaries in their final coalescence stage are not expected to have retained significant matter around them, e.g. in the form

of accretion disks, so no electromagnetic counterpart is expected, at least according to the standard lore. Indeed, no counterpart was found, except for a signal observed by Fermi-GBM, which found a sub-threshold signal 0.4 s after GW150914, with a claimed false-alarm probability of 0.22% (corresponding to 2.9σ), possibly consistent with a short GRB; see Connaughton *et al.* (2016). This statistical significance was computed assigning a larger significance to the events, in the search window, found closest to the time of interest. Using a uniform probability across a 60 s window, the significance of the event reduces to 1.9σ. However, the INTEGRAL γ-ray satellite was covering the full region of GW150914. If the signal was of astrophysical origin, and one assumes that the source spectrum was similar to the template spectrum used in the discovery of the GW150914-GBM candidate event, INTEGRAL should have detected it with a significance between 5σ and 15σ [see Savchenko *et al.* (2016)], but INTEGRAL detected no signal. Nor was any signal detected by the AGILE γ-ray satellite, which covered 65% of the localization region with a sensitivity that would have detected a short GRB of this type even if its flux were about 10 times weaker; see Tavani *et al.* (2016).

It has also been pointed out that the physical requirement for an electromagnetic emission of the type corresponding to the Fermi-GBM signal would exceed by many orders of magnitude what is expected in realistic astrophysical scenarios; see Lyutikov (2016). Given that a statistical significance of 2.9σ is not particularly high, this might already indicate that the Fermi-GBM signal was likely due to noise. Furthermore, the Fermi-GBM data have been re-analyzed in Greiner, Burgess, Savchenko and Yu (2016), where it is stressed that both the SNR and the false-alarm rate reported in Connaughton *et al.* (2016) depend on the assumed hard spectral shape, while a softer spectral shape increases the probability of a background fluctuation. As discussed in Greiner, Burgess, Savchenko and Yu (2016), the spectrum of the candidate event is indeed soft, and softer than typical short-hard GRBs, and is consistent with the background spectrum as measured by Fermi-GBM. The conclusion of Greiner, Burgess, Savchenko and Yu (2016), based also on a more thorough characterization of the background, is that this candidate event was most likely just a statistical fluctuation, and that the 0.2% false-alarm probability was an optimistic lower limit.

- A follow-up search for a high-energy neutrino counterpart gave a negative result; see Adrian-Martinez *et al.* [ANTARES and IceCube and LIGO Scientific and Virgo Collaborations] (2016). No electromagnetic or neutrino counterpart has been found either for the other BH–BH coalescences detected to date.

- In Sesana (2016) it has been pointed out that a coalescence such as GW150914 would appear in the LISA frequency band, with a detectable amplitude, about five years before the signal enters the bandwidth of advanced ground-based interferometers. The signal would stay in the LISA frequency window for thousands of cycles, so that, a few weeks before its appearance in the bandwidth of advanced LIGO or advanced Virgo, the masses would have already been predicted to about 1%, the sky location would already be known to a precision of about $1 \deg^2$ and the coalescence time could be predicted within less than 10 s. This opens the possibility of multi-band GW astronomy, and of alerting electromagnetic telescopes to the search for counterparts of the merger.

- The discovery of the second BH–BH coalescence, GW151226, is presented in [LIGO Scientific Collaboration and Virgo Collaboration], Abbott *et al.* (2016m). In [LIGO Scientific Collaboration and Virgo Collaboration], Abbott *et al.* (2016n), a more detailed analysis of this event is presented, together with a reanalysis of GW150914 using the whole dataset of the O1 run, and a study of the candidate event LVT151012.

  The discovery of the third BH–BH coalescence, GW170104, is presented in [LIGO Scientific Collaboration and Virgo Collaboration], Abbott *et al.* (2017b).

  The first three-detector detection by the LIGO–Virgo network, GW170814, was reported in [LIGO Scientific Collaboration and Virgo Collaboration], Abbott *et al.* (2017c).

- The first GW event due to a NS–NS binary, GW170817, was officially announced on October 16, 2017, and the announcement was immediately followed by tens of papers involving the collaborations that participated in the discovery and the search for the counterpart.

  The paper announcing the GW detection is [LIGO Scientific Collaboration and Virgo Collaboration], Abbott *et al.* (2017c).

  The γ-ray observations are announced in Goldstein *et al.* (2017) for Fermi-GBM and Savchenko *et al.* (2017) for INTEGRAL. The joint observation of

the GW event and the γ-ray burst is described in [LIGO Scientific Collaboration and Virgo Collaboration and Fermi-GBM and INTEGRAL Collaborations], Abbott *et al.* (2017).

The discovery of the electromagnetic counterpart is reported in Coulter *et al.* (2017) [1M2H team], Soares-Santos *et al.* [DES and Dark Energy Camera GW-EM Collaborations] (2017), Valenti *et al.* (2017) [Distance Less Than 40 Mpc (DLT40) Survey], Arcavi *et al.* (2017) [Las Cumbres Observatory (LCO)], Tanvir *et al.* (2017) [VISTA] and Lipunov *et al.* (2017) [MASTER]. An accurate distance to the host galaxy NGC 4993 is given in Hjorth *et al.* (2017).

The full multi-messenger observation of the event is described in Abbott *et al.* [LIGO Scientific Collaboration and Virgo Collaboration and Fermi GBM and INTEGRAL and IceCube and other collaborations] (2017), a paper authored by about 60 collaborations (whose author list is over 10 pages long!).

Follow-up electromagnetic observations and comparison with kilonova models are presented in Cowperthwaite *et al.* (2017) (optical and near-IR observations), Nicholl *et al.* (2017) (optical and UV spectra), Chornock *et al.* (2017) (near-IR), Smartt *et al.* (2017) (optical and near-IR), Arcavi *et al.* (2017a, 2017b) (optical and near-IR), Troja *et al.* (2017) (first detection of the X-ray counterpart), Evans *et al.* (2017) (UV and X-rays), Margutti *et al.* (2017) and Haggard *et al.* (2017) (X-rays), Blanchard *et al.* (2017) (X-ray observations of the host galaxy), Hallinan *et al.* (2017) and Alexander *et al.* (2017) (radio waves), and Andreoni *et al.* (2017) (optical, near-IR and radio). A full listing of the over 80 papers that came out immediately after the discovery of GW170817 can be found at https://lco.global/~iarcavi/kilonovae.html.

Mooley *et al.* (2017) present follow-up radio observations for the next 93 days, which confirm the cocoon scenario, and Ruan, Nynka, Haggard, Kalogera and Evans (2017) present *Chandra* X-ray observations at 110 days after merger, showing that the X-ray emission also brightened.

- Standard references for the theory of GRBs and the subsequent afterglow are Meszaros and Rees (1993) and Piran (2004). The idea that short GRBs are due to NS–NS mergers was advanced in Blinnikov, Novikov, Perevodchikova and Polnarev (1984), and the accumulating evidence prior to GW170817 is reviewed in Nakar (2007) and Berger (2014). The radio signal after a binary merger is studied in

Nakar and Piran (2011), Hotokezaka and Piran (2015) and Hotokezaka *et al.* (2016).

- Lattimer and Schramm (1974) proposed that the decompression of neutron-rich matter in a NS–BH merger could create the condition for r-process nucleosynthesis. For NS–NS mergers, this was studied with numerical simulations in Rosswog *et al.* (1999).

The idea that r-processes after a NS–NS merger would produce an optical transient powered by the decay of many radioactive isotopes was advanced in Li and Paczynski (1998) and further developed in Kulkarni (2005) (which introduced the term "macronova"), while Rosswog (2005) studied nucleosynthesis and GRBs in NS–BH binaries. The term "kilonova" was introduced in Metzger *et al.* (2010), where detailed computations of the light curve are presented. The role of lanthanides in the spectrum of kilonovae has been discussed in Kasen, Badnell and Barnes (2013), and the role of neutrino-driven winds after a NS–NS merger is discussed in Martin *et al.* (2015). A review of the kilonova model is Metzger (2017). A review of the multi-messenger picture that can be carried by NS coalescences, written before the observation of GW170817, is Rosswog (2015).

Studies of r-process nucleosynthesis in GW170817 are given in Cowperthwaite *et al.* (2017), Nicholl *et al.* (2017), Chornock *et al.* (2017), Rosswog *et al.* (2017), Kasliwal *et al.* (2017), Drout *et al.* (2017) and Kasen, Metzger, Barnes, Quataert and Ramirez-Ruiz (2017).

- The dynamics and observational signatures of a "cocoon" surrounding a relativistic jet are discussed in Ramirez-Ruiz, Celotti and Rees (2002), Nakar and Piran (2017), Lazzati, Deich, Morsony and Workman (2017) and Lazzati *et al.* (2017). The cocoon description of the electromagnetic counterpart of GW170717 is proposed in Kasliwal *et al.* (2017), Hallinan *et al.* (2017), Evans *et al.* (2017) and Gottlieb, Nakar, Piran and Hotokezaka (2017).

- Tests of GR from GW150914 are discussed in [LIGO Scientific Collaboration and Virgo Collaboration], Abbott *et al.* (2016e), while the effect of GW151226 on the test of the PN parameters is discussed in [LIGO Scientific Collaboration and Virgo Collaboration], Abbott *et al.* (2016n). Parametrizations of deviations from GR in the waveforms of coalescing binaries have been proposed in Yunes and Pretorius (2009), Mishra, Arun, Iyer and Sathyaprakash (2010) and Li *et al.* (2012).

Limits from GW170817/GRB 170817A on the difference between the speed of light and the speed of GWs, as well as on the difference in Shapiro time delay and on Lorentz violations, are presented in [LIGO Scientific Collaboration and Virgo Collaboration and Fermi-GBM and INTEGRAL Collaborations], Abbott *et al.* (2017). The lower limit on $c_{gw} - c$ from the absence of gravitational Cherenkov radiation in cosmic rays is discussed in Moore and Nelson (2001). The test of the equivalence principe through the Shapiro time delay is discussed in Will (2014). A formalism for parametrizing Lorentz violations is discussed in Colladay and Kostelecky (1998), Kostelecky (2004) and Tasson (2014). The consequences of the limits on the speed of GWs on a class of scalar–tensor and vector–tensor modifications of GR are discussed in Creminelli and Vernizzi (2017), Sakstein and Jain (2017) and Baker *et al.* (2017).

The effect of a graviton mass on the phase of the waveform is computed in Will (1998). Bounds on the graviton mass from GW150914 are presented in [LIGO Scientific Collaboration and Virgo Collaboration], Abbott *et al.* (2016e), while the limits from GW170104, as well as limits on modified dispersion relations, are presented in [LIGO Scientific Collaboration and Virgo Collaboration], Abbott *et al.* (2017b). Using the triple detection of GW170814, the hypothesis of a purely tensor polarization is tested against purely scalar or purely vector polarizations in [LIGO Scientific Collaboration and Virgo Collaboration], Abbott *et al.* (2017c). The interferometer pattern function for the scalar polarization was computed in Maggiore and Nicolis (2000).

- Limits on the graviton mass implicitly assume the existence of a consistent theory of massive gravity. As we mentioned in Section 2.3 of Vol. 1, at the linearized level a specific mass term, the Pauli–Fierz mass term, is required to avoid the situation in which, together with the five degrees of freedom of a massive spin-2 graviton, there also appears a sixth scalar degree of freedom, which would otherwise be a ghost, i.e. would have a negative kinetic energy term. Boulware and Deser (1972) found that, even with the Pauli–Fierz mass term, beyond the linear level the ghost in general reappears. For several decades massive gravity was then abandoned. More recently, a non-linear ghost-free theory of massive gravity, the dRGT theory, has been constructed in de Rham, Gabadadze and Tolley (2011) and Hassan and Rosen (2012a); see de Rham (2014) for a review. This has been extended to ghost-free bigravity theory [where a fixed reference metric that appears in the dRGT theory is promoted to a dynamical metric] in Hassan and Rosen (2012b); see Schmidt-May and von Strauss (2016) for a review. The construction of a consistent theory of massive gravity and of bigravity has an intrinsic field-theoretical interest, allowing us to explore the dynamics of massive spin-2 fields, and provides at least an example of a consistent theoretical framework where the idea of a massive graviton can be tested. One of the original motivations for the construction of these theories was the hope of obtaining a dynamical explanation for dark energy. From this point of view, these theories have not been suc-

cessful so far. The dRGT theory already has difficulty in obtaining a viable background cosmological evolution; see D'Amico *et al.* (2011). In bigravity background FRW solutions exist. However, at the level of cosmological perturbation theory, one must face instabilities in both the scalar and tensor sectors, so that at most one remains with a branch of solutions and a region in parameter space where the background evolution is indistinguishable from that of ΛCDM; see Könnig, Akrami, Amendola, Motta and Solomon (2014), Lagos and Ferreira (2014), Akrami, Hassan, Könnig, Schmidt-May and Solomon (2015) and Cusin, Durrer, Guarato and Motta (2015, 2016).

Supermassive black holes

16

In this chapter we discuss supermassive black holes (SMBHs), with typical masses $(10^5 - 10^{10})M_\odot$. We will consider in particular SMBH binaries, in which both components are SMBHs, as well as binary systems involving a SMBH and a stellar-mass compact object, which give rise to so-called "extreme mass ratio inspirals" (EMRIs). The GWs emitted by SMBH binaries and by EMRIs that fall in the frequency range $10^{-5} - 10^{-1}$ Hz are among the most interesting signals for the space interferometer LISA,[1] while at frequencies of order of a nHz the stochastic background generated by many unresolved SMBH binaries, and possibly the signal from some individually resolvable sources, is already within reach of the present sensitivity of pulsar timing arrays (which we will discuss in detail in Chapter 23). We begin this chapter by reviewing the observational evidence for the SMBH at the center of our Galaxy. Because of its relative closeness, this is the SMBH that we can study in most detail. SMBH binaries will be discussed in Section 16.2. In particular, in Section 16.2.1 we study their possible formation and evolution mechanisms, while in Section 16.2.2 we will see how the GWs emitted by SMBH binaries compare with the planned sensitivity of LISA. EMRIs are discussed in Section 16.3. Their study requires the introduction of new interesting theoretical tools, in particular the self-force approach, which we also will briefly discuss. Finally, in Section 16.4 we will compute the stochastic GW background generated by SMBH binaries.

[1]The LISA space interferometer was initially conceived as a joint ESA-NASA mission. After the withdrawal by NASA in 2011 because of funding problems, it continued under ESA leadership as eLISA (Evolved Laser Interferometer Space Antenna). After the successful performance of the LISA PathFinder a new proposal has been submitted to ESA in 2017 in response to the ESA call for L3 mission concepts, and has been selected by ESA's Science Program Committee for ESA's third large (L3) mission. The project has gone back to the original name LISA, and might also include some new involvement by NASA, see Audley *et al.* (2017). The launch of the mission is currently expected in 2034.

16.1 The central supermassive black hole in our Galaxy

As we zoom into the central arcsecond region of our Galaxy, corresponding to sub-parsec distances from the Galactic center, we find a compact radio source, Sgr A*, with a radio luminosity 2×10^{34} erg/s. Since radiation in the visible spectrum from the Galactic center is completely absorbed by dust, this source is only visible in radio waves, near- and mid-infrared, and X-rays. Radio emission from the Galactic center was first discovered by Jansky in the 1930s, but more detailed studies were only possible after World War II, thanks to advances in radio and microwave astronomy. As we will discuss in this section, there is now very convincing evidence that Sgr A* is associated with a SMBH that sits at the center of the Galaxy. The radiation observed as Sgr A*, in radio waves as well as the infrared, most likely originates from the accretion

Gravitational Waves, Volume 2: Astrophysics and Cosmology. Michele Maggiore.
© Michele Maggiore 2018. Published in 2018 by Oxford University Press.
DOI 10.1093/oso/9780198570899.001.0001

[2]The New Technology Telescope (NTT) is a telescope with a 3.6 m aperture, located in La Silla (Chile), which has also developed advanced techniques for the VLT. The Very Large Telescope (VLT) is a group of four main telescopes each with a diameter of 8.2 m (plus four auxiliary movable telescopes each with diameter 1.8 m) located at Cerro Paranal, in the Atacama Desert, Chile, and is the flagship facility for European ground-based astronomy. Both the NTT and VLT are projects of the European Southern Observatory (ESO). The VLT can perform observation from the mid-infrared (24 μm) to the deep-UV region (300 nm). The telescopes can be used independently or in interferometric mode. In interferometric mode they form the ESO Very Large Telescope Interferometer (VLTI). This allows an angular resolution of the order of a milliarcsecond (mas) to be reached. Such a resolution is equivalent to distinguishing the two headlights of a car at the distance of the Moon! (See http://www.eso.org/public/teles-instr/vlt.html). The Keck Observatory sits on the summit of Mauna Kea, in Hawaii. It is composed of two telescopes, Keck I and Keck II, whose primary mirrors are 10 m in diameter, and which can be operated together in interferometric mode, again reaching mas resolution. It observes in the optical and near-infrared. (See http://www.keckobservatory.org). Both the Keck and the VLT use adaptive optics techniques that measure atmospheric turbulence and correct for it, providing images as sharp as if the telescope were in space.

disk around the SMBH, or from a jet emanating from it.

The evidence for the existence of a central SMBH has grown steadily over the years. The first indication that a large mass is concentrated into this region came in 1980 from the measurement of the velocity of compact clouds of ionized gas in Sgr A West, which suggested the existence of a central "point-like" mass of order a few times $10^6 M_\odot$, in addition to several times $10^6 M_\odot$ in stars within 1 pc from the center. A possible concern with this result was that the gas motion can in principle be sensitive to non-gravitational forces, such as magnetic pressure or frictional forces. However, the result was confirmed in the 1980s and 1990s by the measurement of radial velocities of stars in the innermost arcsecond (the so-called S-stars), and the motion of stars is sensitive only to the gravitational potential. These measurements provided an indication for the presence of a mass of about $3 \times 10^6 M_\odot$, within a radius of order 0.1 pc, corresponding to a lower bound on the mass density of order $10^9 M_\odot/\text{pc}^3$. The main alternative explanation, rather than a SMBH, is the presence of a cluster of dark objects, such as neutron stars or stellar-mass BHs. Such an explanation, however, is only plausible if this hypothetical cluster is dynamically stable over a time-scale that is not too small compared with the age of the Galaxy. A cluster of dark objects with a mass $3 \times 10^6 M_\odot$, concentrated within a radius of 0.1 pc, could still have a lifetime of the order of the age of the Galaxy, so at the time this alternative explanation was not ruled out.

To obtain a more detailed picture of the gravitational potential it was necessary to measure the full velocity field, including both the radial and tangential components. Remarkable progress in this direction came from diffraction-limited studies (i.e. studies using adaptive optics techniques that correct for the distortion induced by turbulence in the atmosphere and reach the theoretical limits of the telescope fixed by diffraction) of the central cluster in the near-infrared. These observations measured the proper-motion velocities of $O(100)$ stars at a distance between 0.01 and 0.3 pc from Sgr A*. This allowed the confinement radius of the central mass to be reduced below 0.006 pc. The corresponding bound on the mass density therefore increased by more than three orders of magnitude, and ruled out a long-lived cluster of dark objects as an alternative explanation.

The next remarkable step was the detection of the deviations from linear motions of a number of stars in the immediate vicinity of Sgr A*, which allowed direct measurements of their orbits. The motion of several S-stars has been monitored since 1992 at the NTT/VLT and since 1995 at the Keck Observatory.[2] At present there are 28 stars whose orbits have been determined. Among these, a star known as S2 (or S0-2 in the Keck nomenclature) has a period as short as 15.9 yr, and has completed a full orbit since monitoring begun. Reconstruction of the orbits of these stars indicates that S2 made in 2002 a closest approach to the central mass at a distance of 6×10^{-4} pc $\simeq 120$ AU, while S16, which has a period of about 36 yr, had a closest approach in 2000, at just 2.2×10^{-4} pc $\simeq 45$ AU, with a speed at periapse of $(12 \pm 2) \times 10^3$ km/s and a rather eccentric orbit, with eccentricity $e \simeq 0.87$. Thus, the central dark mass must be confined at least to within this radius. For the central dark mass, combining the Keck and VLT data, one finds

$$M = (4.30 \pm 0.20|_{\text{stat}} \pm 0.30|_{\text{sys}}) \times 10^6 \, M_\odot \,. \qquad (16.1)$$

A cluster of dark objects with a mass of this order, confined within less than 45 AU, would have a mass density of at least $8 \times 10^{16} \, M_\odot \, \text{pc}^{-3}$, and would either collapse or evaporate on a time-scale smaller than 10^5 yr, making the hypothesis of a central cluster of dark objects highly unlikely. The Schwarzschild radius of a BH with a mass M of the order of that given in eq. (16.1) is

$$R_S \simeq 3.8 \times 10^{-7} \, \text{pc} \left(\frac{M}{4 \times 10^6 M_\odot} \right) \simeq 0.085 \, \text{AU} \,, \qquad (16.2)$$

or, equivalently, $R_S \simeq 18 R_\odot$. The distance of 45 AU reached by the star S16 in its closest approach to the BH corresponds to a distance of about $530 \, R_S$.

The fit to the orbit also provides a measurement of the distance R_0 to the Galactic center,

$$R_0 = (8.28 \pm 0.15|_{\text{stat}} \pm 0.29|_{\text{sys}}) \, \text{kpc} \,. \qquad (16.3)$$

With the present accuracy, all stellar orbits are simultaneously fit extremely well by a single-point-mass potential. Deviation from simple Keplerian orbits can arise either in the presence of an extended mass distribution or because of general-relativistic corrections to the motion in the Schwarzschild potential of the central BH. Within the existing database, S2 is the only star for which one could hope to detect general-relativistic effects, although the present accuracy is not yet sufficient.

The position of the central dark mass is determined from the orbits of the stars with an uncertainty of about 1 mas. This is better than the accuracy on the position of Sgr A*, as determined from its infrared emission. The two positions are offset by just (0.5 ± 6.4) mas West and (9 ± 14) mas South, and are therefore coincident at the 1σ level. This means that, in projection, the central mass and Sgr A* are coincident to an accuracy better than $4 \times 10^{-4} \, \text{pc} \simeq 80 \, \text{AU}$. This strengthens the association of the central dark mass with Sgr A*.

The proper motion of the central dark mass with respect to the central stellar cluster is treated as a free parameter in the orbital fit, and turns out to be $(1.5 \pm 0.5) \, \text{mas yr}^{-1}$, statistically consistent with no motion at all. This limit in turn implies a limit on the mass of any companion BH, to be less than about $5 \times 10^5 (R/16{,}000 \, \text{AU})^{1/2} \, M_\odot$, where R is the distance of a hypothetical companion BH from the central BH (as long as the BH companion is outside the orbits of the stars that are monitored, and its orbital period is long compared with the duration of the study, which is 10–20 yr).

Thus, the study of the orbits of close stars gives remarkable information on the mass and the (absence of) proper motion of the central

BH, and localizes the dark mass within a few hundreds R_S. Remarkably, direct imaging of Sgr A* in the infrared using Very Long Baseline Interferometry (VLBI) can resolve its structure down to even smaller scales. At a distance of about 8 kpc, the Schwarzschild radius R_S given in eq. (16.2) subtends an angle $\theta \simeq 10$ microarcseconds (µas). VLBI studies at wavelengths 7 mm and 3.5 mm suggested an angular size for Sgr A* in the range $20-40$ µas. However, at these wavelengths interstellar scattering dominates, and broadens the image of Sgr A*. Extending VLBI techniques down to $\lambda \simeq 1.3$ mm, where the effect of interstellar scattering is reduced, it has been possible to measure the intrinsic angular diameter of Sgr A*, with the result

$$\theta_{\mathrm{SgrA*}} = 37^{+16}_{-10} \ \mathrm{\mu as}, \tag{16.4}$$

where the errors are at 3σ. If we accept the association of the central SMBH with Sgr A*, which in view of the results from the stellar orbits discussed before seems difficult to dispute, this means that the VLBI measurements are resolving the structure of the SMBH down to about $4\,R_S$. Actually, this measurement is so precise that it even tells us that Sgr A* cannot be exactly centered on the BH. In fact, because of gravitational lensing, radiation originating from a spherical surface of radius R centered around a Schwarzschild BH will have a larger apparent radius R_a, given by

$$R_a = \begin{cases} (3\sqrt{3}/2)R_S & \text{for } R \leqslant (3/2)R_S \ , \\ R(1 - R_S/R)^{-1/2} & \text{for } R \geqslant (3/2)R_S \ . \end{cases} \tag{16.5}$$

Therefore, any spherical emitting region centered around the central SMBH should have a minimum apparent diameter larger than at least $(3\sqrt{3}/2) \times 20\,\mathrm{\mu as} \simeq 52\,\mathrm{\mu as}$. For a Kerr BH, even in the maximally rotating case, the angular diameter in the equatorial plane should still be larger than 45 µas. The fact that the result given in eq. (16.4) is smaller than this, at about 3σ level, indicates that there is an offset between the position of the emission region that we detect as Sgr A* and the SMBH, and that the source of infrared radiation detected as Sgr A* rather occupies a compact portion of the accretion disk or of a jet. VLBI observations at 1.3 mm have also detected variations of the flux from Sgr A*, with a time-scale corresponding to a few times R_S/c. By causality, such a variability indicates again that the source has a size smaller than a few times R_S.

16.2 Supermassive black-hole binaries

16.2.1 Formation and evolution of SMBH binaries

The observational advances due to the use of adaptive optics in ground-based telescopes, mentioned earlier, as well as the Hubble Space Telescope, have also made it possible to study the kinematics of gas or stars in external galaxies down to sub-parsec scales, greatly improving our understanding of the central regions of galaxies. These observations have shown that most, or possibly all, bright galaxies at low redshift harbor SMBHs in their nuclei. As we saw in eq. (14.206), their mass is tightly correlated with the velocity dispersion of the stellar bulge of their host galaxy. Similar correlations can be found between the central BH mass and other properties of the bulge of the host galaxy, such as its stellar mass or its mid-infrared luminosity, and can be written in the form

$$\log_{10} \frac{M_{\rm BH}}{M_\odot} = \alpha + \beta \log_{10} X \,, \qquad (16.6)$$

where we can have $X = \sigma/(200 \, {\rm km/s})$, where σ is the stellar velocity dispersion in the buldge, or $X = L_i/(10^{11} L_{i,\odot})$, where L_i is the mid-infrared luminosity of the bulge, or $X = M_{\rm bulge}/(10^{11} M_\odot)$, where $M_{\rm bulge}$ is the stellar mass of the bulge. The values of α and β for different observables X are given in Table 16.1. The existence of such correlations is understood as evidence of the fact that the central massive BH formed and evolved in symbiosis with the bulge of its host galaxy. The largest BH masses measured in low-redshift galaxies, through measurement of line-of-sight stellar velocities in the central region of the host galaxy, are a few times $10^{10} M_\odot$.[3]

The presence of active galactic nuclei in many distant galaxies, and their interpretation as SMBHs caught during their active periods, indicates that SMBHs in the centers of galaxies were also common in the past. Currently well over 100 quasars are known at redshifts $z > 5.6$, corresponding to objects that formed in the first Gyr of the universe.[4] Currently, the most distant quasar observed, ULAS J1120+0641, is at a redshift $z \simeq 7.085$, and is powered by a central BH with a mass $2 \times 10^9 \, M_\odot$.[5] Thus, SMBHs were already common by that epoch.

A crucial question is how such SMBHs formed and evolved. The current understanding is that SMBHs evolved from initial seeds with masses in the range $(10^2 - 10^5) M_\odot$, at formation redshifts $10 \lesssim z \lesssim 15$. These seeds could be provided by the remnants of Population III stars. As we already mentioned in Chapter 10, these are the first generation of stars, which formed in an environment with essentially zero metallicity, and could reach masses in excess of $100 M_\odot$. Another possibility is that the seeds are formed directly from the collapse of dense gas clouds in the inner regions of gaseous proto-galaxies. These initial BHs further evolved through successive mergers. In CDM cosmology, dark matter halos and galaxies build up hierarchically, by the merger of smaller subunits. The consequences of these mergers on the morphology of the resulting galaxy are commonly observed, for instance with the Hubble Space Telescope. When two galaxies merge, the smaller galaxy sinks toward the center of the larger one because of dynamical friction,[6] while its outer regions are gradually tidally stripped. If the two galaxies harbor a central massive BH, the two massive BHs also sink toward the center of the common gravitational potential. Denoting the core radius of the post-merger galaxy by r_c, the (one-dimensional) velocity dispersions of the stars in the core by σ_c and the number of stars in the core by N_*, each of the two BHs initially sinks independently toward the center, on a time-scale

Table 16.1 Value of the parameters α and β in the $M - X$ relation, for different observables X. We quote the values given in McConnell and Ma (2013). Note that there are many studies of these relations and the values of α and β can vary (also significantly) from study to study.

X	α	β
σ	8.32	5.64
$M_{\rm bulge}$	8.46	1.05
L_i	9.23	1.11

[3]See McConnell *et al.* (2011).

[4]PanSTARRS1 currently has a sample of 124 quasars at $z > 5.6$, see Bañados *et al.* (2016), while SDSS has a sample of 52 quasars at $z > 5.7$, see Jiang *et al.* (2016).

[5]See Mortlock *et al.* (2011).

[6]See e.g. Section 8.1 of Binney and Tremaine (2008) for a discussion of dynamical friction in astrophysics.

of order of the Chandrasekhar dynamical friction time-scale,

$$t_i^{\mathrm{df}} \simeq \left(\frac{4 \times 10^6 \,\mathrm{yr}}{\log N_*}\right)\left(\frac{\sigma_c}{200\,\mathrm{km/s}}\right)\left(\frac{r_c}{100\,\mathrm{pc}}\right)^2\left(\frac{10^8 \, M_\odot}{m_i}\right), \quad (16.7)$$

where m_i is the mass of the i-th BH, $i = 1, 2$, and we define m_2 as the mass of the lighter BH. So, the time-scale for the two BHs to approach each other is given by t_2^{df}, the larger of t_1^{df} and t_2^{df}.[7]

As the lighter BH enters the sphere of influence of the heavier one a bound system begins to form, at the beginning just in the loose sense that the lighter BH is influenced more by the heavier BH than by the cumulative effect of the other stars. When a SMBH binary is formed, its subsequent evolution proceeds through successive stages. At first the semimajor axis $a(t)$ of the binary continues to shrink because of dynamical friction, which extracts orbital energy from the binary and therefore shrinks the orbit.[8] However, as the binary shrinks, its orbital period becomes shorter, and eventually this mechanism shuts off, because a binary cannot lose energy effectively to a star if the effective interaction time (determined from the impact parameter and velocity of the star) becomes longer than one orbital period. Thus, shrinking because of dynamical friction only takes place until the orbital velocity of the BHs in the binary is of the order of the velocity dispersion of the stars in the core. At this stage $a(t)$ has reached a value a_h given by

$$a_h \simeq \frac{Gm_2}{4\sigma_c^2}$$

$$\simeq 3.5\,\mathrm{pc} \left(\frac{180\,\mathrm{km/s}}{\sigma_c}\right)^2\left(\frac{m_2}{10^8 \, M_\odot}\right). \quad (16.8)$$

Note that in the following, we will use for definiteness $180\,\mathrm{km/s}$ as the reference value for σ_c. This is indeed of the order of the value given by the $M - \sigma$ relation (14.206) or (16.6) for a BH with mass $M = 10^8 \, M_\odot$, which we use as the reference value of the BH mass. More precisely, eq. (14.206), with the quoted values for the best-fit parameters, gives $\sigma_c \simeq 176\,\mathrm{km/s}$ for $M = 10^8 \, M_\odot$. If we apply these results to different BH masses, we must also change σ_c according to the $M - \sigma$ relation. For instance, for $M = 10^6 \, M_\odot$, eq. (14.206) gives $\sigma_c \simeq 78\,\mathrm{km/s}$.

After the orbital velocity has become of order σ_c, the binary is called a "hard" binary, because at this stage its binding energy becomes larger than the average kinetic energy of nearby single stars. As a consequence, in three-body encounters it statistically tends to give energy to intruder stars, thus becoming harder.

Using Kepler's law $\omega^2 = Gm/a^3$ and $\omega = 2\pi/P_b$, the binary's orbital period P_b corresponding to the value (16.8) of a is

$$P_b \simeq 6.1 \times 10^4 \,\mathrm{yr} \left(\frac{180\,\mathrm{km/s}}{\sigma_c}\right)^3\left(\frac{10^8 \, M_\odot}{m}\right)^{1/2}\left(\frac{m_2}{10^8 \, M_\odot}\right)^{3/2}, \quad (16.9)$$

where $m = m_1 + m_2$ is the total mass of the binary.

[7] More precisely, during the earlier phase of merger the lighter BH is still accompanied, during its infall, by the stars bound to it. Since the dynamical friction time-scale is inversely proportional to the mass of the bound object inspiralling into the heavier galaxy, in this phase the dynamical time-scale is much shorter than for a "naked" BH, and brings the lighter BH relatively quickly to a distance of order, say, 100 pc, from the center of the common gravitational potential, and hence from the heavier BH. By that time the lighter galaxy has been tidally stripped, and the lighter BH continues to fall toward the center with the longer time-scale given by eq. (16.7), with m_i now equal to the mass of the naked, lighter BH.

[8] Note the the dynamical friction time-scale for this bound object is no longer given by eq. (16.7); see e.g. Dosopoulou and Antonini (2017) for details and references.

Let us now compare these results with the time-scale for orbital shrinking due to GW emission. Using eq. (4.132) of Vol. 1 we see that, for a binary with major semiaxis a, the time to coalescence due to GW emission is

$$\tau_{\rm GW} = \frac{5}{256} \frac{c^5 a^4}{G^3 \mu m^2} \tag{16.10}$$

$$\simeq 5.81 \times 10^6 \,{\rm yr} \left(\frac{a}{0.01\,{\rm pc}}\right)^4 \left(\frac{10^8 M_\odot}{m_1}\right)^3 \frac{m_1^2}{m_2(m_1+m_2)},$$

where $\mu = m_1 m_2/m$ is the reduced mass and we have assumed that the eccentricity e is not too close to 1 [see eqs. (4.136) and (4.140) for the result with generic eccentricity]. Setting $a = 3.5$ pc in this expression, for the above reference values of the masses we get $\tau_{\rm GW} = O(10^{17})$ yr, which is $O(10^7)$ times longer than the age of the Universe. Thus, at the value of the separation where dynamical friction becomes ineffective, GW emission is still completely negligible for shrinking the binary further and bringing it to coalescence. Rather, we see from eq. (16.10) that, for our reference values of the masses, GW back-reaction becomes effective, and leads to further shrinking of the orbit and eventually to coalescence, only once a is of order of 0.01 pc, corresponding to a Keplerian period of order 30 yr. The question is therefore whether some mechanism can bridge the gap, shrinking the binary from $a = O(1)$ pc down to $a = O(0.01)$ pc, when eventually the back-reaction from GW emission can take over.

This issue, known as "the last-parsec problem", is responsible for one of the main uncertainties in the estimate of the abundance of SMBH binaries that are in the regime where their GW emission is significant. Stellar dynamics provides a first possible mechanism that can solve the last-parsec problem. Indeed, once the separation where dynamical friction is no longer effective has been reached, the binary further hardens through three-body interactions, in which it captures a star passing within a distance of order a, and ejects it at much higher velocity through the "slingshot" mechanism. Indeed, stars at a distance from the binary large compared with a see it as a single point-like mass and, when moving in its potential, neither gain nor lose energy. However, stars with low orbital angular momentum can pass very close to the binary. Their effective interaction time, determined by their impact parameter and their speed, can then be shorter than the orbital period of the binary. In this case, from the interaction with the binary, the star can either gain or lose energy, but on the average, at least after several encounters, it will eventually be ejected far from the binary, with a net gain of energy.

This requires stars with low angular momentum, which can get close enough to the binary to extract energy from it. The region in phase space with such small angular momentum is called the "loss cone". The stars that are in this loss cone are, however, gradually captured and ejected, resulting in loss-cone depletion, and to continue the hardening process it is necessary to have efficient mechanisms for refilling the loss cone. In a spherical galaxy, the loss cone can only be repopulated by two-body relaxation, while, if a galaxy is highly flattened or triaxial, the angular momenta of stars can also change because of the non-spherical galactic potential, leading again to a refilling of the loss cone. The efficiency of these mechanisms can be studied through N-body simulations. These indicate that, in spherical galaxies, loss-cone depletion takes place rapidly, while loss-cone refilling by two-body relaxation is too slow and is not sufficient to drive the binary in the GW-dominated regime within a Hubble time (except in the smallest galaxies). Thus, the binary stalls at parsec-scale separation. However, even a small degree of triaxiality, at a level of just a few percent, has a strong effect on loss-cone refilling, and leads to typical coalescence times ranging from a few Gyr for nearly spherical SMBH binaries to $\lesssim 10^8$ yr for very eccentric ones.[9] Thus, the last-parsec problem can be overcome in galaxies that have at least a small triaxiality. This is expected to be the case in most galaxies, because of their evolution through subsequent galaxy mergers. Note that loss-cone refilling could also take place by randomization of the stellar orbits in the galactic core due to the capture of a small satellite galaxy.

[9]See Vasiliev, Antonini and Merritt (2015), and the Further Reading.

A second possible mechanism that can solve the last-parsec problem is that the binary shrinks because of accretion of gas (much as in the phenomenon of planetary migration, which explains why giant gaseous exoplanets can be found close to their stars).

In the following we will assume that the last-parsec problem is indeed solved, and that the SMBH binaries eventually enter in the GW-dominated regime.

16.2.2   SMBH binaries at LISA

Figure 16.1 shows the planned LISA sensitivity curve, for a three-arm configuration, in terms of a dimensionless noise amplitude $h_n(f)$ defined in eq. (16.12) below, and compares it with the signal due to a number of possible sources. The gray shaded area is a "confusion signal" due to the superposition of $O(10^7)$ galactic binaries (mainly white dwarf binaries, but also binaries containing stellar-mass BHs or neutron stars). At higher frequencies one finds galactic binaries that are individually resolvable. With current estimates for the population, one expect to resolve at LISA about 2.5×10^4 galactic binaries. Some of these binaries are already known and characterized from electromagnetic observations, so their GW signal can be predicted, and they will serve as "verification binaries".

The plot also shows the track followed by the signal from SMBH binaries of different total mass M, for equal-mass binaries at a redshift $z = 3$. For coalescing binaries what is actually plotted is a characteristic amplitude *after* matched filtering, defined as follows. Recall from eqs. (15.5) and (15.6) that, after optimal matched filtering, the signal-to-noise ratio S/N is given by

$$\left(\frac{S}{N}\right)^2 = 4 \int_0^\infty df \, \frac{|\tilde{h}(f)|^2}{S_n(f)}, \qquad (16.11)$$

where $\tilde{h}(f) = \tilde{h}_+(f)F_+ + \tilde{h}_\times(f)F_\times$, and $F_{+,\times}(\theta, \phi)$ are the pattern

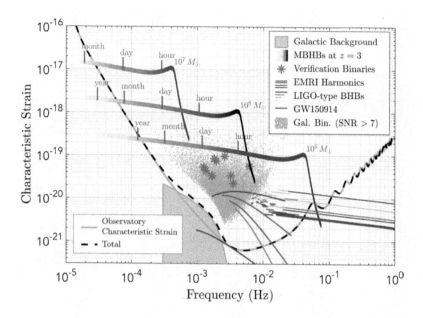

Fig. 16.1 Examples of GW sources in the frequency range of LISA, compared with its sensitivity for a three-arm configuration. From the LISA proposal, Audley *et al.* (2017).

functions, which depend on the source direction.[10] Recalling that the noise spectral density $S_n(f)$ is quadratic in the noise and has dimensions Hz^{-1}, we can define a characteristic dimensionless amplitude of the noise, $h_n(f)$, from

$$h_n^2(f) = f S_n(f).$$ (16.12)

Correspondingly, we can define a characteristic dimensionless amplitude of the signal, $h_c(f)$, from

$$h_c^2(f) = 4f^2 |\tilde{h}(f)|^2,$$ (16.13)

so that

$$\left(\frac{S}{N}\right)^2 = \int_0^\infty d\log f \, \frac{h_c^2(f)}{h_n^2(f)}.$$ (16.14)

For an inspiraling binary we can write

$$h_{+,\times}(t) = A_{+,\times}(t) \cos \Phi_{+,\times}(t),$$ (16.15)

where the amplitudes $A_{+,\times}(t)$ change slowly compared with the phases $\Phi_{+,\times}(t)$. We consider for simplicity a binary with optimal orbital inclination $\iota = 0$. Then, from eq. (4.29), at Newtonian order the plus and cross polarizations have the same amplitude $A_+(t) = A_\times(t) \equiv A(t)$, while $\Phi_+ \equiv \Phi(t)$ and $\Phi_\times = \Phi(t) - \pi/2$. As we discussed in Section 4.1.1 and in Problem 4.1, to compute the Fourier transform one must take into account that eq. (16.15) does not hold on the whole real axis $-\infty < t < \infty$,

[10]For LISA the pattern functions also evolve with time because LISA will be in an Earth-trailing heliocentric orbit between 50 and 65 millions km from the Earth, and along the orbit the orientation of the arms changes.

since eventually the binary coalesces. Nevertheless, its Fourier transform can be computed by a saddle point evaluation of the integral

$$\tilde{h}_{+,\times}(f) = \int_{-\infty}^{t_{\rm coal}} dt\, A(t) \cos \Phi_{+,\times}(t) e^{2\pi i f t}. \qquad (16.16)$$

This is justified as long as the saddle point value is at $t = t_*(f)$ such that $t_* < t_{\rm coal}$, where $t_{\rm coal}$ is the maximum value of t for which we are still in the inspiral phase. We already computed this integral for the two polarizations in eq. (4.366), with the result

$$\tilde{h}_{+,\times}(f) = \frac{1}{2} e^{i\Psi_{+,\times}} A(t_*) \left(\frac{2\pi}{\dot{\Phi}(t_*)} \right)^{1/2}. \qquad (16.17)$$

Using the explicit expression for the inspiral amplitude, we found in eq. (4.369) that (for $\iota = 0$)

$$\frac{1}{2} A(t_*) \left(\frac{2\pi}{\ddot{\Phi}(t_*)} \right)^{1/2} = \alpha_0 \frac{c}{r} \left(\frac{G\mathcal{M}_c}{c^3} \right)^{5/6} f^{-7/6}, \qquad (16.18)$$

[11]In this section we denote $\omega_{\rm gw}$ and $f_{\rm gw}$ simply as ω and f, respectively.

where $\alpha_0 = (5/24)^{1/2} \pi^{-2/3} \simeq 0.21$ and $\mathcal{M}_c = (1+z)M_c$. Alternatively, we can also rewrite this result observing that $\dot{\Phi} = \omega$; see eq. (4.28).[11] Therefore $\dot{\omega} = \ddot{\Phi}$ and $\dot{f} = \ddot{\Phi}/(2\pi)$. We can therefore write eq. (16.17) in the form

$$\tilde{h}_{+,\times}(f) = e^{i\Psi_{+,\times}} \frac{A(f)}{2\dot{f}^{1/2}}, \qquad (16.19)$$

where $A(f) \equiv A[t_*(f)]$. We consider for definiteness a binary with optimal direction with respect to one of the two polarizations, so for example $F_+ = 1$ and $F_\times = 0$.[12] Therefore, in eq. (16.13),

$$|F_+ \tilde{h}_+(f) + F_\times \tilde{h}_\times(f)|^2 = \frac{A^2(f)}{4\dot{f}}. \qquad (16.20)$$

[12]Once again, for LISA, strictly speaking this cannot be the case for a source that one observes for a long time, given that the pattern functions evolve because of the motion of LISA along its orbit. However, this approximation is quite sufficient at the level of Fig. 16.1, where one just wishes to illustrate the typical size of the signals compared to the detector's noise. Rather than taking a source with optimal direction, one could average over the arrival directions, using the fact that, for interferometers, $\langle F_+^2 \rangle = \langle F_\times^2 \rangle = 1/5$ while $\langle F_+ F_\times \rangle = 0$, where the brackets denote the angular average over the arrival direction of the wave; see Section 7.2. This would give an extra factor $\sqrt{2/5}$ in eq. (16.21). The actual search and data analysis will of course take into account the evolution of the pattern functions.

Thus, from eq. (16.13), for inspiraling binaries a useful measure of the signal, to be compared with the dimensionless noise amplitude $h_n(f)$, is given by the quantity

$$h_c(f) = \frac{A(f)\, f}{\dot{f}^{1/2}}, \qquad (16.21)$$

This is the quantity shown in Fig. 16.1 for inspiraling SMBH binaries and for EMRIs, and compared with the dimensionless noise curve $h_n(f)$. It should be stressed that $h_c(f)$ has been defined in terms of the signal-to-noise ration (SNR, or S/N) obtained with optimal matched filtering. It is therefore a measure of the level of the signal that can be obtained after it has been enhanced, with respect to the noise, by optimal filtering (assuming that one has developed accurate enough templates), and not a measure of the GW amplitude itself. As we see from eq. (16.21), it differs from the dimensionless "raw" GW amplitude $A(f)$ by a factor $f/\dot{f}^{1/2}$. This factor can be understood by observing that a signal whose frequency evolves with a time derivative $\dot{f}$ sweeps in a time Δt

a frequency range $\Delta f = \dot{f}\Delta t$. The corresponding number of cycles is $\mathcal{N}_c \simeq f\Delta t \simeq f\Delta f/\dot{f}$; compare with eq. (4.22). The number of cycles that the binary spends at a frequency of order f is obtained by setting $\Delta f \sim f$, which gives

$$\mathcal{N}_c(f) \sim \frac{f^2}{\dot{f}}\,. \tag{16.22}$$

Then eq. (16.21) shows that

$$h_c(f) \sim A(f)\mathcal{N}_c^{1/2}(f)\,. \tag{16.23}$$

This is in agreement with the discussion after eq. (7.186), where we indeed found that, following an inspiral signal for $\mathcal{N}_c(f)$ cycles allows us to increase the SNR by a factor $\mathcal{N}_c^{1/2}(f)$.

Figure 16.1 also shows the evolution in frequency of the SMBH signal in the LISA bandwidth. Using eq. (14.108) with reference values for the masses useful for SMBHs at LISA, and redshifting the frequency by a factor $1/(1+z)$ to obtain the frequency in the observer frame, we see that for an equal-mass spinless binary the inspiral phase terminates when the GW frequency (in the observer frame) reaches the value

$$f_{\rm ISCO} \simeq 4.7 \times 10^{-3}\,{\rm Hz}\,\left(\frac{1}{1+z}\right)\left(\frac{10^6 M_\odot}{M}\right), \tag{16.24}$$

(where here we denote the total BH mass by M). We saw in Chapter 14 that, during the subsequent plunge and merger phase, the GW frequency increases to a value about equal to the frequency of the first quasi-normal mode of the final Kerr BH, given in eq. (14.110). The latter can be rewritten as

$$f_{\rm qnm} \simeq 1.9 \times 10^{-2}\,{\rm Hz}\,\left(\frac{1}{1+z}\right)\left(\frac{10^6 M_\odot}{M}\right). \tag{16.25}$$

In Fig. 16.1, along the tracks of the signals due to SMBH binaries, are marked the frequencies when the time to coalescence is 1 year, 1 month, 1 day or 1 hour. The time to coalescence as a function of the frequency, in the inspiral phase, was given in eq. (4.21), which we can rewrite as

$$\tau \simeq 3.0 \times 10^5 \left(\frac{10^5 M_\odot}{M_c}\right)^{5/3}\left(\frac{1}{1+z}\right)^{5/3}\left(\frac{10^{-3}\,{\rm Hz}}{f}\right)^{8/3}, \tag{16.26}$$

where, as usual, for binaries at cosmological distance we have replaced $M_c \to (1+z)M_c$; see Section 4.1.4. For instance, we see from Fig. 16.1 that an equal-mass binary with $M = 10^5\,M_\odot$ at $z = 3$ enters the LISA bandwidth when $f \simeq 10^{-4}\,{\rm Hz}$. Then, using the fact that for an equal-mass binary of total mass M the chirp mass is $M_c = M/2^{6/5} \simeq 0.43M$, we get $\tau \simeq 1$ yr. Binaries with a larger chirp mass enter the LISA bandwidth at lower frequencies, because the amplitude of their GWs is larger, so they are already visible over the noise when the noise curve is higher. However, they also evolve faster, since $\dot{f} \propto M_c^{5/3}$; see eq. (4.18).

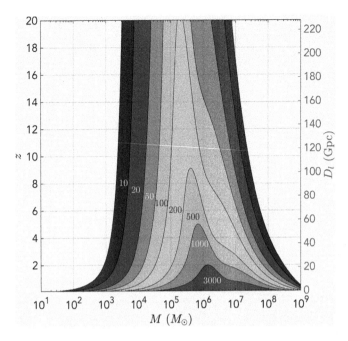

Fig. 16.2 The SNR for SMBH binaries at LISA, in the (z, M) plane. The vertical scale on the right axis gives the corresponding luminosity distance assuming a ΛCDM cosmology with the *Planck* 2015 cosmological parameters. From the LISA proposal, Audley *et al.* (2017).

[13]See the LISA proposal, Audley *et al.* (2017), for a full discussion of the Mission Requirements of LISA.

Then, for instance, an equal-mass binary with $M = 10^6 \, M_\odot$ at $z = 3$ enters the LISA bandwidth $2 - 3$ months before coalescence.

Figure 16.2 shows the SNR after optimal matched filtering in the (z, M) plane, for SMBH binaries at LISA, taking for definiteness a mass ratio $q = 0.2$. Several aspects of this figure are noteworthy, corresponding to different "Mission Requirements" of LISA.[13] For instance, near the lower end of the accessible mass range, SMBH binaries with a total mass M a few times $10^3 M_\odot$ can be visible up to redshift $z \sim 15$ with a SNR of at least 10. This SNR would allow determination of the two component masses and of the luminosity distance with an error of about 20%, which will allow discrimination among different models for the initial seeds. In the mass range $10^4 M_\odot < M < 10^6 M_\odot$ one can detect binaries at $z \lesssim 9$, with an accuracy sufficient to determine the BH masses at the level of 5%, limited by weak lensing, while for sources at $z < 3$ and $10^5 M_\odot < M < 10^6 M_\odot$ the SNR will be better than 200, allowing determination of the dimensionless spin parameter of the largest BH to an accuracy better than 0.1. For binaries with $10^6 M_\odot < M < 10^7 M_\odot$ at $z \sim 2$, the SNR would be sufficiently large to obtain a sky localization of $100 \deg^2$ at least one day before coalescence, allowing observatories to be alerted to search for electromagnetic counterparts. After the merger the sky localization will be significantly improved, down to possibly a few $\deg^2$, allowing for follow-up electromagnetic observations.

16.3   Extreme mass ratio inspirals

An extreme mass ratio inspiral (EMRI) event is the inspiral of a stellar-mass compact object, such as a white dwarf, a neutron star, or a stellar-mass BH, into a SMBH at a center of a galaxy (while a normal star would be tidally disrupted by the central BH).

EMRI events are very interesting because, as we will see in more detail below, depending on the chirp mass of the system the inspiral signal can stay in the LISA bandwidth for years, and the binary may spend many cycles near the ISCO. This would in principle give an accurate mapping of the space-time around the SMBH, allowing us to test its Kerr geometry and providing accurate measurements of its mass and spin. Furthermore, the rate of EMRI events depends on stellar dynamics in the dense stellar environment of the galactic center, where, for instance, in the inner few pc the number density of stars can be in excess of $10^6 \, \mathrm{pc}^{-3}$ and two-body scattering between stars and other dynamical phenomena play a major role. The detection of EMRI events would therefore also give important information on the astrophysics of dense stellar systems.

Another very interesting potential source is from intermediate mass ratio inspirals (IMRI), in which a BH with mass $(10^2 - 10^4) M_\odot$ inspirals into a SMBH. IMRIs events are somewhat more speculative, since the evidence for BHs in this mass range is only indirect. However, if a single IMRI event were to take place within the lifetime of the LISA mission, it would be so strong that, by the end of the inspiral, its signal would be detectable in a time–frequency spectrogram without any matched filtering.

In this section we will discuss the formation mechanism of EMRIs, the potential of LISA in their search, and the challenges to be faced for providing accurate EMRI waveforms.

16.3.1   Formation mechanisms

The formation of a binary made of a central SBMH and a stellar-mass BH follows a different path compared with the formation of a SMBH binary discussed in Section 16.2.1. The sinking of a SMBH from the outer region of a galaxy toward its center because of dynamical friction is specific to its large mass, as we see from eq. (16.7), and this mechanism is irrelevant for the lighter BH, which in interactions with other stars of comparable mass can in principle gain as well as lose energy. However, the nucleus of a galaxy, with a size of a few pc, consists of a cluster of $10^7 - 10^8$ stars, among which there will be in general compact objects such as stellar-mass BHs. In this very dense environment, with stellar densities higher than $10^6 \, \mathrm{pc}^{-3}$ and large relative stellar velocities of order $10^2 - 10^3 \, \mathrm{km/s}$, collisions play an important role. An EMRI can form if successive two-body encounters bring a stellar-mass BH into an orbit sufficiently close to the central SMBH, that GW back-reaction can take over and further shrink the orbit toward coalescence. Actually, even if, because of two-body encounters, the stellar-mass BH is sent on an elliptic orbit sufficiently close that GW back-reaction becomes important, it is still possible that, in such a dense environment, a further two-body encounter (typically taking place near apocenter, in the case of a very eccentric orbit) sends the stellar-mass BH again in an orbit wide enough that GW back-reaction becomes irrelevant. So, what is needed is that two-body relaxation sends a stellar-mass BH on an orbit sufficiently close that the time-scale due to GW back-reaction becomes (much) smaller than the time-scale of further two-body encounters. Actually, a two-body encounter can even send the stellar-mass BH directly on a plunge orbit, in which case it will directly merge with the central SMBH. This case is less interesting for LISA since it would produce a GW burst, rather than a long inspiral signal, and for an extragalactic event the observation of the signal over $O(10^4)$ inspiral cycles is necessary to extract it from the noise.[14]

[14]Such a burst could, however, be detectable at LISA if it originated from a compact object falling into the SMBH in the center of our Galaxy; see Hopman, Freitag and Larson (2007).

Another possible formation channel for EMRIs consists of a binary system that includes a stellar-mass BH among its components, on an orbit that brings it close to the SMBH. In this three-body encounter, one member of the binary is captured by the central BH, while the other is ejected at high speed. A third formation scenario that has been considered is the formation of a massive star in the accretion disk around the SMBH, followed by the rapid evolution of this star, leading to a stellar-mass BH remnant. These formation scenarios differ in their predictions for the typical orbit of the EMRI. In the case in which the EMRI is formed through two-body relaxation, the orbit is expected to have a large eccentricity, with a generic inclination with respect to the spin of the central SMBH. In the capture of a stellar-mass BH from a binary, the orbit would instead be close to circular, but again with a generic inclination. In contrast, massive star formation in a disk around the central BH would give rise to EMRIs with circular orbits in the equatorial plane of the central spinning BH. The theoretical uncertainty in the rate of formation of EMRI is relatively large, and typical estimates are of order $10^{-8} - 10^{-6} \mathrm{yr}^{-1}$ for a central SMBH with a mass $m_1 = 10^6 M_\odot$, corresponding to a detection rate for LISA ranging from a few to hundreds of events per year.

As we have mentioned above, another potentially very interesting source for LISA is from IMRIs, in which the mass m_2 of the lighter BH is intermediate between that of a stellar-mass BH and the mass of a typical central SMBH, say $m_2 \sim (10^2 - 10^4) M_\odot$. As we have discussed in Section 10.3.3, stars lose mass by stellar winds, whose strength increases with the star's metallicity Z. The first generation of stars, Population III stars, which formed in an environment with very low metallicity, were very massive, and could have produced a population of BHs with masses of several hundred solar masses. It has also been suggested that BHs with masses of order $10^3 M_\odot$ can form in compact massive young star clusters. Observationally, IMRIs could be related to ultra-luminous X-ray (ULX) sources. These are variable X-ray sources, with a luminosity $10^{39} - 10^{40}$ erg/s, which (assuming isotropic emission) is higher than the luminosity of any known stellar process, but still much smaller that the luminosity of active galactic nuclei. It has been suggested that ULXs could be due to accretion on intermediate-mass BHs.

16.3.2   EMRIs at LISA

As we see from eq. (16.24), an EMRI into a SMBH with mass $m_1 \simeq 10^5 M_\odot$ at $z \simeq 1$ would produce GWs sweeping in frequency in the inspiral phase up to $f \simeq 2 \times 10^{-2}$ Hz, in the region where LISA has its best sensitivity. We have also seen that EMRIs can have a large eccentricity. As we already discussed in Section 4.1.2, the corresponding signal has a set of harmonics, which in an inspiraling binary evolve simultaneously in time, giving in principle a very specific signature.[15]

Figure 16.1 shows the signals produced by the first five harmonics of an EMRI, as a set of five bending tracks between about 10^{-3} and a few times 10^{-2} Hz, for a system with $(1+z)m_1 = 4 \times 10^5 M_\odot$ and $(1+z)m_2 = 20 M_\odot$ at $z = 1.2$, and eccentricity at plunge time $e = 0.05$, assuming 4 years of observation before the plunge, and zero spin for the SMBH. We see that for this system the first four harmonics fall above the LISA sensitivity curve. Just as for the signal from SMBH binaries, these curves correspond to the quantity $h_c(f)$, which gives a measure of the signal *after* matched filtering. They therefore assume that accurate templates for the sources have been developed, a point that we will discuss further in Section 16.3.3. In contrast, before matched filtering the "raw" GW amplitude of a typical EMRI event at LISA is expected to be below the instrumental noise.

Matched filtering can be particularly effective for EMRIs because they can spend a very large number of inspiral cycles in the LISA bandwidth. Indeed, using eq. (4.23) with reference values for f_{min} and M_c appropriate to LISA, and performing as usual the replacement $M_c \to (1+z)M_c$ for cosmological binaries, the number of cycles spent in the detector bandwidth, for a GW inspiral signal that starts to be detectable at a frequency f_{min}, is

$$\mathcal{N}_{cyc} \simeq 1.0 \times 10^5 \left(\frac{2\,\text{mHz}}{f_{min}}\right)^{5/3} \left(\frac{10^3\,M_\odot}{M_c}\right)^{5/3} \left(\frac{2}{1+z}\right)^{5/3} . \quad (16.27)$$

Note that the number of cycles scales as $M_c^{-5/3}$. In a system with $m_2 \ll m_1$, the chirp mass is much smaller than the total mass,

$$M_c \equiv \frac{(m_1 m_2)^{3/5}}{(m_1 + m_2)^{1/5}} \simeq q^{3/5} m_1 , \quad (16.28)$$

where $q = m_2/m_1 \ll 1$, and therefore the number of cycles is correspondingly larger. For instance, for a binary with $m_1 = 10^6 M_\odot$ and $m_2 = 10 M_\odot$, we have $M_c \simeq 10^3 M_\odot$, and a system at $z \simeq 1.2$, which according to Fig. 16.1 starts to be visible at $f_{min} \sim 2\,\text{mHz}$, spends about 10^5 inspiral cycles in the LISA bandwidth.

[15] More precisely, for a generic orbit one has three fundamental frequencies corresponding to radial, azimuthal and polar motion, and the corresponding harmonics.

This large number of cycles allows in principle extremely precise measurements of the parameters of the system. EMRIs are therefore a potentially very rich source of information. Their orbits are quite relativistic, with generic inclination with respect to the spin of the central BH, and display extreme forms of periastron and orbital precession. They can therefore provide a real "map" of the geometry produced by the central supermassive Kerr BH.

LISA will have the ability to detect EMRIs with a total mass of a few times $10^5 M_\odot$ up to redshift $z = 4$ with a SNR $\gtrsim 20$. This would allow the measurement of the total mass M with an accuracy $\delta M/M < 10^{-4}$, of the mass m_2 of the stellar-mass compact object to $\delta m_2/m_2 < 10^{-3}$, and of the spin of the central SMBH to an accuracy of one part in 10^3. 'Golden' EMRIs, i.e. events detected with SNR > 50 and dimensionless spin parameter $\hat{a}_2 > 0.9$ in a prograde orbit, would allow measurements at the level $\delta M/M \simeq 10^{-5}$, $\delta m_2/m_2 \simeq 10^{-4}$ and $\delta\hat{a}_1/\hat{a}_1 \simeq 10^{-4}$.

At the level of fundamental physics, EMRIs can be used to test the no-hair theorem. In particular, for a Kerr BH with mass M and spin parameter a [as defined in eq. (12.286), so that a has dimensions of length] all higher-mass multipoles M_l and spin multipoles S_l are fixed in terms of M and a by

$$M_l + iS_l = M\,(ia)^l\,. \tag{16.29}$$

EMRIs can be used to detect deviations of the quadrupole moment from that predicted by the Kerr geometry, at the level of 0.1%.

As we discussed in Section 4.1.4, coalescing binaries are the gravitational equivalent of standard candles. To stress the fact that GW detection is more similar to detecting sounds than light, because of the almost omnidirectional nature of the detectors, they are also usually called "standard sirens". Using binaries made of stellar-mass compact objects at $z < 0.2$, EMRIs at $z < 1.5$ and SMBH binaries at $z < 6$, with a four-year mission LISA could measure the Hubble parameter to an accuracy $\delta H_0/H_0$ better than 0.02. This can provide important cosmological information, particularly on the nature of dark energy.[16]

At the level of data analysis the large number of cycles spent in the detector's bandwidth gives the possibility of extracting a small signal from the noise. However, the computational problem becomes formidable. We need templates able to track the signal accurately enough for $10^4 - 10^5$ cycles. This not only requires very accurate theoretical waveforms, but also the scanning of a very fine grid in the parameter space of the binary (where, compared with the discussion in Section 15.1.1, we must now also include the eccentricity of the binary, which in general will be large), although, as discussed in Sections 7.7.1 and 15.1.1, at the level of the search, a number of "extrinsic" parameters can be eliminated analytically.[17] Search techniques have been developed to face these challenges. In particular, one can split the search into shorter time segments where one performs a coherent search, and add up incoherently the power in the coherent segments (similarly to the technique used for blind searches of periodic sources, which we discussed in Section 7.6.3).

---

[16]We will discuss standard sirens in more detail in Section 19.6.

[17]However, in contrast to the case of ground-based interferometers, at LISA the two angles giving the position in the sky of the source are not among the extrinsic parameters, since the motion of LISA in its orbit introduces a Doppler modulation in the signal, which depends on the sky position of the source.

Alternatively, one can use time–frequency techniques, similarly to those used at ground-based interferometers (see the bottom row of Fig. 15.1 for the application to GW150914), or the parameter space of templates can be explored with Markov chain Monte Carlo techniques. The detection of EMRIs at LISA will be further complicated by the fact that $O(10^7)$ galactic white-dwarf binaries generate a stochastic background that is above the LISA noise floor for frequencies below $2 - 3$ mHz, while above a few mHz as many as $O(10^3 - 10^4)$ galactic white-dwarf binaries will be individually resolvable; see Fig. 16.1. One also expects that there will be a stronger signal from the merger of a few SMBH binaries, and all these signals will be superimposed in time and frequency. Therefore, one needs techniques for separating the various signals, removing the stronger source before being able to see the signals due to EMRIs.

16.3.3   Waveforms and the self-force approach

A final highly non-trivial element is the theoretical computation of templates able to follow the actual waveforms with high accuracy over $10^4 - 10^5$ cycles. For EMRIs, the PN expansion is not adequate, since EMRIs are quite relativistic, and the PN expansion converges slowly for $v/c \gtrsim 0.3$. Numerical relativity is also not suitable, since, as we mentioned in Chapter 14, relatively long numerical simulations are currently only possible for mass ratios up to about 4–5, due to the numerical complications introduced by the different length-scales associated with the two BHs (and, even in that case, they cover "only" $O(10^2)$ inspiral cycles).

We can, however, make use of the fact that, in EMRIs, the mass ratio $q = m_2/m_1$ is small, for example of order 10^{-5}, and we can perform an expansion in this parameter. A technique that is particularly suitable in this situation is the "self-force" approach, which has been greatly developed in the last two decades. We already discussed radiation reaction in Section 5.3.5 of Vol. 1, in the context of the coalescence of a compact binary with generic mass ratio. The situation becomes simpler, however, in the limit in which one of the two masses is very light compared with the other, and can be treated as a test mass (even if, as we will see, conceptual and technical issues related to the self-force approach are quite complicated even in this limit). Here we give just a flavor of this method, referring to the Further Reading for more details.[18]

The radiation-reaction problem has a long history, and for electromagnetism it goes back to works by Lorentz, Abrahams, Poincaré and Dirac. It is indeed instructive to start from the back-reaction problem for an electric charge moving in flat space-time, under the influence of some other external force. The typical situation would be an electron moving in a bound orbit in the field created by a positive charge, taken to be very massive. The motion of the electron generates an electric current j^μ, which sources the electromagnetic field. In the Lorentz gauge $\partial_\mu A^\mu = 0$ the electromagnetic field satisfies the equation

$$\Box A^\mu = j^\mu \,.$$
(16.30)

[18]In particular, we closely follow here the very clear reviews by Poisson (2004b) [updated in Poisson, Pound and Vega (2011)] and by Barack (2009).

If we assume that the electron is point-like, the current j^μ has a support only on the electron's world-line. Given the trajectory of the electron, j^μ induces radiative solutions at infinity for A^μ in eq. (16.30). These outgoing electromagnetic waves carry away energy and angular momentum. The electron will therefore lose energy and angular momentum, and will inspiral onto the central massive charge, at least as long as the classical description is still appropriate. In other words, the electromagnetic field generated by the electron acts back on the electron itself, producing a back-reaction force, which should be included in the equations of motion of the electron, at least iteratively.

When solving eq. (16.30) to compute the electromagnetic waves produced at infinity, one uses the retarded Green's function, leading to a solution

$$A^\mu_{\rm ret}(x) = \int d^4x' \, G_{\rm ret}(x; x') j^\mu(x') \, . \tag{16.31}$$

A problem that immediately arises in the computation of the self-force on the electron is that the field $A^\mu_{\rm ret}(x)$ generated by the electron is singular at the position of the electron itself, so one might fear that the self-force is also divergent. One can, however, observe that

$$A^\mu(x) = \frac{1}{2}\left[A^\mu_{\rm ret}(x) + A^\mu_{\rm adv}(x)\right] + \frac{1}{2}\left[A^\mu_{\rm ret}(x) - A^\mu_{\rm adv}(x)\right] \, . \tag{16.32}$$

The first term, which is the symmetric combination of the retarded and advanced solutions, is symmetric under time reversal, which exchanges the retarded and advanced Green's functions. It is therefore the appropriate solution of eq. (16.30) with the time-symmetric boundary condition that there is a flux of electromagnetic radiation coming from infinity that exactly balances the flux radiated by the charge at infinity. In this situation, energy balance is automatically satisfied between the incoming and outgoing radiation, and there is no back-reaction on the electron. This means that the symmetric combination does not contribute to the radiation reaction force, and only the antisymmetric combination contributes. This is very welcome, since the singularities of $A^\mu_{\rm ret}(x)$ and $A^\mu_{\rm adv}(x)$ on the electron world-line are the same. This means that the symmetric combination is singular on the electron world-line, but does not contribute to the self-force. The antisymmetric combination is instead regular on the world-line, and gives the whole contribution to the self-force. Defining $A^\mu_R = (1/2)(A^\mu_{\rm ret} - A^\mu_{\rm adv})$, where the subscript R stands for "regular", and introducing its field strength tensor $F^R_{\mu\nu} = \partial_\mu A^R_\nu - \partial_\nu A^R_\mu$, the equation of motion of a charged particle of mass m and charge e, in the presence of an external force $f^{\rm ext}_\mu$ and including the back-reaction, turns out to be

$$ma_\mu = f^{\rm ext}_\mu + \frac{e}{c} F^R_{\mu\nu} u^\nu \, , \tag{16.33}$$

where $u^\mu = dx^\mu/d\tau$ is the charge's four-velocity. Computing explicitly A^μ_R on the particle world-line, one finally obtains

$$ma^\mu = f^\mu_{\rm ext} + \frac{2e^2}{3mc}\left(\delta^\mu_\nu + u^\mu u_\nu\right)\frac{df^\nu_{\rm ext}}{d\tau} \, . \tag{16.34}$$

The second term on the right-hand side gives the back-reaction force.[19] Note that it is $O(e^2)$ with respect to the first term. A careful analysis shows that the symmetric combination, which is singular on the world-line, does indeed not affect the equation of motion, but gives a divergence that can be reabsorbed into a renormalization of the mass of the particle. As long as the back-reaction force in eq. (16.34) is small compared with the external force, we can use this result to compute for example the classical evolution of a bound electron through a series of quasi-circular orbits.

One would like to repeat a similar analysis for a light mass moving in a gravitational field, which is the typical situation of EMRIs. At first sight, one might think that in general relativity (GR) there is no place for the notion of gravitational "force": in GR, test masses simply move on geodesics. But, in fact, a useful notion of gravitational force can be introduced as follows. Consider a particle moving in a metric $g'_{\mu\nu} = g_{\mu\nu} + h_{\mu\nu}$, where $h_{\mu\nu}$ is a small perturbation. The exact geodesic equation in this metric is

$$\frac{d^2 x^\mu}{d\tau'^2} + \Gamma'^\mu_{\nu\rho} \frac{dx^\nu}{d\tau'} \frac{dx^\rho}{d\tau'} = 0, \qquad (16.35)$$

where $\Gamma'^\mu_{\nu\rho}$ are the Christoffel symbols in the metric $g'_{\mu\nu}$, and τ' is an affine parameter along the trajectory. Expanding to first order in $h_{\mu\nu}$, this can be rewritten as[20]

$$\frac{d^2 x^\mu}{d\tau^2} + \Gamma^\mu_{\nu\rho} \frac{dx^\nu}{d\tau} \frac{dx^\rho}{d\tau} = a^\mu_{\text{grav}}, \qquad (16.36)$$

where $\Gamma^\mu_{\nu\rho}$ are the Christoffel symbols in the metric $g_{\mu\nu}$, and

$$a^\mu_{\text{grav}} = -\frac{1}{2}(g^{\mu\nu} + u^\mu u^\nu)(D_\rho h_{\nu\sigma} + D_\sigma h_{\nu\rho} - D_\nu h_{\rho\sigma})u^\rho u^\sigma, \qquad (16.37)$$

where D_μ is the covariant derivative with respect to $g_{\mu\nu}$. The particle moves along the geodesic of the full metric $g'_{\mu\nu} = g_{\mu\nu} + h_{\mu\nu}$, so of course it does not obey the geodesic equation with respect to $g_{\mu\nu}$. However, we see from eq. (16.36) that, to first order in $h_{\mu\nu}$, its equation of motion can be formally written as a geodesic equation with respect to $g_{\mu\nu}$ supplemented by a "gravitational force" $F^\mu_{\text{grav}} = m a^\mu_{\text{grav}}$, where m is the mass of the particle and a^μ_{grav}, given in eq. (16.37), is $O(h)$. Observe that F^μ_{grav} is gauge-dependent, as can be checked by transforming $h_{\mu\nu}(x) \to h_{\mu\nu}(x) - (D_\mu\xi_\nu + D_\nu\xi_\mu)$; see eq. (1.215). This is to be expected since, because of the equivalence principle, there can be no diffeomorphism-invariant notion of gravitational force in GR.

To make contact with the radiation reaction problem, we can now consider a point-like particle of mass m moving in an external metric $g_{\mu\nu}$, such as the Kerr metric. The particle radiates GWs at infinity, so there will be a back-reaction on it. Since the presence of the particle deforms the total metric into $g_{\mu\nu} + h_{\mu\nu}$, one might be tempted to use $h_{\mu\nu}$ in eq. (16.37) to obtain a gravitational self-force. In this form, this is not correct for several reasons. First of all, we meet the same problem as

[19] Here one has replaced a term $da^\nu/d\tau$ with $m^{-1}df^\nu_{\text{ext}}/d\tau$ on the right-hand side. This reduction-of-order procedure allows one to eliminate the well-known unphysical runaway solution, and can be justified order by order in the back-reaction expansion. See Poisson, Pound and Vega (2011) for a discussion.

[20] See Barack (2009) for the explicit computation.

for the electromagnetic field, namely that $h_{\mu\nu}$ evaluated on the particle world-line is singular. Clearly, the self-force must be determined by a combination in which the singular part is subtracted, similar to A_R^μ for the electromagnetic field. Second, the notion of point-like particle is problematic in GR beyond linearized theory. A first-principles procedure is rather to proceed similarly to what we did in Section 5.2 when we wrote the Einstein equations in the "relaxed" form. In this case it is convenient to use the variable

$$\gamma^{\alpha\beta} = h^{\alpha\beta} - \frac{1}{2} g^{\alpha\beta}(g_{\gamma\delta} h^{\gamma\delta}), \qquad (16.38)$$

rather than the variable $h^{\alpha\beta}$ used in Section 5.2. The Einstein tensor in the metric $g_{\mu\nu} + h_{\mu\nu}$ can be written *exactly* as

$$G^{\alpha\beta}[g+h] = \bar{G}^{\alpha\beta}[g] + \delta G^{\alpha\beta}[g,h] + \Delta G^{\alpha\beta}[g,h], \qquad (16.39)$$

where $\bar{G}^{\alpha\beta}[g]$ is the Einstein tensor computed with the background metric $g_{\mu\nu}$, which is assumed to vanish, $\delta G^{\alpha\beta}[g,h]$ is the part linear in $h^{\alpha\beta}$ (or, equivalently, in $\gamma^{\alpha\beta}$) and $\Delta G^{\alpha\beta}[g,h]$ contains all the terms non-linear in γ. The explicit form of $\delta G^{\alpha\beta}$ is

$$\delta G^{\alpha\beta}[g,h] = -\frac{1}{2} \left(\Box \gamma^{\alpha\beta} + 2 R_\gamma{}^\alpha{}_\delta{}^\beta \gamma^{\gamma\delta} \right)$$
$$+ \frac{1}{2} \left(D^\beta D_\gamma \gamma^{\alpha\gamma} + D^\alpha D_\gamma \gamma^{\beta\gamma} - g^{\alpha\beta} D_\gamma D_\delta \gamma^{\gamma\delta} \right), \quad (16.40)$$

where $\Box \gamma^{\alpha\beta} = g^{\gamma\delta} D_\gamma D_\delta \gamma^{\alpha\beta}$. After imposing the Lorentz gauge condition

$$D_\beta \gamma^{\alpha\beta} = 0, \qquad (16.41)$$

the term in the second line in eq. (16.40) vanishes, and the *exact* Einstein equations can be written as

$$\Box \gamma^{\alpha\beta} + 2 R_\gamma{}^\alpha{}_\delta{}^\beta \gamma^{\gamma\delta} = -\frac{16\pi G}{c^4} T_{\text{eff}}^{\alpha\beta}, \qquad (16.42)$$

where $T_{\text{eff}}^{\alpha\beta}$ contains both the actual energy–momentum tensor of matter, and the non-linearities in γ,

$$T_{\text{eff}}^{\alpha\beta} = T^{\alpha\beta} - \frac{c^4}{8\pi G} \Delta G^{\alpha\beta}[g,h]. \qquad (16.43)$$

Conceptually, this is quite similar to the formulation of the Einstein equation that we discussed in eqs. (5.72) and (5.73), leading to the "relaxed" Einstein equations. Just as we did for the relaxed Einstein equations, we can now observe that eq. (16.42) makes perfect sense independently of the gauge condition (16.41). We can therefore first solve eq. (16.42) for generic $\gamma^{\alpha\beta}$, and after that impose the gauge condition on the solution. Just as for the relaxed Einstein equation, imposing the gauge condition is equivalent to imposing $D_\alpha T_{\text{eff}}^{\alpha\beta} = 0$, which in turn is equivalent to imposing that the particle moves on the exact geodesics of $g_{\alpha\beta} + h_{\alpha\beta}$. In this way one could obtain in principle the exact equations of motion. In practice, one must perform some approximation when integrating the exact, and therefore highly non-linear, wave equation (16.42), and here enters the expansion in powers of the mass of the particle. In this way, one finds that the gravitational self-force is indeed given by eqs. (16.36) and (16.37), together with some prescription for subtracting the singular part from $h_{\alpha\beta}$. If one works within the point-particle approximation, the best one can do is to observe that $h_{\alpha\beta}$ can be split into

$$h_{\alpha\beta} = h_{\alpha\beta}^{\text{dir}} + h_{\alpha\beta}^{\text{tail}}, \qquad (16.44)$$

where $h_{\alpha\beta}^{\text{dir}}(x)$ is a "direct" contribution, propagating *on* the light cone, and coming from the intersection of the past light cone of x with the particle geodesic. In contrast, the tail term, which we already met in the context of the PN expansion (see Section 5.3.4), propagates *inside* the light cone, and therefore its contribution at x comes from all the points in the geodesic in the past of the intersection of the past light cone of x with the particle geodesic. Physically, it corresponds to radiation that scattered one or more times on the background curvature before reaching the observer. While $h_{\alpha\beta}^{\text{dir}}$ is indeed singular on the particle's world-line, $h_{\alpha\beta}^{\text{tail}}$ is regular, and it is therefore quite tempting to postulate that this is the quantity that enters the self-force in eq. (16.37), similarly to A_R^μ in the electromagnetic case. This turns out indeed to be correct, but a full justification is only obtained by going beyond the point-mass limit. Considering both an extended light object of mass m_2, for example a lighter BH, and a heavier object of mass $m_1 \gg m_2$, as in the case of EMRI, one can perform a matched asymptotic expansion, similar to that discussed in Section 5.3.3: near the lighter object the metric will be approximately the Schwarzschild metric associated with the small mass m_2, with small tidal effects due to the heavier BH. In contrast, far away from the light object the metric will be the Schwarzschild (or Kerr) metric of the heavier BH, slightly perturbed by the lighter object, which, in this regime, can indeed be consistently considered as a point mass. Matching the near-zone and far-zone metrics, one obtains the general solution to the problem. The result is that, at linear order in $h_{\alpha\beta}$ and at first order in the small parameter m_2/m_1, the metric perturbation due to an asymptotically small body is that of a point particle moving on a world-line that obeys eq. (16.36), where

$$a_{\text{grav}}^\mu = -\frac{1}{2}(g^{\mu\nu} + u^\mu u^\nu)(D_\rho h_{\nu\sigma}^{\text{tail}} + D_\sigma h_{\nu\rho}^{\text{tail}} - D_\nu h_{\sigma\rho}^{\text{tail}})u^\rho u^\sigma. \qquad (16.45)$$

The equation of motion given by eqs. (16.36) and (16.45) was first found by Mino, Sasaki and Tanaka and by Quin and Wald, and is known as the *MiSaTaQuWa equation*.[21]

One can then in principle compute the change in the particle's world-line generated by the self-force, and derive the evolution of an EMRI through a series of quasi-adiabatic inspiraling orbits. However, the computation of the gravitational self-force at each point of the orbit, through the extraction of the tail part, is computationally expensive, so approximation schemes have been developed. The full gravitational self-force

[21]For a very detailed derivation, clarifying several conceptual points, see Poisson, Pound and Vega (2011).

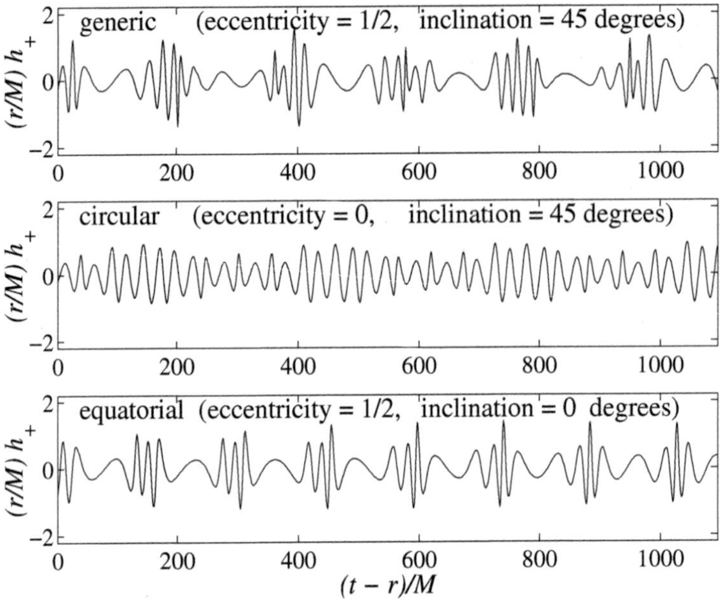

Fig. 16.3 A temporal segment of waveforms produced, in the adiabatic approximation, by an EMRI in a generic orbit and in related circular and equatorial orbits. From Drasco and Hughes (2006).

obtained from the MiSaTaQuWa equation has both a dissipative part, which describes the change in the particle trajectory due to the emission of radiation at infinity and down the BH's event horizon, and a conservative part, which for instance modifies the precession rate of the periastron. The radiative part of the force can be computed more simply, without using the full gravitational self-force formalism, just using the method of BH perturbation theory developed in Chapter 12, in particular the Teukolsky equation in the case of a Kerr BH. One then computes the energy and angular momentum (as well as the Carter constant, which is a third integral of motion in the Kerr geometry) carried away by the radiation field and imposes the respective conservation equations. One can then compute "adiabatic inspiral waveforms", which are obtained neglecting the conservative part of the gravitational self-force, as well as oscillatory terms that average to zero over an orbital period. Figure 16.3 shows an example of the waveforms obtained in this way, for different values of the eccentricity and the inclination of the orbit.

However, neglecting the conservative part of the self-force introduces an error that accumulates over many cycles. While this could still be sufficient for GW detection, particularly for nearly circular orbits, it will probably be less adequate for the subsequent parameter estimation. Computing the full gravitational self-force is, however, rather demanding. Most work has focused on computing the gravitational self-force for a particle moving on a fixed world-line, chosen as a geodesic of the background space-time, although the first computation of the self-force on an orbit, in Schwarzschild space-time, that evolves due to the gravitational self-force itself, has recently been performed. The full program for obtaining the EMRI's waveforms based on the use of the self-force is still under development, but the approach is being actively pursued.

16.4 Stochastic GWs from SMBH binaries

Depending on the frequency range and on the experimental sensitivities, the GW signal due to SMBH binaries could either be dominated by a few individually-resolvable sources or form a stochastic background due to the superposition of many unresolved signals. As we will see in this section, such a stochastic background can be especially relevant for pulsar timing arrays (which, however, could also individually resolve some signals), while at the higher frequencies relevant for LISA the signal will only be made by individual sources. In this section we will see how to compute such a stochastic background, and we will study the different regimes in frequency.

Sum over cosmological sources

Consider a source at a cosmological distance, radiating at an instantaneous GW frequency $f_{\rm gw}^{(r)}$, as measured in the source rest frame. If the source is at redshift z, the GW frequency in the observer rest frame is $f_{\rm gw}^{(\rm obs)} = f_{\rm gw}^{(r)}/(1+z)$ because of the cosmological redshift. Similarly, if $E_{\rm gw}^{(r)}$ is the energy radiated in GWs in the source rest frame, the energy in the observer frame is $E_{\rm gw}^{(\rm obs)} = E_{\rm gw}^{(r)}/(1+z)$. In the following we will denote $f_{\rm gw}^{(\rm obs)} = f$ and $E_{\rm gw}^{(\rm obs)} = E_{\rm gw}$, while we will keep the label (r) on quantities relatives to the rest-frame of the source, and we will write $f_{\rm gw}^{(r)}$ simply as f_r, so in particular

$$f = (1+z)^{-1} f_r \,. \tag{16.46}$$

Suppose at first that there is an ensemble of sources that each produce the same GW energy spectrum $dE_{\rm gw}^{(r)}/d\log f_r$ in their rest frame, and let $n(z)dz$ be the number density of such sources (i.e. the number of sources per unit comoving volume) in the redshift interval $[z, z+dz]$. Given a source at redshift z, the energy that it radiates in the interval $[\log f_r, \log f_r + d\log f_r]$ is

$$\frac{dE_{\rm gw}^{(r)}}{d\log f_r}\, d\log f_r \,. \tag{16.47}$$

After the cosmological redshift, this energy is seen by the observer decreased by an overall factor $(1+z)$, because $E_{\rm gw}^{(\rm obs)} = E_{\rm gw}^{(r)}/(1+z)$. Furthermore, the GWs detected by the observer at a frequency f are those that were emitted by the source at a frequency $f_r = (1+z)f$, and the radiation emitted in the source frame in the interval $[\log f_r, \log f_r + d\log f_r]$

is seen by the observer in the corresponding interval $[\log f, \log f + d \log f]$. According to eq. (16.46), $\log f = \log f_r - \log(1 + z)$ so, given that z is the fixed redshift of the source, $d \log f = d \log f_r$. Therefore, integrating over all sources and denoting by ρ_{gw} the GW energy density in the observer frame,

$$\frac{d\rho_{\text{gw}}}{d \log f}(f) = \int_0^\infty dz \, n(z) \frac{1}{1+z} \left. \left(\frac{dE_{\text{gw}}^{(r)}(f_r)}{d \log f_r} \right) \right|_{f_r=(1+z)f}, \qquad (16.48)$$

where the factor $1/(1+z)$ comes from $E_{\text{gw}}^{(\text{obs})} = E_{\text{gw}}^{(r)}/(1+z)$.

We now define the characteristic GW amplitude $h_c(f)$ of a stochastic background by

$$\langle h_{ij}^{\text{TT}}(t) h_{ij}^{\text{TT}}(t) \rangle = 2 \int_{f=0}^{f=\infty} d \log f \, h_c^2(f), \qquad (16.49)$$

where the angular brackets denote the ensemble average for the stochastic background. The factor of two on the right-hand side is part of the definition, and is motivated by the fact that, in an unpolarized background, the left-hand side is made up of two contributions, $\langle \tilde{h}_+^* \tilde{h}_+ \rangle$ and $\langle \tilde{h}_\times^* \tilde{h}_\times \rangle$, which are equal, while the mixed term $\langle \tilde{h}_+^* \tilde{h}_\times \rangle$ vanishes.

Comparing with eqs. (7.192) and (7.201) we see that

$$\frac{d\rho_{\text{gw}}}{d \log f}(f) = \frac{\pi c^2}{4G} f^2 h_c^2(f). \qquad (16.50)$$

As discussed in Section 7.8, it is also useful to introduce $h_0^2 \Omega_{\text{gw}}$, defined in eq. (7.198) as

$$h_0^2 \Omega_{\text{gw}} = \frac{h_0^2}{\rho_c} \frac{d\rho_{\text{gw}}}{d \log f}, \qquad (16.51)$$

where ρ_c is the critical density of the Universe given in eq. (7.194) and $h_0 = H_0/(100 \, \text{km s}^{-1} \, \text{Mpc}^{-1})$.[22] It is common to express the predictions and the observational results for the stochastic background either in terms of $h_c(f)$ or in terms of $h_0^2 \Omega_{\text{gw}}(f)$. From eqs. (16.50) and (16.51), the relation between the two is

$$h_0^2 \Omega_{\text{gw}}(f) = h_0^2 \frac{2\pi^2}{3H_0^2} f^2 h_c^2(f). \qquad (16.52)$$

[22] As discussed in Section 7.8, multiplying $\Omega_{\text{gw}}(f)$ by h_0^2 allows us to get rid of the uncertainty in the Hubble parameter, since $\rho_c \propto H_0^2$. Theoretical predictions and experimental bounds on stochastic backgrounds of GWs are more conveniently expressed using $h_0^2 \Omega_{\text{gw}}(f)$ rather than $\Omega_{\text{gw}}(f)$, as we will also do in Chapter 22.

From eqs. (16.48) and (16.50) we get

$$h_c^2(f) = \frac{4G}{\pi c^2 f^2} \int_0^\infty \frac{dz}{1+z} n(z) \left. \left(\frac{dE_{\text{gw}}^{(r)}(f_r)}{d \log f_r} \right) \right|_{f_r=(1+z)f}. \qquad (16.53)$$

Actually, even for an ensemble of sources of the same type (for example inspiraling binaries) the GW spectrum $dE_{\text{gw}}^{(r)}/d \log f_r$ will not be the same for each source, as assumed in the derivation of eq. (16.48). Rather, the spectrum will also be characterized by a set of source parameters. For instance, as we saw in Sections 4.1 and 5.6, the GW spectrum $dE_{\rm gw}^{(r)}/d\log f_r$ generated by an inspiraling compact binary, at Newtonian order, depends on the masses of the two bodies through the chirp mass M_c. At 1PN order it depends on both masses separately (or, equivalently, on M_c and on the symmetric mass ratio ν), while at higher PN orders there enters a further dependence on the spins of the two component compact objects. We denote collectively by $\xi = \{\xi_1, \ldots, \xi_m\}$ these parameters. Let

$$\frac{dn}{d\xi_1 \ldots d\xi_m} d\xi_1 \ldots d\xi_m \, dz \equiv \frac{dn}{d\xi} d\xi \, dz \qquad (16.54)$$

be the number density of sources in the redshift interval $[z, z+dz]$ and with source parameters in the interval $[\xi, \xi + d\xi]$. Then eq. (16.48) generalizes to

$$\frac{d\rho_{\rm gw}}{d\log f}(f) = \int_0^\infty \frac{dz}{1+z} \int d\xi \, \frac{dn(z;\xi)}{d\xi} \left(\frac{dE_{\rm gw}^{(r)}(f_r;\xi)}{d\log f_r} \right) \Bigg|_{f_r=(1+z)f}, \qquad (16.55)$$

and similarly

$$h_c^2(f) = \frac{4G}{\pi c^2 f^2} \int_0^\infty \frac{dz}{1+z} \int d\xi \, \frac{dn(z;\xi)}{d\xi} \left(\frac{dE_{\rm gw}^{(r)}(f_r;\xi)}{d\log f_r} \right) \Bigg|_{f_r=(1+z)f}. \qquad (16.56)$$

16.4.1 Regime dominated by GW back-reaction

The above results are very general, and allow us to compute the stochastic background generated by the superposition of signals from many individual sources at cosmological distances. We can now apply it to the GWs produced by SMBH binaries. We first consider the dynamical regime in which the shrinking of the orbit is uniquely determined by the GW back-reaction. We will discuss later what is the range of GW frequencies (in the observer frame) for which this regime is the relevant one. The energy spectrum generated by an inspiraling binary in the source frame, in the Newtonian approximation, is given by eq. (4.41), which we can rewrite as

$$\frac{dE_{\rm gw}^{(r)}}{d\log f_r} = \frac{1}{3G} (GM_c)^{5/3} (\pi f_r)^{2/3}. \qquad (16.57)$$

For computing the portion of the spectrum generated by SMBH binaries in their long inspiral phase the Newtonian approximation is adequate and, assuming for the moment circular orbits, the only parameter ξ is the chirp mass of the binary. Then, inserting eq. (16.57) into eqs. (16.55) and (16.56), we get

$$\frac{d\rho_{\rm gw}}{d\log f}(f) = \frac{(G\pi f)^{2/3}}{3} \int_0^\infty \frac{dz}{(1+z)^{1/3}} \int_0^\infty dM_c \, M_c^{5/3} \frac{dn(z;M_c)}{dM_c} \qquad (16.58)$$

and

$$h_c^2(f) = \frac{4G^{5/3}}{3c^2\pi^{1/3}} f^{-4/3} \int_0^\infty \frac{dz}{(1+z)^{1/3}} \int_0^\infty dM_c \, M_c^{5/3} \frac{dn(z;M_c)}{dM_c}.$$

(16.59)

These results shows that, for SMBH binaries,

$$h_0^2\Omega_{\rm gw}(f) \propto f^{2/3}$$

(16.60)

and

$$h_c(f) \propto f^{-2/3}.$$

(16.61)

Note that this frequency dependence holds independently of the form of the function $dn/dM_c(z; M_c)$, which encodes the details of the formation history of the SMBH binaries.

There are, however, limitations to this result, both on the low-frequency and the high-frequency sides, due to assumptions implicit in the derivation, which we discuss next.

16.4.2 Regime dominated by three-body interactions

The spectrum (16.57) was obtained in eq. (4.41) assuming that the binary system is isolated. In that case, GW back-reaction is the only factor that causes shrinking of the orbit. However, we have already seen that, until the major semiaxis a has reached a value $a_g \ll a_h$ (whose value we will compute more precisely later), the shrinking of the orbit is dominated by other effects; in particular, for $a > a_h$ given in eq. (16.8) it is dominated by dynamical friction, and for $a_g < a < a_h$ by three-body interactions (or by gas accretion). To compute the spectrum of GWs produced in these regimes, we proceed as follows.

We limit ourselves for the moment to quasi-circular Keplerian orbits, and we will come back later to the role of eccentricity. For a quasi-circular evolution of the orbital radius $a(t)$, the orbital frequency of the source $f_s = \omega_s/(2\pi)$ evolves according to Kepler's law, $\omega_s^2(t) = Gm/a^3(t)$. Limiting ourselves to quadrupole emission, which for a circular orbit is fully appropriate in the long inspiral phase, the instantaneous GW frequency in the source rest frame, $f_{\rm gw}^{(r)}(t) \equiv f_r(t)$, is equal to $2f_s(t)$. Therefore

$$f_r(t) = \frac{1}{\pi} \left(\frac{Gm}{a^3(t)} \right)^{1/2}.$$

(16.62)

Since $f_r(t)$ depends on time through $a(t)$, we can write

$$\frac{dE_{\rm gw}^{(r)}}{d\log f_r} = \frac{dE_{\rm gw}^{(r)}}{dt} \frac{dt}{da} \frac{da}{d\log f_r}.$$

(16.63)

From eq. (16.62) we get

$$\frac{da}{d\log f_r} = -\frac{2}{3}a \,. \tag{16.64}$$

The quantity $dE_{\rm gw}^{(r)}/dt$ is the instantaneous radiated power, which we already computed in eq. (4.12), where we used $\omega_{\rm gw} = 2\pi f_{\rm gw}^{(r)} \equiv 2\pi f_r$,

$$\frac{dE_{\rm gw}^{(r)}}{dt} = \frac{32}{5Gc^5}(\pi f_r GM_c)^{10/3} \,. \tag{16.65}$$

Thus, the power spectrum $dE_{\rm gw}^{(r)}/d\log f_r$ of an inspiraling binary in a sequence of quasi-circular orbits is related to the evolution of the radius $a(t)$ by

$$\frac{dE_{\rm gw}^{(r)}}{d\log f_r} = -\frac{64}{15Gc^5}(\pi f_r GM_c)^{10/3}\left(\frac{d\log a}{dt}\right)^{-1} \,. \tag{16.66}$$

As a check of this equation, we can insert the power spectrum (16.57), which is valid when the orbital evolution is governed only by the GW back-reaction, and read the corresponding value of $d\log a/dt$,

$$\left(\frac{d\log a}{dt}\right)_{\rm gw} = -\frac{64}{5c^5}(\pi f_r)^{8/3}(GM_c)^{5/3} \,. \tag{16.67}$$

This gives the radial evolution due to GW back-reaction, in full agreement with the result found from eqs. (4.15) and (4.17) of Vol. 1 [where $a(t)$ for circular orbits was denoted by $R(t)$]. Eliminating f_r through eq. (16.62) we can also rewrite this result as

$$\left(\frac{da}{dt}\right)_{\rm gw} = -\frac{64}{5}\frac{G^3 mm_1 m_2}{c^5 a^3} \,. \tag{16.68}$$

We can now apply eq. (16.66) to the regime where $a_g < a < a_h$, when neither dynamical friction nor GW back-reaction are effective. We assume that in this regime three-body interactions give the dominant contribution to the hardening of the binary, through the slingshot effect, and we neglect the possible role of gas accretion. In this case one can model the radial evolution as

$$\left(\frac{da}{dt}\right)_{\rm 3-body} \simeq -\frac{\kappa\sigma_c^5}{G^2 m^2}a^2 \,, \tag{16.69}$$

where $\kappa \sim 4$ is a numerical constant [related to the so-called hardening rate H by $\kappa = 8H/(9\pi)$], and we have neglected a multiplicative correction $\ln^{-2}(a_h/a)$.[23] Observe that the right-hand side increases with a as a^2, corresponding to the fact that the slingshot effect only takes place if a third star approaches the binary within a distance of order a, so the effective cross-section for such a three-body process is proportional to a^2.

[23]See Sesana, Haardt, Madau and Volonteri (2004).

Inserting eq. (16.69) into eq. (16.66) and again eliminating a using eq. (16.62), we get

$$\frac{dE_{\text{gw}}^{(r)}}{d\log f_r} \simeq (\pi f_r)^4 \frac{G^4 m_1^2 m_2^2 m}{c^5 \sigma_c^5}\,, \tag{16.70}$$

apart from an overall numerical constant $64/(15\kappa) \sim 1$. Observe that in this case the result depends separately on both m_1 and m_2, rather than only on the chirp mass. We therefore insert this result into eq. (16.56), with $d\xi = dm_1 dm_2$, and we get

$$h_c^2(f) \simeq \frac{4\pi^3 G^5}{c^7 \sigma_c^5} f^2 \int_0^\infty dz\, (1+z)^3 \tag{16.71}$$
$$\times \int_0^\infty dm_1 dm_2\, (m_1 + m_2) m_1^2 m_2^2 \frac{dn(z; m_1, m_2)}{dm_1 dm_2}\,.$$

Thus, in this regime

$$h_c(f) \propto f\,, \tag{16.72}$$

and

$$h_0^2 \Omega_{\text{gw}}(f) \propto f^4\,. \tag{16.73}$$

Let us next estimate the frequency range in which these results are valid. Assuming that gas accretion plays a subleading role, the regime dominated by three-body interactions starts at $a = a_h$ given in eq. (16.8) and ends at a value $a = a_g$ determined by requiring that the contribution to $-da/dt$ due to GW back-reaction, given in eq. (16.68), becomes larger than that given by three-body encounters, given in eq. (16.69). This gives

$$a_g \simeq \frac{Gm}{c\,\sigma_c}\nu^{1/5}\,, \tag{16.74}$$

where, as usual, $\nu = m_1 m_2/m^2$ is the symmetric mass ratio. Comparing with eq. (16.8) we see that

$$a_g \simeq 4\frac{\sigma_c}{c}\frac{m}{m_2}\nu^{1/5} a_h\,, \tag{16.75}$$
$$\simeq 6.4 \times 10^{-3}\,\text{pc}\left(\frac{180\,\text{km/s}}{\sigma_c}\right)\left(\frac{m}{10^8\,M_\odot}\right)(4\nu)^{1/5}\,,$$

where the normalization for ν to the value $1/4$ corresponds to the case of equal-mass binaries. Observe that a_g/a_h is proportional to σ_c/c, which is much smaller than 1. Thus, for comparable values of the masses m_1 and m_2, we have $a_g \ll a_h$, and the regime dominated by three-body scattering holds for $a_g < a < a_h$. At a given value of a, a binary on a circular orbit emits mostly at a GW frequency, in the source frame, given by eq. (16.62). The corresponding frequencies in the observer frame are obtained by redshifting the source-frame frequencies. Thus, during the three-body-dominated regime, the binary emits GWs that the observer will see at frequencies f in the range $f_1 < f < f_2$, with

$$f_1 \sim \frac{1}{1+z}\frac{1}{\pi}\left(\frac{Gm}{a_h^3}\right)^{1/2} \tag{16.76}$$

$$\sim 10^{-12}\,\mathrm{Hz} \left(\frac{2}{1+z}\right) \left(\frac{\sigma_c}{180\,\mathrm{km/s}}\right)^3 \left(\frac{10^8\,M_\odot}{m_2}\right) \left(\frac{m}{2m_2}\right)^{1/2},$$

where we have chosen $z = 1$ as a reference value for z, and $m = 2m_2$ for an equal-mass binary. As we will see later, the main contribution to the stochastic GW background from SMBHs is given by binaries with comparable masses, with m_1 in the range $(10^9 - 10^{10})M_\odot$, at $z \simeq 1$. For such values, taking into account also the corresponding change in σ_c from the $M - \sigma$ relation, eq. (16.76) gives $f_1 \sim 10^{-13}$ Hz. Similarly,

$$f_2 \sim \frac{1}{1+z} \frac{1}{\pi} \left(\frac{Gm}{a_g^3}\right)^{1/2} \tag{16.77}$$

$$\sim 7 \times 10^{-9}\,\mathrm{Hz} \left(\frac{2}{1+z}\right) \left(\frac{\sigma_c}{180\,\mathrm{km/s}}\right)^{3/2} \left(\frac{10^8\,M_\odot}{m}\right) (4\nu)^{-3/10},$$

For a stochastic background dominated by comparable-mass binaries with mass $(10^9 - 10^{10})M_\odot$ at $z \simeq 1$, we get f_2 a few times 10^{-10} Hz.

For $f < f_1$, the frequency dependence in $h_0^2 \Omega_{\mathrm{gw}}(f)$ changes further compared with the f^4 behavior in eq. (16.73), and is determined by dynamical friction, and in fact it becomes even steeper than f^4. In any case, on decreasing the frequency from f_2 to f_1, $h_0^2 \Omega_{\mathrm{gw}}(f)$ decreases as f^4, so $h_0^2 \Omega_{\mathrm{gw}}(f_1)$ already has a very small value, and the spectrum at $f < f_1$ is negligible.

16.4.3   High-frequency regime and source discreteness

Putting together the above results, we see that, for $f_1 < f < f_2$, i.e. in the regime dominated by three-body interactions, $h_0^2 \Omega_{\mathrm{gw}}(f) \propto f^4$, while when we enter the regime dominated by GW back-reaction, at $f > f_2$, we have $h_0^2 \Omega_{\mathrm{gw}}(f) \propto f^{2/3}$. Let us next investigate up to what maximum frequency this behavior persists.

First of all, the inspiral phase ends near the innermost stable circular orbit. The corresponding quadrupole radiation at the ISCO, for an equal-mass spinless binary, has already been given in eq. (16.24), which can be rewritten as

$$f_{\mathrm{ISCO}} \simeq 4.7 \times 10^{-7}\,\mathrm{Hz} \left(\frac{1}{1+z}\right) \left(\frac{10^{10}\,M_\odot}{M}\right), \tag{16.78}$$

where we have used $10^{10}\,M_\odot$ as a reference value of the BH mass since, as we will see below, at the frequency of pulsar timing arrays the dominant contribution comes from BHs with masses $(10^9 - 10^{10})M_\odot$. Similarly, the frequency of the first quasi-normal mode of the final Kerr BH, given in eq. (16.25), can be rewritten as

$$f_{\mathrm{max}} \simeq 1.9 \times 10^{-6}\,\mathrm{Hz} \left(\frac{1}{1+z}\right) \left(\frac{10^{10}\,M_\odot}{M}\right). \tag{16.79}$$

Observe that lighter SMBH binaries have higher values of f_{ISCO} and f_{max}, so at higher frequencies their contribution eventually becomes dominant compared with that of the more massive binaries.

To estimate the upper cutoff on the frequency of a stochastic background we must also take into account that the whole formalism that we have used, based on eq. (16.56), assumes the existence of a continuous distribution of sources, with number density $n(z; \xi)$. Of course, actual astrophysical sources form a discrete ensemble, and the continuous approximation is only valid when a large number of sources contribute to each frequency bin. A computation that takes into account the discreteness of the sources can be performed by taking a model for the assembly history of massive BHs, and building an actual population of sources by a Monte Carlo sampling of the distribution $dn/dzdM_c d\log f_r$. One can then compute how many sources contribute to a given bin in observed frequency, for each mass and redshift interval. The result of this analysis[24] is that at frequencies below $f \simeq 10^{-8}$ Hz the Monte Carlo result is in excellent agreement with the result of the continuous approximation. At higher frequencies, however, the situation changes, and the estimate of $h_c(f)$ using the Monte Carlo sampling is significantly smaller than the result of a semi-analytic computation using the corresponding continuous distribution $dn/dzdM_c d\log f_r$. This is due to the fact that, at $f \gtrsim 10^{-8}$ Hz, most of the radiation inferred in the semi-analytic approach turns out to be actually due to less than one source with $M_c \gtrsim 10^8 M_\odot$. This radiation from a "fractional" number of sources is spurious, since a source is either there or not there.

Even when there are several sources, a second question concerns when the GW contributions of the sources are individually resolvable and when they rather contribute to the stochastic background. This depends, of course, on the resolution of the experiment, i.e. on the width of the frequency bins Δf. For pulsar timing arrays, as we will see in Chapter 23, this is basically given by the inverse of the observation time, $\Delta f \sim 1/T$. At a given frequency f, the signal is stochastic only if there are several sources in the frequency bin $[f, f + 1/\Delta T]$. If we have, say, one or a few source per frequency bin, the situation is still very interesting, since these sources are individually resolvable, and would show up as individual continuous sources of GWs. However, their contribution will have to be subtracted from the stochastic background. By removing these sources one finds that, above a frequency f_0, the slope of $h_c(f)$ is steeper, and can be fitted by

$$h_c(f) = h_0 \left(\frac{f}{f_0}\right)^{-2/3} \left(1 + \frac{f}{f_0}\right)^{\gamma}, \qquad (16.80)$$

with $\gamma < 0$. Both f_0 and γ depend on the model chosen for the halo assembly history. Over four models that cover a broad spectrum of predictions for the massive BH assembly history (massive or light BH seeds, and abundant or rare seeds), and a given choice of the $M - \sigma$ relation, the index γ changes relatively little, from -1.11 to -1.04, while f_0 changes from 14.2 nHz to 52.4 nHz; see Table 16.2.

[24]See Sesana, Vecchio and Colacino (2008).

Table 16.2 The values of h_0, f_0 and γ for four models of massive BH assembly, characterized by massive seeds [$M \sim 10^4 M_\odot$, denoted as (m)] or light seeds [$M \sim 10^2 M_\odot$, denoted as (l)] and by rare (r) or abundant (a) seeds. The first two models have different accretion prescriptions. The frequency f_0 is in nHz and h_0 is in units of 10^{-15}. The four models assume the same $M - \sigma$ relation. For different SMBH–galaxy relations the scatter in the parameters of the fit across the four models can actually be larger. From Sesana, Vecchio and Colacino (2008).

Model	f_0	γ	h_0
(l,a)	14.2	-1.09	2.15
(l,a)	42.7	-1.08	0.69
(m,a)	52.4	-1.04	0.67
(m,r)	39.5	-1.11	0.89

16.4.4   Estimates of the SMBH merger rate

The next step in computing $h_c(f)$ is to have a model for the number density of SMBH mergers, i.e. for $dn/dzd\xi$, where $\xi = M_c$ or, at a finer level of description, $\xi = \{m_1, m_2\}$. Equivalently, we can use the total mass m and the mass ratio $q = m_2/m_1$. Note that $0 < q \leqslant 1$ since m_2 is defined as the mass of the lighter BH. Earlier work estimated the number density of SMBH mergers using a combination of dark-matter simulations and semi-analytic techniques, in which the merger histories of dark-matter halos are generated via Monte Carlo methods, and the evolution of the baryonic component is obtained by analytic modeling of the relevant processes such as gas cooling, star formation, supernova feedback and galaxy mergers. More recent work has a more observational approach, taking advantage of the significant progress made with the advent of large-scale surveys.

One defines the merger rate of galaxies $\Phi_{\mathrm{mgr}}(M_G, q_G, z)$ (where M_G is the mass of the heavier galaxy and $q_G \leqslant 1$ is the mass ratio between the two galaxies) as the number of mergers N_{mgr} per unit logarithmic values of M_G, q_G, per unit time t (measured in the observer frame) and per unit redshift z,

$$\Phi_{\mathrm{mgr}}(M_G, q_G, z) \equiv \frac{dN_{\mathrm{mgr}}}{d\log M_G d\log q_G dzdt} . \tag{16.81}$$

Thus $\Phi_{\mathrm{mgr}}(M_G, q_G, z)dzdt$ is the number of events (per logarithmic unit of M_G, q_G) seen by the observer in an observation time dt, and coming from a volume of space given by the shell between redshifts z and $z+dz$.

We can factorize the expression in eq. (16.81) as

$$\frac{dN_{\mathrm{mgr}}}{d\log M_G d\log q_G dzdt} = \frac{dN_{\mathrm{gal}}}{d\log M_G dz} \frac{dn_{\mathrm{mrg}}}{dt} \frac{dP}{d\log q_G} , \tag{16.82}$$

where $dN_{\mathrm{gal}}/d\log M_G dz$ is the number of galaxies per unit of $\log M_G$ and redshift, dn_{mrg}/dt is the merger rate of a *single* galaxy with a lighter galaxy, and dP/dq_G is the probability density of mergers with respect to the variable q_G, i.e. $(dP/dq_G)dq_G$ is the fraction of mergers that have a mass ratio in the interval $[q_G, q_G + dq_G]$.

We next write

$$\frac{dn_{\mathrm{mrg}}}{dt_r} = \frac{f_{\mathrm{gm}}(z, M_G, q_G)}{\tau(z, M_G, q_G)} , \tag{16.83}$$

where t_r is time measured in the rest frame of the galaxy, $\tau(z, M_G, q_G)$ is the characteristic merger time-scale for a pair of galaxies with primary mass M_G and mass ratio q_G, and f_{gm} is the fraction of galaxies that are undergoing mergers (defined for example as the fraction of galaxies with a companion within 30 kpc). From observations, f_{gm} typically has the form $f_{\mathrm{gm}}(z) = a_{\mathrm{gm}}(1+z)^{b_{\mathrm{gm}}}$, with a_{gm} and b_{gm} numerical constants fitted to the data. For instance, for a sample of major mergers with $M_G > 10^{10} M_\odot$ and mass ratio larger than $1/3$, at $z < 1.2$, one finds $a_{\mathrm{gm}} = 0.022 \pm 0.006$ and $b_{\mathrm{gm}} = 1.6 \pm 0.6$.[25]

[25]See Ravi, Wyithe, Shannon and Hobbs (2015) and references therein.

The rest-frame time t_r is related to the observer time t by $(1+z)dt_r = dt$; compare with eq. (16.46). We therefore have

$$\Phi_{\mathrm{mgr}}(M_G, q_G, z) = \frac{f_{\mathrm{gm}}(z, M_G, q_G)}{(1+z)\tau(z, M_G, q_G)} \frac{dN_{\mathrm{gal}}}{d\log M_G dz} \frac{dP}{d\log q_G}. \tag{16.84}$$

In this expression $dN_{\mathrm{gal}}/d\log M_G dz$, $f_{\mathrm{gm}}(z)$ and $dP/d\log q_G$ can be directly determined by observations, while $\tau(z, M_G, q_G)$ can be obtained from numerical simulations of galaxy mergers.

Given an expression for the merger rate, one now has all the elements for computing $h_c(f)$. As we will see in Chapter 23, pulsar timing arrays (PTAs) are sensitive to the region of frequency where the dynamics of SMBH binaries is dominated by GW back-reaction and $h_c(f) \propto f^{-2/3}$ (except for possible effects due to the orbital eccentricity, which we will study in Section 16.4.5). In this context, it is therefore customary to write

$$h_c(f) = A_* \left(\frac{f}{f_*} \right)^{-2/3}, \tag{16.85}$$

where f_* is a reference frequency, which is normally taken to be $f_* = 1\,\mathrm{yr}^{-1} \simeq 31.7$ nHz. Theoretical predictions for the stochastic background from SMBHs, as well as observational limits, are then quoted in terms of A_*. Of course, the value of A_* depends on the choice of f_*, and its value for $f_* = 1/(1\,\mathrm{yr})$ is denoted by $A_{1\mathrm{yr}}$. Taking into account the uncertainties in the quantities that enter eq. (16.84), the theoretical predictions for $A_{1\mathrm{yr}}$ turn out to span the range

$$3.3 \times 10^{-16} < A_{1\mathrm{yr}} < 1.3 \times 10^{-15} \tag{16.86}$$

at 68% c.l., and

$$1.1 \times 10^{-16} < A_{1\mathrm{yr}} < 4.2 \times 10^{-15}, \tag{16.87}$$

at 99.7% c.l.[26] As we will see in Chapter 23, the upper range of this estimate is already excluded by observations with pulsar timing arrays, which currently provide a limit $A_{1\mathrm{yr}} < 1.0 \times 10^{-15}$, see eq. (23.74).

To give an idea of the binary population that makes the main contribution to the signal in the frequency range of PTAs, Fig. 16.4 shows the distribution in the primary mass m_1, mass ratio q and redshift z from a typical model of the SMBH binary population, considering the binaries at a given instant of time that contribute at a level of 0.1 ns or more to the PTA signal in the frequency window $[3 \times 10^{-9} - 10^{-6}]$ Hz (the timing residuals induced by GWs in PTAs will be studied in detail in Chapter 23). We see that the signal comes mostly from comparable-mass binaries at a redshift $z \simeq 1$. The distribution of the number of events is dominated by primary masses $(10^8 - 10^9)M_\odot$. The distribution of energy radiated in GWs with respect to the binary mass is further shifted toward higher masses, since the higher the mass the higher is the total GW energy radiated, as we will see in Fig. 16.7.

---

[26] See Sesana (2013). Similar results are obtained by Ravi, Wyithe, Shannon and Hobbs (2015), that give

$$5.1 \times 10^{-16} < A_{1\mathrm{yr}} < 2.4 \times 10^{-15},$$

at 95% c.l.

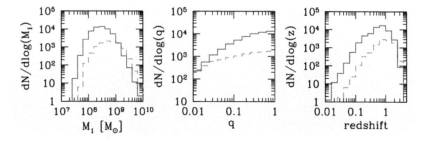

Fig. 16.4 The distribution of the total number of SMBH binaries in the Universe at a given time that contribute at a level of 0.1 ns or more to the timing residuals of PTA in the frequency window $[3 \times 10^{-9} - 10^{-6}]$ Hz, in a typical model of the SMBH binary population. The solid histogram refers to sources still dynamically coupled with their accretion disk and the dashed histogram to sources that are inspiraling after having left their accretion disk behind. From Sesana, Roedig, Reynolds and Dotti (2012).

16.4.5   Effect of the eccentricity

The computation so far has been performed assuming circular orbits. We have seen in Section 4.1.3 that radiation reaction indeed circularizes the orbit. However, the issue is whether the eccentricity is already negligible when we enter the regime dominated by GW back-reaction, after the regimes dominated first by dynamical friction and then by three-body interactions. Actually, it can be shown that in the regime dominated by dynamical friction, while the major semiaxis decreases, the eccentricity increases. The same happens in three-body interactions, as well as if the binary evolves in a massive prograde disk, so eventually the binary can enter the regime dominated by GW back-reaction with a large eccentricity, possibly even larger than 0.9.

We studied the GW emission from eccentric orbits in Vol. 1 in Section 4.1.2, following the classic paper by Peters and Mathews (1963). Before entering into the details of the computation of the stochastic background generated by eccentric binaries, we can understand its main features as follows. As we saw in Section 4.1.2, the radiated power is larger by a factor $F(e)$ compared with the circular case, where $F(e)$ is the Peters–Mathews function

$$F(e) = \frac{1}{(1-e^2)^{7/2}} \left(1 + \frac{73}{24}e^2 + \frac{37}{96}e^4 \right) ; \qquad (16.88)$$

see eq. (4.74).[27] In particular, for $e \to 1^-$, $F(e)$ diverges as $(1-e^2)^{-7/2}$, and the function $F(e)$ is of order $10^2 - 10^3$ for $e \sim 0.8 - 0.9$; see Fig. 4.5. On the other hand, this implies that the time to coalescence is shorter, again by a factor proportional to $(1-e^2)^{7/2}$; see eq. (4.140). Therefore, in a first approximation, the total energy radiated in the inspiral phase is independent of e. However, even if the total radiated energy is comparable to that in the circular case, it has a different distribution in frequency. Indeed, we saw in Section 4.1.2 that, for non-vanishing

[27] In this section we denote the Peters–Mathews function by $F(e)$ rather than $f(e)$, to avoid confusion with the symbol f used for the frequency.

eccentricity, when the source has a frequency ω_s the GWs that are generated are no longer monochromatic at $\omega_{\rm GW} = 2\omega_s$, but are now spread over the frequencies $\omega_{{\rm GW},n} = n\omega_s$, for all integer values of n, and their spectrum is peaked at a value of n that grows with e; see Figs. 4.8–4.10. As a result, compared with the case of circular orbits, power is removed from the $n = 2$ mode and transferred to higher frequencies.

The actual computation is a generalization of that in the circular case. Equation (16.55) is now replaced by

$$\frac{d\rho_{\rm gw}}{d\log f}(f) = \sum_{n=1}^{\infty} \frac{d\rho_{{\rm gw},n}}{d\log f}(f), \qquad (16.89)$$

where

$$\frac{d\rho_{{\rm gw},n}}{d\log f}(f) = \int_0^\infty \frac{dz}{1+z} \int d\xi \, \frac{dn(z;\xi)}{d\xi} \left(\frac{dE_{{\rm gw},n}^{(r)}}{d\log f_{n,r}} \right) \Bigg|_{f_{n,r}=(1+z)f} . \qquad (16.90)$$

Here $E_{{\rm gw},n}^{(r)} = E_{\rm gw}^{(r)}(f_{n,r};\xi)$ is the energy radiated in the n-th harmonic, and $f_{n,r}$ is the frequency of the n-th harmonic in the rest frame. We now observe that $f_{n,r} = (n/2)f_r$, where f_r is the rest-frame GW frequency emitted in the $n = 2$ harmonic, so $d\log f_{n,r} = d\log f_r$. Thus

$$\frac{dE_{{\rm gw},n}^{(r)}}{d\log f_{n,r}} = \frac{dE_{{\rm gw},n}^{(r)}}{d\log f_r}$$

$$= \frac{dE_{{\rm gw},n}^{(r)}}{dt} \frac{dt}{da} \frac{da}{d\log f_r}, \qquad (16.91)$$

similarly to eq. (16.63). The term $dE_{{\rm gw},n}^{(r)}/dt$ is the power P_n radiated in the n-th harmonic, given in eq. (4.107), while Kepler's law again gives eq. (16.64), even when $e \neq 0$. Finally, for non-zero eccentricity, da/dt is given by eq. (4.116). Then, we finally get

$$\frac{dE_{{\rm gw},n}^{(r)}}{d\log f_{n,r}} = \frac{1}{3G}(GM_c)^{5/3} \left(\frac{2\pi f_{n,r}}{n} \right)^{2/3} \frac{g(n,e)}{F(e)}, \qquad (16.92)$$

where $g(n,e)$ was given in eq. (4.108). For circular orbits $g(n, e = 0) = \delta_{n,2}$ and $F(e = 0) = 1$, so we recover eq. (16.57), with the only contribution to the sum over n in eq. (16.89) coming from the $n = 2$ harmonic. Plugging eq. (16.92) into eqs. (16.89) and (16.90), and using $\xi = \{M_c, e\}$ as the relevant parameters in the distribution function $n(z;\xi)$, we find

$$\frac{d\rho_{\rm gw}}{d\log f}(f) = \frac{(G\pi f)^{2/3}}{3} \int_0^\infty \frac{dz}{(1+z)^{1/3}} \int_0^\infty dM_c \int_0^1 de$$

$$\times M_c^{5/3} \frac{dn(z;M_c,e)}{dM_c de} \Phi(e), \qquad (16.93)$$

where

$$\Phi(e) = \sum_{n=1}^{\infty} \left(\frac{n}{2} \right)^{-2/3} \frac{g(n,e)}{F(e)}. \qquad (16.94)$$

Similarly, eq. (16.59) becomes

$$
h_c^2(f) = \frac{4G^{5/3}}{3c^2\pi^{1/3}} f^{-4/3} \int_0^\infty \frac{dz}{(1+z)^{1/3}}
$$
$$
\times \int_0^\infty dM_c \, M_c^{5/3} \int_0^1 de \, \Phi(e) \frac{dn(z; M_c, e)}{dM_c de} . \qquad (16.95)
$$

If the eccentricity e of the binary were a constant, the factor $\Phi(e)$ in these equations would not affect the dependence on the frequency. However, the eccentricity evolves with time. In particular, if we denote by $a_0 = a(t_0)$ and $e_0 = e(t_0)$ the values of the major semiaxis and of the eccentricity at the time t_0 when the binary enters in the regime dominated by GW back-reaction, the subsequent evolution of $a(t)$ and $e(t)$ is determined by eqs. (4.116) and (4.117). We can then invert the relations $a = a(t; a_0, e_0)$ and $e = e(t; a_0, e_0)$ to write the eccentricity at time t as a function of $a(t), a_0$ and e_0. In turn, Kepler's law fixes $a(t)$ in terms of the instantaneous orbital frequency of the source $f_s(t)$, so $e(t) = e[f_s(t); e_0, a_0]$. In the contribution from the n-th harmonics, the observed frequency is $f = n f_s/(1+z)$. Therefore, inside the sum over n in eq. (16.94) we can effectively replace[28]

$$
e \to e(f_s; e_0, a_0) = e\left(\frac{(1+z)f}{n}; e_0, a_0\right) . \qquad (16.96)
$$

This induces a dependence on f in $\Phi(e)$ and therefore an extra dependence in $h_c(f)$ and $h_0^2\Omega_{\rm gw}(f)$. To determine the function $e(f_s; e_0, a_0)$ we recall from our discussion in Section 4.1.3 that the evolution equations for da/dt and de/dt, eqs. (4.116) and (4.117), can be combined into an equation for da/de that can be integrated analytically, giving

$$
a(e) = a_0 \frac{g(e)}{g(e_0)} , \qquad (16.97)
$$

where

$$
g(e) = \frac{e^{12/19}}{1-e^2}\left(1 + \frac{121}{304}e^2\right)^{870/2299} ; \qquad (16.98)
$$

see eqs. (4.127) and (4.128). Finally, eliminating from eq. (16.97) $a(e)$ and a_0 in favor of the corresponding frequencies through Kepler's law, we get

$$
g(e) = g(e_0)\left(\frac{f_{s,0}}{f_s}\right)^{2/3} , \qquad (16.99)
$$

where $f_{s,0}$ is the source frequency at the time t_0 at which the binary enters in the regime dominated by GW back-reaction (or any other reference time, chosen in the regime dominated by GW back-reaction, when a has the value a_0 and the eccentricity has the value e_0). The inverse of the function $g(e)$ exists and is unique, as we see from Fig. 4.11. One can then invert eq. (16.99) numerically, and obtain Φ as a function of

[28] More precisely, the fact that e depends on time means that in principle we must take into account this time dependence in $h_{+,\times}(t)$ *before* performing the Fourier transform to $\tilde{h}_{+,\times}(f)$; compare the steps leading from eqs. (4.31) and (4.32) to eqs. (4.34) and (4.35). However, at least during the long inspiral phase where the binary has a slowly-varying orbital frequency $f_s(t)$, performing this Fourier transform basically amounts to the replacement (16.96).

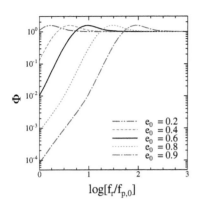

Fig. 16.5 The function $\Phi(f_r/f_{p,0})$ (where the notation $f_{p,0}$ is used for $f_{s,0}$) for different values of e_0. With increasing e_0, the maximum of the curve moves to the right. From Enoki and Nagashima (2007).

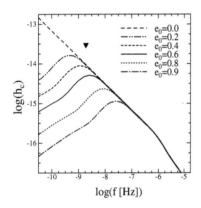

Fig. 16.6 The function $h_c(f)$ for different eccentricities, from Enoki and Nagashima (2007). The black triangle gives the PTA limit as of 2007. Present (2017) limits are at the level of the $e = 0$ curve, as we will see in Section 23.5.

[29]Recall, however, that this value of e_0 was defined, somewhat arbitrarily, as the value at which a_0 was equal to $300R_S$. In general, the impact of eccentricity depends on what is its value at the end of the three-body dominated regime, i.e. when f has the value f_2 given in eq. (16.77). Therefore, the value $e_0 = 0.8$ should be taken as merely illustrative.

$f_r/f_{s,0}$, where, as usual, f_r is related to the observed frequency f by $f_r = (1+z)f$.

Figure 16.5 shows the result for Φ for different values of e_0, defined as the value of e at a time t_0 such that the semimajor axis a_0 is equal to 300 times the Schwarzschild radius of the system. From the figure we see that the eccentricity can indeed significantly suppress the GW background at frequencies low compared with $f_{s,0}$, while it increases the spectrum at intermediate frequencies, reflecting the fact that the eccentricity transfers power from the $n = 2$ to the higher harmonics.

The numerical value of $f_{s,0}$ depends, through Kepler's law, on the total mass of the binary. Therefore, to understand the effect on $h_c(f)$ or on $h_0^2\Omega_{\rm gw}(f)$ one must combine this result for $\Phi(e)$ with a specific model for $dn(z; M_c, e)/dM_cde$ in eq. (16.93). In fact, one must rather use $dn(z; m_1, m_2, e)/dm_1dm_2de$, given that both the chirp mass and the total mass, or equivalently m_1 and m_2, now enter the result.

Figure 16.6 shows $h_c(f)$, using a semi-analytic model in which SMBH formation is related to galaxy formation. Not much is known about the actual distribution of eccentricities, so in Fig. 16.6 one makes the simplifying assumption that all binaries with total mass m had the same eccentricity e_0 at the time when their semimajor axis was $a_0 = 300(2Gm/c^2)$, and compares different choices of e_0. The value $f_* = 1\,{\rm yr}^{-1} \simeq 31.7 \times 10^{-9}$ Hz corresponds to $\log_{10}(f_*/{\rm Hz}) \simeq -7.50$. Thus we see from Fig. 16.6 that this specific model, applied to circular binaries, give a value of $A_{\rm 1yr}$ near the upper limit of the range (16.86). As we will discuss in Section 23.5, PTAs already provide a slightly stronger limit on $A_{\rm 1yr}$. Note, however, that PTAs do not put a limit directly on $h_c(f)$ at $f = 1\,{\rm yr}^{-1} \simeq 31.7\,{\rm nHz}$. Rather, they have their best sensitivity at frequencies of order $2-6$ nHz, depending on the experiment, and conventionally translate their results into the value that $h_c(f)$ would have at $f = 1\,{\rm yr}^{-1}$ *assuming* a behavior $h_c(f) \propto f^{-2/3}$. We actually see from Fig. 16.6 that for, say, $e_0 \gtrsim 0.6$, the value of $h_c(f)$ at the frequencies tested by PTA is significantly below the value for circular orbits, and is below the PTA limits discussed in Section 23.5, even if $h_c(f)$ at $f = 1\,{\rm yr}^{-1}$ is larger than 10^{-15}.

On top of the results shown in this figure, we must also include the suppression in the spectrum at high frequencies computed in eq. (16.80), due to the discreteness in the number of sources. The corresponding knee in the spectrum sets in at a frequency f_0 given in Table 16.2. To compare with the horizontal scale in Fig. 16.6 observe that for the minimum value in this table, $f_0 \simeq 14.2\,{\rm nHz}$, we have $\log_{10}(f_0/{\rm Hz}) \simeq -7.8$ while for the highest value $f_0 \simeq 52.4\,{\rm nHz}$, $\log_{10}(f_0/{\rm Hz}) \simeq -7.3$. Thus, the effect of the eccentricity on the low-frequency side, and the effect of source discreteness on the high-frequency side, significantly reduce the window where $h_c(f) \propto f^{-2/3}$, to the extent that, for $f_0 \simeq 14.2\,{\rm nHz}$ and $e_0 > 0.8$, this window completely disappears.[29]

Figure 16.7 shows the corresponding result for $\Omega_{\rm gw}(f)$, as well as the separated contributions from different intervals of the total mass of the binary (again, the effect of source discreteness is not included in this plot). Independently of the specific details of the model, we see that, in the nHz region relevant for PTAs, the dominant contribution comes from SMBH binaries with a total mass in the range $(10^9 - 10^{10})M_\odot$. For this specific model of the BH assembly history, for $e_0 = 0$ the prediction for $\Omega_{\rm gw}(f)$ in the nHz region is at the level of existing observational limits; compare with eq. (23.76). However, if the typical eccentricity e_0 of binaries in this mass range is large, $e_0 \gtrsim 0.6$, the prediction at PTA frequencies is significantly reduced. We also see from the figure that, as the frequency increases, less massive binaries become more and more important. This is due to the fact that both the ISCO frequency (16.24) that marks the end of the inspiral phase and the maximum frequency (16.25) reached in the merger and ringdown phases, are inversely proportional to the total mass of the binary, so the contribution of lighter binaries extends to higher frequencies. In particular, in the frequency range where LISA has its best sensitivity, around a mHz, the emission is dominated by SMBH binaries with a total mass $(10^6 - 10^7)M_\odot$. However, as already discussed, at frequencies above about 10 nHz, the signal enters in a regime dominated by low-number statistics, so the result at LISA frequency is actually due to just a few individually resolvable binaries, and any residual stochastic background from SMBHs will be well below the LISA sensitivity curve.

Apart from the role of eccentricity, the existence of several other uncertainties must be kept in mind, especially when drawing conclusions from observational upper limits. First of all, as we have mentioned, the computations presented here simply *assume* that the last-parsec problem is solved, either by three-body interactions or by gas accretion. One must also consider the possibility that these effects "overdo" the job, and might dominate over GW back-reaction down to even shorter distances, and possibly up to coalescence, so the binary system will never be in a regime dominated by GW back-reaction. In that case the amplitude of the GW signal and its dependence on the frequency would change, compared with that computed assuming that the dynamics is dominated by GW back-reaction.

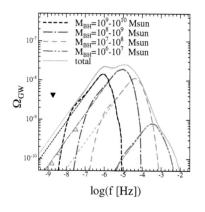

Fig. 16.7 $\Omega_{\rm gw}(f)$, and the separate contributions from different intervals of total mass of the binary, for $e_0 = 0.8$ (solid lines). For each curve corresponding to a different mass range, the dashed lines that prolonge the behavior $\Omega_{\rm gw}(f) \sim f^{2/3}$ toward lower frequencies correspond to the case $e_0 = 0$. The envelope (solid line) is the total contribution from all masses for $e_0 = 0$. From Enoki and Nagashima (2007).

Further reading

- The compact radio source Sgr A* was discovered by Balick and Brown (1974) with the NRAO interferometer at Green Bank. The indication of a central mass of a few times $10^6 M_\odot$ from the motion of gas clouds of ionized gas within Sgr A West is discussed in Lacy, Townes, Geballe and Hollenbach (1980). The motion of stars in the immediate vicinity of Sgr A* has been tracked since 1992 at the NTT/VLT [see Eckart and Genzel (1996)] and since 1995 at the Keck Observatory [see Ghez, Klein, Morris and Becklin (1998)]. Deviation from linear proper motion was detected by Ghez, Morris, Becklin, Tanner and Kremenek (2000) and the first orbit of a star was determined by Schödel *et al.* (2002). Determinations of the central BH mass and distance to the galactic center are given in Ghez *et al.* (2008) using Keck data and Gillessen *et al.* (2009a) using VLT data. The two datasets are combined in Gillessen *et al.* (2009b). Imaging of Sgr A* using VLBI interferometry at 1.3 mm, with a resolution of order $10\,\mu\text{as}$, is reported in Doeleman *et al.* (2008). Time variability of Sgr A* on the scale of a few times the size of the BH horizon was observed by Fish *et al.* (2011). For reviews of the properties of central SMBH see Eckart (2003), Carroll and Ostlie (2007), Melia (2007) and Falcke and Markoff (2013).

- A classic paper on the formation and evolution of massive BHs binaries in galaxy mergers is Begelman, Blandford and Rees (1980). Further studies include Quinlan (1996), Colpi, Mayer and Governato (1999) and Yu (2002); see also Section 8.1 of Binney and Tremaine (2008) and the review by Dotti, Sesana and Decarli (2012). The formation of supermassive black holes is reviewed in Volonteri (2010). A discussion of the time-scales for dynamical friction in different regimes is given in Dosopoulou and Antonini (2017). The role of gas accretion for solving the last-parsec problem is studied in Armitage and Natarajan (2002). Haehnelt and Kauffmann (2002) provided indirect arguments suggesting that SMBH binaries indeed overcome the last-parsec problem, otherwise their ejection rate due to subsequent galaxy mergers would be too large to be consistent with the $M - \sigma$ relation. Studies of the last-parsec problem based on N-body simulations, which reveal the important role of triaxiality of the host galaxy, are presented in Preto, Berentzen, Berczik and Spurzem (2011), Khan, Just and Merritt (2011) and Vasiliev, Antonini and Merritt (2015). A comparison between a semi-analytic model of the hardening rate and N-body simulations is presented in Sesana and Khan (2015).

- The astrophysics and cosmology that can be done with (e)LISA are described in Amaro-Seoane *et al.* (2013). The LISA mission proposal is described in Audley *et al.* (2017).

- Detailed reviews of EMRI are given in Amaro-Seoane *et al.* (2007) and Amaro-Seoane, Gair, Pound, Hughes and Sopuerta (2015). Discussions of EMRIs in the context of LISA are given in Babak, Gair, Petiteau and Sesana (2011) and Amaro-Seoane *et al.* (2013). Efficient search techniques for EMRI at LISA and estimates for the rate of EMRI events at LISA are discussed in Gair *et al.* (2004) and Gair (2009). The effect of mass segregation on the EMRI rate is discussed in Hopman and Alexander (2006). IMRIs are discussed in Miller (2004).

- The MiSaTaQuWa equation was derived in Mino, Sasaki and Tanaka (1997) and Quinn and Wald (1997). The adiabatic approach that only uses the dissipative parts of the self-force was developed in Mino (2003, 2005). Barack and Sago (2010) developed the numerical scheme that allows the computation of the gravitational self-force along any specified, but fixed, bound geodesic orbit of the Schwarzschild geometry. The first computation of the gravitational self-force for a geodesic of the Schwarzschild metric that changes because of the effect of the self-force was performed in Warburton, Akcay, Barack, Gair and Sago (2012). Clear and detailed reviews of the self-force formalism are given in Poisson (2004b) and Poisson, Pound and Vega (2011). The application of the self-force to the study of EMRI is reviewed in Barack (2009); see also Buonanno and Sathyaprakash (2015).

- Approximate waveforms for EMRIs have been proposed in, among others, Barack and Cutler (2004), Hughes, Drasco, Flanagan and Franklin (2005), and Drasco and Hughes (2006). Different approaches to the computation of the EMRI waveforms are reviewed in Drasco (2006).

- The characteristic amplitude used in Section 16.2.2 is introduced in Finn and Thorne (2000). The use of EMRIs to test the multipole moments of Kerr BHs was first discussed in Ryan (1995), and a more recent detailed discussion is given in Barack and Cutler (2007). The possibility of studying the ringdown of BH quasi-normal modes at LISA is studied in Berti, Cardoso and Will (2006b).

- The stochastic background due to SMBH binaries was first studied in Rajagopal and Romani (1995). Equation (16.48) and the behavior $h_c(f) \propto f^{-2/3}$ were first derived in Phinney (2001). Jaffe and Backer (2003), Wyithe and Loeb (2003), Enoki, Inoue, Nagashima and Sugiyama (2004) and Sesana, Haardt, Madau and Volonteri (2004) computed the spectrum with different assumptions on the SMBH binary assembly history. The last of these papers also computed the scaling of $h_c(f)$ with the frequency when stellar dynamics process dominate. Sesana, Vecchio and Colacino (2008) studied a wide range of assembly scenarios, and also showed that the behavior $h_c(f) \sim f^{-2/3}$ changes above 10^{-8} Hz because of the discreteness of the sources. Ravi *et al.* (2012), using semi-analytic prescriptions applied to the Millennium Simulation (a very large N-body simulation of structure formation in ΛCDM), found that the signal in PTAs is not really stochastic, but is dominated by a few sources in each frequency bin. This implies that the search for such a signal using the techniques for stochastic backgrounds is not optimal. In turn, it also means that there are increased chances of detecting individual sources.

- Predictions for the amplitude A_{1yr} that take into account the most recent information on the galaxy merger rate are given in Sesana (2013), McWilliams, Ostriker and Pretorius (2014) and Ravi, Wyithe, Shannon and Hobbs (2015). The time-scale for galaxy merger, which enters into these computations, is estimated in Kitzbichler and White (2008), using the Millennium Simulation.

- The evolution of the eccentricity due to three-body interactions is studied in Mikkola and Valtonen (1992) while its evolution due to dynamical friction is discussed in Colpi, Mayer and Governato (1999) and Dosopoulou and Antonini (2017). The effect of the eccentricity on the stochastic background from SMBH binaries is studied in Enoki, and Nagashima (2007) and Huerta, McWilliams, Gair and Taylor (2015). The effect of gas accretion on the last part of the evolution of the binary is quite sensitive to the details of the modeling of the binary–disk interaction, and can lead to attenuation of the stochastic background but an increase in individually resolvable sources; see Kocsis and Sesana (2011).

Part IV

Cosmology and gravitational waves

Basics of FRW cosmology 17

We now begin Part IV, on the role of GWs in cosmology. Before we get into the heart of the subject, it will be useful to give an overview of Friedmann–Robertson–Walker (FRW) cosmology. We already reviewed some aspects of FRW kinematics in Section 4.1.4. Here we give a more detailed account of FRW cosmology, at the level of background evolution. Of course, FRW cosmology is a subject on which there are many excellent dedicated textbooks (see the Further Reading). The aim of this section is only to give a self-contained account of some basic notions that will be needed in the rest of Part IV and to fix the notations and conventions, referring the reader to the Further Reading for more details.

Starting from this chapter, and for the rest of Part IV, we switch to units $\hbar = c = 1$, which are much more convenient when we deal with classical or quantum field theory.

17.1 The FRW metric

The FRW metric has already been given in eq. (4.141).[1] Observations are consistent with the Universe being spatially flat, to a high degree of accuracy. In the following we will therefore restrict attention to a spatially flat FRW metric, $K = 0$, unless explicitly stated.[2] Then, also setting $c = 1$, the FRW metric reads

$$ ds^2 = -dt^2 + a^2(t)d\mathbf{x}^2 \,, \qquad (17.2) $$

where t is called cosmic time and $a(t)$ is the scale factor. The coordinates $\mathbf{x}$ are called comoving coordinates, while the physical coordinates are defined by

$$ \mathbf{x}_{\rm ph} = a(t)\mathbf{x} \,. \qquad (17.3) $$

For a spatially flat FRW metric the overall normalization of the scale factor can be chosen at will, since it amounts to a rescaling of the coordinates $\mathbf{x}$, and we follow the standard choice of setting $a(t_0) = 1$, where t_0 is the present value of cosmic time. The Hubble parameter is defined by

$$ H = \frac{\dot a}{a} \qquad (17.4) $$

(where the dot denotes the derivative with respect to cosmic time) and its present value is denoted by H_0. Writing

$$ H_0 = h_0 \, 100 \, {\rm km \, s^{-1} \, Mpc^{-1}} \,, \qquad (17.5) $$

[1] From the historical point of view, the name Friedmann–Lemaître–Robertson–Walker (FLRW) is more appropriate. We will, however, follow the most widespread convention, and denote it simply as the FRW metric.

[2] Taking K generic in eq. (4.141), the Friedmann equation that we will derive in eq. (17.48) below is modified into

$$ H^2 = \frac{8\pi G}{3}\rho - \frac{K}{a^2} \,, \qquad (17.1) $$

corresponding to an effective energy density associated to the curvature $\rho_K = -(K/a^2)(3/8\pi G)$. Defining, as in eq. (17.72), $\Omega_K \equiv \rho_{K,0}/\rho_0 = -K/H_0^2$, where the subscript 0 denotes the value at the present value of time, the limit on Ω_K from the *Planck* 2015 data is $|\Omega_K(t_0)| < 5 \times 10^{-3}$ at 95% c.l., see [Planck Collaboration], Ade *et al.* (2016a).

Gravitational Waves, Volume 2: Astrophysics and Cosmology. Michele Maggiore.
© Michele Maggiore 2018. Published in 2018 by Oxford University Press.
DOI 10.1093/oso/9780198570899.001.0001

the measured value of h_0 is close to 0.7, with differences at the level of a few percent, which depend on the cosmological measurements and datasets used.[3] In the following we will keep h_0 explicit, and use $h_0 = 0.7$ as a reference value in the numerical estimates.

The Hubble parameter $H(t)$ defines the relevant cosmological length-scale, which is the horizon scale[4]

$$\lambda_H \equiv H^{-1}\,. \qquad (17.6)$$

Numerically, in our units $c = 1$, the present value of the horizon scale is

$$H_0^{-1} \simeq 4.28275 \left(\frac{0.7}{h_0}\right) \mathrm{Gpc}\,. \qquad (17.7)$$

For the metric (17.2), in Cartesian coordinates the non-vanishing Christoffel symbols are

$$\Gamma^0_{ij} = a^2 H \delta_{ij}\,, \qquad \Gamma^i_{0j} = H \delta^i_j\,, \qquad (17.8)$$

where $i, j = x, y, z$.[5] The Ricci tensor is given by

$$R_{00} = -3(\dot{H} + H^2)\,, \qquad R_{0i} = 0\,, \qquad R_{ij} = a^2(\dot{H} + 3H^2)\delta_{ij}\,, \qquad (17.10)$$

so the Ricci scalar is

$$R = 6(\dot{H} + 2H^2)\,, \qquad (17.11)$$

and the Einstein tensor G^μ_ν is

$$G^0_0 = -3H^2\,, \qquad G^0_i = 0\,, \qquad G^i_j = -(2\dot{H} + 3H^2)\delta^i_j\,. \qquad (17.12)$$

17.1.1 Comoving and physical coordinates

It is interesting to compare the comoving and physical coordinate systems in FRW with the coordinate systems that, in the case of the metric of a GW propagating in flat space, define the TT gauge and the proper detector frame, respectively, as discussed in Section 1.3.3. As we see from eq. (17.9), in comoving coordinates $\Gamma^i_{00} = 0$. This allows us to repeat without any change the derivation given in Section 1.3.3, where we showed that, when we consider a GW propagating in flat space, in the coordinate system that defines the TT gauge a particle initially at rest remains at rest. In the same way, we find that in comoving coordinates a particle initially at rest in the FRW metric remains at rest. This justifies the name "comoving coordinates". The comoving coordinate system in FRW is therefore conceptually analogous to the coordinate system that defines the TT gauge in the case of the metric of a GW traveling in Minkowski space: in both cases, these coordinates are ideally defined by the position of point masses initially at rest. Thus the comoving coordinates of a particle initially at rest in a FRW metric by definition do not evolve with time, just like the TT coordinates of a particle hit by a GW.

In contrast, we see from eq. (17.3) that the physical coordinates of a test mass evolve with the scale factor.[6] In a sense, the name "physical coordinates" is slightly misleading, since it seems to imply that they are the "correct" coordinates; of course, in general relativity (GR) all coordinate systems are on the same footing, and equally legitimate. However, physical coordinates are well suited for using our Newtonian intuition, as long as we are interested in the dynamics at length-scales shorter than the horizon scale λ_H. Indeed, from eq. (17.3),

$$d\mathbf{x}_{\rm ph} = a(d\mathbf{x} + H\mathbf{x}dt)\,, \tag{17.13}$$

which can be inverted to give

$$a d\mathbf{x} = d\mathbf{x}_{\rm ph} - \mathbf{x}_{\rm ph}Hdt\,. \tag{17.14}$$

Then, in physical coordinates the FRW metric (17.2) reads

$$ds^2 = -\left(1 - H^2\mathbf{x}_{\rm ph}^2\right)dt^2 + d\mathbf{x}_{\rm ph}^2 - 2H\mathbf{x}_{\rm ph}{\cdot}d\mathbf{x}_{\rm ph}dt\,. \tag{17.15}$$

We see that, in physical coordinates, at distances $|\mathbf{x}_{\rm ph}| \ll \lambda_H$, the FRW metric reduces to the Minkowski metric, and therefore the Newtonian intuition applies.[7] Comparing with the discussion in Section 1.3.3 we see that physical coordinates in the FRW space-time are conceptually analogous to the coordinates that define the proper detector frame in the case of the metric of a GW traveling in Minkowski space. In both cases, they are the coordinate systems in which we can use Newtonian arguments; and in both cases these Newtonian arguments are anyhow restricted to distances much smaller than the characteristic length-scale of the metric (the wavelength of the GW in the case of GWs propagating over flat space, and the horizon scale λ_H in FRW).

17.1.2   Comoving and physical momenta

Consider a spinless point particle of mass m moving in a background metric $g_{\mu\nu}$. The particle trajectory, in comoving coordinates, is described by a function $x^\mu(t)$. As we saw in eq. (1.65), the point-particle action is given by

$$S = -m \int d\tau\,, \tag{17.19}$$

where (in our units $c = 1$)

$$d\tau^2 = -g_{\mu\nu}dx^\mu(t)dx^\nu(t)\,. \tag{17.20}$$

The corresponding equation of motion is the geodesic equation (1.66),

$$\frac{d^2x^\mu}{d\tau^2} + \Gamma^\mu_{\nu\rho}(x)\frac{dx^\nu}{d\tau}\frac{dx^\rho}{d\tau} = 0\,, \tag{17.21}$$

while from the definition of τ it follows that

$$-g_{\mu\nu}\frac{dx^\mu}{d\tau}\frac{dx^\nu}{d\tau} = 1\,. \tag{17.22}$$

[7] Observe also that, in contrast to comoving coordinates, now $\Gamma^i_{00} \neq 0$. Indeed, in physical coordinates

$$g_{00} = -(1 - H^2 r_{\rm ph}^2)\,,$$
$$g_{0i} = -H x_{{\rm ph},i}\,, \tag{17.16}$$
$$g_{ij} = \delta_{ij}\,.$$

The inverse metric is

$$g^{00} = -1\,,$$
$$g^{0i} = -H x_{\rm ph}^i\,, \tag{17.17}$$
$$g^{ij} = \delta^{ij} - H^2 x_{\rm ph}^i x_{\rm ph}^j\,,$$

and

$$\Gamma^i_{00} = -(\dot{H} + H^2)(1 - H^2 r_{\rm ph}^2)x_{\rm ph}^i\,. \tag{17.18}$$

In the region $Hr_{\rm ph} \ll 1$, Γ^i_{00} is therefore non-vanishing, except at the special point $x_{\rm ph}^i = 0$.

The four-momentum $P^\mu = (P^0, \mathbf{P})$ of a massive particle is defined by

$$P^\mu = m\frac{dx^\mu}{d\tau}, \tag{17.23}$$

and, because of eq. (17.22), satisfies the mass-shell condition $-g_{\mu\nu}P^\mu P^\nu = m^2$. In particular, the energy is given by

$$E = m\frac{dx^0}{d\tau}. \tag{17.24}$$

We now restrict to a FRW space-time. The fact that this space-time is invariant under spatial rotations (or, more generally, the invariance under rotation of the background, when we perturb around FRW) allows us to define some useful quantities, which are intrinsically three-dimensional vectors, and differ by powers of the scale factor from the spatial part $\mathbf{P}$ of the four-vector P^μ. Indeed, using eq. (17.8), the $\mu = i$ component of eq. (17.21) becomes

$$\frac{d^2x^i}{d\tau^2} + 2H\frac{dt}{d\tau}\frac{dx^i}{d\tau} = 0, \tag{17.25}$$

where we have used coordinates $x^\mu = (t, \mathbf{x})$. Then, on a geodesic,

$$\begin{aligned}
\frac{d}{d\tau}\left(a^2\frac{dx^i}{d\tau}\right) &= 2a\frac{da}{d\tau}\frac{dx^i}{d\tau} + a^2\frac{d^2x^i}{d\tau^2} \\
&= 2a\dot{a}\frac{dt}{d\tau}\frac{dx^i}{d\tau} - 2a^2H\frac{dt}{d\tau}\frac{dx^i}{d\tau} \\
&= 0,
\end{aligned} \tag{17.26}$$

where in the second line we have used eq. (17.25). Thus, the *comoving momentum*, defined as

$$p^i \equiv ma^2\frac{dx^i}{d\tau}, \tag{17.27}$$

is constant for a particle moving on a geodesic of FRW. Observe that it is related to the spatial component $\mathbf{P}$ of the four-vector P^μ by $p^i = a^2 P^i$. It is convenient to define its modulus p using as spatial metric δ_{ij} rather than $g_{ij} = a^2\delta_{ij}$,

$$\mathbf{p}^2 \equiv \delta_{ij}p^i p^j, \tag{17.28}$$

so that also $|\mathbf{p}|$ is constant on a geodesic. We will also use the notation $p = |\mathbf{p}|$. Using eqs. (17.24), (17.27) and (17.28), in the frame defined by the coordinates $(t, \mathbf{x})$ eq. (17.22) becomes

$$E^2 = m^2 + \frac{p^2}{a^2}. \tag{17.29}$$

It is then convenient to introduce the "physical" momentum

$$p^i_{\text{ph}} = \frac{1}{a(t)}p^i(t), \tag{17.30}$$

and again define its modulus by

$$\mathbf{p}^2_{\text{ph}} \equiv \delta_{ij}p^i_{\text{ph}}p^j_{\text{ph}}, \tag{17.31}$$

and $p_{\rm ph} = |\mathbf{p}_{\rm ph}|$. In term of the physical momentum the mass-shell relation has the same form as in flat space,

$$E = \sqrt{m^2 + p_{\rm ph}^2}\,. \tag{17.32}$$

On a geodesic, the physical momentum decreases as $1/a$ with the expansion of the Universe. Thus, for a non-relativistic particle $E \simeq m$ is constant, while for a massless particle $E = p_{\rm ph} = p/a(t)$ decreases in time as $1/a$ as the Universe expands. Then the physical wavelength $\lambda_{\rm ph} \equiv 2\pi/p_{\rm ph} = 2\pi a(t)/p$ is stretched by the cosmological expansion as $\lambda(t) \propto a(t)$, while the comoving wavelength $\lambda = 2\pi/p$ stays constant.

The comoving momentum is just the canonical momentum conjugate to the comoving coordinate. Indeed, in FRW and in coordinates $(t, \mathbf{x})$, we have

$$\begin{aligned} d\tau^2 &= dt^2 - a^2(t)\delta_{ij}dx^i(t)dx^j(t) \\ &= dt^2 \left[1 - a^2(t)\dot{\mathbf{x}}^2\right]\,, \end{aligned} \tag{17.33}$$

where we have defined $\dot{\mathbf{x}}^2 = \delta_{ij}\dot{x}^i\dot{x}^j$ and $\dot{x}^i = dx^i/dt$. Therefore the point-particle action in FRW is

$$S = \int dt\, L\,, \tag{17.34}$$

where the Lagrangian L is given by

$$L = -m \left[1 - a^2(t)\dot{\mathbf{x}}^2\right]^{1/2}\,. \tag{17.35}$$

For $a(t) = 1$ and $\dot{\mathbf{x}}^2 \ll 1$ we recover the standard non-relativistic Lagrangian of a free particle, $L = (1/2)m\dot{\mathbf{x}}^2$, apart from a constant rest-mass contribution. The momentum $\pi^i(t)$ canonically conjugate to the comoving coordinate $x^i(t)$ is

$$\begin{aligned} \pi^i &= \frac{\delta L}{\delta \dot{x}^i} \\ &= m \left[1 - a^2(t)\dot{\mathbf{x}}^2\right]^{-1/2} a^2 \dot{x}^i \\ &= ma^2 \frac{dx^i}{d\tau}\,, \end{aligned} \tag{17.36}$$

where in the last line we have again used the relation between dt and $d\tau$ given in eq. (17.33). Thus $\pi^i = p^i$. Note that we are treating the Lagrangian (17.35) as if it were a (time-dependent) Lagrangian in flat space, with metric δ_{ij}, and $\dot{\mathbf{x}}^2 = \delta_{ij}\dot{x}^i\dot{x}^j$. Then $\delta L/\delta \dot{x}^i$, which in general would be equal to π_i, is also the same as π^i.

In correspondence with comoving and physical coordinates we can define a comoving energy density $\rho_{\rm com} = dE/d^3x$ and a physical energy density, that we denote simply by ρ,[8]

$$\rho = \frac{dE}{d^3x_{\rm ph}}\,. \tag{17.37}$$

[8]Observe a possibly confusing aspect of our (rather standard) notation. The coordinate $\mathbf{x}$, without any subscript, is the comoving coordinate, while for physical coordinates we use the notation $\mathbf{x}_{\rm ph}$ (or, when we have long equations involving physical coordinates, and we do not want to clutter the equations with the subscript "ph", we will use the notation $\mathbf{x}_{\rm ph} = \mathbf{r}$, as in Section 17.7). In contrast, for fluid variables such as energy density, pressure, etc., the quantities with no subscript are the physical ones, $\rho \equiv \rho_{\rm phys}$, $p \equiv p_{\rm phys}$, etc., while, when we refer to comoving energy density, etc. we write explicitly $\rho_{\rm com}$, $p_{\rm com}$, etc. This notation turns out to be practical, since we will most often use physical densities, as functions of the comoving coordinates, e.g. $\rho_{\rm phys}(t, \mathbf{x}_{\rm com})$, which in this notation are simply written as $\rho(t, \mathbf{x})$, etc., without cluttering the equations with subscripts.

From eq. (17.3) we see that the volume elements in physical and comoving coordinates are related by $d^3x_{\rm ph} = a^3(t)d^3x$ and therefore

$$\rho(t) = \frac{\rho_{\rm com}(t)}{a^3(t)} . \qquad (17.38)$$

For a non-relativistic particle, the energy is given just by the rest-mass contribution, which does not evolve with time, and therefore for a gas of non-relativistic particles $\rho_{\rm com}$ is constant and $\rho \propto 1/a^3$. For a gas of relativistic particles, taking into account the factor $1/a$ from the redshift of the energy and the factor $1/a^3$ from $d^3x_{\rm ph}$, $\rho \propto 1/a^4$. In the following we will mostly use the physical density ρ. The comoving density $\rho_{\rm com}$ will only enter in Section 17.7, to illustrate the relation between the Lagrangian and the Eulerian points of view in fluid mechanics.

The Fourier transform of a function $f(\mathbf{x})$, with $\mathbf{x}$ the comoving coordinates, is defined by

$$\tilde{f}(\mathbf{k}) = \int d^3x\, f(\mathbf{x})e^{-i\mathbf{k}\cdot\mathbf{x}} , \qquad (17.39)$$

so

$$f(\mathbf{x}) = \int \frac{d^3k}{(2\pi)^3}\, \tilde{f}(\mathbf{k})e^{i\mathbf{k}\cdot\mathbf{x}} , \qquad (17.40)$$

where $\mathbf{k}$ is the comoving wavenumber, and is independent of time. The corresponding physical wavenumber is $\mathbf{k}_{\rm phys}(t) = \mathbf{k}/a(t)$. Since we normalize $a(t)$ so that $a(t_0) = 1$, the comoving wavenumber is the same as the physical wavenumber at the present time, $\mathbf{k} = \mathbf{k}_{\rm phys}(t_0)$.

17.2 Cosmological background equations for a single fluid

The dynamics of the scale factor is governed by the Einstein equations, which it is now convenient to write with an upper and a lower index,

$$G^\mu_\nu = 8\pi G\, T^\mu_\nu . \qquad (17.41)$$

In a FRW space-time, at the level of background evolution, the energy–momentum tensor is taken to be that of a perfect fluid, already given in eq. (10.30),

$$T^\mu_\nu = (\rho + p)\, u^\mu u_\nu + p\,\delta^\mu_\nu , \qquad (17.42)$$

where ρ is the energy density and p the pressure.[9] In the rest frame of the fluid $u^\mu = (1,0,0,0)$, so $u_\mu = (-1,0,0,0)$ and therefore

$$T^0_0 = -\rho(t), \qquad T^i_0 = 0; \qquad T^i_j = p(t)\,\delta^i_j , \qquad (17.43)$$

compare with eq. (10.28).[10] More generally, the form (17.43) of the energy–momentum tensor follows by requiring the symmetries of the

---

[9]Not to be confused, of course, with the comoving momentum p introduced in the previous section. The two quantities will always appear in different contexts.

[10]Observe that, if we write the energy–momentum tensor as in eq. (17.42), ρ and p are scalar under diffeomorphisms, since the tensor structure of T^μ_ν is already carried by $u^\mu u_\nu$ and δ^μ_ν. Then, the identity between the $(0,0)$ component of the tensor $-T^\mu_\nu$ (i.e. the actual energy density) and the scalar ρ only holds in the coordinate system where $u^\mu(x) = (1,0,0,0)$; compare also with the discussion following eq. (10.30) and Note 35 on page 36. This definition of ρ and p applies to a perfect fluid. For a generic fluid one could use eq. (10.48), in which case, in the units $c = 1$ that we are using here,

$$T^0_0 = \rho u^0 u_0 + p(u^0 u_0 + 1) + \Sigma^0_0 . \qquad (17.44)$$

However, for such a generic fluid it can be more practical to change notation, and *define* ρ directly from $T^0_0 = -\rho$ (in any frame), so that ρ will now be the $\mu = 0, \nu = 0$ component of the tensor T^μ_ν. This is the convention that we will use when studying perturbations around flat space as well as cosmological perturbation theory; see eqs. (18.42) and (18.112). As we will see in Section 18.3.2, the difference between the two definitions of ρ (and of p) only shows up at second order in cosmological perturbation theory.

FRW space-time. In the rest frame, the corresponding energy in a co-moving volume V (corresponding to a physical volume $V_{\rm ph} = a^3 V$) is given by

$$
\begin{aligned}
E &= - \int_V d^3x \sqrt{\gamma}\, T^0_0 \\
&= \int_V d^3x\, a^3(t)\rho(t) \\
&= \int_{V_{\rm ph}} d^3x_{\rm ph}\, \rho(t)\,,
\end{aligned}
\tag{17.45}
$$

where $\gamma_{ij} = a^3 \delta_{ij}$ is the spatial metric. This shows that ρ is indeed the physical (rather than comoving) energy density, in the fluid rest frame.[11]

We consider here the case of a single fluid, while multi-component fluids will be discussed in Section 17.3. The Bianchi identity $\nabla_\mu G^\mu_\nu = 0$ imposes energy–momentum conservation, $\nabla_\mu T^\mu_\nu = 0$. Recalling that, for a tensor T^ρ_ν,

$$
\nabla_\mu T^\rho_\nu = \partial_\mu T^\rho_\nu + \Gamma^\rho_{\mu\sigma} T^\sigma_\nu - \Gamma^\sigma_{\mu\nu} T^\rho_\sigma\,,
\tag{17.46}
$$

using the energy–momentum tensor (17.43) and the Christoffel symbols (17.8), we find that the $\nu = 0$ component gives the conservation equation

$$
\dot{\rho} + 3H(\rho + p) = 0\,,
\tag{17.47}
$$

while the equation with $\nu = i$ gives $\partial_i p = 0$, which is automatically satisfied.

Using eq. (17.12) we see that the $\mu = \nu = 0$ component of eq. (17.41) gives the Friedmann equation,

$$
H^2 = \frac{8\pi G}{3}\rho\,,
\tag{17.48}
$$

while the $(0, i)$ components is an identity $0 = 0$, and the (i, j) component gives

$$
2\dot{H} + 3H^2 = -8\pi G p\,.
\tag{17.49}
$$

Equation (17.49) also follows by taking the time derivative of eq. (17.48), using eq. (17.47) to express $\dot{\rho}$ is terms of ρ and p, and then again eq. (17.48) to express ρ in terms of H^2. Thus, we can take eqs. (17.47) and (17.48) as the only independent equations, and their solutions automatically satisfy also eq. (17.49). This redundancy is a consequence of the diffeomorphism invariance of GR or, equivalently, of the Bianchi identities. Finally, the system of equations is closed by the equation of state (EoS) of the cosmological fluid, which is taken to be of the form

$$
p(t) = w(t)\rho(t)\,.
\tag{17.50}
$$

In general, $w(t)$ is a function of time. In a given cosmological epoch, when $w(t) \simeq w$ takes an approximately constant value, w is called the

[11]See also page 425 for a more detailed discussion of conserved quantities in curved space.

EoS parameter. According to standard kinetic theory, for a relativistic gas $w = 1/3$, while for a non-relativistic gas, apart from the factor $1/3$ coming from the spatial average, there is a further suppression $O(v^2/c^2)$ (reinstating c for the moment). This is due to the fact that the pressure of a non-relativistic gas is proportional to the average value of $\mathbf{p} \cdot \mathbf{v} = mv^2$, while the energy density ρ in eq. (17.50) contains the rest-mass contribution, so $E \sim mc^2$ and $w \sim v^2/(3c^2)$. For non-relativistic matter, to a first approximation, we can therefore set $w = 0$.

Adding also a cosmological constant, the Einstein equations become

$$G^\mu_\nu + \Lambda \delta^\mu_\nu = 8\pi G\, T^\mu_\nu\,. \tag{17.51}$$

We see that, formally, at the level of background evolution the cosmological constant term is equivalent to a fluid with energy–momentum tensor

$$(T^\mu_\nu)_\Lambda = -\frac{\Lambda}{8\pi G} \delta^\mu_\nu\,. \tag{17.52}$$

Recalling from the above discussion that the energy–momentum tensor of a perfect fluid has the form $T^\mu_\nu = (-\rho, p, p, p)$, we find that the cosmological constant formally acts as a fluid with energy density

$$\rho_\Lambda = \frac{\Lambda}{8\pi G} \tag{17.53}$$

and pressure $p_\Lambda = -\rho_\Lambda$, and therefore $w_\Lambda = -1$. The epochs of radiation dominance (RD), matter dominance (MD) and cosmological constant dominance (ΛD) are therefore characterized by an EoS of the form $p(t) = w\rho(t)$ with w constant, and equal to $1/3, 0$ and -1, respectively. Setting $p(t) = w(t)\rho(t)$ in eq. (17.47) gives

$$\dot{\rho} + 3(1 + w)H\rho = 0\,. \tag{17.54}$$

Using

$$\begin{aligned} \frac{d\rho}{dt} &= \frac{da}{dt}\frac{d\rho}{da} \\ &= H\frac{d\rho}{d\log a} \end{aligned} \tag{17.55}$$

we can rewrite eq. (17.54) as

$$\frac{d\rho}{d\log a} = -3(1 + w)\rho\,. \tag{17.56}$$

For w constant the integration gives

$$\rho(a) = \rho_0 a^{-3(1+w)}\,, \tag{17.57}$$

where ρ_0 is the present value of the energy density (recall that we set $a(t_0) = 1$). In particular, we have

$$\rho_R(a) = \frac{\rho_{R,0}}{a^4}\,, \tag{17.58}$$

$$\rho_M(a) = \frac{\rho_{M,0}}{a^3}\,, \tag{17.59}$$

$$\rho_\Lambda = \text{constant}\,, \tag{17.60}$$

where the subscripts R, M and Λ refer to radiation, matter and cosmological constant, respectively, and the subscript 0 denotes the present values. This is indeed the behavior expected for the physical energy densities, in agreement with the discussion following eq. (17.38).

Plugging these expressions into the Friedmann equation (17.48) and integrating we find

$$
a(t) \propto \begin{cases} t^{1/2} & (\text{RD}), \\ t^{2/3} & (\text{MD}), \\ e^{Ht} & (\Lambda\text{D}). \end{cases} \tag{17.61}
$$

Thus, for the Hubble parameter we get $H(t) = 1/(2t)$ in RD and $H(t) = 2/(3t)$ in MD, while during ΛD H is constant, and related to Λ by $H^2 = \Lambda/3$, as we see from eqs. (17.48) and (17.53).

More generally, if a fluid has $w = w(t)$, eq. (17.56) gives

$$
\rho(x) = \rho(x_0) \exp\left\{ -3 \int_{x_0}^{x} dx' \, [1 + w(x')] \right\}, \tag{17.62}
$$

where $x \equiv \log a$. Finally, observe also that eq. (17.49), once written in terms of $\ddot{a}$ and combined with the Friedmann equation, gives

$$
\begin{aligned}
\frac{\ddot{a}}{a} &= -\frac{4\pi G}{3}(\rho + 3p) \\
&= -\frac{4\pi G}{3}(1 + 3w)\rho.
\end{aligned} \tag{17.63}
$$

For normal matter $p > 0$ and $\rho > 0$, so $\rho + 3p > 0$ and $\ddot{a}/a < 0$, corresponding to the Newtonian intuition that gravity is attractive, and therefore the expansion of the Universe is decelerating. However, observations of type Ia supernovae have shown that in the recent epoch the Universe has entered a phase of accelerated expansion, corresponding to the presence of a fluid with $w < -1/3$. Such a fluid cannot consist of ordinary matter, and is therefore dubbed "dark energy" (DE). Understanding its physical origin is one of the greatest challenges of modern cosmology. The accelerated expansion of the Universe has also been confirmed to high precision by a combination of supernova data, CMB data and baryon acoustic oscillations (BAO), as well as by several other observations such as the counting of galaxy clusters at different redshifts, lensing, etc.; see the Further Reading. These observations also indicate that the value of w_{DE} is quite close to -1, at a few percent level, and hence consistent with a cosmological constant.[12]

17.3   Multi-component fluids

We now consider a more realistic model with several fluid components. A minimal composition for a realistic model includes photons, neutrinos, baryons,[13] cold dark matter (CDM) and dark energy. We therefore have several fluids, which we label by an index λ, with energy densities ρ_λ

[12]In Section 19.6 we will discuss observational constraints on w_{DE}, as well as possible alternatives to ΛCDM based on modified gravity, that can generate a dynamical DE.

[13]In cosmology one often denotes collectively baryons and leptons as "baryons" (for instance, when one talks of the baryon–photon fluid), so one includes as baryons not only the proton and the neutron, which are indeed baryons in the standard particle physics sense, but even $e^\pm$ or $\mu^\pm$, which, from the particle physics point of view, are of course leptons.

and pressures $p_\lambda = w_\lambda \rho_\lambda$. The total energy density and pressure are given by

$$\rho_{\rm tot}(t) = \sum_\lambda \rho_\lambda(t)\,,$$

$$p_{\rm tot}(t) = \sum_\lambda w_\lambda(t)\rho_\lambda(t)$$

$$\equiv w_{\rm tot}(t)\rho_{\rm tot}(t)\,, \qquad (17.64)$$

where

$$w_{\rm tot}(t) = \frac{\sum_\lambda w_\lambda(t)\rho_\lambda(t)}{\sum_\lambda \rho_\lambda(t)}\,. \qquad (17.65)$$

The Bianchi identities now only require conservation of the total energy–momentum tensor, $\nabla_\mu T^{\mu\nu}_{\rm tot} = 0$. In particular, the $\nu = 0$ component gives

$$\dot\rho_{\rm tot} + 3H(\rho_{\rm tot} + p_{\rm tot}) = 0\,. \qquad (17.66)$$

However, in several situations the energy–momentum tensors of each component can be separately conserved,

$$\dot\rho_\lambda + 3H(1 + w_\lambda)\rho_\lambda = 0 \qquad (\lambda = 1, \dots, n)\,. \qquad (17.67)$$

This happens in particular if the interaction rates Γ of the processes that exchange energy between the components is much smaller than the Hubble rate $H(t)$. In this case the change in the relative energy densities induced by such interactions is negligible compared with the dilution (such as $1/a^4$ for radiation and $1/a^3$ for matter) due to the expansion of the Universe. In other words, on the right-hand side of eq. (17.67) we now have a non-vanishing source term, which is, however, much smaller than the term $3H(1 + w_\lambda)\rho_\lambda$ on the left-hand side, associated with the expansion of the Universe.[14] To go beyond the approximation in which the energy–momentum tensors of different species are separately conserved, the first-principles approach consists in studying the evolution of the phase-space densities of each species, with a Boltzmann equation that includes all the appropriate collision terms.[15]

Once the EoS parameters w_λ have been assigned, eq. (17.67) together with eqs. (17.48) and (17.49) provide a set of $n+2$ equations for the $n+1$ variables $a(t)$ and $\rho_\lambda(t)$. Of course these equations are not all independent, because, once the Einstein equations are satisfied [and therefore, in FRW, once eqs. (17.48) and (17.49) are satisfied], the Bianchi identities force the total energy–momentum conservation (17.66). It is convenient to take the n equations (17.67) plus the Friedmann equation (17.48) as independent, so the system of differential equations is of first-order in $a(t)$.

We now specialize to the case in which the independent components are radiation, matter and a cosmological constant. This is the basic composition of the ΛCDM cosmological model. Before recombination, radiation actually includes two separate non-interacting fluids, the baryon–photon fluid (since, before recombinations, baryons are tightly coupled

---

[14]Note also that, even when two species are tightly coupled, in the sense that $\Gamma \gg H$, we can still have separate conservation equations. An example is provided by the baryon–photon fluid. As we will see in Section 17.4, baryons and photons decouple at a temperature $T = T_{\rm dec} \simeq 0.26$ eV, because the rate of the relevant interactions drops below $H(t)$. Therefore, at $T < T_{\rm dec}$ the separate conservation of the energy–momentum tensors of photons and of baryons is a consequence of $\Gamma \ll H$. However, even in the regime $T_{\rm dec} < T \ll m_B$ the separate conservation holds. This is due to the fact that in this regime baryons are non-relativistic, so $\rho_B \simeq m n_B$, where n_B is the baryon number density. Then the conservation of the baryonic energy density becomes the same as baryon number conservation. Equivalently, in this regime the photons, whose typical energy is $\sim T \ll m_B$, are not able to efficiently transfer energy to the baryons.

[15]See e.g. the discussion in Chapter 3 of Dodelson (2003). We will come back to the Boltzmann equation for photons in Section 20.3.7.

to photons) and the neutrino fluid, while matter is just CDM. After recombination, radiation includes photons and neutrinos, while matter includes CDM and baryons.[16] The total energy density is

$$\rho_{\rm tot} = \rho_R + \rho_M + \rho_\Lambda \,, \tag{17.68}$$

and the total pressure is

$$\begin{aligned} p_{\rm tot} &= p_R + p_M + p_\Lambda \\ &= \frac{1}{3}\rho_R - \rho_\Lambda \,. \end{aligned} \tag{17.69}$$

It is convenient to introduce the critical density today,

$$\rho_0 = \frac{3H_0^2}{8\pi G} \,. \tag{17.70}$$

Numerically, in our units $\hbar = c = 1$, the critical density can be written as[17]

$$\rho_0^{1/4} = 2.99965(3)h_0^{1/2} \times 10^{-3}\,{\rm eV} \,. \tag{17.71}$$

We denote by Ω_λ (with $\lambda = R, M, \Lambda$) the *present* values of the energy fractions,[18]

$$\Omega_R = \frac{\rho_{R,0}}{\rho_0} \,, \qquad \Omega_M = \frac{\rho_{M,0}}{\rho_0} \,, \qquad \Omega_\Lambda = \frac{\rho_\Lambda}{\rho_0} \,. \tag{17.72}$$

Then, the densities $\rho_\lambda(t)$ at a generic time t can be expressed through the scale factor $a(t)$ as

$$\rho_\lambda(a) = \rho_0\Omega_\lambda a^{-3(1+w_\lambda)} \,, \tag{17.73}$$

so, in particular,

$$\rho_R(a) = \rho_0\Omega_R a^{-4} \,, \qquad \rho_M(a) = \rho_0\Omega_M a^{-3} \,, \qquad \rho_\Lambda = \rho_0\Omega_\Lambda \,. \tag{17.74}$$

The Friedmann equation (17.48) then reads

$$\boxed{H(a) = H_0 \left(\Omega_R a^{-4} + \Omega_M a^{-3} + \Omega_\Lambda\right)^{1/2} \,.} \tag{17.75}$$

The critical density at time t is defined by

$$\rho_c(t) = \frac{3H^2(t)}{8\pi G} \,, \tag{17.76}$$

and we denote by $\Omega_\lambda(t)$ (with $\lambda = R, M, \Lambda$) the values of the energy densities $\rho_\lambda(t)$, normalized to the critical density at time t,

$$\Omega_R(t) = \frac{\rho_R(t)}{\rho_c(t)} \,, \qquad \Omega_M(t) = \frac{\rho_M(t)}{\rho_c(t)} \,, \qquad \Omega_\Lambda(t) = \frac{\rho_\Lambda}{\rho_c(t)} \,, \tag{17.77}$$

so that the constants Ω_λ defined in eq. (17.72) are equal to $\Omega_\lambda(t_0)$.

The value of ρ_R is known quite precisely, from measurement of the photon temperature and calculation of the ratio of the temperature of

---

[16] Actually, for the values of the neutrino masses indicated by oscillation experiments, neutrinos eventually become non-relativistic. The current lower limits on the masses m_1, m_2, m_3 of the mass eigenstates are $m_1 \gtrsim m_{\rm sol} \simeq 0.009$ eV and $m_2 \gtrsim m_{\rm atm} \simeq 0.05$ eV, while the third neutrino can be much lighter or even massless. On the other hand, there is a cosmological upper bound on the sum of the neutrino masses, coming from CMB+BAO+SNe which, assuming the validity of ΛCDM, is $\sum_i m_i < O(0.1)$ eV. For $m_1 = m_{\rm sol}$ and $m_2 = m_{\rm atm}$ these mass eigenstates become non-relativistic at $z_1 \simeq 16$ and $z_2 \simeq 94$, respectively, i.e. during MD, afterwards they act as a small hot (or, rather "warm") dark matter component [see the discussion in Gorbunov and Rubakov (2011), Section 8.4].

[17] In normal units, $\rho_0 \propto H_0^2/G$ is a mass per unit volume. Reinstating $\hbar$ and c, the quantity with dimensions of $({\rm energy})^4$ is $\rho_0 c^2 (\hbar c)^3 = (\hbar c)^3 \times (3H_0^2 c^2)/(8\pi G)$. Thus, in principle, the observational error in the numerical value of ρ_0, when it is expressed in $({\rm eV})^4$ (apart from the dependence on h_0, which we keep explicit) comes from the errors in G and in $\hbar$ (of course there is no error in c, which defines the unit of length). However, $\Delta\hbar/\hbar \simeq 6 \times 10^{-9}$, as we see from the relation $\hbar c = 197.3269788(12)$ MeV fm; see e.g. Patrignani *et al.* [Particle Data Group Collaboration] (2016). In contrast, $G = 6.67408(31) \times 10^{-11}\,{\rm m}^3\,{\rm kg}^{-1}\,{\rm s}^{-2}$, so $\Delta G/G \simeq 5 \times 10^{-5}$. Thus, the error in the value of ρ_0/h_0^2, expressed in eV^4 is entirely dominated by the error in G. Obviously, ρ_0 in eq. (17.70) is a purely classical quantity. The dependence on $\hbar$ only enters when, for comparison with the typical scales of particle physics, we wish to express its value in eV^4.

[18] More precisely, we define Ω_R as the value that the radiation fraction would have today if neutrinos were massless; see Note 29 on page 387. Then, we will see that, for values of a such that the neutrinos are relativistic, we have $\rho_R(a) = \rho_0\Omega_R a^{-4}$, as in eq. (17.74), with Ω_R given in eq. (17.79).

neutrinos to that of photons, as we will see in Section 17.5. Dividing ρ_R by the critical density today, which is proportional to H_0^2, introduces an extra uncertainty due to the uncertainty in Hubble parameter. Thus, the quantity that is known more precisely is rather $h_0^2\Omega_R$, where the multiplication by h_0^2 cancels the h_0^2 factor present in ρ_c. The contribution from photons is

$$h_0^2\Omega_\gamma \simeq 2.473 \times 10^{-5}\,, \tag{17.78}$$

as we will see in eq. (17.110). The contribution from neutrinos depends on whether they are relativistic or not. For values of a such that all neutrinos are relativistic, their contribution adds to that of photons so that $\rho_R(a) = \rho_0\Omega_R a^{-4}$ with

$$\Omega_R \simeq 4.184 \times 10^{-5}\, h_0^{-2}\,, \tag{17.79}$$

as we will compute in eq. (17.111). However, this expression no longer holds near the present epoch, when at least two neutrino species become non-relativistic (a third neutrino species could still in principle be sufficiently light, or even massless, to remain relativistic up to the present epoch; see Note 16). Then, at the present epoch they no longer count as relativistic matter, but rather as a distinct "warm" dark matter component.

The exact value of Ω_M obtained from the comparison with cosmological data depends somewhat on the datasets used, as well as on the cosmological model used to fit the cosmological data. Within ΛCDM (with the smallest possible value of neutrino masses consistent with oscillation experiments), the *Planck* 2015 data give[19]

$$\Omega_M = 0.308 \pm 0.012\,. \tag{17.80}$$

In the following we will keep Ω_M explicit in our equations, normalizing it to a reference value $\Omega_M = 0.3$ when we need to give numerical estimates.

By definition $\sum_\lambda \Omega_\lambda(t) = 1$ at all times.[20] Figure 17.1 shows the energy fractions $\Omega_\lambda(t)$ against $x = \log a(t)$. We see that in the far past radiation dominates, as it follows from its behavior $\rho_R \propto a^{-4}$. Matter behaves as $\rho_M \propto a^{-3}$, so it eventually comes to dominate over radiation. Finally, near the present epoch the cosmological term, whose energy density does not even decrease with a, takes over.

Even when the EoS parameters w_λ of the separate fluid components are independent of time, in a multi-component fluid the total EoS parameter $w_{\rm tot}$, defined by $p_{\rm tot} = w_{\rm tot}\rho_{\rm tot}$, is time-dependent,

$$
\begin{aligned}
w_{\rm tot}(a) &= \frac{\sum_\lambda p_\lambda(a)}{\sum_\lambda \rho_\lambda(a)} \\
&= \frac{\sum_\lambda w_\lambda \rho_{\lambda,0}\, a^{-3(1+w_\lambda)}}{\sum_\lambda \rho_{\lambda,0}\, a^{-3(1+w_\lambda)}} \\
&= \frac{\sum_\lambda w_\lambda \Omega_\lambda a^{-3(1+w_\lambda)}}{\sum_\lambda \Omega_\lambda a^{-3(1+w_\lambda)}}\,.
\end{aligned}
\tag{17.81}
$$

In particular, in a model with matter, radiation and cosmological constant,

[19]This value actually refers to the combination of TT, lowP and lensing data. See Table 4 of [Planck Collaboration], Ade *et al.* (2016a), for the result with different combinations of CMB polarization data. Once again, $h_0^2\Omega_M$ is determined more precisely because it does not suffer from the uncertainty on h_0

[20]If we drop the assumption that the Universe is spatially flat, we must further add a term $\Omega_K(t)$ describing the effect of spatial curvature. From the Friedmann equation in the presence of curvature it follows that $\rho_K \propto a^{-2}$; see Note 2.

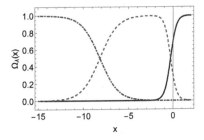

Fig. 17.1 The energy fractions $\Omega_\lambda(x)$, with $\lambda = R$ (dot-dashed), $\lambda = M$ (dashed) and $\lambda = \Lambda$ (solid line), against $x = \log a$. We use the values $\Omega_M = 0.3$, $\Omega_R \simeq 4.18 \times 10^{-5}h_0^{-2}$, $h_0 = 0.7$, $\Omega_\Lambda = 1 - \Omega_M - \Omega_R$. (Observe that near the present epoch the contribution of massive neutrinos should be moved from radiation to non-relativistic matter, or to a new warm dark matter component, but this effect is invisible on this scale).

$$w_{\text{tot}}(a) = \frac{(1/3)\Omega_R a^{-4} - \Omega_\Lambda}{\Omega_R a^{-4} + \Omega_M a^{-3} + \Omega_\Lambda}. \tag{17.82}$$

We plot this function in Fig. 17.2. We see that $w_{\text{tot}}(a)$ smoothly interpolates between the values $w = 1/3$ in RD, $w = 0$ in MD, and $w = -1$ when eventually Λ dominates. Today Ω_M and Ω_Λ are still comparable, and $w_{\text{tot}}(a = 0) \simeq -0.7$.

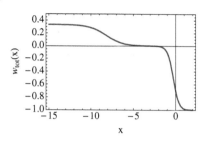

Fig. 17.2 The total EoS parameter w_{tot}, for the same model used in Fig. 17.1

17.4 RD–MD equilibrium, recombination and decoupling

There are several important moments in the thermal history of the Universe. One is the transition between RD and MD. The equilibrium between matter and radiation is given by the condition $\Omega_M a_{\text{eq}}^{-3} = \Omega_R a_{\text{eq}}^{-4}$, so $a_{\text{eq}} = \Omega_R/\Omega_M$ or, in terms of the redshift z,

$$1 + z_{\text{eq}} = \frac{\Omega_M}{\Omega_R}$$

$$= 3513.38 \left(\frac{\Omega_M}{0.3}\right)\left(\frac{4.184 \times 10^{-5}}{\Omega_R h_0^2}\right)\left(\frac{h_0}{0.7}\right)^2. \tag{17.83}$$

In terms of $x = \log a$, for these reference values we have $x_{\text{eq}} \simeq -8.164$. The temperature at matter–radiation equilibrium is given by

$$T_{\text{eq}} = (1 + z_{\text{eq}})T_0, \tag{17.84}$$

where $T_0 \simeq 2.7255(6)$ K is the present photon temperature, measured from the CMB. Numerically,[21]

$$T_{\text{eq}} \simeq 0.825 \,\text{eV} \left(\frac{\Omega_M}{0.3}\right)\left(\frac{4.184 \times 10^{-5}}{\Omega_R h_0^2}\right)\left(\frac{h_0}{0.7}\right)^2. \tag{17.85}$$

Another crucial moment in the thermal history of the Universe is recombination, when the temperature drops sufficiently below the Rydberg, i.e. the binding energy of the hydrogen atom, so that electrons and protons can eventually combine into hydrogen atoms.[22] We define the ionization fraction x_e by

$$x_e = \frac{n_e}{n_e + n_H}, \tag{17.86}$$

where n_e and n_H are the number densities of electrons and hydrogen atoms, respectively. Then, the dependence of x_e on the temperature is governed by the *Saha equation*[23]

$$\frac{x_e^2}{1 - x_e} = \frac{1}{n_B}\left(\frac{m_e T}{2\pi}\right)^{3/2} e^{-\Delta/T}$$

$$= \frac{45\bar{s}}{4\pi^2}\left(\frac{m_e}{2\pi T}\right)^{3/2} e^{-\Delta/T}, \tag{17.87}$$

[21] We use units where the Boltzmann constant $k_B = 1$. The conversion from temperatures to energies is obtained using the fact that, in normal units, $k_B = 8.6173303(50) \times 10^{-5}$ eV/K, see e.g. Patrignani *et al.* [Particle Data Group Collaboration] (2016).

[22] The word "recombination" is somewhat misleading, since it seems to imply that electrons and protons were already combined into atoms at some earlier stage. Rather, this is the first epoch in which electrons and protons combine to form atoms.

[23] See e.g. Section 1.3.1 of Durrer (2008).

where $\Delta = m_p + m_e - m_H \simeq 13.6$ eV is the Rydberg, n_B is the number density of baryons and

$$\bar{s} = \frac{4\pi^2 T^3}{45 n_B} \tag{17.88}$$

is the entropy per baryon; see eq. (17.101) below. Numerically

$$\bar{s} \simeq 7 \times 10^9 \left(\frac{0.2}{\Omega_B h_0^2}\right), \tag{17.89}$$

where Ω_B is the present baryon fraction. Recombination can be defined as the moment at which $x_e = 1/2$ (the precise value is not very important). If $\bar{s}$ were of order one, eq. (17.87) would give $T_{\rm rec} \simeq \Delta$. The fact that $\bar{s}$ is very large (which reflects the fact that there are many more photons than baryons in the Universe) has the effect of lowering $T_{\rm rec}$. Physically, this is due to the fact that even at T below Δ there are still enough photons in the high-energy tail of the thermal distribution, to keep the Universe ionized. Setting $x_e = 1/2$ and using eq. (17.89), eq. (17.87) becomes

$$\left(\frac{T_{\rm rec}}{1\,{\rm eV}}\right)^{-3/2} e^{-\Delta/T_{\rm rec}} \simeq 2.6 \times 10^{-18} \left(\frac{\Omega_B h_0^2}{0.02}\right). \tag{17.90}$$

For $\Omega_B h_0^2 \simeq 0.02$ this give

$$T_{\rm rec} \simeq 0.32\ {\rm eV}, \tag{17.91}$$

corresponding to a redshift

$$z_{\rm rec} \simeq 1376. \tag{17.92}$$

Therefore recombination takes place during MD, slightly after radiation–matter equilibrium.

Shortly after recombination, photons decouple from baryons. Before decoupling, photons are kept in equilibrium with electrons by Compton scattering, and the electrons are tightly coupled to the protons via Coulomb scattering.[24] Therefore photons decouple from electrons (and hence from protons) when the rate for the elastic Thomson scattering of photons on electrons, $\Gamma(T) = \sigma_T n_e$ (where σ_T is the Thomson cross-section and n_e the number density of the remaining free electrons), drops below the Hubble parameter $H(T)$. The condition $\Gamma(T_{\rm dec}) = H(T_{\rm dec})$ gives the decoupling temperature

$$T_{\rm dec} \simeq 0.26\,{\rm eV}, \tag{17.93}$$

corresponding to a redshift

$$z_{\rm dec} \simeq 1090. \tag{17.94}$$

At this point, photons start to propagate freely. This redshift therefore defines the last-scattering surface (LSS) of the CMB photons. Note, however, that, even if photons decouple from electrons, the electrons

[24]In contrast, the direct interaction of photons with protons is much smaller: the Compton scattering cross-section of a photon on a target of mass m, at center-of-mass energies much smaller than m, is given by the Thomson cross-section $\sigma_T = (8\pi/3)\,(\alpha^2/m^2)$. Therefore the cross-section for Compton scattering on protons is suppressed, compared with the scattering on electrons, by a factor $m_e^2/m_p^2 \simeq 3 \times 10^{-7}$.

are still coupled to photons. This is due to the fact that photons are much more numerous than electrons, so the Thompson scattering rate of electrons on photons is still large enough to maintain the electrons in thermal equilibrium with the photons. Therefore the few remaining free electrons maintain the same temperature as the photons, until very low redshift, and the same happens to the few remaining free protons, that are kept in equilibrium with the electrons by Coulomb scattering.

17.5 Effective number of relativistic species

Equation (17.75) is actually over-simplified because it assumes that the species that are relativistic today are the same species that were relativistic in the past, and that for each of them $T \propto 1/a$. If this were the case, the radiation density at time t would be simply related to the radiation density today by $\rho_R(t) = \rho_R(t_0)/a^4(t)$, for all t. However, the real picture is more complicated.

First of all, in the early Universe many more species were relativistic; for instance, at temperatures $T \gg 1$ GeV protons and neutrons are replaced by deconfined quarks with relativistic speeds, at temperatures $T \gg 100$ GeV the W and Z bosons are relativistic, and so on. Thus, at each temperature we must take into account the degrees of freedom that are actually relativistic.

Furthermore when, on lowering the temperature, a species becomes non-relativistic, if it is still coupled to the rest of the primordial plasma then its density drops through annihilation processes, which can no longer be compensated by the inverse process, since the particle is now too heavy to be produced. Then, the entropy of this species is transferred to the particles that are coupled to it, and this modifies the $T \propto 1/a$ behavior of the rest of the plasma. A particularly important example is given by e^+e^- annihilation, which at temperatures below the MeV range produces a difference between the photon and neutrino temperatures. Neutrinos decouple at a temperature of about $2-3$ MeV,[25] when the rate for weak processes such as $\nu_e e^- \to \nu_e e^-$ and $\nu_e \bar{\nu}_e \leftrightarrow e^+e^-$ becomes smaller than the Hubble expansion rate. Shortly afterwards, electrons and positrons become non-relativistic (when T becomes sufficiently smaller than m_e) and their abundance drops. This happens through the annihilation $e^+e^- \to 2\gamma$, which is no longer compensated by the inverse process $2\gamma \to e^+e^-$, since at values of T sufficiently smaller than m_e an e^+e^- pair is too heavy to be produced by the photons.[26] Quantitative computations (using a formalism analogous to the Saha equation discussed in Section. 17.4) show that e^+e^- annihilation takes place at $T \simeq 0.3$ MeV. By that time neutrinos are basically decoupled from electrons and positrons, so processes such as $e^+e^- \to \nu_e\bar{\nu}_e$ are now ineffective and, apart from small corrections due to the fact that the neutrino decoupling is not instantaneous [see eq. (17.108) below], all the entropy of the electrons is transferred to photons, rather than being

[25]More precisely, at temperatures $O(\text{MeV})$, muons and tau have long disappeared from the primordial plasma, and we have only $e^\pm$, photons, and the three neutrino plus antineutrino species. An accurate computation shows that the electron neutrino decouples at a momentum-dependent temperature, whose average over the momenta of the thermal distribution is $T \simeq 1.9$ MeV, while the muon and tau neutrinos, which interact with the electromagnetic plasma only through neutral currents, decouple at a slightly higher temperature, $T \simeq 3.1$ MeV; see Enqvist, Kainulainen and Semikoz (1992) and Section 2.4 of Lesgourgues, Mangano, Miele and Pastor (2013).

[26]Actually, one always has sufficiently energetic photons in the tail of their thermal distributions, and the process $2\gamma \to e^+e^-$ becomes ineffective when the temperature is sufficiently low, so that the photons in the tail become too rare to maintain equilibrium.

shared between photons and neutrinos. As a result, while the neutrino temperature continues to follow its $T_\nu \propto 1/a$ behavior, the photon temperature is raised, compared with the expected $T_\gamma \propto 1/a$ behavior.

To take these effects into account quantitatively, we begin by recalling that the energy density of a species in equilibrium at temperature T is given by

$$\rho = g \int \frac{d^3 k_{\rm ph}}{(2\pi)^3} \frac{E_k}{e^{(E_k - \mu)/T} \mp 1} , \qquad (17.95)$$

where g is the number of spin or helicity states, μ is the chemical potential, $k_{\rm ph}$ is the physical momentum (since ρ is defined as the energy per physical volume $dE/d^3 x_{\rm ph}$), $E_k = (m^2 + k_{\rm ph}^2)^{1/2}$, the $-/+$ signs hold for bosons/fermions, respectively, and we set the Boltzmann constant $k_B = 1$. In the relativistic limit $T \gg m$, $T \gg \mu$ one finds that, for a relativistic boson at temperature T, the energy density is

$$\rho_B = g \frac{\pi^2}{30} T^4 . \qquad (17.96)$$

So, for instance, a gas of photons has $\rho_\gamma = (\pi^2/15) T^4$ because of the two helicity states. For a gas of relativistic fermions the integration gives instead

$$\rho_F = \frac{7}{8} \times g \frac{\pi^2}{30} T^4 . \qquad (17.97)$$

Taking into account the fact that different relativistic species can have different temperatures if some of them have decoupled from the rest of the cosmological fluid, the energy density in relativistic species is given by

$$\rho_R(T) = \sum_{i={\rm bosons}} g_i \frac{\pi^2}{30} T_i^4 + \frac{7}{8} \sum_{i={\rm fermions}} g_i \frac{\pi^2}{30} T_i^4$$

$$= g_*(T) \frac{\pi^2}{30} T^4 , \qquad (17.98)$$

where the function $g_*(T)$ is defined as

$$g_*(T) \equiv \sum_{i={\rm bosons}} g_i \left(\frac{T_i}{T} \right)^4 + \frac{7}{8} \sum_{i={\rm fermions}} g_i \left(\frac{T_i}{T} \right)^4 . \qquad (17.99)$$

Here g_i is the number of spin or helicity states of the i-th species, T_i is its temperature, and we have reserved the notation T for the photon temperature T_γ. The function $g_*(T)$ is called the effective number of relativistic species. After $e^\pm$ have become non-relativistic, and as long as the neutrinos are still relativistic, the contributions to this function come only from photons (two helicity states), the three neutrinos (each with one helicity state) and the three antineutrinos (again one helicity state). As we will show in eq. (17.106), after $e^\pm$ annihilation and while the three neutrinos species remain relativistic, $T_\nu = (4/11)^{1/3} T_\gamma$. Thus the value of $g_*(T)$ at any temperature below the temperature of $e^\pm$

annihilation but above the temperature for which the heaviest neutrino becomes non-relativistic, is given by

$$g_{*,\gamma\nu} = 2 + \frac{7}{8} \times 6 \times \left(\frac{4}{11}\right)^{4/3}$$

$$\simeq 3.363 . \tag{17.100}$$

In the approximation in which the neutrinos are taken to be massless, this would be the same as the value $g_{*,0}$ at the present epoch. With the particle content of the Standard Model, at high temperatures $g_*(T)$ is much higher, rising to a value 106.75 at $T \gtrsim 100$ GeV, where it saturates. Of course, on increasing the temperature further, $g_*(T)$ can become even larger in extensions of the Standard Model, where there can exist further particles with even higher masses, which therefore at sufficiently high temperatures become relativistic and contribute to ρ_R.

The relation between the temperature of photons and that of the neutrinos just after e^+e^- annihilation can be computed using entropy conservation. The entropy density of a single relativistic boson is $s = 2\pi^2 T^3/45$ for each spin state, while that of a relativistic fermions is multiplied by a factor $7/8$. Thus, the total entropy density of a relativistic plasma can be written as

$$s = g_*^S(T)\frac{2\pi^2}{45}T^3 , \tag{17.101}$$

where

$$g_*^S(T) = \sum_{i=\text{bosons}} g_i \left(\frac{T_i}{T}\right)^3 + \frac{7}{8} \sum_{i=\text{fermions}} g_i \left(\frac{T_i}{T}\right)^3 . \tag{17.102}$$

Note that this is different from the function $g_*(T)$ defined in eq. (17.99), except if $T_i = T$ for all species. However, in the early Universe, at temperatures above the MeV range, all particles were in equilibrium at the same temperature, and $g_*(T) = g_*^S(T)$.[27] Just before e^+e^- annihilation the cosmological fluid consists of photons (two helicity states), electrons (with two spin states), positrons (again two spin states), three families of neutrinos (one helicity state each) and three antineutrinos, all at the same temperature T_1, so

$$s_1 = \frac{2\pi^2}{45}T_1^3 \left[2 + \frac{7}{8}(2 + 2 + 6)\right] . \tag{17.104}$$

Just after annihilation we have only photons and neutrinos, with two different temperatures T_γ and T_ν, and

$$s_2 = \frac{2\pi^2}{45}\left[2\,T_\gamma^3(a_2) + \frac{7}{8} \times 6\,T_\nu^3(a_2)\right] . \tag{17.105}$$

Since the entropy $S = sa^3$ is conserved, we have $sa_1^3 = sa_2^3$. If neutrinos receive strictly no energy from e^+e^- annihilation, their temperature

[27]Anticipating, from eq. (17.106) below, that after $e^\pm$ annihilation and before neutrinos become non-relativistic, $T_\nu^3 = (4/11)T_\gamma^3$, we find that the value of $g_*^S(T)$ at a generic temperature after $e^\pm$ annihilation, but before the heaviest neutrino becomes non-relativistic, is

$$g_{*,\gamma\nu}^S = 2 + \frac{7}{8} \times 6 \times \frac{4}{11}$$

$$= \frac{43}{11} \simeq 3.909 . \tag{17.103}$$

Once again, this becomes the same as the present value $g_{*,0}^S$ in the approximation of massless neutrinos.

scales as $1/a$, so $a_1 T_1 = a_2 T_\nu(a_2)$, and we find that, just after annihilation,

$$T_\nu = \left(\frac{4}{11}\right)^{1/3} T_\gamma . \qquad (17.106)$$

After that moment, as long as the neutrinos remain relativistic, both T_ν and T_γ scale as $1/a$, and therefore the relation (17.106) is maintained. Actually, there are two sources of correction to this picture. First of all, neutrino decoupling is not instantaneous, and since it takes place shortly before $e^+ e^-$ annihilation, even neutrinos receive a small amount of entropy from $e^+ e^-$ annihilation. Furthermore, there is an even smaller finite-temperature QED correction to the electromagnetic plasma. This is due to the fact that electromagnetic interactions in the plasma modify the $e^\pm$ and γ dispersion relation, and therefore the energy and pressure of the plasma. These two effects give a small correction, which is usually expressed by leaving eq. (17.106) unaltered, but writing the energy density of the three neutrino species as

$$\rho_\nu = N_{\text{eff}}^{(\nu)} \frac{7}{8} \left(\frac{4}{11}\right)^{4/3} \rho_\gamma , \qquad (17.107)$$

where the "effective number of neutrino species", $N_{\text{eff}}^{(\nu)}$, instead of having the value $N_\nu = 3$ corresponding to the three neutrino families, has a slightly larger value. Detailed computations give[28]

$$N_{\text{eff}}^{(\nu)} \simeq 3.046 . \qquad (17.108)$$

After $e^+ e^-$ annihilation, and as long as the three neutrinos are still relativistic, the total energy density in radiation, including photons and three families of neutrinos, is therefore given by

$$\rho_R(T) = \rho_\gamma(T) \left[1 + N_{\text{eff}}^{(\nu)} \frac{7}{8} \left(\frac{4}{11}\right)^{4/3} \right]$$
$$= \frac{\pi^2}{15} T^4 \left[1 + N_{\text{eff}}^{(\nu)} \frac{7}{8} \left(\frac{4}{11}\right)^{4/3} \right] , \qquad (17.109)$$

where T is the photon temperature. In particular, the energy density in photons is given by $\rho_\gamma = \rho_0 \Omega_\gamma a^{-4}$, where

$$h_0^2 \Omega_\gamma = \frac{8\pi G}{3} \left(\frac{h_0}{H_0}\right)^2 \rho_{\gamma,0}$$
$$= \frac{8\pi G}{3} \left(\frac{1}{100\,\text{km}/(\text{s Mpc})}\right)^2 \frac{\pi^2}{15} T_0^4$$
$$\simeq 2.473 \times 10^{-5} \left(\frac{T_0}{2.7255\,\text{K}}\right)^4 , \qquad (17.110)$$

while, as long as the three neutrinos are still relativistic, the total energy density in radiation is given by $\rho_R(a) = \rho_0 \Omega_R a^{-4}$, with $h_0^2 \Omega_R$ given by

$$h_0^2 \Omega_R = \left[1 + N_{\text{eff}}^{(\nu)} \frac{7}{8} \left(\frac{4}{11}\right)^{4/3} \right] h_0^2 \Omega_\gamma ,$$

[28]See Mangano, Miele, Pastor and Peloso (2002) and Mangano *et al.* (2005), and references therein.

$$\simeq 4.184 \times 10^{-5} \left(\frac{T_0}{2.7255 \, \text{K}} \right)^4 , \qquad (17.111)$$

where we have used the value of $N_{\text{eff}}^{(\nu)}$ given in eq. (17.108). Observe that in the combinations $h_0^2 \Omega_\lambda = h_0^2 \rho_\lambda / \rho_0$ the factor h_0^2 in the numerator cancels the factor h_0^2 present in ρ_0. Therefore these quantities are not affected by the uncertainty in the Hubble parameter.[29]

We can now understand how the evolution of the Hubble parameter in RD is modified by the change in the effective number of relativistic species. Using the expression (17.98) for ρ_R, in RD the Friedmann equation (17.48) becomes

$$H^2 = \frac{8\pi G}{3} g_*(T) \frac{\pi^2}{30} T^4 , \qquad (17.112)$$

where T is the photon temperature. To determine how the temperature scales with a we use again the fact that entropy $S = a^3 s$ is conserved. Let T_r be a reference value of the temperature, below the $e^\pm$ annihilation temperature, but such that the three neutrino species are still relativistic, and let a_r be the corresponding scale factor (in the approximation of massless neutrinos one could take $T_r = T_0$ but, by using T_r, it is not necessary to approximate neutrinos as massless). Then, using eq. (17.101), for a generic temperature $T > T_r$, so in particular for values of T in RD, we have

$$g_*^S(T) T^3 a^3 = g_{*,\gamma\nu}^S T_r^3 a_r^3 , \qquad (17.113)$$

where $g_{*,\gamma\nu}^S$ is given by eq. (17.103). Therefore

$$T = \left(\frac{g_{*,\gamma\nu}^S}{g_*^S(T)} \right)^{1/3} \frac{T_r a_r}{a} . \qquad (17.114)$$

We see that the naive scaling $T \propto 1/a$ is altered by the temperature-dependent factor $g_*^S(T)$, which reflects the fact that the effective number of relativistic species changes. Plugging this into eq. (17.112) we find that, in RD,

$$H^2(a) = \left(\frac{g_*(T)}{g_{*,\gamma\nu}} \right) \left(\frac{g_{*,\gamma\nu}^S}{g_*^S(T)} \right)^{4/3} \frac{a_r^4}{a^4} \frac{8\pi G}{3} \left[g_{*,\gamma\nu} \frac{\pi^2}{30} T_r^4 \right] . \qquad (17.115)$$

The expression in square brackets is just $\rho_R(a_r)$, which is equal to $\rho_0 \Omega_R / a_r^4$ [see eq. (17.74)], and $(8\pi G/3)\rho_0 = H_0^2$. Thus, during RD,

$$H(a) \simeq \left(\frac{g_*(T)}{3.363} \right)^{1/2} \left(\frac{3.909}{g_*^S(T)} \right)^{2/3} \frac{H_0 \Omega_R^{1/2}}{a^2} , \qquad (17.116)$$

where we have used the numerical values of $g_{*,\gamma\nu}$ and $g_{*,\gamma\nu}^S$ given in eqs. (17.100) and (17.103). In principle, in this expression the functions $g_*(T)$ and $g_*^S(T)$ should be expressed in terms of a, inverting $T = T(a)$ from eq. (17.113). In practice, these functions are constant for long periods and then jump to a different constant value when a species becomes

[29] Note that, in agreement with the standard convention in the literature, we are actually defining Ω_R so that it is the value of the radiation fraction today only in the approximation in which the neutrinos are taken to be massless. With the actual values of the neutrino masses, the three neutrinos are relativistic only for $a < a_{\text{NR}}$, where a_{NR} is the value of the scale factor for which the heaviest neutrino becomes non-relativistic (see Note 16). Then, the radiation density is given by $\rho_R(a) = \rho_0 \Omega_R a^{-4}$ only for $a < a_{\text{NR}}$ while it must be corrected for $a > a_{\text{NR}}$, subtracting the contribution from the species that have become non-relativistic. The advantage of this definition is that the precise value of the radiation energy density is only relevant when radiation is important, i.e. in RD or in the early MD, and in this case, according to our definition of Ω_R, we have $\rho_R(a) = \rho_0 \Omega_R a^{-4}$. Thus Ω_R, defined in this way, is the relevant quantity for computing, e.g. the epoch of radiation–matter equilibrium. In contrast, near the present epoch, when radiation is subdominant, the exact value of $\rho_R(a)$ is of little importance, so for most purposes one can use $\rho_R(a) = \rho_0 \Omega_R a^{-4}$ everywhere in analytic computations. Observe also that, when they become non-relativistic, neutrinos are decoupled from photons, so the fact that they become non-relativistic does not increse the photon temperature, in contrast to the case of the electrons computed earlier.

non-relativistic. In particular, in the Standard Model, for values of a corresponding to temperatures $T \gtrsim 100$ GeV, the functions $g_*(T)$ and $g_*^S(T)$ are equal and saturate to the constant value $g_{*,\mathrm{max}} \simeq 106.75$. Then we can write eq. (17.116) as

$$H(a) \simeq 0.62 \left(\frac{106.75}{g_*(T)} \right)^{1/6} \frac{H_0 \Omega_R^{1/2}}{a^2} \qquad (T > 100\,\mathrm{GeV}). \qquad (17.117)$$

We have retained the dependence on $g_*(T)$ because in extensions of the Standard Model, at high temperature it will in general have a different numerical value. For instance in supersymmetric extensions it will be roughly twice the Standard Model value.

17.6   Conformal time and particle horizon

It is convenient to introduce the conformal time η, defined in eq. (4.175) from

$$d\eta = \frac{dt}{a(t)}, \qquad (17.118)$$

so that

$$\eta = \int^t \frac{dt'}{a(t')}, \qquad (17.119)$$

where the lower limit of integration can in principle be chosen arbitrarily. Then, in coordinates $(\eta, \mathbf{x})$, the FRW metric takes the form already given in eq. (4.177),

$$ds^2 = a^2(\eta)\left(-d\eta^2 + d\mathbf{x}^2\right), \qquad (17.120)$$

and is therefore conformal to the Minkowski metric, i.e. of the form $g_{\mu\nu} = \Omega^2(x)\eta_{\mu\nu}$. When using conformal time, one also introduces

$$\mathcal{H} = \frac{a'}{a}, \qquad (17.121)$$

where the prime denotes the derivative with respect to conformal time. From eq. (17.118), for any function $f(t)$ we have $f' = a\dot{f}$, and therefore

$$\mathcal{H} = aH = \dot{a}. \qquad (17.122)$$

In these coordinates the metric is $g_{\mu\nu} = a^2(\eta)\eta_{\mu\nu}$ and the Christoffel symbols are given by

$$\Gamma^\rho_{\mu\nu} = \mathcal{H}\left(\delta^0_\mu \delta^\rho_\nu + \delta^0_\nu \delta^\rho_\mu - \eta^{\rho 0}\eta_{\mu\nu}\right). \qquad (17.123)$$

Explicitly, the non-vanishing components are

$$\Gamma^0_{00} = \mathcal{H}, \qquad \Gamma^0_{ij} = \mathcal{H}\delta_{ij}, \qquad \Gamma^i_{0j} = \mathcal{H}\delta^i_j. \qquad (17.124)$$

The Friedmann equation (17.48) now reads

$$\mathcal{H}^2 = \frac{8\pi G}{3} a^2 \rho_{\mathrm{tot}}, \qquad (17.125)$$

where, having in mind the application to a multi-component fluid, we
have stressed that the energy density on the right-hand side is the total
one. The conservation equation (17.66) becomes

$$\rho'_{\rm tot} + 3\mathcal{H}(1 + w_{\rm tot})\rho_{\rm tot} = 0\,, \tag{17.126}$$

where the expression for $w_{\rm tot}$ in terms of the EoS parameters of the separate fluids is given in eq. (17.65). Taking the derivative of eq. (17.125)
with respect to η and using eq. (17.126) we also get the relation

$$\mathcal{H}' = -\frac{1}{2}(1 + 3w_{\rm tot})\mathcal{H}^2\,. \tag{17.127}$$

In a cosmological model that starts from a big-bang singularity at $t = 0$,
it is natural to define η by choosing $t = 0$ as the lower limit of integration
in eq. (17.119),[30]

$$\eta = \int_0^t \frac{dt'}{a(t')}$$
$$= \int_0^a \frac{d\tilde{a}}{\tilde{a}^2 H(\tilde{a})}\,. \tag{17.128}$$

Since light travels along the curve $ds^2 = 0$, in such a model (without an
earlier inflationary phase) η defined in eq. (17.128) (or more generally
$c\eta$, if we do not set $c = 1$) is equal to the comoving distance traveled
by light from time $t = 0$ up to time t. It therefore represents the maximum size of a region (measured in comoving coordinates $\mathbf{x}$) that can
be causally connected at time t. It is then also called the (comoving)
particle horizon, or the (comoving) causal horizon. The corresponding
physical particle horizon at time t is

$$r_{\rm ph}(t) = a(t)\int_0^t \frac{dt'}{a(t')}\,. \tag{17.129}$$

We next compute the functions $a(\eta)$ and the relation between η and t
in different epochs.

17.6.1 Radiation dominance

Let us assume for the moment that the Universe starts from $t = 0$ in
a RD phase (an assumption that will be crucially modified in inflationary theories; see Section 21.1). As long as we are still deep in RD,
using eq. (17.75) we have $H(a) = H_0\Omega_R^{1/2}a^{-2}$. Then the integration in
eq. (17.128) gives

$$a(\eta) = H_0\Omega_R^{1/2}\eta\,. \tag{17.130}$$

A more accurate expression can be obtained using eq. (17.116). Since
the functions g_* and g_*^S that enter in eq. (17.116) are functions of T, and
hence of a, in principle they appear inside the integral in eq. (17.128).

[30]We will discuss in Section 21.1 how
this changes in the presence of an inflationary phase.

However, one can observe that these functions change in steps, at the epochs when a particle species becomes non-relativistic. The integral in eq. (17.128) gets most of its contribution from values of the integration variable $\tilde{a}$ close to the upper integration limit a and, if the value of the upper limit a is not very close to one of these transition epochs, we can approximate the functions $g_*(\tilde{a})$ and $g_*^S(\tilde{a})$ by their values at $\tilde{a} = a$. Then,

$$a(\eta) \simeq \left(\frac{g_*(T)}{3.363}\right)^{1/2} \left(\frac{3.909}{g_*^S(T)}\right)^{2/3} H_0 \Omega_R^{1/2} \eta \,. \tag{17.131}$$

For $T < O(1)\,\text{MeV}$ the Universe has already been through neutrino decoupling and e^+e^- annihilation, so $g_*(T)$ becomes equal to $g_{*,\gamma\nu} \simeq 3.363$ and $g_*^S(T)$ becomes equal to $g_{*,\gamma\nu}^S \simeq 3.909$, after which they stay constant for the rest of RD (as well as during MD, until neutrinos become non-relativistic). Thus, in the region $T_{\text{eq}} \ll T < O(1)\,\text{MeV}$ (where the condition $T_{\text{eq}} \ll T$ ensures that we are still deep in RD), eq. (17.131) reduces to eq. (17.130). In contrast, just as in eq. (17.117), for temperatures $T \gtrsim 100$ GeV, in the Standard Model $g_*(T) \simeq g_*^S(T) \simeq 106.75$. Then eq. (17.131) becomes

$$a(\eta) \simeq 0.62 \left(\frac{106.75}{g_*(T)}\right)^{1/6} H_0 \Omega_R^{1/2} \eta \qquad (T > 100\,\text{GeV}) \,, \tag{17.132}$$

which is of course equivalent to eq. (17.117), since $H = \dot{a}/a = a'/a^2$.

Integrating the relation $dt = a(\eta)d\eta$ with the initial condition $\eta = 0$ at $t = 0$, we have

$$t = \int_0^\eta d\tilde{\eta}\, a(\tilde{\eta}) \,. \tag{17.133}$$

Using eq. (17.131) and approximating again the functions $g_*(\tilde{\eta})$ and $g_*^S(\tilde{\eta})$ with their values g_* and g_*^S at $\tilde{\eta} = \eta$, we get

$$t \simeq \left(\frac{g_*(T)}{3.363}\right)^{1/2} \left(\frac{3.909}{g_*^S(T)}\right)^{2/3} \frac{1}{2} H_0 \Omega_R^{1/2} \eta^2 \,, \tag{17.134}$$

so, apart from the slowly varying factors $g_*(T)$ and $g_*^S(T)$, in RD we have $t \propto \eta^2$. Combining this with eq. (17.131) we get

$$a(t) \simeq \left(\frac{g_*(T)}{3.363}\right)^{1/4} \left(\frac{3.909}{g_*^S(T)}\right)^{1/3} \Omega_R^{1/4} (2H_0 t)^{1/2} \,. \tag{17.135}$$

In particular, after neutrino decoupling and e^+e^- annihilation (but still as long as we are in RD),

$$t \simeq \frac{1}{2} H_0 \Omega_R^{1/2} \eta^2 \,, \tag{17.136}$$

and

$$a(t) \simeq \Omega_R^{1/4} (2H_0 t)^{1/2} \,, \tag{17.137}$$

while at $T \gtrsim 100$ GeV we get

$$a(t) \simeq 0.79 \left(\frac{106.75}{g_*(T)}\right)^{1/12} \Omega_R^{1/4} (2H_0 t)^{1/2} \,. \tag{17.138}$$

17.6.2   Matter dominance

Consider next the MD phase. Integrating eq. (17.128) up to a value of t in the MD epoch, and separating

$$\int_0^a d\tilde{a} = \int_0^{a_{\rm eq}} d\tilde{a} + \int_{a_{\rm eq}}^a d\tilde{a}\,, \qquad (17.139)$$

where $a_{\rm eq}$ is the value of the scale factor at radiation–matter equilibrium, we get

$$\eta = \eta_{\rm eq} + \int_{a_{\rm eq}}^a \frac{d\tilde{a}}{\tilde{a}^2 H(\tilde{a})}\,. \qquad (17.140)$$

In MD, using eq. (17.75), we can approximate $H(a) \simeq H_0 \Omega_M^{1/2} a^{-3/2}$. Then we get

$$\eta \simeq \eta_{\rm eq} + \frac{2}{H_0 \Omega_M^{1/2}} \left[a^{1/2}(\eta) - a_{\rm eq}^{1/2} \right]\,. \qquad (17.141)$$

Deep in MD $a(\eta) \gg a_{\rm eq}$, $\eta \gg \eta_{\rm eq}$, so

$$\eta \simeq \frac{2}{H_0 \Omega_M^{1/2}} a^{1/2}(\eta)\,, \qquad (17.142)$$

and

$$a(\eta) \simeq \frac{1}{4} H_0^2 \Omega_M \eta^2\,. \qquad (17.143)$$

Observe that in RD $a(\eta) \propto \eta$ while in MD $a(\eta) \propto \eta^2$. The relation between cosmic and conformal time deep in MD is obtained again by integrating eq. (17.133), which gives

$$t = t_{\rm eq} + \int_{\eta_{\rm eq}}^\eta d\tilde{\eta}\, a(\tilde{\eta})$$

$$\simeq t_{\rm eq} + \frac{1}{12} H_0^2 \Omega_M (\eta^3 - \eta_{\rm eq}^3)\,. \qquad (17.144)$$

Again, deep in MD this can be approximated by

$$t \simeq \frac{1}{12} H_0^2 \Omega_M \eta^3\,, \qquad (17.145)$$

so

$$\eta \simeq \left(\frac{12\, t}{H_0^2 \Omega_M} \right)^{1/3}\,. \qquad (17.146)$$

Inserting this into eq. (17.143) we get

$$a(t) \simeq \Omega_M^{1/3} \left(\frac{3}{2} H_0 t \right)^{2/3}\,. \qquad (17.147)$$

17.6.3   Analytic formulas in RD+MD

Using conformal time and including both radiation and matter (but no cosmological constant), the Friedmann equation (17.125) can be integrated analytically. In fact, in this case eq. (17.125) reads

$$\mathcal{H} = H_0 a \left(\frac{\Omega_R}{a^4} + \frac{\Omega_M}{a^3} \right)^{1/2} \tag{17.148}$$

(where $H_0 = \mathcal{H}_0$, since $a_0 = 1$). Using $\mathcal{H} = a'/a$ and recalling that $a_{\rm eq} = \Omega_R/\Omega_M$, this can be rewritten as

$$a' = H_0 \Omega_R^{1/2} \left(1 + \frac{a}{a_{\rm eq}} \right)^{1/2} , \tag{17.149}$$

and therefore, introducing the variable $y = a/a_{\rm eq}$,

$$a_{\rm eq} \int_0^y \frac{dy'}{\sqrt{1+y'}} = H_0 \Omega_R^{1/2} \eta . \tag{17.150}$$

The integral is elementary,

$$\int_0^y \frac{dy'}{\sqrt{1+y'}} = -2 + 2\sqrt{1+y} , \tag{17.151}$$

so in the end we get

$$a(\eta) = a_{\rm eq} \left[\frac{2\eta}{\eta_*} + \left(\frac{\eta}{\eta_*} \right)^2 \right] , \tag{17.152}$$

where

$$\eta_* = \frac{2\Omega_R^{1/2}}{H_0 \Omega_M} . \tag{17.153}$$

It is straightforward to check that, in the limits $\eta \ll \eta_*$ and $\eta \gg \eta_*$, we recover eqs. (17.130) and (17.143), respectively. The value of η at equilibrium, which is by definition the one for which the square bracket in eq. (17.152) is equal to one, is related to η_* by

$$\eta_{\rm eq} = (\sqrt{2} - 1)\eta_* . \tag{17.154}$$

Using eq. (17.133), we can also find an analytic expression for $t(\eta)$,

$$t(\eta) = a_{\rm eq} \left(\frac{\eta^2}{\eta_*} + \frac{\eta^3}{3\eta_*^2} \right) . \tag{17.155}$$

However, since this relation cannot be simply inverted analytically to get $\eta(t)$, there is no correspondingly simple expression for $a(t)$, covering both RD and MD.

Equation (17.152) can be inverted to give

$$\eta = \frac{2}{H_0 \Omega_M^{1/2}} \left[\sqrt{a + a_{\rm eq}} - \sqrt{a_{\rm eq}} \right] . \tag{17.156}$$

17.6.4   Λ dominance

We next consider a phase dominated by a constant vacuum energy, such as an inflationary de Sitter (dS) stage preceding the RD and MD phases, or the future state of the Universe in ΛCDM, where the cosmological constant contribution Ω_Λ will eventually overwhelm the matter density $\Omega_M a^{-3}$. In this case, using eq. (17.53), the Friedmann equation reads simply

$$H^2 = \frac{\Lambda}{3}, \tag{17.157}$$

so H is a constant. Integrating $\dot{a}/a = H$, we get $a(t) = a_* e^{H(t-t_*)}$, with t_* a reference time and $a_* = a(t_*)$ a (positive) integration constant. In the definition of conformal time, eq. (17.119), there is now no special reason for choosing $t = 0$ as the lower integration limit, if we are considering a dS stage that precedes RD. In general, we can define

$$\eta = \eta_1 + \int_{t_1}^{t} \frac{dt'}{a(t')}, \tag{17.158}$$

with η_1 and t_1 some constants that we can choose at our convenience. Then we get

$$\eta = \eta_1 - \frac{1}{a_* H} \left(e^{-H(t-t_*)} - e^{-H(t_1-t_*)} \right). \tag{17.159}$$

It can be convenient to chose η_1 such that $\eta_1 = -e^{-H(t_1-t_*)}/(a_* H)$, so that

$$\eta = -\frac{1}{a_* H} e^{-H(t-t_*)}. \tag{17.160}$$

Observe that, with this choice, the interval $-\infty < t < \infty$ is mapped into $-\infty < \eta < 0$.[31] Then, in terms of η, the scale factor reads

$$\boxed{a(\eta) = -\frac{1}{H\eta}.} \tag{17.161}$$

[31] Of course this can be relevant if one studies an eternal dS for its intrinsic interest. When studying a realistic cosmological model, a dS phase cannot represent the actual state of the Universe for all values of t.

17.6.5   Conformal time at significant epochs

During RD and MD, eq. (17.156) provides a simple analytic expression for conformal time as a function of scale factor a or, equivalently, of redshift z, with $1 + z = 1/a$. In particular, at equilibrium, we have

$$\eta_{\rm eq} = 2(\sqrt{2} - 1) \frac{\Omega_R^{1/2}}{H_0 \Omega_M}, \tag{17.162}$$

in agreement with eqs. (17.153) and (17.154). Inserting the numerical values,

$$\eta_{\rm eq} \simeq 109.28 \, {\rm Mpc} \left(\frac{0.7}{h_0} \right)^2 \left(\frac{0.3}{\Omega_M} \right) \left(\frac{\Omega_R h_0^2}{4.184 \times 10^{-5}} \right)^{1/2}. \tag{17.163}$$

From eq. (17.156), the value of η at decoupling, $\eta_{\rm dec}$, is related to $\eta_{\rm eq}$ by

$$\frac{\eta_{\rm dec}}{\eta_{\rm eq}} = \frac{1}{\sqrt{2}-1}\left[\sqrt{\frac{1+z_{\rm eq}}{1+z_{\rm dec}}+1}-1\right]$$
$$\simeq 2.55\,, \tag{17.164}$$

where in the second line we have used $z_{\rm eq} = 3515$ and $z_{\rm dec} = 1090$. If one uses eq. (17.156) to compute the present value of conformal time, setting $a = 1$ and observing that $a_{\rm eq} \ll 1$, one obtains $\eta_0 = 2/(H_0\Omega_M^{1/2})$. However, in the present epoch we must use the full expression of $H(a)$, including the contribution from Ω_Λ. Using the redshift z instead of a, eq. (17.75) reads

$$H(z) = H_0\sqrt{\Omega_R(1+z)^4 + \Omega_M(1+z)^3 + \Omega_\Lambda}\,. \tag{17.165}$$

From $1/a = 1+z$ it follows that $-da/a^2 = dz$, so eq. (17.128) can be written as

$$\eta(z) = \int_z^\infty \frac{d\tilde{z}}{H(\tilde{z})}\,. \tag{17.166}$$

Thus, in ΛCDM,

$$\boxed{\eta(z) = \frac{1}{H_0}\int_z^\infty \frac{d\tilde{z}}{\sqrt{\Omega_R(1+\tilde{z})^4 + \Omega_M(1+\tilde{z})^3 + \Omega_\Lambda}}\,.} \tag{17.167}$$

Therefore

$$\eta_0 = \frac{1}{H_0}\int_0^\infty \frac{dz}{\sqrt{\Omega_R(1+z)^4 + \Omega_M(1+z)^3 + (1-\Omega_M-\Omega_R)}}$$
$$= \frac{2}{H_0\Omega_M^{1/2}}I(\Omega_M)\,, \tag{17.168}$$

where

$$I(\Omega_M) \equiv \frac{\Omega_M^{1/2}}{2}\int_0^\infty \frac{dz}{\sqrt{\Omega_R(1+z)^4 + \Omega_M(1+z)^3 + (1-\Omega_M-\Omega_R)}}\,. \tag{17.169}$$

The contribution of Ω_R to η_0 only amounts to a small correction at the level of 1%, so we can simply fix it to the value $\Omega_R = 4.184\times 10^{-5}/(0.7)^2$ suggested by eq. (17.111), and we study I as a function of Ω_M only. The result is shown in Fig. 17.3. In particular,

$$I(0.3) \simeq 0.879\,. \tag{17.170}$$

Observe that η_0 is actually determined by $\Omega_M^{-1/2}I(\Omega_M)$. Even if $I(\Omega_M)$ is a growing function of Ω_M, the function $\Omega_M^{-1/2}I(\Omega_M)$ decreases with Ω_M, as shown in Fig. 17.4. This shows that the age of the Universe

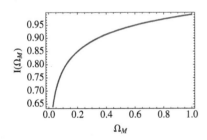

Fig. 17.3 The function $I(\Omega_M)$.

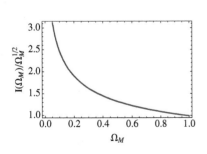

Fig. 17.4 A plot of $I(\Omega_M)\Omega_M^{-1/2}$.

(in conformal time, and therefore also in cosmic time, since one is a monotonic function of the other) is a decreasing function of Ω_M. In particular,

$$\frac{\eta_0(\Omega_M = 0.3)}{\eta_0(\Omega_M = 1)} \simeq 1.6 . \tag{17.171}$$

Thus, introducing a cosmological constant while maintaining the constraints $\Omega_R + \Omega_M + \Omega_\Lambda = 1$ decreases Ω_M and, as a consequences, increases the age of the Universe. Indeed, one of the first observational indications in favor of a cosmological constant was the fact that, with $\Omega_\Lambda = 0$ and the observed value of H_0, the computed age of the Universe turned out to be shorter than the known age of some of the oldest globular clusters in the Milky Way, while the extra factor (17.171) resolves the discrepancy.

Numerically, from eq. (17.168), we get

$$\eta_0 \simeq 13.7 \, \text{Gpc} \left(\frac{0.7}{h_0} \right) \left(\frac{0.3}{\Omega_M} \right)^{1/2} \left(\frac{I(\Omega_M)}{0.879} \right) . \tag{17.172}$$

Comparing with eqs. (17.163) and (17.164), we get

$$\frac{\eta_0}{\eta_{eq}} \simeq 126 \left(\frac{h_0}{0.7} \right) \left(\frac{\Omega_M}{0.3} \right)^{1/2} \left(\frac{I(\Omega_M)}{0.879} \right) \left(\frac{\Omega_R h_0^2}{4.184 \times 10^{-5}} \right)^{-1/2} \tag{17.173}$$

and

$$\frac{\eta_0}{\eta_{dec}} \simeq 49.3 \left(\frac{h_0}{0.7} \right) \left(\frac{\Omega_M}{0.3} \right)^{1/2} \left(\frac{I(\Omega_M)}{0.879} \right) \left(\frac{\Omega_R h_0^2}{4.184 \times 10^{-5}} \right)^{-1/2} . \tag{17.174}$$

In a cosmological model such as the one that we have considered, which has no phase of inflation in the early Universe and starts directly from RD after the big bang, these ratios determine the angle subtended today by the comoving causal horizons at equilibrium and at decoupling, respectively. In particular, the horizon on the last-scattering surface of the CMB would subtend an angle

$$\Delta\theta = \frac{\eta_{dec}}{\eta_0} \simeq 0.02 \, \text{rad} \simeq 1 \, \text{deg} . \tag{17.175}$$

This result will be important when we discuss the effect of GWs on the CMB.

17.6.6   Comoving distance, angular diameter distance and luminosity distance

In cosmology there are several possible notions of distance, whose usefulness depends on the problem at hand. Equation (17.166) gives the comoving horizon at redshift z, i.e. the maximum distance that light might have traveled from the big bang ($\tilde{z} = \infty$) up to a given redshift $\tilde{z} = z$, in a cosmological model that started from RD. More precisely, in a model with an earlier inflationary era, this gives the maximum distance that light might have traveled from the beginning of RD to the given redshift. A related notion is that of comoving distance between us ($\tilde{z} = 0$) and an object at redshift $\tilde{z} = z$, defined again in terms of the light travel time between us and the object. For the spatially flat FRW metric that we are considering, this comoving distance is obtained as in eq. (17.166), with the appropriate change in the integration limit,

$$d_{\rm com}(z) = \int_0^z \frac{d\tilde{z}}{H(\tilde{z})}, \tag{17.176}$$

so in ΛCDM we have

$$d_{\rm com}(z) = \frac{1}{H_0} \int_0^z \frac{d\tilde{z}}{\sqrt{\Omega_R(1+\tilde{z})^4 + \Omega_M(1+\tilde{z})^3 + \Omega_\Lambda}}. \tag{17.177}$$

(If one reinstates the speed of light, the factor $1/H_0$ becomes c/H_0.) This notion of distance is relevant when we are interested in the conformal time that elapsed in the radial propagation of light from the object to us. A different notion of distance is relevant when we know a priori the physical transverse size L of an object. Then the angle $\Delta\theta$ that it subtends on the sky is given (when $\Delta\theta \ll 1$) by $\Delta\theta = L/d_{\rm phys}(z)$, where $d_{\rm phys}(z) = a(z)d_{\rm com}(z)$ is the physical distance associated with the comoving distance $d_{\rm com}(z)$. Because of its relevance to angular measurement, this is also called the *angular diameter distance*, and denoted by $d_A(z)$. Thus

$$d_A(z) \equiv d_{\rm phys}(z) = \frac{1}{1+z} \int_0^z \frac{d\tilde{z}}{H(\tilde{z})} \tag{17.178}$$

and, in ΛCDM,

$$d_A(z) = \frac{1}{H_0} \frac{1}{1+z} \int_0^z \frac{d\tilde{z}}{\sqrt{\Omega_R(1+\tilde{z})^4 + \Omega_M(1+\tilde{z})^3 + \Omega_\Lambda}}. \tag{17.179}$$

Finally, in Section 4.1.4 of Vol. 1 we introduced the notion of luminosity distance $d_L(z)$. This is the notion of distance relevant when we know the intrinsic luminosity of an object (such the electromagnetic luminosity of a type Ia SN, or the GW luminosity of a coalescing binary) since, by definition, the observed flux $\mathcal{F}$ is related to the intrinsic luminosity $\mathcal{L}$ by $\mathcal{F} = \mathcal{L}/(4\pi d_L^2)$; see eq. (4.154). As we saw in eq. (4.160), the luminosity distance is larger than the comoving distance by a factor $(1+z)$, i.e.

$$d_L(z) = (1+z) \int_0^z \frac{d\tilde{z}}{H(\tilde{z})}; \tag{17.180}$$

compare with eq. (4.168) (recalling that we are now using units $c = 1$). Thus, in ΛCDM,

$$d_L(z) = \frac{1+z}{H_0} \int_0^z \frac{d\tilde{z}}{\sqrt{\Omega_R(1+\tilde{z})^4 + \Omega_M(1+\tilde{z})^3 + \Omega_\Lambda}}. \tag{17.181}$$

Figure 17.5 gives a plot of the three functions $d_{\rm com}(z)$, $d_A(z)$ and $d_L(z)$ in ΛCDM, for $\Omega_R = 4.184 \times 10^{-5}/(0.7)^2$, $\Omega_M = 0.3$ and $\Omega_\Lambda = 1 - \Omega_M - \Omega_R \simeq 0.7$. Observe that, for $z \to 0^+$, the three functions become equal, and reproduce the Hubble law $d(z) \simeq H_0^{-1}z$. At large z, however, they are significantly different. In particular, for $z \to \infty$, the comoving distance $d_{\rm com}(z)$ saturates to the finite value (17.168), corresponding to the existence of a finite horizon in a model that started from RD. In contrast, because of its extra factor $1/(1+z)$, for $z \gg 1$ the angular diameter distance eventually decreases as $1/z$, while the luminosity distance increases as z.

Given a measured value of the luminosity distance and a cosmological model (in this case ΛCDM), eq. (17.181) is easily inverted numerically. Observe also that, because of its behavior shown in Fig. 17.5, for a given value of the angular diameter distance (smaller than its maximum value) there are two solutions for z.

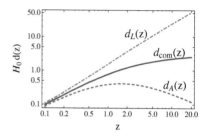

Fig. 17.5 The quantities $H_0 d(z)$ for $d(z)$ equal to $d_{\rm com}(z)$ (solid line), $d_A(z)$ (dashed) and $d_L(z)$ (dot-dashed), on a log-log scale.

17.7   Newtonian cosmology inside the horizon

In this section we give an account of cosmology in the Newtonian limit, both at the background level and at the level of cosmological perturbations. This is a subject that is not treated very much in standard cosmology textbooks, but is quite useful for understanding some aspects of cosmology in simple Newtonian terms, in the limits in which the relevant modes are well inside the horizon. When we study cosmology at the horizon scales, however, the full apparatus of GR becomes necessary. As we will see, the validity of the Newtonian approach is in fact limited to scales well inside the horizon, and to non-relativistic matter with EoS parameter $w = 0$. On the other hand, this approach has the advantage that it allows us to use our Newtonian intuition. Then, the comparison with the full relativistic treatment that will be given in Chapter 18 will allow us to understand which results are really specific to full GR, and which ones are basically already a consequence of Newtonian fluid dynamics in an expanding Universe.

In this section we will use the notation $\mathbf{r} \equiv \mathbf{x}_{\rm ph}$, while we will still denote by $\mathbf{x}$ the comoving coordinates, so

$$\mathbf{r} = a(t)\mathbf{x}. \tag{17.182}$$

When one uses comoving spatial coordinates it is natural to use conformal time η as the time variable, since in coordinates $(\eta, \mathbf{x})$ the metric is conformal to the Minkowski one; see eq. (17.120). In contrast, when using $\mathbf{r}$ it is more natural to use cosmic time t, since in coordinates $(t, \mathbf{r})$ the metric reduces to Minkowski well inside the horizon; see eq. (17.15). In particular, the coordinates $(t, \mathbf{r})$ are the ones in which we can use our Newtonian intuition, as long as we are well inside the horizon. We will also use the notation $\nabla \equiv \partial/\partial \mathbf{x}$, $\nabla_{\mathbf{r}} = \partial/\partial \mathbf{r}$ and, in this section, we will write $\partial/\partial\eta$ and $\partial/\partial t$ explicitly, rather than using the prime and the dot, respectively.

The total velocity field of a fluid is denoted by $\mathbf{V}(t,\mathbf{r})$, or, if we use the pair $(\eta,\mathbf{x})$, as $\mathbf{V}(\eta,\mathbf{x})$. The peculiar velocity field is defined by subtracting the Hubble flow, and is denoted by $\mathbf{v}(t,\mathbf{r})$ or $\mathbf{v}(\eta,\mathbf{x})$, so

$$\mathbf{V}(t,\mathbf{r}) = H\mathbf{r} + \mathbf{v}(t,\mathbf{r})\,. \tag{17.183}$$

Equivalently,

$$\mathbf{V}(\eta,\mathbf{x}) = \mathcal{H}\mathbf{x} + \mathbf{v}(\eta,\mathbf{x})\,, \tag{17.184}$$

since $\mathcal{H} = aH$ [see eq. (17.122)] and $\mathbf{x} = \mathbf{r}/a$.

17.7.1 Newtonian dynamics in expanding backgrounds

The velocity of a particle is defined as $d\mathbf{r}/dt$. As already discussed, this is the definition of velocity for which, inside the horizon, we can apply the usual formulas from Newtonian dynamics, since this intuition is only justified in coordinates $(t,\mathbf{r})$, where the metric takes the form (17.15), which reduces to flat Minkowski space when we are well inside the horizon. Consider a particle evolving on a trajectory $\mathbf{r}(t)$. Writing

$$\begin{aligned}
\frac{d\mathbf{r}}{dt} &= \frac{d(a\mathbf{x})}{dt} \\
&= \frac{da}{dt}\mathbf{x} + a\frac{d\mathbf{x}}{dt} \\
&= H\mathbf{r} + \frac{d\mathbf{x}}{d\eta}\,,
\end{aligned} \tag{17.185}$$

we see that $d\mathbf{r}/dt$ is made of the Hubble flow $H\mathbf{r}$ plus an extra term $d\mathbf{x}/d\eta$, which we therefore identify with the peculiar velocity of the particle. Therefore $d\mathbf{r}/dt$ is the total velocity of the particle, while $d\mathbf{x}/d\eta$ is its peculiar velocity. These are the velocities of a single particle and are functions of t (or η) only, and should not be confused with the total velocity field $\mathbf{V}(\eta,\mathbf{x})$ and with the peculiar velocity field $\mathbf{v}(\eta,\mathbf{x})$. The relation between the single-particle velocities and the corresponding velocity fields will be given in Section 17.7.2.

In the Newtonian limit the equation of motion of a test particle in the gravitational potential generated by a mass density $\rho(\mathbf{r})$ is

$$\frac{d^2\mathbf{r}}{dt^2} = -\frac{\partial\phi}{\partial\mathbf{r}}\,, \tag{17.186}$$

where ϕ is the Newtonian potential,

$$\phi(t,\mathbf{r}) = -G\int d^3\mathbf{r}'\,\frac{\rho(t,\mathbf{r}')}{|\mathbf{r}-\mathbf{r}'|}\,, \tag{17.187}$$

and ρ, as usual, denotes the physical density. Observe that here we have again used the fact that the coordinates $(t,\mathbf{r})$ are the ones in which

Newtonian dynamics is valid, at least inside the horizon. We now transform eq. (17.186) to coordinates $(\eta, \mathbf{x})$, and we split ρ into a spatially homogeneous term plus fluctuations around it,

$$\rho(\eta, \mathbf{x}) = \bar{\rho}(\eta) + \delta\rho(\eta, \mathbf{x}) \,. \tag{17.188}$$

Correspondingly, we also split ϕ as[32]

$$\phi(\eta, \mathbf{x}) = \phi_0(\eta, \mathbf{x}) + \Psi(\eta, \mathbf{x}) \,. \tag{17.189}$$

Using

$$
\begin{aligned}
\frac{d}{dt}\left(\frac{d\mathbf{r}}{dt}\right) &= \frac{d}{dt}\left(H\mathbf{r} + \frac{d\mathbf{x}}{d\eta}\right) \\
&= \frac{1}{a}\frac{d}{d\eta}\left(\mathcal{H}\mathbf{x} + \frac{d\mathbf{x}}{d\eta}\right) \\
&= \frac{1}{a}\frac{d\mathcal{H}}{d\eta}\mathbf{x} + \frac{1}{a}\left(\frac{d^2\mathbf{x}}{d\eta^2} + \mathcal{H}\frac{d\mathbf{x}}{d\eta}\right) \,,
\end{aligned} \tag{17.190}
$$

we get

$$\frac{d\mathcal{H}}{d\eta}\mathbf{x} = -\boldsymbol{\nabla}\phi_0 \,, \tag{17.191}$$

$$\frac{d^2\mathbf{x}}{d\eta^2} + \mathcal{H}\frac{d\mathbf{x}}{d\eta} = -\boldsymbol{\nabla}\Psi \,, \tag{17.192}$$

where $\boldsymbol{\nabla} = \partial/\partial\mathbf{x} = a\partial/\partial\mathbf{r}$. We have used the fact that, in the absence of inhomogeneities, $\mathbf{x}$ is constant, so the terms in which $d\mathbf{x}/d\eta$ and $d^2\mathbf{x}/d\eta^2$ appear must be due to the "peculiar gravitational potential" Ψ only, while the part proportional to $\mathbf{x}$ is present also in the homogeneous case and therefore must be due to ϕ_0.

Friedmann equation from Newtonian dynamics

Equation (17.191) fixes the evolution of the scale factor $a(\eta)$ in terms of the background energy density $\bar{\rho}(\eta)$. Computing $\boldsymbol{\nabla}\phi_0$ from eq. (17.187), and passing to the variable $\mathbf{x}$, we have

$$
\begin{aligned}
\boldsymbol{\nabla}\phi_0(\eta, \mathbf{x}) &= -G\bar{\rho}(\eta)a^2(\eta)\int d^3x' \frac{\partial}{\partial x^i}\frac{1}{|\mathbf{x} - \mathbf{x}'|} \\
&= +G\bar{\rho}(\eta)a^2(\eta)\int d^3x' \frac{\partial}{\partial x'^i}\frac{1}{|\mathbf{x} - \mathbf{x}'|} \,.
\end{aligned} \tag{17.193}
$$

The integral can be computed working first in a sphere with finite volume V, by writing

$$\int_V d^3x' \frac{\partial}{\partial x'^i}\frac{1}{|\mathbf{x} - \mathbf{x}'|} = \int_{\partial V} d^2S'\, n'^i \frac{1}{|\mathbf{x} - \mathbf{x}'|} \,, \tag{17.194}$$

where $d^2S' = |\mathbf{x}'|^2 d\Omega'$ and $\hat{\mathbf{n}}'$ is the exterior unit normal to the boundary. Now $|\mathbf{x}'|$ is on the boundary of V, which in the end we send to infinity,

[32]With the definition of the Bardeen potentials Φ and Ψ that we will introduce in Chapter 18 the perturbation of the potential ϕ will correspond to the Bardeen variable Ψ, which is related to the metric perturbation by $h_{tt} = -2\Psi$; see eq. (18.96).

so we only need to evaluate the right-hand side of eq. (17.194) in the limit $|\mathbf{x}'| \to \infty$ at fixed $\mathbf{x}$, and we can therefore expand

$$\frac{1}{|\mathbf{x} - \mathbf{x}'|} = \frac{1}{|\mathbf{x}'|} + \frac{\mathbf{x} \cdot \mathbf{n}'}{|\mathbf{x}'|^2} + \mathcal{O}\left(\frac{1}{|\mathbf{x}'|^3}\right). \qquad (17.195)$$

The first term gives zero after integration over the solid angle. The second can be integrated using

$$\int d\Omega \, n^i n^j = \frac{4\pi}{3} \delta^{ij}. \qquad (17.196)$$

Equation (17.196) can be easily proved by observing that, for rotational symmetry, the result of the integral must be proportional to δ^{ij}, and the proportionality coefficient can then be obtained by taking the trace of both sides. Finally, the term $\mathcal{O}\left(1/|\mathbf{x}'|^3\right)$ gives zero, after multiplying it by the factor $|\mathbf{x}'|^2$ present in $d^2 S'$, and then taking the limit $|\mathbf{x}'| \to \infty$. In conclusion,

$$\nabla \phi_0(\eta, \mathbf{x}) = G\bar{\rho}(\eta) a^2(\eta) \frac{4\pi}{3} \mathbf{x}. \qquad (17.197)$$

Plugging this into eq. (17.191) and transforming back to the variable t, we get

$$\frac{1}{a} \frac{d^2 a}{dt^2} = -\frac{4\pi G}{3} \bar{\rho}(t), \qquad (17.198)$$

which is indeed the correct general-relativistic equation for the acceleration of the background, in the case of non-relativistic matter (i.e. when the pressure $p = 0$); compare with eq. (17.63). It is remarkable that this general-relativistic equation emerges from a purely Newtonian treatment! The fact that we got the correct Friedmann equation only for non-relativistic matter is of course a consequence of having used the Newtonian expression (17.187), which depends on ρ but not on the pressure p. To take into account other forms of energy density, such as radiation or a cosmological constant, we must resort to a full general-relativistic treatment from the beginning.

Observe also that eq. (17.191) [or, equivalently, eq. (17.197)] integrates to

$$\phi_0(\eta, \mathbf{x}) = -\frac{1}{2} \frac{d\mathcal{H}}{d\eta} |\mathbf{x}|^2 + \text{const}, \qquad (17.199)$$

or, in terms of $(t, \mathbf{r})$,

$$\phi_0(t, \mathbf{r}) = -\frac{1}{2} \left(\frac{1}{a} \frac{d^2 a}{dt^2}\right) |\mathbf{r}|^2 + \text{const}. \qquad (17.200)$$

Actually, if one simply sets $\rho(\eta, \mathbf{x})$ equal to its background value $\bar{\rho}(\eta)$ in eq. (17.187), one finds a divergent integral, so the integration constant in eq. (17.199) is actually infinite. However, a constant term in the Newtonian potential has no physical effect, since in this Newtonian treatment ϕ enters only through its gradient in the equation of motion (17.186), so one can simply redefine this constant, setting it to zero.

Evolution of the peculiar velocity

The second equation, eq. (17.192), gives instead the evolution of the peculiar velocity of a particle,

$$\frac{d^2\mathbf{x}}{d\eta^2} + \mathcal{H}\frac{d\mathbf{x}}{d\eta} = -\boldsymbol{\nabla}\Psi \,, \qquad (17.201)$$

with[33]

$$\Psi(\eta,\mathbf{x}) = -Ga^2(\eta)\int d^3\mathbf{x}' \,\frac{\delta\rho(\eta,\mathbf{x}')}{|\mathbf{x}-\mathbf{x}'|} \,. \qquad (17.202)$$

Recalling that $\boldsymbol{\nabla}^2(1/|\mathbf{x}|) = -4\pi\delta^{(3)}(\mathbf{x})$, we see that Ψ is a solution of the Poisson equation

$$\boldsymbol{\nabla}^2\Psi(\eta,\mathbf{x}) = +4\pi Ga^2(\eta)\delta\rho(\eta,\mathbf{x}) \,. \qquad (17.203)$$

Using

$$\frac{d}{d\eta}\left(a\frac{d\mathbf{x}}{d\eta}\right) = a\left(\frac{d^2\mathbf{x}}{d\eta^2} + \mathcal{H}\frac{d\mathbf{x}}{d\eta}\right) \,, \qquad (17.204)$$

eq. (17.201) can be rewritten as

$$\frac{d\mathbf{p}}{d\eta} = -ma(\eta)\boldsymbol{\nabla}\Psi \,, \qquad (17.205)$$

with

$$\mathbf{p}(\eta) \equiv ma(\eta)\frac{d\mathbf{x}}{d\eta} \,. \qquad (17.206)$$

Thus, in the Newtonian limit, the quantity $\mathbf{p}(\eta)$ is conserved in the absence of the peculiar potential Ψ. Comparison with eq. (17.27) shows that indeed the quantity $\mathbf{p}$ defined in eq. (17.206) is nothing but the Newtonian limit of the comoving momentum. Indeed, to lowest order in the non-relativistic limit the proper time $d\tau$ reduces to dt; see eq. (17.33). Then, using $ad/dt = d/d\eta$, eq. (17.27) reduces to eq. (17.206). In the Newtonian limit, the physical momentum is then

$$\begin{aligned}\mathbf{p}_{\rm ph} &\equiv \frac{1}{a}\mathbf{p} \\ &= m\frac{d\mathbf{x}}{d\eta} \,, \end{aligned} \qquad (17.207)$$

and decreases as $1/a$ when $\Psi = 0$. The action that reproduces the equation of motion (17.201) is

$$S = \int d\eta\, L\left(\mathbf{x}, \frac{d\mathbf{x}}{d\eta}, \eta\right) \,, \qquad (17.208)$$

with

$$L = ma(\eta)\left[\frac{1}{2}\left(\frac{d\mathbf{x}}{d\eta}\right)^2 - \Psi(\eta,\mathbf{x})\right] \,. \qquad (17.209)$$

[33] From the point of view of this Newtonian approach to cosmology, there is now an issue of boundary conditions at infinity, since eq. (17.202) is well-defined only if $\delta\rho$ goes to zero sufficiently fast at spatial infinity, i.e. if the Universe becomes homogeneous on very large spatial scales, $\rho(\eta,\mathbf{x}) \to \bar{\rho}(\eta)$. However, this issue of boundary conditions is a fake one. The Newtonian approach is only valid well inside the horizon, and to discuss what happens on larger scales one needs GR. Actually, since the Universe indeed becomes homogeneous at very large scales, for must purposes at scales comparable to or larger than the horizon it is sufficient to use the linearized GR equations, rather than the full non-linear ones. In Chapter 19 we will present the linearized equations in GR, and we will see that they reduce to the linearized Newtonian equations inside the horizon.

Indeed, the corresponding equation of motion is

$$
0 = \frac{d}{d\eta}\left(\frac{\partial L}{\partial(d\mathbf{x}/d\eta)}\right) - \frac{\partial L}{\partial \mathbf{x}}
$$

$$
= \frac{d}{d\eta}\left(ma\frac{d\mathbf{x}}{d\eta}\right) - \boldsymbol{\nabla}(-ma\Psi)\,, \tag{17.210}
$$

which gives back eq. (17.201), or equivalently eqs. (17.205) and (17.206).[34] The Lagrangian (17.209) is nothing but the first non-trivial term in the non-relativistic expansion of the full relativistic point-particle Lagrangian obtained from eqs. (17.19) and (17.20), with $h_{tt} = -2\Psi$.

We have already seen in eq. (17.36) that, in the full relativistic setting, the comoving momentum p^i is the same as the canonical momentum π^i conjugate to the comoving coordinate $x^i(t)$. Thus, in the Newtonian limit, the comoving momentum, given in eq. (17.206), must be equal to the canonical momentum obtained from the non-relativistic limit of the Lagrangian, i.e. from eq. (17.209). Indeed, the canonical momentum obtained from the Lagrangian (17.209) is

$$
\mathbf{p}_{\text{can}} \equiv \frac{\partial L}{\partial(d\mathbf{x}/d\eta)}
$$

$$
= ma(\eta)\frac{d\mathbf{x}}{d\eta}\,, \tag{17.211}
$$

which agrees with $\mathbf{p}$ defined in eq. (17.206).

The Hamiltonian H_η that generates translations with respect to the time variable η is

$$
H_\eta \equiv \mathbf{p}\cdot\frac{d\mathbf{x}}{d\eta} - L
$$

$$
= \frac{p^2}{2ma(\eta)} + ma(\eta)\Psi\,. \tag{17.212}
$$

Observe that this Hamiltonian is not conserved. For instance, when $\Psi = 0$, p is constant and therefore H_η decreases in time as $1/a(\eta)$. This is a consequence of the fact that the problem is not invariant under time translations, because of the expansion of the Universe. In Hamiltonian form the equations of motion are

$$
\frac{d\mathbf{x}}{d\eta} = \frac{\mathbf{p}}{ma(\eta)}\,, \tag{17.213}
$$

$$
\frac{d\mathbf{p}}{d\eta} = -ma(\eta)\boldsymbol{\nabla}\Psi\,. \tag{17.214}
$$

17.7.2 Newtonian fluid dynamics in an expanding Universe

From single-particle motions to fluid dynamics

We now wish to pass from a single-particle description to a fluid dynamical description, in terms of the density field $\rho(\eta, \mathbf{x})$ and the peculiar

<hr>

[34]Recall that the equation of motion is obtained from the Lagrangian as in eq. (17.210) even when the Lagrangian depends explicitly on time (as is our case); see e.g. Landau and Lifshitz (1976), Section 2.

velocity field $\mathbf{v}(\eta, \mathbf{x})$. The relation between these macroscopic quantities and the single-particle variables is provided by the (comoving) phase-space distribution function $f(\eta, \mathbf{x}, \mathbf{p})$. In particular, we can define the *comoving* number density, n_{com}, by

$$n_{\text{com}}(\eta, \mathbf{x}) \equiv \int \frac{d^3\mathbf{p}}{(2\pi)^3}\, f(\eta, \mathbf{x}, \mathbf{p})\,, \qquad (17.215)$$

where we have reabsorbed into the definition of $f(\eta, \mathbf{x}, \mathbf{p})$ the multiplicative factor g that counts the number of internal degrees of freedom, such as the spin states. The total number of particles is

$$
\begin{aligned}
N &= \int \frac{d^3\mathbf{x}\, d^3\mathbf{p}}{(2\pi)^3}\, f(\eta, \mathbf{x}, \mathbf{p}) \\
&= \int d^3x\, n_{\text{com}}(\eta, \mathbf{x})\,, \qquad (17.216)
\end{aligned}
$$

so $n_{\text{com}}(\eta, \mathbf{x})$ is the number density per comoving volume. From eq. (17.32), the comoving energy density is given by

$$\rho_{\text{com}}(\mathbf{x}) = \int \frac{d^3\mathbf{p}}{(2\pi)^3}\, \sqrt{m^2 + (p^2/a^2)}\, f(\eta, \mathbf{x}, \mathbf{p})\,, \qquad (17.217)$$

where $p = |\mathbf{p}|$. In a non-relativistic fluid $f(\eta, \mathbf{x}, \mathbf{p})$ is sizable only for $p \ll m$, and therefore

$$\rho_{\text{com}}(\mathbf{x}) \simeq m \int \frac{d^3\mathbf{p}}{(2\pi)^3}\, f(\eta, \mathbf{x}, \mathbf{p})\,. \qquad (17.218)$$

Thus, for non-relativistic particles with mass m the comoving energy density is $\rho_{\text{com}}(\mathbf{x}) = m\, n_{\text{com}}(\mathbf{x})$, and $m f(\eta, \mathbf{x}, \mathbf{p})$ represents the energy per cell of the phase space.

As we have seen in eq. (17.185), the peculiar velocity of a particle is given by $d\mathbf{x}/d\eta$. Thus, we define the peculiar velocity field $\mathbf{v}(\eta, \mathbf{x})$ of the fluid at the location $\mathbf{x}$ as the value of $d\mathbf{x}/d\eta$ averaged over momenta, with the phase-space distribution function,

$$
\begin{aligned}
\mathbf{v}(\eta, \mathbf{x}) &\equiv \frac{1}{n_{\text{com}}(\eta, \mathbf{x})} \int \frac{d^3\mathbf{p}}{(2\pi)^3} \frac{d\mathbf{x}}{d\eta} f(\eta, \mathbf{x}, \mathbf{p}) \\
&= \frac{1}{\rho_{\text{com}}(\eta, \mathbf{x})} \int \frac{d^3\mathbf{p}}{(2\pi)^3}\, m \frac{d\mathbf{x}}{d\eta} f(\eta, \mathbf{x}, \mathbf{p}) \\
&= \frac{1}{a(\eta)\rho_{\text{com}}(\eta, \mathbf{x})} \int \frac{d^3\mathbf{p}}{(2\pi)^3}\, \mathbf{p} f(\eta, \mathbf{x}, \mathbf{p})\,, \qquad (17.219)
\end{aligned}
$$

where in the last line we have used the single-particle equation of motion, eq. (17.213). The total velocity field $\mathbf{V}(\eta, \mathbf{x})$ is then obtained using eq. (17.184).

Physical quantities are obtained from the corresponding comoving quantities, recalling that $\mathbf{x}_{\text{ph}} = a(\eta)\mathbf{x}$ and $\mathbf{p}_{\text{ph}}(\eta) = \mathbf{p}/a(\eta)$, so at a given time $d^3\mathbf{x}_{\text{ph}} = a^3 d^3\mathbf{x}$ and $d^3\mathbf{p}_{\text{ph}} = d^3\mathbf{p}/a^3$. The physical number density $n(\eta, \mathbf{x})$ by definition satisfies

$$N = \int d^3x_{\text{ph}} n(\eta, \mathbf{x}) \qquad (17.220)$$

and therefore $n(\eta, \mathbf{x}) = (1/a^3)n_{\text{com}}(\eta, \mathbf{x})$. Similarly, for non-relativistic particles, $\rho(\eta, \mathbf{x}) = mn(\eta, \mathbf{x}) = (1/a^3)\rho_{\text{com}}(\eta, \mathbf{x})$.

One can similarly define other macroscopic quantities using higher-order moments. In particular, the comoving stress tensor $\sigma_{\text{com}}^{ij}(\eta, \mathbf{x})$ is defined by

$$\rho_{\text{com}}(\eta, \mathbf{x})v^i(\eta, \mathbf{x})v^j(\eta, \mathbf{x}) + \sigma_{\text{com}}^{ij}(\eta, \mathbf{x}) \equiv \frac{1}{(am)^2} \int \frac{d^3\mathbf{p}}{(2\pi)^3} \, p^i p^j m f(\eta, \mathbf{x}, \mathbf{p}) \,. \tag{17.221}$$

The corresponding physical stress tensor σ^{ij} is defined again so that $\sigma^{ij} = (1/a^3)\sigma_{\text{com}}^{ij}$, and then

$$\frac{\sigma^{ij}}{\rho} = \frac{\sigma_{\text{com}}^{ij}}{\rho_{\text{com}}} \,. \tag{17.222}$$

For a generic observable $\mathcal{O}(\mathbf{p})$ we introduce the notation

$$\langle \mathcal{O}(\mathbf{p}) \rangle \equiv \frac{1}{n_{\text{com}}(\eta, \mathbf{x})} \int \frac{d^3\mathbf{p}}{(2\pi)^3} \, \mathcal{O}(\mathbf{p}) f(\eta, \mathbf{x}, \mathbf{p}) \,. \tag{17.223}$$

Then we have

$$v^i = \left\langle \frac{dx^i}{d\eta} \right\rangle, \tag{17.224}$$

$$\frac{\sigma^{ij}}{\rho} = \left\langle \frac{dx^i}{d\eta} \frac{dx^j}{d\eta} \right\rangle - \left\langle \frac{dx^i}{d\eta} \right\rangle \left\langle \frac{dx^j}{d\eta} \right\rangle. \tag{17.225}$$

To understand the meaning of σ^{ij} consider first a perfect gas of particles with velocities $d\mathbf{x}/d\eta$. The pressure that they exert is

$$p = \frac{1}{3}n \left\langle \frac{d\mathbf{x}}{d\eta} \cdot \mathbf{p}_{\text{ph}} \right\rangle, \tag{17.226}$$

where we have used the standard derivation from kinetic theory, observing that the momentum that enters is the physical momentum $\mathbf{p}_{\text{ph}} = \mathbf{p}/a$ (again, because the Newtonian results hold with respect to physical coordinates), and p and n are the physical (rather than comoving) pressure and number density. Writing $\mathbf{p}_{\text{ph}} = md\mathbf{x}/d\eta$ [from eq. (17.213)] and $mn = \rho$, we get

$$p = \frac{1}{3}\rho \left\langle \frac{dx^i}{d\eta} \frac{dx^i}{d\eta} \right\rangle \tag{17.227}$$

(summed over i). For an isotropic gas in its rest frame we also have

$$\left\langle \frac{dx^i}{d\eta} \frac{dx^j}{d\eta} \right\rangle \sim \delta^{ij} \,, \qquad \left\langle \frac{dx^i}{d\eta} \right\rangle = 0 \,, \tag{17.228}$$

so we see from eq. (17.225) that σ^{ij} is proportional to δ^{ij}. Its trace, again from eq. (17.225), is $\sigma^{ii} = \rho\langle(dx^i/d\eta)(dx^i/d\eta)\rangle$, which, using eq. (17.227), is equal to $3p$. Therefore

$$\sigma^{ij} = p\delta^{ij} \,. \tag{17.229}$$

In the more general case of a viscous fluid, we can then write

$$\sigma^{ij} = p\delta^{ij} + \Sigma^{ij}\,, \tag{17.230}$$

where the tensor Σ^{ij} contains the effects of the viscosity. Observe, from eq. (17.225), that σ^{ij} characterizes the fluctuations of the particle motion with respect to a single coherent flow.

In the absence of interaction with other species of particles, the phase-space distribution function $f(\eta, \mathbf{x}, \mathbf{p})$ satisfies the collisionless Boltzmann equation, i.e. the continuity equation in phase space,

$$\frac{df}{d\eta} = \left(\frac{\partial f}{\partial \eta}\right)_{\mathbf{x}} + \frac{d\mathbf{x}}{d\eta} \cdot \boldsymbol{\nabla} f + \frac{d\mathbf{p}}{d\eta} \cdot \frac{\partial f}{\partial \mathbf{p}} = 0\,, \tag{17.231}$$

where, for later use, we have stressed that the partial derivative with respect to η is taken at constant comoving coordinate $\mathbf{x}$ (rather than at constant physical coordinate $\mathbf{r}$), since we are considering f as a function of the independent variables η, $\mathbf{x}$ and $\mathbf{p}$. Using eqs. (17.213) and (17.214), the collisionless Boltzmann equation becomes

$$\left(\frac{\partial f}{\partial \eta}\right)_{\mathbf{x}} + \frac{\mathbf{p}}{ma} \cdot \boldsymbol{\nabla} f - am\boldsymbol{\nabla}\Psi \cdot \frac{\partial f}{\partial \mathbf{p}} = 0\,. \tag{17.232}$$

Continuity equation

The continuity equation in comoving coordinates can be obtained from the collisionless Boltzmann equation (17.232), which is a continuity equation in phase space, by integrating it over $d^3\mathbf{p}/(2\pi)^3$. Using

$$\int \frac{d^3p}{(2\pi)^3} \frac{\partial f}{\partial \mathbf{p}} = 0\,, \tag{17.233}$$

(since, on physical grounds, $f(\eta, \mathbf{x}, \mathbf{p})$ vanishes sufficiently fast as $|\mathbf{p}| \to \infty$) and the definitions (17.215) and (17.219), we get

$$\left(\frac{\partial \rho_{\text{com}}}{\partial \eta}\right)_{\mathbf{x}} + \boldsymbol{\nabla} \cdot (\rho_{\text{com}}\mathbf{v}) = 0\,. \tag{17.234}$$

We now write

$$\begin{aligned}\rho_{\text{com}}(\eta, \mathbf{x}) &= \bar{\rho}_{\text{com}}(\eta) + \delta\rho_{\text{com}}(\eta, \mathbf{x}) \\ &= \bar{\rho}_{\text{com}}(\eta)[1 + \delta(\eta, \mathbf{x})]\,, \end{aligned} \tag{17.235}$$

where we have defined the *density contrast*

$$\delta \equiv \frac{\delta\rho_{\text{com}}}{\bar{\rho}_{\text{com}}}\,. \tag{17.236}$$

Since $\bar{\rho}_{\text{com}} = a^{-3}\bar{\rho}$ and $\delta\rho_{\text{com}} = a^{-3}\delta\rho$ (where, as usual, $\bar{\rho}$ and $\delta\rho$ are the physical background density and the fluctuations of the physical density, respectively), we also have

$$\delta \equiv \frac{\delta\rho}{\bar{\rho}}, \tag{17.237}$$

and we do not need to distinguish between a comoving and a physical density contrast. The continuity equation for the background density is obtained by setting $\rho_{\text{com}} = \bar{\rho}_{\text{com}}$ and $\mathbf{v} = 0$ in eq. (17.234), since $\mathbf{v}$ is the peculiar velocity with respect to the Hubble flow, so it is

$$\left(\frac{\partial\bar{\rho}_{\text{com}}}{\partial\eta}\right)_{\mathbf{x}} = 0. \tag{17.238}$$

Therefore, in comoving coordinates, the background density is constant not only in space but also in time, which of course is a consequence of the fact that, when we consider only the background and we set the perturbations to zero, the comoving positions of the particles stay constant in time. Plugging eq. (17.235) into eq. (17.234) and using eq. (17.238) we get

$$\left(\frac{\partial\delta}{\partial\eta}\right)_{\mathbf{x}} + \boldsymbol{\nabla}\cdot[(1+\delta)\mathbf{v}] = 0. \tag{17.239}$$

Observe that we are not assuming δ or $\mathbf{v}$ to be small. The separation between a background part and a part parametrized by δ and $\mathbf{v}$ was uniquely based on the fact that the background density is spatially homogeneous, and the background velocity field is given by the Hubble flow $\mathcal{H}\mathbf{x}$. In principle, the definition (17.184) of $\mathbf{v}$ and the definition (17.235) of δ are valid in general and, within the Newtonian approach, eq. (17.239) is exact. We will discuss on page 410 the corresponding linearized equations.

We now want to rewrite the continuity equation in terms of the variables $(t, \mathbf{r})$, using the physical density $\rho(t, \mathbf{r}) = a^{-3}\rho_{\text{com}}$ and the total velocity field $\mathbf{V}(t, \mathbf{r}) = H\mathbf{r} + \mathbf{v}(t, r)$. To transform to the variables $(t, \mathbf{r})$ we must take into account that, since $\mathbf{r} = a(t)\mathbf{x}$, taking the derivative with respect to t at fixed $\mathbf{r}$ is not the same as taking it at fixed $\mathbf{x}$. Rather, for any differentiable function $g(t, \mathbf{r})$,

$$\begin{aligned} dg &= \left(\frac{\partial g}{\partial t}\right)_{\mathbf{r}} dt + \left(\frac{\partial g}{\partial r^i}\right)_t dr^i \\ &= \left(\frac{\partial g}{\partial t}\right)_{\mathbf{r}} dt + \left(\frac{\partial g}{\partial r^i}\right)_t \left[\left(\frac{\partial r^i}{\partial t}\right)_{\mathbf{x}} dt + \left(\frac{\partial r^i}{\partial x^j}\right)_t dx^j\right] \\ &= \left[\left(\frac{\partial g}{\partial t}\right)_{\mathbf{r}} + \left(\frac{\partial g}{\partial r^i}\right)_t \frac{da}{dt}x^i\right] dt + \left(\frac{\partial g}{\partial r^i}\right)_t a\,dx^i. \end{aligned} \tag{17.240}$$

Therefore

$$\left(\frac{\partial g}{\partial x^i}\right)_t = a\left(\frac{\partial g}{\partial r^i}\right)_t \tag{17.241}$$

and

$$\left(\frac{\partial g}{\partial t}\right)_{\mathbf{x}} = \left(\frac{\partial g}{\partial t}\right)_{\mathbf{r}} + Hr^i\left(\frac{\partial g}{\partial r^i}\right)_t. \tag{17.242}$$

Defining

$$\mathbf{V}_0(t,\mathbf{r}) \equiv H(t)\mathbf{r}, \tag{17.243}$$

we therefore have

$$\left(\frac{\partial}{\partial t}\right)_{\mathbf{x}} = \left(\frac{\partial}{\partial t}\right)_{\mathbf{r}} + \mathbf{V}_0(t,\mathbf{r})\cdot\frac{\partial}{\partial \mathbf{r}}. \tag{17.244}$$

When the peculiar potential $\Psi = 0$, a particle initially at comoving position $\mathbf{x}$ remains at $\mathbf{x}$, so taking the time derivative at constant $\mathbf{x}$ corresponds to following the motion of the particle. This is the Lagrangian point of view. The Eulerian point of view, in contrast, corresponds to looking at the motion of the fluid at a fixed point in physical space, i.e. at fixed $\mathbf{r}$. So, eq. (17.244) gives the relation between the "Lagrangian" and "Eulerian" time derivatives, when $\Psi = 0$. However, when $\Psi \neq 0$, even the comoving coordinates $\mathbf{x}$ change with time, and are no longer exactly equal to the Lagrangian coordinates, which by definition follow the particle's motion.

Using eq. (17.244), eq. (17.234) becomes

$$\left(\frac{\partial(a^3\rho)}{\partial t}\right)_{\mathbf{r}} + \mathbf{V}_0(t,\mathbf{r})\cdot\frac{\partial}{\partial \mathbf{r}}(a^3\rho) + \frac{\partial}{\partial \mathbf{r}}\cdot[a^3\rho(\mathbf{V}-\mathbf{V}_0)] = 0, \tag{17.245}$$

which, dividing by a^3, gives

$$3H\rho + \left(\frac{\partial\rho}{\partial t}\right)_{\mathbf{r}} - \rho\frac{\partial}{\partial \mathbf{r}}\cdot\mathbf{V}_0 + \frac{\partial}{\partial \mathbf{r}}\cdot(\rho\mathbf{V}) = 0. \tag{17.246}$$

Since

$$\frac{\partial}{\partial \mathbf{r}}\cdot\mathbf{V}_0 = H\sum_{i=1}^3\frac{\partial r^i}{\partial r^i} = 3H, \tag{17.247}$$

the first and third terms in the last line of eq. (17.246) cancel, and we get

$$\left(\frac{\partial\rho}{\partial t}\right)_{\mathbf{r}} + \frac{\partial}{\partial \mathbf{r}}\cdot(\rho\mathbf{V}) = 0. \tag{17.248}$$

We see that the continuity equation takes the same form in physical and comoving coordinates [compare eqs. (17.234) and (17.248)], once one uses the expression for the density appropriate to the coordinate system used. Observe also that in eq. (17.234) there appears the peculiar velocity $\mathbf{v}(\eta,\mathbf{x})$, which is the fluid velocity seen by an observer that uses comoving coordinates, while in eq. (17.248) there appears the full velocity field $\mathbf{V}(t,\mathbf{r})$, as appropriate for an Eulerian observer.

The continuity equation for the background is obtained by setting $\rho(t,\mathbf{r}) = \bar\rho(t)$ and $\mathbf{V}(t,\mathbf{r}) = \mathbf{V}_0(t,\mathbf{r}) = H(t)\mathbf{r}$, and therefore

$$\frac{d\bar\rho}{dt} + 3H\bar\rho = 0, \tag{17.249}$$

which gives $\bar{\rho}(t) \sim a^{-3}$. Not surprisingly, for the background, we have recovered eq. (17.47) with $p = 0$, since we are considering a non-relativistic fluid. Plugging $\rho = \bar{\rho}(t)(1+\delta)$ into eq. (17.248) and using this continuity equation for $\bar{\rho}(t)$, the equation for $\delta(t, \mathbf{r})$ becomes

$$\left[\left(\frac{\partial}{\partial t} \right)_{\mathbf{r}} + \mathbf{V}_0 \cdot \frac{\partial}{\partial \mathbf{r}} \right] \delta(t, \mathbf{r}) + \frac{\partial}{\partial \mathbf{r}} \cdot [(1+\delta)\mathbf{v}] = 0 . \tag{17.250}$$

Euler equation

We now take the first moment of the collisionless Boltzmann equation (17.232), i.e. we multiply by $\mathbf{p}/a$, integrate over $d^3\mathbf{p}/(2\pi)^3$, and use eqs. (17.219) and (17.221). This gives

$$\frac{1}{a} \frac{\partial}{\partial \eta} \int \frac{d^3\mathbf{p}}{(2\pi)^3} p_i f + \partial_j \int \frac{d^3\mathbf{p}}{(2\pi)^3} \frac{p_i}{ma} \frac{p_j}{ma} mf$$

$$-a(\partial_j \Psi) \int \frac{d^3\mathbf{p}}{(2\pi)^3} \frac{p_i}{a} m \frac{\partial f}{\partial p_j} = 0 \tag{17.251}$$

(where we have been careful to write the first factor $1/a(\eta)$ outside $\partial/\partial\eta$). In the first term we use

$$\frac{1}{a} \partial_\eta f = \partial_\eta \left(\frac{f}{a} \right) + \frac{1}{a} \mathcal{H} f \tag{17.252}$$

(where $\partial_\eta = \partial/\partial\eta$) and in the third term we integrate $\partial/\partial p_j$ by parts. This gives

$$\partial_\eta(\rho_{\text{com}} v_i) + \mathcal{H}\rho_{\text{com}} v_i + \partial_j(\rho_{\text{com}} v_i v_j + \sigma_{\text{com},ij}) + \rho_{\text{com}} \partial_i \Psi = 0 . \tag{17.253}$$

In this form it is clear that the Euler equation is the equation for evolution of linear momentum: $\rho_{\text{com}} v_i$ is in fact the momentum carried by a unit volume element, the term $\mathcal{H}\rho_{\text{com}} v_i$ describes the decrease in momentum due to the expansion of the Universe, and $\rho_{\text{com}} \partial_i \Psi$ describes the change in momentum due to the potential Ψ. The remaining term is the momentum flux, which is then given by $\partial_j \Pi^{ij}_{\text{com}}$ with

$$\Pi^{ij}_{\text{com}} = \rho_{\text{com}} v^i v^j + \sigma^{ij}_{\text{com}} . \tag{17.254}$$

In particular, when $\sigma^{ij} = p\delta^{ij}$, we have $\Pi^{ij} = \rho v^i v^j + p\delta^{ij}$ (and similarly for the relation between comoving quantities). Expanding the derivatives we get

$$\rho_{\text{com}} \partial_\eta v_i + v_i \partial_\eta \rho_{\text{com}} + \mathcal{H}\rho_{\text{com}} v_i + v_i \partial_j(\rho_{\text{com}} v_j) + \rho_{\text{com}} v_j \partial_j v_i$$

$$+\partial_j \sigma_{\text{com},ij} + \rho_{\text{com}} \partial_i \Psi = 0 . \tag{17.255}$$

Using the continuity equation, the second and fourth terms cancel and, dividing by ρ_{com}, we get the Euler equation in comoving coordinates,

$$\left(\frac{\partial}{\partial \eta} + v_j \frac{\partial}{\partial x^j} \right) v_i + \mathcal{H} v_i = -\frac{\partial \Psi}{\partial x^i} - \frac{1}{\rho} \frac{\partial \sigma_{ij}}{\partial x^j} , \tag{17.256}$$

where we have used eq. (17.222) to express the result in terms of the physical density ρ and the physical stress tensor σ^{ij}. Observe that the result depends on the *convective derivative*

$$\frac{D}{D\eta} \equiv \frac{\partial}{\partial\eta} + \mathbf{v} \cdot \nabla . \qquad (17.257)$$

This can be understood physically by observing that this is the total time derivative computed following the motion of the fluid (the Lagrangian point of view) rather than at a fixed point in space. This is the quantity that is equal to the force exterted on the particles of the fluid, which is given by the right-hand side of eq. (17.256).

In particular, if we can neglect viscosity, then from eq. (17.229) we get

$$\left(\frac{\partial}{\partial\eta} + \mathbf{v} \cdot \nabla\right)\mathbf{v} + \mathcal{H}\mathbf{v} = -\nabla\Psi - \frac{1}{\rho}\nabla p. \qquad (17.258)$$

We can now transform to variables $(t, \mathbf{r})$ using eqs. (17.241) and (17.244), and we get

$$\left(\frac{\partial}{\partial t} + V_j\frac{\partial}{\partial r^j}\right)v_i + Hv_i = -\frac{\partial\Psi}{\partial r^i} - \frac{1}{\rho}\frac{\partial\sigma_{ij}}{\partial r^j}, \qquad (17.259)$$

where, as usual, $\mathbf{V} = \mathbf{V}_0 + \mathbf{v} = H\mathbf{r} + \mathbf{v}$. All quantities are now considered as functions of the independent variables $(t, \mathbf{r})$, and the derivative with respect to t is performed at fixed $\mathbf{r}$. We now write $v_i = V_i - Hr_i$ and we get

$$\left(\frac{\partial}{\partial t} + V_j\frac{\partial}{\partial r^j}\right)V_i = \left(\frac{dH}{dt} + H^2\right)r_i - \frac{\partial\Psi}{\partial r^i} - \frac{1}{\rho}\frac{\partial\sigma_{ij}}{\partial r^j}. \qquad (17.260)$$

Since $dH/dt + H^2 = (1/a)d^2a/dt^2$, we can write

$$\left(\frac{dH}{dt} + H^2\right)r_i = -\frac{\partial}{\partial r_i}\left[-\frac{1}{2a}\frac{d^2a}{dt^2}|\mathbf{r}|^2\right], \qquad (17.261)$$

and we recognize the potential $\phi_0(t, \mathbf{r})$ found in eq. (17.200). Thus, we finally get the Euler equation in physical coordinate

$$\left(\frac{\partial}{\partial t} + V_j\frac{\partial}{\partial r^j}\right)V_i = -\frac{\partial\phi}{\partial r^i} - \frac{1}{\rho}\frac{\partial\sigma_{ij}}{\partial r^j}, \qquad (17.262)$$

where

$$\phi(t, \mathbf{r}) = \phi_0(t, \mathbf{r}) + \Psi(t, \mathbf{r}) \qquad (17.263)$$

is the total gravitational potential. As expected, in Eulerian coordinates the Euler equation is written in terms of the total velocity field $\mathbf{V}$ and the total gravitational potential ϕ, while in comoving coordinates it is written in terms of the peculiar velocity $\mathbf{v}$ and the peculiar gravitational potential Ψ.

Truncation of the BBGKY hierarchy

The continuity equation and the Euler equation provide one scalar equation and one vector equation. However, their unknowns are the scalar ρ, the vector v^i and the tensor σ^{ij} (or, if we neglect viscosity, the pressure p). We therefore need more equations to close the system. Taking the second moment of the collisionless Boltzmann equation gives an equation for σ_{ij}; however, this equation will also involve a correlator of three velocities, $\langle (dx^i/d\eta)(dx^j/d\eta)(dx^k/d\eta) \rangle$. One can find an equation for the three-velocity correlator by taking one more moment of the collisionless Boltzmann equation, which, however, will involve a four-velocity correlator, and so on. This procedure generates an infinity hierarchy of equations, known as the BBGKY hierarchy. At some point it therefore becomes necessary to truncate the hierarchy, assuming that beyond some order the higher-order moments are irrelevant. If this is not the case, then the fluid approximation is simply not adequate. The simplest truncation is obtained by stopping at the level of σ^{ij} and neglecting viscosity, so that the only extra variable is the pressure. One then assumes, on physical grounds, an equation of state, i.e. a relation that expresses p as a function of ρ.

Linearized equations

We can now linearize the continuity and Euler equations. Using comoving coordinates, eqs. (17.239) and (17.256) become

$$\frac{\partial \delta}{\partial \eta} = -\partial_i v^i \,, \tag{17.264}$$

$$\frac{\partial v_i}{\partial \eta} + \mathcal{H} v_i = -\frac{\partial \Psi}{\partial x^i} - \frac{1}{\rho}\frac{\partial \sigma_{ij}}{\partial x^j} \,, \tag{17.265}$$

where it is understood that the derivative with respect to η is performed at constant $\mathbf{x}$. Furthermore the perturbation of the potential, Ψ, is related to the overdensity $\delta\rho$ by the Poisson equation (17.203), which, in terms of $\delta(\eta, \mathbf{x})$, reads

$$\nabla^2 \Psi(\eta, \mathbf{x}) = 4\pi G \bar{\rho} a^2(\eta) \delta(\eta, \mathbf{x}) \,. \tag{17.266}$$

We can also transform eqs. (17.264) and (17.265) so that the derivatives are taken with respect to the physical coordinates $(t, \mathbf{r})$. In this case, we get

$$\left[\frac{\partial}{\partial t} + H(t) r^i \frac{\partial}{\partial r^i} \right] \delta(t, \mathbf{r}) = -\frac{\partial}{\partial r^i} v^i(t, \mathbf{r}) \,, \tag{17.267}$$

$$\frac{\partial V_i}{\partial t} = -\frac{\partial \phi}{\partial r^i} - \frac{1}{\bar{\rho}(t)}\frac{\partial \sigma_{ij}}{\partial r^j} \,, \tag{17.268}$$

where the derivative with respect to t is at constant $\mathbf{r}$. From eq. (17.187), the Poisson equation for the total potential ϕ reads

$$\nabla_{\mathbf{r}}^2 \phi = 4\pi G \rho(t, \mathbf{r}) = 4\pi G \bar{\rho}(t) \left[1 + \delta(t, \mathbf{r}) \right] \,. \tag{17.269}$$

Further reading

- There are many excellent cosmology textbook, including Padmanabhan (2002), Dodelson (2003), Mukhanov (2005), Durrer (2008), Weinberg (2008), Liddle and Lyth (2009), Amendola and Tsujikawa (2010) and Gorbunov and Rubakov (2011).

- Newtonian fluid dynamics in an expanding Universe is discussed in Bertschinger (1996), Bernardeau, Colombi, Gaztanaga and Scoccimarro (2002), Padmanabhan (2002) and Peacock (2007).

18 Helicity decomposition of metric perturbations

[1] At the level of nomenclature, sometimes the expression "degrees of freedom" is reserved for the actual propagating degrees of freedom, which satisfy dynamical equations such as the Klein–Gordon equation. Upon quantization, these degrees of freedom give rise to the actual particle content of the theory. We will rather use the expression "degrees of freedom" for all the components of the fields that appear in the theory, e.g. for the 10 components of $h_{\mu\nu}$ in linearized GR. We will then distinguish between unphysical degrees of freedom (which can be gauged away) and physical degrees of freedom (which cannot). Among the physical degrees of freedom, we will further distinguish between radiative degrees of freedom, which satisfy dynamical equations such as the Klein–Gordon equation and admit wave-like solutions in vacuum (with initial conditions that can be freely specified), and non-radiative degrees of freedom, which satisfy equations such as the Poisson equation, which have no free wave-like solution. Only physical radiative degrees of freedom give rise, upon quantization, to the particle content of the theory.

In this chapter we start to develop the theory of cosmological perturbations, based on the full apparatus of GR. As we will see, cosmological perturbations separate into scalar, vector and tensor perturbations, depending on their properties under spatial rotations (more precisely we will see that, for a Fourier mode with momentum $\mathbf{k}$, this is actually a decomposition in terms of helicity eigenstates, defined with respect to the direction $\hat{\mathbf{k}}$). Tensor perturbations are just gravitational waves, which are the main focus of this book. We will see in Chapter 19 that, at the linearized level, the equations for scalar, vector and tensor perturbations decouple. Nevertheless, in order to appreciate the role of GWs in cosmology, it is quite important to also have a good understanding of the behavior of scalar perturbations (while vector perturbations turn out to be less relevant). We will thus begin in Section 18.1 with a discussion of the helicity decomposition of metric perturbations. It is interesting first of all to see how this separation works when we perturb around flat space. This will allow us to appreciate that the metric perturbation $h_{\mu\nu}$ contains six physical (i.e. gauge-invariant) degrees of freedom: two in the scalar sector, two in the vector sector and two in the tensor sector. However, the degrees of freedom in the scalar and vector sectors satisfy Poisson equations and are therefore non-propagating, while the degrees of freedom in the tensor sector do propagate.[1] This discussion will complement that given in Chapters 1 and 2 of Vol. 1, and will allow us to understand from a different point of view the physical content of linearized GR. We will then discuss how the perturbed energy–momentum tensor can be similarly decomposed, and the relations imposed on the linearized quantities by energy–momentum conservation. Both for the metric perturbations and for the energy–momentum perturbations we will discuss how to construct gauge-invariant variables, to first order in perturbation theory. We will then see how to generalize this construction to perturbations around a FRW background. Thus, the aim of this chapter is basically to construct the variables that are most useful for setting up cosmological perturbation theory, and to understand in some depth their properties. This will greatly facilitate the study of their dynamical evolution, which will be the subject of Chapter 19, and also provides a better understanding of the structure of propagating and non-propagating degrees of freedom in linearized GR.

Gravitational Waves, Volume 2: Astrophysics and Cosmology. Michele Maggiore.
© Michele Maggiore 2018. Published in 2018 by Oxford University Press.
DOI 10.1093/oso/9780198570899.001.0001

18.1   Perturbations around flat space

18.1.1   Helicity decomposition

The first step in the helicity decomposition is to separate metric perturbations according to their properties under spatial rotation. We discuss first the decomposition of perturbations around flat space, and in Section 18.2 we will discuss the decomposition of perturbations around a FRW background. Expanding the metric $g_{\mu\nu}$ around flat space,

$$g_{\mu\nu} = \eta_{\mu\nu} + h_{\mu\nu}\,, \tag{18.1}$$

h_{00} is a scalar under rotations, h_{0i} transforms under rotations as a spatial vector and h_{ij} as a spatial tensor. One can go further, and use the fact that a vector field can be decomposed into a transverse and a longitudinal part. A similar decomposition holds for a spatial tensor field. Thus, one can write the metric perturbation as

$$h_{00} = 2\psi\,, \tag{18.2}$$
$$h_{0i} = \beta_i + \partial_i \gamma\,, \tag{18.3}$$
$$h_{ij} = -2\phi\delta_{ij} + \left(\partial_i\partial_j - \frac{1}{3}\delta_{ij}\mathbf{\nabla}^2\right)\lambda + \frac{1}{2}(\partial_i\epsilon_j + \partial_j\epsilon_i) + h_{ij}^{\rm TT}\,, \tag{18.4}$$

where $\mathbf{\nabla}^2$ is the flat-space Laplacian and the quantities β^i, ϵ^i and $h_{ij}^{\rm TT}$ satisfy the constraints

$$\partial_i\beta^i = 0\,, \tag{18.5}$$
$$\partial_i\epsilon^i = 0\,, \tag{18.6}$$
$$\partial^j h_{ij}^{\rm TT} = 0\,, \tag{18.7}$$
$$\delta^{ij} h_{ij}^{\rm TT} = 0\,, \tag{18.8}$$

i.e. β^i and ϵ^i are transverse vector fields and $h_{ij}^{\rm TT}$ is a symmetric, transverse and traceless tensor field. The combination $[\partial_i\partial_j - (1/3)\delta_{ij}\mathbf{\nabla}^2]$ is chosen so that this term is traceless, $\delta^{ij}[\partial_i\partial_j - (1/3)\delta_{ij}\mathbf{\nabla}^2] = 0$.[2] The quantities $\psi(x), \phi(x), \gamma(x)$ and $\lambda(x)$ are scalar fields under spatial rotations [but, of course, not under full Lorentz transformations of the Minkowski background; for instance, $\psi(x)$ transforms as the $(0,0)$ component of the Lorentz tensor $h_{\mu\nu}$]. Because of eqs. (18.5) and (18.6), the vector fields $\beta^i(x)$ and $\epsilon^i(x)$ are transverse, and therefore contain two degrees of freedom each. The decomposition $h_{0i} = \beta_i + \partial_i\gamma$ corresponds to the decomposition of a vector field into a transverse and a longitudinal part. The meaning of this decomposition becomes clearer when we go to Fourier space. Using the convention on the Fourier transform given in eq. (17.40), we have the correspondence

$$\partial_i \leftrightarrow ik_i\,, \tag{18.9}$$

so

$$\tilde{h}_{0i}(\mathbf{k}) = \tilde{\beta}_i(\mathbf{k}) + ik_i\tilde{\gamma}(\mathbf{k})\,, \tag{18.10}$$

[2]These expressions, as well as all the results that we will discuss in this chapter, are specific to $d = 3$ spatial dimensions. For instance, for generic d the traceless combination is rather

$$[\partial_i\partial_j - (1/d)\delta_{ij}\mathbf{\nabla}^2].$$

As we will discuss, the decomposition into longitudinal and transverse parts is based on the notion of helicity, which in $d = 3$ labels the representations of the group $SO(2)$ of spatial rotations around a fixed axis. For generic d this is replaced by the representations of $SO(d-1)$. Observe also that the variables ψ, ϕ, β_i and $h_{ij}^{\rm TT}$ are dimensionless, while γ and ϵ_i have dimensions of length and λ dimensions of (length)^2.

[3]From now on we will omit the tilde
from quantities defined in momentum
space. Whether we are considering a
function $f(\mathbf{x})$ in $\mathbf{x}$-space or its Fourier
transform $\tilde{f}(\mathbf{k})$ will be clear from the
argument $\mathbf{x}$ or $\mathbf{k}$ (and, if it is clear from
the context, we will also occasionally
drop the argument).

where $\mathbf{k}\cdot\tilde{\boldsymbol{\beta}}(\mathbf{k}) = 0$. So in Fourier space we have a vector field $\tilde{\boldsymbol{\beta}}(\mathbf{k})$ transverse to the direction $\mathbf{k}$, and a longitudinal vector field $\tilde{\gamma}(\mathbf{k})\mathbf{k}$. From a group-theoretical point of view this is a decomposition into irreducible representations of the group $SO(2)$ of rotations around a fixed axis $\mathbf{k}$. These representations are labeled by the helicity h. The fields $\psi(\mathbf{k}), \phi(\mathbf{k}), \gamma(\mathbf{k})$ and $\lambda(\mathbf{k})$ are invariant under rotations around the $\mathbf{k}$ axis, and therefore have zero helicity.[3] A transverse vector field $\boldsymbol{\beta}(\mathbf{k})$ can be written as $\boldsymbol{\beta}(\mathbf{k}) = \beta_1(\mathbf{k})\mathbf{u} + \beta_2(\mathbf{k})\mathbf{v}$, where $\{\mathbf{u}, \mathbf{v}, \hat{\mathbf{k}}\}$ form an orthonormal basis. Under a rotation by an angle α around the $\mathbf{k}$ axis, we have

$$\beta_1(\mathbf{k}) \to \beta_1(\mathbf{k})\cos\alpha - \beta_2(\mathbf{k})\sin\alpha\,,$$
$$\beta_2(\mathbf{k}) \to \beta_1(\mathbf{k})\sin\alpha + \beta_2(\mathbf{k})\cos\alpha\,, \tag{18.11}$$

so $\beta_\pm \equiv \beta_1 \pm i\beta_2$ transform as

$$\beta_\pm(\mathbf{k}) \to e^{\pm i\alpha}\beta_\pm(\mathbf{k})\,. \tag{18.12}$$

By definition, this means that $\beta_+(\mathbf{k})$ is an eigenstate of the helicity with eigenvalue $h = +1$ and β_- is an eigenstate of the helicity with eigenvalue $h = -1$; compare with eq. (2.197) of Vol. 1. Similarly, we have shown in Section 1.2 and in Problem 2.1 in Vol. 1 that the transverse-traceless tensor h_{ij}^{TT} has two independent components h_+ and $h_\times$, and that the combinations $h_+ \pm ih_\times$ are eigenstates of helicity with $h = \pm 2$ (or, equivalently, $h_\times \pm ih_+$ have helicities ∓ 2). Thus, the decomposition (18.2)–(18.4) is a decomposition into helicity eigenstates. The 10 components of the matrix $h_{\mu\nu}$ have been separated into the four functions ψ, ϕ, γ and λ, which are scalars under rotations and therefore have helicity $h = 0$; the transverse vector β^i, with two degrees of freedom with helicities ± 1; and the transverse vector ϵ^i, again with $h = \pm 1$; and, finally, the TT tensor h_{ij}^{TT}, with $h = \pm 2$. Overall, we have four degrees of freedom in the scalar sector, four in the vector sector and two in the tensor sector.

In order to ensure that the helicity decomposition can be inverted (and therefore that there is a one-to-one correspondence between $h_{\mu\nu}$ and the variables $\{\psi, \phi, \gamma, \lambda, \beta^i, \epsilon^i, h_{ij}^{\mathrm{TT}}\}$), we must also impose the boundary conditions that

$$\gamma \to 0\,, \quad \lambda \to 0\,, \quad \nabla^2\lambda \to 0\,, \quad \epsilon_i \to 0\,, \tag{18.13}$$

sufficiently fast at spatial infinity, so that the Laplacian is invertible. These boundary conditions ensure the uniqueness of the decomposition. Then, the inversion can be performed as follows. The variable ψ is of course simply given by eq. (18.2), while ϕ is obtained taking the contraction of eq. (18.4) with δ^{ij}. This gives

$$\psi = \frac{1}{2}h_{00}\,, \tag{18.14}$$

$$\phi = -\frac{1}{6}\delta^{ij}h_{ij}\,. \tag{18.15}$$

Thus, these quantities are local functions of the metric perturbation $h_{\mu\nu}$. All the other variables, in contrast, are constructed from h_{0i} and h_{ij} through applications of the inverse Laplacian, which is a non-local operator with respect to the spatial variables. For instance, to extract γ we take the divergence of eq. (18.3) and use eq. (18.5). This gives $\partial^i h_{0i} = \boldsymbol{\nabla}^2 \gamma$. We then invert the Laplacian [which, with the boundary conditions (18.13), is a well-defined operation] and we get[4]

$$\gamma = \boldsymbol{\nabla}^{-2}(\partial_i h_{0i}) \,. \tag{18.16}$$

To extract $\boldsymbol{\nabla}^2 \lambda$ we apply the operator $\partial^i \partial^j$ to eq. (18.4) and we get

$$\boldsymbol{\nabla}^2 \lambda = -\frac{1}{2}\left(\delta_{ij} - 3\frac{\partial_i \partial_j}{\boldsymbol{\nabla}^2} \right) h_{ij} \,, \tag{18.17}$$

where we have used the boundary condition on $\boldsymbol{\nabla}^2 \lambda$ to invert $\boldsymbol{\nabla}^2(\boldsymbol{\nabla}^2 \lambda)$. Requiring further that λ itself vanishes sufficiently fast at infinity allows us to invert once more the Laplacian in eq. (18.17) and obtain λ. Inserting eq. (18.16) into eq. (18.3) we obtain the explicit expression for β_i in terms of the h_{0i},

$$\beta_i = P_{ij} h_{0j} \,, \tag{18.18}$$

where we have defined the non-local operator

$$P_{ij} = \delta_{ij} - \frac{\partial_i \partial_j}{\boldsymbol{\nabla}^2} \,. \tag{18.19}$$

This operator is transverse, $\partial_i P_{ij} = 0$, and therefore it naturally appears in the expression for the transverse vector β_i. Similarly, taking the divergence of eq. (18.4) and using eqs. (18.15) and (18.17) we get

$$\epsilon_i = 2\boldsymbol{\nabla}^{-2} P_{ik} \partial_l h_{kl} \,. \tag{18.20}$$

Finally, using in eq. (18.4) the explicit expressions for ϕ, λ and ϵ^i in terms of the metric, we get

$$h_{ij}^{\mathrm{TT}} = \Lambda_{ij,kl} h_{kl} \,, \tag{18.21}$$

where

$$\Lambda_{ij,kl} = P_{ik} P_{jl} - \frac{1}{2} P_{ij} P_{kl} \,. \tag{18.22}$$

In Fourier space, using eq. (18.9), $P_{ij} = \delta_{ij} - n_i n_j$, where $\hat{n} = \hat{\mathbf{k}}$ is the unit vector in the direction $\mathbf{k}$. Comparing eq. (18.22) with eqs. (1.35) and (1.36) of Vol. 1, we see that eq. (18.22) gives the Λ tensor that we repeatedly used in Vol. 1 to enforce the TT gauge, except that eqs. (18.19) and (18.22) give its expression as a non-local operator in coordinate space.

It is also important to observe that the relation between $h_{\mu\nu}$ and the variables $\{\psi, \phi, \gamma, \lambda, \beta^i, \epsilon^i, h_{ij}^{\mathrm{TT}}\}$ is non-local in space, because of the inverse Laplacian, but is local in time. This means that assigning initial conditions on a time slice for $h_{\mu\nu}$ is equivalent to assigning them for the variables $\{\psi, \phi, \gamma, \lambda, \beta^i, \epsilon^i, h_{ij}^{\mathrm{TT}}\}$ and therefore the counting of dynamical degrees of freedom can be performed equivalently using $h_{\mu\nu}$ or the variables $\{\psi, \phi, \gamma, \lambda, \beta^i, \epsilon^i, h_{ij}^{\mathrm{TT}}\}$.

[4]Observe that, once we have explicitly separated the $\mu = 0$ and $\mu = i$ indices and we have equations involving only spatial indices, we can use the convention that spatial indices are raised and lowered with δ_{ij}. Thus, in the following we will also often write all the spatial indices as lower indices, with a summation understood over repeated lower spatial indices. However, we will always leave the indices of h_{ij} as lower indices, since raising the indices of h_{ij} with δ^{ij} would create confusion with the $\mu = i, \nu = j$ components of $h^{\mu\nu}$. Indeed, $h^{\mu\nu}$ with $\mu = i, \nu = j$ differs by a sign from $\delta^{ik}\delta^{jl} h_{kl}$, since from $g_{\mu\nu} = \eta_{\mu\nu} + h_{\mu\nu}$ it follows that $g^{\mu\nu} = \eta^{\mu\nu} - h^{\mu\nu} + O(h^2)$.

As we saw in eq. (1.8), the linearized theory is invariant under the gauge transformation

$$h_{\mu\nu}(x) \to h_{\mu\nu}(x) - (\partial_\mu \xi_\nu + \partial_\nu \xi_\mu)\,. \qquad (18.23)$$

When one uses the helicity decomposition of the metric it is useful to perform a similar decomposition of the gauge functions ξ_μ, writing them in the form

$$\xi_0 = A\,, \qquad (18.24)$$
$$\xi_i = B_i + \partial_i C\,, \qquad (18.25)$$

where B_i is a transverse vector, $\partial_i B^i = 0$, while A and C are scalars under rotations. In terms of these variables eq. (18.23) reads

$$\psi \to \psi - \dot{A}\,, \qquad (18.26)$$
$$\phi \to \phi + \frac{1}{3}\mathbf{\nabla}^2 C\,, \qquad (18.27)$$
$$\gamma \to \gamma - A - \dot{C}\,, \qquad (18.28)$$
$$\lambda \to \lambda - 2C\,, \qquad (18.29)$$
$$\beta_i \to \beta_i - \dot{B}_i\,, \qquad (18.30)$$
$$\epsilon_i \to \epsilon_i - 2B_i\,, \qquad (18.31)$$

while $h_{ij}^{\rm TT}$ is gauge-invariant. As dictated by symmetry, the transformation of the scalars ψ, ϕ, γ and λ depends only on the scalar functions A and C, while the transformation of the transverse vector fields β_i and ϵ_i depends only on the transverse vector field B_i. From this point of view, the fact that $h_{ij}^{\rm TT}$ is gauge-invariant is a trivial consequence of the fact that ξ^μ decomposes into a helicity-0 and a helicity-1 part, while a traceless symmetric tensor such as $h_{ij}^{\rm TT}$ is a helicity-2 tensor.

One can now use eqs. (18.26)–(18.31) to fix the gauge. Two natural gauge choices are:

- *The synchronous gauge.* This gauge is defined by $h_{00} = h_{0i} = 0$, so all perturbations are in the spatial part of the metric. In terms of the helicity decomposition this means

$$\psi = \gamma = \beta_i = 0\,. \qquad (18.32)$$

However, this does not fix the gauge uniquely. In fact, we see from eqs. (18.26)–(18.31) that the conditions (18.32) are unchanged by a residual gauge transformation with $A = -f(\mathbf{x})$, $C = f(\mathbf{x})t + g(\mathbf{x})$ and $B_i = B_i(\mathbf{x})$, where $f(\mathbf{x})$, $g(\mathbf{x})$ and $B_i(\mathbf{x})$ are arbitrary functions of the spatial coordinates. A similar problem arises when performing perturbations around the FRW metric, and gives rise to spurious gauge modes in the solutions of the equations for the density perturbations. This is indeed the reason that prompted Bardeen to formulate perturbation theory in terms of gauge-invariant variables, as we will discuss in Section 18.2.

- *The Newtonian gauge* (also called the Poisson gauge, or longitudinal gauge), in which

$$\lambda = \gamma = \beta_i = 0\,. \qquad (18.33)$$

This choice leaves no residual gauge freedom and, as far as scalar perturbations around Minkowski space are concerned, puts the metric in the form

$$ds^2 = -(1 - 2\psi)dt^2 + (1 - 2\phi)d\mathbf{x}^2\,. \qquad (18.34)$$

Comparing with eq. (5.11) we see that, with these sign conventions, the metric generated by a weak and nearly static Newtonian source with energy–momentum tensor $T^{\mu\nu}$ can be written in the form (18.34) with

$$\phi(t, \mathbf{x}) = -\psi(t, \mathbf{x}) = -G \int d^3 x' \frac{T^{00}(t, \mathbf{x}')}{|\mathbf{x} - \mathbf{x}'|}\,. \qquad (18.35)$$

Rather than fixing the gauge, a better approach is to observe that one can form the following gauge-invariant scalar combinations:

$$\Phi = -\phi - \frac{1}{6}\nabla^2\lambda\,, \qquad (18.36)$$

$$\Psi = -\psi + \dot{\gamma} - \frac{1}{2}\ddot{\lambda}\,. \qquad (18.37)$$

Their gauge invariance is immediately checked using eqs. (18.26)–(18.29).[5] In the Newtonian gauge $\Psi = -\psi$ and $\Phi = -\phi$, so eq. (18.34) can be rewritten as

[5] Observe that these quantities are gauge-invariant only at the level of linearized theory.

$$\boxed{ds^2 = -(1 + 2\Psi)dt^2 + (1 + 2\Phi)d\mathbf{x}^2\,.} \qquad (18.38)$$

So, it is often quite convenient to work in the Newtonian gauge. First of all there is no residual gauge freedom, and therefore no spurious gauge mode. Furthermore, since the metric perturbations in the scalar sector are expressed in terms of the gauge-invariant quantities Φ and Ψ, the equations that one obtains for Φ and Ψ are actually valid in any gauge.

The generalization of these definitions to perturbations of FRW spacetime gives the Bardeen variables; see Section 18.2. We will still use the term "Bardeen variables" for the flat-space case. Similarly we can form a gauge-invariant transverse vector

$$\Xi_i = \beta_i - \frac{1}{2}\dot{\epsilon}_i\,. \qquad (18.39)$$

Thus, at the level of linearized theory, we have six gauge-invariant quantities: the two components of the transverse-traceless tensor perturbations h_{ij}^{TT}, the two components of the vector perturbation Ξ_i subject to the condition $\partial_i \Xi^i = 0$, and the two scalar perturbations Φ and Ψ. So, the four gauge functions ξ^μ allow us to eliminate four pure-gauge degrees of freedom from the ten components of $h_{\mu\nu}$, remaining with six gauge-invariant degrees of freedom. We next study the dynamics of these six physical degrees of freedom.

18.1.2 Radiative and non-radiative degrees of freedom

To study the dynamics of these degrees of freedom we consider the linearization of the Einstein action, writing again $g_{\mu\nu} = \eta_{\mu\nu} + h_{\mu\nu}$ and expanding the action to quadratic order. The quadratic part of the Einstein–Hilbert action and the interaction term with an external conserved energy–momentum tensor have already been given in eqs. (2.86) and (2.88). We can rewrite them as

$$S_2 + S_{\rm int} = \int d^4x \left[\frac{1}{2\kappa^2} h_{\mu\nu} \mathcal{E}^{\mu\nu,\rho\sigma} h_{\rho\sigma} + \frac{1}{2} h_{\mu\nu} T^{\mu\nu} \right], \qquad (18.40)$$

where $\kappa \equiv (32\pi G)^{1/2}$.[6] We have introduced the Lichnerowicz operator $\mathcal{E}^{\mu\nu,\rho\sigma}$, defined as[7]

$$\mathcal{E}^{\mu\nu,\rho\sigma} \equiv \frac{1}{2}(\eta^{\mu\rho}\eta^{\nu\sigma} + \eta^{\mu\sigma}\eta^{\nu\rho} - 2\eta^{\mu\nu}\eta^{\rho\sigma})\Box$$
$$- \frac{1}{2}(\eta^{\mu\rho}\partial^\sigma\partial^\nu + \eta^{\nu\rho}\partial^\sigma\partial^\mu + \eta^{\mu\sigma}\partial^\rho\partial^\nu + \eta^{\nu\sigma}\partial^\rho\partial^\mu)$$
$$+ (\eta^{\rho\sigma}\partial^\mu\partial^\nu + \eta^{\mu\nu}\partial^\rho\partial^\sigma), \qquad (18.41)$$

and $\Box = \eta^{\mu\nu}\partial_\mu\partial_\nu$ is the flat-space d'Alembertian. It is useful to perform a similar decomposition also in the energy–momentum tensor, writing

$$T_{00} = \rho, \qquad (18.42)$$
$$T_{0i} = S_i + \partial_i S, \qquad (18.43)$$
$$T_{ij} = p\delta_{ij} + \left(\partial_i\partial_j - \frac{1}{3}\delta_{ij}\boldsymbol{\nabla}^2\right)\sigma + \frac{1}{2}(\partial_i\sigma_j + \partial_j\sigma_i) + \sigma_{ij}^{\rm TT}, \quad (18.44)$$

where

$$\partial_i\sigma^i = 0, \quad \partial_i S^i = 0, \quad \partial^i\sigma_{ij}^{\rm TT} = 0, \quad \delta^{ij}\sigma_{ij}^{\rm TT} = 0. \qquad (18.45)$$

The isotropic part of T_{ij} is $p\delta_{ij}$, where p is the pressure. The remaining terms in T_{ij} define the anisotropic stress tensor and depend on a scalar σ, a transverse vector σ_i and a TT tensor $\sigma_{ij}^{\rm TT}$. As with the analogous decomposition of the metric, the uniqueness of the decomposition is assured if we assume that S, σ, $\boldsymbol{\nabla}^2\sigma$ and σ^i vanish sufficiently rapidly at spatial infinity.

The quantities that appear in the parametrization of $T_{\mu\nu}$ are not all independent, since they are related by energy–momentum conservation $\partial_\mu T^{\mu\nu} = 0$. Imposing $\partial_\mu T^{\mu 0} = 0$ gives

$$\boxed{\dot\rho = \boldsymbol{\nabla}^2 S,} \qquad (18.46)$$

while $\partial_\mu T^{\mu i} = 0$ gives

$$\frac{1}{2}\boldsymbol{\nabla}^2\sigma^i - \dot S^i + \partial^i\left(-\dot S + p + \frac{2}{3}\boldsymbol{\nabla}^2\sigma\right) = 0. \qquad (18.47)$$

[6] Observe that the metric perturbation used in Section 2.2.2 has dimensions of mass, in order to have the canonical field-theoretical normalization of the kinetic term. Here instead the metric perturbation $h_{\mu\nu}$ is adimensional. To compare the formulas of Section 2.2.2 with the result that we will find in this chapter, we must perform the rescaling (2.87), i.e. $h_{\mu\nu} \to (1/\kappa)h_{\mu\nu}$ in the equations of Section 2.2.2.

[7] In the literature there are different conventions for the overall sign of the Lichnerowicz operator. With our convention $\mathcal{E}^{\mu\nu,\rho\sigma}h_{\rho\sigma} = +\Box h^{\mu\nu} - \dots$.

Since σ^i and S^i are transverse [see eq. (18.45)], applying ∂_i to this equation we get the condition

$$\nabla^2 \left(-\dot{S} + p + \frac{2}{3}\nabla^2 \sigma \right) = 0. \qquad (18.48)$$

We impose the boundary condition that the energy–momentum tensor vanishes at infinity. Using the fact that a Poisson equation $\nabla^2 f = 0$ with the boundary condition $f = 0$ at infinity has only the solution $f(\mathbf{x}) = 0$, we get

$$\dot{S} = p + \frac{2}{3}\nabla^2 \sigma. \qquad (18.49)$$

Inserting this into eq. (18.47) we then get

$$\dot{S}^i = \frac{1}{2}\nabla^2 \sigma^i. \qquad (18.50)$$

Thus, the vector equation (18.47) separates into an equation for the transverse vector part and one for the scalars that parametrize the longitudinal vector part. Thus, overall energy–momentum conservation gives two scalar conditions, (18.46) and (18.49), and one condition between transverse vectors, eq. (18.50).

We can now write the linearized action using the decomposition of the metric perturbations given in eqs. (18.2)–(18.4). The quadratic action S_2 decomposes into three terms for the scalar, vector and tensor sectors,

$$S_2 = \frac{1}{32\pi G} \int d^4x \left[\mathcal{L}_{2,\mathrm{s}} + \mathcal{L}_{2,\mathrm{v}} + \mathcal{L}_{2,\mathrm{t}} \right], \qquad (18.51)$$

with

$$\mathcal{L}_{2,\mathrm{s}} = -12\dot{\Phi}^2 + 4\partial_i\Phi\partial^i\Phi + 8\partial_i\Phi\partial^i\Psi, \qquad (18.52)$$
$$\mathcal{L}_{2,\mathrm{v}} = \partial_i\Xi_j\partial^i\Xi^j, \qquad (18.53)$$
$$\mathcal{L}_{2,\mathrm{t}} = -\frac{1}{2}\partial_\mu h_{ij}^{\mathrm{TT}}\partial^\mu h^{ij,\mathrm{TT}}, \qquad (18.54)$$

while the interaction term reads

$$S_{\mathrm{int}} = \int d^4x \left[3\Phi p - \Psi\rho - \Xi_i S^i + \frac{1}{2}h_{ij}^{\mathrm{TT}}\sigma^{\mathrm{TT},ij} \right]. \qquad (18.55)$$

As expected, the action depends only on the gauge-invariant combinations Φ, Ψ, Ξ_i and h_{ij}^{TT}. We can now derive the equations of motion. Consider first the scalar sector. Using $\{\phi, \psi, \lambda, \gamma\}$ as independent variables, the variation with respect to ϕ gives

$$\ddot{\Phi} - \frac{1}{3}\nabla^2(\Phi + \Psi) = -4\pi Gp, \qquad (18.56)$$

while the variation with respect to ψ gives the Poisson equation

$$\nabla^2\Phi = -4\pi G\rho. \qquad (18.57)$$

The variations with respect to λ and γ give a combination of derivatives of these equations. Observe that eqs. (18.56) and (18.57) can also be obtained by taking directly the variation of the action (18.55) with respect to Φ and Ψ, respectively. Plugging eq. (18.57) into eq. (18.56) we can rewrite the latter as

$$\ddot{\Phi} - \frac{1}{3}\boldsymbol{\nabla}^2\Psi = -\frac{4\pi G}{3}(\rho + 3p)\,. \tag{18.58}$$

The term $\ddot{\Phi}$ can be eliminated by observing that eq. (18.57) implies that $\boldsymbol{\nabla}^2\Phi = -4\pi G\rho$. Using eq. (18.46) this becomes $\boldsymbol{\nabla}^2\ddot{\Phi} = -4\pi G\boldsymbol{\nabla}^2\dot{S}$. In flat space it is natural to impose the boundary conditions that $\ddot{\Phi}$ and $\dot{S}$ vanish at infinity. Using again the fact that a Poisson equation $\boldsymbol{\nabla}^2 f = 0$ with $f = 0$ at infinity has only the solution $f(\mathbf{x}) = 0$, we get

$$\ddot{\Phi} = -4\pi G\dot{S}\,. \tag{18.59}$$

Therefore eq. (18.56) can be rewritten as

$$\boldsymbol{\nabla}^2\Psi = 4\pi G(\rho + 3p - 3\dot{S})\,, \tag{18.60}$$

which can be further simplified using eq. (18.49) to write

$$3p - 3\dot{S} = -2\boldsymbol{\nabla}^2\sigma\,. \tag{18.61}$$

Finally, in the vector sector the variation with respect to β_i (or, equivalently, with respect to Ξ^i) gives $\boldsymbol{\nabla}^2\Xi_i = -16\pi GS_i$, and this equation also implies the equation obtained by performing the variation with respect to ϵ_i. In the tensor sector the variation with respect to h_{ij}^{TT} gives $\Box h_{ij}^{\mathrm{TT}} = -16\pi G\sigma_{ij}^{\mathrm{TT}}$. In conclusion, we have

$$\boldsymbol{\nabla}^2\Phi = -4\pi G\rho\,, \tag{18.62}$$
$$\boldsymbol{\nabla}^2\Psi = 4\pi G(\rho - 2\boldsymbol{\nabla}^2\sigma)\,, \tag{18.63}$$
$$\boldsymbol{\nabla}^2\Xi_i = -16\pi GS_i\,, \tag{18.64}$$
$$\Box h_{ij}^{\mathrm{TT}} = -16\pi G\sigma_{ij}^{\mathrm{TT}}\,. \tag{18.65}$$

We see that only the tensor perturbations obey a wave equation. The gauge-invariant scalar and vector perturbations obey a Poisson equation, and therefore represent physical but non-radiative degrees of freedom, which are fully determined by the matter distribution. Adding eqs. (18.62) and (18.63) we also see that $\boldsymbol{\nabla}^2(\Phi + \Psi) = -8\pi G\boldsymbol{\nabla}^2\sigma$, so

$$\Phi + \Psi = -8\pi G\sigma\,. \tag{18.66}$$

Therefore, if the scalar part of the anisotropic stress tensor vanishes, we have $\Psi = -\Phi$. In the absence of matter we have $\Phi = \Psi = 0$ and $\Xi_i = 0$, and we remain with just the two radiative degrees of freedom described by h_{ij}^{TT}, i.e. with the two polarizations of the massless graviton.

18.2   Gauge invariance and helicity decomposition in FRW

18.2.1   Linearized diffeomorphisms and gauge invariance in a curved background

When GR is linearized around some background, the linearized theory inherits from the full theory an invariance under linearized diffeomorphism. This symmetry is typically referred to as the gauge invariance of the linearized theory. In this context the expressions "linearized diffeomorphism" and "gauge transformation" are normally used interchangeably, and indeed a choice of a coordinate system for the linearized theory is usually referred to as a gauge fixing, as we have done in Vol. 1. When studying the linearization around a curved background, such as FRW, it is, however, useful to spell out explicitly the relation between the two. A gauge transformation is defined as a transformation of the fields that does not touch the coordinates. For instance, the $U(1)$ gauge transformation of electromagnetism is defined as the transformation

$$A_\mu(x) \rightarrow A'_\mu(x) = A_\mu(x) - \partial_\mu\theta\,, \tag{18.67}$$

where A_μ is the gauge field and θ the gauge function. A diffeomorphism is instead a transformation of the coordinates and of the fields such that $x^\mu \rightarrow x'^\mu(x)$, while a vector field $V_\mu(x)$ transforms as

$$V_\mu(x) \rightarrow V'_\mu(x') = \frac{\partial x^\rho}{\partial x'^\mu}V_\rho(x)\,, \tag{18.68}$$

a tensor $T_{\mu\nu}$ as

$$T_{\mu\nu}(x) \rightarrow T'_{\mu\nu}(x') = \frac{\partial x^\rho}{\partial x'^\mu}\frac{\partial x^\sigma}{\partial x'^\nu}T_{\rho\sigma}(x)\,, \tag{18.69}$$

and so on, while upper indices transform with $\partial x'/\partial x$ rather than with $\partial x/\partial x'$; for example

$$T^\mu_\nu(x) \rightarrow (T^\mu_\nu)'(x') = \frac{\partial x'^\mu}{\partial x^\rho}\frac{\partial x^\sigma}{\partial x'^\nu}T^\rho_\sigma(x)\,. \tag{18.70}$$

Observe that in the gauge transformation (18.67) the argument of A'_μ is x, while in the diffeomorphisms (18.68) and (18.69) the argument of V'_μ and $T'_{\mu\nu}$ is x'.

For a linearized diffeomorphism we have $x'^\mu = x^\mu + \xi^\mu(x)$, with $\xi^\mu(x)$ infinitesimal. To linear order we can invert it as $x^\mu = x'^\mu - \xi^\mu(x')$. Then eqs. (18.68) and (18.69) become, respectively,

$$V'_\mu(x') = V_\mu(x) - V_\rho\partial_\mu\xi^\rho\,, \tag{18.71}$$
$$T'_{\mu\nu}(x') = T_{\mu\nu}(x) - (T_{\mu\rho}\partial_\nu\xi^\rho + T_{\rho\nu}\partial_\mu\xi^\rho)\,, \tag{18.72}$$

and similarly for all other tensors. Observe that, since $\partial_\mu\xi^\rho$ is already of first order, to linear order the term V_ρ that multiplies it in eq. (18.71) can be written equivalently as $V_\rho(x)$ or $V_\rho(x')$, so we can simply omit its argument. We can now write, to first order,

$$
\begin{aligned}
V'_\mu(x') &= V'_\mu(x + \xi) \\
&= V'_\mu(x) + \xi^\rho \partial_\rho V_\mu(x) \,,
\end{aligned} \tag{18.73}
$$

and similarly for $T'_{\mu\nu}(x')$. Then we get

$$
V'_\mu(x) = V_\mu(x) - (V_\rho \partial_\mu \xi^\rho + \xi^\rho \partial_\rho V_\mu) \,, \tag{18.74}
$$

$$
T'_{\mu\nu}(x) = T_{\mu\nu}(x) - (T_{\mu\rho} \partial_\nu \xi^\rho + T_{\rho\nu} \partial_\mu \xi^\rho + \xi^\rho \partial_\rho T_{\mu\nu}) \,. \tag{18.75}
$$

Now the transformation has taken the form of a gauge transformation, in which the fields are transformed, but the arguments on the left- and right-hand sides are always x. In particular, for the metric $g_{\mu\nu}$, we have

$$
g'_{\mu\nu}(x) = g_{\mu\nu}(x) - (g_{\mu\rho} \partial_\nu \xi^\rho + g_{\rho\nu} \partial_\mu \xi^\rho + \xi^\rho \partial_\rho g_{\mu\nu}) \,. \tag{18.76}
$$

We now expand $g_{\mu\nu}$ around a background, $g_{\mu\nu} = \bar{g}_{\mu\nu} + h_{\mu\nu}$, with $h_{\mu\nu}$ of the same infinitesimal order as $\partial_\mu \xi_\nu$, and by definition we state that, under gauge transformation, $\bar{g}_{\mu\nu}$ is unchanged, and the transformation is totally absorbed by $h_{\mu\nu}$, i.e.

$$
g'_{\mu\nu}(x) = \bar{g}_{\mu\nu}(x) + h'_{\mu\nu}(x) \,. \tag{18.77}
$$

Physically this means that $h_{\mu\nu}(x)$ is considered as a gauge field living on a given background. From eq. (18.76), the gauge transformation of the metric perturbation $h_{\mu\nu}$ is then

$$
\begin{aligned}
h'_{\mu\nu}(x) &= h_{\mu\nu}(x) - (\bar{g}_{\mu\rho} \partial_\nu \xi^\rho + \bar{g}_{\rho\nu} \partial_\mu \xi^\rho + \xi^\rho \partial_\rho \bar{g}_{\mu\nu}) \\
&= h_{\mu\nu}(x) - (\bar{g}_{\mu\rho} \bar{\boldsymbol{\nabla}}_\nu \xi^\rho + \bar{g}_{\rho\nu} \bar{\boldsymbol{\nabla}}_\mu \xi^\rho) \\
&= h_{\mu\nu}(x) - (\bar{\boldsymbol{\nabla}}_\mu \xi_\nu + \bar{\boldsymbol{\nabla}}_\nu \xi_\mu) \,,
\end{aligned} \tag{18.78}
$$

where $\bar{\boldsymbol{\nabla}}_\mu$ is the covariant derivative with respect to the metric $\bar{g}_{\mu\nu}$. When we expand around a flat background, $\bar{\boldsymbol{\nabla}}_\mu$ is equal to ∂_μ, and the gauge transformation takes the form (18.23).

18.2.2 Bardeen variables

We now consider the helicity decomposition in FRW. It is convenient to use coordinates $x^\mu = (\eta, \mathbf{x})$, where η is conformal time and $\mathbf{x}$ are the comoving spatial coordinates, and write

$$
ds^2 = a^2(\eta)(\eta_{\mu\nu} + h_{\mu\nu}) dx^\mu dx^\nu \,. \tag{18.79}
$$

We then write again $h_{\mu\nu}$ as in eqs. (18.2)–(18.4). Just as we have done in flat space (see Note 4 on page 415), for all spatial vectors and tensors that appear in the helicity decomposition, such as β_i, ϵ_i and h_{ij}^{TT}, as well as for the spatial derivative ∂_i acting on them, we use the convention that their indices are raised and lowered with the Kronecker delta δ_{ij}. Therefore, by definition,

$$
\beta_i \equiv \beta^i \,, \qquad \partial_i \gamma \equiv \partial^i \gamma \,, \tag{18.80}
$$

etc. The same convention will be used for the quantities that enter the helicity decomposition of the energy–momentum tensor. Furthermore,

$$\boldsymbol{\nabla}^2 \equiv \delta^{ij}\partial_i\partial_j \qquad (18.81)$$

is still defined as the flat-space Laplacian, and therefore the operator

$$[\partial_i\partial_j - (1/3)\delta_{ij}\boldsymbol{\nabla}^2] \qquad (18.82)$$

that appears in eq. (18.4) still gives zero when contracted with δ^{ij}.[8] Since the flat FRW background is still invariant under spatial rotations, and the time-dependent factor $a(\eta)$ does not affect the transformation properties of the various quantities under spatial rotations, it is still true that the four functions ψ, ϕ, γ and λ have helicity $h = 0$, the transverse vector β^i and ϵ^i each carry helicities $h = \pm 1$, and $h_{ij}^{\rm TT}$ has $h = \pm 2$.[9]

From eq. (18.79) we have $g_{\mu\nu} = \bar{g}_{\mu\nu} + \delta g_{\mu\nu}$ with $\bar{g}_{\mu\nu} = a^2\eta_{\mu\nu}$ and $\delta g_{\mu\nu} = a^2 h_{\mu\nu}$, so with this definition of $h_{\mu\nu}$ the linearized gauge transformation (18.78) reads

$$a^2 h_{\mu\nu} \to a^2 h_{\mu\nu} - (\bar{\boldsymbol{\nabla}}_\mu \xi_\nu + \bar{\boldsymbol{\nabla}}_\nu \xi_\mu)\,, \qquad (18.83)$$

where $\bar{\boldsymbol{\nabla}}_\mu$ is the covariant derivative with respect to the background metric. It is now convenient to write ξ^μ in the form

$$\xi^0 = -A\,, \qquad (18.84)$$
$$\xi^i = B^i + \partial^i C\,, \qquad (18.85)$$

where again $B_i \equiv B^i$ and $\partial^i C \equiv \partial_i C$. In flat space this definition is equivalent to eqs. (18.24) and (18.25). However, in FRW with the background metric $\bar{g}_{\mu\nu} = a^2\eta_{\mu\nu}$, we have

$$\xi_\mu = g_{\mu\nu}\xi^\nu = a^2(A, B_i + \partial_i C)\,. \qquad (18.86)$$

Writing $h_{\mu\nu}$ as in eqs. (18.2)–(18.4) and using the Christoffel symbols given in eq. (17.124), eq. (18.83) gives

$$\psi \to \psi - (A' + \mathcal{H}A)\,, \qquad (18.87)$$
$$\phi \to \phi + \left(\frac{1}{3}\boldsymbol{\nabla}^2 C - \mathcal{H}A\right)\,, \qquad (18.88)$$
$$\gamma \to \gamma - (A + C')\,, \qquad (18.89)$$
$$\lambda \to \lambda - 2C\,, \qquad (18.90)$$
$$\beta_i \to \beta_i - B_i'\,, \qquad (18.91)$$
$$\epsilon_i \to \epsilon_i - 2B_i\,, \qquad (18.92)$$

while $h_{ij}^{\rm TT}$ remains gauge-invariant. In the limit $a = 1$ eqs. (18.87)–(18.92) reduce to eqs. (18.26)–(18.29), as they should. We can now generalize eqs. (18.36) and (18.37), defining the Bardeen variables (also called Bardeen potentials)

$$\Phi = -\phi - \frac{1}{6}\boldsymbol{\nabla}^2\lambda + \mathcal{H}\left(\gamma - \frac{1}{2}\frac{d\lambda}{d\eta}\right)\,, \qquad (18.93)$$

$$\Psi = -\psi + \frac{1}{a}\frac{d}{d\eta}\left[a\left(\gamma - \frac{1}{2}\frac{d\lambda}{d\eta}\right)\right]\,. \qquad (18.94)$$

[8] This convention is very useful, but can occasionally lead to confusion. Consider for instance a four-vector V^μ. Its helicity decomposition can be written as $V^0 = \alpha$, $V^i = \beta^i + \partial^i\gamma$, where α and γ are scalars under rotations and β^i is a transverse vector. With this convention, we then *define* $\beta_i \equiv \beta^i$, $\partial_i\gamma \equiv \partial^i\gamma$. However, V^μ is a four-vector, so $V_\mu = g_{\mu\nu}V^\nu$. Thus, $V_i = g_{i\nu}V^\nu = g_{i0}\alpha + g_{ij}(\beta^j + \partial^j\gamma)$, which in a general metric is of course not the same as $\beta_i + \partial_i\gamma$. So, from $V^i = \beta^i + \partial^i\gamma \equiv \beta_i + \partial_i\gamma$ it does not follow that $V_i = \beta_i + \partial_i\gamma$. The spatial indices can be freely lowered or raised only after having performed the helicity decomposition, i.e. in quantities such as β^i or $\partial^i\gamma$, but not directly in V^i.

[9] There is actually a subtle point concerning the boundary conditions at infinity. As we saw in eq. (18.13), to ensure the uniqueness of the decomposition we must impose that $\gamma, \lambda, \boldsymbol{\nabla}^2\lambda$ and ϵ_i vanish sufficiently fast at spatial infinity. In flat space this is quite natural, since we always deal with localized sources, and we then require that the perturbations go to zero sufficiently far from the sources. In FRW, however, the background density $\bar{\rho}(t)$ is spatially uniform, and therefore there is no notion of "sufficiently far away from the sources". Still, in order to have a mathematically well-defined helicity decomposition, we must assume that in our frame the metric perturbations, and therefore the perturbations of the energy–momentum tensor, go to zero sufficiently fast at spatial infinity, i.e. at spatial distances parametrically larger than the horizon. This is somewhat artificial, since it make reference to a specific spatial origin, but since the conditions refers to distances infinitely larger than the horizon, it can be accepted. Conceptually, this procedure is on the same footing as the procedure of artificially switching off the interaction between particles at times $T \to \pm\infty$ in quantum field theory, in order to have mathematically well-defined in and out asymptotic states.

For $a = 1$ these quantities reduce to the flat-space expressions (18.36) and (18.37). Observing that, under gauge transformation, $\delta[\gamma-(1/2)\lambda'] = -A$, it is straightforward to check that Φ and Ψ are invariant under eqs. (18.87)–(18.90). Similarly, in the vector sector, eq. (18.39) can be generalized simply to

$$\Xi_i = \beta_i - \frac{1}{2}\frac{d\epsilon_i}{d\eta}\,. \tag{18.95}$$

We can now chose again the gauge $\lambda = \gamma = \beta_i = 0$, as in eq. (18.33). In this gauge $\Phi = -\phi$, $\Psi = -\psi$, and in the scalar sector the perturbed metric takes the form

$$ds^2 = a^2(\eta)\left[-(1+2\Psi)d\eta^2 + (1+2\Phi)d\mathbf{x}^2\right]\,. \tag{18.96}$$

This gauge is called the *conformal Newtonian gauge*. As in the flat-space case, it has the advantage that it is a complete gauge fixing, and that the perturbations are expressed in terms of the gauge-invariant Bardeen variables. Therefore the result of a computation in this gauge, when expressed in terms of Φ and Ψ, is valid in any other gauge.

If we also include the vector and tensor sectors, the metric in the conformal Newtonian gauge reads

$$ds^2 = -a^2(1+2\Psi)d\eta^2+a^2\left[(1 + 2\Phi)\delta_{ij} + \frac{1}{2}(\partial_i\epsilon_j + \partial_j\epsilon_i) + h_{ij}^{\rm TT}\right]dx^idx^j\,. \tag{18.97}$$

The inverse metric is immediately obtained by observing that, if $g_{\mu\nu} = a^2(\eta_{\mu\nu} + h_{\mu\nu})$ then, to first order,

$$g^{\mu\nu} = a^{-2}(\eta^{\mu\nu} - h^{\mu\nu})\,. \tag{18.98}$$

For later reference, to first-order in the perturbations the Christoffel symbols in the metric (18.97) are

$$\Gamma^0_{00} = \mathcal{H} + \Psi'\,, \tag{18.99}$$
$$\Gamma^0_{0i} = \partial_i\Psi\,, \tag{18.100}$$
$$\Gamma^0_{ij} = \delta_{ij}\left[\mathcal{H}(1+2\Phi-2\Psi) + \Phi'\right] + u'_{ij} + 2\mathcal{H}u_{ij}\,, \tag{18.101}$$
$$\Gamma^i_{00} = \partial_i\Psi\,, \tag{18.102}$$
$$\Gamma^i_{0j} = (\mathcal{H} + \Phi')\delta_{ij} + u'_{ij}\,, \tag{18.103}$$
$$\Gamma^i_{jk} = (\delta_{ik}\partial_j + \delta_{ij}\partial_k - \delta_{jk}\partial_i)\Phi + \partial_j u_{ik} + \partial_k u_{ij} - \partial_i u_{jk}\,, \tag{18.104}$$

where we have used the shorthand notation

$$u_{ij} = \frac{1}{4}\left(\partial_i\epsilon_j + \partial_j\epsilon_i + 2h_{ij}^{\rm TT}\right)\,. \tag{18.105}$$

These expressions correctly reduce to eq. (17.123) in the unperturbed case.

18.3   Perturbed energy–momentum tensor

A similar treatment can be performed for the energy–momentum tensor. In flat space the most general form of its helicity decomposition has been given in eqs. (18.42)–(18.44). However, in curved space one must be careful that the quantity that is directly related to the energy density ρ is not T_{00} but rather T^0_0, through $T^0_0 = -\rho$, and hence $\delta T^0_0 = -\delta\rho$. In flat space T^0_0 is simply related to T_{00} through a minus sign, so $T_{00} = \rho$ and $\delta T_{00} = \delta\rho$. However, in a generic curved space we rather have $T_{\mu\nu} = g_{\mu\rho}T^\rho_\nu$, so the perturbations of T_{00} involve not only $\delta\rho$ but also the metric perturbations.

At the fundamental field-theoretical level the fact that the basic quantity is T^0_0 is due to the fact that in curved space the relation between the Noether currents j^a_μ and the corresponding Noether charges Q^a (which, in the case of flat space, we discussed in Section 2.1.1) becomes

$$Q^a = \int_\Sigma d\Sigma^\nu \, j^a_\nu \, , \tag{18.106}$$

where Σ is a spacelike hypersurface and $d\Sigma^\nu = n^\nu d\Sigma$, where $d\Sigma$ is the volume element on the surface. The latter is in general given by $d\Sigma = (\det h)^{1/2} \, d^d y$, where h_{ij} is the induced metric on the surface and y^i are coordinates on the surface, with d equal to the dimensionality of the surface, so $d = 3$ in our case. In flat space we found in eqs. (2.12) and (2.13) that the relation between the four-momentum P^μ of a field configuration and the energy–momentum tensor $T_{\mu\nu}$ can be written as

$$E = -\int d^3x \, T^0_0 \, , \tag{18.107}$$

$$P^i = -\int d^3x \, T^i_0 \, , \tag{18.108}$$

since the tensor T^μ_ν actually corresponds to the four conserved currents j^a_ν associated with space-time translations, with $a = \mu$ labeling the current. Its generalization to curved space is therefore

$$P^\mu = -\int d\Sigma \, n^\nu T^\mu_\nu \, . \tag{18.109}$$

In FRW, choosing as spatial hypersurface the one labeled by the comoving coordinates $\mathbf{x}$, we have $n^\mu = (1, 0, 0, 0)$ and

$$P^\mu = -\int d\Sigma \, T^\mu_0 \, . \tag{18.110}$$

Thus, in particular,

$$E = -\int d\Sigma \, T^0_0 \, , \tag{18.111}$$

and the energy density is given by $\rho = -T^0_0$. For this reason, it is conceptually simpler to work with the tensor T^μ_ν with one upper and one lower index.

18.3.1 General decomposition of T^μ_ν

Using T^μ_ν as our basic object, in FRW the decomposition analogous to (18.42)–(18.44) is[10]

$$T^0_0 = -\rho\,, \tag{18.112}$$
$$T^i_0 = S^i + \partial^i S\,, \tag{18.113}$$
$$T^i_j = p\delta^i_j + \Sigma^i_j\,, \tag{18.114}$$

where the anisotropic stress tensor Σ^i_j contains the terms involving σ, σ_i and $\sigma^{\rm TT}_{ij}$ given in eq. (18.44). The components T^0_i are then obtained through $T^0_i = g^{0\mu}g_{i\nu}T^\nu_\mu$. In particular, for the metric in the conformal Newtonian gauge (18.97) we have $g_{0i} = 0$, so $T^0_i = g^{00}g_{ij}T^j_0$. Since in this gauge T^j_0 is already a first-order quantity, to first order we can replace g^{00} and g_{ij} by their background values $\bar{g}^{00} = -1/a^2$ and $\bar{g}_{ij} = a^2\delta_{ij}$, so in this gauge

$$T^0_i = -(S^i + \partial^i S)\,. \tag{18.115}$$

For the helicity decomposition of the energy–momentum tensor, as with any helicity decomposition, we use the convention (18.80) that in the quantities S^i, Σ^{ij}, σ^i and $\sigma^{\rm TT}_{ij}$, as well as in the partial derivative ∂_i acting on these objects, the indices i and j are raised and lowered with δ_{ij}. So, by definition,

$$S^i = S_i\,, \qquad \partial^i S = \partial_i S\,, \qquad \Sigma^{ij} = \Sigma^i_j = \Sigma_{ij}\,, \tag{18.116}$$

and similarly for σ^i and $\sigma^{{\rm TT},ij}$.[11] The decomposition of Σ_{ij} into its scalar, vector and tensor parts can be written as in eq. (18.44),

$$\Sigma_{ij} = \Sigma^s_{ij} + \Sigma^v_{ij} + \Sigma^t_{ij}\,, \tag{18.120}$$

where

$$\Sigma^s_{ij} = \left(\partial_i\partial_j - \frac{1}{3}\delta_{ij}\boldsymbol{\nabla}^2\right)\sigma\,, \tag{18.121}$$
$$\Sigma^v_{ij} = \frac{1}{2}(\partial_i\sigma_j + \partial_j\sigma_i)\,, \tag{18.122}$$
$$\Sigma^t_{ij} = \sigma^{\rm TT}_{ij}\,. \tag{18.123}$$

The vector σ^i is still defined so as to satisfy $\partial_i\sigma^i = 0$, and $\sigma^{\rm TT}_{ij}$ is by definition transverse and traceless again with respect to the metric δ^{ij}, i.e. $\partial_i\sigma^{\rm TT}_{ij} = 0$ and $\delta^{ij}\sigma^{\rm TT}_{ij} = 0$.

At the background level FRW cosmology is obtained assuming a perfect fluid form for the energy–momentum tensor, in which only ρ and p are non-vanishing. Therefore, up to first order in the perturbation, the most general form of the energy–momentum tensor (for a single fluid) is,

$$T^0_0 = -(\bar{\rho} + \delta\rho)\,, \tag{18.124}$$
$$T^i_0 = S^i + \partial^i S\,, \tag{18.125}$$
$$T^i_j = (\bar{p} + \delta p)\delta^i_j + \Sigma^i_j\,, \tag{18.126}$$

[10]Observe that, in contrast to what we have done for a perfect fluid in eq. (17.42), we are now *defining* ρ from eq. (18.112), which is valid in any frame, so that ρ is now the component of a tensor, rather than a diff-invariant quantity; compare with Note 10 on page 374. The relation to the corresponding definition for perfect fluids is worked out in detail in Section 18.3.2, where we will see that the difference only shows up at second order in cosmological perturbation theory.

[11]Similarly to what we have already discussed in Note 8, one should, however, be aware that, with the definitions $S^i = S_i$, $\partial^i = \partial_i$ and $\Sigma^i_j = \Sigma_{ij}$, even if $T^i_j = p\delta^i_j + \Sigma^i_j$, in general $T_{ij} \neq p\delta_{ij} + \Sigma_{ij}$. In fact, for $T_{\mu\nu}$ the indices are still raised and lowered with the metric $g_{\mu\nu}$, so the correct expression is

$$T_{ij} = g_{i\mu}T^\mu_j$$
$$= g_{0i}T^0_j + g_{ik}T^k_j\,. \tag{18.117}$$

For the metric in the conformal Newtonian gauge, (18.97), we have $g_{0i} = 0$, even including the perturbations. However, $g_{ij} = a^2(\delta_{ij} + h_{ij})$, with

$$h_{ij} = 2\Phi\delta_{ij} + \frac{1}{2}(\partial_i\epsilon_j + \partial_j\epsilon_i) + h^{\rm TT}_{ij}\,. \tag{18.118}$$

Then, to first order,

$$T_{ij} = a^2[p(\delta_{ij} + h_{ij}) + \Sigma_{ij}]\,, \tag{18.119}$$

which differs from $p\delta_{ij} + \Sigma_{ij}$ both in the term $p\,h_{ij}$ and in the overall factor a^2. In practice the simplest way to avoid these complications is to use only T^μ_ν when developing cosmological perturbation theory, as we will do in the following, and never use $T_{\mu\nu}$.

where the overbar denotes the background values and the quantities $\{\delta\rho, \delta p, S, S^i, \Sigma^i_j\}$ are first-order in the perturbation. The background values $\bar{\rho}$ and $\bar{p}$ are related by the EoS

$$\bar{p} = w\bar{\rho}, \tag{18.127}$$

where w can in principle be a function of time. However for radiation, matter and cosmological constant it simply has a constant value $w = 1/3, 0$ and -1, respectively. However, a dynamical dark energy will in general have a time-dependent w. For the perturbations, one introduces the density contrast

$$\delta \equiv \frac{\delta\rho}{\rho} \tag{18.128}$$

(which we already introduced in a Newtonian context in Chapter 17.7) and the speed of sound squared,

$$c_s^2 \equiv \frac{\delta p}{\delta\rho}. \tag{18.129}$$

For a generic fluid, the pressure p is a function of ρ and of internal properties of the fluid, such as its entropy density s, so $p = p(\rho, s)$. Then

$$c_s^2 \equiv \left(\frac{\partial p}{\partial\rho}\right)_s + \left(\frac{\partial p}{\partial s}\right)_\rho \frac{\partial s}{\partial\rho}. \tag{18.130}$$

The first term defines the *adiabatic speed of sound*,

$$c_{s,(a)}^2 \equiv \left(\frac{\delta p}{\delta\rho}\right)_s, \tag{18.131}$$

while the second term is the non-adiabatic speed of sound. So, in the most general case, the pressure perturbation δp and the density perturbation $\delta\rho$ are given by two independent functions. However, to a good approximation both the relativistic fluid that dominates in RD and the CDM fluid that dominates in MD can be taken as barotropic fluids, which are defined by the fact that $p = p(\rho)$ (even at the level of perturbations). In this case the non-adiabatic sound speed vanishes, and

$$\delta p = \frac{dp}{d\rho}\delta\rho. \tag{18.132}$$

Since $\delta\rho$ is already of first order, in this expression we can replace $dp/d\rho$ by the background expression $d\bar{p}/d\bar{\rho}$. In FRW $\bar{p}$ and $\bar{\rho}$ depend only on time, and therefore $d\bar{p}/d\bar{\rho} = \dot{\bar{p}}/\dot{\bar{\rho}}$. Then, using eq. (18.127), we get

$$c_s^2 = w + \frac{\bar{\rho}\dot{w}}{\dot{\bar{\rho}}}. \tag{18.133}$$

So, for a barotropic fluid in FRW, c_s^2 is not an independent quantity, and pressure perturbations are entirely specified by density perturbations, together with the background values $\bar{\rho}(t)$ and $\bar{p}(t)$. If furthermore w is constant, we have the very simple relation

$$c_s^2 = w. \tag{18.134}$$

However, this holds only if the fluid is barotropic and w is constant. We have seen in Section 17.81 that in a fluid with multiple components the total EoS parameter depends on time, even if the EoS parameters of the separate components do not.

18.3.2   Perturbations of perfect fluids

Equations (18.124)–(18.126) give the most general form of a perturbed energy–momentum tensor. A simpler expression can, however, be obtained for a perfect fluid. In this case, the energy–momentum tensor has been given in eq. (17.42), which we rewrite here as

$$T^\mu_\nu = (\rho_{\rm pf} + p_{\rm pf})\, u^\mu u_\nu + p_{\rm pf}\, \delta^\mu_\nu \,. \tag{18.135}$$

We have added the subscript "pf" (for "perfect fluid") to ρ and p to stress that these are the standard definitions used in the case of perfect fluids, while we will reserve ρ and p without subscripts for the definitions given for a generic fluid in eqs. (18.112)–(18.114). When applied to a perfect fluid, the definition of ρ in eq. (18.112) does not reduce to the quantity $\rho_{\rm pf}$ defined by eq. (18.135). Indeed, comparing eq. (18.135) with eq. (18.112) we see that

$$\rho = -\rho_{\rm pf} u^0 u_0 - p_{\rm pf}(u^0 u_0 + 1)\,. \tag{18.136}$$

As we have already pointed out in Note 10 on page 374 in Chapter 17, and in Note 10 on page 426, $\rho_{\rm pf}$ and $p_{\rm pf}$ are scalars under diffeomorphisms, while ρ and p are only scalars under spatial rotations. Under diffeomorphisms, $\rho = -T^0_0$ and $p = (1/3)\delta^i_j T^j_i$ transform as dictated by the transformation of T^μ_ν. In a given frame, the two definitions agree only at space-time points where $u^0 u_0 = -1$.

Let us compute the form of the perturbations for a perfect fluid. If the perfect fluid form is maintained also at the perturbation level, then to first order

$$\delta T^\mu_\nu = (\delta\rho_{\rm pf} + \delta p_{\rm pf})\, \bar{u}^\mu \bar{u}_\nu + (\bar\rho_{\rm pf} + \bar p_{\rm pf})(\delta u^\mu \bar{u}_\nu + \bar{u}^\mu \delta u_\nu) + \delta p_{\rm pf}\, \delta^\mu_\nu \,. \tag{18.137}$$

To compute δu^μ we recall from Section 1.3.1 that, for a particle moving on a trajectory $x_0(\tau)$,

$$u^\mu \equiv \frac{dx_0^\mu}{d\tau}\,, \tag{18.138}$$

where $d\tau$ is the proper time of the particle, defined by

$$d\tau^2 = -g_{\mu\nu} dx_0^\mu dx_0^\nu \,. \tag{18.139}$$

By definition, u^μ satisfies $g_{\mu\nu} u^\mu u^\nu = -1$. We use coordinates $(\eta, \mathbf{x})$, so the metric, including the scalar perturbations, is given by eq. (18.96). Then

$$d\tau^2 = a^2 d\eta^2 \left[(1 + 2\Psi) - (1 + 2\Phi)(v_0^i)^2 \right] \,, \tag{18.140}$$

where

$$v_0^i = \frac{dx_0^i}{d\eta} \tag{18.141}$$

is the peculiar velocity of the particle; compare with eq. (17.185). Consider first the spatial component $u^i = dx_0^i/d\tau$. We write

$$\frac{dx_0^i}{d\tau} = \frac{dx_0^i}{d\eta} \frac{d\eta}{d\tau}$$

$$= v_0^i \frac{d\eta}{d\tau}. \tag{18.142}$$

At the background level there is no peculiar velocity, so v_0^i is already a first-order quantity in the perturbation. Then, in $d\eta/d\tau$, we can replace $d\tau$ by its lowest-order expression, $d\tau = a\,d\eta$ and, to first order in the perturbation,

$$u^i = \frac{1}{a} v_0^i. \tag{18.143}$$

The component u_0 can instead be determined from $g_{\mu\nu} u^\mu u^\nu = -1$. To first order, using eq. (18.97), this gives[12]

$$u^0 = \frac{1}{a}(1 - \Psi). \tag{18.144}$$

These results carry over from the single-particle motion to the fluid motion, as discussed in Section 17.7.2, so to first order in the perturbations

$$u^\mu = \frac{1}{a}\left(1 - \Psi, v^i\right), \tag{18.145}$$

where $u^\mu = u^\mu(\eta, \mathbf{x})$ is the four-velocity field of the fluid and $v^i = v^i(\eta, \mathbf{x})$ is its peculiar velocity field. So, the background value is $\bar{u}^\mu = a^{-1}(1, 0)$ and the first-order perturbation is $\delta u^\mu = a^{-1}(-\Psi, v^i)$. From this we also get

$$u_\mu = g_{\mu\nu} u^\nu$$

$$= a[-(1 + \Psi), v^i], \tag{18.146}$$

so $\bar{u}_\mu = a(-1, 0)$ and $\delta u_\mu = a(-\Psi, v^i)$. For v^i we use the same convention as for S^i, eq. (18.116), defining $v_i = v^i$.

To go back to the variables ρ and p we now observe that, from eqs. (18.145) and (18.146), $u^0 u_0 = -1 + O(\Psi^2)$. Therefore, to first order in perturbation theory, we can set $u^0 u_0 = -1$ and we see from eq. (18.136) that ρ becomes the same as $\rho_{\rm pf}$, and similarly for p. The difference only shows up at second order in cosmological perturbation theory. Thus, to first order we replace $\rho_{\rm pf}$ and $p_{\rm pf}$ with ρ and p and, plugging eqs. (18.145) and (18.146) into eq. (18.137), we get

$$\delta T^0_{\ 0} = -\delta\rho, \tag{18.147}$$

$$\delta T^i_{\ 0} = -(\rho + p)v^i, \tag{18.148}$$

$$\delta T^0_{\ i} = (\rho + p)v_i, \tag{18.149}$$

$$\delta T^i_{\ j} = c_s^2 \delta\rho\, \delta^i_j. \tag{18.150}$$

[12]Observe that the inclusion of vector and tensor perturbations in the metric does not affect this result, since they do not contribute to g_{00}.

The peculiar velocity field v^i can be further decomposed into a transverse and a longitudinal part,

$$v^i(\eta, \mathbf{x}) = v^{\mathrm{T},i}(\eta, \mathbf{x}) + \partial^i v(\eta, \mathbf{x}),\qquad(18.151)$$

where $\partial_i v^{\mathrm{T},i} = 0$ and v is called the *velocity potential*. It is also useful to define the *velocity divergence*

$$\theta = \partial_i v^i = \boldsymbol{\nabla}^2 v.\qquad(18.152)$$

Comparing with eqs. (18.124)–(18.126) we find that, for a perfect fluid, the pressure perturbations are related to the density perturbations by $\delta p = c_s^2 \delta\rho$ and the anisotropic stress tensor vanishes,

$$\Sigma_{ij} = 0.\qquad(18.153)$$

We also see that the quantities S^i and S are related to the velocity field by

$$S^i = -(\rho + p) v^{\mathrm{T},i},\qquad(18.154)$$
$$S = -(\rho + p) v.\qquad(18.155)$$

Actually, eqs. (18.154) and (18.155) are not specific to perfect fluids, since already for a perfect fluid T_0^i is given by the most general vector field, so they rather relate, in general, the quantities S^i and S used in the general parametrization (18.124)–(18.126) to the velocity fields v^i and v, which have a more immediate physical meaning. We will therefore use interchangeably the pair (S, S^i) or the pair $(v, v^{\mathrm{T},i})$ also for the most general fluid, with the two set of variables being related by eqs. (18.154) and (18.155).

Observe that, for perfect fluids, the helicity-2 part of the anisotropic stress tensor, $\Sigma_{ij}^t \equiv \sigma_{ij}^{\mathrm{TT}}$ vanishes. In eq. (18.65) we found that, in flat space, this quantity is the only source for the tensor part of the metric perturbations. The same is true in FRW; we will see this explicitly in Section 19.5, where we will write the perturbed Einstein equations in the tensor sector, but this is already clear from the fact that, at the linearized level, only a helicity-2 tensor can source the helicity-2 quantity h_{ij}^{TT}. Thus, we can already conclude that a perfect fluid cannot be a source for GW production. All phenomena of GW production in cosmology come from departures from the perfect fluid assumption.

Finally, we give the perturbations of $T_{\mu\nu}$ with both lower indices. To linear order we have

$$\begin{aligned}T_{\mu\nu} &= (\bar{g}_{\mu\rho} + \delta g_{\mu\rho})(\bar{T}_\nu^\rho + \delta T_\nu^\rho)\\&= \bar{g}_{\mu\rho}\bar{T}_\nu^\rho + \left(\bar{g}_{\mu\rho}\delta T_\nu^\rho + \delta g_{\mu\rho}\bar{T}_\nu^\rho\right),\end{aligned}\qquad(18.156)$$

where we have split T_ν^μ into a background term plus the perturbation,

$$T_\nu^\mu = \bar{T}_\nu^\mu + \delta T_\nu^\mu.\qquad(18.157)$$

Then $\bar{T}_{\mu\nu} = \bar{g}_{\mu\rho}\bar{T}^{\rho}_{\nu}$ and

$$\delta T_{\mu\nu} = \bar{g}_{\mu\rho}\delta T^{\rho}_{\nu} + \delta g_{\mu\rho}\bar{T}^{\rho}_{\nu}\,. \tag{18.158}$$

We see that $\delta T_{\mu\nu}$ involves both the matter perturbations δT^{μ}_{ν} and the metric perturbations $\delta g_{\mu\nu}$, as we anticipated at the beginning of this section. For a perfect fluid in the metric (18.97), the background values of $T_{\mu\nu}$ are given by

$$\bar{T}_{00} = a^2\bar{\rho}\,, \tag{18.159}$$
$$\bar{T}_{0i} = 0\,, \tag{18.160}$$
$$\bar{T}_{ij} = a^2\bar{p}\,\delta_{ij}\,, \tag{18.161}$$

while for the perturbations we get

$$\delta T_{00} = a^2(\delta\rho + 2\bar{\rho}\Psi)\,, \tag{18.162}$$
$$\delta T_{0i} = -a^2(\bar{\rho} + \bar{p})v_i\,, \tag{18.163}$$
$$\delta T_{ij} = a^2 c_s^2 \delta\rho\,\delta_{ij} + a^2\bar{p}\left[2\Phi\delta_{ij} + \frac{1}{2}(\partial_i\epsilon_j + \partial_j\epsilon_i) + h_{ij}^{\mathrm{TT}}\right]\,. \tag{18.164}$$

We see that δT_{00} depends not only on the matter perturbation $\delta\rho$ but also on the scale factor $a(\eta)$ of the background and on the metric perturbation Ψ. Similarly, δT_{ij} depends on $a(\eta)$, Φ, ϵ_i and h_{ij}^{TT}. Observe also that, while T^{μ}_{ν} is given in eqs. (18.124)–(18.126) without making specific reference to a gauge, the expressions for $\delta T_{\mu\nu}$ given in eqs. (18.162)–(18.164) are valid only in the conformal Newtonian gauge, since to compute eq. (18.158) we made use of eq. (18.97).

Comparing with eqs. (18.147)–(18.150), we see that in cosmological perturbation theory it is in general more convenient to work with T^{μ}_{ν} than with $T_{\mu\nu}$.

18.3.3   Linearized energy–momentum conservation

We now compute the consequences of energy–momentum conservation,

$$\nabla_{\mu}T^{\mu}_{\nu} = 0\,, \tag{18.165}$$

for the perturbed energy–momentum tensor of a general fluid (without restricting to the perfect fluid approximation). We write T^{μ}_{ν} in the form (18.124)–(18.126) and we use eq. (18.97) for the metric. The Christoffel symbols are given in eqs. (18.99)–(18.104). The equation $\nabla_{\mu}T^{\mu}_{0} = 0$ has a zeroth-order term, involving only the background,

$$\boxed{\bar{\rho}' + 3\mathcal{H}(\bar{\rho} + \bar{p}) = 0\,.} \tag{18.166}$$

Of course this is just the usual conservation equation (17.47), except that it is now written using conformal time, and with the background quantities denoted by overbars. Collecting the terms of first order in the perturbation we get

$$(\delta\rho)' + 3\mathcal{H}(\delta\rho + \delta p) + 3(\bar{\rho} + \bar{p})\Phi' - \nabla^2 S = 0\,. \tag{18.167}$$

In flat space the background values are $\bar{\rho} = \bar{p} = 0$, so $\delta\rho$ is the same as ρ, and conformal time is the same as coordinate time t, since $a = 1$. Then eq. (18.167) correctly reduces to eq. (18.46). Using eqs. (18.155) and (18.152) we see that

$$\nabla^2 S = -(\bar{\rho} + \bar{p})\theta \,, \tag{18.168}$$

and therefore eq. (18.167) can be rewritten as[13]

$$(\delta\rho)' + 3\mathcal{H}(\delta\rho + \delta p) = -(\bar{\rho} + \bar{p})(\theta + 3\Phi') \,. \tag{18.169}$$

Finally, we divide by $\bar{\rho}$ and use

$$\left(\frac{\delta\rho}{\bar{\rho}}\right)' = \frac{(\delta\rho)'}{\bar{\rho}} + 3\mathcal{H}(1 + w)\frac{\delta\rho}{\bar{\rho}} \,, \tag{18.170}$$

where $\bar{\rho}'$ has been traded for $\bar{\rho}$ and $\bar{p}$ using eq. (18.166) and we have also used $\bar{p} = w\bar{\rho}$. Then, in terms of $\delta = \delta\rho/\rho$ (which, to linear order, is the same as $\delta\rho/\bar{\rho}$), eq. (18.169) reads

$$\boxed{\delta' + 3\mathcal{H}(c_s^2 - w)\delta = -(1 + w)(\theta + 3\Phi') \,.} \tag{18.171}$$

For non-relativistic matter, $c_s^2 = w = 0$, and this simplifies to

$$\delta'_M \simeq -(\theta + 3\Phi') \,. \tag{18.172}$$

This expresses the fact that the matter density contrast δ_M can increase at a point $\mathbf{x}$ either because there is a velocity flow toward $\mathbf{x}$, or because of a time-varying gravitational potential Φ. Comparing with eq. (17.264) we see that the equation $\delta'_M = -\theta$ is just the linearized Newtonian equation. However, the Newtonian approximation misses the term Φ'. This is not surprising, since we saw in eq. (5.10) that the Newtonian limit for a massive particle is sensitive only to h_{00}, and therefore misses the potential Φ that enters in h_{ij}. We will, however, see from the linearized Einstein equations, that well inside the horizon Φ' is negligible compared with θ, so in this limit the Newtonian approximation is indeed correct.

We next consider eq. (18.165) with $\nu = i$. At the background level, as we have already mentioned, it just gives $\partial_i \bar{p} = 0$, which is automatically satisfied in FRW. At first order, we get

$$\left(S'_i + 4\mathcal{H}S_i - \frac{1}{2}\nabla^2\sigma_i\right) - \partial_i\left[\delta p + \frac{2}{3}\nabla^2\sigma + (\bar{\rho} + \bar{p})\Psi - (S' + 4\mathcal{H}S)\right] = 0. \tag{18.173}$$

We see that the vector equation $\nabla_\mu T^\mu_i = 0$ splits into a longitudinal and a transverse part. The uniqueness of the helicity decomposition (which is ensured by imposing the appropriate boundary conditions at infinity; see the discussion in Note 9 on page 423), allows us to set both parts separately to zero, so we get a scalar equation,

$$S' + 4\mathcal{H}S = \delta p + \frac{2}{3}\nabla^2\sigma + (\bar{\rho} + \bar{p})\Psi \,, \tag{18.174}$$

[13]Observe, however, that the flat-space limit is clearer using S rather than θ. Since in flat space $\bar{\rho} = \bar{p} = 0$, the limit of a perturbation around Minkowski space with a small but non-zero value of S (and therefore a non-vanishing perturbation T^0_i) corresponds to $\theta \to \infty$ in such a way that $(\bar{\rho} + \bar{p})\theta$ stays small (since it must be a perturbation) but non-zero.

and an equation for the transverse vectors,

$$S_i' + 4\mathcal{H}S_i = \frac{1}{2}\boldsymbol{\nabla}^2\sigma_i\,. \tag{18.175}$$

In the flat-space limit we recover eqs. (18.49) and (18.50) (recalling that in flat space $\bar{p} = 0$, so δp is the same as the total pressure p). Taking the Laplacian of eq. (18.174) (recall that $\boldsymbol{\nabla}^2 \equiv \partial_i\partial^i \equiv \partial_i\partial_i$ is the flat-space Laplacian) and using eqs. (18.168) and (18.166) we get

$$\boxed{\theta' + \left[\mathcal{H}(1 - 3w) + \frac{w'}{1+w}\right]\theta = -\boldsymbol{\nabla}^2\left[\Psi + \frac{\delta p}{\bar{\rho}(1+w)} + \frac{2}{3\bar{\rho}(1+w)}\boldsymbol{\nabla}^2\sigma\right].}$$

$$\tag{18.176}$$

Observe that $\boldsymbol{\nabla}^2\delta p = \boldsymbol{\nabla}^2 p$, since $p = \bar{p} + \delta p$ and in FRW $\bar{p}$ depends only on time. For non-relativistic matter, $c_s^2 = w = 0$, and eq. (18.176) becomes

$$\theta' + \mathcal{H}\theta = -\boldsymbol{\nabla}^2\left[\Psi + \frac{1}{\bar{\rho}}\left(p + \frac{2}{3}\boldsymbol{\nabla}^2\sigma\right)\right]\,. \tag{18.177}$$

We can compare this with the divergence of the linearized Euler equation (17.265). Recalling that the tensor σ_{ij} that appears in eq. (17.265) was defined in eq. (17.230) as $\sigma_{ij} = p\delta_{ij} + \Sigma_{ij}$, we see that

$$\partial_i\partial_j\sigma_{ij} = \boldsymbol{\nabla}^2 p + \partial_i\partial_j\Sigma_{ij}\,. \tag{18.178}$$

From eqs. (18.120)–(18.123), $\partial_i\partial_i\Sigma_{ij} = (2/3)\boldsymbol{\nabla}^2(\boldsymbol{\nabla}^2\sigma)$. Thus, we find that eq. (18.177) is just the divergence of the linearized Euler equation that we obtained directly in Newtonian cosmology.

Similarly, eq. (18.175) can be written in terms of v_i^{T} as

$$\frac{\partial}{\partial\eta}[(\bar{\rho} + \bar{p})v_i^{\mathrm{T}}] + 4\mathcal{H}[(\bar{\rho} + \bar{p})v_i^{\mathrm{T}}] = -\frac{1}{2}\boldsymbol{\nabla}^2\sigma_i\,, \tag{18.179}$$

or, developing the derivative in the first term,

$$\boxed{\frac{\partial v_i^{\mathrm{T}}}{\partial\eta} + \left[\mathcal{H}(1 - 3w) + \frac{w'}{1+w}\right]v_i^{\mathrm{T}} = -\frac{1}{2\bar{\rho}(1+w)}\boldsymbol{\nabla}^2\sigma_i\,.} \tag{18.180}$$

For non-relativistic matter $w = 0$ and we get

$$\frac{\partial v_i^{\mathrm{T}}}{\partial\eta} + \mathcal{H}v_i^{\mathrm{T}} = -\frac{1}{2\bar{\rho}}\boldsymbol{\nabla}^2\sigma_i\,, \tag{18.181}$$

which is the transverse part of the linearized Euler equation (17.265) obtained from Newtonian cosmology. Of course, within the Newtonian approach we can only obtain the equation for non-relativistic matter, $w = 0$. The equations with generic w given by eqs. (18.176) and (18.180) are out of reach of the Newtonian approach.

Finally, we observe that for a perfect fluid the terms $\boldsymbol{\nabla}^2\sigma$ and $\boldsymbol{\nabla}^2\sigma_i$, which come from the anisotropic stress tensor, vanish.

18.3.4   Gauge-invariant combinations

Just as with metric perturbations, also with the perturbation of the energy–momentum tensor we can construct quantities that are invariant under linearized diffeomorphisms. The procedure is quite similar to that discussed for the metric perturbations. In order not to mix matter perturbations with metric perturbations it is, however, convenient to work with T^μ_ν. The transformation of T^μ_ν under diffeomorphism is given in eq. (18.70) and, from the discussion in Section 18.2.1, it follows that under a gauge transformation of the linearized theory

$$T^\mu_\nu(x) \to (T^\mu_\nu)'(x)$$
$$= T^\mu_\nu(x) - \left(\bar{T}^\mu_\rho \partial_\nu \xi^\rho - \bar{T}^\rho_\nu \partial_\rho \xi^\mu + \xi^\rho \partial_\rho \bar{T}^\mu_\nu\right). \quad (18.182)$$

Observe that, for linearized diffeomorphisms, $\partial_\mu \xi_\nu$ is by definition of first order in the perturbation, and of the same order as $h_{\mu\nu}$. Thus, in the second line of eq. (18.182) we could replace the occurrences of $T_{\mu\nu}$ multiplied by derivatives of ξ^μ by their background values, since terms such as $\delta T^\mu_\rho \partial_\nu \xi^\rho$ are of second order. We write T^μ_ν as in eqs. (18.124)–(18.126), but with S^i and S expressed in terms of $v^{\mathrm{T},i}$ and v, using eqs. (18.154) and (18.155). For ξ^μ we use eqs. (18.84) and (18.85). The background quantities $\bar\rho$ and $\bar p$ are gauge-invariant by definition since, just as for the case of the gauge transformation of the metric discussed in Section 18.2.1, by definition the infinitesimal gauge transformation is entirely reabsorbed into a gauge transformation of the perturbations. Then we get

$$\delta\rho \to \delta\rho + A\bar\rho' = \delta\rho - 3\mathcal{H}\bar\rho(1+w)A, \quad (18.183)$$
$$\delta p \to \delta p + A\bar p' = \delta p - 3\mathcal{H}\bar\rho(1+w)c^2_{s,(a)}A, \quad (18.184)$$
$$v \to v + C', \quad (18.185)$$
$$v^{\mathrm{T},i} \to v^{\mathrm{T},i} + B'^i, \quad (18.186)$$
$$\Sigma_{ij} \to \Sigma_{ij}. \quad (18.187)$$

where $c^2_{s,(a)} = \bar p'/\bar\rho'$ is the adiabatic speed of sound [compare with eqs. (18.129)–(18.133)], and in eqs. (18.183) and (18.184) we have made use of eq. (18.166).

We see that Σ_{ij} is gauge-invariant, and therefore also its scalar, vector and tensor components σ, σ^i and $\sigma^{\mathrm{TT}}_{ij}$, defined in eqs. (18.121)–(18.123), are gauge-invariant. In contrast, $\delta\rho$, δp and v are not gauge-invariant. With $\delta\rho$ and δp we can form the combination

$$\delta p - c^2_{s,(a)}\delta\rho, \quad (18.188)$$

which is gauge-invariant. This quantity vanishes for adiabatic perturbations, since in this case $c^2_{s,(a)} = c^2_s$, and by definition $\delta p = c^2_s\delta\rho$.

Further gauge-invariant combinations can, however, be formed combining the transformation properties of the energy–momentum perturbations with the transformation properties of the metric perturbations

(18.87)–(18.87).   In particular, in the scalar sector we can form the gauge-invariant combinations

$$(\delta\rho)_* \equiv \delta\rho - 3\mathcal{H}(\bar{\rho}+\bar{p})(v+\gamma)\,, \qquad (18.189)$$

$$(\delta p)_* \equiv \delta p - 3\mathcal{H}(\bar{\rho}+\bar{p})c_{s,(a)}^2(v+\gamma) \qquad (18.190)$$

and

$$v_* \equiv v + \frac{1}{2}\lambda'\,. \qquad (18.191)$$

Equivalently, we can define the gauge-invariant density contrast

$$\delta_* = \delta - 3\mathcal{H}(1+w)(v+\gamma) \qquad (18.192)$$

and a gauge-invariant sound speed

$$c_{s,*}^2 = \frac{(\delta p)_*}{(\delta\rho)_*}\,. \qquad (18.193)$$

Of course, one can take different linear combinations of these gauge-invariant perturbations. For instance, instead of δ_* defined in eq. (18.192), one could use

$$\delta_g \equiv \delta - 3\mathcal{H}(1+w)\left(\gamma - \frac{1}{2}\lambda'\right) \qquad (18.194)$$

which, because of eq. (18.191), differs from δ_* by a term that depends only on the gauge-invariant quantity v_*. However, we will see in Chapter 19 that the gauge-invariant density contrast δ_* is the source for the gauge-invariant Bardeen potential Φ, and therefore provides a gauge-invariant generalization of Poisson law. From eq. (18.177) we see that, in order of magnitude, for non-relativistic matter $\mathcal{H}\tilde{\theta}(\mathbf{k})/k^2 \sim \tilde{\Psi}(\mathbf{k})$. As we will see in the next chapter, for modes well inside the horizon $\tilde{\Psi} \ll \tilde{\delta}(\mathbf{k})$, so $\tilde{\delta}_*(\mathbf{k}) \simeq \tilde{\delta}(\mathbf{k})$. Therefore, well inside the horizon the density contrast δ is approximately gauge-invariant. This is no longer true at scales comparable to the horizon, where the gauge-invariant quantity (again, at linear order) is δ_*.

Finally, in the vector sector we can define a gauge-invariant combination

$$v_*^{\mathrm{T},i} \equiv v^{\mathrm{T},i} + \beta^i\,, \qquad (18.195)$$

while Σ_{ij}^v (and therefore σ^i) is already gauge-invariant. In the tensor sector Σ_{ij}^t is gauge-invariant.

In the conformal Newtonian gauge we have $\lambda = \gamma = \beta_i = 0$ and the expressions for the gauge-invariant combinations simplify. Performing the Fourier transform with respect to the comoving coordinates, as in eq. (17.39), eq. (18.152) becomes $\tilde{\theta}(\mathbf{k}) = -k^2\tilde{v}(\mathbf{k})$, where $k = |\mathbf{k}|$. Then, in the conformal Newtonian gauge, we have

$$\tilde{\delta}_*(\mathbf{k}) = \tilde{\delta}(\mathbf{k}) + \frac{3\mathcal{H}(1+w)\tilde{\theta}(\mathbf{k})}{k^2}\,, \qquad (18.196)$$

while $v_*^{\mathrm{T},i} \equiv v^{\mathrm{T},i}$ and $v_* = v$. So, in particular, the gauge-invariant velocity field is the same as the velocity field in the conformal Newtonian gauge.

Further reading

- For FRW the helicity decomposition is discussed e.g. in the textbooks by Amendola and Tsujikawa (2010) and by Gorbunov and Rubakov (2011) and in Chapter 14 of Liddle and Lyth (2000). The helicity decomposition in flat space is discussed in Flanagan and Hughes (2005) and Jaccard, Maggiore and Mitsou (2013), which we have followed in our presentation. See also Section 5.5 of Poisson and Will (2014).

- The gauge-invariant perturbations of the metric and of the energy–momentum tensor were introduced in Bardeen (1980). Cosmological perturbation theory in both the synchronous and the conformal Newtonian gauges was developed in Ma and Bertschinger (1995). Gauge-invariant cosmological perturbation theory is reviewed in Kodama and Sasaki (1984, 1987), Mukhanov, Feldman and Brandenberger (1992) and Durrer (2008).

Evolution of cosmological perturbations

19

We now study the evolution of cosmological perturbations during radiation dominance (RD), matter dominance (MD) and the recent dark energy (DE) dominated epoch. We will first discuss scalar perturbations and then move to tensor perturbations. We will then study how tensor perturbations can be used as probes of dark energy and modified gravity.

19.1 Evolution equations in the scalar sector

Perturbing the Einstein equations (17.41) we get

$$\delta G^\mu_\nu = 8\pi G \, \delta T^\mu_\nu \,. \tag{19.1}$$

In this section we study the scalar sector. We use the conformal Newtonian gauge (18.96). The Christoffel symbols in this gauge have already been given in eqs. (18.99)–(18.104). Restricting to the scalar sector, we set $u_{ij} = 0$ in these equations, and for the perturbed Einstein tensor we get

$$\delta G^0_0 = 2a^{-2} \left[\boldsymbol{\nabla}^2 \Phi - 3\mathcal{H}(\Phi' - \mathcal{H}\Psi) \right] , \tag{19.2}$$

$$\delta G^i_0 = -2a^{-2} \partial^i (\Phi' - \mathcal{H}\Psi) , \tag{19.3}$$

$$\delta G^i_j = 2a^{-2} \left[(\mathcal{H}^2 + 2\mathcal{H}')\Psi + \mathcal{H}\Psi' - \Phi'' - 2\mathcal{H}\Phi' + \frac{1}{3}\boldsymbol{\nabla}^2(\Phi + \Psi) \right] \delta^i_j$$
$$- a^{-2} \left(\partial^i \partial_j - \frac{1}{3}\delta^i_j \boldsymbol{\nabla}^2 \right) (\Phi + \Psi) \,. \tag{19.4}$$

We recall that we are using the convention that all spatial indices of the quantities that appears in the helicity decomposition, as well as the index of the spatial derivative ∂_i acting on them, are raised and lowered with the Kronecker delta δ_{ij}; see eq. (18.80). Therefore, on the right-hand sides of eqs. (19.3) and (19.4) we can freely lower the spatial indices. Recall also that $\boldsymbol{\nabla}^2 = \partial^i \partial_i = \delta^{ij} \partial_i \partial_j$ is the flat-space Laplacian.[1]

The most general perturbed energy–momentum tensor was given in eqs. (18.124)–(18.126). Restricting to the scalar sector, we have

$$\delta T^0_0 = -\delta\rho \,, \tag{19.5}$$

$$\delta T^i_0 = -(\bar\rho + \bar p)\partial^i v \,, \tag{19.6}$$

[1]Of course, these formulas are specific to the fact that we are limiting ourselves to a spatially flat FRW metric. For open or closed FRW space-times δ_{ij} is replaced by the spatial metric γ_{ij}, the spatial derivative ∂_i becomes the covariant derivative with respect to this spatial metric, and similarly for the Laplacian.

Gravitational Waves, Volume 2: Astrophysics and Cosmology. Michele Maggiore.
© Michele Maggiore 2018. Published in 2018 by Oxford University Press.
DOI 10.1093/oso/9780198570899.001.0001

$$\delta T^i_j = \delta p\, \delta^i_j + \left(\partial^i\partial_j - \frac{1}{3}\delta^i_j \boldsymbol{\nabla}^2\right)\sigma\,, \qquad (19.7)$$

where we have used eq. (18.122) for the scalar part of the anisotropic stress tensor, and we have written $S = -(\rho + p)v$; see eq. (18.155). Since v is already a first-order perturbation, this is equivalent to $S = -(\bar\rho + \bar p)v$ and, since the background values $\bar\rho$ and $\bar p$ are independent of $\mathbf{x}$, $\partial_i[(\bar\rho + \bar p)v] = (\bar\rho + \bar p)\partial_i v$.

Using these expressions, the $(0,0)$ component of eq. (19.1) gives

$$\boxed{\boldsymbol{\nabla}^2\Phi - 3\mathcal{H}(\Phi' - \mathcal{H}\Psi) = -4\pi G a^2 \delta\rho\,.} \qquad (19.8)$$

We recognize here a generalization to FRW of the flat-space Poisson equation (18.62).[2] Note also the presence of a factor a^2 on the right-hand side of eq. (19.8), which can be reabsorbed (in this equations, as well as in all other equations that follow) using physical coordinates $\mathbf{r} = a\mathbf{x}$ and cosmic time t, in which case eq. (19.8) becomes

$$\boldsymbol{\nabla}^2_{\mathbf{r}}\Phi - 3H(\dot\Phi - H\Psi) = -4\pi G \delta\rho\,. \qquad (19.9)$$

The $(i,0)$ equation gives[3]

$$\boxed{\Phi' - \mathcal{H}\Psi = 4\pi G a^2(\bar\rho + \bar p)v\,.} \qquad (19.12)$$

Inserting $(\Phi' - \mathcal{H}\Psi)$ from eq. (19.12) into eq. (19.8) we get

$$\boldsymbol{\nabla}^2\Phi = -4\pi G a^2\left[\delta\rho - 3\mathcal{H}(\bar\rho + \bar p)v\right]\,. \qquad (19.13)$$

Recalling that we are working in the conformal Newtonian gauge, where $\gamma = 0$, we see that the term in brackets on the right-hand side is just the expression for the gauge-invariant density perturbation $(\delta\rho)_*$, defined in eq. (18.189), in this gauge. Thus, eq. (19.13) can be rewritten as

$$\boxed{\boldsymbol{\nabla}^2\Phi = -4\pi G a^2 \bar\rho\, \delta_*\,,} \qquad (19.14)$$

where $\delta_* = (\delta\rho)_*/\bar\rho$ is the gauge-invariant density contrast. This result is quite natural, since it tells us that the gauge-invariant potential Φ is sourced by a gauge-invariant generalization of the density perturbation. Even though we have found this result working in the conformal Newtonian gauge, eq. (19.14) is clearly valid in any gauge, since it is a relation between gauge-invariant quantities.

We next consider the (i,j) equation. Defining $V^i_j \equiv G^i_j - 8\pi G T^i_j$, we see from eqs. (19.4) and (19.7) that V^i_j has the generic structure

$$V^i_j = A\delta^i_j + \left(\partial^i\partial_j - \frac{1}{3}\delta^i_j\boldsymbol{\nabla}^2\right)B\,. \qquad (19.15)$$

Taking the trace of the equation $V^i_j = 0$ kills the terms proportional to $[\partial_i\partial_j - (1/3)\delta_{ij}\boldsymbol{\nabla}^2]$, and we get $3A = 0$, while applying the traceless

[2]Observe that, in flat space, the background value $\bar\rho$ vanishes, so ρ in eq. (18.62) was the same as the first-order perturbation $\delta\rho$.

[3]Note that, defining $V^i \equiv G^i_0 - 8\pi G T^i_0$, when we include also the vector sector V^i separates as usual as $V^i = V^{\mathrm{T},i} + \partial^i V$, with $\partial_i V^{\mathrm{T},i} = 0$. To extract the equation for the scalar sector one can then take the divergence of the equation $V^{\mathrm{T},i} + \partial^i V = 0$, obtaining $\boldsymbol{\nabla}^2 V = 0$. In this sense, one might think that the correct $(i,0)$ equation in the scalar sector is rather

$$\boldsymbol{\nabla}^2(\Phi' - \mathcal{H}\Psi) = 4\pi G a^2(\bar\rho + \bar p)\boldsymbol{\nabla}^2 v$$
$$= 4\pi G a^2(\bar\rho + \bar p)\theta\,. \qquad (19.10)$$

However, the use of the helicity decomposition is only meaningful if we impose boundary conditions such that the decomposition is invertible; see eq. (18.13) and the discussion in Note 9 on page 423. This also implies that the Laplacian is invertible and the boundary conditions are such that the only solution to an equation of the form $\boldsymbol{\nabla}^2 V = 0$ is $V = 0$. Therefore eqs. (19.12) and (19.10) are completely equivalent. Alternatively, one can perform the spatial Fourier transform, and rewrite eq. (19.10) in the form

$$\tilde\Phi'(\mathbf{k}) - \mathcal{H}\tilde\Psi(\mathbf{k}) = -4\pi G a^2(\bar\rho + \bar p)\tilde\theta(\mathbf{k})/k^2. \qquad (19.11)$$

Observe that this way of solving the equation in momentum space is completely equivalent to setting to zero the solution of the homogeneous equation $\boldsymbol{\nabla}^2(\Phi' - \mathcal{H}\Psi) = 0$ when one works in coordinate space.

operator $[\partial^i \partial_j - (1/3)\delta^i_j \nabla^2]$ kills the term δ^i_j in V^i_j, and we remain with $\nabla^2 \nabla^2 B = 0$. Requiring again the appropriate boundary conditions at infinity on B and on $\nabla^2 B$ (see Note 3), this is actually equivalent to $B = 0$. In other words, the invertibility of the helicity decomposition ensures that, in the scalar sector, the equation $V^i_j = 0$ separates into the two equations $A = 0$ and $B = 0$. Then, we get

$$\boxed{\Phi + \Psi = -8\pi G a^2 \sigma} \qquad (19.16)$$

and

$$\Phi'' - \frac{1}{3}\nabla^2(\Phi + \Psi) + 2\mathcal{H}\Phi' - \mathcal{H}\Psi' - (\mathcal{H}^2 + 2\mathcal{H}')\Psi = -4\pi G a^2 \delta p, \quad (19.17)$$

which correctly reduce to eqs. (18.56) and (18.66) in the flat-space limit. Using eq. (19.16) to write $\nabla^2(\Phi + \Psi)$ in terms of $\nabla^2 \sigma$, we can also rewrite eq. (19.17) as

$$\boxed{\Phi'' + 2\mathcal{H}\Phi' - \mathcal{H}\Psi' - (\mathcal{H}^2 + 2\mathcal{H}')\Psi = -4\pi G a^2 \left(\delta p + \frac{2}{3}\nabla^2 \sigma\right),}$$

$$(19.18)$$

Since the Einstein equation (17.41) implies the energy–momentum conservation $\nabla_\mu T^\mu_\nu = 0$, the linearized Einstein equations imply the linearized energy–momentum conservation equations. In the scalar sector these are given by eqs. (18.171) and (18.176).

As we will see, different combinations of the four linearized Einstein equations and of the two linearized energy–momentum conservation equations can be used to study the evolution in the scalar sector. In any case, there are only four independent equations [for example the four Einstein equations (19.8), (19.12), (19.16) and (19.18)], for the two metric variables Φ, Ψ and the four fluid variables $\delta\rho, \delta p, v, \sigma$, so we have overall six variables. To close the system, we need further relations between the fluid variables. To proceed further, it is then useful to distinguish between the case when only one type of fluid contributes to the energy–momentum tensor and the case of a fluid with several different components.

19.1.1   Single-component fluid

We consider first a single-component fluid. In this case one relation is given by the equation of state (EoS), which expresses the properties of the particular fluid medium that we are considering. The simplest possibility is to restrict to a perfect fluid. In this case, at the background level, p and ρ are related by $\bar{p} = w\bar{\rho}$ and, at the level of perturbations, $\delta p = c_s^2 \delta\rho$, where c_s^2 is the adiabatic speed of sound; see eq. (18.131). This allows us to eliminate δp from the set of independent fluid variables.

The second relation corresponds to the truncation of the BBGKY hierarchy, as discussed on page 410. For a perfect fluid, by definition the BBGKY hierarchy is truncated at the level of the anisotropic stress tensor, which in the scalar sector amounts to setting $\sigma = 0$. Then, for a perfect fluid eq. (19.16) gives

$$\Psi = -\Phi\,. \tag{19.19}$$

Inserting this in the other equations, performing the Fourier transform with respect to the spatial variables, and using $\delta = \delta\rho/\rho$ and $\theta = \boldsymbol{\nabla}^2 v$ in place of $\delta\rho$ and v, we get

$$k^2\tilde{\Phi} + 3\mathcal{H}(\tilde{\Phi}' + \mathcal{H}\tilde{\Phi}) = 4\pi Ga^2\bar{\rho}\,\tilde{\delta}\,, \tag{19.20}$$

$$\tilde{\Phi}' + \mathcal{H}\tilde{\Phi} = -4\pi Ga^2\bar{\rho}(1+w)\tilde{\theta}/k^2\,, \tag{19.21}$$

$$\tilde{\Phi}'' + 3\mathcal{H}\tilde{\Phi}' + (\mathcal{H}^2 + 2\mathcal{H}')\tilde{\Phi} = -4\pi Ga^2 c_s^2\bar{\rho}\,\tilde{\delta}\,. \tag{19.22}$$

where $\tilde{\Phi} = \tilde{\Phi}(\eta,\mathbf{k})$, $\tilde{\delta} = \tilde{\delta}(\eta,\mathbf{k})$, etc. Thus, we remain with three equations for the three independent variables $\tilde{\Phi}$, $\tilde{\delta}$ and $\tilde{\theta}$. In terms of these variables, the linearized energy–momentum conservation equations (18.171) and (18.176) become, in momentum space,

$$\tilde{\delta}' + 3\mathcal{H}(c_s^2 - w)\tilde{\delta} = -(1+w)(\tilde{\theta} + 3\tilde{\Phi}') \tag{19.23}$$

and

$$\tilde{\theta}' + \left[\mathcal{H}(1-3w) + \frac{w'}{1+w}\right]\tilde{\theta} = k^2\left(\frac{c_s^2}{1+w}\,\tilde{\delta} - \tilde{\Phi}\right)\,. \tag{19.24}$$

Even if only three of the five equations (19.20)–(19.24) are independent, it is useful to write down all of them, and to take different independent combinations depending on the situation. To solve these equations, a useful strategy is to multiply eq. (19.20) by c_s^2 and add it to eq. (19.22). In this way the source terms cancel, and we get a homogeneous equation involving only Φ,

$$\tilde{\Phi}'' + 3\mathcal{H}(1+c_s^2)\tilde{\Phi}' + \left[\mathcal{H}^2(1+3c_s^2) + 2\mathcal{H}' + c_s^2 k^2\right]\tilde{\Phi} = 0\,. \tag{19.25}$$

Using eq. (17.127) we can rewrite this as

$$\boxed{\tilde{\Phi}'' + 3\mathcal{H}(1+c_s^2)\tilde{\Phi}' + \left[3\mathcal{H}^2(c_s^2 - w) + c_s^2 k^2\right]\tilde{\Phi} = 0\,,} \tag{19.26}$$

where w is the EoS parameter in this single-fluid model. Observe that, if $w = c_s^2$ (as is the case for a perfect fluid), the term $\mathcal{H}^2(c_s^2 - w)$ vanishes.

Equation (19.25) or (19.26) is called the "master equation" for Φ. To solve it we must assign the initial values $\tilde{\Phi}(\eta_{\rm in},\mathbf{k})$ and $\tilde{\Phi}'(\eta_{\rm in},\mathbf{k})$ for the mode $\mathbf{k}$ of interest. Note that, at the linearized level, different momentum modes evolve independently. Plugging the resulting solution for $\tilde{\Phi}(\eta,\mathbf{k})$ into eq. (19.20) we get the time evolution of $\tilde{\delta}(\eta,\mathbf{k})$, while eq. (19.21) determines $\tilde{\theta}(\eta,\mathbf{k})$.

19.1.2 Multi-component fluid

We next consider the linearized equations for a set of n fluids, labeled by an index $\lambda = 1, \ldots, n$. The linearized Einstein equations are still given by eqs. (19.8), (19.12), (19.16) and (19.18) except that on the right-hand sides there now appears the perturbation of the total energy–momentum tensor. So, writing the equations directly in momentum space, we have

$$k^2\tilde{\Phi} + 3\mathcal{H}(\tilde{\Phi}' - \mathcal{H}\tilde{\Psi}) = 4\pi G a^2 \sum_\lambda \widetilde{\delta\rho}_\lambda , \qquad (19.27)$$

$$\tilde{\Phi}' - \mathcal{H}\tilde{\Psi} = 4\pi G a^2 \sum_\lambda (\bar{\rho}_\lambda + \bar{p}_\lambda)\tilde{v}_\lambda , \qquad (19.28)$$

$$\tilde{\Phi} + \tilde{\Psi} = -8\pi G a^2 \sum_\lambda \tilde{\sigma}_\lambda , \qquad (19.29)$$

$$\tilde{\Phi}'' + 2\mathcal{H}\tilde{\Phi}' - \mathcal{H}\tilde{\Psi}' - (\mathcal{H}^2 + 2\mathcal{H}')\tilde{\Psi} = -4\pi G a^2 \sum_\lambda \left(\widetilde{\delta p}_\lambda - \frac{2}{3}k^2\tilde{\sigma}_\lambda \right) . \qquad (19.30)$$

We assume that the energy–momentum tensors of the various fluids are separately conserved; see the discussion in Section 17.3. In this case, the linearization of the n conservation equations gives, in the scalar sector, n copies of eqs. (18.167) and (18.174),

$$(\widetilde{\delta\rho})'_\lambda + 3\mathcal{H}(\widetilde{\delta\rho}_\lambda + \widetilde{\delta p}_\lambda) + (\bar{\rho}_\lambda + \bar{p}_\lambda)(3\tilde{\Phi}' - k^2 v_\lambda) = 0 , \qquad (19.31)$$

$$[(\bar{\rho}_\lambda + \bar{p}_\lambda)\tilde{v}_\lambda]' + 4\mathcal{H}(\bar{\rho}_\lambda + \bar{p}_\lambda)\tilde{v}_\lambda + \widetilde{\delta p}_\lambda - \frac{2}{3}k^2\tilde{\sigma}_\lambda + (\bar{\rho}_\lambda + \bar{p}_\lambda)\tilde{\Psi} = 0 . \qquad (19.32)$$

We now define

$$\delta_\lambda = \frac{\delta\rho_\lambda}{\rho_\lambda} , \qquad (19.33)$$

which to linear order is of course the same as $\delta_\lambda = \delta\rho_\lambda/\bar{\rho}_\lambda$, and the total density contrast

$$\delta_{\text{tot}} = \frac{\delta\rho_{\text{tot}}}{\rho_{\text{tot}}}$$
$$= \frac{\sum_\lambda \delta\rho_\lambda}{\sum_\lambda \rho_\lambda} , \qquad (19.34)$$

which of course to linear order is the same as

$$\delta_{\text{tot}} = \frac{1}{\bar{\rho}_{\text{tot}}} \sum_\lambda \bar{\rho}_\lambda \delta_\lambda . \qquad (19.35)$$

We also assume that all fluids are perfect fluids with $\bar{p}_\lambda = w_\lambda \bar{\rho}_\lambda$ and $(\delta p)_\lambda = c^2_{s,\lambda}(\delta\rho)_\lambda = c^2_{s,\lambda}\bar{\rho}_\lambda\delta_\lambda$, and $\sigma_\lambda = 0$. Then eq. (19.29) gives again $\Psi = -\Phi$. Substituting this in the other equations, and using as independent variables $\tilde{\Phi}$, $\tilde{\delta}_\lambda$ and $\tilde{\theta}_\lambda = -k^2\tilde{v}_\lambda$ we get

$$k^2\tilde{\Phi} + 3\mathcal{H}(\tilde{\Phi}' + \mathcal{H}\tilde{\Phi}) = 4\pi G a^2 \bar{\rho}_{\text{tot}}\tilde{\delta}_{\text{tot}} , \qquad (19.36)$$

$$\tilde{\Phi}' + \mathcal{H}\tilde{\Phi} = -\frac{4\pi G a^2}{k^2} \sum_\lambda \bar{\rho}_\lambda(1 + w_\lambda)\tilde{\theta}_\lambda , \qquad (19.37)$$

$$\tilde{\Phi}'' + 3\mathcal{H}\tilde{\Phi}' + (\mathcal{H}^2 + 2\mathcal{H}')\tilde{\Phi} = -4\pi G a^2 \sum_\lambda c^2_{s,\lambda}\bar{\rho}_\lambda\tilde{\delta}_\lambda , \qquad (19.38)$$

while eqs. (19.31) and (19.32) become

$$\tilde{\delta}_\lambda' + 3\mathcal{H}(c_{s,\lambda}^2 - w_\lambda)\tilde{\delta}_\lambda = -(1 + w_\lambda)(\tilde{\theta}_\lambda + 3\tilde{\Phi}'), \quad (19.39)$$

$$\tilde{\theta}_\lambda' + \left[\mathcal{H}(1 - 3w_\lambda) + \frac{w_\lambda'}{1 + w_\lambda}\right]\tilde{\theta}_\lambda = k^2 \left(\frac{c_{s,\lambda}^2}{1 + w_\lambda}\tilde{\delta}_\lambda - \tilde{\Phi}\right). \quad (19.40)$$

Equations (19.36) and (19.40), with $\lambda = 1, \ldots, n$, are $2n + 3$ equations for the $2n + 1$ variables $\tilde{\delta}_\lambda$, $\tilde{\theta}_\lambda$ and $\tilde{\Phi}$. Of course, they are not all independent, since the linearized Einstein equations imply the linearized conservation of the total energy–momentum tensor. Therefore, in the scalar sector, once eqs. (19.36)–(19.38) are satisfied, it is sufficient to satisfy eqs. (19.39) and (19.40) for $n - 1$ fluids, and the conservation laws for δ_λ and for θ_λ for the n-th fluid is then automatically satisfied. Thus, in the end we have $2n + 1$ independent equations for $2n + 1$ variables. In many situations, however, it can be convenient to take as independent equations the $2n$ conservation equations (19.39) and (19.40), plus eq. (19.36), so the system of equations is only of first order in the time derivative.

Note also that, in the multi-fluid case, in general we can no longer obtain a single master equation for $\tilde{\Phi}$. Indeed, if we combine eqs. (19.36) and (19.38) so to eliminate the source term, we again get eq. (19.25), but with c_s replaced by an "effective speed of sound" $c_{s,\text{eff}}$, given by

$$c_{s,\text{eff}}^2(\tilde{\delta}_1, \ldots, \tilde{\delta}_n) = \frac{\sum_\lambda c_{s,\lambda}^2 \bar{\rho}_\lambda \tilde{\delta}_\lambda}{\sum_\lambda \bar{\rho}_\lambda \tilde{\delta}_\lambda}. \quad (19.41)$$

Since $c_{s,\text{eff}}^2$ depends on the variables $\tilde{\delta}_\lambda$, in this case we do not have an equation involving $\tilde{\Phi}$ only.

In general, the single-fluid approximation is adequate when we study the perturbations in the dominant component, for example the perturbation of radiation during RD or the perturbation of cold dark matter (CDM) during MD. In contrast, the multi-fluid equations must be used when we study the perturbations in a subdominant component, such as the CDM perturbations during RD. Even in this case, however, one can simplify the problem by retaining, on the right-hand side of eq. (19.41), only the contribution of the dominant component. Observe that this requires not only that a species λ has $\bar{\rho}_\lambda$ much larger than the energy density of the other species, but also that its perturbations δ_λ and its speed of sound $c_{s,\lambda}$ are such that its value of $c_{s,\lambda}^2 \bar{\rho}_\lambda \delta_\lambda$ in the numerator of eq. (19.41), as well as its value of $\bar{\rho}_\lambda \delta_\lambda$ in the denominator, dominates over the respective contributions of the other species. If this is the case, then $c_{s,\text{eff}}^2(\tilde{\delta}_1, \ldots, \tilde{\delta}_n)$ becomes approximately the same as $c_{s,\lambda}^2$ for the dominant component, so the potential Φ is again determined by the master equation (19.25) in which c_s^2 is replaced by the value of $c_{s,\lambda}^2$ of the dominant component. We can then plug the value of Φ so determined into eqs. (19.39) and (19.40) for the subdominant components, so the problem simplifies considerably.

In the following we will apply these results to a multi-component fluid made of radiation, non-relativistic matter and dark energy. A more general question then arises, concerning under what conditions radiation can be treated as a fluid. Actually, the fluid approximation is valid as long as the photons are tightly coupled to the baryons, which is true until recombination. Therefore, the results that we will present are in principle valid only until recombination. However, recombination takes place during MD, when radiation is already subdominant. Thus, at least in a first approximation, the subsequent evolution of the matter perturbations is not affected by the treatment of radiation and, as far as matter perturbations are concerned, the results that we shall discuss are valid even after recombination. In contrast, the correct treatment of the radiation perturbations after recombination is different and will be discussed in Section 20.3, when we study CMB anisotropies.

19.1.3   Super-horizon and sub-horizon regimes

In the equations that we have been discussing two crucial scales appear. One is the comoving Hubble parameter $\mathcal{H}$. The other is the comoving momentum k (and, as we shall see, we further have the dimensionless numbers c_s^2 and w). Equivalently, we can think in terms of the physical Hubble parameter $H = \mathcal{H}/a(t)$ and the physical momentum $k_{\rm ph}(t) = k/a(t)$. Depending on the ratio between these two scales we can identify two limiting regimes, corresponding to super-horizon and sub-horizon modes.[4]

The super-horizon regime

This is given by the limit

$$k \ll \mathcal{H}(\eta)\,, \tag{19.42}$$

or, equivalently, $k_{\rm ph}(t) \ll H(t)$. In terms of the comoving wavelength $\lambda = 2\pi/k$, this means[5]

$$\frac{\lambda}{2\pi} \gg \mathcal{H}^{-1}(\eta)\,, \tag{19.43}$$

i.e. the reduced comoving wavelength $\lambdabar$ is much larger than the comoving horizon $\mathcal{H}^{-1}(\eta)$. For the physical wavelength $\lambda_{\rm ph} = a(t)\lambda$, in this regime

$$\frac{\lambda_{\rm ph}(t)}{2\pi} \gg H^{-1}(t)\,, \tag{19.44}$$

so the reduced physical wavelength is much larger than the physical horizon. In this regime, to lowest order we can formally set $k = 0$ in the equations. Then eq. (19.36) becomes

$$3\mathcal{H}(\tilde{\Phi}' + \mathcal{H}\tilde{\Phi}) \simeq 4\pi G a^2 \bar{\rho}_{\rm tot}\tilde{\delta}_{\rm tot}\,. \tag{19.45}$$

Writing $\tilde{\theta}_\lambda = -k^2\tilde{v}_\lambda$ the remaining independent equations (19.39) and (19.40) become

$$\tilde{\delta}_\lambda' + 3\mathcal{H}(c_{s,\lambda}^2 - w_\lambda)\tilde{\delta}_\lambda \simeq -3(1 + w_\lambda)\tilde{\Phi}'\,, \tag{19.46}$$

$$\tilde{v}_\lambda' + \left[\mathcal{H}(1 - 3w_\lambda) + \frac{w_\lambda'}{1 + w_\lambda}\right]\tilde{v}_\lambda \simeq -\frac{c_{s,\lambda}^2}{1 + w_\lambda}\tilde{\delta}_\lambda + \tilde{\Phi}\,. \tag{19.47}$$

[4]As we mentioned in Note 4 on page 370, we use the expression "horizon scale" as a synonym of "Hubble scale". This should not be confused with the particle horizon (or causal horizon), which gives the actual size of the causally connected regions. The particle horizon rather depends on the whole past cosmological history, and will be much larger than the local Hubble scale if there was an earlier inflationary phase. With the same (standard) abuse of language we will use the expressions "super-horizon" and "sub-horizon" to denote modes with reduced comoving wavelength larger or, respectively, smaller that $\mathcal{H}^{-1}(\eta)$. A more careful terminology would be to call them "super-Hubble" and "sub-Hubble" modes, respectively.

[5]Observe that in the cosmological equations there enter the parameters k and $\mathcal{H}$, without a further factor of 2π, so the relevant ratio is $\mathcal{H}/k$ or equivalently $\lambdabar/\mathcal{H}^{-1}$, where $\lambdabar = 1/k$ is the reduced wavelength, while the wavelength λ is defined by $\lambda = 2\pi/k = 2\pi\lambdabar$.

Thus, $\tilde{v}_\lambda$ disappears from the equation for $\tilde{\delta}_\lambda$. We can first solve the $n+1$ equations (19.45) and (19.46) for the $n+1$ variables $\tilde{\delta}_\lambda$ and Φ, and then plug the solution into the right-hand side of eq. (19.47) and solve for $\tilde{v}_\lambda$.

It is also useful to write eq. (19.37) explicitly in the super-horizon regime, in the form

$$4\pi G a^2 \sum_\lambda \bar{\rho}_\lambda (1+w_\lambda)\tilde{v}_\lambda = \Phi' + \mathcal{H}\tilde{\Phi}. \tag{19.48}$$

The sub-horizon regime

This is given by the opposite limit

$$k \gg \mathcal{H}(\eta), \tag{19.49}$$

so that the wavelength is now well inside the horizon, $\lambdabar \ll \mathcal{H}^{-1}(\eta)$ or $\lambdabar_{\rm ph}(t) \ll H^{-1}(t)$. In this regime we recover the limit of Newtonian cosmology discussed in Section 17.7. In particular, eq. (19.36) reduces to the Poisson equation,

$$k^2 \tilde{\Phi} \simeq 4\pi G a^2 \sum_\lambda \bar{\rho}_\lambda \tilde{\delta}_\lambda. \tag{19.50}$$

Observe, however, that k and $\mathcal{H}$ are not the only scales in the problem, due to the presence of the dimensionless numbers c_s^2 and w. In particular, we see from eq. (19.25) that another important scale is given by the combination $c_s k$. We will discuss its role later.

Typical scales of horizon entry

To understand the evolution of the physical wavelength of a mode with respect to the horizon scale, we consider first a model with a RD phase followed by a MD phase. Using eq. (17.152) we see that

$$\mathcal{H} = \frac{2(\eta + \eta_*)}{\eta(\eta + 2\eta_*)}, \tag{19.51}$$

where η_* is related to the value of conformal time at RD–MD equality by $\eta_{\rm eq} = (\sqrt{2}-1)\eta_*$; see eq. (17.154). Deep in RD we have $\eta \ll \eta_*$, and eq. (19.51) gives $\mathcal{H} \simeq 1/\eta$, while deep in MD $\eta \gg \eta_*$, and eq. (19.51) gives $\mathcal{H} \simeq 2/\eta$. Of course, these limiting behaviors also follow immediately from the fact that in RD $a(\eta) \propto \eta$ while in MD $a(\eta) \propto \eta^2$. Thus, $\mathcal{H}^{-1} \simeq \eta$ in RD and $\mathcal{H}^{-1} \simeq \eta/2$ in MD. Therefore, compared with the fixed comoving scale $\lambda = 2\pi/k$, it is clear that, as we go backward in time toward $\eta = 0$, eventually every modes is outside the horizon.

We illustrate this in Fig. 19.1, where we plot the comoving horizon size $\mathcal{H}^{-1}$ obtained from eq. (19.51). Comoving wavelengths are constant in time, so they are just horizontal lines on this plot. We show in particular the comoving (reduced) wavelength $\lambdabar_{\rm eq} = \lambda_{\rm eq}/(2\pi) = k_{\rm eq}^{-1}$ that enters

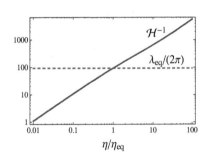

Fig. 19.1 The comoving horizon size $\mathcal{H}^{-1}$ in a model including only an RD+MD phase (solid line), compared with the constant comoving wavelength that enters the horizon at equilibrium. The vertical axis is in units of Mpc and we have used $\Omega_M = 0.3$, $h_0 = 0.7$ in eq. (17.163).

the horizon at the RD–MD transition. The value of this comoving momentum is obtained by requiring $k_{\rm eq} = \mathcal{H}(\eta_{\rm eq})$, which, using eq. (19.51), gives

$$k_{\rm eq} = \frac{2(2-\sqrt{2})}{\eta_{\rm eq}} \simeq \frac{1.172}{\eta_{\rm eq}}. \qquad (19.52)$$

Inserting the numerical value of $\eta_{\rm eq}$ from eq. (17.163) we get

$$k_{\rm eq} \simeq 1.53 \times 10^{-2} \frac{h_0}{\rm Mpc} \left(\frac{h_0}{0.7}\right)\left(\frac{\Omega_M}{0.3}\right), \qquad (19.53)$$

where we have fixed $h_0^2\Omega_R$ to its standard reference value. Observe that at cosmological scales the Hubble parameter fixes the reference length, so the uncertainty in the Hubble parameter affects all other quantities. For this reason, it is customary to express quantities with dimension of length in units of Mpc/h_0, rather than just Mpc, and momenta in units of h_0/Mpc, rather than just Mpc^{-1}. The corresponding wavelength is

$$\lambda_{\rm eq} = \frac{2\pi}{k_{\rm eq}}$$

$$\simeq 409 \frac{\rm Mpc}{h_0}\left(\frac{0.7}{h_0}\right)\left(\frac{0.3}{\Omega_M}\right), \qquad (19.54)$$

so for $h_0 = 0.7$ and $\Omega_M = 0.3$ we have $\lambda_{\rm eq} \simeq 584$ Mpc, and

$$\lambdabar_{\rm eq} \equiv \frac{\lambda_{\rm eq}}{(2\pi)} \simeq 93\,{\rm Mpc}. \qquad (19.55)$$

It is instructive to compare also the evolution of the reduced physical wavelengths $\lambdabar_{\rm ph}(t) = a(t)\lambdabar$ with the physical Hubble scale $H^{-1}(t)$. Furthermore, beside radiation and matter, we now include also the cosmological constant, which begins to dominate near the recent epoch. Both $\lambdabar_{\rm ph}$ and H^{-1} evolve with time, but with a different dependence of the scale factor. While $\lambdabar_{\rm ph}(t) \propto a$, we see from the Friedmann equation (17.75) that[6]

$$H^{-1} \simeq H_0^{-1}\Omega_R^{-1/2}a^2 \quad (\text{RD}) \qquad (19.56)$$

and

$$H^{-1} \simeq H_0^{-1}\Omega_M^{-1/2}a^{3/2} \quad (\text{MD}). \qquad (19.57)$$

In Fig. 19.2 the solid line is H^{-1}, obtained from eq. (17.75), as a function of $x = \log a$. Radiation–matter equilibrium is at $x \simeq -8.1$; the present time corresponds to $x = 0$, and we also plot for illustration part of the future evolution, $x > 0$, when Λ-domination (ΛD) sets in. H^{-1} changes from the behavior $H^{-1} \propto a^2$ at $x < -8.1$ to the slightly slower growth $\propto a^{3/2}$ when it enters MD, and eventually flattens and becomes constant during ΛD. The three dashed lines show the function $\lambdabar_{\rm ph}(a) = a\lambdabar$ for three different values of the reduced comoving wavelength $\lambdabar$.[7] For illustration, we choose (1) $\lambda = \lambda_{\rm eq}$, i.e. $\lambdabar = \lambda_{\rm eq}/(2\pi) \simeq 93$ Mpc (with the values $h_0 = 0.7$ and $\Omega_M = 0.3$), which is the mode that enters at RD–MD equilibrium; (2) a smaller wavelength $\lambdabar = 10$ Mpc, which is among

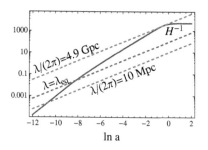

Fig. 19.2 The physical horizon size H^{-1} (solid line) compared with the physical reduced wavelengths $a(t)\lambda/(2\pi)$ for different values of the comoving reduced wavelength $\lambda/(2\pi)$ (dashed lines), plotted against $x = \log a$, in a model including RD, MD and a ΛD phase. The vertical axis is in Mpc. We use our standard reference values $\Omega_M = 0.3$, $\Omega_R \simeq 4.18 \times 10^{-5}h_0^{-2}$, $h_0 = 0.7$ and $\Omega_\Lambda = 1 - \Omega_M - \Omega_R$.

[6]We neglect here the effect due to the change in the number of effective relativistic species during RD. As we saw in eq. (17.117), this just introduces a prefactor in eq. (19.56) which ranges from one at the present epoch to a value which, in the Standard Model, saturates at $\simeq 0.6$ at $T > 100$ GeV.

[7]Observe that, since we normalize the scale factor so that $a(t_0) = 1$, the value of the comoving wavelength is the same as the value of the physical wavelength at the present time t_0.

the smallest wavelengths relevant for structure formation, and enters the horizon deep in RD; (3) a very long wavelength with $\lambdabar \simeq 4.94$ Gpc. As we see from the figure, wavelengths longer than this never enter the horizon.[8]

The value $a = a_*$ of the scale factor at which a reduced comoving wavelength $\lambdabar$ crosses the horizon is given by

$$\lambdabar a_* = H_0^{-1} \left(\Omega_R a_*^{-4} + \Omega_M a_*^{-3} + \Omega_\Lambda \right)^{-1/2}, \tag{19.58}$$

and the corresponding redshift z_* is given by $1 + z_* = 1/a_*$. For modes that enter the horizon deep in RD eq. (19.58) gives

$$1 + z_* \simeq \frac{1}{\lambdabar H_0 \Omega_R^{1/2}}$$

$$\simeq 4.63 \times 10^5 \left(\frac{1\,\text{Mpc}}{\lambdabar} \right) \left(\frac{4.184 \times 10^{-5}}{h_0^2 \Omega_R} \right)^{1/2}. \tag{19.59}$$

For modes entering during MD, in the intermediate epoch when we can neglect both Ω_R and Ω_Λ, eq. (19.58) gives instead

$$1 + z_* \simeq \frac{1}{(\lambdabar H_0)^2 \Omega_M}$$

$$\simeq 1.53 \times 10^3 \left(\frac{200\,\text{Mpc}}{\lambdabar} \right)^2 \left(\frac{0.7}{h_0} \right)^2 \left(\frac{0.3}{\Omega_M} \right). \tag{19.60}$$

Finally, if we keep both Ω_M and Ω_Λ, we get

$$\left(\frac{H_0^{-1}}{\lambdabar} \right)^2 = \frac{\Omega_M}{a_*} + \Omega_\Lambda a_*^2. \tag{19.61}$$

The right-hand side, as a function of a_*, goes to infinity for $a_* \to 0^+$ and for $a_* \to \infty$, and has a non-zero minimum value at $a_* = (\Omega_M/2\Omega_\Lambda)^{1/3}$. From this we find that the maximum value of $\lambdabar$ for which eq. (19.61) admits a solution is given by

$$\lambdabar_{\text{max}} = H_0^{-1} \frac{2^{1/3}}{\sqrt{3}} \Omega_M^{-1/3} \Omega_\Lambda^{-1/6}, \tag{19.62}$$

which for our reference values gives $\lambdabar_{\text{max}} \simeq 4.94$ Gpc.

To get a physical sense for the value of different comoving scales, consider the mass of non-relativistic matter contained today in a sphere of radius $\lambdabar/2$. Assuming that $\lambdabar$ is sufficiently large that the average density of the material inside it is given by the matter density of the Universe $\rho_{M,0}$ (which of course is no longer true if we go to sufficiently small scales, such as the scale of a Galaxy), the enclosed mass is

$$M = \frac{4\pi}{3} \rho_{M,0} \left(\frac{\lambdabar}{2} \right)^3. \tag{19.63}$$

Using $\rho_{M,0} = \Omega_M \rho_0$ and

$$\rho_0 = \frac{3H_0^2}{8\pi G} = 2.77537(13) \times 10^{11} h_0^2 \frac{M_\odot}{\text{Mpc}^3}, \tag{19.64}$$

we get

$$M \simeq 2.14 \times 10^{10} M_\odot \left(\frac{\lambdabar}{1\,\text{Mpc}}\right)^3 \left(\frac{h_0}{0.7}\right)^2 \left(\frac{\Omega_M}{0.3}\right). \qquad (19.65)$$

The scale $\lambdabar_{\text{eq}} \simeq 93$ Mpc therefore contains a mass $M \simeq 2 \times 10^{16} M_\odot$. This is larger than the mass of even the heaviest and rarest dark matter halos, which reach at most $M \sim (1-2) \times 10^{15} M_\odot$, while the mass of a typical galaxy like ours, including its DM halo, is of order $M \sim (10^{11}-10^{12})\, M_\odot$. So, all comoving reduced wavelengths that correspond to fluctuations that collapsed to form a single DM halo are smaller that $\lambdabar_{\text{eq}}$, and therefore entered the horizon in RD. For instance, the mode with $\lambdabar = 4$ Mpc, according to eq. (19.65), encloses the typical mass of a galaxy, $M \sim 10^{12}\, M_\odot$. As we see from eq. (19.59), it entered the horizon at a redshift $z_* \simeq 1 \times 10^5$. Structure formation is nevertheless sensitive also to longer wavelengths, that entered near the RD–MD transition or during MD, through correlations between density perturbations.[9] The CMB is also sensitive both to modes that enter the horizon during RD and to modes that enter in MD, as we will see in Chapter 20.

[9]The measure of correlations on such large scales is one of the target of future surveys such as EUCLID and the SKA.

19.2 Initial conditions

The next step is to set up the initial conditions for the perturbations. The initial conditions are given deep into RD, when all cosmologically relevant modes are well outside the horizon. To integrate eqs. (19.45) and (19.46) we need in principle to assign the initial values of $\tilde{\Phi}$, $\tilde{\theta}_\lambda$ and $\tilde{\delta}_\lambda$, for all the different fluids. However, due to the existence of decaying modes, the initial conditions on the physically relevant solutions for $\tilde{\Phi}$ and $\tilde{\theta}_\lambda$ are in fact fixed in terms of the initial conditions on $\tilde{\delta}_\lambda$, as we next show.

Deep into RD we can neglect CDM and baryons, and use the single-fluid equations. The evolution of Φ is then governed by the master equation (19.25). In RD the term $[\mathcal{H}^2(1 + 3c_s^2) + 2\mathcal{H}']$ vanishes, as we see using $\mathcal{H} = 1/\eta$ and $c_s^2 = 1/3$, and we are left with

$$\tilde{\Phi}'' + \frac{4}{\eta}\tilde{\Phi}' + \frac{k^2}{3}\tilde{\Phi} = 0. \qquad (19.66)$$

For modes well outside the horizon we have $k\eta \ll 1$. The derivatives of Φ are parametrically of order $1/\eta$, so we can neglect the term $k^2\Phi$, and we get a homogeneous equation

$$\tilde{\Phi}'' + \frac{4}{\eta}\tilde{\Phi}' \simeq 0, \qquad (19.67)$$

valid in RD, and for modes well outside the horizon. This equation has the constant solution

$$\tilde{\Phi} = c_1, \qquad (19.68)$$

and the solution $\tilde{\Phi} = c_2/\eta^3$, with c_1 and c_2 constants. The latter is a decaying mode, and diverges as we go back in time toward $\eta = 0$.

Therefore, if this mode is present at the epoch in which we fix the initial condition, and comparable to the constant mode (19.68), we must conclude that, going further backward in time, close to the big bang (or, more precisely, near the end of the inflationary period and at the very beginning of the RD era) this mode should have been extraordinarily large compared with the constant mode (19.68). In the absence of a physical mechanism that could justify this, it is much more natural to assume that, when the Universe emerged from an early phase of inflation, these modes were of comparable size; then, by the time we fix the initial conditions, the decaying mode has practically vanished. Therefore, we will systematically neglect the decaying modes. In this case, plugging the constant solution $\tilde{\Phi} = \tilde{\Phi}_{\rm in}$ into eq. (19.45), which holds for super-horizon modes, we get

$$3\mathcal{H}^2 \tilde{\Phi}_{\rm in} \simeq 4\pi G a^2 \bar{\rho}_{\rm tot} (\tilde{\delta}_{\rm tot})_{\rm in} \,. \tag{19.69}$$

Using the Friedmann equation (17.125), we find that

$$\boxed{(\tilde{\delta}_{\rm tot})_{\rm in} = 2\tilde{\Phi}_{\rm in} \,.} \tag{19.70}$$

Therefore, once we discard the decaying mode, the initial condition on $\tilde{\Phi}$ is fixed in terms of the initial conditions on $\tilde{\delta}_{\rm tot}$. In turn, deep in RD, $\tilde{\delta}_{\rm tot}$ is the same as the density contrast of radiation, $\tilde{\delta}_R$; see eq. (19.34).

Using eq. (19.46) we see that, since $c_{s,R}^2 = w_R = 1/3$, for super-horizon modes we have

$$\tilde{\delta}_R' = -4\tilde{\Phi}' \,, \tag{19.71}$$

so

$$\tilde{\delta}_R(\eta, \mathbf{k}) = -4\tilde{\Phi}(\eta, \mathbf{k}) + 6\tilde{\Phi}_{\rm in}(\mathbf{k}) \,, \tag{19.72}$$

where the integration constant $6\tilde{\Phi}_{\rm in}(\mathbf{k})$ is fixed by eq. (19.70), observing that deep in RD $\tilde{\delta}_{\rm tot} \simeq \tilde{\delta}_R$. Since $\tilde{\Phi}$ has a constant and a decaying mode, the same is true for $\tilde{\delta}_R$. So, neglecting the decaying mode, $\tilde{\delta}_R$ also is constant deep in RD, for modes outside the horizon, and is equal to $2\tilde{\Phi}_{\rm in}$.

The same analysis can be performed for $\tilde{v}_\lambda$. Using eq. (19.47) together with eq. (19.70) we find that, during RD and for super-horizon modes, the velocity potential for radiation satisfies

$$\tilde{v}_R' \simeq \frac{1}{2}\tilde{\Phi} \,. \tag{19.73}$$

Since $\tilde{\Phi}(\eta) = \tilde{\Phi}_{\rm in}$ is a constant, this equation can be integrated to give

$$\tilde{v}_R \simeq \frac{1}{2}\eta\tilde{\Phi}_{\rm in} + c_1 \,. \tag{19.74}$$

The velocity potential for matter satisfies instead (again during RD and for super-horizon modes)

$$\tilde{v}_M' + \frac{1}{\eta}\tilde{v}_M \simeq \tilde{\Phi}_{\rm in} \,, \tag{19.75}$$

which has the solution

$$\tilde{v}_M \simeq \frac{1}{2}\eta\tilde{\Phi}_{\rm in} + \frac{c_2}{\eta}\,. \tag{19.76}$$

Thus, for both $\tilde{v}_R$ and $\tilde{v}_M$ we have a solution growing linearly in η plus a subdominant term (the constant c_1 and the decaying mode c_2/η, respectively). By the same argument used for Φ, we retain only the growing term. Thus,

$$\boxed{\tilde{v}_{R,\rm in} = \tilde{v}_{M,\rm in} = \frac{1}{2}\eta_{\rm in}\,\tilde{\Phi}_{\rm in}\,.} \tag{19.77}$$

For super-horizon modes $k\eta \ll 1$, so $k\tilde{v}_{\lambda,\rm in} \ll \tilde{\Phi}_{\rm in}$. In conclusion, the initial conditions on Φ and of v_λ are fixed in terms of that on $\delta_{\rm tot}$, and in the end we only need to assign the initial conditions on the separate $\tilde{\delta}_\lambda$, as we now discuss.

19.2.1 Adiabatic and isocurvature perturbations

Consider first a cosmic fluid with two components, radiation (R) and matter (M). According to the definitions (19.33) and (19.34), the respective density contrasts are $\delta_R = \delta\rho_R/\rho_R$ and $\delta_M = \delta\rho_M/\rho_M$, while the total density contrast for this two-component fluid is

$$\delta_{\rm tot} = \frac{\rho_R\delta_R + \rho_M\delta_M}{\rho_R + \rho_M}\,, \tag{19.78}$$

and $\rho_R \gg \rho_M$ since we are deep in RD. The simplest option, to span the space of initial conditions, could be to use δ_R and δ_M as independent quantities. So, a generic perturbation could be decomposed in a radiation perturbation with $\{\delta_R \neq 0, \delta_M = 0\}$ and a matter perturbation, defined by $\{\delta_R = 0, \delta_M \neq 0\}$. However, there is a physically more convenient choice of basis. Namely, we fix the conformal Newtonian gauge, and we consider first perturbations that, in this gauge, are due only to temperature fluctuations. For radiation $\rho_R \propto T^4$ while for matter $\rho_M = mn_M$, where n_M is the number density of matter, and $n_M \propto T^3$. Therefore $\rho_M \propto T^3$, and we have $\delta_R = 4\delta T/T$ and $\delta_M = 3\delta T/T$. Thus, for such perturbations,

$$\delta_M = \frac{3}{4}\delta_R\,. \tag{19.79}$$

As a second independent component, we can use perturbations δ_R and δ_M such that the total density contrast vanishes, $\delta = 0$. To understand the physical meaning of this separation, consider the entropy per matter particle $\bar{s} \propto T^3/n_M$; compare with eq. (17.88). Since $\rho_M = mn_M$ we also have $\bar{s} \propto T^3/\rho_M$. Then

$$\frac{\delta\bar{s}}{\bar{s}} = 3\frac{\delta T}{T} - \frac{\delta\rho_M}{\rho_M}$$

$$= \frac{3}{4}\delta_R - \delta_M\,. \tag{19.80}$$

Thus, we see that temperature perturbations conserve the entropy, and for this reason are called *adiabatic perturbations*. In contrast, the quantity that drives the metric perturbation Φ is the total density contrast δ_{tot}, which appears on the right-hand side of eq. (19.45). Thus, perturbations with an initial value $\delta_{\text{tot}} = 0$ do not excite Φ; so, in this case there are no metric perturbations, and therefore no perturbations in the curvature either. Thus, perturbations with $\delta_{\text{tot}} = 0$ are called *isocurvature perturbations*. Observe that, since we are deep in RD, isocurvature perturbations are approximately the same as perturbations with $\delta_M \neq 0$ and $\delta_R = 0$.

Observe also that we have specified that the definition of adiabatic and isocurvature perturbations is performed in the conformal Newtonian gauge. In fact, as we discussed in Section 18.3.4, the density contrast is not gauge-invariant, and so, if $\delta_{\text{tot}} = 0$ in one gauge, then in a different gauge we have in general $\delta_{\text{tot}} \neq 0$. Equivalently, the notion of temperature fluctuations is not gauge-invariant; on super-horizon scales, given a frame where δT is non-zero, we can always find a different coordinate system where $\delta T = 0$. Therefore, the definitions that we have given of adiabatic and isocurvature modes only make sense with respect to a given gauge, which is chosen to be the conformal Newtonian gauge.[10]

These definitions can be extended to a fluid with several components; consider for instance a cosmic fluid with three components: radiation (R), cold dark matter (CDM) and baryons (B). We can define CDM isocurvature modes from the condition $\{\delta_{\text{CDM}} \neq 0, \delta_B = 0\}$, while δ_R is such that $\delta_{\text{tot}} = 0$, i.e. $\rho_{\text{CDM}}\delta_{\text{CDM}} + \rho_R\delta_R = 0$. Deep in RD, where ρ_{CDM} is negligible compared with ρ_R, this is in practice equivalent to requiring $\delta_R = 0$. Thus, in practice we can define CDM isocurvature modes from the condition $\{\delta_{\text{CDM}} \neq 0, \delta_B = 0, \delta_R = 0\}$. Similarly baryon isocurvature modes are defined by $\{\delta_{\text{CDM}} = 0, \delta_B \neq 0, \delta_R = 0\}$. As a third independent initial condition we take the adiabatic initial condition, which is still defined as being due to temperature fluctuations only (in the conformal Newtonian gauge), and therefore satisfies

$$\delta_{\text{CDM}} = \delta_B = \frac{3}{4}\delta_R. \tag{19.81}$$

The usefulness of the separation between adiabatic and isocurvature initial conditions is that they are produced by quite distinct physical mechanisms. Adiabatic initial conditions are generated by mechanisms that act in the same way on all species and everywhere in space. In particular, the mechanism of generation of initial perturbations from quantum fluctuations during inflation, that we will study in Section 21.3, generates adiabatic perturbations. Isocurvature perturbations need a source that does not act homogeneously in space. A typical example is the accretion of matter onto cosmic strings or other cosmological defects. Adiabatic and isocurvature perturbations have very distinct signatures in the spectrum of the CMB. While adiabatic perturbation produce a series of acoustic peaks (as we will review in Chapter 20), cosmic strings generate only a single broad Doppler peak. Comparison with observation shows that the data are entirely consistent with adiabatic perturbation.

[10]It is also possible to define adiabatic perturbations in terms of quantities that are gauge-invariant (at the level of linearized theory). This can be done by using the quantity $\delta p - c_{s,(a)}^2 \delta\rho$ introduced in eq. (18.188). As we saw, this quantity is gauge-invariant at the linearized level, and vanishes for adiabatic perturbations. Therefore the condition that it vanishes provides a gauge-invariant definition of adiabatic perturbations. It can also be shown that the four-divergence of the entropy flux is proportional to this gauge-invariant combination; see Appendix A5.2 of Durrer (2008).

Any admixture of CDM isocurvature perturbations in the initial conditions, if present at all, is below a few percent. In the following we will then only consider adiabatic perturbations. Observe that, as long as we are in the linear regime, an initially adiabatic mode will remain adiabatic, and isocurvature modes remain isocurvature. When we enter the non-linear regime, however, the two modes will mix.

To summarize, we assign adiabatic initial conditions for all cosmologically relevant modes. Writing

$$\tilde{\Phi}_{\text{in}}(\mathbf{k}) = A(k)\,, \tag{19.82}$$

eq. (19.70) (with $\delta_{\text{tot}} \simeq \delta_R$) gives

$$\tilde{\delta}_{R,\text{in}}(\mathbf{k}) = 2A(k)\,, \tag{19.83}$$

while

$$\tilde{\delta}_{B,\text{in}}(\mathbf{k}) = \tilde{\delta}_{\text{CDM},\text{in}}(\mathbf{k}) = \frac{3}{2}A(k) \tag{19.84}$$

and

$$k\tilde{v}_{R,\text{in}}(\mathbf{k}) = k\tilde{v}_{M,\text{in}}(\mathbf{k}) = \frac{1}{2}k\eta_{\text{in}}A(k)\,. \tag{19.85}$$

We will discuss in Section 19.4 the dependence of $A(k)$ on k.[11]

19.2.2 The variables ζ and $\mathcal{R}$

In this subsection we introduce two variables, ζ and $\mathcal{R}$, that for adiabatic modes have the very useful property that they are time-independent, in the limit of modes well outside the horizon.[12] We have defined adiabatic initial conditions by the property that, in the conformal Newtonian gauge, they are only due to temperature fluctuations. This means that, in this gauge, for each species λ at the initial time $\eta = \eta_{\text{in}}$ we have

$$\begin{aligned}
\delta\rho_\lambda(\eta_{\text{in}},\mathbf{x}) &= \frac{\partial\rho_\lambda}{\partial T}(\eta_{\text{in}},\mathbf{x})\delta T(\eta_{\text{in}},\mathbf{x}) \\
&= \frac{\partial\rho_\lambda}{\partial\eta}(\eta_{\text{in}},\mathbf{x})\frac{d\eta}{dT}\delta T(\eta_{\text{in}},\mathbf{x}) \\
&= \bar{\rho}'_\lambda(\eta_{\text{in}})\frac{\delta T(\eta_{\text{in}},\mathbf{x})}{T'(\eta_{\text{in}})}\,,
\end{aligned} \tag{19.86}$$

where, in the second line, we have observed that to first order we can replace ρ'_λ by $\bar{\rho}'_\lambda$, which is independent of $\mathbf{x}$. Similarly,

$$\delta p_\lambda(\eta_{\text{in}},\mathbf{x}) = \bar{p}'_\lambda(\eta_{\text{in}})\frac{\delta T(\eta_{\text{in}},\mathbf{x})}{T'(\eta_{\text{in}})}\,. \tag{19.87}$$

This means that adiabatic initial conditions have the form

$$\delta\rho_\lambda(\eta_{\text{in}},\mathbf{x}) = \bar{\rho}'_\lambda(\eta_{\text{in}})\,\epsilon(\eta_{\text{in}},\mathbf{x})\,, \tag{19.88}$$
$$\delta p_\lambda(\eta_{\text{in}},\mathbf{x}) = \bar{p}'_\lambda(\eta_{\text{in}})\,\epsilon(\eta_{\text{in}},\mathbf{x})\,, \tag{19.89}$$

for some function $\epsilon(\eta_{\text{in}},\mathbf{x})$, independent of λ, in which all the dependence on $\mathbf{x}$ resides. Actually, this definition assumes thermal equilibrium, since

[11] As we will discuss in Section 19.4.1, the initial values, such as $\tilde{\Phi}_{\text{in}}(\mathbf{k})$, are actually stochastic variables with zero mean and a variance given by $A(k)$; see eq. (19.175).

[12] We follow the clear presentation in Section 5.2 of Gorbunov and Rubakov (2011).

we have assumed that $\delta\rho_\lambda$ and δp_λ are functions of the temperature only. We can generalize the definition of adiabatic initial conditions, dropping the assumption of thermal equilibrium, by simply stating that adiabatic initial conditions are *defined* (in the conformal Newtonian gauge) by eqs. (19.88) and (19.89), for some function $\epsilon(\eta_{\rm in}, \mathbf{x})$.

Consider now the adiabatic mode, i.e. the solution of the equations with adiabatic initial condition. We can show that it satisfies eqs. (19.88) and (19.89) not just at the initial time, but at all times, as long as it is well outside the horizon. In principle this is far from obvious, since we are trying to solve the $n + 1$ equations (19.45) and (19.46) using only two functions Φ and ϵ. To show that this ansatz works, we observe that in momentum space, and for generic η, eq. (19.88) becomes simply

$$\tilde{\delta}_\lambda(\eta, \mathbf{k}) = \frac{\bar{\rho}'_\lambda(\eta)}{\bar{\rho}_\lambda(\eta)}\tilde{\epsilon}(\eta, \mathbf{k}), \tag{19.90}$$

$$= -3\mathcal{H}(\eta)[1 + w_\lambda(\eta)]\tilde{\epsilon}(\eta, \mathbf{k}), \tag{19.91}$$

where we have assumed the validity of the separate conservation equations (17.67). Equation (19.89) states that the perturbations have a speed of sound given by

$$c_{s,\lambda}^2 \equiv \frac{\delta p_\lambda}{\delta\rho_\lambda} = \frac{\bar{p}'_\lambda}{\bar{\rho}'_\lambda} = \frac{(w_\lambda\bar{\rho}_\lambda)'}{\bar{\rho}'_\lambda}$$

$$= w_\lambda - \frac{w'_\lambda}{3\mathcal{H}(1 + w_\lambda)}. \tag{19.92}$$

Plugging eqs. (19.91) and (19.92) into eq. (19.46), which holds for super-horizon modes, we get

$$-3(1 + w_\lambda)(\mathcal{H}\tilde{\epsilon})' = -3(1 + w_\lambda)\tilde{\Phi}'. \tag{19.93}$$

Thus, the n equations for $\lambda = 1, \ldots, n$ collapse to a single equation

$$(\mathcal{H}\tilde{\epsilon})' = \tilde{\Phi}', \tag{19.94}$$

whose solution is

$$\tilde{\Phi}(\eta, \mathbf{k}) = \mathcal{H}\tilde{\epsilon}(\eta, \mathbf{k}) + \tilde{\zeta}(\mathbf{k}), \tag{19.95}$$

where $\tilde{\zeta}(\mathbf{k})$ is an integration constant, independent of time. We can now plug this expression for Φ into eq. (19.45), and we get the equation

$$\tilde{\epsilon}' + \left(\frac{\mathcal{H}'}{\mathcal{H}} + \mathcal{H}\right)\tilde{\epsilon} + \tilde{\zeta} = \frac{1}{2}\tilde{\delta}_{\rm tot}. \tag{19.96}$$

We next observe that, with our ansatz, and to linear order,

$$\tilde{\delta}_{\rm tot} \equiv \frac{\sum_\lambda \delta\rho_\lambda}{\sum_\lambda \rho_\lambda} = \frac{\tilde{\epsilon}\sum_\lambda \bar{\rho}'_\lambda}{\sum_\lambda \bar{\rho}_\lambda}$$

$$= \tilde{\epsilon}\frac{\bar{\rho}'_{\rm tot}}{\bar{\rho}_{\rm tot}}. \tag{19.97}$$

Using $\bar{\rho}_{\rm tot} \propto \mathcal{H}^2$, we get

$$\tilde{\delta}_{\rm tot} = \frac{2\tilde{\epsilon}\mathcal{H}'}{\mathcal{H}} \,. \tag{19.98}$$

Therefore eq. (19.96) becomes

$$\tilde{\epsilon}' + \mathcal{H}\tilde{\epsilon} + \tilde{\zeta} = 0 \,, \tag{19.99}$$

which fixes ϵ as a function of ζ. Therefore, the ansatz (19.88), (19.89) indeed provides a solution of the $n+1$ equations (19.45) and (19.46). Finally, plugging these solutions into the right-hand side of eq. (19.47), we can solve it for the variables $\tilde{v}_\lambda$.

An important consequence of this derivation is that it shows that the combination

$$\tilde{\zeta}(\mathbf{k}) = \tilde{\Phi}(\eta, \mathbf{k}) - \mathcal{H}\tilde{\epsilon}(\eta, \mathbf{k}) \tag{19.100}$$

is constant for super-horizon adiabatic modes, as we see from eq. (19.94). Using eqs. (19.97) and (18.166) we have

$$\tilde{\epsilon} = \tilde{\delta}_{\rm tot} \frac{\bar{\rho}_{\rm tot}}{\bar{\rho}'_{\rm tot}}$$

$$= -\frac{\bar{\rho}_{\rm tot}\tilde{\delta}_{\rm tot}}{3\mathcal{H}(\bar{\rho}_{\rm tot} + \bar{p}_{\rm tot})} \,. \tag{19.101}$$

Therefore

$$\tilde{\zeta}(\mathbf{k}) = \tilde{\Phi}(\eta, \mathbf{k}) + \frac{\bar{\rho}_{\rm tot}(\eta)\tilde{\delta}_{\rm tot}(\eta, \mathbf{k})}{3[\bar{\rho}_{\rm tot}(\eta) + \bar{p}_{\rm tot}(\eta)]} \,. \tag{19.102}$$

Observe that, in the proof that ζ is constant for super-horizon adiabatic modes, we have never used any specific form of the background. So, ζ is constant independently of whether we are in RD, in MD or at the transition between RD and MD, as long as the mode is super-horizon and the time dependence of $\tilde{\Phi}(\eta, \mathbf{k})$ and $\tilde{\delta}_{\rm tot}(\eta, \mathbf{k})$ is that of the adiabatic mode.

Instead of $\tilde{\zeta}$, one can use

$$\tilde{\mathcal{R}} \equiv \tilde{\Phi} + \mathcal{H}\tilde{v}_{\rm tot} \,, \tag{19.103}$$

where

$$\tilde{v}_{\rm tot} = \frac{\sum_\lambda (\bar{\rho}_\lambda + \bar{p}_\lambda)\tilde{v}_\lambda}{\sum_\lambda (\bar{\rho}_\lambda + \bar{p}_\lambda)} \,. \tag{19.104}$$

In fact, using eqs. (19.36) and (19.37),

$$\tilde{\zeta} - \tilde{\mathcal{R}} = \frac{k^2\tilde{\Phi}}{12\pi G a^2(\bar{\rho}_{\rm tot} + \bar{p}_{\rm tot})}$$

$$= \frac{2}{9(1 + w_{\rm tot})}\frac{k^2}{\mathcal{H}^2}\tilde{\Phi} \,. \tag{19.105}$$

Outside the horizon the right-hand side, which is proportional to $k^2/\mathcal{H}^2$, is negligible, so for super-horizon modes $\tilde{\mathcal{R}}$ is the same as $\tilde{\zeta}$. It can be shown that the three-dimensional curvature scalar of surfaces of constant $\rho_{\rm tot}$ is determined by ζ, through

$$R^{(3)} = -\frac{4}{a^2}\boldsymbol{\nabla}^2\zeta\,, \tag{19.106}$$

while for surfaces of constant $v_{\rm tot}$ (i.e. surfaces of constant time for a comoving observer) we have

$$R^{(3)} = -\frac{4}{a^2}\boldsymbol{\nabla}^2\mathcal{R}\,. \tag{19.107}$$

The adiabatic initial conditions can be expressed in terms of the initial value of $\mathcal{R}_{\rm in}$, observing that eq. (19.103), deep in RD, gives

$$\tilde{\mathcal{R}}_{\rm in} \equiv \tilde{\Phi}_{\rm in} + \frac{1}{\eta_{\rm in}}\tilde{v}_{R,\rm in}\,. \tag{19.108}$$

Then, using eq. (19.85), we get

$$\tilde{\mathcal{R}}_{\rm in} = \frac{3}{2}\tilde{\Phi}_{\rm in}\,. \tag{19.109}$$

Therefore the initial conditions (19.82)–(19.85) can be written equivalently in terms of $\tilde{\mathcal{R}}_{\rm in}(k)$ instead of $A(k)$.

19.3 Solutions of the equations for scalar perturbations

19.3.1 Numerical integration

Given the equations and their initial conditions, the most straightforward approach is to integrate them numerically. We have not yet specified the function $A(k)$, that enters in eqs. (19.82)–(19.85). However, at the level of the linearized equations, different momentum modes do not mix, so this is just an overall normalization factor for each mode, and we can get rid of it by computing the evolution of the perturbations normalized to their initial values, $\tilde{\Phi}(\eta,\mathbf{k})/\tilde{\Phi}(\eta_{\rm in},\mathbf{k})$, etc., which is equivalent to setting $A(k) = 1$. Given that the equations are linear, the final output of the integration can be multiplied back by $A(k)$ to obtain the correct momentum dependence of the different modes. It is convenient to introduce the functions

$$\hat{\theta}_\lambda(\eta,\mathbf{x}) \equiv \frac{\theta_\lambda(\eta,\mathbf{x})}{\mathcal{H}(\eta)}\,, \tag{19.110}$$

$$\hat{k}(\eta) \equiv \frac{k}{\mathcal{H}(\eta)}\,, \tag{19.111}$$

and to use as time-evolution variable $x = \log a$ instead of η. We also define

$$\xi(x) \equiv \frac{1}{H}\frac{dH}{dx}\,. \tag{19.112}$$

Observe that, in terms of $\mathcal{H}$, we have the relations

$$1 + \xi = \frac{1}{\mathcal{H}} \frac{d\mathcal{H}}{dx}$$
$$= \frac{\mathcal{H}'}{\mathcal{H}^2}, \tag{19.113}$$

where, as usual, the prime denotes $d/d\eta$ while the derivative with respect to x will be written explicitly. We can also relate $\xi(x)$ to the total EoS parameter $w_{\text{tot}}(x)$ defined in eq. (17.65), which therefore satisfies

$$\dot{\rho}_{\text{tot}} + 3H(1 + w_{\text{tot}})\rho_{\text{tot}} = 0. \tag{19.114}$$

The Friedmann equation gives $\rho_{\text{tot}} \propto H^2$, and therefore

$$\frac{\dot{\rho}_{\text{tot}}}{\rho_{\text{tot}}} = \frac{2\dot{H}}{H} = 2\frac{dH}{dx}. \tag{19.115}$$

Thus

$$w_{\text{tot}}(x) = -1 - \frac{2}{3}\xi(x). \tag{19.116}$$

We consider a model with radiation, matter and a cosmological constant. Using eq. (17.75) we have the explicit expressions[13]

$$\xi(x) = -\frac{4\Omega_R e^{-4x} + 3\Omega_M e^{-3x}}{2(\Omega_R e^{-4x} + \Omega_M e^{-3x} + \Omega_\Lambda)} \tag{19.117}$$

and

$$\hat{k}^2(x) = \left(\frac{H_0^{-1}}{\lambdabar}\right)^2 \frac{1}{e^{2x}(\Omega_R e^{-4x} + \Omega_M e^{-3x} + \Omega_\Lambda)}. \tag{19.118}$$

For the perturbations we neglect anisotropic stresses, so that $\Psi = -\Phi$. There are no perturbations associated with the cosmological constant term, as we see on setting $c_{s,\lambda}^2 = w_\lambda = -1$ in eq. (19.39).[14] In momentum space the perturbation variables are therefore $\tilde{\delta}_R$, $\tilde{\delta}_M$, $\hat{\theta}_R$, $\hat{\theta}_M$ and $\tilde{\Phi}$ (in order not to excessively burden the notation, we do not put a tilde over the Fourier modes of $\hat{\theta}$). The conservation equations (19.39) and (19.40) become

$$\partial_x \tilde{\delta}_M = -3\partial_x \tilde{\Phi} - \hat{\theta}_M, \tag{19.119}$$

$$\partial_x \tilde{\delta}_R = -\frac{4}{3}(3\partial_x \tilde{\Phi} + \hat{\theta}_R), \tag{19.120}$$

$$\partial_x \hat{\theta}_M = -(2 + \xi)\hat{\theta}_M - \hat{k}^2 \tilde{\Phi}, \tag{19.121}$$

$$\partial_x \hat{\theta}_R = -(1 + \xi)\hat{\theta}_R + \hat{k}^2 \left(\frac{1}{4}\tilde{\delta}_R - \tilde{\Phi}\right), \tag{19.122}$$

where, in $\theta' = \mathcal{H}\hat{\theta}' + \mathcal{H}'\hat{\theta}$, we have replaced $\mathcal{H}'$ using (19.113), and we have used $d/d\eta = \mathcal{H}d/dx$. To close the system of equations we choose eq. (19.36), which in these variables reads

$$\partial_x \tilde{\Phi} + \left(1 + \frac{\hat{k}^2}{3}\right)\tilde{\Phi} = \frac{1}{2}\tilde{\delta}_{\text{tot}}, \tag{19.123}$$

[13]Observe that here we are not including the change in the number of effective relativistic species, described in Section 17.5, and we are also neglecting neutrino masses.

[14]This is an important difference between ΛCDM and models where dark energy is dynamical. For a dynamical dark energy, in general w_{DE} is not exactly equal to -1, and therefore there are dark-energy perturbations.

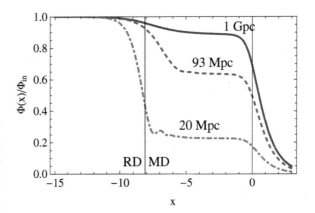

Fig. 19.3 The evolution of $\tilde{\Phi}(\eta, k)$, normalized to its initial value, against $x = \log a$, for three different values of the reduced wavelength $\tilde{\lambda} = 20$ Mpc, 93 Mpc and 1 Gpc. We use $\Omega_M = 0.3$, $\Omega_R h_0^2 = 4.18 \times 10^{-5}$, $h_0 = 0.7$ and $\Omega_\Lambda = 1 - \Omega_M - \Omega_R$.

[15]Note that the adiabatic initial conditions (19.82)–(19.85) have been computed assuming that the modes are well outside the horizon, and keeping only the leading terms in a small-k expansion. When performing accurate numerical work it can be useful to compute also the next corrections in an expansion in powers of $\hat{k}_{\rm in}^2$. See Durrer (2008), Section 2.4.3, for the explicit formulas.

where

$$\tilde{\delta}_{\rm tot} = \tilde{\delta}_R \frac{\bar{\rho}_R}{\bar{\rho}_{\rm tot}} + \tilde{\delta}_M \frac{\bar{\rho}_M}{\bar{\rho}_{\rm tot}}$$

$$= \frac{\tilde{\delta}_R \Omega_R e^{-4x} + \tilde{\delta}_M \Omega_M e^{-3x}}{\Omega_R e^{-4x} + \Omega_M e^{-3x} + \Omega_\Lambda}. \quad (19.124)$$

Factoring out the common function $A(k)$, the adiabatic initial conditions (19.82)–(19.85) become[15]

$$\tilde{\Phi}_{\rm in} = 1, \quad \tilde{\delta}_{R,\rm in} = 2, \quad \tilde{\delta}_{M,\rm in} = \frac{3}{2}, \quad \hat{\theta}_{R,\rm in} = \hat{\theta}_{M,\rm in} = -\frac{1}{2}\hat{k}_{\rm in}^2. \quad (19.125)$$

Observe that, since the initial conditions are assigned deep in RD, where $\mathcal{H} = 1/\eta$, we have $\hat{k}_{\rm in} \simeq k\eta_{\rm in}$. The numerical integration of eqs. (19.119)–(19.123) is then straightforward, and we plot the results in Figs. 19.3–19.8. We use our reference values $\Omega_M = 0.3$, $h_0 = 0.7$, $\Omega_R h_0^2 = 4.18 \times 10^{-5}$ and $\Omega_\Lambda = 1 - \Omega_M - \Omega_R \simeq 0.7$. The behavior of the solutions depends on whether the mode enters the horizon during RD or during MD. In Figs. 19.3 and 19.4 we therefore plot the result for three cases: (1) a mode with a reduced wavelength $\tilde{\lambda} = 20$ Mpc, which enters the horizon during RD (as we see from Fig. 19.2 on page 445); (2) the mode with $\tilde{\lambda}_{\rm eq} = 93$ Mpc, which (with our choice of parameters) enters the horizon just at the RD–MD transition; see eq. (19.53); (3) a mode with $\tilde{\lambda}_{\rm eq} = 1$ Gpc that, according to eq. (19.60), enters during MD, at a redshift $z_* \simeq 460$ (i.e. $x_* \simeq -6.1$), so deep in MD, but still well before the present epoch when the cosmological constant begins to dominate.

From Fig. 19.3 we see that, for modes with very large wavelength (which therefore enter the horizon in MD), $\tilde{\Phi}$ stays constant at its initial value during RD, decreases somewhat at the RD–MD transition and then deep in MD (say at $x \gtrsim -5$, i.e. $z \lesssim 150$) stabilizes to another constant value. The numerical integration indicates that, in the limit of

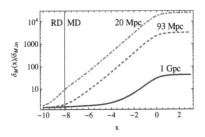

Fig. 19.4 The same as in Fig. 19.3 for the matter density contrast δ_M, normalized to its initial value, using a logarithmic vertical scale.

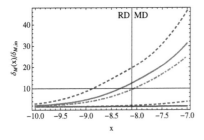

Fig. 19.5 The matter density contrast $\delta_M/\delta_{M,\rm in}$. The lines correspond, from bottom to top, to $\tilde{\lambda} = 1$ Gpc, 93, 20, 16 and 10 Mpc. We plot them on a horizontal scale that enlarges the region near the RD–MD transition.

very small k, this second plateau is at the value $\tilde{\Phi} \simeq 0.9\tilde{\Phi}_{\rm in}$, a result that we will prove analytically. In contrast, the small-wavelength mode with $\lambdabar = 20$ Mpc, which enters in RD, stays constant until is outside the horizon, and starts to decrease as soon as it enters the horizon; at the RD–MD transition it therefore already has a smaller value, and then in MD, after some oscillations, it stabilizes again to a second plateau, which of course is now smaller than 0.9. In other words, during MD all modes are constant, while in RD they are constant when they are outside the horizon and decrease when they are inside the horizon. So, the earlier the mode enters the horizon in RD, the smaller is its constant value in MD. The mode that enters at equilibrium has of course an intermediate behavior between these two cases. Another change of regime takes place near the present epoch, when the cosmological constant begins to dominate. In the plot we have extended the horizontal axis up to $x = 3$, and thus in the future, to better appreciate this part of the evolution. We see that during ΛD the value of Φ starts decreasing again for all modes, and goes asymptotically to zero in the future.

Figure 19.4 illustrates the behavior of $\tilde{\delta}_M$ for these three modes, which shows that the low-wavelength modes start to grow earlier. We also see that, in MD, the growth is linear with the scale factor a, and then stops during ΛD. Figure 19.5 is an enlargement of the region near the RD–MD transition, and shows that, for the low-$\lambdabar$ modes, the growth already begins in RD, with the lowest $\lambdabar$ starting to grow earlier. We will obtain an analytical understanding of these results in Sections 19.3.2 and 19.3.3.

Finally, in Figs. 19.6–19.8 we show the results for $\tilde{\theta}_M$, $\tilde{\delta}_R$ and $\tilde{\theta}_R$. In order not to clutter the plots, we only show the case $\lambda = \lambda_{\rm eq}$. We see that $\tilde{\theta}_M$ has a behavior similar to $\tilde{\delta}_M$, staying constant in RD and starting to grow (in absolute value) deep in MD. However, it again goes to zero in ΛD. The density contrast and the velocity potential of radiation, on the other hand, never grow significantly (compare the vertical scales in Figs. 19.4 and 19.7), and rather display a set of oscillations in MD, which then decay in ΛD. We hasten to add that, after recombination, radiation cannot be treated as a fluid characterized only by density perturbations and a velocity potential. The correct treatment involves all moments of the distribution function, and the interaction with baryons, and will be discussed in Section 20.3.7.

Of course, the fact that $\tilde{\delta}_M$ and $\tilde{\theta}_M$ grow so much, at least for short wavelengths, raises a question about the validity of the linearized approximation. Linear perturbation theory is only valid for $|\tilde{\delta}_M| \ll 1$ and $|\tilde{\theta}_M| \ll 1$. Remember, however, that the quantities shown in the plots are the functions normalized to their initial value, i.e. to the function $A(k)$ that appears in eq. (19.82). We will discuss in Section 19.4 the experimental information that we have on $A(k)$, and we will then be able to understand which modes stay in the linear regime up to the present epoch. Actually, since the initial conditions on $\tilde{\theta}_M$ are suppressed by a factor $\hat{k}_{\rm in}^2 \ll 1$ compared with the initial conditions on $\tilde{\delta}_M$, we will only have to check if $\tilde{\delta}_M$ stays small.

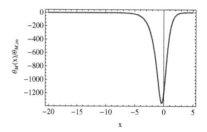

Fig. 19.6 The function $\tilde{\theta}_M$ for $\lambdabar = \lambda_{\rm eq}/(2\pi) = 93$ Mpc, normalized to its initial value, vs. $x = \log a$. Same parameters as in Fig. 19.3, in this and in the following figures.

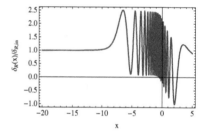

Fig. 19.7 The function $\tilde{\delta}_R$ for $\lambdabar = \lambda_{\rm eq}/(2\pi) = 93$ Mpc, normalized to its initial value, vs. $x = \log a$.

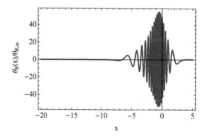

Fig. 19.8 The function $\tilde{\theta}_R$ for $\lambdabar = \lambda_{\rm eq}/(2\pi) = 93$ Mpc, normalized to its initial value, vs. $x = \log a$.

19.3.2 Analytic solutions in RD

Several aspects of the numerical solutions discussed in Section 19.3.1 can be easily understood analytically. We consider first the evolution during RD. We examine separately the metric perturbations and the perturbations in the energy density of radiation and of matter.

Metric perturbations

In RD $c_s^2 = w = 1/3$ and $\mathcal{H} = 1/\eta$. The master equation (19.25) then simplifies to (19.66). As we have already seen, outside the horizon this has a constant solution and a decaying solution. Furthermore, the initial conditions (19.125) gives $\tilde{\Phi}_{\rm in} = (1/2)\tilde{\delta}_{R,\rm in}$ which, since we are deep in RD, is equivalent to $\tilde{\Phi}_{\rm in} = (1/2)\tilde{\delta}_{\rm tot}$. Then eq. (19.123) gives $\tilde{\Phi}'_{\rm in} = -\hat{k}_{\rm in}^2/3 \ll 1$, so we are already basically selecting the constant mode, and we do not need to wait a long transient before it disappears. This explains the fact that, in Fig. 19.3, $\tilde{\Phi}$ is constant outside the horizon. If the mode enters the horizon during RD, however, the behavior changes. Indeed, the master equation (19.26) in RD becomes

$$\tilde{\Phi}'' + \frac{4}{\eta}\tilde{\Phi}' + c_s^2 k^2 \tilde{\Phi} = 0 \tag{19.126}$$

(where, in RD, $c_s^2 = 1/3$). With the initial conditions $\tilde{\Phi}(\eta_{\rm in}, \mathbf{k}) = \tilde{\Phi}_{\rm in}(\mathbf{k})$ and $\tilde{\Phi}'(\eta_{\rm in}, \mathbf{k}) = 0$, set at a value of $\eta_{\rm in}$ such that $k\eta_{\rm in} \to 0$, this equation has the exact solution

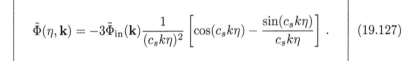

$$\tilde{\Phi}(\eta, \mathbf{k}) = -3\tilde{\Phi}_{\rm in}(\mathbf{k}) \frac{1}{(c_s k\eta)^2}\left[\cos(c_s k\eta) - \frac{\sin(c_s k\eta)}{c_s k\eta}\right]. \tag{19.127}$$

This solution is shown in Fig. 19.9. We see that, to be precise, the relevant parameter is actually $c_s k\eta$, rather than $k\eta$, but since in RD $c_s = \sqrt{1/3} \simeq 0.57$, the two are close to each other. In the limit $c_s k\eta \ll 1$ the solution goes to a constant, while for $c_s k\eta \gg 1$ eq. (19.127) gives

$$\tilde{\Phi}(\eta, \mathbf{k}) \simeq -3\tilde{\Phi}_{\rm in}(\mathbf{k})\frac{1}{(c_s k\eta)^2}\cos(c_s k\eta), \tag{19.128}$$

so it goes to zero, with some oscillations. This agrees with the RD part of the result found numerically in Fig. 19.3: for modes with very low k, we have $k\eta \ll 1$ for all η up to the RD–MD transition, so the mode stays constant during the whole RD phase. For modes with larger k, eventually $c_s k\eta$ becomes larger than 1 already during RD, and at that point the mode starts decreasing (when comparing with Fig. 19.3, observe that in Fig. 19.9 we plot the solution against $c_s k\eta$, while in Fig. 19.9 we plot it against $\log a$, which in RD is proportional to $\log \eta$).

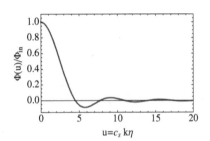

Fig. 19.9 The solution (19.127), normalized to its initial value, against $u \equiv c_s k\eta$.

Perturbations of radiation density

We next consider δ_R. For super-horizon modes deep in RD we have already found in eq. (19.72) that $\tilde{\delta}_R$ is constant, at a value

$$\tilde{\delta}_R(\eta, \mathbf{k}) \simeq 2\tilde{\Phi}_{\rm in}(\mathbf{k}) \,. \tag{19.129}$$

Consider now the evolution of a mode well inside the horizon in RD, so that $\hat{k} \gg 1$. Deep in RD we have $\delta_{\rm tot} \simeq \delta_R$, so eq. (19.123) gives

$$\tilde{\delta}_R(\eta, \mathbf{k}) \simeq \frac{2}{3}\hat{k}^2 \tilde{\Phi}(\eta, \mathbf{k}) \,, \tag{19.130}$$

which is just the Poisson equation of Newtonian cosmology, in these variables. We now use the fact that, in RD,

$$\hat{k} \equiv k/\mathcal{H} \simeq k\eta \,, \tag{19.131}$$

and we plug in $\tilde{\Phi}(\eta, \mathbf{k})$ from eq. (19.128), using eq. (19.70) to express $\tilde{\Phi}_{\rm in}$ as $\tilde{\delta}_{R,\rm in}/2$. Then we get

$$\tilde{\delta}_R(\eta, \mathbf{k}) \simeq -3\tilde{\delta}_{R,\rm in}(\mathbf{k}) \cos(c_s k\eta) \tag{19.132}$$

(where again $c_s = 1/\sqrt{3}$), which indeed reproduces the oscillating behavior that can be found numerically inside the horizon. The value of $\tilde{\theta}_R$ can be obtained from eq. (19.37), which deep in RD reads

$$\tilde{\Phi}' + \frac{1}{\eta}\tilde{\Phi} = -\frac{16\pi G a^2}{3k^2}\bar{\rho}_R\tilde{\theta}_R$$

$$= -\frac{2}{\eta^2 k^2}\tilde{\theta}_R \,. \tag{19.133}$$

Plugging eq. (19.127) into the left-hand side gives the general analytic form of $\tilde{\theta}_R$ in RD. In particular, for modes inside the horizon the leading contribution to the right-hand side comes from the term $\tilde{\Phi}'$, when one takes the derivative of the cosine in eq. (19.128), while the other terms are suppressed by a factor $1/(k\eta) \ll 1$. Thus, for sub-horizon modes, we get

$$\tilde{\theta}_R \simeq -\frac{3\sqrt{3}}{2}\tilde{\Phi}_{\rm in} k \sin k\eta \,, \tag{19.134}$$

or, in terms of the dimensionless quantity $\hat{\theta}_R = \tilde{\theta}_R/\mathcal{H} \simeq \tilde{\theta}_R\eta$,

$$\hat{\theta}_R \simeq -\frac{3\sqrt{3}}{2}\tilde{\Phi}_{\rm in} k\eta \sin k\eta \,. \tag{19.135}$$

Comparing with eq. (19.132) we see that, for radiation, the oscillations of the velocity potential are shifted by $\pi/2$ from those of the density contrast. This is typical of acoustic oscillations, and is easy to visualize in coordinate space. At a given point $\mathbf{x}$, when the density contrast of a fluid element has reached its maximum compression, its velocity field has gone to zero, and the fluid is ready to rebound.

Perturbations of CDM density

We next consider matter perturbations during RD (recall that in this section we are neglecting baryons, so non-relativistic matter is only made of CDM). Since in RD non-relativistic matter is subdominant, we can simply take the value of $\tilde{\Phi}$ from eq. (19.127) [which has been obtained from the master equation (19.25), in the single-fluid approximation in which the only fluid considered was radiation] and use it as a source in eqs. (19.119) and (19.121). This is equivalent to approximating $\delta_{\rm tot} \simeq \delta_R$ in eq. (19.123), so that eqs. (19.120), (19.122) and (19.123) form a closed system for $\tilde{\Phi}, \tilde{\delta}_R$ and $\tilde{\theta}_R$; this gives in particular the solution (19.127) for $\tilde{\Phi}$, which we can now plug into eqs. (19.119) and (19.121).

Using η instead of x and $\tilde{\theta}$ instead of $\hat{\theta}$, eqs. (19.39) and (19.40) give

$$\tilde{\delta}'_M + \tilde{\theta}_M = -3\tilde{\Phi}', \qquad (19.136)$$

$$\tilde{\theta}'_M + \frac{1}{\eta}\tilde{\theta}_M = -k^2\tilde{\Phi}. \qquad (19.137)$$

Writing

$$\tilde{\theta}'_M + \frac{1}{\eta}\tilde{\theta}_M = \frac{1}{\eta}\frac{d(\eta\tilde{\theta}_M)}{d\eta}, \qquad (19.138)$$

eq. (19.137) integrates to

$$\tilde{\theta}_M(\eta, \mathbf{k}) = -\frac{k^2}{\eta}\int_0^\eta d\eta_1\, \eta_1 \tilde{\Phi}(\eta_1, \mathbf{k}), \qquad (19.139)$$

where the integration constant has been chosen so that $\tilde{\theta}_M(\eta = 0, \mathbf{k}) = 0$. For modes far outside the horizon, we can formally set $k = 0$ in eq. (19.139), which gives $\tilde{\theta}_M = 0$. Then eq. (19.136), together with the fact that $\tilde{\Phi}$ is constant outside the horizon, tells us that $\tilde{\delta}_M$ is also constant outside the horizon, as expected for a density contrast.

The behavior well inside the horizon is obtained by plugging the explicit expression for $\tilde{\Phi}$ given in eq. (19.127) into eqs. (19.136) and (19.139) and integrating. The solution for $\tilde{\delta}_M(\eta)$, in the limit $c_s k\eta \gg 1$, can be found analytically and is

$$\boxed{\tilde{\delta}_M = \tilde{\delta}_{M,\rm in} + 9\tilde{\Phi}_{\rm in}\left[\log(c_s k\eta) + \gamma_E - \frac{2}{3}\right],} \qquad (19.140)$$

where $\gamma_E \simeq 0.577$ is the Euler–Mascheroni constant, and $c_s = 1/\sqrt{3}$ since we are in RD. This shows that, during RD, the density contrast of CDM grows logarithmically in η for modes well inside the horizon, which is know as the Mészáros effect. This logarithmic growth is very important for structure formation, as we will see later. This should be contrasted with the behavior of $\tilde{\delta}_R$ which, as we have seen, during RD and inside the horizon performs acoustic oscillations, and does not grow in amplitude.

19.3.3 Analytic solutions in MD

We next consider the solution in MD. We examine separately the metric and matter perturbations, and we will finally examine the solution across the RD–MD transition for super-horizon modes.

Metric perturbations

In MD, to determine $\tilde{\Phi}$ we can again use a single-fluid approximation in which we retain only the contribution of matter. Then $\tilde{\Phi}$ satisfies the master equation (19.25), where now $c_s^2 = 0$ and $\mathcal{H} = 2/\eta$. This gives

$$\tilde{\Phi}'' + \frac{6}{\eta}\tilde{\Phi}' = 0\,, \qquad (19.141)$$

which has a constant solution and a decaying mode $\tilde{\Phi} \propto \eta^{-5}$. Observe that, in contrast to the RD case (19.126), now the homogeneous term $c_s^2 k^2 \tilde{\Phi}$ vanishes simply because in MD $c_s = 0$, without the need to take the small-k limit. Thus, neglecting as usual the decaying mode, we find that in MD $\tilde{\Phi}$ is constant for all modes, and not only for super-horizon modes. This confirms the numerical result shown in Fig. 19.3, where we see that all modes show a plateau when they are well inside the MD phase. However, we also see from the numerical solution that the transient period between RD and MD is relatively long, and the modes settle to a plateau only well after RD–MD equilibrium.

Matter perturbations

For super-horizon modes, the behavior of $\tilde{\delta}_M$ can be read from eq. (19.45), just as we did for $\tilde{\delta}_R$ in RD. Deep in MD $\tilde{\Phi}$ is again constant, at a value $\tilde{\Phi}_{MD}$, while now $\tilde{\delta}_{tot} \simeq \tilde{\delta}_M$. Thus for super-horizon modes in MD we get

$$\tilde{\delta}_M(\eta, \mathbf{k}) \simeq 2\tilde{\Phi}_{MD}(\mathbf{k}) \qquad (k\eta \ll 1)\,. \qquad (19.142)$$

In the sub-horizon regime and in MD eq. (19.36) reduces instead to the Poisson equation

$$k^2 \tilde{\Phi} \simeq 4\pi G a^2 \bar{\rho}_M \tilde{\delta}_M\,. \qquad (19.143)$$

Writing $\bar{\rho}_M = \rho_0 \Omega_M / a^3$ and using $H_0^2 = (8\pi G/3)\rho_0$ we get

$$\tilde{\delta}_M(\eta, \mathbf{k}) = \frac{2\tilde{\Phi}_{MD}(\mathbf{k})}{3\Omega_M}\left(\frac{k}{H_0}\right)^2 a(\eta) \qquad (k\eta \gg 1)\,, \qquad (19.144)$$

and therefore in MD, on sub-horizon scales, $\tilde{\delta}_M$ grows linearly with the scale factor, as we see indeed from Fig. 19.4.

Super-horizon modes across the RD–MD transition

We see from Fig. 19.3 that, even in the limit $k \to 0$, the constant value of $\tilde{\Phi}$ during MD is different from the constant value during RD. To understand the transition between RD and MD analytically, we take the

$k \to 0$ limit of eqs. (19.119)–(19.123), without assuming that either radiation or matter dominate. First of all, we observe that eqs. (19.121) and (19.122) become homogeneous in $\hat{\theta}_M$ and $\hat{\theta}_R$, respectively; furthermore, in the limit $k \to 0$ the initial conditions (19.125) give $\hat{\theta}_{R,\text{in}} = \hat{\theta}_{M,\text{in}} = 0$. Therefore, we find that

$$\hat{\theta}_R(\eta, k = 0) = \hat{\theta}_M(\eta, k = 0) = 0 \qquad (19.145)$$

over the whole RD+MD evolution. Plugging this into eqs. (19.119), (19.120) and (19.123) we get the equations

$$\partial_x \tilde{\delta}_M = -3\partial_x \tilde{\Phi}, \qquad (19.146)$$

$$\partial_x \tilde{\delta}_R = -4\partial_x \tilde{\Phi}, \qquad (19.147)$$

$$\partial_x \tilde{\Phi} + \tilde{\Phi} = \frac{\bar{\rho}_R \tilde{\delta}_R + \bar{\rho}_M \tilde{\delta}_M}{2\bar{\rho}_{\text{tot}}}, \qquad (19.148)$$

which are valid only for the $k = 0$ mode. Integration of eqs. (19.146) and (19.147) with the adiabatic initial conditions (19.82)–(19.85) gives

$$\tilde{\delta}_M(\eta, k = 0) = -3\tilde{\Phi}(\eta, k = 0) + \frac{9}{2}\tilde{\Phi}_{\text{in}}(k = 0), \qquad (19.149)$$

$$\tilde{\delta}_R(\eta, k = 0) = -4\tilde{\Phi}(\eta, k = 0) + 6\tilde{\Phi}_{\text{in}}(k = 0), \qquad (19.150)$$

and therefore (again for the $k = 0$ mode only)

$$\tilde{\delta}_R(\eta, k = 0) = \frac{4}{3}\tilde{\delta}_M(\eta, k = 0). \qquad (19.151)$$

Plugging this into eq. (19.148) we get the equation, for $\tilde{\Phi}(\eta, k = 0)$,

$$\partial_x \tilde{\Phi} + \tilde{\Phi} = \left(\frac{4}{3}\bar{\rho}_R + \bar{\rho}_M\right)\frac{1}{2\bar{\rho}_{\text{tot}}}\tilde{\delta}_M. \qquad (19.152)$$

Taking the derivative ∂_x of this equation and replacing $\partial_x \tilde{\delta}_M$ from eq. (19.146) we get a second-order equation for $\tilde{\Phi}$. Using $y = a/a_{\text{eq}}$ as the time-evolution variable, it reads

$$\frac{d^2\tilde{\Phi}}{dy^2} + \frac{21y^2 + 54y + 32}{2y(y+1)(3y+4)}\frac{d\tilde{\Phi}}{dy} + \frac{1}{y(y+1)(3y+4)}\tilde{\Phi} = 0. \qquad (19.153)$$

[16] As first found by Kodama and Sasaki (1984) with some convenient change of variables. Nowadays the solution is easily found with any symbolic manipulation program.

Despite the complicated appearance of this equation, its solutions can be found analytically.[16] The two independent solutions are proportional to $\sqrt{1+y}/y^3$ and $(-16 - 8y + 2y^2 + 9y^3)/y^3$, respectively. The Taylor expansion of $16\sqrt{1+y}$ is $16 + 8y - 2y^2 + y^3 + O(y^4)$, so there is a solution $\tilde{\Phi}(y)$ that has a finite limit for $y \to 0^+$, given by

$$\tilde{\Phi}(y) = \frac{\tilde{\Phi}_{\text{in}}}{10y^3}\left(16\sqrt{1+y} - 16 - 8y + 2y^2 + 9y^3\right), \qquad (19.154)$$

where $\tilde{\Phi}_{\text{in}} = \tilde{\Phi}(y = 0)$. This function goes to $\tilde{\Phi}_{\text{in}}$ for $y \to 0$ and to $(9/10)\tilde{\Phi}_{\text{in}}$ for $y \to \infty$. Thus, for very large wavelengths, the plateau in

MD is at $9/10$ of the value in RD, consistently with the numerical result shown in Fig. 19.3.

Actually, this result can be obtained in a more elegant way using the fact that the variable ζ is constant for super-horizon modes with adiabatic initial conditions, as we have shown in Section 19.2.2. Recall that this is valid independently of the background evolution, so is also valid throughout the RD–MD transition; therefore, for these super-horizon modes, ζ has the same value in RD and in MD. In contrast, Φ is constant in RD and in MD, but not across the transition. Thus, from eq. (19.102), we see that deep in RD, where $\tilde{\Phi} = \tilde{\Phi}_{\rm in}$,

$$\zeta = \tilde{\Phi}_{\rm in} + \frac{1}{4}\tilde{\delta}_R$$
$$= \frac{3}{2}\tilde{\Phi}_{\rm in}\,, \tag{19.155}$$

where in the second line we have used eq. (19.129). Deep in MD, denoting by $\tilde{\Phi}_{\rm MD}$ the constant value of $\tilde{\Phi}$ at the MD plateau, eq. (19.102) gives

$$\zeta = \tilde{\Phi}_{\rm MD} + \frac{1}{3}\tilde{\delta}_M$$
$$= \frac{5}{3}\tilde{\Phi}_{\rm MD}\,, \tag{19.156}$$

where in the second line we have used eq. (19.142). Since ζ is constant for super-horizon modes, we get $(3/2)\tilde{\Phi}_{\rm in} = (5/3)\tilde{\Phi}_{\rm MD}$, and hence

$$\tilde{\Phi}_{\rm MD} = \frac{9}{10}\tilde{\Phi}_{\rm in}\,. \tag{19.157}$$

19.3.4 Analytic solutions during dark-energy dominance

Finally, we consider the evolution deep into Λ-dominance. In this case there are no energy or pressure perturbations, since by definition Λ is a constant. Thus, to determine the evolution of $\tilde{\Phi}$ we can use eq. (19.22), with the right-hand side set to zero [or, equivalently, we can use the master equation (19.26) with $w = -1$ and $c_s^2 = 0$]. Furthermore, using eq. (17.161), we see that $\mathcal{H} = -1/\eta$. Then eq. (19.22) becomes

$$\tilde{\Phi}'' - \frac{3}{\eta}\tilde{\Phi}' + \frac{3}{\eta^2}\tilde{\Phi} = 0\,. \tag{19.158}$$

One solution is

$$\tilde{\Phi} \propto \eta \propto \frac{1}{a}\,, \tag{19.159}$$

while the second solution decays even faster, $\Phi \propto \eta^3 \propto 1/a^3$. This reproduces the decay of $\tilde{\Phi}$ with the scale factor that we observe in Fig. 19.3 in the ΛD phase. For $\tilde{\delta}_M$, even inside the horizon, we then find that when we enter ΛD it no longer grows. Indeed, the Poisson equation still has the form (19.143), because in the absence of perturbations of Λ the dominant contribution on the right-hand side is still given by $\tilde{\delta}_M$. Then we get

$$\tilde{\delta}_M \propto \frac{\tilde{\Phi}}{a^2 \bar{\rho}_M} \propto \text{const}, \qquad (19.160)$$

since $\bar{\rho}_M \propto 1/a^3$ and $\tilde{\Phi} \propto 1/a$. This agrees with the result of the numerical integration shown in Fig 19.4. Therefore the growth of structure terminates in ΛD.

19.4   Power spectra for scalar perturbations

19.4.1   Definitions and conventions

When discussing cosmological perturbation theory, it is convenient to change the definition of the Fourier transform from our standard conventions, given in the Notation section in Vol. 1 and in eq. (17.39). This standard definition of the Fourier transform is appropriate for functions that go to zero sufficiently fast at infinity, so that the integral can be extended to an infinite volume. In cosmology, however, we rather measure an observable, such as the density contrast, in coordinate space, over the finite spatial volume V accessible to the observations, so the Fourier modes must be defined by integrating over this finite volume. In order to simplify the expressions for the power spectrum that we will introduce later, a convenient normalization of the Fourier transform of any observable $f(\mathbf{x})$ is

$$\tilde{f}(\mathbf{k}) = \frac{1}{V^{1/2}} \int_V d^3x \, f(\mathbf{x}) e^{-i\mathbf{k}\cdot\mathbf{x}}. \qquad (19.161)$$

The inverse of eq. (19.161) is[17]

$$f(\mathbf{x}) = V^{1/2} \int \frac{d^3k}{(2\pi)^3} \tilde{f}(\mathbf{k}) e^{i\mathbf{k}\cdot\mathbf{x}}. \qquad (19.162)$$

Observe that, with these definitions, if $f(\mathbf{x})$ is dimensionless [as is the case for instance for $\Phi(\mathbf{x})$ and $\delta(\mathbf{x})$], its Fourier transform has dimensions of $k^{-3/2}$. Therefore quantities such as $k^{3/2}\tilde{\delta}(\mathbf{k})$ are dimensionless.

Until now we have studied how scalar modes evolve, given their initial values. The initial values are specified by the function $A(k)$ for adiabatic scalar perturbation; see eqs. (19.82)–(19.85). Similarly, we will see in Section 19.5 that in the tensor sector we must assign the initial values $h_{A,\text{in}}(\mathbf{k})$. The important point about these initial values is that they are stochastic variables. No theory of initial conditions can predict for instance the initial value of the Bardeen variable $\Phi_{\text{in}}(\mathbf{x})$ at a specific point $\mathbf{x}$ in space, in the specific realization of the Universe in which we live. The best that we can hope to predict (and to measure) is statistical information about the correlators of these quantities. This is particularly

[17]We consider a volume V with size much larger than the wavelength of the modes under considerations. For such modes the ensemble average over several realizations of the Universe, that we will introduce below, can be replaced by a spatial average over several independent volumes, assuming a form of ergodic hypothesis. In turn, for these modes, we can still treat the momenta as a continuous set (even if, strictly speaking, in a finite volume with vanishing boundary condition, momenta become discrete), and we can use the same inverse Fourier transform as in the infinite-volume limit, except that our choice of inserting a normalization factor $V^{-1/2}$ in eq. (19.161) induces a factor $V^{1/2}$ in eq. (19.162).

evident when the initial fluctuations originate from quantum fluctuations in the early Universe, as in inflationary theories. Thus $\Phi_{\text{in}}(\mathbf{x})$ [or, in momentum space, $\tilde{\Phi}_{\text{in}}(\mathbf{k})$] is a stochastic variable. In the case of Gaussian fluctuations, all non-trivial information about a stochastic variable is contained in its two-point correlator $\langle \Phi_{\text{in}}(\mathbf{x})\Phi_{\text{in}}(\mathbf{x}') \rangle$. Any sign of non-Gaussianity in the primordial initial conditions would be of great interest, since it would give us indications on the specific mechanisms responsible for the generation of the initial perturbations in the primordial Universe. However, at present all cosmological data are consistent with Gaussian initial conditions, and in the following we will only consider Gaussian initial conditions.

Some information on these correlators comes first of all from the fact that the FRW background is invariant under spatial translations and under rotations. The invariance under spatial translation ensures that $\langle \Phi_{\text{in}}(\mathbf{x})\Phi_{\text{in}}(\mathbf{x}') \rangle = f(\mathbf{x} - \mathbf{x}')$, and invariance under spatial rotation further implies that $f(\mathbf{x} - \mathbf{x}')$ is actually a function of $|\mathbf{x} - \mathbf{x}'|$ only. For the correlator between the momentum modes we then have

$$\langle \tilde{\Phi}_{\text{in}}(\mathbf{k})\tilde{\Phi}_{\text{in}}^*(\mathbf{k}') \rangle = \frac{1}{V} \int_V d^3x \, d^3x' \langle \Phi_{\text{in}}(\mathbf{x})\Phi_{\text{in}}(\mathbf{x}') \rangle e^{-i\mathbf{k}\cdot\mathbf{x}} e^{i\mathbf{k}'\cdot\mathbf{x}'}$$

$$= \frac{1}{V} \int_V d^3X \, d^3y \, f(y) e^{-i(\mathbf{k}-\mathbf{k}')\cdot\mathbf{X}} e^{-i(\mathbf{k}+\mathbf{k}')\cdot\mathbf{y}/2} \,, \quad (19.163)$$

where $\mathbf{X} = (\mathbf{x} + \mathbf{x}')/2$ and $\mathbf{y} = \mathbf{x} - \mathbf{x}'$. The integral over $\mathbf{X}$ factorizes and gives $(2\pi)^3 \delta^{(3)}(\mathbf{k} - \mathbf{k}')$, and the remaining integral depends only on $k = |\mathbf{k}|$ since in it there is no preferred spatial direction. Thus, in a FRW background the two-point correlator in momentum space has the general form

$$\langle \tilde{\Phi}_{\text{in}}(\mathbf{k})\tilde{\Phi}_{\text{in}}^*(\mathbf{k}') \rangle = \frac{1}{V}(2\pi)^3 \delta^{(3)}(\mathbf{k} - \mathbf{k}') P_{\Phi,\text{in}}(k) \,. \quad (19.164)$$

All non-trivial information is in the function $P_{\Phi,\text{in}}(k)$, which is called the *power spectrum* of Φ_{in}. The label "in" attached to $P_{\Phi,\text{in}}(k)$ stresses that this is actually the *primordial* power spectrum, i.e. the power spectrum of the initial fluctuations. Since in a finite volume

$$(2\pi)^3 \delta^{(3)}(\mathbf{k} = 0) = V \,, \quad (19.165)$$

we also have the equality

$$P_{\Phi,\text{in}}(k) = \langle |\tilde{\Phi}_{\text{in}}(\mathbf{k})|^2 \rangle \,. \quad (19.166)$$

Observe that, with our definition (19.161), $\tilde{\Phi}(\eta, \mathbf{k})$ has dimensions $k^{-3/2}$ and therefore its power spectrum has dimensions of volume. The factor $V^{-1/2}$ in eq. (19.161) was chosen so as to get rid of explicit factors of V in eq. (19.166). The same holds for the power spectra of all other quantities, such as a density contrast; $\tilde{\delta}(\eta, \mathbf{k})$ has dimensions $k^{-3/2}$, and its primordial power spectrum

$$P_{\delta,\text{in}}(k) = \langle |\tilde{\delta}_{\text{in}}(\mathbf{k})|^2 \rangle \quad (19.167)$$

has again dimensions of a volume. In terms of the function $P_{\Phi,\mathrm{in}}(k)$ the two-point correlator in coordinate space reads

$$
\langle \Phi_{\mathrm{in}}(\mathbf{x})\Phi_{\mathrm{in}}(\mathbf{x}') \rangle = V \int \frac{d^3 k}{(2\pi)^3} \frac{d^3 k'}{(2\pi)^3} \langle \tilde{\Phi}_{\mathrm{in}}(\mathbf{k})\tilde{\Phi}_{\mathrm{in}}^*(\mathbf{k}') \rangle e^{i\mathbf{k}\cdot\mathbf{x} - i\mathbf{k}'\cdot\mathbf{x}'}
$$

$$
= \int \frac{d^3 k}{(2\pi)^3} P_{\Phi,\mathrm{in}}(k) e^{i\mathbf{k}\cdot(\mathbf{x}-\mathbf{x}')} \,. \tag{19.168}
$$

In particular, for the variance of the field, we get

$$
\langle \Phi_{\mathrm{in}}^2(\mathbf{x}) \rangle = \int \frac{d^3 k}{(2\pi)^3} P_{\Phi,\mathrm{in}}(k)
$$

$$
= \frac{1}{2\pi^2} \int_0^\infty \frac{dk}{k} k^3 P_{\Phi,\mathrm{in}}(k) \,. \tag{19.169}
$$

It is therefore convenient to define the function

$$
\boxed{\mathcal{P}_{\Phi,\mathrm{in}}(k) = \frac{k^3}{2\pi^2} P_{\Phi,\mathrm{in}}(k) \,,} \tag{19.170}
$$

which is also usually called the power spectrum, and gives the power per logarithmic interval of k. It is also common to denote $\mathcal{P}_{\Phi,\mathrm{in}}(k)$ as $\Delta_{\Phi,\mathrm{in}}^2(k)$. Note that $\mathcal{P}_{\Phi,\mathrm{in}}(k)$ is dimensionless. Similarly, for the matter density contrast we define

$$
P_{M,\mathrm{in}}(k) = \langle |\tilde{\delta}_{M,\mathrm{in}}(\mathbf{k})|^2 \rangle \,, \tag{19.171}
$$

$$
\mathcal{P}_{M,\mathrm{in}}(k) = \frac{k^3}{2\pi^2} P_{M,\mathrm{in}}(k) \,, \tag{19.172}
$$

and

$$
\Delta_{M,\mathrm{in}}(k) = [\mathcal{P}_{M,\mathrm{in}}(k)]^{1/2} \,, \tag{19.173}
$$

and so on for all other quantities that appear in cosmological perturbation theory.

We can now better clarify the statement in eq. (19.82) that the initial value of $\tilde{\Phi}_{\mathrm{in}}(\mathbf{k})$ is equal to a function $A(k)$. Actually, $\tilde{\Phi}_{\mathrm{in}}(\mathbf{k})$ is a stochastic variable with zero average,

$$
\langle \tilde{\Phi}_{\mathrm{in}}(\mathbf{k}) \rangle = 0 \,, \tag{19.174}
$$

while

$$
\langle |\tilde{\Phi}_{\mathrm{in}}(\mathbf{k})|^2 \rangle = A^2(k) = P_{\Phi,\mathrm{in}}(k) \,. \tag{19.175}
$$

19.4.2   The primordial power spectrum

In the scalar sector the initial conditions can be assigned giving the primordial power spectrum of Φ. An equivalent possibility, which in fact is nowadays more common, is to work in terms of the primordial power spectrum of $\mathcal{R}$, $\mathcal{P}_{\mathcal{R},\mathrm{in}}(k)$, which is usually called the (primordial) curvature power spectrum because of the relation of $\mathcal{R}$ with the three-dimensional curvature of surfaces in the comoving frame; see eq. (19.107). Because of eq. (19.109), the two are related by

$$\mathcal{P}_{\mathcal{R},\text{in}}(k) = \frac{9}{4}\mathcal{P}_{\Phi,\text{in}}(k) \, . \tag{19.176}$$

The simplest possible form for these power spectra, at least in a range of values of k around a reference value k_* (called the "pivot scale"), is just a power law,[18]

$$\mathcal{P}_{\mathcal{R},\text{in}}(k) = A_{\mathcal{R}}\left(\frac{k}{k_*}\right)^{n_s - 1} \, . \tag{19.177}$$

This form, as we will see, is suggested in particular by the inflationary mechanism of generation of initial conditions. In terms of $\mathcal{P}_{\Phi,\text{in}}(k)$ we therefore have

$$\mathcal{P}_{\Phi,\text{in}}(k) = A_{\Phi}\left(\frac{k}{k_*}\right)^{n_s - 1} \, , \tag{19.178}$$

with

$$A_{\Phi} = \frac{4}{9}A_{\mathcal{R}} \, . \tag{19.179}$$

The corresponding expression for $P_{\Phi,\text{in}}(k)$ is[19]

$$P_{\Phi,\text{in}}(k) = \frac{2\pi^2}{k^3}A_{\Phi}\left(\frac{k}{k_*}\right)^{n_s - 1} \, . \tag{19.181}$$

The scalar amplitude $A_{\mathcal{R}}$ and the scalar spectrum power-law index n_s are taken among the parameters of the cosmological model, and are fitted to the data. The quantity $n_s - 1$ is also called the spectral tilt, and parametrizes the deviations from a flat spectrum. In general, a power spectrum with $n_s - 1 > 0$ is called "blue", while a power spectrum with $n_s - 1 < 0$ is called "red". The flat spectrum with $n_s = 1$ is called the Harrison–Zeldovich spectrum. The pivot scale k_* is a parameter chosen by the experimentalist, typically in the middle of the range of scales probed by the experiment.[20]

A more detailed parametrization of the power spectrum that includes small deviations from a power-law behavior introduces on more parameter, $dn_s/d\log k$, called the *running index*. This is defined assuming that, around the pivot scale k_*, $\log[\mathcal{P}_{\mathcal{R},\text{in}}(k)]$ is a slowly varying function of $\log k/k_*$,

$$\mathcal{P}_{\mathcal{R},\text{in}}(k) \simeq A_{\mathcal{R}} \exp\left[(n_s - 1)\log\frac{k}{k_*} + \frac{1}{2}\left(\frac{dn_s}{d\log k}\right)\log^2\frac{k}{k_*}\right] \, , \tag{19.182}$$

or, equivalently,

$$\boxed{\mathcal{P}_{\mathcal{R},\text{in}}(k) \simeq A_{\mathcal{R}}\left(\frac{k}{k_*}\right)^{n_s - 1 + \frac{1}{2}\left(\frac{dn_s}{d\log k}\right)\log(k/k_*)}} \, . \tag{19.183}$$

[18]Observe that both $A_{\mathcal{R}}$ and n_s depend on the pivot scale chosen in the definition, i.e. $A_{\mathcal{R}} = A_{\mathcal{R}}(k_*)$ and $n_s = n_s(k_*)$. Typically, we will not write this dependence explicitly, in order not to burden the notation.

[19]In the literature, instead of A_{Φ}, the quantity δ_H defined by $A_{\Phi} = (5\delta_H/3)^2$ is sometimes used. Then

$$P_{\Phi,\text{in}}(k) = \frac{50\pi^2}{9k^3}\delta_H^2\left(\frac{k}{k_*}\right)^{n_s - 1} \, . \tag{19.180}$$

[20]As we already mentioned, the value of $A_{\mathcal{R}}$ depends on the choice made for the pivot scale and, when comparing results from different experiments, one must pay attention to the values chosen. Typical pivot scales often used by CMB experiment such as WMAP and *Planck* are $k_* = 0.002\,\text{Mpc}^{-1}$, or $k_* = 0.05\,\text{Mpc}^{-1}$, see also Note 40 on page 489.

The current best-fit values from the 2015 data release of the *Planck* mission are (at 68% c.l.)

$$n_s = 0.9677 \pm 0.0060 \tag{19.184}$$

and

$$\log(10^{10} A_{\mathcal{R}}) = 3.062 \pm 0.029 \tag{19.185}$$

(defined with a pivot scale $k_* = 0.05 \, \text{Mpc}^{-1}$) so

$$A_{\mathcal{R}} \simeq 2.14 \times 10^{-9} \tag{19.186}$$

and $A_\Phi \simeq 0.95 \times 10^{-9}$. The running index is currently consistent with zero: $dn_s/d \log k = -0.0057 \pm 0.0071$ at 95% c.l.[21]

From eq. (19.181) we see that the function $A(k)$ defined in eq. (19.175) [and which enters in the adiabatic initial conditions (19.82)–(19.85), although in the stochastic sense described by eqs. (19.174) and (19.175)] is related to A_Φ by

$$A^2(k) = \frac{2\pi^2}{k^3} A_\Phi \left(\frac{k}{k_*} \right)^{n_s - 1} . \tag{19.187}$$

A good measure of the typical size of the initial matter density contrast is given by the quantity $\Delta_{M,\text{in}}(k)$ defined in eq. (19.173). Using eq. (19.84) we get

$$\Delta_{M,\text{in}}(k) = \frac{3}{2} \left[\mathcal{P}_{\Phi,\text{in}}(k) \right]^{1/2}$$

$$= \frac{3}{2} A_\Phi^{1/2} \left(\frac{k}{k_*} \right)^{(n_s - 1)/2} . \tag{19.188}$$

Neglecting the tilt, from eqs. (19.179) and (19.186) we see that the typical initial value of the matter density contrast is then given by

$$\Delta_{M,\text{in}}(k) \simeq 4.6 \times 10^{-5} , \tag{19.189}$$

independent of k. Therefore, a given momentum mode leaves the linear regime when it is amplified by a factor of order a few times 10^4. From the numerical integration of the evolution equations we find that (with our values of the cosmological parameters) the mode that is amplified so that, today, $\Delta_M(k) \simeq 1$, is given by $\lambdabar_{\text{nl}} \simeq 16.4$ Mpc; compare also with Fig. 19.4. In terms of momenta, measured in units of $h_0/$Mpc (taking into account that this numerical result was obtained setting $h_0 = 0.7$), for the value of the cosmological parameters that we have used, the non-linear modes are those with $k > k_{\text{nl}}$, where,

$$k_{\text{nl}}(z = 0) \simeq 0.09 \, \frac{h_0}{\text{Mpc}} . \tag{19.190}$$

The argument $z = 0$ reminds us that this is the mode that becomes non-linear today, at $z = 0$. Actually, for cosmological structure formation it is more relevant to know which is mode that enters the non-linear regime at the redshift at which observations are performed. Taking $z = 1$ as a typical redshift probed by these observations, we get

$$k_{\text{nl}}(z = 1) \simeq 0.16 \, \frac{h_0}{\text{Mpc}} . \tag{19.191}$$

[21]See Table 4 of [Planck Collaboration], Ade *et al.* (2016a). Observe that the values of the cosmological parameters, such as n_s and $A_{\mathcal{R}}$, depend on the cosmological model used to fit the data, and the values given here are obtained assuming ΛCDM with its minimal set of six free parameters. A possible choice of these six independent parameters is given by the baryon density $\omega_b = \Omega_b h_0^2$, the cold dark matter density $\omega_c = \Omega_c h_0^2$, the Hubble parameter today $H_0 = h_0(100 \, \text{km s}^{-1}\text{Mpc}^{-1})$, the amplitude of scalar perturbation $A_{\mathcal{R}}$, the scalar spectrum index n_s and the redshift at which the Universe is half-reionized z_{re}, while neutrino masses are fixed to the minimal value allowed by oscillation experiments, $\sum_\nu m_\nu = 0.06$ eV. Extensions of this minimal model including new parameters describing for example more general neutrino masses, extra neutrino families, running of the spectral indices, GWs, etc. give slightly different best-fit values and in general larger errors in the parameter determination, because of the extra degeneracies induced by the new parameters. The value of the parameters also depends on the choice of datasets that are combined. For instance, the value of n_s given in eq. (19.184) is obtained combining the *Planck TT* spectrum with *Planck* lowP data (i.e. polarization data at low multipoles, $l < 29$) and lensing. Using only TT+lowP one obtains for instance $n_s = 0.9655 \pm 0.0062$, while TT+lowP+BAO gives $n_s = 0.9673 \pm 0.0045$. This gives an idea of how the numbers can change depending on the datasets used.

19.4.3 Transfer function and growth rate

The result of the numerical integration of the cosmological perturbation equations for the different modes can be conveniently summarized in terms of simple fitting formulas. Going back to the evolution of $\tilde{\Phi}(a, k)$ shown in Fig. 19.3, we can first of all ask how the value of $\tilde{\Phi}(a, k)$ at the plateau during MD is related to the initial value deep in RD, at least in linear theory. In this case each mode evolves separately, and the relation between the initial and final values is linear, so one can write $\tilde{\Phi}(a_{\rm MD}, k) = T(k; a_{\rm MD}, a_{\rm in})\tilde{\Phi}(a_{\rm in}, k)$, where $T(k; a_{\rm MD}, a_{\rm in})$ is called the *transfer function*. We already encountered this concept in Vol. 1, in the context of the theory of the linear response of a detector to an input signal. Here $a_{\rm MD}$ is a value of the scale factor, chosen on the plateau in MD; see Fig. 19.3. Its precise value is not very important. For definiteness, one can fix for instance $a_{\rm MD} = 0.03$, corresponding to $x_{\rm MD} \simeq -3.51$. The exact value of $a_{\rm in}$ is irrelevant, as long as it is deep in RD. We will denote $T(k; a_{\rm MD}, a_{\rm in})$ simply by $T(k)$. We further observe that, in the limit $k \to 0$, the value of $T(k)$ was computed analytically in eq. (19.157), and is equal to $9/10$. It is convenient to extract this factor explicitly, and define the transfer function from

$$\tilde{\Phi}(a_{\rm MD}, k) = \frac{9}{10} T(k)\tilde{\Phi}(a_{\rm in}, k), \qquad (19.192)$$

so that now, by definition, $T(0) = 1$. The subsequent evolution, from $a = a_{\rm MD}$ until the present time, depends on the dark-energy content of the theory. In ΛCDM the evolution is that shown in Fig. 19.3. In any case, it is convenient to separate the evolution until the end of the plateau, which basically depends only on the matter and radiation content of the theory, which is by now relatively well established, from the subsequent evolution, which in contrast is sensitive to the specific dark-energy model used.[22] To parametrize the subsequent evolution one introduces a function $D(a)$, called the *growth function* [the reason for this name will become clear from eq. (19.195)], and writes the value of $\tilde{\Phi}$ at $a > a_{\rm MD}$ as

$$\boxed{\tilde{\Phi}(a, k) = \frac{9}{10} T(k)\frac{D(a)}{a}\,\tilde{\Phi}(a_{\rm in}, k).} \qquad (19.193)$$

[22]The evolution during RD and MD can, however, be affected in so-called early dark-energy models, in which dark energy can still provide a fraction of the total energy density of the Universe even in the far past. This is of course not the case for ΛCDM; compare with Fig. 17.1 on page 380.

By definition, in a model with $\Omega_\Lambda = 0$, we have $D(a) = a$, so that the extra factor $D(a)/a = 1$, and $\tilde{\Phi}$ remains constant from $a_{\rm MD} \simeq 0.03$ up to the present time. In principle, the evolution at $a > a_{\rm MD}$ might also depend on k. However, the numerical integration shows that, in ΛCDM, the function $D(a)$ is basically independent of k, since all relevant modes are now well inside the horizon, and we can compute their subsequent evolution in the $k \to \infty$ limit. The parametrization (19.193) therefore allows us to encode the evolution into a function $T(k)$ that describes the evolution from the initial time until a value of a deep in the MD phase, which depends on k but not on a, since in this region $\tilde{\Phi}$ has reached its MD plateau and is therefore independent of time, and a subsequent evolution from the end of the MD phase to the present epoch, which depends on a but is the same for all modes, and is therefore described by a function $D(a)$.

This parametrization also allows us to express $\delta_M(a, k)$ in terms of the initial conditions. Indeed, as we see from Fig. 19.2 on page 445 and the discussion around it, all modes of interest for cosmology, except for the largest observable ones, are already well inside the horizon for $a > a_{\rm MD}$. Therefore, for $a > a_{\rm MD}$ these modes satisfy the Poisson equation (19.50). In ΛCDM the leading contribution to the right-hand side comes from the matter density contrast, since there are no dark-energy perturbations (and the contribution of radiation, $\bar{\rho}_R \delta_R$ today is negligible compared to $\bar{\rho}_M \delta_M$). Therefore for $a > a_{\rm MD}$ we have

$$
\begin{aligned}
k^2 \tilde{\Phi}(a, k) &\simeq 4\pi G \bar{\rho}_M(a) a^2 \tilde{\delta}_M(a, k) \\
&= \frac{3}{2a} H_0^2 \Omega_M \tilde{\delta}_M(a, k) ,
\end{aligned} \tag{19.194}
$$

where we have used $\bar{\rho}_M(a) = \bar{\rho}_{M,0}/a^3 = \rho_0 \Omega_M/a^3$. Then,

$$
\begin{aligned}
\tilde{\delta}_M(a, k) &\simeq \frac{2k^2 a}{3 H_0^2 \Omega_M} \tilde{\Phi}(a, k) \\
&= \frac{3}{5} \frac{k^2 T(k)}{H_0^2 \Omega_M} D(a) \tilde{\Phi}_{\rm in}(k) \\
&= \frac{2}{5} \frac{k^2 T(k)}{H_0^2 \Omega_M} D(a) \tilde{\delta}_{M,\rm in}(k) ,
\end{aligned} \tag{19.195}
$$

where in the last line we have eliminated $\tilde{\Phi}_{\rm in}$ in favor of $\tilde{\delta}_{M,\rm in}$ using the adiabatic initial condition (19.84). Observe that $D(a)$ describes the growth of the matter density contrast. When the cosmological constant is negligible, $D(a) = a$ and we recover the result given in eq. (19.144) for sub-horizon modes in MD. When the cosmological constant starts to become important, $D(a)$ grows more slowly than a and eventually goes to a constant deep in ΛD, as we see from the numerical integration in Fig. 19.3 and the analytic result of Section 19.3.4. Thus, in ΛCDM the growth of $\tilde{\delta}_M$ eventually shuts off.

The values of $\tilde{\Phi}(a, k)$ and of $\tilde{\delta}_M(a, k)$ in the recent epoch are therefore known, as functions of the initial values, if we know the functions $T(k)$ and $D(a)$. In principle $T(k)$ can be obtained, for each k, by the numerical integration of the evolution equations. The result can, however, be compactly summarized by a fitting formula known, after Bardeen, Bond, Kaiser and Szalay, as the BBKS transfer function, and given by[23]

[23]The BBKS transfer function is also often written in terms of the variable $q = k/(h_0^2 \Omega_M \, {\rm Mpc}^{-1})$ instead of $k/k_{\rm eq}$, in which case eq. (19.197) reads

$$
\begin{aligned}
T(q) = &\frac{\log(1 + 2.34q)}{2.34q} [1 + 3.89q \\
&+ (16.1q)^2 + (5.46q)^3 + (6.71q)^4]^{-1/4} .
\end{aligned} \tag{19.196}
$$

This transfer function does not include the effect of baryons, which is studied in Eisenstein and Hu (1998).

$$
\begin{aligned}
T(\kappa) = &\frac{\log(1 + 0.171\kappa)}{0.171\kappa} \\
&\times \left[1 + 0.284\kappa + (1.18\kappa)^2 + (0.399\kappa)^3 + (0.490\kappa)^4\right]^{-1/4} ,
\end{aligned} \tag{19.197}
$$

where $\kappa = k/k_{\rm eq}$ and $k_{\rm eq} \simeq 0.073\, h_0^2 \Omega_M\, {\rm Mpc}^{-1}$; see eq. (19.53). We plot this function in Fig. 19.10. Observe that $T(0) = 1$, as it should, while in the limit $k \gg k_{\rm eq}$ we have

$$T(k) \propto \frac{\log k}{k^2}\,. \qquad (19.198)$$

This can be understood observing that the modes with $k \gg k_{\rm eq}$ enter the horizon deep in RD. By the time they reach the RD–MD transition, according to eq. (19.128) $\tilde{\Phi}(k)$ has decreased by a factor of order $1/(k\eta_{\rm eq})^2 = (k_{\rm eq}/k)^2$. However, this result is only correct far from the RD–MD transition. Close to the transition there is a further logarithmic enhancement, as we have found for $\tilde{\delta}_M$ in eq. (19.140).

The growth function can be computed combining eqs. (19.119) and (19.121) into a second order equation for $\tilde{\delta}_M$,

$$\begin{aligned}
\partial_x^2 \tilde{\delta}_M &= -3\partial_x^2 \tilde{\Phi} - \partial_x \hat{\theta}_M \\
&= -3\partial_x^2 \tilde{\Phi} + (2+\xi)\hat{\theta}_M + \hat{k}^2 \tilde{\Phi} \\
&= -3\partial_x^2 \tilde{\Phi} - (2+\xi)(\partial_x \tilde{\delta}_M + 3\partial_x \tilde{\Phi}) + \hat{k}^2 \tilde{\Phi}\,, \quad (19.199)
\end{aligned}$$

so

$$\partial_x^2 \tilde{\delta}_M + (2+\xi)\partial_x \tilde{\delta}_M = -3\partial_x^2 \tilde{\Phi} - 3(2+\xi)\partial_x \tilde{\Phi} + \hat{k}^2 \tilde{\Phi}\,. \qquad (19.200)$$

For modes well inside the horizon, on the right-hand side the term $\hat{k}^2 \tilde{\Phi}$ dominates. Then, using the Poisson equation (19.194), which is valid for sub-horizon modes during the whole evolution from MD to ΛD, we get a single second-order equation for $\tilde{\delta}_M$,

$$\partial_x^2 \tilde{\delta}_M + (2+\xi)\partial_x \tilde{\delta}_M - \frac{3\Omega_M}{2a^3 h^2(x)}\tilde{\delta}_M = 0\,, \qquad (19.201)$$

where $h(x) = H(x)/H_0$. It is convenient to express the derivatives with respect to a, which gives

$$\frac{d^2\tilde{\delta}_M}{da^2} + (3+\xi)\frac{1}{a}\frac{d\tilde{\delta}_M}{da} - \frac{3\Omega_M}{2a^5 h^2(a)}\tilde{\delta}_M = 0\,. \qquad (19.202)$$

Denoting by $D(a)$ the solution that satisfies the initial condition $D(a) = a$ deep in MD, when the cosmological constant term is negligible, one finds

$$\boxed{D(a) = \frac{5\Omega_M}{2} h(a) \int_0^a \frac{da'}{[a' h(a')]^3}\,.} \qquad (19.203)$$

It can be seen that this is indeed a solution by direct substitution into eq. (19.202), while the behavior $D(a) \propto a$ in MD can be checked using the fact, that, in MD, $h(a) = \Omega_M^{1/2} a^{-3/2}$. The function $D(a)$ given by eq. (19.203) is easily computed numerically (actually, neglecting radiation, as we have already done in arriving at this equation, the integral can even be performed analytically; the result is, however, given

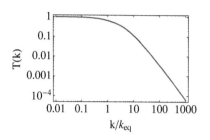

Fig. 19.10 The BBKS transfer function.

by a not very illuminating combination of hypergeometric functions). In Fig. 19.11 we plot the function $D(a)/a$ against the redshift z [so $a = (1+z)^{-1}$], setting $\Omega_M = 0.3$. With this value of Ω_M, the present value of the growth function is

$$D(a_0) \simeq 0.779 . \tag{19.204}$$

In Fig. 19.12 we show the dependence of $D(a_0)$ on Ω_M, with $\Omega_\Lambda = 1 - \Omega_M$. For $\Omega_M = 1$ we have $D(a_0) = 1$, as it should, i.e. we get back the result of a pure MD phase. As the relative importance of the cosmological constant is increased, $D(a_0)$ decreases toward zero, so the growth of structures is slowed down, as we indeed saw from the result of the numerical integration shown in Fig. 19.4.

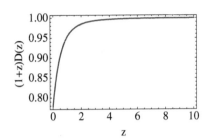

Fig. 19.11 The function $D(a)/a$, plotted against the redshift z.

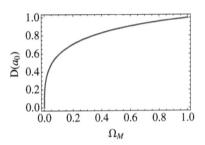

Fig. 19.12 The present value of the growth function, $D(a_0)$, as a function of Ω_M, with $\Omega_\Lambda = 1 - \Omega_M$.

19.4.4 The linearly processed power spectrum

The transfer function allows us to connect the primordial power spectrum to the linearly evolved power spectrum today. Actually, we have seen in eq. (19.191) that at, say, $z = 1$, for comoving momenta above $k_{\rm nl} \simeq 0.16 h_0/{\rm Mpc}$ the linear approximation breaks down. In principle, for higher momenta, i.e shorter length-scales, what one should do is to compute the power spectrum of matter, as a function of momenta and redshifts, with a full non-linear theory, for example using N-body simulations, and then compare this with the observational data. It is, however, common to shift the burden of dealing with the non-linear theory from the predictions to the data, in the following sense. Given an observation of an inhomogeneity at some redshift (say, CMB perturbations, or galaxy clustering, etc.) we can evolve it backward in time to infer the primordial spectrum that generated it. If the mode in question remained in the linear regime from the primordial epoch until the redshift at which the measurement refers (as is the case for the CMB), this can be done simply using cosmological perturbation theory. If in contrast the mode goes non-linear, more sophisticated methods are needed. In any case, evolving the measurement backward (with a full non-linear theory, if needed), we can find the value of $\tilde{\delta}_M(k)$ of this mode at the primordial epoch. Now we can formally evolve it again forward in time, up to the present epoch, using the *linear* transfer function, and present the result of the measurement in terms of the value that the power spectrum would have had at the present epoch if the evolution from the primordial epoch to the present time had been linear. Of course, this represents the true matter power spectrum of the present Universe only for modes with $k < k_{\rm nl}$, which indeed have remained linear. However, it is a useful way of displaying in a single plot measurements made at widely different redshifts. For instance, the CMB anisotropies give us information on the density fluctuations at $z \simeq 1100$, while measurements of galaxy clusters give information on redshifts, say, $0.1 \lesssim z \lesssim 1.6$, and so on.

The primordial power spectrum of Φ is given by eq. (19.166). Similarly, the power spectrum at a subsequent epoch, when the scale factor

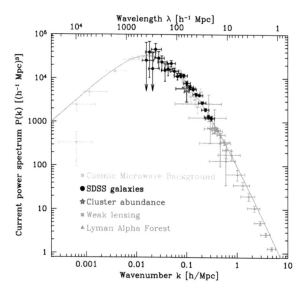

Fig. 19.13 The linear matter power spectrum today, compared with data from CMB, SDSS galaxies, cluster abundance, weak lensing and the Lyman alpha forest. The plot is obtained in a model with $\Omega_M = 0.28$, $h_0 = 0.72$, $n_s = 1$. From Tegmark *et al.* (2004).

has the value a, is

$$P_\Phi(a; k) = \langle |\tilde{\Phi}(a, \mathbf{k})|^2 \rangle . \tag{19.205}$$

Using eq. (19.193) we therefore have, at a generic epoch well after the RD–MD transition,

$$P_\Phi(a; k) = \left(\frac{9}{10} T(k) \right)^2 \left(\frac{D(a)}{a} \right)^2 P_{\Phi,\text{in}}(k) . \tag{19.206}$$

Similarly, for the matter power spectrum, using eq. (19.195) we get

$$P_M(a; k) = \left(\frac{3}{5} \frac{k^2 T(k) D(a)}{H_0^2 \Omega_M} \right)^2 P_{\Phi,\text{in}}(k) . \tag{19.207}$$

Using eq. (19.181) we get the linearly-processed matter power spectrum,

$$P_M(a; k) = \frac{18\pi^2}{25} A_\Phi \left(\frac{D(a)}{H_0^2 \Omega_M} \right)^2 k\, T^2(k) \left(\frac{k}{k_*} \right)^{n_s - 1} . \tag{19.208}$$

The function $P_M(k)$ is shown in Fig. 19.13, together with several observational data, reported on this plot as described above. Apart from the weak momentum dependence in the tilt factor, the dependence on k is given by the term $k T^2(k)$. Therefore we have the limiting behaviors

$$P_M(k) \propto k \qquad (k \ll k_{\text{eq}}) \tag{19.209}$$

and, from eq. (19.198),

$$P_M(k) \propto \frac{\log^2 k}{k^3} \qquad (k \gg k_{\rm eq}), \qquad (19.210)$$

and $P_M(k)$ has a maximum near $k \simeq k_{\rm eq}$.

In terms of the function $\mathcal{P}_M$ given in eq. (19.172), the asymptotic behaviors are then $\mathcal{P}_M(k) \propto k^4$ at low k and $\mathcal{P}_M(k) \propto \log^2 k$ at large k. When discussing the matter power spectrum, in the literature P_M rather than $\mathcal{P}_M$ is typically used, particularly when comparing with observations, as in Fig. 19.13. For the power spectrum of Φ the opposite is the case, and $\mathcal{P}_\Phi$ is more commonly used, particularly when computing the theoretical predictions, for example from inflationary models.

19.5   Tensor perturbations

In the previous sections we have studied the behavior of scalar perturbations. We should next consider vector and tensor perturbations. However, in most situations vector perturbations are of limited interest in cosmology. Writing the perturbation equations in the vector sector one finds that they have only decaying modes, in RD, MD and ΛD, and therefore any initial fluctuation in the vector sector quickly disappears. We therefore now turn to the study of tensor perturbations of the FRW metric. As we will see, for modes well inside the horizon, tensor perturbations correspond to GWs propagating in a FRW background.

19.5.1   Cosmological evolution

We work again in the conformal Newtonian gauge, and we keep only $h_{ij}^{\rm TT}$ in eq. (18.97), so that now

$$ds^2 = a^2 \left[-d\eta^2 + \left(\delta_{ij} + h_{ij}^{\rm TT} \right) dx^i dx^j \right]. \qquad (19.211)$$

The Christoffel symbols have already been given in eqs. (18.99)–(18.104). Then, for the Einstein tensor we get $\delta G_0^0 = 0$, $\delta G_0^i = 0$ and

$$\delta G_j^i = \frac{1}{2a^2} \left[(h_{ij}^{\rm TT})'' + 2\mathcal{H}(h_{ij}^{\rm TT})' - \boldsymbol{\nabla}^2 h_{ij}^{\rm TT} \right]. \qquad (19.212)$$

For the energy–momentum tensor we again use the decomposition given in eqs. (18.124)–(18.126), with Σ_{ij} given in eqs. (18.121)–(18.123). In the tensor sector, the perturbed Einstein equations (19.1) then give

$$(h_{ij}^{\rm TT})'' + 2\mathcal{H}(h_{ij}^{\rm TT})' - \boldsymbol{\nabla}^2 h_{ij}^{\rm TT} = 16\pi G a^2 \sigma_{ij}^{\rm TT}. \qquad (19.213)$$

It is convenient to go into momentum space and expand $h_{ij}^{\rm TT}$ in the basis of the polarization tensors. Then we write

$$\tilde{h}_{ij}^{\rm TT}(\eta, \mathbf{k}) = \sum_{A=+,\times} e_{ij}^A(\hat{\mathbf{k}}) \tilde{h}_A(\eta, \mathbf{k}), \qquad (19.214)$$

and similarly

$$\tilde{\sigma}_{ij}^{\mathrm{TT}}(\eta,\mathbf{k}) = \sum_{A=+,\times} e_{ij}^A(\hat{\mathbf{k}})\tilde{\sigma}_A(\eta,\mathbf{k})\,. \tag{19.215}$$

The polarization tensors $e_{ij}^A(\hat{\mathbf{k}})$ were given in eqs. (1.54)–(1.56) of Vol. 1, which we recall here,

$$e_{ij}^+(\hat{\mathbf{k}}) = \hat{\mathbf{u}}_i\hat{\mathbf{u}}_j - \hat{\mathbf{v}}_i\hat{\mathbf{v}}_j\,, \tag{19.216}$$

$$e_{ij}^\times(\hat{\mathbf{k}}) = \hat{\mathbf{u}}_i\hat{\mathbf{v}}_j + \hat{\mathbf{v}}_i\hat{\mathbf{u}}_j\,, \tag{19.217}$$

where $\hat{\mathbf{u}}$ and $\hat{\mathbf{v}}$ are unit vectors orthogonal to $\hat{\mathbf{k}}$ and to each other. Observe that these polarization tensors are normalized as[24]

$$e_{ij}^A(\hat{\mathbf{k}})e_{ij}^{A'}(\hat{\mathbf{k}}) = 2\delta^{AA'}\,. \tag{19.218}$$

In the frame where $\hat{\mathbf{k}}$ is along the $\hat{z}$ direction, we can choose $\hat{\mathbf{u}} = \hat{\mathbf{x}}$ and $\hat{\mathbf{v}} = \hat{\mathbf{y}}$, so

$$e_{ab}^+ = \begin{pmatrix} 1 & 0 \\ 0 & -1 \end{pmatrix}_{ab}, \qquad e_{ab}^\times = \begin{pmatrix} 0 & 1 \\ 1 & 0 \end{pmatrix}_{ab}, \tag{19.219}$$

with $a,b = 1,2$ spanning the (x,y) plane. Equation (19.213) then splits into two independent equations for the functions $h_A(\eta,k)$,

$$\boxed{\tilde{h}_A'' + 2\mathcal{H}\tilde{h}_A' + k^2\tilde{h}_A = 16\pi Ga^2\tilde{\sigma}_A\,.} \tag{19.220}$$

Before performing the numerical integration, we can extract analytically some general features of the solution. It is convenient to introduce a field $\tilde{\chi}_A(\eta,\mathbf{k})$ from

$$\tilde{h}_A(\eta,\mathbf{k}) = \frac{1}{a(\eta)}\tilde{\chi}_A(\eta,\mathbf{k})\,. \tag{19.221}$$

Then eq. (19.220) becomes

$$\tilde{\chi}_A'' + \left(k^2 - \frac{a''}{a}\right)\tilde{\chi}_A = 16\pi Ga^3\tilde{\sigma}_A\,. \tag{19.222}$$

We first consider the case $\sigma_A = 0$. This study is relevant to the situation in which GWs have been previously generated and we want to understand how such GWs evolve in the FRW background, after the source that generated them has switched off.[25] In this case

$$\tilde{\chi}_A'' + \left(k^2 - \frac{a''}{a}\right)\tilde{\chi}_A = 0\,. \tag{19.223}$$

In a generic cosmological epoch $a''/a \sim 1/\eta^2$. If the modes are well inside the horizon, $k\eta \gg 1$, the term a''/a can be neglected.[26] Then, inside the horizon, the solution for $\tilde{\chi}_A$ is just a simple oscillating function

$$\tilde{\chi}_A(\eta,\mathbf{k}) \propto \sin(k\eta + \alpha) \qquad (k\eta \ll 1) \tag{19.224}$$

[24]Recall that, after having performed the helicity decomposition, and therefore having separated the temporal index from the spatial indices, our convention is that the spatial indices are raised and lowered by δ_{ij}, and repeated lower indices, or repeated upper indices are summed over.

[25]However, we will see in Section 19.5.2 that relativistic particles such as neutrinos, moving on geodesics of the metric that includes the tensor perturbations, generate an anisotropic stress $\sigma_A = O(h)$, and therefore affect eqs. (19.220) and (19.222).

[26]In RD the estimate $a''/a \sim 1/\eta^2$ does not hold since $a \propto \eta$ and $a'' = 0$. However, in this case setting a''/a to zero is exact, rather than just an approximation valid inside the horizon.

and therefore

$$\tilde{h}_A(\eta, \mathbf{k}) \propto \frac{1}{a(\eta)} \sin(k\eta + \alpha) \qquad (k\eta \ll 1), \qquad (19.225)$$

with the overall amplitude and the phase α fixed by the initial conditions. The scaling $\tilde{h}_A(\eta, \mathbf{k}) \propto 1/a(\eta)$ times a oscillating factor is just what one expects from the fact that gravitons are massless particles. Indeed, for modes well inside the horizon we can apply the procedure discussed in Section 1.4 of Vol. 1 for computing the energy density of GWs, i.e. we can average the energy–momentum tensor over several wavelengths to obtain a gauge-invariant quantity (an operation that no longer makes sense if the wavelength is greater than the horizon size). Therefore their energy-density is given by the standard expression (1.136), and is proportional to $\sum_A \langle \dot{h}_A^2 \rangle$.[27] We use $\dot{h}_A = (1/a)h'_A$ and we observe that, if

$$h_A(\eta, \mathbf{k}) \propto \frac{\sin(k\eta + \alpha)}{a(\eta)}, \qquad (19.226)$$

then

$$h'_A(\eta, \mathbf{k}) \propto \frac{k \cos(k\eta + \alpha)}{a(\eta)} + O\left(\frac{1}{a^2}\right), \qquad (19.227)$$

so

$$\dot{h}_A(\eta, \mathbf{k}) \propto \frac{k \cos(k\eta + \alpha)}{a^2(\eta)} + O\left(\frac{1}{a^3}\right). \qquad (19.228)$$

In $\langle \dot{h}_A^2 \rangle$ the term $\cos^2(k\eta + \alpha)$, averaged over several periods, simply gives a factor $1/2$, and it then follows that $\rho_{\rm gw} \propto a^{-4}$, as indeed we expect for any form of radiation. Observe that only when the tensor perturbations are well inside the horizon can they be described as a collection of gravitons, i.e. as a collection of particles with well-defined energy and momentum, whose energy density scales as $1/a^4$.

We can now perform the numerical integration of eq. (19.220) with $\sigma_A = 0$. Since the two polarizations satisfy the same equation, and the evolution equation, as well as the initial conditions in the early Universe, depend on $\mathbf{k}$ only through $k = |\mathbf{k}|$, we will denote $\tilde{h}_A(\eta, \mathbf{k})$ simply by $\tilde{h}(\eta, k)$. To study eq. (19.220) we proceed as we have done for scalar perturbations. First of all, we set initial conditions deep in RD, for super-horizon modes. In this case eq. (19.220) becomes

$$\tilde{h}'' + \frac{2}{\eta}\tilde{h}' \simeq 0, \qquad (19.229)$$

which has a constant mode and a decaying mode $\tilde{h}(\eta, k) \propto 1/\eta$. Again, we neglect the decaying mode. Thus, at an initial time $\eta_{\rm in}$ deep in RD, we set the initial conditions $\tilde{h}(\eta_{\rm in}, k) = \tilde{h}_{\rm in}(k)$ and $\tilde{h}'(\eta_{\rm in}, k) = 0$. The value of $\tilde{h}_{\rm in}$ depends of course on the specific mechanism that generated GWs in the primordial Universe. However, since eq. (19.220) is linear in $\tilde{h}$, different momentum modes do not mix, and to study how a given mode evolves we can simply set $\tilde{h}_{\rm in}(k) = 1$, i.e. we study the evolution of $\tilde{h}(\eta, k)$ normalized to its initial value.

[27] We are considering here the energy density with respect to the physical coordinates, so in eq. (19.211) the explicit factor a^2 is reabsorbed by $a^2 dx^i dx^j = dx^i_{\rm phys} dx^j_{\rm phys}$. In this case the background metric takes the form (17.15), which at subhorizon scales approaches the Minkowski metric, and we can then use eq. (1.136), which was obtained by specifying the general expression (1.125) to a Minkowski background metric.

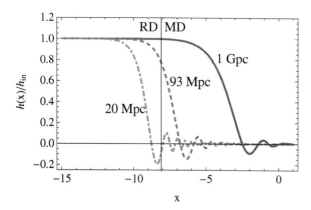

Fig. 19.14 The evolution of $\tilde{h}(\eta, k)$, normalized to its initial value, against $x = \log a(\eta)$, for three different values of the reduced wavelength $\lambdabar = 20$ Mpc, 93 Mpc and 1 Gpc.

To perform the numerical integration we pass again to the dimensionless variables $x = \log a$ and $\hat{k} = k/\mathcal{H}$. Then eq. (19.220) becomes

$$\partial_x^2 \tilde{h} + (3 + \xi)\partial_x \tilde{h} + \hat{k}^2 \tilde{h} = 0 \,, \qquad (19.230)$$

where ξ was defined in eq. (19.112), and the explicit expressions for $\xi(x)$ and $\hat{k}(x)$ in ΛCDM were given in eqs. (19.117) and (19.118). We neglect here the change in the number of effective relativistic species or the effect of neutrinos; compare with Note 13 on page 455. We will come back to this when we discuss the transfer function for tensor modes in Section 19.5.2. We use our reference cosmological model with matter, radiation and cosmological constant, with $\Omega_M = 0.3$ and $\Omega_\Lambda = 1 - \Omega_M - \Omega_R \simeq 0.7$. The result of the numerical integration is shown in Fig. 19.14, for the same reduced wavelengths that we plotted in the scalar sector in Fig. 19.3.

We see that $\tilde{h}(\eta, k)$ remains constant as long as the mode is well outside the horizon, i.e. as long as $k\eta \ll 1$. When it enters the horizon it quickly decays to zero, independently of whether the horizon entry is in RD or MD, in agreement with the $1/a$ behavior of sub-horizon modes discussed before. This has important implications for understanding the effect of primordial GWs on the CMB. By the epoch of decoupling, which according to eq. (17.94) takes place at a redshift $z_{\text{dec}} \simeq 1090$ and therefore $x_{\text{dec}} \simeq -7.0$, all short-wavelength GWs generated in the primordial Universe have essentially disappeared. Only long-wavelength modes can still have a sizable amplitude. Therefore, primordial GWs can affect the CMB on large angular scales, but not on small angular scales.[28] We will discuss their effect quantitatively in Chapter 20.

In Figs. 19.15 and 19.16 we plot the evolution of the modes with $\lambdabar = 20$ Mpc and $\lambdabar = 1$ Gpc, respectively, from a value of x for which they are already well within the horizon (we take $x > -10$ for the former and $x > -5$ for the latter) up to the cosmological future, deep in the

[28]More precisely, we will see in Section 20.3.5 [see in particular eq. (20.126) and the discussion following it] that the GW effect on the CMB is given by a line-of-sight integral, from decoupling up to the present time, of the time derivative of $\tilde{h}(\eta, \mathbf{k})$. For modes entering in RD, this becomes the integral over a function that oscillates rapidly.

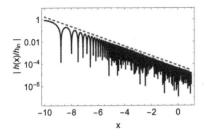

Fig. 19.15 The numerical evolution of the mode with $\bar{\lambda} = 20$ Mpc. At $x = -10$ this mode is already well inside the horizon, and we show its subsequent evolution in the RD, MD and ΛD phases (solid line). The dashed line is the curve $\propto 1/a = e^{-x}$, with amplitude at $x = -10$ chosen equal to $2\tilde{h}(x = -10)$ (the factor of 2 is introduced for visual reasons).

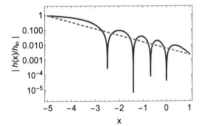

Fig. 19.16 The numerical evolution of the mode with $\bar{\lambda} = 1$ Gpc, from the value of $x = -5$ when it is already well inside the horizon, and its subsequent evolution in the MD and ΛD phases (solid line), compared with the behavior $1/a = e^{-x}$(dashed), with amplitude normalized to $\tilde{h}(x = -5)$.

[29]Note that we have not included the variation of the effective number of relativistic species, which could be done by using eq. (17.131). This modified expression for the scale factor should be used consistently both when computing $\mathcal{H}$ in eq. (19.220) and when expressing the amplitude of the solution $\tilde{h}(\eta, k)$ in terms of $a(\eta)$. In any case, we already know from the general argument following eq. (19.223) that the two effects compensate so that eventually, inside the horizon, $\tilde{h}(\eta, k) \propto 1/a(\eta)$ times an oscillating function.

ΛD phase. We now use a logarithmic vertical scale that allows us to follow in more detail the evolution when $|\tilde{h}(x)/\tilde{h}_{\rm in}| \ll 1$. The dashed line shows the behavior $\tilde{h}(x) \propto 1/a = e^{-x}$. We see that, in agreement with eq. (19.226), once a mode is inside the horizon it decays as $1/a$ times an oscillating factor, in all three phases, RD, MD and ΛD.

We can understand in more detail this behavior of $\tilde{h}(\eta, k)$ analytically. We will examine separately the different cosmological epochs, and we will also discuss how to get some analytic understanding of the transition from MD to ΛD.

Analytic solution in RD

Deep in RD eq. (19.220) (with $\tilde{\sigma}_A = 0$) reads

$$\tilde{h}'' + \frac{2}{\eta}\tilde{h}' + k^2\tilde{h} = 0\,, \tag{19.231}$$

This can be solved exactly, and has the solutions $\sin(k\eta)/(k\eta)$ and $\cos(k\eta)/(k\eta)$. We now impose the initial conditions $\tilde{h}(\eta_{\rm in}, k) = \tilde{h}_{\rm in}(k)$ and $\tilde{h}'(\eta_{\rm in}, k) = 0$, at an initial value $\eta_{\rm in}$ such that $k\eta_{\rm in} \to 0$. This selects the former solution, so

$$\boxed{\tilde{h}(\eta, k) \simeq \tilde{h}_{\rm in}(k)\frac{\sin(k\eta)}{k\eta} \qquad \text{(RD)}\,.} \tag{19.232}$$

For $k\eta \ll 1$ this solution is constant, i.e. it reduces to the constant mode that we identified above in the limit $k = 0$. In contrast, for $k\eta \ll 1$ the solution proportional to $\cos(k\eta)/(k\eta)$ goes as $1/\eta$, and is therefore the decaying mode.

Using eq. (17.130) for the scale factor we see that eq. (19.232) can be written as

$$\tilde{h}(\eta, k) = \tilde{h}_{\rm in}(k)\frac{a_{\rm in}}{a(\eta)}\frac{\sin(k\eta)}{k\eta_{\rm in}}\,. \tag{19.233}$$

Therefore, while outside the horizon $\tilde{h}(\eta, k)$ is constant, once inside the horizon $(k\eta \gg 1)$, $\tilde{h}(\eta, k)$ scales as $1/a(\eta)$ times an oscillating function, in agreement with eq. (19.226). Equation (19.232) fixes the overall amplitude and phase, which were left generic in eq. (19.226), for the solution with initial conditions $\tilde{h}(\eta_{\rm in}, k) = \tilde{h}_{\rm in}(k)$ and $\tilde{h}'(\eta_{\rm in}, k) = 0$.[29]

Analytic solution in MD

We consider next the MD phase. In MD we have $\mathcal{H} = 2/\eta$ and eq. (19.220) becomes

$$\tilde{h}'' + \frac{4}{\eta}\tilde{h}' + k^2\tilde{h} = 0\,. \tag{19.234}$$

Again, this can be solved exactly. Two independent solutions are

$$h_1(\eta, k) = \frac{1}{(k\eta)^2}\left[\frac{\sin(k\eta)}{k\eta} - \cos(k\eta)\right] \tag{19.235}$$

and

$$h_2(\eta, k) = \frac{1}{(k\eta)^2}\left[\frac{\cos(k\eta)}{k\eta} + \sin(k\eta)\right]. \qquad (19.236)$$

For $k\eta \to 0$ the solution $h_1(\eta, k)$ goes to a constant, $h_1(\eta, k) \to 1/3$, so it is a constant mode, while $h_2(\eta) \propto 1/\eta^3$ is the decaying mode. Again, their existence could have been derived directly by setting $k^2 = 0$ in eq. (19.234).[30] For a generic value of k the RD solution (19.232) will be matched to a linear combination of these two solutions. However, if we consider a mode k that is well outside the horizon at RD–MD equilibrium, $k\eta_{\rm eq} \ll 1$, the solution during RD will just be the constant solution. Since the function $h_1(\eta)$ satisfies $h_1(0) = 1/3$ and $h'(0) = 0$, a super-horizon RD solution will just match to $h_1(\eta)$, with only a small admixture of the decaying mode $h_2(\eta)$, with a coefficient that goes to zero as $k\eta_{\rm eq} \to 0$. Then, for modes such that $k\eta_{\rm eq} \ll 1$, during MD the solution is

$$\boxed{\tilde{h}(\eta, k) \simeq \frac{3\tilde{h}_{\rm in}(k)}{(k\eta)^2}\left[\frac{\sin(k\eta)}{k\eta} - \cos(k\eta)\right] \quad (\text{MD}, k\eta_{\rm eq} \ll 1),}$$

$$(19.240)$$

where $\tilde{h}_{\rm in}(k)$ is the initial value assigned deep in RD. This can also be written in terms of the Bessel function

$$J_{3/2}(z) = \sqrt{\frac{2}{\pi z}}\left(\frac{\sin z}{z} - \cos z\right) \qquad (19.241)$$

as

$$\tilde{h}(\eta, k) = \tilde{h}_{\rm in}(k)\frac{3(\pi/2)^{1/2}}{(k\eta)^{3/2}}J_{3/2}(k\eta). \qquad (19.242)$$

Equivalently, we can write it in terms of the spherical Bessel function $j_1(x) = x^{-2}(\sin x - x\cos x)$ as

$$\boxed{\tilde{h}(\eta, k) = \tilde{h}_{\rm in}(k)\frac{3j_1(k\eta)}{k\eta} \quad (\text{MD}, k\eta_{\rm eq} \ll 1).}$$

$$(19.243)$$

Similarly, the solution (19.236) is proportional to $(k\eta)^{-3/2}J_{-3/2}(k\eta)$.[31] If the value of k is not too small, eventually $k\eta$ will become large during MD, and the mode enters the horizon. In this regime the solution (19.240) becomes

$$\tilde{h}(\eta, k) \simeq -\frac{3\tilde{h}_{\rm in}(k)}{(k\eta)^2}\cos(k\eta) \quad (\text{MD}, k\eta_{\rm eq} \ll 1, k\eta \gg 1). \quad (19.245)$$

Recalling from eq. (17.143) that in MD the scale factor $a(\eta) \propto \eta^2$, we recover the result that, when the mode is well inside the horizon, $\tilde{h}(\eta, k) \propto 1/a(\eta)$, times an oscillating factors. Once again, this is the behavior expected for a collection of gravitons. Observe that this is true

[30]The fact that, as long as a GW is outside the horizon, there is a constant and a decaying mode, is not specific to RD or MD, but holds more generally. Indeed, for super-horizon modes eq. (19.220) (with the source term set to zero) becomes

$$\tilde{h}'' + 2\mathcal{H}\tilde{h}' \simeq 0. \qquad (19.237)$$

One solution is of course $\tilde{h}' = 0$ and therefore $\tilde{h}$ constant. The other solution is $\tilde{h}' \propto a^{-2}$. Writing $\tilde{h}' = ad\tilde{h}/dt = a\dot{a}d\tilde{h}/da$, for this solution we have

$$\frac{d\tilde{h}}{da} \propto \frac{1}{a^3\dot{a}}. \qquad (19.238)$$

If $a \propto t^n$ (with $n > 0$), $a^3\dot{a} \propto t^{4n-1} \propto a^{(4n-1)/n}$, so in this case

$$\frac{d\tilde{h}}{da} \propto \frac{1}{a^{(4n-1)/n}}. \qquad (19.239)$$

This is a decreasing mode (i.e. $|\tilde{h}|$ is a decreasing function of a, where $a \propto t^n$ increases with time), as long as $(4n-1)/n > 1$. This implies $n > 1/3$, which is always the case in a normal cosmological context. Indeed, repeating the analysis of Section 17.2, we see that the condition $n \leqslant 1/3$ requires an unphysical fluid with EoS parameter $w \geqslant 1$. The same happens for $a \propto e^{Ht}$, as in de Sitter expansion, since now $\dot{a} = Ha$ and $d\tilde{h}/da \propto a^{-4}$, or in quasi-de Sitter expansion. So, the solution $\tilde{h}' \propto a^{-2}$ describes a super-horizon decaying mode.

[31]More generally, if we take $a(\eta) \propto \eta^\alpha$, so that $\mathcal{H} = \alpha/\eta$, the solution of $\tilde{h}'' + 2\mathcal{H}\tilde{h}' + k^2\tilde{h} = 0$ that reduces to a constant $\tilde{h}_{\rm in}(k)$ outside the horizon is

$$\tilde{h}(\eta, k) = \tilde{h}_{\rm in}(k)\frac{\Gamma(\nu+1)}{(k\eta/2)^\nu}J_\nu(k\eta),$$

$$(19.244)$$

where $\nu = \alpha - (1/2)$. In MD $\alpha = 2$ and, using $\Gamma(5/2) = 3\sqrt{\pi}/4$, we get back eq. (19.242). In RD $\alpha = 1$ and, recalling that $\Gamma(3/2) = \sqrt{\pi}/2$ and $J_{1/2}(z) = [2/(\pi z)]^{1/2}\sin z$, we get back eq. (19.232).

also for the solution (19.236), with the only difference being that the oscillating factor is $\sin(k\eta)$ rather than $-\cos(k\eta)$. So, even if the condition $k\eta_{\rm eq} \ll 1$ is not satisfied, and therefore the matching with RD involves both functions $h_1(\eta, k)$ and $h_2(\eta, k)$, once the mode enters the horizon during MD it evolves as $1/a$ times oscillating factors [as indeed we already know from the discussion after eq. (19.223)], so its energy density scales as $1/a^4$, as it should.

Analytic solution in the ΛD phase

We next consider a de Sitter phase, such as the ΛD phase obtained when the cosmological constant begins to dominate. We use for the moment the scale factor written in the form $a(\eta) = -1/(H\eta)$. This is a convenient choice when studying a de Sitter phase in isolation. Actually, to understand the evolution across the MD–ΛD transition, one should be careful about the convention chosen for the overall constant that can always be added to η; see eqs. (17.158)–(17.161). In eq. (17.130) this constant is chosen so that, during RD and MD, η is positive, while, if one writes the de Sitter scale factor as $a(\eta) = -1/(H\eta)$, one has implicitly chosen a different constant, such that η is negative. When studying the whole evolution across RD, MD and ΛD, it is in general more convenient to use a variable η that evolves continuously, as we will do below when we study the MD–ΛD transition analytically. For the moment, using $a(\eta) = -1/(H\eta)$ we have $\mathcal{H} = -1/\eta$, so eq. (19.220) becomes

$$\tilde{h}'' - \frac{2}{\eta}\tilde{h}' + k^2\tilde{h} = 0. \tag{19.246}$$

The most general solution is

$$\tilde{h}(\eta, k) = \tilde{h}_{\rm in}(k)\Big\{\alpha_k\left[\cos(k\eta) + k\eta\sin(k\eta)\right] \\ + \beta_k\left[\sin(k\eta) - k\eta\cos(k\eta)\right]\Big\}, \tag{19.247}$$

with generic coefficients α_k, β_k. We consider first the wavelengths that are well inside the horizon during ΛD. Then we see from eq. (19.243) that, in the limit $|k\eta| \gg 1$,

$$\tilde{h}(\eta, k) \simeq \tilde{h}_{\rm in}(k)\, k\eta\left[\alpha_k\sin(k\eta) - \beta_k\cos(k\eta)\right]. \tag{19.248}$$

Recalling that this result has been obtained choosing the definition of η such that $a(\eta) = -1/(H\eta)$, we see that, once again, inside the horizon $\tilde{h}(\eta, k) \propto 1/a$ times an oscillating factor, in agreement with the discussion after eq. (19.223) and the numerical results shown in Figs. 19.15 and 19.16. We then obtain again the result $\rho_{\rm gw} \propto 1/a^4$, and therefore the interpretation in terms of a collection of gravitons.

Consider next a mode that is super-horizon at the end of the MD era and at the beginning of the ΛD era, i.e. a mode that, at that epoch, satisfies $k/\mathcal{H} \ll 1$. In eqs. (19.246) and (19.243) we have defined the additive constant in η so that $a(\eta) = -1/(H\eta)$, so in this case the condition $k/\mathcal{H} \ll 1$ becomes $|k\eta| \ll 1$. In this limit the term in eq. (19.243) proportional to α_k goes to a constant, so it is the constant mode, while the term proportional to β_k goes to $(k\eta)^3/3 \propto 1/a^3$, and is therefore a decaying mode (equivalently, since in a de Sitter expansion asymptotically $\eta \to 0^-$, the term $(k\eta)^3$ goes to zero during the de Sitter phase). Thus, for asymptotically small values of k, the mode stays constant even during ΛD, matching continuously the value $\tilde{h}(\eta, k) \simeq \tilde{h}_{\rm in}(k)$ of a constant super-horizon mode during MD. It should, however, be observed that the MD–ΛD transition takes place near the present epoch. Thus, the regime just described only takes place for modes for which λ is asymptotically large compared with H_0, and which are therefore unobservable. A mode that is super-horizon during MD, but still with a reduced wavelength that is not too large compared to H_0, will rather match at the MD–ΛD transition to a superposition of the constant and the decaying mode. This is illustrated in Fig. 19.17, where we show the behavior of two modes with long reduced wavelengths, $\lambda = 6$ Gpc and $\lambda = 10$ Gpc, obtained from the numerical integration of eq. (19.230). We plot the result up to $x = 5$ to emphasize the behavior in the phase dominated by the cosmological constant. These modes are already too large to be observable, since they are always outside the horizon, as we see from Fig. 19.2. Nevertheless, we see that their constant value during ΛD is different from that in MD.

To get an analytic understanding of this, let us perform explicitly the analytic matching between the MD and ΛD solutions.[32] As already mentioned, when one writes the MD scale factor as in eq. (17.143) and the ΛD scale factor as $a(\eta) = -1/(H\eta)$, one is implicitly using a different convention for the additive constant for η in the MD and ΛD phases, and therefore η is discontinuous at the transition. It can be useful to have also an approximate analytic treatment, in which we use a variable η that is continuous across the different epochs, such as RD, MD and ΛD, and match the approximate analytic solutions in the different epochs by requiring continuity of $\tilde{h}(\eta, k)$ and $\tilde{h}'(\eta, k)$ across the transitions.[33] In particular, to study the evolution across MD and ΛD, we can approximate

$$a(\eta) = \begin{cases} (1/4)H_0^2 \Omega_M \eta^2 & (\eta < \eta_*), \\ -1/[H(\eta - \eta_1)] & (\eta_* < \eta < \eta_1), \end{cases} \qquad (19.249)$$

where we have used eq. (17.143) for the scale factor during MD. We further require that the constant H in the ΛD phase be given by $H = H_0 \Omega_\Lambda^{1/2}$; see eq. (17.75). Requiring continuity of a and a' then fixes the matching point η_* and the additive constant η_1. We get $\eta_1 = (3/2)\eta_*$ and

$$\eta_* = 2H_0^{-1} \Omega_\Lambda^{-1/6} \Omega_M^{-1/3} . \qquad (19.250)$$

Of course, this way of matching abruptly two different phases is very crude, and is only useful to get a first analytic understanding. With this definition of η, the solution (19.243) reads

$$\tilde{h}(\eta, k) = \tilde{h}_{\rm in}(k)[\alpha_k(\cos u + u \sin u) + \beta_k(\sin u - u \cos u)], \qquad (19.251)$$

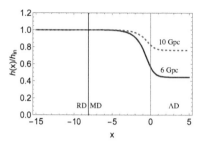

Fig. 19.17 The evolution of the super-horizon modes with $\lambda = 6, 10$ Gpc, on a scale that emphasizes the future ΛD phase.

[32]This exercise will also be useful later, when we compute the production of GWs during inflation through the Bogoliubov coefficients; see Section 21.3.7.

[33]Observe that in a numerical treatment, such as that based on eq. (19.230), the solution provides a unique function $\mathcal{H}(x) = dx/d\eta$, where $x = \log a$, and therefore implicitly defines a unique continuous function $\eta(x)$ obtained by integrating $d\eta = dx/\mathcal{H}(x)$, with a choice of the additive constant for η fixed, once and for all, for the whole evolution.

where $u = k(\eta - 3\eta_*/2)$. Note that now the variable η, in a de Sitter phase, tends asymptotically to $3\eta_*/2$ (i.e. $u \to 0^-$), while before we had $\eta \to 0^-$. Using in eq. (19.250) our reference values $h_0 = 0.7, \Omega_M = 0.3$ and $\Omega_\Lambda = 0.7$ we find that the matching, for a mode with $\lambdabar = 1/k$, is performed at $u = u_* \simeq -6.79\,\mathrm{Gpc}/\lambdabar$. We see that, for both $\lambdabar = 6$ Gpc and $\lambdabar = 10$ Gpc, $|u_*|$ is not small, so we are not in the regime where the the matching would give $\alpha_k \simeq 1$. Evolving these solutions further into ΛD (i.e. decreasing u toward zero) the decaying mode proportional to β_k gradually vanishes, while the solution $(\cos u + u \sin u) \to 1$. Therefore the constant plateau in ΛD differs by a factor α_k, compared with the value in MD. Requiring continuity of $\tilde{h}(\eta, k)$ and $\tilde{h}'(\eta, k)$ at $\eta = \eta_*$, for $\lambdabar = 6$ Gpc we get $\{\alpha_k \simeq 0.35, \beta_k \simeq -0.13\}$, while for $\lambdabar = 10$ Gpc we get $\{\alpha_k \simeq 0.71, \beta_k \simeq 0.32\}$. By comparison, the plateau values in Fig. 19.17 are at $0.44\tilde{h}_{\mathrm{in}}$ and $0.76\tilde{h}_{\mathrm{in}}$, respectively. Thus, this crude matching procedure give results that are qualitatively correct and, for these modes, are quantitatively accurate at the $(10-20)\%$ level.

19.5.2   Transfer function for tensor modes

We can now summarize the precedings results in terms of a transfer function $T_{\mathrm{GW}}(k)$ for tensor modes, defined by

$$\tilde{h}_A(\eta_0, k) = T_{\mathrm{GW}}(k)\tilde{h}_A(\eta_{\mathrm{in}}, k). \tag{19.252}$$

More generally, in Chapter 20 we will also make use of the transfer function $\mathsf{h}(k\eta)$, defined by

$$\tilde{h}_A(\eta, \mathbf{k}) = \mathsf{h}(k\eta)\,\tilde{h}_A(\eta_{\mathrm{in}}, \mathbf{k}), \tag{19.253}$$

so $T_{\mathrm{GW}}(k) = \mathsf{h}(k\eta_0)$. Observe that the free evolution in FRW, given by eq. (19.220) with $\sigma_A = 0$, is independent of A, so $T_{\mathrm{GW}}(k)$ is independent of the polarization. Furthermore, different polarizations do not mix. In the scalar case, in eq. (19.193) we explicitly extracted from the definition of the transfer function a numerical factor $9/10$ and a factor $D(a)/a$, which were convenient to factor out because of the specific evolution of scalar perturbations. For GWs we will rather use the definition (19.252).

Depending on its comoving momentum, a mode will re-enter the horizon during RD or during MD.[34] From the results of Section 19.5.1 it is clear that the transfer function has different behaviors depending on whether the mode re-enters the horizon in RD or in MD, i.e. on whether $k > k_{\mathrm{eq}}$ or $k < k_{\mathrm{eq}}$. The value k_{eq} of the comoving momentum of the mode that enters at the RD–MD transition was computed in eq. (19.53). It is useful to translate this into the corresponding value of the frequency $f_{\mathrm{eq}} = k_{\mathrm{eq}}/(2\pi)$. Reinstating the appropriate factor of c and expressing the result in Hz, we get

$$f_{\mathrm{eq}} \simeq 1.66 \times 10^{-17}\,\mathrm{Hz}\left(\frac{h_0}{0.7}\right)^2\left(\frac{\Omega_M}{0.3}\right). \tag{19.254}$$

Observe that in eq. (19.53) we computed the comoving momentum; since we are using units $a_0 = 1$, this is the same as the physical momentum

[34]Even larger modes never re-enter the horizon; see Fig. 19.2 on page 445. However, modes that are still super-horizon today are not observables. Note also that, as we will see in Section 21.3, during a phase of primordial inflation all relevant modes were initially inside the horizon and then left the horizon during inflation; see Fig. 21.4. Therefore, rather than "horizon entry", we use the expression "horizon re-entry".

today. So, gravitons whose physical frequency today is larger than the value f_{eq} given in eq. (19.254) re-entered the horizon during RD.

As we discussed in Vol. 1, ground-based interferometers are sensitive to GWs whose frequency today is in the range $O(10 - 10^3)$ Hz, while planned space-borne interferometers could operate in the range $10^{-4} - 10^{-1}$ Hz, and the pulsar timing arrays that we will discuss in Chapter 23 operate around 10^{-9} Hz. Beside astrophysical sources, these detectors could in principle detect a stochastic GW background generated in the early Universe by production mechanisms such as those that we will discuss in Chapters 21 and 22. In that case, since these detectors all operate in the regime $f \gg f_{eq}$, the corresponding GWs would correspond to primordial tensor modes that re-entered the horizon deep in the RD era.

In contrast, we will see in Section 20.1 [see in particular the discussion following eq. (20.17)] that the multipoles of the CMB temperature with $l < O(150)$ are affected by Fourier modes of scalar and tensor perturbations that re-entered during MD, while multipoles with $l > O(150)$ are affected by modes that re-entered during RD. So, the CMB is sensitive both to scalar modes entering in RD and to a range of scalar modes entering in MD. However, given the existing bounds on the amplitude of the tensor modes, which we will discuss in Chapter 20, GWs give a subdominant contribution compared with scalar modes. Furthermore, we will also find in Chapter 20 that the effect of GWs on the temperature–temperature correlator quickly disappears for multipoles above $l > O(50)$ (see Fig. 20.14 on page 538), while the effect of GWs on the correlator of the B-mode of the CMB polarization could be potentially visible only up to $l = O(150)$, unless one is able to subtract the lensing of the scalar contribution (see Fig. 20.28 on page 568). Thus, CMB experiments are mostly sensitive to primordial GWs that re-entered during MD, i.e. $f < f_{eq}$.

In principle, the transfer function can be accurately computed numerically, given a cosmological model. However, relatively simple approximate analytic expressions can be obtained using the results of Section 19.5.1. We have seen that, independently of whether the mode re-enters in RD, MD or ΛD, once inside the horizon it scales as $1/a$ times oscillating factors. These oscillations are very fast, particularly near the recent epoch. Indeed, at $\eta = \eta_0$, in the factor $\sin(k\eta)$ the frequency difference between neighboring peaks is $\Delta f \sim \eta_0^{-1} \sim 10^{-18}$ Hz. Thus, for computing the transfer function at frequencies such as those probed by pulsar timing arrays or interferometers, we can simply average these oscillations, replacing the factor $\sin^2(k\eta)$ (which appears in expressions quadratic in the GW, such as the energy density) by $1/2$.[35] In this approximation,

$$\tilde{h}^2(\eta_0, \mathbf{k}) \simeq \frac{1}{2}\tilde{h}^2(\eta_{in}, \mathbf{k}) \left(\frac{a_*(k)}{a_0}\right)^2 , \qquad (19.255)$$

where $\tilde{h}(\eta_{in}, \mathbf{k}) = \tilde{h}_{k,in}$ is the primordial value of the tensor perturbation, determined when the perturbation is well outside the horizon, $a_*(k)$ is

[35] In contrast, replacing $\sin(k\eta) \to 1$ amounts to replacing $\tilde{h}(\eta_0, k)$ by its envelope; see Figs. 19.15 and 19.16. In the end, accurate numerical factors are only obtained by numerical evaluation of the transfer function.

the value of the scale factor when the mode with comoving momentum k re-enters the horizon, and the factor of $1/2$ comes from $\langle \sin^2(k\eta) \rangle$. Of course, eq. (19.255) is approximate because it assumes that a mode is exactly constant as long as it is outside the horizon, and suddenly starts to evolve exactly as $1/a$ times trigonometric functions as it re-enters the horizon. To get an accurate matching between these asymptotic super-horizon and sub-horizon regimes, one should rather rely on the numerical integration.

Setting as usual $a_0 = 1$, the transfer function $T_{\rm GW}(k)$ is therefore approximately given by

$$T_{\rm GW}^2(k) \simeq \frac{1}{2} a_*^2(k) \,. \tag{19.256}$$

Horizon crossing is determined by the condition $\mathcal{H} = k$, i.e. $H = k/a$. Thus, the scale factor $a_*(k)$ is determined by

$$H[a_*(k)] = \frac{k}{a_*(k)} \,. \tag{19.257}$$

As we have seen, the primordial modes of interest for direct GW detection enter the horizon deep in RD. We must then use eq. (17.116) for the Hubble parameter, which takes into account the change in the number of relativistic degrees of freedom with temperature. Then, for modes that re-enter deep in RD we get

$$a_*(k) \simeq \left(\frac{g_*(T_k)}{3.363} \right)^{1/2} \left(\frac{3.909}{g_*^S(T_k)} \right)^{2/3} \frac{H_0 \Omega_R^{1/2}}{k} \,, \tag{19.258}$$

where $g_*(T_k)$ and $g_*^S(T_k)$ are the values of $g_*(T)$ and $g_*^S(T)$ when the mode with comoving momentum k enters the horizon. At $T \gg 100$ GeV this becomes

$$a_*(k) \simeq 0.62 \left(\frac{106.75}{g_*(T_k)} \right)^{1/6} \frac{H_0 \Omega_R^{1/2}}{k} \,; \tag{19.259}$$

see eq. (17.117). In contrast, after neutrino decoupling, when only photons and neutrinos are the relativistic degrees of freedom (but still deep in RD, where $k \gg k_{\rm eq}$), eq. (19.258) becomes

$$a_*(k) \simeq \frac{H_0 \Omega_R^{1/2}}{k} \,. \tag{19.260}$$

For a mode with comoving momentum k that re-enters during RD the temperature T_k when the mode re-enters the horizon is obtained using eq. (17.113) in the form

$$g_*^S(T_k) T_k^3 a_*^3(k) \simeq 3.909 \, T_0^3 \tag{19.261}$$

(where T_0 is the present photon temperature), and using eq. (19.258) for $a_*(k)$. This gives

$$T_k \simeq 7.04 \times 10^7 \, {\rm GeV} \, c(T_k) \left(\frac{f}{1 \, {\rm Hz}} \right) \,, \tag{19.262}$$

where $f = k/(2\pi)$ is the physical frequency of the mode today, and

$$c(T_k) = \left(\frac{g_*(T_k)}{3.363}\right)^{-1/2} \left(\frac{g_*^S(T_k)}{3.909}\right)^{1/3} \qquad (19.263)$$

is a slowly varying function that is equal to one for temperature smaller than the temperature of neutrino decoupling, around 1 MeV, and slowly decreases with increasing temperature saturating, in the Standard Model, to a value $c(T_k) \simeq 0.53$ at $T \gg 100$ GeV.

Thus, GWs whose frequency today is $f \sim 10^2$ Hz, in the range of ground-based interferometers, re-entered when $T_k \sim 3 \times 10^9$ GeV [taking into account that $c(T_k) \simeq 0.53$ at these temperatures]. GWs with $f \sim 0.1$ Hz, which is in the range of the proposed advanced space interferometer DECIGO, re-entered when $T_k \sim 3 \times 10^6$ GeV, while a GW with $f \sim 10^{-3}$ Hz, in the range where LISA has its best sensitivity (see Fig. 16.1 on page 335), re-entered when the temperature was of order 3×10^4 GeV. Taking $f \simeq 10^{-9}$ Hz, relevant for pulsar timing arrays, we get $T_k = O(100)$ MeV, while GWs with $f \sim 10^{-12} - 10^{-11}$ Hz re-entered the horizon around the time of big-bang nucleosynthesis (which starts when $T \sim 0.1$ MeV; see the discussion on page 712). If we extrapolate eq. (19.262) close to matter–radiation equilibrium (which is correct within about a factor of 2), we find that GWs with frequencies $f \sim f_{eq}$ re-entered when $T_k \sim$ eV.

In contrast, for the modes that re-enter in MD well after the RD–MD transition, i.e. $k \ll k_{eq}$, the Hubble parameter is given by $H(a) \simeq H_0 \Omega_M^{1/2} a^{-3/2}$, and therefore eq. (19.257) gives

$$\begin{aligned} a_*(k) &\simeq \frac{H_0^2 \Omega_M}{k^2} \\ &= \frac{1}{\sqrt{2}} \left(\frac{k_{eq}}{k}\right) \frac{H_0 \Omega_R^{1/2}}{k}, \end{aligned} \qquad (19.264)$$

where we have used eqs. (17.162) and (19.52) to write

$$k_{eq} = \sqrt{2} \frac{H_0 \Omega_M}{\Omega_R^{1/2}}. \qquad (19.265)$$

Note that $a_*(k)$ now scales as $1/k^2$ rather than $1/k$. Of course, in order to accurately follow the transfer function across the RD–MD transition, we need to integrate eq. (19.220) numerically for the momenta of interest.[36] The temperature of the photons at the time of re-entry is again computed using eq. (19.261), where now $a_*(k)$ is given by eq. (19.264), and observing that in MD $g_*^S(T_k)$ is constant and equal to its present value $\simeq 3.909$. This gives

$$T_k \sim 1 \,\text{eV} \left(\frac{f}{f_{eq}}\right)^2. \qquad (19.266)$$

Once again, more accurate computations require numerical integration of the evolution equations. In Figs. 19.14–19.16 on pages 477–478 we

[36] Observe that eq. (19.264) holds only for $k \ll k_{eq}$ and eq. (19.260) holds only for $k \gg k_{eq}$. To find continuity at $k = k_{eq}$ between eqs. (19.260) and (19.264) we must include both matter and radiation in $H(a)$, which gives an extra factor $\sqrt{2}$ in $H(a_{eq})$.

have shown the result of the integration of the evolution equations, both for modes re-entering in RD and for those re-entering in MD, using the simplified model (17.75) for the cosmological evolution of the background. More accurate results (including for example a first-principles treatment of recombination and decoupling, and realistic neutrino masses) can be obtained using publicly available Boltzmann codes, such as CAMB and CLASS.

Observe that, for GWs that re-entered in MD, $T_k \lesssim O(1)$ eV. Thus, the transfer function depends only on well-understood physics, and can be accurately and reliably computed numerically. In contrast, for modes observable at space-borne or ground-based interferometers, the transfer function also depends on the cosmological evolution at temperatures T much higher than the electroweak scale. In principle, once inside the horizon their evolution could be affected by any phenomenon (for example a phase transition in the early Universe) that would alter the subsequent cosmological expansion. Their transfer function also depends on the effective number of relativistic degrees of freedom near the time of re-entry, which in extensions of the Standard Model would be different from the asymptotic value $g_*(T \gg 100\,\text{GeV}) \simeq 106.75$. Thus, in principle the GW transfer function at these frequencies could carry very interesting information on high-energy physics.

19.5.3   GW damping from neutrino free-streaming

The preceding discussion assumed that, in eq. (19.220), the anisotropic stress σ_A vanishes. However relativistic free-streaming particles, moving on the geodesics of the metric perturbed by the presence of the tensor modes $\tilde{h}_A$, generate an anisotropic stress tensor $O(h)$, which must therefore be accounted for in eq. (19.220). In particular, as we saw on page 383, neutrinos decouple from the primordial plasma at temperatures of order of MeV, and then start to free-stream as relativistic particles. The anisotropic stress that they generate can be computed using the collisionless Boltzmann equation for relativistic particles, which we will introduce in Section 20.3.7 [see eq. (20.162)], and computing $d\mathbf{p}/d\eta$ using the geodesic of massless particles perturbed by the presence of the GW, as in the computations that we will perform in Sections 20.3.1 and 20.3.5.[37] The result is that, after decoupling, free-streaming neutrinos generate an anisotropic stress tensor

[37]See the Further Reading for the explicit computation.

$$\tilde{\sigma}_A(\eta, k) = -4\rho_\nu(\eta) \int_{\eta_{\nu\text{dec}}}^{\eta} d\eta' \, \frac{j_2[k(\eta - \eta')]}{k^2(\eta - \eta')^2} \tilde{h}'_A(\eta', k), \qquad (19.267)$$

where $\eta_{\nu\text{dec}}$ is the value of conformal time at which neutrinos decouple, $\rho_\nu(\eta)$ is the neutrino energy density and

$$j_2(x) = -\frac{\sin x}{x} - \frac{3\cos x}{x^2} + \frac{3\sin x}{x^3}, \qquad (19.268)$$

is a spherical Bessel function. We insert this into eq. (19.220) and write

$$16\pi G a^2 \rho_\nu(\eta) = 6\Omega_\nu(\eta)\,\mathcal{H}^2, \qquad (19.269)$$

where

$$\Omega_\nu(\eta) = \frac{\rho_\nu(\eta)}{\rho_c(\eta)} \tag{19.270}$$

is the energy fraction in neutrinos at time η, and we have used the Friedmann equation. Then, for $\eta > \eta_{\nu\text{dec}}$ we get the integro-differential equation

$$\tilde{h}_A'' + 2\mathcal{H}\tilde{h}_A' + k^2\tilde{h}_A = -24\Omega_\nu(\eta)\,\mathcal{H}^2(\eta) \int_{\eta_{\nu\text{dec}}}^\eta d\eta' \, \frac{j_2[k(\eta - \eta')]}{k^2(\eta - \eta')^2}\tilde{h}_A'(\eta', k)\,. \tag{19.271}$$

The neutrino energy fraction at time η can be written in terms of the present energy fractions of neutrinos, photons and matter as

$$\begin{aligned}
\Omega_\nu(\eta) &= \frac{\Omega_\nu a^{-4}}{\Omega_M a^{-3} + \Omega_R a^{-4}} \\
&= \frac{\Omega_\nu}{\Omega_R[1 + a(\eta)/a_{\text{eq}}]}\,,
\end{aligned} \tag{19.272}$$

where $\Omega_R = \Omega_\gamma + \Omega_\nu$.[38] Deep in RD, $a(\eta) \ll a_{\text{eq}}$ and $\Omega_\nu(\eta)$ has the constant value Ω_ν/Ω_R. For three neutrino species, using eqs. (17.107) and (17.108), we get

$$f_\nu \equiv \frac{\Omega_\nu}{\Omega_R} \simeq 0.40890\,. \tag{19.273}$$

Observe that the term on the right-hand side of eq. (19.271) does not alter the fact that super-horizon modes stay constant. Indeed, as η' ranges from $\eta_{\nu\text{dec}}$ to η the variable $k(\eta - \eta')$ inside the integral ranges from $k(\eta - \eta_{\nu\text{dec}})$ to zero. Thus, its maximum value is smaller than $k\eta$. For a mode that at time η is super-horizon we have $k\eta \ll 1$, so for these modes the argument $k(\eta - \eta') \to 0$ for all η' in the integration range. In the limit $x \to 0$ the function $j_2(x)/x^2$ goes to a finite value, $j_2(x)/x^2 \to 1/15$. Thus, we see that for super-horizon modes we still have a constant solution for $\tilde{h}_A$. Similarly, it is still true that the other solution is a decreasing mode.

Thus, the effect of the anisotropic stress term in eq. (19.271) can be summarized as follows. (1) For $\eta < \eta_{\nu\text{dec}}$ neutrinos are still coupled to the primordial plasma, and this term is absent. (2) For $\eta_{\nu\text{dec}} < \eta \ll \eta_{\text{eq}}$ neutrinos are free-streaming, so this term is now present, and the energy fraction $\Omega_\nu(\eta)$ is constant and sizable, $\Omega_\nu(\eta) = f_\nu \simeq 0.4$. Still, modes that during RD are super-horizon and only re-enter in MD are not affected, and remain constant. Modes that re-enter the horizon during RD are instead affected by this term. Numerical integration of eq. (19.271) shows that, for the modes that re-entered after neutrino decoupling but before matter–radiation equilibrium, i.e. modes with $\eta_{\text{eq}}^{-1} < k < \eta_{\nu\text{dec}}^{-1}$, the solution given in eq. (19.232) is replaced by

$$\tilde{h}(\eta, k) \simeq \tilde{h}_{\text{in}}(k)A\frac{\sin k\eta}{k\eta}\,, \tag{19.274}$$

[38] Recall, from Note 29 on page 387, that Ω_ν is actually defined as the energy fraction that the neutrinos would have today if they were massless, so that the expression $\Omega_\nu a^{-4}$ is anyhow correct even in the massive case, as long as the neutrinos are relativistic, but the true present value of the neutrino energy fraction is actually slightly different from Ω_ν.

with $A \simeq 0.80$. (3) Modes that re-enter the horizon deep in MD only feel the effect of the anisotropic stress term when the term $\Omega_\nu(\eta)$ has become small, $\Omega_\nu(\eta) \simeq f_\nu a_{\rm eq}/a \ll f_\nu$ so, the later they re-enter, the less they are affected. However, for the modes that re-enter near the beginning of MD, which are relevant for CMB observations, the term $\Omega_\nu(\eta)$ is not yet negligible, and they do experience suppression, although to a lesser degree than the modes that re-entered in RD.

Similar considerations could be made for the free-streaming of photons, which also generates an anisotropic stress. However, photons decouple when we are already in MD. As we saw in eqs. (17.83) and (17.94), for photons $1+z_{\rm dec} \simeq 1090$ while (for $\Omega_M = 0.3$, $h_0 = 0.7$) $1+z_{\rm eq} \simeq 3513$, so $a_{\rm dec}/a_{\rm eq} \simeq 3.22$. At decoupling, the energy fraction in photons is then

$$\Omega_\gamma(\eta_{\rm dec}) = \frac{\Omega_\gamma}{\Omega_R[1 + a_{\rm dec}/a_{\rm eq}]}$$

$$\simeq 0.14, \tag{19.275}$$

and then decreases further, eventually as $1/a$, as we go deeper into MD. Thus, photon free-streaming has no effect on the evolution of tensor modes during RD, and only start to have an effect in MD, just after decoupling, when, however, $\Omega_\gamma(\eta)$ is already small. Thus, the damping due to photon free-streaming is less important.

It is also interesting to observe that, as we discussed at the end of the Section 19.5.2, when computing the transfer function for GWs with frequencies relevant for interferometers, the result depends on physics at temperatures well beyond the electroweak scale, where in principle new relativistic free-streaming particles could appear.

19.5.4   The tensor power spectrum, $\Omega_{\rm gw}(f)$ and $h_c(f)$

The definition of the power spectrum for tensor perturbations is completely analogous to that for scalar perturbations. Similarly to eq. (19.164), the primordial tensor power spectrum is defined by[39]

$$\langle \tilde{h}_{A,{\rm in}}(\mathbf{k})\tilde{h}^*_{A',{\rm in}}(\mathbf{k}')\rangle = \frac{1}{V}(2\pi)^3\delta^{(3)}(\mathbf{k}-\mathbf{k}')\delta_{AA'}\frac{1}{4}P_{T,{\rm in}}(k), \tag{19.276}$$

where $\tilde{h}_{A,{\rm in}}(\mathbf{k}) = \tilde{h}_A(\eta_{\rm in},\mathbf{k})$. Using eq. (19.218) we find that $\tilde{h}_{ij}^{\rm TT}(\eta,\mathbf{k})$, given in eq. (19.214), satisfies

$$\langle \tilde{h}_{ij}^{\rm TT}(\eta_{\rm in},\mathbf{k})(\tilde{h}_{ij}^{\rm TT})^*(\eta_{\rm in},\mathbf{k}')\rangle = \frac{1}{V}(2\pi)^3\delta^{(3)}(\mathbf{k}-\mathbf{k}')P_{T,{\rm in}}(k). \tag{19.277}$$

As in eq. (19.170), one further defines

$$\mathcal{P}_{T,{\rm in}}(k) = \frac{k^3}{2\pi^2}P_{T,{\rm in}}(k). \tag{19.278}$$

In coordinate space, using eq. (19.162), we get

$$\langle h_{ij}^{\rm TT}(\eta_{\rm in},\mathbf{x})h_{ij}^{\rm TT}(\eta_{\rm in},\mathbf{x}')\rangle = \int \frac{d^3k}{(2\pi)^3}\,e^{i\mathbf{k}\cdot(\mathbf{x}-\mathbf{x}')}P_{T,{\rm in}}(k), \tag{19.279}$$

[39]The factor $1/4$ in eq. (19.276) is related to our choice of normalization of the polarization tensors in eq. (19.218). If one rather uses the definitions $e_{ij}^+ = (\hat{\mathbf{u}}_i\hat{\mathbf{u}}_j - \hat{\mathbf{v}}_i\hat{\mathbf{v}}_j)/\sqrt{2}$ and $e_{ij}^\times = (\hat{\mathbf{u}}_i\hat{\mathbf{v}}_j + \hat{\mathbf{v}}_i\hat{\mathbf{u}}_j)/\sqrt{2}$, so that $e_{ij}^A e_{ij}^{A'} = \delta^{AA'}$, then the factor $1/4$ in eq. (19.276) becomes $1/2$, so that in any case the coefficient on the right-hand side of eq. (19.277) is unchanged.

so in particular

$$\langle h_{ij}^{\mathrm{TT}}(\eta_{\mathrm{in}},\mathbf{x})h_{ij}^{\mathrm{TT}}(\eta_{\mathrm{in}},\mathbf{x})\rangle = \int \frac{d^3k}{(2\pi)^3}\, P_{T,\mathrm{in}}(k)$$
$$= \int_0^\infty \frac{dk}{k}\, \mathcal{P}_{T,\mathrm{in}}(k)\,. \qquad (19.280)$$

The primordial tensor power spectrum is parametrized as

$$\mathcal{P}_{T,\mathrm{in}}(k) = A_T(k_*)\left(\frac{k}{k_*}\right)^{n_T}, \qquad (19.281)$$

where k_* is the same pivot scale as used in the scalar spectrum (19.178), A_T is the amplitude and n_T is the tilt of the tensor spectrum.[40] Note that, for historical reasons, there is a different convention compared with eq. (19.178): in the scalar case a flat spectrum corresponds to $n_s = 1$, while for the tensor case the definition of n_T is such that the flat spectrum corresponds to $n_T = 0$. Similarly to eq. (19.183), one could also introduce in principle a running of the tensor index.

The tensor-to-scalar ratio r is defined by

$$\boxed{r(k) = \frac{\mathcal{P}_{T,\mathrm{in}}(k)}{\mathcal{P}_{\mathcal{R},\mathrm{in}}(k)}\,.} \qquad (19.282)$$

In particular, setting k equal to the pivot scale k_* used in the tensor and scalar spectra, we get

$$r(k_*) = \frac{A_T(k_*)}{A_{\mathcal{R}}(k_*)}\,. \qquad (19.283)$$

CMB results are often expressed in terms of $r_{0.05}$, which is defined as the value of $r(k_*)$ for a pivot scale $k_* = 0.05\,\mathrm{Mpc}^{-1}$, or of $r_{0.002}$, which is defined at a pivot scale $k_* = 0.002\,\mathrm{Mpc}^{-1}$; see Note 40. This gives the overall amplitude of the tensor-to-scalar ratio at the pivot scale, while its evolution with k is contained in the spectral indices n_s and n_T and possibly in their running. We will discuss in detail in the following chapters the existing observational information on r, and the theoretical predictions.

The quantity $r(k)$ characterizes the spectrum of a primordial background of GWs, in the language that it is typically used in CMB experiments. This primordial background, once evolved to the present time, will manifest itself as a stochastic GW background that, depending on how far it extends in frequency, could in principle be detectable at GW interferometers (as we discussed in Vol. 1) or at pulsar timing arrays (which will be discussed in Chapter 23). In that case the stochastic background is typically described by a spectral density $S_h(f)$, and its energy density is described by the quantity $\Omega_{\mathrm{gw}}(f)$, which we introduced in Section 7.8. It is useful to work out the relation between these quantities, providing a bridge between the language used in the context of the CMB and the language used in the context of direct GW detection.

[40] The pivot scale is conveniently chosen roughly in the middle of the range of momenta explored by an experiment. In eq. (20.17) we will see that the multipole l of the CMB anisotropies corresponds to a comoving momentum $k_l \simeq 10^{-4}l\,h_0/\mathrm{Mpc}$. As we will see in Chapter 20, the effect of scalar perturbations on the CMB is detected from $l = 4$ up to $l = O(2500)$, corresponding to momenta from about $4\times10^{-4}\,h_0/\mathrm{Mpc}$ up to about $0.3\,h_0/\mathrm{Mpc} \simeq 0.2\,\mathrm{Mpc}^{-1}$. A typical pivot scale chosen by CMB experiments is $k_* = 0.050\,\mathrm{Mpc}^{-1}$, which is about in the middle of this range. However, the effect of tensor perturbations on the CMB could be visible at most in the range $4 < l < O(100)$, corresponding to a maximum comoving momentum of order $0.010\,h_0/\mathrm{Mpc} \simeq 0.007\,\mathrm{Mpc}^{-1}$. Thus, when one studies both scalar and tensor perturbations, a more convenient conventional choice is $k_* = 0.002\,\mathrm{Mpc}^{-1}$.

Let us recall that a stochastic background of GWs, assumed to be stationary, Gaussian, isotropic and unpolarized, is determined uniquely by the spectral density $S_h(f)$, defined by eq. (7.190). According to eq. (7.192),

$$\langle h_{ij}^{\mathrm{TT}}(\eta_0, \mathbf{x}) h_{ij}^{\mathrm{TT}}(\eta_0, \mathbf{x}) \rangle = 4 \int_0^\infty df \, S_h(f), \qquad (19.284)$$

where we are now using conformal time, and η_0 is as usual the present value. Observe that $S_h(f)$ is defined as the spectral density of the GWs observed *today*. It is therefore related to the power spectrum $\mathcal{P}_{T,0}(k)$ today, which, in analogy to eq. (19.280), is defined by

$$\langle h_{ij}^{\mathrm{TT}}(\eta_0, \mathbf{x}) h_{ij}^{\mathrm{TT}}(\eta_0, \mathbf{x}) \rangle = \int_0^\infty \frac{dk}{k} \mathcal{P}_{T,0}(k). \qquad (19.285)$$

Comparing eqs. (19.284) and (19.285) and using the physical frequency f as the argument of $\mathcal{P}_{T,0}$, we see that[41]

$$S_h(f) = \frac{1}{4f} \mathcal{P}_{T,0}(f). \qquad (19.287)$$

Then, using eq. (7.202),

$$\Omega_{\mathrm{gw}}(f) = \frac{\pi^2}{3H_0^2} f^2 \mathcal{P}_{T,0}(f). \qquad (19.288)$$

We can relate $\Omega_{\mathrm{gw}}(f)$ to the primordial tensor power spectrum using the transfer function (19.253), which gives

$$\mathcal{P}_{T,0}(f) = |T_{\mathrm{GW}}(f)|^2 \mathcal{P}_{T,\mathrm{in}}(f). \qquad (19.289)$$

where we have used $f = k/(2\pi)$ as the argument of T_{GW}.[42] Therefore,

$$\boxed{\Omega_{\mathrm{gw}}(f) = \frac{\pi^2}{3H_0^2} f^2 |T_{\mathrm{GW}}(f)|^2 \mathcal{P}_{T,\mathrm{in}}(f).} \qquad (19.290)$$

This relation between $\Omega_{\mathrm{gw}}(f)$ and $\mathcal{P}_{T,\mathrm{in}}(f)$ is exact. We can now obtain simple approximate expressions using the analytic approximation (19.256) for the transfer function. In particular, for modes with $f \gg f_{\mathrm{eq}}$, that re-entered in RD, i.e. for modes relevant for pulsar timing arrays and GW interferometers, we can use eq. (19.258), and we get

$$h_0^2 \Omega_{\mathrm{gw}}(f) \simeq \frac{1}{24} h_0^2 \Omega_R \left(\frac{g_*(T_k)}{3.363} \right) \left(\frac{3.909}{g_*^S(T_k)} \right)^{4/3} \mathcal{P}_{T,\mathrm{in}}(f) \qquad (f \gg f_{\mathrm{eq}}). \qquad (19.291)$$

For the frequencies relevant for space-borne and ground-based interferometers we have $T_k \gg 100$ GeV and, in the Standard Model, $g_*(T_k) = g_*^S(T_k) \simeq 106.75$. We can then rewrite

$$h_0^2 \Omega_{\mathrm{gw}}(f) \simeq 6.73 \times 10^{-7} \left(\frac{106.75}{g_*(T_k)} \right)^{1/3} \mathcal{P}_{T,\mathrm{in}}(f), \qquad (f \gtrsim 10^{-4} \, \mathrm{Hz}), \qquad (19.292)$$

[41] We can obtain the same result by comparing eq. (7.190) directly with the analog of eq. (19.276), written for $\tilde{h}_A(\eta_0, \mathbf{k})$, i.e.

$$\langle \tilde{h}_A(\eta_0, \mathbf{k}) \tilde{h}_{A'}^*(\eta_0, \mathbf{k}') \rangle \qquad (19.286)$$
$$= \frac{1}{V} (2\pi)^3 \delta^{(3)}(\mathbf{k} - \mathbf{k}') \delta_{AA'} \frac{1}{4} P_{T,0}(k).$$

However, in this case we must be careful that the quantity $\tilde{h}_A(f, \hat{\mathbf{n}})$ that appears in eq. (7.190) is not the same as the quantity $\tilde{h}_A(\mathbf{k})$ (with $\mathbf{k} = 2\pi f \hat{\mathbf{n}}$) that appears in eq. (19.286). Indeed, $\tilde{h}_A(f, \hat{\mathbf{n}})$ was defined in eq. (7.188) using $\int df \int d^2 \hat{\mathbf{n}}$, while $\tilde{h}_A(\mathbf{k})$ was defined using eq. (19.161), where rather appears $d^3 k = k^2 dk d^2 \hat{\mathbf{n}}$ (and we also have a different convention for the volume factors). In practice, it is much simpler to compare directly the correlator of $h_{ij}^{\mathrm{TT}}(\eta_0, \mathbf{x})$, since for the GW amplitude $h_{ij}^{\mathrm{TT}}(\eta_0, \mathbf{x})$ in coordinate space we are using the same conventions.

[42] Note that, in the approximation (19.256), the transfer function $T_{\mathrm{GW}}(f)$ is real. However, more generally, the transfer function defined in eq. (19.252) relates two generic complex Fourier modes, and is therefore in general complex.

where we have used the value (17.111) for $h_0^2\Omega_R$. Observe that the f^2 factor in eq. (19.290) has canceled against a similar factor coming from the transfer function. As a result, for modes that re-entered in RD, the frequency dependence of $h_0^2\Omega_{\mathrm{gw}}(f)$ is the same as that of the primordial spectrum $\mathcal{P}_{T,\mathrm{in}}(f)$. In particular, we will see in Chapter 21 that single-field slow-roll inflation predicts an almost flat primordial spectrum (up to a maximum cutoff). Then, $h_0^2\Omega_{\mathrm{gw}}(f)$ will be almost flat for all frequencies $f \gtrsim f_{\mathrm{eq}}$ up to the cutoff.

For modes that re-entered in MD the situation is different, since the transfer function obtained from eq. (19.264) is proportional to $1/k^2 \propto 1/f^2$. Thus, now

$$|T_{\mathrm{GW}}(f)|^2 \propto f^{-4} \tag{19.293}$$

and, from eq. (19.290),

$$h_0^2\Omega_{\mathrm{gw}}(f) \propto f^{-2}\mathcal{P}_{T,\mathrm{in}}(f) . \tag{19.294}$$

Thus, in single-field slow-roll inflation $h_0^2\Omega_{\mathrm{gw}}(f)$ goes approximately as $1/f^2$ for $f < f_{\mathrm{eq}}$. More precisely, inserting eqs. (19.256) and (19.264) into eq. (19.290) we get

$$h_0^2\Omega_{\mathrm{gw}}(f) \simeq \frac{1}{48}\, h_0^2\Omega_R \left(\frac{f_{\mathrm{eq}}}{f}\right)^2 \mathcal{P}_{T,\mathrm{in}}(f) \qquad (f \ll f_{\mathrm{eq}})$$

$$\simeq 8.72 \times 10^{-7} \left(\frac{f_{\mathrm{eq}}}{f}\right)^2 \mathcal{P}_{T,\mathrm{in}}(f) . \tag{19.295}$$

As we already saw in eq. (16.49), for stochastic GW backgrounds it is also convenient to introduce a characteristic GW amplitude $h_c(f)$, defined by

$$\langle h_{ij}^{\mathrm{TT}}(t) h_{ij}^{\mathrm{TT}}(t) \rangle = 2 \int_{f=0}^{f=\infty} d\log f \, h_c^2(f) , \tag{19.296}$$

where $h_{ij}^{\mathrm{TT}} = h_{ij}^{\mathrm{TT}}(\eta_0, \mathbf{x})$ is the GW amplitude at the present epoch. Comparing with eq. (19.287), we get

$$h_c^2(f) = \frac{1}{2}\mathcal{P}_{T,0}(f) , \tag{19.297}$$

or, from eq. (19.287),

$$h_c^2(f) = 2f S_h(f) . \tag{19.298}$$

The relation between $h_c(f)$ and $\Omega_{\mathrm{gw}}(f)$ is then obtained from eq. (19.288),

$$\Omega_{\mathrm{gw}}(f) = \frac{2\pi^2}{3H_0^2}\, f^2 h_c^2(f) , \tag{19.299}$$

in agreement with eq. (16.52).

19.6 Standard sirens, dark energy and modified gravity

In Section 4.1.4 of Vol. 1 we studied coalescing binaries at cosmological distances and we found that their detection can provide a direct measurement of their luminosity distance, with an absolute calibration that does not need any intermediate "distance ladder". We have indeed seen in Chapter 15 that, by applying Bayesian parameter estimation to the observed BH–BH or NS–NS coalescences, one obtains a measurement of their luminosity distances. The GW signal, by itself, does not give direct information on the redshift z. If, however, the GW source has an identified electromagnetic counterpart, like the NS–NS binary GW170817 that we discussed in detail in Section 15.3, this provides a measurement of the redshift and allows us to obtain both d_L and z. Otherwise, if an electromagnetic counterpart is not observed, statistical techniques can be used to extract cosmological information from a large number of detections, as we will discuss. In this sense, coalescing binaries are the gravitational analog of standard candles. To stress that GW detection is more similar to the detection of sound than to that of light (because of the almost 4π sensitivity of the detectors, and because the signal is produced by coherent motions of the sources), in cosmological applications it has become common to refer to coalescing binaries as "standard sirens".

In Section 4.1.4 of Vol. 1 we did not yet have all the tools necessary to study GW propagation in a curved background, and we instead used the propagation of a scalar field as a proxy for our problem; see eqs. (4.177)–(4.180). Now, having studied the evolution of tensor perturbations over a FRW background, and having developed more extensively some elements of cosmology in Chapter 17, we can come back to the problem in more detail. The equation for tensor perturbations over a FRW background, in GR, is given by eq. (19.220), neglecting neutrino free-streaming [which can be included as in eq. (19.271)]. Introducing $\tilde{\chi}_A(\eta, \mathbf{k})$ from eq. (19.221), $\tilde{h}_A = (1/a)\tilde{\chi}_A$, we found that the free propagation is governed by eq. (19.223). In this equation, the coefficient $(k^2 - a''/a)$ determines the effective speed of propagation of tensor perturbations. The term $a''/a \sim 1/\eta^2$ is small compared with k^2 for sub-horizon modes. We saw in Section 15.4.3 that the bound on $|c_{\text{gw}} - c|/c$ from GW170817/GRB 170817A is $O(10^{-15})$. Even with respect to such an accuracy, for the GWs observed at ground or space-based interferometers the term a''/a is utterly negligible compared with k^2. For instance, a GW frequency $f \sim 10^2$ Hz corresponds to a reduced wavelength $\lambdabar \sim 500$ km, so $(k\eta)^{-2} \sim (\lambdabar/H_0^{-1})^2 \sim 10^{-41}$. Then, we can write simply

$$\tilde{\chi}_A'' + k^2 \tilde{\chi}_A = 0. \tag{19.300}$$

This shows that the dispersion relation of GWs in FRW is $\omega = k$, i.e. GWs propagate at the speed of light (recall that we are using units $c = 1$). The factor $1/a$ in eq. (19.221), by contrast, tells us how the

GW amplitude decreases in the propagation over cosmological distances from the source to the observer. As we saw in Section 4.1.4, for inspiraling binaries it combines with several factors $(1 + z)$ coming from the transformation between quantities computed in the source frame and the corresponding quantities in the observer frame, together with a redefinition of the chirp mass, to eventually produce the dependence of the GW amplitude on the luminosity distance, $\tilde{h}_A(\eta, \mathbf{k}) \propto 1/d_L(z)$.

An important consequence of a direct measurement of $d_L(z)$ over cosmological distances is that it allows us to probe whether dark energy is given by a simple cosmological constant, as in the ΛCDM model that we have been discussing, or by some other mechanism. As we will see for an explicit example in Section 19.6.4, it is possible to obtain an accelerated expansion of the Universe without introducing a cosmological constant, but rather by modifying GR on cosmological scales. In this case the Friedmann equation (17.75) takes the form

$$ H(a) = H_0 \left[\Omega_R a^{-4} + \Omega_M a^{-3} + \rho_{\mathrm{DE}}(a)/\rho_0 \right]^{1/2} , \qquad (19.301) $$

where, as usual, ρ_0 is the critical density today [see eq. (17.70)] and $\rho_{\mathrm{DE}}(a)$ is a function, determined by the modification of GR considered, which formally plays the role of a DE density. If there is no energy exchange between this term and the usual energy densities of matter and radiation (as is indeed the case in typical modified gravity models), ρ_{DE} will satisfy its own conservation equation, analogous to eq. (17.67),

$$ \dot{\rho}_{\mathrm{DE}} + 3H(1 + w_{\mathrm{DE}})\rho_{\mathrm{DE}} = 0 , \qquad (19.302) $$

where w_{DE} is some function of time, or equivalently of redshift, that has the meaning of the dark-energy EoS. From eq. (17.63) we see that to have accelerated expansion in the recent epoch, we must have $w_{\mathrm{DE}}(z) < -1/3$. Actually, from the fact that ΛCDM works so well in fitting the data, we expect that in any viable dynamical DE model $w_{\mathrm{DE}}(z)$ will be relatively close to -1, at least in the recent cosmological epoch when DE becomes relevant. From eq. (17.62), the function $w_{\mathrm{DE}}(z)$ determines $\rho_{\mathrm{DE}}(z)$ according to

$$ \rho_{\mathrm{DE}}(z) = \rho_{\mathrm{DE}}(0) \exp \left\{ 3 \int_0^z \frac{d\tilde{z}}{1 + \tilde{z}} \left[1 + w_{\mathrm{DE}}(\tilde{z}) \right] \right\} , \qquad (19.303) $$

where we have transformed the integral over $dx = d \log a$ into an integral over dz. The corresponding comoving distance is obtained from eq. (17.176),

$$ d_{\mathrm{com}}(z) = \frac{1}{H_0} \int_0^z \frac{d\tilde{z}}{\sqrt{\Omega_R(1 + \tilde{z})^4 + \Omega_M(1 + \tilde{z})^3 + \rho_{\mathrm{DE}}(\tilde{z})/\rho_0}} , \qquad (19.304) $$

to be compared with eq. (17.177) for ΛCDM, while the luminosity distance is obtained from eq. (17.180),

$$ d_L(z) = \frac{1 + z}{H_0} \int_0^z \frac{d\tilde{z}}{\sqrt{\Omega_R(1 + \tilde{z})^4 + \Omega_M(1 + \tilde{z})^3 + \rho_{\mathrm{DE}}(\tilde{z})/\rho_0}} , $$
$$ (19.305) $$

and, similarly, the angular diameter distance is obtained from $d_A(z) = (1 + z)^{-1}d_{\text{com}}(z)$. In the recent cosmological epoch, when DE becomes important, the contribution of Ω_R in eq. (19.305) is very small, and can be neglected, and the luminosity distance depends on H_0 and on $\rho_{\text{DE}}(z)$ [with $\Omega_M + \Omega_R + \rho_{\text{DE}}(0)/\rho_0 = 1$, i.e. $\Omega_M \simeq 1 - \rho_{\text{DE}}(0)/\rho_0$].

19.6.1   Testing cosmological models against observations

To test a model, such as ΛCDM or a modified gravity model, against cosmological observations one must first of all compute both the background evolution and the cosmological perturbations of the given model. The latter are then implemented in a Boltzmann code that integrates the Boltzmann equations for the multipoles of the radiation field (see Section 20.3.7) and, given the initial conditions (in general assumed to be the adiabatic initial conditions discussed in Section 19.2.1) and the cosmological parameters of the model, produces the evolution of the power spectra of the temperature and polarization of the CMB, as well as the linear matter power spectrum. These can then be compared with several cosmological datasets, in particular (1) the observed multipoles of CMB temperature and polarization (including their cross-correlations), which we will discuss in detail in Chapter 20; (2) observations of type Ia SNe, which are standard candles (or, rather, as we discussed in Section 10.2.2, can be used as standard candles after including empirical corrections correlating the peak luminosity with the shape and color of the luminosity curve); (3) observations of baryon acoustic oscillations (BAO), which originate from the coupled acoustic oscillations of baryons and photons before decoupling and which leave an imprint in the observed matter power spectrum; (4) observations of cosmological structure formation in the linear and quasi-linear regime (while comparison with structure formation in the full non-linear regime eventually requires the use of N-body simulations).

These observations are nicely complementary to each other and, within a given model, allow us to fix with great accuracy the cosmological parameters. As we mentioned in Note 21 on page 468, in its simplest form ΛCDM has only six free parameters: the amplitude $A_{\mathcal{R}}$ and tilt n_s of the initial power spectrum (19.177), the Hubble parameter today H_0, the fraction of baryonic matter, Ω_B [or rather the combination $\omega_b = h_0^2\Omega_b$, where the factor h_0^2 cancels a factor $1/h_0^2$ implicit in the definitions of the energy fractions Ω_i; see eqs. (17.70) and (17.72)], the fraction $\omega_M = h_0^2\Omega_M$ of total non-relativistic matter, including both baryons and dark matter, and the redshift z_{re} at which the Universe becomes half-reionized after the formation of the first generation of stars. Simple extensions include a non-vanishing curvature, neutrino masses, etc. With vanishing curvature, the energy fraction Ω_Λ associated with the cosmological constant in ΛCDM [or, more generally, the energy fraction $\Omega_{\text{DE}} \equiv \rho_{\text{DE}}(z = 0)/\rho_0$] is fixed in terms of Ω_M by the condition $\Omega_\Lambda + \Omega_M + \Omega_R = 1$, while $\omega_R = h_0^2\Omega_R$ is fixed very accurately by the

CMB temperature, as we saw in eq. (17.79). Thus, the simplest version of ΛCDM is specified by the parameters

$$\{A_s, n_s, H_0, \omega_M, \omega_b, z_{\rm re}\}\,. \tag{19.306}$$

It is quite remarkable that, with just six free parameters, the model fits extremely well thousands of CMB multipoles (see Fig. 20.1 on page 509), the BAO and SNe observations at several different redshifts, as well as structure formation data. For this reason, ΛCDM is currently considered the standard cosmological paradigm, and any cosmological model that one might present as an alternative to ΛCDM must, first of all, pass with equal success these very stringent tests, which crucially involve both its background evolution and its cosmological perturbations.

In general, to test alternatives to ΛCDM and in particular its DE sector, one can proceed in two complementary ways. Either one writes down an explicit alternative theory, works out its specific consequences and compares how well the theory fits the data with respect to ΛCDM (we will see an explicit example of a viable alternative theory in Section 19.6.4), or one develops phenomenological parametrizations of deviations from ΛCDM, both at the background level and at the level of cosmological perturbations. At the background level, a simple example of the latter strategy is given by the so-called wCDM model, in which the EoS of DE is assumed to be a constant w, taken to be a new free parameter, to be fitted to the data together with the other parameters of ΛCDM. The introduction of w changes the background evolution of the model, while cosmological perturbations are taken to be the same as in ΛCDM. In this case, the observational constraint that one obtains on w is quite strong, with the current limit given by[43] $w = -1.00^{+0.04}_{-0.05}$. At first sight, such a strong limit on $|w+1|$ might seem a stringent confirmation of ΛCDM. However, any fundamental theory that produces a dynamical DE with $w_{\rm DE}$ different from -1 will also in general give rise to a time-dependent DE EoS $w_{\rm DE}(z)$. Furthermore, it will also unavoidably generate differences in the behavior of the cosmological perturbations, and this will be relevant for fitting the data at the present level of accuracy. In this sense, wCDM does not correspond to any actual physical model. The correct interpretation of an analysis based on wCDM is that if within wCDM one finds deviations from $w = -1$, this can be an indication of deviations from ΛCDM. However, if within wCDM the fit to the data drives us back toward $w = -1$, this a priori does not exclude the existence of other dynamical DE models, not well approximated by this artificial family of models, and that might fit the data as well as, or better than, ΛCDM. Indeed, the model that we will discuss in Section 19.6.4 has a time-dependent DE with $w_{\rm DE}(z = 0) \simeq -1.15$, and still fits the current cosmological data at the same level as ΛCDM.

A better parametrization, which at least addresses the issue of the time dependence of the EoS, is obtained by expanding the dependence of $w_{\rm DE}(a)$ on a to first order around the present value $a = 1$. Thus, at least in the recent cosmological epoch, one can write

$$w_{\rm DE}(a) = w_0 + w_a(a - 1)\,, \tag{19.307}$$

[43]This value is obtained by combining the first-year results from the Dark Energy Survey (DES) on galaxy clustering and weak lensing with *Planck* CMB data, BAOs from the SDSS, 6dF and BOSS surveys and type Ia SNe from the JLA dataset; see Abbott *et al.* [DES Collaboration] (2017).

In terms of redshift, this can be rewritten as

$$w_{\rm DE}(z) = w_0 + w_a \frac{z}{1+z}.$$ (19.308)

From eq. (19.303), the corresponding expression for the DE density is

$$\rho_{\rm DE}(z) = \rho_{\rm DE}(0)(1+z)^{3(1+w_0+w_a)} e^{-3w_a \frac{z}{1+z}}.$$ (19.309)

The current limits in the (w_0, w_a) plane are less stringent, allowing for values of $|w_0+1|$ of order 0.2 or larger, and values of $|w_a| = O(1)$. Otherwise, one can try to estimate the constraints on $w_{\rm DE}(z)$ in independent redshift bins, without assuming a functional form. This unavoidably results in even less stringent constraints.[44]

[44]See in particular Figs. 4 and 5 of the *Planck* dark energy paper, [Planck Collaboration], Ade *et al.* (2016b).

19.6.2   Cosmology with standard sirens

We can now discuss the impact on cosmology, and in particular on the study of DE, of measurements of luminosity distances with standard sirens. One can distinguish two situations, depending on the redshift.

Local measurements

[45]Which follows from the flatness condition; if we want to include a spatial curvature Ω_K, the argument of the square root in eq. (19.305) becomes $\Omega_R(1 + \tilde{z})^4 + \Omega_M(1 + \tilde{z})^3 + \Omega_K(1 + \tilde{z})^2 + \rho_{\rm DE}(\tilde{z})/\rho_0$ and, by definition of Ω_K, we have $\Omega_R+\Omega_M+\Omega_K+\Omega_{\rm DE} = 1$; see Note 2 on page 369. Here, given the strong observational limits $|\Omega_K| < 5 \times 10^{-3}$, we neglect spatial curvature.

If the source is at a relatively short distance, the luminosity distance is insensitive to the specific form of $\rho_{\rm DE}(z)$ in eq. (19.305). Using the condition $\Omega_R + \Omega_M + \Omega_{\rm DE} = 1$,[45] we see that, in the limit $z \ll 1$,

$$d_L(z) = \frac{z}{H_0} + O(z^2).$$ (19.310)

[46]The first measurement of H_0 from GW170817 is reported in [LIGO Scientific Collaboration and Virgo Collaboration and 1M2H and Dark Energy Camera GW-E and DES and DLT40 and Las Cumbres Observatory and VINROUGE and MASTER Collaborations], Abbott *et al.* (2017), and gives

$$H_0 = 70.0^{+12.0}_{-8.0} \ {\rm km\,s^{-1}\,Mpc^{-1}}.$$ (19.311)

In eq. (19.312) we quote the value given in Guidorzi *et al.* (2017), where the original analysis is improved by modeling the broadband X-ray to radio emission to constrain the inclination of the source (which, as we have seen, is partly degenerate with the luminosity distance, and is therefore responsible for an important contribution to the error in d_L). This gives a significantly higher value $H_0 = 74.0^{+11.5}_{-7.5} \ {\rm km\,s^{-1}\,Mpc^{-1}}$, which rises further to the value (19.312) with an improved estimate of the peculiar velocity.

Indeed, to first non-trivial order in z, all notions of distance in cosmology (comoving distance, luminosity distance or angular diameter distance) become equivalent, and reproduce the Hubble law $d = z/H_0$. Thus, a measurement of a standard siren at low z does not give direct information on $\rho_{\rm DE}(z)$, but rather gives a measurement of H_0.

The first example of such a measurement has been provided by the NS–NS binary GW170817, which we studied in great detail in Section 15.3. As we saw in eq. (15.21), this source was at a redshift $z = 0.009680 \pm 0.00079$. From the measured value of the luminosity distance, given in Table 15.6, one could then infer H_0. For a source at such a low redshift one must, however, correct for the contribution to the redshift given by the radial component of the "peculiar velocity", i.e. of the random velocity of the host galaxy with respect to us. In the case of GW170817 the host galaxy, NGC 4993, is part of a group of galaxies with a measured coherent bulk flow towards a large concentration of mass known as the Great Attractor, and the peculiar velocity of NGC 4993 gives a contribution of the order 10% to the observed redshift. After correcting for this effect, one obtains[46]

$$H_0 = 75.5^{+11.6}_{-9.6} \ {\rm km\,s^{-1}\,Mpc^{-1}}.$$ (19.312)

To understand the cosmological significance of this measurement we observe that current information on H_0 from electromagnetic probes basically comes from two distinct sets of observations. On the one hand, the measurements of the CMB, combined with other cosmological datasets such as BAO and SNe, as we have discussed. On the other hand, one can perform "local" measurements of the expansion rate based for instance on observations of Cepheid variables in nearby galaxies. The CMB, BAO and SNe observations are performed at large redshifts. In particular, as we will discuss in great detail in Chapter 20, the CMB signal comes from the decoupling epoch that, as we saw in eq. (17.94), takes place near $z = 1090$. These observations can be translated into a measurement of the value of $H(z)$ today, H_0, only by assuming a cosmological model, such as ΛCDM. In contrast, local measurements are largely independent of the cosmological model. Currently the *Planck* 2015 observations, assuming ΛCDM, give[47]

$$H_0 = 66.93 \pm 0.62 \,\mathrm{km\,s^{-1}\,Mpc^{-1}} . \qquad (19.313)$$

In contrast, local measurements give[48]

$$H_0 = 73.24 \pm 1.74 \,\mathrm{km\,s^{-1}\,Mpc^{-1}} . \qquad (19.314)$$

The difference between these two values is at the level of $(73.24 - 66.93)/\sqrt{1.74^2 + 0.62^2} \simeq 3.4$ standard deviations. This tension[49] could be due to unaccounted systematic errors, or else it could be a signal of deviations from ΛCDM, particularly in the dark-energy sector. Indeed, a dark-energy density with a "phantom" equation of state, i.e. $w(z) < -1$, could reconcile the two measurements.[50]

A measurement of H_0 based on standard sirens at low redshifts is again a local measurement, independent of the cosmological model, but with completely different systematic errors compared with type Ia SNe, and would contribute to clarifying the issue. The single measurement (19.312) provided by GW170817/GRB 170817A has an error that is too large to discriminate between the values in eqs. (19.313) and (19.314). However, with N measurements from standard sirens with approximately the same error, we would obtain a measurement of H_0 with an error smaller by approximately a factor $1/\sqrt{N}$. This means that with $O(50)$ standard sirens with measured redshift we should be able to discriminate between the values of H_0 in eqs. (19.313) and (19.314). Agreement with eq. (19.313) would confirm ΛCDM and would point toward unaccounted systematics in type Ia SNe, while agreement with eq. (19.314) would be a strong indication in favor of phantom DE. Advanced LIGO/Virgo, at their target sensitivities, are expected to detect several more events like GW170817/GRB 170817A, while we have seen in Section 16.3.2 that the space interferometer LISA, in a four-year mission, could measure the Hubble parameter to an accuracy $\delta H_0/H_0 \lesssim 0.02$.

To use coalescing binaries as standard candles, the measurement of the luminosity distance must be complemented by information on the redshift. This can be obtained from the observation of an electromagnetic counterpart, as in GW170817/GRB 170817A. Otherwise, with a

[47]See [Planck Collaboration], Ade *et al.* (2016a), together with the more recent constraints on the reionization redshift obtained in [Planck Collaboration], Aghanim *et al.* (2016).

[48]See Riess *et al.* (2016).

[49]In cosmology, it has become customary to use the word "tension" to designate the intermediate situation in which the difference between two results is not sufficiently significant, statistically, to conclusively prove the existence of a disagreement, but is certainly significant enough to warrant further investigation.

[50]For example, one can fit the *Planck* CMB data (plus other combination of datasets such as BAO, SNe and weak lensing) and adding also the local H_0 measurement as a prior, using wCDM rather than ΛCDM. The result is that the tension is resolved, and the best fit provides values of w below -1 at about the 2σ level, typically in the range $[-1.3, -1.1]$, depending on the datasets included; see Di Valentino, Melchiorri and Silk (2016). In Section 19.6.4 we will discuss a better motivated modification of gravity with a phantom DE EoS, which also produces a higher value of H_0.

large number of events, a statistical approach could be used, in which for each event one considers all galaxies within the localization volume as the possible host. Each galaxy has a different redshift, and therefore provides a different measurement of H_0 and, if there are many galaxies in the localization volume, little information can be obtained on H_0 from a single event. However, with many events of this type, one can form a likelihood function in which each potential host galaxy, for each detected event, is treated as a measurement of H_0; with a large number of events, the likelihood function would converge toward the correct value of H_0. Further information could be folded in the method, for example by weighting galaxies according to their mass content or by including priors on H_0, which automatically exclude galaxies that would give values of H_0 too far away from those obtained from CMB or from local measurements.

Another statistical approach could exploit the fact that GW observation actually measure the combination $m_i^{(\mathrm{obs})} = (1+z)m_i$; see eq. (15.19). If one knew the actual masses m_i, one could then determine z. In the case of NS–NS binaries, one can make use of the fact that the NS mass distribution is relatively narrow, and try to exploit this to obtain statistical information on the redshift. The usefulness of these statistical techniques will depend on how much larger will be the sample of standard sirens without detected counterpart, compared with the sample of those with counterparts.

Standard sirens at cosmological redshifts

Third-generation GW detectors, such as the space interferometer LISA or ground-based interferometers such as the Einstein Telescope (ET), could detect standard sirens at cosmological distances, actually probing the function $w_{\mathrm{DE}}(z)$, rather than just the Hubble parameter.

At LISA, in the merger of supermassive black hole (SMBH) binaries at $z \lesssim 6$, the luminosity distance d_L can be determined to an accuracy that, if dominated by the instrumental error, could be better than 1%. However, this accuracy is degraded by the fact that, during propagation over cosmological distances, the signal is affected by gravitational lensing from masses along the line of sight. Weak lensing produces a random magnification or demagnification, inducing a systematic error (while strong lensing, which generates multiple gravitational "images", could in principle give a further handle to decode the signal). Unless suitable delensing techniques are used, by correlating the signal with lensing maps, weak lensing is expected to introduce an error that can be of order

$$\frac{\Delta d_L}{d_L} \sim (0.03 - 0.05)z\,, \tag{19.315}$$

and that will then largely dominate the uncertainty at $z \gtrsim 0.4$. In contrast, at these redshifts the effect of the peculiar velocity of the host galaxy becomes negligible.

To determine the redshift one needs either an electromagnetic counterpart or a sufficiently accurate localization of the source that might allow identification of the host galaxy. LISA can determine the sky localization of a long-duration inspiral signal thanks to the modulation of the signal due to LISA's motion around the Sun. Furthermore, the inclusion of sub-dominant harmonics in the analysis of the GW signal significantly improves the sky localization and therefore the chances of identifying the host galaxy.

The results that can be obtained from standard sirens at cosmological distances have been most often expressed using wCDM as a benchmark, i.e. in terms of limits on w, leading in general to the estimate that w could be measured to an accuracy of order $\Delta w \sim 0.1$ both at LISA and at ET (after taking into account the error induced by lensing). Using wCDM is practical, because of the simplicity of the model, although one should be aware of the limitations of this parametrization discussed on page 495. Furthermore, forecasts of the accuracy that can be obtained for w have regularly been given assuming that the other cosmological parameters of wCDM [in particular, Ω_M and H_0, which are those that enter in the luminosity distance (19.305)] are kept fixed at their best-fit values obtained from ΛCDM. This is, however, incorrect. In general, in each model the cosmological parameters must be determined self-consistently, by comparing the model with a set of cosmological observations, such as CMB, BAO and SNe, and performing Bayesian parameter estimation. In essence, this amounts to requiring that each model adjust itself so to reproduce a set of fixed measured distances at large redshifts (such as the scales implied by the positions of the peaks in the CMB, or the angular diameter distance fixed by the BAO scale, or the luminosity distances of a set of type Ia SNe). Thus, Bayesian parameter estimation automatically adjusts the parameters in each model so to reproduce these distance measurements at large redshift, and therefore goes in the direction of compensating the differences in luminosity distance (or in comoving distance or in angular diameter distance) induced by the DE parameter w, or more generally by a non-trivial functional form of $w_{\rm DE}(z)$. For a given w, at $z \gtrsim 0.5$ the actual difference between the luminosity distances in wCDM and in ΛCDM, when using the respective best-fit parameters, can be one order of magnitude smaller than the difference obtained by keeping Ω_M and H_0 fixed to the ΛCDM best-fit values. We will see an explicit example in Section 19.6.4. This is the same effect that allows us to investigate DE by performing a measurement of H_0 from standard sirens at low redshift, since parameter estimations in ΛCDM and in modified gravity give different values of H_0; see Note 50 on page 497.

In the next subsection we will see, however, that the effect on the luminosity distance due to the DE EoS is not necessarily the strongest signature of modified gravity that could be detected with standard sirens. Modified gravity models that predict a non-trivial DE EoS at the same time generically predict also a modified propagation equation of the tensor perturbations. This effect is not compensated by parameter estimation and, as we will see, can provide the dominant signature of modified gravity, in a measurement of luminosity distances from standard sirens.

19.6.3   Tensor perturbations in modified gravity

The preceeding discussion focused on the effect of a non-trivial DE EoS. The most natural motivation for a non-trivial dark energy EoS is the assumption that gravity is modified at cosmological scales. However, in a generic modified gravity theory, one can have an even more significant effect due to the fact that the propagation equation (19.220) of tensor perturbations can be different from that in GR. As we will see, this effect is not partially compensated by Bayesian parameter estimation, and can dominate by one order of magnitude the effect from a non-trivial $w_{\rm DE}(z)$.

In a generic modified gravity theory, both the coefficient of the k^2 term and that of the $2\mathcal{H}$ term in eq. (19.220) (as well as the inhomogeneous term on the right-hand side) can in principle be modified, and this is indeed what happens in many of the theories that have been put forward in recent years. A change in the coefficient of the k^2 term gives a propagation speed of GWs different from the speed of light. The limit $|c_{\rm gw} - c|/c < O(10^{-15})$ from GW170817/GRB 170817A now puts a very stringent limit on such a modification, and this has ruled out a large class of scalar–tensor and vector–tensor modifications of GR; see the Further Reading.

Other modified gravity theories, however, lead to a propagation equation where the coefficient of the k^2 term is unchanged, while the coefficient of the $2\mathcal{H}$ term is modified, i.e. they give a propagation equation of the form

$$\tilde{h}_A'' + 2\mathcal{H}[1 - \delta(\eta)]\tilde{h}_A' + k^2\tilde{h}_A = 0 \,, \tag{19.316}$$

with $\delta(\eta)$ some function, and we have restricted to free propagation, setting the source term to zero. We will study in Section 19.6.4 an example of a model where the free propagation of tensor perturbations has this form. In this case we introduce $\tilde{\chi}_A(\eta, \mathbf{k})$ from

$$\tilde{h}_A(\eta, \mathbf{k}) = \frac{1}{\tilde{a}(\eta)}\tilde{\chi}_A(\eta, \mathbf{k}) \,, \tag{19.317}$$

where

$$\frac{\tilde{a}'}{\tilde{a}} = \mathcal{H}[1 - \delta(\eta)] \,, \tag{19.318}$$

and we get

$$\tilde{\chi}_A'' + \left(k^2 - \frac{\tilde{a}''}{\tilde{a}}\right)\tilde{\chi}_A = 0 \,. \tag{19.319}$$

Once again, inside the horizon the term $\tilde{a}''/\tilde{a}$ is totally negligible, so GWs propagate at the speed of light. However, over cosmological distances $\tilde{h}_A$ now decreases as $1/\tilde{a}$ rather than as $1/a$, so, in the propagation from the source to the observer, it will be multiplied by a factor $\tilde{a}_{\rm em}/\tilde{a}_{\rm obs} \equiv \tilde{a}(z)/\tilde{a}(0)$ rather than by a factor $a_{\rm em}/a_{\rm obs} = a(z)/a(0)$, where the labels refer to the emission time (at redshift z) and the observation time (at redshift zero). Since the ratio $\tilde{a}(z)/\tilde{a}(0)$ is independent of the normalization of $\tilde{a}$, we can choose $\tilde{a}(0) = 1$, just as we choose $a(0) = 1$. So, while in GR the GW signal from an inspiraling binary at redshift z has the form

$$h_{+,\times}(t; z) = \frac{1}{d_L(z)} g_{+,\times}(t; z), \tag{19.320}$$

where the functions $g_{+,\times}(t; z)$ can be obtained from eqs. (4.170), (4.171) and (4.194) of Vol. 1, in such a modified gravity model we rather have

$$h_{+,\times}(t; z) = \frac{\tilde{a}(z)}{a(z)} \frac{1}{d_L(z)} g_{+,\times}(t; z). \tag{19.321}$$

Thus, standard sirens actually measure a "GW luminosity distance" $d_L^{\rm gw}(z)$ defined by

$$d_L^{\rm gw}(z) = \frac{a(z)}{\tilde{a}(z)} d_L^{\rm em}(z)$$

$$= \frac{1}{(1+z)\tilde{a}(z)} d_L^{\rm em}(z), \tag{19.322}$$

where $d_L^{\rm em}(z) \equiv d_L(z)$ is the standard luminosity distance (19.305) for electromagnetic signals. Equation (19.318) can be rewritten as

$$\frac{d}{d\eta} \log \frac{a}{\tilde{a}} = \delta(\eta)\mathcal{H}(\eta), \tag{19.323}$$

which upon integration, passing from $d\eta$ to dz, gives

$$d_L^{\rm gw}(z) = d_L^{\rm em}(z) \exp\left\{ -\int_0^z \frac{dz'}{1+z'} \delta(z') \right\}. \tag{19.324}$$

The fact that in modified gravity the GW luminosity distance is in general different from that for electromagnetic signals opens the possibility of model-independent tests of GR, by comparing the results obtained from standard candles with those from standard sirens.

We also see that, in general, at the level of background evolution a modified gravity model is characterized by two functions: $w_{\rm DE}(z)$, which determines the evolution of the background, and $\delta(z)$, which determines the propagation of GWs over such a background.[51] In the study of standard sirens, the modification due to $\delta(z)$ in eq. (19.324) can be much more significant, numerically, than the modification of the electromagnetic luminosity distance due to a non-trivial EoS $w_{\rm DE}(z)$. Indeed, in general both $\delta(z)$ and $|w_{\rm DE}(z) + 1|$ will be of the same order, i.e. of the order of the deviation of the modified gravity model from ΛCDM. However, as we have discussed and as we will check with an explicit example in Section 19.6.4, after performing Bayesian parameter estimation in both ΛCDM and in the modified gravity model, the differences in $d_L^{\rm em}(z)$ will be smoothed out, since in each model the respective best-fit values of H_0 and Ω_M are determined by fitting the theory to electromagnetic observations, such as CMB, BAO and SNe, which basically

[51] Furthermore, at the level of cosmological perturbations, the behavior of the Fourier modes of the perturbations in a modified gravity model will be different from that in ΛCDM, and can be parametrized phenomenologically in terms of some functions of redshift and wavenumber; see e.g. Kunz (2012) for a review.

amounts to fixing the values of some distance scales at large redshifts. Thus, for distinguishing a modified gravity model from ΛCDM with standard sirens, the extra factor due to $\delta(z)$ in eq. (19.324) will eventually make the dominant contribution.

19.6.4   An explicit example: non-local gravity

Motivations for non-local gravity

As an example of a successful modified gravity model, we discuss a non-local modification of gravity that has been introduced and developed in recent years.[52] The Einstein–Hilbert action of GR and the actions of its typical extensions such as scalar–tensor theories are local. However, as a matter of principle, once one includes quantum fluctuations the relevant quantity becomes the quantum effective action, which is a quantity whose variation determines the equations of motion of the vacuum expectation values of the quantum fields. A crucial property of the quantum effective action is that whenever in the fundamental theory there are massless particles, in the quantum effective action there are also non-local terms. In gravity, because of the massless graviton, non-local terms are unavoidably present, and could in principle significantly affect the infrared (IR) behavior of the theory. These non-localities are well understood in the ultraviolet regime, where standard techniques have been developed for computing them, but are much less so in the IR, which is indeed the regime where they could in principle give rise to important cosmological effects. Several hints of strong IR effects come for instance from computations of quantum field theory in de Sitter space. However, the whole issue of IR corrections in de Sitter space is unsettled, because of the intrinsic difficulty of the problem. Given the difficulty of a pure top-down approach, one can take an alternative and more phenomenological strategy. In general, strong IR effects manifest themselves through the generation of non-local terms, proportional to inverse powers of the d'Alembertian operator, in the quantum effective action. For instance, in QCD it has been suggested that IR fluctuations generate in the quantum effective action a term

$$\frac{m_g^2}{2} \text{Tr} \int d^4x \, F_{\mu\nu} \frac{1}{\Box} F^{\mu\nu} , \tag{19.325}$$

where $F_{\mu\nu}$ is the non-abelian field strength. The introduction of this term has been shown to reproduce the results on the non-perturbative gluon propagator in the IR, obtained from operator product expansions and from lattice QCD. Physically, this non-local term corresponds to giving a mass m_g to the gluons: indeed, choosing the Lorentz gauge and expanding in powers of the gauge field A^μ, the above term reduces to a gluon mass term $m_g^2 \text{Tr}(A_\mu A^\mu)$, plus extra non-local interactions. Note that the use of a non-local operator such as $\Box^{-1}$ allows us to write a mass term without violating gauge invariance. However, this only makes sense at the level of quantum effective actions, where non-localities are allowed, and indeed are unavoidably generated by quantum loops of

[52]We closely follow here the discussion in Belgacem, Dirian, Foffa and Maggiore (2017a), to which we refer the reader for a detailed review of the conceptual aspects and the cosmological consequences of the model.

massless particles. The fundamental action of a quantum field theory, by contrast, must be local. Thus, non-local terms of this form describe dynamical mass generation by quantum fluctuations, at the level of the quantum effective action.

In a similar spirit, one can study a model of gravity based on the quantum effective action

$$\Gamma_{\rm RR} = \frac{m_{\rm Pl}^2}{2} \int d^4x \sqrt{-g} \left[R - \frac{1}{6} m^2 R \frac{1}{\Box^2} R \right] , \qquad (19.326)$$

where $m_{\rm Pl}$ is the reduced Planck mass and m is a new mass parameter that replaces the cosmological constant of ΛCDM. Writing the metric in the form $g_{\mu\nu} = e^{2\sigma} \bar{g}_{\mu\nu}$, where σ is the conformal mode and $\bar{g}_{\mu\nu}$ a reference metric, one can check that the non-local term in eq. (19.326) corresponds to a mass term for the conformal mode of the metric. Thus, the physical assumption underlying eq. (19.326) is that IR effects induce a dynamical mass generation for the conformal mode. Independently of its motivations, which can be quite attractive from a field-theoretical point of view, we will use this model as an example of a modified gravity theory, thanks to the fact that, as we will see, it produces a viable cosmological evolution with a dynamical DE, fits the cosmological data very well, and also gives a non-trivial GW propagation, of the type discussed in Section 19.6.3. We will refer to the model defined by eq. (19.326) as the "RR" model. Observe that, strictly speaking, the RR model does not even belong to the class of "modified gravity" theories. We are not modifying Einstein–Hilbert gravity as a fundamental theory, but are rather trying to capture its leading quantum effects in the IR.

Comparison with cosmological observations

To study the cosmological consequences of the model it is useful to put it into a local form, which can be done by introducing two auxiliary fields $U = -\Box^{-1}R$ and $S = -\Box^{-1}U$. Then, the equations of motion of the theory become formally equivalent to those of a scalar–tensor theory.[53] It is then straightforward to integrate the equations of motion numerically, assuming a FRW form for the background metric. One then finds that the Friedmann equations takes the form (19.301) with an effective DE density determined by the evolution of the auxiliary fields, shown in Fig. 19.18. The corresponding DE EoS, determined through eq. (19.302), is shown in Fig. 19.19. We see that the non-local term induces a dynamical DE, which generates accelerated expansion in the recent cosmological epoch, without the need to introduce a cosmological constant. We also observe that $w_{\rm DE}(z) < -1$, i.e. the model naturally produces a "phantom" DE.

One can next study the cosmological perturbations of the model. One finds that the cosmological perturbations are well-behaved. This is already a non-trivial results, since most of the modified gravity models proposed in the literature have been shown to suffer from fatal instabilities at the level of perturbations. One can then implement the perturbations in a Boltzmann code and compare with CMB, BAO, SNe and

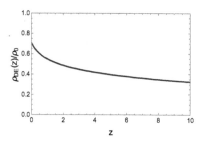

Fig. 19.18 The evolution of $\rho_{\rm DE}(z)$ in the RR model with $u_0 = 0$. From Belgacem, Dirian, Foffa and Maggiore (2017a).

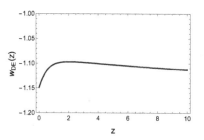

Fig. 19.19 The corresponding evolution of $w_{\rm DE}(z)$. From Belgacem, Dirian, Foffa and Maggiore (2017a).

[53]However, the fields U and S are not genuine degrees of freedom. They are just auxiliary quantities, whose initial conditions would in principle be fixed in terms of the initial conditions of the metric, if one had an explicit derivation of the non-local term in the quantum effective action. In particular, at the quantum level, there are no creation/annihilation operator associated with them; see the discussion in Section 2.4 of Belgacem, Dirian, Foffa and Maggiore (2017a). In practice, three of the four initial conditions on the auxiliary fields parametrize irrelevant directions, while the initial value u_0 of the U field is such that in the large-u_0 limit the results reduce to those of ΛCDM; here we discuss the results of the "minimal model" with $u_0 = 0$.

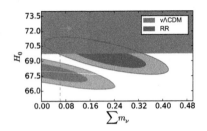

Fig. 19.20 The 1σ and 2σ contours in the plane $(H_0, \sum_\nu m_\nu)$, obtained from the *Planck*+BAO+JLA dataset, for the minimal RR model (upper, rightmost elliptical contours) and $\nu\Lambda$CDM, i.e. ΛCDM with varying neutrino masses (lower, leftmost). The horizontal dashed line is the central value of the local H_0 measurement (19.314), and the gray areas are the corresponding 1σ and 2σ limits (whose upper parts extend above the scale of the figure). The dashed vertical line marked by an arrow is the lower limit on the the sum of neutrino masses from oscillation experiments. From Belgacem, Dirian, Foffa and Maggiore (2017a).

structure formation data, along the lines discussed in Section 19.6.1. The result is that the model fits the data at a level statistically indistinguishable from ΛCDM, and with the same number of parameters, given that the cosmological constant has been replaced by the mass scale m. Assuming flatness, both Λ and m can be taken as derived quantities in the respective models.

Bayesian parameter estimation also reveals some interesting features. First of all, the Hubble parameter comes out higher, reducing to just 2.0σ the tension with the local measurement. A second interesting result concerns the sum of the neutrino masses. In principle, this is also a parameter that affects cosmological predictions, and oscillation experiments provide a lower bound $\sum_\nu m_\nu \gtrsim 0.06$ eV. However, in the *Planck* baseline analysis of ΛCDM, one fixes it to the lower limit, $\sum_\nu m_\nu = 0.06$ eV. If one rather leaves $\sum_\nu m_\nu$ as a free parameter when analyzing the cosmological data, in ΛCDM one finds that the one-dimensional marginalized likelihood for $\sum_\nu m_\nu$ has its maximum at a value of zero, and this marginalized likelihood only implies an upper bound $\sum_\nu m_\nu \lesssim 0.23$ eV (at 95% c.l.). The situation is different in the RR non-local model (with u_0 small). In that case, letting the sum of neutrino masses be a free parameter, one finds a one-dimensional marginalized likelihood for $\sum_\nu m_\nu$ that is peaked at a nonzero value, higher than the lower limit set by oscillation experiments. Thus, the RR model provides a prediction for the value of the sum of the neutrino masses. In contrast, in ΛCDM, for the best-fit value one simply gets back the value of the prior that one has put in, and with no prior one can only obtain an upper limit on $\sum_\nu m_\nu$. The predictions of the RR model and of ΛCDM for H_0 and for the sum of neutrino masses are shown in Fig. 19.20, where we also compare with the local H_0 measurement and with the lower limit on the sum of neutrino masses from oscillation experiments.

The GW luminosity distance

The RR model also provides a nice example of a modified gravity model with a non-trivial propagation equation for GWs. Indeed, the equation governing the free propagation of tensor perturbations over FRW turns out to be of the form (19.316), with

$$\delta(\eta) = \frac{3\gamma(d\bar{V}/d\log a)}{2(1-3\gamma\bar{V})}, \tag{19.327}$$

where $\bar{V}$ is the background evolution of the dimensionless auxiliary field $V \equiv H_0^2 S$ and $\gamma = m^2/(9H_0^2)$. Inserting the numerical solution for $\bar{V}$ and using eq. (19.324) one then finds the ratio $d_L^{\rm gw}/d_L^{\rm em}$ in the RR model, which is shown in Fig. 19.21 as a function of redshift. We see that the relative difference between the gravitational and electromagnetic luminosity distances, at $z \gtrsim 1.5$, is about 3%.

In Fig. 19.22 we show the relative difference $\Delta d_L/d_L$ for three different cases. The upper curve (dot-dashed) is the relative difference between the electromagnetic luminosity distance in the RR model and the luminosity distance of ΛCDM, $(d_L^{\rm RR,em} - d_L^{\Lambda\rm CDM})/d_L^{\Lambda\rm CDM}$, when we use the

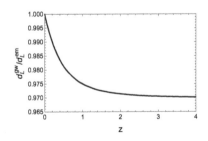

Fig. 19.21 The ratio $d_L^{\rm gw}(z)/d_L^{\rm em}(z)$ in the RR model. From Belgacem, Dirian, Foffa and Maggiore (2017b).

same fiducial values for h_0 and Ω_M, taken here for definiteness equal to the values for the RR model, $h_0 = 0.701$ and $\Omega_M = 0.292$. In this case we see that, over a range of redshifts relevant for third-generation GW interferometers, the relative difference is of order 2%. However, as we have discussed, this is not the quantity relevant to observations. For each model, the actual predictions are those obtained by using its own values of the cosmological parameters. The corresponding result is given by the middle curve (dashed) in Fig. 19.22, using $\{h_0 = 0.701, \Omega_M = 0.292\}$ for RR and $\{h_0 = 0.681, \Omega_M = 0.305\}$ for ΛCDM (obtained, in both cases, by fitting the models to *Planck*+BAO+JLA SNe). We see that, at redshifts $z \gtrsim 1$, $|\Delta d_L|/d_L$ becomes one order of magnitude smaller, as anticipated in the discussion on page 499.

This relative difference in the predictions of the electromagnetic luminosity distances is the one relevant for standard candles. For standard sirens we rather need to compare the GW luminosity distance of the RR model, $d_L^{\mathrm{RR,gw}}$, with the luminosity distance $d_L^{\Lambda\mathrm{CDM}}$ of ΛCDM (which, in contrast, is the same for GWs and for electromagnetic signals). Since we see from Fig. 19.21 that $d_L^{\mathrm{RR,gw}}$, at $z \gtrsim 1$, is smaller than $d_L^{\mathrm{RR,em}}$ by about 3%, it is clear that this will induce a variation of the same order in the comparison with ΛCDM. Indeed, the result for

$$\frac{\Delta d_L^{\mathrm{gw}}}{d_L} \equiv \frac{d_L^{\mathrm{RR,gw}} - d_L^{\Lambda\mathrm{CDM}}}{d_L^{\Lambda\mathrm{CDM}}}, \tag{19.328}$$

using again the respective values of $\{h_0, \Omega_M\}$ for the RR model and for ΛCDM, is given by the lower curve (solid line) in Fig. 19.22. We see that the difference, in absolute value, now rises again to values of order 3%, and the sign of the difference is opposite, compared with the upper curve.

This clearly shows the importance of accounting for the possibility of modified GW propagation, when using standard sirens to study DE and modified gravity.

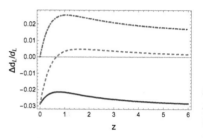

Fig. 19.22 $\Delta d_L^{\mathrm{em}}/d_L$ using the same values of h_0 and Ω_M (upper dot-dashed curve) and using for each model its own values of h_0 and Ω_M (middle, dashed curve). Lower solid line: $\Delta d_L^{\mathrm{gw}}/d_L$ using again for each model its own values of h_0 and Ω_M. From Belgacem, Dirian, Foffa and Maggiore (2017b).

Further reading

- Clear presentations of the equations of cosmological perturbations and of their solutions can be found in Dodelson (2003), Amendola and Tsujikawa (2010) and Gorbunov and Rubakov (2011). Note that our notation for the Bardeen potentials (as well as the metric signature) is the same as in Amendola and Tsujikawa, while Gorbunov and Rubakov exchange the definitions of Φ and Ψ with respect to our conventions, and also have the opposite metric signature. Thus, in the absence of anisotropic stress,

when $\Psi = -\Phi$, our equations for the cosmological perturbations reduce to those of Gorbunov and Rubakov on replacing $\Phi \to -\Phi$. The BBKS transfer function is given in Appendix G of Bardeen, Bond, Kaiser and Szalay (1986); see also Hu and Sugiyama (1996). The extension to include the effect of baryons is presented in Eisenstein and Hu (1998). The transfer function for the tensor modes is discussed in Turner, White and Lidsey (1993) (where, however, a model with $\Omega_M = 1$ is studied), Pritchard and Kamionkowski (2005), Smith, Kamionkowski and Cooray (2006), Boyle (2006) and Boyle and Steinhardt (2008).

- The damping of tensor modes due to neutrino free-streaming was first studied in Rebhan and Schwarz (1994), Durrer and Kahniashvili (1998) and Weinberg (2004), and has been further discussed in Bashinsky (2005), Dicus and Repko (2005), Watanabe and Komatsu (2006), Boyle and Steinhardt (2008) and Dent, Krauss, Sabharwal and Vachaspati (2013). The numerical integration of the integro-differential equation (19.271) and the resulting modifications of the TT and BB CMB correlators are explicitly discussed in Appendix A of Pritchard and Kamionkowski (2005). The effect of neutrino free-streaming to second order in cosmological perturbation theory is computed in Mangilli, Bartolo, Matarrese and Riotto (2008). The free-streaming of neutrinos is clearly discernible in the CMB data; see Sellentin and Durrer (2015).

- The parametrization of the DE EoS in terms of (w_0, w_a) has been proposed in Chevallier and Polarski (2001) and Linder (2003).

- The possibility of measuring the Hubble constant with coalescing binaries was pointed out by Schutz (1986). Its use for constraining the EoS of DE is discussed in Holz and Hughes (2005), Kocsis, Frei, Haiman and Menou (2006), Dalal, Holz, Hughes and Jain (2006), Arun, Iyer, Sathyaprakash, Sinha and Van Den Broeck (2007), Cutler and Holz (2009), Nissanke, Holz, Hughes, Dalal and Sievers (2010), Van Den Broeck, Trias, Sathyaprakash and Sintes (2010) and Zhao, Van Den Broeck, Baskaran and Li (2011). Statistical methods for extracting information on the cosmological parameters without determining the redshift from an electromagnetic counterpart were already discussed in the original paper by Schutz (1986). More recent studies include Del Pozzo (2012), Taylor, Gair and Mandel (2012), Taylor and Gair (2012) and Messenger and Read (2012). Techniques for reducing the error due to lensing are discussed in Shapiro, Bacon, Hendry and Hoyle (2010) and Hilbert, Gair and King (2011).

- The measurement of H_0 from the joint GW170817 and GRB 170817A detection is presented in [LIGO Scientific Collaboration and Virgo Collaboration and 1M2H and Dark Energy Camera GW-E and DES and DLT40 and Las Cumbres Observatory and VINROUGE and MASTER Collaborations], Abbott *et al.* (2017). In Guidorzi *et al.* (2017) a higher value of H_0 is obtained by constraining the inclination of the source with a model of the broadband X-ray to radio emission and a different estimate of the peculiar velocity.

- Modified propagation of GWs has been observed in several specific models. For instance, in the now ruled out DGP model [Dvali, Gabadadze and Porrati (2000)], at cosmological scales gravity leaks into extra dimensions, affecting the $1/d_L(z)$ behavior; see Deffayet and Menou (2007). Modified GW propagation can be included in the general effective field theory approach to dark energy reviewed in Gleyzes, Langlois and Vernizzi (2015) [see also Nishizawa (2017)], and has been discussed in some scalar–tensor theories of the Horndeski class in Saltas, Sawicki, Amendola and Kunz (2014), Lombriser and Taylor (2016) and Arai and Nishizawa (2017). The discussion of the GW luminosity distance that we have presented is from Belgacem, Dirian, Foffa and Maggiore (2017b), where it is also realized that, in theories where GW propagation is modified, after parameter estimation this effect largely dominates over the effect from the DE EoS.

- The "RR" non-local gravity model was proposed in Maggiore and Mancarella (2014), following earlier work in Maggiore (2014). The interpretation of the non-local term as a dynamical mass generation for the conformal mode is discussed in Maggiore (2015, 2016). A detailed comparison with cosmological data has been carried out in Dirian, Foffa, Khosravi, Kunz and Maggiore (2014), Dirian, Foffa, Kunz, Maggiore and Pettorino (2015, 2016), Dirian (2017) and Belgacem, Dirian, Foffa and Maggiore (2017a). Detailed reviews of conceptual aspects and cosmological consequences of the model are Maggiore (2017) and Belgacem, Dirian, Foffa and Maggiore (2017a).

The imprint of GWs on the CMB

We now turn to the effect of primordial GWs on the cosmic microwave background (CMB). We will see that the effect of GWs on CMB temperature anisotropies and on CMB polarization is potentially a powerful tool for detecting GWs of primordial origin, such as GWs generated by an early inflationary phase. In this chapter we will study how scalar and tensor perturbations affect the CMB, while in the next chapter we will study how inflation generates primordial GWs and in Chapter 22 we will discuss several other cosmological production mechanisms.

20.1 The CMB multipoles

The CMB is characterized by a temperature T that is remarkably uniform over the sky, with small fluctuations. These tiny anisotropies, however, carry extraordinary information about the early Universe. We denote by $T_0(\hat{\mathbf{n}})$ the CMB temperature measured today, by a detector located on the Earth (or on a satellite), looking in the direction $\hat{\mathbf{n}}$. Its average over all directions of the sky is

$$T_0 \equiv \int \frac{d^2\hat{\mathbf{n}}}{4\pi} T_0(\hat{\mathbf{n}}) \simeq 2.7255(6)\,\mathrm{K}\,. \tag{20.1}$$

We denote by $\delta T_0(\hat{\mathbf{n}})$ the fluctuations with respect to this average value, $\delta T_0(\hat{\mathbf{n}}) = T_0(\hat{\mathbf{n}}) - T_0$. This is a scalar function of the unit vector $\hat{\mathbf{n}}$, or equivalently of its polar angles (θ, ϕ), and can therefore be expanded in spherical harmonics,

$$\frac{\delta T_0(\hat{\mathbf{n}})}{T_0} = \sum_{l=1}^{\infty} \sum_{m=-l}^{l} a_{lm}\, Y_{lm}(\hat{\mathbf{n}})\,. \tag{20.2}$$

This expansion defines the coefficients a_{lm}. The fact that $\delta T_0(\hat{\mathbf{n}})/T_0$ is real, together with $Y_{lm}^* = (-1)^m Y_{l,-m}$, implies that

$$a_{lm}^* = (-1)^m a_{l,-m}\,. \tag{20.3}$$

Recalling that the spherical harmonics satisfy the orthogonality condition (3.228), the inverse of eq. (20.2) is

$$a_{lm} = \int d^2\hat{\mathbf{n}}\, \frac{\delta T_0(\hat{\mathbf{n}})}{T_0}\, Y_{lm}^*(\hat{\mathbf{n}})\,. \tag{20.4}$$

Gravitational Waves, Volume 2: Astrophysics and Cosmology. Michele Maggiore.
© Michele Maggiore 2018. Published in 2018 by Oxford University Press.
DOI 10.1093/oso/9780198570899.001.0001

Observe that the sum in eq. (20.2) starts from $l = 1$ rather than from $l = 0$ since, by definition, in $\delta T_0(\hat{\mathbf{n}})$ the constant mode T_0 has already been subtracted, so the CMB anisotropies start from the dipole, $l = 1$. Furthermore, the cosmological contribution to the dipole is masked by a much larger contribution due to the motion of the Earth with respect to the CMB rest frame. Thus, the dipole is just used to compute the Earth's velocity in the CMB rest frame, and the remaining expansion starts from $l = 2$.[1]

The coefficients a_{lm} are stochastic variables, which are determined from the stochastic initial conditions in the early Universe. In principle, if we had an ensemble of universes, we could define an ensemble average for them. Assuming that the process that generates the initial conditions is statistically spatially isotropic, this ensemble average would take the form

$$\langle a_{lm} a^*_{l'm'} \rangle = C_l \, \delta_{ll'} \, \delta_{mm'} \,. \tag{20.5}$$

However, we only have one realization of our Universe. Nevertheless, assuming statistical isotropy, we can consider the different values of m for a given l as independent quantities extracted from the same statistical distribution. We can then replace the average over an ensemble of universes by an average over m, defining

$$C_l = \frac{1}{2l+1} \sum_{m=-l}^{l} |a_{lm}|^2 \,. \tag{20.6}$$

This, however, implies that, for a given l, we have only $2l + 1$ quantities at our disposal in order to sample the statistical property of the distribution. This intrinsic uncertainty is called *cosmic variance*, and represents a fundamental limitation to our knowledge.[2] Cosmic variance is particularly visible for very small l, say $l < 10$, as the shaded band in Fig. 20.1.

The coefficients C_l determine the correlation function of the temperature fluctuations,

$$\langle \delta T_0(\hat{\mathbf{n}}) \delta T_0(\hat{\mathbf{n}}') \rangle = T_0^2 \sum_{lm,l'm'} \langle a_{lm} a^*_{l'm'} \rangle Y_{lm}(\hat{\mathbf{n}}) Y^*_{l'm'}(\hat{\mathbf{n}}')$$

$$= T_0^2 \sum_{lm} C_l \, Y_{lm}(\hat{\mathbf{n}}) Y^*_{lm}(\hat{\mathbf{n}}') \,. \tag{20.7}$$

Using the addition theorem for spherical harmonics,

$$\sum_{m=-l}^{l} Y_{lm}(\hat{\mathbf{n}}) Y^*_{lm}(\hat{\mathbf{n}}') = \frac{2l+1}{4\pi} P_l(\hat{\mathbf{n}} \cdot \hat{\mathbf{n}}') \,, \tag{20.8}$$

where the P_l are Legendre polynomials, we get

$$\langle \delta T_0(\hat{\mathbf{n}}) \delta T_0(\hat{\mathbf{n}}') \rangle = \frac{T_0^2}{4\pi} \sum_l (2l+1) C_l P_l(\hat{\mathbf{n}} \cdot \hat{\mathbf{n}}') \,. \tag{20.9}$$

[1] Actually, the Earth's motion also induces an aberration effect, i.e. an apparent deflection of the observed CMB photons. This effect has been measured [see [Planck Collaboration], Aghanim *et al.* (2014)], and one could then in principle subtract it from the total dipole contribution to isolate the cosmological contribution. However, the error in the measurement of the velocity from the aberration is presently too large to extract the much smaller cosmological contribution to the dipole.

[2] The error in the *Planck* 2015 measurements of the multipoles of the temperature anisotropies is dominated by cosmic variance up to $l \simeq 1600$, which means that, up to these multipoles, the experiment has reached the best attainable accuracy. This holds for the multipoles of the temperature–temperature correlator. For the correlators involving the CMB polarization, which will be discussed in Section 20.4, in the *Planck* 2015 data the fundamental limitation due to cosmic variance is only attained for lower values of l.

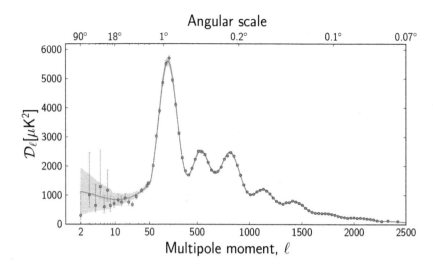

Fig. 20.1 The coefficients $\mathcal{D}_l$ that give the CMB anisotropy power spectrum, measured by the *Planck* collaboration. The vertical band visible at low l is the cosmic variance. From [Planck Collaboration], Ade *et al.* (2014a) "Planck 2013 results. XV". Reproduced with permission @ ESO.

When one wants to distinguish the coefficients a_{lm} and C_l from analogous coefficients that determine the correlation between temperature and polarizations, or between polarizations, which we will introduce in Section 20.4, one rather denotes them by a_{lm}^{TT} and C_l^{TT}, where the superscript TT stresses that these coefficients determine the temperature–temperature correlation function. In this section we will denote them simply by a_{lm} and C_l.

In general, observational data are displayed by plotting the related quantity

$$\mathcal{D}_l = T_0^2 \frac{l(l+1)}{2\pi} C_l \,. \tag{20.10}$$

This quantity is usually referred as the CMB temperature power spectrum. Figure 20.1 shows the coefficients $\mathcal{D}_l$, as measured by the *Planck* collaboration. Three main features stand out from this plot.

- The low-l region, say $l \lesssim 40$, is approximately flat when plotted in terms of $\mathcal{D}_l$. This is the region of the Sachs–Wolfe plateau, which we will discuss in Section 20.3.2. We will see that, in the limit in which the initial power spectrum has the Harrison–Zeldovich form, i.e. $n_s = 1$ in eq. (19.178), for small l one gets $C_l \propto 1/l(l+1)$, so in this limit $\mathcal{D}_l$ is exactly flat (apart from a smaller contribution from the ISW effect; see below). This is indeed the reason why $\mathcal{D}_l$ is defined as proportional to $l(l+1)C_l$ in eq. (20.10). Actually, n_s turns out to be close but not exactly equal to 1, as we saw in eq. (19.184), so $\mathcal{D}_l$ is not exactly flat even for small l.

- An intermediate region, where $\mathcal{D}_l$ displays a set of peaks. These are acoustic peaks, reflecting the acoustic oscillations in the radiation energy density at last scattering.
- A large-l region, where damping effects set in, and the power spectrum decreases.

The expansion in spherical harmonics on the sphere is analogous to the Fourier transform in the plane, and for sufficiently small angular scales, where one cannot appreciate the curvature of the sphere, it reduces to the Fourier transform. At large l, the Legendre polynomial $P_l(\cos\theta)$ oscillates as $\cos[(l+1/2)\theta - \pi/4]$, so the characteristic angles θ that contribute mostly to multipole l are of order π/l: multipoles with large l capture features at small angular scales, while small l corresponds to large angular scales. It is conventional to define a characteristic angular scale θ_l associated with the multipole moment l,

$$\theta_l \equiv \frac{\pi}{l}. \tag{20.11}$$

This quantity labels the top horizontal axis in Fig. 20.1. In the definition (20.11) the numerical coefficient has been chosen according to the number of nodes of $P_l(\cos\theta)$. So, for instance, $P_1(\cos\theta)$ vanishes at $\theta = -\pi/2, \pi/2$, so these nodes separate the whole circle $-\pi \leqslant \theta \leqslant \pi$ into two parts of length $\theta_{l=1} = \pi$ each; for the quadrupole $l = 2$, $P_l(\cos\theta)$ has four nodes, separating the circle into regions of approximate size $2\pi/4 = \pi/2$ each (although now the nodes are not exactly equally spaced), and so on for all the other multipoles. Of course, the precise numerical factor in eq. (20.11) is just a conventional definition, and a given multipole receives contributions from a whole range of angles of order θ_l.

The computation of CMB anisotropies can be separated into two parts, which correspond to two quite distinct physical regimes. At $z < z_{\rm dec}$ the photons are strongly coupled to the baryons. Therefore, together they form a single photon–baryon fluid. This means in particular that radiation can indeed by described as a fluid with density and velocity perturbations, whose evolution is given by the equations that we have studied in Chapter 19. After decoupling, in contrast, photons freely propagate toward the observer. Correspondingly (at least in the approximation in which decoupling is instantaneous, see below), the CMB anisotropies can be split into two parts: (1) the perturbations to $\delta T/T$ coming from the perturbations of the photon energy density, the photon velocity potential and the metric perturbations at the time of last scattering; (2) the effects that perturb the subsequent free propagation of photons from last scattering to us. Let us stress that this separation into a regime where the photon can be treated as a fluid and a free-streaming regime is an approximation. The exact first-principles approach involves the Boltzmann equation for the photons, and leads to a set of equations that can be integrated numerically through so-called Boltzmann codes, as we will discuss in Section 20.3.7. However, the simplified treatment allows us to obtain analytic results that are still relatively accurate, and to get an understanding of the physics involved. Therefore we will use this approximation in the following sections.

Before entering into the details of the computations, it is useful to understand qualitatively some basic features of the effect of GWs on the CMB. In particular, for GWs the most interesting region is the low-l region. This is due to the fact that GWs do not affect the primordial CMB anisotropies at the time of last scattering, as we will prove in Section 20.3.5. Thus, the only effect of GWs comes from their influence on the propagation of photons from last scattering to the present time. As we have seen in Section 19.5 (see in particular Fig. 19.14), the amplitude of a primordial GW (i.e. a GW produced in the early Universe, much before last scattering) freely propagating in a FRW background starts to decrease when it enters the horizon. Thus, GWs that entered the horizon well before last scattering, i.e. modes with $k\eta_{\rm dec} \gg 1$, by the time of last scattering already have a small amplitude, and in the subsequent evolution they are rapidly oscillating functions. As we will see in this Section 20.3.5, the effect of GWs on CMB temperature multipoles is given by a line-of-sight integral of the derivative of h. For modes with $k\eta_{\rm dec} \gg 1$ this becomes an integral over a rapidly oscillating function, which averages to zero. Thus, tensor modes with $k\eta_{\rm dec} \gg 1$ have very little effect on CMB anisotropies, and the most interesting region for attempting to detect the effect of primordial GWs on the CMB is $k\eta_{\rm dec} \lesssim 1$.

A GW with comoving momentum $k \lesssim 1/\eta_{\rm dec}$ imprints anisotropies on the CMB on a length-scale $\lambda = 2\pi/k \gtrsim 2\pi\eta_{\rm dec}$. As we can see from Fig. 20.2, if such a GW acts soon after last scattering, an observer at the Earth sees the corresponding perturbations under an angle

$$\theta \simeq \frac{(\lambda/2)}{\eta_0 - \eta_{\rm dec}} \simeq \frac{\lambda}{2\eta_0} \, , \qquad (20.12)$$

since $\eta_0 \gg \eta_{\rm dec}$.[3] Using $\lambda \gtrsim 2\pi\eta_{\rm dec}$, this gives

$$\theta \gtrsim \pi \frac{\eta_{\rm dec}}{\eta_0} \, . \qquad (20.13)$$

If the GW perturbation acts long after last scattering, its wavelength subtends an even larger angle; see again Fig. 20.2. Thus, primordial GWs can only affect large-angle anisotropies. Using the value of $\eta_{\rm dec}/\eta_0$ from eq. (17.174), we see that they can only affect angles

$$\theta \gtrsim \frac{\pi}{50} \, {\rm rad} \simeq 3.6° \, . \qquad (20.14)$$

Therefore, from eq. (20.11), primordial GWs basically only affect multipoles with

$$l < O(50) \, . \qquad (20.15)$$

We will later confirm this intuitive argument with a detailed explicit computation. The result, shown in Fig. 20.14 on page 538, will confirm that the effect of GWs quickly falls for multipoles with l larger than a value of order 50–100. Similar arguments hold for the effect of GWs on

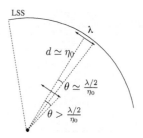

Fig. 20.2 GWs propagating in a direction transverse to the line of sight leave perturbations on the scale λ of their comoving wavelength. The corresponding angle seen by the observer is $\theta \simeq \lambda/(2\eta_0)$ if the GW acts very close to the last scattering surface (LSS), and larger than that if the GW is closer to the observer.

[3]In Fig. 20.2 we are assuming that the direction of propagation of the GW is orthogonal to the direction $\hat{\mathbf{n}}$ of the line of sight. Indeed, we will see in eq. (20.123) that the effect is proportional to $n^i n^j (h_{ij}^{\rm TT})'$, and therefore vanishes when $\hat{\mathbf{n}}$ coincides with the propagation direction of the GW.

the CMB polarization. The detailed computation will show that in this case the effect of GWs quickly decreases beyond l of order 100–150, as we will see in Figs. 20.27 and 20.28.

It is also useful to give the approximate relation between the comoving momentum k of a perturbation and the typical multipole moment l of the CMB anisotropies produced by such a mode. Combining eqs. (20.11) and (20.12) and using $\lambda = 2\pi/k$ we find that the multipole moment l probes typical momenta of the order of a scale k_l defined by

$$k_l = \frac{l}{\eta_0} \,. \tag{20.16}$$

Using the numerical value of η_0 in eq. (17.172) and setting for simplicity $\Omega_M = 0.3$ gives

$$k_l \simeq 1 \times 10^{-4}\, l \,\frac{h_0}{\text{Mpc}} \,. \tag{20.17}$$

As we discussed in Note 40 on page 489, typical pivot scales often used by CMB experiments are $k_* = 0.002\,\text{Mpc}^{-1}$ and $k_* = 0.05\,\text{Mpc}^{-1}$. Setting $h_0 = 0.7$, the former corresponds to a multipole moment $l_* \simeq 27$ and the latter to $l_* \simeq 690$.[4]

In eq. (19.190) we found that scalar modes become non-linear today at $k_{\text{nl}} \simeq 0.08\, h_0/\text{Mpc}$. However, at the epoch of last scattering all the scalar modes corresponding to even the highest multipoles probed by CMB were still fully linear. Thus, the perturbations on the last scattering surface are very well described by cosmological perturbation theory at linear order. The situation is slightly different for the propagation of the CMB photons from last scattering to us. At the level of sensitivity of the *Planck* 2015 data, the CMB lensing, which is a higher-order effect, becomes relevant for the temperature–temperature correlator. Furthermore, scalar modes with $k \gtrsim k_{\text{nl}}$, which affect CMB multipoles with $l \gtrsim 800$, start to become non-linear near the present cosmological epoch. Therefore, CMB multipoles with $l \gtrsim 800$ could in principle be affected by this non-linear evolution, through a modification of the late ISW effect, which we will introduce in Section 20.3.2. However, we will see that for such multipoles the ISW effect is very small.[5] Thus, the physics of the CMB is mostly described by linear cosmological perturbation theory.

20.2   Null geodesics

In order to understand how scalar and tensor perturbations affect CMB multipoles, we must study the propagation of photons in a perturbed FRW metric. In this section we begin by recalling the notions of null geodesic and affine parameter.

In Section 1.3.1 we introduced the geodesic equation for a massive particle, using the particle proper time τ to parametrize the trajectory. To study the massless limit it is useful first of all to be more general and describe the trajectory of a massive particle by a curve $x^\mu(\lambda)$, where λ is for the moment an arbitrary parameter along the particle's world-line.

[4]Of course, in general, a Fourier mode k contributes to a range of multipoles. The value of l to which it contributes most is equal to l/η_0 times a constant of order 1, which is different for the various effects that contribute to the CMB. We will see for instance in eq. (20.111) that in the so-called Sachs–Wolfe effect, which we will introduce in Section 20.3.2, the contribution to the C_l is given by an integral over dk/k of the primordial power spectrum of scalar perturbations, weighted with a factor $j_l^2(k\eta_0)$. The spherical Bessel function $j_l(k\eta_0)$ has its maximum at $k\eta_0 \simeq l$ (for example the maximum is at $k\eta_0 \simeq \{12, 104, 507\}$ for $l = \{10, 100, 500\}$, respectively), but of course the result for C_l comes from the integration over a full range of momenta, determined in a first approximation by the width of the function $j_l^2(k\eta_0)$ at its maximum; see Fig. 20.4 on page 529.

[5]Actually, we will see in Figs. 20.8 and 20.9 that, for multipoles $l \gtrsim 800$, the late ISW effect is completely negligible in the temperature–temperature (TT) correlator. The best way to measure the ISW is to perform a correlation between the temperature perturbations and a tracer of the matter distribution such as the galaxy density field; see the Further Reading. A separate detection of the ISW effect is interesting because, as we will see, in ΛCDM the late ISW effect is due to the fact that the cosmological constant starts to dominate, and, more generally, the late ISW effect is a probe of dark energy.

For a massive particle the equation of motion for $x^\mu(\lambda)$ can be obtained from the action (see also the discussion in Section 22.5.2)

$$S = -m \int d\lambda \left(-g_{\mu\nu}[x(\lambda)] \frac{dx^\mu}{d\lambda} \frac{dx^\nu}{d\lambda} \right)^{1/2}, \qquad (20.18)$$

by taking the variation with respect to $x^\mu(\lambda)$, with $x^\mu(\lambda)$ kept fixed at the boundaries. The constant m just ensures that the action is dimensionless, as it should be. Being an overall factor, it does not affect the equations of motion. Nevertheless, it only makes sense for a massive particle, and its presence already makes it clear that the limit $m \to 0$ cannot be taken in a trivial manner. The action (20.18) is invariant under reparametrizations, under which the D fields $x^\mu(\lambda)$ (in D space-time dimensions), living in the one-dimensional world-line parametrized by λ, transform as scalars,

$$\lambda \to \lambda'(\lambda), \qquad (20.19)$$
$$x^\mu(\lambda) \to (x^\mu)'(\lambda') = x^\mu(\lambda). \qquad (20.20)$$

To understand how to take the massless limit, it is useful to rewrite the action (20.18) by introducing an auxiliary field $e(\lambda)$, as

$$S = \frac{1}{2} \int d\lambda \left[\frac{1}{e(\lambda)} g_{\mu\nu}[x(\lambda)] \frac{dx^\mu}{d\lambda} \frac{dx^\nu}{d\lambda} - e(\lambda)m^2 \right]. \qquad (20.21)$$

The two actions (together with the condition $e > 0$) are equivalent. Indeed, taking the variation of the action (20.21) with respect to $e(\lambda)$ we get

$$e^2(\lambda) = -\frac{1}{m^2} g_{\mu\nu} \frac{dx^\mu}{d\lambda} \frac{dx^\nu}{d\lambda}. \qquad (20.22)$$

For a timelike geodesic we have $g_{\mu\nu}(dx^\mu/d\lambda)(dx^\nu/d\lambda) < 0$. Therefore the solution of eq. (20.22) with $e > 0$ is

$$e(\lambda) = \frac{1}{m} \left(-g_{\mu\nu} \frac{dx^\mu}{d\lambda} \frac{dx^\nu}{d\lambda} \right)^{1/2}. \qquad (20.23)$$

Plugging this into the action (20.21) we get back the action (20.18). The introduction of the auxiliary field $e(\lambda)$ allows us to get rid of the square root in the original action.[6] The new action (20.21) is still reparametrization invariant, if we supplement eqs. (20.19) and (20.20) with the following transformation for $e(\lambda)$:

$$e(\lambda) \to e'(\lambda') = e(\lambda) \frac{d\lambda}{d\lambda'}, \qquad (20.24)$$

i.e. $e(\lambda)d\lambda = e(\lambda')d\lambda'$. Variation of the action (20.21) with respect to $x^\mu(\lambda)$ gives

$$\frac{d^2 x^\mu}{d\lambda^2} + \Gamma^\mu_{\nu\rho}(x) \frac{dx^\nu}{d\lambda} \frac{dx^\rho}{d\lambda} = \frac{1}{e} \frac{de}{d\lambda} \frac{dx^\mu}{d\lambda}. \qquad (20.25)$$

This is not yet the same as the geodesic equation (17.21), because of the non-vanishing term on the right-hand side. This is due to the fact

[6] The reader familiar with string theory will recognize that the analogous procedure allows one to eliminate the square root of the Nambu–Goto action, to obtain the Polyakov form of the bosonic string action; see for example Section 1.2 of Polchinski (1998). We will review this construction in Section 22.5.2.

that the parameter λ is for the moment totally arbitrary. However, we can now use the local invariance (20.24) to fix the gauge $e(\lambda) = 1/m$. Because of eq. (20.23), this amounts to choosing λ so that $d\lambda = (-g_{\mu\nu}dx^\mu dx^\nu)^{1/2}$, i.e. λ is chosen to be equal to the particle proper time τ. With this choice $de/d\lambda = 0$ and the right-hand side of eq. (20.25) vanishes. We therefore get back the standard form (1.66) of the geodesic equation. We see that the choice of proper time to parametrize the world-line of a massive particle, while not compulsory, is useful because it simplifies the geodesic equation.

Once the gauge has been fixed so that $de/d\lambda = 0$, we still have the possibility of performing a residual transformation of the form

$$\lambda' = \alpha\lambda + \beta. \tag{20.26}$$

Under this transformation $e(\lambda) \to e'(\lambda') = e(\lambda)/\alpha = 1/(m\alpha)$, so the condition $de/d\lambda = 0$ is preserved, and the right-hand side of eq. (20.25) still vanishes. A transformation of the form (20.26) is called an affine transformation. Correspondingly, a parameter λ for which the right-hand side of eq. (20.25) vanishes is called an *affine parameter*.

Apart from the nice technical aspect of allowing us to get rid of the square root, the form (20.21) of the action is also useful because it has a well-defined massless limit. Indeed, setting $m = 0$ there, we get the action of a massless point particle,

$$S_{\text{massless}} = \frac{1}{2} \int d\lambda \, \frac{1}{e(\lambda)} g_{\mu\nu}[x(\lambda)] \frac{dx^\mu}{d\lambda} \frac{dx^\nu}{d\lambda}. \tag{20.27}$$

Variation of the action (20.27) with respect to x^μ gives again the geodesic equation (20.25), since the mass term in eq. (20.21) did not contribute to the variation with respect to x^μ. Variation with respect to $e(\lambda)$ now no longer determines $e(\lambda)$ itself, as in the massive case [see eq. (20.22)]. Rather, it gives

$$g_{\mu\nu} \frac{dx^\mu}{d\lambda} \frac{dx^\nu}{d\lambda} = 0, \tag{20.28}$$

i.e. independently of the definition of λ, it imposes the condition that the photon evolves along a null geodesic,

$$g_{\mu\nu}dx^\mu(\lambda)dx^\nu(\lambda) = 0. \tag{20.29}$$

Thus, for time-like geodesics we can determine the form of the affine parameter explicitly, by gauge-fixing $e(\lambda)$ to a constant and solving eq. (20.22). In particular, the proper time τ can be defined from

$$g_{\mu\nu} \frac{dx^\mu}{d\tau} \frac{dx^\nu}{d\tau} = -1, \tag{20.30}$$

which can be solved for $d\tau$ as $d\tau^2 = -g_{\mu\nu}dx^\mu dx^\nu$.

For a massless particle, instead of eq. (20.30) we rather have eq. (20.28), which cannot be used to determine λ. Nevertheless, we can still fix λ, modulo the transformation (20.26), by requiring it to be an affine parameter. Indeed, starting with a generic choice of the parameter λ such that $e(\lambda)$ is some generic function, we can perform a reparametrization transformation $\lambda \to \lambda'(\lambda)$ with $\lambda'(\lambda)$ defined by

$$d\lambda' = e(\lambda)d\lambda \,. \qquad (20.31)$$

According to eq. (20.24) we then have $e'(\lambda') = 1$, so in this new parametrization the right-hand side of eq. (20.25) vanishes, and the geodesic equation becomes

$$\frac{d^2 x^\mu}{d\lambda'^2} + \Gamma^\mu_{\nu\rho}(x)\frac{dx^\nu}{d\lambda'}\frac{dx^\rho}{d\lambda'} = 0 \,, \qquad (20.32)$$

as we can also directly check by replacing $d/d\lambda$ by $e\,d/d\lambda'$ in eq. (20.25). At the same time eq. (20.28) becomes

$$g_{\mu\nu}\frac{dx^\mu}{d\lambda'}\frac{dx^\nu}{d\lambda'} = 0 \,. \qquad (20.33)$$

We henceforth drop the prime and write λ' simply as λ, with the understanding that λ is now an affine parameter.

Equations (20.32) and (20.33) determine the dynamics of a (point-like) massless particle in curved space. In terms of this affine parameter λ we define the photon four-momentum P^μ similarly to what we did in eq. (17.23) for massive particles,

$$P^\mu = \frac{dx^\mu}{d\lambda} \,. \qquad (20.34)$$

Requiring that $dx^0/d\lambda$ is equal to the energy P^0 of the photon fixes the parameter α in the affine transformation (20.26), while, of course, the choice of the parameter β remains undetermined, but it has no effect since only $d\lambda$ enters the equations. Then eq. (20.28) becomes the mass-shell condition

$$g_{\mu\nu}[x^\mu(\lambda)]P^\mu P^\nu = 0 \,, \qquad (20.35)$$

while eq. (20.32) becomes

$$\frac{dP^\mu}{d\lambda} + \Gamma^\mu_{\nu\rho}[x^\mu(\lambda)]P^\nu P^\rho = 0 \,, \qquad (20.36)$$

where we have stressed that the Christoffel symbols and the metric in eqs. (20.35) and (20.36) must be evaluated on the particle trajectory, i.e. at the space-time points $x^\mu(\lambda)$. Observe that the mass-shell condition (20.35) is consistent with the equation of motion (20.36). Indeed, using eq. (20.36),

$$\frac{d}{d\lambda}\left(g_{\mu\nu}[x^\mu(\lambda)]P^\mu P^\nu\right) = (\partial_\rho g_{\mu\nu} - g_{\mu\sigma}\Gamma^\sigma_{\nu\rho} - g_{\nu\sigma}\Gamma^\sigma_{\mu\rho})P^\mu P^\nu P^\rho$$

$$= 0 \,, \qquad (20.37)$$

since the term in parentheses is just the covariant derivative of the metric, $\nabla_\rho g_{\mu\nu}$. Thus, the value of $g_{\mu\nu}[x^\mu(\lambda)]P^\mu P^\nu$ is conserved along a geodesic, and therefore if we set it to zero at an initial space-time point of a geodesic, it remains zero along the geodesic. In this sense, eq. (20.35) can be seen as a constraint on the initial data, preserved by the equations of motion.

In practice, the affine parameter λ can be eliminated using the $\mu = 0$ component of eq. (20.34),

$$P^0 \equiv \frac{dx^0}{d\lambda}, \tag{20.38}$$

i.e. $d\lambda \equiv dx^0/P^0$, where $P^0 = E$ is the photon energy. This allows us to eliminate the derivatives with respect to λ, according to

$$\frac{d}{d\lambda} = \frac{dx^0}{d\lambda}\frac{d}{dx^0} = P^0 \frac{d}{dx^0}. \tag{20.39}$$

As a check, let us verify that in a FRW metric the evolution of P^0 with the scale factor, determined from eq. (20.36), is the correct one for the energy of a massless particle. With this purpose in mind we work in coordinates $(t, \mathbf{x})$, where the correct scaling is $P^0 \propto 1/a$. The non-vanishing Christoffel symbols were given in eq. (17.8). The $\mu = 0$ component of eq. (20.36) gives

$$\frac{dP^0}{d\lambda} + a^2 H \delta_{ij} P^i P^j = 0. \tag{20.40}$$

The condition (20.35) gives $a^2 \delta_{ij} P^i P^j = (P^0)^2$. In the coordinates $(t, \mathbf{x})$ that we are using here, eq. (20.39) gives $d/d\lambda = P^0 d/dt$. Then eq. (20.40) becomes

$$P^0 \frac{dP^0}{dt} = -H(P^0)^2, \tag{20.41}$$

which, apart from the irrelevant solution $P^0 = 0$, has the solution $P^0 \propto a^{-1}$. Thus, $dt/d\lambda$ indeed scales as $1/a$, as required for the energy of a massless particle.

The same computation as in eq. (17.26), with $\tau \to \lambda$, shows that, in FRW, $a^2 dx^i/d\lambda$ is conserved on the solutions of eq. (20.32). Thus, in this case (or, more generally, when perturbing around FRW) we define the comoving photon momentum p^i by

$$p^i = a^2 \frac{dx^i}{d\lambda}, \tag{20.42}$$

and the physical photon momentum as

$$p^i_{\rm ph} = \frac{1}{a}p^i = a\frac{dx^i}{d\lambda}. \tag{20.43}$$

We then define $\mathbf{p}^2 = \delta_{ij}p^i p^j$ and $\mathbf{p}^2_{\rm ph} = \delta_{ij}p^i_{\rm ph}p^j_{\rm ph}$, as in eqs. (17.28) and (17.31), respectively. Then eq. (20.35), specialized to FRW and coordinates $(t, \mathbf{x})$, reads $E = |\mathbf{p}_{\rm ph}|$, so in terms of these variables we have the same dispersion relation as in flat space. The spatial components P^i of the photon four-momentum P^μ are then related to the comoving momentum by $p^i = a^2 P^i$, just as we found for massive particles in Section 17.1.2.

20.3   Temperature anisotropies at large angles

We now specify this general formalism to photons propagating around a perturbed FRW background, and we compute the resulting anisotropies in the CMB, focusing on the effects that are important in the large-angle, low-multipole regime, which is the most relevant one for studying the effect of GWs on the CMB.

20.3.1   Photon geodesics in a perturbed FRW metric

When studying cosmological perturbations we find it convenient to use conformal coordinates $(\eta, \mathbf{x})$. In this case eq. (20.39) becomes

$$\frac{d}{d\lambda} = P^0 \frac{d}{d\eta}, \tag{20.44}$$

where $P^0 = d\eta/d\lambda$ is the value of the photon energy in the conformal frame. When we need to distinguish between the value of P^0 in the conformal frame, with coordinates $(\eta, \mathbf{x})$, and the value in the frame with coordinates $(t, \mathbf{x})$, we will use the notations $P^\eta = d\eta/d\lambda$ and $P^t = dt/d\lambda$, respectively; otherwise in this section P^0 denotes P^η. From $dt = a d\eta$ it follows that $P^\eta = P^t/a$, and therefore, on a geodesic,

$$P^\eta \propto a^{-2}. \tag{20.45}$$

The non-vanishing Christoffel symbols for a FRW metric plus perturbations have been given, in conformal coordinates, in eqs. (18.99)–(18.104). Making use of eq. (20.44), and considering for the moment only scalar perturbations, the $\mu = 0$ component of eq. (20.36) gives

$$P^0 \frac{dP^0}{d\eta} = -\left(\mathcal{H} + \Psi'\right)\left(P^0\right)^2 - 2P^0 P^i \partial_i \Psi$$
$$- \left[\mathcal{H}(1 + 2\Phi - 2\Psi) + \Phi'\right]\mathbf{P}^2, \tag{20.46}$$

where, as usual, the prime denotes the partial derivative $\partial/\partial\eta$, and $\mathbf{P}^2 = \delta_{ij} P^i P^j$. We must now be careful to distinguish between partial and total derivatives with respect to η, since the potentials Φ and Ψ, just like the metric and the Christoffel symbols in eqs. (20.35) and (20.36), are evaluated on the photon trajectory, and therefore are functions both of η and of $\mathbf{x}(\eta)$. Therefore

$$\frac{d}{d\eta}\Psi[\eta, \mathbf{x}(\eta)] = \frac{\partial\Psi}{\partial\eta} + \frac{dx^i(\eta)}{d\eta}\partial_i\Psi$$
$$= \Psi' + \frac{1}{P^0} P^i \partial_i \Psi, \tag{20.47}$$

where we have used eq. (20.44) to write $dx^i/d\eta$ in terms of $dx^i/d\lambda = P^i$. We can rewrite eq. (20.47) as

$$P^i \partial_i \Psi = P^0 \left(\frac{d\Psi}{d\eta} - \Psi'\right), \tag{20.48}$$

which can be used to transform the second term on the right-hand side of eq. (20.46). The term $\mathbf{P}^2$ in eq. (20.46) can be transformed using eq. (20.35). To linear order in the perturbations this gives

$$\mathbf{P}^2 = (1 - 2\Phi + 2\Psi)\left(P^0\right)^2 . \tag{20.49}$$

Then eq. (20.46) becomes

$$\frac{dP^0}{d\eta} = -\left[2\mathcal{H} + \Phi' - \Psi' + 2\frac{d\Psi}{d\eta}\right]P^0 . \tag{20.50}$$

We can now split P^0 into a background term and a first-order term,

$$P^0 = \bar{P}^0 + \delta P^0 . \tag{20.51}$$

At the background level we have

$$\frac{d\bar{P}^0}{d\eta} = -2\mathcal{H}\bar{P}^0 , \tag{20.52}$$

which gives $\bar{P}^0 \propto a^{-2}$, as we have already found in eq. (20.45). It is then convenient to introduce an auxiliary variable

$$\varepsilon = a^2 P^0 \tag{20.53}$$

and write $\varepsilon = \bar{\varepsilon} + \delta\varepsilon$, with $\bar{\varepsilon} = a^2\bar{P}^0$ and $\delta\varepsilon = a^2\delta P^0$. Then

$$\frac{d\varepsilon}{d\eta} = \left(\Psi' - \Phi' - 2\frac{d\Psi}{d\eta}\right)\varepsilon , \tag{20.54}$$

which splits into $d\bar{\varepsilon}/d\eta = 0$, so that $\bar{\varepsilon}$ is constant, and

$$\frac{d(\delta\varepsilon)}{d\eta} = \left(\Psi' - \Phi' - 2\frac{d\Psi}{d\eta}\right)\bar{\varepsilon} . \tag{20.55}$$

Since $\bar{\varepsilon}$ is constant, this can be rewritten as

$$\frac{d(\delta\varepsilon/\bar{\varepsilon})}{d\eta} = \Psi' - \Phi' - 2\frac{d\Psi}{d\eta} . \tag{20.56}$$

In turn $\delta\varepsilon/\bar{\varepsilon} = \delta P^0/\bar{P}^0$, so in the end

$$\frac{d}{d\eta}\left(\frac{\delta P^0}{\bar{P}^0}\right) = \Psi' - \Phi' - 2\frac{d\Psi}{d\eta} . \tag{20.57}$$

Integrating this equation over time, from the emission point E, at time $\eta = \eta_{\rm dec}$ and position $\mathbf{x}(\eta_{\rm dec})$, to the observation point O, at the present time η_0 and position $\mathbf{x} = 0$, we get

$$\boxed{\left.\frac{\delta P^0}{\bar{P}^0}\right|_E^O = -2\Psi[\eta, \mathbf{x}(\eta)]\Big|_E^O + \int_E^O d\eta\left\{\Psi'[\eta, \mathbf{x}(\eta)] - \Phi'[\eta, \mathbf{x}(\eta)]\right\} .}$$

$$\tag{20.58}$$

Observe that the total derivative term $d\Psi/d\eta$, upon integration, produces a term that depends only on the values of Ψ at the emission and observation points, while the terms involving the partial derivatives $\Psi' \equiv \partial\Psi/\partial\eta$ and $\Phi' \equiv \partial\Phi/\partial\eta$ depend on the whole path. Recall also, from eq. (19.16), that in the absence of anisotropic stress we have $\Phi = -\Psi$.

20.3.2   Sachs–Wolfe, ISW and Doppler contributions

Equation (20.58) gives the perturbation of P^0, which is the energy of a photon measured in the conformal frame. From this we can obtain the ratio of the temperatures at the observation and emission points, again in the conformal frame,

$$\left(\frac{T_O}{T_E}\right)^{\text{conf}} = \frac{(P^0)_O}{(P^0)_E}, \tag{20.59}$$

where the label "conf" on the left-hand side stresses that these are the temperatures in the conformal frame. To linear order this implies that

$$\left(\frac{\delta T}{T}\right)^{\text{conf}}_O = \left(\frac{\delta T}{T}\right)^{\text{conf}}_E + \left(\frac{\delta P^0}{\bar{P}^0}\right)_O - \left(\frac{\delta P^0}{\bar{P}^0}\right)_E. \tag{20.60}$$

The term

$$\left(\frac{\delta P^0}{\bar{P}^0}\right)_O - \left(\frac{\delta P^0}{\bar{P}^0}\right)_E = \frac{\delta P^0}{\bar{P}^0}\bigg|_E^O \tag{20.61}$$

can then be read from eq. (20.58). Apparently, we have separated the temperature fluctuations at the observation point O into an "intrinsic" part, given by the temperature fluctuations at the emission point E, plus the term (20.61), which came from integration of the geodesic equation (20.50) and which therefore describes the perturbations generated during the propagation from the source to the observer. However, we will see that this separation is specific to our use of the conformal frame, and different choices of coordinates can mix the local term $-2\Psi(\eta_E, \mathbf{x}_E)$ in eq. (20.58) with the term $(\delta T/T)^{\text{conf}}_E$ in eq. (20.60). Furthermore, this is not yet the most convenient form for the temperature fluctuations. The reason is that the temperatures that appear in this equation are computed in the conformal frame. In this frame, the baryon–photon fluid on the LSS has velocity perturbations, which are described by the velocity potential u^μ. As we found in eq. (18.145), in conformal coordinates this is given by

$$u^0 = \frac{1}{a}(1 - \Psi), \tag{20.62}$$

$$u^i = \frac{1}{a}v^i, \tag{20.63}$$

where $v^i = v^i(\eta, \mathbf{x})$ is the peculiar velocity field. As a consequence, photons coming from different directions have different Doppler shifts and therefore different energies. Furthermore, the "intrinsic" contribution $(\delta T/T)^{\text{conf}}_E$ is also affected by the fact that different points on the LSS have different values of the potential Ψ, and therefore experience different gravitational redshifts. Therefore, for the temperatures in the conformal frame we cannot use the simple relation $\rho_\gamma \propto T^4$, which would give $\delta T/T = (1/4)\delta_\gamma$.

We can extract explicitly these effects using, instead of the temperature in the conformal frame, the temperature computed in the rest frame of the baryon–photon fluid, as follows. We introduce the variable

$$\mathcal{E}(x) = -g_{\mu\nu}(x)P^\mu u^\nu(x), \qquad (20.64)$$

[7]Of course, the LSS is not a mathematical surface, but rather a region of space with a finite thickness, determined as discussed in Section 20.3.6, so it makes sense to consider a small volume element within it.

and we consider a small volume element within the LSS.[7] By the equivalence principle, in this small region we can always choose the metric $g_{\mu\nu}$ so that $g_{\mu\nu} \simeq \eta_{\mu\nu}$. Furthermore, without spoiling the condition $g_{\mu\nu} \simeq \eta_{\mu\nu}$, we can perform locally a Lorentz boost, so that in this infinitesimal region the velocity field of the baryon–photon fluid takes the form $u^\mu = (1,0,0,0)$, i.e. we can go to the rest frame of this infinitesimal volume element of the baryon–photon fluid. In this frame eq. (20.64) gives $\mathcal{E}_E = (P^0)_E$, where the subscript E denotes that the quantities are evaluated at the emission space-time point. Therefore $\mathcal{E}_E = \mathcal{E}(\eta_E, \mathbf{x}_E)$ is a scalar under diffeomorphisms that is equal to the physical photon energy, at the emission time η_E, in the rest frame of the infinitesimal baryon–photon fluid element located at the point $\mathbf{x}_E$ on the LSS. The explicit relation between $\mathcal{E}$ and P^μ in the conformal frame is obtained using eqs. (18.96), (20.62) and (20.63), which give

$$\mathcal{E} = a\left[(1+\Psi)P^0 - (1+2\Phi)\mathbf{P}\cdot\mathbf{v}\right]. \qquad (20.65)$$

To first order, $\mathbf{P}^2$ is given by eq. (20.49). However, since $\mathbf{v}$ is already first-order in the perturbation, in $(1+2\Phi)\mathbf{P}\cdot\mathbf{v}$ we can replace $(1+2\Phi)$ by 1 and for $\mathbf{P}$ we can use the lowest-order expression $\mathbf{P} = P^0\hat{\mathbf{P}}$, where $\hat{\mathbf{P}}$ is the unit vector in the propagation direction. Since the photon propagates from the LSS toward us, it is convenient to write $\hat{\mathbf{P}} = -\hat{\mathbf{n}}$, so $\hat{\mathbf{n}}$ is the unit vector in the direction in which the observer is pointing to see the photon. Then, to first order,

$$\mathcal{E} = aP^0(1 + \Psi + \hat{\mathbf{n}}\cdot\mathbf{v}). \qquad (20.66)$$

[8]Recall indeed that, for a clock following a time-like geodesic, the proper time τ is related to the coordinate time t by

$$d\tau = (-g_{\mu\nu}dx^\mu dx^\nu)^{1/2}$$
$$= \left(-g_{\mu\nu}\frac{dx^\mu}{dt}\frac{dx^\nu}{dt}\right)^{1/2} dt. (20.67)$$

For a clock at rest this gives

$$d\tau = [-g_{00}(x)]^{1/2}dt. \qquad (20.68)$$

Since $d\tau$ is invariant, the coordinate time t_1 at position x_1 is related to the coordinate time t_2 at position x_2 by

$$[-g_{00}(x_1)]^{1/2}dt_1 = [-g_{00}(x_2)]^{1/2}dt_2. \qquad (20.69)$$

In the cosmological metric (18.96), using cosmic time, we have $-g_{00} = 1 + 2\Psi$, and therefore

$$\frac{dt_1}{dt_2} = 1 + \Psi(x_2) - \Psi(x_1), \qquad (20.70)$$

so the frequency of a photon observed at a point x_2 is related to the frequency of the photon emitted at a point x_1 by

$$\frac{\omega_O}{\omega_E} = 1 + \Psi_O - \Psi_E, \qquad (20.71)$$

which correctly reproduces the term $1+\Psi$ in eq. (20.66).

Recall that P^0 is actually P^η, which, as we saw in eq. (20.45), scales as a^{-2}. Thus, the term aP^0 in eq. (20.66) scales as $1/a$. This agrees with the fact that $\mathcal{E}$ is the value of the photon energy at the emission point x_E, in the rest frame of the infinitesimal baryon–photon fluid element located at x_E and in coordinates such that $g_{00}(x_E) = -1$. The terms Ψ and $\hat{\mathbf{n}}\cdot\mathbf{v}$ in eq. (20.66) represent the gravitational Doppler effect[8] and the kinematic Doppler effect, respectively. We next separate P^0 as in eq. (20.51), so $\mathcal{E} = \bar{\mathcal{E}} + \delta\mathcal{E}$, where

$$\bar{\mathcal{E}} = a\bar{P}^0, \qquad (20.72)$$

and

$$\delta\mathcal{E} = a\left[\delta P^0 + \bar{P}^0(\Psi + \hat{\mathbf{n}}\cdot\mathbf{v})\right]. \qquad (20.73)$$

Rather than considering the ratio (20.59), we now consider the ratio $\mathcal{E}_O/\mathcal{E}_E$. Because of the physical meaning of $\mathcal{E}$ that we have discussed, we have

$$\left(\frac{T_O}{T_E}\right) = \frac{\mathcal{E}_O}{\mathcal{E}_E}, \qquad (20.74)$$

where the temperature T_E on the left-hand side is the temperature measured in a locally Minkowskian frame in which the baryon–photon fluid is at rest. From eq. (20.74) we get

$$\left(\frac{\delta T}{T}\right)_O = \left(\frac{\delta T}{T}\right)_E + \left(\frac{\delta \mathcal{E}}{\overline{\mathcal{E}}}\right)_O - \left(\frac{\delta \mathcal{E}}{\overline{\mathcal{E}}}\right)_E. \qquad (20.75)$$

Using eqs. (20.72) and (20.73),

$$\frac{\delta \mathcal{E}}{\overline{\mathcal{E}}} = \frac{\delta P^0}{\overline{P}^0} + \Psi + \hat{\mathbf{n}} \cdot \mathbf{v}. \qquad (20.76)$$

Therefore, using eq. (20.58), we get

$$\left(\frac{\delta T}{T}\right)_O = \left(\frac{\delta T}{T}\right)_E - \Psi_O + \Psi_E + \hat{\mathbf{n}} \cdot \mathbf{v}_O - \hat{\mathbf{n}} \cdot \mathbf{v}_E$$
$$+ \int_E^O d\eta \left\{ \Psi'[\eta, \mathbf{x}(\eta)] - \Phi'[\eta, \mathbf{x}(\eta)] \right\}. \qquad (20.77)$$

Observe that, because of the factor -2Ψ in eq. (20.58), the term Ψ in eq. (20.77) has the opposite sign compared with the contribution from the gravitational Doppler effect. We next observe that the term Ψ_O is a constant for all directions of propagation $\hat{\mathbf{n}}$ and only contributes to the monopole, i.e. it is already taken into account in the value of the average temperature given in eq. (20.1). Similarly, the term $\hat{\mathbf{n}} \cdot \mathbf{v}_O$ represents the motion of the observer in the CMB rest frame, and has no cosmological significance. The value of $\mathbf{v}_O$ is obtained from a measurement of the dipole, and this term is then subtracted. We can therefore just discard these terms, and write

$$\left(\frac{\delta T}{T}\right)_O = \left(\frac{\delta T}{T}\right)_E + \Psi_E - \hat{\mathbf{n}} \cdot \mathbf{v}_E + \int_E^O d\eta \, (\Psi' - \Phi'). \qquad (20.78)$$

In the frame where the infinitesimal fluid element at position x_E is at rest and g_{00} at the location x_E of the fluid element is equal to -1 we have no Doppler or gravitational redshift, so $\rho_\gamma \propto T^4$, and therefore $\delta_\gamma = 4\delta T/T$. Then, writing all arguments explicitly,

$$\boxed{\begin{aligned} \frac{\delta T}{T}(\eta_0, \hat{\mathbf{n}}) = &\frac{1}{4}\delta_\gamma(\eta_{\text{dec}}, r_L\hat{\mathbf{n}}) + \Psi(\eta_{\text{dec}}, r_L\hat{\mathbf{n}}) - \hat{\mathbf{n}} \cdot \mathbf{v}(\eta_{\text{dec}}, r_L\hat{\mathbf{n}}) \\ &+ \int_{\eta_{\text{dec}}}^{\eta_0} d\eta \left\{ \Psi'[\eta, (\eta_0 - \eta)\hat{\mathbf{n}}] - \Phi'[\eta, (\eta_0 - \eta)\hat{\mathbf{n}}] \right\}, \end{aligned}}$$
$$(20.79)$$

where $r_L = \eta_0 - \eta_{\text{dec}}$ is the distance to the LSS in conformal coordinates, so $\mathbf{x}_E = r_L\hat{\mathbf{n}}$ is the position of the fluid element. Inside the integral we have also used

$$\mathbf{x}(\eta) = (\eta_0 - \eta)\hat{\mathbf{n}}. \qquad (20.80)$$

This is the unperturbed trajectory of a photon coming from the direction $\hat{\mathbf{n}}$ and that at $\eta = \eta_{\rm dec}$ is on the LSS at the position $\mathbf{x} = (\eta_0 - \eta_{\rm dec})\hat{\mathbf{n}}$ and at $\eta = \eta_0$ is at the observer location $\mathbf{x} = 0$. Since $\mathbf{x}(\eta)$ appears as the argument of Ψ' or Φ', which already are first-order quantities, the use of the unperturbed photon trajectory is here appropriate.

Observe that we could write $(\delta_\gamma)_{\rm E} \propto T_{\rm E}^4$, and therefore $(\delta T/T)_{\rm E} = (1/4)\delta_\gamma$, only because $T_{\rm E}$ is the emission temperature in the baryon–photon rest frame. In the conformal frame the expression for $\delta T/T$ in terms of δ_γ would be more complicated, since the fluid in different directions $\hat{\mathbf{n}}$ has different velocities. This would generate again the gravitational and kinematic Doppler terms that appear in eq. (20.79), and which are not explicitly present in eq. (20.60). This shows that the separation between "intrinsic" perturbations on the LSS and perturbations generated during the propagation to the observer is gauge-dependent.[9]

The term

$$\left[\frac{\delta T}{T}(\eta_0, \hat{\mathbf{n}})\right]_{\rm SW} \equiv \frac{1}{4}\delta_\gamma(\eta_{\rm dec}, r_L\hat{\mathbf{n}}) + \Psi(\eta_{\rm dec}, r_L\hat{\mathbf{n}}) \tag{20.81}$$

is called the *Sachs–Wolfe* (SW) effect. The term $-\hat{\mathbf{n}}\cdot\mathbf{v}$ describes, as we have seen, the special-relativistic Doppler effect (in the small-velocity limit). Finally, the term

$$\left[\frac{\delta T}{T}(\eta_0, \hat{\mathbf{n}})\right]_{\rm ISW} \equiv \int_{\eta_{\rm dec}}^{\eta_0} d\eta \left\{\Psi'[\eta, (\eta_0 - \eta)\hat{\mathbf{n}}] - \Phi'[\eta, (\eta_0 - \eta)\hat{\mathbf{n}}]\right\} \tag{20.82}$$

is called the *integrated Sachs–Wolfe* (ISW) effect. Physically, we can understand it as follows. If, during its journey from the emission point to the observer, the photon passes through a constant potential well, it initially gains some energy by falling into the well, but then it loses exactly the same amount when it comes out of it. However, if the potential well evolves with time, the two effects in general no longer compensate, and the photon can in the end gain or lose energy. Thus, the ISW effect vanishes for constant Φ and Ψ, but is non-zero in the presence of time-dependent scalar [or, as we will see in eq. (20.123), tensor] perturbations.

Observe from Fig. 19.3 that the potential Φ is constant during MD. The same happens for Ψ, which in the absence of anisotropic stresses is simply equal to $-\Phi$. Therefore the ISW effect receives contribution only from two regimes. One is the epoch after the RD–MD transition, when Φ has not yet settled to its constant value. As we see from Fig. 19.3, it indeed takes quite some time after the RD–MD transition before Φ becomes constant. This gives rise to the *early ISW effect*. The second epoch is the recent regime when dark energy (possibly in the form of a cosmological constant, or more generally in the form of a dynamical dark energy) begins to dominate. Then, we see again from Fig. 19.3 that Φ is no longer constant. This gives rise to the *late ISW effect*. This effect is a potentially interesting probe of the properties of dark energy.

[9]Indeed, as an extreme example, one can consider a reference frame in which the foliation of equal-time surfaces is defined by the surfaces of constant energy density. In this case, by definition, the "intrinsic" $\delta T/T$ vanishes, and all the effect is moved to the gravitational and kinematic Doppler shifts. For instance, suppose that in the conformal frame in a direction $\hat{\mathbf{n}}$ we have $\delta T < 0$ on the LSS. We can then choose the foliation in this direction so that the equal-time surface corresponds, from the point of view of the conformal frame, to moving toward earlier times, so the temperature in this direction rises because of its $1/a$ scaling. Now $\delta T = 0$ by definition on such a surface, but the Doppler shift will exactly undo this, so in the end the observed value $(\delta T/T)_{\rm O}$ will be the same, as it should be, since it is an observable quantity.

20.3.3   Expression of the C_l in terms of the $\Theta_l(k)$

As a next step, we translate the result (20.79) into the effect of the scalar perturbations on the coefficients C_l. We use the notation

$$\Theta(\eta_0, \hat{\mathbf{n}}, \mathbf{x}) \equiv \frac{\delta T}{T}(\eta_0, \hat{\mathbf{n}}, \mathbf{x}) \tag{20.83}$$

for the temperature fluctuations measured today, i.e. at $\eta = \eta_0$, by an observer located at the position $\mathbf{x}$, and looking into the direction $\hat{\mathbf{n}}$ (so that the photons propagate toward the observer, with momentum in the direction $-\hat{\mathbf{n}}$). In eq. (20.79) the observer was located at $\mathbf{x} = 0$. We then write

$$\Theta(\eta_0, \hat{\mathbf{n}}, \mathbf{x}) = \int \frac{\sqrt{V} d^3k}{(2\pi)^3}\, \tilde{\Theta}(\eta_0, \hat{\mathbf{n}}, \mathbf{k})\, e^{i\mathbf{k}\cdot\mathbf{x}}, \tag{20.84}$$

where the conventions on the volume factor are the same as in eq. (19.162). Then

$$\Theta(\eta_0, \hat{\mathbf{n}}, \mathbf{x}=0) = \int \frac{\sqrt{V} d^3k}{(2\pi)^3}\, \tilde{\Theta}(\eta_0, \hat{\mathbf{n}}, \mathbf{k}). \tag{20.85}$$

Using eq. (20.79) we see that $\tilde{\Theta}(\eta_0, \hat{\mathbf{n}}, \mathbf{k})$ actually depends on $\hat{\mathbf{n}}$ only through the combination $\hat{\mathbf{n}}\cdot\mathbf{k}$. Consider for instance the term $\Psi(\eta_{\rm dec}, r_L\hat{\mathbf{n}})$ in eq. (20.79). Writing

$$\Psi(\eta, \mathbf{x}) = \int \frac{\sqrt{V} d^3k}{(2\pi)^3}\, \tilde{\Psi}(\eta, \mathbf{k})\, e^{i\mathbf{k}\cdot\mathbf{x}}, \tag{20.86}$$

we have

$$\Psi(\eta_{\rm dec}, r_L\hat{\mathbf{n}}) = \int \frac{\sqrt{V} d^3k}{(2\pi)^3}\, \tilde{\Psi}(\eta_{\rm dec}, \mathbf{k})\, e^{i r_L \mathbf{k}\cdot\hat{\mathbf{n}}}, \tag{20.87}$$

so indeed the integrand is not a function of the two vectors $\mathbf{k}$ and $\hat{\mathbf{n}}$ separately, but only of $\mathbf{k}$ and $\mathbf{k}\cdot\hat{\mathbf{n}}$. For the ISW term we have

$$\int_{\eta_{\rm dec}}^{\eta_0} d\eta\, \Psi'[\eta, (\eta_0-\eta)\hat{\mathbf{n}}] = \int_{\eta_{\rm dec}}^{\eta_0} d\eta \int \frac{\sqrt{V} d^3k}{(2\pi)^3}\, \tilde{\Psi}'(\eta, \mathbf{k})\, e^{i(\eta_0-\eta)\mathbf{k}\cdot\hat{\mathbf{n}}}, \tag{20.88}$$

so again the integrand on the right-hand side is a function only of $\mathbf{k}$ and $\mathbf{k}\cdot\hat{\mathbf{n}}$. Similarly, the velocity field for scalar perturbations is a gradient, $\mathbf{v}(\eta, \mathbf{x}) = \nabla v$ [see eq. (18.151)], so again the Doppler term $\hat{\mathbf{n}}\cdot\mathbf{v}$ in eq. (20.79) produces a factor proportional to $\mathbf{k}\cdot\hat{\mathbf{n}}$ in momentum space.

The dependence on $\cos\theta \equiv \hat{\mathbf{k}}\cdot\hat{\mathbf{n}}$ can be expanded in Legendre polynomials $P_l(\cos\theta)$, since the functions $P_l(z)$ are a complete set on the interval $z \in [-1, 1]$. In particular, plane waves have an expansion in terms of Legendre polynomials given by

$$e^{i\mathbf{k}\cdot\mathbf{x}} = \sum_{l=0}^{\infty} (2l+1) i^l\, j_l(kr) P_l(\hat{\mathbf{k}}\cdot\hat{\mathbf{x}}), \tag{20.89}$$

where $j_l(z) = (\pi/2z)^{1/2} J_{l+(1/2)}(z)$ are spherical Bessel functions and $r = |\mathbf{x}|$, $k = |\mathbf{k}|$. It is therefore convenient to define the multipoles

$\tilde{\Theta}_l(\eta, \mathbf{k})$ from

$$\tilde{\Theta}(\eta, \hat{\mathbf{n}}, \mathbf{k}) = \tilde{\Theta}(\eta, \cos\theta, \mathbf{k})$$
$$= \sum_{l=0}^{\infty} (2l+1) i^l \, \tilde{\Theta}_l(\eta, \mathbf{k}) P_l(\cos\theta) \,. \qquad (20.90)$$

Using the orthogonality of the Legendre polynomials,

$$\int_{-1}^{1} dx \, P_l(x) P_{l'}(x) = \frac{2}{2l+1} \delta_{ll'} \,, \qquad (20.91)$$

this relation can be inverted as

$$\tilde{\Theta}_l(\eta, \mathbf{k}) = \frac{1}{2i^l} \int_{-1}^{1} d\cos\theta \, P_l(\cos\theta) \tilde{\Theta}(\eta, \cos\theta, \mathbf{k}) \,. \qquad (20.92)$$

From eqs. (20.85) and (20.90) we have in particular

$$\Theta(\eta_0, \hat{\mathbf{n}}, \mathbf{x}=0) = \sum_{l=0}^{\infty} (2l+1) i^l \int \frac{\sqrt{V} d^3k}{(2\pi)^3} \, \tilde{\Theta}_l(\eta_0, \mathbf{k}) P_l(\hat{\mathbf{k}} \cdot \hat{\mathbf{n}}) \,. \qquad (20.93)$$

Since we will always be interested in the $\tilde{\Theta}_l$ at the present time η_0, to keep the notation light we henceforth suppress the argument η_0 and write $\tilde{\Theta}_l(\eta_0, \mathbf{k})$ simply as $\tilde{\Theta}_l(\mathbf{k})$.

In order to express the C_l in terms of the $\tilde{\Theta}_l(\mathbf{k})$ we proceed as follows. We write eq. (20.4) as

$$a_{lm} = \int d^2\hat{\mathbf{n}} \, \Theta(\eta_0, \hat{\mathbf{n}}, \mathbf{x}=0) \, Y^*_{lm}(\hat{\mathbf{n}}) \qquad (20.94)$$
$$= \sum_{l_1=0}^{\infty} (2l_1+1) i^{l_1} \int \frac{\sqrt{V} d^3k}{(2\pi)^3} \, \tilde{\Theta}_{l_1}(\mathbf{k}) \int d^2\hat{\mathbf{n}} \, Y^*_{lm}(\hat{\mathbf{n}}) P_{l_1}(\hat{\mathbf{n}} \cdot \hat{\mathbf{k}}) \,.$$

We now use the identity

$$\int d^2\hat{\mathbf{n}} \, Y^*_{lm}(\hat{\mathbf{n}}) P_{l_1}(\hat{\mathbf{n}} \cdot \hat{\mathbf{k}}) = \delta_{ll_1} \frac{4\pi}{2l+1} Y^*_{lm}(\hat{\mathbf{k}}) \,, \qquad (20.95)$$

which is easily proved by inserting on the left-hand side the expression for $P_{l_1}(\hat{\mathbf{n}} \cdot \hat{\mathbf{k}})$ obtained from the addition theorem for spherical harmonics, written in the form

$$P_{l_1}(\hat{\mathbf{n}} \cdot \hat{\mathbf{k}}) = \frac{4\pi}{2l_1+1} \sum_{m_1=-l_1}^{l_1} Y^*_{l_1 m_1}(\hat{\mathbf{k}}) Y_{l_1 m_1}(\hat{\mathbf{n}}) \,, \qquad (20.96)$$

and using the orthogonality of spherical harmonics, eq. (3.228). Then

$$a_{lm} = i^l \, 4\pi \int \frac{\sqrt{V} d^3k}{(2\pi)^3} \, \tilde{\Theta}_l(\mathbf{k}) Y^*_{lm}(\hat{\mathbf{k}}) \,. \qquad (20.97)$$

We next observe that eq. (20.6) gives the definition of the C_l, as they are measured experimentally. The corresponding quantity that can be computed theoretically is

$$C_l = \frac{1}{2l+1} \sum_{m=-l}^{l} \langle |a_{lm}|^2 \rangle , \qquad (20.98)$$

where the angle brackets denotes the stochastic average over the initial conditions, performed using the primordial power spectra introduced in Section 19.4.2. This gives

$$C_l = \frac{(4\pi)^2 V}{2l+1} \sum_{m=-l}^{l} \int \frac{d^3k}{(2\pi)^3} \frac{d^3k'}{(2\pi)^3} Y_{lm}^*(\hat{\mathbf{k}}) Y_{lm}(\hat{\mathbf{k}}') \langle \tilde{\Theta}_l(\mathbf{k}) \tilde{\Theta}_l^*(\mathbf{k}') \rangle$$

$$= 4\pi V \int \frac{d^3k}{(2\pi)^3} \frac{d^3k'}{(2\pi)^3} P_l(\hat{\mathbf{k}} \cdot \hat{\mathbf{k}}') \langle \tilde{\Theta}_l(\mathbf{k}) \tilde{\Theta}_l^*(\mathbf{k}') \rangle , \qquad (20.99)$$

where in the second line we have used again eq. (20.96). As discussed in Section 19.4, quantities such as $\tilde{\Theta}_l(\mathbf{k})$ or $\tilde{\Psi}(\mathbf{k})$ are stochastic variables because their initial conditions are stochastic. For given initial conditions, their evolution is deterministic. Furthermore, the initial conditions are imposed deep in the radiation era, when all relevant modes are well outside the horizon and in the linear regime. Thus, restoring for a moment the time variable, we can write

$$\tilde{\Theta}_l(\eta_0, \mathbf{k}) = T_{\Theta_l}(k) \tilde{\Theta}_l(\eta_{\rm in}, \mathbf{k}) . \qquad (20.100)$$

The transfer function $T_{\Theta_l}(k)$ is determined by the perturbation equations in the FRW background and, because of the spatial isotropy of the background, it depends only on the modulus k of $\mathbf{k}$, as we have indeed seen from the explicit equations of Chapter 19. Furthermore, the initial conditions for different quantities are proportional to each other [see for example eqs. (19.82)–(19.85) for the case of adiabatic initial conditions], so $\tilde{\Theta}_l(\eta_{\rm in}, \mathbf{k})/\tilde{\Psi}_{\rm in}(\mathbf{k})$ is just a numerical (l-dependent) constant. This means that a quantity such as $\tilde{\Theta}_l(\eta_0, \mathbf{k})/\tilde{\Psi}_{\rm in}(\mathbf{k})$ is a deterministic rather than a stochastic variable, because the initial conditions cancel in the ratio, and furthermore it depends only on the modulus $k = |\mathbf{k}|$ and on l. We define

$$\Theta_l(\eta_0, k) \equiv \frac{\tilde{\Theta}_l(\eta_0, \mathbf{k})}{\tilde{\Psi}_{\rm in}(\mathbf{k})} . \qquad (20.101)$$

Then (again suppressing the argument η_0)

$$V \langle \tilde{\Theta}_l(\mathbf{k}) \tilde{\Theta}_l^*(\mathbf{k}') \rangle = \Theta_l(k) \Theta_l^*(k') V \langle \tilde{\Psi}_{\rm in}(\mathbf{k}) \tilde{\Psi}_{\rm in}^*(\mathbf{k}') \rangle$$

$$= |\Theta_l(k)|^2 (2\pi)^3 \delta^{(3)}(\mathbf{k} - \mathbf{k}') P_{\Psi,\rm in}(k) , \qquad (20.102)$$

where in the second line we have used eq. (19.164), and $P_{\Psi,\rm in}(k)$ is the primordial power spectrum of Ψ. Inserting this into eq. (20.99) and using $P_l(1) = 1$ we get

$$C_l = \frac{2}{\pi} \int_0^\infty dk\, k^2\, |\Theta_l(k)|^2\, P_{\Psi,\rm in}(k) . \qquad (20.103)$$

Equivalently, in terms of $\mathcal{P}_\Psi$ defined as in eq. (19.170), we have

$$C_l = 4\pi \int_0^\infty \frac{dk}{k} \, |\Theta_l(k)|^2 \, \mathcal{P}_{\Psi,\text{in}}(k) \,. \qquad (20.104)$$

This is the final expression for the C_l in terms of the Θ_l. Observe that we have chosen to normalize $\tilde{\Theta}_l(\eta_0, \mathbf{k})$ to $\tilde{\Psi}_{\text{in}}(\mathbf{k})$. We could as well have chosen to introduce a variable $\tilde{\Theta}_l(\eta_0, \mathbf{k})/\tilde{\delta}_M(\eta_{\text{in}}, \mathbf{k})$, where δ_M is the matter density contrast, and then eq. (20.103) would be equivalently written as

$$C_l = \frac{2}{\pi} \int_0^\infty dk \, k^2 \left| \frac{\tilde{\Theta}_l}{\tilde{\delta}_{M,\text{in}}} \right|^2 (k) \, P_{M,\text{in}}(k) \,, \qquad (20.105)$$

where $P_{M,\text{in}}(k)$ is the primordial matter power spectrum, or we could have normalized to the matter density contrast today. It is important to observe that the coefficients C_l depend on the initial conditions, which determine the initial power spectrum, as well as on the cosmological model, which determines the transfer functions in eq. (20.100) and therefore the quantity $\Theta_l(k)$.

20.3.4 Scalar contribution to the C_l

We can now finally compute the effect of the scalar perturbations on the C_l. We will begin with an analytic treatment of the large-angle regime, i.e. angles larger than a few degrees, corresponding to modes that were outside the horizon at decoupling, $k\eta_{\text{dec}} < 1$. Indeed, we saw in eq. (20.14) that this is the only regime where GWs can potentially affect the temperature anisotropies, and our aim is to understand how to detect GWs from the CMB, disentangling them from the effect of scalar perturbations. In term of multipoles, this means $l < O(50)$; see eq. (20.15). We also limit ourselves to adiabatic perturbations. We will then show the full result obtained numerically for generic multipoles, using Boltzmann codes.

The SW effect

We consider first the SW effect (20.81), which, as we will see, in the large-angle regime gives the dominant contribution. At decoupling we are in MD, so for super-horizon modes eq. (19.142) applies. Recall also that in the absence of anisotropic stresses we have $\Psi = -\Phi$. Then, for super-horizon modes, $\tilde{\Psi}(\eta, \mathbf{k})$ and $\tilde{\delta}_M(\eta, \mathbf{k})$ are actually constant in time and related by $\tilde{\delta}_M(\mathbf{k}) = -2\tilde{\Psi}(\mathbf{k})$. The adiabatic initial conditions fix $\tilde{\delta}_{M,\text{in}} = (3/4)\tilde{\delta}_{\gamma,\text{in}}$; see eqs. (19.83) and (19.84). Therefore, for adiabatic modes that are outside the horizon at decoupling, we have

$$\frac{1}{4}\tilde{\delta}_\gamma(\mathbf{k}) = -\frac{2}{3}\tilde{\Psi}(\mathbf{k}) \,. \qquad (20.106)$$

Then eqs. (20.81) and (20.87) give

$$
\Theta(\eta_0, \hat{\mathbf{n}}, \mathbf{x} = 0)\Big|_{k < \eta_{\rm dec}^{-1}} = \frac{1}{3}\Psi(\eta_{\rm dec}, r_L\hat{\mathbf{n}})\Big|_{k < \eta_{\rm dec}^{-1}}
$$

$$
= \frac{1}{3}\int_{k < \eta_{\rm dec}^{-1}} \frac{\sqrt{V}d^3k}{(2\pi)^3}\,\tilde{\Psi}(\eta_{\rm dec}, \mathbf{k})\,e^{ir_L\mathbf{k}\cdot\hat{\mathbf{n}}}\,, \quad (20.107)
$$

where the subscript $k < \eta_{\rm dec}^{-1}$ in $\Theta(\eta_0, \hat{\mathbf{n}}, \mathbf{x} = 0)$ and $\Psi(\eta_{\rm dec}, r_L\hat{\mathbf{n}})$ means that the relation holds only for the Fourier modes of these quantities that are outside the horizon at decoupling. Inserting eq. (20.93) into the left-hand side of eq. (20.107) (and observing that a similar relation holds for generic $\mathbf{x}$) we find

$$
\sum_{l=0}^{\infty}(2l + 1)i^l\,\tilde{\Theta}_l(\eta_0, \mathbf{k})P_l(\cos\theta) = \frac{1}{3}\tilde{\Psi}(\eta_{\rm dec}, \mathbf{k})\,e^{ir_L\mathbf{k}\cdot\hat{\mathbf{n}}}\,, \quad (20.108)
$$

which, once again, is valid only for the modes with $k < \eta_{\rm dec}^{-1}$. Finally, using the expansion (20.89) for the plane wave in Legendre polynomials and comparing the different multipoles, we get

$$
\boxed{\left[\tilde{\Theta}_l(\eta_0, \mathbf{k})\right]_{\rm SW} = \frac{1}{3}\tilde{\Psi}(\eta_{\rm dec}, \mathbf{k})j_l(kr_L)\,,} \quad (20.109)
$$

where we recall that $r_L = \eta_0 - \eta_{\rm dec}$ is the distance to the LSS. We next observe that, for super-horizon modes, the constant value during MD is related to the constant initial value deep in RD by eq. (19.157).[10] Therefore $\tilde{\Psi}(\eta_{\rm dec}, \mathbf{k}) = (9/10)\tilde{\Psi}_{\rm in}(\mathbf{k})$ and, for the quantity $\Theta_l(\mathbf{k})$ defined in eq. (20.101) we get

$$
\left[\Theta_l(\eta_0, \mathbf{k})\right]_{\rm SW} = \frac{3}{10}j_l[k(\eta_0 - \eta_{\rm dec})]\,. \quad (20.110)
$$

Plugging this into eq. (20.104) and observing that $\eta_0 - \eta_{\rm dec} \simeq \eta_0$, we get

$$
\boxed{C_l^{\rm SW} \simeq \frac{9\pi}{25}\int_0^\infty \frac{dk}{k}\,j_l^2(k\eta_0)\mathcal{P}_{\Psi,\rm in}(k)\,,} \quad (20.111)
$$

which, as we saw in eq. (20.15), is valid for low multipoles, $l < O(50)$, since in the derivation we used $k\eta_{\rm dec} < 1$. If we take the power-law form (19.178) for the primordial power spectrum, the integral in eq. (20.111) can be performed analytically. For the flat spectrum $n_s = 1$ the result is particularly simple,

$$
C_l^{\rm SW}(n_s = 1) \simeq \frac{9\pi}{50}\frac{A_\Psi}{l(l + 1)}\,. \quad (20.112)
$$

This means that the coefficients $\mathcal{D}_l$ defined in eq. (20.10) at low l become independent of l (for $n_s = 1$),

$$
\mathcal{D}_l^{\rm SW}(n_s = 1) \simeq \frac{9}{100}T_0^2 A_\Psi
$$

$$
= \frac{1}{25}T_0^2 A_\mathcal{R}\,, \quad (20.113)
$$

[10] As we saw in eq. (17.94), $z_{\rm dec} \simeq 1090$, so $x_{\rm dec} = \log a_{\rm dec} \simeq -7.0$. From Fig. 19.3 we see that for typical modes the plateau during MD has not yet been fully reached at $x \simeq -7.0$. In the spirit of the analytic estimates that we are presenting here, we neglect these corrections. Accurate numerical values are obtained through Boltzmann codes, as we will discuss later.

where in the second line we have used eq. (19.179). This is called the *Sachs–Wolfe plateau*, and we approximately see it in the low-l region of Fig. 20.1. The integral in eq. (20.111) can also be performed analytically for generic n_s although the result, given by a combination of Gamma functions, is not particularly illuminating. For n_s close to 1, which is the physically relevant case, the result is, however, quite simple,

$$C_l^{\text{SW}}(n_s) \simeq \frac{9\pi}{50} \frac{A_\Psi}{l(l+1)} \left(\frac{l}{k_* \eta_0}\right)^{n_s-1}, \qquad (20.114)$$

where k_* is the pivot scale defined in eq. (19.178). The parameter n_s is determined by a global fit to the CMB spectrum, also in combination with other cosmological observations (in order to break the degeneracy between various parameters). We see, however, that n_s has quite a distinctive effect on the CMB multipoles. A value $n_s < 1$ tilts the Sachs–Wolfe plateau, enhancing the anisotropies in the low-l region and decreasing them in the high-l region. This is called a "red" tilt (while $n_s > 1$ is called a "blue" tilt). As we saw in eq. (19.184), the fit to the most recent data from the *Planck* collaboration, performed assuming ΛCDM as the cosmological model, gives $n_s = 0.9677(60)$ at 68% c.l., so the tilt is indeed red, a fact that has important implications for inflation, as we will see. The corresponding best-fit line is shown in Fig. 20.1. Observe that as we move toward the lowest-l values, we are more and more affected by cosmic variance.

The ISW effect

We now turn to the contribution of scalar perturbations to the ISW effect, given by eq. (20.82). Proceeding just as we have done for the SW effect, and neglecting anisotropic stresses, so that in eq. (20.82) we can write $\Psi' - \Phi' = 2\Psi'$, we now find

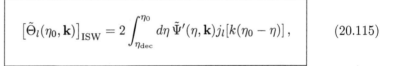

$$\left[\tilde{\Theta}_l(\eta_0, \mathbf{k})\right]_{\text{ISW}} = 2 \int_{\eta_{\text{dec}}}^{\eta_0} d\eta \, \tilde{\Psi}'(\eta, \mathbf{k}) j_l[k(\eta_0 - \eta)], \qquad (20.115)$$

and therefore for $\Theta_l(\eta_0, \mathbf{k})$ defined in eq. (20.101) we get

$$\left[\Theta_l(\eta_0, \mathbf{k})\right]_{\text{ISW}} = 2 \int_{\eta_{\text{dec}}}^{\eta_0} d\eta \, \frac{\tilde{\Psi}'(\eta, \mathbf{k})}{\tilde{\Psi}_{\text{in}}(\mathbf{k})} j_l[k(\eta_0 - \eta)]. \qquad (20.116)$$

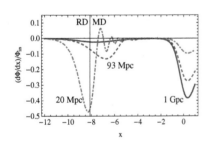

Fig. 20.3 The evolution of $d\tilde{\Phi}/dx$, normalized to its initial value, against $x = \log a$, for $\lambda = 20$ Mpc, 93 Mpc and 1 Gpc, as in Fig. 19.3.

The evolution of $\tilde{\Phi} = -\tilde{\Psi}$ for different comoving momenta was shown in Fig. 19.3. The corresponding function $d\tilde{\Phi}/d\log a = \mathcal{H}^{-1}\tilde{\Phi}'$ is shown in Fig. 20.3. As we have already remarked, the derivative $d\tilde{\Phi}/d\log a$ is very small (in units of $\tilde{\Phi}_{\text{in}}$) during MD, and the only contributions come from the region close to the RD–MD transition, where $\tilde{\Phi}$ has not yet reached its constant MD value, and from the recent epoch when the transition occurs between MD and the epoch where the cosmological constant begins to dominate. Choosing some value $\bar{\eta}$ deep in MD, we can therefore distinguish between an *early ISW effect*, given by

$$\left[\tilde{\Theta}_l(\eta_0, \mathbf{k})\right]_{\text{early ISW}} = 2 \int_{\eta_{\text{dec}}}^{\bar{\eta}} d\eta \, \tilde{\Psi}'(\eta, \mathbf{k}) j_l[k(\eta_0 - \eta)], \qquad (20.117)$$

and a *late ISW effect*, given by

$$\left[\tilde{\Theta}_l(\eta_0, \mathbf{k})\right]_{\text{late ISW}} = 2 \int_{\bar{\eta}}^{\eta_0} d\eta \, \tilde{\Psi}'(\eta, \mathbf{k}) j_l[k(\eta_0 - \eta)]. \qquad (20.118)$$

Choosing $\bar{\eta}$ in the region where Ψ' is very small, the separation is basically independent of the exact value of $\bar{\eta}$.

In Fig. 20.4 we plot the spherical Bessel functions $j_l(u)$ for $l = 5, 10$ and 20. We see that $j_l(u)$ peaks at $u \simeq l$; for $u < l$ it goes to zero quite fast, while for $u > l$ it has damped oscillations, and therefore peaks again at a few overtones. In the early ISW effect the term $\tilde{\Psi}'(\eta, \mathbf{k})$ is non-vanishing only for modes with $k \sim 1/\eta_{\text{eq}}$, and when $\eta \sim \eta_{\text{eq}}$. Then we can approximate $j_l[k(\eta_0 - \eta)] \simeq j_l[k(\eta_0 - \eta_{\text{eq}})] \simeq j_l(k\eta_0)$. We then find that in the early ISW effect the Bessel function gives the largest contribution for $l \sim k\eta_0 \sim \eta_0/\eta_{\text{eq}} = O(100)$, where we have used eq. (17.173). In contrast, in the late ISW effect we see from Fig. 20.3 that there is a significant contribution only from the modes with $k \sim 1/\eta_0$. Therefore the Bessel function gives the largest contribution for very small l, say $l < O(5 - 10)$. The late ISW effect can be a useful probe of the properties of dark energy. However, in the region $l \lesssim 5 - 10$ we have to fight against a large cosmic variance.

The Doppler effect given by the term $-\hat{\mathbf{n}} \cdot \mathbf{v}$ in eq. (20.79) can be treated similarly to the SW term. As we saw in eq. (19.135), in RD the oscillations of the velocity are out of phase by $\pi/2$ with respect to the oscillations of the density perturbation. This remains true also until decoupling, being a generic property of acoustic oscillations, so the maxima of the SW term correspond to minima of the Doppler term.

In Figs. 20.5–20.8 we show the separate contributions to $\mathcal{D}_l$ from the SW effect, the Doppler effect, the early ISW effect and the late ISW effect, respectively, in ΛCDM, computed numerically with a Boltzmann code (see Section 20.3.7). For the primordial scalar spectrum we have used the values $A_{\mathcal{R}} = 2.3 \times 10^{-9}$ and $n_s = 0.96$. Observe that the SW, ISW and Doppler contributions add up linearly in Θ_l, but $\mathcal{D}_l$ depends on $|\Theta_l|^2$. Therefore the total $\mathcal{D}_l$ is not merely the sum of the separate SW, Doppler and ISW contributions, but there are also interference terms. So, the plots in these figures give a first understanding of the relative importance and shape of the various contributions, but it is also important whether these different terms add up in phase or out of phase. In particular, we have seen that the Doppler term is out of phase with the SW term (as can indeed be seen also from Figs. 20.5 and 20.6), so in the end its overall contribution to the $\mathcal{D}_l$ is less important than what one could have inferred from the value $\mathcal{D}_l^{\text{Doppler}}$ obtained by inserting only the Doppler effect into the computation of the $\mathcal{D}_l$. In contrast, the

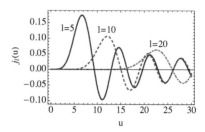

Fig. 20.4 The spherical Bessel functions $j_l(u)$ for $l = 5$ (solid line), $l = 10$ (dashed) and $l = 20$ (dot-dashed).

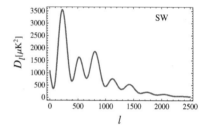

Fig. 20.5 The contribution of the SW effect to $\mathcal{D}_l$, obtained using the CLASS Boltzmann code. Data points in Figs. 20.5–20.9 courtesy of S. Foffa.

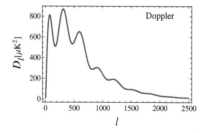

Fig. 20.6 The Doppler contribution.

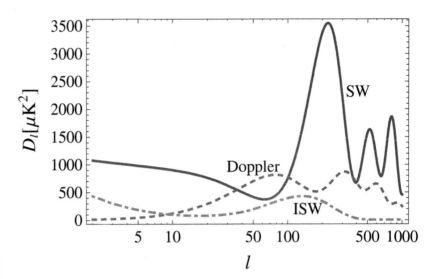

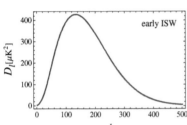

Fig. 20.7 The contribution of the early ISW effect.

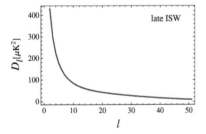

Fig. 20.8 The contribution of the late ISW effect.

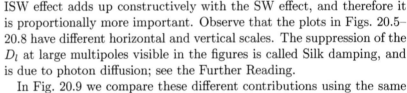

Fig. 20.9 The SW, Doppler and ISW contributions to $\mathcal{D}_l$ on a log–linear scale. Observe that the total $\mathcal{D}_l$, shown in Fig. 20.1, also involve the interference terms between the SW, Doppler and ISW contributions.

ISW effect adds up constructively with the SW effect, and therefore it is proportionally more important. Observe that the plots in Figs. 20.5–20.8 have different horizontal and vertical scales. The suppression of the $\mathcal{D}_l$ at large multipoles visible in the figures is called Silk damping, and is due to photon diffusion; see the Further Reading.

In Fig. 20.9 we compare these different contributions using the same scale, and a logarithmic horizontal scale. The SW plateau corresponds to the fact that $\mathcal{D}_l^{\rm SW}$ changes by less than a factor of 2 as l is varied from $l = O(1)$ to $l \simeq 30$, and it actually appears as a plateau only on this logarithmic scale; compare Figs. 20.5 and 20.9. The deviations from an exactly constant plateau are partly due to the fact that we set $n_s \neq 1$, which tilts the spectrum according to eq. (20.114) (in these figures we have used $n_s = 0.96$). Furthermore, as we stressed in eq. (20.107), eq. (20.111) is actually an approximation valid only in the limit $k\eta_{\rm dec} \ll 1$. For finite k there are corrections, so the contribution to $\mathcal{D}_l$ from the SW effect (20.81) is not exactly flat even for $n_s = 1$.

20.3.5   Tensor contribution to the C_l

GW contribution to $\delta T/T$

We now come to the effect of GWs on CMB anisotropies. We first compute how the photon trajectories are affected when there are both scalar and tensor perturbations. This effect is obtained from eq. (20.36), using for the Christoffel symbols the expression given in eqs. (18.99)–(18.104), in which we now include also the tensor perturbations. We still neglect vector perturbations, so u_{ij} in eq. (18.105) is equal to $(1/2)h_{ij}^{\text{TT}}$. When computing the geodesic equation for P^0 the only extra contribution due to h_{ij}^{TT} comes from Γ_{ij}^0, Then eq. (20.46) is replaced by

$$P^0 \frac{dP^0}{d\eta} = -\left(\mathcal{H} + \Psi'\right)\left(P^0\right)^2 - 2P^0 P^i \partial_i \Psi - \left[\mathcal{H}(1 + 2\Phi - 2\Psi) + \Phi'\right]\mathbf{P}^2$$
$$- \left[\frac{1}{2}(h_{ij}^{\text{TT}})' + \mathcal{H}h_{ij}^{\text{TT}}\right]P^i P^j . \tag{20.119}$$

At the same time, using the metric (18.97) with scalar and tensor perturbations, eq. (20.35) gives

$$(1 + 2\Psi)\left(P^0\right)^2 = (1 + 2\Phi)\mathbf{P}^2 + h_{ij}^{\text{TT}} P^i P^j . \tag{20.120}$$

To zeroth order in the perturbations this equation gives the unperturbed relation $\mathbf{P}^i = P^0 \hat{\mathbf{P}}^i$, where $\hat{\mathbf{P}} = -\hat{\mathbf{n}}$ is the propagation direction. Since h_{ij}^{TT} is already first-order, in eq. (20.120) the term $h_{ij}^{\text{TT}} P^i P^j$ can be replaced by $(P^0)^2 h_{ij}^{\text{TT}} n^i n^j$. We can then solve eq. (20.120) for $\mathbf{P}^2$, obtaining, to first order,

$$\mathbf{P}^2 = (1 - 2\Phi + 2\Psi - h_{ij}^{\text{TT}} n^i n^j)\left(P^0\right)^2 , \tag{20.121}$$

which replaces eq. (20.49) when there are also tensor perturbations. Inserting this into eq. (20.119) we get

$$\frac{dP^0}{d\eta} = -\left[2\mathcal{H} + \Phi' - \Psi' + 2\frac{d\Psi}{d\eta} + \frac{1}{2}(h_{ij}^{\text{TT}})' n^i n^j\right] P^0 , \tag{20.122}$$

which generalizes eq. (20.50) to scalar and tensor perturbations.[11] Observe that n^i is the propagation direction of the photon (and not of the GW), so $h_{ij}^{\text{TT}} n^i$ is non-zero, except if the propagation direction of the GW coincides with the radial direction. The GWs that have the greatest effect on the photon trajectory are those that propagate in the direction perpendicular to the photon.

To first order, the expression (20.66) for the parameter $\mathcal{E}$ is unaffected by h_{ij}^{TT}, since its presence in $g_{\mu\nu}$ only generates a term proportional to $h_{ij}^{\text{TT}} P^i v^j$, which is second-order because it involves h_{ij}^{TT} and the fluid velocity v^j, which are both first-order quantities. Furthermore, to first order, the expression (18.145) for u^μ is also unaffected, as we see on repeating the steps leading to eq. (18.143), which gives $u^i = (1/a)v^i$, and observing that, in the determination of u^0 through $g_{\mu\nu} u^\mu u^\nu = -1$, h_{ij}^{TT} gives only a higher-order contribution proportional to $h_{ij}^{\text{TT}} v^i v^j$. In conclusion, the rest of the arguments leading to eq. (20.79) go through without any change and we conclude that, in the presence of both scalar and tensor perturbations,

[11] Observe that all terms involving $\mathcal{H}$ times a first-order perturbation, i.e. $\mathcal{H}\Phi$, $\mathcal{H}\Psi$ and $\mathcal{H}h_{ij}^{\text{TT}}$, canceled upon inserting eq. (20.121) into eq. (20.119). This can be understood as a consequence of the fact that the massless action (20.27) is invariant under the conformal rescaling $g_{\mu\nu} \to e^{2\Omega(x)}g_{\mu\nu}$, $d\lambda \to e^{2\Omega(x)}d\lambda$. Choosing $e^{2\Omega(x)} = 1/a^2$ we can transform the metric $g_{\mu\nu} = a^2\eta_{\mu\nu}$ into the flat metric $\eta_{\mu\nu}$, while at the same time $P^\mu = dx^\mu/d\lambda \to a^2 P^\mu$. Thus, to zeroth order, in conformal coordinates $a^2 P^\mu$ satisfies the geodesic equation in the metric $\eta_{\mu\nu}$, so $d(a^2 P^\mu)/d\tau = 0$, which is equivalent to $dP^\mu/d\eta = -2\mathcal{H}P^\mu$. This is another way of understanding the zeroth order term $-2\mathcal{H}P^0$ in eq. (20.122). To first order, no further term involving $\mathcal{H}$ can appear, because the overall factor a^2 has been removed from the metric by the conformal transformation, so the geodesic equation for $a^2 P^0$ must be of the form $d(a^2 P^0)/d\eta = f(\Psi, \Phi, h_{ij}^{\text{TT}})a^2 P^0$, for some function f of the metric perturbations. Furthermore, for dimensional reasons f can depend only on the derivatives of the metric perturbations.

$$\frac{\delta T}{T}(\eta_0, \hat{\mathbf{n}}) = \frac{1}{4}\delta_\gamma(\eta_{\text{dec}}, r_L\hat{\mathbf{n}}) + \Psi(\eta_{\text{dec}}, r_L\hat{\mathbf{n}}) - \hat{\mathbf{n}}\cdot\mathbf{v}(\eta_{\text{dec}}, r_L\hat{\mathbf{n}})$$
$$+ \int_{\eta_{\text{dec}}}^{\eta_0} d\eta \left\{\Psi'[\eta, (\eta_0 - \eta)\hat{\mathbf{n}}] - \Phi'[\eta, (\eta_0 - \eta)\hat{\mathbf{n}}]\right\}$$
$$- \frac{1}{2}n^i n^j \int_{\eta_{\text{dec}}}^{\eta_0} d\eta \, (h_{ij}^{\text{TT}})'[\eta, (\eta_0 - \eta)\hat{\mathbf{n}}].$$

$$(20.123)$$

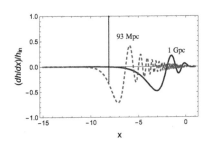

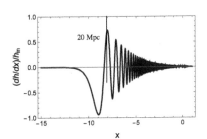

Fig. 20.10 The evolution of $[d\tilde{h}(x,k)/dx]/\tilde{h}_{\text{in}}(k)$ vs. $x = \log a$, for $\lambda = 93$ Mpc and 1 Gpc.

Fig. 20.11 The evolution of $[d\tilde{h}(x,k)/dx]/\tilde{h}_{\text{in}}(k)$ vs. $x = \log a$, for $\lambda = 20$ Mpc.

We see that GWs contribute only to the ISW effect. Just as for the scalar contribution to the ISW effect, the tensor contribution to the ISW depends only on the time derivative $(h_{ij}^{\text{TT}})'$, for the same physical reason as discussed after eq. (20.82). We now pass to the Fourier modes, writing

$$h_{ij}^{\text{TT}}(\eta, \mathbf{x}) = \int \frac{\sqrt{V}d^3k}{(2\pi)^3}\tilde{h}_{ij}^{\text{TT}}(\eta, \mathbf{k})e^{i\mathbf{k}\cdot\mathbf{x}},$$

$$(20.124)$$

and we expand, as usual,

$$\tilde{h}_{ij}^{\text{TT}}(\eta, \mathbf{k}) = \sum_{A=+,\times} e_{ij}^A(\hat{\mathbf{k}})\tilde{h}_A(\eta, \mathbf{k}).$$

$$(20.125)$$

Then

$$\int_{\eta_{\text{dec}}}^{\eta_0} d\eta \, (h_{ij}^{\text{TT}})'[\eta, (\eta_0 - \eta)\hat{\mathbf{n}}]$$

$$(20.126)$$

$$= \sum_{A=+,\times} \int \frac{\sqrt{V}d^3k}{(2\pi)^3}e_{ij}^A(\hat{\mathbf{k}})\int_{x_{\text{dec}}}^{0} dx \, \frac{d\tilde{h}_A(x,\mathbf{k})}{dx}e^{i[\eta_0 - \eta(x)]\mathbf{k}\cdot\hat{\mathbf{n}}},$$

where we used $x = \log a(\eta)$ as the first argument of $\tilde{h}_A$. Figures 20.10 and 20.11 show $d\tilde{h}/dx$, computed numerically from eq. (19.230), for our reference values of λ, where $\tilde{h} = \tilde{h}_A$ is the polarization amplitude that we have already plotted in Fig. 19.14. In Fig. 20.10 we plot the result for the mode $\lambda = 93$ Mpc, which enters the horizon at RD–MD equilibrium, and for $\lambda = 1$ Mpc, which enters in MD, while in Fig. 20.11 we show $d\tilde{h}/dx$ for the mode $\lambda = 20$ Mpc, which enters during RD. We see that, for modes that enter in RD, after decoupling (i.e. at $x > x_{\text{dec}} \simeq -7.0$), $d\tilde{h}/dx$ oscillates very rapidly. Therefore the integral over dx in eq. (20.126), which is evaluated from $x = x_{\text{dec}}$ to $x = 0$, is the integral of a rapidly oscillating function, and is negligibly small. In contrast, for longer modes, the integral gives a larger result.[12]

As in eq. (19.253), we write $\tilde{h}_A(\eta, \mathbf{k}) = \mathsf{h}(k\eta)\,\tilde{h}_A(\eta_{\text{in}}, \mathbf{k})$, where $\mathsf{h}(k\eta)$ is the appropriate transfer function. Then, we can write the contribution of GWs to the temperature fluctuation $\Theta = \delta T/T$, seen today by an observer located at $\mathbf{x} = 0$, looking in the direction $\hat{\mathbf{n}}$, as

$$\Theta^{\text{GW}}(\eta_0, \hat{\mathbf{n}}, \mathbf{x} = 0) = -\frac{1}{2}n^i n^j \int_{\eta_{\text{dec}}}^{\eta_0} d\eta \, (h_{ij}^{\text{TT}})'[\eta, (\eta_0 - \eta)\hat{\mathbf{n}}]$$

[12]Of course, this assumes that the tensor perturbations have been generated at a primordial epoch, and simply evolve freely according to eq. (19.220) with $\sigma_A = 0$. Further tensor perturbations might be generated after decoupling or in the recent epoch, for example by astrophysical sources.

$$= -\frac{1}{2} n^i n^j \int_{\eta_{\text{dec}}}^{\eta_0} d\eta \int \frac{\sqrt{V} d^3 k}{(2\pi)^3} \, (\tilde{h}_{ij}^{\text{TT}})'(\eta, \mathbf{k}) \, e^{i(\eta_0 - \eta)\mathbf{k}\cdot\hat{\mathbf{n}}} \,, \quad (20.127)$$

$$= -\frac{1}{2} n^i n^j \int_{\eta_{\text{dec}}}^{\eta_0} d\eta \int \frac{\sqrt{V} d^3 k}{(2\pi)^3} \, \mathsf{h}'(k\eta) \tilde{h}_{ij}^{\text{TT}}(\eta_{\text{in}}, \mathbf{k}) \, e^{i(\eta_0 - \eta)\mathbf{k}\cdot\hat{\mathbf{n}}} \,.$$

Computation of the coefficients C_l^{GW}

Having Θ^{GW}, we can first of all compute the GW contribution to the coefficients a_{lm}. The total value of the a_{lm} is then given by $a_{lm} = a_{lm}^{\text{S}} + a_{lm}^{\text{GW}}$, where a_{lm}^{S} is the scalar contribution computed in Section 20.3.4. Compared with the scalar case, in the tensor case an extra complication is that now $\hat{\mathbf{n}}$ appears not only in the combination $\hat{\mathbf{n}}\cdot\mathbf{k}$, but also explicitly, in the term $n^i n^j \tilde{h}_{ij}^{\text{TT}}(\eta_{\text{in}}, \mathbf{k})$. Thus, we cannot use eq. (20.97) directly, but must start from eq. (20.4). Inserting eq. (20.127) into eq. (20.4) and writing

$$\tilde{h}_{ij}^{\text{TT}}(\eta_{\text{in}}, \mathbf{k}) = \sum_{A=+,\times} e_{ij}^A(\hat{\mathbf{k}}) \tilde{h}_{A,\text{in}}(\mathbf{k}) \,, \quad (20.128)$$

we get

$$a_{lm}^{\text{GW}} = \int d^2 \hat{\mathbf{n}} \, Y_{lm}^*(\hat{\mathbf{n}}) \, \Theta^{\text{GW}}(\eta_0, \hat{\mathbf{n}}, \mathbf{x} = 0)$$

$$= -\frac{1}{2} \int d^2 \hat{\mathbf{n}} \, Y_{lm}^*(\hat{\mathbf{n}}) \int_{\eta_{\text{dec}}}^{\eta_0} d\eta \int \frac{\sqrt{V} d^3 k}{(2\pi)^3} \, \mathsf{h}'(k\eta)$$

$$\times \sum_{A=+,\times} e_{ij}^A(\hat{\mathbf{k}}) \tilde{h}_{A,\text{in}}(\mathbf{k}) \, n^i n^j e^{i(\eta_0 - \eta)\mathbf{k}\cdot\hat{\mathbf{n}}} \,. \quad (20.129)$$

The presence of the factor $n^i n^j$ significantly complicates the evaluation of the integral over $d^2 \hat{\mathbf{n}}$. It is then convenient to use the identity

$$\frac{\partial}{\partial \hat{k}_i} \frac{\partial}{\partial \hat{k}_j} e^{i(\eta_0 - \eta)\mathbf{k}\cdot\hat{\mathbf{n}}} = -(\eta_0 - \eta)^2 k^2 n^i n^j e^{i(\eta_0 - \eta)\mathbf{k}\cdot\hat{\mathbf{n}}} \,, \quad (20.130)$$

where $\hat{k}_i = k_i/|\mathbf{k}|$, and rewrite eq. (20.129) as

$$a_{lm}^{\text{GW}} = \frac{1}{2} \int d^2 \hat{\mathbf{n}} \, Y_{lm}^*(\hat{\mathbf{n}}) \int_{\eta_{\text{dec}}}^{\eta_0} d\eta \int \frac{\sqrt{V} d^3 k}{(2\pi)^3} \frac{1}{k^2 (\eta_0 - \eta)^2} \mathsf{h}'(k\eta)$$

$$\times \sum_{A=+,\times} e_{ij}^A(\hat{\mathbf{k}}) \tilde{h}_{A,\text{in}}(\mathbf{k}) \frac{\partial}{\partial \hat{k}_i} \frac{\partial}{\partial \hat{k}_j} e^{i(\eta_0 - \eta)\mathbf{k}\cdot\hat{\mathbf{n}}} \,. \quad (20.131)$$

We next expand the plane wave as in eq. (20.89), so

$$a_{lm}^{\text{GW}} = \frac{1}{2} \int d^2 \hat{\mathbf{n}} \, Y_{lm}^*(\hat{\mathbf{n}}) \int_{\eta_{\text{dec}}}^{\eta_0} d\eta \int \frac{\sqrt{V} d^3 k}{(2\pi)^3} \frac{1}{k^2 (\eta_0 - \eta)^2} \mathsf{h}'(k\eta)$$

$$\times \sum_{A=+,\times} e_{ij}^A(\hat{\mathbf{k}}) \tilde{h}_{A,\text{in}}(\mathbf{k}) \sum_{l'=0}^{\infty} i^{l'} (2l'+1) j_{l'}[k(\eta_0 - \eta)] \frac{\partial}{\partial \hat{k}_i} \frac{\partial}{\partial \hat{k}_j} P_{l'}(\hat{\mathbf{k}}\cdot\hat{\mathbf{n}}) \,.$$

$$(20.132)$$

The integral over $d^2\hat{\mathbf{n}}$ can now be performed using eq. (20.95),

$$
a_{lm}^{\mathrm{GW}} = 2\pi\, i^l \int_{\eta_{\mathrm{dec}}}^{\eta_0} d\eta \int \frac{\sqrt{V}d^3k}{(2\pi)^3} \frac{1}{k^2(\eta_0-\eta)^2}\, \mathsf{h}'(k\eta)
$$

$$
\times \sum_{A=+,\times} e_{ij}^A(\hat{\mathbf{k}})\tilde{h}_{A,\mathrm{in}}(\mathbf{k})\, j_l[k(\eta_0-\eta)]\frac{\partial}{\partial\hat{k}_i}\frac{\partial}{\partial\hat{k}_j}Y_{lm}^*(\hat{\mathbf{k}})\,. \quad (20.133)
$$

When computing the C_l using eq. (20.98) with $a_{lm} = a_{lm}^{\mathrm{S}} + a_{lm}^{\mathrm{GW}}$ we have the purely scalar term, which is the one that we have computed in Section 20.3.4, a purely GW term and a mixed term. The mixed term depends on the correlator $\langle\tilde{\Phi}(\eta_{\mathrm{in}},\mathbf{k})\tilde{h}_{ij}^{\mathrm{TT}}(\eta_{\mathrm{in}},\mathbf{k})\rangle$. If the mechanism that generates the primordial perturbations preserves invariance under rotations, there is no privileged spatial direction, so this correlator vanishes. We are then left with the purely scalar term from Section 20.3.4, plus the purely GW term,

$$
C_l^{\mathrm{GW}} = \frac{1}{2l+1}\sum_{m=-l}^{l}\langle|a_{lm}^{\mathrm{GW}}|^2\rangle\,. \quad (20.134)
$$

Inserting eq. (20.133) into this expression and computing the stochastic average using eqs. (19.276) and (19.278) we get

$$
C_l^{\mathrm{GW}} = \frac{\pi\kappa_l}{4}\int_0^\infty \frac{dk}{k}\,\mathcal{P}_{T,\mathrm{in}}(k)\left[\int_{\eta_{\mathrm{dec}}}^{\eta_0} d\eta\,\frac{j_l[k(\eta_0-\eta)]}{k^2(\eta_0-\eta)^2}\,\mathsf{h}'(k\eta)\right]^2\,,
$$
$$(20.135)$$

where

$$
\kappa_l = \int d^2\hat{\mathbf{k}} \sum_{A=+,\times} e_{ij}^A(\hat{\mathbf{k}})e_{kl}^A(\hat{\mathbf{k}})
$$

$$
\times \frac{1}{2l+1}\sum_{m=-l}^{l}\left[\frac{\partial}{\partial\hat{k}_i}\frac{\partial}{\partial\hat{k}_j}Y_{lm}^*(\hat{\mathbf{k}})\right]\left[\frac{\partial}{\partial\hat{k}_k}\frac{\partial}{\partial\hat{k}_l}Y_{lm}(\hat{\mathbf{k}})\right]\,. \quad (20.136)
$$

Observe that the integral over dk in eq. (20.135) formally goes up to $k=\infty$, but the presence of the derivative of the transfer function $\mathsf{h}'(k\eta)$, which appears in an integral over $\eta > \eta_{\mathrm{dec}}$, suppresses the contribution of the modes with $k > 1/\eta_{\mathrm{dec}}$. To compute κ_l we begin by observing that

$$
\sum_{A=+,\times} e_{ij}^A(\hat{\mathbf{k}})e_{kl}^A(\hat{\mathbf{k}}) = P_{ik}P_{jl} + P_{jk}P_{il} - P_{ij}P_{kl}\,, \quad (20.137)
$$

where $P_{ij}(\hat{\mathbf{k}}) = \delta_{ij} - \hat{k}_i\hat{k}_j$ is the projector that we already introduced in eq. (1.35). The above equality can be easily checked by choosing, without loss of generality, $\hat{\mathbf{k}} = (0,0,1)$: in this case both $e_{ij}^A(\hat{\mathbf{k}})$ and $P_{ij}(\hat{\mathbf{k}})$ are non-vanishing only when all their indices are in the $(1,2)$ plane, and eq. (20.137) can be easily checked for all components. Actually, the structure on the right-hand side of eq. (20.137) is also uniquely selected by the fact that it is symmetric in (i,j), symmetric in (k,l), symmetric under exchange $(i, j) \leftrightarrow (k, l)$, transverse in all indices (as follows from $\hat{k}^i P_{ij}(\hat{\mathbf{k}}) = 0$) and vanishes upon contraction with δ^{ij} or δ^{kl}, all properties that are obeyed by the left-hand side. Thus, in fact, one only has to check the proportionality constant between the left- and right-hand sides, which can be done by selecting a single combination of indices. In terms of the Lambda tensor defined in eq. (1.36) [see also eq. (18.22)], we can rewrite this as

$$\sum_{A=+,\times} e_{ij}^A(\hat{\mathbf{k}}) e_{kl}^A(\hat{\mathbf{k}}) = \Lambda_{ij,kl}(\hat{\mathbf{k}}) + \Lambda_{ij,lk}(\hat{\mathbf{k}}) . \tag{20.138}$$

Then

$$\kappa_l = \frac{2}{2l+1} \sum_{m=-l}^{l} \int d^2\hat{\mathbf{k}} \, \Lambda_{ij,kl}(\hat{\mathbf{k}}) \left[\frac{\partial}{\partial \hat{k}_i} \frac{\partial}{\partial \hat{k}_j} Y_{lm}^*(\hat{\mathbf{k}}) \right] \left[\frac{\partial}{\partial \hat{k}_k} \frac{\partial}{\partial \hat{k}_l} Y_{lm}(\hat{\mathbf{k}}) \right] . \tag{20.139}$$

We now recall the property (1.37) of the Lambda tensor, $\Lambda_{ij,i'j'} \Lambda_{i'j',kl} = \Lambda_{ij,kl}$, as well as the fact that it is symmetric under exchange of the first and second pair of indices, and we rewrite eq. (20.139) as

$$\kappa_l = \frac{2}{2l+1} \sum_{m=-l}^{l} \int d^2\hat{\mathbf{k}} \left[\Lambda_{i'j',ij} \frac{\partial}{\partial \hat{k}_i} \frac{\partial}{\partial \hat{k}_j} Y_{lm}^*(\hat{\mathbf{k}}) \right]$$
$$\times \left[\Lambda_{i'j',kl}(\hat{\mathbf{k}}) \frac{\partial}{\partial \hat{k}_k} \frac{\partial}{\partial \hat{k}_l} Y_{lm}(\hat{\mathbf{k}}) \right] . \tag{20.140}$$

Comparing with eq. (3.271) we recognize that the expressions in square bracket are just proportional to the E2 tensor spherical harmonics, which in the present notation read

$$(\mathbf{T}_{lm}^{E2})_{ij} = c_l^{(2)} \Lambda_{ij,i'j'}(\hat{\mathbf{k}}) \frac{\partial}{\partial \hat{k}_{i'}} \frac{\partial}{\partial \hat{k}_{j'}} Y_{lm}(\hat{\mathbf{k}}) , \tag{20.141}$$

with $c_l^{(2)} = [2(l-2)!/(l+2)!]^{1/2}$. Thus

$$\kappa_l = \frac{2}{2l+1} \frac{1}{[c_l^{(2)}]^2} \sum_{m=-l}^{l} \int d^2\hat{\mathbf{k}} \, (\mathbf{T}_{lm}^{E2})_{ij} (\mathbf{T}_{lm}^{E2})_{ij}^*$$
$$= \frac{(l+2)!}{(l-2)!} , \tag{20.142}$$

where in the last line we have used the orthonormality property of tensor spherical harmonics, eq. (3.277). Then, we finally get[13]

$$\boxed{C_l^{\mathrm{GW}} = \frac{\pi}{4} \frac{(l+2)!}{(l-2)!} \int_0^\infty \frac{dk}{k} \, \mathcal{P}_{T,\mathrm{in}}(k) \left[\int_{\eta_{\mathrm{dec}}}^{\eta_0} d\eta \, \frac{j_l[k(\eta_0 - \eta)]}{k^2(\eta_0 - \eta)^2} \, \mathrm{h}'(k\eta) \right]^2 .}$$
$$\tag{20.143}$$

[13] Our result agrees with that given in Durrer and Kahniashvili (1998) and in eq. (2.256) of Durrer (2008), after taking into account our different conventions (in particular, our h_{ij}^{TT} is denoted there as $2H_{ij}$, the result is written in terms of the power spectrum today rather than factorizing explicitly the transfer function, and the normalization of the power spectrum is different). Our result differs instead from eq. (9.55) of Gorbunov and Rubakov (2011), where the overall coefficient is $9\pi/2$ rather than $\pi/4$. Their factor 9 is just a typo due to the fact that they used $\mathrm{h}^2 = [3j_1(k\eta)/k\eta]^2$ but they counted twice the factor of 3^2 in this expression, factorizing it but also leaving it in h^2. The remaining factor $1/2$ is due to an inconsistent factor $1/\sqrt{2}$ in the normalization of the polarization tensors, between their eqs. (2.44) and (9.52).

This is the basic result that allows us to compute the effect on the CMB, given the primordial spectrum of GWs and the transfer function. We will see in Section 21.1 how this formula translates into limits on the ratio r for GWs produced during inflation. As we stressed at the beginning of Section 20.1, the analytic treatment in these sections, and therefore also eq. (20.143), is still approximate, because of the sharp separation that we are making between a regime where the photons can be treated as a fluid, before decoupling, and a free-streaming regime after decoupling. The exact numerical approach based on the Boltzmann equation will be discussed in Section 20.3.7, and involves the integration of a very large number of coupled differential equation, one for each multipole considered. The analytic results that we present here are, however, very useful for physical understanding, and reasonably accurate numerically. Let us therefore further manipulate eq. (20.143), to obtain some simple analytic formulas depending only on some constant to be evaluated numerically.

Since the integral in $d\eta$ in eq. (20.143) is over $\eta > \eta_{\text{dec}}$, the function $h'(k\eta)$ that appears in the integral suppresses the modes with $k\eta_{\text{dec}} \gg 1$, so the main contribution comes from modes with $k\eta_{\text{dec}} \lesssim 1$. These modes re-entered the horizon after decoupling, so in MD, where the transfer function can be approximated as in eq. (19.240),

$$h(k\eta) \simeq \frac{3}{(k\eta)^2} \left[\frac{\sin(k\eta)}{k\eta} - \cos(k\eta) \right] . \tag{20.144}$$

We neglect the small modification of the transfer function due to the recent dark-energy-dominated epoch. Then, retaining only the modes with $k < k_{\text{dec}}$ and writing the primordial tensor power spectrum as in eq. (19.281), we obtain

$$C_l^{\text{GW}} \simeq \frac{9\pi A_T}{4} \frac{(l+2)!}{(l-2)!} \int_0^{k_{\text{dec}}} dk \left(\frac{k}{k_*} \right)^{n_T} k\, \mathcal{U}_l^2(k) , \tag{20.145}$$

where

$$\mathcal{U}_l(k) \equiv \int_{\eta_{\text{dec}}}^{\eta_0} d\eta \frac{j_l[k(\eta_0 - \eta)]}{k^2(\eta_0 - \eta)^2} \left[\frac{3\cos(k\eta)}{(k\eta)^3} - \frac{3\sin(k\eta)}{(k\eta)^4} + \frac{\sin(k\eta)}{(k\eta)^2} \right] . \tag{20.146}$$

Passing to the dimensionless variables $\kappa = k/k_{\text{dec}}$, $\kappa_* = k_*/k_{\text{dec}}$ and $y = k_{\text{dec}}\eta$, the corresponding contribution to $\mathcal{D}_l$ can then be written as

$$\mathcal{D}_l^{\text{GW}} \simeq \frac{9 A_T T_0^2}{8} \int_0^1 d\kappa \left(\frac{\kappa}{\kappa_*} \right)^{n_T} v_l(\kappa) , \tag{20.147}$$

where

$$v_l(\kappa) \equiv l(l+1) \frac{(l+2)!}{(l-2)!} \kappa\, u_l^2(\kappa) \tag{20.148}$$

and

$$u_l(\kappa) = \int_1^{y_0} dy \frac{j_l[\kappa(y_0 - y)]}{\kappa^2(y_0 - y)^2} \left[\frac{3\cos(\kappa y)}{(\kappa y)^3} - \frac{3\sin(\kappa y)}{(\kappa y)^4} + \frac{\sin(\kappa y)}{(\kappa y)^2} \right] , \tag{20.149}$$

where $y_0 = k_{\rm dec}\eta_0 = \eta_0/\eta_{\rm dec}$. From eq. (17.174) we have $y_0 \simeq 49.3$, using our reference values for the cosmological parameters. It is then straightforward to compute the integral in eq. (20.149) numerically for different values of κ, and obtain $v_l(\kappa)$. In Fig. 20.12 we plot $v_l(\kappa)$ for $l = 4, 20$ and 40. We see that $v_l(\kappa)$ is a rather sharply peaked function, with a maximum at a value of κ that increases as l. For $n_T = 0$ the integral

$$\bar{v}_l \equiv \int_0^1 d\kappa\, v_l(\kappa) \qquad (20.150)$$

in eq. (20.147) is also easily computed numerically, and the result is shown in Fig. 20.13. We see that, for $4 \leqslant l \leqslant 40$, the integral is a slowly varying function of l.

For larger values of l the peak of $v_l(\kappa)$ progressively moves into the region $\kappa > 1$, which is excluded from the integration domain in eq. (20.147) (or, more generally, would be suppressed using the appropriate transfer function in this regime), and then the value of the integral quickly drops.

For $n_T \neq 0$ the integral in eq. (20.147) can be evaluated by observing that, if $|n_T|$ is not too large, the factor κ^{n_T} is slowly varying compared with the sharply peaked function $v_l(\kappa)$. We can therefore replace it by $\bar{\kappa}_l^{n_T}$, where $\bar{\kappa}_l$ is the value of κ where $v_l(\kappa)$ is maximum. The latter can be computed numerically for several l and we find that, in the region $4 < l < 40$, it is well fitted by $\bar{\kappa}_l \simeq 0.022\, l + 0.055$. Writing $\kappa_* = k_*/k_{\rm dec} = k_*\eta_{\rm dec}$ and using the reference value $\eta_{\rm dec} \simeq \eta_0/49.3$ from eq. (17.174), we finally get

$$\mathcal{D}_l^{\rm GW} \simeq \frac{9 A_T T_0^2}{8}\, \bar{v}_l \left(\frac{l+2.7}{k_*\eta_0}\right)^{n_T}. \qquad (20.151)$$

Equivalently, in this regime

$$l(l+1) C_l^{\rm GW} \simeq 0.16 A_T \left(\frac{\bar{v}_l}{0.023}\right) \left(\frac{l+2.7}{k_*\eta_0}\right)^{n_T}, \qquad (20.152)$$

where we have taken 0.023 as a reference value for the almost constant function $\bar{v}_l$. In Fig. 20.14 we show the GW contribution to $\mathcal{D}_l$, for a tensor-to-scalar ratio $r = 0.1$ and $n_T = 0$ (dashed curve), and we compare it with the scalar SW term (solid curve) that we have already shown in Figs. 20.5 and 20.9. We see that, for $r = 0.1$, at $l < O(100)$, the tensor contribution is at least one order of magnitude smaller than the scalar contribution, and then quickly becomes several orders of magnitude smaller for $l > O(100)$.[14] Both curves have been obtained by numerical integration of the full set of equations through a Boltzmann code (using the formalism that we will discuss in Section 20.3.7), and therefore represent the "exact" numerical solution to the problem. The dot-dashed line that stops at $l = 100$ is the analytic approximation (20.151), in which we have used the values of $\bar{v}_l(\kappa)$ computed numerically for different l and shown in Fig. 20.13. We see that for $l < O(40)$ the result from the full Boltzmann code is very well reproduced by the simple analytic approximation given by eqs. (20.148)–(20.151).

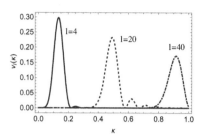

Fig. 20.12 The function $v_l(\kappa)$ for $l = 4, 20, 40$.

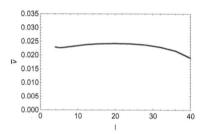

Fig. 20.13 $\bar{v}_l \equiv \int_0^1 d\kappa\, v_l(\kappa)$ as a function of l.

[14] As we will discuss in detail in Chapter 21, values of r of order 0.1 are of the order of the existing observational upper bounds.

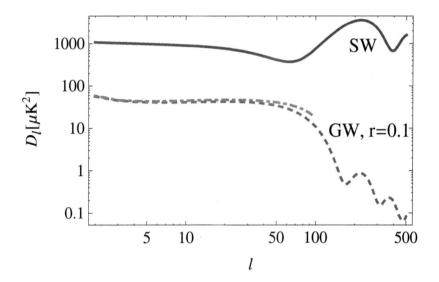

Fig. 20.14 The contribution to $\mathcal{D}_l$ from the SW effect of the scalar modes (solid curve) compared with the GW contribution (dashed), on a log–log scale. We have used $A_{\mathcal{R}} = 2.3 \times 10^{-9}$ and $n_s = 0.96$ for the primordial scalar spectrum, and $r = 0.1$ and $n_T = 0$ for the primordial tensor spectrum. Both curves have been obtained with the Boltzmann code CLASS. The dot-dashed line that stops at $l = 100$ is the analytic approximation (20.151), with $\bar{v}_l$ as in Fig. 20.13.

20.3.6   Finite thickness of the LSS

The analytic results of the previous sections have been obtained assuming an instantaneous transition between a fluid regime and the free-streaming regime. One can go beyond this approximation of instantaneous decoupling by recalling that, if a particle scatters with a cross-section σ off an ensemble of targets with number density n, its mean free path is $\ell = 1/(n\sigma)$. In our case the photons scatter off the electrons, with number density n_e, and the relevant cross-section is the Thomson cross-section σ_T. One can then define the optical depth τ by

$$\tau(\eta_1, \eta_2) \equiv \int_{t_2}^{t_1} dt\, n_e(t)\sigma_T$$
$$= \int_{\eta_2}^{\eta_1} d\eta\, n_e(\eta)\sigma_T a(\eta)\,, \qquad (20.153)$$

where $\eta_1 > \eta_2$. The probability that a photon did not scatter off an electron in the conformal time interval $[\eta_2, \eta_1]$ is

$$P(\eta_1, \eta_2) = e^{-\tau(\eta_1, \eta_2)}\,. \qquad (20.154)$$

The probability density that a photon observed at time η_1 had undergone its last scattering in the interval $[\eta_2, \eta_2 + d\eta_2]$ is given by $P(\eta_1, \eta_2 + d\eta_2) -$

$P(\eta_1, \eta_2)$. We then define the *visibility function* $V(\eta_1, \eta_2)$ from

$$V(\eta_1, \eta_2)d\eta_2 = P(\eta_1, \eta_2 + d\eta_2) - P(\eta_1, \eta_2) \,. \tag{20.155}$$

We are mostly interested in $V(\eta) \equiv V(\eta_0, \eta)$, since we observe the photons today. Using eqs. (20.154) and (20.155),

$$V(\eta) = \frac{d}{d\eta}e^{-\tau(\eta_0, \eta)}$$

$$= -e^{-\tau(\eta_0, \eta)}\frac{d\tau(\eta_0, \eta)}{d\eta} \,. \tag{20.156}$$

Observe that

$$\frac{d\tau(\eta_0, \eta)}{d\eta} = -n_e(\eta)\sigma_T a(\eta) \tag{20.157}$$

is negative. Since $\tau(\eta_0, \eta_0) = 0$, while at very early times (say, at an initial time $\eta = 0$) $\tau(\eta_0, 0) \to \infty$, the visibility function satisfies

$$\int_0^{\eta_0} d\eta\, V(\eta_0, \eta) = e^{-\tau(\eta_0, \eta)}\Big|_0^{\eta_0} = 1 \,, \tag{20.158}$$

so $V(\eta)$ is the normalized probability density that a photon observed today last scattered at conformal time η. Much before decoupling τ is very large and $V(\eta)$ is exponentially suppressed. Much later $d\tau(\eta_0, \eta)/d\eta$ is small because the density of free electrons, $n_e(\eta)$, is small. Thus, $V(\eta)$ is peaked at the decoupling epoch. Its maximum can be used to define the position, in conformal time, of the LSS, i.e. the value of conformal time at decoupling, $\eta_{\rm dec}$, and the width at this maximum gives the thickness of the LSS, $\Delta\eta_{\rm dec}$. Inserting the numerical value of σ_T and the known expressions for the functions $n_e(\eta)$ and $a(\eta)$ near decoupling one finds[15]

$$\frac{\Delta\eta_{\rm dec}}{\eta_{\rm dec}} \simeq 0.04 \,. \tag{20.159}$$

[15]See for example eq. (8.166) of Gorbunov and Rubakov (2011).

The instantaneous decoupling approximation that we have been using corresponds to neglecting $\Delta\eta_{\rm dec}$ and taking $V(\eta) = \delta(\eta - \eta_{\rm dec})$. To go beyond this approximation we must perform a convolution of our previous results for $\delta T/T$ with the visibility function. Then eq. (20.123) must be replaced by

$$\frac{\delta T}{T}(\eta_0, \hat{\mathbf{n}}) = \int_0^{\eta_0} d\eta'\, V(\eta')\frac{\delta T}{T}(\eta_0, \eta', \hat{\mathbf{n}}) \,, \tag{20.160}$$

where

$$\frac{\delta T}{T}(\eta_0, \eta', \hat{\mathbf{n}}) = \frac{1}{4}\delta_\gamma[\eta', (\eta_0 - \eta')\hat{\mathbf{n}}] + \Psi[\eta', (\eta_0 - \eta')\hat{\mathbf{n}}] - \hat{\mathbf{n}}\cdot\mathbf{v}[\eta', (\eta_0 - \eta')\hat{\mathbf{n}}]$$

$$+ \int_{\eta'}^{\eta_0} d\eta\, \{\Psi'[\eta, (\eta_0 - \eta)\hat{\mathbf{n}}] - \Phi'[\eta, (\eta_0 - \eta)\hat{\mathbf{n}}]\}$$

$$- \frac{1}{2}n^i n^j \int_{\eta'}^{\eta_0} d\eta\, (h_{ij}^{\rm TT})'[\eta, (\eta_0 - \eta)\hat{\mathbf{n}}] \,. \tag{20.161}$$

20.3.7   The Boltzmann equation for photons

The treatment of the CMB that we have given until now is based on approximations that have allowed us to obtain analytic results. Basically, we have split the problem into two parts. Before recombination, the photons are tightly coupled to the baryons and we have treated them as a perfect fluid. The corresponding formalism for computing cosmological perturbations was developed in Chapter 19, where photon perturbations were described in terms of just two variables, their density perturbation $\delta_\gamma(\eta, \mathbf{x})$ and their velocity potential $v_\gamma(\eta, \mathbf{x})$, or equivalently by their Fourier transforms $\tilde{\delta}_\gamma(\eta, \mathbf{k})$ and $\tilde{v}_\gamma(\eta, \mathbf{k})$.

After decoupling, we have described photons as particles freely streaming in a perturbed background, according to eqs. (20.58) and (20.79). The temperature perturbations seen by an observer located at $\mathbf{x}$, looking in the direction $\hat{\mathbf{n}}$, are described by a function $\Theta(\eta, \hat{\mathbf{n}}, \mathbf{x})$. After Fourier-transforming the $\mathbf{x}$ dependence, we have then found that the photon perturbations are described by a full set of multipoles $\tilde{\Theta}_l(\eta, \mathbf{k})$; see eq. (20.90).[16]

The transition between the fluid regime and the free-streaming regime has been assumed at first to be instantaneous, at $z = z_{\text{dec}}$, and the density and velocity perturbations computed at $z = z_{\text{dec}}$ from the fluid description have been used as initial conditions for the subsequent free streaming evolution. Including the visibility function, as in eq. (20.156), allows us to go beyond the instantaneous decoupling approximation. Nevertheless, this treatment is still intrinsically approximate, since before decoupling we are anyhow treating the photon field as a fluid characterized only by its density and velocity perturbations, while after decoupling all multipoles of the temperature field appear. Given the accuracy of the CMB measurement, it is necessary to go beyond this approximation and develop a first-principles approach to the computation of the C_l. In this section we give an overview of the main aspects of such an approach.

In general, an ensemble of photons is described by a distribution function in phase space. As in Section 20.3.1, we use conformal coordinates $x^\mu = (\eta, \mathbf{x})$. The distribution function will in general be a function of the coordinates x^μ and of the photon four-momentum $P^\mu = dx^\mu/d\lambda$. However, the four-momentum satisfies the mass-shell condition (20.35), which allows us to eliminate P^0 from the list of independent variables. Equivalently, instead of $P^i = dx^i/d\lambda$ we can use the comoving photon momentum $p^i = a^2 dx^i/d\lambda$ [see eq. (20.42)], or the physical momentum $p^i_{\text{ph}} = a dx^i/d\lambda$ [see eq. (20.43)]. In the following we will use for definiteness $\mathbf{p}_{\text{ph}}$. The distribution function can then be taken to be a function $f(\eta, \mathbf{x}, \mathbf{p}_{\text{ph}})$. In the absence of interactions with other particles the photons would satisfy the collisionless Boltzmann equation, which we have already written down for non-relativistic particles; see eq. (17.231). In terms of $\mathbf{p}_{\text{ph}}$, it reads

$$\frac{\partial f}{\partial \eta} + \frac{d\mathbf{x}}{d\eta} \cdot \boldsymbol{\nabla} f + \frac{d\mathbf{p}_{\text{ph}}}{d\eta} \cdot \frac{\partial f}{\partial \mathbf{p}_{\text{ph}}} = 0 \,. \tag{20.162}$$

[16]Of course, from the point of view of the observation all we need is $\Theta(\eta, \hat{\mathbf{n}}, \mathbf{x} = 0)$, where $\mathbf{x} = 0$ is the location of the observer. This function of $\hat{\mathbf{n}}$ can then be just expanded in spherical harmonics, as in eq. (20.2). However, for computing the photon perturbations in terms of the scalar perturbations of the metric, we need the full function $\Theta(\eta, \hat{\mathbf{n}}, \mathbf{x})$, since the equations relating $\Theta(\eta, \hat{\mathbf{n}}, \mathbf{x})$ to $\Psi(\eta, \mathbf{x})$ and $\Phi(\eta, \mathbf{x})$ are differential equations with respect to the $\mathbf{x}$ coordinates. After Fourier-transforming, at the level of linear perturbations, we then have a linear set of equations between the Fourier modes $\tilde{\Theta}(\eta, \hat{\mathbf{n}}, \mathbf{k})$ and the Fourier modes $\tilde{\Psi}(\eta, \mathbf{k}), \tilde{\Phi}(\eta, \mathbf{k})$. For scalar perturbations we saw that, in $\tilde{\Theta}(\eta, \hat{\mathbf{n}}, \mathbf{k})$, the dependence on $\hat{\mathbf{n}}$ is only through $\cos\theta = \hat{\mathbf{n}} \cdot \hat{\mathbf{k}}$, so it can be expanded on the basis of Legendre polynomials; see eq. (20.90). Then the function $\tilde{\Theta}_l(\eta, \mathbf{k})$ describes the contribution to the l-th temperature multipole, due to the Fourier modes of scalar perturbations with momentum $\mathbf{k}$.

For tensor perturbations we have the further complication that the dependence on $\hat{\mathbf{n}}$ is not only through $\hat{\mathbf{n}} \cdot \mathbf{k}$, since $n^i n^j$ also appears contracted with $(h_{ij}^{\text{TT}})'$, as we see from eq. (20.161). To illustrate the Boltzmann-equation approach we shall restrict to scalar perturbations. See Durrer (2008), Sect. 4.5.3, for the Boltzmann equation with tensor perturbations.

The interaction with the other particles generate a non-vanishing term $C[f]$ on the right-hand side, called the collision term, so that

$$\frac{\partial f}{\partial \eta} + \frac{d\mathbf{x}}{d\eta} \cdot \nabla f + \frac{d\mathbf{p}_{ph}}{d\eta} \cdot \frac{\partial f}{\partial \mathbf{p}_{ph}} = C[f] \,. \qquad (20.163)$$

In our case the collision term describes the change in the distribution function of photons with momentum $\mathbf{p}$, due to the process

$$e^-(\mathbf{q}) + \gamma(\mathbf{p}) \leftrightarrow e^-(\mathbf{q}') + \gamma(\mathbf{p}') \,. \qquad (20.164)$$

The explicit expression for the collision term is[17]

$$C[f(\mathbf{p})] = \frac{1}{|\mathbf{p}|} \int \frac{d^3q}{(2\pi)^3 2E_e(q)} \frac{d^3q'}{(2\pi)^3 2E_e(q')} \frac{d^3p'}{(2\pi)^3 2E_\gamma(p')} |\mathcal{M}|^2$$
$$\times (2\pi)^4 \delta^{(4)}(p + q - p' - q') \left[f_e(\mathbf{q}')f(\mathbf{p}') - f_e(\mathbf{q})f(\mathbf{p}) \right] \,, \qquad (20.165)$$

where $\mathcal{M}$ is the amplitude for the process (20.164) and $f_e(\mathbf{q})$ is the electron distribution function.

The equilibrium distribution is defined by the fact that, on it, the collision term vanishes, i.e. the rate of the process $e^-(\mathbf{q}) + \gamma(\mathbf{p}) \to e^-(\mathbf{q}') + \gamma(\mathbf{p}')$ is balanced by the rate of the inverse process. Thus, in an unperturbed FRW background the equilibrium distribution f_0 satisfies

$$\frac{\partial f_0}{\partial \eta} + \frac{d p_{ph}}{d\eta} \frac{\partial f_0}{\partial p_{ph}} = 0 \,, \qquad (20.166)$$

where $p_{ph} = |\mathbf{p}_{ph}|$,[18] and we have taken into account the fact that the FRW metric is spatially homogeneous and isotropic, so $f_0(\eta, \mathbf{x}, \mathbf{p}_{ph}) = f_0(\eta, p_{ph})$. To lowest order, $p_{ph} \propto a^{-1}$, so

$$\frac{d p_{ph}}{d\eta} = -\mathcal{H} p_{ph} \qquad (20.167)$$

and

$$\frac{\partial f_0}{\partial \eta} - \mathcal{H} p_{ph} \frac{\partial f_0}{\partial p_{ph}} = 0 \,. \qquad (20.168)$$

Looking for a solution of the Bose–Einstein form

$$f_0(\eta, p_{ph}) = \frac{1}{\exp\{p_{ph}/T(\eta)\} - 1} \,, \qquad (20.169)$$

eq. (20.168) gives $T' = -\mathcal{H}T$ and therefore $T \propto a^{-1}$, which is the expected result for the time evolution of the photon temperature in an unperturbed FRW background. We can now go to first order in the perturbation. It is convenient to write the photon distribution function as

$$f(\eta, \mathbf{x}, \mathbf{p}_{ph}) = \left[\exp\left\{ \frac{p_{ph}}{T(\eta)[1 + \Theta(\eta, \mathbf{x}, \mathbf{p}_{ph})]} \right\} - 1 \right]^{-1} \,, \qquad (20.170)$$

and consider $\Theta(\eta, \mathbf{x}, \mathbf{p}_{ph})$ as a first-order perturbation. On the left-hand side of eq. (20.163) the term $d\mathbf{x}/d\eta$ is related to the momentum by

[17]See Chapter 4 of Dodelson (2003) for an explicit derivation.

[18]Note that the label "ph" on the momentum stands for "physical" (and not for "photon"!), to stress that this is the physical, rather than the comoving momentum. We reserve here the letters p and p' for the photon momenta, and q and q' for the electron momenta.

eqs. (20.43) and (20.44), while $d\mathbf{p}_{\rm ph}/d\eta$ is computed by expanding the geodesic equation to first order. The collision term can be computed using the matrix element for Compton scattering. If, for simplicity, we neglect for a moment the angular dependence of Compton scattering and its dependence on the photon polarization (which are of course included when one wants to perform accurate computations), we have

$$|\mathcal{M}|^2 = 8\pi\sigma_T m_e^2 , \qquad (20.171)$$

where σ_T is the Thomson cross-section.[19] Then, putting together all first-order terms, one finds

$$\frac{\partial\Theta}{\partial\eta} + \hat{p}^i\partial_i\Theta + \frac{\partial\Phi}{\partial\eta} + \hat{p}^i\partial_i\Psi = n_e\sigma_T a(\eta)\left(\Theta_0 - \Theta + \hat{\mathbf{p}}\cdot\mathbf{v}_b\right) , \quad (20.172)$$

where $\mathbf{v}_b$ is the baryon velocity field and

$$\Theta_0(\eta,\mathbf{x}) \equiv \int \frac{d^2\hat{\mathbf{p}}}{4\pi} \, \Theta(\eta,\mathbf{x},\hat{\mathbf{p}}) . \qquad (20.173)$$

We have also made use of the fact that Θ depends on $\hat{\mathbf{p}}$ but not on $|\mathbf{p}|$, as we shall see shortly. Observe that, for the unit vector, the distinction between physical and comoving momentum is irrelevant, so we write $\hat{p}^i_{\rm ph} = \hat{p}^i$. The angular dependence of Compton scattering, as well as the dependence on the photon polarization vectors, produces further terms on the right-hand side, which typically give corrections at the level of 1% (the polarization term does, however, play a very important role since the CMB polarization carries important informations, as we will discuss in Section 20.4). We can rewrite the precedings equations as

$$\frac{\partial(\Theta+\Phi)}{\partial\eta} + \hat{p}^i\partial_i(\Theta+\Psi) = -\tau'(\eta)\left(\Theta_0 - \Theta + \hat{\mathbf{p}}\cdot\mathbf{v}_b\right) , \qquad (20.174)$$

where

$$\tau(\eta) \equiv \tau(\eta_0,\eta) = \int_\eta^{\eta_0} d\eta' \, n_e(\eta')\sigma_T a(\eta') \qquad (20.175)$$

is the optical depth from time η to the present time; compare with eq. (20.153). An important aspect of eq. (20.174) is that it depends on $\hat{\mathbf{p}} = \hat{\mathbf{p}}_{\rm ph}$ but not on the modulus $p_{\rm ph} = |\mathbf{p}_{\rm ph}|$. This originates from the fact that the physical photon momentum $p_{\rm ph}$ is much smaller than the electron momentum $q_{\rm ph}$, which in turn is much smaller than m_e (see Note 19), so the collision term becomes independent of the modulus of $\mathbf{p}_{\rm ph}$. In turn this means that Θ, which in principle is a function of $\eta,\mathbf{x}$ and $\mathbf{p}_{\rm ph}$ (since it is related to the distribution function is phase space), is actually a function only of $\eta, \mathbf{x}$ and the unit vector $\hat{\mathbf{p}}$. We can now make contact with the formalism developed in the preceding sections. The quantity $\Theta(\eta,\mathbf{x},\hat{\mathbf{n}})$ that we used before, and which was defined as the temperature perturbations seen by an observer located at the position

x and looking into the direction $\hat{\mathbf{n}}$, is just the same as the phase-space distribution function $\Theta(\eta, \mathbf{x}, \hat{\mathbf{p}})$, for photons propagating in the direction $\hat{\mathbf{p}} = -\hat{\mathbf{n}}$.

So, eq. (20.174) (once supplemented also with the contribution from the angular and polarization dependence of the matrix element) provides a first-principles approach to the computation of the temperature anisotropies, to first order in cosmological perturbation theory. The term involving the baryon velocity field can be simplified, recalling, from eq. (18.151), that a velocity field can be decomposed into a longitudinal and a transverse part. The transverse part is a vector perturbation and, as we mentioned at the beginning of Section 19.5, it has only decaying modes and can be neglected. We therefore retain only the longitudinal part, so $v_b^i = \partial^i v_b$ and, in momentum space,

$$\tilde{\mathbf{v}}_b(\eta, \mathbf{k}) = i\mathbf{k}\tilde{v}_b(\eta, \mathbf{k}),\qquad(20.176)$$

where $\tilde{v}_b(\eta, \mathbf{k})$ is the baryon velocity potential. Then, recalling from eq. (20.90) that $\tilde{\Theta}(\eta, \mathbf{k}, \hat{\mathbf{p}})$ actually depends on $\hat{\mathbf{p}}$ only through $\cos\theta = \hat{\mathbf{p}}\cdot\hat{\mathbf{k}}$, in Fourier space eq. (20.174) becomes

$$\frac{\partial}{\partial\eta}\left[\tilde{\Theta}(\eta, \cos\theta, \mathbf{k}) + \tilde{\Phi}(\eta, \mathbf{k})\right] + ik\cos\theta\left[\tilde{\Theta}(\eta, \cos\theta, \mathbf{k}) + \tilde{\Psi}(\eta, \mathbf{k})\right]$$
$$= -\tau'(\eta)\left[\tilde{\Theta}_0(\eta, \mathbf{k}) - \tilde{\Theta}(\eta, \cos\theta, \mathbf{k}) + ik\cos\theta\,\tilde{v}_b(\eta, \mathbf{k})\right].\quad(20.177)$$

Observe the different roles of $\hat{\mathbf{p}}$, which enters only as a unit vector, and represents the direction of the photon momentum, and of $\mathbf{k}$, which enters instead because of the Fourier transform, so that the dependence on $\mathbf{x}$ and $\hat{\mathbf{p}}$ is now replaced by a dependence on $\mathbf{k}$ and $\hat{\mathbf{p}}$. Observe also that, obviously, $\tilde{\Phi}$, $\tilde{\Psi}$ and $\tilde{v}_b$ do not carry any dependence on $\hat{\mathbf{p}}$, and therefore on $\cos\theta$. We now project $\tilde{\Theta}(\eta, \cos\theta, \mathbf{k})$ onto the basis of Legendre polynomials using eq. (20.92), multiplying both sides of eq. (20.177) by $P_l(\cos\theta)$ and integrating over $d\cos\theta$. For $l \geqslant 1$ we also use the recursion relation for the Legendre polynomials,

$$(2l + 1)xP_l(x) = (l + 1)P_{l+1}(x) + lP_{l-1}(x).\qquad(20.178)$$

We then obtain a set of coupled equations for the multipoles $\tilde{\Theta}_l(\eta, \mathbf{k})$.[20] The equations for the first two moments are

$$\tilde{\Theta}_0' - k\tilde{\Theta}_1 = -\tilde{\Phi}',\qquad(20.179)$$
$$3\tilde{\Theta}_1' + k(\tilde{\Theta}_0 - 2\tilde{\Theta}_2) = -k\tilde{\Psi} + \tau'(3\tilde{\Theta}_1 + ik\tilde{v}_b),\qquad(20.180)$$

while for $l > 2$ one finds

$$\tilde{\Theta}_l' + \frac{l}{2l + 1}k\tilde{\Theta}_{l-1} - \frac{l + 1}{2l + 1}k\tilde{\Theta}_{l+1} = \tau'\tilde{\Theta}_l.\qquad(20.181)$$

This set of equations can be truncated at a very large l and integrated numerically, through so-called Boltzmann codes.[21] This provides very accurate numerical solutions for the coefficients C_l. From eq. (20.181) we can also understand that the fluid description becomes appropriate

[20] Our definition (20.92) of $\tilde{\Theta}_l(\eta, \mathbf{k})$ is the same as in Gorbunov and Rubakov (2011), and differs by a factor $(-1)^l$ from the convention used in eq. (4.99) of Dodelson (2003). Observe also that we define the velocity potential from $v_b^i = \partial^i v_b$, so $\tilde{\mathbf{v}}_b = ik\tilde{v}_b$, while Dodelson defines v_b from $\tilde{\mathbf{v}}_b = i(\mathbf{k}/k)\tilde{v}_b$.

[21] The first publicly available Boltzmann code was CMBfast. Nowadays the most widely used public codes are CAMB and CLASS; see the Further Reading.

in the limit $\tau \gg k\eta$, which is called the *tight-coupling* limit. Indeed, in order of magnitude $\tilde{\Theta}'_l \sim (1/\eta)\tilde{\Theta}_l$. Then, we see by inspection that eq. (20.181) admits two kinds of hierarchies among the $\tilde{\Theta}_l$, namely

$$\tilde{\Theta}_l \sim \frac{k\eta}{\tau}\tilde{\Theta}_{l-1} \qquad (20.182)$$

and

$$\tilde{\Theta}_l \sim \frac{\tau}{k\eta}\tilde{\Theta}_{l-1}. \qquad (20.183)$$

For modes such that $k\eta \ll \tau$ the hierarchy (20.183) is clearly unphysical, and the corresponding solution is eliminated by requiring that the solution does not diverge at large multipoles. The second hierarchy is instead the one that is physically meaningful, and tells us that the contribution of the higher multipoles becomes smaller and smaller. We can then keep only the lowest multipoles. This corresponds precisely to the truncation of the BBGKY hierarchy discussed in Section 17.7.2 for a Newtonian fluid. In particular, if we keep only $\tilde{\Theta}_0$ and $\tilde{\Theta}_1$, the photon fluid is entirely described by eqs. (20.179) and (20.180), with $\tilde{\Theta}_2$ set to zero. The physical meaning of $\tilde{\Theta}_0$ and $\tilde{\Theta}_1$ is easily understood. From $\rho_\gamma \propto T^4$ we see that

$$\tilde{\Theta}_0 = \frac{1}{4}\tilde{\delta}_\gamma. \qquad (20.184)$$

The meaning of $\tilde{\Theta}_1$ can instead be understood from eq. (19.39), which for photons reads

$$\tilde{\delta}'_\gamma = -\frac{4}{3}\left(\tilde{\theta}_\gamma + 3\tilde{\Phi}'\right). \qquad (20.185)$$

Using eq. (20.184) and comparing with eq. (20.179) we find

$$k\tilde{\Theta}_1 = -\frac{1}{3}\tilde{\theta}_\gamma. \qquad (20.186)$$

We see that the first two multipoles, $\tilde{\Theta}_0$ and $\tilde{\Theta}_1$ are related, respectively, to the density perturbation and velocity potential of the photons. In the tight-coupling limit, all higher multipoles can be neglected and the fluid description is appropriate. As we discussed in Section 17.7.2, if we also include the multipole $\tilde{\Theta}_2$, we are adding viscosity terms, i.e. we are taking into account dissipation in the system.

For $z \gg z_{\rm dec}$ the optical depth is very large and the fluid description is appropriate. Observe that even in this case, for a given large but finite τ, there will be sufficiently short modes for which $k\eta$ becomes of order of or larger than τ. This corresponds to the fact that a fluid description is anyhow a coarse-grained macroscopic description, which breaks down at sufficiently short scales. For $z < z_{\rm dec}$, τ drops to a very small value, the fluid description is no longer valid and we are rather in the free-streaming regime. The analytic results presented in this chapter assume an instantaneous decoupling and a sharp transition between these two regimes, while the numerical integration of the full set of equations (20.179)–(20.181) provides the correct solution to the problem, with a smooth transition between these two regimes.

20.4 CMB polarization

In the previous sections we have only been concerned with the intensity of the CMB, as encoded in its temperature fluctuations. However, electromagnetic radiation can be polarized, and it turns out that the CMB polarization carries rich and very important information, which can also be especially important from the point of view of the imprint left by GWs. In this section we discuss the formalism needed to describe the CMB polarization, the generation of CMB polarization by scalar and tensor perturbations, and the present experimental situation.

20.4.1 Stokes parameters

We begin by developing the formalism needed to discuss the CMB polarization. Consider first a nearly monochromatic electromagnetic wave, with frequency ω, propagating in the z direction. At a given point in space, the components of the electric field are given by

$$E_x(t) = A_x(t) \cos\left[\omega t - \theta_x(t)\right] , \tag{20.187}$$
$$E_y(t) = A_y(t) \cos\left[\omega t - \theta_y(t)\right] , \tag{20.188}$$

with A_i and θ_i real. The condition that the wave is nearly monochromatic means that $A_i(t)$ and $\theta_i(t)$ are slowly-varying on the time-scale ω^{-1}, i.e. have Fourier modes with typical frequencies much smaller than ω. It is convenient to introduce the complex quantities

$$\mathcal{E}_a(t) = A_a(t) e^{i[\omega t - \theta_a(t)]} , \tag{20.189}$$

where $a = \{x, y\}$ is an index labeling the directions in the transverse plane. Then

$$E_a = \mathrm{Re}\,\mathcal{E}_a . \tag{20.190}$$

The four real functions $\{A_x, A_y, \theta_x, \theta_y\}$ that fully characterize the electromagnetic wave can be conveniently assembled into the four independent real components of a 2×2 Hermitian tensor $\langle \mathcal{E}_a \mathcal{E}_b^* \rangle$, where the angle brackets denote a temporal average over several periods of the wave. As for any 2×2 Hermitian matrix, we can expand it in the basis of the identity matrix and the three Pauli matrices

$$\sigma_1 = \begin{pmatrix} 0 & 1 \\ 1 & 0 \end{pmatrix}, \qquad \sigma_2 = \begin{pmatrix} 0 & -i \\ i & 0 \end{pmatrix} \qquad \sigma_3 = \begin{pmatrix} 1 & 0 \\ 0 & -1 \end{pmatrix} ,$$
$$\tag{20.191}$$

so we can write

$$\langle \mathcal{E}_a \mathcal{E}_b^* \rangle = \frac{1}{2} I \delta_{ab} + \frac{1}{2} \left(U\sigma_1 + V\sigma_2 + Q\sigma_3 \right)_{ab}$$
$$= \frac{1}{2} \begin{pmatrix} I+Q & U-iV \\ U+iV & I-Q \end{pmatrix}_{ab} , \tag{20.192}$$

with real coefficients I, U, V and Q. The parameter I, proportional to the identity matrix, is given in terms of the complex electric field by

$$I = \langle |\mathcal{E}_x|^2 \rangle + \langle |\mathcal{E}_y|^2 \rangle$$
$$= \langle A_x^2 \rangle + \langle A_y^2 \rangle , \tag{20.193}$$

as we immediately see on taking the trace of eq. (20.192). Therefore I is proportional to the intensity of the wave. The quantities I, U, V and Q are called the Stokes parameters. Equation (20.192) is inverted by $U = \mathrm{Tr}(\tilde{P}\sigma_1)$, $V = \mathrm{Tr}(\tilde{P}\sigma_2)$ and $Q = \mathrm{Tr}(\tilde{P}\sigma_3)$, where $\tilde{P}_{ab} \equiv \langle \mathcal{E}_a \mathcal{E}_b^* \rangle$. This gives

$$Q = \langle |\mathcal{E}_x|^2 \rangle - \langle |\mathcal{E}_y|^2 \rangle, \tag{20.194}$$

$$U = 2\,\mathrm{Re}\,\langle \mathcal{E}_x^* \mathcal{E}_y \rangle, \tag{20.195}$$

$$V = 2\,\mathrm{Im}\,\langle \mathcal{E}_x^* \mathcal{E}_y \rangle. \tag{20.196}$$

Equivalently, in terms of A_a and θ_a,

$$Q = \langle A_x^2 \rangle - \langle A_y^2 \rangle, \tag{20.197}$$

$$U = 2\langle A_x A_y \cos(\theta_x - \theta_y)\rangle, \tag{20.198}$$

$$V = 2\langle A_x A_y \sin(\theta_x - \theta_y)\rangle. \tag{20.199}$$

From eqs. (20.193) and (20.197)–(20.199) it follows that

$$I^2 \geqslant Q^2 + U^2 + V^2, \tag{20.200}$$

which becomes an equality if $A_a(t)$ and $\theta_a(t)$ are independent of time (i.e. for a purely monochromatic wave) and we can then omit the averaging procedure.[22] The Stokes parameters Q and U describe linear polarization. Consider for instance a wave that, with respect to the chosen (x, y) axes in the transverse plane, is linearly polarized along the x axis. Then $\mathcal{E}_x \neq 0$ and $\mathcal{E}_y = 0$, and therefore $Q \neq 0$, while $U = V = 0$. Similarly, a wave that is linearly polarized at $\pi/4$ with respect to the x axis has $\mathcal{E}_x = \mathcal{E}_y$ (i.e $A_x = A_y$, $\theta_x = \theta_y$), and therefore $Q = V = 0$ and $U \neq 0$. A wave linearly polarized in any other direction can be written as a superposition of waves polarized along the x axis and at $\pi/4$ with respect to the x axis, and therefore has Q and U non-zero, while still $V = 0$. In contrast, V describes circular polarization. Indeed, circular polarization corresponds to $\theta_y(t) = \theta_x(t) \pm \pi/2$ and $A_x = A_y$, so that eqs. (20.187) and (20.188) become

$$E_x(t) = A_x(t)\cos\left[\omega t - \theta_x(t)\right], \tag{20.204}$$

$$E_y(t) = \pm A_x(t)\sin\left[\omega t - \theta_x(t)\right], \tag{20.205}$$

and the electric field sweeps in time a circle in the transverse plane. In terms of $\mathcal{E}_a$ the condition $\theta_y(t) = \theta_x(t) \pm \pi/2$ and $A_x = A_y$ becomes $\mathcal{E}_x = \pm i\mathcal{E}_y$, and we see from eqs. (20.194)–(20.196) that for such a wave $Q = U = 0$, while $V \neq 0$. Finally, the wave is unpolarized if $Q = U = V = 0$.

While the intensity I is positive-definite, Q, U and V are not, and their sign carries physical information. We see from eq. (20.194) that a wave with $Q > 0, U = V = 0$ has linear polarization along the x axis, while $Q < 0, U = V = 0$ gives a wave linearly polarized along the y axis; see Fig. 20.15. Similarly, the polarizations corresponding to $U > 0, Q = V = 0$ and to $U < 0, Q = V = 0$ are shown in Fig. 20.16.

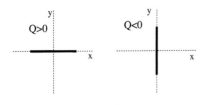

Fig. 20.15 The polarizations corresponding to $Q > 0, U = V = 0$ (left) and to $Q < 0, U = V = 0$ (right).

For circular polarization, the sign of V corresponds to counterclockwise ($V > 0$) or clockwise ($V < 0$) rotation.

The definition of Q and U depends on the choice of axes in the transverse plane. Consider a rotation of the axes by an angle ψ, so that

$$E_x \to E_x \cos\psi - E_y \sin\psi \,, \tag{20.206}$$
$$E_y \to E_x \sin\psi + E_y \cos\psi \,. \tag{20.207}$$

The components of the complex vector $\mathcal{E}_a$ transform in the same way, and then, from eqs. (20.194)–(20.196), we find that V is invariant, while

$$Q \to Q \cos 2\psi - U \sin 2\psi \,, \tag{20.208}$$
$$U \to Q \sin 2\psi + U \cos 2\psi \,. \tag{20.209}$$

In particular,

$$(Q \pm iU) \to e^{\pm 2i\psi}(Q \pm iU) \,. \tag{20.210}$$

These are the transformation properties of fields with helicities ± 2, respectively.[23] A rotation by $\pi/4$ transforms a U-polarization into a Q-polarization, as we see from eqs. (20.208) and (20.209), and also by comparing Figs. 20.15 and 20.16. The quantity $Q^2 + U^2$ is invariant under rotations around the propagation direction.

The trace part of the tensor $\langle \mathcal{E}_a \mathcal{E}_b^* \rangle$ in eq. (20.192) just gives the intensity of the wave, while the polarization properties are encoded in the traceless part. It is then convenient to define the traceless tensor

$$\mathcal{P}_{ab} \equiv \frac{1}{I}\left[\langle \mathcal{E}_a \mathcal{E}_b^* \rangle - \frac{1}{2}\delta_{ab}\langle \mathcal{E}^c \mathcal{E}_c^* \rangle \right] \,, \tag{20.211}$$

where $\langle \mathcal{E}^c \mathcal{E}_c^* \rangle = I$ is the trace, and we have normalized this tensor to I in order to deal with a dimensionless quantity. In terms of the Stokes parameters,

$$\mathcal{P}_{ab} = \frac{1}{2I}\begin{pmatrix} Q & U - iV \\ U + iV & -Q \end{pmatrix}_{ab} \,. \tag{20.212}$$

As we will see in Section 20.4.4, the mechanism that generates the CMB polarization produces only linear polarization, and no circular polarization. In the following we therefore set $V = 0$. The Hermitian tensor $\mathcal{P}_{ab}$ then becomes real and symmetric. In the context of CMB studies, it is conventional to reabsorb the normalization factor I defining the dimensionless Stokes parameters,

$$\mathcal{Q} = \frac{Q}{4I} \,, \qquad \mathcal{U} = \frac{U}{4I} \,, \tag{20.213}$$

so that

$$\mathcal{P}_{ab} = 2\begin{pmatrix} \mathcal{Q} & \mathcal{U} \\ \mathcal{U} & -\mathcal{Q} \end{pmatrix}_{ab} \,, \tag{20.214}$$

and therefore $\mathcal{Q} = (\mathcal{P}_{11} - \mathcal{P}_{22})/4$ and $\mathcal{U} = \mathcal{P}_{12}/2$.[24] In general, in a polarization map involving linearly polarized waves, the polarization at a point is represented by a rod whose length is

$$P = \sqrt{\mathcal{Q}^2 + \mathcal{U}^2} \,, \tag{20.215}$$

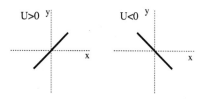

Fig. 20.16 The polarization corresponding to $U > 0, Q = V = 0$ (left) and to $U < 0, Q = V = 0$ (right).

[23] We already met the same transformation property in eq. (2.194) for the combinations $(h_+ \pm i h_\times)$, where h_+ and $h_\times$ are the plus and cross polarizations of a GW.

[24] The factor $1/4$ in the definition (20.213) is motivated by the fact that the energy density in the CMB is $\rho = I/(8\pi)$, so $\delta\rho/\rho = \delta I/I$. Since $\rho \propto T^4$, we have $\delta T/T = \delta I/(4I)$. As we will see, Q and U are already first-order quantities, generated by the density perturbations. Then the perturbation variables are $\delta T/T = \delta I/(4I)$, $\mathcal{Q} = Q/(4I)$ and $\mathcal{U} = U/(4I)$. Note also that different conventions exist in the literature. For instance Gorbunov and Rubakov (2011) rescale Q and U by a factor I rather than $4I$. Normally in the CMB literature $\mathcal{Q}$ and $\mathcal{U}$ are actually denoted simply as Q and U, which, however, can generate confusion about the factor $1/4$. Here we follow the notation in Durrer (2008).

and whose angle α to the x axis is given by

$$Q = P\cos 2\alpha\,, \qquad \mathcal{U} = P\sin 2\alpha\,. \tag{20.216}$$

This definition is motivated by the fact that, for $V = 0$, the matrix $\mathcal{P}_{ab}/2$, with $\mathcal{P}_{ab}$ given in eq. (20.214), has eigenvalues $\lambda_\pm = \pm P$. The eigenvector $\mathbf{v}$ with eigenvalue $+P$, apart from an arbitrary normalization, has components

$$(v_1, v_2) = (Q + P, \mathcal{U})\,. \tag{20.217}$$

Using eq. (20.216) these components become

$$(v_1, v_2) = (2\cos^2\alpha, 2\sin\alpha\cos\alpha)\,, \tag{20.218}$$

and therefore $v_2/v_1 = \tan\alpha$. Therefore α, defined from eq. (20.216), is the angle of this eigenvector of $\mathcal{P}_{ab}$ with the x axis, while P is its eigenvalue.

Note that the rod used to represent the polarization has no directionality, i.e. it is not an arrow; it simply represents the axis along which the electric field of a linearly polarized wave oscillates. Observe also from eqs. (20.208) and (20.209) that the Stokes parameters Q and U are invariant under a rotation by an angle $\psi = \pi$.

20.4.2 Polarization maps. E and B modes

The variables that we have just introduced describe the polarization at a point in space. We next promote the polarization tensor in eq. (20.214) to a tensor field, and we want to understand the spatial pattern of polarization maps, in a flat two-dimensional plane.[25] We begin by observing that for a vector field, in any number of dimensions, we have the usual decomposition into a transverse and a longitudinal part, $v_i = v_i^T + \partial_i v$ with $\partial_i v_i^T = 0$. In two spatial dimensions (denoting as usual the two-dimensional indices as $a, b, \ldots = 1, 2$) this further simplifies because a transverse vector field can always be written as $v_a^T = \epsilon_{ab}\partial_b w$ for some function w. Here ϵ_{ab} is the Levi-Civita antisymmetric tensor in a flat two-dimensional space, with $\epsilon_{12} = -\epsilon_{21} = 1$, and we keep for simplicity all indices as lower indices, since they are raised and lowered with the two-dimensional flat metric δ_{ab}. Thus, in two dimensions a generic vector field $v_a(x)$ can be written in terms of two scalar fields $v(x)$ and $w(x)$ as

$$v_a(x) = \partial_a v + \epsilon_{ab}\partial_b w\,. \tag{20.219}$$

Similarly, in a flat two-dimensional space, a symmetric traceless tensor field such as $\mathcal{P}_{ab}$ can be uniquely decomposed as

$$\mathcal{P}_{ab} = -\left(\partial_a\partial_b - \frac{1}{2}\delta_{ab}\boldsymbol{\nabla}^2\right)\mathcal{P}_E - \frac{1}{2}\left(\epsilon_{ac}\partial_b\partial_c + \epsilon_{bc}\partial_a\partial_c\right)\mathcal{P}_B\,, \tag{20.220}$$

where $\mathcal{P}_E$ and $\mathcal{P}_B$ are scalar functions. By analogy with the decomposition of the electromagnetic field into a divergence and a curl, the Fourier

[25]For the CMB we are actually interested in the polarization pattern on the celestial sphere. The discussion in this subsection will be useful for understanding the issue in a simpler setting, and also provides the correct description of the polarization pattern around a point, in a region sufficiently small that the curvature of the sphere can be neglected. We will perform the generalization to a sphere in the Section 20.4.3.

modes generated by $\mathcal{P}_E$ are called E modes, while those generated by $\mathcal{P}_B$ are called B modes.

Equation (20.220) allows us to express the two independent components of the symmetric traceless tensor field $\mathcal{P}_{ab}(x)$ [or, equivalently, $Q(x)$ and $U(x)$] in terms of two fields $\mathcal{P}_E$ and $\mathcal{P}_B$ that are scalars under rotations. Dealing with scalars rather than with the components of a tensor is in principle a simplification, in particular when in Section 20.4.3 we will generalize this construction to the sphere, since a scalar on the sphere can be expanded in spherical harmonics, while the expansion of a tensor involves the more complicated tensor spherical harmonics. The price that we pay, however, is that $\mathcal{P}_E$ and $\mathcal{P}_B$ are non-local functions of $\mathcal{P}_{ab}$. Indeed, applying $\partial_a\partial_b$ to eq. (20.220) we get

$$2\partial_a\partial_b\mathcal{P}_{ab} = -\boldsymbol{\nabla}^2\boldsymbol{\nabla}^2\mathcal{P}_E \,, \tag{20.221}$$

while applying $\epsilon_{ac}\partial_b\partial_c$ we get

$$2\epsilon_{ac}\partial_b\partial_c\mathcal{P}_{ab} = -\boldsymbol{\nabla}^2\boldsymbol{\nabla}^2\mathcal{P}_B \,, \tag{20.222}$$

where we have used

$$\epsilon_{ab}\epsilon_{cd} = \delta_{ac}\delta_{bd} - \delta_{ad}\delta_{bc} \,. \tag{20.223}$$

and $\epsilon_{ab}\epsilon_{ac} = \delta_{bc}$. We see that $\mathcal{P}_{E,B}$ can be obtained in terms of $\mathcal{P}_{ab}$ by inverting $\boldsymbol{\nabla}^2$ twice, i.e. performing twice a convolution with the Green's function of the Laplace operator. Therefore $\mathcal{P}_{E,B}$ are given in terms of $\mathcal{P}_{ab}$ by an expression that is non-local in space (but local in time). The fact that the Laplacian is invertible (assuming that $\mathcal{P}_{ab}$ vanishes at infinity, so that the convolution with the Green's function of the Laplacian is well-defined), and that its inversion is unique, proves that existence and the uniqueness of the decomposition (20.220). From eqs. (20.214) and (20.220), the explicit relation between $\{Q,\mathcal{U}\}$ and $\{\mathcal{P}_E,\mathcal{P}_B\}$ is given by

$$4Q = -(\partial_x^2 - \partial_y^2)\mathcal{P}_E - 2\partial_x\partial_y\mathcal{P}_B \,, \tag{20.224}$$

$$4\mathcal{U} = -2\partial_x\partial_y\mathcal{P}_E + (\partial_x^2 - \partial_y^2)\mathcal{P}_B \,. \tag{20.225}$$

The functions $\mathcal{P}_E$ and $\mathcal{P}_B$ determine the spatial pattern of the E and B modes, respectively. An especially characteristic pattern appears near the minima or maxima of one of these functions. We consider first the E modes, so we set $\mathcal{P}_B = 0$, and we assume that, at some point $\mathbf{x} = \mathbf{x}_0$ in the transverse plane, $\mathcal{P}_E$ has an extremum. We use polar coordinates (r, θ) around this point (with the angle θ measured counterclockwise from the x axis) and we study the polarization pattern in a small region around $\mathbf{x}_0$. We transform ∂_x and ∂_y in eqs. (20.224) and (20.225) into derivatives with respect to r and θ and we set $(\partial_r\mathcal{P}_E)_0 = (\partial_\theta\mathcal{P}_E)_0 = 0$, as appropriate for an extremum (where the subscript $(\ldots)_0$ reminds us that the derivative is taken at $\mathbf{x}_0$). We also assume that in a sufficiently

Fig. 20.17 The polarization pattern corresponding to an E mode around a maximum of $\mathcal{P}_E$.

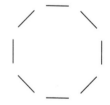

Fig. 20.18 The polarization pattern corresponding to an E mode around a minimum of $\mathcal{P}_E$.

Fig. 20.19 The polarization pattern corresponding to a B mode around a maximum of $\mathcal{P}_B$.

Fig. 20.20 The polarization pattern corresponding to a B mode around a minimum of $\mathcal{P}_B$.

[26]Recall that, in two dimensions, the transformation $\mathbf{x} \to -\mathbf{x}$ that flips the sign of both component is not a parity transformation: it has determinant $+1$, and is just the same as a rotation by π. Parity transformations in two dimensions are defined as the flip of sign of a single coordinate. Under such a transformation ϵ_{ab} is a pseudotensor, since it remains unchanged, rather than changing sign as a "good" tensor should do. In contrast, under a transformation, $\{x \to -x, y \to -y\}$, which is just a rotation by π, ϵ_{ab} transforms as a normal tensor, i.e. gets a $(-1)^2 = +1$ sign, and indeed under $\mathbf{x} \to -\mathbf{x}$ all four of Figs. (20.17)–(20.20) are invariant.

small neighborhood of $\mathbf{x}_0$ the pattern is symmetric under rotation, so we neglect the term $(\partial_\theta^2 \mathcal{P}_E)_0$. Then eqs. (20.224) and (20.225) give

$$4\mathcal{Q} = -(\partial_r^2 \mathcal{P}_E)_0 \cos 2\theta \,, \tag{20.226}$$

$$4\mathcal{U} = -(\partial_r^2 \mathcal{P}_E)_0 \sin 2\theta \,, \tag{20.227}$$

Comparing with eqs. (20.215) and (20.216) we see that $P = |(\partial_r^2 \mathcal{P}_E)_0|/4$ while the relation between α and θ depends on the sign of $(\partial_r^2 \mathcal{P}_E)_0$, i.e. on whether $\mathbf{x}_0$ is a minimum or a maximum. At a maximum $(\partial_r^2 \mathcal{P}_E)_0 < 0$, so $P = -(\partial_r^2 \mathcal{P}_E)_0/4$ and $\alpha = \theta$. Therefore the rod representing the polarization is in the radial direction. The polarization pattern is therefore the one shown in Fig. 20.17. At a minimum $(\partial_r^2 \mathcal{P}_E)_0 > 0$ and we can rewrite eqs. (20.226) and (20.227) in the form (20.216) with $P = (\partial_r^2 \mathcal{P}_B)_0/4$ and $\alpha = \theta - \pi/2$. So, the polarization rod is now orthogonal to the radial direction, and the corresponding pattern is shown in Fig. 20.18.

We can proceed similarly for the B modes. Setting now $\mathcal{P}_E = 0$ and expanding near an extremum $\mathbf{x}_0$ of $\mathcal{P}_B$, again assuming spherical symmetry in a small neighborhood of $\mathbf{x}_0$, we get

$$4\mathcal{Q} = -(\partial_r^2 \mathcal{P}_B)_0 \sin 2\theta \,, \tag{20.228}$$

$$4\mathcal{U} = (\partial_r^2 \mathcal{P}_B)_0 \cos 2\theta \,. \tag{20.229}$$

At a maximum, $(\partial_r^2 \mathcal{P}_B)_0 < 0$ and we can rewrite $\mathcal{Q}$ and $\mathcal{U}$ in the form (20.216) with $P = |(\partial_r^2 \mathcal{P}_B)_0|/4$ and $\alpha = \theta - \pi/4$. The polarization rod therefore makes an angle $-\pi/4$ with the radial direction. Recalling that θ is measured counterclockwise from the x axis, the polarization pattern is then the one shown in Fig. 20.19. Conversely, if $(\partial_r^2 \mathcal{P}_B)_0 > 0$, we can rewrite $\mathcal{Q}$ and $\mathcal{U}$ in the form (20.216) with $P = (\partial_r^2 \mathcal{P}_B)_0/4$ and $\alpha = \theta + \pi/4$, and the pattern is the one shown in Fig. 20.20.

Observe that, if we make a reflection around a single axis, for example the y axis, the patterns in Figs. 20.17 and 20.18 are unchanged. In contrast, under such a transformation the patterns in Figs. 20.19 and 20.20 are interchanged. This results from the presence of the tensor ϵ_{ab} in the term involving $\mathcal{P}_B$ in eq. (20.220), since ϵ_{ab} transforms as a pseudotensor under a parity transformation that flips the sign of a single axis, for example $\{x \to x, y \to -y\}$ or $\{x \to -x, y \to y\}$. As a consequence, while $\mathcal{P}_E$ is a true scalar, $\mathcal{P}_B$ is a pseudoscalar, i.e. it changes sign under transformations such as such as $\{x \to x, y \to -y\}$. This flip of sign transforms the maximum of $\mathcal{P}_B$ into a minimum, so it exchanges the pattern of Fig. 20.19 with that of Fig. 20.20.[26]

The transformation properties of $\mathcal{Q}$ and $\mathcal{U}$ under $\mathbf{x} \to \mathbf{x}' = \{x, -y\}$ are obtained from eqs. (20.224) and (20.225) using $\partial_x \to \partial_x$, $\partial_y \to -\partial_y$, $\mathcal{P}_E(\mathbf{x}) \to \mathcal{P}_E(\mathbf{x}')$ and $\mathcal{P}_B(\mathbf{x}) \to -\mathcal{P}_B(\mathbf{x}')$. This gives

$$\mathcal{Q}(\mathbf{x}) \to \mathcal{Q}(\mathbf{x}') \,, \tag{20.230}$$

$$\mathcal{U}(\mathbf{x}) \to -\mathcal{U}(\mathbf{x}') \,, \tag{20.231}$$

as we can also see from Figs. 20.15 and 20.16. The same holds for the transformation $\mathbf{x} \to \mathbf{x}' = \{-x, y\}$.

20.4.3   Polarization and tensor spherical harmonics

The discussion in Section 20.4.2 is appropriate for an infinite plane wave propagating along the z axis. For the CMB we are interested instead in photons coming radially from all directions, i.e. we are interested in the polarization tensor defined on a sphere, rather than on a plane. It is therefore convenient to use polar coordinates (r, θ, ϕ) and to introduce the metric on the unit sphere,

$$ds^2 = d\theta^2 + \sin^2\theta \, d\phi^2 \,, \tag{20.232}$$

i.e.

$$g_{ab} = (1, \sin^2\theta) \,, \tag{20.233}$$

where the indices a and b take the values $\{\theta, \phi\}$. As in the previous sections, we denote by $\hat{\mathbf{n}}$ the unit vector from the observer toward the observation direction, so the propagation direction of the photon is $\hat{\mathbf{p}} = -\hat{\mathbf{n}}$. The polarization properties of an electromagnetic wave arriving from the direction $\hat{\mathbf{n}}$ are encoded in the tensor $\langle \mathcal{E}_a \mathcal{E}_b^* \rangle$. Its trace is now taken with respect to the metric g_{ab} given in eq. (20.233), so is given by $\langle \mathcal{E}^a \mathcal{E}_a^* \rangle = g^{ab} \langle \mathcal{E}_a \mathcal{E}_b^* \rangle$. Thus the traceless tensor $\mathcal{P}_{ab}$ is now defined as

$$\mathcal{P}_{ab} = \frac{1}{\langle \mathcal{E}^c \mathcal{E}_c^* \rangle} \left[\langle \mathcal{E}_a \mathcal{E}_b^* \rangle - \frac{1}{2} g_{ab} \langle \mathcal{E}^c \mathcal{E}_c^* \rangle \right] \,, \tag{20.234}$$

which generalizes eq. (20.211) to the sphere. To define the Stokes parameters in a way that reduces to eq. (20.214) in the flat limit, we begin by observing that for each direction $\hat{\mathbf{n}}$ (except for the north pole), we can construct an orthonormal Cartesian coordinate system $\{\hat{\mathbf{x}}, \hat{\mathbf{y}}, \hat{\mathbf{n}}\}$, in which $\hat{\mathbf{x}}$ points in the $\hat{\boldsymbol{\theta}}$ direction and $\hat{\mathbf{y}}$ in the $-\hat{\boldsymbol{\phi}}$ direction, as shown in Fig. 20.21. We denote by V_x and V_y the components of a vector $\mathbf{V}$, tangent to the sphere, with respect to these axes. Since $dx = d\theta$ and $dy = -\sin\theta d\phi$, we have $V_x dx + V_y dy = V_\theta d\theta + V_\phi d\phi$, with $V_\theta = V_x$ and $V_\phi = -\sin\theta \, V_y$. Similarly, for a tensor, the Cartesian and polar coordinates are related by $T_{\theta\theta} = T_{xx}$, $T_{\theta\phi} = -\sin\theta \, T_{xy}$ and $T_{\phi\phi} = \sin^2\theta \, T_{yy}$. Thus, in polar coordinates the polarization tensor (20.214) reads

$$\mathcal{P}_{ab} = 2 \begin{pmatrix} \mathcal{Q} & -\mathcal{U}\sin\theta \\ -\mathcal{U}\sin\theta & -\mathcal{Q}\sin^2\theta \end{pmatrix}_{ab} \,, \tag{20.235}$$

where now a and b take the values $\{\theta, \phi\}$. This tensor is traceless with respect to the metric (20.233) of the unit sphere, $g^{ab}\mathcal{P}_{ab} = 0$. As in Section 20.4.2, we have set $V = 0$ since, as we will see, circular polarization is not generated in the CMB.

The generalization of eq. (20.220) to the sphere is obtained by writing

$$\mathcal{P}_{ab}(\theta, \phi) = -\frac{1}{2} K_{ab} \mathcal{P}_E(\theta, \phi) - \frac{1}{2} L_{ab} \mathcal{P}_B(\theta, \phi) \,, \tag{20.236}$$

where

$$K_{ab} = \boldsymbol{\nabla}_a \boldsymbol{\nabla}_b + \boldsymbol{\nabla}_b \boldsymbol{\nabla}_a - g_{ab} \boldsymbol{\nabla}^2 \,, \tag{20.237}$$

$$L_{ab} = \epsilon_a{}^c \boldsymbol{\nabla}_b \boldsymbol{\nabla}_c + \epsilon_b{}^c \boldsymbol{\nabla}_a \boldsymbol{\nabla}_c \,. \tag{20.238}$$

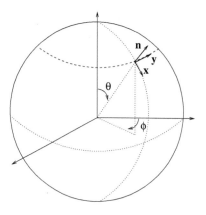

Fig. 20.21 The $\{\hat{\mathbf{x}}, \hat{\mathbf{y}}, \hat{\mathbf{n}}\}$ reference frame discussed in the text.

As usual, ∇_a is the covariant derivative with respect to the metric g_{ab}, $\nabla^2 = \nabla_a \nabla^a$ and ϵ_{ab} is the two-dimensional Levi-Civita symbol,

$$\epsilon_{ab} = \sqrt{g} \begin{pmatrix} 0 & 1 \\ -1 & 0 \end{pmatrix}. \tag{20.239}$$

The indices are raised and lowered with g_{ab} rather than with δ_{ab}, so we now have been careful in writing the indices properly as upper or lower. Since $g = \sin^2 \theta$ and $0 \leqslant \theta \leqslant \pi$, so $\sin \theta \geqslant 0$, we have $\sqrt{g} = \sin \theta$,

$$\epsilon_{ab} = \begin{pmatrix} 0 & \sin \theta \\ -\sin \theta & 0 \end{pmatrix} \tag{20.240}$$

and

$$\epsilon^a{}_b = \begin{pmatrix} 0 & \sin \theta \\ -1/\sin \theta & 0 \end{pmatrix}. \tag{20.241}$$

The two-dimensional Levi-Civita symbol satisfies the identities

$$\epsilon_{ca} \epsilon^c{}_b = g_{ab}, \tag{20.242}$$
$$\epsilon_{ab} \epsilon_{cd} = g_{ac} g_{bd} - g_{ad} g_{bc}, \tag{20.243}$$

and is covariantly conserved,

$$\nabla_c \epsilon^{ab} = 0. \tag{20.244}$$

The existence and uniqueness of the decomposition (20.236) can be proved, as in the flat-space case, by displaying its inversion explicitly. We perform the computation in some detail.[27]

[27]The reader not interested in these technical steps can simply skip the part in smaller type and jump to the result, eqs. (20.256) and (20.257).

We begin by observing that the operators K and L are orthogonal in the sense that, on any regular function $f(\theta, \phi)$,

$$K^{ab} L_{ab} f = 0, \tag{20.245}$$

and similarly $L^{ab} K_{ab} f = 0$.[28] This can be proved using the fact that, on a scalar, the covariant derivatives commute, $[\nabla_a, \nabla_b] f = 0$, while on a vector

$$[\nabla_a, \nabla_b] B^c = R^c{}_{dab} B^d, \tag{20.246}$$

and on a tensor

$$[\nabla_a, \nabla_b] T^{cd} = R^c{}_{eab} T^{ed} + R^d{}_{eab} T^{ec}. \tag{20.247}$$

[28]Observe that, on a scalar function f, we have $\nabla_a \nabla_b f = (\partial_a \partial_b - \Gamma^c_{ab} \partial_c) f = \nabla_b \nabla_a f$, so $K_{ab} f = (2 \nabla_a \nabla_b - g_{ab} \nabla^2) f$. However, on a tensor such as $L_{ab} f$, applying $\nabla_a \nabla_b$ is different from applying $\nabla_b \nabla_a$, so the general definition of the symmetric operator K_{ab} is the one in eq. (20.237).

For the metric of the unit sphere $g_{ab} = (1, \sin^2 \theta)$, the Riemann tensor is simply given by

$$R_{abcd} = g_{ac} g_{bd} - g_{ad} g_{bc}. \tag{20.248}$$

Inserting the explicit expression for K^{ab} into eq. (20.245), we get a term proportional to

$$g^{ab} L_{ab} f = 2 \epsilon^{bc} \nabla_b \nabla_c f = 0 \tag{20.249}$$

and a term

$$(\nabla^a \nabla^b + \nabla^b \nabla^a) L_{ab} f = 2 \nabla_a \nabla_b (\epsilon^a{}_c \nabla^b \nabla^c + \epsilon^b{}_c \nabla^a \nabla^c) f. \tag{20.250}$$

Defining $T^{ab} = \epsilon^a{}_c \nabla^b \nabla^c f$ and using eqs. (20.247) and (20.248) we get

$$[\nabla_a, \nabla_b]T^{ab} = 2g_{ab}T^{ab} = 0 \,. \tag{20.251}$$

Thus, the two terms on the right-hand side of eq. (20.250) are the same, and

$$2\nabla_a\nabla_b(\epsilon^a{}_c\nabla^b\nabla^c + \epsilon^b{}_c\nabla^a\nabla^c)f = 4\epsilon^b{}_c\nabla^a\nabla_b\nabla_a\nabla^c f \,, \tag{20.252}$$

where we have also used eq. (20.244). Finally, defining $B^c = \nabla^c f$ and using eq. (20.246), we get

$$4\epsilon^b{}_c\nabla^a\nabla_b\nabla_a\nabla^c f = 4(1+\nabla^2)(\epsilon^{bc}\nabla_b\nabla_c f) \,, \tag{20.253}$$

which vanishes because $\epsilon^{bc}\nabla_b\nabla_c f = 0$. This proves eq. (20.245). With similar manipulations we can compute $K^{ab}K_{ab}$ and $L^{ab}L_{ab}$. In particular

$$\begin{aligned} K^{ab}K_{ab}f &= 2\nabla^a\nabla^b K_{ab} \\ &= 2(2\nabla^a\nabla^b\nabla_a\nabla_b - \nabla^2\nabla^2)f \\ &= 2\nabla^2(\nabla^2 + 2)f \,, \end{aligned} \tag{20.254}$$

where in the first line we have used $g^{ab}K_{ab} = 0$, in the second $\nabla_b\nabla_a f = \nabla_a\nabla_b f$, and in the third eqs. (20.246) and (20.248). Similarly, also $L^{ab}L_{ab}f = 2\nabla^2(\nabla^2+2)f$. Therefore

$$\begin{aligned} K^{ab}\mathcal{P}_{ab} &= -(1/2)K^{ab}(K_{ab}\mathcal{P}_E + L_{ab}\mathcal{P}_B) \\ &= -\nabla^2(\nabla^2+2)\mathcal{P}_E \,. \end{aligned} \tag{20.255}$$

On the other hand, $g^{ab}\mathcal{P}_{ab} = 0$ and $(\nabla^a\nabla^b + \nabla^b\nabla^a)\mathcal{P}_{ab} = 2\nabla^a\nabla^b\mathcal{P}_{ab}$, so $K^{ab}\mathcal{P}_{ab} = 2\nabla^a\nabla^b\mathcal{P}_{ab}$. Then we conclude that

$$\boxed{2\nabla_a\nabla_b\mathcal{P}^{ab} = -\nabla^2(\nabla^2+2)\mathcal{P}_E \,,} \tag{20.256}$$

and similarly

$$\boxed{2\epsilon_a{}^c\nabla_b\nabla_c\mathcal{P}^{ab} = -\nabla^2(\nabla^2+2)\mathcal{P}_B \,.} \tag{20.257}$$

The "2" is present in (∇^2+2) because the Ricci scalar for the unit sphere is $R = 2$, as we can check from eq. (20.248). Thus, eqs. (20.256) and (20.257) reduce to the flat-space results (20.221) and (20.222) for high-frequency modes for which the curvature of the sphere can be neglected, as they should do.

The inversion of eqs. (20.256) and (20.257) is unique modulo the kernels of the operators ∇^2 and (∇^2+2). Consider a generic function on the sphere $f(\theta, \phi)$ and expand it in spherical harmonics,

$$f(\theta, \phi) = \sum_{l=0}^{\infty}\sum_{m=-l}^{l} c_{lm}Y_{lm}(\theta, \phi) \,. \tag{20.258}$$

Using

$$\nabla^2 Y_{lm} = -l(l+1)Y_{lm} \,, \tag{20.259}$$

we see that the most general solution of the equation $\nabla^2 f = 0$ is $f =$ constant, while the most general solution of $(\nabla^2 + 2)f = 0$ is a linear combination of the spherical harmonics $Y_{l=1,m}$, since $(\nabla^2 + 2)Y_{l=1,m} = 0$. Therefore, the inversion of eq. (20.256) is unique except for the fact that, in an expansion of $\mathcal{P}_E$ is spherical harmonics, the coefficients of the $l = 0$ and $l = 1$ spherical harmonics are not fixed. The same holds for $\mathcal{P}_B$. However, $\mathcal{P}_E$ and $\mathcal{P}_B$ are just auxiliary quantities that determine the observable quantity $\mathcal{P}_{ab}$ through eq. (20.236), i.e. through $K^{ab}\mathcal{P}_E$ and $L^{ab}\mathcal{P}_B$. The operators K^{ab} and L^{ab}, acting on the spherical harmonics with $l = 0, 1$ give zero, so the coefficients of the spherical harmonics with $l = 0, 1$ can be set to zero without loss of generality. Therefore, in this sense the decomposition (20.236) is well-defined and unique.

The advantage of introducing $\mathcal{P}_E$ and $\mathcal{P}_B$ is indeed that these quantities are scalars under rotation, and therefore can be expanded in ordinary spherical harmonics. In contrast Q and U are components of the tensor $\mathcal{P}_{ab}$ and expanding them directly on the celestial sphere requires the use of tensor spherical harmonics, as we will see. We then write the expansion of $\mathcal{P}_E$ and $\mathcal{P}_B$ as

$$\mathcal{P}_E(\theta, \phi) = -\sum_{l=2}^{\infty}\sum_{m=-l}^{l} N_l a_{lm}^E Y_{lm}(\theta, \phi), \tag{20.260}$$

$$\mathcal{P}_B(\theta, \phi) = -\sum_{l=2}^{\infty}\sum_{m=-l}^{l} N_l a_{lm}^B Y_{lm}(\theta, \phi), \tag{20.261}$$

where

$$N_l \equiv \left[2\frac{(l-2)!}{(l+2)!}\right]^{1/2}, \tag{20.262}$$

and $a_{lm}^{E,B}$ are the expansion coefficients, analogous to the coefficients a_{lm} in the expansion of the temperature field, eq. (20.2). The factor N_l in eqs. (20.260) and (20.261), as well as the overall minus sign, are a normalization of the coefficients $a_{lm}^{E,B}$, introduced for later convenience and, as already discussed, the coefficients $a_{lm}^{E,B}$ with $l = 0$ and with $l = 1$ can be set to zero without loss of generality. When we consider both the temperature and polarization multipoles, we will denote the temperature multipoles a_{lm} by a_{lm}^T.

Inserting eqs. (20.260) and (20.261) into eq. (20.236) we get

$$\mathcal{P}_{ab}(\theta, \phi) = \sum_{l=2}^{\infty}\sum_{m=-l}^{l} \left[a_{lm}^E (Y_{lm}^E)_{ab}(\theta, \phi) + a_{lm}^B (Y_{lm}^B)_{ab}(\theta, \phi)\right],$$

$$\tag{20.263}$$

where

$$(Y_{lm}^E)_{ab} = \frac{N_l}{2} K_{ab} Y_{lm}$$

$$= \frac{N_l}{2}\left(\nabla_a \nabla_b + \nabla_b \nabla_a - g_{ab}\nabla_c \nabla^c\right) Y_{lm} \tag{20.264}$$

and

$$(Y_{lm}^B)_{ab} = \frac{N_l}{2} L_{ab} Y_{lm}$$
$$= \frac{N_l}{2} \left(\epsilon^c{}_b \nabla_a \nabla_c + \epsilon^c{}_a \nabla_b \nabla_c \right) Y_{lm} \,. \qquad (20.265)$$

The functions $(Y_{lm}^E)_{ab}$ and $(Y_{lm}^B)_{ab}$ are the tensor spherical harmonics on the sphere. As we can see from the derivation leading to eq. (20.263), they provide a basis for the expansion of a traceless symmetric tensor $\mathcal{P}_{ab}(\theta, \phi)$ living on the sphere. Observe that $(Y_{lm}^E)_{ab}$ and $(Y_{lm}^B)_{ab}$ are traceless with respect to the metric g_{ab}, i.e.

$$g^{ab}(Y_{lm}^E)_{ab} = 0 \,, \qquad g^{ab}(Y_{lm}^B)_{ab} = 0 \,. \qquad (20.266)$$

A parity transformation on the sphere is defined as the transformation $\hat{\mathbf{n}} \to -\hat{\mathbf{n}}$. Since the polarization tensor $\mathcal{P}_{ab}$ is a true tensor, and

$$Y_{lm}(-\hat{\mathbf{n}}) = (-1)^l Y_{lm}(\hat{\mathbf{n}}) \,, \qquad (20.267)$$

we find from eqs. (20.264) and (20.265) that, under parity,

$$a_{lm}^E \to (-1)^l a_{lm}^E \,, \qquad a_{lm}^B \to (-1)^{l+1} a_{lm}^B \,, \qquad (20.268)$$

corresponding to electric and magnetic parities, respectively. Thus $\mathcal{P}_E$ is a true scalar and $\mathcal{P}_B$ is a pseudoscalar.

To get more explicit expressions for $(Y_{lm}^E)_{ab}$ and $(Y_{lm}^B)_{ab}$ we observe that, for the metric g_{ab}, the only non-vanishing Christoffel symbols are

$$\Gamma^\theta_{\phi\phi} = -\sin\theta\cos\theta \,, \qquad \Gamma^\phi_{\theta\phi} = \Gamma^\phi_{\phi\theta} = \mathrm{ctg}\,\theta \,. \qquad (20.269)$$

On a scalar such as Y_{lm}, we have $\nabla_a Y_{lm} = \partial_a Y_{lm}$, while

$$(\nabla_a \nabla_b + \nabla_b \nabla_a) Y_{lm} = 2(\partial_a \partial_b - \Gamma^c_{ab} \partial_c) Y_{lm} \,, \qquad (20.270)$$

and

$$\nabla_c \nabla^c Y_{lm} = \frac{1}{\sqrt{-g}} \partial_a \left(\sqrt{-g}\, g^{ab} \partial_b \right) Y_{lm}$$
$$= \left[\partial_\theta^2 + \mathrm{ctg}\,\theta\, \partial_\theta + \frac{1}{\sin^2\theta} \partial_\phi^2 \right] Y_{lm} \,. \qquad (20.271)$$

Then we get

$$(Y_{lm}^E)_{ab}(\theta, \phi) = \frac{N_l}{2} \begin{pmatrix} W & X \\ X & -\sin^2\theta\, W \end{pmatrix}_{ab} Y_{lm}(\theta, \phi) \qquad (20.272)$$

and

$$(Y_{lm}^B)_{ab}(\theta, \phi) = \frac{N_l}{2} \begin{pmatrix} -(1/\sin\theta)X & \sin\theta\, W \\ \sin\theta\, W & \sin\theta\, X \end{pmatrix}_{ab} Y_{lm}(\theta, \phi) \,, \qquad (20.273)$$

where

$$X = 2\partial_\theta \partial_\phi - 2\,\mathrm{ctg}\,\theta\, \partial_\phi \,, \qquad (20.274)$$
$$W = \partial_\theta^2 - \mathrm{ctg}\,\theta\, \partial_\theta - \frac{1}{\sin^2\theta} \partial_\phi^2 \,. \qquad (20.275)$$

[29]Our definition of X is the same as
in the original Zerilli (1970) paper, and
is the one commonly used in black-hole
perturbation theory. In the study of
CMB polarization, following the paper
by Kamionkowski, Kosowsky and Stebbins (1997), it is costumary to redefine
X by extracting a factor $\sin\theta$, so that in
eq. (20.272) we rather have $\sin\theta\, X$ instead of X, and in eq. (20.273) the (11)
and (22) matrix elements read $-X$ and
$\sin^2\theta\, X$, respectively.

We have already met the differential operators X and W in Chapter 12
in the context of black-hole perturbation theory; see eqs. (12.42) and
(12.43).[29] Using

$$\partial_\phi Y_{lm} = im Y_{lm} \qquad (20.276)$$

and

$$\left(\partial_\theta^2 + \operatorname{ctg}\theta - \frac{m^2}{\sin^2\theta}\right) Y_{lm} = -l(l+1)Y_{lm}\,, \qquad (20.277)$$

we can also write

$$XY_{lm} = 2im\left(\partial_\theta - \operatorname{ctg}\theta\right) Y_{lm}\,, \qquad (20.278)$$
$$WY_{lm} = \left[2\partial_\theta^2 + l(l+1)\right] Y_{lm}\,. \qquad (20.279)$$

It follows from eq. (20.245), upon integration by parts, that

$$\int d^2\hat{\mathbf{n}} \left[(Y_{l'm'}^E)^{ab}(\hat{\mathbf{n}})\right]^* (Y_{lm}^B)_{ab}(\hat{\mathbf{n}}) = \frac{N_l^2}{4} \int d^2\hat{\mathbf{n}} \, (K^{ab} Y_{l'm'}^*) L_{ab} Y_{lm}$$
$$= \frac{N_l^2}{4} \int d^2\hat{\mathbf{n}} \, Y_{l'm'}^* K^{ab} L_{ab} Y_{lm} = 0\,, \qquad (20.280)$$

while, using eq. (20.254),

$$\int d^2\hat{\mathbf{n}} \left[(Y_{l'm'}^E)^{ab}(\hat{\mathbf{n}})\right]^* (Y_{lm}^E)_{ab}(\hat{\mathbf{n}}) = \frac{N_l^2}{4} \int d^2\hat{\mathbf{n}} \, (K^{ab} Y_{l'm'}^*) K_{ab} Y_{lm}$$
$$= \frac{N_l^2}{4} \int d^2\hat{\mathbf{n}} \, Y_{l'm'}^* K^{ab} K_{ab} Y_{lm}$$
$$= \frac{N_l^2}{2} \int d^2\hat{\mathbf{n}} \, Y_{l'm'}^* \boldsymbol{\nabla}^2(\boldsymbol{\nabla}^2 + 2) Y_{lm}$$
$$= \frac{N_l^2}{2} l(l+1)[l(l+1) - 2] \int d^2\hat{\mathbf{n}} \, Y_{l'm'}^* Y_{lm}$$
$$= \delta_{ll'}\delta_{mm'}\,, \qquad (20.281)$$

and similarly for the integral of $(Y_{l'm'}^B)_{ab}^*(Y_{lm}^B)^{ab}$. Thus

$$\int d^2\hat{\mathbf{n}} \left[(Y_{lm}^A)_{ab}(\hat{\mathbf{n}})\right]^* (Y_{l'm'}^{A'})^{ab}(\hat{\mathbf{n}}) = \delta^{AA'}\delta_{ll'}\delta_{mm'}\,, \qquad (20.282)$$

where $A = \{E, B\}$. The constant N_l has been introduced in eqs. (20.264)
and (20.265) in order to ensure this normalization. Equation (20.263) is
therefore inverted by

$$a_{lm}^A = \int d^2\hat{\mathbf{n}} \left[(Y_{lm}^A)_{ab}(\hat{\mathbf{n}})\right]^* \mathcal{P}^{ab}(\hat{\mathbf{n}})\,. \qquad (20.283)$$

Inserting eq. (20.264) into eq. (20.283), using $g_{ab}\mathcal{P}^{ab} = 0$ and integrating
by parts we get

$$\boxed{a_{lm}^E = N_l \int d^2\hat{\mathbf{n}} \, Y_{lm}^*(\hat{\mathbf{n}})\boldsymbol{\nabla}_a\boldsymbol{\nabla}_b\mathcal{P}^{ab}(\hat{\mathbf{n}})\,.} \qquad (20.284)$$

Similarly, inserting eq. (20.265) into eq. (20.283), integrating by parts and using (20.244) we get

$$a_{lm}^B = N_l \epsilon^c{}_b \int d^2\hat{n}\, Y_{lm}^*(\hat{n}) \boldsymbol{\nabla}_a \boldsymbol{\nabla}_c \mathcal{P}^{ab}(\hat{n})\,. \qquad (20.285)$$

This formalism has allowed us to expand in spherical harmonics a symmetric and traceless tensor $\mathcal{P}_{ab}$ living on the sphere, with the indices a and b taking the two values $\{\theta, \phi\}$. In Section 3.5.2 of Vol. 1 we developed the expansion in tensor harmonics of a symmetric and traceless tensor $\mathcal{P}_{ij}$, with indices $i, j = 1, 2, 3$, that lives in three-dimensional space, and we were interested in particular in the case in which the tensor is also transverse, $\partial^i \mathcal{P}_{ij} = 0$. We applied that formalism to the expansion of the metric perturbation $h_{ij}^{\rm TT}$ in tensor spherical harmonics; see eq. (3.275). Of course the two formalisms are related, and it is instructive to work out this relation explicitly.

For a generic symmetric and traceless tensor, expressed in Cartesian coordinates with indices $i, j = 1, 2, 3$, we found that its five independent components can be expanded in the basis of the five tensor spherical harmonics that we denoted by $(\mathbf{T}_{lm}^{S0})_{ij}$, $(\mathbf{T}_{lm}^{E1})_{ij}$, $(\mathbf{T}_{lm}^{B1})_{ij}$, $(\mathbf{T}_{lm}^{E2})_{ij}$ and $(\mathbf{T}_{lm}^{B2})_{ij}$, which are given explicitly in eqs. (3.268)–(3.272). Of these, only $(\mathbf{T}_{lm}^{E2})_{ij}$ and $(\mathbf{T}_{lm}^{B2})_{ij}$ are transverse, so the expansion of a transverse and traceless symmetric tensor $\mathcal{P}_{ij}$ involves only these two quantities, and has the form

$$\mathcal{P}_{ij}(r, \theta, \phi) = \sum_{l=2}^{\infty} \sum_{m=-l}^{l} \left[u_{lm}(r)(\mathbf{T}_{lm}^{E2})_{ij}(\theta, \phi) + v_{lm}(r)(\mathbf{T}_{lm}^{B2})_{ij}(\theta, \phi) \right];$$

$$(20.286)$$

compare with eq. (3.275).[30] Observe that the sum starts from $l = 2$. As we discussed after eq. (3.273), $\mathbf{T}_{lm}^{E2}$ and $\mathbf{T}_{lm}^{B2}$ vanish for $l = 0, 1$, while $\mathbf{T}_{lm}^{E1}$ and $\mathbf{T}_{lm}^{B1}$ vanish for $l = 0$. A complete set of $s = 2$ spherical harmonic is given by $\mathbf{T}_{lm}^{S0}$ with $l \geqslant 0$, $\mathbf{T}_{lm}^{E1}$ and $\mathbf{T}_{lm}^{B1}$ with $l \geqslant 1$, and $\mathbf{T}_{lm}^{E2}$ and $\mathbf{T}_{lm}^{B2}$ with $l \geqslant 2$.

Equation (20.286) gives the Cartesian components of the tensors, i.e. it gives $\mathcal{P}_{ij}$ with $i, j = \{x, y, z\}$. We rather need the polar components $\mathcal{P}_{\alpha\beta}$ with $\alpha = \{r, \theta, \phi\}$. We have already transformed the tensor harmonics to polar components in eq. (12.41), when studying black-hole perturbation theory.[31] Using eqs. (12.37)–(12.41) we have

$$(\mathbf{T}_{lm}^{E2})_{\alpha\beta} = \frac{N_l}{2} r^2 \begin{pmatrix} 0 & 0 & 0 \\ 0 & W & X \\ 0 & X & -\sin^2\theta\, W \end{pmatrix}_{\alpha\beta} Y_{lm} \qquad (20.287)$$

and

$$(\mathbf{T}_{lm}^{B2})_{\alpha\beta} = \frac{N_l}{2} r^2 \begin{pmatrix} 0 & 0 & 0 \\ 0 & -(1/\sin\theta)X & \sin\theta\, W \\ 0 & \sin\theta\, W & \sin\theta\, X \end{pmatrix}_{\alpha\beta} Y_{lm}\,, \qquad (20.288)$$

[30] In principle, there can be also a time dependence in $\mathcal{P}_{ij}$, which results in a time dependence of the coefficients $u_{lm}(r)$ and $v_{lm}(r)$. This time dependence plays no role in the structure of the expansion in tensor spherical harmonics, and we do not write it explicitly.

[31] Actually in Section 12.2.1 we were even more general, since we studied tensor harmonics with Lorentz indices, for example $(\mathbf{T}_{lm}^{E2})_{\mu\nu}$. Here we need only the spatial part, since we are studying the expansion of a tensor $\mathcal{P}_{ij}$ with only spatial indices, rather than the expansion of a Lorentz tensor $h_{\mu\nu}$.

[32] As for the r dependence, the overall r^2 factor in eqs. (20.287) and (20.288), which arose in the passage from Cartesian to polar coordinates, combines with the r dependence of $u_{lm}(r)$ and $v_{lm}(r)$ in eq. (20.286) to give the desired dependence of $\mathcal{P}_{ij}$ on r. In our case, if we want to make contact with the expansion of a tensor $\mathcal{P}_{ab}(\theta, \phi)$ that does not depend on r, the dependence of $u_{lm}(r)$ and $v_{lm}(r)$ on r must be chosen such that $r^2 u_{lm}(r) \equiv a^E_{lm}$ and $r^2 v_{lm}(r) \equiv a^B_{lm}$ are independent of r.

[33] This unusual labeling of the initial and final variables is due to the fact that we will eventually integrate over the direction of initial momenta $\hat{\mathbf{p}}'$, and we will then remain with a function of $\mathbf{p}$. Observe also that we reserve the notation $\mathbf{k}$ for labeling the Fourier modes of the perturbations.

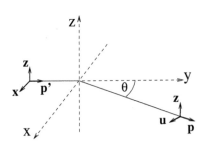

Fig. 20.22 The geometry of the photon–electron scattering discussed in the text

where the indices α and β run over the values $\{r, \theta, \phi\}$. We explicitly see the transverse nature of these spherical harmonics, since the (r, α) components in eqs. (20.287) and (20.288) vanish, and we remain with only a tensor in the (θ, ϕ) plane. Comparing with eqs. (20.272) and (20.273) we see that the remaining structure in the (θ, ϕ) plane is just given by the tensor harmonics on the sphere, $(Y^E_{lm})_{ab}$ and $(Y^B_{lm})_{ab}$.[32]

20.4.4 Generation of CMB polarization

In the generation of CMB polarization there are two key elements: first, the fact that Compton scattering of photons off free electrons depends on the polarization of the incoming and outgoing photons; second, the fact that, because of the perturbation $\delta T/T$, the photon flux arriving on an given electron from all directions is anisotropic, and in particular has a non-vanishing quadrupole moment. We will examine the roles of these two elements in turn.

Polarization dependence of Compton scattering

We consider the process $\gamma e^- \to \gamma e^-$ in which an incoming photon with momentum $\mathbf{p}'$ and polarization vector $\boldsymbol{\epsilon}'$ scatters off a electron at rest, and we denote by $\mathbf{p}$ and $\boldsymbol{\epsilon}$ the momentum and the polarization vector of the outgoing photon.[33] As we will see, we are interested in this process at the epoch of last scattering (and during the subsequent propagation of the photons toward us). The energy of a CMB photon is then much smaller than $m_e c^2$, and therefore the scattering is elastic, $|\mathbf{p}| = |\mathbf{p}'| \equiv p$. Thus, Compton scattering reduces to the classical Thomson scattering, so the differential cross section is independent of p and is given by

$$\frac{d\sigma}{d\Omega} = \frac{3\sigma_T}{8\pi} |\boldsymbol{\epsilon}^* \cdot \boldsymbol{\epsilon}'|^2 , \qquad (20.289)$$

where σ_T is the Thomson cross-section. If we have a flux of unpolarized radiation coming from a given fixed direction, this dependence of the cross-section on the polarization vectors generates an outgoing polarized radiation. To understand this point, consider the geometry of the scattering process shown in Fig. 20.22, where we have set the momentum of the incoming photons along the y axis, $\mathbf{p}' = p(0, 1, 0)$, and the outgoing photon is scattered in the (x, y) plane with scattering angle θ, so $\mathbf{p} = p(\sin\theta, \cos\theta, 0)$. Since the polarization vector is orthogonal to the propagation direction, using the linear polarization vectors as a basis, the polarization $\boldsymbol{\epsilon}'$ of the incoming photon is a superposition $\boldsymbol{\epsilon}' = \epsilon'_x \hat{\mathbf{x}} + \epsilon'_z \hat{\mathbf{z}}$, while for the outgoing photon $\boldsymbol{\epsilon} = \epsilon_u \hat{\mathbf{u}} + \epsilon_z \hat{\mathbf{z}}$, where $\hat{\mathbf{u}} = (\cos\theta, -\sin\theta, 0)$.

Consider first the case in which the incoming photon has polarization along the z axis. If the outgoing photon is also polarized along the z axis, we have $\boldsymbol{\epsilon}^* \cdot \boldsymbol{\epsilon}' = 1$ and $d\sigma/d\Omega = 3\sigma_T/(8\pi)$. In contrast, if the outgoing photon is polarized along the $\hat{\mathbf{u}}$ direction, we have $\boldsymbol{\epsilon}^* \cdot \boldsymbol{\epsilon}' = 0$ and the differential cross-section vanishes. Thus, an incoming photon with polarization perpendicular to the scattering plane produces an outgoing photon again with polarization perpendicular to the scattering plane, with a differential cross-section $d\sigma/d\Omega = 3\sigma_T/(8\pi)$.

Consider next an incoming photon with polarization parallel to the scattering plane, so along the x axis in the geometry of Fig. 20.22. Now, if the outgoing photon is polarized along the $\hat{\mathbf{z}}$ direction, we have $\boldsymbol{\epsilon}^*\cdot\boldsymbol{\epsilon}' = 0$ and the cross-section vanishes, while, if the outgoing photon is polarized along the $\hat{\mathbf{u}}$ direction, we have $\boldsymbol{\epsilon}^*\cdot\boldsymbol{\epsilon}' = \hat{\mathbf{x}}\cdot\hat{\mathbf{u}} = \cos\theta$ and $d\sigma/d\Omega = [3\sigma_T/(8\pi)]\cos^2\theta$. Therefore if the incoming photon has a polarization parallel to the scattering plane, the outgoing photon also has polarization in the scattering plane but $d\sigma/d\Omega$ is now suppressed by a factor $\cos^2\theta$. Thus, if the initial beam of photons is unpolarized, the radiation observed at an angle θ is preferentially polarized in the direction perpendicular to the scattering plane, as shown in Fig. 20.23. The effect become maximum for a scattering angle $\theta = \pi/2$. In this case an initial photon with polarization along the $\hat{\mathbf{x}}$ direction simply cannot scatter at $\theta = \pi/2$, because in this case $\hat{\mathbf{u}}$ becomes equal to $-\hat{\mathbf{y}}$ and $\boldsymbol{\epsilon}'$ is necessarily orthogonal to $\boldsymbol{\epsilon}$. Thus, all photons that scattered at $\theta = \pi/2$ had an initial polarization along $\hat{\mathbf{z}}$, and therefore their final polarization is also along $\hat{\mathbf{z}}$. Thus, the polarization observed at $\theta = \pi/2$ is fully polarized, while at a generic angle it is partially polarized, in the direction perpendicular to the scattering plane. For forward scattering, $\theta = 0$, the effect disappears and the outgoing radiation is unpolarized.

Integration over incoming photon directions

In the preceding discussion, photons arrive at a given electron from a fixed direction. In the cosmic fluid, however, photons arrive at an electron from all directions. If this incoming photon flux were isotropic, there would be no net polarization of the outgoing radiation. Consider for instance the radiation scattered, say, in the $\hat{\mathbf{z}}$ direction. It receives contributions from photons coming from all directions, which happened to be scattered in the $\hat{\mathbf{z}}$ direction. From the preceding discussion, it follows for instance that the radiation coming from the $\hat{\mathbf{y}}$ direction and scattered into the $\hat{\mathbf{z}}$ direction is fully polarized perpendicular to the scattering plane, i.e. along the $\hat{\mathbf{x}}$ direction; see Fig. 20.24. For radiation coming from the $\hat{\mathbf{x}}$ direction, and scattered again in the $\hat{\mathbf{z}}$ direction, the scattering plane is instead the (x, z) plane, and the polarization is fully along the $\hat{\mathbf{y}}$ direction; see Fig. 20.25. If the intensity of the radiation coming from the $\hat{\mathbf{x}}$ direction is the same as that coming from the $\hat{\mathbf{y}}$ direction, the sum of their contributions generates no net polarizations for the radiation outgoing in the $\hat{\mathbf{z}}$ direction. Thus, the second crucial element in the generation of CMB polarization is the anisotropy of the radiation impinging on a given electron. So, CMB polarization anisotropies emerge as a consequence of Thomson scattering plus CMB density perturbations. In particular, comparing Figs. 20.24 and 20.25 we see that we need a quadrupole anisotropy in the incoming radiation, since the cancellation takes place between radiation coming from an angle θ and radiation coming from an angle $\theta + (\pi/2)$. A dipole anisotropy is not

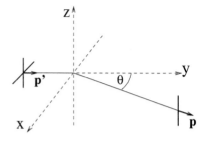

Fig. 20.23 Unpolarized incoming radiation generates outgoing radiation linearly polarized in the transverse plane

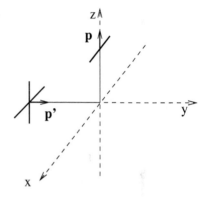

Fig. 20.24 The polarization generated in the $\hat{\mathbf{z}}$ direction by radiation coming from the $\hat{\mathbf{y}}$ direction.

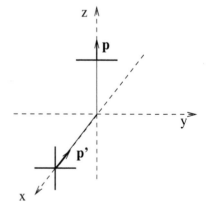

Fig. 20.25 The polarization generated in the $\hat{\mathbf{z}}$ direction by radiation coming from the $\hat{\mathbf{x}}$ direction.

sufficient. Observe that, in any case, Thomson scattering generates only linear polarization, and does not generate a net circular polarization. For this reason, we can neglect the Stokes parameter V.

The necessity of having both Thomson scattering and anisotropies considerably restricts the epochs at which the CMB polarization can be generated. The photons that impinge on a given electron from different directions underwent their previous scattering at a distance of the order of the photon mean free path λ_γ. To generate significant polarization, it is therefore necessary that the plasma is sufficiently inhomogeneous on the scale λ_γ. Before decoupling, photons and electrons are tightly coupled, and the mean free path λ_γ is very small. On the scale λ_γ the anisotropies are therefore negligible, and no significant net CMB polarization is generated. Near decoupling, λ_γ becomes large, and for a while there are still enough free electrons around that Thomson scattering is sufficiently frequent, so the conditions for generation of CMB polarization are met. However, decoupling takes place soon after recombination (defined as the epoch where the electron fraction $x_e = 1/2$; see page 382); thus, shortly after recombination, there are basically no more free electrons around, so even if now λ_γ is large, no scattering takes place and no polarization is produced. Thus, the CMB polarization is produced only during a short interval of time near recombination.[34] In Section 20.3 we computed the CMB anisotropies and we found that the leading-order result could be obtained by approximating the visibility function (20.156) by a Dirac delta. The same approximation cannot be used for the CMB polarization, since this effectively amounts to separating the plasma evolution into two epochs: one, before recombination, in which the optical depth is taken to be infinite and $\lambda_\gamma = 0$, and one, after recombination, when photons stream out freely, without interacting with electrons. In both limits, there is no generation of CMB polarization. To compute the CMB polarization it is necessary to take into account the finite width $\Delta\eta$ of the visibility function, and the result goes to zero as $\Delta\eta \to 0$. The polarization anisotropies in the CMB are therefore suppressed with respect to the temperature anisotropies.

To be more quantitative, we proceed as follows. As a first step we compute the net polarization generated when a radial flux of photons whose intensity has a given angular distribution $f(\theta', \phi')$ scatters off an electron at the origin. We consider an incoming photon whose propagation direction $\hat{\mathbf{p}}'$ is

$$\hat{\mathbf{p}}' = (\sin\theta' \sin\phi', \sin\theta' \cos\phi', \cos\theta'), \qquad (20.290)$$

i.e $\hat{\mathbf{p}}'$ is the unit radial vector with polar angles (θ', ϕ'). In the plane transverse to $\hat{\mathbf{p}}'$ we introduce a basis of two unit vectors $\mathbf{e}'_1$ and $\mathbf{e}'_2$ orthogonal to each other. We choose

$$\mathbf{e}'_1 = (\cos\theta' \sin\phi', \cos\theta' \cos\phi', -\sin\theta'), \qquad (20.291)$$

$$\mathbf{e}'_2 = (\cos\phi', -\sin\phi', 0), \qquad (20.292)$$

which corresponds to unit vectors in the $\hat{\boldsymbol{\theta}}$ and $\hat{\boldsymbol{\phi}}$ directions, respectively. We use $\mathbf{e}'_1$ and $\mathbf{e}'_2$ as a basis for the polarization vector of the initial

[34]Except that, much later, after formation of the first stars, the UV radiation from these early stars partially reionizes the Universe, and there are again free electrons around. This takes place around $z \simeq 8 - 10$ [see [Planck Collaboration], Aghanim *et al.* (2016)], and produces an extra "reionization bump" in the CMB polarization spectrum, see Fig. 20.27 on page 568 and the discussion around it.

photon.[35] Consider first an incoming photon with propagation direction $\hat{\mathbf{p}}'$ and polarization vector $\mathbf{e}_1'$, which is scattered into the $\hat{\mathbf{z}}$ direction, as in Fig. 20.24 (which illustrates the situation in which $\theta' = \pi/2$, $\phi' = 0$) or in Fig. 20.25 (which corresponds to $\theta' = \pi/2$, $\phi' = -\pi/2$). The polarization vector of the outgoing photon is in the (x, y) plane, and we take $\mathbf{e}_1 = \hat{\mathbf{x}}$ and $\mathbf{e}_2 = \hat{\mathbf{y}}$ as a basis for the final polarization. According to eq. (20.289), the probability that the final photon has polarization along $\hat{\mathbf{x}}$ is proportional to

$$|\mathbf{e}_1' \cdot \hat{\mathbf{x}}|^2 = \cos^2 \theta' \sin^2 \phi' \,, \tag{20.293}$$

while the probability that it has polarization along $\hat{\mathbf{y}}$ is proportional to

$$|\mathbf{e}_1' \cdot \hat{\mathbf{y}}|^2 = \cos^2 \theta' \cos^2 \phi' \,. \tag{20.294}$$

If instead the incoming photon has polarization vector $\mathbf{e}_2'$, these probabilities are proportional, respectively, to

$$|\mathbf{e}_2' \cdot \hat{\mathbf{x}}|^2 = \cos^2 \phi' \,, \tag{20.295}$$

and

$$|\mathbf{e}_2' \cdot \hat{\mathbf{y}}|^2 = \sin^2 \phi' \,. \tag{20.296}$$

If the initial radiation is unpolarized, there is an equal probability that the incoming photon has polarization $\mathbf{e}_1'$ or $\mathbf{e}_2'$. Thus, apart from a common proportionality constant C, which will cancel when we compute $\mathcal{P}_{ab}$ defined in eq. (20.212), we have

$$\mathcal{E}_x^2 = C \left(\cos^2 \theta' \sin^2 \phi' + \cos^2 \phi' \right), \tag{20.297}$$
$$\mathcal{E}_y^2 = C \left(\cos^2 \theta' \cos^2 \phi' + \sin^2 \phi' \right). \tag{20.298}$$

As a check, for $\theta' = \pi/2, \phi' = 0$ we get $\mathcal{E}_y = 0$ and the polarization is entirely along the x axis, in agreement with the result shown in Fig. 20.24, while for $\theta' = \pi/2$, $\phi' = -\pi/2$ we get $\mathcal{E}_x = 0$, in agreement with Fig. 20.25.

The Stokes parameter I of the radiation scattered in the $\hat{\mathbf{z}}$ direction is obtained by integrating $\mathcal{E}_x^2 + \mathcal{E}_y^2$ over all directions of the incoming radiation, weighted with the intensity of the incoming radiation $f(\theta', \phi')$,

$$I = C \int d\Omega' \, f(\theta', \phi')(\cos^2 \theta' \sin^2 \phi' + \cos^2 \phi' + \cos^2 \theta' \cos^2 \phi' + \sin^2 \phi')$$
$$= C \int d\Omega' \, f(\theta', \phi') \left(1 + \cos^2 \theta' \right), \tag{20.299}$$

where $d\Omega' = d\cos\theta' d\phi'$. Similarly, Q is obtained by integrating $\mathcal{E}_x^2 - \mathcal{E}_y^2$ over all directions of the incoming radiation,

$$Q = C \int d\Omega' \, f(\theta', \phi')(\cos^2 \theta' \sin^2 \phi' + \cos^2 \phi' - \cos^2 \theta' \cos^2 \phi' - \sin^2 \phi')$$
$$= C \int d\Omega' \, f(\theta', \phi') \sin^2 \theta' \cos 2\phi' \,. \tag{20.300}$$

[35] As usual, at $\theta' = 0$ the angle ϕ' is undefined. However, we will later integrate over all arrival direction, and in the resulting integral this ambiguity at a single point is irrelevant.

As discussed in Section 20.4.1, the U component can be derived by projecting on axes rotated by $\pi/4$, i.e. on $\mathbf{e}_1 = (1/\sqrt{2})(\hat{\mathbf{x}} + \hat{\mathbf{y}})$ and $\mathbf{e}_2 = (1/\sqrt{2})(\hat{\mathbf{x}} - \hat{\mathbf{y}})$, which gives

$$U = -C \int d\Omega'\, f(\theta', \phi')\, \sin^2\theta' \sin 2\phi' \,. \tag{20.301}$$

The angular factors in Q and U can be expressed in terms of the spherical harmonics with $l = 2$ and $m = \pm 2$, given on page 140 of Vol. 1. In particular,

$$Y_{2,2}(\theta, \phi) + Y_{2,-2}(\theta, \phi) = \left(\frac{15}{8\pi}\right)^{1/2} \sin^2\theta \cos 2\phi\,, \tag{20.302}$$

$$Y_{2,2}(\theta, \phi) - Y_{2,-2}(\theta, \phi) = i\left(\frac{15}{8\pi}\right)^{1/2} \sin^2\theta \sin 2\phi\,. \tag{20.303}$$

This shows that, on expanding the intensity of the incoming radiation $f(\theta', \phi')$ in spherical harmonics, the only contributions to Q and U come from the $l = 2$ term, i.e. from the quadrupole, as we already anticipated from the discussion of Figs. 20.24 and 20.25.

We can get rid of our specific choice of $\hat{\mathbf{p}}$, which we have set along the $\hat{\mathbf{z}}$ direction, by observing that the preceding results can be written in a rotationally invariant form as

$$I(\hat{\mathbf{n}}) = C \int d\hat{\mathbf{n}}'\, f(\hat{\mathbf{n}}')\, \left[1 + (\hat{\mathbf{n}} \cdot \hat{\mathbf{n}}')^2\right]\,, \tag{20.304}$$

$$Q(\hat{\mathbf{n}}) = -C \int d\hat{\mathbf{n}}'\, f(\hat{\mathbf{n}}')\, \left[(\hat{\mathbf{n}}' \cdot \mathbf{e}_1)(\hat{\mathbf{n}}' \cdot \mathbf{e}_1) - (\hat{\mathbf{n}}' \cdot \mathbf{e}_2)(\hat{\mathbf{n}}' \cdot \mathbf{e}_2)\right]\,, \tag{20.305}$$

$$U(\hat{\mathbf{n}}) = -C \int d\hat{\mathbf{n}}'\, f(\hat{\mathbf{n}}')\, 2(\hat{\mathbf{n}}' \cdot \mathbf{e}_1)(\hat{\mathbf{n}}' \cdot \mathbf{e}_2)\,, \tag{20.306}$$

where we have used $\hat{\mathbf{n}} = -\hat{\mathbf{p}}$ and $\hat{\mathbf{n}}' = -\hat{\mathbf{p}}'$, while $\mathbf{e}_1(\hat{\mathbf{n}})$ and $\mathbf{e}_2(\hat{\mathbf{n}})$ are the unit vectors orthogonal to the $\hat{\mathbf{n}}$ direction, which for $\hat{\mathbf{p}} = \hat{\mathbf{z}}$ reduce to $\mathbf{e}_1 = \hat{\mathbf{x}}$ and $\mathbf{e}_2 = \hat{\mathbf{y}}$. This result can be rewritten compactly as[36]

$$\mathcal{P}_{ab}(\hat{\mathbf{n}}) = \frac{\int d\hat{\mathbf{n}}'\, f(\hat{\mathbf{n}}')\, S_{ab}(\hat{\mathbf{n}}, \hat{\mathbf{n}}')}{2 \int d\hat{\mathbf{n}}'\, f(\hat{\mathbf{n}}')\, \left[1 + (\hat{\mathbf{n}} \cdot \hat{\mathbf{n}}')^2\right]}\,. \tag{20.307}$$

where we have defined

$$S_{ab}(\hat{\mathbf{n}}, \hat{\mathbf{n}}') = \left[1 - (\hat{\mathbf{n}} \cdot \hat{\mathbf{n}}')^2\right] g_{ab} - 2(\hat{\mathbf{n}}' \cdot \mathbf{e}_a)(\hat{\mathbf{n}}' \cdot \mathbf{e}_b)\,, \tag{20.308}$$

and $\mathbf{e}_a = \mathbf{e}_a(\hat{\mathbf{n}})$ (with $a = 1, 2$). Observe that in the ratio the overall proportionality constant C has canceled.

To relate the function $f(\hat{\mathbf{n}}')$ to the temperature anisotropies we consider the geometry shown in Fig. 20.26. The observer at the origin detects an ensemble of photons arriving from the direction $\hat{\mathbf{n}}$. We denote by η_1 the conformal time at which a photon of this ensemble underwent its last scattering. Thus, this photon last scattered at the point $\mathbf{x}_1 = (\eta_0 - \eta_1)\hat{\mathbf{n}}$, and then propagated toward the observer, with a propagation direction $\hat{\mathbf{p}} = -\hat{\mathbf{n}}$ (in Figs. 20.24 and 20.25 the situation $\hat{\mathbf{p}} = \hat{\mathbf{z}}$

[36]The computation is easier if we first work in Cartesian coordinates, so that a and b take the values $\{x, y\}$. Then it can immediately be seen that eq. (20.307), with g_{ab} replaced by δ_{ab}, reproduces eq. (20.214). Once written in this tensorial form, the identity is automatically true in polar coordinates, where a and b take the values $\{\theta, \phi\}$, once we replace δ_{ab} by g_{ab}.

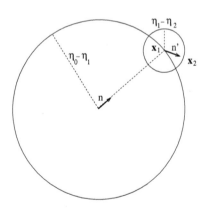

Fig. 20.26 The geometry of the last two scattering events.

was illustrated). The probability density that a photon observed today, at conformal time η_0, last scattered in the time interval $[\eta_1, \eta_1 + d\eta_1]$ is given by $V(\eta_0, \eta_1)d\eta_1$, where $V(\eta_0, \eta)$ is the visibility function introduced in eq. (20.155). In turn, a photon that last scattered in $\mathbf{x}_1$ underwent its previous scattering at an earlier time η_2 with a probability density given by $V(\eta_1, \eta_2)d\eta_2$, and this scattering event took place at the point $\mathbf{x}_2 = \mathbf{x}_1 + (\eta_1 - \eta_2)\hat{\mathbf{n}}'$, as illustrated in Fig. 20.26.

To compute the polarization of the photon flux propagating toward the observer in the $-\hat{\mathbf{n}}$ direction, we assume that a net polarization was generated only in the last scattering, by the anisotropy of the flux impinging on an electron located at $\mathbf{x}_1$.[37] The temperature anisotropies of the photons arriving at time η_1 at the electron at $\mathbf{x}_1$ from the various directions $\hat{\mathbf{n}}'$ can be computed exactly as in eqs. (20.160) and (20.161), except that the observation time is now η_1, and we also keep track of the fact that these are the fluctuations seen by an electron at $\mathbf{x}_1$ (while in eqs. (20.160) and (20.161) the observer was at $\mathbf{x} = 0$), so

$$\frac{\delta T}{T}(\eta_1, \mathbf{x}_1, \hat{\mathbf{n}}') = \int_0^{\eta_1} d\eta_2\, V(\eta_1, \eta_2)\frac{\delta T}{T}(\eta_1, \eta_2, \mathbf{x}_1, \hat{\mathbf{n}}'),\qquad(20.309)$$

where

$$\begin{aligned}\frac{\delta T}{T}(\eta_1, \eta_2, \mathbf{x}_1, \hat{\mathbf{n}}') &= \frac{1}{4}\delta_\gamma[\eta_2, \mathbf{x}_1 + (\eta_1 - \eta_2)\hat{\mathbf{n}}'] + \Psi[\eta_2, \mathbf{x}_1 + (\eta_1 - \eta_2)\hat{\mathbf{n}}'] \\ &\quad - \hat{\mathbf{n}}'\!\cdot\!\mathbf{v}[\eta_2, \mathbf{x}_1 + (\eta_1 - \eta_2)\hat{\mathbf{n}}'] \\ &\quad + \int_{\eta_2}^{\eta_1} d\eta\, \{\Psi'[\eta, \mathbf{x}_1 + (\eta_1 - \eta)\hat{\mathbf{n}}'] - \Phi'[\eta, \mathbf{x}_1 + (\eta_1 - \eta)\hat{\mathbf{n}}']\} \\ &\quad - \frac{1}{2}n'^i n'^j \int_{\eta_2}^{\eta_1} d\eta\, (h_{ij}^{\rm TT})'[\eta, \mathbf{x}_1 + (\eta_1 - \eta)\hat{\mathbf{n}}'],\end{aligned}\qquad(20.310)$$

and $\mathbf{x}_1 = (\eta_0 - \eta_1)\hat{\mathbf{n}}$. The intensity $I(\eta_1, \mathbf{x}_1, \hat{\mathbf{n}}')$ of the radiation arriving at the electron at $\mathbf{x}_1$ at time η_1 from a direction $\hat{\mathbf{n}}'$ can be split into its background value plus perturbations,

$$I(\eta, \mathbf{x}_1, \hat{\mathbf{n}}') = I_0(\eta) + \delta I(\eta, \mathbf{x}_1, \hat{\mathbf{n}}').\qquad(20.311)$$

The background value $I_0(\eta)$ is homogeneous and isotropic, while, since $I \propto T^4$,

$$\frac{\delta I}{I}(\eta_1, \mathbf{x}_1, \hat{\mathbf{n}}') = 4\frac{\delta T}{T}(\eta_1, \mathbf{x}_1, \hat{\mathbf{n}}').\qquad(20.312)$$

As a final step, we must integrate over $V(\eta_0, \eta_1)d\eta_1$, to take into account all possible values of η_1 at which the last scattering took place. Thus, for an electron located at $\mathbf{x}_1$, the function $f(\hat{\mathbf{n}}')$ in eq. (20.307) is given by

$$f(\mathbf{x}_1, \hat{\mathbf{n}}') = \int_0^{\eta_0} d\eta_1\, V(\eta_0, \eta_1)\left[I_0(\eta_1) + \delta I(\eta_1, \mathbf{x}_1, \hat{\mathbf{n}}')\right],\qquad(20.313)$$

where $\delta I(\eta_1, \mathbf{x}_1, \hat{\mathbf{n}}')$ is obtained from eqs. (20.309), (20.310) and (20.312). Since $I_0(\eta_1)$ is isotropic, it does not contribute to the numerator in

[37] This is of course an approximation, useful for obtaining an analytic understanding of the generation of CMB polarization. Actually, the radiation arriving at the electron can in general already be polarized, because of previous Compton scatterings. As we have seen, scattering processes at an epoch much before decoupling are ineffective in generating a net polarization, because the mean free path of the photons is too small, while much later than decoupling there are not enough free electrons around, and multiple scattering events do not take place. However, near the recombinations epochs, a fraction of the radiation that scatters on an electron has already been partially polarized by previous scatterings. The correct way to take into account these effects is with a Boltzmann equation approach, as we shall discuss.

eq. (20.307). Indeed, we have already seen that only the quadrupole part of $f(\hat{\mathbf{n}}')$ gives a vanishing integral. Then, the numerator is of first order in the perturbation, and to first order we can replace the denominator in eq. (20.307) by its zeroth-order term. Then

$$
\begin{aligned}
\mathcal{P}_{ab}(\hat{\mathbf{n}}) &= \frac{\int_0^{\eta_0} d\eta_1\, V(\eta_0,\eta_1) \int d\hat{\mathbf{n}}'\, \delta I(\eta_1,\mathbf{x}_1,\hat{\mathbf{n}}') S_{ab}(\hat{\mathbf{n}},\hat{\mathbf{n}}')}{2\int_0^{\eta_0} d\eta_1\, V(\eta_0,\eta_1) I_0(\eta_1) \int d\hat{\mathbf{n}}'\,[1+(\hat{\mathbf{n}}\cdot\hat{\mathbf{n}}')^2]} \\
&= \frac{3}{8\pi}\frac{\int_0^{\eta_0} d\eta_1\, V(\eta_0,\eta_1) I_0(\eta_1) \int d\hat{\mathbf{n}}'\,[\delta T(\eta_1,\mathbf{x}_1,\hat{\mathbf{n}}')/T(\eta_1)] S_{ab}(\hat{\mathbf{n}},\hat{\mathbf{n}}')}{\int_0^{\eta_0} d\eta_1\, V(\eta_0,\eta_1) I_0(\eta_1)},
\end{aligned}
$$
(20.314)

where we have used

$$
\int d\hat{\mathbf{n}}'\,[1+(\hat{\mathbf{n}}\cdot\hat{\mathbf{n}}')^2] = \frac{16\pi}{3}. \tag{20.315}
$$

Since $V(\eta)$ is sharply peaked around the value $\eta = \eta_{\mathrm{dec}}$ while $I_0(\eta)$ is a smooth function, we can replace $I_0(\eta) \simeq I_0(\eta_{\mathrm{dec}})$ and extract it from the integrals, both in the numerator and in the denominator. Then, recalling that $\int_0^{\eta_0} d\eta\, V(\eta_0,\eta) = 1$, we can approximate

$$
\boxed{\;\mathcal{P}_{ab}(\hat{\mathbf{n}}) \simeq \frac{3}{8\pi}\int_0^{\eta_0} d\eta_1\, V(\eta_0,\eta_1) \int d\hat{\mathbf{n}}'\, \frac{\delta T(\eta_1,\mathbf{x}_1,\hat{\mathbf{n}}')}{T(\eta_1)} S_{ab}(\hat{\mathbf{n}},\hat{\mathbf{n}}').\;}
$$
(20.316)

Since $\mathbf{x}_1 = (\eta_0 - \eta_1)\hat{\mathbf{n}}$, the dependence on $\mathbf{x}_1$ in $\delta T/T$ on the right-hand side contributes to the dependence on $\hat{\mathbf{n}}$ of $\mathcal{P}_{ab}(\hat{\mathbf{n}})$ [while η_1 is integrated over, and we do not write explicitly the dependence of $\mathcal{P}_{ab}(\hat{\mathbf{n}})$ on η_0].

Contribution of scalar and tensor anisotropies

The polarization tensor is now determined by inserting the temperature anisotropies obtained from eqs. (20.309) and (20.310) into eq. (20.316). In particular, we can compute the separate contributions from scalar perturbations and from tensor perturbations. From eq. (20.316) we can now understand a very important property of the CMB polarization. Namely, *scalar perturbations only generate the polarization E mode, while tensor perturbations generate both E and B modes.*

We consider first the SW term in eq. (20.310). Expanding in Fourier modes as in eq. (20.86), we get

$$
\begin{aligned}
\left(\frac{\delta T}{T}\right)_{\mathrm{SW}}(\eta_1,\mathbf{x}_1,\hat{\mathbf{n}}') &= \int_0^{\eta_1} d\eta_2\, V(\eta_1,\eta_2)\left(\frac{\delta_\gamma}{4}+\Psi\right)[\eta_2,\mathbf{x}_1+(\eta_1-\eta_2)\hat{\mathbf{n}}'] \\
&= \int_0^{\eta_1} d\eta_2\, V(\eta_1,\eta_2)\int \frac{\sqrt{V}d^3k}{(2\pi)^3}\left(\frac{1}{4}\tilde{\delta}_\gamma+\tilde{\Psi}\right)(\eta_2,\mathbf{k})\, e^{i[\mathbf{k}\cdot\mathbf{x}_1+(\eta_1-\eta_2)\mathbf{k}\cdot\hat{\mathbf{n}}']}.
\end{aligned}
$$
(20.317)

The crucial point is that this expression depends on $\hat{\mathbf{n}}'$ only through $\mathbf{k}\cdot\hat{\mathbf{n}}'$. The same is true for the Doppler term in the scalar contribution since, for scalar perturbations, $v_i = \partial_i v$ and therefore in the Fourier transform the term $\hat{\mathbf{n}}' \cdot \mathbf{v}$ becomes $i\mathbf{k} \cdot \hat{\mathbf{n}}'$. Also in the ISW term, after Fourier transformation, $\hat{\mathbf{n}}'$ enters only through $\mathbf{k} \cdot \hat{\mathbf{n}}'$, as we see from eq. (20.310). This is an unavoidable consequence of the fact that, for scalar perturbations, a Fourier mode is characterized only by its momentum $\mathbf{k}$, and has no polarization vector or polarization tensor in its transverse plane. Thus, the contribution of scalar perturbations with Fourier mode $\mathbf{k}$ to the angular integral in eq. (20.316) is of the form

$$\int d\hat{\mathbf{n}}' \, F(\hat{\mathbf{k}} \cdot \hat{\mathbf{n}}') S_{ab}(\hat{\mathbf{n}}, \hat{\mathbf{n}}') \,, \tag{20.318}$$

for some function F. To evaluate this integral we use polar coordinates with $\hat{\mathbf{k}}$ as polar axis, so $\hat{\mathbf{k}} = (0, 0, 1)$. In this frame we write

$$\hat{\mathbf{n}} = (\sin\theta\cos\phi, \sin\theta\sin\phi, \cos\theta) \,,$$
$$\hat{\mathbf{n}}' = (\sin\theta'\cos\phi', \sin\theta'\sin\phi', \cos\theta') \,, \tag{20.319}$$

while the polarization vectors $\mathbf{e}_\theta$ and $\mathbf{e}_\phi$ that enter in eq. (20.308), and which by definition are orthogonal to $\hat{\mathbf{n}}$, can be chosen as

$$\mathbf{e}_\theta = (\cos\theta\sin\phi, \cos\theta\cos\phi, -\sin\theta) \,, \tag{20.320}$$
$$\mathbf{e}_\phi = (\cos\phi, -\sin\phi, 0) \,, \tag{20.321}$$

which correspond to unit vectors in the $\hat{\boldsymbol{\theta}}$ and $\hat{\boldsymbol{\phi}}$ directions, respectively. Then $\hat{\mathbf{k}} \cdot \hat{\mathbf{n}}' = \cos\theta'$, while

$$2(\hat{\mathbf{n}}' \cdot \mathbf{e}_\theta)(\hat{\mathbf{n}}' \cdot \mathbf{e}_\phi) = \cos\theta\sin^2\theta'\sin(2\phi + 2\phi') - \sin\theta\sin 2\theta'\cos(2\phi + 2\phi') \,. \tag{20.322}$$

From this we see that

$$\int d\hat{\mathbf{n}}' \, F(\hat{\mathbf{k}} \cdot \hat{\mathbf{n}}') S_{\theta\phi}(\hat{\mathbf{n}}, \hat{\mathbf{n}}') = -\int_{-1}^{1} d\cos\theta \, F(\cos\theta) \int_0^{2\pi} d\phi \, 2(\hat{\mathbf{n}}' \cdot \mathbf{e}_\theta)(\hat{\mathbf{n}}' \cdot \mathbf{e}_\phi)$$
$$= 0 \,. \tag{20.323}$$

Therefore $\mathcal{P}_{\theta\phi}$ vanishes for scalar perturbations. Similarly we find that $(\hat{\mathbf{n}}' \cdot \mathbf{e}_\theta)^2$ and $(\hat{\mathbf{n}}' \cdot \mathbf{e}_\phi)^2$ depend on ϕ only through the combination $(\phi + \phi')$, so after integrating over ϕ' the dependence of ϕ disappears. Therefore, for scalar perturbations, $\mathcal{P}_{\theta\theta}$ and $\mathcal{P}_{\phi\phi}$ are functions of θ only. From this, using the fact that the only non-vanishing Christoffel symbols on the sphere are those given in eq. (20.269), we find that $\epsilon_a{}^c \nabla_b \nabla_c \mathcal{P}^{ab} = 0$. Then eq. (20.257) gives $\mathcal{P}_B = 0$. Scalar perturbations do not generate the B polarization.

The situation is different for tensor modes. In this case $\hat{\mathbf{n}}'$ enters also through the combination $n'^i n'^j (h_{ij}^{\mathrm{TT}})'$ in eq. (20.310), which reflects the fact that tensor modes have a preferential direction in the transverse plane, given by their polarization tensor. Therefore the preceding argument no longer goes through, and in particular the tensor contribution is not described just by a function $F(\hat{\mathbf{k}} \cdot \hat{\mathbf{n}}') = F(\cos\theta')$, which in the scalar case could be factored out from the integral over ϕ'. Now this function has a generic dependence of ϕ', so $\mathcal{P}_{\theta\phi}$ in general does not vanish and $\mathcal{P}_{\theta\theta}$ and $\mathcal{P}_{\phi\phi}$ are functions of both θ' and ϕ'. Thus, for tensor perturbations both the E and B modes are non-vanishing.

The analytic expression that we have obtained also allows us to understand that the finite thickness of the LSS is necessary for obtaining a non-vanishing polarization. Consider first the SW term. Inserting eq. (20.317) into eq. (20.316) and recalling that $\mathbf{x}_1 = (\eta_0 - \eta_1)\hat{\mathbf{n}}$, we get

$$
\mathcal{P}_{ab,\mathrm{SW}}(\hat{\mathbf{n}}) \simeq \frac{3}{8\pi} \int_0^{\eta_0} d\eta_1\, V(\eta_0, \eta_1) \int_0^{\eta_1} d\eta_2\, V(\eta_1, \eta_2) \int \frac{\sqrt{V} d^3 k}{(2\pi)^3}
$$
$$
\times e^{i(\eta_0 - \eta_1)\mathbf{k}\cdot\hat{\mathbf{n}}} \left(\frac{1}{4}\tilde{\delta}_\gamma + \tilde{\Psi} \right)(\eta_2, \mathbf{k}) \int d\hat{\mathbf{n}}'\, S_{ab}(\hat{\mathbf{n}}, \hat{\mathbf{n}}')\, e^{i(\eta_1 - \eta_2)\mathbf{k}\cdot\hat{\mathbf{n}}'}. \tag{20.324}
$$

Neglecting the thickness of the LSS, as we did in a first approximation when we studied the temperature anisotropies, amounts to approximating $V(\eta_0, \eta_1)$ by $\delta(\eta_1 - \eta_{\mathrm{dec}})$ and then $V(\eta_1, \eta_2)$ by $\delta(\eta_2 - \eta_{\mathrm{dec}})$. The factor $e^{i(\eta_1 - \eta_2)\mathbf{k}\cdot\hat{\mathbf{n}}'}$ in eq. (20.324) then becomes equal to 1, and the integral over $d\hat{\mathbf{n}}'$ vanishes. As we have already seen, $S_{ab}(\hat{\mathbf{n}}, \hat{\mathbf{n}}')$ has a quadrupole structure with respect to $\hat{\mathbf{n}}'$, and to obtain a non-vanishing integral we need to expand the exponential $e^{i(\eta_1 - \eta_2)\mathbf{k}\cdot\hat{\mathbf{n}}'}$ to second order, so as to bring down two factors of $\hat{\mathbf{n}}'$. Thus, the SW contribution to the polarization is proportional to $(k\Delta\eta_{\mathrm{dec}})^2$, where $\Delta\eta_{\mathrm{dec}}$ is the thickness of the LSS.

From eq. (20.310) the Doppler contribution is given by

$$
\mathcal{P}_{ab,\mathrm{Dop}}(\hat{\mathbf{n}}) \simeq - \int_0^{\eta_0} d\eta_1\, V(\eta_0, \eta_1) \int_0^{\eta_1} d\eta_2\, V(\eta_1, \eta_2) \int \frac{\sqrt{V} d^3 k}{(2\pi)^3}
$$
$$
\times e^{i(\eta_0 - \eta_1)\mathbf{k}\cdot\hat{\mathbf{n}}} \int d\hat{\mathbf{n}}'\, S_{ab}(\hat{\mathbf{n}}, \hat{\mathbf{n}}')\, \hat{\mathbf{n}}'\cdot\tilde{\mathbf{v}}(\eta_2, \mathbf{k})\, e^{i(\eta_1 - \eta_2)\mathbf{k}\cdot\hat{\mathbf{n}}'}. \tag{20.325}
$$

Observe that the Doppler term $-\hat{\mathbf{n}}'\cdot\tilde{\mathbf{v}}$ already carries one factor $\hat{\mathbf{n}}'$, and to have a non-vanishing integral over $d\hat{\mathbf{n}}'$ we only need to expand the exponential to first order. Thus, the Doppler term is proportional to $k\Delta\eta_{\mathrm{dec}}$. Recall from the discussion on page 529 that the Fourier modes with momentum k contribute mostly to multipoles with $l \sim k\eta_0$. Thus, the condition $k\Delta\eta_{\mathrm{dec}} \simeq 1$ corresponds to

$$
l \simeq \frac{\eta_0}{\eta_{\mathrm{dec}}} \frac{\eta_{\mathrm{dec}}}{\Delta\eta_{\mathrm{dec}}} \simeq 1200\,, \tag{20.326}
$$

where we have used the numerical values given in eqs. (17.174) and (20.159). For such high multipoles the temperature anisotropies that source the polarization anisotropies are already strongly suppressed, as we see from Fig. 20.1.[38] We are therefore mostly interested in multipoles such that $k\Delta\eta_{\mathrm{dec}} \ll 1$. In this regime, the dominant contribution of the scalar perturbations to the polarization anisotropies therefore comes from the Doppler effect, which is suppressed only by $k\Delta\eta_{\mathrm{dec}}$, while the SW effect is suppressed by $(k\Delta\eta_{\mathrm{dec}})^2$. The contribution to the polarization from the scalar ISW is even smaller. Indeed, we see from eq. (20.310) that the scalar ISW carries no explicit factor of $\hat{\mathbf{n}}'$, just like the scalar

[38]The physical mechanism that suppresses them is also related to the finite width of the LSS, and is called Silk damping. It is due to the fact that around decoupling the mean free path of the photons starts to increase, while the photons are still partly coupled to the electrons, so photon diffusion smooths out anisotropies on small scales.

SW, so again we need to expand $e^{i(\eta_1-\eta_2)\mathbf{k}\cdot\hat{\mathbf{n}}'}$ to second order, which gives the same $(k\Delta\eta_{\mathrm{dec}})^2$ suppression as in the SW case. Furthermore, we have an integral over η from η_1 to η_2. This will give two distinct contribution, one near decoupling and one near reionization; see Note 34 on page 560. In each case, this gives a further suppression proportional to the width of the LSS and to the duration of the reionization epoch, respectively. In particular, the contribution near decoupling will be suppressed by an extra factor $\Delta\eta_{\mathrm{dec}}$. On dimensional grounds, the integral over $d\eta$ in the ISW is compensated by the partial derivative with respect to conformal time in Ψ', which, near decoupling, is of order $(1/\eta_{\mathrm{dec}})\Psi$. Thus, the integral appearing in the ISW effect carries an extra suppression $\Delta\eta_{\mathrm{dec}}/\eta_{\mathrm{dec}}$, and overall the contribution of the scalar ISW to the polarization anisotropies is suppressed by $(\Delta\eta_{\mathrm{dec}}/\eta_{\mathrm{dec}})(k\Delta\eta_{\mathrm{dec}})^2$.

For the contribution to polarization from the tensor ISW effect, given by the last line of eq. (20.310), the situation is different, since there are two explicit factors $\hat{\mathbf{n}}'$ in the combination $n'^i n'^j (h_{ij}^{\mathrm{TT}})'$. Therefore, in this case we can replace $e^{i(\eta_1-\eta_2)\mathbf{k}\cdot\hat{\mathbf{n}}'}$ by 1, and we do not pick up any suppression proportional to $k\Delta\eta_{\mathrm{dec}}$. However, just as for the scalar ISW, the integration over $d\eta$ from η_1 to η_2 brings a suppression factor $\Delta\eta_{\mathrm{dec}}/\eta_{\mathrm{dec}}$.

In summary, all contributions to the polarization vanish in the limit $\Delta\eta_{\mathrm{dec}} \to 0$, in agreement with the physical argument presented on page 560. For scalar perturbations, the dominant contribution comes from the Doppler effect and is suppresed by a factor of order $k\Delta\eta_{\mathrm{dec}}$ while the contribution of tensor perturbations is suppressed by a factor of order $\Delta\eta_{\mathrm{dec}}/\eta_{\mathrm{dec}}$. Most importantly, tensor perturbations contribute to both E and B modes, while scalar perturbations contribute only to E modes. Thus, looking at the B modes, one can hope to see the effect of GWs, disentangled from the effect of scalar perturbations. We will see, however, that a B mode can also be generated by lensing of an E mode, as well as by scattering on polarized dust.

Polarization correlators

Explicit expressions for the polarization multipoles can now be obtained by inserting into eq. (20.316) the temperature anisotropies given by eq. (20.310) and carrying out the integrals numerically, or with some analytic approximation.[39] In practice, however, to obtain precise quantitative results to be compared with the data it is necessary to make use of Boltzmann codes for the polarization multipoles (see the Further Reading), which take into account the exact form of $\delta T/T$, as well as effects such as the multiple scatterings mentioned in Note 37. In eq. (20.263) we have introduced the multipoles a_{lm}^E and a_{lm}^B, while we now denote the temperature multipoles by a_{lm}^T. Just as in eq. (20.5), with these quantities we can form the diagonal correlators

$$\langle a_{lm}^T a_{l'm'}^{T,*}\rangle = C_l^{TT}\,\delta_{ll'}\delta_{mm'}\,, \qquad (20.327)$$

$$\langle a_{lm}^E a_{l'm'}^{E,*}\rangle = C_l^{EE}\,\delta_{ll'}\delta_{mm'}\,, \qquad (20.328)$$

[39] Several useful analytic approximations are discussed in Section 10.3 of Gorbunov and Rubakov (2011).

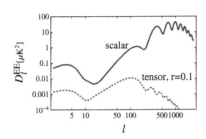

Fig. 20.27 The contribution to $\mathcal{D}_l^{EE}$ from scalar perturbations (solid line), compared to the contribution of tensor perturbations with $r = 0.1$ and $n_t = 0$ (dotted). Figs. 20.27–20.29 obtained with the CLASS Boltzmann code, data courtesy of S. Foffa.

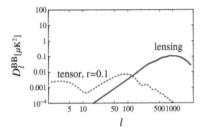

Fig. 20.28 The contribution to $\mathcal{D}_l^{BB}$ from the lensing of scalar perturbations (solid line), compared to the contribution of tensor perturbations with $r = 0.1$ and $n_t = 0$ (dotted).

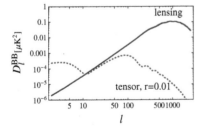

Fig. 20.29 As in Fig. 20.28 for $r = 0.01$ and $n_t = 0$. Note the different vertical scale compared with Fig. 20.28.

$$\langle a_{lm}^B a_{l'm'}^{B,*} \rangle = C_l^{BB} \delta_{ll'} \delta_{mm'} . \tag{20.329}$$

Furthermore, the polarization multipoles are correlated with the temperature multipoles, since they both originated from the temperature perturbations. We can therefore form all possible off-diagonal correlators, writing more generally

$$\langle a_{lm}^X a_{l'm'}^{Y,*} \rangle = C_l^{XY} \delta_{ll'} \delta_{mm'} , \tag{20.330}$$

where X and Y take the values $\{T, E, B\}$, and we define the corresponding $\mathcal{D}_l^{XY}$ as in eq. (20.10). The off-diagonal correlators involving the B polarization, C^{TB} and C^{EB}, are, however, constructed with a scalar and a pseudoscalar, and therefore vanish if the mechanism that generates the primordial perturbation conserves parity. The non-vanishing ones are then the TT, TE, EE and BB correlators.

In Fig. 20.27 we show $\mathcal{D}_l^{EE}$, plotting both the contribution from scalar perturbations and that from tensor perturbations, with a tensor-to-scalar ratio $r = 0.1$ (and a tensor tilt $n_t = 0$). For scalar perturbations the generation of polarization at the LSS gives a contribution that grows with l, reaching a maximum around $l = O(1000)$. The "bump" seen at $l \lesssim 5$ is called the "reionization bump". As we mentioned in Note 34, this is generated at $z = 8 - 10$, after the first stars have formed. These first-generation stars are very massive and emit strong UV radiation, which partially reionizes the Universe, producing free electrons. Thomson scattering of the CMB photons on these free electrons then produces an extra contribution to the polarization. The difference between the typical value of l for the polarization generated at the LSS and that at reionization reflects the difference between the corresponding values of conformal time. This plot shows that, even for a value $r = 0.1$ (which, as we will discuss later, is of the order of the largest value allowed by the observations) in the EE correlator, the GW contribution is masked by the scalar contribution, at all multipoles.

The BB correlator is therefore in principle more promising for detecting a tensor contribution. However, even in the BB correlator the scalar contribution is actually non-vanishing. This is due to gravitational lensing of the E mode from the matter distribution in which the CMB photons propagate after last scattering, and which can partially convert an E mode into a B mode. The effect of lensing on the CMB correlators can be computed in terms of the statistical properties of the lensing potential, which is determined by the power spectrum of Ψ; see the Further Reading. In Fig. 20.28 we plot the contribution to $\mathcal{D}_l^{BB}$ from the lensing of scalar perturbations, and we compare it with the contribution of tensor perturbations with $r = 0.1$ and $n_t = 0$. We see that at $l \gtrsim O(100)$ lensing dominates, but at smaller l the GW contribution is more important. Note also that we have used the same vertical scale in Figs. 20.27 and 20.28. This allows us to appreciate that the tensor contributions to the EE and BB correlators are quite similar. In Fig. 20.29 we show the result for $r = 0.01$ (but now with a different vertical scale).

From these figures we see that, in the BB correlator, primordial GWs at a level $r = 0.1$ dominate over lensing at $l \lesssim O(100)$, while for $r = 0.01$ there is still a useful region at $l \lesssim O(10)$. Furthermore, if the lensing potential is known from other sources, such as large-scale structure observations, one can apply "delensing" techniques to subtract, to some extent, the contribution of the lensed E mode, and improve the sensitivity to the B mode.

20.4.5   Experimental situation

The first evidence for a non-zero E polarization was reported in 2002 by the Degree Angular Scale Interferometer (DASI), a telescope installed at the South Pole. The TE spectrum at large angles, due to reionization, was seen by the Wilkinson Microwave Anisotropy Probe (WMAP) in the release of its first year of data in 2003, while the second, third and fourth peaks of the EE spectrum were detected by the Cosmic Background Imager (CBI) in 2004 (the first peak, as we see from Fig. 20.30, is weaker and was harder to detect). Accurate spectra were then obtained by WMAP in its three-year data release in 2006. Several other experiments have provided measurements of the E polarization, including CAPMAP, BOOMERANG03, MAXIPOL, QUAD, BICEP1 and QUIET. Figures 20.30 and 20.31 show the data from the 2015 release of the *Planck* collaboration. The solid lines in the upper panels represents the prediction from ΛCDM, with its best-fit parameters determined using the TT data only. Therefore, these curves are pure predictions for the EE and TE correlators. The bottom panels show the residuals with respect to this best-fit model. Note that these figures are on a linear–linear scale, while the theoretical prediction shown in Fig. 20.27 was on a log–log scale.

The B mode at subdegree scales, i.e. at values of l where the signal is due to lensing of the E mode, was detected in 2013 by the South Pole Telescope (SPTpol) and confirmed by the POLARBEAR experiment. In 2014 the BICEP2 (Background Imaging of Cosmic Extragalactic Polarization) experiment announced the detection of B-mode polarization over the range $30 < l < 150$. At $l \gtrsim 100$ the data were consistent with the prediction from the lensing of the E mode. The data at $l < 100$, however, clearly received an additional contribution from some extra component. Modeling the signal uniquely with lensing plus tensor modes, and no foreground subtraction, the best value for r was found to be $r = 0.20^{+0.07}_{-0.05}$, with $r = 0$ excluded at the 7.0σ level. The BICEP2 collaboration used dust models that predicted that polarized dust emission was about a factor $5 - 10$ lower than the observed signal. Under-estimating the large theoretical uncertainty in these models, the result was initially presented as the detection of primordial GWs, and caused great excitement. This claim was somewhat in tension with the upper bound on r previously found by the *Planck* collaboration from temperature anisotropies,

$$r_{0.002} < 0.11 \quad (95\% \text{ c.l.}), \quad (20.331)$$

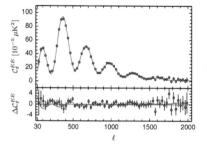

Fig. 20.30 The C_l^{EE} multipoles of the EE correlator and, in the bottom panel, the residual with respect to the best-fit ΛCDM prediction. From [Planck Collaboration], Adam, *et al.* (2016b) "Planck 2015 results. I. Overview of products and scientific results". Reproduced with permission @ ESO.

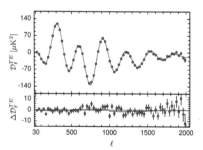

Fig. 20.31 The D_l^{TE} multipoles of the TE correlator and, in the bottom panel, the residual with respect to the best-fit ΛCDM prediction. From [Planck Collaboration] Adam, *et al.* (2016b) "Planck 2015 results. I. Overview of products and scientific results". Reproduced with permission @ ESO.

where the subscript is a reminder that this value has been obtained setting $k_* = 0.002\,\mathrm{Mpc}^{-1}$ as the pivot scale in eq. (19.281). The crucial issue is, however, the contribution of galactic foregrounds. These depend on the frequency of observation. The typical window for CMB observations is in the range 100–150 GHz, where the black-body spectrum peaks. At lower frequencies, say $\lesssim 100$ GHz, the dominant foreground for the B-mode measurement comes from polarized Galactic synchrotron emission. At higher frequency the dominant contribution is from emission due to polarized Galactic dust, in which asymmetric dust grains of size about 0.1 µm align with the Galactic magnetic field, producing polarized emission.

For estimating the polarized synchrotron emission from the Galaxy one can use low-frequency polarized maps from WMAP, which show that this effect is small compared with the B polarization measured by BICEP2. In contrast, subsequent studies of the B modes generated by polarized dust indicated that it fully accounts for the BICEP2 measurement. Indeed, shortly after the BICEP2 announcement the *Planck* experiment provided full-sky maps of polarized Galactic dust. At frequencies $\gtrsim 350$ GHz the B-mode signal from Galactic dust grains becomes dominant, even at high Galactic latitude. One can therefore measure it directly at these high frequencies. To estimate its intensity in the 100–150 GHz range relevant for CMB observations involves some modeling and extrapolation, which introduce uncertainties. For good modeling of foregrounds, it is very important to perform measurements at different frequencies, and *Planck* observed the full polarized sky at seven different frequencies from 30 to 353 GHz. *Planck* data also included the direct dust measurement in the small patch of sky observed by BICEP2, which had been chosen by BICEP2 because it was believed to be particularly clean from dust (assuming that a small level of dust contamination in the temperature anisotropies corresponds to a small level of contamination in the polarization anisotropies). The maps provided by *Planck* showed that this assumption, and especially the dust modeling used by BICEP2, were not valid. The polarized dust contribution in the BICEP2 field of view is significantly larger than that predicted by the dust models that were used by BICEP2, and is of the same order as the signal from tensor modes with $r = 0.2$, so it can fully account for the observed signal, without invoking GWs (as had already been warned about in some theoretical works; see the Further Reading).

After that, a joint analysis of the BICEP2, Keck Array (an array of BICEP2-like receivers, also located at the South Pole, that observed the same field of view as BICEP2 in the seasons 2012 and 2013) and *Planck* data was performed. This study, which correlated the *Planck* measurement of dust at 353 GHz with the BICEP2/Keck Array maps at 150 GHz, showed that the two signals are strongly correlated. Modeling the signal using ΛCDM plus lensing, polarized dust and primordial GWs, the result was that there is no statistically significant evidence for GWs. The upper limit on r from this analysis is, setting $n_T = 0$,

$$r_{0.05} < 0.12 \quad (95\% \text{ c.l.})\,. \tag{20.332}$$

Transforming to the pivot scale $k_* = 0.002 \, \mathrm{Mpc}^{-1}$ by using eq. (19.177) with n_s from eq. (19.184) gives

$$r_{0.002} < 0.09 \quad (95\% \text{ c.l.}), \tag{20.333}$$

which is of the same order as the limit obtained from the *Planck* 2013 data release using temperature anisotropies; see eq. (20.331). It should also be observed that the precise value of the limit on r depends on the cosmological model used. In particular the running of the spectral scalar index n_s or the addition of extra relativistic degrees of freedom introduces degeneracies in the data analysis, which allows the bound on r to become somewhat less tight.

While at present the limits on r from temperature anisotropies and from B polarization are quite comparable, in the future significant improvements are expected to come essentially from observations of the B polarization, which can potentially provide much more stringent limits. Eventually, the level at which we can detect a contribution of primordial GWs from measurements of B mode polarization will depend on the ability to characterize and subtract the polarized foregrounds. This requires complex techniques of component separation, based on the multi-frequency capability of the instruments. Using B mode polarization, future satellite experiments could detect values of r as low as $r = O(10^{-3})$.

Further reading

- The CMB was accidentally discovered by Penzias and Wilson in 1964–65, from accurate residual noise measurements with the Bell Labs twenty-foot horn reflector at Crawford Hill, New Jersey. For this discovery they were awarded the Nobel Prize in 1978. The Nobel Lecture by Penzias (1978) traces the history of the experimental discovery as well as of the theoretical ideas, mostly motivated by the aim of understanding nucleosynthesis, that led to the idea of a hot early Universe and the prediction of a relic radiation. The first paper that recognized explicitly the black-body nature of the radiation, and that it was potentially detectable at microwave frequencies, was Doroshkevich and Novikov (1964). The significance of the Penzias and Wilson discovery as a relic radiation from the big bang was immediately recognized by Dicke, Peebles, Roll and Wilkinson (1965), who at that time were preparing a search for relic microwave radiation from the big bang. The paper with the experimental dis-

covery Penzias and Wilson (1965) and the Dicke *et al.* paper with the theoretical interpretation were submitted to and published jointly in *Astrophysical Journal Letters*. The history of the discovery and subsequent studies of the CMB is described in the collection of essays edited by Peebles, Page and Partridge (2009).

- The temperature anisotropies of the CMB were discovered in 1992 by the FIRAS instrument on the COBE satellite; see Smoot *et al.* (1992), Bennett *et al.* (1994) and Bennett *et al.* (1996). The angular resolution of about 7° allowed determination of the multipoles with $l < 20$. For this discovery Smoot and Mather were awarded the Nobel Prize in 2006. Several other subsequent experiments with improved angular resolution measured the spectrum of the CMB multipoles. The first acoustic peak was clearly detected by the balloon-borne experiments Boomerang (de Bernardis *et al.*, 2000) and Maxima (Hanany *et al.*, 2000). The second peak was definitively detected by the satellite experiment WMAP. In its nine years of data taking WMAP produced high-precision maps of the CMB anisotropies, which have opened the era of high-precision cosmology; see for example Spergel *et al.* (2003, 2007), Komatsu *et al.* (2011) and Hinshaw *et al.* (2013). Currently the most accurate CMB data are those obtained by the *Planck* mission, in its 2015 data release. These data, in combination with other cosmological datasets (such as type Ia supernovae, baryon acoustic oscillations and cosmological structure formation), give the presently most accurate estimates of the cosmological parameters, as well as a wealth of other cosmological information. The results from the intermediate 2013 *Planck* data release are presented in a series of over 40 papers from the collaboration, while the *Planck* 2015 results are presented in a set of 28 papers. Among them, an overview of the 2015 results is presented in [Planck Collaboration], Adam *et al.* (2016b), the estimation of cosmological parameters is presented in [Planck Collaboration], Ade *et al.* (2016a), while constraints on inflation are presented in [Planck Collaboration], Ade *et al.* (2016c).

- The SW and ISW effects were first computed by Sachs and Wolfe (1967), while the damping due to photon diffusion, or Silk damping, was discussed in Silk (1968). The modern approach to the computation of the CMB anisotropies also evolved from the work of Peebles and Yu (1970), Doroshkevich, Zel'dovich and Sunyaev (1978), Wilson and Silk (1981), Bond and Efstathiou (1984, 1987), Hu and Sugiyama (1995, 1996), and many others. For a history of theoretical and experimental advances in the early CMB studies see White, Scott and Silk (1994). Excellent textbooks discussing the CMB are Dodelson (2003), Durrer (2008), and Gorbunov and Rubakov (2011).

- Modification of the late ISW effect by non-linear clustering on small scales (for example by quasars) was first discussed in Rees and Sciama (1968), and is called the Rees–Sciama effect. More generally, the late ISW effect is affected by the non-linear mode evolution for all scales $k \gtrsim k_{\rm nl}$ (which is also often generically called the Rees–Sciama effect). The late ISW contribution is a probe of dark energy, and can be detected separately by performing correlations between the galaxy density field and the CMB temperature perturbations, as suggested by Crittenden and Turok (1996). It was indeed first detected by correlating WMAP data with galaxy

surveys in Boughn and Crittenden (2004) and in several subsequent works; see for example Granett, Neyrinck and Szapudi (2008) and Nadathur and Crittenden (2016). The direct effect of non-linear clustering on the TT correlation, through the late ISW effect, is below the error due to cosmic variance; see Cooray (2002). Weak lensing of the CMB is reviewed in Lewis and Challinor (2006).

- Accurate numerical predictions can be obtained from Boltzmann codes. Early codes integrated numerically the full set of $O(2000)$ equations, one for each multipole, and were very slow, requiring about 2 CPU hours for a single spectrum. This is not very practical, since the runs must be repeated on a whole grid in the multi-dimensional space of cosmological parameters. The breakthrough came with the use of a line-of-sight integration method, with the publicly available code CMBfast, by Seljak and Zaldarriaga (1996), which required about 5 minutes for a single spectrum. More recent and even faster public codes are CAMB by Lewis, Challinor and Lasenby (2000) (see also http://camb.info), and CLASS, described in Lesgourgues (2011) and Blas, Lesgourgues and Tram (2011).

- The effect of relic GWs on the CMB anisotropies was studied in Rubakov, Sazhin and Veryaskin (1982), Fabbri and Pollock (1983), Abbott and Wise (1984), Starobinskii (1985), and Allen and Koranda (1994). The problem of disentangling the scalar from the tensor contributions to the CMB was studied with Boltzmann codes in Crittenden, Bond, Davis, Efstathiou and Steinhardt (1993).

- The possibility of detecting primordial GWs from their effect on the CMB polarization was proposed by Bond and Efstathiou (1984) and Polnarev (1985), and further studied by Crittenden, Davis and Steinhardt (1993). The decomposition of the CMB polarization into E and B modes was proposed in Seljak (1997), Zaldarriaga and Seljak (1997) and Kamionkowski, Kosowsky and Stebbins (1997). See also Stebbins (1996) for a related analysis in the context of weak lensing. For reviews on CMB polarization see Kosowsky (1999), Zaldarriaga (2003) and Baumann *et al.* [CMBPol Study Team Collaboration] (2009).

- The first evidence for the E polarization was reported in Kovac *et al.* [DASI experiment] (2002). The lensed B-mode was detected by Hanson *et al.* [SPTpol Collaboration] (2013).

- The BICEP2 result for the B polarization was presented in [BICEP2 Collaboration], Ade *et al.*

(2014), where it was claimed that it was due to primordial GWs. Flauger, Hill and Spergel (2014) pointed out that the uncertainty in the polarization power spectrum is such that polarized dust plus lensing can fully account for the observed signal, while Mortonson and Seljak (2014) reached the same conclusion by fitting the data with lensing, tensor modes and polarized dust, assuming no prior on the amplitude of the dust polarization, and using only its known dependence on l. In [Planck Collaboration], Adam *et al.* (2016a), the *Planck* Collaboration provided full-sky maps of dust, at 353 GHz, that, extrapolated to the BICEP2 frequency, indicated again that the BICEP2 signal is entirely consistent with that from polarized dust. The joint BICEP2/Keck Array/*Planck* analysis is presented in [BICEP2 and Planck Collaborations], Ade *et al.* (2015), and concluded that there is no statistical evidence in favor of a detection of primordial GWs.

21 Inflation and primordial perturbations

In this chapter we study the generation of a primordial spectrum of scalar and tensor perturbations during inflation. We will begin in Section 21.1 with an introduction to inflationary cosmology, limiting ourselves to single-field slow-roll inflation. We will then study in Section 21.2 the quantization of a scalar field in curved space. These results will then allow us, in Section 21.3, to compute the spectrum of scalar and tensor perturbations generated in single-field slow-roll inflationary models. We will then compare these results with CMB observations, and we will discuss their implications for the stochastic background of GWs at the present epoch.

21.1 Inflationary cosmology

Inflation was initially conceived as a solution to shortcomings of standard cosmology, such as the flatness and horizon problems. On top of this, its great virtue is that it provides a testable theory of initial conditions for the cosmological evolution studied in Chapter 19. In this section we begin by examining the main features of typical inflationary models. Inflation is by now a topic on which there are many excellent dedicated reviews and textbooks. We will therefore limit ourselves to summarizing the main aspects that will be relevant in the following for our discussion, referring the reader to the Further Reading for more details.

21.1.1 The flatness problem

In Chapter 19 we limited ourselves to a spatially flat FRW metric (17.2). More generally, the background metric consistent with spatial translation and rotations is given by eq. (4.141),

$$ds^2 = -dt^2 + a^2(t) \left[\frac{dr^2}{1 - Kr^2} + r^2(d\theta^2 + \sin^2\theta \, d\phi^2) \right], \qquad (21.1)$$

where K is a constant with dimensions (length)$^{-2}$.[1] Flat space corresponds to $K = 0$, while positive K corresponds to a (spatially) closed model and negative K to an open model. The spatial radius of curvature is $R_{\text{curv}} = a/|K|^{1/2}$. In the presence of curvature the Friedmann

[1]In contrast to eq. (4.141), we now denote curvature by a capital K, reserving the notation k for the momentum of Fourier modes.

Gravitational Waves, Volume 2: Astrophysics and Cosmology. Michele Maggiore.
© Michele Maggiore 2018. Published in 2018 by Oxford University Press.
DOI 10.1093/oso/9780198570899.001.0001

equation (17.48) becomes

$$H^2 + \frac{K}{a^2} = \frac{8\pi G}{3}\rho \,. \tag{21.2}$$

We can formally rewrite this as

$$H^2 = \frac{8\pi G}{3}(\rho + \rho_K)\,, \tag{21.3}$$

where

$$\rho_K(t) = -\frac{3}{8\pi G}\frac{K}{a^2(t)}\,. \tag{21.4}$$

The origin of the curvature problem is that ρ_K decreases with the scale factor only as $1/a^2$, while $\rho_M \propto 1/a^3$ and $\rho_R \propto 1/a^4$. Therefore, unless in the past it was extremely small, today curvature should dominate over radiation and matter, which observationally is not the case. To be more quantitative, we define $\Omega_K(t) = \rho_{\rm curv}(t)/\rho_c(t)$, where $\rho_c(t)$ is the critical density at time t defined in eq. (17.76). Then

$$\Omega_K(t) = -\frac{K}{a^2(t)H^2(t)}\,. \tag{21.5}$$

During RD we have $a(t) \propto t^{1/2}$, so $\Omega_K \propto 1/(aH)^2 \propto t \propto a^2$. Similarly, during MD $a(t) \propto t^{2/3}$, and $\Omega_K \propto t^{2/3} \propto a$. Thus, if t_i is a value of time deep in RD and t_0 is as usual the present value of time, we have

$$\Omega_K(t_0) = \Omega_K(t_i)\left(\frac{a_{\rm eq}}{a_i}\right)^2\left(\frac{a_0}{a_{\rm eq}}\right)\,, \tag{21.6}$$

where, as usual, we denote by $a_{\rm eq}$ the scale factor at matter–radiation equilibrium. For clarity, in this and the next subsection we write the present value of the scale factor a_0 explicitly, rather than setting $a_0 = 1$. For this order-of-magnitude estimate we have also neglected the recent epoch when dark energy begins to dominate. Writing $(a_{\rm eq}/a_i)^2(a_0/a_{\rm eq}) = (a_{\rm eq}/a_0)(a_0/a_i)^2$ and using $a_{\rm eq}/a_0 = (1+z_{\rm eq})^{-1} \simeq 1/3500$ and $\rho_R \propto 1/a^4$, we can rewrite this as

$$\Omega_K(t_0) \simeq \frac{\Omega_K(t_i)}{3500}\left(\frac{\rho_{R,i}}{\rho_{R,0}}\right)^{1/2}\,. \tag{21.7}$$

If we take t_i of order of the Planck time $t_{\rm Pl}$, when $\rho_{R,i}^{1/4}$ was of order $M_{\rm Pl}$, using the numerical value of the radiation energy density today $\rho_{R,0}^{1/4} \simeq 2.41 \times 10^{-4}{\rm eV}$ we find $\Omega_K(t_0) \sim 10^{60}\Omega_K(t_i)$. Cosmological observation, in particular based on analysis of the CMB, show that today $\Omega_K(t_0)$ is consistent with zero, and its upper bound at 95% c.l. is $|\Omega_K(t_0)| < 5 \times 10^{-3}$; see Note 2 on page 369. Then, we should conclude that the initial conditions in the Universe were such that $|\Omega_K(t_i)| < 10^{-62}$. It is certainly not natural to appeal to such extremely fine-tuned initial conditions. One might object that the extrapolation back to the Planckian epoch requires entering a regime where the laws of physics and the cosmological model are uncertain. However, we can simply go back in time to an epoch, such as the nucleosynthesis epoch $T = O(1)$ MeV, where the standard cosmological model is still known to work well, and the energies are such that the fundamental laws of physics are well understood. The bound on Ω_K at the present epoch implies that at nucleosynthesis $|\Omega_K|$ was smaller than 10^{-18}. Such a small value should have a dynamical explanation, rather than coming from extremely fine-tuned initial conditions.

The origin of the problem is that in the RD and MD epochs the expansion is decelerated, so $\dot{a}$ decreases with time. Since $a^2 H^2 = \dot{a}^2$, $|\Omega_K(t)|$ in eq. (21.5) increases with time. The idea of inflation is that the RD phase was preceded by a phase where the expansion of the Universe was accelerating, rather than decelerating. Then, $\dot{a}^2 = a^2 H^2$ was increasing with time and, according to eq. (21.5), independently of the initial conditions $|\Omega_K(t)|$ was dynamically driven toward very small values. Inflation is defined as such an accelerating phase. As we discussed following eq. (17.63), this requires that during inflation the effective equation of state (EoS) parameter $w < -1/3$. The simplest example is a de Sitter expansion, where the scale factor evolves exponentially with cosmic time, $a(t) \propto e^{Ht}$, where $H = \dot{a}/a$ is a constant. As discussed following eq. (17.53), this is the evolution obtained from a constant vacuum energy, and corresponds to $w = -1$. Power-like inflation corresponds instead to an expansion $a(t) \propto t^n$ with $n > 1$, which still gives $\ddot{a} > 0$. It is useful to define

$$N(t_2, t_1) \equiv \int_{t_1}^{t_2} dt\, H(t)\,. \tag{21.8}$$

For a de Sitter expansion H is constant and $a(t) \propto e^{Ht}$, so

$$N(t_2, t_1) = \log \frac{a(t_2)}{a(t_1)} \qquad \text{(de Sitter)}\,. \tag{21.9}$$

The quantity $N(t_2, t_1)$ is called the number of "e-folds" in the evolution from t_1 to t_2. Let us estimate the minimum duration of the inflationary phase necessary to solve the flatness problem. We denote by t_i and a_i the cosmic time and scale factor, respectively, when inflation begins (where we set the initial conditions), and by t_{end} and a_{end} the value of cosmic time and scale factor when inflation terminates.

If there was a previous hot plasma at the beginning of inflation, its temperature would be quickly redshifted away (exponentially, for de Sitter inflation) to negligibly small values. Thus, to generate the hot RD phase that corresponds to the beginning of the standard big-bang picture, the Universe must reheat after inflation, and the reheating temperature actually corresponds to the maximum temperature of the Universe of the standard hot big-bang picture. Thus, when $a = a_{\text{end}}$, inflation terminates and reheating begins. We denote by a_{reh} the value of the scale factor when reheating terminates and RD begins. As usual, we also denote by a_{eq} the value at RD–MD equilibrium, and by a_0 the present value (which we will eventually fix to $a_0 = 1$).

To estimate the minimum number of e-folds necessary to solve the flatness problem, let us make the crude approximation that reheating completes in a very short time, so that $a_{\rm end} \simeq a_{\rm reh}$ ("instantaneous reheating"). We assume that after the end of inflation the usual RD epoch starts, followed by the MD epoch. We also assume for simplicity a de Sitter inflation. During de Sitter inflation H is constant, so $\Omega_K \propto 1/a^2$. Then eq. (21.6) is replaced by

$$\Omega_K(t_0) = \Omega_K(t_i) \left(\frac{a_i}{a_{\rm end}}\right)^2 \left(\frac{a_{\rm eq}}{a_{\rm end}}\right)^2 \left(\frac{a_0}{a_{\rm eq}}\right), \qquad (21.10)$$

where now t_i is an initial time at the beginning of inflation. Assuming generically $\Omega_K(t_i) = O(1)$, which is what we expect in the absence of any fine tuning, and requiring $|\Omega_K(t_0)| < 10^{-2}$ gives

$$\frac{a_{\rm end}}{a_i} \simeq |\Omega_K(t_0)|^{-1/2}(1+z_{\rm eq})^{-1/2}\frac{a_0}{a_{\rm end}}$$
$$\gtrsim 0.17\,\frac{a_0}{a_{\rm end}}, \qquad (21.11)$$

where we have used $a_{\rm eq}/a_0 = (1+z_{\rm eq})^{-1} \simeq 1/3500$. Thus, in order of magnitude, the minimum amount of expansion during inflation must be almost as much as from the end of inflation up to the present time. Denoting by $\rho_{\rm reh}$ the energy density at the end of reheating, we have $\rho_{R,0} \simeq \rho_{\rm reh}(a_{\rm reh}/a_0)^4$ (neglecting also the change in the number of effective relativistic species; see Section 17.5). Thus, in the approximation $a_{\rm end} \simeq a_{\rm reh}$,

$$\frac{a_0}{a_{\rm end}} \simeq \left(\frac{\rho_{\rm reh}}{\rho_{R,0}}\right)^{1/4}. \qquad (21.12)$$

Then, using the numerical value of the radiation energy density today, $\rho_{R,0}^{1/4} \simeq 2.41 \times 10^{-4}\,{\rm eV}$, eq. (21.11) gives

$$\Delta N \gtrsim 64 - \log\left(\frac{10^{16}\,{\rm GeV}}{\rho_{\rm reh}^{1/4}}\right), \qquad (21.13)$$

where $\Delta N = N(t_e, t_i) = \log(a_{\rm end}/a_i)$ is the number of e-folds from the beginning to the end of inflation. In the approximation of instantaneous reheating we have $\rho_{\rm reh}^{1/4} \simeq \rho_{\rm inf}^{1/4}$, where $\rho_{\rm inf}^{1/4}$ is the inflationary energy scale. As we will see in eq. (21.327), the inflationary scale must be below $O(10^{16})$ GeV, otherwise the effect of GWs produced during inflation would have already been detected in the CMB temperature anisotropies. For such an inflationary scale the minimum amount of inflation necessary to solve the flatness problem is $\Delta N_{\rm min} \simeq 64$. The lowest conceivable reheating temperature is around 1 MeV, in order not to spoil the predictions of primordial nucleosynthesis.[2] In this case the minimum number of e-folds becomes $\Delta N_{\rm min} \simeq 20$. Of course, it would be quite a coincidence if the number of inflationary e-folds were just the minimum number required for solving the flatness problem. In general, we could have $\Delta N \gg \Delta N_{\rm min}$. Correspondingly, the prediction for Ω_K today would become much smaller than the existing bound.

[2]One might also be worried that inflation with a reheating temperature below the TeV range would wash out any baryon asymmetry generated earlier. However, this is more model-dependent, and there are indeed models where baryogenesis can proceed below the electroweak scale; see for example Katz and Riotto (2016) and references therein.

In more realistic models of reheating, $\rho_{\rm reh}^{1/4}$ is a few orders of magnitudes lower than $\rho_{\rm inf}^{1/4}$, in a model-dependent way. One should also take into account the evolution of $\Omega_K(t)$ during a realistic reheating phase of finite duration. Typical models of reheating proceed through decay of the scalar field that drives inflation into massive particles, resulting in a matter-dominated reheating phase. In that case, in eq. (21.10) one should also take into account that $\Omega_K(t)$ would further increase by a factor $a_{\rm reh}/a_{\rm end}$ from the end of inflation to the end of reheating. This would change somewhat the exact numbers in eq. (21.13).

21.1.2   The horizon problem

The horizon problem stems from the fact that, as we have seen in Chapter 20, the anisotropies in the CMB temperature are at the level $\delta T/T \sim 10^{-5}$, even for low multipoles, such as the quadrupole, which correspond to large angular scales of the order of the whole sky. This means that the whole of the LSS must have been in causal contact in the past. Regions that never were in causal contact would necessarily have relative temperatures differences with at least $\delta T/T = O(1)$. The maximum comoving distance that can be in causal contact is fixed by the comoving horizon, defined in eq. (17.128). To be slightly more accurate, if we consider two particles that interacted at an initial time t_1 and then moved at the speed of light in opposite directions, then, generalizing eq. (17.128), we see that their comoving distance at time t_2 is

$$d_H(t_2,t_1) = 2\int_{a_1}^{a_2} \frac{da}{a^2 H(a)} \,, \tag{21.14}$$

where $a_1 = a(t_1)$ and $a_2 = a(t_2)$. In a cosmological model that starts at $t = t_1$ deep in a RD phase, we can take as a lower limit of the integral $a_1 = 0$. The integral in eq. (21.14) converges near the lower limit, since in RD $H^2 \propto a^{-4}$ so $a^2 H(a)$ is a constant. The integral is then dominated by its upper integration limit. We have already computed the value of the comoving horizon at decoupling in eqs. (17.172) and (17.174) in a model where an initial RD phase is followed by MD and then by the present DE-dominated era, and we found in eq. (17.175) that the comoving horizon at decoupling subtends, today, an angle $\Delta\theta \simeq 1°$ in the sky; see eq. (17.175). Even accounting for the factor of 2 in eq. (21.14), there should be no correlation of the CMB temperature on angular scales larger than about 2°. The fact that the CMB temperature is instead remarkably uniform, at the level $\delta T/T \sim 10^{-5}$, over the whole sky, clearly shows that the size of the region in causal contact is much larger than that computed using a cosmological model that starts from RD at the initial time. Again, the solution to the problem is provided by inflation. Setting $a_1 = 0$ and writing eq. (21.14) as

$$d_H(a_2, a_1 = 0) = 2\int_0^{a_2} \frac{da}{a\dot{a}} \,, \tag{21.15}$$

we see that the convergence of the integral at the lower limit depends on whether the expansion is accelerated or decelerated. The integral diverges if $\dot{a}$ is constant, which is the limiting case between acceleration and deceleration, and more generally it diverges if the expansion in the early Universe, near $a = 0$, was accelerated, i.e. $\dot{a} \propto a^n$ with $n > 0$, while it converges if it was decelerated. An early phase of accelerated expansion therefore has the consequence of increasing the size of the comoving horizon. In order to determine the minimum number of e-folds required to solve the horizon problem, we assume again for simplicity a de Sitter inflation, with a constant Hubble parameter H. Then eq. (21.14), computed between the time a_i at which inflation begins and the time $a_{\rm end}$ at which it ends, gives

$$d_H(t_e, t_i) = \frac{2}{H} \int_{a_i}^{a_{\rm end}} \frac{da}{a^2}$$

$$\simeq \frac{2}{H a_i}, \tag{21.16}$$

where in the second line we have used $a_{\rm end} \gg a_i$. This expression diverges if $a_i \to 0$, so by taking a_i sufficiently small we can obtain a sufficiently large horizon scale. To this quantity we should in principle add the contribution obtained by integrating from the end of inflation to the present time, but we have seen that this quantity is small compared with the comoving diameter of the LSS, so basically the whole contribution must come from the inflationary phase. The physical radius of the LSS today is $(\eta_0 - \eta_{\rm dec}) \simeq \eta_0$. Using eq. (17.168), together with the numerical values $\Omega_M \simeq 0.3$ and eq. (17.170), gives $\eta_0 \simeq 3.2/H_0$, so the comoving radius of the LSS today is of order $3.2/(H_0 a_0)$, where again, for clarity, we are not setting $a_0 = 1$. Then, the horizon problem is solved if

$$H_0 a_0 \gtrsim 3.2 H a_i . \tag{21.17}$$

Since H is the constant value of the Hubble parameter during inflation, it is also the same as the value of the Hubble parameter at the beginning of RD (assuming, again, instantaneous reheating). Then the Friedmann equation (17.75), after reinstating a_0, gives[3]

$$H \simeq H_0 \Omega_R^{1/2} \left(\frac{a_0}{a_{\rm end}} \right)^2 , \tag{21.18}$$

and eq. (21.17) becomes

$$\frac{a_{\rm end}}{a_i} \gtrsim 3.2 \, \Omega_R^{1/2} \frac{a_0}{a_{\rm end}}$$

$$\simeq 0.04 \frac{a_0}{a_{\rm end}} . \tag{21.19}$$

This is parametrically the same as the flatness condition (21.11), and somewhat less restrictive numerically, leading to

$$\Delta N \gtrsim 62.8 - \log \left(\frac{10^{16} \, \text{GeV}}{\rho_{\rm reh}^{1/4}} \right) . \tag{21.20}$$

[3] Observe that we are neglecting the change in the number of relativistic degrees of freedom with temperature during RD, which is correctly described by eq. (17.116). Of course this approximation, as well as the fact that we have assumed instantaneous reheating, is only appropriate at the level of order-of-magnitude estimates.

Then, the amount of e-folds that solves the flatness problem also solves the horizon problem. A more accurate formula, including the effect of reheating, will be given in eq. (21.330).

21.1.3   Single-field slow-roll inflation

A simple model of inflation is obtained by adding to Einstein gravity a scalar field ϕ, the inflaton, slowly rolling in a potential $V(\phi)$. The action is then given by $S = S_E + S_{\text{infl}}$, where S_E is the Einstein–Hilbert action,

$$S_E = \frac{1}{16\pi G} \int d^4x \sqrt{-g}\, R \,, \tag{21.21}$$

and the inflaton action is[4]

$$S_{\text{infl}} = \int d^4x \sqrt{-g} \left[-\frac{1}{2} g^{\mu\nu} \partial_\mu \phi \partial_\nu \phi - V(\phi) \right]. \tag{21.22}$$

We consider for simplicity a potential $V(\phi)$ that is everywhere non-negative. Using

$$\delta\sqrt{-g} = -\frac{1}{2}\sqrt{-g}\, g_{\mu\nu} \delta g^{\mu\nu} \,, \tag{21.23}$$

the energy–momentum tensor of the scalar field is

$$T_{\mu\nu} \equiv -\frac{2}{\sqrt{-g}} \frac{\delta S_\phi}{\delta g^{\mu\nu}}$$

$$= \partial_\mu \phi \partial_\nu \phi - g_{\mu\nu} \left[\frac{1}{2} g^{\rho\sigma} \partial_\rho \phi \partial_\sigma \phi + V(\phi) \right]. \tag{21.24}$$

For a spatially homogeneous field in the FRW metric (17.2) this gives

$$\rho_{\text{infl}} = \frac{1}{2}\dot{\phi}^2 + V(\phi) \,, \tag{21.25}$$

$$p_{\text{infl}} = \frac{1}{2}\dot{\phi}^2 - V(\phi) \,. \tag{21.26}$$

Thus, such a field effectively acts as a fluid with a time-dependent EoS parameter

$$w = \frac{(1/2)\dot{\phi}^2 - V(\phi)}{(1/2)\dot{\phi}^2 + V(\phi)} \,. \tag{21.27}$$

We see that, when the evolution is such that $(1/2)\dot{\phi}^2 \ll V(\phi)$, the effective EoS parameter $w \simeq -1$, and we have an accelerated expansion, with a behavior close to de Sitter. Let us then discuss in more detail the evolution equations, and explore under what conditions the kinetic term of the field is much smaller than the potential energy. For the background evolution we assume a flat FRW metric with scale factor $a(t)$, and a spatially homogeneous field $\phi(t)$, where t is cosmic time. The (00) component of the Einstein equation gives the usual Friedmann equation

$$3H^2 = 8\pi G\, \rho_{\text{infl}} \,, \tag{21.28}$$

where $\rho_{\rm infl}$ is given by eq. (21.25). Variation of the action with respect to ϕ gives the Klein–Gordon equation in curved space, which, specialized to FRW and to a homogeneous field ϕ, reads

$$\ddot{\phi} + 3H\dot{\phi} + V_\phi = 0 \,, \tag{21.29}$$

where $V_\phi \equiv dV/d\phi$ (we will also use the notation $V_{\phi\phi} \equiv d^2V/d\phi^2$) and, as usual, a dot indicates the derivative with respect to cosmic time. Energy–momentum conservation implies the conservation equation

$$\dot{\rho}_{\rm infl} + 3H(\rho_{\rm infl} + p_{\rm infl}) = 0 \,, \tag{21.30}$$

which can also be derived as a consequence of the Klein–Gordon equation (21.29), using the definitions of $\rho_{\rm infl}$ and $p_{\rm infl}$ given in eqs. (21.25) and (21.26). As usual, the $(0i)$ Einstein equation, in a homogeneous background, vanishes identically, and the (ij) equation is a consequence of eqs. (21.28) and (21.30).

Taking the time derivative of eq. (21.28) and using eq. (21.30), together with $\rho_{\rm infl} + p_{\rm infl} = \dot{\phi}^2$, which follows from eqs. (21.25) and (21.26), we get the useful relation

$$\dot{H} = -4\pi G\dot{\phi}^2 \,, \tag{21.31}$$

which can also be derived taking the difference of the (ii) and (00) Einstein equations.

In order to have a period of accelerated expansion we impose the *first slow-roll condition*,

$$\frac{1}{2}\dot{\phi}^2 \ll V(\phi) \,. \tag{21.32}$$

As we have seen, when this condition is satisfied, the evolution of the scale factor is of the inflationary type. We also require that this condition be satisfied over an extended period of time, which is obtained by imposing that, in absolute value, the time derivative of the left-hand side of eq. (21.32) should be much smaller than the time-derivative of the right-hand side. This gives the *second slow-roll condition*,

$$|\ddot{\phi}| \ll |V_\phi(\phi)| \,. \tag{21.33}$$

When the slow-roll conditions are satisfied, eqs. (21.28) and (21.29) become

$$3H^2 \simeq 8\pi G V \,, \tag{21.34}$$
$$3H\dot{\phi} \simeq -V_\phi \,. \tag{21.35}$$

Using eq. (21.35), the second slow-roll condition can be rewritten as

$$|\ddot{\phi}| \ll 3H|\dot{\phi}| \,, \tag{21.36}$$

which means that $\dot{\phi}$ changes slowly on the typical time-scale of the problem, which is given by the Hubble parameter. Similarly, using eqs. (21.31), (21.32) and (21.34) we get

$$-\dot{H} \ll 3H^2 \,, \tag{21.37}$$

which means that also H changes slowly on this time-scale, and the evolution is almost de Sitter-like.

From eq. (21.34), $H \simeq (8\pi G V/3)^{1/2}$. Inserting this into eq. (21.35) we get

$$\dot\phi \simeq -\frac{1}{\sqrt{24\pi G}} \frac{V_\phi(\phi)}{V^{1/2}(\phi)} , \qquad (21.38)$$

which determines the evolution of ϕ in terms of ϕ only, in the slow-roll limit. Inserting this into eq. (21.32), the first slow-roll condition can be written as a condition on the potential,

$$\frac{1}{48\pi G} \left(\frac{V_\phi}{V}\right)^2 \ll 1 . \qquad (21.39)$$

In order to write the second slow-roll condition also as a condition on V we take the time derivative of eq. (21.38), which gives

$$\ddot\phi \simeq -\frac{1}{\sqrt{24\pi G}} \dot\phi V^{1/2} \left[\frac{V_{\phi\phi}}{V} - \frac{(V_\phi)^2}{2V^2}\right]$$

$$\simeq -\frac{1}{8\pi G} H\dot\phi \left[\frac{V_{\phi\phi}}{V} - \frac{(V_\phi)^2}{2V^2}\right] , \qquad (21.40)$$

where in the second line we have used eq. (21.34). Then eq. (21.36) becomes

$$\frac{1}{24\pi G} \left|\frac{V_{\phi\phi}}{V} - \frac{(V_\phi)^2}{2V^2}\right| \ll 1 . \qquad (21.41)$$

Using eq. (21.39), we see that the second term in eq. (21.41) is already much smaller than 1, so the second slow-roll condition is equivalent to

$$\frac{1}{24\pi G} \left|\frac{V_{\phi\phi}}{V}\right| \ll 1 . \qquad (21.42)$$

It is conventional to define the dimensionless slow-roll parameters ε_V and η_V as

$$\varepsilon_V \equiv \frac{1}{16\pi G} \left(\frac{V_\phi}{V}\right)^2 , \qquad (21.43)$$

$$\eta_V \equiv \frac{1}{8\pi G} \frac{V_{\phi\phi}}{V} . \qquad (21.44)$$

Note that $\varepsilon_V > 0$ by definition, while η_V can be positive or negative. We have added the subscript V to ε_V and η_V to stress that this is the definition of the slow-roll parameters made in terms of the potential. A different definition of these parameters, also commonly used in the literature, is given by

$$\varepsilon_H \equiv -\frac{\dot H}{H^2} , \qquad (21.45)$$

$$\eta_H \equiv \varepsilon_H - \frac{\ddot H}{2H\dot H} , \qquad (21.46)$$

so that

$$\frac{\ddot{H}}{H\dot{H}} = 2(\varepsilon_H - \eta_H).$$ (21.47)

The subscript H stresses that this is the definition of the slow-roll parameters constructed using the Hubble parameter. Combining eqs. (21.31), (21.34) and (21.35) we find that, to lowest order in the slow-roll parameters, $\varepsilon_H \simeq \varepsilon_V$ and $\eta_H \simeq \eta_V$. More generally,

$$\varepsilon_H = \varepsilon_V + O(\varepsilon^2),$$ (21.48)
$$\eta_H = \eta_V + O(\varepsilon^2),$$ (21.49)

where, for the purpose of organizing the slow-roll expansion, the second slow-roll parameter η_V will be taken to be parametrically of order ε. Thus, when working beyond lowest order, one should be careful about the definition used.[5]

Alternatively, when performing a systematic expansion in the slow-roll approximation, it can be convenient to introduce a hierarchy of slow-roll parameters ε_i from $\varepsilon_1 \equiv \varepsilon_H$ and

$$\varepsilon_{i+1} \equiv \frac{\dot{\varepsilon}_i}{H\varepsilon_i}$$ (21.50)

for $i \geqslant 1$. In particular,

$$\varepsilon_2 = 4\varepsilon_H - 2\eta_H.$$ (21.51)

In terms of ε_V and η_V the slow-roll conditions (21.39) and (21.42) can be rewritten as

$$\frac{\varepsilon_V}{3} \ll 1, \qquad \frac{|\eta_V|}{3} \ll 1,$$ (21.52)

and, to lowest order, we can also write

$$\frac{\dot{H}}{H^2} \simeq -\varepsilon_V,$$ (21.53)

$$\frac{\ddot{H}}{H\dot{H}} \simeq 2(\varepsilon_V - \eta_V),$$ (21.54)

while the derivatives of ϕ are given by

$$\frac{3\dot{\phi}^2}{2V} \simeq \varepsilon_V,$$ (21.55)

$$\frac{\ddot{\phi}}{H\dot{\phi}} \simeq \varepsilon_V - \eta_V.$$ (21.56)

It is also common to define another slow-roll parameter δ by

$$\delta \equiv \frac{\ddot{\phi}}{H\dot{\phi}}.$$ (21.57)

Thus,

$$\delta = \varepsilon_V - \eta_V + O(\varepsilon^2).$$ (21.58)

[5] In the literature, when working to lowest order in the slow-roll expansion, the two slow-roll parameters are usually denoted simply by ε and η. However, since we are already using the letter η to denote conformal time, to avoid confusion we will always keep a generic subscript, in general V, on the second slow-roll parameter, and correspondingly we will also in general write ϵ_V in equations in which both slow-roll parameters appear. Of course, in such first-order equations, we could equivalently write ε_H, η_H. When only the first slow-roll parameter appears in equations valid to first order, we will occasionally write it simply as ε.

Observe that, in the slow-roll approximation, the original second-order equation (21.29) has been approximated by a first-order equation. One might wonder about the mapping of the initial conditions between the original equation, which requires giving the initial values of ϕ and $\dot{\phi}$, and eq. (21.38), which requires only the initial values of ϕ. The important point, which underlines the usefulness of the slow-roll approximation, is that the solution of eq. (21.38) is an attractor of the solution of eq. (21.29), for a broad range of initial conditions. This can be proved using the Hamilton–Jacobi formalism; see the Further Reading.

Finally, using the slow-roll equations (21.34) and (21.35), the number of e-folds between times t_1 and t_2, defined in eq. (21.8), can be written as

$$N(t_2,t_1) = \int_{\phi_1}^{\phi_2} d\phi \, \frac{H}{\dot{\phi}}$$
$$\simeq 8\pi G \int_{\phi_2}^{\phi_1} d\phi \, \frac{V(\phi)}{V_\phi(\phi)} . \tag{21.59}$$

Given the form of the potential, we can compute the value of the field ϕ_e when the inflationary phase ends. This happens when one of the slow-roll conditions is no longer satisfied. Of course, the precise numerical values of $|\varepsilon_V|$ and $|\eta_V|$ for which we declare that the slow-roll regime terminates are slightly arbitrary. It is conventional to say that the slow-roll regime terminates when either $|\epsilon_V|$ or $|\eta_V|$ becomes equal to 1, whichever happens first (although at this point $|\epsilon_V|/3$ and $|\eta_V|/3$ are still smaller than 1, so the slow-roll approximation has not yet gone astray). Denoting by ϕ_i the initial value of the field, the total number of e-folds during inflation is then given by

$$\Delta N \simeq 8\pi G \int_{\phi_e}^{\phi_i} d\phi \, \frac{V(\phi)}{V_\phi(\phi)} . \tag{21.60}$$

Since ϕ_e is fixed once the potential $V(\phi)$ has been given, the condition that the number of e-folds should be large enough to solve the flatness and horizon problems gives a condition on the initial field ϕ_i. We will examine this separately for the two cases $\{V_\phi > 0, \phi_i > \phi_e\}$ (corresponding to large-field inflation) and $\{V_\phi < 0, \phi_i < \phi_e\}$ (corresponding to small-field inflation). More generally, for single-field inflation there is a one-to-one correspondence between the value of the field ϕ during slow-roll and the number of e-folds $N_e(\phi)$ to the end of inflation,

$$N_e(\phi) \simeq 8\pi G \int_{\phi_e}^{\phi} d\phi_1 \, \frac{V(\phi_1)}{V_\phi(\phi_1)} . \tag{21.61}$$

21.1.4   Large-field and small-field inflation

In the literature there exist a great variety of inflationary models. A useful classification distinguishes between large-field and small-field inflationary models. Let us first recall that (in the units $\hbar = c = 1$ that we are using) the Planck mass $M_{\rm Pl}$ is defined by

$$G = \frac{1}{M_{\rm Pl}^2}, \qquad (21.62)$$

or, more generally, $M_{\rm Pl}^2 = \hbar c/G$. We will also use the *reduced Planck mass* $m_{\rm Pl}$, defined from

$$M_{\rm Pl}^2 = 8\pi m_{\rm Pl}^2, \qquad (21.63)$$

so that $8\pi G = 1/m_{\rm Pl}^2$.[6] Numerically[7]

$$M_{\rm Pl} = 1.220910(29) \times 10^{19}\,{\rm GeV}\,, \qquad (21.64)$$
$$m_{\rm Pl} = 2.435361(58) \times 10^{18}\,{\rm GeV}\,. \qquad (21.65)$$

The use of the reduced Planck mass usually simplifies a number of 8π factors in the equations.

Large-field models are defined by the fact that the inflationary behavior is obtained only for super-Planckian values of the field, i.e. at $\phi \gtrsim m_{\rm Pl}$.[8] In contrast, in small-field models inflation is realized at $\phi \ll m_{\rm Pl}$. As we will see, these two classes of models have rather different predictions, in particular concerning GW production.[9]

Typical large-field models are characterized by a potential $V(\phi)$ growing as a power ϕ^n around a minimum, which, with a shift of the field, can be set without loss of generality at $\phi = 0$. The simplest example is $V(\phi) = (1/2)m^2\phi^2$, which corresponds just to a massive scalar field without self-interaction. For this potential, eqs. (21.43) and (21.44) give

$$\varepsilon_V = \eta_V = \frac{2m_{\rm Pl}^2}{\phi^2}\,. \qquad (21.66)$$

The slow-roll conditions are therefore satisfied only for very large values of the field. In order to have inflation, we must begin with an initial value ϕ_i sufficiently larger than $m_{\rm Pl}$ (where, for definiteness, we have taken $\phi_i > 0$). The field ϕ will then slowly roll down the potential, and inflation will stop approximately when $\varepsilon_V = \eta_V \simeq 1$, i.e. when $\phi \simeq \phi_e = \sqrt{2}\,m_{\rm Pl}$. According to eq. (21.60),

$$\Delta N \simeq \frac{1}{4m_{\rm Pl}^2}(\phi_i^2 - \phi_e^2)$$
$$\simeq \frac{\phi_i^2}{4m_{\rm Pl}^2}\,, \qquad (21.67)$$

where we have neglected the second term since $\phi_e^2/(4m_{\rm Pl}^2) = 1/2$, while we need $\Delta N \gg 1$. This gives

$$\phi_i \simeq 15.5\,m_{\rm Pl}\left(\frac{\Delta N}{60}\right)^{1/2}\,. \qquad (21.68)$$

Thus, to have at least 60 e-folds of inflation we need an initial value $\phi_i \gtrsim 15 m_{\rm Pl} \simeq 3 M_{\rm Pl}$. The need for super-Planckian values of the field raises an issue of consistency of the approach. A first requirement is

[6]Unfortunately, the notation $M_{\rm Pl}$ is sometimes used in the literature for the Planck mass and sometimes for the reduced Planck mass, so one must be careful about the conventions used.

[7]See http://www.physics.nist.gov/cgi-bin/cuu/Value?plkmc2gev

[8]One can argue that $m_{\rm Pl}$ is a more natural unit than $M_{\rm Pl}$ to define the scale of super-Planckian physics. Indeed, when expanding the Einstein–Hilbert action around flat space, the factor $8\pi G$ is reabsorbed into the canonical normalization of the graviton field, and the gravitational vertex is proportional to $\kappa = (32\pi G)^{1/2} = 2/m_{\rm Pl}$; see eq. (2.117) of Vol. 1. Of course, precise numbers can only be obtained by computing explicitly the gravitational quantum corrections to the specific process considered.

[9]It should be observed, however, that the distinction is not so clear-cut. In particular, we will see below that the Starobinsky model formally belongs to the class of large-field models, but it predicts a small value of the tensor-to-scalar ratio r, which is typical of small-field models.

that the potential $V(\phi_i)$ should be much smaller than $m_{\rm Pl}^4$, otherwise quantum gravity effects cannot be neglected. With $\phi_i \simeq 15 m_{\rm Pl}$, however, this is obtained simply by requiring $m \ll 0.1 m_{\rm Pl}$, which is not a stringent requirement. Still, this is a necessary but not sufficient condition to ensure the consistency of an approach involving super-Planckian field excursions, as we will see in a moment. First, let us observe that a similar analysis can be performed for a more general potential of the form

$$V(\phi) = g_n \phi^n . \tag{21.69}$$

For this potential

$$\varepsilon_V = \frac{n^2 m_{\rm Pl}^2}{2\phi^2} , \tag{21.70}$$

$$\eta_V = n(n-1) \frac{m_{\rm Pl}^2}{\phi^2} . \tag{21.71}$$

For $n > 2$ we have $\eta_V > \epsilon_V$, so the second slow-roll condition is the first to be violated and inflation terminates when $\eta_V \simeq 1$, i.e. $(\phi_e/m_{\rm Pl})^2 \simeq n(n-1)$. From eq. (21.60),

$$\Delta N \simeq \frac{1}{2n} \left(\frac{\phi_i}{m_{\rm Pl}} \right)^2 - \frac{n-1}{2} . \tag{21.72}$$

If n is not too large, we can neglect $(n-1)/2$ with respect to $\Delta N = O(60)$, and we get

$$\phi_i \simeq 11.0 \, n^{1/2} m_{\rm Pl} \left(\frac{\Delta N}{60} \right)^{1/2} , \tag{21.73}$$

so again we need super-Planckian excursions of the field. Dimensionally $g_n = (\text{mass})^{4-n}$, so we can write $g_n = c_n m_{\rm Pl}^{4-n}$, with c_n dimensionless. The condition $V(\phi_i) \ll m_{\rm Pl}^4$ then gives

$$c_n \lesssim \frac{1}{(11.0 \, n^{1/2})^n} \left(\frac{60}{\Delta N} \right)^{n/2} . \tag{21.74}$$

Consider for instance a theory with a potential $V(\phi) = (1/4)\lambda\phi^4$, i.e. $n = 4$ and $c_4 = \lambda/4$. In order to have $V(\phi_i) \ll m_{\rm Pl}^4$, with ϕ_i sufficiently large to ensure 60 e-folds of inflation, we need $\lambda \ll 10^{-5}$. By itself such a requirement is not too strong, and basically means that the theory is in the weak-coupling regime.

However, the condition $V(\phi_i) \ll m_{\rm Pl}^4$ is not sufficient to ensure the consistency of a dynamics involving super-Planckian field values. A more subtle requirement comes from the fact that, even if we start from a classical theory with a potential $V(\phi)$, the dynamics will actually be governed by the effective potential $V_{\rm eff}(\phi)$ that includes the quantum loop corrections. Consider for definiteness the theory of a scalar field (21.22), with potential $V(\phi) = (1/2)m^2\phi^2$, coupled to gravity. Feynman graphs with $n \geq 4$ external ϕ lines generate terms in the effective potential proportional to ϕ^n. Since $V_{\text{eff}}(\phi)$ has dimensions of (mass)^4, on dimensional grounds these corrections have the form

$$\Delta V = m_{\text{Pl}}^4 \left[c_4 \left(\frac{\phi}{m_{\text{Pl}}} \right)^4 + c_6 \left(\frac{\phi}{m_{\text{Pl}}} \right)^6 + c_8 \left(\frac{\phi}{m_{\text{Pl}}} \right)^8 + \ldots \right] . \quad (21.75)$$

If the coefficients c_i are $O(1)$, the conditions (21.74) that ensure that we are in the slow-roll regime are violated. Even more importantly, if $c_i = O(1)$, then for super-Planckian excursions of the field the energy density in the effective potential becomes of order m_{Pl}^4 and we are outside the limit of validity of semiclassical gravity. In this regime a full theory of quantum gravity would be needed to make any statement. The issue is therefore whether the coefficients c_i can be of order 1.

If we assume that our "fundamental" theory is just the theory of a scalar field (21.22), with potential $V(\phi) = (1/2)m^2\phi^2$, coupled to gravity, then the coefficients c_i generated by loop corrections are indeed small. We can understand this even without doing any explicit one-loop computation, just by observing that in the limit $m \to 0$ this theory acquires a shift symmetry $\phi \to \phi + \text{const}$, which forbids the generation of the terms shown in eq. (21.75) in the one-loop effective potential. Therefore, the coefficients c_i in eq. (21.75) must be proportional to powers of the mass m. For dimensional reasons, this means that the c_i, which are dimensionless, are suppressed by powers of m/m_{Pl}, which we have already taken to be much smaller than 1 to satisfy the slow-roll conditions. More explicitly, computing for instance the one-loop graph in which four external scalar field lines are attached to a graviton loop, we find that $c_4 = O(m^4/m_{\text{Pl}}^4)$.[10] The same conclusion applies for a potential $V(\phi) = g_n\phi^n$.

However, in a realistic fundamental theory, the inflaton field ϕ will be coupled not only to gravity, but also to all other fields in the theory,[11] and these couplings in general will not be small. Furthermore, it could also be coupled to states with masses at the Planck scale, which would appears in a UV completion such as string theory. In these cases, coefficients $c_i = O(1)$ will be generated, unless some symmetry protects them. Thus, despite its apparent simplicity, it is non-trivial to embed large-field inflation in a consistent fundamental field-theoretical model. A possibility is to invoke an exact shift symmetry in the fundamental theory, giving rise for instance to an inflationary potential of the form

$$V(\phi) = \Lambda^4 \left[1 + \cos \left(\frac{\phi}{f} \right) \right] , \quad (21.76)$$

where Λ and f are mass scales. This model is called *natural inflation*.

So, embedding large-field inflation in a fundamental theory is possible but quite non-trivial. In contrast, initial conditions giving rise to large-field inflation can be devised quite naturally. We can imagine that, in the Planckian epoch, the Universe was highly inhomogeneous and strongly curved on the scale of the Planck length. It is quite natural to expect

---

[10] Again, the explicit computation is not needed to understand this behavior. Each of the four vertices in this graph brings a factor $1/m_{\text{Pl}}$. Thanks to the four graviton propagators running in the loop, the loop integral is UV-convergent, so it can depend only on the momenta of the external fields. For spatially homogeneous and slowly evolving fields the only scale associated with the external lines is then the mass m.

[11] This coupling to matter is also necessary in order to produce the reheating from the inflaton decay, when the slow-roll regime terminates and the inflaton oscillates around the minimum of the potential.

that the typical fluctuations of the inflaton field would also be $O(m_{\mathrm{Pl}})$. Statistically, there would be some patch of about Planckian size where the gradient term of the inflaton field ϕ and the spatial curvature term were smaller than the potential term, while ϕ had fluctuated to super-Planckian values. Such a patch would satisfy the slow-roll conditions and would begin to inflate. Then, curvature and gradient terms would be reduced further, since during inflation $\Omega_K \propto 1/a^2$ and also the gradient terms decrease as $1/a^2$, while the potential term stays constant, so this patch would continue to inflate. The present observable Universe would then have emerged from the inflation of such a patch, and one can similarly imagine several other patches, outside our horizon, that had undergone a similar evolution. Such a scenario, emerging from "chaotic" initial conditions, as well as large-field inflation in general, is also called *chaotic inflation.*

Small-field inflation is instead characterized by sub-Planckian excursions of the field. A typical example is obtained with a potential of the type shown in Fig. 21.1, known as a hill-top potential. It has a maximum at $\phi = 0$ and a long, almost flat, and monotonically decreasing part near $\phi = 0$, and then a minimum at some larger value of ϕ. The corresponding inflationary model is also referred as hill-top inflation. If the initial condition for ϕ is close to zero (or, because of quantum fluctuations, even if we initially set ϕ exactly equal to zero), then ϕ will evolve across the long plateau, and there it will be in the slow-roll regime. It is typically assumed that, near the origin, the potential has the form

$$V(\phi) \simeq V_0 - \frac{\lambda}{4}\phi^4 \,. \tag{21.77}$$

$V(\phi)$

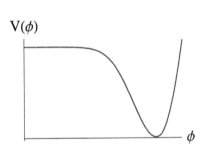

Fig. 21.1 A typical potential for small-field inflation

Then, as long as $(\lambda/4)\phi^4 \ll V_0$, eqs. (21.43) and (21.44) give

$$\eta_V = -\frac{3\lambda m_{\mathrm{Pl}}^2 \phi^2}{V_0} \,. \tag{21.78}$$

Therefore $\eta_V < 0$, while $\varepsilon_V = -\eta_V(\lambda\phi^4/V_0)$, so $\varepsilon_V \ll |\eta_V|$. Thus, the slow-roll conditions are satisfied for sufficiently small ϕ, and inflation ends when $|\eta_V| \simeq 1$, i.e. when

$$\phi_e^2 \simeq \frac{V_0}{3\lambda m_{\mathrm{Pl}}^2} \,. \tag{21.79}$$

Since we have assumed $(\lambda/4)\phi^4 \ll V_0$, this is consistent as long as $(\lambda/4)\phi_e^4 \ll V_0$, i.e.

$$V_0 \ll 36\,\lambda m_{\mathrm{Pl}}^4 \,. \tag{21.80}$$

We have seen that, for single-field inflation, there is a one-to-one correspondence between the value of the field ϕ during slow-roll and the number of e-folds $N_e(\phi)$ to the end of inflation. For hill-top inflation, using eq. (21.61), we have

$$N_e(\phi) \simeq \frac{V_0}{2\lambda m_{\mathrm{Pl}}^2 \phi^2} \,, \tag{21.81}$$

where we have neglected $1/\phi_e^2$ compared with $1/\phi^2$, which is a good approximation when $N_e(\phi) \gg 1$. Observe that N_e formally diverges if the initial value $\phi = 0$, corresponding to the fact that, classically, if we set the initial value of ϕ on the maximum of the potential at $\phi = 0$, it remains there. However, this is an unstable situation, and quantum fluctuations will destabilize it. In this case we must replace ϕ^2 by the (renormalized) quantum expectation value $\langle \phi^2 \rangle$.

Initial conditions for small-field inflation can be naturally obtained by assuming that in the early Universe, at high temperatures, the effective potential has a single minimum at $\phi = 0$. As the temperature decreases, a phase transition occurs so that the effective potential now acquires the form of Fig. 21.1. At high temperatures the field ϕ is in the minimum at $\phi = 0$, and when the form of the potential changes it will be pushed out from the value $\phi = 0$ by quantum fluctuations and will start to roll toward the new minimum.

A yet different type of inflationary model is *hybrid inflation*. In this case there are two fields: the inflaton field ϕ and a multi-component Higgs-like field Φ. The prototype example is given by the potential

$$\mathcal{V}(\Phi, \phi) = \frac{\lambda}{4} \left(\Phi^\dagger \Phi - v^2 \right)^2 + \frac{1}{2} g^2 \phi^2 \Phi^\dagger \Phi + U(\phi) \,, \qquad (21.82)$$

where $U(\phi)$ is a typical potential of chaotic inflation, for example $U(\phi) = (1/2)m^2\phi^2$, while $\Phi^\dagger \Phi = \sum_a \sigma_a^2$, where $a = 1, \ldots, N$ labels the components of the Higgs field. The case of a single-component field, $N = 1$, is ruled out because it would lead to over-production of dangerous sub-horizon domain walls. Using the notation $\sigma^2 = \sum_a \sigma_a^2$, we can rewrite eq. (21.82) as

$$
\begin{aligned}
\mathcal{V}(\sigma, \phi) &= \frac{\lambda}{4} \left(\sigma^2 - v^2 \right)^2 + \frac{1}{2} g^2 \phi^2 \sigma^2 + U(\phi) \\
&= \frac{1}{4}\lambda v^4 + \frac{1}{2}(g^2\phi^2 - \lambda v^2)\sigma^2 + \frac{1}{4}\lambda\sigma^4 + U(\phi) \,. \quad (21.83)
\end{aligned}
$$

The idea behind this model is that, as long as $g^2\phi^2 > \lambda v^2$, the effective mass term for σ is positive, so σ oscillates around $\sigma = 0$, and then ϕ rolls down the potential

$$V(\phi) \simeq V_0 + U(\phi) \,, \qquad (21.84)$$

where $V_0 = (1/4)\lambda v^4$. As soon as ϕ goes below the critical value given by $\phi_c^2 = \lambda v^2/g^2$, the extremum $\sigma = 0$ become unstable, and the field σ acquires a tachyonic mass term and rapidly rolls down toward the new minimum, which for a fixed $\phi < \phi_c$ is at $\sigma^2 = v^2(1 - \phi^2/\phi_c^2)$. Inflation ends either when the slow-roll conditions for $V(\phi)$ are violated or at $\phi = \phi_c$, whichever comes first. In both cases, when $\phi < \phi_c$, the fields $\{\sigma, \phi\}$ rapidly evolve toward the global minimum at $\{\sigma = v, \phi = 0\}$. Despite the fact that $U(\phi)$ is a typical potential used in chaotic inflation, i.e. in large-field models, depending on the value of the parameters this multi-field dynamics can be significantly different from large-field inflation. In particular, we can have sufficient inflation without super-Planckian excursions of the field. This happens if we choose the parameters so that, at $\phi = \phi_c$, we are still in the slow-roll regime, and we further require that $U(\phi_c) \ll V_0$. Then, the slow-roll parameter η_V is given by

$$\eta_V = \frac{1}{8\pi G} \frac{V_{\phi\phi}}{V}$$
$$\simeq m_{\rm Pl}^2 \frac{U''}{V_0} \,. \tag{21.85}$$

Thanks to the requirement $U(\phi_c) \ll V_0$, this can be much smaller than 1 without requiring $\phi \gg m_{\rm Pl}$. For instance, if we take $U(\phi) = (1/2)m^2\phi^2$, then, instead of eq. (21.66), we get $m_{\rm Pl}^2 m^2 \ll V_0$, independently of ϕ. For ε, from eq. (21.43) we see that in this model

$$\varepsilon_V = \eta_V \frac{2U(\phi)}{V_0} \,. \tag{21.86}$$

In particular, at $\phi = \phi_c$, the condition $U(\phi_c) \ll V_0$ implies that $\varepsilon_V \ll \eta_V$. So in this model ε_V and η_V are both small for sub-Planckian field values, and the slow-roll conditions are satisfied down to $\phi = \phi_c$. At this point, inflation ends not because ε_V or η_V reach values of order unity, but because the value $\sigma = 0$ becomes unstable.

21.1.5   Starobinsky model

In the literature there is a huge variety of inflationary models, which are obtained not only by playing with the form of the potential, but also by considering multi-field inflation or non-canonical kinetic terms. Detailed discussions can be found in the references in the Further Reading. Here we will focus on a model, due to Starobinsky, that, besides being historically the first inflationary model proposed, is also arguably the simplest and most elegant and, as we will see in Section 21.3.5 (see in particular Fig. 21.6), is also the one that fits best the current cosmological data. As we will see, it can be reformulated in terms of a scalar–tensor theory, and then it will belong to the class of large-field inflationary models that we have discussed. However, in its most economical and elegant formulation it does not involve a scalar field but only gravity, and is obtained by adding a term proportional to the square of the Ricci scalar to the Einstein–Hilbert action. The action of the Starobinsky model (also called the R^2 model) is[12]

$$S = \frac{1}{16\pi G} \int d^4x \sqrt{-g} \left[R + \frac{R^2}{6M^2} \right] , \tag{21.87}$$

where M is a new mass scale that characterizes the model, and fixes the energy scale at which inflation takes place, while the factor $1/6$ is a conventional normalization of M^2. Actually, eq. (21.87) is a member of larger class of modified-gravity theories, known as $f(R)$ gravity, whose action has the form

$$S = \frac{1}{16\pi G} \int d^4x \sqrt{-g} \, f(R) , \tag{21.88}$$

[12]Recall that our signature is $(-,+,+,+)$. If one rather uses the signature $(+,-,-,-)$, the Ricci scalar changes sign and the term in square brackets becomes $[-R + R^2/(6M^2)]$, i.e. the Einstein–Hilbert term changes sign while the R^2 term does not.

and which have been widely studied both with motivation coming from inflation in the early Universe and as a possible explanation for dark energy. It is instructive, and takes no extra effort, to work out the equations of motion and some of their consequences, for a generic function $f(R)$. We perform the variation using the usual result (21.23) for the variation of $\sqrt{-g}$. The variation of $f(R)$ is given by $\delta f(R) = f'(R)\delta R$, where $f'(R) = df/dR$, and the variation of the Ricci scalar is

$$\delta R = R_{\mu\nu}\delta g^{\mu\nu} + \left(g_{\mu\nu}\Box - \nabla_\mu\nabla_\nu\right)\delta g^{\mu\nu}. \tag{21.89}$$

In GR, the first term in eq. (21.89) combines with the variation of $\sqrt{-g}$ to give the Einstein tensor, while the second term gives a total derivative in the action and does not contribute to the equations of motion. In $f(R)$ gravity, in contrast, integrating by parts twice, we get the operator $(g_{\mu\nu}\Box - \nabla_\mu\nabla_\nu)$ acting on $f'(R)$. Thus, the equation of motion derived from the action (21.88) is[13]

$$f'(R)R_{\mu\nu} - \frac{1}{2}f(R)g_{\mu\nu} + \left(g_{\mu\nu}\Box - \nabla_\mu\nabla_\nu\right)f'(R) = 8\pi G T_{\mu\nu}, \tag{21.90}$$

where $T_{\mu\nu}$ is the usual energy–momentum tensor of matter. For $f(R) = R$ we have $f'(R) = 1$ and we recover the Einstein equations. An important and rather general property of $f(R)$ gravity can be derived by taking the trace of eq. (21.90), which gives

$$f'(R)R - 2f(R) + 3\Box f'(R) = 8\pi G T, \tag{21.91}$$

In GR this equation becomes $R = -8\pi G T$, so the Ricci scalar is determined algebraically in terms of the trace of the energy–momentum tensor. In $f(R)$ theory, in contrast, this can be taken as a dynamical equation for the field

$$\varphi \equiv f'(R). \tag{21.92}$$

If we denote by $R = R_0(\varphi)$ the inverse of eq. (21.92) (if it exists, which of course is not the case in GR), then eq. (21.91) reads

$$\Box\varphi + V_\phi(\varphi) = \frac{8\pi G}{3}T, \tag{21.93}$$

where

$$V_\phi(\varphi) = \frac{1}{3}\left\{\varphi R_0(\varphi) - 2f[R_0(\varphi)]\right\}. \tag{21.94}$$

We see from this expression that $f(R)$ gravity has an extra scalar propagating degree of freedom compared to GR.

We next study under what conditions eq. (21.91) admits a de Sitter solution in the absence of matter. For a de Sitter space R is constant and $R_{\mu\nu} = (1/4)Rg_{\mu\nu}$. For such an ansatz the full equation (21.90) is equivalent to its trace, eq. (21.91). Furthermore, when R is constant, $\Box f'(R) = 0$. Then, in vacuum, eq. (21.91) becomes

$$f'(R)R - 2f(R) = 0. \tag{21.95}$$

[13] Note that in GR we can either use the metric as the only independent variable in the variation, or treat the metric and the Christoffel symbols as independent variables. In GR the latter formalism, known as the Palatini formalism, gives the same equations of motion. This is no longer the case in $f(R)$ gravity. In the following, we always consider the action derived using the metric as the only independent variable in the variational principle.

For $f(R) = \alpha R^2$, with α an arbitrary constant, this equation is satisfied identically. Thus, the Starobinsky model has an approximate vacuum de Sitter solution when the term $R^2/(6M^2) \gg R$, i.e. at large curvatures, $R \gg 6M^2$. Note that we are taking M smaller than the Planck mass, so it make sense to consider the regime $R \gg M$ without worrying about super-Planckian curvatures. We will indeed see that the model provides a primordial spectrum of density perturbations in agreement with the CMB normalization for $M \simeq 10^{13}$ GeV, which is well below $m_{\rm Pl}$.

To study the cosmological solutions of the Starobinsky model more quantitatively we now specialize to $f(R) = R + R^2/(6M^2)$ and we consider a spatially flat FRW metric. Then, using the fact that in FRW $R = 6(\dot{H} + 2H^2)$, the (00) component of eq. (21.90) in vacuum reads

$$\dot{H}\left(1 - \frac{\dot{H}}{6H^2}\right) + \frac{\ddot{H}}{3H} = -\frac{M^2}{6}. \tag{21.96}$$

As we already know from the discussion following eq. (21.95), in the regime $R \gg 6M^2$ (i.e. $H^2 \gg M^2$) there is an approximate de Sitter solution. We can therefore search for a slow-roll solution of eq. (21.96). In the slow-roll regime, $|\ddot{H}/H| \ll |\dot{H}|$ and $|\dot{H}/H^2| \ll 1$. Then, to lowest order, the second and third terms on the left-hand side of eq. (21.96) can be neglected, and we get $\dot{H} \simeq -M^2/6$. So, if we denote by t_i an initial time during the slow-roll phase,

$$H(t) \simeq H(t_i) - \frac{M^2}{6}(t - t_i). \tag{21.97}$$

Comparing with eq. (21.45) we see that the slow-roll parameter ε_H at time t is given by

$$\varepsilon_H = \frac{M^2}{6H^2(t)}. \tag{21.98}$$

If we start from $H^2 \gg M^2$, we have $\varepsilon_H \ll 1$ and we are in the slow-roll regime. However, $H(t)$ decreases, so eventually ϵ_H becomes of order 1. At this point the slow-roll condition is no longer satisfied and inflation ends.

To better understand the existence of an extra scalar degree of freedom in $f(R)$ gravity, and to make contact with the scalar-field inflation discussed in Sections 21.1.3 and 21.1.4, we can rewrite the $f(R)$ theory as a scalar–tensor theory as follows. The action (21.88) is equivalent to the action obtained by introducing an auxiliary scalar field χ and given by

$$S[g_{\mu\nu}, \chi] = \frac{1}{16\pi G} \int d^4x \sqrt{-g} \left[f(\chi) + (R - \chi)f'(\chi)\right], \tag{21.99}$$

where $f'(\chi) \equiv df/d\chi$. Indeed, the variation with respect to χ gives

$$f''(\chi)(R - \chi) = 0, \tag{21.100}$$

so, as long as $f''(\chi) \neq 0$, we have $\chi = R$. Plugging this back into eq. (21.99) we get eq. (21.88). We now introduce the field $\varphi = f'(\chi)$. Since χ is fixed algebraically by its equation of motion to the value $\chi = R$, this is actually the same as the field φ defined in eq. (21.92). We assume that the function $f(\chi)$ is such that we can invert this definition to obtain $\chi = \chi(\varphi)$ (which is the case in particular for the Starobinsky model). Then eq. (21.99) takes the form

$$S[g_{\mu\nu}, \varphi] = \frac{1}{16\pi G} \int d^4x \sqrt{-g}\, [\varphi R - U(\varphi)] \,, \qquad (21.101)$$

where

$$U(\varphi) = \chi(\varphi)\varphi - f[\chi(\varphi)] \,. \qquad (21.102)$$

We now define a conformally rescaled metric

$$\tilde{g}_{\mu\nu} = \varphi g_{\mu\nu} \,. \qquad (21.103)$$

Denoting by a tilde the quantities constructed with the metric $\tilde{g}_{\mu\nu}$, we have

$$R = \varphi \left[\tilde{R} + 3\tilde{\Box} \log \varphi - \frac{3}{2}\tilde{g}^{\mu\nu} \partial_\mu \log \varphi \, \partial_\nu \log \varphi \right] \,, \qquad (21.104)$$

while $\sqrt{-g} = \varphi^{-2}\sqrt{-\tilde{g}}$. Then

$$\sqrt{-g}\, \varphi R = \sqrt{-\tilde{g}} \left[\tilde{R} + 3\tilde{\Box} \log \varphi - \frac{3}{2}\tilde{g}^{\mu\nu} \partial_\mu \log \varphi \, \partial_\nu \log \varphi \right] \,. \qquad (21.105)$$

The term $\tilde{\Box} \log \varphi$ is a total covariant derivative and drops from the action. We finally introduce a new scalar field ϕ from

$$\log \varphi = \sqrt{\frac{2}{3}} \frac{\phi}{m_{\rm Pl}} \,. \qquad (21.106)$$

Then the action (21.101) becomes

$$S[\tilde{g}_{\mu\nu}, \varphi] = \int d^4x \sqrt{-\tilde{g}} \left[\frac{1}{16\pi G}\tilde{R} - \frac{1}{2}\tilde{g}^{\mu\nu} \partial_\mu \phi \partial_\nu \phi - V(\phi) \right] \,, \qquad (21.107)$$

where

$$V(\phi) = \frac{m_{\rm Pl}^2}{2} \frac{U(\varphi)}{\varphi^2} \,, \qquad (21.108)$$

with $\varphi(\phi)$ given by eq. (21.106). In particular, for the Starobinsky model $f(\chi) = \chi + \chi^2/(6M^2)$, so $\chi(\varphi) = 3M^2(\varphi - 1)$, $U(\varphi) = (3/2)M^2(\varphi - 1)^2$, and

$$V(\phi) = \frac{3}{4}m_{\rm Pl}^2 M^2 \left(1 - e^{-\sqrt{\frac{2}{3}}\frac{\phi}{m_{\rm Pl}}} \right)^2 \,. \qquad (21.109)$$

This potential is shown in Fig. 21.2. We see that it has a long, flat part at large values of the field, where the slow-roll conditions can be realized. We use the notation

$$x = \sqrt{\frac{2}{3}} \frac{\phi}{m_{\rm Pl}} \,, \qquad (21.110)$$

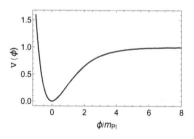

Fig. 21.2 The potential of the Starobinsky model. $V(\phi)$ is in units of $(3/4)m_{\rm Pl}^2 M^2$.

so $x = \log \varphi$. The slow-roll parameters in the scalar–tensor formulation can be computed using eqs. (21.43) and (21.44), which give

$$\varepsilon_V(x) = \frac{4}{3} \frac{1}{(e^x - 1)^2}, \tag{21.111}$$

$$\eta_V(x) = -\frac{4}{3} \frac{e^{-x}(1 - 2e^{-x})}{(1 - e^{-x})^2}. \tag{21.112}$$

The functions $\varepsilon_V(x)$ and $-\eta_V(x)$ are shown in Fig. 21.3. In the limit of large x, $\varepsilon_V \simeq (4/3)e^{-2x}$ and $\eta_V \simeq -(4/3)e^{-x}$, so the slow-roll parameters satisfy $\varepsilon_V \ll |\eta_V| \ll 1$. However, we see from the plot that, evolving toward smaller x, the condition $|\varepsilon_V| < 1$ is violated before the condition $|\eta_V| < 1$, and slow roll terminates when $x \simeq 0.8$, i.e.

$$\phi_e \simeq 1.0 \, m_{\mathrm{Pl}}. \tag{21.113}$$

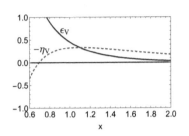

Fig. 21.3 The slow-roll parameters ε_V and $-\eta_V$ versus $x \equiv \log \varphi$.

The number of e-folds to the end of inflation can be computed using eq. (21.61), and we get

$$N_e(x) \simeq \frac{3}{2} \int_{x_e}^{x} dx_1 \frac{V(x_1)}{dV/dx_1}$$

$$\simeq \frac{3}{4} e^x, \tag{21.114}$$

where we have again used the fact that the integral is dominated by the upper integration limit. Thus

$$N_e(\phi) \simeq \frac{3}{4} \exp\left\{ \sqrt{\frac{2}{3}} \frac{\phi}{m_{\mathrm{Pl}}} \right\}, \tag{21.115}$$

and the relation between the value ϕ_i of the field ϕ when inflation begins, and the total number ΔN of inflationary e-folds is

$$\phi_i \simeq m_{\mathrm{Pl}} \left(5.36 + 1.20 \log \frac{\Delta N}{60} \right). \tag{21.116}$$

In this sense, the Starobinsky model is a large-field model, since inflation takes place for super-Planckian values of the field. Nevertheless, we will see that, in contrast to the large-field models discussed in Section 21.1.4, it predicts a small value of the tensor-to-scalar ratio r. Using eq. (21.114) we can rewrite the slow-roll parameters (21.111) and (21.112) in terms of N_e. In particular, for $x \gg 1$

$$\varepsilon_V \simeq \frac{3}{4N_e^2}, \tag{21.117}$$

$$\eta_V \simeq -\frac{1}{N_e}. \tag{21.118}$$

These relations will be useful when we compute the spectra of primordial scalar and tensor perturbations predicted by the model.

21.2   Quantum fields in curved space

A crucial feature of inflation is that it provides a framework for predicting the primordial spectrum of scalar and tensor perturbations. We will see that, for both scalar and tensor perturbations, the problem can be reduced to the study of a quantum scalar field in the FRW background. In Section 21.2.1 we begin by studying the quantization of a scalar field in a generic curved space. This is a subject that is of great interest in itself, and some of these more general results will also be useful later. In Section 21.2.2 we will then specialize to quantum fields in FRW space-time.

21.2.1   Field quantization in curved space

Let us consider a real massive scalar field ϕ in curved space. With our signature $(-,+,+,+)$ its action is

$$S = -\frac{1}{2} \int d^4x \sqrt{-g} \left[g^{\mu\nu} \partial_\mu \phi \partial_\nu \phi + m^2 \phi^2 \right] , \qquad (21.119)$$

and the equation of motion is the Klein–Gordon (KG) equation

$$(\Box - m^2)\phi = 0 , \qquad (21.120)$$

where

$$\Box \phi = \nabla_\mu \nabla^\mu \phi = \frac{1}{\sqrt{-g}} \partial_\mu \left(\sqrt{-g} \, g^{\mu\nu} \partial_\nu \right) \phi . \qquad (21.121)$$

In flat space the solutions are the plane waves $e^{\pm ikx}$, with $k_0^2 \equiv \omega_k^2 = \mathbf{k}^2 + m^2$. We can then define the functions

$$u_\mathbf{k}(x) = \frac{1}{\sqrt{2\omega_k}} e^{ikx} = \frac{1}{\sqrt{2\omega_k}} e^{-i\omega_k t + i\mathbf{k}\cdot\mathbf{x}} , \qquad (21.122)$$

and write the classical field as

$$\phi(x) = \int \frac{d^3k}{(2\pi)^3} \left[a_\mathbf{k} u_\mathbf{k}(x) + a_\mathbf{k}^* u_\mathbf{k}^*(x) \right] . \qquad (21.123)$$

Given two classical (and in general complex) solutions u_1 and u_2 of the KG equation in flat space, we can define a scalar product

$$\langle u_1 | u_2 \rangle = i \int d^3x \left[u_1^* \partial_0 u_2 - (\partial_0 u_1^*) u_2 \right] . \qquad (21.124)$$

Since the integral on the right-hand side is only over d^3x, while $u_{1,2}$ are functions of t and $\mathbf{x}$, the left-hand side is in principle still a function of time. However, if u_1 and u_2 are solutions of the KG equation, the integral is actually time-independent, as we see immediately on taking its time derivative and using the KG equation.[14] This conserved scalar product can be used to express the various conserved quantities of the theory, that are obtained from the Noether theorem, in terms of the generators of the corresponding symmetries.[15]

[14] And assuming that the fields decay sufficiently fast at spatial infinity that we can integrate the spatial derivatives by parts. In practice this can be done also for plane waves, with the understanding that they must be convolved to wave-packets.

[15] See Maggiore (2005), Section 3.3.1.

This construction can be generalized to curved space, where the use of the conserved scalar product is particularly convenient, especially in the context of the Bogoliubov transformation that we will introduce. Let $u_i(x)$ be a complete set of solutions of the curved-space KG equation, where the index i labels the solutions. We use for simplicity a discrete notation, so we expand the classical field as

$$\phi(x) = \sum_i [a_i u_i(x) + a_i^* u_i^*(x)] .$$

$$(21.125)$$

In spatially homogeneous space-times (such as FRW) the index i is again the spatial momentum $\mathbf{k}$, and we will pass from this discrete notation to a continuum notation using the convention

$$\sum_i \to \int \frac{d^3k}{(2\pi)^3} ,$$

$$(21.126)$$

which corresponds to a choice of normalization of the $a_\mathbf{k}$ coefficients. Correspondingly,

$$\delta_{ij} \to (2\pi)^3 \delta^{(3)}(\mathbf{k} - \mathbf{k}') .$$

$$(21.127)$$

The construction of the scalar product can be generalized to curved space by choosing a spatial hypersurface Σ (taken to be a Cauchy surface in a globally hyperbolic space-time) with unit normal n^μ, and writing

$$\langle u_1 | u_2 \rangle = i \int_\Sigma d\Sigma^\mu \, J_\mu ,$$

$$(21.128)$$

where

$$J_\mu = u_1^* \partial_\mu u_2 - (\partial_\mu u_1^*) u_2 .$$

$$(21.129)$$

The analogue of the fact that in Minkowski space the scalar product (21.124) does not depend on t is now the statement that the scalar product (21.128) is independent of the hypersurface Σ (whose position in flat space is labeled by the time coordinate t). This statement can be proved[16] by taking another spatial hypersurface Σ' and observing that (if the fields $u_{1,2}$ decay sufficiently fast at spatial infinity on both Σ and Σ')

$$\int_\mathcal{M} d^4x \sqrt{-g} \, \nabla^\mu J_\mu = \int_\Sigma d\Sigma^\mu \, J_\mu - \int_{\Sigma'} d\Sigma^\mu \, J_\mu ,$$

$$(21.130)$$

[16]See Hawking and Ellis (1973), Section 2.8.

where $\mathcal{M}$ is the four-dimensional space bounded by the three-dimensional spatial hypersurfaces Σ and Σ'. If u_1 and u_2 satisfy the KG equation, we have

$$\nabla^\mu J_\mu = u_1 \Box u_2 - (\Box u_1) u_2 = 0 ,$$

$$(21.131)$$

and therefore the scalar product is indeed independent of Σ. We see from the definition (21.124) that the scalar product satisfies

$$\langle u_1 | u_2 \rangle^* = \langle u_2 | u_1 \rangle ,$$

$$(21.132)$$

$$\langle u_1^* | u_2^* \rangle = -\langle u_2 | u_1 \rangle .$$

$$(21.133)$$

Furthermore, it is linear in the second variable and anti-linear in the first,

$$\langle u_1|\alpha u_2 + \beta u_3\rangle = \alpha\langle u_1|u_2\rangle + \beta\langle u_1|u_3\rangle\,, \tag{21.134}$$
$$\langle \alpha u_1 + \beta u_2|u_3\rangle = \alpha^*\langle u_1|u_3\rangle + \beta^*\langle u_2|u_3\rangle\,. \tag{21.135}$$

The modes $u_i(x)$ can be chosen so that they are orthonormal with respect to this scalar product,

$$\langle u_i|u_j\rangle = \delta_{ij}\,, \tag{21.136}$$
$$\langle u_i|u_j^*\rangle = 0\,. \tag{21.137}$$

From eq. (21.133) it also follows that

$$\langle u_i^*|u_j^*\rangle = -\delta_{ij}\,. \tag{21.138}$$

The solutions $u_i(x)$ of the KG equations that obey eqs. (21.136) and (21.137) are called the *mode functions*. Passing to a quantum description, the field is expanded as

$$\phi(x) = \sum_i \left[a_i u_i(x) + a_i^\dagger u_i^*(x)\right]\,, \tag{21.139}$$

where the creation and annihilation operators satisfy

$$[a_i, a_j^\dagger] = \delta_{ij}\,, \tag{21.140}$$

together with $[a_i, a_j] = 0$, $[a_i^\dagger, a_j^\dagger] = 0$. Using eqs. (21.122), (21.126) and (21.127) we see that in the flat-space limit we recover the standard flat-space quantization, with the usual normalization factors. The orthonormality properties (21.136) and (21.138) of the modes allow us to invert eq. (21.139) very simply,

$$a_i = \langle u_i|\phi\rangle\,, \tag{21.141}$$
$$a_i^\dagger = -\langle u_i^*|\phi\rangle\,. \tag{21.142}$$

A crucial difference of QFT in curved space, compared with flat space, is that the choice of the modes used to perform the quantization procedure is not unique. In Minkowski space one requires that the modes are of "positive energy" with respect to time t, i.e. that they are positive eigenvalues of the operator $i\partial/\partial t$, where $\partial/\partial t$ is the Killing vector associated with the invariance of Minkowski space under time translations. This selects the mode proportional to $e^{-i\omega_k t}$ (where $\omega_k > 0$). In curved space there is no such unambiguous definition, since the invariance under diffeomorphisms does not allow us to select a preferred definition of time. Modes that are "positive frequency" with respect to a variable t are in general a mixture of positive and negative frequencies with respect to another variables $t'(t)$. Indeed, the expansion of $e^{-i\omega t'(t)}$ in Fourier modes involves in general both positive and negative frequencies with respect to t, and in a general space-time (and in particular in FRW) there is no Killing vector associated with time translations that could provide a privileged definition of time. As a consequence, there is no unique choice of the mode functions. Given a complex mode function $u_i(x)$ we can construct the quantities

$$U_i(x) = \sum_j \left[\alpha_{ij} u_j(x) + \beta_{ij} u_j^*(x) \right] . \tag{21.143}$$

Clearly, $U_i(x)$ still satisfy the KG equation. We now require that also the $U_i(x)$ are mode functions, i.e. that

$$\langle U_i | U_j \rangle = \delta_{ij} , \tag{21.144}$$
$$\langle U_i | U_j^* \rangle = 0 . \tag{21.145}$$

We then find from eqs. (21.136), (21.137) and (21.143) that the coefficients α_{ij} and β_{ij} must satisfy

$$\sum_k (\alpha_{ik} \alpha_{jk}^* - \beta_{ik} \beta_{jk}^*) = \delta_{ij} \tag{21.146}$$

and

$$\sum_k (\alpha_{ik} \beta_{jk} - \alpha_{jk} \beta_{ik}) = 0 . \tag{21.147}$$

The transformation (21.143), with the coefficients α_{ij} and β_{ij} chosen to satisfy eqs. (21.146) and (21.147), is called a *Bogoliubov transformation*. We can then expand the field with respect to the U_i modes,

$$\phi(x) = \sum_i \left[A_i U_i(x) + A_i^\dagger U_i^*(x) \right] . \tag{21.148}$$

This defines new creation and annihilation operators A_i and $A_i^\dagger$. Equations (21.144) and (21.145) allow us to invert eq. (21.148) just as we have done in eqs. (21.141) and (21.142) for a_i and $a_i^\dagger$,

$$A_i = \langle U_i | \phi \rangle , \tag{21.149}$$
$$A_i^\dagger = -\langle U_i^* | \phi \rangle . \tag{21.150}$$

Inserting here the expression (21.143) for U_i, and using eq. (21.135) we get

$$A_i = \sum_j \left(\alpha_{ij}^* a_j - \beta_{ij}^* a_j^\dagger \right) , \tag{21.151}$$

$$A_i^\dagger = \sum_j \left(\alpha_{ij} a_j^\dagger - \beta_{ij} a_j \right) . \tag{21.152}$$

Then, computing the commutator and making use of eq. (21.146), we get

$$[A_i, A_j^\dagger] = \delta_{ij} . \tag{21.153}$$

Similarly, eq. (21.147) ensures that $[A_i, A_j] = 0$, $[A_i^\dagger, A_j^\dagger] = 0$. Thus, the orthonormality of the mode functions, or equivalently eqs. (21.146) and

(21.147), ensures that A_i and $A_j^\dagger$ satisfy the standard commutation relations. It is useful to write down explicitly also the inverse of eq. (21.143), which is

$$u_i(x) = \sum_j \left[\alpha_{ji}^* U_j(x) - \beta_{ji} U_j^*(x) \right] , \qquad (21.154)$$

as can be verified by direct substitution into eq. (21.143), making use of eqs. (21.146) and (21.147). Similarly, inversion of eqs. (21.151) and (21.152) gives

$$a_i = \sum_j \left(\alpha_{ji} A_j + \beta_{ji}^* A_j^\dagger \right) , \qquad (21.155)$$

$$a_i^\dagger = \sum_j \left(\alpha_{ji}^* A_j^\dagger + \beta_{ji} A_j \right) . \qquad (21.156)$$

Finally, we observe that, using eq. (21.143) together with eqs. (21.136)–(21.138), the Bogoliubov coefficients α_{ij} and β_{ij} can be compactly written as

$$\alpha_{ij} = \langle u_j | U_i \rangle , \qquad (21.157)$$

$$\beta_{ij} = -\langle u_j^* | U_i \rangle . \qquad (21.158)$$

We see that the introduction of the scalar product (21.128) allows us to write in a rather compact manner the Bogoliubov coefficients and to provide an elegant derivation of relations such as eq. (21.151).

The modes a_i define a vacuum state, which we call $|0_a\rangle$, by the condition that, for all i,

$$a_i |0_a\rangle = 0 . \qquad (21.159)$$

We can then construct the Fock space by acting on $|0_a\rangle$ with $a_i^\dagger$. Similarly, using the operators A_i we can define another vacuum state, which we call $|0_A\rangle$, requiring that for all i

$$A_i |0_A\rangle = 0 . \qquad (21.160)$$

We can then construct the Fock space by acting on $|0_A\rangle$ with $A_i^\dagger$. The vacuum states $|0_a\rangle$ and $|0_A\rangle$ and the corresponding definitions of one-particle states, $a_i^\dagger |0_a\rangle$ and $A_i^\dagger |0_A\rangle$, are physically different if the Bogoliubov coefficients β_{ij} are non-vanishing. Indeed, eq. (21.151) gives

$$A_i |0_a\rangle = -\sum_k \beta_{ik}^* a_k^\dagger |0_a\rangle . \qquad (21.161)$$

If $\beta_{ij} \neq 0$ the right-hand side is non-zero, so the vacuum with respect to the a_i operator is not a vacuum state with respect to A_i. Rather, $|0_a\rangle$ is a multi-particle state with respect to the particles defined by $A_i^\dagger |0_A\rangle$, with occupation number

$$\langle 0_a | A_i^\dagger A_i | 0_a \rangle = \sum_{jk} \beta_{ij} \beta_{ik}^* \langle 0_a | a_j a_k^\dagger | 0_a \rangle$$

$$= \sum_j |\beta_{ij}|^2 , \qquad (21.162)$$

where in the second line we used eq. (21.140). Similarly, $|0_A\rangle$ is a multi-particle state with respect to the particle states defined by $a_i^\dagger |0_a\rangle$. The explicit relation between $|0_A\rangle$ and $|0_a\rangle$ can be obtained using

$$
\begin{aligned}
0 &= a_i |0_a\rangle \\
&= \sum_j \left(\alpha_{ji} A_j + \beta_{ji}^* A_j^\dagger \right) |0_a\rangle \,.
\end{aligned}
\tag{21.163}
$$

This equation is solved by

$$
|0_a\rangle = \mathcal{N} e^{-\frac{1}{2} \sum_{ij} C_{ij} A_i^\dagger A_j^\dagger} |0_A\rangle \,,
\tag{21.164}
$$

where

$$
C_{ij} = \sum_k \beta_{ik}^* (\alpha^{-1})_{kj} \,,
\tag{21.165}
$$

and $\mathcal{N}$ is a normalization constant.[17] We see that, in QFT in curved space, there is no unique choice of the vacuum state. The choice of vacuum is not just a formal mathematical step. The appropriate vacuum state depends on the physics of the problem, as well as on the observer. We will see some explicit examples in the following discussion.

For future reference, it is also useful to collect explicitly the formulas for the Bogoliubov transformation when the index i actually corresponds to momentum $\mathbf{k}$, using eqs. (21.126) and (21.127). Then eq. (21.143) becomes

$$
U_{\mathbf{k}}(x) = \int \frac{d^3 k'}{(2\pi)^3} \left[\alpha_{\mathbf{k}\mathbf{k}'} u_{\mathbf{k}'}(x) + \beta_{\mathbf{k}\mathbf{k}'} u_{\mathbf{k}'}^*(x) \right] \,,
\tag{21.167}
$$

while eqs. (21.146) and (21.147) become

$$
\int \frac{d^3 k}{(2\pi)^3} \left(\alpha_{\mathbf{k}_1 \mathbf{k}} \alpha_{\mathbf{k}_2 \mathbf{k}}^* - \beta_{\mathbf{k}_1 \mathbf{k}} \beta_{\mathbf{k}_2 \mathbf{k}}^* \right) = (2\pi)^3 \delta^{(3)}(\mathbf{k}_1 - \mathbf{k}_2) \,,
\tag{21.168}
$$

$$
\int \frac{d^3 k}{(2\pi)^3} \left(\alpha_{\mathbf{k}_1 \mathbf{k}} \beta_{\mathbf{k}_2 \mathbf{k}} - \beta_{\mathbf{k}_1 \mathbf{k}} \alpha_{\mathbf{k}_2 \mathbf{k}} \right) = 0 \,.
\tag{21.169}
$$

Finally, eqs. (21.151) and (21.155) become, respectively,

$$
A_{\mathbf{k}} = \int \frac{d^3 k'}{(2\pi)^3} \left[\alpha_{\mathbf{k}\mathbf{k}'}^* a_{\mathbf{k}'} - \beta_{\mathbf{k}\mathbf{k}'}^* a_{\mathbf{k}'}^\dagger \right]
\tag{21.170}
$$

and

$$
a_{\mathbf{k}} = \int \frac{d^3 k'}{(2\pi)^3} \left[\alpha_{\mathbf{k}'\mathbf{k}} A_{\mathbf{k}'} + \beta_{\mathbf{k}'\mathbf{k}}^* A_{\mathbf{k}'}^\dagger \right] \,,
\tag{21.171}
$$

while eq. (21.162) reads

$$
\langle 0_a | A_{\mathbf{k}}^\dagger A_{\mathbf{k}} | 0_a \rangle = \int \frac{d^3 k'}{(2\pi)^3} |\beta_{\mathbf{k}\mathbf{k}'}|^2 \,.
\tag{21.172}
$$

[17] This can be proved by inserting this expression into eq. (21.163) and expanding the exponentials order by order. A simpler and more elegant derivation can be obtained by representing the commutation relation (21.153) on a space of functions $\psi(\xi)$, with ξ_i auxiliary variables, identifying $A_i^\dagger = \xi_i$ and $A_i = \partial/\partial \xi_i$. Equation (21.163) is then transformed into a differential equation

$$
\sum_j \left(\alpha_{ji} \frac{\partial}{\partial \xi_j} + \beta_{ji}^* \xi_j \right) \psi(\xi) = 0 \,,
\tag{21.166}
$$

where $\psi(\xi) = \langle \xi | 0_a \rangle$ is the wavefunction in the ξ representation, and whose solution gives eq. (21.164). In the case of FRW this computation is performed in detail in the Solved Exercises section of Mukhanov and Winitzki (2007), but the same trick can also be used for the general case.

21.2.2   Quantum fields in a FRW background

We now specialize the above general considerations to the case of a massless real scalar field in a spatially-flat FRW metric. Using cosmic time t, the action (21.119) becomes

$$S = \frac{1}{2} \int d^3x\, dt\, a^3 \left(\dot{\phi}^2 - a^{-2}\partial_i\phi\partial_i\phi \right),$$ (21.173)

and the corresponding equation of motion is

$$\ddot{\phi} + 3H\dot{\phi} - \frac{1}{a^2}\nabla^2\phi = 0,$$ (21.174)

where $\nabla^2 = \partial_i\partial_i$ is the flat-space Laplacian. Performing the spatial Fourier transform we get

$$\ddot{\phi}_{\mathbf{k}} + 3H\dot{\phi}_{\mathbf{k}} + \frac{k^2}{a^2}\phi_{\mathbf{k}} = 0,$$ (21.175)

where we use the notation $\phi_{\mathbf{k}}(t)$ for the Fourier modes $\tilde{\phi}(t,\mathbf{k})$. Using conformal time, eq. (21.175) becomes

$$\phi_{\mathbf{k}}'' + 2\mathcal{H}\phi_{\mathbf{k}}' + k^2\phi_{\mathbf{k}} = 0.$$ (21.176)

As in the tensor case [compare with eq. (19.221)] it is useful to introduce the field

$$\chi(\eta,\mathbf{x}) = a(\eta)\phi(\eta,\mathbf{x}).$$ (21.177)

Then (after an integration by parts)

$$S = \frac{1}{2} \int d^3x\, d\eta \left(\chi'^2 - \partial_i\chi\partial_i\chi + \frac{a''}{a}\chi^2 \right),$$ (21.178)

where, as usual, the prime denotes the derivative with respect to η. The corresponding equation of motion is

$$\chi_{\mathbf{k}}'' + \left(k^2 - \frac{a''}{a} \right)\chi_{\mathbf{k}} = 0,$$ (21.179)

where again we have performed the spatial Fourier transform. In terms of the variable χ and conformal time η, the action is formally the same as that of a scalar field in Minkowski space, with a time-dependent mass $m_\chi^2(\eta) = -a''/a$. At the level of the equation of motion, there is therefore no friction term in the equation for $\chi_{\mathbf{k}}(\eta)$, while in the equation for $\phi_{\mathbf{k}}(t)$ the term $3H\dot{\phi}_{\mathbf{k}}$ appears.

In FRW the mode functions $u_{\mathbf{k}}(\eta,\mathbf{x})$ can be written in a separable form

$$u_{\mathbf{k}}(\eta,\mathbf{x}) = f_k(\eta)e^{i\mathbf{k}\cdot\mathbf{x}},$$ (21.180)

thanks to the invariance of the FRW space-time under spatial transformations. Furthermore, the invariance under spatial rotations ensures that $f_k(\eta)$ depends on the momentum $\mathbf{k}$ only through $k = |\mathbf{k}|$. Choosing

602 *Inflation and primordial perturbations*

the hypersurface Σ as a hypersurface of constant η, the scalar product (21.128) reads

$$\langle u_{\mathbf{k}} | u_{\mathbf{k}'} \rangle = i \int d^3x \, e^{-i(\mathbf{k}-\mathbf{k}')\cdot\mathbf{x}} \left(f_k^* f_{k'}' - f_k'^* f_{k'} \right)$$
$$= -i(2\pi)^3 \delta^{(3)}(\mathbf{k} - \mathbf{k}') \left(f_k f_k'^* - f_k' f_k^* \right), \qquad (21.181)$$

where inside the parentheses we have been able to replace $f_{k'}$ by f_k because of the delta function multiplying the expression. Thus, the condition that $u_{\mathbf{k}}$ be mode functions reduces to the condition that

$$f_k f_k'^* - f_k' f_k^* = i. \qquad (21.182)$$

In FRW we will use the term "mode functions" also to denote the functions $f_k(\eta)$ normalized as in eq. (21.182). Recall that the Wronskian of two differentiable functions $g_1(\eta)$ and $g_2(\eta)$ is defined by

$$W[g_1, g_2] = g_1 g_2' - g_1' g_2. \qquad (21.183)$$

If g_1 and g_2 are solutions of a linear second-order differential equation, their Wronskian is time-independent, and if the solutions are linearly independent, the Wronskian is non-vanishing. Then, the condition (21.182) can be written as

$$W[f_k, f_k^*] = i. \qquad (21.184)$$

According to the discussion in Section 21.2.1, given some mode functions $u_{\mathbf{k}}(\eta)$ of the form (21.180), we can associate with them creation and annihilation operators $a_{\mathbf{k}}$ and $a_{\mathbf{k}}^\dagger$, expanding the field operator as

$$\chi(\eta, \mathbf{x}) = \int \frac{d^3k}{(2\pi)^3} \left(f_k(\eta) e^{i\mathbf{k}\cdot\mathbf{x}} a_{\mathbf{k}} + f_k^*(\eta) e^{-i\mathbf{k}\cdot\mathbf{x}} a_{\mathbf{k}}^\dagger \right), \qquad (21.185)$$

with

$$[a_{\mathbf{k}}, a_{\mathbf{k}'}^\dagger] = (2\pi)^3 \delta^{(3)}(\mathbf{k} - \mathbf{k}'). \qquad (21.186)$$

In FRW it is particularly interesting to consider a Bogoliubov transformation (21.167) with

$$\alpha_{\mathbf{k}\mathbf{k}'} = \alpha_k (2\pi)^3 \delta^{(3)}(\mathbf{k} - \mathbf{k}'), \qquad (21.187)$$
$$\beta_{\mathbf{k}\mathbf{k}'} = \beta_k (2\pi)^3 \delta^{(3)}(\mathbf{k} + \mathbf{k}'). \qquad (21.188)$$

Then, the new modes $U_{\mathbf{k}}(x)$ still have a separable form

$$U_{\mathbf{k}}(\eta, \mathbf{x}) = F_k(\eta) e^{i\mathbf{k}\cdot\mathbf{x}}, \qquad (21.189)$$

and, from eq. (21.167), we find that the relation between F_k and f_k is given by a particularly simple Bogoliubov transformation,

$$F_k(\eta) = \alpha_k f_k(\eta) + \beta_k f_k^*(\eta). \qquad (21.190)$$

The relation (21.168) between the Bogoliubov coefficients reduces to

$$|\alpha_k|^2 - |\beta_k|^2 = 1, \qquad (21.191)$$

and eq. (21.169) is identically satisfied. Equation (21.172) becomes

$$N_{\mathbf{k}}^{(A)} \equiv \langle 0_a | A_{\mathbf{k}}^{\dagger} A_{\mathbf{k}} | 0_a \rangle = V |\beta_k|^2 \,, \qquad (21.192)$$

where V is the spatial volume and we have used $(2\pi)^3 \delta^{(3)}(\mathbf{k} = 0) = V$. The number density of particles of type A and momentum $\mathbf{k}$, in the vacuum of the particles of type a, is then given by

$$n_{\mathbf{k}}^{(A)} \equiv \frac{N_{\mathbf{k}}^{(A)}}{V} = |\beta_k|^2 \,. \qquad (21.193)$$

Observe that $n_{\mathbf{k}}^{(A)}$ actually depends only on $k = |\mathbf{k}|$. The total number of particles of type A in the state $|0_a\rangle$ is

$$N^{(A)} = V \int \frac{d^3 k}{(2\pi)^3} |\beta_k|^2 \qquad (21.194)$$

$$= \int \frac{d^3 x \, d^3 k}{(2\pi)^3} |\beta_k|^2 \,. \qquad (21.195)$$

The latter expression makes it clear that $|\beta_k|^2$ is the occupation number per cell of phase space. Finally, the relation between vacua given in eq. (21.164) takes the form

$$|0_a\rangle = \mathcal{N} \exp \left\{ -\int \frac{d^3 k}{(2\pi)^3} \frac{\beta_k^*}{2\alpha_k} A_{\mathbf{k}}^{\dagger} A_{-\mathbf{k}}^{\dagger} \right\} |0_A\rangle \,, \qquad (21.196)$$

where $\mathcal{N}$ is a normalization factor.[18] In quantum optics a state of this form, given by the exponential of an expression quadratic in the creation operators, is called a *squeezed state*.

21.2.3 Vacuum fluctuations in de Sitter inflation

The preceding considerations hold for a generic FRW space-time. We consider now in particular a de Sitter-like FRW metric. Let us begin by recalling that the full four-dimensional de Sitter space can be defined starting from five-dimensional Minkowski space, with coordinates $X_0, X_1, \ldots, X_4$, and imposing the condition

$$-X_0^2 + \sum_{i=1}^{4} X_i^2 = H^{-2} \,, \qquad (21.197)$$

where H is a constant. This condition describes a four-dimensional hyperboloid embedded in a five-dimensional Minkowski space. More generally, D-dimensional de Sitter space-time is defined as the hyperboloid

$$-X_0^2 + \sum_{i=1}^{D} X_i^2 = H^{-2} \,, \qquad (21.198)$$

embedded in Minkowski space with D spatial dimensions plus time,

$$ds^2 = -dX_0^2 + \sum_{i=1}^{D} dX_i^2 \,. \qquad (21.199)$$

[18]The normalization factor is finite only if, for $k \to \infty$, $|\beta_k|^2$ goes to zero faster than k^{-3}, so that

$$\int \frac{d^3 k}{(2\pi)^3} \frac{|\beta_k|^2}{|\alpha_k|^2} = \int \frac{d^3 k}{(2\pi)^3} \frac{|\beta_k|^2}{1 + |\beta_k|^2}$$

converges. Otherwise, the state $|0_a\rangle$ is not part of the Fock space generated by the $A_{\mathbf{k}}^{\dagger}$ operators acting on $|0_A\rangle$. From eq. (21.195) we see that the condition that $|\beta_k|^2$ goes to zero faster than k^{-3} is also necessary to ensure the convergence of the total number density of particles, $N^{(A)}/V$. As we shall see, in the physical situations that we will consider $|\beta_k|^2$ decreases exponentially for k above a critical cutoff value, so there will be no problem of convergence.

This description makes it manifest that de Sitter space is invariant under the Lorentz group $O(1, D)$ of the embedding Minkowski space. There are several ways to choose coordinates on de Sitter space. A global system of coordinates, i.e. a system of coordinates that covers the full D-dimensional manifold, is obtained by writing

$$X_0 = H^{-1}\sinh(H\tau)\,, \tag{21.200}$$

$$X_i = H^{-1}\xi_i\cosh(H\tau)\,, \tag{21.201}$$

with the condition $\sum_{i=1}^{D}\xi_i^2 = 1$, so the ξ_i coordinates describe a $(D-1)$-dimensional unit sphere embedded in D-dimensional Euclidean space. Writing the interval eq. (21.199) in these coordinates one finds

$$ds^2 = -d\tau^2 + H^{-2}\cosh^2(H\tau)d\Omega_d^2\,, \tag{21.202}$$

where $d = D-1$ and $d\Omega_d^2$ is the metric of the d-dimensional unit sphere. This choice of coordinates has the advantages of providing a global coordinate system and of displaying clearly the topology of de Sitter space. Another useful choice is obtained by introducing coordinates (t, x_i), where $i = 1, \ldots, d$, writing

$$X_0 = H^{-1}\sinh(Ht) + \frac{1}{2}H\mathbf{x}^2 e^{Ht}\,, \tag{21.203}$$

$$X_i = e^{Ht}x_i \quad (i = 1, \ldots, d) \tag{21.204}$$

$$X_D = H^{-1}\cosh(Ht) - \frac{1}{2}H\mathbf{x}^2 e^{Ht}\,, \tag{21.205}$$

which indeed satisfy the constraint (21.198). Since

$$X_0 + X_D = H^{-1}e^{Ht}\,, \tag{21.206}$$

which is positive, these coordinates do not cover all of de Sitter space, but only one half of it. In these coordinates

$$ds^2 = -dt^2 + e^{2Ht}d\mathbf{x}^2\,. \tag{21.207}$$

This is just what we have called the "de Sitter metric" in the previous chapters. We see that it is actually just one patch of the full de Sitter space. In cosmology we are actually interested in FRW metrics that, for some limited period of time, during inflation, have the form (21.207) or, more precisely, are approximately of this form (as in the case of slow-roll inflation). Following standard usage, we will simply refer to (21.207) as the de Sitter metric, although we have seen that, more precisely, technically it only describes a patch of the full de Sitter space. In any case, for applications to inflation, it is not the "eternal" de Sitter metric that is relevant, but only a FRW metric that, for some limited time span, is close to eq. (21.207).

As usual we introduce conformal time η. Then for de Sitter space $a(\eta) = -1/(H\eta)$ with $\eta < 0$, as we saw in eq. (17.161), so $a''/a = 2/\eta^2$ and eq. (21.179) becomes

$$\chi_k'' + \left(k^2 - \frac{2}{\eta^2}\right)\chi_k = 0 \qquad \text{(de Sitter)}\,. \tag{21.208}$$

Two independent solutions of this equations are

$$f_k(\eta) = \frac{1}{\sqrt{2k}} e^{-ik\eta} \left(1 - \frac{i}{k\eta}\right) \qquad \text{(de Sitter)} \qquad (21.209)$$

and its complex conjugate. It is straightforward to verify that $f_k(\eta)$ satisfies eq. (21.182), so it is a mode function.

The quantization of a *massive* scalar field in de Sitter space is also intrinsically interesting from a field-theoretical point of view, and the result will also be useful when we compute the corrections to the leading slow-roll approximation. For a massive field the action (21.119) gives the massive KG equation $(-\Box + m^2)\phi$. Specializing it to de Sitter space and introducing again $\chi = a\phi$, one finds that eq. (21.208) is replaced by

$$\chi_k'' + \left(k^2 + \frac{m^2}{H^2\eta^2} - \frac{2}{\eta^2}\right)\chi_k = 0. \qquad (21.210)$$

We introduce

$$\nu = \left(\frac{9}{4} - \frac{m^2}{H^2}\right)^{1/2}. \qquad (21.211)$$

For $m = 0$ we have $\nu = 3/2$. Then eq. (21.210) can be rewritten as

$$\chi_k'' + \left(k^2 - \frac{\nu^2 - \frac{1}{4}}{\eta^2}\right)\chi_k = 0. \qquad (21.212)$$

The solution of this equation that reduces to eq. (21.209) for $m = 0$ is

$$f_k(\eta) = -\frac{1}{2}\sqrt{\frac{\pi}{k}} (-k\eta)^{1/2} H_\nu^{(1)}(-k\eta), \qquad (21.213)$$

where $H_\nu^{(1)}(x)$ is the Hankel function of the first kind, $H_\nu^{(1)}(x) = J_\nu(x) + iY_\nu(x)$, and $J_\nu(x)$ and $Y_\nu(x)$ are the Bessel functions of the first and second kinds, respectively. Using

$$J_{n+\frac{1}{2}}(x) = \sqrt{\frac{2x}{\pi}} j_n(x), \qquad (21.214)$$

$$Y_{n+\frac{1}{2}}(x) = \sqrt{\frac{2x}{\pi}} y_n(x), \qquad (21.215)$$

and the expressions for the spherical Bessel functions with $n = 1$,

$$j_1(x) = -\frac{d}{dx}\frac{\sin x}{x}, \qquad (21.216)$$

$$y_1(x) = +\frac{d}{dx}\frac{\cos x}{x}, \qquad (21.217)$$

we can check that, in the massless limit, the solution (21.213) reduces to the solution (21.209). The action (21.119) can also be generalized by introducing a non-minimal coupling to the curvature, replacing

$$m^2\phi^2 \to (m^2 + \xi R)\phi^2, \qquad (21.218)$$

where ξ is a parameter. In de Sitter R is constant, and its value is given by $R = 12H^2$; see eq. (17.11). Thus, the solution (21.213) also holds for the non-minimal case, with

$$\begin{aligned} \nu &= \left(\frac{9}{4} - \frac{m^2 + 12\xi H^2}{H^2} \right)^{1/2} \\ &= \left(\frac{9}{4} - 12\xi - \frac{m^2}{H^2} \right)^{1/2} . \end{aligned} \tag{21.219}$$

The vacuum defined with respect to the modes (21.209) in the massless case, or more generally with respect to the modes (21.213) for a massive non-minimally coupled field, is called the *Bunch–Davies vacuum*. It is selected physically by the fact that, well inside the horizon, where $|k\eta| \gg 1$, it reduces to the purely positive-frequency mode of Minkowski space. Thus, the Bogoliubov coefficients that connect the definition of particle selected by the modes (21.209) and the definition used in Minkowski space, reduce to $\alpha_k = 1$ and $\beta_k = 0$ in the limit $|k\eta| \gg 1$. In other words, the Bunch–Davies vacuum defined by the modes (21.209) or (21.213) coincides with the Minkowski vacuum for momenta such that the mode is well inside the horizon.

To understand quantitatively under what conditions the Bunch–Davies vacuum is the most natural choice at the beginning of an inflationary epoch, we observe that the value of k that appears in eq. (21.209) is the comoving momentum or, equivalently, the physical momentum today, setting $a_0 = 1$. Because of the cosmological redshift, the corresponding physical momentum at the beginning of inflation was very large. To get a numerical estimate, we observe, from Fig. 19.13 on page 473, that the typical values of comoving momenta relevant for cosmology range from $k \sim 10^{-3} h_0/\text{Mpc}$ for CMB scales up to, say, $10\, h_0/\text{Mpc}$ for structure formation, and even larger. Indeed, reduced wavelengths as small as a few times 10^{-2} Mpc can still be of direct cosmological relevance. According to eq. (19.65), they contain a mass of order $10^5 M_\odot$, and correspond to the first gravitationally bound objects.

The corresponding physical momenta at the beginning of inflation are given by

$$\begin{aligned} k_{\text{phys}}(t_i) &= k\, \frac{a_0}{a_i} \\ &= k\, \frac{a_0}{a_{\text{end}}} \frac{a_{\text{end}}}{a_i} , \end{aligned} \tag{21.220}$$

where as usual we denote by a_i and a_{end} the scale factors at the beginning and end of inflation, respectively. We have seen in eq. (21.11) that, to solve the flatness problem, the minimum number of e-folds must be such that $(a_{\text{end}}/a_i) \sim 0.17(a_0/a_{\text{end}})$ (assuming instantaneous reheating). If the number of total inflationary e-folds ΔN is larger than $(\Delta N)_{\text{min}}$, then (a_{end}/a_i) is even much larger, by a factor $\exp\{\Delta N - (\Delta N)_{\text{min}}\}$. Then the physical momentum at the beginning of inflation, corresponding to the comoving momentum k, is of order

$$k_{\text{phys}}(t_i) \sim 0.17 \, k \left(\frac{a_0}{a_{\text{end}}}\right)^2 e^{\Delta N - (\Delta N)_{\min}} \,. \tag{21.221}$$

To get this order-of-magnitude estimate we have assumed for simplicity an instantaneous reheating. In the same approximation the energy density at the end of inflation is the same as the radiation energy density at the beginning of RD, and $(a_0/a_{\text{end}})^2 = (\rho_{\inf}/\rho_{R,0})^{1/2}$.[19] Using the present value of the radiation energy density,

$$\rho_{R,0}^{1/4} \simeq 2.41 \times 10^{-4} \text{eV} \,, \tag{21.222}$$

and transforming from Mpc^{-1} to eV using (in units $\hbar = c = 1$)

$$1 \, \text{Mpc}^{-1} \simeq 6.4 \times 10^{-30} \, \text{eV} \,, \tag{21.223}$$

eq. (21.221) gives

$$k_{\text{phys}}(t_i) \sim 2 \times 10^{18} \text{GeV} \left(\frac{k}{1 \, \text{Mpc}^{-1}}\right) \left(\frac{\rho_{\inf}^{1/4}}{10^{16} \, \text{GeV}}\right)^2 e^{\Delta N - (\Delta N)_{\min}} \,. \tag{21.224}$$

Using the notation $\rho_{\inf}^{1/4} = M$, we can rewrite eq. (21.224) as

$$\frac{k_{\text{phys}}(t_i)}{M} \sim 200 \left(\frac{M}{10^{16} \, \text{GeV}}\right) \left(\frac{k}{1 \, \text{Mpc}^{-1}}\right) e^{\Delta N - (\Delta N)_{\min}} \,. \tag{21.225}$$

We see that, for $\Delta N \gg (\Delta N)_{\min}$, the physical momenta $k_{\text{phys}}(t_i)$ at the beginning of inflation, corresponding to the typical comoving momenta relevant for cosmology, are much larger than the energy scale of inflation itself, and they can even become super-Planckian. It is quite natural to assume that, at the beginning of inflation, the modes corresponding to these huge momenta were not populated. These considerations select the Bunch–Davies vacuum.

It should be observed, however, that if ΔN is not much larger than $(\Delta N)_{\min}$ and if the inflationary scale is low, $M \ll 10^{16} \, \text{GeV}$, then for modes of cosmological relevance we can have $k_{\text{phys}}(t_i)/M \ll 1$. Then, modes with these momenta could in principle be populated, for instance if an earlier RD phase had preceded inflation. In such cases the initial state will not necessarily be the Bunch–Davies vacuum.[20]

Observe that the solution (21.209) for χ_k diverges as $\eta \to 0^-$. This can be traced to the fact that in eq. (21.208) the term $-2/\eta^2$ formally plays the role of an effective time-dependent mass squared, $m_\chi^2(\eta) = -2/\eta^2$, which is tachyonic. However, χ is just an auxiliary field introduced in order to write the equations of motions in an easily solvable form. The physical field is $\phi(x) = \chi(x)/a(\eta)$, so in de Sitter $\phi(x) = -H\eta \chi(x)$. Therefore, writing again $\phi(x) = \phi_k(\eta)e^{i\mathbf{k}\cdot\mathbf{x}}$, the solution for $\phi_k(\eta)$ corresponding to eq. (21.209) is

$$\phi_k(\eta) = -\frac{H}{\sqrt{2k}} e^{-ik\eta} \left(\eta - \frac{i}{k}\right) \qquad \text{(de Sitter)} \,. \tag{21.226}$$

[19]It should be stressed that neglecting reheating is only useful for a first crude estimate of the orders of magnitudes involved. For instance, we will see in eq. (21.352) that, in the Starobinsky model of inflation, in order to reproduce the correct normalization for the spectrum of scalar amplitudes, the mass scale M must be of order 3×10^{13} GeV. However, with typical reheating mechanisms, the reheating temperature is rather of order 3×10^9 GeV; see for example Mijic, Morris and Suen (1986). Therefore, in this case, at the beginning of the RD phase $\rho_R^{1/4} \sim 10^{-4}\rho_{\inf}^{1/4}$.

[20]The choice of Bunch–Davies vacuum is also sometimes motivated by the fact that, when one studies quantum field theory in a full de Sitter space-time, choices of the vacuum different from the Bunch–Davies vacuum may lead to conceptual problems, such as difficulties in having well-defined loop amplitudes for interacting fields; see the Further Reading. While theoretically interesting from the point of view of QFT in curved space, these considerations are largely irrelevant for the study of inflation. As we discussed below eq. (21.207), what we call a "de Sitter" phase of inflation has nothing to do with the full de Sitter space-time. Rather, it is just an epoch in which, for a limited interval of time, the FRW metric approaches the form (21.207), which is the same as the metric in the coordinates that cover one patch of the full de Sitter space-time. The potential problems of non-Bunch–Davies vacua in de Sitter are related to the global properties of the full de Sitter space-time (for example the presence of non-local singularities in some Green's functions due to the existence of antipodal points in de Sitter space, or the failure of some analyticity requirements on the Green's function in the Euclidean continuation of de Sitter space-time) which have no relevance to a finite phase of exponential expansion in FRW. From the point of view of FRW, there is obviously no conceptual problem in assuming that, when the inflationary expansion begins, after say an earlier RD phase, the modes corresponding to the Bunch–Davies definition of particles had non-vanishing occupation number.

We see that, in the limit $\eta \to 0^-$, $\phi_k(\eta)$ goes to a constant,

$$\phi_k(\eta) \to \frac{iH}{\sqrt{2k^3}}. \qquad (21.227)$$

Even if the solution (21.209) is specific to the de Sitter metric, the existence of a mode that reduces to the flat-space mode at $|k\eta| \gg 1$, i.e. for wavelengths well inside the horizon, is a general property of the FRW background. Indeed, for dimensional reasons, the term a''/a in eq. (21.179) is $O(1/\eta^2)$, and therefore for $|k\eta| \gg 1$ we can neglect the term a''/a with respect to k^2. Then eq. (21.179) becomes a free wave equation, with solutions $\chi(\eta) \propto e^{\pm ik\eta}$.[21]

[21] The dimensional estimate $a''/a = O(1/\eta^2)$ does not hold in RD, where $a \propto \eta$ and $a''/a = 0$. In this case, however, the wave equation reduces to a free wave equation exactly.

Let us now compute the quantum expectation value $\langle 0|\chi^2(x)|0\rangle$. Writing the quantum field χ as in eq. (21.185), in a generic FRW metric we get

$$\langle 0|\chi^2(\eta, \mathbf{x})|0\rangle = \int \frac{d^3k}{(2\pi)^3} |f_k(\eta)|^2$$
$$= \frac{1}{2\pi^2} \int_0^\infty \frac{dk}{k} k^3 |f_k(\eta)|^2, \qquad (21.228)$$

where for the moment $f_k(\eta)$ are the mode functions in a generic FRW metric, associated with the (yet unspecified) vacuum $|0\rangle$ used in $\langle 0|\chi^2|0\rangle$. Recalling the definition of the power spectrum of a scalar field given by eqs. (19.169) and (19.170), we can rewrite this result as

$$\langle 0|\chi^2(\eta, \mathbf{x})|0\rangle = \int_0^\infty \frac{dk}{k} \mathcal{P}_\chi(k; \eta), \qquad (21.229)$$

where

$$\mathcal{P}_\chi(k; \eta) = \frac{1}{2\pi^2} k^3 |f_k(\eta)|^2. \qquad (21.230)$$

Since $\phi = \chi/a$, it also follows that

$$\langle 0|\phi^2(\eta, \mathbf{x})|0\rangle = \int_0^\infty \frac{dk}{k} \mathcal{P}_\phi(k; \eta), \qquad (21.231)$$

where

$$\mathcal{P}_\phi(k; \eta) = \frac{1}{2\pi^2 a^2(\eta)} k^3 |f_k(\eta)|^2. \qquad (21.232)$$

Observe that, because of the time dependence of the FRW background, the expectation value $\langle 0|\phi^2(\eta, \mathbf{x})|0\rangle$, and hence the power spectrum, in general depends on η, but is of course independent of $\mathbf{x}$, because of the spatial homogeneity of FRW. In particular in Minkowski space, where $a(\eta) = 1$ and $|f_k| = (2k)^{-1/2}$, we get

$$\mathcal{P}_\phi(k) = \frac{k^2}{4\pi^2} \qquad \text{(Minkowski)}. \qquad (21.233)$$

Consider now the vacuum fluctuations in de Sitter space, with a Bunch–Davies vacuum. Plugging into eq. (21.232) the expression for $f_k(\eta)$ given in eq. (21.209) gives

$$\mathcal{P}_\phi(k; \eta) = \frac{H^2 k^2}{4\pi^2} \left(\eta^2 + \frac{1}{k^2}\right). \qquad (21.234)$$

We see that there are two quite distinct regimes. For any given k, at sufficiently early time we have $|k\eta| \gg 1$, so the mode is well inside the horizon. In this limit the function $f_k(\eta)$ that gives the fluctuations of the field χ approaches the Minkowski value. However in the definition of the field ϕ there is an extra factor $1/a$, so $\mathcal{P}_\phi(k; \eta)$ is smaller than the Minkowski value by a factor $a^2(\eta) = H^2\eta^2$. In the opposite limit, when the mode is well outside the horizon, we have instead

$$\mathcal{P}_\phi(k) \simeq \frac{H^2}{4\pi^2} \qquad (|k\eta| \ll 1), \qquad (21.235)$$

so the power spectrum approaches a constant value, independent both of η and of k.

Observe that, given the expression (21.234) for the power spectrum, the integral in eq. (21.231) is divergent both in the ultraviolet and in the infrared. The UV divergence is cured by the standard techniques of renormalization of QFT in curved space,[22] and is not a great cause of concern. The IR divergence (which is logarithmic since, as $k \to 0$, $\mathcal{P}_\phi(k)$ goes to a constant) is more subtle, and is an indication that we have not fully understood the physics involved. In particular, in this case the IR divergence is a signal of the fact that, at scales very large compared with the horizon, the separation into a classical field satisfying eq. (21.175) plus quantum fluctuations around this configuration is no longer adequate, given that quantum fluctuations are not small. One should in principle resum the quantum contributions (including higher loops, which are also in principle large when the first loop correction becomes large) and find a new background, i.e. a new solution of the quantum-corrected equations of motion, around which to expand.

It is also interesting to compute the spectrum of perturbations, assuming a non-Bunch–Davies vacuum. During the cosmological epoch when the FRW expansion is approximately of the de Sitter type (21.207), the modes $F_k(\eta)$ associated with a generic vacuum state can be written as in eq. (21.190), so

$$F_k(\eta) = \frac{\alpha_k}{\sqrt{2k}} e^{-ik\eta} \left(1 - \frac{i}{k\eta}\right) + \frac{\beta_k}{\sqrt{2k}} e^{+ik\eta} \left(1 + \frac{i}{k\eta}\right), \qquad (21.236)$$

with α_k and β_k related by eq. (21.191). These modes reduces to the Bunch–Davies modes for $\alpha_k = 1, \beta_k = 0$. For generic values of α_k and β_k, the power spectrum is

$$\mathcal{P}_\phi(k; \eta) = \frac{1}{2\pi^2 a^2(\eta)} k^3 |F_k(\eta)|^2$$

$$= \frac{H^2 k^2 \eta^2}{4\pi^2} \qquad (21.237)$$

$$\times \left\{ \left(1 + 2|\beta_k|^2\right) \left(1 + \frac{1}{k^2\eta^2}\right) + 2\,\mathrm{Re}\left[\alpha_k \beta_k^* e^{-2ik\eta} \left(1 - \frac{2i}{k\eta} - \frac{1}{k^2\eta^2}\right)\right] \right\},$$

where we have used eq. (21.191). For modes well inside the horizon, $|k\eta| \gg 1$, the oscillating factor $e^{-2ik\eta}$ averages to zero, and we get

$$\mathcal{P}_\phi(k; \eta) \simeq \frac{H^2 k^2 \eta^2}{4\pi^2} \left(1 + 2|\beta_k|^2\right) \qquad (|k\eta| \gg 1). \qquad (21.238)$$

[22]See for example the textbook by Birrell and Davies (1982).

In the opposite limit of super-horizon modes, $|k\eta| \ll 1$, we rather get

$$\mathcal{P}_\phi(k) \simeq \frac{H^2}{4\pi^2} |\alpha_k - \beta_k|^2 \qquad (|k\eta| \ll 1), \qquad (21.239)$$

which generalizes eq. (21.235). Note that, for super-horizon modes, even in this general case $\mathcal{P}_\phi(k;\eta)$ becomes time-independent. However, $\mathcal{P}_\phi(k)$ is no longer scale-invariant, and rather has a dependence on k determined by $|\alpha_k - \beta_k|^2$.

21.3 Primordial perturbations in single-field slow-roll inflation

To compute the primordial spectrum of scalar and tensor perturbations generated in single-field slow-roll inflation, let us first examine the evolution of the physical horizon size H^{-1}, compared with the evolution of the physical wavelength of a given mode. For a cosmological model that begins from a RD phase we showed this in Fig. 19.2 on page 445. We now extend this cosmological model into the past, assuming an earlier period of slow roll inflation. During inflation the physical horizon size H^{-1} is almost constant (or, more precisely, during slow roll it slowly increases since, to lowest order in the slow-roll expansion, $\varepsilon_H \simeq \varepsilon_V > 0$ so, according to eq. (21.53), $\dot{H} < 0$). In contrast, the physical wavelength is proportional to $a(\eta)$, so eventually all physical wavelengths are inside the horizon in the far past, if the inflationary phase is sufficiently long. This behavior is shown in Fig. 21.4. In this figure we show the evolution of the modes with $\lambda/(2\pi) = 4.9$ Gpc and $\lambda/(2\pi) = 10$ Mpc. The former wavelength is so large that (for our usual reference values $h_0 = 0.7, \Omega_M = 0.3$) modes with even higher wavelengths never re-enter the horizon, because of the flattening of H^{-1} in the recent DE-dominated era. This mode therefore represents an upper limit on the wavelengths of cosmological relevance. The mode with $\lambda/(2\pi) = 10$ Mpc, in contrast, is of the order of the shortest wavelengths shown in Fig. 19.13 on page 473. By definition, the minimum number of e-folds necessary to solve the horizon problem is such that even the longest modes probed by the CMB were well inside the horizon during part of the inflationary evolution, so a causal contact could be established between perturbations separated by this comoving distance. Thus, independently of the value of the inflationary scale and of the details of reheating (in the figure we have used for definiteness $\rho_{\rm infl}^{1/4} = 10^{16}$ GeV and assumed instantaneous reheating), in a successful inflationary model the minimum number of e-folds during inflation, $\Delta N_{\rm min}$, is such that all cosmologically relevant modes were well inside the horizon at some time during inflation, which we can take as the "initial" time for the subsequent evolution. Then, they exit from the horizon at some later time, still during the inflationary phase, and re-enter much later, either during RD or MD, depending on their wavelength (as shown in more detail in Fig. 19.2 on page 445), except for modes with reduced wavelength greater than about 4.9 Gpc,

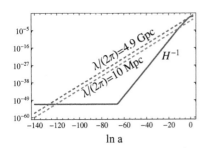

Fig. 21.4 The physical horizon size H^{-1} (solid line) in Mpc, compared with the physical reduced wavelengths $a(t)\lambda/(2\pi)$, for typical values of λ relevant for cosmology (dashed lines), plotted agains $x = \log a$, in a model including inflation followed by RD, MD and a ΛD phase. We consider for illustration a de Sitter inflation with inflationary scale $\rho_{\rm infl}^{1/4} = 10^{16}$ GeV and instantaneous reheating.

which never re-enter the horizon because they are still outside when the recent DE-dominated phase sets in.

We now have all the elements for computing the primordial spectrum of scalar and tensor perturbations predicted by single-field slow-roll inflation.

21.3.1 Mukhanov–Sasaki equation

We first consider the scalar sector. In this case the metric perturbations are described by the Bardeen variables Ψ and Φ. We will see that during inflation there are no anisotropic stresses, so $\Phi = -\Psi$ and the metric perturbations are fully described by Ψ. Furthermore we have the inflaton field ϕ, which we also split into a spatially independent background plus perturbations,

$$\phi(\eta, \mathbf{x}) = \phi_0(\eta) + \delta\phi(\eta, \mathbf{x}) \,. \tag{21.240}$$

The perturbations in the scalar sector are therefore described by $\Psi(\eta, \mathbf{x})$ and $\delta\phi(\eta, \mathbf{x})$. The full system of perturbation equations has been written down in Section 19.1 for a generic energy–momentum tensor. In the present setting the energy–momentum tensor is that of the inflaton field. Since we have only two independent variables $\Psi(\eta, \mathbf{x})$ and $\delta\phi(\eta, \mathbf{x})$ we only need two independent equations. A convenient choice is given by the (00) equation and the scalar part of the $(0i)$ equation, eqs. (19.8) and (19.12), since these are first-order in the time derivative. In contrast, the scalar part of the (ij) equations is second-order in time, and is implied by the (00) and $(0i)$ equations through the Bianchi identities.

To compute the right-hand sides of eqs. (19.8) and (19.12) in terms of $\delta\phi$ we use the energy–momentum tensor $T_{\mu\nu}$ given in eq. (21.24). As discussed in Section 18.3, when developing perturbation theory in curved space it is more convenient to use T^μ_ν rather than $T_{\mu\nu}$ since, in a generic curved space, we still have $T^0_0 = -\rho$ and therefore $\delta T^0_0 = -\delta\rho$. From eq. (21.24),

$$T^\mu_\nu = g^{\mu\rho}\partial_\rho\phi\partial_\nu\phi - \delta^\mu_\nu\left[\frac{1}{2}g^{\rho\sigma}\partial_\rho\phi\partial_\sigma\phi + V(\phi)\right] \,. \tag{21.241}$$

From this, using eq. (18.96) with $\Phi = -\Psi$, to linear order we get

$$\delta T^0_0 = -\frac{1}{a^2}\left[-(\phi'_0)^2\Psi + \phi'_0\delta\phi'\right] - \frac{dV(\phi_0)}{d\phi_0}\delta\phi \,, \tag{21.242}$$

$$\delta T^i_0 = \frac{1}{a^2}\phi'_0\partial^i\delta\phi \,. \tag{21.243}$$

Comparing with eqs. (19.5) and (19.6) we can now find the energy density perturbation and the velocity potential of the inflaton field. From eqs. (21.25) and (21.26), $\bar\rho_{\rm infl} + \bar{p}_{\rm infl} = (\phi'_0)^2/a^2$. Then

$$\delta\rho_{\rm infl} = \frac{1}{a^2}\left[-(\phi'_0)^2\Psi + \phi'_0\delta\phi'\right] + \frac{dV(\phi_0)}{d\phi_0}\delta\phi \,, \tag{21.244}$$

$$v_\phi = -\frac{\delta\phi}{\phi'_0} \,. \tag{21.245}$$

Computing T^i_j we see that the anisotropic stress term comes only from $\partial_i\delta\phi\partial_j\delta\phi$, which is of second order in the perturbation. Therefore to linear order the inflaton generates no anisotropic stress and $\Psi = -\Phi$, as we already anticipated. Inserting eqs. (21.244) and (21.245) into eqs. (19.8) and (19.12) we get a system of two equations for the two perturbation variables Ψ and $\delta\phi$,

$$\nabla^2\Psi - 3\mathcal{H}(\Psi' + \mathcal{H}\Psi) = -4\pi G\left[(\phi_0')^2\Psi - \phi_0'\delta\phi' - a^2\frac{dV(\phi_0)}{d\phi_0}\delta\phi\right],$$
(21.246)

$$\Psi' + \mathcal{H}\Psi = 4\pi G\phi_0'\delta\phi.$$
(21.247)

We next observe that, according to eq. (19.103), in our case the variable $\mathcal{R}$ is given by

$$\mathcal{R} = -\Psi - \frac{\mathcal{H}\delta\phi}{\phi_0'},$$
(21.248)

since $\Phi = -\Psi$ and the only source for the velocity potential is the inflaton field, so $v_{\rm tot} = v_\phi$. It is convenient to introduce the *Mukhanov–Sasaki* variable u,

$$u \equiv -z\mathcal{R},$$
(21.249)

where the variable z, defined by

$$z \equiv \frac{a\phi_0'}{\mathcal{H}} = \frac{a\dot\phi_0}{H},$$
(21.250)

depends only on the background evolution. Then (making use also of the classical equation of motion for the background), the two first-order equations (21.246) and (21.247) can be combined into a single second-order equation for u, the Mukhanov–Sasaki equation,

$$u'' - \frac{z''}{z}u - \nabla^2 u = 0.$$
(21.251)

After Fourier transformation, we get

$$u_{\mathbf{k}}'' + \left(k^2 - \frac{z''}{z}\right)u_{\mathbf{k}} = 0.$$
(21.252)

Analogous (but much longer) manipulations can be performed directly at the level of the quadratic action for the fluctuations. Using $\{u,\Psi\}$ as independent fields instead of $\{\delta\phi,\Psi\}$, eliminating Ψ with its own equation of motion, and dropping total derivatives, one finds that the action for u is

$$S = \frac{1}{2}\int d^3x\, d\eta\left(u'^2 - \partial_i u\partial_i u + \frac{z''}{z}u^2\right),$$
(21.253)

which is the same as the action (21.178), with a''/a replaced by z''/z. Observe that the field $u(x)$ has the standard normalization for a scalar field.

Rewriting the Mukhanov–Sasaki equation in terms of the curvature perturbations $\mathcal{R} = -u/z$ we get

$$\mathcal{R}_{\mathbf{k}}'' + \frac{2z'}{z}\mathcal{R}_{\mathbf{k}}' + k^2\mathcal{R}_{\mathbf{k}} = 0\,. \qquad (21.254)$$

In the limit $|k\eta| \to 0$, i.e. for super-horizon modes, we can drop the k^2 term. Then the equation has a solution with $\mathcal{R}_{\mathbf{k}}$ constant and a solution with $\mathcal{R}_{\mathbf{k}} \propto z^{-2}$. The latter is a decaying mode.[23] Indeed, to lowest order in the slow-roll expansion, $\phi_0'/\mathcal{H}$ is constant and $z(\eta)$ evolves as $a(\eta)$. Then the second solution is proportional to

$$\int^{\eta} \frac{d\tilde{\eta}}{z^2(\tilde{\eta})} \propto \int^{t} \frac{d\tilde{t}}{a^3(\tilde{t})} = \int^{a} \frac{d\tilde{a}}{H\tilde{a}^4}$$

$$\simeq -\frac{1}{3Ha^3} + \text{const}\,, \qquad (21.256)$$

where we have taken H constant. The constant term coming from the lower integration limit can be reabsorbed into the constant mode, so this solution represents a decaying mode $\mathcal{R}_{\mathbf{k}} \propto a^{-3}$. This result has been obtained to lowest order in the slow-roll expansion. At generic order the numerical coefficients and the exact dependence on a will change somewhat, but this will not affect the fact that the solution $\mathcal{R}_{\mathbf{k}}' \propto z^{-2}$ is a decaying mode. Thus, in the end, apart from a quickly disappearing decaying mode, outside the horizon the curvature $\mathcal{R}_{\mathbf{k}}$ is constant.[24]

21.3.2 Scalar perturbations to lowest order in slowroll

Equation (21.252) depends in a non-trivial manner on the background solution for $a(\eta)$ and $\phi_0'(\eta)$, through the variable z. The equation, however, greatly simplifies in the slow-roll regime.[25] We begin by observing that

$$z^2(\eta) = -\frac{a^2(\eta)}{4\pi G}\frac{\dot{H}}{H^2}$$

$$= \frac{a^2(\eta)}{4\pi G}\,\varepsilon_H(\eta)\,, \qquad (21.257)$$

where in the first line we have used $z = a\dot{\phi}/H$ together with eq. (21.31), and in the second line we have used the definition (21.45). In the slow-roll approximation each derivative with respect to conformal time carries one more power of ε.[26] In particular, $\varepsilon' = O(\varepsilon^2)$. Thus, when computing z''/z, the result at lowest non-trivial order is obtained by taking ε constant. This gives $z''/z = a''/a$, and eq. (21.252) becomes the same as eq. (21.179), with $\chi_{\mathbf{k}}(\eta)$ replaced by $u_{\mathbf{k}}(\eta)$. The power spectrum of $u_{\mathbf{k}}(\eta)$ can then be read immediately from eq. (21.230),

$$\mathcal{P}_u(k;\eta) = \frac{1}{2\pi^2}k^3|f_k(\eta)|^2\,. \qquad (21.258)$$

[23]There is here a subtle point concerning the $k \to 0$ limit. Indeed, if we compute the time derivative $\mathcal{R}'$ using eq. (21.248) for $\mathcal{R}$, taking Ψ' and $\delta\phi'$ from eqs. (21.246) and (21.247), and using for the background eqs. (21.29) and (21.31) (which are exact, i.e. do not assume the slow-roll expansion), we get

$$\mathcal{R}_{\mathbf{k}}' = \frac{1}{4\pi G}\frac{\mathcal{H}}{(\phi_0')^2}k^2\Psi_{\mathbf{k}}\,. \qquad (21.255)$$

Thus, $\mathcal{R}_{\mathbf{k}=0}' = 0$, so $\mathcal{R}_{\mathbf{k}=0}$ is constant. In contrast, eq. (21.254) for $k = 0$ is a second-order differential equation that, beside the constant solution, also has a decaying mode. When $k = 0$ the latter is a spurious solution that is introduced when, in the manipulations performed to get the Mukhanov–Sasaki equation, one takes a Laplacian of eq. (21.247) (or, in momentum space, multiplies by $-k^2$). This operation does not introduce spurious solutions as long as $k^2 \neq 0$, but is does for the zero mode $k = 0$.

[24]We stress that this result holds only for single-field inflation. For multi-field inflation the situation is more complex; see for example Section 7.9 of Liddle and Lyth (2000).

[25]Of course, for a given background evolution, one can obtain more accurate results by integrating numerically eq. (21.252) for a large set of values of k. Both the slow-roll approximation and the direct numerical integration are used in the comparison with the data, see [Planck Collaboration], Ade *et al.* (2014b).

[26]For the purpose of organizing the slow-roll expansion, the second slow-roll parameter η_V will be taken to be parametrically of order ε_V. Observe also that, when counting the order of the expansion we will use generically the letter ε, without distinguishing between ε_H and ε_V.

Assuming a Bunch–Davies vacuum, $f_k(\eta)$ is given in eq. (21.209). From eq. (21.249),

$$\mathcal{R}_\mathbf{k}(\eta) = -\frac{H}{a\dot\phi_0}\, u_\mathbf{k}(\eta)\,. \tag{21.259}$$

Therefore, the primordial spectrum of $\mathcal{R}$ is given by

$$\mathcal{P}_\mathcal{R}(k;\eta) = \left(\frac{H}{a\dot\phi_0}\right)^2 \mathcal{P}_u(k;\eta)\,. \tag{21.260}$$

For super-horizon modes $|f_k(\eta)|^2 = 1/(2k^3\eta^2)$, so

$$\mathcal{P}_u(k;\eta) = \frac{1}{4\pi^2\eta^2} \qquad (|k\eta| \ll 1)\,. \tag{21.261}$$

Then, using the fact that to lowest order in the slow-roll expansion $a(\eta) = -1/(H\eta)$ and H is a constant,

$$\boxed{\mathcal{P}_\mathcal{R}(k) \simeq \left(\frac{H^2}{2\pi\dot\phi_0}\right)^2 \qquad (|k\eta| \ll 1)\,.} \tag{21.262}$$

This result is valid to lowest order in the slow-roll expansion, i.e. assuming that H and $\dot\phi_0$ are exactly constant, and we see that in this approximation the spectrum on super-horizon scales is time-independent. This does not come as a surprise, since we saw following eq. (21.254) that $\mathcal{R}_\mathbf{k}$, and therefore its power spectrum, is constant on super-horizon scales. We also found there that this result holds in general, independently of the slow-roll expansion. Equation (21.262) provides an explicit check of this general argument to lowest order in the slow-roll expansion.

Furthermore, we see that the power spectrum is also flat with respect to k, i.e. to this order the spectral index $n_s = 0$. As we already mentioned, this is referred to as a Harrison–Zeldovich spectrum. Observe from eq. (21.239) that a flat spectrum is a peculiarity of the choice of the Bunch–Davies vacuum. Different choices of the vacuum will in general give $n_s \neq 0$ already at this zeroth order in the slow-roll expansion.

21.3.3 Scalar perturbations to first order. Spectral tilt

To compute the power spectrum to next order in the slow-roll expansion, we must compute z''/z in eq. (21.252) beyond leading order. In the intermediate steps of the computation it is convenient to use as the two independent slow-roll parameters ε_H and δ. We can then evaluate z'' by making use of eqs. (21.45) and (21.57), after transforming the derivatives from cosmic to conformal time. This gives the exact results

$$\frac{z'}{z} = \mathcal{H}(\eta)\,(1 + \varepsilon_H + \delta)\,, \tag{21.263}$$

$$\frac{z''}{z} = \mathcal{H}^2(\eta)\,\big[2 + 2\varepsilon_H + 3\delta + \delta(\delta + \varepsilon_H) + \mathcal{H}^{-1}(\delta' + \varepsilon_H')\big]\,. \tag{21.264}$$

In particular, to first order in the slow-roll parameters,

$$\frac{z''}{z} \simeq \mathcal{H}^2(\eta)(2 + 2\varepsilon_H + 3\delta)\,. \tag{21.265}$$

In this expression we must also take into account the modification of the zeroth order relation $a(\eta) = -1/(H\eta)$, i.e. of $\mathcal{H} = -1/\eta$, which is valid only in an exact de Sitter metric. This can be obtained by rewriting eq. (21.45) in terms of $\mathcal{H}$ and of derivatives with respect to conformal time, which gives

$$\mathcal{H}' = (1 - \varepsilon_H)\mathcal{H}^2\,. \tag{21.266}$$

In the expansion to first order in the slow-roll parameter, ε_H must be taken non-zero but constant, since its derivative is $O(\varepsilon^2)$. Then eq. (21.266) integrates to

$$\eta = -\frac{1}{(1 - \varepsilon_H)\mathcal{H}}\,, \tag{21.267}$$

or, again to first order in ε,

$$\mathcal{H} \simeq -\frac{1 + \varepsilon_H}{\eta}\,. \tag{21.268}$$

Writing $\mathcal{H} = d\log a/d\eta$, we see that for constant ε_H this integrates further to

$$a(\eta) \propto \eta^{-(1+\varepsilon_H)}\,. \tag{21.269}$$

Inserting eq. (21.268) into eq. (21.265), and eliminating δ in favor of η_H using $\delta = \varepsilon_H - \eta_H + O(\varepsilon^2)$ [see eq. (21.58)], we can rewrite the result as

$$\frac{z''}{z} = \frac{\nu^2 - 1/4}{\eta^2}\,, \tag{21.270}$$

where

$$\nu = \frac{3}{2} + (3\varepsilon_H - \eta_H) + O(\varepsilon^2)\,. \tag{21.271}$$

Inserting eq. (21.270) into the Mukhanov–Sasaki equation (21.252) and comparing with eq. (21.212) we see that, at first order in the slow-roll expansion, the zeroth order solution (21.209) is replaced by eq. (21.213), with the value of ν given in eq. (21.271). Then

$$\mathcal{P}_u(k; \eta) = \frac{k^2}{8\pi}(-k\eta)\,|H_\nu^{(1)}(-k\eta)|^2\,, \tag{21.272}$$

and, using eq. (21.249), the curvature power spectrum is given by

$$\mathcal{P}_\mathcal{R}(k; \eta) = \frac{k^2}{8\pi}\frac{(-k\eta)}{z^2(\eta)}\,|H_\nu^{(1)}(-k\eta)|^2\,. \tag{21.273}$$

We next write explicitly the dependence of $z(\eta)$ on conformal time, again to first order in the slow-roll parameters. To this purpose it is convenient to denote by η_k the value of conformal time when the mode with comoving momentum k exits the horizon. By definition, this is given by the solution of

$$H(\eta_k) = \frac{k}{a(\eta_k)}, \tag{21.274}$$

which expresses the condition that the physical momentum $k/a(\eta)$ becomes equal to $H(\eta)$. We also define

$$H_k \equiv H(\eta_k), \tag{21.275}$$

and, similarly, we denote $\dot\phi_0(\eta_k) \equiv (\dot\phi_0)_k$ and $z(\eta_k) \equiv z_k$. The dependence of $z(\eta)$ on conformal time is obtained, to first order in the slow-roll parameters, by observing that to this order (hence taking ε_V and η_V constant) eq. (21.263) integrates to

$$z(a) \propto a^{1+2\varepsilon_V - \eta_V}, \tag{21.276}$$

where we have used $\delta = \varepsilon_V - \eta_V + O(\varepsilon^2)$. Together with eq. (21.269), to first order in ε we get

$$z(\eta) \propto \eta^{-(1+\delta\nu)}, \tag{21.277}$$

where $\delta\nu = 3\varepsilon_V - \eta_V$. For a given k, the proportionality constant can be fixed by normalizing $z(\eta)$ to the value that it has when this mode crosses the horizon, so[27]

$$z(\eta) = z_k \times \left(\frac{\eta}{\eta_k}\right)^{-(1+\delta\nu)}. \tag{21.278}$$

[27]Of course, $z(\eta)$ by itself is independent of k. Therefore the k dependence of z_k in eq. (21.278) is such as to cancel the k dependence in $(-\eta_k)^{-(1+\delta\nu)}$. However, this way of writing $z(\eta)$ is quite convenient, as we will see.

To zeroth order, η_k is simply determined by the condition $k\eta_k = -1$. To obtain η_k to first order we observe that, in terms of $\mathcal{H}$, eq. (21.274) reads $\mathcal{H}(\eta_k) = k$, which, using eq. (21.268), gives

$$\eta_k = -\frac{1 + \varepsilon_H}{k}. \tag{21.279}$$

Then, to first order, eq. (21.278) can be rewritten as

$$z(\eta) = (1 + \varepsilon_H) z_k \times (-k\eta)^{-(1+\delta\nu)}. \tag{21.280}$$

We insert this into eq. (21.273) and use

$$z_k^2 = a^2(\eta_k) \left(\frac{\dot\phi_0}{H}\right)_k^2$$

$$= k^2 \left(\frac{\dot\phi_0}{H^2}\right)_k^2, \tag{21.281}$$

where in the second line we have eliminated $a(\eta_k)$ using eq. (21.274). Then, we finally get

$$\mathcal{P}_{\mathcal{R}}(k;\eta) = (1 - 2\varepsilon_V) \frac{1}{8\pi} \left(\frac{H^2}{\dot\phi_0}\right)_k^2 (-k\eta)^{3+2\delta\nu} |H_{\frac{3}{2}+\delta\nu}^{(1)}(-k\eta)|^2, \tag{21.282}$$

where, in the first-order terms, we use for definiteness ε_V and η_V both in the prefactor and in $\delta\nu$. This is the power spectrum of $\mathcal{R}$ to first order in the slow-roll parameters, which means that we have kept the terms of first order in the slow-roll parameters ε_V and η_V, and we have neglected terms $O(\varepsilon^2)$ and $O(\varepsilon')$. It is otherwise exact as far as the dependence on $-k\eta$ is concerned.

We can now take the limit $|k\eta| \to 0$. From the general argument discussed following eq. (21.254) we know that $\mathcal{R}_{\mathbf{k}}$, and hence its power spectrum $\mathcal{P}_{\mathcal{R}}(k;\eta)$, must become independent of η on super-horizon scales. In eq. (21.262) we checked this to zeroth order in the slow-roll parameters, i.e. setting $\varepsilon_V = \eta_V = 0$. Using eq. (21.273) we can now check it explicitly to first order in the slow-roll parameters. In the limit $(-k\eta) \to 0$ we have[28]

$$H_\nu^{(1)}(-k\eta) \simeq -\frac{i}{\pi} 2^\nu \Gamma(\nu) (-k\eta)^{-\nu}. \qquad (21.283)$$

So, the dependence on η indeed cancels and, on super-horizon scales,

$$\mathcal{P}_{\mathcal{R}}(k;\eta) \simeq (1 - 2\varepsilon_V) \left(\frac{2^{\nu-\frac{3}{2}}\Gamma(\nu)}{\Gamma(3/2)} \right)^2 \frac{1}{4\pi^2} \left(\frac{H^2}{\dot{\phi}_0} \right)_k^2 \qquad (|k\eta| \ll 1). \qquad (21.284)$$

To first order in ε, we can also expand

$$\frac{2^{\delta\nu}\Gamma(\frac{3}{2} + \delta\nu)}{\Gamma(3/2)} = 1 + C\delta\nu + O(\delta\nu^2), \qquad (21.285)$$

where

$$C = 2 - \log 2 - \gamma_E \simeq 0.7296, \qquad (21.286)$$

and γ_E is the Euler–Mascheroni constant. Thus, we finally get

$$\mathcal{P}_{\mathcal{R}}^{1/2}(k;\eta)\Big|_{|k\eta| \ll 1} = [1 + (3C - 1)\varepsilon_V - C\eta_V] \left(\frac{H^2}{2\pi|\dot{\phi}_0|} \right)_k. \qquad (21.287)$$

The term $O(\varepsilon)$ in the prefactor is independent of k and numerically small, so it is often neglected. The most important aspect of this result is that the spectrum now has a dependence on k even in the super-horizon limit, which enters through the $(H^2/|\dot{\phi}_0|)_k$ term.[29]

We can also rewrite the result (21.287) in terms of the inflationary potential, using eqs. (21.34) and (21.38). If we also want to keep track correctly of the $O(\varepsilon)$ terms in the prefactor, we also need the first-order correction to eqs. (21.34) and (21.38). The first correction to eq. (21.34) is obtained keeping also the term $\dot{\phi}^2/2$ in eq. (21.25). Together with eq. (21.55), this gives

$$H^2 = \frac{8\pi G V}{3} \left(1 + \frac{\varepsilon_V}{3} + O(\varepsilon^2) \right), \qquad (21.288)$$

[28]See for example Gradshteyn and Ryzhik (1980), eqs. 8.443 and 8.334.3.

[29]This result was first derived in Stewart and Lyth (1993); see also Section 7.6 of Liddle and Lyth (2000). To compare with Section 7.6 of Liddle and Lyth (2000), observe that what is called η_H there corresponds to our parameter $-\delta$, i.e. to $-\varepsilon_V + \eta_V + O(\varepsilon^2)$ in our notation, and also what we call C is called $-C$ in Liddle and Lyth (2000) (so our C is positive). Thus, our eq. (21.287) agrees with their eq. (7.134) [and our eq. (21.290) agrees with their eq. (7.139), after taking into account a correction $2C \to 3C$ mentioned in the errata of their book]. Observe also that, to this order, there is no need to distinguish between (ε_H, η_H) and (ε_V, η_V). We only keep the subscript on the slow-roll parameters to avoid confusion between the second slow-roll parameter and conformal time η.

while, using eqs. (21.56) and (21.288), eq. (21.38) becomes

$$\dot\phi \simeq -\frac{1}{\sqrt{24\pi G}}\frac{V_\phi(\phi)}{V^{1/2}(\phi)}\left(1 - \frac{1}{2}\varepsilon_V + \frac{1}{3}\eta_V + O(\varepsilon^2)\right). \qquad (21.289)$$

Plugging these expressions into eq. (21.287) we get

$$\mathcal{P}_{\mathcal{R}}^{1/2}(k;\eta)\Big|_{|k\eta|\ll1} = \frac{1}{\sqrt{12}\,\pi m_{\rm Pl}^3}\left[1 + \left(3C - \frac{1}{6}\right)\varepsilon_V - \left(C + \frac{1}{3}\right)\eta_V\right]$$
$$\times\left(\frac{V^{3/2}}{|V_\phi|}\right)_k. \qquad (21.290)$$

Equivalently, we can express the result in terms of the slow-roll parameter ε_V using eq. (21.43), and then eq. (21.290) can be written as

$$\mathcal{P}_{\mathcal{R}}(k;\eta)\Big|_{|k\eta|\ll1} = \left[1 + 2\left(3C - \frac{1}{6}\right)\varepsilon_V - 2\left(C + \frac{1}{3}\right)\eta_V\right]$$
$$\times\frac{1}{24\pi^2 m_{\rm Pl}^4}\left(\frac{V}{\varepsilon_V}\right)_k. \qquad (21.291)$$

From eq. (21.290) we can read both the amplitude at a pivot scale k_* and the tilt of the primordial spectrum. Comparing with eq. (19.177) we see that the amplitude $A_{\mathcal{R}}$ is simply given by

$$A_{\mathcal{R}} = \left[1 + 2\left(3C - \frac{1}{6}\right)\varepsilon_V - 2\left(C + \frac{1}{3}\right)\eta_V\right]$$
$$\times\frac{1}{12\pi^2 m_{\rm Pl}^6}\left(\frac{V^3}{V_\phi^2}\right)_{k_*}. \qquad (21.292)$$

For computing the tilt, we first write eq. (19.177) in the form

$$\log\mathcal{P}_{\mathcal{R}}(k) = \log A_{\mathcal{R}} + (n_s - 1)\log(k/k_*), \qquad (21.293)$$

from which it follows that, to this order,

$$n_s(k) - 1 = \frac{d\log\mathcal{P}_{\mathcal{R}}(k)}{d\log k}$$
$$= \frac{d\log(V^3/V_\phi^2)_k}{d\log k}. \qquad (21.294)$$

In principle $n_s(k)$ is a function of k. We define the tilt $n_s \equiv n_s(k_*)$. As we saw in eq. (19.183), the scale dependence of the tilt (its "running") can in principle be encoded in further parameters such as $(dn_s/d\log k)_{k_*}$ (and possibly higher-order derivatives, although these are at present beyond experimental reach).

Let us recall that the dependence of V and V_ϕ on k follows from the fact that they are functions of $\phi(\eta)$, and $\phi(\eta)$ must be evaluated at the time η_k at which the mode with comoving momentum k crosses the horizon, given by eq. (21.274). We can then transform the derivative with respect to k into a derivative with respect to $\phi_k \equiv \phi(\eta_k)$, as follows.

Let t_k be the cosmic time corresponding to conformal time η_k, and t_* be the value of t_k when k is equal to the pivot scale k_*, i.e. t_* is the cosmic time when the pivot scale crosses the horizon. Integrating eq. (21.38) from $t_k = t_*$ to $t_k = t_* + dt_k$ we get

$$d\phi_k = -\frac{m_{\rm Pl}}{\sqrt{3}} \left(\frac{V_\phi}{V^{1/2}}\right)_{k_*} dt_k \,, \tag{21.295}$$

where $d\phi_k = \phi(t_k) - \phi(t_*)$. We next use eq. (21.274), observing that in slow-roll inflation $H(t)$ is to a first approximation constant, while $a(t)$ evolves exponentially with t. Therefore, to lowest order, we can neglect the variation of H and replace it with its constant value $H_* \equiv H(t_*)$. Then, t_k is approximately determined by the equation $a(t_k) \simeq k/H_*$, so

$$\left(\frac{da}{dt}\right)_{t_*} dt_k \simeq \frac{dk}{H_*} \,. \tag{21.296}$$

Therefore

$$(aH)_{t_*} \, dt_k \simeq \frac{dk}{H_*} \,, \tag{21.297}$$

which, again using eq. (21.274), gives

$$dt_k = \frac{1}{H_*} \frac{dk}{k} \,. \tag{21.298}$$

Putting this into eq. (21.295) and again using eq. (21.34) to eliminate H_* in favor of $V(k_*)$, we get

$$d\phi_k = -m_{\rm Pl}^2 \left(\frac{V_\phi}{V}\right)_{k_*} d\log k \,. \tag{21.299}$$

Using this result to replace in eq. (21.294) the derivative with respect to $d\log k$ by the derivative with respect to $d\phi_k$ we get

$$n_s - 1 = m_{\rm Pl}^2 \left[\left(\frac{V_\phi}{V}\right)\left(\frac{2V_{\phi\phi}}{V_\phi} - \frac{3V_\phi}{V}\right)\right]_{k_*}$$
$$= m_{\rm Pl}^2 \left[2\frac{V_{\phi\phi}}{V} - 3\left(\frac{V_\phi}{V}\right)^2\right]_{k_*} . \tag{21.300}$$

Recalling the definitions (21.43) and (21.44) of the slow-roll parameters, we finally find

$$\boxed{n_s - 1 = 2\eta_V - 6\epsilon_V \,,} \tag{21.301}$$

where the slow-roll parameters η_V and ϵ_V are evaluated at the conformal time $\eta(k_*)$. Equation (21.301) is valid only to lowest order in the slow-roll expansion, i.e. the right-hand side receives corrections $[1 + O(\varepsilon)]$. Of course, to lowest order, we do not need to distinguish between (ϵ_V, η_V) and (ε_H, η_H), and one commonly uses the generic notation (ε, η) (where the second slow-roll parameter η should not be confused with conformal time).

21.3.4   Tensor perturbations

The computation of the spectrum of GWs is quite similar to that in the scalar case. According to eq. (19.220), in the absence of anisotropic stress the two polarizations of the GW satisfy the equation

$$\tilde{h}_A'' + 2\mathcal{H}\tilde{h}_A' + k^2 \tilde{h}_A = 0 \,. \tag{21.302}$$

Comparing with eq. (21.176) we see that this is just the same as the classical equation of motion of a massless scalar field ϕ, and the quantization therefore proceeds in the same way. We must, however, take into account that h_A is not a canonically normalized field (and, in fact, is even dimensionless rather than having the canonical dimensions of mass of a scalar field). To get the correct normalization we expand the Einstein action to quadratic order around a FRW background. This requires expanding the Christoffel symbols [which we computed to linear order in eqs. (18.99)–(18.104)] to second order in the perturbation. The second-order corrections in the Christoffel symbol come from the expansion of the inverse metric $g^{\mu\nu}$.[30] If we write $g_{\mu\nu} = a^2(-1, \delta_{ij} + h_{ij}^{\rm TT})$, we have $g^{00} = -a^{-2}$ and $g^{0i} = 0$ exactly, while g^{ij} is given by an infinite series,

$$g^{ij} = \frac{1}{a^2}\left[1 - h_{ij}^{\rm TT} + h_{ik}^{\rm TT}h_{kj}^{\rm TT} + \ldots\right] \tag{21.303}$$

(where, on $h_{ij}^{\rm TT}$, we raise and lower indices with δ_{ij}). Similarly, we must expand the determinant of the metric to second order, which gives

$$\sqrt{-g} = a^4\left(1 - \frac{1}{2}h_{ij}^{\rm TT}h_{ij}^{\rm TT}\right) \,. \tag{21.304}$$

The computation is then long but straightforward, and gives

$$S_2[h] = \frac{1}{64\pi G}\int d^3x\, d\eta\, a^2\left[\partial_\eta h_{ij}^{\rm TT}\partial_\eta h_{ij}^{\rm TT} - \partial_k h_{ij}^{\rm TT}\partial_k h_{ij}^{\rm TT}\right] \,. \tag{21.305}$$

We now pass to the Fourier modes, using the convention (17.40),

$$S_2[h] = \frac{1}{64\pi G}\int \frac{d^3k}{(2\pi)^3}\, d\eta\, a^2 \tag{21.306}$$
$$\times \left[\partial_\eta \tilde{h}_{ij}^{\rm TT}(\eta, \mathbf{k})\partial_\eta \tilde{h}_{ij}^{\rm TT,*}(\eta, \mathbf{k}) - \partial_k \tilde{h}_{ij}^{\rm TT}(\eta, \mathbf{k})\partial_k \tilde{h}_{ij}^{\rm TT,*}\right] \,.$$

We can now substitute the expansion (19.214). Taking into account the normalization (19.218) of the polarization tensors, we get

$$S_2[h] = \frac{1}{32\pi G}\sum_A\int \frac{d^3k}{(2\pi)^3}\, d\eta\, a^2 \tag{21.307}$$
$$\times \left[\partial_\eta \tilde{h}_A(\eta, \mathbf{k})\partial_\eta \tilde{h}_A^*(\eta, \mathbf{k}) - \partial_k \tilde{h}_A(\eta, \mathbf{k})\partial_k \tilde{h}_A^*(\eta, \mathbf{k})\right] \,.$$

We finally transform back to coordinate space,[31] and we get

---

[30] In contrast, when we write $g_{\mu\nu} = \eta_{\mu\nu} + h_{\mu\nu}$, $h_{\mu\nu}$ is by definition the *exact* form of the metric perturbations, to all orders. It is sometime convenient to further separate $h_{\mu\nu} = h_{\mu\nu}^{(1)} + h_{\mu\nu}^{(2)} + \ldots$ where $h_{\mu\nu}^{(1)}$ is a first-order perturbation, $h_{\mu\nu}^{(2)}$ a second-order perturbations, etc.

[31] Observe that going first to Fourier space was necessary in order to use the decomposition (19.214), which only makes sense for the Fourier modes, since the polarization tensors are defined with respect to a momentum $\mathbf{k}$. One can then use eq. (19.218) to eliminate the polarization tensors and finally transform back to coordinate space. In contrast, writing directly $h_{ij}^{\rm TT} = \sum_A e_{ij}^A h_A$ in coordinate space makes no sense.

$$S_2[h] = \frac{1}{32\pi G} \sum_A \int d^3x\, d\eta\, a^2 \left[\partial_\eta h_A \partial_\eta h_A - \partial_k h_A \partial_k h_A \right]$$

$$= -\frac{1}{32\pi G} \sum_A \int d^4x \sqrt{-\bar{g}}\, \bar{g}^{\mu\nu} \partial_\mu h_A \partial_\nu h_A$$

$$= -\frac{1}{2} \sum_A \int d^4x \sqrt{-\bar{g}}\, \bar{g}^{\mu\nu} \partial_\mu \varphi_A \partial_\nu \varphi_A \,, \qquad (21.308)$$

where $\bar{g}_{\mu\nu} = a^2 \eta_{\mu\nu}$ is the background FRW metric and

$$\varphi_A(\eta, \mathbf{x}) = \frac{1}{\sqrt{16\pi G}}\, h_A(\eta, \mathbf{x}) \,. \qquad (21.309)$$

We see that the action governing the two polarizations h_A is the same as the curved-space action of two canonically-normalized scalar fields φ_A.[32] From eq. (19.276),

$$\frac{1}{4} P_{T,\mathrm{in}}(k; \eta) = \langle |\tilde{h}_{A,\mathrm{in}}(\mathbf{k}; \eta)|^2 \rangle \qquad (21.310)$$

where A is any of the two polarization states, $A = +$ or $A = \times$ (i.e. there is no sum over A). Then

$$\frac{1}{4} P_{T,\mathrm{in}}(k; \eta) = 16\pi G \langle |\tilde{\varphi}_{A,\mathrm{in}}(\mathbf{k})|^2 \rangle$$

$$= 16\pi G\, P_\varphi(k; \eta) \,, \qquad (21.311)$$

so $P_{T,\mathrm{in}}(k; \eta) = 64\pi G\, P_\varphi(k; \eta)$ and similarly

$$\mathcal{P}_{T,\mathrm{in}}(k; \eta) = 64\pi G\, \mathcal{P}_\varphi(k; \eta) \,. \qquad (21.312)$$

As in the computation of the spectrum of a scalar field, we next introduce the field $\chi = \varphi/a$ (we drop the index A). Then eq. (21.302) becomes the same as eq. (21.179). We compute to first order in slow-roll expansion. Inserting $a(\eta)$ from eq. (21.269), we have

$$\frac{a''}{a} = \frac{2 + 3\varepsilon_V + O(\varepsilon^2)}{\eta^2} \,, \qquad (21.313)$$

where, as before, in the first-order term we use for definiteness ε_V rather than ε_H. Then eq. (21.179) becomes the same as eq. (21.212), with

$$\nu = \frac{3}{2} + \varepsilon_V + O(\varepsilon^2) \,. \qquad (21.314)$$

Assuming that also for gravitons the appropriate vacuum state is the Bunch–Davies vacuum, the mode functions are given by eq. (21.213), and the power spectrum of φ is

$$\mathcal{P}_\varphi(k; \eta) = \frac{1}{2\pi^2 a^2(\eta)} k^3 |f_k(\eta)|^2$$

$$= \frac{k^2}{8\pi} \frac{(-k\eta)}{a^2(\eta)} |H_\nu^{(1)}(-k\eta)|^2 \,, \qquad (21.315)$$

[32] The normalization factor $(16\pi G)^{-1/2}$ in eq. (21.309) depends of course on our convention (19.218) for the normalization of the polarization tensor. Different conventions exists in the literature. In particular, if one rather uses $e_{ij}^A(\hat{\mathbf{k}}) e_{ij}^{A'}(\hat{\mathbf{k}}) = \delta^{AA'}$, the normalization factor in eq. (21.309) becomes $(32\pi G)^{-1/2}$. The metric perturbation h_{ij}^{TT} does not depend on the normalization chosen for the polarization tensor, which instead affects the definitions of h_A. Thus, the numerical factors in the definition of the power spectrum of h_A, eq. (19.276), depend on the normalization choice of the e_{ij}^A, in such a way that the power spectrum of h_{ij}^{TT} given in eq. (19.279) is always the same; see Note 39 on page 488.

with ν given in eq. (21.314). In order to write $a^2(\eta)$ explicitly in terms of η we proceed just as we did for $z^2(\eta)$ in eqs. (21.274)–(21.280). Using eq. (21.269) we get

$$a(\eta) = a_k \left(\frac{\eta}{\eta_k}\right)^{-(1+\varepsilon_V)}$$

$$= (1 + \varepsilon_V)\frac{k}{H_k}(-k\eta)^{-(1+\varepsilon_V)} \qquad (21.316)$$

(where all expansions are valid only to first order in ε_V) and therefore

$$\mathcal{P}_{T,\mathrm{in}}(k;\eta) = (1 - 2\varepsilon_V)\frac{H_k^2}{\pi m_{\mathrm{Pl}}^2}(-k\eta)^{3+2\varepsilon_V}|H^{(1)}_{\frac{3}{2}+\varepsilon_V}(-k\eta)|^2. \qquad (21.317)$$

This gives the tensor power spectrum during a quasi-de Sitter epoch of inflation, to first order in the slow-roll parameter, for generic values of $(-k\eta)$. We now take the super-horizon limit, using eq. (21.283). Then the dependence on η cancels and, again using eq. (21.285), we get

$$\left.\mathcal{P}_{T,\mathrm{in}}^{1/2}(k;\eta)\right|_{|k\eta|\ll 1} = [1 - (1 - C)\varepsilon_V]\frac{\sqrt{2}}{\pi}\frac{H_k}{m_{\mathrm{Pl}}}, \qquad (21.318)$$

where the constant C was defined in eq. (21.286). Once again, apart from a k-independent $O(\varepsilon)$ correction to the amplitude, which is small and is typically omitted, we see that the result at first order in the slow-roll expansion induces a dependence on k, this time through H_k, and therefore a spectral tilt.

Similarly to what we did for scalar perturbations, we can trade H for the potential V using eq. (21.288), and we get

$$\left.\mathcal{P}_{T,\mathrm{in}}^{1/2}(k;\eta)\right|_{|k\eta|\ll 1} = \left[1 - \left(\frac{5}{6} - C\right)\varepsilon_V\right]\left(\frac{2}{3}\right)^{1/2}\frac{V_k^{1/2}}{\pi m_{\mathrm{Pl}}^2}. \qquad (21.319)$$

The amplitude of the tensor perturbations is therefore given, up to corrections $O(\varepsilon^2)$, by

$$A_T = [1 - 2(1 - C)\varepsilon_V]\frac{2H_*^2}{\pi^2 m_{\mathrm{Pl}}^2}$$

$$= \left[1 - 2\left(\frac{5}{6} - C\right)\varepsilon_V\right]\frac{2V_*}{3\pi^2 m_{\mathrm{Pl}}^4}, \qquad (21.320)$$

where $H_* = H_{k_*}$ and $V_* = V_{k_*}$. The tilt is obtained by proceeding as in the scalar case. Equation (21.294) is now replaced by

$$n_T = \left(\frac{d\log V}{d\log k}\right)_{k_*}. \qquad (21.321)$$

Using eq. (21.299) again, we get

$$n_T = -2\varepsilon_V, \qquad (21.322)$$

where ε_V must in principle be evaluated at the conformal time $\eta(k_*)$ (in any case, to this order in the slow-roll expansion ε_V is constant). Computing to the next-order would produce corrections $[1 + O(\varepsilon)]$ on the right-hand side. Slow-roll inflation therefore predicts a negative tilt for the primordial GW spectrum. We will come back later to the consequences for attempts at detecting these primordial GWs with space-borne or ground-based experiments, which are sensitive to GW wavelengths much shorter than the typical cosmological wavelengths, and therefore to much higher momenta.

From eqs. (21.287) and (21.318), on super-horizon scales the tensor-to-scalar ratio r defined in eq. (19.282) is then given by

$$r(k) = \left[1 + 2C(\eta_V - 2\varepsilon_V) + O(\varepsilon^2)\right] \frac{8}{m_{\rm Pl}^2} \left(\frac{\dot\phi_0^2}{H^2}\right)_k , \qquad (21.323)$$

where in the $O(\varepsilon)$ term in the amplitude we can use equivalently (ε_V, η_V) or (ε_H, η_H). Using eqs. (21.292) and (21.320) this can also be rewritten in terms of the potential as

$$r(k) = \left[1 + 2\left(C + \frac{1}{3}\right)(\eta_V - 2\varepsilon_V) + O(\varepsilon^2)\right] 8m_{\rm Pl}^2 \left(\frac{V_\phi}{V}\right)_k^2 , \qquad (21.324)$$

and, comparing with the definition (21.43) of the slow-roll parameter ε_V, we finally get

$$\boxed{r = 16\varepsilon_V \left[1 + 2\left(C + \frac{1}{3}\right)(\eta_V - 2\varepsilon_V) + O(\varepsilon^2)\right] ,} \qquad (21.325)$$

which is valid for generic k, and so in particular when both sides are evaluated at the pivot scale k_*.

Comparing eqs. (21.325) and (21.322) we see that, to lowest order in the slow-roll parameters, single-field slow-roll inflation predicts a relation between the ratio r and the tilt of the tensor spectrum,

$$\boxed{r = -8n_T .} \qquad (21.326)$$

Since to lowest order ε_V no longer appears, this relation is independent of the specific choice of the inflationary potential. Its experimental confirmation would then provide very convincing evidence in favor of single-field slow-roll inflation.

We finally observe that, since the amplitude of scalar perturbations $A_{\mathcal{R}}$ is known [see eq. (19.186)], and the amplitude of tensor perturbations is given in terms of the potential V_* by eq. (21.320), if one measures the tensor-to-scalar ratio $r(k_*) = A_T(k_*)/A_{\mathcal{R}}(k_*)$ at the pivot scale k_*, one can determine the scale of the inflationary potential. To lowest order in ε,

$$V_*^{1/4} \simeq m_{\rm Pl} \left(\frac{3\pi^2}{2} r(k_*) A_{\mathcal{R}}(k_*)\right)^{1/4}$$

$$\simeq 1.83 \times 10^{16} \, \text{GeV} \left(\frac{r(k_*)}{0.1} \right)^{1/4} \left(\frac{A_{\mathcal{R}}(k_*)}{2.14 \times 10^{-9}} \right)^{1/4} . \quad (21.327)$$

As we saw in Section 20.4.5 (and we will see also from Fig. 21.6 on page 627), the experimental upper limit on the tensor-to-scalar ratio is $r(k_*) \lesssim 0.1$, obtained typically at a pivot scale $k_* = 0.002 \, \text{Mpc}^{-1}$ (or a similar limit at $k_* = 0.05 \, \text{Mpc}^{-1}$; see also Note 40 on page 489).[33] This implies that the inflationary scale must be below about $2 \times 10^{16} \, \text{GeV}$. In turn, for the Hubble parameter H_* during inflation, using $H_*^2 \simeq (8\pi G/3)V_* = V_*/(3m_{\text{Pl}}^2)$, this implies

$$\frac{H_*}{m_{\text{Pl}}} = \frac{V_*^{1/2}}{\sqrt{3}\, m_{\text{Pl}}^2}$$

$$\simeq 3.25 \times 10^{-5} \left(\frac{r(k_*)}{0.1} \right)^{1/2} \left(\frac{A_{\mathcal{R}}(k_*)}{2.14 \times 10^{-9}} \right)^{1/2} , \quad (21.328)$$

so a measurement of r would fix the value of H_*. In particular, the bound $r < 0.1$ implies $H_* \lesssim 8 \times 10^{13} \, \text{GeV}$.

21.3.5   Predictions from a sample of inflationary models

We can now use the results from Sections 21.3.2–21.3.4 to work out the explicit predictions of single-field slow-roll inflation for a sample of typical choices of the potential V, and we can then compare them with observations. Given V, using eqs. (21.292) and (21.294) for scalar perturbations and eqs. (21.320) and (21.321) for tensor perturbations, we obtain the predictions for the amplitudes and tilts $(A_{\mathcal{R}}, n_s)$ and (A_T, n_T), respectively. These are expressed in terms of the potential V_* and its derivatives, i.e. of the potential $V(\phi)$ evaluated when its argument ϕ is equal to $\phi(\eta_*)$, where $\eta_* = \eta(k_*)$ is the conformal time at which the mode with k equal to the pivot scale k_* left the horizon during inflation. To have an explicit prediction, we therefore need $\phi(\eta_*)$. As we saw in eq. (21.61), in the slow-roll approximation we can trade $\phi(\eta_*)$ for the value N_* that measures the number of e-folds from the moment at which the scale k_* crosses the horizon to the end of inflation.

If we knew exactly the expansion history of a given inflationary model, including the details of the reheating mechanism, then for each mode with momentum k we could compute the time when it exits the horizon during inflation. Thus, in particular, given the pivot scale k_*, the corresponding value of N_* would be fixed. The computation is quite similar to that performed in Section 21.1.2, where we computed the minimum number of e-folds required to solve the horizon problem. In that case we required that inflation was sufficiently long that the mode with $1/k = \eta_0$ was inside the horizon at some sufficiently early time during inflation. In the approximation of instantaneous reheating this led to the condition (21.20).

One can perform the same computation for a mode with a generic value of k. For the purpose of comparing with the data, we will see that

[33]The reference value $A_{\mathcal{R}}(k_*) = 2.14 \times 10^{-9}$ that we are using here is actually the value measured at $k_* = 0.05 \, \text{Mpc}^{-1}$; see eq. (19.186). Using the spectral tilt $n_s = 0.9677$ one finds that, at $k_* = 0.002 \, \text{Mpc}^{-1}$, $A_{\mathcal{R}}(k_*) = 2.37 \times 10^{-9}$.

it is also important to include an intermediate reheating phase, which can be modeled with a generic equation-of-state parameter $w_{\rm reh}$. Then one finds[34]

$$N(k) \simeq 67 - \log\left(\frac{k}{H_0}\right) + \frac{1}{4}\log\left(\frac{V_k^2}{m_{\rm Pl}^4 \rho_{\rm end}}\right) - \frac{1 - 3w_{\rm reh}}{12(1 + w_{\rm reh})}\log\left(\frac{\rho_{\rm end}}{\rho_{\rm reh}}\right),$$
$$(21.330)$$

where, as usual, V_k is the value of the potential when the mode k crosses the horizon during inflation, $\rho_{\rm end}$ is the energy density when inflation ends and reheating begins, $\rho_{\rm reh}$ is the energy density when reheating ends and RD begins, and $m_{\rm Pl}$ is the reduced Planck mass.

The uncertainty regarding the reheating phase is then modeled in terms of two parameters, $w_{\rm reh}$ and $\rho_{\rm reh}$. The latter must be less than or equal to $\rho_{\rm end}$, with equality holding only if reheating is instantaneous and all the energy density of inflation is immediately transformed into radiation. On the other hand, in order not to spoil big-bang nucleosynthesis, the reheating temperature must be larger than the MeV scale, so $\rho_{\rm reh}^{1/4} \gtrsim O(1)$ MeV; see Note 2 on page 577. The other parameter, $w_{\rm reh}$, depends on the specific reheating mechanism. Several reheating mechanisms proceed through the decay of the inflaton into massive modes, resulting in an approximately matter-dominated phase, with $w_{\rm reh} \simeq 0$, although this value is affected by averaging over the inflaton oscillations around its minimum. Because of the uncertainty in $\rho_{\rm reh}^{1/4}$ and $w_{\rm reh}$, for a given inflationary potential there is a whole range of values of $N_* = N(k_*)$ that are a priori possible, and this parametrizes our ignorance of the reheating stage. For typical inflationary models and reheating mechanisms, N_* is in the range $50-60$, although in some models the range can be larger. Thus, in general one keeps N_* as a free parameter and gives the predictions of a given inflationary model as a function of N_*. We will now illustrate the procedure for several inflationary models.

Large-field inflation

Consider first a potential of the form $V(\phi) = g_n \phi^n$, as in eq. (21.69). This is a typical potential of large-field inflation. Note that g_n has dimensions of $({\rm mass})^{4-n}$. Using eq. (21.292) we get

$$A_{\mathcal{R}}(k_*) = \frac{1}{12\pi^2 m_{\rm Pl}^6}\frac{g_n}{n^2}\phi^{n+2}(\eta_*),$$
$$(21.331)$$

where $\eta_* = \eta(k_*)$ is the conformal time at which the mode with k equal to the pivot scale k_* crosses the horizon (and, here and in the following, we no longer write explicitly the factor $1 + O(\varepsilon)$ in the amplitude). Proceeding as in eq. (21.72) [with $\phi(\eta_*)$ and N_* replacing ϕ_i and ΔN], in this model

$$\phi(\eta_*) \simeq m_{\rm Pl}(2nN_*)^{1/2},$$
$$(21.332)$$

where we have neglected the term $(n-1)/2$ in eq. (21.72), which is $O(1)$, since $N_* \gg 1$. Then, trading $\phi(\eta_*)$ for N_*, eq. (21.331) reads

$$A_{\mathcal{R}} = \frac{1}{12\pi^2 n^2}\frac{g_n}{m_{\rm Pl}^{4-n}}(2nN_*)^{(n+2)/2}.$$
$$(21.333)$$

[34] See for example Martin and Ringeval (2010) or the *Planck* paper on inflation, [Planck Collaboration], Ade *et al.* (2014b). To make contact with eq. (21.20) observe that, when studying the solution of the horizon problem, we were considering a mode with $k = \eta_0^{-1} \simeq H_0/3.2$, so $\log(k/H_0) \simeq -1.2$. Furthermore, we assumed instantaneous reheating, so $\rho_{\rm end} = \rho_{\rm reh}$. Thus, the last term in eq. (21.330) was absent. We also assumed a de Sitter expansion neglecting the evolution of V, so in that approximation V_k was the same as the constant energy density during inflation, $V_k = \rho_{\rm end}$, which is turn was the same as $\rho_{\rm reh}$ in the approximation of instantaneous reheating. Thus, in these approximations, eq. (21.330) becomes

$$N(k_0) \simeq 68.2 - \log(m_{\rm Pl}/\rho_{\rm reh}^{1/4}),$$
$$(21.329)$$

Using the numerical value (21.65) of $m_{\rm Pl}$ we recover eq. (21.20). The condition that the number of e-folds ΔN is such that the mode with $k = 1/\eta_0$ was inside the horizon at some early time during inflation implies that $\Delta N \geqslant N(k_0)$. Observe also that, to be more accurate, we should also include the change in the effective number of relativistic degrees of freedom. This results in an extra term $-(1/12)\log g_{*,{\rm reh}}$ on the right-hand side of eq. (21.330), where $g_{*,{\rm reh}}$ is the effective number of relativistic degrees of freedom during reheating. However, with the Standard Model value $g_{*,{\rm reh}} \sim 100$, this only gives an extra factor $-(1/12)\log g_{*,{\rm reh}} \simeq -0.4$, and even in the extreme case $g_{*,{\rm reh}} \sim 1000$ it only becomes $\simeq -0.6$.

As we saw in eq. (19.186), the value of $A_\mathcal{R}$ is known quite accurately from observation. We can therefore use this result to fix the value of the dimensionless coupling $g_n/M_{\rm Pl}^{4-n}$ as a function of N_*,

$$\frac{g_n}{m_{\rm Pl}^{4-n}} = 2.54 \times 10^{-7} \frac{n^2}{(120n)^{(n+2)/2}} \left(\frac{60}{N_*}\right)^{(n+2)/2} \left(\frac{A_\mathcal{R}}{2.14 \times 10^{-9}}\right).$$

$$(21.334)$$

In particular, for a potential $V(\phi) = (1/2)m^2\phi^2$ we have $n = 2$ and $g_2 = m^2/2$, and we get

$$\frac{m}{m_{\rm Pl}} \simeq 5.9 \times 10^{-6} \left(\frac{60}{N_*}\right) \left(\frac{A_\mathcal{R}}{2.14 \times 10^{-9}}\right)^{1/2}, \qquad (21.335)$$

while for a potential $V(\phi) = (1/4)\lambda\phi^4$ we get

$$\lambda \simeq 1.47 \times 10^{-13} \left(\frac{60}{N_*}\right)^3 \left(\frac{A_\mathcal{R}}{2.14 \times 10^{-9}}\right). \qquad (21.336)$$

This fixes the coupling g_n, from the condition of reproducing the amplitude of the scalar perturbations at a given pivot scale, in terms of the number of e-folds from the moment when the pivot scale crosses the horizon to the end of inflation.

The scalar index n_s is obtained as a function of N_* by plugging eq. (21.332) into eqs. (21.70) and (21.71), which gives

$$\varepsilon_V = \frac{n}{4N_*}, \qquad (21.337)$$

$$\eta_V = \frac{n-1}{2N_*}. \qquad (21.338)$$

Therefore eq. (21.301) gives

$$n_s - 1 = -\frac{n+2}{2N_*}. \qquad (21.339)$$

Finally, eqs. (21.325) and (21.337) give

$$r = \frac{4n}{N_*}. \qquad (21.340)$$

A particularly useful way of plotting the predictions of the theory and comparing them with the data is to show the prediction in the plane (n_s, r), as a function of N_*. This is shown in Fig. 21.5, for the three cases $n = 4, n = 2$ and $n = 1$. For each choice of the power n in the potential, eqs. (21.339) and (21.340) describe a straight line in the (r, n_s) plane, expressed in parametric form with N_* as the parameter. As $N_* \to \infty$ all lines approach asymptotically the point $(n_s = 1, r = 0)$. The most relevant range $N_* \in [50, 60]$ is shown as a solid segment in Fig. 21.5, while the rest of each line is shown dashed.

Figure 21.6 shows the marginalized 68% and 95% c.l. regions for n_s and $r_{0.002}$ from the *Planck* 2015 data in combination with other datasets,

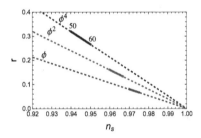

Fig. 21.5 The prediction in the (n_s, r) plane of the theory $V(\phi) = g_n \phi^n$ with $n = 4$ (upper line), $n = 2$ (middle line) and $n = 1$ (lower line). The position on the lines is parametrized by N_*, and the solid part of each line describes the range $N_* \in [50, 60]$.

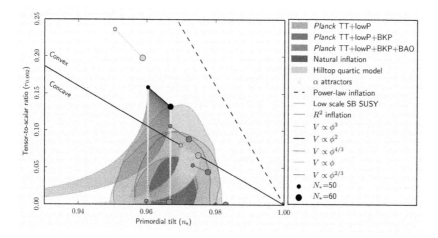

Fig. 21.6 Marginalized 68% and 95% c.l. regions for n_s and $r_{0.002}$ from the Planck 2015 data in combination with other datasets, compared to the theoretical predictions of selected inflationary models. From top to bottom, the segments shows the predictions, for $N_* \in [50, 60]$, for $V \propto \phi^n$ with $n = 3, 2, 1, 2/3$, respectively. The segment with $r \simeq 0$ that best fits the data is the prediction of R^2 inflation, i.e. of the Starobinsky model. From [Planck Collaboration], Ade *et al.* (2016bc). Reproduced with permission @ ESO.

compared to the theoretical predictions of various inflationary models. We see that the model with $V(\phi) = (1/4)\lambda\phi^4$, which was once among the most popular inflationary models, is ruled out with great statistical significance (actually, it was already ruled out to a significant confidence level by WMAP), and in fact for $N_* \in [50, 60]$ its prediction, shown in Fig. 21.5, is even outside the scale of Fig. 21.6. Even the model with $V(\phi) = (1/2)m^2\phi^2$ is excluded at about 95% c.l. The line corresponding to a potential $V(\phi) \propto \phi$ separates convex potentials, $V(\phi) \propto \phi^n$ with $n > 1$, from concave potentials, with $n < 1$. Concave potentials are compatible with the data, while convex potentials with $n > 2$ are excluded. The bounds can, however, be somewhat relaxed by allowing for a running of the scalar spectral index, since the introduction of one more parameters introduces degeneracies.

Lyth bound

In single-field slow-roll inflation, in order to generate large values of r we must have large excursions of the inflation field ϕ. This can be understood from eq. (21.323). Using

$$\frac{1}{H}\frac{d}{dt} = \frac{d}{dN}, \qquad (21.341)$$

where $N = \log a$, we can rewrite eq. (21.323), as

$$r(N) = \frac{8}{m_{\rm Pl}^2}\left(\frac{d\phi_0}{dN}\right)^2. \qquad (21.342)$$

Therefore the excursion of the inflaton field from the time when the pivot scale leaves the horizon to the end of inflation is

$$
\begin{aligned}
\Delta\phi_0 &= \int_{\phi_*}^{\phi_e} d\phi_0 \\
&= \int_{N(\phi_*)}^{N(\phi_e)} dN \, \frac{d\phi_0}{dN} \\
&= \frac{m_{\mathrm{Pl}}}{\sqrt{8}} \int_{N(\phi_*)}^{N(\phi_e)} dN \, r^{1/2}(N) \,.
\end{aligned}
\tag{21.343}
$$

Neglecting the weak dependence of r on N, and denoting as usual by N_* the number of e-folds from the time when the pivot scale leaves the horizon to the end of inflation, $N_* = |N(\phi_e) - N(\phi_*)|$, we get

$$
\begin{aligned}
|\Delta\phi_0| &\simeq \frac{m_{\mathrm{Pl}}}{\sqrt{8}} r^{1/2} N_* \\
&\simeq 6.7 \, m_{\mathrm{Pl}} \left(\frac{r}{0.1}\right)^{1/2} \left(\frac{N_*}{60}\right) \,.
\end{aligned}
\tag{21.344}
$$

If we require that $\Delta\phi_0$ remains sub-Planckian, eq. (21.344) gives a bound on r known as the *Lyth bound*. For instance, requiring $|\Delta\phi_0| < m_{\mathrm{Pl}}$ gives

$$
r < 2.2 \times 10^{-3} \left(\frac{60}{N_*}\right)^2 \,.
\tag{21.345}
$$

Starobinsky inflation

We next consider the Starobinsky model, or R^2 model, as it is also called. Plugging eqs. (21.117) and (21.118) into eqs. (21.301) and (21.325) and evaluating as usual n_s and r at the pivot scale, we find that the Starobinsky model predicts

$$
n_s - 1 \simeq -\frac{2}{N_*}
\tag{21.346}
$$

$$
r \simeq \frac{12}{N_*^2} \,.
\tag{21.347}
$$

These predictions are consistent with the *Planck* 2015 data (at 95% c.l.) for $54 < N_* < 62$; see Fig. 21.6. We see that r is $O(1/N_*^2)$, while we saw in eq. (21.340) that in typical large-field models $r = O(1/N_*)$. Numerically, eq. (21.347) gives

$$
r \simeq 3.3 \times 10^{-3} \left(\frac{60}{N_*}\right)^2 \,.
\tag{21.348}
$$

To reach values of this order is the target of proposed future space missions.

The value of the mass scale M that appears in the Starobinsky model [see eq. (21.87)] can be fixed as usual in terms of N_* by comparing with the numerical value of the scalar amplitude $A_{\mathcal{R}}$. Using eq. (21.292) with the potential (21.109), in the limit of large ϕ, where inflation takes place, we get

$$A_{\mathcal{R}}(k_*) \simeq \frac{3}{128\pi^2} \left(\frac{M}{m_{\mathrm{Pl}}}\right)^2 \exp\left\{2\sqrt{\frac{2}{3}}\frac{\phi_*}{m_{\mathrm{Pl}}}\right\} . \tag{21.349}$$

In the same approximation, eq. (21.60) [with $\phi(\eta_*)$ and N_* replacing ϕ_i and ΔN] gives

$$N_* \simeq \frac{3}{4} \exp\left\{\sqrt{\frac{2}{3}}\frac{\phi_*}{m_{\mathrm{Pl}}}\right\} . \tag{21.350}$$

Therefore

$$A_{\mathcal{R}} \simeq \frac{1}{24\pi^2} \left(\frac{M}{m_{\mathrm{Pl}}}\right)^2 N_*^2 . \tag{21.351}$$

Using as usual the value given in eq. (19.186) as the reference for $A_{\mathcal{R}}$ we get

$$M \simeq 1.19 \times 10^{-5} m_{\mathrm{Pl}} \left(\frac{A_{\mathcal{R}}}{2.14 \times 10^{-9}}\right)^{1/2} \left(\frac{60}{N_*}\right)$$

$$\simeq 2.89 \times 10^{13} \,\mathrm{GeV} \left(\frac{A_{\mathcal{R}}}{2.14 \times 10^{-9}}\right)^{1/2} \left(\frac{60}{N_*}\right) . \tag{21.352}$$

21.3.6 The relic inflationary GW background today

In the preceding sections we have learned how to compute the primordial GW spectrum generated by a given inflationary model. As we discussed in detail in Chapter 20, these primordial GWs can in principle be detected through their effect on the multipole moments of the temperature and of the polarization of the CMB photons. In particular, GWs affect temperature and polarization anisotropies through their contribution to the ISW effect [see eqs. (20.123) and (20.316)], i.e. they affect the propagation of the CMB photons in their journey from the last scattering surface to us.

A different reason why the GWs produced during inflation (or, more generally, in the early Universe) are interesting is the fact that today they form a relic stochastic background of GWs. We can therefore investigate the possibility of detecting such a stochastic background directly with ground-based or space-borne interferometers, using the techniques discussed in Section 7.8 of Vol. 1, as well as with pulsar timing arrays, which we will discuss in Chapter 23. In this case we are interested in the power spectrum today, which is related to the primordial power spectrum through the transfer function, as in eq. (19.289).[35] The relation between the primordial tensor power spectrum and the energy fraction $\Omega_{\mathrm{gw}}(f)$, which is the quantity normally used in the context of interferometers and pulsar timing array experiments, has been worked out in eq. (19.290).

We can now insert in eq. (19.290) a generic primordial tensor power spectrum of the form (19.281). It is convenient to write $A_T(k_*) =$

[35]Of course the transfer function from the time when we set initial condition up to a generic subsequent time also enters in the comparison with CMB data, given that, to compute the ISW effect, we need the GW amplitude at a generic moment between last scattering and the present time. This is automatically taken into account in the codes, such as CAMB or CLASS, that give the CMB predicitons of a given cosmological model, such as ΛCDM.

$r(k_*)A_{\mathcal{R}}(k_*)$, and also $(k/k_*) = (f/f_*)$, where $f = k/(2\pi)$ is the physical frequency today and $f_* = k_*/(2\pi)$ is the pivot frequency,

$$\mathcal{P}_{T,\text{in}}(f) = r(k_*)A_{\mathcal{R}}(k_*)\left(\frac{f}{f_*}\right)^{n_T}. \qquad (21.353)$$

For instance, if $k_* = 0.002\,\text{Mpc}^{-1}$,

$$f_* \simeq 3.09 \times 10^{-18}\,\text{Hz}, \qquad (21.354)$$

while for $k_* = 0.05\,\text{Mpc}^{-1}$ we get $f_* \simeq 7.73\times10^{-17}$ Hz. Then eq. (19.290) becomes

$$\Omega_{\text{gw}}(f) = \frac{\pi^2}{3H_0^2}\,f^2|T_{\text{GW}}(f)|^2\,r(k_*)A_{\mathcal{R}}(k_*)\left(\frac{f}{f_*}\right)^{n_T}. \qquad (21.355)$$

As we saw in eq. (19.258), for modes that re-entered the horizon deep in RD, i.e. for $f \gg f_{\text{eq}}$ [with $f_{\text{eq}} \simeq 3 \times 10^{-17}$ Hz; see eq. (19.254)], the transfer function is proportional to $1/f$, apart from a weak dependence due to the change in the effective number of relativistic degrees of freedom, so the factor $f^2|T_{\text{GW}}(f)|^2$ in eq. (21.355) becomes constant, apart from this weak dependence. Furthermore, we saw that in the Standard Model the function $g_*(T)$ saturates to a constant value $g_* \simeq 106.75$ at $T \gg 100$ GeV. Thus, for modes that re-entered the horizon at temperatures $T \gtrsim 100$ GeV, corresponding to modes with frequencies today $f \gtrsim 10^{-4}$ Hz, there is not even this weak dependence and $f^2|T_{\text{GW}}(f)|^2$ is constant. In this limit the relation between $h_0^2\Omega_{\text{gw}}(f)$ and $\mathcal{P}_{T,\text{in}}(f)$ is given by eq. (19.292). Writing the primordial tensor power spectrum as in eq. (21.353) and normalizing $r(k_*)$ and $A_{\mathcal{R}}(k_*)$ to their usual reference values, eq. (19.292) becomes[36]

$$h_0^2\Omega_{\text{gw}}(f) \simeq 1.44 \times 10^{-16}\left(\frac{r(k_*)}{0.1}\right)\left(\frac{A_{\mathcal{R}}(k_*)}{2.14 \times 10^{-9}}\right) \qquad (21.357)$$
$$\times \left(\frac{f}{f_*}\right)^{n_T}\left(\frac{106.75}{g_*(T_k)}\right)^{1/3} \qquad (f \gtrsim 10^{-4}\,\text{Hz}).$$

This is the expression for $h_0^2\Omega_{\text{gw}}(f)$ relevant for pulsar timing arrays and for space-borne and ground-based interferometers. The expression in the region $f_{\text{eq}} \ll f \lesssim 10^{-4}$ Hz is quite similar, with just a small correction due to the separate evolution of $g_*(T_k)$ and $g_*^S(T_k)$, and can be obtained from eq. (19.291). In the opposite limit $f \ll f_{\text{eq}}$ we use eq. (19.295), which gives

$$h_0^2\Omega_{\text{gw}}(f) \simeq 1.87 \times 10^{-16}\left(\frac{r(k_*)}{0.1}\right)\left(\frac{A_{\mathcal{R}}(k_*)}{2.14 \times 10^{-9}}\right) \qquad (21.358)$$
$$\times \left(\frac{f_{\text{eq}}}{f}\right)^2\left(\frac{f}{f_*}\right)^{n_T} \qquad (f \ll f_{\text{eq}}).$$

Notice the extra factor $(f_{\text{eq}}/f)^2$ compared with eq. (21.357). These expressions are not specific to inflation, but hold for any model for which

[36]In these estimates we use the simple analytic approximation to the transfer function discussed in Section 19.5.2. More accurate results are obtained through a numerical evaluation of the transfer function. This provides more accurate numbers and a smooth transition between the regimes $f \ll f_{\text{eq}}$ and $f \gg f_{\text{eq}}$, which can also be encoded in simple fitting formulas. For instance, Turner, White and Lidsey (1993) (TWL) express $\Omega_{\text{gw}}(f)$ by multiplying the result computed analytically for $f \ll f_{\text{eq}}$ by the transfer function

$$T^2(f) = 1+1.57(f/f_{\text{eq}}) +3.43(f/f_{\text{eq}})^2, \qquad (21.356)$$

which provides a good fit to the numerically determined transfer function. Observe that the coefficients in eq. (21.356) are different from those given in eq. (16) of TWL because their definition of k_{eq} is different from ours, $k_{\text{eq}} = 2(2 - \sqrt{2})k_{\text{eq}}^{\text{TWL}}$, where k_{eq} denotes our definition (which is nowadays the usual one) and $k_{\text{eq}}^{\text{TWL}}$ is their definition [see also eq. (15) of Nakayama, Saito, Suwa and Yokoyama (2008)]. Observe also that the transfer function of TWL is computed setting $\Omega_M + \Omega_R = 1$, i.e. with no dark energy. One should also include the damping from neutrino free streaming discussed in Section 19.5.3. As we will discuss, a further transfer function is required to take into account the effect of reheating.

the primordial power spectrum can be written as a simple power law, as in eq. (21.353), over the whole range of frequencies in which one is interested. We observe that the overall coefficient in $h_0^2 \Omega_{\rm gw}(f)$ is very small, of order 10^{-16}. As we will see in Section 22.7, this is several orders of magnitudes below the sensitivity of ground-based interferometers, as well as of the planned LISA space interferometer. The proposed Japanese mission DECIGO aims at reaching the sensitivity necessary for detecting this inflationary GW background. It would observe GWs in the range between 1 mHz and 100 Hz, with its best sensitivity in the region $0.1 - 10$ Hz. Ideas for a possible successor to LISA, the Big Bang Observer (BBO), had also been discussed earlier in the literature, although BBO has not yet taken the form of a concrete project.

The value of $h_0^2 \Omega_{\rm gw}$ at the frequencies of space-borne or ground-based interferometers depends crucially also on the tilt n_T of the spectrum. For a blue spectrum, i.e. $n_T > 0$, the factor $(f/f_*)^{n_T}$ can provide an important enhancement factor, given that f_* has the very small value given in eq. (21.354), while $f \sim 10^{-3}$ Hz for LISA and $f \sim 10^2$ Hz for LIGO/Virgo. Unfortunately, as we saw in eq. (21.322), for single-field slow-roll inflation the tilt is rather slightly negative, $n_T = -r/8$. The prediction of single-field slow-roll inflation can therefore be written in terms of a single parameter r, which depends on the specific inflationary potential considered. In particular in the Standard Model, where at $T > 100$ GeV the factor $g_*(T)$ saturates to the value $\simeq 106.75$, at frequencies $f \gtrsim 10^{-4}$ Hz we have

$$ h_0^2 \Omega_{\rm gw}(f) \simeq 1.44 \times 10^{-16} \left(\frac{r(k_*)}{0.1} \right) \left(\frac{A_{\mathcal{R}}(k_*)}{2.14 \times 10^{-9}} \right) \left(\frac{f}{f_*} \right)^{-r(k_*)/8} . $$
(21.359)

Given the observational bound $r < 0.1$, we see that $|n_T|$ is very small, so $\Omega_{\rm gw}(f)$ is nearly flat for $f \gg f_{\rm eq}$. Alternatively, we can write the tensor amplitude $A_T(k_*)$ in terms of H_* (the value of the Hubble parameter during inflation evaluated at the time when the pivot scale crosses the horizon) using eq. (21.320). This gives, setting again $g_*(T) \simeq 106.75$,

$$ \Omega_{\rm gw}(f) \simeq 1.36 \times 10^{-17} \left(\frac{H_*}{10^{-5} m_{\rm Pl}} \right)^2 \left(\frac{f}{f_*} \right)^{-r(k_*)/8} , $$
(21.360)

which again is valid for $f \gtrsim 10^{-4}$ Hz, and can be easily extended to $f_{\rm eq} \ll f \lesssim 10^{-4}$ Hz by taking into account the evolution of the effective number of relativistic species. The equivalence between eqs. (21.359) and (21.360) can also be obtained directly using eq. (21.328).

Finally, at very high frequencies the spectrum is cut off by the fact that, if the wavelength is sufficiently small, the mode never exits the horizon during inflation, and therefore is never in the super-horizon regime. This means that there is no amplification for modes whose comoving momentum k satisfies $k > a_{\rm end} H_{\rm end}$, where $a_{\rm end}$ and $H_{\rm end}$ are the scale factor and Hubble parameters at the end of inflation. The comoving momentum k is equal to the physical momentum today, since we set $a_0 = 1$. So (in our units $\hbar = c = 1$) $k = 2\pi f$, where f is the physical frequency today, and the maximum frequency for which there is amplification is given by $2\pi f_{\max} \simeq a_{\mathrm{end}} H_{\mathrm{end}}$. To obtain an order-of-magnitude estimate of $f_{\max}$ we assume at first instantaneous reheating and use eq. (21.12), together with $H_*^2 \simeq \rho_{\mathrm{infl}}/(3m_{\mathrm{Pl}}^2)$. This gives

$$f_{\max} \simeq 1.4 \times 10^8 \, \mathrm{Hz} \left(\frac{H_*}{10^{-5} m_{\mathrm{Pl}}} \right)^{1/2}, \qquad (21.361)$$

or, using eq. (21.328),

$$f_{\max} \simeq 2.5 \times 10^8 \, \mathrm{Hz} \left(\frac{r}{0.1} \right)^{1/4} \left(\frac{A_{\mathcal{R}}}{2.14 \times 10^{-9}} \right)^{1/4}. \qquad (21.362)$$

Beyond $f_{\max}$, $h_0^2 \Omega_{\mathrm{gw}}(f)$ is cut off exponentially. This is due to the fact that the typical frequency scale of the inflationary background is given by the value of H during inflation, and in a time-dependent background oscillating at this frequency the production of quanta with $f \gg H$ is exponentially suppressed. This will also be apparent from the full quantum computation that we will discuss in Section 21.3.7. In Fig. 21.7 we plot the function $h_0^2 \Omega_{\mathrm{gw}}(f)$ for $r = 0.1$ (solid curve), $r = 0.01$ (dashed) and $r = 0.001$ (dot-dashed).

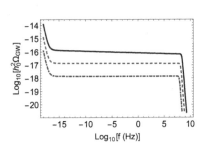

Fig. 21.7 The inflationary prediction for $h_0^2 \Omega_{\mathrm{gw}}(f)$ for $r = 0.1$ (solid curve), $r = 0.01$ (dashed) and $r = 0.001$ (dot-dashed), assuming instantaneous reheating.

[37] Reheating can itself be a powerful extra source of GWs, not related to the mechanism of amplification of vacuum fluctuations that we are discussing here. The corresponding production of GWs will be studied in detail in Section 22.3. Here we study the effect that the existence of a reheating phase has on the spectrum of GWs produced by the amplification of vacuum fluctuations.

Effect of reheating

Including a realistic reheating phase can change the high-frequency behavior in an important manner.[37] This can be understood as follows. Let us compare two cosmological models that have the same value of H during inflation (which we take for simplicity as a de Sitter phase), and also the same function $H(a)$ once reheating is completed in both models, so that we are comparing two models that have the same cosmological evolution in the late RD and in the MD and present epochs. However, the first model has instantaneous reheating, i.e. H^{-1} switches suddenly from a constant value during inflation to a behavior $H^{-1}(a) \propto a^2$ during RD. The second model rather has an intermediate reheating phase where $H^{-1}(a)$ grows more slowly with a. As we have already mentioned, typical reheating mechanisms proceed through decay of the inflaton into massive states, resulting in an approximately matter-dominated reheating phase. In this case $H^{-1}(a) \propto a^{3/2}$ during reheating, but our discussion is more general, and holds for any reheating phase where $H^{-1}(a)$ grows more slowly than a^2. The evolution of $H^{-1}(a)$ for the two models is shown in Fig. 21.8, where the first model is represented by a dashed line and the second model by a solid line. By construction, for $a < a_1$ and for $a > a_3$ the two models coincide, so in the plot we can only see the solid line, while in the region $a_1 < a < a_3$ the functions $H^{-1}(a)$ for the two models are different. Notice that, if we require that the two models have the same value of H during inflation and the same expansion history $H(a)$ after reheating, inflation must end at different values of a in the two models, marked as a_1 in the model with an extended reheating phase and a_2 in the model with instantaneous reheating. In the former model reheating terminates when $a = a_3$, and after that,

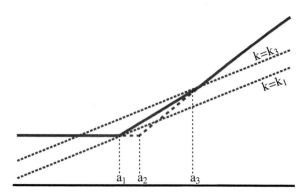

Fig. 21.8 The evolution of $H^{-1}(a)$ in a model with an extended reheating phase (solid line) and in a model with instantaneous reheating (dashed line, partly covered by the solid line). The evolution of the physical reduced wavelength for two different values of k is also shown.

by construction, the evolutions in the two models coincide again. In the plot we also show the evolution of the physical reduced wavelength $\lambdabar_{\rm phys}(a) = \lambdabar a = k^{-1}a$, for two different modes. One is the mode with $k = k_1$ such that $\lambdabar_{\rm phys}(a_1) = H(a_1)$. In the model with instantaneous reheating this mode exits the horizon when $a = a_1$ and re-enter later, during RD, while in the model with extended reheating the modes with $k > k_1$ never exit the horizon. The second mode that we show in the plot is the one with a value $k = k_3$ such that it re-enters the horizon when $a = a_3$. Note that $k_3 < k_1$, i.e. for the corresponding comoving reduced wavelength we have $\lambdabar_3 > \lambdabar_1$.

We can now easily compare the behavior of $h_0^2\Omega_{\rm gw}(f)$ for the two models. The modes with $k < k_3$, i.e. $\lambdabar > \lambdabar_3$, exit the horizon during inflation at a value of a when the two models coincide, and re-enter the horizon at $a > a_3$, when again the two models coincide. As long as it is outside the horizon a mode stays constant independently of the form of $H(a)$, so for these modes there is no difference between the two models. For $k > k_3$ the situation is different. First of all, we see that the two models have their cutoffs at different values of k, given by the largest value of k for which a mode never exits the horizon. For the model with extended reheating this cutoff is at $k = k_1$, while for the model with instantaneous reheating the cutoff is at a value $k_2 > k_1$, determined by $k_2/a_2 = H_{\rm dS}$, where $H_{\rm dS}$ is the constant value of H during the de Sitter phase.

More interestingly, even the modes with $k_3 < k < k_1$ undergo a different evolution. Indeed we see from the figure that, in the model with extended reheating, these modes re-enter the horizon earlier, compared with the model with instantaneous reheating. More quantitatively, in the model with instantaneous reheating the Hubble parameter in RD (i.e. for all $a > a_2$) can be written, using as a reference the value at

$a = a_3$, as

$$H(a) = H(a_3) \left(\frac{a_3}{a}\right)^2 . \tag{21.363}$$

The value $a = a_*(k)$ when a mode re-enters the horizon is determined as usual by $k/a_* = H(a_*)$. Using eq. (21.363) we get

$$a_*(k) = a_3 \left(\frac{k_3}{k}\right) , \tag{21.364}$$

where we have used $k_3/a_3 = H(a_3)$. For the model with extended reheating, assuming a matter-dominated expansion, for $a_1 < a < a_3$ we rather have

$$H(a) = H(a_3) \left(\frac{a_3}{a}\right)^{3/2} , \tag{21.365}$$

and therefore

$$a_*(k) = a_3 \left(\frac{k_3}{k}\right)^2 . \tag{21.366}$$

Therefore there is an extra suppression factor (k_3/k) compared with the model with instantaneous reheating. Recalling from eq. (19.256) that the transfer function of the tensor modes is proportional to $a_*^2(k)$, we see that, in the model with extended reheating, for modes with $k > k_3$ (but still smaller than the cutoff at $k = k_1$) $\Omega_{\rm gw}(f)$ is suppressed by an extra factor $(k_3/k)^2$, compared with the result for the model with instantaneous reheating that we showed in Fig. 21.7. Since in the model with extended reheating $a_3 = a_{\rm reh}$ is the value of the scale factor when reheating terminates, in this model we will denote k_3 by $k_{\rm reh}$. Thus, in the model with an extended reheating phase, taken to be matter-dominated, for $f > f_{\rm reh}$ we find that $\Omega_{\rm gw}(f)$ is suppressed by an extra factor $(f_{\rm reh}/f)^2$, compared with the case that we studied before. Of course, this is is the same result that we have already obtained by computing $\Omega_{\rm gw}(f)$ for the modes that re-enter during RD, i.e. $f > f_{\rm eq}$, and the modes that re-enter during MD, at $f < f_{\rm eq}$, with $f_{\rm reh}$ instead of $f_{\rm eq}$.

It is also easy to generalize the result to a generic evolution during reheating, with a general equation-of-state parameter $w_{\rm reh} < 1/3$. In this case, from eq. (17.57), $H^2(a) \propto a^{-3(1+w_{\rm reh})}$. Then we get

$$a_*(k) = a_3 \left(\frac{k_3}{k}\right)^{1+\frac{1-3w_{\rm reh}}{1+3w_{\rm reh}}} . \tag{21.367}$$

Thus, compared with the instantaneous reheating case $a_*(k) = a_3(k_3/k)$, there is an extra suppression factor in a_* equal to $(k_3/k)^{(1-3w_{\rm reh})/(1+3w_{\rm reh})}$. The extra suppression factor in $\Omega_{\rm gw}(f)$ is the square of this, so for $f > f_{\rm reh}$ eq. (21.359) is replaced by

$$h_0^2 \Omega_{\rm gw}(f) \simeq 1.44 \times 10^{-16} \left(\frac{r(k_*)}{0.1}\right) \left(\frac{A_{\mathcal{R}}(k_*)}{2.14 \times 10^{-9}}\right) \tag{21.368}$$

$$\times \left(\frac{f}{f_*}\right)^{-r(k_*)/8} \left(\frac{f}{f_{\rm reh}}\right)^{-2(1-3w_{\rm reh})/(1+3w_{\rm reh})} , \quad (f \gg f_{\rm reh}) .$$

apart, as usual, from overall numerical factors of order 1 that can only be obtained by evaluating the transfer function numerically.[38] The value of $f_{\rm reh}$ can be written in terms of the reheating temperature using eq. (19.262). Using, as usual, the fact that at $T \gg 100$ GeV in the Standard Model $g_*(T) = g_*^S(T) \simeq 106.75$, and taking 10^6 GeV as a reference value for the reheating temperature, eq. (19.262) gives

$$f_{\rm reh} \simeq 0.027 \text{ Hz} \left(\frac{T_{\rm reh}}{10^6 \text{GeV}} \right) \left(\frac{g_{*,\rm reh}}{106.75} \right)^{1/6} . \qquad (21.370)$$

The result for $h_0^2 \Omega_{\rm gw}(f)$ is shown in Fig. 21.9, setting for definiteness $r = 0.1$ and $w_{\rm reh} = 0$, and for three different reheating temperatures, $T_{\rm reh} = 10^6, 10^9$ and 10^{12} GeV. We see that, for a reheating temperature of order 10^6 GeV the signal at ground-based interferometers, say at $f \sim 10^2$ Hz, which would already be too small to be observed with present and near-future detectors even without this extra suppression, would be further suppressed by an extra factor of order $(0.027/10^2)^2 \sim 10^{-7}$, and therefore would be totally below any future hope of detection. The knee in the spectrum would be just in the band of ground-based interferometers for $T_{\rm reh} \sim 10^{10}$ GeV.

As we have mentioned, the space-borne interferometer DECIGO has been proposed as a future mission for reaching a sensitivity to $h_0^2 \Omega_{\rm gw}(f)$ of order 10^{-17} in the region of frequencies around, say, $0.1-10$ Hz. In this case a reheating temperature larger than about $10^8 - 10^9$ GeV would be necessary in order not to suppress the signal in the most sensitive region.

What could be learned from the observation of the inflationary GW background?

A detection of the inflationary GW background would constitute a major discovery. Besides providing further clear evidence for inflation, it would also provide direct observational confirmation of the fact that gravity must be quantized. While at a purely theoretical level there is probably little doubt that this should be so, in particular if one treats gravity as a low-energy effective QFT, observationally there is yet no evidence for the quantum nature of gravity. However, the computation of the primordial spectrum that we have discussed is based on writing the TT part of the metric perturbation, $h_{ij}^{\rm TT}$, as a quantum field, when we computed the quantum vacuum expectation value $\langle 0| \tilde{h}_{A,\rm in}^*(\mathbf{k}; \eta) \tilde{h}_{A,\rm in}(\mathbf{k}; \eta) |0\rangle$ in eq. (21.310). Confirmation of this prediction would therefore provide direct evidence of the fact that the graviton is described by a quantum field with its usual expansion in creation and annihilation operators.[39]

An observation of the inflationary GW spectrum across several frequencies, from the very low frequencies to which CMB experiments are sensitive to the higher frequencies of space-borne and ground-based interferometers, would be a mine of information. The overall amplitude of the spectrum would fix the Hubble parameter during inflation [see eq. (21.360)], and therefore the inflationary scale. Thanks to the long

[38]Indeed, Nakayama, Saito, Suwa and Yokoyama (2008), in a numerical study of a model where the evolution during reheating is basically matter-dominated, find that reheating induces an extra suppression in $\Omega_{\rm gw}(f)$ described by the transfer function

$$T^2(f) = \qquad (21.369)$$
$$[1 - 0.32(f/f_{\rm reh}) + 0.99(f/f_{\rm reh})^2]^{-1}.$$

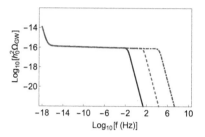

Fig. 21.9 The inflationary prediction for $h_0^2 \Omega_{\rm gw}(f)$ for $r = 0.1$ including a matter-dominated reheating phase with $T_{\rm reh} = 10^6$ GeV (solid curve), 10^9 GeV (dashed) and 10^{12} GeV (dot-dashed).

[39]One might wonder whether the observation of the spectrum of scalar perturbations already tells us something about the quantum nature of gravity. Indeed, scalar perturbations are obtained from the quantum vacuum expectation value of the curvature perturbation $\mathcal{R}$, which is a superposition of the scalar metric perturbation Ψ and of the perturbation $\delta\phi$ of the inflaton field; see eq. (21.248). However, the answer is negative. The scalar metric perturbations Ψ and Φ in GR are non-radiative fields that obey a Poisson equation [see eqs. (19.14) and (19.16)], rather than a Klein–Gordon equation. They do not have plane-wave solutions in vacuum, whose coefficients, in the quantum theory, would be promoted to creation and annihilation operators, and there are no physical quanta associated with them. The quantum nature of the field $\mathcal{R}$ is fully inherited from the quantum nature of the inflaton field.

lever arm one could measure the spectral tilt n_T, which in turn would give information on the specific inflationary model. Furthermore, as we discussed in Section 19.5.2, GWs with frequencies relevant for LISA re-entered the horizon when $T \sim 3 \times 10^4$ GeV, while GWs with frequencies relevant for LIGO/Virgo re-entered the horizon when $T \sim 3 \times 10^9$ GeV. They would therefore carry an imprint of physics at these energy scales, through the transfer function, which depends on the effective number of degrees of freedom at the time of re-entry. Finally, we have seen that reheating induces a knee in the spectrum, at a frequency that depends on the reheating temperature, and whose subsequent slope depends on the equation of state during reheating.

We have seen that CMB observations have already put very meaningful limits $r < 0.1$ in their frequency range, i.e. for the extremely small frequencies of the order of the pivot scale given in eq. (21.354). In contrast, detection of the inflationary background (21.368) is not possible with the current and near-future sensitivity of pulsar timing arrays or ground-based interferometers, or with LISA.[40] As already mentioned, the proposed Japanese project DECIGO aims to reach the sensitivity necessary for detecting the inflationary GW background, with its best sensitivity in the range $0.1 - 10$ Hz.

The situation could, however, be different if the vacuum is not the Bunch–Davies one. In that case we see from eq. (21.239) that one could in principle obtain a blue spectrum, i.e. a spectrum with a positive tilt $n_T > 0$, which would therefore give a much larger values of $h_0^2 \Omega_{\rm gw}(f)$ at higher frequencies. We will also see in Section 22.6 that some alternatives to inflation can predict a blue spectrum of GWs. In any case, the computation of the inflationary GW spectrum clearly illustrates how cosmological GWs can in principle open a window on the very early Universe and on the corresponding high-energy physics.

21.3.7 A full quantum computation of $\Omega_{\rm gw}(f)$

A peculiar aspect of the computation of scalar and tensor perturbations that we have illustrated so far is that the starting point is the quantum fluctuations of a field, whether this be a scalar field or the gravitational field. Using the methods of QFT in curved space discussed in Section 21.2 we have been able to compute the spectrum of quantum fluctuations of the field when the corresponding mode has left the horizon during inflation. When the mode later re-enters the horizon, during RD or MD, the fluctuations associated with it are then interpreted as classical fluctuations. The validity of such a mixed quantum/classical approach is not obvious, but can actually be justified by showing that, when the modes are outside the horizon, their canonical momentum shrinks to zero, and therefore only the expectation values of the fields, rather than those involving also the canonical momenta, are relevant; see the Further Reading. Thus, once the fluctuations have re-entered the horizon they behave as classical fluctuations. Alternatively, in order to clarify this issue it is very instructive to perform a purely quantum

---

[40]Basically, the difficulty in obtaining a significant sensitivity to $h_0^2 \Omega_{\rm gw}(f)$ at high frequency is due to the fact that $h_0^2 \Omega_{\rm gw}(f) \propto f^3 S_h(f)$; see eq. (7.202) in Vol. 1. The spectral density of the signal, $S_h(f)$, is the quantity to be compared with the spectral density of the noise, $S_n(f)$ [possibly in a non-trivial way, as when one performs the correlation between two detectors, as in eq. (7.239)]. Of course, at widely different frequencies the technological problems that one has to address to have a small noise spectral density are completely different. However, a crucial point is the factor f^3 in $h_0^2 \Omega_{\rm gw}(f) \propto f^3 S_h(f)$, which makes it more and more difficult to reach interesting values of $h_0^2 \Omega_{\rm gw}(f)$ as the frequency increases.

computation of the tensor perturbations, using the Bogoliubov coefficients. We illustrate the computation in a model in which a phase of de Sitter inflation is followed by RD and MD.[41]

The physics of the problem is completely analogous to that of adiabatic and sudden perturbations, discussed in most elementary textbooks of non-relativistic quantum mechanics. Consider a non-relativistic quantum-mechanical system in the Schrödinger representation. Suppose that at $t < 0$ the system is described by a time-independent Hamiltonian $\mathcal{H} = \mathcal{H}_0$ and, at $t = 0$, we slowly switch on a perturbation $\mathcal{H}_1(t)$, so the total Hamiltonian becomes $\mathcal{H}(t) = \mathcal{H}_0 + \mathcal{H}_1(t)$. In this case one can prove a theorem (valid to all orders in perturbation theory) known as the *adiabatic theorem*, which states that if at $t = 0$ the system is an eigenstate $|n^0\rangle$ of the initial Hamiltonian $\mathcal{H}_0$, then the state of the system will evolve smoothly into the corresponding eigenstate $|n(t)\rangle$ of the Hamiltonian $\mathcal{H}(t)$.[42] The opposite case is that of sudden perturbations. Suppose that, at some time $t = 0$, the Hamiltonian changes suddenly from $\mathcal{H}$ to $\mathcal{H}'$, so that $\mathcal{H}(t) = \mathcal{H}$ for $t < 0$ and $\mathcal{H}(t) = \mathcal{H}'$ for $t > 0$. This is an idealization of the situation in which the Hamiltonian changes on a time-scale much shorter than the characteristic time-scale of the system. In this case integrating the Schrödinger equation from $t = -\epsilon$ to $t = +\epsilon$ we get

$$|\psi(t = \epsilon)\rangle - |\psi(t = -\epsilon)\rangle = -\frac{i}{\hbar} \int_{-\epsilon}^{\epsilon} dt\, \mathcal{H}(t)|\psi(t)\rangle. \qquad (21.371)$$

If $\mathcal{H}(t)$ is finite (i.e. it is not proportional to a Dirac delta), the right-hand side goes to zero for $\epsilon \to 0$, and therefore

$$\lim_{t \to 0^+} |\psi(t)\rangle - \lim_{t \to 0^-} |\psi(t)\rangle = 0. \qquad (21.372)$$

In other words, the state of the system is unchanged under a sudden perturbation. Physically, the system has had no time to adjust to the change in the Hamiltonian. However, if $|\psi\rangle$ was an eigenstate of $\mathcal{H}$, in general it will no longer be an eigenstate of the new Hamiltonian $\mathcal{H}'$. We can rather expand it in the basis $|n'\rangle$ of the new eigenstates, by computing the probability amplitudes $\langle n'|\psi\rangle$.

The same approach can be applied to our cosmological context. Consider for instance the transition from a de Sitter to the RD phase. The time-scale over which this transition is completed is given by the inverse of the Hubble parameter $H^{-1}(t)$, evaluated at the time t_1 at which the transition takes place; i.e. the typical frequency of the background variation is given by $H(t_1)$. The modes whose physical frequencies at time t_1 are much larger than $H(t_1)$ (i.e. the modes whose wavelengths are much smaller than $H^{-1}(t_1)$, and which therefore are well inside the horizon at that time) feel the transition as very slow, and have time to adjust to it. In contrast, for modes well outside the horizon the transition is sudden. Suppose for instance that, during the de Sitter phase, the system was in the ground state, for example with respect to the Bunch–Davies definition of particle, so all the momentum modes were empty.

[41] Of course, in an exactly de Sitter phase, $\dot{\phi}_0 = 0$ and eq. (21.262) gives a diverging result for the scalar perturbation. In other words, in a purely de Sitter inflation $r = 0$ since A_T is finite but $A_{\mathcal{R}}$ diverges. Here, however, we are just interested in showing that, for the tensor perturbations, the full quantum computation of the tensor amplitude A_T gives the same result (21.320) that we obtained with the mixed quantum/classical approach, and we can therefore work to lowest order in the slow-roll expansion, assuming an exact de Sitter scale factor, followed by a transition to RD and then MD. To perform the same exercise for scalar perturbations, even to lowest order, we should instead consider a slow-roll phase with a small but non-zero $\dot{\phi}_0$.

[42] For the proof of the theorem see for example Griffiths (1995), Section 10.1.2. The subtle point is to decide when the perturbation can be considered slow enough. The result is that the typical frequency of the perturbation must be much smaller than the typical frequency *differences* $\omega_n^0 - \omega_m^0$ between the unperturbed energy levels. Therefore, in the presence of degeneracy, we are never in the condition where the adiabatic theorem can be applied.

The modes inside the horizon at the time of the transition have time to adjust themselves smoothly to the transition, and the state evolves so that it is still empty in these modes, with respect to the new definition of particle appropriate to RD. In contrast, for modes well outside the horizon at the transition epoch the state does not have time to evolve and remains unchanged. However, as we have seen in Section 21.2.1, a vacuum state with respect to some modes appropriate to de Sitter (such as the modes that define the Bunch–Davies vacuum) is a multi-particle state with respect to a different set of modes, such as the modes appropriate to RD. The net effect is that, from the point of view of an observer in RD, there is now a spectrum of particles. Physically, we might picture this phenomenon as particle creation by the time-varying gravitational field at the transition epoch, in analogy to particle creation by a time-varying electric field.

More quantitatively, suppose that just before the transition the quantum state was $|s\rangle$, and let a_k and $a_k^\dagger$ be the annihilation and creation operators appropriate to a given observer in de Sitter (i.e. appropriate to a given choice of vacuum in de Sitter, such as the Bunch–Davies vacuum). In general, we can express $|s\rangle$ in terms of the occupation numbers $n_k = a_k^\dagger a_k$. Consider the modes well outside the horizon. For these modes the transition is sudden and the physical state does not have time to change during the transition, so that it is still given by $|s\rangle = |\{n_k\}\rangle$ just after the transition. However, we now must express the occupation numbers with respect to $N_k = A_k^\dagger A_k$. Using the Bogoliubov transformation (21.171) we see that

$$N_k = n_k + 2|\beta_k|^2 \left(n_k + \frac{1}{2} \right). \qquad (21.373)$$

This shows, first of all, that any preexisting value of n_k is amplified, with an amplification factor $1 + 2|\beta_k|^2$. Furthermore, even the "half-quantum" due to vacuum fluctuations is amplified. Even if the system was in the vacuum state before the transition, after the transition it is a multi-particle state with respect to the new definition of particles.

The preceding discussion is valid for modes that are super-horizon at the time of transition. Sub-horizon modes are not amplified, so there is a cutoff in the spectrum at a comoving momentum k given by $k/a(t_1) \simeq H(t_1)$. For the de Sitter–RD transition, the corresponding maximum value of the physical frequency today is the one already computed in eq. (21.361).

To perform the explicit computation of the Bogoliubov coefficients we consider first the particle production at the de Sitter–RD transition. We take an FRW model in which $a(t) \propto e^{Ht}$ for $-\infty < t < t_1$, and $a(t) \propto t^{1/2}$ for $t_1 < t < t_{eq}$, where t_{eq} is the time when MD sets in. We are therefore approximating the de Sitter–RD transition as instantaneous at $t = t_1$, which is valid for the modes that are well outside the horizon at time t_1. In terms of conformal time η, we write

$$a(\eta) = -\frac{1}{H_{dS}\eta} \qquad (-\infty < \eta < \eta_1), \qquad (21.374)$$

where $\eta_1 < 0$, and

$$a(\eta) = \frac{1}{H_{\mathrm{dS}}\eta_1^2}(\eta - 2\eta_1) \qquad (\eta_1 < \eta < \eta_{\mathrm{eq}}). \tag{21.375}$$

The choice of scale factor in eq. (21.375) neglects the change in the number of relativistic degrees of freedom during RD. For an accurate quantitative estimate this must be included, as we have done in Section 21.3.6; see for example eq. (21.357). In this section we use this simplified model just to show, conceptually, how a stochastic GW background emerges in this purely quantum approach. We have fixed the constants so that $a(\eta)$ and $a'(\eta)$ are continuous across the transition. During the de Sitter phase we assume that the state is in the Bunch–Davies vacuum, so the mode functions are given by eq. (21.209), i.e.

$$f_k(\eta) = \frac{1}{\sqrt{2k}}e^{-ik\eta}\left(1 - \frac{i}{k\eta}\right) \qquad (\eta < \eta_1). \tag{21.376}$$

In RD, in contrast, $a''/a = 0$ and the solutions of eq. (21.179) are simply $e^{\pm ik\eta}$, so

$$f_k(\eta) = \frac{1}{\sqrt{2k}}\left(\alpha_k e^{-ik\eta} + \beta_k e^{ik\eta}\right) \qquad (\eta_1 < \eta < \eta_{\mathrm{eq}}). \tag{21.377}$$

Requiring that the $f_k(\eta)$ and its derivative are continuous across the transition gives

$$\alpha_k = 1 - \frac{i}{k\eta_1} - \frac{1}{2k^2\eta_1^2}, \qquad \beta_k = \frac{e^{-2ik\eta_1}}{2k^2\eta_1^2}. \tag{21.378}$$

Note that $|\alpha_k|^2 - |\beta_k|^2 = 1$, as it should. The number of particles produced at the de Sitter–RD transition, per cell of phase space, is given by eq. (21.373); in particular, if before the transition the system was in the vacuum state, $n_k = 0$, after the transition

$$N_k = |\beta_k|^2 = \frac{1}{4k^4\eta_1^4}. \tag{21.379}$$

Denoting by a_{end} the value of the scale factor at the end of inflation, eq. (21.374) gives $a_{\mathrm{end}} = -1/(H_{\mathrm{dS}}\eta_1)$, so eq. (21.374) can be rewritten as

$$N_k = \frac{a_{\mathrm{end}}^4 H_{\mathrm{dS}}^4}{4k^4}. \tag{21.380}$$

In the simplified cosmological model given by eqs. (21.374) and (21.375), which does not take into account the change in the number of degrees of freedom in RD at high temperatures, the relation between the Hubble parameter at the end of inflation (which is also the beginning of the RD era in this simplified model that assumes instantaneous reheating) is obtained from eq. (17.75), which gives

$$H_{\mathrm{dS}}^2 = \frac{H_0^2\Omega_R}{a_{\mathrm{end}}^4}, \tag{21.381}$$

where we have set as usual $a_0 = 1$. Then we can rewrite N_k as

$$N_k = \frac{H_0^2 H_{\mathrm{dS}}^2}{4k^4} \Omega_R . \qquad (21.382)$$

Using eq. (7.204) of Vol. 1 and writing $k = 2\pi f$ we get

$$\Omega_{\mathrm{gw}}(f) = \frac{N_k k^4}{\pi^2 \rho_0}$$
$$= \frac{1}{12\pi^2} \left(\frac{H_{\mathrm{dS}}}{m_{\mathrm{Pl}}} \right)^2 \Omega_R , \qquad (21.383)$$

where we have used, as usual, $\rho_0 = 3H_0^2/(8\pi G) = 3H_0^2 m_{\mathrm{Pl}}^2$. Equation (21.383) can also be rewritten in the form

$$\Omega_{\mathrm{gw}}(f) = \frac{1}{36\pi^2} \frac{\rho_{\mathrm{infl}}}{m_{\mathrm{Pl}}^4} \Omega_R , \qquad (21.384)$$

where $\rho_{\mathrm{infl}} = 3H_{\mathrm{dS}}^2 m_{\mathrm{Pl}}^2$.[43] We see that k cancels, and $\Omega_{\mathrm{gw}}(f)$ is indeed independent of f, i.e. of k, for the particles created at the de Sitter–RD transition. The modes whose frequency f today is much larger than the value f_{eq} given in eq. (19.254) are well inside the horizon at the RD–MD transition, and for them this second transition is adiabatic. According to the discussion at the beginning of this section, they are not further amplified at the RD–MD transition. Then, eq. (21.383) is the final result for $f \gg f_{\mathrm{eq}}$.

We can now compare with the result obtained from the computation of the primordial power spectrum performed in the preceding sections. For a phase of exact de Sitter inflation $H = H_{\mathrm{dS}}$ is constant and $n_T = 0$. Then eq. (21.318) gives

$$\mathcal{P}_{T,\mathrm{in}} = \frac{2}{\pi^2} \frac{H_{\mathrm{dS}}^2}{m_{\mathrm{Pl}}^2} . \qquad (21.388)$$

Inserting this into eq. (19.291) and neglecting, as we have done here, the variation of the effective number of degrees of freedom with temperature, i.e. setting $(g_*/3.363)(3.909/g_*^S)^{4/3} = 1$, we find, in the limit $f \gg f_{\mathrm{eq}}$,

$$\Omega_{\mathrm{gw}}(f) = \frac{1}{24} \Omega_R \mathcal{P}_{T,\mathrm{in}}(f)$$
$$= \frac{1}{12\pi^2} \left(\frac{H_{\mathrm{dS}}}{m_{\mathrm{Pl}}} \right)^2 \Omega_R , \qquad (21.389)$$

in full agreement with eq. (21.383). Using the expression (17.116) for the Hubble parameter in RD, which takes into account the change in the number of relativistic degrees of freedom, provides a further factor $(g_*/3.363)(3.909/g_*^S)^{4/3}$ in both computations. Thus, the result of the computation using the Bogoliubov coefficients for a pure de Sitter inflation correctly reproduces eq. (19.291) with $n_T = 0$ and H_* replaced by H_{dS}, as expected.

Modes with $f \lesssim f_{\mathrm{eq}}$ are still outside the horizon at the RD–MD transition, and therefore undergo further amplification. The computation is

[43]In some of the earlier literature the relation between H_{dS} and a_{end}, which in the instantaneous reheating approximation is correctly given by eq. (21.381), was rather computed by writing

$$H_0 \simeq H_{\mathrm{dS}} \left(\frac{a_{\mathrm{end}}}{a_{\mathrm{eq}}} \right)^2 \left(\frac{a_{\mathrm{eq}}}{a_0} \right)^{3/2} , \qquad (21.385)$$

This would give (setting again $a_0 = 1$) $H_{\mathrm{dS}} = H_0 a_{\mathrm{eq}}^{1/2}/a_{\mathrm{end}}^2$, which is different from eq. (21.381) because $a_{\mathrm{eq}} = \Omega_R/\Omega_M$. Then eq. (21.384) was written as

$$\Omega_{\mathrm{gw}}(f) = \frac{1}{36\pi^2} \frac{\rho_{\mathrm{infl}}}{m_{\mathrm{Pl}}^4} (1 + z_{\mathrm{eq}})^{-1}$$
$$= \frac{1}{36\pi^2} \frac{\rho_{\mathrm{infl}}}{m_{\mathrm{Pl}}^4} \frac{\Omega_R}{\Omega_M} , \qquad (21.386)$$

or equivalently

$$\Omega_{\mathrm{gw}}(f) = \frac{16}{9} \frac{\rho_{\mathrm{infl}}}{\rho_{\mathrm{Planck}}} (1 + z_{\mathrm{eq}})^{-1} , \qquad (21.387)$$

where $\rho_{\mathrm{Planck}} = M_{\mathrm{Pl}}^4 = 64\pi^2 m_{\mathrm{Pl}}^4$. However, eq. (21.385) implicitly assumes $\Omega_M = 1 - \Omega_R \simeq 1$; i.e. it is an approximation to the correct Friedmann equation (17.75) only when $\Omega_\Lambda = 0$. Thus, if one uses that expression, then by consistency one must also set $a_{\mathrm{eq}} = \Omega_R$. The correct result for generic Ω_M is given by eq. (21.384) and is in fact independent of Ω_M, rather than proportional to $(1 + z_{\mathrm{eq}})^{-1} \equiv a_{\mathrm{eq}} = \Omega_R/\Omega_M$.

completely analogous to the one that we have just performed. We can write the scale factor during MD as

$$a(\eta) = a_{\rm eq} \frac{(\eta_{\rm eq} - 4\eta_1 + \eta)^2}{4(\eta_{\rm eq} - 2\eta_1)^2},\qquad (21.390)$$

which solves the Friedmann equation with non-relativistic matter sources, and is chosen so that the scale factor and its derivative are continuous across the RD–MD transition, with the RD scale factor given in eq. (21.375) and with $a_{\rm eq} = (\eta_{\rm eq} - 2\eta_1)/(H_{\rm dS}\eta_1^2)$. Solving eq. (21.179) with this scale factor one finds the mode function

$$F_k(\eta) = \frac{1}{\sqrt{2k}}\left[1 - \frac{i}{2k(\eta_{\rm eq} - 2\eta_1)}\left(\frac{a_{\rm eq}}{a(\eta)}\right)^{1/2}\right]e^{-ik\eta},\qquad (21.391)$$

and its complex conjugate. Matching the positive-frequency RD mode $(2k)^{-1/2}e^{-ik\eta}$ to the the general MD solution $\tilde{\alpha}_k F_k + \tilde{\beta}_k F_k^*$ by imposing the continuity of the function and of its derivative, one obtains the Bogoliubov coefficients $\tilde{\alpha}_k$ and $\tilde{\beta}_k$ for the RD–MD transition. The overall Bogoliubov transformation from de Sitter to MD is given by composition of the two Bogoliubov transformations, i.e.

$$\begin{pmatrix} \alpha_{k,{\rm tot}} & \beta_{k,{\rm tot}} \\ \beta_{k,{\rm tot}}^* & \alpha_{k,{\rm tot}}^* \end{pmatrix} = \begin{pmatrix} \alpha_k & \beta_k \\ \beta_k^* & \alpha_k^* \end{pmatrix}\begin{pmatrix} \tilde{\alpha}_k & \tilde{\beta}_k \\ \tilde{\beta}_k^* & \tilde{\alpha}_k^* \end{pmatrix}.\qquad (21.392)$$

The computation is straightforward, and provides the extra factor in $\Omega_{\rm gw}(f)$, proportional to $(f_{\rm eq}/f)^2$ for modes with $f \ll f_{\rm eq}$, that we already found in eq. (21.358).

Further reading

- The first inflationary model was proposed by Starobinsky (1979, 1980). The model provides both inflation and a successful reheating. The paper by Guth (1981) was the first where the advantages of inflationary cosmology were appreciated. It was based on a scalar field tunneling from a metastable state of a potential $V(\phi)$ with non-vanishing vacuum energy V_0 to the ground state with $V(\phi) = 0$. The tunneling proceeded through nucleation of bubbles of true vacuum at different points of space, and reheating was proposed to be due to collisions between bubbles. This model, later called "old inflation", did not work because to have sufficient inflation the bubble nucleation rate must be sufficiently small, but in this case it turns out that the bubbles never get in causal contact with each other, so reheating does not take place. Small-field inflation (at the epoch also called "new inflation") was proposed by Linde (1982) and Albrecht and Steinhardt (1982), and provided a second successful mechanism of inflation, after that of Starobinsky. Chaotic inflation was proposed in Linde (1983a) and hybrid inflation in Linde (1994). Natural inflation was proposed in Freese, Frieman and Olinto (1990).

- There are several excellent textbooks and reviews dealing with inflation; see for example Lyth and Riotto (1999), Liddle and Lyth (2000, 2009), Mukhanov (2005), Weinberg (2008), Baumann (2009) and Gorbunov and Rubakov (2011). A compilation of inflationary models is given in Martin, Ringeval and Vennin (2014). The Starobinsky model and $f(R)$ gravity are discussed in detail in De Felice and Tsujikawa (2010). The fact that the slow-roll solution is an attractor of the solutions of the full second-order equation is shown using the Hamilton–Jacobi formalism by Salopek and Bond (1990); see also Section 3.7 of Liddle and Lyth (2000). A comparison of the inflationary predictions with the *Planck* 2013 data is performed in [Planck Collaboration], Ade *et al.* (2014b), where one can also find a review of the relevant inflationary results and a detailed discussion of the methodology for comparing inflationary predictions with the data. Comparison with the *Planck* 2015 data is performed in [Planck Collaboration], Ade *et al.* (2016c).

- A classic textbook on quantum field theory in curved space is Birrell and Davies (1982). A more recent introduction is Mukhanov and Winitzki (2007). Discussions of de Sitter space, including different coordinate systems, can be found for example in Hawking and Ellis (1973) and in Birrell and Davies (1982).

- Early work on particle creation in an expanding universe was done by Parker (1969, 1971), Zel'dovich (1970) and Zel'dovich and Starobinsky (1971). However, it was incorrectly assumed that the equation governing graviton production is conformally invariant, so it was concluded that no graviton production occurred. The generation of gravitons by amplification of vacuum fluctuations was first proposed, without reference to inflation, by Grishchuk (1974). The fact that an inflationary de Sitter model would result in the production of relic GWs was first discussed in Starobinsky (1979, 1980). The amplification of scalar perturbations was first considered, without reference to inflation, in Lukash (1980). The generation of inhomogeneities from quantum fluctuations amplified by inflation was discussed by Mukhanov and Chibisov (1981, 1982), Hawking (1982), Starobinsky (1982), Guth and Pi (1982, 1985) and Bardeen, Steinhardt and Turner (1983). The Mukhanov–Sasaki equation was derived in Mukhanov (1985) and in Sasaki (1986). The corresponding action (21.253) was derived in Mukhanov (1989). A first systematic discussion of the theory of cosmological perturbations was presented in Mukhanov, Feldman and Brandenberger (1992). The computation of the inflationary primordial power spectra beyond the lowest order in the slow-roll expansion is performed in Stewart and Lyth (1993).

- The effect of the GW spectrum generated by de Sitter inflation on the multipole moments of the CMB was computed for $l = 2, 3$ in Rubakov, Sazhin and Veryaskin (1982) and for general l in Fabbri and Pollock (1983) and Starobinsky (1985). Abbott and Wise (1984) extended the computation to power-law inflation. The quantum computation using the Bogoliubov coefficients was performed in Abbott and Harari (1986) and Allen (1988); see also the reviews by Allen (1997) and Maggiore (2000). Further early works include Sahni (1990), Souradeep and Sahni (1992), Liddle and Lyth (1992), Davis, Hodges, Smoot, Steinhardt and Turner (1992) and Lucchin, Matarrese and Mollerach (1992).

- The effect of an extended phase of reheating on the high-frequency portion of the spectrum is studied in Nakayama, Saito, Suwa and Yokoyama (2008) and in Kuroyanagi, Nakayama and Saito (2011). Accurate numerical evaluation of the inflationary GW spectrum, including the effect of the change in the effective number of relativistic species, the damping due to neutrino free-streaming studied in Section 19.5.3, and numerical evolution beyond the slow-roll approximation are presented in Watanabe and Komatsu (2006) and Kuroyanagi, Chiba and Sugiyama (2009). See the Further Reading of Chapter 20 for references on the transfer function of GWs, which enters into these computations. A review of GW production during inflation is Guzzetti, Bartolo, Liguori and Matarrese (2016).

- The DECIGO project is described in Kawamura *et al.* (2011). For discussion of the Big Bang Observer (BBO) see for example Crowder and Cornish (2005) and Harry, Fritschel, Shaddock, Folkner and Phinney (2006).

- The decoherence of quantum fluctuations when they re-enter the horizon is discussed in Guth and Pi (1985), Starobinsky (1986), Grishchuk and Sidorov (1990) Starobinsky and Yokoyama (1994), Albrecht, Ferreira, Joyce and Prokopec (1994) and Polarski and Starobinsky (1996). A recent discussion from a different perspective is given in Burgess, Holman, Tasinato and Williams (2015).

- The choice of vacuum state in inflation and in de Sitter space has been subject to much work. At the historical level, the Bunch–Davies vacuum actually first appeared in Chernikov and Tagirov (1968) and then in Bunch and Davies (1978). It is also called the Euclidean vacuum, because the correlation functions computed in this vacuum state can be obtained by analytic continuations from the Euclidean version of de Sitter space, i.e. a sphere. As discussed in Mottola (1985) and Allen (1985), the symmetries of de Sitter space only fix the vacuum up to a one-parameter family, called α-vacua, which includes the Bunch–Davies vacuum as a special case. Problems of consistency of the α-vacua in de Sitter space have been addressed in Banks and Mannelli (2003), Einhorn and Larsen (2003)

and Danielsson (2002). As we discussed in Note 20 on page 607, it should, however, be observed that consistency problems that might arise in the definition of quantum field theory in the full de Sitter space do not necessarily apply to inflation, where the FRW metric only approaches a (quasi)-de Sitter form for just a finite part of the evolution, and in fact even there corresponds to only one patch of de Sitter space, as discussed following eq. (21.207).

22 Stochastic backgrounds of cosmological origin

In Chapter 21 we have studied in detail the production of a stochastic GW background due to the amplification of vacuum fluctuations during inflation. Because of its potential importance, and of the many technical steps needed in the relevant computations, we devoted a full chapter to this topic. In this chapter we will examine several other cosmological production mechanisms that have been proposed in the literature, and we will then summarize the present knowledge on stochastic GW backgrounds.

22.1 Characteristic frequency of relic GWs

We first discuss what is the typical frequency today of a GW signal of cosmological origin. Part of the answer depends on the dynamics of the production mechanism, and therefore is model-dependent, and part is kinematic, depending on the redshift from the production era to the present time. First of all, it is useful to separate the kinematics from the dynamics.

We consider a graviton produced with a frequency f_*, at a production time $t = t_*$, when the temperature had a value T_*, within the RD or MD phases. Its frequency today is redshifted to a value $f_0 = f_* a(t_*)/a(t_0)$. To compute the ratio $a(t_*)/a(t_0)$ we use eq. (17.113), setting $a_r = a_0 = 1$ and $T_r = T_0 \simeq 2.7255(6)$ K. This gives

$$f_0 = f_* \frac{a(t_*)}{a(t_0)}$$

$$\simeq 7.80 \times 10^{-14} f_* \left(\frac{106.75}{g_*^S(T_*)} \right)^{1/3} \left(\frac{1\,\text{GeV}}{T_*} \right), \qquad (22.1)$$

where, as usual, we have normalized the effective number of relativistic species to its high-energy value in the Standard Model; see Section 17.5. We now want to estimate the characteristic value of the frequency f_* of a graviton produced in the early Universe at time t_*, when the temperature was T_*. Here of course the dynamics of the specific production mechanism enters, but still some general considerations are possible. One of the relevant parameters in this estimate is certainly the Hubble parameter at the time of production, $H(t_*) \equiv H_*$. It can then be useful to write the reduced wavelength $\lambdabar_*$ of a graviton at the time of

Gravitational Waves, Volume 2: Astrophysics and Cosmology. Michele Maggiore.
© Michele Maggiore 2018. Published in 2018 by Oxford University Press.
DOI 10.1093/oso/9780198570899.001.0001

production as

$$\lambda_* = \epsilon_* H_*^{-1} . \qquad (22.2)$$

Since H_*^{-1} is the size of the horizon at time t_*, $\epsilon_* > 1$ corresponds to super-horizon modes at the time of production. Because of causality, the generation of super-horizon modes will be suppressed. This does not mean that the production of modes with $\epsilon_* > 1$ is strictly forbidden by causality. The spectrum of modes produced by causal mechanisms is continuous with frequency, and does not drop suddenly to zero below a critical frequency. Thus, to some extent, even super-horizon modes are produced. However, at sufficiently low frequencies, such that the corresponding wavelengths are well outside the horizon, causality requires that their spectrum goes to zero, as $f \to 0$, as

$$h_0^2 \Omega_{\text{gw}}(f) \propto f^3 . \qquad (22.3)$$

This follows from the fact, for super-horizon modes, the source term $\tilde{\sigma}_A^{\text{TT}}(\mathbf{k})$ in eq. (19.220) should have a flat power spectrum, i.e. $\langle |\tilde{\sigma}_A^{\text{TT}}(\mathbf{k})|^2 \rangle$ will be independent of $\mathbf{k}$. This corresponds to white noise and reflects the absence of causal correlations. The same will then be true for $\tilde{h}_A(\mathbf{k})$. This means that, for super-horizon modes, the power spectrum $\mathcal{P}_T(k)$ at the time of production satisfies

$$\mathcal{P}_T(k) \propto k \qquad (22.4)$$

[see eq. (19.280)] or, equivalently, $S_h(f)$ in eq. (19.284) is independent of the frequency (recall that $k = 2\pi f$). Then, from eq. (19.288) or from eq. (7.202) of Vol. 1, we get eq. (22.3).

Thus, for typical gravitons produced at time t_*, we expect $\epsilon_* = O(1)$ or smaller. More precisely, in the absence of any other length-scale in the problem, we can indeed expect that the peak of the spectrum will be at $\epsilon_* = O(1)$, while if other length-scales smaller than H_*^{-1} enter the problem (or small dimensionless parameters such as a speed of sound $c_s^2 < 1$), ϵ_* will in general be some orders of magnitude smaller than 1.

During RD, $H_*^2 = (8\pi/3)G\rho_{\text{rad}}$. Then, from eq. (17.98),

$$H_*^2 = \frac{\pi^2 g_* T_*^4}{90 m_{\text{Pl}}^2} , \qquad (22.5)$$

and, using $2\pi f_* \equiv H_*/\epsilon_*$, eq. (22.1) can be written as

$$f_0 \simeq 2.65 \times 10^{-8} \frac{1}{\epsilon_*} \left(\frac{T_*}{1 \, \text{GeV}} \right) \left(\frac{g_*}{106.75} \right)^{1/6} \text{Hz} . \qquad (22.6)$$

The effect of the dynamics on the value of f_0 has therefore been isolated into the parameter ϵ_*.[1]

We will typically be interested in production mechanisms taking place at temperatures $T \gtrsim 100$ GeV. Then the relation between time and temperature during the RD phase can be obtained using eqs. (17.114) and

[1] Here we have assumed that the temperature is sufficiently high that $g_*^S = g_*$. Otherwise, in general we should replace $g_*^{1/6}$ by $[g_*^3 (g_*^S)^{-2}]^{1/6}$ in eq. (22.6); compare with eqs. (17.98) and (17.114). Because of the power 1/6, however, this has little effect.

(17.138). Inserting the numerical values, this gives

$$t_* \simeq 0.23 \left(\frac{106.75}{g_*}\right)^{1/2} \left(\frac{1\,\text{MeV}}{T_*}\right)^2 \text{s}. \tag{22.7}$$

This tells us how far back in time we are exploring the Universe, when we observe a graviton produced at temperature T_*. Using eq. (22.6) we see that, by detecting a stochastic background of relic GWs that today has a typical frequency f_0, we are looking back at the Universe at time

$$t_* \simeq 1.6 \times 10^{-22} \frac{1}{\epsilon_*^2} \left(\frac{1\,\text{Hz}}{f_0}\right)^2 \left(\frac{106.75}{g_*}\right)^{1/6} \text{s}. \tag{22.8}$$

Table 22.1 The production time t_* and the production temperature T_* for GWs observed today at frequency f_0, if at the time of production they had a reduced wavelength equal to the horizon scale.

f_0 (Hz)	t_* (s)	T_* (GeV)
10^{-4}	1.6×10^{-14}	3.8×10^3
1	1.6×10^{-22}	3.8×10^7
10^2	1.6×10^{-26}	3.8×10^9
10^3	1.6×10^{-28}	3.8×10^{10}

Table 22.1 gives some representative values for the production time t_* and the temperature of the Universe T_* corresponding to frequencies relevant to space-borne and to ground-based interferometers, setting $\epsilon_* = 1$. One should, however, keep in mind that the estimate $\epsilon_* \sim 1$ can sometimes be incorrect, even as an order of magnitude. We will later illustrate this point with some specific examples. The values in Table 22.1 are more correctly interpreted as a starting point for understanding how the result is affected by the dynamics of the specific production mechanism.

The highest possible value of the frequency that a relic GW frequency could have today can be obtained assuming that at the Planck epoch a graviton is produced with energy of order of the reduced Planck mass m_{Pl}, and redshifting it to the present epoch, assuming no inflationary phase and only RD followed by MD. This would give

$$f_0 \simeq 64 \left(\frac{g_*}{106}\right)^{1/6} \text{GHz} \qquad (T_* \simeq m_{\text{Pl}}). \tag{22.9}$$

The dependence on g_* is rather weak because of the power $1/6$ in eq. (22.6). For $g_* = 1000$, f_0 only increases by a factor ~ 1.5 relative to $g_* = 100$. More realistically, we have by now very convincing evidence for the existence of an inflationary phase in the early Universe, so the maximum production temperature for primordial gravitons should rather be determined by the reheating temperature. Indeed, a stochastic background produced by classical mechanisms before inflation would be diluted and completely washed away by the subsequent inflationary phase. As we already mentioned in Chapter 21, the reheating temperature is at most equal to the inflationary scale (if reheating is almost instantaneous and all the energy density in the inflaton field is immediately converted into radiation), and in typical reheating models is several orders of magnitude lower. With the existing bounds on the scale of inflation discussed in Section 21.3.6, the highest possible frequency today of a relic GW background will rather be $O(0.1)$ GHz [see eq. (21.361)], and possibly much smaller, depending on the reheating temperature.

22.2   GW production by classical fields

22.2.1   General formalism

We now study the GWs produced by a classical field. The subject is conceptually interesting by itself, and covers a broad range of situations.[2]

We consider a generic inhomogeneous, time-evolving scalar field in an expanding universe. Such a field has a time-varying and spatially dependent energy–momentum tensor and is therefore in general a source for a stochastic background of GWs. The simplest approach for computing the resulting GW production, which indeed was used in early papers on the subject, is to use the formalism reviewed in Section 3.1 of Vol. 1 [and originally developed in Section 10.4 of the textbook by Weinberg (1972)]. In particular, in this formalism the GW amplitude in the far zone and the GW energy spectrum are given, in terms of the energy–momentum tensor of the source, by eqs. (3.14) and (3.16), which we recall here,[3]

$$
h_{ij}^{\rm TT}(t,\mathbf{x}) = \frac{4G}{r}\,\Lambda_{ij,kl}(\hat{\mathbf{n}}) \int_{-\infty}^{\infty} \frac{d\omega}{2\pi}\,\tilde{T}_{kl}(\omega,\omega\hat{\mathbf{n}})e^{-i\omega(t-r)}\,, \qquad (22.10)
$$

$$
\frac{dE_{\rm gw}}{d\Omega} = \frac{G}{2\pi^2}\Lambda_{ij,kl}(\hat{\mathbf{n}}) \int_{0}^{\infty} d\omega\,\omega^2 \tilde{T}_{ij}(\omega,\omega\hat{\mathbf{n}})\tilde{T}_{kl}^{*}(\omega,\omega\hat{\mathbf{n}})\,, \qquad (22.11)
$$

where $\Lambda_{ij,kl}$ is the Lambda tensor that projects onto the TT gauge, defined in eq. (1.36), and $\hat{\mathbf{n}}$ is the propagation direction of the wave. One could then take an ensemble average of $\langle \tilde{T}_{ij}(\omega,\omega\hat{\mathbf{n}})\tilde{T}_{kl}^{*}(\omega,\omega\hat{\mathbf{n}})\rangle$, to compute the resulting stochastic background. However, in the present cosmological setting such an approach would be inadequate for several reasons. First of all, in this case the source is not localized and therefore there is no wave zone where eq. (22.10) could hold. Second, this expression has been obtained by expanding the metric perturbation around flat space, rather than around the FRW background. The correct approach is to use cosmological perturbation theory, as we have developed in Chapter 19. In particular, the dynamics of the tensor perturbations is given by eq. (19.220), where the source term σ_A is defined by eqs. (18.114), (18.120), (18.123) and (19.215). In Section 19.5 we were mostly interested in the evolution of the GWs, given some initial value $h_{A,\rm in}$, so in eq. (19.220) we set the source term to zero (except when we studied the effect of neutrino free-streaming). Here we are considering the production mechanism, so the role of σ_A is crucial.

Lowering the upper index in eq. (18.114), the helicity decomposition of a generic energy–momentum tensor T_{ij} reads

$$
T_{ij} = a^2(p\delta_{ij} + \Sigma_{ij})\,, \qquad (22.12)
$$

where Σ_{ij} is given in eqs. (18.120)–(18.123). The tensor $T_{ij}^{\rm TT} = a^2\sigma_{ij}^{\rm TT}$ is the TT part of T_{ij}. It follows from the discussion in Section 1.2 that it can be extracted from T_{ij} by taking the Fourier transform and multiplying the momentum mode $\tilde{T}_{ij}(\mathbf{k})$ by the Lambda tensor $\Lambda_{ij,kl}(\hat{\mathbf{k}})$.

[2]We follow in particular the formalism developed by Dufaux, Bergman, Felder, Kofman and Uzan (2007).

[3]Recall that in Part IV we have switched to units $c = 1$.

It is convenient to introduce the field $h_{ij} = ah_{ij}$ and, correspondingly, $\mathsf{h}_A = a\mathsf{h}_A$, similarly to what we did in Section 21.3.4 [where we introduced φ_A from eq. (21.309) and, as in eq. (21.177), $\chi_A = a\varphi_A$]. Then eq. (19.220) becomes

$$\tilde{\mathsf{h}}_A''(\mathbf{k}) + \left(k^2 - \frac{a''}{a} \right) \tilde{\mathsf{h}}_A(\mathbf{k}) = 16\pi G a^3 \tilde{\sigma}_A^{\mathrm{TT}}(\mathbf{k}) \,. \tag{22.13}$$

This equation further simplifies if the typical momenta $\mathbf{k}$ for which the source term is important are well inside the horizon. For such momenta $k^2 \gg a''/a \sim 1/\eta^2$, and we can solve eq. (22.13) using the retarded Green's function of the operator $\partial_\eta^2 + k^2$. This is just the well-known retarded Green's function of the harmonic oscillator,

$$G(\eta - \eta') = \frac{\sin[k(\eta - \eta')]}{k} \theta(\eta - \eta') \,. \tag{22.14}$$

In particular, if we assume that before an initial conformal time η_i the source was not active, and as initial condition we set $\tilde{\mathsf{h}}_A(\eta_i, \mathbf{k}) = \tilde{\mathsf{h}}_A'(\eta_i, \mathbf{k}) = 0$, the solution is

$$\tilde{\mathsf{h}}_A(\eta, \mathbf{k}) = 16\pi G \int_{-\infty}^{\infty} d\eta' \, G(\eta - \eta') a^3(\eta') \tilde{\sigma}_A^{\mathrm{TT}}(\eta', \mathbf{k})$$

$$= \frac{16\pi G}{k} \int_{\eta_i}^{\eta} d\eta' \, \sin[k(\eta - \eta')] a^3(\eta') \tilde{\sigma}_A^{\mathrm{TT}}(\eta', \mathbf{k}) \,. \tag{22.15}$$

If the source turns off at some final conformal time η_f, then

$$\tilde{\mathsf{h}}_A(\eta_f, \mathbf{k}) = \frac{16\pi G}{k} \int_{\eta_i}^{\eta_f} d\eta' \, \sin[k(\eta_f - \eta')] a^3(\eta') \tilde{\sigma}_A^{\mathrm{TT}}(\eta', \mathbf{k}) \tag{22.16}$$

is the initial value of the subsequent free evolution. During RD and MD, modes that were generated as sub-horizon remain inside the horizon, so their subsequent evolution is determined by eq. (22.13) with the source term set to zero and again $|a''/a| \ll k^2$, i.e. by the free wave equation

$$\tilde{\mathsf{h}}_A'' + k^2 \tilde{\mathsf{h}}_A \simeq 0 \,, \tag{22.17}$$

whose solution is simply

$$\tilde{\mathsf{h}}_A(\eta, \mathbf{k}) = \mathcal{A}_A(\mathbf{k}) \sin[k(\eta - \eta_f)] + \mathcal{B}_A(\mathbf{k}) \cos[k(\eta - \eta_f)] \qquad (\eta \geqslant \eta_f) \,. \tag{22.18}$$

Matching at $\eta = \eta_f$ the values of $\tilde{\mathsf{h}}_A$ and $\tilde{\mathsf{h}}_A'$ obtained from eqs. (22.16) and (22.18) one finds

$$\mathcal{A}_A(\mathbf{k}) = \frac{16\pi G}{k} \int_{\eta_i}^{\eta_f} d\eta' \, \cos[k(\eta_f - \eta')] a^3(\eta') \tilde{\sigma}_A^{\mathrm{TT}}(\eta', \mathbf{k}) \,, \tag{22.19}$$

$$\mathcal{B}_A(\mathbf{k}) = \frac{16\pi G}{k} \int_{\eta_i}^{\eta_f} d\eta' \, \sin[k(\eta_f - \eta')] a^3(\eta') \tilde{\sigma}_A^{\mathrm{TT}}(\eta', \mathbf{k}) \,. \tag{22.20}$$

The corresponding GW energy density is given by eq. (1.136). Expressed in terms of h and of conformal time, it reads

$$\rho_{\rm gw}(\eta) = \frac{1}{16\pi G a^4} \sum_{A=+,\times} \langle (\mathsf{h}'_A - \mathcal{H}\mathsf{h}_A)^2 (\eta, \mathbf{x}) \rangle , \qquad (22.21)$$

where $\langle \ldots \rangle$ represents a spatial average over several wavelengths and, as usual, $f' \equiv df/d\eta$. For sub-horizon modes, $\mathsf{h}'_A \sim k\mathsf{h}_A \gg \mathcal{H}\mathsf{h}_A$, so the second term in parentheses can be neglected, and, in this limit,

$$\rho_{\rm gw}(\eta) = \frac{1}{16\pi G a^4} \sum_{A=+,\times} \langle \mathsf{h}'^2_A (\eta, \mathbf{x}) \rangle . \qquad (22.22)$$

We now expand $\mathsf{h}'_A(\eta, \mathbf{x})$ in Fourier modes. Then

$$\langle \mathsf{h}'^2_A(\eta, \mathbf{x}) \rangle = \int \frac{d^3k}{(2\pi)^3} \frac{d^3k'}{(2\pi)^3} \mathsf{h}'_A(\eta, \mathbf{k}) \mathsf{h}'^*_A(\eta, \mathbf{k}') \langle e^{i(\mathbf{k}-\mathbf{k}')\cdot\mathbf{x}} \rangle . \qquad (22.23)$$

The spatial average over a volume V of size much larger than the reduced wavelengths $1/k$ and $1/k'$ gives

$$\langle e^{i(\mathbf{k}-\mathbf{k}')\cdot\mathbf{x}} \rangle = \frac{(2\pi)^3}{V} \delta^{(3)}(\mathbf{k} - \mathbf{k}') , \qquad (22.24)$$

so in the end we get

$$\rho_{\rm gw}(\eta) = \frac{1}{16\pi G a^4} \frac{1}{V} \sum_{A=+,\times} \int \frac{d^3k}{(2\pi)^3} |\mathsf{h}'_A(\eta, \mathbf{k})|^2 , \qquad (22.25)$$

where $\mathsf{h}_A(\eta, \mathbf{k})$ is given by eq. (22.15) for $\eta \leqslant \eta_f$, and by eq. (22.18) for $\eta \geqslant \eta_f$. Observe that this formalism allows us to follow the time evolution of the energy density in the tensor perturbations, not only during the period in which the source is no longer active and the tensor perturbations evolve freely in a FRW background, but even in the earlier epoch when the source was active and the tensor perturbations were building up. For $\eta \geqslant \eta_f$ eq. (22.25) can be further simplified if we are not interested in the details of the time evolution of the signal on time-scales of the order of $1/k$ or smaller. In this case we can further perform a temporal average of $\rho_{\rm gw}$ over several periods, retaining only the coarser features of the temporal evolution. Then, using eq. (22.18), together with $\langle \sin^2[k(\eta - \eta_f)] \rangle = \langle \cos^2[k(\eta - \eta_f)] \rangle = 1/2$ and $\langle \sin[k(\eta - \eta_f)] \cos[k(\eta - \eta_f)] \rangle = 0$, we get[4]

$$\rho_{\rm gw}(\eta \geqslant \eta_f) = \frac{1}{32\pi G a^4(\eta) V} \sum_{A=+,\times} \int \frac{d^3k}{(2\pi)^3} k^2 \left[|\mathcal{A}_A(\mathbf{k})|^2 + |\mathcal{B}_A(\mathbf{k})|^2 \right]$$

$$= \frac{4\pi G}{a^4(\eta) V} \sum_{ij} \int \frac{d^3k}{(2\pi)^3} \left[\left| \int_{\eta_i}^{\eta_f} d\eta' \cos(k\eta') a(\eta') \tilde{T}_{ij}^{\rm TT}(\eta', \mathbf{k}) \right|^2 \right.$$

$$\left. + \left| \int_{\eta_i}^{\eta_f} d\eta' \sin(k\eta') a(\eta') \tilde{T}_{ij}^{\rm TT}(\eta', \mathbf{k}) \right|^2 \right]$$

[4]See eq. (26) of Dufaux, Bergman, Felder, Kofman and Uzan (2007). The different factor $(2\pi)^3$ follows from our different conventions for the Fourier transform.

$$= \frac{4\pi G}{a^4(\eta)V} \sum_{ij} \int \frac{d^3k}{(2\pi)^3} \int_{\eta_i}^{\eta_f} d\eta' d\eta'' \, a(\eta')a(\eta'') \cos[k(\eta' - \eta'')]$$

$$\times \tilde{T}_{ij}^{\mathrm{TT}}(\eta', \mathbf{k})\tilde{T}_{ij}^{*,\mathrm{TT}}(\eta'', \mathbf{k}) \,, \tag{22.26}$$

where in the second equality we have expanded $\cos[k(\eta_f - \eta')]$ and $\sin[k(\eta_f - \eta')]$ in terms of sin and cos of $k\eta_f$ and $k\eta'$ and recombined the various terms. We have also made use of

$$\sum_A |\tilde{\sigma}_A|^2 = \frac{1}{2} \sum_{ij} |\tilde{\sigma}_{ij}|^2 \,, \tag{22.27}$$

which follows from eqs. (19.215) and (19.218), and we have used $T_{ij}^{\mathrm{TT}} = a^2 \sigma_{ij}^{\mathrm{TT}}$ instead of $\sigma_{ij}^{\mathrm{TT}}$. The overall factor $1/a^4$ correctly gives the cosmological redshift of the energy density of the GWs, as it should for sub-horizon modes; compare with the discussion in Section 19.5.1.

These results allow us to compute the energy density of the GWs generated by a given classical source in a FRW background. In the special case of a localized source in flat space, we must therefore be able to recover eq. (22.11). Observe that eq. (22.11) gives the total energy radiated by the source, up to time $+\infty$. To recover it, we therefore set $a(\eta) = 1$ in eq. (22.26) (so η becomes the same as t) and we extend the integration up to $\eta_f = \infty$. Recall that the lower limit of integration was actually $\eta = -\infty$, which we replaced by η_i because the source was not active before η_i; see eq. (22.15). So, the total GW energy density produced by the source in the flat-space limit, as predicted by eq. (22.26), is

$$\rho_{\mathrm{gw}} = \frac{2\pi G}{V} \sum_{ij} \int \frac{d^3k}{(2\pi)^3} \int_{-\infty}^{+\infty} dt' dt''$$

$$\times \left[e^{ik(t'-t'')} \tilde{T}_{ij}^{\mathrm{TT}}(t', \mathbf{k})\tilde{T}_{ij}^{*,\mathrm{TT}}(t'', \mathbf{k}) + e^{-ik(t'-t'')} \tilde{T}_{ij}^{\mathrm{TT}}(t', \mathbf{k})\tilde{T}_{ij}^{*,\mathrm{TT}}(t'', \mathbf{k}) \right]$$

$$= \frac{4\pi G}{V} \sum_{ij} \int \frac{d^3k}{(2\pi)^3} \tilde{T}_{ij}^{\mathrm{TT}}(k, \mathbf{k})\tilde{T}_{ij}^{*,\mathrm{TT}}(k, \mathbf{k}) \,, \tag{22.28}$$

where, in the last line, $\tilde{T}_{ij}^{\mathrm{TT}}(\omega, \mathbf{k})$ (with $\omega = k$) is the Fourier transform of $\tilde{T}_{ij}^{\mathrm{TT}}(t, \mathbf{k})$ with respect to the time variable. We next write (with the summation over repeated indices understood)

$$\tilde{T}_{ij}^{\mathrm{TT}}(k, \mathbf{k})\tilde{T}_{ij}^{*,\mathrm{TT}}(k, \mathbf{k}) = \Lambda_{ij,kl}(\hat{\mathbf{k}})\tilde{T}_{kl}(k, \mathbf{k}) \, \Lambda_{ij,mn}(\hat{\mathbf{k}})\tilde{T}_{mn}^*(k, \mathbf{k})$$

$$= \Lambda_{kl,mn}(\hat{\mathbf{k}})\tilde{T}_{kl}(k, \mathbf{k})\tilde{T}_{mn}^*(k, \mathbf{k}) \,, \tag{22.29}$$

where we have used $\Lambda_{ij,kl}(\hat{\mathbf{k}})\Lambda_{ij,mn}(\hat{\mathbf{k}}) = \Lambda_{kl,mn}(\hat{\mathbf{k}})$; see eq. (1.37). Then the total GW energy $E_{\mathrm{gw}} = V\rho_{\mathrm{gw}}$ is given by

$$E_{\mathrm{gw}} = 4\pi G\Lambda_{ij,kl} \int \frac{d^3k}{(2\pi)^3} \, \tilde{T}_{ij}(k, \mathbf{k})\tilde{T}_{kl}^*(k, \mathbf{k})$$

$$= \frac{G}{2\pi^2} \Lambda_{ij,kl} \int_0^\infty k^2 dk \int d\Omega \, \tilde{T}_{ij}(k, \mathbf{k})\tilde{T}_{kl}^*(k, \mathbf{k}) \,. \tag{22.30}$$

Therefore, in the flat-space limit, $dE_{\mathrm{gw}}/d\Omega$ is indeed given by eq. (22.11).

22.2.2   GW generation by a stochastic scalar field

If the source is given by random fields, a stochastic average over the field configurations is further required. Thus, the energy density becomes

$$
\rho_{\mathrm{gw}}(\eta) = \frac{4\pi G}{a^4(\eta)V} \int \frac{d^3k}{(2\pi)^3} \int_{\eta_i}^{\eta_f} d\eta' d\eta''\, a(\eta')a(\eta'') \cos[k(\eta'-\eta'')]
$$
$$
\times \sum_{ij} \langle \tilde{T}_{ij}^{\mathrm{TT}}(\eta',\mathbf{k})\tilde{T}_{ij}^{*,\mathrm{TT}}(\eta'',\mathbf{k})\rangle\,, \tag{22.31}
$$

where now $\langle\ldots\rangle$ denotes the ensemble average over the field configurations. We consider for illustration a single scalar field ϕ with action (21.22). The energy–momentum tensor is then given by eq. (21.24), so in particular

$$
T_{ij} = \partial_i\phi\partial_j\phi - g_{ij}\left[\frac{1}{2}g^{\rho\sigma}\partial_\rho\phi\partial_\sigma\phi + V(\phi)\right]. \tag{22.32}
$$

In eq. (22.32), writing $g_{ij} = a^2(\delta_{ij}+h_{ij})$, the term h_{ij} gives a second-order contribution, in which the GWs produced at first order become themselves sources for further GW production. To first order, we can therefore replace g_{ij} by $a^2\delta_{ij}$. Then, recalling that $\Lambda_{ij,kl}\delta_{kl}=0$, to first order the term in square brackets in eq. (22.32) does not contribute to the TT part, and we obtain

$$
\tilde{T}_{ij}^{\mathrm{TT}}(\eta,\mathbf{k}) = \Lambda_{ij,kl}(\hat{\mathbf{k}})\tilde{T}_{kl}(\eta,\mathbf{k})
$$
$$
= \Lambda_{ij,kl}(\hat{\mathbf{k}}) \int d^3x\, e^{-i\mathbf{k}\cdot\mathbf{x}}\partial_k\phi\partial_l\phi
$$
$$
= \Lambda_{ij,kl}(\hat{\mathbf{k}}) \int \frac{d^3p}{(2\pi)^3}\, p_k(p-k)_l\tilde{\phi}(\eta,\mathbf{p})\tilde{\phi}(\eta,\mathbf{k}-\mathbf{p})\,. \tag{22.33}
$$

Recalling further from Section 1.2 that $\Lambda_{ij,kl}(\hat{\mathbf{k}})k_l=0$, we get

$$
\tilde{T}_{ij}^{\mathrm{TT}}(\eta,\mathbf{k}) = \Lambda_{ij,kl}(\hat{\mathbf{k}}) \int \frac{d^3p}{(2\pi)^3}\, p_k p_l\tilde{\phi}(\eta,\mathbf{p})\tilde{\phi}(\eta,\mathbf{k}-\mathbf{p})\,. \tag{22.34}
$$

Then

$$
\sum_{ij} \langle \tilde{T}_{ij}^{\mathrm{TT}}(\eta',\mathbf{k})\tilde{T}_{ij}^{*,\mathrm{TT}}(\eta'',\mathbf{k})\rangle = \Lambda_{kl,mn}(\hat{\mathbf{k}}) \int \frac{d^3p}{(2\pi)^3}\frac{d^3p'}{(2\pi)^3}
$$
$$
\times p_k p_l p'_m p'_n \,\langle \tilde{\phi}(\eta',\mathbf{p})\tilde{\phi}(\eta',\mathbf{k}-\mathbf{p})\tilde{\phi}^*(\eta'',\mathbf{p}')\tilde{\phi}^*(\eta'',\mathbf{k}-\mathbf{p}')\rangle\,, \tag{22.35}
$$

where we have again used $\Lambda_{ij,kl}(\hat{\mathbf{k}})\Lambda_{ij,mn}(\hat{\mathbf{k}}) = \Lambda_{kl,mn}(\hat{\mathbf{k}})$. If the random field ϕ obeys Gaussian statistics,[5] the four-point correlator can be computed in terms of the two-point functions using Wick's theorem,

$$
\langle \tilde{\phi}(\eta',\mathbf{p})\tilde{\phi}(\eta',\mathbf{k}-\mathbf{p})\tilde{\phi}^*(\eta'',\mathbf{p}')\tilde{\phi}^*(\eta'',\mathbf{k}-\mathbf{p}')\rangle
$$
$$
= \langle \tilde{\phi}(\eta',\mathbf{p})\tilde{\phi}^*(\eta',\mathbf{p}-\mathbf{k})\rangle\,\langle \tilde{\phi}(\eta'',-\mathbf{p}')\tilde{\phi}^*(\eta'',\mathbf{k}-\mathbf{p}')\rangle
$$
$$
+\langle \tilde{\phi}(\eta',\mathbf{p})\tilde{\phi}^*(\eta'',\mathbf{p}')\rangle\,\langle \tilde{\phi}(\eta',\mathbf{k}-\mathbf{p})\tilde{\phi}^*(\eta'',\mathbf{k}-\mathbf{p}')\rangle
$$
$$
+\langle \tilde{\phi}(\eta',\mathbf{p})\tilde{\phi}^*(\eta'',\mathbf{k}-\mathbf{p}')\rangle\,\langle \tilde{\phi}(\eta',\mathbf{k}-\mathbf{p})\tilde{\phi}^*(\eta'',\mathbf{p}')\rangle\,, \tag{22.36}
$$

[5]In the case of preheating that we will discuss in Section 22.3, this is only true in the initial linear phase. At the end of the preheating phase the dynamics becomes highly non-linear, and the fields strongly non-Gaussian.

where we have used $\tilde{\phi}(\eta, -\mathbf{k}) = \tilde{\phi}^*(\eta, \mathbf{k})$ for real fields. Assuming invariance under spatial translations and rotations, the two-point function has the general form

$$\langle \tilde{\phi}(\eta', \mathbf{p})\tilde{\phi}^*(\eta'', \mathbf{p}')\rangle = (2\pi)^3 \delta^{(3)}(\mathbf{p} - \mathbf{p}')F(\eta', \eta''; p)\,, \qquad (22.37)$$

where the proportionality to $\delta^{(3)}(\mathbf{p} - \mathbf{p}')$ follows from invariance under spatial translations, and the dependence of the function $F(\eta', \eta''; p)$ on $p = |\mathbf{p}|$ follows from invariance under rotations.[6]

Using eq. (22.37) we see that the first contraction on the right-hand side of eq. (22.36) is proportional to $\delta^{(3)}(\mathbf{k})$. After integration over $d^3k = k^2 dk\, d\Omega$ in eq. (22.31), it therefore gives a vanishing contribution.[7] Using eq. (22.37) to compute the other two terms in eq. (22.36) and recalling that $(2\pi)^3\delta^{(3)}(\mathbf{p} = 0) = V$, we get

$$\sum_{ij}\langle \tilde{T}_{ij}^{\rm TT}(\eta', \mathbf{k})\tilde{T}_{ij}^{*,\rm TT}(\eta'', \mathbf{k})\rangle = V\Lambda_{kl,mn}(\hat{\mathbf{k}}) \qquad (22.38)$$

$$\times \int \frac{d^3p}{(2\pi)^3}\, p_k p_l [p_m p_n + (k-p)_m(k-p)_n]F(\eta', \eta''; p)F(\eta', \eta''; |\mathbf{k} - \mathbf{p}|)\,.$$

The terms containing k_m or k_n give zero when contracted with $\Lambda_{kl,mn}(\hat{\mathbf{k}})$, while, using the explicit expression (1.39), we get

$$\Lambda_{kl,mn}(\hat{\mathbf{k}})p_k p_l p_m p_n = \frac{1}{2}p^4 \sin^4\theta(\hat{\mathbf{k}}, \hat{\mathbf{p}})\,, \qquad (22.39)$$

where $\theta(\hat{\mathbf{k}}, \hat{\mathbf{p}})$ denotes the angle between $\hat{\mathbf{k}}$ and $\hat{\mathbf{p}}$ [compare with the computation leading to eq. (4.346) in Vol. 1]. Plugging this result into eq. (22.31) we finally obtain

$$\rho_{\rm gw}(\eta \geqslant \eta_f) = \frac{4\pi G}{a^4(\eta)}\int \frac{d^3k}{(2\pi)^3}\int_{\eta_i}^{\eta_f} d\eta' d\eta''\, a(\eta')a(\eta'') \cos[k(\eta' - \eta'')]$$

$$\times \int \frac{d^3p}{(2\pi)^3}p^4 \sin^4\theta(\hat{\mathbf{k}}, \hat{\mathbf{p}})F(\eta', \eta''; p)F(\eta', \eta''; |\mathbf{k} - \mathbf{p}|)\,. \; (22.40)$$

The integral over d^3p in the second line depends on $\mathbf{k}$, through $|\mathbf{k} - \mathbf{p}|$ and $\theta(\hat{\mathbf{k}}, \hat{\mathbf{p}})$. However, since it is a scalar, and there is no other vector with which we can contract k_i, it can depend on $\mathbf{k}$ only through the modulus k, and not on the direction $\hat{\mathbf{k}}$. This means that the integral over d^3k can be written as $d^3k = 4\pi k^2 dk$. As in eq. (7.197), we write

$$\rho_{\rm gw}(\eta) = \int_{k=0}^{k=\infty} d(\log k)\, \frac{d\rho_{\rm gw}}{d\log k}(\eta, k)\,. \qquad (22.41)$$

Then, for $\eta \geqslant \eta_f$,

$$\frac{d\rho_{\rm gw}}{d\log k}(\eta, k) = \frac{2G}{\pi a^4(\eta)}k^3 \int_{\eta_i}^{\eta_f} d\eta' d\eta''\, a(\eta')a(\eta'') \cos[k(\eta' - \eta'')]$$

$$\times \int \frac{d^3p}{(2\pi)^3}p^4 \sin^4\theta(\hat{\mathbf{k}}, \hat{\mathbf{p}})F(\eta', \eta''; p)F(\eta', \eta''; |\mathbf{k} - \mathbf{p}|)\,.$$

$$(22.42)$$

[6] Observe that in all computations in this section we use the normalization (17.39) of the Fourier transform, rather than the normalization (19.161). Therefore there is no explicit factor $1/V$ in eq. (22.37), in contrast to eq. (19.164), where the normalization (19.161) was used.

[7] Observe that this contraction corresponds to a disconnected diagram, in which the two fields at time η are contracted among themselves, and also the two fields at time η' are contracted among them. So, the disconnected diagram does not contribute, as we know from experience with quantum field theory. Technically, disconnected diagrams can just be reabsorbed into a redefinition of $\tilde{T}_{ij}^{\rm TT}(\eta, \mathbf{k})$, which is equivalent to a renormalization of the corresponding operator in the field-theoretical language.

This is the final result of this section, and gives the energy spectrum of the stochastic background of GWs generated by a random Gaussian scalar field, for $\eta \geqslant \eta_f$, i.e. after the source has switched off, in terms of its two-point function $F(\eta', \eta''; p)$, defined by eq. (22.37).

22.3   GWs from preheating after inflation

If a thermalized hot plasma ever existed before inflation, at the end of inflation its temperature would be redshifted away by a factor $e^{\Delta N}$, where the number of e-folds during inflation, ΔN, is larger than a minimum value $(\Delta N)_{\min}$ typically of order 50–60, as we saw in Section 21.1. In order to generate the hot thermalized relativistic plasma that we identify as the initial state of the hot big-bang theory, the Universe must therefore be reheated. For single-field slow-roll inflation the first successful reheating mechanism that was studied is the perturbative decay of the inflaton field into other matter fields after the end of the slow-roll phase. It was later realized that the energy transfer from the inflaton to normal matter at the end of inflation can be much more efficient, involving non-linear physics and explosive production of particles, which of course is also potentially interesting from the point of view of GW generation. This non-linear mechanism is referred to as "preheating". In particular, two interesting preheating mechanisms have emerged, namely parametric resonance in single-field inflation and tachyonic preheating in hybrid inflation. We discuss them in the following subsections.

22.3.1   Parametric resonance in single-field inflation

We consider a single-field slow-roll inflationary model, of the type discussed in Section 21.1.4, with an inflaton field ϕ and an inflationary potential $V(\phi)$, and we couple it for simplicity to a single scalar matter field χ with a coupling $\chi^2 \phi^2$,

$$S = \int d^4x \sqrt{-g} \left[\frac{1}{16\pi G} R - \frac{1}{2} g^{\mu\nu} \partial_\mu \phi \partial_\nu \phi - V(\phi) \right.$$
$$\left. - \frac{1}{2} g^{\mu\nu} \partial_\mu \chi \partial_\nu \chi - \frac{1}{2} g^2 \chi^2 \phi^2 \right], \qquad (22.43)$$

where g is a coupling constant. The background solution is $\phi = \phi(t)$, $\chi = 0$ so, including perturbations around it, we write

$$\phi(t, \mathbf{x}) = \phi(t) + \delta\phi(t, \mathbf{x}), \qquad (22.44)$$
$$\chi(t, \mathbf{x}) = \delta\chi(t, \mathbf{x}). \qquad (22.45)$$

At the end of inflation the field ϕ performs oscillations around the minimum of $V(\phi)$ with a frequency given by $m_\phi = \sqrt{d^2V/d\phi^2}$, with the second derivative evaluated at the minimum of $V(\phi)$. Therefore the solution of the equation of motion for $\phi(t)$ is approximately

$$\phi(t) = \Phi(t) \sin m_\phi t, \qquad (22.46)$$

where $\Phi(t)$ is a function that changes slowly on the time-scale m_ϕ^{-1}, and which we will approximate just by a constant. The equation for the perturbations is [compare with eq. (21.175)]

$$\delta\ddot{\phi}_{\mathbf{k}} + 3H\delta\dot{\phi}_{\mathbf{k}} + \left(\frac{k^2}{a^2} + m_\phi^2\right)\delta\phi_{\mathbf{k}} = 0, \qquad (22.47)$$

$$\delta\ddot{\chi}_{\mathbf{k}} + 3H\delta\dot{\chi}_{\mathbf{k}} + \left(\frac{k^2}{a^2} + g^2\phi^2(t)\right)\delta\chi_{\mathbf{k}} = 0. \qquad (22.48)$$

The equation for $\delta\phi_{\mathbf{k}}$ is the usual equation for the perturbations of a scalar field in FRW, damped by the friction term due to the expansion of the Universe, and furthermore has a mass term with the "correct", non-tachyonic sign. This equation has no unstable mode, as we studied in Section 19.3. In contrast, the equation for $\delta\chi_{\mathbf{k}}$ has a time-dependent frequency

$$\begin{aligned} \omega_k^2(t) &= k^2 + g^2 a^2(t)\phi^2(t) \\ &= k^2 + g^2\Phi^2 a^2(t)\sin m_\phi t. \end{aligned} \qquad (22.49)$$

We can rewrite eq. (22.48) in the form

$$\frac{d^2}{dz^2}(\delta\chi_k) + 3\frac{H}{m_\phi}\frac{d}{dz}(\delta\chi_k) + [A(k) - 2q\cos(2z)]\,\delta\chi_k, \qquad (22.50)$$

where $z = m_\phi t$,

$$q = \frac{g^2\Phi^2}{4m_\phi^2} \qquad (22.51)$$

is called the *resonance parameter* and

$$A(k) = \frac{k^2}{m_\phi^2 a^2} + 2q. \qquad (22.52)$$

Equation (22.50), with the friction term proportional to H/m_ϕ neglected (which is a good approximation during the rapid phase of parametric resonance that we will find below), is a Mathieu equation. Depending on the values of $A(k)$ and of q, it can have either oscillatory or exponential solutions, with a band structure corresponding to sets of stable or unstable modes. In general, if q is large, a broad range of values of k can be in resonance, and one is then in the broad-resonance regime. In this situation energy is efficiently pumped from the inflaton field into the resonant modes $\delta\chi_k$, leading to large time-dependent and inhomogeneous field configurations. Once the resonant modes reach an amplitude $\chi^2 \sim \Phi^2$, the perturbative approach in which we expanded around $\chi = 0$ breaks down, and preheating ends. At this point the field χ dissipates its energy via interaction with the other fields of the theory, eventually leading to thermalization.

The scalar field dynamics can be computed numerically, solving the equation of motion of the interacting scalar fields in FRW. The numerical study shows that there is an initial linear phase in which the field χ

grows and its statistics is still Gaussian. In this regime some analytic checks of the numerical results for GW production can be made using the results of Section 22.2.2. This phase evolves into a highly non-linear phase, where the statistics of the field χ is highly non-Gaussian, characterized by the formation, expansion and collision of bubble-like field inhomogeneities. This is followed by a stage of turbulence and, finally, thermalization. Most of the GW production takes place in the violent bubble-like phase. The resulting GW production can then be computed from eq. (22.13), either dropping the term a''/a and using eq. (22.26), or directly integrating eq. (22.13) numerically. The fact that most of the GWs are produced in the bubble collisions makes the GW production mechanism somewhat similar to that operating in strongly first-order phase transitions, which we will discuss in Section 22.4.

The details of the numerical results depend of course on the specific inflationary model used. Most numerical results have been obtained for the inflationary potential $V(\phi) = (1/4)\lambda\phi^4$, which, as we have seen in Section 21.3.5, is now ruled out, or for $V(\phi) = (1/2)m^2\phi^2$, which is presently very close to being ruled out. The result depends of course also on the coupling g between the inflaton and the field χ or, more generally, on the form assumed for the interaction between ϕ and χ. Thus, the predictions for the GW spectrum are necessarily strongly dependent on the details of the model used.

The typical frequency of the GWs produced can, however, be estimated as in Section 22.1. The characteristic scale H_* in this case is provided by the Hubble parameter at the peak of GW production. In a first approximation, we can identify H_* with the value of the Hubble parameter at the end of the reheating phase and the beginning of RD. In this approximation we can use eq. (22.5) with T_* equal to the reheating temperature. However, the numerical studies typically show that the spectrum peaks at wavelengths one or two orders of magnitude smaller than H_*^{-1}, so the peak frequency is shifted toward correspondingly larger values. We see from eq. (22.6) (taking ϵ_* of order, say, 0.1) that, to have a signal at $f \sim 200 - 300$ Hz, near the best sensitivity region of ground-based interferometers such as LIGO and VIRGO, one should have $T_* \sim 10^9$ GeV. The spectrum is quite peaked around the maximum; on the low-frequency side its tails must decrease as f^3 by causality [see eq. (22.3)], while at frequencies higher than the peak frequency it declines even more rapidly, corresponding to the fact that modes with larger values of k are not excited by parametric resonance. The numerical studies also indicate that, for inflationary models of the type $V(\phi) = (1/2)m^2\phi^2$, the typical peak value of $\Omega_{\rm gw}$ today is roughly independent of the inflationary scale, and of order

$$h_0^2\Omega_{\rm gw} \sim 10^{-11}\,. \tag{22.53}$$

This is considerably higher than the value obtained in inflation from the amplification of vacuum fluctuations which, as we saw in eq. (21.368), with the current bounds on the tensor-to-scalar tensor ratio r cannot exceed a value of order 10^{-16}.

22.3.2 Tachyonic preheating in hybrid inflation

In hybrid inflationary models, discussed in Section 21.1.4, preheating occurs when the inflaton field ϕ decreases below the critical value ϕ_c for which the Higgs-like field Φ acquires a tachyonic mass; see the discussion following eq. (21.84).

Naively, one might think that the process in which a Higgs field Φ evolves from the local maximum $\Phi = 0$ of its Mexican hat potential, toward a spontaneous symmetry-breaking minimum, could be described in terms of the evolution of a spatially homogeneous component of the Higgs field [or, equivalently, in terms of a spatially homogeneous configuration of the field $\sigma \equiv (\Phi^\dagger \Phi)^{1/2}$, see the definitions after eq. (21.82)], which rolls down from the top of the potential to the symmetry-breaking minimum (which, for hybrid inflation, is at $\phi = 0$ and $\sigma = v$). The field should then starts performing oscillations around this minimum. Initially the oscillations would have an amplitude comparable to the vacuum expectation value itself, and would then be damped by the production of particles coupled to σ, so that σ eventually settles into the minimum of the potential, with amplitude of oscillations much smaller than v. This can take place after a large number of oscillations.

However, numerical investigations have revealed that the actual dynamical evolution leading to the final spontaneous symmetry-breaking state is more complex and violent, and in particular the approximation of homogenous field is inadequate. Indeed, when the field σ acquires a tachyonic mass $-(1/2)m^2\sigma^2$, all momentum modes with $k < m$ become unstable, and start to grow exponentially as

$$\tilde{\sigma}_k(t) \propto e^{t\sqrt{m^2 - k^2}} . \tag{22.54}$$

When these fluctuations become large they correspond to quantum states with large occupation numbers and can then be described as classical waves. As soon as the oscillations in the various momentum modes become large, the process becomes very complicated and non-linear, with these classical waves scattering among them because of the $\lambda\sigma^4$ interaction in eq. (21.83). The full dynamics can, however, be studied numerically by solving relativistic wave equations on a lattice, and one finds that the field sets into its minimum after only $O(1)$ oscillations. The rapid transfer of energy from the potential $V(0)$ into the amplitude of the inhomogeneous oscillations due to the tachyonic instability has been called *tachyonic preheating*, and is substantially different from the parametric resonance phenomenon that drives preheating in single-field inflation.

Again, the process of tachyonic preheating can efficiently create GWs. The specific form of the potential $U(\phi)$ in eq. (21.83) mostly affects the dynamics during preheating by setting the velocity with which ϕ reaches the critical value ϕ_c, so we can just use $\dot{\phi}_c$ as one of the parameters of the model or, equivalently, we can use the dimensionless quantity $V_c = g\dot{\phi}_c/(\lambda v^2)$ [not to be confused with the potential $V(\phi)$ defined in eq. (21.84)]. The parameters of the model are therefore $\{\lambda, g, v, V_c\}$. The dependence on v can be extracted analytically, since by passing to the dimensionless variables ϕ/v, σ/v and vx^μ (and neglecting the expansion of the Universe during the short preheating era) the dependence on v drops from the equations of motion derived from eq. (21.83). In particular, one finds that the peak frequency is independent of v while the value of $h_0^2\Omega_{\rm gw}$ at the peak scales as v^2. A detailed analysis of the dependence of the results on the model parameters can be found in Dufaux, Felder, Kofman and Navros (2009). For instance analytic estimates (confirmed within one order of magnitude by the numerical simulations) indicate that, when V_c is not too low, so that preheating is actually initiated by the classical rolling of the inflaton, the peak frequency today, f_0, is given by

$$f_0 \sim 7 \times 10^{10} \, \lambda^{1/4} V_c^{1/3} \, {\rm Hz} \qquad (22.55)$$

(except if $g^2/\lambda \ll 1$, in which case, after the first tachyonic growth, the typical momenta are significantly shifted toward lower frequencies), while the corresponding peak value of the GW energy density is given by

$$h_0^2\Omega_{\rm gw}(f_0) \sim 10^{-6} V_c^{-2/3} \left(\frac{v}{M_{\rm Pl}}\right)^2 . \qquad (22.56)$$

As an example, in Fig. 22.1 we show the predictions for $h_0^2\Omega_{\rm gw}(f)$ from tachyonic preheating for $\lambda = 10^{-5}$ and different values of V_c, while in Fig. 22.2 we show the result for fixed $V_c = 10^{-3}$, two different values $\lambda = 10^{-14}$ (left) and 10^{-5} (right), and several different values of λ/g^2. All spectra are shown for $v = 10^{-3}M_{\rm Pl}$. Lowering v leaves the shape of the spectra unchanged but reduces their amplitude proportionally to v^2, as we see from eq. (22.56). Observe from eq. (22.55) that for sufficiently small values of λ the peak of the spectrum can be in the region accessible to ground-based or even space-borne interferometers.

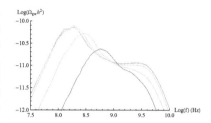

Fig. 22.1 Gravity wave spectra after preheating for $\lambda = 10^{-5}$ and different values of V_c. From Dufaux, Felder, Kofman and Navros (2009).

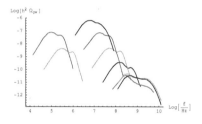

Fig. 22.2 $h_0^2\Omega_{\rm gw}(f)$ for fixed $V_c = 10^{-3}$, two different values $\lambda = 10^{-14}$ (left) and 10^{-5} (right), and several different values of λ/g^2. From Dufaux, Felder, Kofman and Navros (2009).

22.4 GWs from first-order phase transitions

22.4.1 Crossovers and phase transitions

During its cosmological evolution the Universe has undergone a number of changes of regime that, depending on the precise dynamics of the model, can be either phase transitions or smooth crossovers. The most recent one took place at the QCD scale, at a critical temperature T_c whose value is estimated from lattice QCD simulations to be about 150 MeV. Above this temperature quarks and gluons are deconfined and form a quark–gluon plasma, while below they are confined into hadrons. At earlier times and higher temperature, $T_c \sim 100$ GeV, there takes place the electroweak phase transition (or, as we will see, crossover). As the Universe cools and T decreases below T_c, the Higgs field acquires a non-vanishing expectation value, breaking $SU(2) \times U(1)$ to $U(1)_{\rm em}$, and the W and Z bosons become massive. Further phase transitions could have occurred earlier, for instance at the grand unification scale.

From the point of view of GW production, the interesting situation is when there is a first-order phase transition. A first-order phase transition is a violent event, and can be a significant source of GWs, as we will see in this section. In contrast, if the transition is second-order (or just a smooth crossover), no significant GW production is expected to take place.

Crossover/phase transition at the electroweak scale

The order of the electroweak phase transition depends on the value of the parameters of the underlying particle physics model, and in particular on the Higgs mass. Unfortunately, in the Standard Model, with the observed value of the Higgs mass $m_H = 125.09 \pm 0.21(\text{stat.}) \pm 0.11(\text{syst.})$ GeV, at the electroweak scale there is no first-order phase transition. Rather, there is just a smooth crossover at a temperature $T = 150.5 \pm 1.5$ GeV (a first-order phase transition would have taken place if the Higgs field had been much lighter, $m_H \lesssim 80$ GeV) and no GW production. This is unfortunate since, as we will see in more detail in eq. (22.135) (and we already understand from the approximate estimate in Table 22.1 on page 646), the GW signal from a first-order phase transition at the electroweak scale can fall within the LISA frequency band.

However, a strong first-order phase transition at the electroweak scale would also be welcome because it would make possible electroweak baryogenesis. Thus, the search for extensions of the Standard Model which feature a strong first-order phase transition is well motivated. In the minimal supersymmetric Standard Model (MSSM), a strong first-order phase transition requires $m_H < 115$ GeV, so again with the actual Higgs mass there is no appreciable GW signal. One can, however, construct other extensions of the Standard Model where the phase transition is first-order. In particular, the addition of an extra real singlet field or of a second Higgs doublet can induce a strong first-order phase transition (in the latter case, if the extra Higgs states are heavier than about 300 GeV). Another example of a model that leads to a first-order phase transition is obtained by adding to the Standard Model a dimension-6 operator, which could emerge as an effective low-energy description of some strongly-coupled dynamics at the TeV scale.

Crossover/phase transition at the QCD scale

For the QCD deconfinement transition the situation is somewhat similar. On the one hand, for a first-order phase transition the GW signal could be in the frequency range accessible to pulsar timing arrays [see eq. (22.135)], so it is potentially very interesting. On the other hand, after many years of studies, lattice QCD simulations with three dynamical quark flavors have conclusively shown that in QCD there is no phase transition but just a crossover.

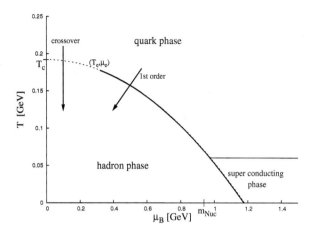

Fig. 22.3 A conjectured form of the phase diagram of QCD in the (μ_B, T) plane. Solid lines indicate a first-order phase transition, while the dashed line indicates a crossover. From Schwarz and Stuke (2009). ©SISSA Medialab Srl. Reproduced by permission of IOP Publishing. All rights reserved. [Observe that the exact phase diagram, including the existence of the critical endpoint (μ_e, T_e), and the value of the critical temperature, are still under debate. In particular, more recent lattice QCD calculations give a critical temperature $T_c \simeq 155$ MeV].

These studies have, however, been performed assuming that the baryonic chemical potential μ_B (i.e. the energy needed to add a baryon to a thermalized state at fixed volume and entropy) is small, $\mu_B \ll T$. The situation for larger values of μ_B is still uncertain. Figure 22.3 shows the conjectured QCD phase diagram in the (μ_B, T) plane, using results from perturbative computations, lattice QCD and nuclear physics. The solid line indicates a first-order phase transition, which terminates at a critical endpoint (μ_e, T_e). The correctness of this picture is still under debate. In the standard cosmological setting, μ_B/T is very small. Indeed, CMB and nucleosynthesis data show that the baryon-to-photon ratio

$$\frac{n_B}{n_\gamma} \sim \frac{\mu_B}{T} \sim 10^{-9}\,. \tag{22.57}$$

This quantity is conserved after baryogenesis and the last first-order phase transition, when the evolution conserves entropy. Thus, in the standard scenario, at the QCD scale $\mu_B/T \sim 10^{-9}$, and in this case the cosmological transition at the QCD scale is a crossover and does not generate GWs.

Non-standard scenarios have been investigated that could lead to a large value of μ_B/T and then, possibly, to a first-order phase transition. A possibility is that, at temperatures above the QCD phase transition, the value of μ_B/T was $O(1)$ thanks to some mechanism that generates a high baryon number. Then the QCD phase transition could be of first order. At the same time, during this first-order phase transition, the QCD vacuum could become trapped in a false vacuum state, generating a short period of inflation that dilutes μ_B/T to the current value. A different attempt at obtaining a large value of μ_B/T is based on the observation that μ_B depends both on the specific baryon asymmetry b (i.e. the baryon asymmetry per entropy) and on the specific lepton asymmetry l. The specific baryon asymmetry is fixed by cosmological observations and is very small, $b = (8.85 \pm 0.24) \times 10^{-11}$. Direct measurement of the specific lepton asymmetry l are more difficult, because the cosmic neutrino background escapes direct detection and is only indirectly constrained. The combined bounds from CMB and nucleosynthesis are only of order $|l| \lesssim 0.02$. The standard cosmological scenario predicts that the lepton asymmetry should be of the same order as the baryon asymmetry. This is due to the fact that after inflation any pre-existing baryon or lepton asymmetry has been washed away. They must therefore be re-generated after inflation, by baryon- and lepton-number-violating processes. In the Standard Model this can happen somewhat above the electroweak scale, due to sphaleron processes. In this case in each single sphaleron process the baryon-number violation is equal to the lepton-number violation, $\Delta(B - L) = 0$. However, there are extensions of the Standard Model, for example with heavy (s)neutrinos, where a large lepton asymmetry can be produced. For $|l| \gg b$, μ_B grows linearly with l, reaching values of order 0.2 GeV for $|l| \sim 0.02$, so not far from the critical point (μ_e, T_e) in Fig. 22.3. Thus, considering also the theoretical uncertainties in Fig. 22.3, it is not yet completely excluded that in extensions of the Standard Model leading to a large lepton asymmetry the QCD transition could be first-order.

The bottom-line is that, both at the electroweak scale and at the QCD scale, using the Standard Model (and, for the QCD scale, standard scenarios for the baryonic chemical potential) there is no first-order phase transition, so no GW production. We also see from eq. (22.6) that a phase transition at the grand-unification scale would give a signal at frequencies to high to be observed, even at ground-based interferometers. There is, however, at present still room for extensions of the Standard Model, or of standard scenarios, that could lead to a first-order phase transition, either at the electroweak or at the QCD scale. This is an example of the fact that the predictions for cosmological stochastic backgrounds suffer from uncertainties in the particle physics model at high energies, as well as from uncertainties in the cosmological model at early times. On the other hand, this means that the detection of a primordial stochastic background due to a first-order cosmological phase transition would reveal features of high-energy particle physics that would hardly be accessible otherwise, and would therefore be a very remarkable result. In Section 22.4.2 we will therefore study the GW production in a first-order phase transition in the early Universe. We will see that this study is also quite interesting from a conceptual point of view.

22.4.2   First-order phase transitions in cosmology

The dynamics of a phase transition is governed by the finite-temperature effective potential.[8] A typical example of a finite-temperature effective potential $V(\phi, T)$ inducing a first-order phase transition is given by

$$V(\phi, T) = \frac{1}{2}M^2(T)\phi^2 - \frac{1}{3}\delta(T)\phi^3 + \frac{1}{4}\lambda\phi^4 \,, \tag{22.58}$$

where $\phi \geqslant 0$ (for example ϕ is the modulus of a field) and, at least in some range of temperatures near a value T_0,

$$M^2(T) = \gamma(T^2 - T_0^2) \,, \tag{22.59}$$
$$\delta(T) = AT \,, \tag{22.60}$$

where the parameters γ, A and λ are strictly positive. The potential $V(\phi, T)$ is shown in Fig. 22.4 for different values of T, in the case $A^2 < 4\lambda\gamma$. For $T > T_0$ the field has a minimum at $\phi = 0$, where $V = 0$. If ϕ is the order parameter of a symmetry, then at this minimum the symmetry is unbroken. At $T \gg T_0$ this is the only minimum of the potential. If $A^2 < 4\lambda\gamma$, then, as we lower the temperature, a second extremum of $V(\phi)$ appears, at a temperature T_1 given by

$$T_1^2 = \frac{T_0^2}{1 - A^2/(4\lambda\gamma)} \,. \tag{22.61}$$

This extremum is at $\phi = \phi_1 = AT_1/(2\lambda)$, and $V(\phi_1) > 0$. At $T < T_1$ this extremum splits into a local maximum and a local minimum. Lowering the temperature further, the value of the potential at the local minimum gets lower and lower, until is becomes degenerate with the minimum at $\phi = 0$, at a critical temperature T_c given by

$$T_c^2 = \frac{T_0^2}{1 - 2A^2/(9\lambda\gamma)} \,. \tag{22.62}$$

At $T < T_c$ this minimum becomes the true minimum of the potential. Finally, as the temperature is lowered further to the value T_0, the extremum in $\phi = 0$ becomes a local maximum.

Before entering into a more technical analysis, let us first understand qualitatively, from Fig. 22.4, how a first-order phase transition takes place in an expanding Universe. At large temperatures the effective potential $V(\phi, T)$ has only the minimum at $\phi = 0$, and the field sits at this minimum. As the Universe cools down and its temperature becomes lower than T_c, this minimum becomes metastable, so the Universe is now trapped in a false vacuum. The transition to the true vacuum can take place via either thermal or quantum fluctuations, which nucleate bubbles of true vacuum. At temperature T, the nucleation of a vacuum bubble results in an energy gain equal to $(4\pi/3)r^3\Delta\rho_{\rm vac}(T)$, where r is the bubble radius and $\Delta\rho_{\rm vac}(T)$ is the energy density difference between the metastable state and the true vacuum. On the other hand, one loses an energy equal to $4\pi r^2 \sigma$, where σ is the surface tension of the bubble

[8] For a review of the definition of the finite-temperature effective potential and the relevant computational techniques in thermal field theory see for example Quiros (1999).

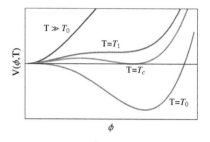

Fig. 22.4 The potential $V(\phi, T)$ for a first-order phase transition, for different values of T. We have set $A^2/(4\lambda\gamma) = 0.5$.

(which is calculable given the effective potential, as we will see). For a given value of $T < T_c$, corresponding to a given value of $\Delta\rho_{\rm vac}(T)$, there is therefore a critical radius $r_c(T) = 3\sigma/\Delta\rho_{\rm vac}(T)$ (a more precise computation, using the exact profile of the scalar field, will be given below) such that bubbles nucleated with $r < r_c(T)$ shrink and disappear because of the surface tension, while bubbles with $r > r_c(T)$ survive and expand.

Of course, the probability of nucleating a large bubble is small. In standard equilibrium thermodynamics, if the system is in a metastable state and we wait long enough, it will eventually undergo the transition to the true vacuum state, even if the transition rate is small. In the cosmological setting, however, there is an explicit time-scale due to the expansion of the Universe, and we must compare the transition rate from the false to the true vacuum with the Hubble expansion rate. If, at a given temperature, the nucleation rate $\Gamma(T)$ of supercritical bubbles, i.e. bubbles with $r > r_c(T)$, is much smaller than the Hubble parameter $H(T)$, the Universe will have cooled down significantly further before nucleation of supercritical bubbles has a significant chance of taking place. When T is smaller than T_c but close to it, $\Delta\rho_{\rm vac}(T)$ is small, so $r_c(T)$ is large and the nucleation rate of bubbles with $r > r_c(T)$ is small. As the Universe cools down further, $\Delta\rho_{\rm vac}(T)$ increases, $r_c(T)$ decreases, and the nucleation rate $\Gamma(T)$ of supercritical bubbles increases. A first-order phase transition will take place if, at some temperature T_* in the range $T_0 < T_* < T_c$, $\Gamma(T_*)$ becomes as large as $H(T_*)$.[9] If this never happens, then, once the system reaches the temperature T_0, the field ϕ smoothly rolls down for $\phi = 0$, which is now a maximum, toward the true minimum, and we have a second-order phase transition.

Denoting by $\eta(T)$ the vacuum expectation value of the scalar field at the true vacuum at temperature T and observing that the effective potential $V(\phi, T)$ is actually the free energy density of the system, the vacuum energy density associated with the transition is

$$\rho_* = \left[-V[\eta(T), T] + T\frac{d}{dT}V[\eta(T), T] \right]_{T=T_*} . \qquad (22.63)$$

Since we are interested in transitions taking place during RD, it is convenient to normalize this quantity to the radiation energy density (17.98), defining

$$\alpha = \frac{30\rho_*}{\pi^2 g_*(T_*)T_*^4} . \qquad (22.64)$$

Another important parameter is the bubble nucleation rate at the transition temperature, which is modeled as an exponential,

$$\Gamma(t) = \Gamma_* e^{\beta(t-t_*)} , \qquad (22.65)$$

where t_* is the time at which the transition takes place [defined more quantitatively in eq. (22.123)], and at which the rate has a value Γ_*. We will see later how this exponential form emerges, and how to compute β from the effective potential. Since β has dimensions of inverse time, the

[9]A more precise quantitative estimate will be given in eq. (22.124).

relevant dimensionless quantity is actually the ratio β/H_*, where $H_* = H(T_*)$, which indeed will appear in the computation of GW production.

22.4.3 Thermal tunneling theory

We now discuss more quantitatively how bubbles of true vacuum are nucleated and how their nucleation rate can be computed, first reviewing the bounce computation in quantum mechanics, then studying QFT at zero temperature and finally turning to the relevant case of tunneling in QFT at finite temperature. Apart from the application to GW production in first-order phase transitions, the theory has several applications in QFT, and is of great elegance in itself.

Tunneling and the bounce solution

Let us consider a quantum-mechanical system with one degree of freedom $q(t)$ and Hamiltonian $H = p^2/2 + V(q)$, where

$$V(q) = \frac{1}{2}\omega^2 q^2 + \frac{1}{2}gq^4 . \tag{22.66}$$

We have set the particle mass $m = 1$, and g is a coupling constant. For $g > 0$ the potential is bounded from below and the Hamiltonian H is positive definite. The partition function of the system at temperature $T = 1/\beta$ can be computed using the basic result relating the operator formalism of quantum mechanics to the Euclidean path-integral formulation[10]

$$\operatorname{tr} e^{-\beta H/\hbar} = \int_{q(-\beta/2)=q(\beta/2)} [dq] e^{-S_E/\hbar} , \tag{22.67}$$

where the trace is over the Hilbert space of the system and $[dq]$ denotes the integration measure in the path integral.[10] Since we will use a semiclassical approximation, which is formally an expansion in $\hbar$, it is convenient for the moment to keep $\hbar$ explicit rather than setting $\hbar = 1$. The path integral in eq. (22.67) is performed over trajectories $q(t)$ with $-\beta/2 \leqslant t \leqslant \beta/2$, and periodic boundary conditions. In eq. (22.67) we have performed the rotation to Euclidean space, and the Euclidean action is

$$S_E = \int_{-\beta/2}^{\beta/2} dt \left[\frac{1}{2}\dot{q}^2 + V(q)\right] . \tag{22.68}$$

In the limit $\beta \to \infty$ the left-hand side of eq. (22.67) is dominated by the contribution of the ground state, whose energy we denote by E_0, so $\operatorname{tr} e^{-\beta H/\hbar} \simeq e^{-\beta E_0/\hbar}$, while on the right-hand side the only trajectories that contribute are those that satisfy $q(-\beta/2) = q(\beta/2) = 0$, since otherwise the Euclidean action diverges as $\beta \to \infty$. Thus, the ground-state energy can be computed from

$$E_0(g) = - \lim_{\beta \to \infty} \frac{\hbar}{\beta} \log \int_{q(-\beta/2)=q(\beta/2)=0} [dq] e^{-S_E/\hbar} . \tag{22.69}$$

As long as $g > 0$, this allows us to compute the energy $E_0(g)$, which will have the form $E_0(g) = (1/2)\hbar\omega[1 + O(g)]$, corresponding to a harmonic

[10] See for example Section 9.6 of Maggiore (2005).

oscillator with frequency ω, plus anharmonic corrections parametrized by g. Observe that, in the limit $\beta \to \infty$, rescaling $q \to q/\sqrt{g}$,

$$\frac{1}{\hbar} \int_{-\infty}^{\infty} dt \left[\frac{1}{2}\dot{q}^2 + \frac{1}{2}\omega^2 q^2 + \frac{1}{2}gq^4 \right] \to \frac{1}{g\hbar} \int_{-\infty}^{\infty} dt \left[\frac{1}{2}\dot{q}^2 + \frac{1}{2}\omega^2 q^2 + \frac{1}{2}q^4 \right],$$

(22.70)

so the perturbative expansion in powers of g is the same as the semi-classical expansion in powers of $\hbar$.

We now want to consider the situation $g < 0$. The corresponding potential is shown in Fig. 22.5. Now the potential is unbounded from below, and $E_0(g)$ can be obtained by analytic continuation from the expression valid for $g > 0$, given in eq. (22.69).[11] As is physically clear, and as we will show below, this analytic continuation generates an imaginary part for E_0, so that now the wavefunction $\psi(t)$ of the corresponding eigenstate evolves as

$$\psi(t) \propto e^{-i(\mathrm{Re}E_0)t - |\mathrm{Im}E_0|t},$$

(22.71)

corresponding to the fact that a state localized near the minimum at $q = 0$ eventually decays by tunneling.[12] Thus, to find E_0 we still have to compute the path-integral on the right-hand side of eq. (22.69), except that now $g < 0$ so the potential is the one given shown in Fig. 22.5. Observe also that, from the procedure of analytic continuation, we inherit the boundary conditions $q(-\beta/2) = q(\beta/2) = 0$.

We compute this path integral in the semiclassical approximation. Thus, we look for the classical solutions $q_{\mathrm{cl}}(t)$ of the equation of motion derived from the action S_E, that satisfy the boundary conditions $q(-\beta/2) = q(\beta/2) = 0$, with $\beta \to \infty$. We then write

$$q(t) = q_{\mathrm{cl}}(t) + \hbar^{1/2}\varphi(t),$$

(22.72)

where φ are the quantum fluctuations, and we integrate over φ. Then, to quadratic order in φ

$$\int_{q(-\beta/2)=q(\beta/2)=0} [dq]e^{-S_E/\hbar} =$$

$$= e^{-S[q_{\mathrm{cl}}]/\hbar} \int [d\varphi]\, e^{-\frac{1}{2} \int_{-\infty}^{\infty} dt_1 dt_2\, \varphi(t_1)M(t_1,t_2)\varphi(t_2)},$$

(22.73)

where the operator $M(t_1, t_2)$ is given by

$$M(t_1, t_2) = \left. \frac{\delta^2 S}{\delta q(t_1)\delta q(t_2)} \right|_{q=q_{\mathrm{cl}}}$$

$$= \left[-\frac{d^2}{dt_1^2} + V''[q_{\mathrm{cl}}(t_1)] \right] \delta(t_1 - t_2).$$

(22.74)

Let $\varphi_n(t)$ be a complete set of (in general complex) normal modes of the operator M,

$$\int_{-\infty}^{\infty} dt'\, M(t, t')\varphi_n(t') = \lambda_n \varphi_n(t).$$

(22.75)

[11] For a detailed discussion of how analytic continuation works in this problem, see Chapter 36 of Zinn-Justin (2002).

[12] Observe that for $g < 0$ the Hamiltonian does not have a real spectrum, because it is unbounded from below and the boundary conditions on the wavefunction at spatial infinity are not real.

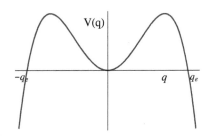

Fig. 22.5 The potential $V(q)$ for $g < 0$. The points $q = \pm q_e$ where $V(q) = 0$ correspond to the escape points of the tunneling process in Minkowski space.

We take these normal modes to be orthogonal with respect to the scalar product

$$\langle f|g \rangle = \int_{-\infty}^{\infty} dt\, f^*(t) g(t)\,, \qquad (22.76)$$

but we find it convenient not to require them to be orthonormal, so

$$\langle \varphi_n | \varphi_m \rangle = \delta_{nm} u_n\,, \qquad (22.77)$$

where the u_n give the norms of the modes. We then expand

$$\varphi(t) = \sum_n \xi_n \varphi_n(t)\,, \qquad (22.78)$$

where the ξ_n are real coefficients, ranging from $-\infty$ to $+\infty$. The measure of the path integral is defined as

$$[d\varphi] = \mathcal{N} \prod_n \sqrt{\frac{u_n}{2\pi\hbar}}\, d\xi_n\,, \qquad (22.79)$$

where $\mathcal{N}$ is a normalization constant that we will fix later. We then have

$$\int_{-\infty}^{\infty} dt_1 dt_2\, \varphi(t_1) M(t_1, t_2) \varphi(t_2) = \sum_n \lambda_n u_n \xi_n^2\,. \qquad (22.80)$$

The integration over the modes with non-zero eigenvalues λ_n is Gaussian and gives a result proportional to

$$e^{-S[q_{\mathrm{cl}}]/\hbar} (\det{}' M)^{1/2}\,, \qquad (22.81)$$

where the prime means that we have excluded the eigenvectors with $\lambda_n = 0$.[13] We will discuss later the treatment of the zero modes. If $S[q_{\mathrm{cl}}]$ is non-zero, the factor $e^{-S[q_{\mathrm{cl}}]/\hbar}$ is non-perturbative in $\hbar$. If the non-zero eigenvalues of M are all positive, or more generally if there are an even number of negative eigenvalues, $\det{}' M > 0$ and the contribution (22.81) is real, so it contributes to the real part of E_0. However, in this case we have no right to retain it, since this term is exponentially small as $\hbar \to 0^+$. It is then negligible compared with the terms obtained by performing perturbation theory around the trivial solution $q_{\mathrm{cl}}(t) = 0$, since these contributions are given by an expansion in powers of the coupling g, which, in view of the comment after eq. (22.70), is the same as an expansion in powers of $\hbar$. However, if there is one or, more generally, an odd number of negative eigenvalues, then $\det{}' M < 0$ and the contribution (22.81) is imaginary. In this case, it gives the leading correction to $\mathrm{Im}\, E_0$.

The equation of motion obtained from the Euclidean action (22.68) is

$$\ddot{q} = V'\,, \qquad (22.82)$$

which is formally the same as that obtained from a Minkowskian action with a potential $-V(q)$, shown in Fig. 22.6. One obvious solution satisfying the boundary conditions $q(-\beta/2) = q(\beta/2) = 0$ is the trivial

[13]The integration over a mode with $\lambda_n < 0$ is defined by analytic continuation and again gives a factor proportional to $\lambda_n^{1/2}$, so eq. (22.81) holds for both positive and negative eigenvalues. For negative eigenvalues, however, the proportionality constant differs by a factor $1/2$ because of the procedure of analytic continuation; see the detailed discussion in Callan and Coleman (1977), after their eq. (2.22). In the following we will not need the explicit proportionality factors, and we will not keep track of them, limiting ourselves to quoting the final result in eq. (22.98). These factors are accurately computed in Coleman (1977a,b) and Callan and Coleman (1977).

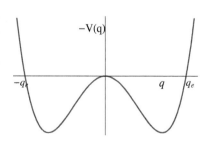

Fig. 22.6 The potential $-V(q)$ for $g < 0$.

one, $q_{cl}(t) = 0$. However, in this case $V''(q_{cl}) = V''(0) \equiv \omega^2$ and the operator M is the one corresponding to the harmonic oscillator, with only positive eigenvalues. So, it does not contribute to the imaginary part. Thus, we have to look for non-trivial solutions. The first integral of eq. (22.82) is

$$\frac{1}{2}\dot{q}_{cl}^2 - V(q_{cl}) = E \,. \tag{22.83}$$

The solutions with $E < 0$ corresponds to oscillations around a minimum of $-V$, say the minimum at $q > 0$, with inversion points q_1 and q_2. From eq. (22.83), the period β of this solution is given by

$$\beta = 2 \int_{q_1}^{q_2} dq \, \frac{1}{\sqrt{2(E+V)}}$$

$$= 2 \int_{q_1}^{q_2} dq \, \frac{1}{\sqrt{2E + \omega^2 q^2 + (1/2)gq^4}} \,. \tag{22.84}$$

In the limit $\beta \to \infty$ the integral on the right-hand side must diverge. For $E < 0$ but close to zero, the inversion point q_1 is at $q_1 \simeq 2|E|/\omega$, and the integrand near the extremum $q = q_1$ is approximately proportional to $1/\sqrt{q^2 - q_1^2}$. For $q_1 \neq 0$, $\sqrt{q^2 - q_1^2} \simeq \sqrt{(q - q_1)2q_1}$ and this singularity is integrable. In contrast, for $q_1 = 0$, we have $1/\sqrt{q^2 - q_1^2} = 1/q$, whose integral diverges at $q = 0$. This means that the solution whose period $\beta \to \infty$ is the one whose inversion point $q_1 \to 0^+$, and therefore with $E \to 0$. Thus, in the limit $\beta \to \infty$, on the solution satisfying the boundary conditions $q(-\infty) = q(\infty) = 0$ the first integral of eq. (22.82) is

$$\frac{1}{2}\dot{q}_{cl}^2 - V(q_{cl}) = 0 \,. \tag{22.85}$$

In the limit $\beta \to \infty$ we understand from Fig. 22.6 that this solution starts from $q \to 0^+$ at $t \to -\infty$, goes to the point $q = q_e$ where it bounces, and then goes back to $q = 0$ as $t \to \infty$. This solution is called the *bounce*. Actually, in the limit $\beta \to \infty$ the system is invariant under time translations, so if $q_{cl}(t)$ is a solutions of the equations of motion with these boundary conditions, then $q_{cl}(t; t_0) = q_{cl}(t - t_0)$ is also a solution, for an arbitrary value of the parameter t_0. Therefore, there is a whole family of classical bounce solutions around which we must expand the path integral, parametrized by t_0, which is called a "collective coordinate". Using eq. (22.85), we see that the action (22.68) evaluated on this solution is positive-definite and, in the limit $\beta \to \infty$, has the value

$$S_{cl} = \int_{-\infty}^{\infty} dt \, \dot{q}_{cl}^2 \,, \tag{22.86}$$

independent of t_0. We next consider the fluctuations around a bounce solution. It is easy to see that there is a zero mode, i.e. an eigenvector with zero eigenvalue, given by

$$\varphi_0(t) = \dot{q}_{cl}(t) \,. \tag{22.87}$$

Indeed, differentiating eq. (22.82) with respect to time, we get

$$\left[-\frac{d^2}{dt^2} + V''[q_{cl}(t)] \right] \dot{q}_{cl}(t) = 0 , \qquad (22.88)$$

which shows that $\dot{q}_{cl}(t)$ is an eigenvector of the operator M, with zero eigenvalue. More generally we can observe that, whenever the classical solution $q_{cl}(t; \gamma)$ depends on a set of collective coordinates γ_i, the functions $\partial q_{cl}/\partial \gamma_i$ are zero modes. This can be shown starting from the classical equation of motion, $(\delta S/\delta q)_{q=q_{cl}} = 0$, and taking its derivative with respect to γ_i,

$$0 = \frac{\partial}{\partial \gamma_i} \left(\frac{\delta S}{\delta q(t)} \right)_{q=q_{cl}}$$

$$= \int dt' \left(\frac{\delta^2 S}{\delta q(t) \delta q(t')} \right)_{q=q_{cl}} \frac{\partial q_{cl}(t'; \gamma)}{\partial \gamma_i} . \qquad (22.89)$$

The operator $[\delta^2 S/\delta q(t)\delta q(t')]_{q=q_{cl}}$ is just $M(t, t')$, and we see that $\partial q_{cl}/\partial \gamma_i$ are zero modes. In our case the collective coordinate is t_0 and q_{cl} is a function of $t - t_0$, so $\partial q_{cl}/\partial t_0 = -\partial q_{cl}/\partial t$, and hence $\dot{q}_{cl}(t)$ is a zero mode.

From the fact that $\dot{q}_{cl}(t)$ is a zero mode it also follows that the operator M in eq. (22.74) has one and only one negative eigenvalue. In fact, M has the form of a Hermitian Hamiltonian in a one-dimensional quantum-mechanics problem. As we know from quantum mechanics, its spectrum is such that the eigenfunction with the lowest eigenvalue (i.e. the ground state in the Hamiltonian analogy) has no node, the first excited state has one node, etc. Since the zero mode $\dot{q}_{cl}(t)$ has one node, corresponding to the point $q = q_e$ where the solution bounces back, it follows that the zero mode is the first excited state of the operator M, and M has one and only one eigenvector with lower eigenvalue, i.e. with $\lambda < 0$. Thus, $(\det'M)^{1/2} = i|\det'M|^{1/2}$, and we get a contribution to $\mathrm{Im}\, E_0$.

We must next understand how to treat the zero mode, i.e. the integration over the variable ξ_0 associated with the mode $\varphi_0(t) = \dot{q}_{cl}(t)$. In the path integral the integration over ξ_0 runs from $-\infty$ to $+\infty$, and nothing else depends on ξ_0, so this is a formally divergent term. However, we must recall that we are working in the limit $\beta \to \infty$, and that there is already a factor of $1/\beta$ on the right-hand side of eq. (22.69), which eventually must cancel. So, while for the other occurrences of β in our computation to lowest order it was legitimate to set directly $\beta = \infty$ [for example in the integration limits in eq. (22.86)], here we have to be more careful. The infinitesimal fluctuation in the direction parametrized by ξ_0 actually corresponds to moving in the space of classical solutions. Indeed, consider an infinitesimal fluctuation $\delta \xi_0 \, \varphi_0$ around the classical solution $q_{cl}(t; t_0)$. Using the fact that $\varphi_0(t) = (\partial/\partial t)q_{cl}(t; t_0) = -(\partial/\partial t_0)q_{cl}(t; t_0)$, we see that

$$q_{cl}(t; t_0) + \delta \xi_0 \varphi_0(t) = q_{cl}(t; t_0) - \delta \xi_0 \frac{\partial}{\partial t_0} q_{cl}(t; t_0)$$

$$= q_{cl}(t; t_0 - \delta \xi_0) . \qquad (22.90)$$

[14]More precisely, we can transform the integral over $d\xi_0$ into an integral over dt_0 using a simpler version of the Faddeev–Popov trick used in the quantization of gauge theories. First of all observe that, given a classical configuration $q_{cl}(t; t_0)$, we define the fluctuations around it as

$$\varphi(t; t_0) \equiv q(t) - q_{cl}(t; t_0). \quad (22.91)$$

The generic field configuration $q(t)$ knows nothing about t_0, so φ depends on both t and t_0. We start from the identity

$$1 = \int dt_0 \, \delta[f(t_0)] \left| \frac{\partial f}{\partial t_0} \right|, \quad (22.92)$$

which is valid for any (differentiable) function $f(t_0)$, and we chose

$$f(t_0) = \int dt \, \varphi(t; t_0) \dot{q}_{cl}(t; t_0), \quad (22.93)$$

i.e. the projection of the quantum fluctuations $\varphi(t; t_0)$ onto the zero mode. Using eq. (22.78), $f(t_0) = \xi_0(t_0) u_0$ while, using eq. (22.91),

$$\frac{\partial f}{\partial t_0} = \int dt \left[-\dot{q}_{cl}(t; t_0) \frac{\partial}{\partial t_0} q_{cl}(t; t_0) \right.$$
$$\left. + \varphi(t; t_0) \frac{\partial}{\partial t_0} \dot{q}_{cl}(t; t_0) \right]. \quad (22.94)$$

In the first term we use $-\partial q_{cl}/\partial t_0 = +\dot{q}_{cl}$, where the dot is as usual the derivative with respect to t, so the first term is just u_0. Then we get

$$1 = \int dt_0 \, \delta(\xi_0) \left[1 - \frac{1}{u_0} \varphi(t; t_0) \ddot{q}_{cl} \right]. \quad (22.95)$$

The second term depends on the quantum field φ. Exponentiating the square bracket in the form $\exp\{-(1/u_0)\varphi(t; t_0)\ddot{q}_{cl}\}$, this term can be reabsorbed into the action. However, the action appears in the exponent in the form $S_E/\hbar$, while this term has no factor of $\hbar$, and is therefore a loop correction (actually, it can be shown that it only contributes at the two-loop level). To lowest order in the semiclassical expansion we can therefore neglect it, and write eq. (22.95) as

$$1 \simeq \int dt_0 \, \delta(\xi_0). \quad (22.96)$$

Inserting this into eq. (22.79) eliminates the integral over ξ_0 and replaces it with an integral over the collective coordinate t_0, with a Jacobian equal to 1.

Thus, moving by $\delta\xi_0$ in the direction of the zero mode $\varphi_0(t)$ is equivalent to changing the classical solution around which we are expanding, i.e. it is equivalent to performing a shift of the collective coordinate $t_0 \to t_0 + \delta t_0$, with $\delta t_0 = -\delta\xi_0$. Thus, the integration over ξ_0 is equivalent to the integration over the collective coordinate t_0, and corresponds to summing, in the path integral, over all possible bounce solutions.[14] For finite β, t_0 ranges between $-\beta/2$ and $\beta/2$, so this finally give an overall factor of β, which is very welcome since it matches the factor $1/\beta$ on the right-hand side of eq. (22.69).

The final step of this (long!) computation is to observe that, besides the bounce solutions, we have multi-bounces, corresponding to solutions bouncing back and forth several times between $q = 0$ and $q = q_e$. Their effect is to exponentiate the one-bounce contribution, which compensates for the log in eq. (22.69).

Thus, writing $\text{Im}\, E_0 = \Gamma/2$, the tunneling rate Γ is given by

$$\Gamma = A e^{-S_b}, \quad (22.97)$$

where S_b is the value of the Euclidean action, evaluated on the bounce solution, and A is a proportionality constant, which is given by

$$A = \hbar \left(\frac{S_b}{2\pi\hbar} \right)^{1/2} \left| \frac{\det'[-\partial_t^2 + V''(\phi_b)]}{\det[-\partial_t^2 + V''(0)]} \right|^{-1/2}. \quad (22.98)$$

The factor $(S_b/2\pi\hbar)^{1/2}$ is related to the integration over the zero mode, and comes from the factor $(u_0/2\pi\hbar)^{1/2}$ in eq. (22.79) on observing, from eq. (22.86), that the norm u_0 of the zero mode $\varphi_0 = \dot{q}_{cl}$ is just S_b. The factor $\mathcal{N}$ in eq. (22.79) is fixed by requiring that the path integral computed with the integration measure (22.79) reproduces the standard result for the transition amplitude of the harmonic oscillator, and provides the factor $\det^{1/2}[-\partial_t^2 + V''(0)]$ in eq. (22.98). Finally, the factor of two from $\Gamma = 2\, \text{Im}\, E_0$ is compensated by an extra factor $1/2$ from the integration over the negative eigenvalue; see Note 13.

The same computation can be performed for a potential $V(q)$ such as that shown in Fig. 22.7. Now the system has a "false vacuum" in $q = 0$ and a true vacuum in $q = q_0$. The equation of motion (22.82) derived from the Euclidean action is again formally the same as that of a particle moving in the potential $-V(q)$, which is shown in Fig. 22.8, and again has a bounce solution, from $q = 0$ to $q = q_e$ and back. Observe, however, that, in this case, what we are computing in this way is not the imaginary part of the energy of a state in the spectrum of the Hamiltonian. The potential shown in Fig. 22.7 is bounded from below, and the corresponding Hamiltonian has a spectrum of purely real eigenvalues. The rate Γ must simply interpreted as the tunneling rate of a state initially localized near the false vacuum.

Tunneling in QFT at zero temperature

We next consider a scalar field theory in flat space, with Minkowskian action

$$S_M = \int d^4x \left[-\frac{1}{2}\eta^{\mu\nu}\partial_\mu\phi\partial_\nu\phi - V(\phi) \right], \qquad (22.99)$$

[recall that our signature is $\eta_{\mu\nu} = (-,+,+,+)$], and we consider first QFT at zero temperature. We chose a potential $V(\phi)$ that has a false vacuum in $\phi = 0$, chosing the zero of the potential such that $V(\phi) = 0$, and a true vacuum at $\phi = \phi_0 > 0$, with $V(\phi_0) < 0$, as in Fig. 22.7 (with $q \to \phi$, $q_0 \to \phi_0$).[15] The rotation to Euclidean signature $(+,+,+,+)$ is obtained by performing the Wick rotation, i.e. introducing Euclidean time τ from $\tau = it$.[16] Then $d^4x = -id\tau d^3x \equiv -i(d^4x)_E$. Defining the Euclidean action S_E from $e^{iS_M} = e^{-S_E}$, i.e. $S_E = -iS_M$, we therefore find

$$S_E = \int d^4x_E \left[\frac{1}{2}\partial_\mu\phi\partial_\mu\phi + V(\phi) \right], \qquad (22.100)$$

where the sum over repeated lower indices is performed with the Euclidean metric, $\partial_\mu\phi\partial_\mu\phi \equiv (\partial\phi/\partial\tau)^2 + (\partial_i\phi)^2$. We can now repeat the computation performed in the quantum-mechanical case, with minor modifications. Namely, we look for solutions of the equations of motion derived from the Euclidean action, subject to the boundary condition

$$\lim_{\tau\to\pm\infty} \phi(\tau,\mathbf{x}) = 0. \qquad (22.101)$$

Furthermore, we must now also impose

$$\lim_{|\mathbf{x}|\to\infty} \phi(\tau,\mathbf{x}) = 0, \qquad (22.102)$$

otherwise the action evaluated on the solution would be infinite. Again, the fluctuations around the trivial solution $\phi(\tau,\mathbf{x}) = 0$ have no negative eigenvalue and do not contribute to the imaginary part, so we must rather look for a non-trivial classical solution. In the zero-temperature case that we are considering here, it is reasonable to look for a solution that respects the $O(4)$ symmetry of Euclidean space. Namely, we chose an arbitrary point of four-dimensional Euclidean space, and we use it as the origin for the coordinate system. With respect to this origin, we define a radial coordinate $\rho = (\tau^2 + \mathbf{x}^2)^{1/2}$, and we look for a solution $\phi = \phi(\rho)$. Of course, the origin of the coordinate system can be chosen arbitrarily, which means that now the solution has a set of collective coordinates $\{\tau_0, \mathbf{x}_0\}$ corresponding to the choice of the origin, reflecting the invariance under (Euclidean) time translation as well as spatial translation. As in the quantum-mechanical tunneling computation of the previous subsection, we let τ run over the finite interval $-\beta/2 < \tau < \beta$, letting $\beta \to \infty$ at the end, and similarly we work in a large but finite spatial volume V. Then, the integration over the collective coordinate in the path integral gives a factor βV. The factor β is canceled by the factor $1/\beta$ in eq. (22.69), as in the quantum-mechanical case. The result

[15]More precisely, we take $V(\phi)$ to be the effective potential, which also takes into account the loop corrections. As first shown in a classic paper by Coleman and Weinberg (1973), radiative corrections can change the qualitative shape of the potential, and develop a true minimum at $\phi = \phi_0 > 0$, thus inducing spontaneous symmetry breaking.

[16]The plus sign in the definition $\tau = +it$ is fixed by the condition that the pole structure of the propagator obtained from the Euclidean action is the same as that of the $-i\epsilon$ prescription of the Feynman propagator in Minkowski space.

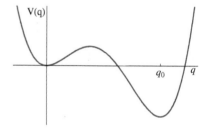

Fig. 22.7 The potential $V(q)$.

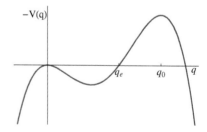

Fig. 22.8 The function $-V(q)$.

is therefore that in QFT the rate Γ is proportional to the volume V, and the quantity that is finite is the rate per unit volume, Γ/V. This corresponds to the fact that the nucleation of bubbles of true vacuum can take place anywhere in space.

On a function $\phi = \phi(\rho)$, the action (22.100) reads

$$S_E = 2\pi^2 \int_0^\infty d\rho\, \rho^3 \left[\frac{1}{2} \left(\frac{d\phi}{d\rho} \right)^2 + V(\phi) \right] \tag{22.103}$$

(where $2\pi^2$ is the solid angle in four dimensions), so the equation of motion becomes

$$\frac{d^2\phi}{d\rho^2} + \frac{3}{\rho} \frac{d\phi}{d\rho} = V'(\phi). \tag{22.104}$$

On the $O(4)$-symmetric solution, the two boundary conditions (22.101) are unified into a single boundary condition

$$\lim_{\rho \to \infty} \phi(\rho) = 0. \tag{22.105}$$

Furthermore, requiring that the solution be regular at the origin of the polar coordinate system gives

$$\left. \frac{d\phi}{d\rho} \right|_{\rho=0} = 0. \tag{22.106}$$

Equation (22.104) can be formally interpreted as the motion of a particle of unit mass in one dimension, described by the coordinate ϕ, with ρ playing the role of time, moving in the potential $-V$ shown in Fig. 22.8, and further subject to a damping force $-\eta(\rho)v$, where $v = d\phi/d\rho$ is the velocity of this fictitious particle and $\eta(\rho) = 3/\rho$ a rather peculiar friction coefficient, depending on "time" ρ. This fictitious particle is subject to the boundary conditions (22.105) and (22.106). This mechanical analogy allows us to understand the qualitative features of the solution. The "particle" must start at "time" $\rho = 0$ with zero velocity [see eq. (22.106)], from some initial position ϕ_i chosen so that, at time $\rho \to \infty$, it reaches the position $\phi = 0$. Because of the friction term, its initial energy must therefore be greater than zero, so the initial position must be in the region $\phi_e < \phi_i < \phi_0$, see Fig. 22.9. Its precise value can be determined by integrating eq. (22.104) numerically, assigning a value of $\phi(\rho = 0)$ in the range $\phi_e < \phi(\rho = 0) < \phi_0$, setting $d\phi/d\rho|_{\rho=0} = 0$ and using an under-shooting/over-shooting procedure. If we chose a value of $\phi(\rho = 0)$ smaller than ϕ_i, then, because of the friction term, the solution will not have enough energy to reach $\phi = 0$, and will bounce back and perform oscillations around the minimum of $-V$. If we choose a value of $\phi(\rho = 0)$ such that $\phi_i < \phi(\rho = 0) < \phi_0$, then the solution over-shoots, and reaches the region $\phi < 0$, so eventually $\phi \to -\infty$, as we see from the form of $-V$ in Fig. 22.9 (observe that the friction term goes to zero as $\rho \to +\infty$, so it does not significantly damp the evolution at large values of "time" ρ).

In this way we can determine the solution, $\phi_b(\rho)$, numerically. We still call this solution a bounce since, if we follow its evolution as a

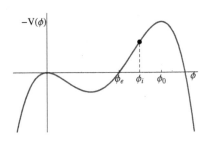

Fig. 22.9 The function $-V(\phi)$, and the initial position ϕ_i.

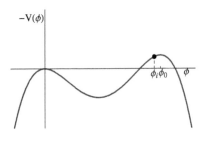

Fig. 22.10 As in Fig. 22.9, with a smaller energy difference between the true and the false vacuua

function of Euclidean time τ at $\mathbf{x} = 0$, it corresponds to a solution that for $\tau \to -\infty$ (which means $\rho \to +\infty$) goes to zero, increases to the maximum value ϕ_i, which is reached at $\tau = 0$, and then for positive τ bounces back, reaching $\phi = 0$ again as $\tau \to +\infty$. When we continue back to Minkowski space, the solution $\phi = \phi_b(\rho)$ becomes $\phi = \phi_b(\sqrt{-t^2 + r^2})$, where $r = |\mathbf{x}|$. If we look at it as a function of r for fixed t, for example at $t = 0$, we see that at small r it is near the value $\phi = \phi_i$, and thus close to the true vacuum ϕ_0. As r increases, ϕ evolves toward the value $\phi = 0$, corresponding to the false vacuum. It therefore represents a "bubble" in which the field configuration in the inner region is close to the true vacuum, and the field in the exterior is close to the false vacuum. Since the full dependence of the solution on t and r is through the combination $-t^2 + r^2$, with time this bubble of true vacuum expands into the false vacuum at the speed of light.[17] Actually, the description in terms of a bubble is especially appropriate if the transition between the inner and outer region is sharp. This is the case when the energy difference between the true and the false vacuum is small, which is the situation depicted in Fig. 22.10. In this case we see from Fig. 22.10 that the value ϕ_i must be quite close to the maximum ϕ_i of the function $-V$. From the mechanical analogy we see that in this case, in the Euclidean solution, the "particle" with coordinate ϕ will stay for a long "time" ρ near the value ϕ_i, since here the potential is close to flat, up to a "time" $\rho = R$ when this slow-roll regime terminates and the particle quickly rolls toward the "position" $\phi = 0$. Translated in terms of the Minkowskian solution, at fixed time $t = 0$, this describes a field configuration in which, for $r < R$, ϕ is very close to the value $\phi = \phi_0$ corresponding to the true vacuum; it remains almost constant up to $r = R$, and then it quickly goes to the false vacuum value $\phi = 0$. The transition between the region where $\phi \simeq \phi_0$ and the region where $\phi \simeq 0$ takes place in a layer of width $\delta r \ll R$. This corresponds to the *thin-wall* solution, in which a bubble of true vacuum is separated by a thin wall from the surrounding false vacuum. Again, because of the dependence on $-t^2 + r^2$, with time this bubble expands into the false vacuum at the speed of light.

We now have all the elements to readily adapt the computation of the transition rate that we performed in the quantum-mechanical case.[18] The result is

$$\frac{\Gamma}{V} = A\, e^{-S_b/\hbar}\,, \qquad (22.107)$$

where S_b is the Euclidean action (22.103) evaluated on the bounce solution $\phi_b(\rho)$.[19] As in the quantum-mechanical computation, the prefactor A in eq. (22.107) can be computed by keeping track of the normalization factors u_n for the four zero modes [which give a factor $(S_b/2\pi)^{1/2}$ each], as well as of the normalization factor $\mathcal{N}$ in the measure (22.79). This gives[20]

$$A = \left(\frac{S_b}{2\pi}\right)^2 \left|\frac{\det'[-\partial_\mu\partial_\mu + V''(\phi_b)]}{\det[-\partial_\mu\partial_\mu + V''(0)]}\right|^{-1/2}. \qquad (22.111)$$

[17] Observe, however, that this Minkowskian description in terms of an expanding bubble only holds after the bubble has been nucleated, and therefore only for the expanding branch of the hyperboloid $r^2(t) = R^2 + t^2$. The situation is analogous to the description of barrier penetration by a particle in quantum mechanics, in which the WKB approximation is used to compute the tunneling probability, and only after the particle escapes can it be described as freely propagating.

[18] It can again be proved that the fluctuations around the bounce solution have one and only one negative eigenvalue; see Section 6.5 of Coleman (1977b).

[19] Observe that, even if V is negative for some values of the field, it can be shown that $S_b > 0$. Indeed, consider the family of functions $\phi_\lambda(x) = \phi_b(x/\lambda)$. Rescaling the integration variable $x \to \lambda x$ in the action, we get

$$S[\phi_\lambda] = \int d^4x \left[\frac{\lambda^2}{2}(\partial_\mu\phi_b)^2 + \lambda^4 V(\phi_b)\right]. \qquad (22.108)$$

Since ϕ_b is a solution of the equations of motion, it is an extremum in the space of all functions, so in particular within the family of function $\phi_\lambda(x)$. Thus, the action (22.108) must be stationary at $\lambda = 1$. This gives

$$\int d^4x\, V(\phi_b) = -(1/4)\int d^4x\, (\partial_\mu\phi_b)^2\,, \qquad (22.109)$$

and therefore

$$S_b = \frac{1}{4}\int d^4x\, (\partial_\mu\phi_b)^2 > 0\,, \quad (22.110)$$

see Section 6.2 of Coleman (1977b). Observe that this is a generalization of the result (22.85), (22.86) valid in the quantum-mechanical case.

[20] See Coleman (1977b), eq. (6.47).

Tunneling in QFT at finite temperature

We can finally generalize this formalism to finite-temperature QFT, which is the situation relevant to a phase transition in the early Universe. A first difference, with respect to the previous computation, is that the potential $V(\phi)$ must now be replaced by the effective potential at finite temperature, $V(\phi, T)$. Furthermore, we should not take the limit $\beta \to \infty$ in eq. (22.67), or in its field-theoretical generalization, since $\beta = 1/T$ is the temperature of the system. Thus, the tunneling rate is now obtained from the finite-temperature path integral

$$\int \mathcal{D}\phi \, e^{-S_4}, \tag{22.112}$$

where

$$S_4[\phi] = \int_0^{1/T} d\tau \int d^3x \left[\frac{1}{2}\partial_\mu\phi\partial_\mu\phi + V(\phi, T)\right]. \tag{22.113}$$

The boundary conditions in the path integral are $\phi(0, \mathbf{x}) = \phi(1/T, \mathbf{x})$, and we have finally set $\hbar = 1$. In the limit $T \to 0$ we recover the results of the previous subsection. Here we are rather interested in the opposite limit, in which T is large compared with $1/R$, where R is the typical bubble radius. In this limit the dependence of ϕ on Euclidean time becomes negligible, since the overall range of Euclidean times becomes small, and we can write

$$S_4[\phi] \simeq \frac{1}{T}S_3[\phi], \tag{22.114}$$

where

$$S_3[\phi] = \int d^3x \left[\frac{1}{2}(\partial_i\phi)^2 + V(\phi, T)\right] \tag{22.115}$$

and $\phi = \phi(\mathbf{x})$. In this limit it is natural to look for a solution of the equation of motion with $O(3)$ symmetry, $\phi = \phi_b(r)$, where $r = |\mathbf{x}|$.[21] For an $O(3)$-symmetric solution $\phi(r)$, the action S_3 reduces to

$$S_3 = 4\pi \int_0^\infty dr \, r^2 \left[\frac{1}{2}\left(\frac{d\phi}{dr}\right)^2 + V(\phi, T)\right], \tag{22.116}$$

and the corresponding equation of motion is

$$\frac{d^2\phi}{dr^2} + \frac{2}{r}\frac{d\phi}{dr} = V'(\phi, T). \tag{22.117}$$

The boundary conditions are now

$$\lim_{r\to\infty} \phi(r) = 0, \tag{22.118}$$

in order to have a finite action, and

$$\left.\frac{d\phi}{dr}\right|_{r=0} = 0, \tag{22.119}$$

[21]More precisely, at a given T one can have both solutions of the equations of motion with $O(4)$ symmetry and solutions with $O(3)$ symmetry. The solution with the lowest action dominates the path integral. At low T the $O(4)$-symmetric solution will dominate, and will reproduce the result of the previous subsection, while above a critical temperature the $O(3)$-symmetric solution will dominate.

in order to ensure regularity at the origin. There are now only three zero modes, corresponding to the invariance under spatial translation, so Γ is proportional to V. The factor $1/\beta = T$ in eq. (22.69) is no longer canceled, but this is not a problem since now T is finite. Furthermore, each of the three zero modes contributes to the prefactor with a factor $[S_4/(2\pi)]^{1/2} = [S_3/(2\pi T)]^{1/2}$, so overall they give a factor $[S_3/(2\pi T)]^{3/2}$. Thus, the tunneling rate Γ is given by

$$\frac{\Gamma}{V} = A\, e^{-S_{3,b}(T)/T}\,, \tag{22.120}$$

where $S_{3,b}(T)$ is the action $S_3[\phi, T]$ evaluated on the bounce solution $\phi_b(r)$, and the prefactor A is given by

$$A = T \left(\frac{S_3(T)}{2\pi T} \right)^{3/2} \left| \frac{\det'[-\boldsymbol{\nabla}^2 + V''(\phi_b)]}{\det[-\boldsymbol{\nabla}^2 + V''(0)]} \right|^{-1/2}. \tag{22.121}$$

A precise evaluation of the determinants in eq. (22.121) is in general difficult, even numerically. However, for most purposes it is sufficient to observe that (in units $\hbar = c = 1$) A has dimensions of $(\text{mass})^4$, as is clear from the fact that $\Gamma/V \sim (\text{mass})^4$.[22] In the high-temperature limit the only relevant mass scale becomes T, so we must have $A = cT^4$, with c a number of order 1. Thus, finally,

$$\frac{\Gamma}{V} \simeq cT^4\, e^{-S_{3,b}(T)/T}\,. \tag{22.122}$$

At time t the rate of bubble nucleation in a Hubble volume $H^{-3}(t)$ is $(\Gamma/V) \times H^{-3}(t)$. The time t_* at which the phase transition takes place can then be obtained by requiring that the total number of bubbles nucleated from time $t = 0$ to $t = t_*$ is of order 1,

$$\int_0^{t_*} dt\, \frac{\Gamma}{V H^3(t)} = O(1)\,. \tag{22.123}$$

We can transform the integral with respect to time into an integral with respect to temperature using $T \propto 1/a$, which gives $dT/T = -H dt$. For transitions taking place in RD, which is the case of interest, using eq. (17.112) for $H^2(t_*)$ and setting $c = O(1)$, we find that the temperature T_* at which the phase transition takes place is determined by the condition

$$\int_{T_*}^{\infty} \frac{dT}{T} \left(\frac{45}{4\pi^3 g_*(T)} \right)^2 \left(\frac{M_{\mathrm{Pl}}}{T} \right)^4 e^{-S_3(T)/T} = O(1)\,. \tag{22.124}$$

Once T_* has been determined in this way, the parameter α is given by eqs. (22.63) and (22.64). The information on the transition rate is in principle fully contained in eq. (22.122). However, it can be useful to

---

[22] Equivalently, we can directly observe that, in eq. (22.121), $\det'$ differs from $\det$ by the omission of three eigenvalues corresponding to the three zero modes, so

$$\frac{\det'[-\boldsymbol{\nabla}^2 + V''(\phi_b)]}{\det[-\boldsymbol{\nabla}^2 + V''(0)]}$$

has dimensions $1/\lambda^3$, where λ is the dimension of the eigenvalues. The latter is the same as the dimension of the operator $\boldsymbol{\nabla}^2$, i.e. $(\text{mass})^2$, so

$$\left| \frac{\det'[-\boldsymbol{\nabla}^2 + V''(\phi_b)]}{\det[-\boldsymbol{\nabla}^2 + V''(0)]} \right|^{-1/2} \sim (\text{mass})^3.$$

Taylor-expand this expression around the transition time t_*, expressing T as a function of time, $T = T(t)$, and writing

$$\frac{S_{3,b}(T)}{T} \simeq \left(\frac{S_{3,b}(T)}{T}\right)_{t=t_*} + (t - t_*)\frac{d}{dt}\left(\frac{S_{3,b}(T)}{T}\right)_{t=t_*}. \qquad (22.125)$$

We then define

$$\beta = -\frac{d}{dt}\left(\frac{S_{3,b}(T)}{T}\right)_{t=t_*}, \qquad (22.126)$$

so, in a small interval of times near $t = t_*$, eq. (22.122) becomes

$$\Gamma(t) \simeq \Gamma_* e^{\beta(t-t_*)}. \qquad (22.127)$$

Again using $dT/T = -Hdt$, eq. (22.126) becomes

$$\boxed{\frac{\beta}{H_*} = T_*\frac{d}{dT}\left(\frac{S_{3,b}(T)}{T}\right)_{T=T_*}.} \qquad (22.128)$$

Notice that, in order to compute α, it is sufficient to know the transition temperature T_* and the potential $V(\phi, T)$, while to compute β one must also know how the action $S_{3,b}(T)$ depends on T in a neighborhood of T_*.[23]

Equations (22.124) and (22.128) allows one in principle to perform an accurate numerical computation of T_* and of β. We can, however, also obtain an analytic understanding of typical orders of magnitude that one would obtain for β. The integral in eq. (22.124) will in general be dominated by its value at $T = T_*$, so in a first approximation

$$\left(\frac{45}{4\pi^3 g_*(T_*)}\right)^2 \left(\frac{M_{\rm Pl}}{T_*}\right)^4 e^{-S_3(T_*)/T_*} = O(1). \qquad (22.129)$$

Setting $g_*(T_*)$ to its high-temperature value in the Standard Model, $g_* \simeq 106.75$, this gives

$$\frac{S_3(T_*)}{T_*} \simeq 4\log\frac{M_{\rm Pl}}{T_*} - 11.4. \qquad (22.130)$$

For a transition at the electroweak scale, $T_* = O(100\,{\rm GeV})$ and we get $S_3(T_*)/T_* \simeq 145$, while, for the QCD scale, $S_3(T_*)/T_* \simeq 173$. In general, barring fine tuning, we expect that $T_* d/dT[S_3(T)/T]_{T_*}$ will be of the same order of magnitude as $S_3(T_*)/T_*$, so as a rough estimate we expect $\beta/H_* = O(100)$, both at the electroweak scale and at the QCD scale. This estimate is roughly confirmed by the explicit numerical computations. Of course, precise numbers can only obtained with an accurate numerical evaluation of eq. (22.128), given the potential $V(\phi, T)$, and values of β/H_* as low as $O(10)$ or as high as $O(10^3)$ can be obtained in different models. In general α, β and T_* are correlated and, in typical models, as α increases, so that the energy released in the transition increases, β and T_* decrease.

[23]It should also be stressed that the whole formalism that we have discussed computes the nucleation rate assuming that quantum fluctuations in the thermal environment are responsible for the bubble nucleation. It is in principle possible that different mechanisms are at work, in which bubbles are nucleated by various forms of impurities and inhomogeneities of non-thermal origin. For instance, cosmic strings could act as nucleation sites, as could inhomogeneities due to some pre-existing turbulence; see Witten (1984). In this case β could be very different, and very difficult to predict.

Finally we observe that, in the thin-wall limit, eq. (22.116) allows us to formalize the physical intuition, discussed at the beginning of this section, that the dynamics of true vacuum bubbles is determined by a competition between the volume energy gained by tunneling to the true vacuum, and the effect of surface tension, which tends to shrink the bubbles to zero. Indeed, in the thin-wall limit the bubble configuration has $\phi \simeq \phi_0$ for $r < R$, and $\phi \simeq 0$ and $V(\phi) \simeq 0$ for $r > R$, with the transition taking place in a small range δr near $r = R$. Then the term proportional to the potential in eq. (22.116) is negative and proportional to the bubble volume,

$$4\pi \int_0^\infty dr\, r^2\, V(\phi, T) \simeq 4\pi \int_0^R dr\, r^2\, V(\phi, T)$$

$$= -|\bar{V}(T)|\frac{4\pi R^3}{3}, \qquad (22.131)$$

where $\bar{V}(T) < 0$ is the average of the potential inside the bubble volume, which is of order $V(\phi_0(T), T)$. The term proportional to $(d\phi/dr)^2$ in eq. (22.116) can instead be simplified by observing that, in the thin-wall limit, $(d\phi/dr)$ is sharply peaked around $r = R$, so

$$4\pi \int_0^\infty dr\, r^2 \frac{1}{2}\left(\frac{d\phi}{dr}\right)^2 \simeq 4\pi R^2 \int_0^\infty dr\, \frac{1}{2}\left(\frac{d\phi}{dr}\right)^2$$

$$\equiv 4\pi R^2 S_1(T). \qquad (22.132)$$

So, this term is positive and proportional to the bubble surface, and we recognize the contribution from the surface tension. Thus, in the thin-wall approximation, on a bubble of radius R,

$$S_3(R) = -|\bar{V}(T)|\frac{4\pi R^3}{3} + 4\pi R^2 S_1(T). \qquad (22.133)$$

The bounce solution is an extremum of this action in the whole space of field configurations, so in particular with respect to the parameter R. The bubble radius R in the thin-wall approximation can therefore be determined by minimizing eq. (22.133) with respect to R, which gives

$$R(T) = \frac{2S_1(T)}{|\bar{V}(T)|}. \qquad (22.134)$$

22.4.4   Bubble dynamics and GW production

In a first-order phase transition, once these supercritical bubbles have nucleated, their subsequent dynamics depends on their environment. Conceptually, the simplest situation is that of vacuum bubbles, i.e. bubbles that are assumed to expand in the vacuum, neglecting their interaction with the cosmic fluid. In that case, the bubbles expand until their walls move at a speed close to the speed of light. The energy gained in the transition from the metastable state to the ground state is transferred to the kinetic energy of the bubble wall. As the bubble expands, more and more regions of space convert to the ground state, and the wall becomes more and more energetic. At the same time, it also becomes thinner, and therefore the energy density stored in the wall increases very rapidly. Of course, as long as we have a single spherical bubble, this large energy cannot be converted to GWs. However, when two bubbles collide, spherical symmetry is broken and GW production can take place.

More realistically, in the cosmological setting in which we are eventually interested, the bubbles interact with the cosmic fluid. This interaction can in principle be quite complex, and results in friction over the bubble wall, which slows down and changes the bubble evolution, while at the same time the cosmic fluid is stirred by the expanding bubbles, with the generation of turbulent motions and acoustic waves. Indeed, three main mechanisms for GW production have been identified, namely GWs from bubble collisions, from turbulence and from colliding acoustic waves in the fluid. We examine them in turn.

GWs from bubble collisions

Before discussing the results obtained in the literature from numerical computations, let us first estimate the orders of magnitude involved, for the peak frequency and the amplitude of the stochastic background produced by bubble collisions.

At the time of production, the characteristic frequency scale of the process is given by the parameter β that appears in eq. (22.127); thus the typical frequency f_* at the time of production can be estimated as $2\pi f_* \sim \beta$.[24] Inserting this value into eq. (22.6) and setting $g_* \simeq 106.75$, as is appropriate at the electroweak scale, the frequency of the GWs produced, redshifted to the present epoch, can then be estimated to be peaked at

$$f_{\text{peak}} \sim \text{a few} \times 10^{-6} \left(\frac{\beta}{H_*} \right) \left(\frac{T_*}{100\,\text{GeV}} \right) \text{Hz} \,. \qquad (22.135)$$

Thus, if there is a first-order phase transition at the electroweak scale $T_* = O(100)$ GeV, for values of β/H_* in the typical range $10 - 10^3$ we find that f_0 is in the range $10^{-4} - 10^{-2}$ Hz, which is just in the frequency band of the LISA space-interferometer.

Similarly, if there is a first-order phase transition in QCD at $T_* \simeq 150$ MeV, we get

$$f_{\text{peak}} \sim \text{a few} \times 10^{-9} \left(\frac{\beta}{H_*} \right) \left(\frac{T_*}{150\,\text{MeV}} \right) \text{Hz} \,. \qquad (22.136)$$

If β/H_* is not too large, this could potentially be within the frequency window of pulsar timing arrays, as we will see in Chapter 23.

To get a first rough order-of-magnitude estimate of the value of $\Omega_{\text{gw}}(f)$ at the peak frequency, we can observe that the typical wavelength λ_* of the radiation, at the time of production, is of the order of the typical radius R_* of the bubbles when they collide. Then, the energy liberated in

[24]Actually, when the bubble velocity v_b is not close to the speed of light, there are two different time- or length-scales in the problem. One is the typical time-scale of the phase transition, β^{-1}, which can also identified with the duration of the phase transition. The other is the typical size of the bubbles near the end of the phase transition, when most bubble collisions take place, which is rather given by $R_* = v_b\beta^{-1}$. One should then ask which of these two scales is more appropriate for estimating the peak frequency through eq. (22.6). The answer is quite subtle, and depends on the detailed time evolution of the energy–momentum tensor of the source. It turns out that, in a first approximation, β^{-1} is appropriate for GWs from bubble collisions, while $v_b\beta^{-1}$ is appropriate for turbulence; see the Further Reading. A fit to the dependence on v_b for bubble collisions will be given in eqs. (22.144) and (22.145).

GWs in the collision of two bubbles is of order $E_{GW} \sim GM_B^2/\lambda_*$, where $M_B \sim \rho_{\rm vac}\lambda_*^3$ is the energy of a typical bubble and $\rho_{\rm vac}$ is the difference in energy density from the metastable state to the ground state. The fraction of the false-vacuum energy that goes into GWs instead of going into radiation is given by

$$\frac{\Omega_{\rm gw}}{\Omega_R} \sim \frac{E_{\rm GW}}{M_B}.\tag{22.137}$$

This is, however, valid only for bubble collisions in vacuum, where the energy gained in the transition to the true vacuum can only go into the kinetic energy of the bubble walls, i.e., at the fundamental level, into the kinetic energy of the scalar field. For bubbles evolving in a primordial plasma, the underlying physics is quite different. Indeed, since the energy gained in the transition grows with the bubble radius as R^3, while the energy stored into the bubble walls only grows as R^2, in the presence of other channels only a tiny fraction of the vacuum energy will go into the scalar field. Most of the energy goes into the primordial fluid, where part of it is dissipated. We must therefore insert an efficiency factor κ, which gives the fraction of the available energy that actually contributes to GW generation, and replace M_B by κM_B. Then

$$E_{GW} \sim \kappa^2\frac{GM_B^2}{\lambda_*}.\tag{22.138}$$

Therefore

$$\frac{E_{\rm GW}}{M_B} \sim \kappa^2\frac{GM_B}{\lambda_*} \sim \kappa^2 G\rho_{\rm vac}\lambda_*^2.\tag{22.139}$$

Using $G\rho_{\rm vac} \sim H_*^2$, we get, at the time of production

$$\Omega_{\rm gw} \sim \kappa^2\lambda_*^2 H_*^2\Omega_R.\tag{22.140}$$

For a phase transition with characteristic time-scale β we have $\lambda_* = 1/f_* \sim 1/\beta$, and therefore we can also rewrite it as

$$\Omega_{\rm gw} \sim \kappa^2\left(\frac{H_*}{\beta}\right)^2\Omega_R.\tag{22.141}$$

Despite the crudeness of the approximation, this relation allows us to get a first feeling for the numbers involved. Equation (22.141) suggests that, even in the most optimistic case $\kappa = 1$ (which, as we will see, is far from being realized in realistic situations), for typical values of β/H_* in the range range $10 - 10^3$, $\Omega_{\rm gw}$ at the peak ranges from a (very optimistic) $10^{-2}\Omega_R \sim 10^{-7}$ to $10^{-6}\Omega_R \sim 10^{-11}$. As we will see, however, the efficiency factor κ and other effects can significantly lower these estimates. It is also interesting to observe that the dependence on $\epsilon_* = \lambdabar_* H_*$ in eqs. (22.6) and (22.140) is such that, as ϵ_* decreases, the peak frequency increases as $1/\epsilon_*$ while the amplitude at the peak decreases as ϵ_*^2.

To compute the GW spectrum more quantitatively, one can begin by considering the collision of two vacuum bubbles. This can be simulated

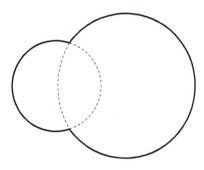

Fig. 22.11 The envelope approximation for bubble collision.

[25]See Huber and Konstandin (2008a).

numerically, studying the dynamics of the underlying scalar field and computing the GW production from the time-evolving energy–momentum tensor of the scalar fields. For bubbles evolving in vacuum, it turns out that the numerical results are well reproduced, with an accuracy of about 20%, by the so-called *envelope approximation*. This is an analytic approximation in which all the energy is assumed to be stored in the bubble walls, assumed to be thin, and, as the two bubbles collide, one uses as a source for GW production the energy–momentum tensor of the uncollided part of the bubble walls, represented by the thick lines in Fig. 22.11, while the complicated overlap region between the colliding bubbles is simply neglected. This approximation can then be used to study the GWs produced by nucleating at random points in space an ensemble of bubbles. The number of bubbles nucleated in a time step δt is drawn from a Poisson distribution whose mean is determined by the bubble nucleation rate $\Gamma(t)$ computed in Section 22.4.3, and the results are then averaged over several realization of the nucleation process. The envelope approximation allows one to compute the cumulative GWs produced by collisions in a system with hundreds of bubbles, which would be unfeasible if one had to evolve numerically the profile of the scalar field of each bubble. With this combination of analytic and numerical techniques, Kamionkowski, Kosowsky and Turner (1994) found that the peak value of $h_0^2 \Omega_{\rm gw}$ is well fitted by

$$h_0^2 \Omega_{\rm gw}(f_{\rm peak}) \simeq 1.1 \times 10^{-6} \kappa^2 \left(\frac{H_*}{\beta}\right)^2 \left(\frac{\alpha}{1+\alpha}\right)^2 \left(\frac{v_b^3}{0.24 + v_b^3}\right) \left(\frac{100}{g_*}\right)^{1/3},$$
(22.142)

where α, defined in eq. (22.64), determines the strength of the phase transition, and v_b is the average velocity of the bubble walls at collision. Comparing with eq. (22.141) we see that the crude estimate leading to eq. (22.141) correctly reproduces the dependence on κ and on H_*/β, and even gives the correct order of magnitude of the overall numerical factor, in the limit of strong first-order phase transitions, $\alpha \gg 1$, and relativistic bubble velocities. If α or v_b are small, however, there are further suppression factors. Equation (22.142) has been widely used in the literature. Later, numerical work with increased accuracy[25] has confirmed the basic features of this result and provided a more accurate fitting formula,

$$h_0^2 \Omega_{\rm gw}(f_{\rm peak}) \simeq 1.84 \times 10^{-6} \kappa^2 \left(\frac{H_*}{\beta}\right)^2 \left(\frac{\alpha}{1+\alpha}\right)^2 \left(\frac{v_b^3}{0.42 + v_b^2}\right) \left(\frac{100}{g_*}\right)^{1/3},$$
(22.143)

while the peak frequency is well fitted by

$$f_{\rm peak} \simeq 3.83 \times 10^{-6}\,{\rm Hz}\, \left(\frac{\beta}{H_*}\right) \left(\frac{T}{100\,{\rm GeV}}\right) \left(\frac{g_*}{106.75}\right)^{1/6} g(v_b),$$
(22.144)

where

$$g(v_b) \simeq \frac{2.7}{1.8 - 0.1v_b + v_b^2} \qquad (22.145)$$

is a function of v_b, which monotonically decreases from $g(0) = 1.5$ to $g(1) = 1$. We see that the approximate estimate (22.135) is consistent with the numerical result (22.144).[26]

For QCD we rather use as a reference value $g_* = 64.75$, which takes into account all degrees of freedom that are relativistic just above the transition (including the strange quark),[27] and setting $T_* = 150$ MeV. Then eq. (22.144) reads

$$f_{\text{peak}} \simeq 5.28 \times 10^{-9} \, \text{Hz} \left(\frac{\beta}{H_*}\right) \left(\frac{T}{150 \, \text{MeV}}\right) \left(\frac{g_*}{64.75}\right)^{1/6} g(v_b),$$

$$(22.146)$$

The latter numerical study also allowed determination of the frequency spectrum in the region near the peak. In general, at sufficiently low frequency, such that the corresponding wavelength is well outside the horizon, causality requires that $h_0^2 \Omega_{\text{gw}}(f) \propto f^3$; see eq. (22.3) and the discussion after it. Existing numerical studies of GW production in bubble collisions in the envelope approximation only explore the frequency region not very far from the peak, and do not reach the regime of super-horizon modes. For frequencies lower than the peak, but not yet in the super-horizon regime, the frequency dependence is well fitted by a power $h_0^2 \Omega_{\text{gw}}(f) \propto f^{2.8}$, not far from the f^3 behavior expected for the even lower frequencies corresponding to super-horizon modes. At frequencies higher than the peak, in contrast, the numerical study shows that $h_0^2 \Omega_{\text{gw}}(f)$ decays only as $1/f$. This relatively slow decay can be quite interesting from the observational point of view, since it means that a sizable signal could still be present in the bandwidth of a detectors, such as LISA, even when the peak of the spectrum lies at a lower frequency.

The next step is to compute κ and v_b in terms of the parameters of the phase transition. In the physically relevant case of bubbles evolving in the cosmic fluid this is a complicated problem, and the result depends on the detailed dynamics of the bubbles. In general, there are two possible "combustion" modes in the collision of two bubbles: detonation, which basically takes place when the boundaries propagate faster than the speed of sound, and deflagration, when instead they move slower. Only in the first case can we expect a large production of GWs. In deflagration there will be no large concentration of kinetic energy near the bubble walls, so GW production is suppressed. The bubble velocity has a relatively simple form in the case of Chapman–Jouguet detonation (i.e. when the detonation propagates at a speed such that, in the frame of the shock wave, the fluid velocity is exactly equal to the speed of sound). In this case one can show that v_b is fixed in terms of α by

$$v_b(\alpha) = \frac{\sqrt{1/3} + \sqrt{\alpha^2 + 2\alpha/3}}{1 + \alpha}. \qquad (22.147)$$

[26]See also Jinno and Takimoto (2017) for related analytic results in the thin-wall and envelope approximation.

[27]A listing of the degrees of freedom as a function of mass can be found in Table I of Watanabe and Komatsu (2006). As we saw in eq. (17.99), the contribution of fermions must be multiplied by a factor 7/8.

Observe also that, while for bubbles propagating and colliding in vacuum the source of GW production is the energy–momentum tensor of the scalar field, for bubbles propagating in the cosmic fluid by far the dominant contribution is given by the energy–momentum tensor $T_{ij} = (\rho + p)v_i v_j$ of the fluid displaced and stirred by the bubbles. In the envelope approximation one assumes again that only the fluid at the envelope contributes (an assumption to which we will return later). Assuming Jouguet detonation and the envelope approximation, one can estimate[28] that

$$\kappa(\alpha) = \frac{1}{1 + A\alpha} \left(A\alpha + \frac{4}{27}\sqrt{\frac{3\alpha}{2}} \right), \qquad (22.148)$$

[28]See again Kamionkowski, Kosowsky and Turner (1994)

where $A \simeq 0.715$. Equations (22.147) and (22.148) have been widely used in the literature, because of their relative simplicity. However, to model bubble evolution in a plasma more realistically it is also necessary to include terms representing friction between the bubble walls and the plasma. These have been parametrized phenomenologically by adding to the equation of motion of the scalar field a term proportional to $\eta u^\mu \partial_\mu \phi$, where u^μ is the plasma four-velocity and η a friction coefficient. Then, κ and v_b are also functions of η. In general, only for a limited range of values of the parameters α and η does the solution correspond to detonation, while the most probable outcome is either deflagration or a runaway solution where the walls continuously accelerate.[29]

[29]See Espinosa, Konstandin, No and Servant (2010) for more recent estimates of $\kappa(\alpha)$ in different regimes.

Finally, the values of α and β should in principle be derived from the fundamental theory under consideration, using the nucleation theory discussed in Section 22.4.3. This has been done for some selected models that could give a first-order phase transition at the electroweak scale, such as non-minimal supersymmetric extensions of the Standard Model; see the Further Reading. It is clear that, in the end, the computation of GW production from bubble collisions suffers from many uncertainties, first of all because at present these extensions of the Standard Model are unavoidably somewhat speculative and, on a more technical side, because of the uncertainties regarding bubble evolution in a realistic thermal environment. In any case, existing predictions at the electroweak scale in general do not exceed $h_0^2 \Omega_{\rm gw} \sim 10^{-12}$ at the peak frequency. Figure 22.12 shows the signal obtained for different typical values of the parameters α and β/H_*, comparing the result from the collision of two bubbles (left panel) with the result for the nucleation and collision of an ensemble of bubbles (right panel). The slower decay with frequency in the latter case is due to the cumulative effect from the late nucleation of small bubbles, which is absent in the two-bubble case. Observe that the signal is too small for a space interferometer such as LISA but, thanks to the slow $1/f$ decrease at high frequencies, for some values of the parameters it could be detectable at an advanced space interferometer such as DECIGO or BBO.

It is important to observe that, when the envelope approximation is applied to bubbles evolving in the primordial plasma, rather than in vacuum, we are actually assuming that the main source of GW gener-

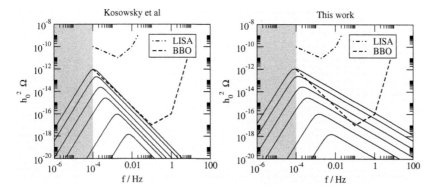

Fig. 22.12 $h_0^2\Omega_{\rm gw}(f)$ for several values of α and β/H_*. Left panel, using the numerical results from the collision of two bubbles studied in Kosowsky, Turner and Watkins (1992a, 1992b). Right panel, using the envelope approximation for an ensemble of bubbles, from the numerical study of Huber and Konstandin (2008b). The curves from top to bottom correspond to values of $(\alpha, \beta/H_*)$ equal to $(0.20, 30)$, $(0.15, 40)$, $(0.10, 60)$, $(0.07, 100)$, $(0.05, 300)$ and $(0.03, 1000)$, respectively. The corresponding transition temperature decreases from 130 GeV (top) to 70 GeV (bottom). From Huber and Konstandin (2008b). ©SISSA Medialab Srl. Reproduced by permission of IOP Publishing. All rights reserved.

ation is the energy–momentum tensor of the fluid near the envelope of the colliding bubbles. This is an assumption that can actually miss a potentially dominant contribution to GW production, as we will discuss later.

GWs from turbulence in the cosmic fluid

Collision of bubbles injects a large amount of energy into the cosmic fluid. At the electroweak epoch this fluid has a very high Reynolds number, of order 10^{13}, and therefore turbulent motion rapidly sets in. Furthermore, even the magnetic Reynolds number is very large, so magnetic fields generated during the phase transition are amplified and full magnetohydrodynamic (MHD) turbulence develops. When the phase transition terminates, so does the energy injection. However, because of the very high Reynolds number, MHD turbulence decays quite slowly, over several Hubble times, and therefore provides a long-lasting source of GWs. The peak frequency $f_{\rm peak}$ turns out to be determined by the size R_* of the bubbles at collisions, $k_{\rm peak} = 2\pi f_{\rm peak} \simeq 1/R_*$; see the discussion in Note 24 on page 676. At low frequencies one has a spectrum $h_0^2\Omega_{\rm gw}(k) \propto k^3$, as required by causality (at least in the limit of super-horizon modes). On the high-frequency side the GW spectrum decays as k^{-n}, but the value of n depends strongly on how turbulence is modeled, in particular on the power spectrum of the turbulent fluid motion and on whether the source is modeled as short-lasting or long-lasting. In the more appropriate case of a long-lasting source analytical estimates suggest $n = 5/3$ if the turbulent fluid motion has a Kolmogorov spectrum, and $n = 3/2$ for a Iroshnikov–Kraichnan spectrum. Probably, only full MHD numerical simulations can address the issue precisely. A fit to the numerical results for MHD turbulence is given by[30]

$$h_0^2 \Omega_{\mathrm{gw}}(f_{\mathrm{peak}})\big|_{\mathrm{turb}} \simeq 3.3 \times 10^{-5} \left(\frac{\kappa \alpha}{1+\alpha} \right)^{3/2} \left(\frac{H_*}{\beta} \right) v_b$$
$$\times \frac{(f/f_{\mathrm{peak}})^3}{([1 + (f/f_{\mathrm{peak}})]^{11/3}(1 + 4k_*/H_*)}, \qquad (22.149)$$

where k_* is the physical wavenumber at the time of production, and

$$f_{\mathrm{peak}} \simeq 2.7 \times 10^{-5}\,\mathrm{Hz}\,\frac{1}{v_b} \left(\frac{\beta}{H_*} \right) \left(\frac{T_*}{100\,\mathrm{GeV}} \right) \left(\frac{g_*}{100} \right)^{1/6}. \qquad (22.150)$$

These formulas include both the contribution to GW production from the turbulent kinetic motion of the fluid, and that due to the magnetic field, which are generated by the turbulent motion itself, and, for simplicity, assumes equipartition, i.e. the same amount of energy density in the turbulent kinetic motion and in magnetic fields.

In general, the GW spectrum produced by turbulence peaks at a higher frequency than that due to bubble collisions, because of the factor $1/v_b$ in eq. (22.150), and the overall amplitude can be somewhat larger. The two contributions are shown separately in Fig. 22.13, and their sum is shown in Fig. 22.14. We see that the lower peak from bubble collisions causes the hump on the left of the true peak of the spectrum, which could be a distinctive feature of this signal.

GWs from sound waves in the cosmic fluid

The picture discussed in the preceding subsections basically assumes that the primordial plasma can be described as an incompressible and turbulent fluid. Recently, it has become possible to test this assumption using fully three-dimensional simulations of bubble nucleation in the coupled system of the scalar field and the fluid, and to compute the resulting GW production. The numerical simulations indicate that sound waves in the fluid, created by the bubble collisions, are important. These are completely missed in the envelope approximation, where the fluid on the colliding part of the bubble walls (the dashed lines in Fig. 22.11) is simply neglected. These sound waves, which describe coherent motions of the fluid, are, however, a potentially important source of GWs. Furthermore, they last for a long time after the phase transition has been completed. This boosts their GW signal by the ratio of their lifetime, which can be of the order of the Hubble time, to the duration of the transition, i.e. by a factor H_*^{-1}/β^{-1}. For a typical value $\beta/H_* \sim 10^2$, the resulting GW signal can then be two orders of magnitudes larger than that due to bubble collisions.

To describe the coupled field–fluid system one starts from the energy–

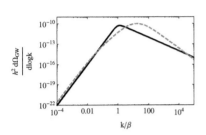

Fig. 22.13 The GW spectra from bubble collisions (solid) and MHD turbulence (dashed) for $v_b = 0.7$, $\kappa \alpha/(1 + \alpha) = 0.1$, $\beta/H_* = 10$ and $T_* = 100$ MeV (as appropriate for a phase transition at the QCD scale). From Caprini, Durrer and Siemens (2010).

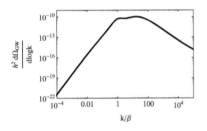

Fig. 22.14 The total GW spectrum resulting from the sum of the two contributions in Fig. 22.13. From Caprini, Durrer and Siemens (2010).

momentum tensor of the total system,

$$T_{\text{tot}}^{\mu\nu} = T_{\text{field}}^{\mu\nu} + T_{\text{fluid}}^{\mu\nu} \tag{22.151}$$

where

$$T_{\text{field}}^{\mu\nu} = \partial^\mu \phi \partial^\nu \phi - \frac{1}{2} \eta^{\mu\nu} (\partial \phi)^2 \,, \tag{22.152}$$

and

$$T_{\text{fluid}}^{\mu\nu} = (\rho + p)u^\mu u^\nu + \eta^{\mu\nu} p \,, \tag{22.153}$$

where p is the pressure in the rest frame, ρ is the energy density, and u^μ the four-velocity of the fluid; compare with eq. (10.30). Over the relatively short time-scale of the phase transition one can in a first approximation neglect the expansion of the Universe and use the flat-space metric. The fundamental interaction between the field ϕ responsible for the bubble profile and the matter fields composing the primordial plasma results in an effective friction on the bubble. To describe this field–fluid interaction phenomenologically it is common to split the energy–momentum conservation equation $\partial_\mu T_{\text{tot}}^{\mu\nu} = 0$ as

$$\partial_\mu T_{\text{field}}^{\mu\nu} = \delta^\nu \,, \tag{22.154}$$

$$\partial_\mu T_{\text{fluid}}^{\mu\nu} = -\delta^\nu \,, \tag{22.155}$$

where

$$\delta^\nu = \eta u^\mu \partial_\mu \phi \partial^\nu \phi \tag{22.156}$$

and η is an adjustable friction parameter. Of course this splitting, and the precise choice of δ^ν, are somewhat arbitrary, but this latter term, proportional to the fluid four-velocity, is believed to give a good phenomenological description of the friction-induced interaction between the field and the fluid. Starting from a system in the symmetric phase, one models as usual bubble nucleation with the rate given in eq. (22.65), and from this one generates a set of nucleation times and locations, at each of which one inserts a static bubble with a Gaussian profile for the scalar field. One can then evolve the system numerically on a lattice, and compute the time evolution of the energy–momentum tensor, as well as the resulting GW production. In this way one finds[31] that the GW energy Ω_{GW} from acoustic waves is related to the value $\Omega_{\text{GW}}^{\text{ea}}$ obtained in the envelope approximation by

[31]See Hindmarsh, Huber, Rummukainen and Weir (2014, 2015).

$$\frac{\Omega_{\text{GW}}}{\Omega_{\text{GW}}^{\text{ea}}} \gtrsim 60 \, C_{\text{GW}} \left(\frac{\beta}{H_*} \right) \,, \tag{22.157}$$

where C_{GW} is a dimensionless parameter. The numerical simulation gives $C_{\text{GW}} \sim 0.04$. Since β/H_* can easily be of order 10^2, this "acoustic" production mechanism can give a signal orders of magnitude larger than the standard estimates based on the envelope approximation. The peak frequency is estimated to be comparable to the value (22.150) obtained from turbulence. One should keep in mind, however, that these results were obtained in simulations with values of $\alpha \lesssim 0.1$ and small fluid velocities. The extrapolation to higher values of α might be non-trivial, with the possible conversion of part of this acoustic signal into the turbulent one discussed earlier.[32]

[32]See Section 2.1.2 of Caprini *et al* (2016).

22.5   Cosmic strings

22.5.1   Global and local strings

Cosmic strings are configurations of scalar and/or gauge fields whose energy density is concentrated along a one-dimensional line, either an infinite line or a closed loop. Such objects might have formed in the early Universe during a symmetry-breaking phase transitions (independently of the order of the phase transition) in extensions of the Standard Model. The prototype example is given by the breaking of a $U(1)$ symmetry, due to a complex scalar field Φ whose potential, in the broken phase, has the standard Mexican hat potential,

$$V(\Phi) = \frac{1}{4}\lambda \left(|\Phi|^2 - \eta^2 \right)^2 . \tag{22.158}$$

If we write $\Phi = |\Phi|e^{i\psi}$, then at the minimum of the potential we have $|\Phi| = \eta$, but the phase ψ is undefined. Therefore there is not just a single vacuum state, but a whole manifold $\mathcal{M}$ of vacua, which in this case is just a circle, parametrized by an angle ψ, with $0 \leqslant \psi \leqslant 2\pi$. Since Φ is single-valued, as we go around a closed path in physical space, the phase ψ must change by $\Delta\psi = 2\pi n$, where $n = 0, \pm 1, \pm 2, \dots$. If $n \neq 0$, as we go around the circle in physical space, the phase ψ wraps one or more times around this circle in the field space. We then have a configuration that is topologically non-trivial, and cannot be deformed continuously into the topologically trivial configuration $\psi = 0$.[33] Consider now a circle $\mathcal{C}_1$ in physical space, such that along it $\Delta\psi = 2\pi$, and study what happens if we shrink it to a point or, more precisely, to an infinitesimally small circle $\mathcal{C}_2$ with $\Delta\psi = 0$. In order for n to change discontinuously from $n = 1$ to $n = 0$ it is necessary that, in the process, a singularity be encountered, i.e. we must encounter at least a point where the phase ψ is undefined, which happens when the modulus $|\Phi| = 0$. This means that at least one string of false vacuum must be caught inside a loop with $\Delta\psi = 2\pi$; see Fig. 22.15. Furthermore, the string must either be infinitely long or be a loop that winds around the circle $\mathcal{C}_1$, otherwise we would be able to unwind the circle $\mathcal{C}_1$ from this line of singularities. Of course, in a cosmological setting, an infinite loop should rather be seen as an idealization of a closed loop with size much larger than the horizon. At the string location we have $\Phi = 0$, so we are in the false vacuum, and there is an associated energy density. Thus, we have a concentration of energy density along a string.

To be more quantitative, we have to specify the Lagrangian, and in particular we will see that crucial differences emerge depending on whether the $U(1)$ symmetry that is broken is a global or a local symmetry. The corresponding strings are called global or local strings, respectively. Consider first a theory with a global $U(1)$ symmetry, with just a complex scalar field Φ and Lagrangian density

$$\mathcal{L} = -\partial_\mu \Phi \partial^\mu \Phi^* - V(\Phi), \tag{22.159}$$

[33]More formally, the existence of this topologically non-trivial configuration is due to the fact that the first homotopy group of the vacuum manifold, $\pi_1(\mathcal{M})$, is non-trivial. The first homotopy group classifies the mappings $S^1 \to \mathcal{M}$ from a circle to the vacuum manifold $\mathcal{M}$, which for a $U(1)$ group is itself a circle S^1.

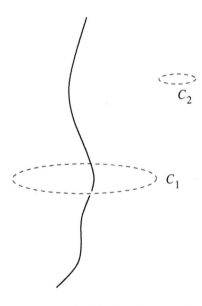

Fig. 22.15 A infinite string caught inside the circle $\mathcal{C}_1$.

with $V(\Phi)$ given by eq. (22.158) (we limit ourselves for the moment to flat space). For $\eta \neq 0$ we can write

$$\Phi(x) = \left(\eta + \frac{\varphi(x)}{\sqrt{2}}\right) e^{i\psi(x)}, \qquad (22.160)$$

and we get

$$\mathcal{L} = -\frac{1}{2}\partial_\mu\varphi\partial^\mu\varphi - \frac{\lambda}{4}\left(\sqrt{2}\,\eta\varphi + \frac{1}{2}\varphi^2\right)^2 - \left(\eta + \frac{\varphi}{\sqrt{2}}\right)^2 \partial_\mu\psi\partial^\mu\psi$$
$$= \mathcal{L}_2 + \mathcal{L}_{\text{int}}, \qquad (22.161)$$

where the quadratic term is

$$\mathcal{L}_2 = -\frac{1}{2}\partial_\mu\varphi\partial^\mu\varphi - \frac{1}{2}m_s^2\varphi^2 - \eta^2\partial_\mu\psi\partial^\mu\psi, \qquad (22.162)$$

with $m_s^2 = \lambda\eta^2$, and $\mathcal{L}_{\text{int}}$ contains the interaction terms. We therefore have a massless scalar field ψ, which is the Goldstone boson associated with the breaking of the global $U(1)$ symmetry, and a massive scalar field φ with mass m_s.

The equation of motion derived from the Lagangian (22.159) is

$$\Box\Phi = \frac{\lambda}{2}\left(|\Phi|^2 - \eta^2\right)\Phi. \qquad (22.163)$$

We can look for a static solution representing a string along the z axis, of the form

$$\Phi(\rho, \theta) = \eta f(m_s\rho)e^{in\theta}, \qquad (22.164)$$

where (ρ, θ) are polar coordinates in the (x, y) plane, and the integer n is the winding number of the string. Observe that, in terms of the phase ψ of the field Φ, this ansatz means $\psi(\rho, \theta) = n\theta$. Substituting into the equations of motion, one finds that near the origin $f(m_s\rho) \propto \rho^{|n|}$ while, for $m_s\rho \gg 1$,

$$f(m_s\rho) \simeq 1 - O\left(\frac{1}{m_s^2\rho^2}\right). \qquad (22.165)$$

The energy density of this configuration is

$$\mathcal{E} = \partial_i\Phi\partial_i\Phi^* + V(\Phi)$$
$$= |\partial_\rho\Phi|^2 + \frac{1}{\rho^2}|\partial_\theta\Phi|^2 + V(\Phi). \qquad (22.166)$$

Using the asymptotic form of $f(m_s\rho)$ we see that the energy density is peaked around $\rho \sim m_s^{-1}$, and vanishes at large distances as $1/\rho^2$. The $1/\rho^2$ behavior at infinity is due to the $\partial_\theta\Phi$ term in eq. (22.166). Indeed, $\partial_\theta\Phi = in\eta f(m_s\rho)e^{in\theta}$, so at large distances $|\partial_\theta\Phi| \to n\eta$. Then, as $\rho \to \infty$

$$\mathcal{E}(\rho, \theta) \to \frac{n^2\eta^2}{\rho^2}. \qquad (22.167)$$

Since the energy density is localized in a region of width $\sim m_s^{-1}$ (which is the Compton wavelength of the massive state described by φ) and goes to zero as $1/\rho^2$ at infinity, this configuration describes a localized string. However, when we integrate $\mathcal{E}$ over $dx dy = \rho \, d\rho \, d\theta$ to get the energy of the configuration per unit length, the $1/\rho^2$ behavior is not integrable, and the energy per unit length diverges logarithmically. The string tension μ, i.e. the energy per unit length, is given by

$$\mu \simeq 2\pi n^2 \eta^2 \log(m_s R) \,, \tag{22.168}$$

where R is the infrared cutoff, and we have taken m_s^{-1} as the order of magnitude of the distance where the asymptotic behavior is no longer valid and the behavior $f(m_s \rho) \propto \rho^{|n|}$ sets in. This infrared divergence is a reflection of the existence in the spectrum of the Goldstone boson θ, which, being massless, induces long-range interactions between different parts of the string. In a condensed matter system the infrared divergence is cut off by the finite volume of the system. Indeed, global strings of this type are observed in condensed matter, in particular as quantized vortex lines in liquid helium. More generally, in a system with several global strings the cutoff will be provided by the average distance between strings, and for a string forming a loop this will the loop size itself.

Consider now a model with spontaneous breaking of a *local $U(1)$* symmetry. The prototype example is the *Abelian Higgs model*, with Lagrangian density

$$\mathcal{L} = -|D_\mu \Phi|^2 - V(\Phi) - \frac{1}{4} F_{\mu\nu} F^{\mu\nu} \,, \tag{22.169}$$

where again $V(\Phi)$ is given by eq. (22.158) and the covariant derivative is given by $D_\mu \Phi = (\partial_\mu - i g A_\mu)\Phi$. The particle spectrum now consists of a massive Higgs field with mass $m_s = \lambda^{1/2}\eta$, as before, plus a massive gauge boson with mass $m_v = g\eta$. As usual in the Higgs mechanism, the Goldstone boson ψ that appears in the model with global symmetry breaking has been turned into the longitudinal component of the massive gauge field.

This system again has static string-like solutions, which have the form

$$\Phi(\rho, \theta) = \eta f(\rho) e^{in\theta} \,, \tag{22.170}$$

$$A_A = -\epsilon_{AB} x_B \frac{n}{g\rho^2} \alpha(\rho) \,, \tag{22.171}$$

where $A, B = 1, 2$ correspond to the x and y directions, (ρ, θ) are again the polar coordinates in the (x, y) plane and $f(\rho)$ and $\alpha(\rho)$ are two functions. Now there are two mass scales in the problem, m_s and m_v, so $f(\rho)$ no longer depends on the single combination $m_s \rho$, as in the global case. Inserting this ansatz into the equations of motion one finds that at large distances the field Φ tends asymptotically to the value

$$\Phi \to \eta e^{in\theta} \,, \tag{22.172}$$

as in the global case, while, at the same time,

$$A_\mu \to \frac{n}{g} \partial_\mu \theta \,. \tag{22.173}$$

Observe that this configuration carries a quantized magnetic flux,

$$\int_S d\mathbf{S} \cdot \mathbf{B} = \int_C d\mathbf{l} \cdot \mathbf{A} = \frac{2\pi n}{g}, \qquad (22.174)$$

where C is a closed curve bounding the surface S. This configuration is called the Nielsen–Olesen vortex and is the prototype example of a *local string*, i.e. a string-like solution in a theory with a spontaneously broken local symmetry. As $\rho \to \infty$ we have $D_\mu \Phi \to 0$, since the gauge field contribution cancels the scalar field contribution. Thus, the term $\partial_\theta \Phi$, which in the global case goes to a constant, leading to a $1/\rho^2$ behavior in the energy density and to a logarithmic divergence in the string tension, is now replaced by $D_\theta \Phi$, which goes to zero as $\rho \to \infty$. Furthermore, also $F_{\mu\nu} \to 0$ as $\rho \to \infty$. Thus, the energy per unit length of a local string is finite and, on setting $n = 1$, is given by

$$\mu = 2\pi \eta^2 g(\beta), \qquad (22.175)$$

where $\beta = (m_s/m_v)^2 = \lambda/g^2$ and $g(\beta)$ is a slowly varying function such that $g(1) = 1$.

Beside the fact that local strings have a finite string tension, while a single straight global string has a logarithmically divergent string tension, the other crucial difference between global and local strings is in the mechanisms by which they can radiate away energy. Both global and local strings can in principle emit gravitational waves. We will discuss in section 22.5.4 the GW emission by loops of cosmic strings, and we will see that the power radiated in GWs is $P_{\mathrm{GW}} = \Gamma G \mu^2$, where Γ is a numerical coefficient that depends on the loop shape, with typically $\Gamma \sim 50 - 100$. For a global string there is, however, another channel of massless particles available, i.e. it can emit the Goldstone particles associated with the broken global symmetry. In this case a computation similar to the one that we will discuss for the GW emission gives a power emitted in scalar radiation $P_S = \Gamma_S \eta^2$, where again $\Gamma_S = O(100)$. Apart from the logarithmic enhancement in eq. (22.168), which, as we have discussed, for a loop is actually cut off by the loop size, even for a global string we have $\mu \sim \eta^2$, so $G\mu^2 \sim (\eta/M_{\mathrm{Pl}})^2 \eta^2$, and the radiation in GWs is suppressed, compared with the radiation in Goldstone bosons, by a factor $(\eta/M_{\mathrm{Pl}})^2$, which even for a GUT theory is at least of order 10^{-8}. Thus, a global string loop mainly decays by emission of Goldstone bosons. This channel does not exist for local strings, which rather radiate GWs. In the following, we will therefore restrict ourselves to the study of local strings.

Finally, we observe that for quite some time cosmic strings have been considered as an alternative to inflation as a mechanism for producing the initial density perturbations from which cosmological structure formation evolved. However, as discussed in Section 19.2.1, they would seed isocurvature perturbations and, as the main mechanism for structure formation, they have been ruled out by cosmological observations, in particular those of the CMB. The spectrum of the C_l predicted by cosmic strings has only a single broad peak, covering values of l in the range

$\sim 150 - 400$, rather than the set of much sharper acoustic peaks beautifully observed by CMB experiments. As we have already mentioned in Section 19.2.1, the present cosmological data are entirely consistent with adiabatic perturbations, and any admixture of CDM isocurvature perturbations in the initial conditions, if present at all, is below a few percent. As we will discuss later, the abundance of strings is fixed by their string tension. Writing

$$G\mu = A \times 10^{-7}, \qquad (22.176)$$

and requiring that cosmic strings do not spoil the agreement with CMB data, gives a bound on the string tension. The most recent analysis[34] gives (at 95% c.l.) $A < 2.4$ for strings described by the Abelian Higgs model, and $A < 1.3$ for strings modeled with the Nambu–Goto action, which we will describe in the next subsection.

[34]See [Planck Collaboration], Ade *et al.* (2016a), "Planck 2015 results. XIII. Cosmological parameters".

22.5.2 Effective description and Nambu–Goto action

At a fundamental level the dynamics of Nielsen–Olesen vortices, i.e. their motion in flat or curved space as well as their interaction with other vortices, is determined by the dynamics of the underlying scalar and gauge fields. However, if we are only interested in the dynamics on scales much larger than the characteristic thickness of the vortex (which is determined by whichever is the largest between the Compton wavelength of the scalar, m_s^{-1}, and that of the massive vector field, m_v^{-1}), we can use an effective description in which we neglect the internal structure and regard the string as a one-dimensional object in space (defined by the line of zeros of the Higgs field), which spans a two-dimensional surface in space-time.

The dynamics of such a string is a generalization of the dynamics of a relativistic point particle, which may be useful to recall here (see also the discussion in Section 20.2). In Newtonian mechanics a point particle moving in three-dimensional space is described by three functions $X^i(t)$ with $i = 1, 2, 3$. This description is not convenient in the relativistic case, since it breaks Lorentz invariance. Rather, it is convenient to introduce a variable τ that parametrizes the world-line swept by the particle, and describe the particle using four functions $X^\mu(\tau)$, with $\mu = 0, \dots, 3$. Apparently, we have introduced one extra degree of freedom, $X^0(\tau)$. However, the way we decide to parametrize the world-line is arbitrary, so we also impose invariance under reparametrizations: if we transform $\tau \to \tilde{\tau}(\tau)$, the four functions $X^\mu(\tau)$ transform into four new functions $\tilde{X}^\mu(\tilde{\tau})$, such that $\tilde{X}^\mu(\tilde{\tau}) = X^\mu(\tau)$, and we require that the action governing $X^\mu(\tau)$ be invariant under such a transformation. In field-theoretical language, we can view the four functions $X^\mu(\tau)$ as four fields living in a $(0 + 1)$-dimensional space-time, scalar under reparametrizations of the world-line, i.e. under one-dimensional diffeomorphisms. With respect to the description in terms of $X^i(t)$ we have therefore added one more field, X^0, but we also have one "gauge" symmetry, so the number of degrees of freedom is the same. We could for instance chose the gauge $X^0(\tau) = \tau$, and we would be back to the description in terms of $X^i(t)$, with $t \equiv X^0(\tau) = \tau$. However, it is in general more useful to keep the redundant description in terms of $X^\mu(\tau)$, which explicitly preserves Lorentz invariance. The action of a point-like massive particle in a curved space with metric $g_{\mu\nu}$ is given by

$$S_{\text{pp}} = -m \int ds \,, \tag{22.177}$$

where $ds^2 = -g_{\mu\nu}(X)dX^\mu(\tau)dX^\nu(\tau)$. This action is explicitly invariant under reparametrizations, since ds is the invariant length. Observe that, for a time-like path, $ds^2 > 0$. Using $dX^\mu(\tau) = \dot{X}^\mu d\tau$, where the dot denotes the derivative with respect to τ, we can rewrite eq. (22.177) as

$$S_{\text{pp}} = -m \int d\tau \left(-g_{\mu\nu}\dot{X}^\mu \dot{X}^\nu\right)^{1/2} ; \tag{22.178}$$

see also eq. (20.18). As we have already discussed [see eqs. (20.21)–(20.23)], we can get rid of the square root by introducing an auxiliary field $e(\tau)$ and writing

$$S_{\text{pp}} = \frac{1}{2} \int d\tau \left(e^{-1}g_{\mu\nu}\dot{X}^\mu \dot{X}^\nu - em^2\right). \tag{22.179}$$

Indeed, the variation with respect to $e(\tau)$ gives

$$e(\tau) = \frac{1}{m}\left(-g_{\mu\nu}\dot{X}^\mu \dot{X}^\nu\right)^{1/2}, \tag{22.180}$$

which, inserted into eq. (22.179), gives back eq. (22.178). The action (22.179) is invariant under the reparametrization $\tau \to \tilde{\tau}(\tau)$, with $X^\mu(\tau) \to \tilde{X}^\mu(\tilde{\tau})$ and $e(\tau) \to \tilde{e}(\tilde{\tau})$ such that

$$\tilde{X}^\mu(\tilde{\tau}) = X^\mu(\tau),$$
$$\tilde{e}(\tilde{\tau})d\tilde{\tau} = e(\tau)d\tau. \tag{22.181}$$

Observe also that eq. (22.179) has a well-defined massless limit, which is not the case for the action (22.177).

The generalization to the string dynamics is obtained by parametrizing the world-sheet of the string by a time-like parameter τ (analogous to the parameter τ for the world-line of a point-like particle) and a space-like parameter σ. The position of the string in space-time (whose coordinates we denote by x^μ) is then described by the four functions $X^\mu(\tau, \sigma)$, so that the position in space-time of the point of the world-sheet labeled by (τ, σ) is given by $x^\mu = X^\mu(\tau, \sigma)$. For cosmic strings we will consider a four-dimensional space-time, while the theory of a fundamental bosonic string requires for consistency $D = 26$ space-time dimensions and super-strings live in 10 dimensions (or, non-perturbatively, in 11 dimensions). The generalization of the point-particle action (22.178) is given by the Nambu–Goto action

$$S_{\text{NG}} = -\mu \int d^2\zeta \sqrt{-h} \,, \tag{22.182}$$

where $\zeta^a = (\tau, \sigma)$, the indices a, b take the values 1 and 2 corresponding to τ and σ, and μ is the string tension. We have also introduced $h = \det h_{ab}$, where

$$h_{ab} = g_{\mu\nu}(X)\partial_a X^\mu \partial_b X^\nu \qquad (22.183)$$

is the induced metric, i.e. the metric induced on the world-sheet by its embedding into a space-time with metric $g_{\mu\nu}$, through the functions $X^\mu(\tau, \sigma)$. Under diffeomorphisms of the world-sheet, $\zeta \to \tilde{\zeta}(\zeta)$, the four fields X^μ transform as scalars,

$$X^\mu(\zeta) \to \tilde{X}^\mu(\tilde{\zeta}) = X^\mu(\zeta). \qquad (22.184)$$

Then h_{ab} transforms as a metric tensor on the world-sheet, and the Nambu–Goto action is invariant. Just as in the point-particle case, we can get rid of the square root by introducing an auxiliary field, which in this case is a tensor γ_{ab}, with Lorentzian signature $(-, +)$. The action (22.182) is in fact equivalent to the Polyakov action

$$S_P = -\frac{\mu}{2}\int d^2\zeta \sqrt{-\gamma}\, \gamma^{ab} g_{\mu\nu}(X)\partial_a X^\mu \partial_b X^\nu, \qquad (22.185)$$

[35] See Section 1.2 of Polchinski (1998), which we are following here.

where $\gamma = \det \gamma_{ab}$. This can be shown by taking the variation with respect to γ_{ab}, which gives[35] $\gamma_{ab} = \kappa h_{ab}$, where κ is an arbitrary constant. Inserting this back into eq. (22.185), κ cancels between $\sqrt{-\gamma}$ and γ^{ab}, while $h^{ab}g_{\mu\nu}\partial_a X^\mu \partial_b X^\nu = h^{ab}h_{ab} = 2$, and we get back eq. (22.182). In this formulation, the theory is invariant under two-dimensional diffeomorphisms, under which $X^\mu(\zeta)$ transforms as in eq. (22.184), while γ_{ab} transforms as a metric,

$$\gamma_{ab}(\zeta) \to \tilde{\gamma}_{ab}(\tilde{\zeta}) = \gamma_{cd}(\zeta)\frac{\partial \zeta^c}{\partial \tilde{\zeta}^a}\frac{\partial \zeta^d}{\partial \tilde{\zeta}^b}. \qquad (22.186)$$

In the Polyakov formulation we further have an invariance under Weyl transformation, in which the world-sheet coordinates ζ^a, as well as the fields X^μ, are not changed, while

$$\gamma_{ab} \to e^{2\omega(\zeta)}\gamma_{ab}. \qquad (22.187)$$

The Nambu–Goto action can indeed be obtained from the Nielsen–Olesen vortex solution, as follows. The solution (22.170), (22.171) was obtained considering a static string along the z axis. Consider now a more general string solution, in general neither static nor a straight line. The location of the string, identified by the line of points where the Higgs field vanishes, defines a world-sheet in space-time $X^\mu(\zeta)$, so that a point labeled by ζ^a on the world-sheet has space-time coordinates $x^\mu = X^\mu(\zeta)$. At each point on the world-sheet we can define two space-like vectors $n_A^\mu(\zeta)$, $A = 1, 2$, which are orthogonal to the world-sheet, i.e.

$$g_{\mu\nu}n_A^\mu(\zeta)\partial_a X^\nu(\zeta) = 0, \qquad (22.188)$$

for $a = 1, 2$. We also normalize them so that

$$g_{\mu\nu}n_A^\mu n_B^\nu = \delta_{AB}. \qquad (22.189)$$

Then, at least infinitesimally close to the world-sheet, a generic space-time point x^μ can be written as

$$x^\mu(\xi) \equiv x^\mu(\zeta^a, \rho^A) = X^\mu(\zeta) + \rho^A n_A^\mu(\zeta), \qquad (22.190)$$

and we can therefore trade the four coordinates x^μ for the four coordinates $\xi \equiv (\zeta^a, \rho^A)$.[36]

Sufficiently close to the world-sheet (i.e. at spatial distances much smaller than its typical radius of curvature) we can still approximate the solution as a straight line in space, so the Nielsen–Olesen vortex solution at the space-time point $x^\mu(\xi)$ is still given by eqs. (22.170) and (22.171), in which now the transverse plane is the one spanned by the (n_1^μ, n_2^μ) vectors, and the role of the (x, y) coordinates is taken by (ρ_1, ρ_2). Thus, sufficiently close to the world-sheet, Φ is still given by (22.170), where now (ρ, θ) are the polar coordinates corresponding to the Cartesian coordinates (ρ_1, ρ_2), i.e.

$$\Phi[x(\xi)] = \eta f(\rho)e^{in\theta}, \qquad (22.191)$$

while the gauge field can be written as

$$A^\mu[x(\xi)] = -n_A^\mu(\zeta)\epsilon_{AB}\rho_B \frac{n}{g\rho^2}\alpha(\rho). \qquad (22.192)$$

We can now reduce the action of the Abelian Higgs model,

$$S = -\int d^4x \sqrt{-g}\left[|D_\mu\Phi|^2 + V(\Phi) + \frac{1}{4}F_{\mu\nu}F^{\mu\nu}\right], \qquad (22.193)$$

to an effective action for the variables $X^\mu(\zeta)$, observing that

$$d^4x\sqrt{-g} = d^4\xi\sqrt{-M}, \qquad (22.194)$$

where $d^4\xi = d^2\zeta d^2\rho$ and $M = \det M_{\alpha\beta}$, with

$$M_{\alpha\beta} = g_{\mu\nu}\frac{\partial x^\mu}{\partial\xi^\alpha}\frac{\partial x^\nu}{\partial\xi^\beta}, \qquad (22.195)$$

and α runs over the four values $a = 1, 2$ and $A = 1, 2$. Sufficiently close to the string world-sheet we can use eq. (22.190). Then, for the matrix elements where $\alpha = a$ and $\beta = b$ we get

$$M_{ab} \simeq g_{\mu\nu}\left(\partial_a X^\mu + \rho^A\partial_a n_A^\mu\right)\left(\partial_b X^\nu + \rho^B\partial_b n_B^\nu\right)$$
$$= h_{ab} + O(\rho/R), \qquad (22.196)$$

where R is the typical radius of curvature of the string. Similarly,

$$M_{AB} \simeq g_{\mu\nu}n_A^\mu n_B^\nu = \delta_{AB}, \qquad (22.197)$$
$$M_{Aa} \simeq g_{\mu\nu}n_A^\mu\left(\partial_a X^\nu + \rho^A\partial_a n_A^\nu\right) = O(\rho/R), \qquad (22.198)$$

where we have used eqs. (22.188) and (22.189). Therefore, to lowest order in ρ/R, $\det M_{\alpha\beta} \simeq \det h_{ab}$, and the action (22.193) becomes

$$S \simeq -\int d^2\zeta\sqrt{-h}\int d^2\rho\left[|D_\mu\Phi|^2 + V(\Phi) + \frac{1}{4}F_{\mu\nu}F^{\mu\nu}\right](\zeta, \rho). \qquad (22.199)$$

[36] Actually, in a curved space-time with metric $g_{\mu\nu}$, to make the construction generally covariant we must pass to a system of Riemann normal coordinates, as discussed in Section 1.3.2 of Vol. 1, based on the geodesics leaving from the point $X^\mu(\zeta)$ with a given tangent vector n^μ. However, in a sufficiently small region close to the world-sheet, the geodesics in the direction n_A^μ parametrized by a proper-time parameter ρ_A becomes the same as the straight line $\rho^A n_A^\mu$. Furthermore, in the cosmological context in which we are eventually interested, the metric $g_{\mu\nu}$ only changes on a cosmological scale, while the string thickness is a microscopic quantity.

[37] Observe that the derivation that we have given does not go through for global strings, as signaled by the fact that their energy per unit length diverges. The reason is that in this case the theory also has a Goldstone particle, which, being massless, cannot be integrated out, but must be kept in the low-energy effective theory. The corresponding effective string action in this case is the Kalb-Ramond action, which besides the Nambu–Goto term also involves an antisymmetric tensor field, whose dual is related to the Goldstone boson; see Section 4.4.2 of Vilenkin and Shellard (2000).

The expression in square brackets is evaluated over the fields (22.191) and (22.192). On this static configuration, this expression coincides with the component $-T_0^0$ of the energy–momentum tensor derived from the Lagrangian (22.169), and its integral over $d^2\rho$ is therefore the energy per unit length of the string, i.e. the string tension μ. We have therefore recovered the Nambu–Goto action (22.182). Keeping also higher-order terms in ρ/R gives a further contribution, which depends for instance on the extrinsic curvatures $K_{ab}^A = -g_{\mu\nu}\partial_a n_A^\mu \partial_b X^\nu$ associated with the directions n_A^μ. These terms, in the language of effective actions, are higher-derivative operators, and are suppressed at scales $r \ll R$ with respect to the Nambu–Goto term. Their presence, however, is a signal of the fact that the string description is here only an effective one. A fundamental (bosonic) string, or the bosonic sector of superstring theory, taken as a fundamental theory, has only the Nambu–Goto term.[37]

22.5.3   String dynamics. Cusps and kinks

The use of the Nambu–Goto action allows us to study the dynamics of a local string much more easily, compared with what could be done by studying the dynamics of the vortex solution directly in the Abelian Higgs model. Furthermore, this effective action approach makes it clear that the dynamics on scales much larger than the thickness of the string is independent of the details of the underlying fundamental theory. We consider for simplicity the motion of strings in flat space-time, $g_{\mu\nu} = \eta_{\mu\nu}$, and we use the Polyakov action (22.185). As we have already mentioned, the variation with respect to γ_{ab} gives $\gamma_{ab} = \kappa h_{ab}$, where κ is an arbitrary constant, and h_{ab} is defined in eq. (22.183). Using the Weyl symmetry (22.187) we can always set $\kappa = 1$. Setting $\gamma_{ab} = h_{ab}$, the equation of motion derived by performing the variation of the action (22.185) with respect to X^μ is

$$\partial_a\left(\sqrt{-h}\,h^{ab}\partial_b X^\mu\right) = 0\,. \qquad (22.200)$$

This is the same as $\Box X^\mu = 0$, where $\Box = (-h)^{-1/2}\partial_a(\sqrt{-h}\,h^{ab}\partial_b)$ is the two-dimensional d'Alembertian operator with respect to the metric h^{ab}, and acting on fields, such as X^μ, that are scalars under diffeomorphisms of the world-sheet. Of course, one obtains the same equation of motion by directly taking the variation with respect to X^μ of the Nambu–Goto action (22.182), using $\delta h = h h^{ab}\delta h_{ab}$.

In general, eq. (22.200) has a highly non-linear dependence on X^μ, since h_{ab} is itself a functional of X^μ given by eq. (22.183). However, we can now make use of the two-dimensional diffeomorphism invariance to fix the gauge to simplify this equation. The two-dimensional diffeomorphisms allow us to fix two conditions. It is convenient to require

$$h_{01} = 0 \qquad (22.201)$$

and

$$h_{00} + h_{11} = 0\,, \qquad (22.202)$$

so that h_{ab} can be written in the form

$$h_{ab} = e^{\phi}\eta_{ab}, \qquad (22.203)$$

i.e. the world-sheet metric becomes conformally flat.[38] This gauge is therefore called the *conformal gauge*. The advantage of this gauge choice is that $-h = e^{2\phi}$ while $h^{ab} = e^{-\phi}\eta^{ab}$, so in the combination $\sqrt{-h}\,h^{ab}$ that appears in eq. (22.200) the conformal factor cancels, and we are left with a simple linear equation

$$(-\partial_{\tau}^2 + \partial_{\sigma}^2)X^{\mu} = 0, \qquad (22.204)$$

which is just the Klein–Gordon wave equation in two-dimensional flat space. Furthermore, using eq. (22.183) with $g_{\mu\nu} = \eta_{\mu\nu}$, the gauge conditions (22.201) and (22.202) give, respectively,

$$\eta_{\mu\nu}\partial_{\sigma}X^{\mu}\partial_{\tau}X^{\nu} = 0, \qquad (22.205)$$
$$\eta_{\mu\nu}(\partial_{\sigma}X^{\mu}\partial_{\sigma}X^{\nu} + \partial_{\tau}X^{\mu}\partial_{\tau}X^{\nu}) = 0. \qquad (22.206)$$

These two equations do not involve second derivative with respect to τ, and are therefore constraints on the initial values of $\partial_{\tau}X^{\mu}$ and $\partial_{\sigma}X^{\mu}$, which are preserved by the time evolution.

The dynamics of the string can be further simplified as follows.[39] The conditions (22.201) and (22.202) do not remove the gauge freedom completely. Indeed, consider a diffeomorphism $\zeta^a \to \tilde{\zeta}^a(\zeta)$, i.e. $\tau \to \tilde{\tau}(\tau,\sigma)$ and $\sigma \to \tilde{\sigma}(\tau,\sigma)$. The metric h_{ab} transforms in the same way as γ_{ab} in eq. (22.186). Using eqs. (22.201) and (22.202) and requiring that $\tilde{h}_{01} = 0$ gives

$$\partial_{\tau}\tilde{\tau}\partial_{\sigma}\tilde{\tau} = \partial_{\tau}\tilde{\sigma}\partial_{\sigma}\tilde{\sigma}, \qquad (22.207)$$

while requiring $\tilde{h}_{00} + \tilde{h}_{11} = 0$ gives

$$(\partial_{\tau}\tilde{\tau})^2 + (\partial_{\sigma}\tilde{\tau})^2 = (\partial_{\tau}\tilde{\sigma})^2 + (\partial_{\sigma}\tilde{\sigma})^2. \qquad (22.208)$$

We see by inspection that a solution of eqs. (22.207) and (22.208) is obtained using functions $\tilde{\sigma}(\tau,\sigma)$ and $\tilde{\tau}(\tau,\sigma)$ that satisfy the two conditions

$$\partial_{\tau}\tilde{\sigma} = \partial_{\sigma}\tilde{\tau}, \qquad (22.209)$$
$$\partial_{\sigma}\tilde{\sigma} = \partial_{\tau}\tilde{\tau}. \qquad (22.210)$$

Applying ∂_{σ} to the first equation, and using the second one, we also get

$$(-\partial_{\tau}^2 + \partial_{\sigma}^2)\tilde{\tau} = 0, \qquad (22.211)$$

and similarly $(-\partial_{\tau}^2 + \partial_{\sigma}^2)\tilde{\sigma} = 0$. Having at our disposal a function $\tilde{\tau}$ that satisfies eq. (22.211) but is otherwise arbitrary, we can use it to fix the gauge $X^0(\tau,\sigma) = \tau$, since X^0 also satisfies eq. (22.204). Once $\tilde{\tau}(\tau,\sigma)$ has been fixed in this way, the derivatives of the function $\tilde{\sigma}$ are also fixed, so the gauge fixing is now complete, apart from the irrelevant possibility of shifting σ by a constant. In this gauge, therefore, X^0 drops from the set of dynamical variables and, using the notation $X^0 \equiv t$, we

[38]Locally, i.e. in an open neighborhood of a point on the world-sheet, it is always possible to impose these conditions. The same is true globally, if the world-sheet has the topology of a plane or, for closed string loops, of a cylinder. These are the cases of relevance for the effective string that we are discussing here.

[39]We follow Section 6.2 of Vilenkin and Shellard (2000), to which we refer for more details.

have $\tau = t$. Thus, the string is now described by a vector $\mathbf{X}(t, \sigma)$, and eqs. (22.204)–(22.206) become

$$\ddot{\mathbf{X}} - \mathbf{X}'' = 0, \tag{22.212}$$
$$\dot{\mathbf{X}} \cdot \mathbf{X}' = 0, \tag{22.213}$$
$$\dot{\mathbf{X}}^2 + \mathbf{X}'^2 = 1, \tag{22.214}$$

where the dot denotes the derivative with respect to t and the prime that with respect to σ. The solution of eq. (22.212) is a superposition of generic left-moving and right-moving waves,

$$\mathbf{X}(t, \sigma) = \frac{1}{2} \left[\mathbf{a}(\sigma - t) + \mathbf{b}(\sigma + t) \right], \tag{22.215}$$

with $\mathbf{a}$ and $\mathbf{b}$ arbitrary vector functions of one variable. The constraints (22.213) and (22.214) then give

$$\mathbf{a}'^2 = \mathbf{b}'^2 = 1. \tag{22.216}$$

The energy–momentum tensor of the string is obtained from eq. (22.185),

$$T^{\mu\nu}(x) = \frac{2}{\sqrt{-g}} \frac{\delta S}{\delta g_{\mu\nu}(x)}$$
$$= -\mu \int d^2\zeta \sqrt{-h}\, h^{ab} \frac{1}{\sqrt{-g}} \delta^{(4)}(x - X) \partial_a X^\mu \partial_b X^\nu, \tag{22.217}$$

where, after taking the variation with respect to $g_{\mu\nu}$, we have used the equation of motion $\gamma^{ab} = h^{ab}$. In flat space, and using the conformal gauge (22.203), we get

$$T^{\mu\nu}(x) = \mu \int d^2\zeta\, \delta^{(4)}(x - X) \left[\partial_\tau X^\mu \partial_\tau X^\nu - \partial_\sigma X^\mu \partial_\sigma X^\nu \right]. \tag{22.218}$$

If we fix the residual gauge freedom by further imposing $X^0(\tau, \sigma) = \tau$, we finally get for the energy density $\rho(x) = T^{00}$,

$$\rho(t, \mathbf{x}) = \mu \int d\sigma d\tau\, \delta^{(3)}[\mathbf{x} - \mathbf{X}(\tau, \sigma)]\, \delta(t - \tau)$$
$$= \mu \int d\sigma\, \delta^{(3)}[\mathbf{x} - \mathbf{X}(t, \sigma)]. \tag{22.219}$$

The total energy is therefore

$$E = \int d^3x\, \rho(x)$$
$$= \mu \int d\sigma. \tag{22.220}$$

Thus, in this gauge σ provides a measure of length on the string, given by $d\sigma = dE/\mu$. Similarly, for the momentum we get

$$P^i = \int d^3x\, T^{0i}$$
$$= \mu \int d\sigma\, \dot{X}^i. \tag{22.221}$$

Consider now a closed loop. According to eq. (22.220), the parameter σ can be taken to be in the range $0 \leqslant \sigma \leqslant L$, where $L = E/\mu$ is called the invariant length of the loop. For a closed loop we require periodic boundary conditions $\mathbf{X}(t, \sigma + L) = \mathbf{X}(t, \sigma)$. In terms of the functions $\mathbf{a}$ and $\mathbf{b}$ introduced in eq. (22.215), this implies

$$-\mathbf{a}(\sigma + L - t) + \mathbf{a}(\sigma - t) = \mathbf{b}(\sigma + L + t) - \mathbf{b}(\sigma + t)$$
$$\equiv \mathbf{\Delta} , \qquad (22.222)$$

and, because of invariance under translations in the world-sheet, the vector $\mathbf{\Delta}$ is independent of σ. From eq. (22.221), the momentum $\mathbf{P}$ is given by

$$\begin{aligned}
\mathbf{P} &= \frac{\mu}{2} \int_0^L d\sigma \, [\partial_t \mathbf{a}(\sigma - t) + \partial_t \mathbf{b}(\sigma + t)] \\
&= \frac{\mu}{2} \int_0^L d\sigma \, [-\partial_\sigma \mathbf{a}(\sigma - t) + \partial_\sigma \mathbf{b}(\sigma + t)] \\
&= \frac{\mu}{2} [-\mathbf{a}(L - t) + \mathbf{a}(-t) + \mathbf{b}(L + t) - \mathbf{b}(t)] \\
&= \mu \mathbf{\Delta} .
\end{aligned} \qquad (22.223)$$

In the center-of-mass frame of the string, which is defined by $\mathbf{P} = 0$, we therefore have $\mathbf{\Delta} = 0$, and eq. (22.222) shows that $\mathbf{a}$ and $\mathbf{b}$ are separately periodic,

$$\mathbf{a}(\sigma + L - t) = \mathbf{a}(\sigma - t) , \qquad (22.224)$$
$$\mathbf{b}(\sigma + L + t) = \mathbf{b}(\sigma + t) . \qquad (22.225)$$

With this formalism it is easy to understand the emergence of cusps in the generic motion of a string. Indeed, from eq. (22.215) we have

$$(\partial_t \mathbf{X})^2 = \frac{1}{4} \left(-\mathbf{a}' + \mathbf{b}' \right)^2 . \qquad (22.226)$$

Thus, using eq. (22.216), we see that at a point of the world-sheet where $\mathbf{a}' = -\mathbf{b}'$, the world-sheet moves at the speed of light. Equation (22.216) shows that, as σ runs from zero to L, the functions $\mathbf{a}'(\sigma - t)$ and $\mathbf{b}'(\sigma + t)$ sweep curves on the unit sphere. Because of eqs. (22.224) and (22.225), these are actually closed curves on the sphere. The periodic boundary conditions on these vector functions also imply that

$$\int_0^L d\sigma \, \mathbf{a}' = \int_0^L d\sigma \, \mathbf{b}' = 0 , \qquad (22.227)$$

so these curves cannot be confined to a single hemisphere. This means that, as σ runs from zero to L, the curves swept by $\mathbf{a}'(\sigma - t)$ and by $-\mathbf{b}'(\sigma + t)$ will generically intersect, and therefore in a typical motion of the string there will be points on the world-sheet that travel at the speed of light. Since this happens just at the intersection point, the corresponding point of the string travels at the speed of light only at some instant of time. When this happens, the spatial configuration of the string takes the form of a cusp. Indeed, let us choose the origin of the world-sheet coordinates so that the point of the world-sheet travelling at the speed of light corresponds to $\tau = \sigma = 0$. We also choose the origin of the spatial coordinates so that, at $\tau = 0$, this point is at $\mathbf{x} = 0$, and $\mathbf{a} = \mathbf{b} = 0$, and we again use the gauge $t = \tau$. Expanding $\mathbf{a}(\sigma_-)$ and $\mathbf{b}(\sigma_+)$ in powers of $\sigma_\pm \equiv \sigma \pm t$ near the point $t = \sigma = 0$, i.e. $\sigma_+ = \sigma_- = 0$, we get

$$\mathbf{a}(\sigma_-) = \mathbf{a}_0'\sigma_- + \frac{1}{2}\mathbf{a}_0''\sigma_-^2 + \frac{1}{6}\mathbf{a}_0'''\sigma_-^3 + \dots , \tag{22.228}$$

$$\mathbf{b}(\sigma_+) = \mathbf{b}_0'\sigma_+ + \frac{1}{2}\mathbf{b}_0''\sigma_+^2 + \frac{1}{6}\mathbf{b}_0'''\sigma_+^3 + \dots . \tag{22.229}$$

so that

$$\mathbf{a}'(\sigma_-) = \mathbf{a}_0' + \mathbf{a}_0''\sigma_- + \dots , \tag{22.230}$$

$$\mathbf{b}'(\sigma_+) = \mathbf{b}_0' + \mathbf{b}_0''\sigma_+ + \dots . \tag{22.231}$$

The condition that the curves spanned by $\mathbf{a}'(\sigma - t)$ and $-\mathbf{b}'(\sigma + t)$ intersect at $\sigma = t = 0$ gives $\mathbf{a}_0' = -\mathbf{b}_0'$. Thus, from eq. (22.215), at $t = 0$ we have

$$\mathbf{X}(0, \sigma) = \frac{1}{4}(\mathbf{a}_0'' + \mathbf{b}_0'')\sigma^2 + \frac{1}{6}(\mathbf{a}_0''' + \mathbf{b}_0''')\sigma^3 + \dots . \tag{22.232}$$

If $(\mathbf{a}_0'' + \mathbf{b}_0'') \neq 0$, we can chose the axes so that $(\mathbf{a}_0'' + \mathbf{b}_0'')$ points, say, along the x axis. Then, in the (x, y) plane, we have

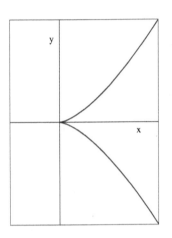

Fig. 22.16 The cusp developed by a cosmic string.

$$X_x(0, \sigma) \simeq c_x\sigma^2 , \tag{22.233}$$

$$X_y(0, \sigma) \simeq c_y\sigma^3 , \tag{22.234}$$

where $c_x = (1/4)|\mathbf{a}_0'' + \mathbf{b}_0''|$ and $c_y = (1/6)(\mathbf{a}_0''' + \mathbf{b}_0''')_y$ (assuming that $c_y \neq 0$). This is the parametric description of a cusp, as shown in Fig. 22.16 and, near $\sigma = 0$, the shape of the string is given by $X_x^3(0, \sigma) \propto X_y^2(0, \sigma)$. The velocity of the string at this point is orthogonal to the direction of the cusp. This can be shown by inserting eqs. (22.230) and (22.231) into eq. (22.216) and requiring that it be satisfied order by order in σ, which gives

$$|\mathbf{a}_0'| = |\mathbf{b}_0'| = 1 , \tag{22.235}$$

$$\mathbf{a}_0' \cdot \mathbf{a}_0'' = \mathbf{b}_0' \cdot \mathbf{b}_0'' = 0 . \tag{22.236}$$

Thus the direction of the cusp, which is given by $(\mathbf{a}_0'' + \mathbf{b}_0'')$, is orthogonal to the direction of the velocity, which is given by $(1/2)(-\mathbf{a}_0' + \mathbf{b}_0') = \mathbf{b}_0'$, since $\mathbf{a}_0' = -\mathbf{b}_0'$.

Of course, the cusp singularity is an artifact of the zero-thickness approximation implicit in the use of the Nambu–Goto action. For a Nielsen–Olesen vortex the singularity will be smoothed by its inner structure, on the scales m_s^{-1} or m_v^{-1}, whichever is larger, and the velocity at the cusp will be strictly smaller than the speed of light. However, this smoothing occurs on scales that are much smaller than the horizon-scale length typical of the cosmic strings in which we are interested, and the idealized cusp obtained in the Nambu–Goto approximation is an exceedingly good description of the dynamics. Clearly, the presence of such a singularity where formally the string reaches the speed of light is very interesting for GW production, as we will discuss further in Section 22.5.4.

Another significant structure that can develop on cosmic strings is a kink, i.e. a point where the functions $\mathbf{a}'$ and $\mathbf{b}'$ are discontinuous, so that there is a discontinuity in the vector $\mathbf{X}'$ tangent to the string. While cusps, as we have seen, are a rather generic product of the evolution of a single string, kinks are formed when two strings interact with each other, exchanging parts as in Fig. 22.17, or when a single string self-intersects, as in Fig. 22.18. In the latter case there is both formation of a kink and emission of a smaller string loop. Numerical simulations for the Abelian Higgs model show that such processes, also referred as reconnections, are quite frequent. A kink implies that both $\mathbf{a}'$ and $\mathbf{b}'$ are discontinuous. Since $\mathbf{a}$ is a function of $\sigma - t$ and $\mathbf{b}$ is a function of $\sigma + t$, after the kink has been generated it splits into two kinks, moving in opposite directions along the string, at the speed of light. The argument that we have presented regarding the generic intersection of the curves spanned by $\mathbf{a}'$ and $\mathbf{b}'$ on the unit sphere indicates that cusps are generically expected to appear on smooth strings a few times per oscillation period. However, if a string develops kinks, the corresponding curve on the unit sphere has discontinuities, and in this case there are more chances that the two curves do not intersect. In that case the string has kinks, traveling on it at the speed of light, but does not develop cusps.

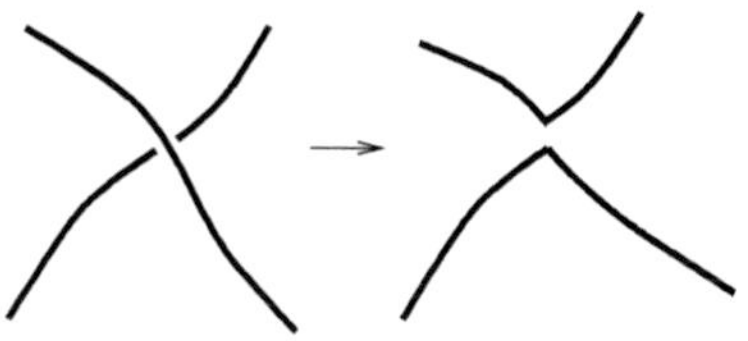

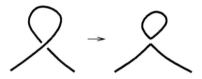

Fig. 22.17 The kinks generated by the reconnection of two cosmic strings.

Fig. 22.18 The kink and the closed loop generated by the self intersection of a cosmic string.

22.5.4 Gravitational radiation from cosmic strings

In the Standard Model these is no spontaneously broken local $U(1)$ symmetry that could give rise to cosmic strings.[40] However, cosmic strings could emerge in extensions of the Standard Model based on larger grand-unified groups. If cosmic strings exist, they would be very remarkable objects. For $G\mu \simeq 10^{-7}$, which is about the largest value consistent with the CMB [see eq. (22.176)], we have $\mu^{1/2} \simeq 4 \times 10^{15}$ GeV, in our units $\hbar = c = 1$. Reinstating c, a mass per unit length μ such that $G\mu/c^2 = 10^{-7}$ has the huge numerical value $\mu \simeq 10^{21}$ g/cm. A cosmic string loop with this tension and a length equal to one astronomical unit (and thus much smaller than the cosmological scale that most cosmic string loops would be expected to have, as we shall see) would already have a mass $\sim 10 M_\odot$. This energy is concentrated in a tube of width $\sim m_s^{-1}$ or $\sim m_v^{-1}$. As we saw in Section 22.5.1, $m_s = \lambda^{1/2}\eta$ and $m_v = g\eta$, so their value is basically determined by the Higgs vacuum expectation value η. For a phase transition at a GUT scale, and any reasonable values of the couplings λ and g, the thickness of the string is much smaller than, say, the proton radius. Thus, we have a huge energy per unit length, concentrated in an extremely narrow tube. Because of their huge tension, these objects oscillate relativistically and we have

[40]The spontaneous breaking of the $SU(2) \times U(1)$ gauge symmetry of the Standard Model does not give rise to stable string configurations; see Section 4.2.6 of Vilenkin and Shellard (2000).

even seen that, during their evolutions, cusps are generically formed, where a point of the string instantaneously moves at the speed of light. Cosmic strings are therefore potentially very interesting sources of GWs.

Stochastic GW background from a cosmic string network

To study GW generation by a network of cosmic strings, we must address two questions. First, what is the radiation generated by a cosmic string loop and, second, what are the statistical properties of a cosmic string network. Let us begin with the radiation from a single loop. A first naive estimate can be obtained using the quadrupole formula (3.75). The mass of a string loop of length L is $M \sim \mu L$, so its quadrupole moment is $Q \sim \mu L^3$. The typical oscillation frequency of a loop of length L oscillating relativistically is $\omega \sim 1/L$, so $\dddot{Q} \sim Q\omega^3 \sim \mu$, and eq. (3.75) gives $P_{\rm quad} \sim G\mu^2$. However, the use of the quadrupole formula is obviously inappropriate, given that the loop moves relativistically. One must rather use the formalism discussed in Section 3.1 of Vol. 1, valid for (weak-field) sources with arbitrary velocity. An oscillating string loop of length L emits at a discrete set of frequencies ω_n. Observe that, for a closed loop of length L in its rest frame, the functions $\mathbf{a}$ and $\mathbf{b}$ are periodic with respect to σ period L; see eqs. (22.224) and (22.225). From eq. (22.215) this means that

$$\mathbf{X}\left(\sigma + \frac{L}{2}, t + \frac{L}{2}\right) = \mathbf{X}(\sigma, t)\,, \tag{22.237}$$

so the string configuration is periodic in time with period $L/2$, modulo a shift in the parameter σ. Thus, the oscillation frequencies of a cosmic string loop of length L are $\omega_n = 2\pi f_n$, with $f_n = n/(L/2) = 2n/L$. Adapting to this case eqs. (3.16)–(3.20), which were written for a simply periodic source, we see that the energy radiated per unit solid angle is still given by eq. (3.16), where the Fourier transform of the energy–momentum tensor of the source is

$$\tilde{T}_{ij}(\omega, \mathbf{k}) = \sum_n \theta_{ij}(\omega_n, \mathbf{k})\, 2\pi\delta(\omega - \omega_n)\,, \tag{22.238}$$

and the radiated power is

$$P = \sum_n P_n\,, \tag{22.239}$$

where (in our present units $c = 1$)

$$\frac{dP_n}{d\Omega} = \frac{G\omega_n^2}{\pi} \Lambda_{ij,kl}(\hat{\mathbf{n}})\, \theta_{ij}(\omega_n, \omega_n\hat{\mathbf{n}})\theta_{kl}^*(\omega_n, \omega_n\hat{\mathbf{n}})\,. \tag{22.240}$$

One can now consider some explicit exact solutions for the motion of the loop, and compute numerically the power radiated in the various harmonics, and hence the total power. The result is of the form

$$P = \Gamma G\mu^2\,, \tag{22.241}$$

where the proportionality to $G\mu^2$ is the same as in the quadrupole formula, and is fixed by the facts that eq. (22.240) is proportional to G and that each factor $\theta_{ij}(\omega_n, \omega_n \hat{\mathbf{n}})$ is proportional to μ. The numerical factor Γ depends on the shape of the loop, and turns out to be a large number, $\Gamma \sim 50 - 100$, because many harmonics contribute. A value of Γ of this order is obtained for loops with cusps, for loops with kinks, and also for loops with neither cusps nor kinks. However, the dependence of P_n on n is different in these cases. In particular, cusps or kinks correspond to very small-scale structures on the string, and contribute to harmonics with very large n. Thus, for loops with cusps or with kinks, P_n decreases quite slowly with n. For loops with cusps, at large n, one has

$$P_n^{\text{cusps}} \sim G\mu^2 n^{-4/3}, \qquad (22.242)$$

while for loops with kinks

$$P_n^{\text{kinks}} \sim G\mu^2 n^{-5/3}. \qquad (22.243)$$

The energy radiated in GWs is taken from the energy $E = \mu L$ of the loop. Thus, as the loop loses energy, it must shrink and eventually disappear. From eq. (22.241) we get we have

$$\mu \frac{dL}{dt} = -\Gamma G\mu^2. \qquad (22.244)$$

Since Γ, to a first approximation, is independent of the loop size, a loop created initially with a length L_0 at time t_0 evolves as

$$L(t) = L_0 - \Gamma G\mu(t - t_0). \qquad (22.245)$$

Thus, in the absence of interaction with other strings, the loop lifetime is due to GW emission and is given by $T = L_0/(\Gamma G\mu)$. For $G\mu \sim 10^{-7}$ and $\Gamma \sim 10^2$ we have $T \sim 10^5 L_0$. Therefore, the loop is long-lived compared with the natural length-scale given by its size, and performs $O(10^5)$ oscillations before disappearing.

The other crucial aspect for computing the GW production from an ensemble of cosmic strings is the distribution of loop sizes. Several analytical and numerical investigations have shown that a cosmic string network evolves toward an attractor solution characterized by a scaling regime, in which the average distance between strings remains a fixed fraction of the Hubble length $H^{-1}(t)$, i.e. it scales linearly with cosmic time t, and the energy density of the network remains a constant fraction of the total energy density of the Universe. If cosmic strings were absolutely stable, the energy of a single cosmic string $E = \mu L$ would scale with the expansion as $a(t)$, since $L \propto a(t)$. A physical volume V scales as a^3, so the energy density of cosmic strings would scale only as $1/a^2$, i.e. much more slowly than radiation, so during RD the ratio of the energy density in cosmic strings to the energy density in radiation would increase as a^2. If this were the case, cosmic strings would quickly reach an unacceptably large density and would be dangerous relics to be washed out by an inflationary epoch, just like monopoles or domain walls. However, cosmic strings lose energy by GW emission, and this saves them from becoming dangerous relics. What happens in a cosmic string network is that long strings cross each other and produce smaller loops, as in Fig. 22.18, which then shrink and eventually disappear by emitting GWs. Numerical simulations give, for the number density (per physical volume) of loops of length L,[41]

[41] See Blanco-Pillado, Olum and Shlaer (2014).

$$\frac{n(L,t)}{a^3(t)} \simeq \frac{0.18}{t^{3/2}(L+\Gamma G\mu t)^{5/2}} \qquad \text{(RD)} \qquad (22.246)$$

which is valid for L smaller than a cutoff value $L_{\max} \simeq 0.1t$, and

$$\frac{n(L,t)}{a^3(t)} \simeq \frac{0.27 - 0.45(L/t)^{0.31}}{t^2(L+\Gamma G\mu t)^2} \qquad \text{(MD)}, \qquad (22.247)$$

with a cutoff $L_{\max} \simeq 0.18t$. This describes a population of loops with a broad range of values of L, from $L = 0$ up to values not much smaller than the horizon radius. Writing $L = \alpha t$, in this distribution there are loops up to $\alpha \sim 0.1$. It is still under debate whether there exists also a separate population of small loops, with typical values $\alpha \sim \Gamma G\mu$, as was assumed in earlier works. The scaling property of these distributions is reflected in the fact that they have the form $n(L,t)/a^3(t) = t^{-4}N(L/t)$, for some function N.

Given the energy radiated by a loop and the loop distribution, the resulting GW background can be computed as follows. The power radiated by a single loop of length L can be written as in eq. (22.239). It is convenient to extract explicitly the factor $G\mu^2$, writing $P_n = G\mu^2 p_n$, so

$$\frac{dE_{\rm GW}}{dt} = G\mu^2 \sum_n p_n$$

$$= G\mu^2 L \int_0^\infty df_e\, S(Lf_e)\,, \qquad (22.248)$$

where

$$S(x) \equiv \sum_n p_n \delta(x - x_n)\,. \qquad (22.249)$$

The subscript e in f_e is a reminder that this is the frequency at the time of emission, before any redshifting, $x = Lf_e$, $x_n = Lf_{e,n}$, and $f_{e,n} = 2n/L$ are the frequencies at which the loop of length L emits. Comparing with eq. (22.241) we have

$$\int_0^\infty dx\, S(x) = \Gamma\,. \qquad (22.250)$$

This continuous notation is useful because is allows us to write the frequency spectrum of the GWs emitted by a single loop, in the interval between the emission time t_e and $t_e + dt_e$, as

$$\frac{dE_{\rm GW}(t_e)}{df_e} = G\mu^2 LS(Lf_e)dt_e\,. \qquad (22.251)$$

The energy density of the radiation emitted by a network of loops with number density $n(L,t)dL$ in the interval between time t_e and $t_e + dt_e$ is therefore

$$\frac{d\rho_{\mathrm{GW}}(t_e)}{df_e} = G\mu^2 dt_e \int_0^{\alpha t_e} dL\, n(L, t_e) L S(Lf_e)\,, \qquad (22.252)$$

where we have taken into account the fact that, at time t_e, the maximum length of the loop is αt_e. We now observe that ρ_{GW} redshifts as $1/a^4$, so $\rho_{\mathrm{GW}}(t_e) = a^{-4}(t_e)\rho_{\mathrm{GW}}(t_0)$, where t_0 is the present value of cosmic time and we have set $a(t_0) = 1$. The emission frequency is related to the frequency f_{obs} observed today by $f_e = f_{\mathrm{obs}}/a(t_e)$, so

$$\frac{d\rho_{\mathrm{GW}}(t_e)}{df_e} = \frac{1}{a^3(t_e)}\frac{d\rho_{\mathrm{GW}}(t_0)}{df_{\mathrm{obs}}}\,. \qquad (22.253)$$

Integrating over the emission time t_e from the primordial epoch $t_e \simeq t_*$ when cosmic strings first formed to the present time t_0, we get

$$\left(\frac{d\rho_{\mathrm{GW}}}{df}\right)_0 = G\mu^2 \int_{t_*}^{t_0} dt\, a^3(t) \int_0^{\alpha t} dL\, n(L, t) L\, S\left(\frac{Lf}{a(t)}\right)\,, \qquad (22.254)$$

where we henceforth use simply f to denote f_{obs}, and we have denoted the integration variable t_e simply by t. Therefore, using the definition (7.198), $\Omega_{\mathrm{gw}}(f) = \rho_c^{-1} f d\rho_{\mathrm{GW}}/df$, together with $\rho_c = 3H_0^2/(8\pi G)$, we get

$$h_0^2 \Omega_{\mathrm{gw}}(f) = \frac{8\pi}{3}\left(\frac{h_0}{H_0}\right)^2 (G\mu)^2 f \int_{t_*}^{t_0} dt\, a^3(t) \int_0^{\alpha t} dL\, n(L, t) L\, S\left(\frac{Lf}{a(t)}\right).$$

$$(22.255)$$

Given the function $S(x)$ that models the loop emission and the number density of loops $n(L,t)$, this integral can then be computed numerically. Currently the typical value of the power P_n in a realistic string network is not known. One can, however, study various limiting cases. In particular, as a model for the loop radiation we can use the radiation from loops with cusps, as given in eq. (22.242), or that for loops with kinks, as given in eq. (22.243). For the number density we can use the distribution for large loops given in eqs. (22.246) and (22.247), which is the case favored by the most recent numerical studies. Alternatively, one can study the contribution from a population of small loops, with $\alpha \sim \Gamma G\mu$, which was suggested by earlier analytic and numerical studies that assumed that the loop size is determined by the smallest length-scale on which gravitational back-reaction washes out inhomogeneities on the string. Actually, more recent studies have suggested that such small-scale structures should be even smaller, with $\alpha \sim (\Gamma G\mu)^n$, with $n = 3/2$ in RD and $n = 5/2$ in MD. It is conventional to parametrize this uncertainty by introducing $\varepsilon \equiv \alpha/(\Gamma G\mu)$, so $\varepsilon = 1$ corresponds to the "old" scenario, while $\varepsilon \ll 1$ in the scenario with $\alpha \sim (\Gamma G\mu)^n$ with $n > 1$; for instance, if $\Gamma = 50$, $G\mu = 10^{-7}$ and $n = 5/2$ we get $\varepsilon \sim 10^{-8}$.

A final parameter that must be taken into account is the recombi-nation probability p, which measures the probability that two strings

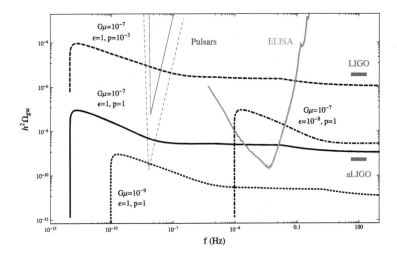

Fig. 22.19 $h_0^2\Omega_{\mathrm{gw}}(f)$ for cosmic strings with cusps, with a small initial loop size $\varepsilon \equiv \alpha/(\Gamma G\mu)$, for different choices of ε, $G\mu$ and the reconnection probability p, assuming $\Gamma = 50$. From Binetruy, Bohe, Caprini and Dufaux (2012). ©SISSA Medialab Srl. Reproduced by permission of IOP Publishing. All rights reserved.

crossing each other exchange their segments, as in Fig. 22.17. For the effective Nambu–Goto strings that describe Nielsen–Olesen vortices one basically has $p = 1$. However, it has also been suggested that fundamental superstrings can expand to cosmological scales. For fundamental strings in typical string models, where the string scale is close to the Planck scale, this is quite unlikely, since such strings would be unstable to break-up into smaller Planckian-size strings, and even if they were stable, they would anyhow be removed by inflation. However, in string models with a lower scale, based for example on large extra dimensions, fundamental strings (F-strings) or one-dimensional D-branes (D-strings) of cosmological size could exist, and could be produced after inflation. For such strings the reconnection probability p is suppressed by a factor of order g_s^2, where g_s is the string coupling. Furthermore, strings moving in a higher-dimensional space-time can avoid intersections more easily. As a result, one estimates $10^{-3} \lesssim p \lesssim 1$ for F-strings and $0.1 \lesssim p \lesssim 1$ for D-strings. In turn, this gives a larger number density of long strings, since a small reconnection probability means that processes such as those in Fig. 22.18 are suppressed, and long strings are much more abundant. Indeed, $n(L,t)$ scales possibly as $1/p$ (although the exact dependence is still debated) and the values in eqs. (22.246) and (22.247) correspond to $p = 1$.

When computing the stochastic background of GWs one must also take into account that, depending on the detector sensitivity, some of the strongest bursts coming from cusps could be detected individually (a point that we will discuss in more detail later), and therefore must be subtracted from the contribution of the stochastic background.

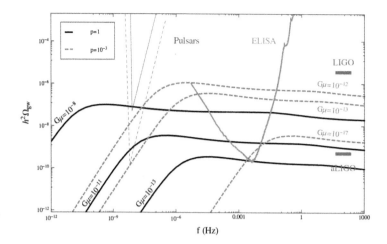

Fig. 22.20 $h_0^2\Omega_{\rm gw}(f)$ for cosmic strings with cusps, with a large initial loop size ($\alpha = 0.1$) and different choices of $G\mu$ and the reconnection probability p, assuming $\Gamma = 50$. From Binetruy, Bohe, Caprini and Dufaux (2012). ©SISSA Medialab Srl. Reproduced by permission of IOP Publishing. All rights reserved.

Figure 22.19 shows $h_0^2\Omega_{\rm gw}(f)$ generated by strings with cusps, taking for $n(L,t)$ a distribution of small loops with a fixed value $\varepsilon \equiv \alpha/(\Gamma G\mu)$, for different choices of ε, $G\mu$, and the reconnection probability p, assuming $\Gamma = 50$. The theoretical prediction is compared with the limit set by the LIGO S5 run, with the design sensitivity of advanced LIGO, with the sensitivity expected for LISA and with pulsar timing experiments. Pulsar timing arrays (PTAs) will be discussed in Chapter 23, and provide very interesting limits in the frequency range between 10^{-9} and 10^{-7} Hz. The solid wedge-shaped line corresponds to the current sensitivities, and the dashed line to expected future sensitivities.

Figure 22.20 shows $h_0^2\Omega_{\rm gw}(f)$ generated again by strings with cusps, but taking for $n(L,t)$ a distribution of large loops with $\alpha = 0.1$ and various values of $G\mu$ and p. For large loop size and recombination parameter $p = 1$, pulsar timing arrays impose a very strong bound. The nine-year data set released in 2015 by the pulsar timing array NANOGrav provides a limit

$$G\mu < 3.3 \times 10^{-8}\,, \tag{22.256}$$

at 95% c.l., while the European Pulsar Timing Array (EPTA) currently provides a limit $G\mu < 1.3 \times 10^{-7}$. These bounds are stronger than the CMB limit (22.176), although they are more model-dependent since various parameters, such as Γ and α, enter the computation. The space interferometer LISA is expected to reach values as small as $G\mu \sim 10^{-13}$, for $p = 1$ and large loops.

GW bursts from cusps and kinks

Beside generating a stochastic background, cusps and kinks on strings can also generate GW bursts so strong that, potentially, they could be detectable individually. To compute the burst generated by a cusp one can use the expansion (22.228), (22.229) for the time evolution of the cosmic string near the cusp, and plug it into the energy–momentum tensor of the string, eq. (22.218). The resulting waveform is obtained using the formalism of Section 3.1 of Vol. 1, valid for weak-field sources with arbitrary velocity. This gives the result in the local wave zone, i.e. at distances large compared with the size of the source, but small compared with cosmological scales. One must then evolve the GWs to the present time through a FRW metric. The computation has been performed in detail in Damour and Vilenkin (2000, 2001). The result is that the amplitude in Fourier space is

$$\tilde{h}(f) \sim \frac{G\mu L}{[(1+z)fL]^{1/3}} \frac{1+z}{t_0 z}, \qquad (22.257)$$

where t_0 is the present value of cosmic time and z the redshift of the source. The Fourier mode $\tilde{h}(f)$ is focused in a narrow cone in the direction of the cusp velocity, with an opening angle

$$\theta_m \simeq [(1+z)fL/2]^{-1/3}. \qquad (22.258)$$

Writing the loop length as $L = \alpha t$, and expressing t in terms of the redshift, one finds

$$\tilde{h}(f) \sim G\mu\alpha^{2/3}(ft_0)^{-1/3}\phi_h(z), \qquad (22.259)$$

where

$$\phi_h(z) = z^{-1}(1+z)^{-1/3}(1+z/z_{\rm eq})^{-1/3}. \qquad (22.260)$$

The rate of bursts due to cusps can be estimated assuming that C cusps are formed per oscillation cycle of the string, with $C = O(1)$. A string with length L has on oscillation period $T(L) = L/2$ [see eq. (22.237)], so the number of cusp events per unit space-time volume is

$$\nu(t) \simeq \frac{Cn(L,t)}{T(L)}$$
$$= \frac{2Cn(L,t)}{L}. \qquad (22.261)$$

Then, the rate of GW bursts in the direction of the observer due to cusps in the redshift interval dz, observed at frequency f, is

$$d\dot{N} \sim \frac{\theta_m^2}{4}(1+z)^{-1}\nu(z)dV(z), \qquad (22.262)$$

where the factor $\theta_m^2/4$ is the beaming angle of the burst, the factor $(1+z)^{-1}$ comes from the cosmological time dilatation between the time $t_{\rm obs}$ measured by the observer and the time t of the source, $dt_{\rm obs} =$

$(1+z)dt$, and $dV(z)$ is the proper volume between redshifts z and $z+dz$. In MD

$$dV(z) = 54\pi t_0^3 \left[(1+z)^{1/2} - 1\right]^2 (1+z)^{-11/2} dz, \qquad (22.263)$$

while in RD

$$dV(z) = 72\pi t_0^3 (1+z_{\rm eq})^{1/2} (1+z)^{-5} dz. \qquad (22.264)$$

Given the amplitude of a single burst (22.257) and the burst rate (22.262) as a function of redshift, one can compute the typical amplitude of a burst generated by cusps over a period, say, of one year, and compare with the experimental sensitivities. A similar analysis can be done for kinks. Similarly to the stochastic background case discussed earlier, the result depends on the population of loops considered, in particular on whether one considers loops with typical length $L = \alpha t$ with $\alpha \sim \Gamma G\mu$, which was once considered the standard scenario, or whether $\alpha \sim 0.1$, which is favored by the most recent simulations, or even values of $\varepsilon = \alpha/(\Gamma G\mu) \ll 1$, as discussed following eq. (22.255). As we see from eq. (22.259), the signal is not sharply peaked at a specific frequency, but rather has a smooth power-law behavior, so it can be relevant both for ground-based and for space-borne interferometers.

The result for the scenario with $\varepsilon = 1$ is shown in Fig. 22.21, in the frequency range of LIGO/Virgo, as a function of $\alpha = 50G\mu$. We see that, for some values of the parameters, the signal is potentially detectable.

22.6   Alternatives to inflation

We have seen in Chapter 21 that inflation provides an elegant solution to the homogeneity and flatness problems of FRW cosmology (as well as diluting unwanted relics such as monopoles that could appear in extensions of the Standard Model and would otherwise dominate over radiation since their density scales slower than $1/a^4$), and furthermore provides the initial adiabatic perturbations that seed structure formation. As such, it is a very successful paradigm. The simplest scenario for obtaining inflation is the single-field slow-roll inflation that we discussed a length in Section 21.1, or relatively simple modifications of it, such as the hybrid inflationary model [see eq. (21.82)]. There have, however, been several attempts at incorporating inflation into a more fundamental theory, such as string theory, that could at the same time address fundamental problems such as the big-bang singularity. Such models have in general distinctive predictions for the stochastic GW background, which differ from that of standard slow-roll inflation computed in Section 21.3.[42] In the following we will briefly discuss some of these models, just to give a flavor of how the prediction of standard slow-roll inflation can be modified in alternative theories of inflation. Of course, all these models suffer from uncertainties due to the difficulty of dealing with the regime of Planckian curvatures, so the results must be taken only as examples of possible scenarios.

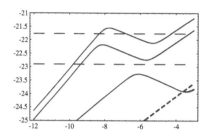

Fig. 22.21 GW amplitude of bursts emitted by cosmic string cusps (upper and middle curves) and kinks (lower curve), in the LIGO/Virgo frequency band, as a function of $\alpha = 50G\mu$, in a base-10 log–log plot. The upper curve assumes that the average number of cusps per loop oscillation is $C = 1$. The middle curve assumes $C = 0.1$. The lower curve gives the kink signal (assuming only one kink per loop). The horizontal dashed lines indicate the 1-sigma noise levels (after optimal filtering) of initial LIGO and advanced LIGO. The short-dashed line indicates the "confusion" amplitude noise of the stochastic GW background. From Damour and Vilenkin (2001).

[42] It should also be observed that in principle these models could also be joined, at sub-Planckian curvature, to standard slow-roll inflation, providing a way of addressing the initial singularity problem while still matching later to standard inflationary cosmology.

As an example of such alternatives to standard inflation, we consider the pre-big-bang model. The starting point is the low-energy effective action from string theory, which, besides the metric, also includes a scalar field, the dilaton ϕ.[43] The effective action in the metric–dilaton sector, to lowest order in the derivatives and in the coupling e^ϕ, is given by

$$S_{\text{eff}} \sim \int d^D x \sqrt{-g} \left[e^{-\phi} \left(R - \partial_\mu \phi \partial^\mu \phi \right) \right] , \qquad (22.265)$$

where ϕ is the dilaton field. This action receives two kinds of perturbative corrections: corrections parametrized by e^ϕ, which are just higher loops in the field-theory sense, and corrections proportional to α', where $\lambda_s = \sqrt{2\alpha'}$ is the string length. The latter are genuinely string effects, and take into account the finite size of the string. Since α' has dimensions of $(\text{length})^2$, it is associated with higher-derivative operators such as the square of the Riemann tensor, $R^2_{\mu\nu\rho\sigma}$. Remarkably, there exists a regime in the early Universe evolution of this model where the perturbative approach is well justified, and in a first approximation one can use eq. (22.265), neglecting the corrections. This regime occurs if we take as initial condition a Universe at weak coupling and low curvatures. Then, a generic inhomogeneous string vacuum shows a gravitational instability. Specializing for simplicity to a homogeneous model, the solution of the equations of motion derived from the action (22.265) reads, for generic anisotropic FRW scale factors $a_i(t)$, $i = 1, \ldots, D-1$,

$$a_i(t) = (-t)^{c_i} ,$$
$$\phi_i(t) = \phi_0 + c_0 \log(-t) , \qquad (22.266)$$

with

$$\sum_{i=1}^{D-1} c_i^2 = 1 , \qquad \sum_{i=1}^{D-1} c_i = 1 + c_0 . \qquad (22.267)$$

This is just a generalization in the presence of the dilaton of the Kasner solution of general relativity. Considering for simplicity an isotropic model, one has in particular a solution with $c_i = -1/\sqrt{D-1}$ and $c_0 = -1 - \sqrt{D-1}$. This corresponds to a super-inflationary evolution: the Hubble parameter $H = \dot{a}/a$ grows until it reaches the string scale $1/\lambda_s$, and eventually formally diverges at a singularity. When $H \sim 1/\lambda_s$, of course, the description based on the lowest-order action is no longer adequate. The inclusion of α' corrections, at least to first order, i.e. terms $O(R^2_{\mu\nu\rho\sigma})$, has been shown to turn the unbounded growth of the curvature into a de Sitter phase with a linearly growing dilaton. Of course, in this regime the curvature is anyhow of order one in string units, so a full non-perturbative study would in principle be necessary, but the fact that the $O(\alpha')$ corrections already regularize the big-bang singularity is encouraging.

Since in this new regime the dilaton keeps growing linearly, finally even e^ϕ becomes large, even if the evolution started at very weak coupling; at this stage loop corrections are important, and they can trigger the exit from the de Sitter phase. An example of this behavior, using both α' corrections to first order and loop corrections derived from orbifold compactifications of string theory (using modular-invariant loop corrections to the Kähler potential known to all orders, as well as one-loop corrections to the gravitational couplings of the higher-derivative terms), has been studied explicitly, and is shown in Fig. 22.22.

Using this type of evolution as an example of what could be obtained from string theory, the corresponding stochastic GW background can be computed as follows. We consider a model with three phases: first an isotropic superinflationary evolution described by eq. (22.266), with the value of the constants c_i and c_0 determined by eq. (22.267). Then this phase is matched to a de Sitter phase with linearly growing dilaton, which may be considered as representative of a typical solution in the large-curvature regime; the constant values of H and $\dot\phi$ in this phase are taken as free parameters, of the order of the string scale. Finally, we match this phase to the standard RD era. Of course, this model can only be taken as indicative of possible behavior in string cosmology, since what happens in the large-curvature phase is certainly much more complicated, and a full understanding would require a non-perturbative approach. However, we will see that some important characteristics of the GW spectrum depend only on the low-curvature regime, where we can use the lowest-order effective action and which is therefore under good theoretical control.

We specialize for illustration to a isotropic model compactified to four dimensions. Writing the scale factors and the dilaton in terms of conformal time, from eq. (22.266) we find that for $-\infty < \eta < \eta_s$ (with $\eta_s < 0$) we have a dilaton-dominated regime with

$$a(\eta) = -\frac{1}{H_s\eta_s} \left(\frac{\eta - (1-\alpha)\eta_s}{\alpha\eta_s} \right)^{-\alpha} , \qquad (22.268)$$

$$\phi(\eta) = \phi_s - \gamma \log \frac{\eta - (1-\alpha)\eta_s}{\eta_s} , \qquad (22.269)$$

and the Kasner conditions eq. (22.267) give $\alpha = 1/(1+\sqrt{3})$ and $\gamma = \sqrt{3}$.

At a value $\eta = \eta_s$ the curvature becomes of the order of the string scale, and there we assume a de Sitter expansion with linearly growing dilaton. In terms of conformal time, this means

$$a(\eta) = -\frac{1}{H_s\eta} , \qquad \phi(\eta) = \phi_s - 2\beta \log \frac{\eta}{\eta_s} . \qquad (22.270)$$

The parameter H_s is of the order of the string scale, while β is a free parameter of the model. Note that $2\beta = \dot\phi/H_s$, where $\dot\phi$ is the derivative with respect to cosmic time, and in this model $\beta > 0$.

The stringy phase lasts for $\eta_s < \eta < \eta_1$ (again $\eta_1 < 0$), and at η_1 we match this phase to a standard RD era. This gives, for $\eta_1 < \eta < \eta_r$ (with $\eta_r > 0$),

$$a(\eta) = \frac{1}{H_s\eta_1^2} (\eta - 2\eta_1) , \qquad \phi = \phi_0 . \qquad (22.271)$$

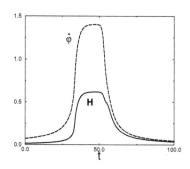

Fig. 22.22 The evolution of the Hubble parameter H and of $\dot\phi$ in string cosmology with α' and loop corrections. From Foffa, Maggiore and Sturani (1999).

Then, the standard MD era follows. We have chosen the additive and multiplicative constants in $a(\eta)$ in such a way that $a(\eta)$ and $da/d\eta$ (and therefore also da/dt) are continuous across the transitions. The equation for the tensor perturbations in these three regimes can be solved exactly (see the Further Reading), and matched at the transition points. The matching gives the Bogoliubov coefficients, similarly to the computation in Section 21.3.7. The quantity η_1 can be estimated by observing that, since it marks the end of the de Sitter phase, $|\eta_1| = 1/[H_s a(\eta_1)]$. On the other hand, assuming no intermediate inflationary phase and a direct matching to RD, η_1 also marks the beginning of the RD phase. Therefore, the Friedmann equation (17.117) gives

$$H_s \simeq 0.62 \left(\frac{106.75}{g_*(T)}\right)^{1/6} \frac{H_0 \Omega_R^{1/2}}{a^2(\eta_1)}. \qquad (22.272)$$

Then, writing

$$k|\eta_1| = \frac{2\pi f}{H_s a(\eta_1)}, \qquad (22.273)$$

the parameter η_1 can be traded for a parameter f_1 defined by $k|\eta_1| = f/f_1$. Numerically,

$$f_1 \simeq 2.5 \times 10^{10} \, \text{Hz} \, \left(\frac{H_s}{0.1 M_{\text{pl}}}\right)^{1/2} \left(\frac{106.75}{g_*}\right)^{1/12}, \qquad (22.274)$$

where we have chosen a reference value H_s of the order of a typical string scale. Similarly, we can introduce a parameter f_s instead of η_s, from $k|\eta_s| = f/f_s$. However, this parameter depends on the duration of the string phase, and therefore, in contrast to f_1, is totally unknown, even as an order of magnitude. However, since $|\eta_1| < |\eta_s|$, we have $f_s < f_1$.

To summarize, the model has a parameter f_s with dimensions of a frequency, which can have any value in the range $0 < f_s < f_1$, with $f_1 = O(1-10)$ GHz, and a dimensionless parameter β. It is useful to further trade the latter for the parameter $\mu = |\beta - (3/2)|$.

Performing the matching between the solutions in the three phases one obtains the Bogoliubov coefficients and then one finally gets the spectrum. The computation can be done fully analytically. From the exact expressions one finds that, in the regime $f \ll f_s$ we have $h_0^2 \Omega_{\text{gw}}(f) \propto f^3$, while for $f_s \ll f < f_1$ the slope changes to $h_0^2 \Omega_{\text{gw}}(f) \propto f^{3-2\mu}$. Thus, the exact analytic solution obtained from the computation of the Bogoliubov coefficients is well approximated by

$$h_0^2 \Omega_{\text{gw}}(f) \simeq h_0^2 \Omega_{\text{gw}}(f_1) \times \begin{cases} \left(\frac{f_1}{f_s}\right)^{2\mu} \left(\frac{f}{f_1}\right)^3 & (f \ll f_s), \\ \\ \left(\frac{f}{f_1}\right)^{3-2\mu} & (f_s \ll f < f_1), \end{cases}$$
$$(22.275)$$

where the factor $(f_1/f_s)^{2\mu}$ is determined by continuity in $f = f_s$. The overall normalization is determined by the fact that the value of $h_0^2 \Omega_{\text{gw}}(f)$

at the cutoff value f_1 is fixed by the scale H_s reached by the Hubble parameter during the string phase. Writing $H_s = g_1 M_{Pl}$, one finds $h_0^2 \Omega_{gw}(f_1) \simeq g_1^2 h_0^2 \Omega_R$, where Ω_R is the present fraction of the energy density in radiation. Assuming that $H_s = g_1 M_{Pl}$ is of the order of the string scale, one gets, in typical string models, $g_1 \sim 0.1$. Thus, the value of the peak frequency f_1, as well as that $h_0^2 \Omega_{gw}(f_1)$, is fixed, at least within some order of magnitude, and assuming typical values for the string mass scale. In contrast, the parameters (μ, f_s) must be taken as free parameters of the model, except for the bounds $\mu \geqslant 0$ and $f_s < f_1$ that follow from their definition. Furthermore we must have $\mu \leqslant 3/2$, otherwise, moving from its value at $f = f_s$ toward lower frequencies, $h_0^2 \Omega_{gw}(f)$ would increase and would eventually violate observational limits, such as the nucleosynthesis bound, that we will discuss in Section 22.7. Depending on the values of the two parameters μ and f_s, a broad range of possibilities for the GW spectrum is possible. If f_s is very close to f_1, or if μ is very close to zero, when moving from its peak at f_1 toward lower frequencies the spectrum decreases as f^3 (just like a black-body spectrum), and is totally unobservable at the frequencies at which ground-based interferometers, space interferometers and pulsar timing array operate. This corresponds to the solid line in Fig. 22.23. In contrast, if $\mu \simeq 3/2$, the spectrum is almost flat in the region $f_s \ll f < f_1$, so, if f_s is sufficiently small, it can be sizable at the frequencies of interferometers, or even pulsar timing, and in fact possibly even at the very small frequencies probed by the CMB, as shown by the two dashed lines in Fig. 22.23.

The preceding computation only takes into account the matching between the string phase and the RD phase. For waves that enter during RD this is all that is needed. Otherwise, in general, the spectrum is multiplied by the transfer function $T^2(f/f_{eq})$, just as we discussed for the slow-roll inflationary spectrum. For $f < f_{eq}$ we then have an extra factor of order $(f_{eq}/f)^2$. Thus, just as in slow-roll inflation $h_0^2 \Omega_{gw}(f)$ changes from an (almost) flat spectrum at $f \gg f_{eq}$ to a $1/f^2$ behavior at $f < f_{eq}$, in the pre-big-bang scenario $h_0^2 \Omega_{gw}(f)$ changes from an f^3 to an f behavior. Of course, this is relevant only if the knee frequency f_s is so low as to be itself comparable to f_{eq}, otherwise, at $f < f_{eq}$, $h_0^2 \Omega_{gw}(f)$ is anyhow utterly negligible. However, this is the case that would be interesting for detection of this stochastic background from its effect on the CMB. The $\propto f$ behavior of $h_0^2 \Omega_{gw}(f)$ at $f \lesssim 10^{-17}$ Hz is visible in the left-most curve of Fig. 22.23.

Despite this wide range of possibilities, the low-frequency behaviors $\propto f$ at $f < f_{eq}$ and $\propto f^3$ at $f_{eq} < f < f_s$ are general properties of this model that depend only on the form of the low-energy effective action, and would therefore be a specific signature of the model. Performing the exact computation one finds that some oscillations are also superimposed on the asymptotic behaviors (22.275), and are visible when the spectrum is plotted on a more narrow range of frequencies; see Fig. 22.24.

The main phenomenological difficulty that must be addressed in this model, to make it a viable competitor with standard slow-roll inflation,

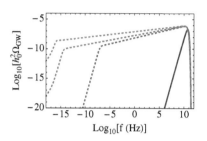

Fig. 22.23 $\Omega_{gw}(f)$ versus f for different values of f_s and μ. From left to right, the curves correspond to $\{f_s = 10^{-16} \text{ Hz}, \mu = 1.45\}$, $\{f_s = 10^{-14.5} \text{ Hz}, \mu = 1.42\}$ and $\{f_s = 10^{-7} \text{ Hz}, \mu = 1.40\}$ (all three dashed), and $f_s = f_1$ (solid line). We set $f_1 = 2.5 \times 10^{10}$ Hz and $h_0^2 \Omega_{gw}(f_1) = 10^{-6}$.

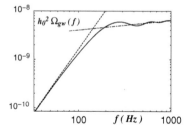

Fig. 22.24 $\Omega_{gw}(f)$ versus f for $\mu = 1.4$, $f_s = 100$ Hz and $f_1 = 2.5 \times 10^{10}$ Hz (solid line). The dashed lines are the low- and high-frequency limits. From Buonanno, Maggiore and Ungarelli (1997).

is the generation of a nearly scale-invariant spectrum of adiabatic perturbations. The simplest mechanism give an unacceptable blue-tilted spectrum of perturbations. However, mechanisms based on the decay of massive axions ("curvaton mechanism") could give a correct spectrum of primordial perturbations; see the Further Reading. In any case, the pre-big-bang model provides an interesting explicit example of how the spectrum of relic GWs can be a signature of alternative early-universe scenarios, as we can appreciate by comparing the pre-big-bang prediction in Fig. 22.23 with the predictions from standard single-field slow-roll inflation in Figs. 21.7 and 21.9.

Several other alternative models addressing the issue of the big-bang singularity have been proposed. Many of them are characterized by an initial contracting phase, leading to Planckian curvatures. At this point the universe bounces back, and the standard phase of expansion follows. Similarly to the pre-big-bang model, the main theoretical problem is the lack of control on the Planckian regime. An example of this family is the ekpyrotic model, which can be described in terms of collision of branes moving in an extra spatial dimension, as well as related cyclic models where the Universe undergoes subsequents phases of contraction, bounce and expansion; see the Further Reading. These models generically predict a strongly blue spectrum, similar to the $f_s = f_1$ case of the pre-big-bang model, shown as a solid line in Fig. 22.23. Once their peak value at high frequency has been normalized so that it does not exceed the nucleosynthesis bound (which we will discuss in Section 22.7), the GW spectrum in these models is therefore completely unobservable at all frequencies at which existing or planned detectors operate.

22.7 Bounds on primordial GW backgrounds

We next investigate the observational limits on stochastic GW backgrounds of primordial origin that can be obtained from various cosmological observations.

22.7.1 The nucleosynthesis bound

Big-bang nucleosynthesis (BBN) successfully predicts the primordial abundances of deuterium (D), ^{3}He, ^{4}He and ^{7}Li (while heavier elements, up the iron group, are synthesized in stars, and even heavier elements are produced through neutron capture, as we discussed in Section 15.3.4). In the simplest scenario the only cosmological parameter that enters into these predictions is the baryon-to-photon ratio $\eta_B = n_B/n_\gamma$.[44] The agreement with observation of the prediction for the primordial abundances of these four nuclei (whose abundances span more than nine of orders of magnitude) in terms of a single parameter η_B has long been one of the cornerstones of standard big-bang cosmology. In more recent years, η_B has also been accurately determined by CMB experiments

[44]There are in principle analogous neutrino asymmetry parameters η_{ν_l}, which, however, have an effect only if they are several orders of magnitude larger than η_B. In standard BBN one assume that this is not the case, for example that η_{ν_l} is of order η_B, in which case they have a negligible effect, and can be omitted altogether.

such as WMAP and *Planck*, through their determination of the baryon fraction Ω_B. The agreement between the CMB and the BBN determinations of η_B is another remarkable success of the standard cosmological model. In turn, this agreement puts constraints on possible deviations from the standard cosmological model at the BBN epoch. In particular it constraints the existence of extra forms of energy density at the BBN epoch, such as the energy density in GWs or in more exotic components such as primordial black holes, which are not included in the primordial nucleosynthesis computations, and which would spoil the agreement if they were too large.

To physically understand how such a limit comes out, we begin by observing that a crucial quantity in the computations of BBN is the ratio of the number density of neutrons, n_n, to that of protons, n_p. As long as thermal equilibrium is maintained we have (for non-relativistic nucleons, as is appropriate at $T \sim$ MeV, when BBN takes place)

$$\frac{n_n}{n_p} = e^{-Q/T}, \qquad (22.276)$$

where $Q = m_n - m_p \simeq 1.3$ MeV. Equilibrium is maintained by the process $pe \leftrightarrow n\nu$, as long as its rate $\Gamma_{pe \to n\nu}$ is much larger than the Hubble parameter $H(t)$. When the rate drops below $H(t)$ the process can no longer compete with the expansion of the Universe and the ratio n_n/n_p remains frozen at the value $\exp(-Q/T_{\rm f})$, where $T_{\rm f}$ is the value of the temperature at the time of freeze-out. This number therefore determines the density of neutrons available for nucleosynthesis. Since practically all neutrons available will eventually form ^{4}He, the final primordial abundance of ^{4}He is exponentially sensitive to the freeze-out temperature $T_{\rm f}$.

Precise computations of BBN involve detailed numerical studies of the relevant kinetic equations, including a whole series of thermonuclear reactions. However, the main effect of extra contributions to the energy density on $T_{\rm f}$ can be easily understood analytically by restricting to the process $pe \leftrightarrow n\nu$ and taking for simplicity

$$\Gamma_{pe \to n\nu} \sim G_F^2 T^5 \qquad (22.277)$$

(which is really appropriate only in the limit $T \gg Q$). The Hubble parameter is given as usual by the Friedmann equation (17.48), $H^2 = (8\pi/3)G\rho$, where ρ includes all forms of energy density at the time of BBN. BBN takes place deep into the RD phase, but slightly below the MeV scale, when ρ is made up of photons and neutrinos plus any types of extra energy density $\rho_{\rm extra}$ that might be present, such as, indeed, GWs,

$$\rho = \rho_\gamma + \rho_\nu + \rho_{\rm extra}. \qquad (22.278)$$

The principle of the nucleosynthesis bound is that, since the computation performed with $\rho_{\rm extra} = 0$ correctly reproduces the primordial abundances of the light elements, any extra form of energy density present at the time of nucleosynthesis cannot be too large, since otherwise it would spoil the agreement. To understand the physics involved we must take into account that there are several events that happen at temperatures around an MeV.

First of all, the electron neutrino ν_e decouples from the primordial plasma at $T \simeq 1.9$ MeV (while ν_μ and ν_τ already decoupled at $T \simeq 3.1$ MeV; see Note 25 on page 383). Subsequently, at $T \simeq 0.3$ MeV, e^+e^- annihilation takes place. As we have discussed in Section 17.5, in a first approximation the entropy of electrons and positrons is transferred uniquely to the photons, whose temperature is thereby raised, compared with the neutrino temperature, by a factor $(11/4)^{1/3}$, as we saw in eq. (17.106). However, neutrino decoupling is not instantaneous, and given that the temperature corresponding to an ideal instantaneous neutrino decoupling is close to the temperature of e^+e^- annihilation, a small fraction of the $e^\pm$ entropy actually goes into neutrinos. This, together with a finite-temperature QED effect due to the fact that electromagnetic interactions in the plasma modify the $e^\pm$ and γ dispersion relation, actually increases the ratio T_ν/T_γ, compared with the value $(4/11)^{1/3}$. It is customary to take this effect into account by introducing an effective number of neutrino species [see eq. (17.107)], which in the Standard Model with three neutrino families turns out to have the value $N_{\rm eff}^{(\nu)} \simeq 3.046$ [see eq. (17.108)].

A third relevant temperature is given by the temperature at which BBN starts. Naively, one would imagine that nucleosynthesis begins, with the formation of deuterium through the process $p + n \to \mathrm{D} + \gamma$, when the temperature drops below the deuterium binding energy, $B_\mathrm{D} \simeq 2.2$ MeV. However, the value of η_B is of order 10^{-9}, so the number density of photons exceeds that of baryons by a factor $\eta_B^{-1} \sim 10^9$. This means that, even at temperatures somewhat below B_D, there are still enough photons in the high-energy tail of the thermal distribution to destroy the newly formed deuterium nuclei through the inverse process $\mathrm{D} + \gamma \to p + n$. Deuterium production is then delayed (the so-called deuterium bottleneck) until $T \simeq 100$ keV, when the number density of the photons in the energy tail of the thermal distribution has been sufficiently suppressed. Once deuterium begins to form, the whole chain of BBN reactions starts. Thus, because of the deuterium bottleneck, BBN takes place after neutrino decoupling and e^+e^- annihilation, when the neutrino energy density has the form (17.107), with $N_{\rm eff}^{(\nu)} \simeq 3.046$.

To take into account the possible existence of extra energy components, it is convenient to introduce an "effective number of neutrino species" $N_{\rm eff}$, defined so that before e^+e^- annihilation

$$\rho_\nu + \rho_{\rm extra} \equiv \frac{7}{8} N_{\rm eff} \rho_\gamma \,. \tag{22.279}$$

Thus, the addition of an extra component with an energy density equal to that of a massless neutrino would correspond to increasing $N_{\rm eff}$ by one unit, compared with its Standard Model value, but of course in general this is just a way of parametrizing any extra form of energy. To obtain the expression for $\rho_\nu + \rho_{\rm extra}$ just after e^+e^- annihilation we assume

that, similarly to the neutrinos, ρ_{extra} describes a relativistic component, decoupled from the primordial plasma (as would indeed be the
case for GWs). Then both ρ_ν and ρ_{extra} scale as $1/a^4$, while the photon temperature evolves according to $g_*^S(T)T^3 a^3 = \text{const}$, i.e. constant
entropy, so

$$\rho_\gamma \sim T^4 \sim \frac{1}{a^4 (g_*^S)^{4/3}} \,. \tag{22.280}$$

As we saw in Section 17.5, before e^+e^- annihilation the active degrees
of freedoms are the photon (with two helicities), $e^\pm$ with two helicities
each, and the three neutrinos and three antineutrinos, all with the same
temperature. Then, at this epoch, $g_*^S(1\,\text{MeV}) = 2+(7/8)(4+6) = 43/4$,
while, after e^+e^- annihilation, $(g_*^S)_0 = 43/11$ [see eq. (17.103)], so

$$\frac{(g_*^S)_0}{g_*^S(1\,\text{MeV})} = \frac{4}{11} \,. \tag{22.281}$$

Then, just after e^+e^- annihilation (and up to the present epoch, since,
neglecting neutrino masses, no other particle becomes non-relativistic in
the subsequent evolution),

$$\rho_\nu + \rho_{\text{extra}} \equiv N_{\text{eff}} \frac{7}{8} \left(\frac{4}{11}\right)^{4/3} \rho_\gamma \,. \tag{22.282}$$

We can then write, after e^+e^- annihilation,

$$\rho_{\text{extra}} = (N_{\text{eff}} - N_{\text{eff}}^{(\nu)}) \frac{7}{8} \left(\frac{4}{11}\right)^{4/3} \rho_\gamma$$

$$\simeq (N_{\text{eff}} - 3.046) \frac{7}{8} \left(\frac{4}{11}\right)^{4/3} \rho_\gamma \,. \tag{22.283}$$

The Hubble parameter during BBN can be written as

$$H^2 = \frac{8\pi G}{3} \left[1 + N_{\text{eff}} \frac{7}{8} \left(\frac{4}{11}\right)^{4/3}\right] \rho_\gamma$$

$$= \frac{8\pi G}{3} \left[1 + N_{\text{eff}} \frac{7}{8} \left(\frac{4}{11}\right)^{4/3}\right] \frac{\pi^2}{15} T^4 \,. \tag{22.284}$$

The freeze-out temperature T_{f} is determined by the condition $\Gamma \sim H$,
i.e.

$$G_F^2 T_{\text{f}}^5 \sim \left(\frac{8\pi^3}{45}\right)^{1/2} \left[1 + N_{\text{eff}} \frac{7}{8} \left(\frac{4}{11}\right)^{4/3}\right]^{1/2} \frac{T_{\text{f}}^2}{M_{\text{Pl}}} \,, \tag{22.285}$$

which gives

$$T_{\text{f}} \sim \left(\frac{8\pi^3}{45}\right)^{1/6} \left[1 + N_{\text{eff}} \frac{7}{8} \left(\frac{4}{11}\right)^{4/3}\right]^{1/6} \frac{1}{(G_F^2 M_{\text{Pl}})^{1/3}} \,. \tag{22.286}$$

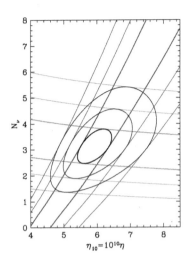

Fig. 22.25 BBN-only constraints on η_B and $N_{\rm eff}$. The nearly vertical curves are the 1σ, 2σ and 3σ limits from D/H, while the nearly horizontal curves are from ^{4}He, and the closed contours combine both. From Cyburt, Fields, Olive and Skillman (2005).

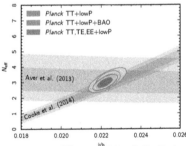

Fig. 22.26 The combined limit in the plane $(N_{\rm eff}, \omega_b = h_0^2 \Omega_B)$ from *Planck*+BAO data and BBN. From [Planck Collaboration], Ade *et al.* (2016a). "Planck 2015 results. XIII. Cosmological parameters". Reproduced with permission ©ESO.

We see that any extra form of energy, and in particular an energy density in relic gravitons, results in a larger freeze-out temperature, so more available neutrons and then overproduction of ^{4}He. However, since the density of ^{4}He also increases with the baryon-to-photon ratio η_B, we could compensate for an increase in $N_{\rm eff}$ with a decrease in η_B. Thus, the comparison with ^{4}He alone is not sufficient to put a bound on $N_{\rm eff}$. However, lowering the value of η_B results in an increase in the abundance of D, so this provides a lower limit on η_B. Figure 22.25 shows in particular that the measurements of the abundances of ^{4}He and D/H are nicely complementary for the determination of η_B and $N_{\rm eff}$. Indeed, the ^{4}He fraction has only a very weak (actually, logarithmic) sensitivity to η_B, but is very sensitive to $N_{\rm eff}$, while for D/H the opposite is the case. Combined together, they already provide a useful bound $N_{\rm eff} < 4.6$ at 95% c.l.

The limit is further improved by adding the value of η_B determined directly by the CMB, as was first done by WMAP and then, with improved accuracy, by *Planck*. In particular, combining *Planck* 2015 data from the TT, TE and EE correlators and low-multipole polarization (lowP) data, together with BBN computations of the deuterium abundance, gives

$$N_{\rm eff} = 2.91 \pm 0.37 \,, \tag{22.287}$$

at 95% c.l. (while using helium abundance give a similar value, $N_{\rm eff} = 2.99 \pm 0.39$). The combined contours are shown in Fig. 22.26.

If we use as upper limit the value $N_{\rm eff} = 2.91 + 0.37 = 3.28$, we have $N_{\rm eff} - 3.046 = 0.234$, which we use as the reference value for this quantity. Then, eq. (22.283) reads

$$\rho_{\rm extra}(t_{\rm BBN}) < 0.053 \left(\frac{N_{\rm eff} - 3.046}{0.234} \right) \rho_\gamma(t_{\rm BBN}) \,, \tag{22.288}$$

and the energy density in GWs at time of BBN is bounded by $\rho_{\rm gw}(t_{\rm BBN}) \leqslant \rho_{\rm extra}(t_{\rm BBN})$. To obtain a bound on the GW energy density at the present time we observe that, since we have defined $N_{\rm eff}$ in eq. (22.282) in terms of the energy density of neutrinos after $e^+ e^-$ annihilation, from this epoch to the present time there is no further change in the effective number of degrees of freedom $g_*(T)$, so the GW energy density scales exactly as the photon energy density. Thus, we get the nucleosynthesis bound at the present time,

$$\frac{\rho_{\rm gw}}{\rho_\gamma} < 0.053 \left(\frac{N_{\rm eff} - 3.046}{0.234} \right) \,. \tag{22.289}$$

Of course this bound holds only for GWs that were already produced at the time of nucleosynthesis, $T \sim 0.1$ MeV. It does not apply to stochastic backgrounds produced later, such as backgrounds of astrophysical origin.

Note that this is a bound on the total energy density in GWs, integrated over all frequencies. However, only the frequencies that were inside the horizon at the time of BBN contribute to the expansion of the Universe. The corresponding GW frequency today is obtained from eq. (22.6). We take for definiteness $T \sim 0.1$ MeV (corresponding to the onset of BBN after the deuterium bottleneck), so $g_*^S(T_*) = 43/11$, and we set for definiteness $\epsilon_* = 0.1$, to require that the (reduced) wavelengths are well inside the horizon scale at production.[45] Then this corresponds to GWs that today have a frequency

$$f_{\rm BBN} \simeq 1.5 \times 10^{-11}\,{\rm Hz}\,. \tag{22.290}$$

Writing, as in eq. (7.198) of Vol. 1,

$$\Omega_{\rm gw}(f) = \frac{1}{\rho_c}\frac{d\rho_{\rm gw}}{d\log f}\,, \tag{22.291}$$

and inserting the numerical value $h_0^2\rho_\gamma/\rho_c \simeq 2.473\times10^{-5}$ [see eq. (17.110)], eq. (22.289) gives[46]

$$\int_{f=f_{\rm BBN}}^{f=\infty} d(\log f)\, h_0^2\Omega_{\rm gw}(f) < 1.3 \times 10^{-6}\left(\frac{N_{\rm eff}-3.046}{0.234}\right)\,, \tag{22.292}$$

where $\log f$ denotes as usual the natural logarithm, and we have restricted the integral over f to $f > f_{\rm BBN}$. The BBN bound is therefore a bound on the integral of $h_0^2\Omega_{\rm gw}(f)$ for $f > f_{\rm BBN}$, and not directly on $h_0^2\Omega_{\rm gw}(f)$. From the mathematical point of view one could of course imagine a function $h_0^2\Omega_{\rm gw}(f)$ with a very narrow peak at some frequency f, with a peak value larger than the right-hand side of eq. (22.292), but sufficiently narrow that its contribution to the integral was small enough. However, all cosmological mechanisms that we have discussed rather give a spectrum that, even in the most peaked cases, such as phase transitions, still covers at least one decade in frequency. For such spectra, we can transform the bound in eq. (22.292) into a bound on $h_0^2\Omega_{\rm gw}(f)$, requiring that, if the integral cannot exceed a given value, even its positive-definite integrand $h_0^2\Omega_{\rm gw}(f)$ cannot exceed it over an appreciable interval of frequencies $\Delta\log f \sim 1$, so we can impose

$$h_0^2\Omega_{\rm gw}(f) < 1.3 \times 10^{-6}\left(\frac{N_{\rm eff}-3.046}{0.234}\right) \quad {\rm (for\ } f > f_{\rm BBN}). \tag{22.293}$$

This is the meaning of the bound marked as "BBN" in Fig. 22.27 (where we set $N_{\rm eff} - 3.046 = 0.234$). Of course, the precise bound on the spectrum depends on the actual shape of $h_0^2\Omega_{\rm gw}(f)$. For instance if $h_0^2\Omega_{\rm gw}(f)$ has the form

$$h_0^2\Omega_{\rm gw}(f) = (h_0^2\Omega_{\rm gw})_*\left(\frac{f}{f_*}\right)^\alpha\,, \tag{22.294}$$

with $\alpha > 0$, for $f_{\rm min} < f < f_{\rm max} \equiv f_*$, and vanishes beyond the cutoff $f_{\rm max}$, then the bound on the peak value $(h_0^2\Omega_{\rm gw})_*$ of the spectrum is

$$(h_0^2\Omega_{\rm gw})_* < 1.3 \times 10^{-6}\,\alpha\left[1-\left(\frac{f_{\rm min}}{f_{\rm max}}\right)^\alpha\right]^{-1}\left(\frac{N_{\rm eff}-3.046}{0.234}\right)$$
$$\simeq 1.3 \times 10^{-6}\,\alpha\left(\frac{N_{\rm eff}-3.046}{0.234}\right)\,, \tag{22.295}$$

[45] There is of course some arbitrariness in the choice $\epsilon_* = 0.1$. In any case ϵ_* must be sufficiently smaller than 1 in order for the bound to apply. Recall in fact from Section 1.4 of Vol. 1 that the energy of GWs is defined only as an average over a volume larger than several wavelengths, so the wavelength must be sufficiently small that the volume used for averaging is at least well inside the Hubble volume. Indeed, we saw in Section 19.5.1 that only when the modes are well inside the horizon can we associate with them an energy density that scales as $1/a^4$. There is also some arbitrariness in the choice $T \sim 0.1$ MeV. BBN is an extended process that lasts from temperatures of a few MeV until temperatures of about 10 keV. Certainly GWs that were still outside the horizon at $T = 10$ keV, corresponding to $f \simeq 10^{-12}$ Hz, are not bounded by BBN, and those that entered slightly earlier have less of an effect on BBN. So, rather than a sharp cutoff at $f \simeq 10^{-11}$ Hz, one has more precisely a smooth decrease of the effect of GWs, and therefore a less and less stringent bound, as f decreases from $\sim 10^{-10}$ to $\sim 10^{-12}$ Hz.

[46] Observe, from the definition

$$h_0^2\Omega_{\rm gw} = \frac{h_0^2}{\rho_c}\frac{d\rho_{\rm gw}}{d\log f}\,,$$

that $h_0^2\Omega_{\rm gw}$ is independent of h_0, since the explicit factor h_0^2 cancels with the factor h_0^2 that is present in ρ_c. For this reason we prefer to give the bounds in terms of $h_0^2\Omega_{\rm gw}$ rather than $\Omega_{\rm gw}$. This avoids the need to rescale the values of the bounds whenever a more accurate value of h_0 is determined. Furthermore, the determination of h_0 from CMB data assumes the validity of the ΛCDM cosmological model; see Note 16 on page 342. Thus, quoting a result for $\Omega_{\rm gw}(f)$ assuming a value of h_0, such as $h_0 = 0.68$, rather than giving $h_0^2\Omega_{\rm gw}(f)$, introduces a spurious dependence on the cosmological model that, even if numerically small, has nothing to do with the measurement being performed.

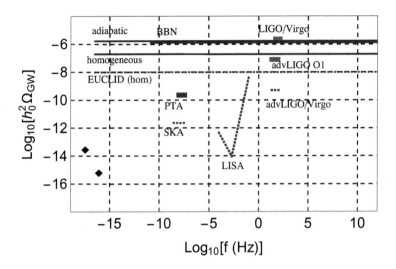

Fig. 22.27 The limits on $h_0^2\Omega_{\rm gw}(f)$ discussed in the text. The solid lines refer to existing (2017) limits, while dashed lines are forecasts for future experiments. The BBN limit is the solid line at $f \gtrsim 1.5 \times 10^{-11}$ Hz. The bound on extra radiation in the CMB from *Planck* 2015 plus structure formation, for adiabatic initial conditions, is almost superimposed but holds for $f \gtrsim 3 \times 10^{-17}$ Hz, and is labeled "adiabatic". The same bound for homogeneous initial conditions (extrapolated from the bound on adiabatic initial conditions as discussed in Note 47) is labeled "homogeneous". Expectations for EUCLID for homogeneous initial conditions are shown dotted, assuming an improvement by a factor 20 compared with *Planck*. The CMB limits on r at the pivot scales $k_* = 0.002\,{\rm Mpc}^{-1}$ and $k_* = 0.05\,{\rm Mpc}^{-1}$ (corresponding to $f_* = 3.09 \times 10^{-18}$ Hz and 7.73×10^{-17} Hz, respectively) are shown as filled diamonds. The short segments at $f = O(10-100)$ Hz show the limit set by initial LIGO/Virgo and the limit set by advanced LIGO in the O1 run. The expectations for the final design of advanced LIGO/Virgo and for LISA are also shown. The present PTA limit (discussed in Section 23.5) is compared with future expectations for the SKA.

where the second equality holds in the limit $(f_{\rm min}/f_{\rm max})^\alpha \ll 1$. Thus the bound on the peak value of the spectrum can be stronger or weaker than eq. (22.293), depending on whether $\alpha < 1$ or $\alpha > 1$, respectively. If $h_0^2\Omega_{\rm gw}(f)$ has a constant value $(h_0^2\Omega_{\rm gw})_*$ between the frequencies $f_{\rm min}$ and $f_{\rm max}$, and is small outside this interval, then eq. (22.292) rather gives

$$(h_0^2\Omega_{\rm gw})_* < \frac{1.3 \times 10^{-6}}{\log(f_{\rm max}/f_{\rm min})}\left(\frac{N_{\rm eff} - 3.046}{0.234}\right). \qquad (22.296)$$

As we have seen in this chapter, the spectrum of some cosmological signals, such as those due to cosmic strings, indeed extends over many decades in frequencies, and in this case the BBN bound is correspondingly more stringent.

22.7.2 Bounds on extra radiation from the CMB

CMB observations provide another sensitive probe of the presence of extra relativistic components, in this case at the time of last scattering. Indeed, extra radiation, such as that due to a fourth relativistic neutrino or to GWs, affects the rate of expansion of the Universe and the epoch of matter–radiation equilibrium, shifting it to a later time, and as a consequence it shifts the position of the acoustic peaks of the CMB. Furthermore, since structure formation can only start in MD, extra radiation delays the start of structure formation. Then, the horizon is larger when structures begin to collapse, and this shifts the peak of the matter power spectrum to larger scales, which can be tested using probes of structure formation, such as galaxy surveys and the Lyman-α forest.

If the GW perturbations are adiabatic, their effect on the CMB is indistinguishable from that of extra neutrinos, and one can therefore immediately apply the bounds on extra massless neutrino families obtained from the CMB. In this case the *Planck* 2015 data (combining TT, TE, EE and lowP data) together with BAO give $N_{\rm eff} < 3.5$ at 95% c.l. Proceeding as in the derivation of eq. (22.292) we then obtain

$$\int_{f=f_{\rm CMB}}^{f=\infty} d(\log f)\, h_0^2 \Omega_{\rm gw}(f) < 2.5 \times 10^{-6}\,. \qquad (22.297)$$

Combining *Planck* data with BAO, lensing and the deuterium BBN abundance tightens the bound to

$$\int_{f=f_{\rm CMB}}^{f=\infty} d(\log f)\, h_0^2 \Omega_{\rm gw}(f) < 1.7 \times 10^{-6}\,, \qquad (22.298)$$

which is the value that we plot in Fig. 22.27. The numerical value is therefore of the same order as that from the BBN bound (22.292). However, now the lower limit of the integral is lower, since all modes that were inside the horizon at the epoch of CMB decoupling contribute to the expansion. Putting into eq. (22.6) the value $T_{\rm dec} \simeq 0.26\,{\rm eV}$ from eq. (17.93) and the corresponding value of g_*, and again setting for definiteness $\epsilon_* = 0.1$, we get

$$f_{\rm CMB} \simeq 3 \times 10^{-17}\,{\rm Hz}\,. \qquad (22.299)$$

The bound therefore extends to much lower frequencies, and is shown as the dashed line (marked "adiabatic") in Fig. 22.27. Just as in the case of the BBN bound, it must be recalled that this is actually an integrated bound, and the corresponding bound on $h_0^2 \Omega_{\rm gw}(f)$ becomes stronger for a spectrum that is spread over several decades in frequency, as in eq. (22.296).

As discussed in Section 19.2.1, adiabatic perturbations are defined by the fact that, in the conformal Newtonian gauge, we only have temperature perturbations, so the perturbations in radiation (and therefore, in this case, the perturbations in GWs) are generated by the same temperature fluctuations responsible for matter perturbation. This is not the case for any of the GW production mechanism that we have discussed, such as inflation, phase transitions, cosmic strings or pre-big-bang cosmology. For instance, in inflation the initial matter perturbations originate from the decay of the inflaton field at reheating after inflation, while GWs are generated by the amplification of quantum fluctuations during inflation, as discussed in Section 21.3, so the two kind of perturbations are totally unrelated. The same holds for example for the radiation emitted by cosmic string cusps, for which again there is no reason why δ_M should be equal to $(3/4)\delta_{\rm gw}$, or why $\delta_{\rm gw}$ should be equal to δ_γ. As discussed in Section 19.2.1, a second independent type of initial perturbation are isocurvature perturbations, defined by the fact that, in the conformal Newtonian gauge, the total density contrast vanishes. Alternatively, as far as GW perturbations are concerned, we can consider as a second independent component the perturbations whose initial value $(\delta_{\rm gw})_{\rm in} = 0$, i.e. homogeneous initial conditions in the GW field. Because of the source term in eq. (19.39), non-vanishing GW perturbations will anyhow be generated at later times. For such initial conditions the degeneracy with the effect of extra massless neutrinos is broken and, since the CMB is fully consistent with adiabatic initial condition, the resulting bound for homogeneous initial conditions is stronger than that for adiabatic initial conditions, by almost an order of magnitude; see the line labeled "homogeneous" in Fig. 22.27.[47] Future CMB experiments, such as EUCLID, could improve these limits by about a factor of 20.

22.7.3 Bounds from the CMB at large angles

We have seen in Chapter 20 that the low-l multipoles of the temperature and polarization anisotropies of the CMB provide an upper bound on the tensor-to-scalar ratio r at a given pivot scale $k_* = 2\pi f_*$. Using eq. (19.290) this translates into an upper bound on $h_0^2 \Omega_{\rm gw}(f_*)$. In particular, we saw that CMB experiments typically express their results in terms of an upper bound on $r(k_*)$ at a pivot scale $k_* = 0.05\,{\rm Mpc}^{-1}$, corresponding to $f_* \simeq 7.73 \times 10^{-17}$ Hz; see eq. (21.354). This is slightly larger than $f_{\rm eq} \simeq 3 \times 10^{-17}$ Hz [see eq. (19.254)], and we can use the approximation to the transfer function leading to eq. (19.291), apart from a numerical factor close to 1.

At the pivot scale the tensor primordial spectrum (19.281) is simply given by $\mathcal{P}_{T,{\rm in}}(f_*) = A_T(f_*)$, and we write as usual $A_T(f_*) = r(f_*)A_{\mathcal{R}}(f_*)$. Then we get

$$h_0^2 \Omega_{\rm gw}(f = 7.73 \times 10^{-17}\,{\rm Hz}) \simeq 4 \times 10^{-16} \left(\frac{r_{0.05}}{0.1} \right) \left(\frac{A_{\mathcal{R}}(k_{0.05})}{2.14 \times 10^{-9}} \right),$$

$$(22.300)$$

apart from a numerical factor close to 1, which can be obtained from a numerical evaluation of the transfer function. As we saw in eq. (20.332), the current limit on $r_{0.05}$ is $r_{0.05} < 0.12$.

Alternatively, we can use the bound on $r(k_*)$ obtained at the pivot scale $k_* = 0.002\,{\rm Mpc}^{-1}$, corresponding to $f_* \simeq 3.09 \times 10^{-18}$ Hz; see

[47]See Smith, Pierpaoli and Kamionkowski (2006) and Sendra and Smith (2012). More precisely, using WMAP data the limit is 8.7×10^{-6} for adiabatic perturbations and 1.0×10^{-6} for homogeneous perturbations. With *Planck* 2015 data the limit on adiabatic perturbations becomes the one in eq. (22.297), while the limit on homogeneous perturbations has not been recomputed. We have drawn the line in Fig. 22.27 assuming that the same factor $\simeq 8.7$ between adiabatic and homogeneous perturbations remains with *Planck* 2015 data, so this line should only be taken as indicative.

eq. (21.354). This pivot scale is smaller than $f_{\rm eq}$, so we can use the approximation to the transfer function leading to eq. (19.295), which gives

$$h_0^2 \Omega_{\rm gw}(f_*) \simeq 1.87 \times 10^{-16} \left(\frac{f_{\rm eq}}{f_*}\right)^2 \left(\frac{r(k_*)}{0.1}\right) \left(\frac{A_{\mathcal{R}}(k_*)}{2.14 \times 10^{-9}}\right) ;$$

$$(22.301)$$

compare with eq. (21.358). In particular, inserting the numerical value $f_* \simeq 3.09 \times 10^{-18}$ Hz, we get[48]

$$h_0^2 \Omega_{\rm gw}(f = 3.09 \times 10^{-18} \, {\rm Hz}) \simeq 2.4 \times 10^{-14} \left(\frac{r_{0.002}}{0.1}\right) \qquad (22.302)$$
$$\times \left(\frac{h_0}{0.7}\right)^4 \left(\frac{\Omega_M}{0.3}\right)^2 \left(\frac{A_{\mathcal{R}}(k_{0.002})}{2.37 \times 10^{-9}}\right) .$$

The current *Planck* limit from temperature anisotropies is $r_{0.002} < 0.11$ at 95% c.l. [see eq. (20.331)], while the combined BICEP2/Keck Array/*Planck* limit from polarization is $r_{0.002} < 0.09$ [see eq. (20.333)].

Observe that the bound on $h_0^2 \Omega_{\rm gw}(f)$ at $f = 3.09 \times 10^{-18}$ Hz is much weaker than the bound at $f = 7.73 \times 10^{-17}$ Hz, despite the fact that the bounds on $r_{0.002}$ and on $r_{0.05}$ are very similar. This is due to the factor $(f_{\rm eq}/f)^2$ that appears in the transfer function at $f \lesssim f_{\rm eq}$. These limits are marked as filled diamonds in Fig. 22.27.

22.7.4 Limits on stochastic backgrounds from interferometers

Correlating the output of different detectors, as discussed in detail in Section 7.8, upper limits on stochastic GW background have been put by the initial LIGO and Virgo interferometers, and have been further improved by advanced LIGO. Assuming a flat spectrum for $\Omega_{\rm gw}(f)$, the limit set in the first advanced LIGO run, in the frequency band $20 - 86$ Hz, is

$$h_0^2 \Omega_{\rm gw} < 7.9 \times 10^{-8} \qquad (95\% \, {\rm c.l.}). \qquad (22.303)$$

In this frequency range this is better than the bounds from BBN and from the CMB plus structure formation, and also has a different meaning, since it comes from a direct search for GWs. Thus, it applies not only to stochastic backgrounds that were already present at cosmological epochs, such as those probed by CMB or by BBN, but even to stochastic backgrounds of more recent astrophysical origin, such as those generated by a large number of unresolved BH–BH binaries. Advanced LIGO/Virgo are expected to probe further down to $h_0^2 \Omega_{\rm gw} \simeq 5 \times 10^{-9}$ (for a flat spectrum), on a band centered around 30 Hz.

The space-borne interferometer LISA can search for stochastic backgrounds which are not flat in frequency, but rather a broken power-law, such as those due to first-order phase transitions or cosmic strings. For such spectra, LISA aims at reaching,[49]

[48] Recall, from eq. (19.185) and Note 33 on page 624, that $A_{\mathcal{R}} \simeq 2.14 \times 10^{-9}$ is the value measured at the pivot scale $k_* = 0.05 \, {\rm Mpc}^{-1}$. Using the spectral tilt $n_s = 0.9677$ one finds that, at $k_* = 0.002 \, {\rm Mpc}^{-1}$, $A_{\mathcal{R}}(k_*) \simeq 2.37 \times 10^{-9}$. Therefore, in eq. (22.302) we use this as a reference value for $A_{\mathcal{R}}(k_*)$.

[49] See the LISA proposal, Audley *et al.* (2017).

$$h_0^2 \Omega_{\rm gw}(f) \simeq 1.3 \times 10^{-11} \left(\frac{10^{-4}\,{\rm Hz}}{f} \right) , \qquad 0.1\,{\rm mHz} < f < 20\,{\rm mHz}$$
(22.304)

and

$$h_0^2 \Omega_{\rm gw}(f) \simeq 4.5 \times 10^{-12} \left(\frac{f}{10^{-2}\,{\rm Hz}} \right)^3 , \qquad 20\,{\rm mHz} < f < 0.2\,{\rm Hz} .$$
(22.305)

These existing and forecast limits are shown in Fig. 22.27. Finally, we have already mentioned limits on stochastic backgrounds of GWs that can be obtained by pulsar timing arrays (PTAs). We will discuss them in more detail in the next chapter, after having discussed the basic principles of PTAs.

Further reading

- For reviews of stochastic backgrounds of GWs see Allen (1997) and Maggiore (2000). Reviews of cosmological backgrounds focused on LISA are Binetruy, Bohe, Caprini and Dufaux (2012) and, for cosmological phase transitions, Caprini *et al.* (2016).

- Parametric resonance and out-of-equilibrium particle production at the end of inflation are discussed by Traschen and Brandenberger (1990), Kofman, Linde and Starobinsky (1994, 1997), Shtanov, Traschen and Brandenberger (1995) and Boyanovsky, de Vega, Holman, Lee and Singh (1995). GW production from preheating in chaotic inflation was first studied by Khlebnikov and Tkachev (1997a, 1997b) and considered again by Easther and Lim (2006). The general formalism for GW production by inhomogeneous random fields and accurate numerical studies of GW production during preheating are presented in Dufaux, Bergman, Felder, Kofman and Uzan (2007), which we have followed in our presentation. Further works include Felder and Kofman (2007) and Easther, Giblin and Lim (2007). Tachyonic preheating in hybrid inflation has been studied in Felder *et al.* (2001) and Felder, Kofman and Linde (2001). GW production from preheating in hybrid inflation has been studied in Garcia-Bellido and Figueroa (2007), Garcia-Bellido, Figueroa and Sastre (2008) and Dufaux, Felder, Kofman and Navros (2009), which also contains a useful critical review of earlier literature and a detailed study of parameter space. The production of GWs from gauge fields at preheating is discussed in Dufaux, Figueroa and Garcia-Bellido (2010).

- The (lack of) electroweak phase transition in the Standard Model for realistic values of the Higgs mass is studied in Kajantie, Laine, Rummukainen and Shaposhnikov (1996a, 1996b). The corresponding crossover temperature is accurately determined, with lattice simulations, in D'Onofrio and Rummukainen (2016). The addition of a second Higgs doublet or of an extra real scalar singlet can, however, lead to a strong first-order phase transition; see Fromme, Huber and Seniuch (2006) and Barger, Langacker, McCaskey, Ramsey-Musolf and Shaughnessy (2008), respectively. The crossover nature of the QCD phase transition with dynamical quarks (and zero baryon chemical potential) is determined, using lattice QCD, in Aoki, Endrodi, Fodor, Katz and Szabo (2006). The possibility that a large lepton asymmetry could induce a first-order QCD phase transition is discussed in Schwarz and Stuke (2009). The possibility of having a first-order QCD phase transition thanks to a large pre-existing value of μ_B/T, which is later diluted by a short inflationary period at the QCD scale, is discussed in Boeckel and Schaffner-Bielich (2010). Schwaller (2015) discusses the GW production in a first-order phase transition in extensions of the Standard Model with a confining dark sector.

- The theory of decay of false vacuum at finite temperature was developed in classic papers by Coleman (1977a) and Callan and Coleman (1977), elaborating on previous work by Voloshin, Kobzarev and Okun (1974), and is reviewed in Coleman (1977b). Instantons and bounces are also discussed in detail in the textbook by Zinn-Justin (2002) and in the review by Vandoren and van Nieuwenhuizen (2008). The extension to finite temperature and the cosmological consequences are studied in particular by Linde (1977, 1981, 1983b). The theory of relativistic bubble combustions is developed in Steinhardt (1982). Bubble nucleation in first-order cosmological phase transitions is studied in Hogan (1983), Turner, Weinberg and Widrow (1992), Anderson and Hall (1992), Enqvist, Ignatius, Kajantie and Rummukainen (1992) and Ignatius, Kajantie, Kurki-Suonio and Laine (1994), and is reviewed in Quiros (1999).

- The fact that a first-order cosmological phase transition can be a powerful source of GWs was first pointed out by Witten (1984) in the context of the QCD phase transition, and was further investigated by Hogan (1986), while Turner and Wilczek (1990) estimated GW production in cosmological bubble collisions in the context of inflation.

The theory of GWs production from bubble collisions in a first-order phase transition was developed in a series of papers by Kosowsky, Turner and Watkins (1992a, 1992b), Kosowsky and Turner (1993) and Kamionkowski, Kosowsky and Turner (1994), using a combination of analytic and numerical techniques. In particular the envelope approximation was proposed in Kosowsky and Turner (1993). The numerical results in the envelope approximation were subsequently refined by Huber and Konstandin (2008b), who also determined the shape of the spectrum near the peak. Jinno and Takimoto (2017) show how to obtain analytic results in the envelope approximation by computing the two-point correlator of the energy–momentum tensor, using the formalism that we have discussed in Section 22.2. Different regimes of bubble evolution are studied in Espinosa, Konstandin, No and Servant (2010). The determination of the parameters α and β in supersymmetric extensions of the Standard Model that could give a first-order phase transition is discussed in Apreda, Maggiore, Nicolis and Riotto (2002) and Huber and Konstandin (2008a, 2008b). The dependence of the peak frequency on the bubble velocity v_b is discussed in Caprini, Durrer, Konstandin and Servant (2009), where it is also pointed out that causality implies $h_0^2 \Omega_{\rm gw}(f) \propto f^3$ for super-horizon modes.

In Caprini, Durrer and Servant (2009) it is found that the relevant time-scale is β^{-1} for bubble collisions and $v_b \beta^{-1}$ for turbulence. GW production from bubble collisions and MHD turbulence at the QCD phase transition has been studied in Caprini, Durrer and Siemens (2010). Detailed reviews of the literature are given in Binetruy, Bohe, Caprini and Dufaux (2012), Caprini *et al.* (2016) and Weir (2017).

- GW production from turbulence was discussed in Kamionkowski, Kosowsky and Turner (1994) and subsequently studied in Kosowsky, Mack and Kahniashvili (2002), Dolgov, Grasso and Nicolis (2002), Caprini and Durrer (2006), Caprini, Durrer and Servant (2009) and Caprini, Durrer and Siemens (2010). GW production from sound waves generated in first-order phase transitions is studied in Hindmarsh, Huber, Rummukainen and Weir (2014, 2015) and Giblin and Mertens (2014).

- Cosmic strings are reviewed in Vilenkin (1985), Hindmarsh and Kibble (1995) and the textbook by Vilenkin and Shellard (2000). The relation of topological defects (such as domain walls and strings) to the topology of the vacuum manifold was discussed in a classic paper by Kibble (1976). The idea that cosmic strings could seed structure formation was proposed by Zeldovich (1980) and Vilenkin (1981a). Our discussion of the string dynamics and Nambu–Goto action follows Polchinski (1998) and Vilenkin and Shellard (2000). The development of cusps is discussed in Turok (1984). The fact that oscillating loops emit GWs efficiently was first pointed out by Vilenkin (1981b), and has been much studied in the literature; see for example Hogan and Rees (1984), Vachaspati and Vilenkin (1985), Bennett and Bouchet (1988), Caldwell and Allen (1992) and Caldwell, Battye and Shellard (1996). Early work on GWs from cosmic strings is reviewed in Allen (1997). A detailed review of the stochastic GW background from cosmic strings is given in Binetruy, Bohe, Caprini and Dufaux (2012). Numerical simulations of the cosmic string network are discussed in Blanco-Pillado, Olum and Shlaer (2014), where it is found that the network is dominated by large loops with $\alpha \sim 0.1$. The emission of GW bursts from cosmic string cusps was suggested in Berezinsky, Hnatyk and Vilenkin (2000) and studied in detail by Damour and Vilenkin (2000, 2001, 2005). For comparison with experimental sensitivities and existing bounds see also Siemens, Mandic and Creighton (2007) and Olmez, Mandic and Siemens (2010).

- The possibility of production of fundamental strings and D-strings of cosmological size during brane inflation is discussed in Majumdar and Christine-Davis (2002), Jones, Stoica and Tye (2002, 2003), Sarangi and Tye (2002) and Dvali and Vilenkin (2004). Interactions and reconnections between such cosmological-size F- and D-strings is studied in Jackson, Jones and Polchinski (2005).

- The pre-big-bang model was developed in Veneziano (1991), Gasperini and Veneziano (1993a, 1993b, 1994) and Brustein and Veneziano (1994) and is reviewed in Gasperini and Veneziano (2003) and Gasperini (2007). The possibility of obtaining a smooth evolution from the pre-big-bang phase to the post-big-bang era using the α' correction of string theory as well as the string loop corrections is discussed in Gasperini, Maggiore and Veneziano (1997), Brustein and Madden (1998) and Foffa, Maggiore and Sturani (1999). The generation of adiabatic density perturbations in this model can take place through the decay of massive axions, leading to the "curvaton" mechanism, as discussed in Enqvist and Sloth (2002), Lyth and Wands (2002) and Lyth, Ungarelli and Wands (2003); see also Mollerach (1990) and Copeland, Easther and Wands (1997). The spectrum of GWs is computed in Brustein, Gasperini, Giovannini and Veneziano (1995), Brustein, Gasperini and Veneziano (1997) and Buonanno, Maggiore and Ungarelli (1997). Our discussion of the computation of the GW background closely follows Maggiore (2000).

- The ekpyrotic model was proposed in Khoury, Ovrut, Steinhardt and Turok (2001) and Khoury, Ovrut, Seiberg, Steinhardt and Turok (2002). Its cyclic extension is discussed in Steinhardt and Turok (2002). The GW spectrum is computed in Khoury, Ovrut, Steinhardt and Turok (2001) and in Boyle, Steinhardt and Turok (2004). For reviews see Lehners (2008) and Appendix B of Baumann *et al.* [CMBPol Study Team Collaboration] (2009).

- For reviews of nucleosynthesis and discussions of the BBN bound see Sarkar (1996), Mangano, Miele, Pastor and Peloso (2002), Cyburt, Fields, Olive and Skillman (2005) and Iocco, Mangano, Miele, Pisanti and Serpico (2009). The bound on N_{eff} from *Planck* is given in Ade *et al.*, [Planck Collaboration] (2016a). The limit of N_{eff} from the effect of extra radiation on the CMB was proposed in Smith, Pierpaoli and Kamionkowski (2006) and updated in Sendra and Smith (2012) and Pagano, Salvati and Melchiorri (2016).

- Limits on stochastic GW backgrounds from the initial LIGO/Virgo interferometers are presented in Aasi *et al.* [LIGO Scientific and VIRGO Collaborations] (2014). The limit from the O1 run of advanced LIGO is presented in [LIGO Scientific Collaboration and Virgo Collaboration], Abbott *et al.* (2017a).

- For reviews of stochastic GW background of astrophysical origin see Section 11 of Maggiore (2000) and references therein and, for a more recent discussion, Rosado (2011).

Stochastic backgrounds and pulsar timing arrays

In this final chapter we discuss pulsar timing arrays (PTAs), and in particular how they can be used to search for stochastic GW backgrounds. As we have seen in Chapters 6 and 11, pulsars are clocks with remarkable stability. A GW passing between the Earth and a pulsar affects the timing residuals, and the idea of using the Earth and a pulsar as the end-point masses of a GW detector was already put forward in the late 1970s. For a single Earth–pulsar system, the effect of a GW will be masked by many other effects, such as intrinsic period noise of the pulsar, uncorrected interstellar delays or inaccuracies in Solar System ephemerides. However, these sources of noise can be fought by correlating the signals from a network of several widely spaced pulsars. This has led to the PTA concept. This method is being implemented by several international collaborations:

- The European Pulsar Timing Array (EPTA) uses five 100 m class telescopes across Europe: the Effelsberg Radio Telescope in Germany, the Lovell Telescope in the UK, the Nançay Decimetric Radio Telescope in France, the Sardinia Radio Telescope in Italy and the Westerbork Synthesis Radio Telescope in the Netherlands. It currently monitors 41 pulsars at several frequencies (the use of different frequencies allows one to partially correct for the effects of the interstellar medium). The five telescopes also operate by combining their signals coherently, as a part of the Large European Array of Pulsars (LEAP) project. This results in a sensitivity equivalent to that of a telescope with a diameter of 194 m, with the advantage (compared with the Arecibo Telescope, see the next bullet point) of being fully steerable. Using them as single telescopes gives the highest overall sampling rate of pulsars (the so-called "cadence") among the existing PTAs, of about one data point per week for each pulsar.

- NANOGrav (the North American Nanohertz Observatory for Gravitational Waves) is composed of several institutions across the US and Canada, and uses the Arecibo Telescope in Puerto Rico and the Green Bank Telescope in West Virginia, which are among the world's largest single-dish telescopes. In particular Arecibo has been for many years the largest and most sensitive single radio telescope in the world, with a diameter of 305 m.[1] However, the dish is so large that it must remain fixed in position, limiting the

[1]The largest radio telescope is now the Five-hundred-meter Aperture Spherical Radio Telescope (FAST), in southwest China, with a fixed spherical dish of diameter 500 m, built in a natural depression. It is operative and taking data since 2017.

Gravitational Waves, Volume 2: Astrophysics and Cosmology. Michele Maggiore.
© Michele Maggiore 2018. Published in 2018 by Oxford University Press.
DOI 10.1093/oso/9780198570899.001.0001

accessible portion of the sky. On the other hand, the Green Bank Telescope can cover 85% of the sky. NANOGrav currently monitors 49 pulsars, each at at least two frequencies, of which 27 are monitored at Arecibo.

- The Parkes Pulsar Timing Array (PPTA) uses the Parkes Observatory in Australia, and it therefore has access to the pulsars in the Southern Hemisphere. It currently monitors 25 pulsars at three frequencies, including one of the most accurately timed pulsars, PSR J0437-4715.

These collaborations monitor those millisecond pulsars that are known to be the most stable clocks, looking for effects on the timing residuals that could eventually be due to GWs, after elimination of the various noise sources. The different collaborations also combine their data in the framework of the International Pulsar Timing Array (IPTA).

In frequency space, PTAs are sensitive to GWs whose frequency is larger than the inverse of the time span T of the data, $f \gtrsim 1/T$. For the existing datasets, which span a period of more than 10 yr, this implies a lower frequency $f_{\min}$ slightly below 3 nHz. Modulations of the signal with $f < f_{\min}$, and hence with periods longer than about 10 yr, cannot be seen as oscillating terms but rather as secular variations. These are fitted to polynomials and reabsorbed, through standard pulsar-timing techniques, into a fit to the pulsar spindown period [see eq. (6.43)], and other pulsar parameters. Therefore, PTAs are not sensitive to GWs with $f \lesssim 1/T$. On the other hand, pulsars are typically observed once every few weeks (or at most once per week for EPTA), so data sampling limits the maximum detectable GW frequency to about $50 \, \mathrm{yr}^{-1} \simeq 1 \times 10^{-6} \, \mathrm{Hz}$.

In this frequency range a particularly interesting target is the stochastic GW background generated by supermassive BH binaries, which we have discussed in Chapter 16. We will indeed see that the predictions discussed in Section 16.4 are well within the present and near-future sensitivity of PTAs, to the extent that the models of SBMH binary population and evolution that give the largest prediction for the signal have already been ruled out. Furthermore, PTAs can possibly detect even single supermassive BH binaries at low redshift. Other, more hypothetical, sources in this frequency range include some of the backgrounds of cosmological origin discussed in Chapter 22, such as the background generated by cosmic strings, or by a strong first-order QCD phase transition.

23.1 GW effect on the timing of a single pulsar

We first compute how a GW passing between the Earth and a pulsar affects the observed periodicity of the pulses as observed on the Earth. We choose the reference frame such that the Earth is at the origin of the coordinate system, and we denote by $\hat{\mathbf{n}}$ the propagation direction of the GW and by $\hat{\mathbf{n}}_a$ the unit vector from the Earth toward the pulsar (the label a will be useful later for generalizing to several pulsars). As long as we deal with a single pulsar, without loss of generality we can set $\hat{\mathbf{n}}_a = \hat{\mathbf{x}}$, so for the moment we take the pulsar to be at a distance d_a along the x axis. We denote by ν_a the rotational frequency of the pulsar, so $T_a = 1/\nu_a$ is the rotational period, while we reserve the letter $f = \omega_{\mathrm{gw}}/(2\pi)$ for the frequency of the GW.

To study how a passing GW affects the periodicity of the pulses we adapt the computation of the light travel time in a single arm of a GW interferometer; see eq. (9.9) or eq. (9.138) of Vol. 1. We work in the TT gauge, so the positions of the two endpoints of the arm, in this case the Earth and the pulsar, are fixed, and are not affected by the passage of a GW. In the units $c = 1$ that we are using here the space-time interval in the TT frame is

$$ds^2 = -dt^2 + [\delta_{ij} + h_{ij}^{\mathrm{TT}}(t, \mathbf{x})dx^i dx^j]. \tag{23.1}$$

Consider a bunch of photons emitted by the pulsar at an emission time t_{em}, when the beam points toward the Earth. For a light beam propagating along the x axis, the condition that it travels along null geodesics, $ds^2 = 0$, gives

$$dx^2 = \frac{dt^2}{1 + h_{xx}^{\mathrm{TT}}[t, \mathbf{x}(t)]}, \tag{23.2}$$

where $\mathbf{x}(t)$ is the photon trajectory. We are considering the propagation from the pulsar, at $\mathbf{x} = d_a\hat{\mathbf{x}}$, toward the observer, at $\mathbf{x} = 0$, and so in the $-\hat{\mathbf{x}}$ direction. Then eq. (23.2) gives

$$\begin{aligned} dx &= -\frac{dt}{\left\{1 + h_{xx}^{\mathrm{TT}}[t, \mathbf{x}(t)]\right\}^{1/2}} \\ &= -\left\{1 - \frac{1}{2}h_{xx}^{\mathrm{TT}}[t, \mathbf{x}(t)]\right\}dt, \end{aligned} \tag{23.3}$$

where the second equality holds to first order in h. Thus, the photons will arrive at the observer at a time t_{obs} such that

$$d_a = t_{\mathrm{obs}} - t_{\mathrm{em}} - \frac{1}{2}\int_{t_{\mathrm{em}}}^{t_{\mathrm{obs}}} dt'\, h_{xx}^{\mathrm{TT}}[t', \mathbf{x}(t')]. \tag{23.4}$$

Since h_{xx}^{TT} is a small quantity, to first order in the integral we can replace t_{obs} by its zeroth-order value $t_{\mathrm{obs}} = t_{\mathrm{em}} + d_a$, and for the photon trajectory we can take the unperturbed path $\mathbf{x}(t) = (t_{\mathrm{obs}} - t)\hat{\mathbf{n}}_a$. To generalize to the case in which $\hat{\mathbf{n}}_a$ is not necessarily along the x axis, we must also replace h_{xx}^{TT} by $n_a^i n_a^j h_{ij}^{\mathrm{TT}}$. Thus, to first order in h,

$$t_{\mathrm{obs}} = t_{\mathrm{em}} + d_a + \frac{n_a^i n_a^j}{2}\int_{t_{\mathrm{em}}}^{t_{\mathrm{em}}+d_a} dt'\, h_{ij}^{\mathrm{TT}}[t', (t_{\mathrm{em}} + d_a - t')\hat{\mathbf{n}}_a]. \tag{23.5}$$

Observe that d_a is the distance between the Earth and the pulsar in the TT gauge. This quantity is independent of h_{ij}^{TT}, since in the TT frame the coordinates $x = 0$ and $x = d_a$ are *defined* by the position of the Earth and of the pulsar, respectively, even when the GW is passing (recall the discussion in Section 1.3.3 of Vol. 1). Thus, the dependence of t_{obs} on the GW is fully given explicitly by the factor h_{ij}^{TT} in the integrand on the right-hand side of eq. (23.5).

Consider now the photons emitted after one rotational period of the pulsar, i.e. at an emission time $t'_{\text{em}} = t_{\text{em}} + T_a$ when again the beam of electromagnetic radiation emitted by the pulsar points toward the Earth. By the same computation, they will be observed at an observation time t'_{obs} obtained by replacing t_{em} by t'_{em} in eq. (23.5), i.e.

$$t'_{\text{obs}} = t_{\text{em}} + T_a + d_a \tag{23.6}$$
$$+ \frac{n_a^i n_a^j}{2} \int_{t_{\text{em}}+T_a}^{t_{\text{em}}+T_a+d_a} dt'\, h_{ij}^{\text{TT}}[t', (t_{\text{em}} + T_a + d_a - t')\hat{\mathbf{n}}_a]$$
$$= t_{\text{em}} + T_a + d_a + \frac{n_a^i n_a^j}{2} \int_{t_{\text{em}}}^{t_{\text{em}}+d_a} dt'\, h_{ij}^{\text{TT}}[t' + T_a, (t_{\text{em}} + d_a - t')\hat{\mathbf{n}}_a],$$

where in the second equality we have written $t' = t'' + T_a$ and we have finally renamed the integration variable t'' as t'. Then, subtracting eq. (23.5) from eq. (23.6), we get

$$t'_{\text{obs}} - t_{\text{obs}} = T_a + \Delta T_a, \tag{23.7}$$

where

$$\Delta T_a = \frac{n_a^i n_a^j}{2} \int_{t_{\text{em}}}^{t_{\text{em}}+d_a} dt' \left\{ h_{ij}^{\text{TT}}[t' + T_a, \mathbf{x}_0(t')] - h_{ij}^{\text{TT}}[t', \mathbf{x}_0(t')] \right\}, \tag{23.8}$$

and

$$\mathbf{x}_0(t') = (t_{\text{em}} + d_a - t')\hat{\mathbf{n}}_a. \tag{23.9}$$

Equation (23.7) shows that the time interval at which the pulses are seen by the observer is equal to the rotational period of the pulsar plus an extra term ΔT_a induced by the GW. We next recall that T_a is the rotation period of the pulsar, typically of the order of a few milliseconds. In contrast, the GWs that can be detected with pulsar timing have periods $T_{\text{gw}} = 2\pi/\omega_{\text{gw}}$ of the order of the observation time, $1 - 10$ yr. This means that $\omega_{\text{gw}} T_a$ is an extremely small quantity, say of order 10^{-10}. Therefore, in eq. (23.8) we can Taylor-expand the dependence on T_a in $h_{ij}^{\text{TT}}[t' + T_a, \mathbf{x}_0(t')]$ to first order, and we get

$$\frac{\Delta T_a}{T_a} = \frac{1}{2} n_a^i n_a^j \int_{t_{\text{em}}}^{t_{\text{em}}+d_a} dt' \left[\frac{\partial}{\partial t'} h_{ij}^{\text{TT}}(t', \mathbf{x}) \right]_{\mathbf{x}=\mathbf{x}_0(t')}. \tag{23.10}$$

Observe that $\partial/\partial t'$ is a partial derivative acting only on the first argument of h_{ij}^{TT}, and not on the time dependence in $\mathbf{x}(t')$, exactly as we found for the gravitational Doppler shift of the frequency of a photon (which in the CMB context is described by the ISW effect); see eq. (20.123). We will discuss later the relation between these two apparently different computations.

For a monochromatic GW propagating along the $\hat{\mathbf{n}}$ direction we write

$$h_{ij}^{\mathrm{TT}}(t, \mathbf{x}) = \mathcal{A}_{ij}(\hat{\mathbf{n}}) \cos\left[\omega_{\mathrm{gw}}(t - \hat{\mathbf{n}} \cdot \mathbf{x})\right] , \qquad (23.11)$$

where $n^i \mathcal{A}_{ij}(\hat{\mathbf{n}}) = 0$. Substituting this into eq. (23.10) and carrying out the integral we get

$$\frac{\Delta T_a}{T_a} = \frac{n_a^i n_a^j \mathcal{A}_{ij}}{2(1 + \hat{\mathbf{n}} \cdot \hat{\mathbf{n}}_a)} \left\{ \cos(\omega_{\mathrm{gw}} t_{\mathrm{obs}}) - \cos\left[\omega_{\mathrm{gw}} t_{\mathrm{em}} - \omega_{\mathrm{gw}} \tau_a \hat{\mathbf{n}} \cdot \hat{\mathbf{n}}_a\right] \right\} , \qquad (23.12)$$

where $\tau_a \equiv t_{\mathrm{obs}} - t_{\mathrm{em}}$ is the light travel time between the Earth and the pulsar (which, in our units $c = 1$, is the same as d_a). It is conventional to define the quantity

$$z_a(t) \equiv \left(\frac{\nu_0 - \nu(t)}{\nu_0} \right)_a . \qquad (23.13)$$

Thus $z_a(t) = -(\Delta \nu_a / \nu_a)(t) = \Delta T_a / T_a$, since $T_a = 1/\nu_a$.[2] Equation (23.12) can then be rewritten as

$$z_a(t) = \frac{n_a^i n_a^j}{2(1 + \hat{\mathbf{n}} \cdot \hat{\mathbf{n}}_a)} \left[h_{ij}^{\mathrm{TT}}(t, \mathbf{x} = 0) - h_{ij}^{\mathrm{TT}}(t - \tau_a, \mathbf{x}_a) \right] , \qquad (23.14)$$

where we now write t_{obs} simply as t and we denote the pulsar's position by $\mathbf{x}_a = d_a \hat{\mathbf{n}}_a$, while $\mathbf{x} = 0$ is the observer's position.[3] The timing residuals R_a of the a-th pulsar, measured with respect to a reference time $t = 0$, are defined by

$$R_a(t) = \int_0^t dt' \, z_a(t') . \qquad (23.15)$$

It can also be useful to write the result in terms of the two polarizations of the GW. To this purpose, let us chose the reference frame such that $\hat{\mathbf{n}} = (0, 0, 1)$. Then the GW amplitude in the TT gauge has the form (1.33), which we recall here,

$$h_{ij}^{\mathrm{TT}} = \begin{pmatrix} h_+ & h_\times & 0 \\ h_\times & -h_+ & 0 \\ 0 & 0 & 0 \end{pmatrix} , \qquad (23.16)$$

and $h_{+,\times} = h_{+,\times}(t - z)$. With respect to such a choice of axes, we write the unit vector in the direction of the a-th pulsar as

$$\hat{\mathbf{n}}_a = (\sin\theta_a \cos\phi_a, \sin\theta_a \sin\phi_a, \cos\theta_a) . \qquad (23.17)$$

Then

$$n_a^i n_a^j h_{ij}^{\mathrm{TT}} = \sin^2\theta_a \left(h_+ \cos 2\phi_a + h_\times \sin 2\phi_a \right) , \qquad (23.18)$$

[2] One should be aware of the fact that occasionally in the literature $z_a(t)$ is defined with the opposite sign.

[3] More precisely, it is convenient to use as t_{obs} the time of arrival at the Solar System barycenter (and similarly $\mathbf{x} = 0$ is the position of the solar system barycenter), and then take into account separately the orbital motion of the Earth, which gives rise the Roemer time delay, as well as to general-relativistic effects such as those discussed in Section 6.2.2 of Vol. 1.

and eq. (23.14) gives

$$z_a(t) = \frac{1}{2}(1 - \cos\theta_a)\cos 2\phi_a \left[h_+(t) - h_+(t - \tau_a - \tau_a \cos\theta_a)\right]$$

$$+ \frac{1}{2}(1 - \cos\theta_a)\sin 2\phi_a \left[h_\times(t) - h_\times(t - \tau_a - \tau_a \cos\theta_a)\right]. \quad (23.19)$$

Observe that the right-hand side of eq. (23.14) is regular for $\hat{\mathbf{n}}\cdot\hat{\mathbf{n}}_a = -1$. Indeed, $(1 + \hat{\mathbf{n}}\cdot\hat{\mathbf{n}}_a)$ is equal to $(1 + \cos\theta_a)$. However, we see from eq. (23.18) that $n_a^i n_a^j h_{ij}^{\mathrm{TT}}$ produces a factor $\sin^2\theta_a$, so the overall dependence on θ_a is given by $\sin^2\theta_a/(1 + \cos\theta_a) = 1 - \cos\theta_a$.

It is instructive to compare eq. (23.10) with the result obtained by computing the Doppler shift of the photons emitted by the pulsar labeled by a, $(\Delta\nu_\gamma/\nu_\gamma)_a$. This can be done by studying how a photon geodesic is affected by the passing GW. Actually, in Section 20.3.1 we have already solved this problem in a more general setting, namely in a FRW background, and considering both scalar and tensor perturbations. We can then read the result from eq. (20.123), specializing it to flat space and keeping the tensor perturbations only. Observe also that the temperature perturbations $\delta T/T$ are equal to the perturbations in the photon frequency $\Delta\nu_\gamma/\nu_\gamma$; see eq. (20.60). Furthermore, we set to zero the intrinsic fluctuations at the emission point, since we are now only interested in the fluctuations generated at the observation point during the propagation.[4] Then, eq. (20.123) gives

$$\left(\frac{\Delta\nu_\gamma}{\nu_\gamma}\right)_a = -\frac{1}{2}n_a^i n_a^j \int_{t_{\mathrm{em}}}^{t_{\mathrm{obs}}} dt' \left[\frac{\partial}{\partial t'}h_{ij}^{\mathrm{TT}}(t',\mathbf{x})\right]_{\mathbf{x}=\mathbf{x}_0(t')}, \quad (23.20)$$

where again $\mathbf{x}_0(t)$ is given by eq. (23.9), and we have made use of the fact that in flat space conformal time becomes the same as cosmic time. Comparing with eqs. (23.8) and (23.10) and recalling that $(\Delta T/T)_a = -(\Delta\nu/\nu)_a$ we see that the Doppler shift of the photons $\Delta\nu_\gamma/\nu_\gamma$ is the same as the relative change in the rate of the pulses, $\Delta\nu/\nu$, when the latter is computed in the limit $\omega_{\mathrm{gw}}T \to 0$, but the two quantities are different for generic $\omega_{\mathrm{gw}}T$. This can be understood as follows. In our computation of $\Delta\nu/\nu$ we have considered two different geodesics: one leaving from the pulsar in the direction of the observer at time t_{em}, and one leaving after one rotational period, at time $t_{\mathrm{em}} + T$. Each geodesic is affected by the passing GW, and the observed fluctuation in the periodicity of the pulses is given by the difference of the effects on the two geodesics. The computation of the gravitational Doppler shift can instead be seen as a computation of the difference in the time of arrival between two successive crests (or, more generally, between two points with phases φ and $\varphi + d\varphi$, with $d\varphi$ infinitesimal) of an electromagnetic wave, belonging to the same geodesic. Since the relative Doppler shift $\Delta\nu_\gamma/\nu_\gamma$ is independent of the frequency, the result is the same as that obtained for $\Delta\nu/\nu$ from the difference of the two geodesics in the limit of T infinitesimal, but is different for finite T.[5]

---

[4]Of course the source, in this case the pulsar, also has its own intrinsic fluctuations. In this context these fluctuations just contribute to the observational error.

[5]Indeed, the original derivation of the response of the Earth–pulsar system to a GW, given in Detweiler (1979) and regularly quoted or repeated in the literature, is actually a computation of the Doppler shift $\Delta\nu_\gamma/\nu_\gamma$ of the photon frequency, based on an earlier computation by Estabrook and Wahlquist (1975) which was performed in the context of GW searches using Doppler tracking of spacecraft. In Detweiler (1979), and in all the subsequent literature, this result is then used, without any comment, as the result for $\Delta\nu/\nu$. However, these two quantities are logically distinct: ν_γ is the frequency of the photons received by the radio telescope, while ν is the frequency at which the observer receives the pulses from the pulsar because of the lighthouse effect. The original derivation that we have given in eqs. (23.2)–(23.10) has the advantage of computing the relevant quantity, $\Delta\nu/\nu$, directly. It also makes it clear that the argument suggesting that $\Delta\nu_\gamma/\nu_\gamma = \Delta\nu/\nu$ is only correct in the limit $\omega_{\mathrm{gw}}T_a \to 0$, otherwise the correct result for pulsar timing is given by eq. (23.8), and not by eq. (23.10). In practice, given that the rotational period T_a of the pulsar is of the order of milliseconds, while $\omega_{\mathrm{gw}} = 2\pi/T_{\mathrm{GW}}$ with T_{GW} of order 10 yr, the limit $\omega_{\mathrm{gw}}T_a \to 0$ is fully appropriate in this situation.

23.2   Response to a continuous signal

A first application of eq. (23.14) is to a single continuous source of GWs, such as a SMBH binary in its long quasi-circular inspiral phase. For PTAs, relevant sources must have an orbital frequency $f_s \sim 10^{-9}$ Hz. According to Kepler's law, the corresponding major semiaxis is

$$a \simeq 0.1 \, \mathrm{pc} \left(\frac{m}{10^{10} \, M_\odot} \right)^{1/3} \left(\frac{1 \, \mathrm{nHz}}{f_s} \right)^{2/3} , \qquad (23.21)$$

where m is the total mass of the binary. For such a binary, the terms $h_{ij}(t, \mathbf{x} = 0)$ and $h_{ij}(t - \tau_a, \mathbf{x}_a)$ in eq. (23.14), over the time-scale $T = O(10)$ yr of PTA observations, would be seen as continuous GWs. Observe, however, that these two terms correspond to different emission times. The term $h_{ij}(t, \mathbf{x} = 0)$ is the GW that arrives at time t at the observer's location, and that therefore was emitted by the SMBH binary at a time $t_{\mathrm{ret}} = t - r/c$, where r is the distance from the observer to the SMBH, and, for clarity, we have reinserted the speed of light c. In contrast, $h_{ij}(t - \tau_a, \mathbf{x}_a)$ is the GW that hit the pulsar at time $t - \tau_a$, where, as usual, τ_a is the light travel time between the pulsar and the Earth. Therefore, this GW was emitted at a retarded time $t'_{\mathrm{ret}} = t - \tau_a - r'/c$, where r' is the distance from the pulsar to the SMBH binary. Since the SMBH will in general be at a cosmological distance, whereas the pulsar is still within the Galaxy, we have $r, r' \gg c\tau_a$ and therefore $r' \simeq r - c\tau_a \cos\vartheta$, where ϑ is the angle between the directions of the SMBH and of the pulsar, as seen by the observer. Therefore,

$$t'_{\mathrm{ret}} \simeq t_{\mathrm{ret}} - \tau_a (1 - \cos\vartheta) . \qquad (23.22)$$

Thus, the two terms in eq. (23.14) correspond to GWs that were emitted at different times by the source and one should ask if, during this time interval, which is of order τ_a, the source has appreciably evolved. The evolution of the GW frequency of an inspiraling binary is given by eq. (4.18), which can be rewritten as

$$\frac{\dot{f}_r}{f_r} = \frac{96}{5} \left(\frac{GM_c}{c^3} \right)^{5/3} (\pi f_r)^{8/3} , \qquad (23.23)$$

where, as in Chapter 16, $f_r \equiv f_{\mathrm{gw}}^{(r)}$ is the GW frequency in the source rest frame, related to the observed frequency f by $f = (1+z)^{-1} f_r$. In the inspiral phase the characteristic time-scale over which the observed signal evolves is therefore

$$\tau_{\mathrm{gw}} \equiv \frac{f}{\dot{f}} \simeq 420 \, \mathrm{pc}/c \left(\frac{10^{10} \, M_\odot}{M_c} \right)^{5/3} \left(\frac{2}{1+z} \right)^{8/3} \left(\frac{6.3 \, \mathrm{nHz}}{f} \right)^{8/3} . \qquad (23.24)$$

Thus, for the heaviest SMBH binaries, $M_c \sim 10^{10} \, M_\odot$, in the stage of their evolution when they radiate at an observed GW frequency $f_s \simeq$ 6.3 nHz (which, as we shall see, is the frequency where PPTA has its best sensitivity) and at a typical redshift $z \simeq 1$, the source evolves on a time-scale $\tau_{\rm gw}$ such that $c\tau_{\rm gw} = O(0.4)$ kpc. For $M_c \sim 10^9 M_\odot$ we rather get $c\tau_{\rm gw} = O(20)$ kpc. The range of distances of the pulsars that are monitored by PTAs is typically $0.4 - 3$ kpc, with a few pulsars up to about 5 kpc; thus, depending on the mass of the SMBH binary, on the exact frequency at which GWs are searched and on the distance of the pulsar, the source might have undergone a non-negligible evolution during the light travel time from the pulsar to us. Note in particular the strong dependence on frequency, so that at higher frequencies the time $\tau_{\rm gw}$ over which the source evolves is shorter and therefore the effect of source evolution becomes more important.

When there is significant source evolution during the light travel time from the source to us, the two terms in eq. (23.14) effectively produce two continuous waves with two different frequencies and phases. Since during the inspiral the GW frequency increases with time, the lower-frequency sinusoid corresponds to the pulsar term $h_{ij}(t - \tau_a, \mathbf{x}_a)$, and is specific to the each pulsar, while the higher-frequency sinusoid comes from the Earth term $h_{ij}(t, \mathbf{x} = 0)$, and is the same for all pulsars.

23.3   Response to a stochastic GW background

We now apply eq. (23.14) to a stochastic background of GWs. The GW amplitude corresponding to a superposition of waves of all frequencies and coming from all directions can be written as in eq. (7.188) of Vol. 1, which we recall here,

$$h_{ij}(t, \mathbf{x}) = \sum_{A=+,\times} \int_{-\infty}^{\infty} df \int d^2\hat{\mathbf{n}}\, \tilde{h}_A(f, \hat{\mathbf{n}})\, e_{ij}^A(\hat{\mathbf{n}})\, e^{-2\pi i f(t - \hat{\mathbf{n}} \cdot \mathbf{x})}. \quad (23.25)$$

We work in the TT gauge, so $h_i^i = \partial^j h_{ij} = 0$. The polarization tensors $e_{ij}^A(\hat{\mathbf{n}})$ are given in eq. (1.54). Inserting this expression into eq. (23.14) we get

$$z_a(t) = \sum_{A=+,\times} \int_{-\infty}^{\infty} df \int d^2\hat{\mathbf{n}}\, \tilde{h}_A(f, \hat{\mathbf{n}})\, F_a^A(\hat{\mathbf{n}}) e^{-2\pi i f t}$$
$$\times \left[1 - e^{2\pi i f \tau_a (1 + \hat{\mathbf{n}} \cdot \hat{\mathbf{n}}_a)}\right], \quad (23.26)$$

where

$$F_a^A(\hat{\mathbf{n}}) = \frac{n_a^i n_a^j e_{ij}^A(\hat{\mathbf{n}})}{2(1 + \hat{\mathbf{n}} \cdot \hat{\mathbf{n}}_a)}. \quad (23.27)$$

Comparing with eqs. (7.21) and (7.218) of Vol. 1, and observing that the GW signal $h_k(t)$ in the k-detector corresponds in this case to $z_a(t)$ for the a-th pulsar, we see that $F_a^A(\hat{\mathbf{n}})$ is the equivalent, for the a-th pulsar, of the pattern function of the k-th detector $F_k^A(\hat{\mathbf{n}})$.

For a stochastic background we must compute $\langle z_a(t) z_b(t) \rangle$, where the angle brackets denote the ensemble average over the stochastic variables $\tilde{h}_A(f, \hat{\mathbf{n}})$. Assuming that the stochastic GW background is stationary, isotropic and unpolarized, the ensemble average is computed using eq. (7.190) of Vol. 1,

$$\langle \tilde{h}_A^*(f, \hat{\mathbf{n}})\tilde{h}_{A'}(f', \hat{\mathbf{n}}')\rangle = \delta(f - f') \frac{\delta^2(\hat{\mathbf{n}}, \hat{\mathbf{n}}')}{4\pi} \delta_{AA'} \frac{1}{2} S_h(f) \,. \qquad (23.28)$$

This gives

$$\langle z_a(t)z_b(t)\rangle = \frac{1}{2} \int_{-\infty}^{\infty} df \, S_h(f) \int \frac{d^2\hat{\mathbf{n}}}{4\pi} \, \mathcal{K}_{ab}(f; \hat{\mathbf{n}}) \sum_{A=+,\times} F_a^A(\hat{\mathbf{n}})F_b^A(\hat{\mathbf{n}}) \,, \qquad (23.29)$$

where

$$\mathcal{K}_{ab}(f; \hat{\mathbf{n}}) = \left[1 - e^{-2\pi i f \tau_a(1+\hat{\mathbf{n}}\cdot\hat{\mathbf{n}}_a)}\right] \left[1 - e^{2\pi i f \tau_b(1+\hat{\mathbf{n}}\cdot\hat{\mathbf{n}}_b)}\right] \,. \qquad (23.30)$$

Comparison with the results presented in Section 7.8.3, in particular eq. (7.226), shows that the integral over $d^2\hat{\mathbf{n}}$ in eq. (23.29) is the equivalent of the overlap reduction function in the case of two GW interferometers. Observe that, in $\mathcal{K}_{ab}(f; \hat{\mathbf{n}})$, in each square bracket there are two terms, which originated from the terms $h_{ij}(t, \mathbf{x} = 0)$ and $h_{ij}(t - \tau_a, \mathbf{x}_a)$ in eq. (23.14). Thus, the four contributions obtained by expanding the products in eq. (23.30) correspond to terms obtained by computing separately

$$\langle h_{ij}(t, \mathbf{x} = 0)h_{kl}(t, \mathbf{x} = 0)\rangle \,, \qquad \langle h_{ij}(t, \mathbf{x} = 0)h_{kl}(t - \tau_b, \mathbf{x}_b)\rangle \,,$$
$$\langle h_{ij}(t - \tau_a, \mathbf{x}_a)h_{kl}(t, \mathbf{x} = 0)\rangle \,, \qquad \langle h_{ij}(t - \tau_a, \mathbf{x}_a)h_{kl}(t - \tau_b, \mathbf{x}_b)\rangle \,, \qquad (23.31)$$

respectively. However, terms at different space-time points, such as $\langle h_{ij}(t, \mathbf{x} = 0)h_{kl}(t - \tau_b, \mathbf{x}_b)\rangle$, are negligible. This is due to the fact that, as we have seen, PTAs are sensitive to GWs with a period $T = O(10)$ yr. This is very small compared with the time τ_b taken by the electromagnetic signal to go from the b-th pulsar to the observer. The closest known millisecond pulsar, PSR J0437-4715, is at a distance of 156 pc, and, as we have mentioned, typical millisecond pulsars used in PTAs are at distances from us of about $0.4-3$ kpc.[6] This means that $h_{ij}(t, \mathbf{x} = 0)$ and $h_{kl}(t - \tau_b, \mathbf{x}_b)$ are uncorrelated, and similarly for the term at the space-time location of the two pulsars, $h_{ij}(t - \tau_a, \mathbf{x}_a)$ and $h_{kl}(t - \tau_b, \mathbf{x}_b)$. Thus, the last three terms in eq. (23.31) are negligible with respect to $\langle h_{ij}(t, \mathbf{x} = 0)h_{kl}(t, \mathbf{x} = 0)\rangle$.

This physical argument is made more precise by the explicit expression in eq. (23.30). Indeed, even for a distance $c\tau$ as small as 156 pc and f of order of the minimum value in the PTA frequency range, $f \simeq 3 \times 10^{-9}$ Hz, $f\tau_a$ is of order 50, which becomes about 300 for a millisecond pulsar at a more typical distance of 1 kpc. Thus, we can replace

$$\mathcal{K}_{ab}(f; \hat{\mathbf{n}}) \to 1 \,, \qquad (23.32)$$

since the terms neglected are rapidly oscillating exponentials and contribute negligibly to the integral over frequencies in eq. (23.29).[7]

[6] See Table 2 of Verbiest *et al.* (2016) for a list of the pulsars and their properties used in the first IPTA data release.

[7] To be more precise observe that, even if $f\tau_a$ is large, $f\tau_a(1+\hat{\mathbf{n}}\cdot\hat{\mathbf{n}}_a)$ goes to zero when $\hat{\mathbf{n}}\cdot\hat{\mathbf{n}}_a \to -1$. This corresponds to a GW propagating exactly from the direction of the pulsar toward us, i.e. exactly parallel to the propagation direction of the photon. In this limit of course the GW must have no effect on the photon, since GWs act on the transverse plane. Indeed, in this case $1 - e^{-2\pi i f \tau_a(1+\hat{\mathbf{n}}\cdot\hat{\mathbf{n}}_a)} \to 0$ and $\mathcal{K}_{ab}(f; \hat{\mathbf{n}})$ vanishes, as it should [while, for $\hat{\mathbf{n}}\cdot\hat{\mathbf{n}}_a \to +1$, $z_a(t)$ vanishes because of $F_a^A(\hat{\mathbf{n}})$, which is proportional to $1 - \hat{\mathbf{n}}\cdot\hat{\mathbf{n}}_a$; see eq. (23.19)]. Thus, if $f\tau_a \gtrsim 50$, the approximation of neglecting the term $e^{-2\pi i f \tau_a(1+\hat{\mathbf{n}}\cdot\hat{\mathbf{n}}_a)}$ is valid everywhere, except in a small solid angle with $|1+\hat{\mathbf{n}}\cdot\hat{\mathbf{n}}_a| \lesssim 1/50$. This corresponds to a cone with opening angle $\Delta\theta$ given by $(\Delta\theta)^2/2 \simeq 1/50$, and therefore to a fraction of the solid angle $\pi(\Delta\theta)^2/(4\pi) \simeq 1/100$. For a pulsar at 1 kpc the approximation is valid everywhere except in a region with $|1 + \hat{\mathbf{n}}\cdot\hat{\mathbf{n}}_a| \lesssim 1/300$, corresponding to a fraction $1/600$ of the full solid angle. The approximation (23.32) therefore introduces an error of about 1% or 0.2%, respectively, in the integral over $d^2\hat{\mathbf{n}}$.

In this approximation,

$$\langle z_a(t) z_b(t) \rangle = \frac{1}{2} \int_{-\infty}^{\infty} df \, S_h(f) \int \frac{d^2\hat{n}}{4\pi} \sum_{A=+,\times} F_a^A(\hat{n}) F_b^A(\hat{n}) \,. \quad (23.33)$$

The final step is to compute the angular integral. This can be done by writing the polarization tensors explicitly as in eq. (1.54),

$$e_{ij}^+(\hat{n}) = \hat{u}_i \hat{u}_j - \hat{v}_i \hat{v}_j \,, \quad (23.34)$$

$$e_{ij}^\times(\hat{n}) = \hat{u}_i \hat{v}_j + \hat{v}_i \hat{u}_j \,, \quad (23.35)$$

and choosing an explicit form for the unit vectors $\hat{u}$ and $\hat{v}$, which are orthogonal to $\hat{n}$ and to each other. We write $\hat{n}$ in the form

$$\hat{n}(\theta, \phi) = (\sin\theta \cos\phi, \sin\theta \sin\phi, \cos\theta) \,. \quad (23.36)$$

Then a convenient choice for $\hat{u}$ and $\hat{v}$ is

$$\hat{u} = (\sin\phi, -\cos\phi, 0) \,, \quad (23.37)$$

$$\hat{v} = (\cos\theta \cos\phi, \cos\theta \sin\phi, -\sin\theta) \,. \quad (23.38)$$

The angular integral in eq. (23.33) can then be performed analytically, and the result is[8]

$$C(\theta_{ab}) \equiv \int \frac{d^2\hat{n}}{4\pi} \sum_{A=+,\times} F_a^A(\hat{n}) F_b^A(\hat{n})$$

$$= x_{ab} \log x_{ab} - \frac{1}{6} x_{ab} + \frac{1}{3} \,, \quad (23.39)$$

where

$$x_{ab} = \frac{1}{2}(1 - \cos\theta_{ab}) \,, \quad (23.40)$$

and θ_{ab} is the relative angle between the two pulsars. This expression for $C(\theta_{ab})$ was first found by Hellings and Downs, and is known as the Hellings–Downs curve. The function $C(\theta)$ is shown in Fig. 23.1. Observe also that, in the limit $\mathcal{K}_{ab}(f; \hat{n}) \to 1$, in the correlation in eq. (23.29) the dependences on the frequency and on θ_{ab} factorize. Using $S_h(-f) = S_h(f)$, we finally get[9]

$$\langle z_a(t) z_b(t) \rangle = C(\theta_{ab}) \int_0^\infty df \, S_h(f) \,. \quad (23.41)$$

It is also useful to write the result in terms of the correlator of the timing residuals $R_a(t)$ defined in eq. (23.15),

$$r_{ab}(t) \equiv \langle R_a(t) R_b(t) \rangle \,. \quad (23.42)$$

Inserting eq. (23.26) [with the square bracket in the second line set equal to 1] in eq. (23.15) we get

$$R_a(t) = \sum_{A=+,\times} \int_{-\infty}^{\infty} df \int d^2\hat{n} \, \tilde{h}_A(f, \hat{n}) \, F_a^A(\hat{n}) \frac{1}{(-2\pi i f)} \left(e^{-2\pi i f t} - 1 \right) \,. \quad (23.43)$$

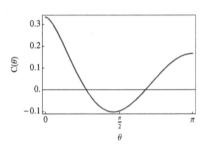

Fig. 23.1 The correlation function $C(\theta)$.

[8]The integral is elementary but its computation is somewhat laborious. It is performed in detail in Appendix C of Anholm, Ballmer, Creighton, Price and Siemens (2009).

[9]Equation (23.41) is actually the result of a theoretical computation that ignores the fact that PTAs, like any GW detector, are not really sensitive to a function of time, in this case of $z_a(t)$, but only to its Fourier modes $\tilde{z}_a(f)$ in a given range of frequencies. The integral over frequencies from $f = 0$ to $f = \infty$ in eq. (23.41) therefore does not correspond to anything observable, and the quantities actually observed correspond to the integral from a minimum frequency f_l to a maximum frequency f_h. More precisely, one could form a correlation weighted with the detector sensitivity, similarly to the two-detector correlation denoted by Y in eq. (7.223).

Computing again the correlator using eq. (23.28) we get

$$r_{ab}(t) = 2C(\theta_{ab}) \int_0^\infty df \, \frac{S_h(f)}{(2\pi f)^2} \, [1 - \cos(2\pi f t)] \,. \qquad (23.44)$$

Restricting the interval to the useful frequency bandwidth (see Note 9) we actually have

$$r_{ab}(t) = C(\theta_{ab}) \int_{f_l}^{f_h} df \, \frac{S_h(f)}{(2\pi f)^2} \, 2\,[1 - \cos(2\pi f t)] \,. \qquad (23.45)$$

The dependence of this correlator on time is due to the fact that the definition of the timing residuals, eq. (23.15), arbitrarily picks an origin of time, $t = 0$, thereby breaking the invariance under time translations. Observe that, for t sufficiently small, so that $2\pi f_h t \ll 1$, we can expand the term $\cos(2\pi f t)$ and we get $r_{ab}(t) \simeq t^2 \langle z_a(t) z_b(t) \rangle$. This correctly reflects the fact that, for such small values of the observation time, eq. (23.15), with the Fourier modes of $z_a(t)$ restricted to the observable frequencies $f < f_h$, gives $R_a(t) \simeq t z_a(t)$. However, the upper limit on the observation frequency, f_h, is given by the cadence of the pulsar's observation. Thus, the timing residuals are only measured at sampling times and frequencies such that $2\pi f t \gtrsim 1$, or $2\pi f t \gg 1$, and the factor $2[1 - \cos(2\pi f t)]$ in eq. (23.45) is generically of order 1.

It is conventional to define

$$P_g(f) = \frac{2}{3} \frac{S_h(f)}{(2\pi f)^2} \,. \qquad (23.46)$$

Then, apart from the factor $2[1 - \cos(2\pi f t)] = O(1)$, and possibly the contribution of an optimal filter function (see Note 9),

$$r_{ab} \simeq \zeta(\theta_{ab}) \int_{f_l}^{f_h} df \, P_g(f) \,, \qquad (23.47)$$

where $\zeta(\theta_{ab}) = (3/2)C(\theta_{ab})$, i.e. (for $\theta \neq 0$),

$$\zeta(\theta_{ab}) = \frac{3}{2} x_{ab} \log x_{ab} - \frac{1}{4} x_{ab} + \frac{1}{2} \,. \qquad (23.48)$$

Similarly to what we did in eqs. (7.197) and (7.198) for the energy density, we also define $dr_{ab}/d\log f$ from

$$r_{ab} = \int_0^\infty d\log f \, \frac{dr_{ab}}{d\log f}(f) \,, \qquad (23.49)$$

so

$$\frac{dr_{ab}}{d\log f}(f) \simeq \zeta(\theta_{ab}) f P_g(f) \,. \qquad (23.50)$$

It is useful to rewrite the result in term of the characteristic amplitude $h_c(f)$ defined in eq. (19.296). From eqs. (19.298) and (23.46),

$$P_g(f) = \frac{h_c^2(f)}{12\pi^2 f^3} \,, \qquad (23.51)$$

and therefore

$$\frac{dr_I}{d\log f} \simeq \zeta(\theta_{ab}) \frac{h_c^2(f)}{12\pi^2 f^2} \,. \qquad (23.52)$$

23.4   Extracting the GW signal from noise

In the preceding sections we have seen how to compute the GW signal in PTAs. Of course, in an actual experiment, on top of the GW signal there will be a noise $n_a(t)$. This might be due to intrinsic timing irregularities in the spindown of the different pulsars, as well as to timing residuals due to irregularities in terrestrial time standards, or, for example, fluctuations in the refractive index of the interstellar medium. The output $s_a(t)$ corresponding to the timing residuals of the a-th pulsar will then be of the form

$$s_a(t) = R_a(t) + n_a(t) \,. \tag{23.53}$$

In the regime where the noise is much larger than the GW signal we can extract the GW signal from the noise by applying correlation techniques similar to the matched filtering discussed in detail in Sections 7.3 and 7.8.3 of Vol. 1. We write

$$\langle s_a(t)s_b(t)\rangle = \langle R_a(t)R_b(t)\rangle + \langle R_a(t)n_b(t)\rangle$$
$$+\langle n_a(t)R_b(t)\rangle + \langle n_a(t)n_b(t)\rangle\,, \tag{23.54}$$

where the ensemble average can be replaced by a temporal average over the observation time T,

$$\langle s_a(t)s_b(t)\rangle = \frac{1}{T}\int_0^T dt\, s_a(t)s_b(t)\,. \tag{23.55}$$

In practice this is computed as a discrete sum over the times at which the pulsars are sampled. Equation (23.55) corresponds to the simplest implementation of matched filtering; see eq. (7.38). As discussed in Sections 7.3, this suppresses an uncorrelated term such as $\langle n_a(t)n_b(t)\rangle$, with respect to the correlated signal $\langle R_a(t)R_b(t)\rangle$, by a factor $(\tau_{\rm noise}/T)^{1/2}$, where $\tau_{\rm noise}$ is the typical autocorrelation time of the noise. Basically this emerges because $n_a(t)n_b(t)$ performs a random walk and its time integral grows only as $T^{1/2}$, while for a positive-definite quantity it grows as T.[10]

Furthermore, we see from eq. (23.52) that we have two further handles for discriminating a true GW signal from noise. One is the dependence on the frequency, which for a GW signal is proportional to $f^{-3}h_c^2(f)$, and the second is the dependence on the angles θ_{ab} among the (a,b) pulsars. Concerning the dependence on the frequency, it is usually assumed that, within the frequency window available to a PTA, $h_c(f)$ can be modeled as a power law,

$$h_c(f) = A_* \left(\frac{f}{f_*}\right)^\alpha \,. \tag{23.56}$$

In Section 16.4.1 we found for instance that, for SMBH binaries with negligible eccentricity, $\alpha = -2/3$. Observe, however, that α can become positive in the PTA window if the typical eccentricities when the binary enters the regime dominated by GW back-reaction are large, $e_0 \gtrsim 0.6$; see Fig. 16.6. As in eq. (16.85), f_* is a reference frequency, normally

[10]The overall scaling of the signal-to-noise ratio with the observation time T is, however, more complicated, since one must also take into account that, by increasing T, one lowers the minimum frequency $f_l = 1/T$ available to PTA observations; see eq. (23.71).

fixed in the context of PTAs at $f_* = 1\,\mathrm{yr}^{-1}$. From eqs. (23.52) and (23.56)

$$\frac{dr_{ab}}{d\log f}(f) \propto f^{2\alpha-2}. \qquad (23.57)$$

For $\alpha < 1$ (which is the case even for SMBHs with large eccentricity where, from Fig. 16.6, $\alpha \simeq 0.8$ in the nHz region) this is a red spectrum. Given α, this behavior can in principle be tested with optimal matched filtering as discussed in Section 7.8.3, or more generally by looking for excess power in the low-frequency bins.

If we monitor a large number $\mathcal{N}_p$ of pulsars, we have at our disposal $N_p = \mathcal{N}_p(\mathcal{N}_p - 1)/2$ independent pulsar pairs. We can then further enhance the GW signal with respect to the noise, using as a template the very specific angular dependence of the correlation in eq. (23.41). Observe that the Hellings–Downs curve is a specific consequence of the quadrupolar nature of GWs, reflected in the geometric factors in eq. (23.19).[11]

Correlated noise such as that due to clock noise at the Earth will produce a different pattern, related to the monopole nature of this effect, while inaccuracies in the Solar System ephemerides will lead to a dipole spatial correlation between pulsars. The precision of the method increases with the number of pulsars, which is indeed the reason that led to the concept of PTAs. As of 2016, the first IPTA data release contained 49 millisecond pulsars. However, it should be stressed that with present datasets the existing limits are dominated by just the few pulsars with the highest timing precision. In the future, the Square Kilometer Array (SKA) is expected to carry out the timing of several hundreds of pulsars, at the accuracy level currently achieved for just a few pulsars.

A simple way of searching for a correlation of the data with the Hellings–Downs curve is to introduce the statistic

$$\rho = \frac{1}{\sigma_r \sigma_\zeta}\frac{1}{N_p}\sum_{a<b}(r_{ab} - \bar{r})(\zeta_{ab} - \bar{\zeta}), \qquad (23.58)$$

where the sum over (ab) runs over the N_p distinct pulsar pairs, $\zeta_{ab} = \zeta(\theta_{ab})$, $\bar{r}, \bar{\zeta}$ are the averages computed over all pulsar pairs, and σ_r^2 and σ_ζ^2 are the respective variances. For large N_p, if there is no correlation in the data, ρ will be distributed as a Gaussian with zero mean and a variance $\sigma_\rho^2 = 1/N_p$. Therefore, the statistical significance of a measured value ρ_* can be expressed in terms of the number of standard deviations, ρ_*/σ_ρ.

A further improvement, with respect to the use of eq. (23.58), can be obtained by performing a full Bayesian analysis, which combines both the information on the temporal correlations induced by GWs with a given frequency spectrum, and the angular correlations given by the Hellings–Down curve.[12] To this purpose we observe first of all that, by the very nature of pulsar timing measurements, the timing residuals are measured at a discrete set of values of time. Furthermore, each pulsar is in general monitored at a different set of times, which can be represented

[11]Note, however, that the derivation of the Hellings–Downs curve also relies upon the assumption that the stochastic background is unpolarized, i.e. that $\langle \tilde{h}_A^*(f, \hat{n})\tilde{h}_{A'}(f', \hat{n}')\rangle$ in eq. (23.28) is proportional to $\delta_{AA'}$. Otherwise, more generally, the factor $\delta_{AA'}S_h(f)$ in eq. (23.28) must be replaced by a Hermitian matrix $S_{AA'}(f)$ and, in eq. (23.29), the factor $S_h(f)\sum_{A=+,\times}F_a^A(\hat{n})F_b^A(\hat{n})$ on the right-hand side must be replaced by $\sum_{A,A'=+,\times}S_{AA'}(f)F_a^A(\hat{n})F_b^{A'}(\hat{n})$.

[12]We follow here the discussion in van Haasteren, Levin, McDonald and Lu (2009).

as a matrix t_{ai} that gives the values of the times t_i at which the timing residuals of the a-th pulsar are measured. We then write the timing residuals for the a-th pulsar, measured at the i-th time, as R_{ai}. The timing residual are made of a "deterministic" part $R_{ai}^{\det}$ and a stochastic part. The deterministic part includes for example the effect due to the pulsar spindown, which can be fitted to the data by performing a quadratic fit as in eq. (6.41), as well as several other effects that can in principle be modeled. The stochastic part is due both to intrinsic pulsar noise (as well as other effects such as clock and receiver noise and fluctuations in the refractive index of the interstellar medium) and, hopefully, to GWs. We can then write

$$R_{ai} = R_{ai}^{\det} + R_{ai}^{\rm GW} + R_{ai}^{\rm N}, \qquad (23.59)$$

where $R_{ai}^{\rm GW}$ is the stochastic GW contribution and $R_{ai}^{\rm N}$ is due to all other stochastic noise sources. A stochastic Gaussian processes is fully characterized by its two-point correlator, which, in this discrete notation, is represented by coherence matrices,

$$\langle R_{ai}^{\rm GW} R_{bj}^{\rm GW} \rangle = C_{(ai)(bj)}^{\rm GW}, \qquad (23.60)$$

$$\langle R_{ai}^{\rm N} R_{bj}^{\rm N} \rangle = C_{(ai)(bj)}^{\rm N}, \qquad (23.61)$$

while (setting to zero the correlation between the GW and the noise) the total coherence matrix is

$$C_{(ai)(bj)} = C_{(ai)(bj)}^{\rm GW} + C_{(ai)(bj)}^{\rm N}. \qquad (23.62)$$

The deterministic contribution $R_{ai}^{\det}$ will depend on a set of model parameters denoted collectively by $\boldsymbol{\eta}$, while the stochastic noise contribution depends on a set of model parameters denoted collectively by $\boldsymbol{\xi}$, and the GW contribution is parametrized by A_* and α as in eq. (23.56). We can then construct the likelihood function, as in eq. (7.63). If both the noise and the GW background are Gaussian, the timing residuals are distributed as a multi-dimensional Gaussian, so the likelihood is

$$\Lambda(\{R_{ai}\}|\boldsymbol{\theta}, A_*, \alpha) = \frac{1}{\sqrt{(2\pi)^n \det C}} \qquad (23.63)$$
$$\times e^{-\frac{1}{2}\sum_{(ai)(bj)}(R_{ai}-R_{ai}^{\det})C_{(ai)(bj)}^{-1}(R_{bj}-R_{bj}^{\det})},$$

where $\boldsymbol{\theta} = \{\boldsymbol{\eta}, \boldsymbol{\xi}\}$ and n is the dimension of the covariance matrix. The covariance matrix due to GWs can be obtained by repeating for the correlator at different times, $\langle R_a(t)R_b(t')\rangle$, the computation performed for $\langle R_a(t)R_b(t)\rangle$. We assume a GW spectrum of the form (23.56). Then one obtains[13]

[13]See van Haasteren, Levin, McDonald and Lu (2009).

$$C_{(ai)(bj)}^{\rm GW} = \alpha_{ab} \frac{A_*^2 f_*^\gamma}{(2\pi)^2 f_L^{\gamma+1}} \left\{ \Gamma(-1-\gamma)\sin\left(\frac{-\pi\gamma}{2}\right) [f_L \tau_{(ai)(bj)}]^{\gamma+1} \right.$$
$$\left. -\sum_{k=0}^\infty (-1)^k \frac{[f_L \tau_{(ai)(bj)}]^{2k}}{(2k)!(2k-1-\gamma)} \right\}. \qquad (23.64)$$

where $\gamma = 1 - 2\alpha$, $\Gamma(z)$ is the Euler gamma function and $\tau_{(ai)(bj)} = 2\pi(t_{ai} - t_{bj})$. For $a \neq b$ the term α_{ab} is the same as the function $\zeta(\theta_{ab})$ in eq. (23.48). For $a = b$ we have instead $\alpha = 1$, so for generic a and b we can write

$$\alpha_{ab} = \frac{3}{2}x_{ab}\log x_{ab} - \frac{1}{4}x_{ab} + \frac{1}{2} + \frac{1}{2}\delta_{ab}, \qquad (23.65)$$

and x_{ab} is given in eq. (23.40). The quantity f_L in eq. (23.64) is a low-frequency cutoff, necessary for the convergence of the integral. However, the effect of GWs with frequencies smaller than the total observation time of PTAs does not show up as an oscillating term, but as a secular variation, which is automatically reabsorbed in the fit to the spindown parameters. Therefore, in the end one can show that the result does not depend on f_L as long as f_L is sufficiently small compared with the inverse of the observation time.

This formalism is in principle much more powerful than the separate use of eq. (23.58) and of the spectral information. Indeed, the quantity r_{ab} in eq. (23.42) takes into account only the correlation between pulsars at equal time, while the covariance matrix $C^{\text{GW}}_{(ai)(bj)}$ contains the information on all correlators $\langle R_a(t_i)R_b(t_j)\rangle$ at different times. Furthermore, in this formalism we can also take into account the fact that different pulsars are monitored at a different set of times, which are represented as a matrix t_{ai} that gives the values of time t_i at which the timing residuals of the a-th pulsar are measured.

The possible drawback of the method is that in eq. (23.63) we need the covariance matrix of the noise. This requires careful modeling of the various noise sources, both those intrinsic to the detector (receiver noise, etc.) or to the observation process (such as fluctuations in the interstellar medium) and those intrinsic to the pulsars. However, if one is able to model the noise carefully, one has the full likelihood function (23.63), which gives the distribution probability of all parameters $\{\boldsymbol{\theta}, A_*, \alpha\}$. Eventually we are interested mostly in the probability distribution for A_* and α, marginalized over the model parameters $\boldsymbol{\theta}$. A direct integration over this multi-dimensional space is out of question owing to its computational cost. However one can perform it via Markov chain Monte Carlo (MCMC), apart from some possible analytic maximization over some parameter of the timing model.[14] In particular, one is interested in the one-dimensional marginalized likelihood for the parameter A_* in eq. (23.56), to see if there is statistical evidence for a non-vanishing value of A_*, or else to put an upper limit on it.

In this Bayesian analysis the overall evidence for a model [such as a model of the noise plus a non-vanishing GW contribution with a given value of α in eq. (23.56)] can be quantified in terms of the Bayes factor. To define the Bayes factor one starts from Bayes' theorem, which we have already discussed in Section 7.4.1. Let $\mathcal{M}_i$ and $\mathcal{M}_j$ be two models. For instance the model $\mathcal{M}_i$ has a given description of noise plus a GW contribution, and the model $\mathcal{M}_j$ has the same description of the noise but no GW contribution (and therefore corresponds to the null hypothesis). We can use eq. (7.55) where $A = \mathcal{M}$ is the model and d

[14]Alternatively, one can use some computationally less expensive indicator, such as the maximum likelihood estimator, as discussed in Section 7.4.2.

are the data. Thus, $P(\mathcal{M}_i|d)$ is the probability of the model $\mathcal{M}_i$ given the data (a strictly Bayesian concept), while $P(d|\mathcal{M}_i)$ is the probability of the data given the model $\mathcal{M}_i$ (a concept that makes sense also in the frequentist approach), and is just the likelihood function, which in our case is given by eq. (23.63). Finally, $P(\mathcal{M}_i)$ is the prior probability of the model (which, in the absence of extra information, can be taken as flat; see, however, the discussion on page 350 of Vol. 1). Then eq. (7.55) gives

$$\frac{P(\mathcal{M}_i|d)}{P(\mathcal{M}_j|d)} = \frac{P(d|\mathcal{M}_i)}{P(d|\mathcal{M}_j)}\frac{P(\mathcal{M}_i)}{P(\mathcal{M}_j)}$$
$$\equiv B_{ij}\frac{P(\mathcal{M}_i)}{P(\mathcal{M}_j)}, \tag{23.66}$$

which defines the Bayes factor B_{ij}. The Bayes factor gives the relative probabilities of models i and j, up to the ratio of the prior probabilities $P(\mathcal{M})$. As we already mentioned on page 296, conventionally, its numerical value is translated into a statement about the evidence of model i with respect to model j using the Jeffreys scale. Model i is favored with respect to model j if $B_{ij} > 1$. For $1 < B_{ij} < 3$ the evidence is deemed "weak", for $3 < B_{ij} < 20$ is "definite", for $20 < B_{ij} < 150$ is "strong", and for $B_{ij} > 150$ is "very strong".

However, this scale is an oversimplification, and gives only a first approximate guidance. A better way to assess the significance of a Bayes factor in a specific problem is to determine empirically the distribution of the Bayes factor by creating several noise simulations, which, however, requires a detailed understanding of all noise sources. Alternatively, one could apply to PTA techniques conceptually analogous to the shifting technique discussed on page 371 of Vol. 1, where, in the search for coincidences between GW interferometers, one artificially shifts the data stream of one detector with respect to the other, to estimate the background. As we saw in Section 15.1.1, this technique has been successfully applied to the discovery of compact binary coalescences. In the present case an example of how one can form a similar ensemble of mock noise data is by scrambling, at the level of data analysis, the position of the pulsars in the sky, i.e. attributing the timing residual of the a-th pulsar to another pulsar, at angle θ_b. In principle, this should destroy all spatial correlations. One can then compute the Bayes factor on this scrambled pulsar set, assuming the presence of GWs, with respect to the hypothesis of no GWs. Of course, because of the scrambling, any correlation with the Hellings–Downs curve is now due to chance. By performing the scrambling a large number of times we can empirically determine the distribution of the Bayes factor in the absence of any spatial correlation, and use it to assess the statistical significance of a given measured Bayes factor.

Finally, it is interesting to see how the signal-to-noise ratio (SNR) scales with the observation time T and with the other parameters of the search.[15] Using eq. (23.47) and proceeding as in the derivation of eq. (7.239), we see that the SNR is given by

$$\text{SNR} = \left[2T \sum_{a<b} \zeta_{ab}^2 \int_{f_l}^{f_h} df \, \frac{P_g^2(f)}{P_a(f)P_b(f)} \right]^{1/2} , \qquad (23.67)$$

where $P_a(f)$ is the noise spectral density of the a-th pulsar.[16] The factor $T^{1/2}$ in eq. (23.67) is the standard scaling factor gained by the correlated signal with respect to an uncorrelated noise, as discussed after eq. (23.55). However, a further dependence on T is hidden in the fact that the lowest frequency f_l accessible to observation is $f_l = 1/T$, so by increasing T we also increase the available bandwidth. Let us assume for simplicity that the pulsars all have a white noise (which is only true in a first approximation) and that all pulsars are sampled with a uniform cadence $C = 1/\Delta t$. Then the intrinsic pulsar noise is $P_a(f) = 2\sigma_a^2 \Delta t$, where σ_a is the root mean square of the white noise of the a-th pulsar. For the pulsars in the IPTA dataset, σ_a ranges from values smaller than $0.5\,\mu s$ for the pulsars with the highest timing precision, to values of order $5\,\mu s$. On top of this, $P_a(f)$ contains the GW signal. In the regime where the intrinsic pulsar noise is much larger than the GW signal, the latter can be neglected when evaluating $P_a(f)$, but as we will see this is not true for all frequencies with present PTA data. Thus, we write

$$P_a(f) = 2\sigma_a^2 \Delta t + P_g(f) \qquad (23.68)$$

and

$$\text{SNR} = \left[2T \sum_{ab} \zeta_{ab}^2 \int_{f_l}^{f_h} df \, \frac{P_g^2(f)}{[2\sigma_a^2 \Delta t + P_g(f)][2\sigma_b^2 \Delta t + P_g(f)]} \right]^{1/2} . \qquad (23.69)$$

Consider first the regime where, for all frequencies between f_l and f_h, the GW signal is much smaller than the intrinsic noise, i.e. $P_g(f) \ll 2\sigma_a^2 \Delta t$. We write $P_g(f)$ using eqs. (23.51) and (23.56). For the values of α expected for typical PTA sources, the frequency integral is dominated by the lower limit f_l. Then

$$\int_{f_l}^{f_h} df \, \frac{P_g^2(f)}{[2\sigma_a^2 \Delta t + P_g(f)][2\sigma_b^2 \Delta t + P_g(f)]} \propto \frac{A_*^4}{(2\sigma_a^2 \Delta t)(2\sigma_b^2 \Delta t)} f_l^{1-2\beta} , \qquad (23.70)$$

where $\beta = 3 - 2\alpha$. Thus, setting $f_l = 1/T$, we get

$$\text{SNR} \propto A_*^2 \left(\sum_{ab} \frac{\zeta_{ab}^2}{\sigma_a^2 \sigma_b^2} \right)^{1/2} \frac{T^\beta}{\Delta t} . \qquad (23.71)$$

This expression clearly shows the various dependences. Since in the weak-signal limit the SNR is quadratic in the signal, it is of course proportional to A_*^2. The expression in parentheses shows that the SNR is strongly dominated by the pulsars with the best timing precisions. If for simplicity we assume that we have $\mathcal{N}_p$ pulsars with high timing

[16] To simplify the notation, we assume that the observation time T is the same for all pulsar pairs. However, all the following formulas go through on replacing T by T_{ab}, where T_{ab} is the observation time for which the observations of pulsars a and b overlap [to be taken, of course, inside the sum over (a, b)].

precision and comparable values of $\sigma_a \sim \sigma$, then the SNR is proportional to $1/\sigma^2$, so of course the SNR improves when the timing precision of the best pulsars is increased. Furthermore, the sum over (a, b) runs effectively over the $\mathcal{N}_p(\mathcal{N}_p - 1)/2$ best pulsar pairs, so for large $\mathcal{N}_p$ the SNR is proportional to $\mathcal{N}_p$. It also increases linearly with the cadence $1/\Delta t$ of the observations. Finally, the less trivial aspect of the result is the dependence on the observation time T, which combines the $T^{1/2}$ factor typical of matched filtering with the fact that, by increasing T, we extend the useful bandwidth toward lower frequencies. How much we gain therefore depends on the slope of the GW spectrum. In particular, for SMBH with negligible eccentricity, $\alpha = -2/3$, so $\beta = 13/3$ and the SNR increases very rapidly with T, SNR $\propto T^{13/3}$.

In the opposite limit, when the GW signal is stronger than the pulsar noises, $2\sigma_a^2 \Delta t \ll P_g(f)$,

$$\int_{f_l}^{f_h} df \, \frac{P_g^2(f)}{[2\sigma_a^2 \Delta t + P_g(f)][2\sigma_b^2 \Delta t + P_g(f)]} \simeq \Delta f = \frac{1}{\Delta t}$$

and

$$\text{SNR} \propto T^{1/2} \mathcal{N}_p \frac{1}{\Delta t} . \tag{23.72}$$

Thus, the SNR now grows only as $T^{1/2}$ and is independent of A_*, as long as A_* is sufficiently large to be in the strong-signal limit. Interestingly, assuming a stochastic GW background with $A_* = 10^{-15}$ and $\alpha = -2/3$, and a typical cadence $1/\Delta t = 20 \, \text{yr}^{-1}$, corresponding to the fact that a pulsar is typically observed every few weeks, one finds that the noise of several pulsars is already in the intermediate regime between these two limits. In this intermediate regime, setting $\alpha = -2/3$, one finds

$$\text{SNR} \propto \mathcal{N}_p \left(\frac{A_*}{\sigma} \right)^{3/13} \frac{T^{1/2}}{(\Delta t)^{3/26}} . \tag{23.73}$$

We see that it is possible to improve the PTA sensitivity to stochastic backgrounds in several ways: (1) discovering new millisecond pulsars amenable to observations with high timing precision; (2) increasing the observation time; (3) reducing the value of σ_a by improving the timing precision and (4) increasing the cadence, for instance by combining data from multiple telescopes. The preceding scaling laws give a quantitative estimate of the improvement in sensitivity obtained in each case.

Increasing the cadence also has the effect of raising the maximum frequency available to PTAs. From eq. (16.25) we see that, to reach the frequency range $1 - 10 \, \mu\text{s}$ where one could in principle study the radiation due to the merger and ringdown of SMBHs, we need a cadence better than one observation per day. Ultimately there are high-frequency limitations, due to the need to add a large number of pulses to get a stable pulse profile, as we have already mentioned in Section 6.2. This limitation is not significant for current observations, but might become relevant at the precision level of the SKA.

23.5   Searches for stochastic backgrounds with PTAs

The sensitivity of pulsar timing arrays has significantly improved in the last few years, and further significant improvements are expected in the near future. On a longer time-scale, further remarkable advances are expected with the SKA. Here we summarize some of the most recent limits obtained, as of 2017, although these results will (hopefully!) soon become obsolete.

Writing the characteristic amplitude $h_c(f)$ as in eq. (23.56), and setting $f_* = 1\,\text{yr}^{-1} \simeq 31.7$ nHz and $\alpha = -2/3$, PPTA (2015) finds the limit

$$A_{1\,\text{yr}} < 1.0 \times 10^{-15}\,, \tag{23.74}$$

at 95% c.l. This limit is determined by the four pulsars with the highest timing precision. Similar limits have been obtained by the other collaborations: $A_{1\,\text{yr}} < 1.5 \times 10^{-15}$ from the NANOGrav nine-year data release, $A_{1\,\text{yr}} < 3.0 \times 10^{-15}$ from EPTA and $A_{1\,\text{yr}} < 1.7 \times 10^{-15}$ from the first IPTA data release.[17]

Comparing with the range of predictions for $A_{1\,\text{yr}}$ given in eq. (16.86), we see that PTAs have already entered the region where detection of the stochastic background from SMBH binaries is possible, and a detection could take place in the next few years. As we have already discussed in Chapter 16, to correctly interpret the implications of a lack of detection on the modeling of SMBH binaries, one must be aware, first of all, of the fact that all predictions simply *assume* that the last-parsec problem is solved. Furthermore, after evolving through the regimes dominated by dynamical friction and later by three-body interaction or accretion (see the discussion in Section 16.2.1) the binaries might have acquired a large eccentricity, which could significantly suppress the signal in the PTA frequency range, and would in any case change the functional form of $h_c(f)$, with respect to the template $h_c(f) \propto f^{-2/3}$ used in the search; compare with Fig. 16.6 on page 362.

To translate eq. (23.74) into a limit on $h_0^2\Omega_{\text{gw}}(f)$ observe that the pivot frequency f_* is just an auxiliary quantity used to parametrize the spectrum, and is chosen arbitrarily. The actual limit on $h_0^2\Omega_{\text{gw}}(f)$ rather refers to a frequency f corresponding to the mean sensitivity frequency of the detector. For PPTA (2015) this turns out to be $f \simeq 0.2\,\text{yr}^{-1} \simeq 6.3$ nHz.[18] For $\alpha = -2/3$ we have

$$\Omega_{\text{gw}}(f) = \frac{2\pi^2}{3H_0^2} f^2 A_*^2 \left(\frac{f}{f_*}\right)^{-4/3}\,. \tag{23.75}$$

Then the PPTA (2015) limit on $A_{1\,\text{yr}}$ translates into a limit

$$h_0^2\Omega_{\text{gw}}(f = 6.3\,\text{nHz}) < 2.2 \times 10^{-10}\,. \tag{23.76}$$

In contrast, the limit reported by EPTA refers to a slightly lower frequency $f = 2.8$ nHz, and is given by

$$h_0^2\Omega_{\text{gw}}(f = 2.8\,\text{nHz}) < 1.1 \times 10^{-9}\,. \tag{23.77}$$

[17]The first IPTA data release is mostly concerned with methodological issues on the combination of data from the various PTAs, and does not use the most updated datasets of its individual PTA.

[18]See Shannon *et al.* (2015), Supplement Section S2.2, for a discussion of the determination of the mean frequency. Previous results obtained by PPTA in 2013 had a lower mean frequency, $f \simeq 2.8$ nHz, and translated into $h_0^2\Omega_{\text{gw}}(f = 2.8\,\text{nHz}) < 1.3 \times 10^{-9}$; see Shannon *et al.* (2013).

These bounds were shown in Fig. 22.27 on page 716, together with the other existing limits on stochastic GW backgrounds.

Targeting a spectrum with the shape expected from cosmic strings one finds similar bounds on $h_0^2\Omega_{\rm gw}$, which can be translated into bounds on the string tension, which we have already mentioned in eq. (22.256). In particular, the limit from NANOGrav can be translated into a limit on the string tension $G\mu < 3.3 \times 10^{-8}$, which is better by a factor of 4 than the limit obtained from the CMB, while EPTA gives the limit $G\mu < 1.3 \times 10^{-7}$.

Similarly, one can search for a flat spectrum in $h_0^2\Omega_{\rm gw}(f)$, such as that predicted by inflation (apart from a small red tilt), as we saw in Section 21.3.6. This corresponds to $\alpha = -1$. In this case NANOGrav gives

$$h_0^2\Omega_{\rm gw}(f = 1\,{\rm yr}^{-1}) < 4.2 \times 10^{-10}\,. \tag{23.78}$$

It should, however, be stressed again that PTAs have basically no sensitivity at the frequency $f = 1\,{\rm yr}^{-1} \simeq 31.7$ nHz. The best sensitivity is in the range of a few nHz, as in eqs. (23.76) and (23.77). The result (23.78) is then conventionally extrapolated to a reference frequency $f = 1\,{\rm yr}^{-1}$ assuming the given value of α in the spectrum. As we saw in eq. (21.358), for single-field slow-roll inflation at PTA frequencies the CMB limit on the tensor-to-scalar ratio r corresponds to $h_0^2\Omega_{\rm gw}(f) \lesssim 2 \times 10^{-16}$, so this inflationary background is out of reach of present PTA sensitivities.

Finally, we observe that the limit on $h_0^2\Omega_{\rm gw}$ from pulsar timing holds for all sorts of GW backgrounds, not necessarily of primordial origin, in contrast to the BBN bound discussed in Section 22.7, which holds only for GWs that were already present at the time of BBN.

Further reading

- The response of the Earth–pulsar system to a GW was given in Detweiler (1979), using the result of a computation by Estabrook and Wahlquist (1975) that was performed in the context of GW searches from Doppler tracking of spacecrafts. It is, however, a computation of the Doppler shift $\Delta\nu_\gamma/\nu_\gamma$ of the photons received, and not of $\Delta\nu/\nu$, where ν is the observed frequency of arrival of the pulses due to the lighthouse effect of the pulsar. The equality between these quantities has been implicitly assumed both in Detweiler (1979) and in all the subsequent literature. However, as we have discussed in Section 23.1 and in Note 5 on page 728, these two quantities are logically distinct, and the equality $\Delta\nu_\gamma/\nu_\gamma = \Delta\nu/\nu$ holds only in the limit $\omega_{\rm gw}T \to 0$. Note also that the overall sign of our

result, obtained either through the direct computation of $\Delta\nu/\nu$ in eq. (23.14) or from the computation of $\Delta\nu_\gamma/\nu_\gamma$ in eq. (23.20), agrees with eq. (19) of Estabrook and Wahlquist (1975) but differs from that given in Anholm, Ballmer, Creighton, Price and Siemens (2009), which we trace to an incorrect sign in going from eq. (A11) to eq. (A13) of that paper. Our sign agrees with that in eqs. (3) and (4) of Sesana and Vecchio (2010), keeping into account their opposite-sign definition of $z_a(t)$.

- The possibility of using the Earth–pulsar system to detect GWs was discussed in Sazhin (1978), Detweiler (1979), Mashhoon (1982) and Bertotti, Carr and Rees (1983). The use of PTAs for detecting stochastic GWs was suggested by Hellings and Downs (1983), and the method was further developed by Foster and Backer (1990), Kaspi, Taylor and Ryba (1994), Jenet, Hobbs, Lee and Manchester (2005), Jenet *et al.* (2006) and Hobbs *et al.* (2009).

- The European Pulsar Timing Array (EPTA) is described in Kramer and Champion (2013), while the coherent use of its five telescopes within the LEAP project is described in Bassa *et al.* (2016) and Liu *et al.* (2016). The North American Nanohertz Observatory for Gravitational Waves (NANOGrav) is described in McLaughlin (2013), the Parkes Pulsar Timing Array (PPTA) in Manchester *et al.* (2013), and the International Pulsar Timing Array (IPTA) in Hobbs *et al.* (2010). The potential of the Square Kilometer Array (SKA) for pulsar timing and GW detection is discussed in Kramer (2007).

- Fully Bayesian analysis of PTA data is developed in van Haasteren, Levin, McDonald and Lu (2009), van Haasteren *et al.* (2011) and van Haasteren and Levin (2013). Data analysis techniques for PTAs based on the methods used for GW interferometers are discussed in Anholm, Ballmer, Creighton, Price and Siemens (2009). Pulsar scrambling is discussed

in Cornish and Sampson (2016) and Taylor *et al.* (2017). The scaling of the SNR with the parameters of the search is discussed in Siemens, Ellis, Jenet and Romano (2013). The possibility of using PTAs to study stochastic backgrounds of longitudinal components of GWs in alternative theories of gravity is discussed in Chamberlin and Siemens (2012).

- The possibility of detecting single sources with PTAs is discussed in Jenet, Lommen, Larson and Wen (2004), Sesana and Vecchio (2010), Finn and Lommen (2010), Lee *et al.* (2011) and Madison *et al.* (2016).

- As of 2017, the best limits on stochastic backgrounds are presented in Shannon *et al.* (2015) for PPTA, Lentati *et al.* (2015) for EPTA, and Arzoumanian *et al.* (2016) for NANOGrav. The first data release of IPTA is presented in Verbiest *et al.* (2016). EPTA limits on individual continuous GW sources are presented in Babak *et al.* (2016).

- Reviews on PTAs include Burke-Spolaor (2015) and Lommen (2015).

Bibliography

Abbott, L. F. and M. B. Wise (1984). "Constraints on generalized inflationary cosmologies," *Nucl. Phys. B* **244**, 541.

Abbott, L. F. and D. D. Harari (1986). "Graviton production in inflationary cosmology," *Nucl. Phys. B* **264**, 487.

Abbott, T. M. C. *et al.* [DES Collaboration] (2017). "Dark Energy Survey year 1 results: Cosmological constraints from galaxy clustering and weak lensing," arXiv:1708.01530 [astro-ph.CO].

Abrahams, A. M. and G. B. Cook (1994). "Collisions of boosted black holes: Perturbation theory prediction of gravitational radiation," *Phys. Rev. D* **50**, 2364.

Ahlers, M., P. Mertsch and S. Sarkar (2009). "On cosmic ray acceleration in supernova remnants and the FERMI/PAMELA data," *Phys. Rev. D* **80**, 123017. [arXiv:0909.4060 [astro-ph.HE]].

Ajith, P. *et al.* (2007). "Phenomenological template family for black-hole coalescence waveforms," *Class. Quant. Grav.* **24**, S689. [arXiv:0704.3764 [gr-qc]].

Ajith, P. *et al.* (2008). "A template bank for gravitational waveforms from coalescing binary black holes. I. Non-spinning binaries," *Phys. Rev. D* **77**, 104017. Erratum: *Phys. Rev. D* **79** (2009) 129901. [arXiv:0710.2335 [gr-qc]].

Ajith, P. *et al.* (2011). "Inspiral-merger-ringdown waveforms for black-hole binaries with non-precessing spins," *Phys. Rev. Lett.* **106**, 241101. [arXiv:0909.2867 [gr-qc]].

Akgun, T. and I. Wasserman (2008). "Toroidal magnetic fields in type II superconducting neutron stars," *Mon. Not. Roy. Astron. Soc.* **383**, 1551. [arXiv:0705.2195 [astro-ph]].

Akrami, Y., S. F. Hassan, F. Könnig, A. Schmidt-May and A. R. Solomon (2015). "Bimetric gravity is cosmologically viable," *Phys. Lett. B* **748**, 37. [arXiv:1503.07521 [gr-qc]].

Albrecht, A. and P. J. Steinhardt (1982). "Cosmology for Grand Unified Theories with radiatively induced symmetry breaking," *Phys. Rev. Lett.* **48**, 1220.

Albrecht, A., P. Ferreira, M. Joyce and T. Prokopec (1994). "Inflation and squeezed quantum states," *Phys. Rev. D* **50**, 4807. [astro-ph/9303001].

Alcubierre, M. (2008). *Introduction to 3+1 Numerical Relativity.* Oxford University Press, Oxford.

Alexander, K. D. *et al.* (2017). "The electromagnetic counterpart of the binary neutron star merger LIGO/VIRGO GW170817. VI. Radio constraints on a relativistic jet and predictions for late-time emission from the kilonova ejecta," *Astrophys. J.* **848**, L21. [arXiv:1710.05457 [astro-ph.HE]].

Alic, D., C. Bona-Casas, C. Bona, L. Rezzolla and C. Palenzuela (2012). "Conformal and covariant formulation of the Z4 system with constraint-violation damping," *Phys. Rev. D* **85**, 064040. [arXiv:1106.2254 [gr-qc]].

Allen, B. (1985). "Vacuum states in de Sitter space," *Phys. Rev. D* **32**, 3136.

Allen, B. (1988). "The stochastic gravity wave background in inflationary Universe models," *Phys. Rev. D* **37**, 2078.

Allen, B. (1997). "The stochastic gravity wave background: Sources and detection," in *Relativistic Gravitation and Gravitational Radiation. Proceedings of 1995 Les Houches School of Physics.* J.-A. Marck and J.-P. Lasota eds., Cambridge University Press, Cambridge. [gr-qc/9604033].

Allen, B. and S. Koranda (1994). "Temperature fluctuations in the cosmic background radiation from inflationary cosmological models," *Phys. Rev. D* **50**, 3713. [astro-ph/9404068].

Allen, B., W. G. Anderson, P. R. Brady, D. A. Brown and J. D. E. Creighton (2012). "FINDCHIRP: An algorithm for detection of gravitational waves from inspiraling compact binaries," *Phys. Rev. D* **85**, 122006. [gr-qc/0509116].

Allen, G., N. Andersson, K. D. Kokkotas and B. F. Schutz (1998). "Gravitational waves from pulsating stars: Evolving the perturbation equations for a relativistic star," *Phys. Rev. D* **58**, 124012. [gr-qc/9704023].

Amaro-Seoane, P. *et al.* (2007). "Astrophysics, detection and science applications of intermediate- and extreme mass-ratio inspirals," *Class. Quant. Grav.* **24**, R113. [astro-ph/0703495].

Amaro-Seoane, P. *et al.* (2013). "eLISA/NGO: Astrophysics and cosmology in the gravitational-wave millihertz regime," *GW Notes* **6**, 4. [arXiv:1201.3621 [astro-ph.CO]].

Amaro-Seoane, P., J. R. Gair, A. Pound, S. A. Hughes and C. F. Sopuerta (2015). "Research update on extreme-mass-ratio inspirals," *J. Phys. Conf. Ser.* **610**, 012002. [arXiv:1410.0958 [astro-ph.CO]].

Amendola, L. and S. Tsujikawa (2010). *Dark Energy. Theory and Observations*, Cambridge University Press, Cambridge.

Anderson, A. and R. H. Price (1991). "Intertwining of the equations of black hole perturbations," *Phys. Rev. D* **43**, 3147.

Anderson, G. W. and L. J. Hall (1992). "The electroweak phase transition and baryogenesis," *Phys. Rev. D* **45**, 2685.

Anderson, M. *et al.* (2008). "Magnetized neutron star mergers and gravitational wave signals," *Phys. Rev. Lett.* **100**, 191101. [arXiv:0801.4387 [gr-qc]].

Andersson, N. (1992). "A numerically accurate investigation of black-hole normal modes," *Proc. Roy. Soc. Lond. A* **439**, 47.

Andersson, N. (1993). "On the asymptotic distribution of quasinormal-mode frequencies for Schwarzschild black holes," *Class. Quant. Grav.* **10**, L61.

Andersson, N. (1995). "Excitation of Schwarzschild black hole quasinormal modes," *Phys. Rev. D* **51**, 353.

Andersson, N. (1997). "Evolving test fields in a black hole geometry," *Phys. Rev. D* **55**, 468. [gr-qc/9607064].

Andersson, N. (1998). "A new class of unstable modes of rotating relativistic stars," *Astrophys. J.* **502**, 708. [gr-qc/9706075].

Andersson, N. (2003). "Gravitational waves from instabilities in relativistic stars," *Class. Quant. Grav.* **20**, R105. [astro-ph/0211057].

Andersson, N. and S. Linnaeus (1992). "Quasinormal modes of a Schwarzschild black hole: Improved phase-integral treatment," *Phys. Rev. D* **46**, 4179.

Andersson, N., Y. Kojima and K. D. Kokkotas (1996). "On the oscillation spectra of ultracompact stars: An extensive survey of gravitational wave modes," *Astrophys. J.* **462**, 855. [gr-qc/9512048].

Andersson, N., K. D. Kokkotas and B. F. Schutz (1996). "Space-time modes of relativistic stars," *Mon. Not. Roy. Astron. Soc.* **280**, 1230.

Andersson, N. and K. D. Kokkotas (1998). "Towards gravitational wave asteroseismology," *Mon. Not. Roy. Astron. Soc.* **299**, 1059. [gr-qc/9711088].

Andersson, N., K. D. Kokkotas and B. F. Schutz (1999). "Gravitational radiation limit on the spin of young neutron stars," *Astrophys. J.* **510**, 846. [astro-ph/9805225].

Andersson, N. and K. D. Kokkotas (2001). "The r-mode instability in rotating neutron stars," *Int. J. Mod. Phys. D* **10**, 381. [gr-qc/0010102].

Andersson, N. *et al.* (2011). "Gravitational waves from neutron stars: Promises and challenges," *Gen. Rel. Grav.* **43**, 409. [arXiv:0912.0384 [astro-ph.SR]].

Andreoni, I. *et al.* (2017). "Follow up of GW170817 and its electromagnetic counterpart by Australian-led observing programs," [arXiv:1710.05846 [astro-ph.HE]].

Anholm, M., S. Ballmer, J. D. E. Creighton, L. R. Price and X. Siemens (2009). "Optimal strategies for gravitational wave stochastic background searches in pulsar timing data," *Phys. Rev. D* **79**, 084030. [arXiv:0809.0701 [gr-qc]].

[ANTARES and IceCube and LIGO Scientific Collaboration and Virgo Collaboration] Adrian-Martinez, S. *et al.* (2016). "High-energy neutrino follow-up search of gravitational wave event GW150914 with ANTARES and IceCube," *Phys. Rev. D* **93**, 122010. [arXiv:1602.05411 [astro-ph.HE]].

Antoniadis, J. *et al.* (2013). "A massive pulsar in a compact relativistic binary," *Science* **340**, 6131. [arXiv:1304.6875 [astro-ph.HE]].

Aoki, W., N. Tominaga, T. C. Beers, S. Honda and Y. S. Lee (2014). "A chemical signature of first-generation very massive stars," *Science* **345**, 912.

Aoki, Y., G. Endrodi, Z. Fodor, S. D. Katz and K. K. Szabo (2006). "The order of the quantum chromodynamics transition predicted by the Standard Model of particle physics," *Nature* **443**, 675. [hep-lat/0611014].

Apreda, R., M. Maggiore, A. Nicolis and A. Riotto (2002). "Gravitational waves from electroweak phase transitions," *Nucl. Phys. B* **631**, 342. [gr-qc/0107033].

Arai, S. and A. Nishizawa (2017). "Generalized framework for testing gravity with gravitational-wave propagation. II. Constraints on Horndeski theory," arXiv:1711.03776 [gr-qc].

Arcavi, I. *et al.* (2017a). "Optical follow-up of gravitational-wave events with Las Cumbres observatory," *Astrophys. J.* **848**, L33. [arXiv:1710.05842 [astro-ph.HE]].

Arcavi, I. *et al.* (2017b). "Optical emission from a kilonova following a gravitational-wave-detected neutron-star merger," *Nature* **551**, 64. [arXiv:1710.05843 [astro-ph.HE]].

Armitage, P. J. and P. Natarajan (2002). "Accretion during the merger of supermassive black holes," *Astrophys. J.* **567**, L9. [astro-ph/0201318].

Arnowitt, R. L., S. Deser and C. W. Misner (1962). "The dynamics of general relativity," in *Gravitation: An Introduction to Current Research*, L. Witten ed., Wiley, New York. Also available as [arXiv:gr-qc/0405109].

Arras, P. *et al.* (2003). "Saturation of the r-mode instability," *Astrophys. J.* **591**, 1129. [astro-ph/0202345].

Arun, K. G., B. R. Iyer, B. S. Sathyaprakash, S. Sinha and C. Van Den Broeck (2007). "Higher signal harmonics, LISA's angular resolution and dark energy," *Phys. Rev. D* **76**, 104016. Erratum: *Phys. Rev. D* **76**, 129903 (2007). [arXiv:0707.3920 [astro-ph]].

Arzoumanian, Z. *et al.* [NANOGrav Collaboration] (2016). "The NANOGrav nine-year data set: Limits on the isotropic stochastic gravitational wave background," *Astrophys. J.* **821**, 13. [arXiv:1508.03024 [astro-ph.GA]].

Ashtekar, A., S. Fairhurst, and B. Krishnan (2000). "Isolated horizons: Hamiltonian evolution and the first law", *Phys. Rev. D* **62** 104025.

Ashtekar, A., C. Beetle, and J. Lewandowski (2001). "Mechanics of rotating isolated horizons", *Phys. Rev. D* **64**, 044016.

Ashtekar, A. and B. Krishnan (2004). "Isolated and dynamical horizons and their applications," *Living Rev. Rel.* **7**, 10.

Astier, P. *et al.* [The SNLS Collaboration] (2006). "The Supernova Legacy Survey: Measurement of Ω_m, Ω_Λ and w from the first year data set," *Astron. Astrophys.* **447**, 31. [astro-ph/0510447].

Audley, H. *et al.*, "Laser Interferometer Space Antenna," [arXiv:1702.00786 [astro-ph.IM]].

Babak, S., J. R. Gair, A. Petiteau and A. Sesana (2011). "Fundamental physics and cosmology with LISA," *Class. Quant. Grav.* **28**, 114001. [arXiv:1011.2062 [gr-qc]].

Babak, S. *et al.* (2013). "Searching for gravitational waves from binary coalescence," *Phys. Rev. D* **87**, 024033. [arXiv:1208.3491 [gr-qc]].

Babak *et al.*, S. (2016). "European Pulsar Timing Array limits on continuous gravitational waves from individual supermassive black hole binaries," *Mon. Not. Roy. Astron. Soc.* **455**, 1665. [arXiv:1509.02165 [astro-ph.CO]].

Bachelot, A. and A. Motet-Bachelot (1992). "Les résonances d'un trou noir de Schwarzschild," *Ann. Inst. H. Poincaré A* **59**, 3.

Backer, D.C., S.R. Kulkarni, C. Heiles, M.M. Davis and W. M. Goss (1982). "A millisecond pulsar," *Nature* **300**, 615.

Baiotti, L. and L. Rezzolla (2006). "Challenging the paradigm of singularity excision in gravitational collapse," *Phys. Rev. Lett.* **97**, 141101. [gr-qc/0608113].

Baiotti, L., R. De Pietri, G. M. Manca and L. Rezzolla (2007). "Accurate simulations of the dynamical bar-mode instability in full general relativity," *Phys. Rev. D* **75**, 044023. [astro-ph/0609473].

Baiotti, L., B. Giacomazzo and L. Rezzolla (2008). "Accurate evolutions of inspiralling neutron-star binaries: Prompt and delayed collapse to black hole," *Phys. Rev. D* **78**, 084033. [arXiv:0804.0594 [gr-qc]].

Baiotti, L. and L. Rezzolla (2017). *Rept. Prog. Phys.* **80**, 096901. "Binary neutron-star mergers: A review of Einstein's richest laboratory," arXiv:1607.03540 [gr-qc].

Baker, J. G., J. Centrella, D. I. Choi, M. Koppitz and J. van Meter (2006). "Gravitational wave extraction from an inspiraling configuration of merging black holes," *Phys. Rev. Lett.* **96**, 111102. [gr-qc/0511103].

Baker, J. G. *et al.* (2006). "Getting a kick out of numerical relativity," *Astrophys. J.* **653** L93. [astro-ph/0603204].

Baker, J. G. *et al.* (2007a). "Modeling kicks from the merger of non-precessing black-hole binaries," *Astrophys. J.* **668** 1140. [astro-ph/0702390].

Baker, J. G. *et al.* (2007b). *Phys. Rev. Lett.* **99**, 181101. "Consistency of post-Newtonian waveforms with numerical relativity," [gr-qc/0612024].

Baker, T. *et al.* (2017). "Strong constraints on cosmological gravity from GW170817 and GRB 170817A," [arXiv:1710.06394 [astro-ph.CO]].

Balick, B. and R. L. Brown (1974). "Intense sub-arcsecond structure in the galactic center," *Astrophys. J.* **194**, 265.

Bambi, C., J. Jiang and J. F. Steiner (2016). *Class. Quant. Grav.* **33**, 064001. "Testing the no-hair theorem with the continuum-fitting and the iron line methods: A short review," [arXiv:1511.07587 [gr-qc]].

Bañados, E. *et al.* (2016). "The Pan-STARRS1 distant $z > 5.6$ quasar survey: more than 100 quasars within the first Gyr of the Universe," *Astrophys. J. Suppl.* **227**, 11. [arXiv:1608.03279 [astro-ph.GA]].

Banks, T. and L. Mannelli (2003). "De Sitter vacua, renormalization and locality," *Phys. Rev. D* **67**, 065009. [hep-th/0209113].

Barack, L. (2009). "Gravitational self force in extreme mass-ratio inspirals," *Class. Quant. Grav.* **26**, 213001. [arXiv:0908.1664 [gr-qc]].

Barack, L. and C. Cutler (2004). "LISA capture sources: Approximate waveforms, signal-to-noise ratios, and parameter estimation accuracy," *Phys. Rev. D* **69**, 082005. [gr-qc/0310125].

Barack, L. and C. Cutler (2007). "Using LISA EMRI sources to test off-Kerr deviations in the geometry of massive black holes," *Phys. Rev. D* **75** 042003. [gr-qc/0612029].

Barack, L. and N. Sago (2010). "Gravitational self-force on a particle in eccentric orbit around a Schwarzschild black hole," *Phys. Rev. D* **81**, 084021. [arXiv:1002.2386 [gr-qc]].

Barausse, E. and L. Rezzolla (2009). "Predicting the direction of the final spin from the coalescence of two black holes," *Astrophys. J.* **704**, L40. [arXiv:0904.2577 [gr-qc]].

Barausse, E., E. Racine and A. Buonanno (2009). "Hamiltonian of a spinning test-particle in curved spacetime," *Phys. Rev. D* **80**, 104025. [arXiv:0907.4745 [gr-qc]].

Barausse, E. and A. Buonanno (2010). "An improved effective-one-body Hamiltonian for spinning black-hole binaries," *Phys. Rev. D* **81**, 084024. [arXiv:0912.3517 [gr-qc]].

Barausse, E., V. Cardoso and G. Khanna (2010). "Test bodies and naked singularities: Is the self-force the cosmic censor?," *Phys. Rev. Lett.* **105**, 261102. [arXiv:1008.5159 [gr-qc]].

Barausse, E., V. Cardoso and G. Khanna (2011). "Testing the cosmic censorship conjecture with point particles: The effect of radiation reaction and the self-force," *Phys. Rev. D* **84**, 104006. [arXiv:1106.1692 [gr-qc]].

Barausse, E., A. Buonanno and A. Le Tiec (2012). "The complete non-spinning effective-one-body metric at linear order in the mass ratio," *Phys. Rev. D* **85**, 064010. [arXiv:1111.5610 [gr-qc]].

Barausse, E., V. Morozova and L. Rezzolla (2012). "On the mass radiated by coalescing black-hole binaries," *Astrophys. J.* **758** 63. Erratum: *Astrophys. J.* **786** (2014) 76. [arXiv:1206.3803 [gr-qc]].

Barbon R., V. Buondi, E. Cappellaro and M. Turatto (1999). "The Asiago supernova catalogue — 10 years after," *Astron. Astrophys. Suppl.* **139**, 531. [astro-ph/9908046].

Bardeen, J. M. (1980). "Gauge invariant cosmological perturbations," *Phys. Rev. D* **22**, 1882.

Bardeen, J. M. and W. H. Press (1973). "Radiation fields in the Schwarzschild background," *J. Math. Phys.* **14**, 7.

Bardeen, J. M., P. J. Steinhardt and M. S. Turner (1983). "Spontaneous creation of almost scale-free density perturbations in an inflationary universe," *Phys. Rev. D* **28**, 679.

Bardeen, J. M., J. R. Bond, N. Kaiser and A. S. Szalay (1986). "The statistics of peaks of Gaussian random fields," *Astrophys. J.* **304** 15.

Barger, V., P. Langacker, M. McCaskey, M. J. Ramsey-Musolf and G. Shaughnessy (2008). "LHC phenomenology of an extended Standard Model with a real scalar singlet," *Phys. Rev. D* **77** 035005. [arXiv:0706.4311 [hep-ph]].

Barker, B.M. and R.F. O'Connell (1975). "Gravitational two-body problem with arbitrary masses, spins, and quadrupole moments," *Phys. Rev. D* **12**, 329.

Barker, B.M. and R.F. O'Connell (1979). "The gravitational interaction: Spin, rotation, and quantum effects — A review," *Gen. Rel. Grav.* **11**, 149.

Bashinsky, S. (2005). "Coupled evolution of primordial gravity waves and relic neutrinos," [astro-ph/0505502].

Bassa, C. G. *et al.* (2016). "LEAP: The Large European Array for Pulsars," *Mon. Not. Roy. Astron. Soc.* **456**, 2196. [arXiv:1511.06597 [astro-ph.IM]].

Baumann, D. *et al.* [CMBPol Study Team Collaboration] (2009). "CMBPol mission concept study: probing inflation with CMB polarization," *AIP Conf. Proc.* **1141**, 10. [arXiv:0811.3919 [astro-ph]].

Baumann, D. (2009). "Inflation," [arXiv:0907.5424 [hep-th]].

Baumgarte, T. W. and S. L. Shapiro (2003). "Numerical relativity and compact binaries," *Phys. Rept.* **376**, 41. [gr-qc/0211028].

Baumgarte, T. W. and S. L. Shapiro (2010). *Numerical Relativity: Solving Einstein's Equations on the Computer*, Cambridge University Press, Cambridge.

Bauswein, A. and H.-T. Janka (2012). "Measuring neutron-star properties via gravitational waves from binary mergers," *Phys. Rev. Lett.* **108**, 011101. [arXiv:1106.1616 [astro-ph.SR]].

Bauswein, A., T. W. Baumgarte and H.-T. Janka (2013). "Prompt merger collapse and the maximum mass of neutron stars," *Phys. Rev. Lett.* **111**, 131101. [arXiv:1307.5191 [astro-ph.SR]].

Bauswein, A. and N. Stergioulas (2015). "Unified picture of the post-merger dynamics and gravitational wave emission in neutron star mergers," *Phys. Rev. D* **91**, 124056. [arXiv:1502.03176 [astro-ph.SR]].

Bauswein, A., N. Stergioulas and H.-T. Janka (2016). "Exploring properties of high-density matter through remnants of neutron-star mergers," *Eur. Phys. J. A* **52**, 56. [arXiv:1508.05493 [astro-ph.HE]].

Bauswein, A., J. Clark, N. Stergioulas and H.-T. Janka (2016). "Dynamics and gravitational-wave emission of neutron-star merger remnants," [arXiv:1602.00950 [astro-ph.HE]].

Begelman, M. C., R. D. Blandford and M. J. Rees (1980). "Massive black hole binaries in active galactic nuclei," *Nature* **287**, 307.

Bekenstein, J. D. (1973). "Black holes and entropy," *Astrophys. J.* **183** 657.

Bekenstein, J. D. (1974). "The quantum mass spectrum of the Kerr black hole," *Lett. Nuovo Cim.* **11**, 467.

Belgacem, E., Y. Dirian, S. Foffa and M. Maggiore (2017a). "Nonlocal gravity. Conceptual aspects and cosmological predictions," arXiv:1712.07066 [hep-th].

Belgacem, E., Y. Dirian, S. Foffa and M. Maggiore (2017b). "The gravitational-wave luminosity distance in modified gravity theories," arXiv:1712.08108 [astro-ph.CO].

Benacquista, M. (2006). "Relativistic binaries in globular clusters," *Living Rev. Rel.* **9**, 2.

Benhar, O., V. Ferrari and L. Gualtieri (2004). "Gravitational wave asteroseismology revisited," *Phys. Rev. D* **70**, 124015. [astro-ph/0407529].

Bennett, C. L. *et al.* (1994). "Cosmic temperature fluctuations from two years of COBE differential microwave radiometers observations," *Astrophys. J.* **436**, 423 [astro-ph/9401012].

Bennett, C. L. *et al.* (1996). "Four year COBE DMR cosmic microwave background observations: Maps and basic results," *Astrophys. J.* **464**, L1. [astro-ph/9601067].

Bennett, D. P. and F. R. Bouchet (1988). "Evidence for a scaling solution in cosmic string evolution," *Phys. Rev. Lett.* **60**, 257.

Berezinsky, V., B. Hnatyk and A. Vilenkin (2000). "Superconducting cosmic strings as gamma-ray burst engines," [astro-ph/0001213].

Berger, E. (2014). "Short-duration gamma-ray bursts," *Annu. Rev. Astron. Astrophys.* **52**, 43. [arXiv:1311.2603 [astro-ph.HE]].

Bernardeau, F., S. Colombi, E. Gaztanaga and R. Scoccimarro (2002). "Large-scale structure of the Universe and cosmological perturbation theory," *Phys. Rept.* **367**, 1. [astro-ph/0112551].

Bernuzzi, S. and D. Hilditch (2010). "Constraint violation in free evolution schemes: Comparing BSSNOK with a conformal decomposition of Z4," *Phys. Rev. D* **81**, 084003. [arXiv:0912.2920 [gr-qc]].

Bernuzzi, S., A. Nagar, T. Dietrich and T. Damour (2015). "Modeling the dynamics of tidally interacting binary neutron stars up to the merger," *Phys. Rev. Lett.* **114**, 161103. [arXiv:1412.4553 [gr-qc]].

Bernuzzi, S., T. Dietrich, and A. Nagar (2015). "Modeling the complete gravitational wave spectrum of neutron star mergers," *Phys. Rev. Lett.* 115, 091101. [arXiv:1504.01764 [gr-qc]].

Berti, E. (2004). 'Black hole quasinormal modes: Hints of quantum gravity?," [gr-qc/0411025].

Berti, E., M. Cavaglia and L. Gualtieri (2004). "Gravitational energy loss in high-energy particle collisions: Ultrarelativistic plunge into a multidimensional black hole," *Phys. Rev. D* **69**, 124011. [hep-th/0309203].

Berti, E. and V. Cardoso (2006). "Quasinormal ringing of Kerr black holes. I. The excitation factors," *Phys. Rev. D* **74**, 104020. [gr-qc/0605118].

Berti, E., V. Cardoso and C. M. Will (2006a). "Considerations on the excitation of black hole quasinormal modes," *AIP Conf. Proc.* **848**, 687 [gr-qc/0601077].

Berti, E., V. Cardoso and C. M. Will (2006b). "On gravitational-wave spectroscopy of massive black holes with the space interferometer LISA," *Phys. Rev. D* **73**, 064030. [gr-qc/0512160].

Berti, E. *et al.* (2007). "Inspiral, merger and ringdown of unequal mass black hole binaries: A multipolar analysis," *Phys. Rev. D* **76**, 064034. [gr-qc/0703053].

Berti, E., V. Cardoso and A. O. Starinets (2009). "Quasinormal modes of black holes and black branes," *Class. Quant. Grav.* **26**, 163001. [arXiv:0905.2975 [gr-qc]].

Berti, E. *et al.* (2010). "Semianalytical estimates of scattering thresholds and gravitational radiation in ultrarelativistic black hole encounters," *Phys. Rev. D* **81** 104048. [arXiv:1003.0812 [gr-qc]].

Bertotti, B., B.J. Carr and M.J. Rees (1983). "Limits from the timing of pulsars on the cosmic gravitational wave background," *Mon. Not. Roy. Astron. Soc.* **203**, 945.

Bertschinger, E. (1996). "Cosmological dynamics," in *Cosmology and Large Scale Structure. Proceedings of the 1993 Les Houches Summer School*, R. Schaeffer *et al.* eds., Elsevier, Amsterdam. [astro-ph/9503125].

Bibby, J. L., P.A. Crowther, J. O. Furness and J. S. Clark (2008). "A downward revision to the distance of the 1806-20 cluster and associated magnetar from Gemini near-infrared spectroscopy," *Mon. Not. Roy. Astron. Soc.* **386**, L23. [arXiv:0802.0815].

[BICEP2 Collaboration], Ade, P. A. R. *et al.* (2014). "Detection of B-mode polarization at degree angular scales by BICEP2," *Phys. Rev. Lett.* **112**, 241101. [arXiv:1403.3985 [astro-ph.CO]].

[BICEP2 and Planck Collaborations], Ade, P. A. R. *et al.* (2015). "A joint analysis of BICEP2/Keck Array and *Planck* Data," *Phys. Rev. Lett.* **114**, 101301. [arXiv:1502.00612 [astro-ph.CO]].

Bietenholz, M. F. (2006). "Radio images of 3C 58: Expansion and motion of its wisp," *Astrophys. J.* **645**, 1180. [astro-ph/0603197].

Bietenholz, M. F. and R. L. Nugent (2015). "New expansion rate measurements of the Crab Nebula in radio and optical," *Mon. Not. Roy. Astron. Soc.* **454**, 2416. [arXiv:1509.04687 [astro-ph.HE]].

Bildsten L. and G. Ushomirsky (2000). "Viscous boundary layer damping of r-modes in neutron stars," *Astrophys. J.* **529**, L33. [astro-ph/9911155].

Binetruy, P., A. Bohe, C. Caprini and J. F. Dufaux (2012). "Cosmological backgrounds of gravitational waves and eLISA/NGO: Phase transitions, cosmic strings and other sources," *JCAP* **1206**, 027. [arXiv:1201.0983 [gr-qc]].

Bini, D. and T. Damour (2013). "Analytical determination of the two-body gravitational interaction potential at the fourth post-Newtonian approximation," *Phys. Rev. D* **87**, 121501. [arXiv:1305.4884 [gr-qc]].

Binney, J. and M. Merrifield (1998). *Galactic Astronomy*, Princeton University Press, Princeton, NJ.

Binney, J. and S. Tremaine (2008). *Galactic Dynamics*, 2nd edition, Princeton University Press, Princeton, NJ.

Binnington, T. and E. Poisson (2009). "Relativistic theory of tidal Love numbers," *Phys. Rev. D* **80**, 084018. [arXiv:0906.1366 [gr-qc]].

Birrell, N. D. and P. C. W. Davies (1982). *Quantum Fields in Curved Space*, Cambridge University Press, Cambridge.

Blair, W. P. *et al.* (2007). "Spitzer Space Telescope observations of Kepler's supernova remnant: A detailed look at the circumstellar dust component," *Astrophys. J.* **662**, 998. [astro-ph/0703660].

Blanchard, P. K. *et al.* (2017). "The electromagnetic counterpart of the binary neutron star merger LIGO/VIRGO GW170817. VII. Properties of the host galaxy and constraints on the merger timescale," *Astrophys. J.* **848**, L22. [arXiv:1710.05458 [astro-ph.HE]].

Blanchet, L. (2006). "Gravitational radiation from post-Newtonian sources and inspiralling compact binaries," *Living Rev. Rel.* **9**, 4.

Blanchet, L., T. Damour and G. Schäfer (1990). "Postnewtonian hydrodynamics and postnewtonian gravitational wave generation for numerical relativity," *Mon. Not. Roy. Astron. Soc.* **242**, 289.

Blanchet L., M. S. S. Qusailah and C. M. Will (2005). "Gravitational recoil of inspiralling black-hole binaries to second post-Newtonian order," *Astrophys. J.* **635** 508. [astro-ph/0507692].

Blanchet, L., A. Buonanno and G. Faye (2006). "Higher-order spin effects in the dynamics of compact binaries. II. Radiation field," *Phys. Rev. D* **74**, 104034. Errata: [*Phys. Rev. D* **75** (2007) 049903 and **81** (2010) 089901]. [gr-qc/0605140].

Blanchet, L., S. L. Detweiler, A. Le Tiec and B. F. Whiting (2010). "High-order post-newtonian fit of the gravitational self-force for circular orbits in the Schwarzschild geometry," *Phys. Rev. D* **81**, 084033. [arXiv:1002.0726 [gr-qc]].

Blanco-Pillado, J. J., K. D. Olum and B. Shlaer (2014). "The number of cosmic string loops," *Phys. Rev. D* **89**, 023512 [arXiv:1309.6637 [astro-ph.CO]].

Blas, D., J. Lesgourgues and T. Tram (2011). "The Cosmic Linear Anisotropy Solving System (CLASS) II: Approximation schemes," *JCAP* **1107**, 034. [arXiv:1104.2933 [astro-ph.CO]].

Blinnikov, S. I., I. D. Novikov, T. V., Perevodchikova and A. G. Polnarev (1984). "Exploding neutron stars in close binaries," *Soviet Astronomy Letters* **10**, 177.

Boeckel, T. and J. Schaffner-Bielich (2010). "A little inflation in the early universe at the QCD phase transition," *Phys. Rev. Lett.* **105**, 041301. Erratum: *Phys. Rev. Lett.* **106** (2011) 069901. [arXiv:0906.4520 [astro-ph.CO]].

Bogdanov, S. *et al.* (2006). "Chandra X-ray observations of nineteen millisecond pulsars in the globular cluster 47 Tucanae," *Astrophys. J.* **646** 1104. [astro-ph/0604318].

Bona, C., C. Palenzuela-Luque, and C. Bona-Casas (2009). *Elements of Numerical Relativity and Relativistic Hydrodynamics: From Einstein's Equations to Astrophysical Simulations*, Lecture Notes in Physics, Vol. 783, Springer-Verlag, Berlin, Heidelberg.

Bond J. R. and G. Efstathiou (1984). "Cosmic background radiation anisotropies in universes dominated by nonbaryonic dark matter," *Astrophys. J. Lett.* **285**, L45.

Bond J. R. and G. Efstathiou (1987). "The statistics of cosmic background radiation fluctuations," *Mon. Not. Roy. Astron. Soc.* **226**, 655.

Bondi, H., M. G. J. van der Burg and A. W. K. Metzner (1962). "Gravitational waves in general relativity. 7. Waves from axisymmetric isolated systems," *Proc. Roy. Soc. Lond. A* **269**, 21.

Boughn, S. and R. Crittenden (2004). "A correlation of the cosmic microwave sky with large scale structure," *Nature* **427**, 45. [astro-ph/0305001].

Boyanovsky, D., H. J. de Vega, R. Holman, D. S. Lee and A. Singh (1995). "Dissipation via particle production in scalar field theories," *Phys. Rev. D* **51**, 4419. [hep-ph/9408214].

Boylan-Kolchin, M., C.-P. Ma and E. Quataert (2004). "Core formation in galactic nuclei due to recoiling black holes," *Astrophys. J.* **613** L37. [astro-ph/0407488].

Boyle, L. A. (2006). *Gravitational Waves and the Early Universe*, PhD Thesis, Princeton University.

Boyle, L. A., P. J. Steinhardt and N. Turok (2004). "The cosmic gravitational wave background in a cyclic universe," *Phys. Rev. D* **69**, 127302. [hep-th/0307170].

Boyle, L. A. and P. J. Steinhardt (2008). "Probing the early universe with inflationary gravitational waves," *Phys. Rev. D* **77**, 063504. [astro-ph/0512014].

Boyle, M. *et al.* (2007). "High-accuracy comparison of numerical relativity simulations with post-Newtonian expansions," *Phys. Rev. D* **76**, 124038. [arXiv:0710.0158 [gr-qc]].

Boyle, M. *et al.* (2008). 'High-accuracy numerical simulation of black-hole binaries: Computation of the gravitational-wave energy flux and comparisons with post-Newtonian approximants," *Phys. Rev. D* **78**, 104020. [arXiv:0804.4184 [gr-qc]].

Brenneman, L. W. and C. S. Reynolds (2006). "Constraining black hole spin via X-ray spectroscopy," *Astrophys. J.* **652**, 1028. [astro-ph/0608502].

Brezin, E., C. Itzykson and J. Zinn-Justin (1970). "Relativistic Balmer formula including recoil effects," *Phys. Rev. D* **1**, 2349.

Brown, J. D. and J. W. York, Jr. (1993). "Quasilocal energy and conserved charges derived from the gravitational action," *Phys. Rev. D* **47**, 1407 [gr-qc/9209012].

Brown, J. D., S. R. Lau and J. W. York (1997). "Energy of isolated systems at retarded times as the null limit of quasilocal energy," *Phys. Rev. D* **55**, 1977. [gr-qc/9609057].

Brown, J. D., S. R. Lau and J. W. York (2000). "Action and energy of the gravitational field," [gr-qc/0010024].

Brügmann, B., W. Tichy and N. Jansen (2004). "Numerical simulation of orbiting black holes," *Phys. Rev. Lett.* **92**, 211101. [gr-qc/0312112].

Brügmann, B. , J. A. Gonzalez, M. Hannam, S. Husa and U. Sperhake (2008). "Exploring black hole superkicks," *Phys. Rev. D* **77**, 124047 [arXiv:0707.0135 [gr-qc]].

Brustein, R. and G. Veneziano (1994). "The graceful exit problem in string cosmology," *Phys. Lett. B* **329**, 429. [hep-th/9403060].

Brustein, R., M. Gasperini, M. Giovannini and G. Veneziano (1995). "Relic gravitational waves from string cosmology," *Phys. Lett. B* **361**, 45. [hep-th/9507017].

Brustein, R., M. Gasperini and G. Veneziano (1997). "Peak and endpoint of the relic graviton background in string cosmology," *Phys. Rev. D* **55**, 3882. [hep-th/9604084].

Brustein, R. and R. Madden (1998). "A model of graceful exit in string cosmology," *Phys. Rev. D* **57**, 712. [hep-th/9708046].

Bunch, T. S. and P. C. W. Davies (1978). "Quantum field theory in de Sitter space: Renormalization by point splitting," *Proc. Roy. Soc. Lond. A* **360**, 117.

Buonanno, A., M. Maggiore and C. Ungarelli (1997). "Spectrum of relic gravitational waves in string cosmology," *Phys. Rev. D* **55**, 3330. [gr-qc/9605072].

Buonanno, A. and T. Damour (1999). "Effective one-body approach to general relativistic two-body dynamics," *Phys. Rev. D* **59** 084006. [gr-qc/9811091].

Buonanno, A. and T. Damour (2000). "Transition from inspiral to plunge in binary black hole coalescences," *Phys. Rev. D* **62** 064015. [gr-qc/0001013].

Buonanno, A., Y. Chen and T. Damour (2006). "Transition from inspiral to plunge in precessing binaries of spinning black holes," *Phys. Rev. D* **74** 104005. [gr-qc/0508067].

Buonanno, A., G. B. Cook and F. Pretorius (2007). "Inspiral, merger and ring-down of equal-mass black-hole binaries," *Phys. Rev. D* **75** 124018. [gr-qc/0610122].

Buonanno, A. *et al.* (2007). "Toward faithful templates for non-spinning binary black holes using the effective-one-body approach," *Phys. Rev. D* **76**, 104049. [arXiv:0706.3732 [gr-qc]].

Buonanno, A., L. E. Kidder and L. Lehner (2008). "Estimating the final spin of a binary black hole coalescence," *Phys. Rev. D* **77** 026004. [arXiv:0709.3839 [astro-ph]].

Buonanno, A. *et al.* (2009). "Effective-one-body waveforms calibrated to numerical relativity simulations: Coalescence of non-spinning, equal-mass black holes," *Phys. Rev. D* **79**, 124028. [arXiv:0902.0790 [gr-qc]].

Buonanno, A. and B. S. Sathyaprakash (2015). "Sources of Gravitational Waves: Theory and Observations," in *General Relativity and Gravitation: A Centennial Perspective*, A. Ashtekar *et al.* eds., Cambridge University Press, Cambridge. [arXiv:1410.7832 [gr-qc]].

Burgess, C. P., R. Holman, G. Tasinato and M. Williams (2015). "EFT beyond the horizon: Stochastic inflation and how primordial quantum fluctuations go classical," *JHEP* **1503**, 090. [arXiv:1408.5002 [hep-th]].

Burgio, G. F., V. Ferrari, L. Gualtieri and H.-J. Schulze (2011). "Oscillations of hot, young neutron stars: Gravitational wave frequencies and damping times," *Phys. Rev. D* **84**, 044017. [arXiv:1106.2736 [astro-ph.SR]].

Burke-Spolaor, S. (2015). "Gravitational-wave detection and astrophysics with pulsar timing arrays," [arXiv:1511.07869 [astro-ph.IM]].

Burnham, R. Jr. (1978). *Burnham's Celestial Handbook: An Observer's Guide to the Universe Beyond the Solar System*, Vol. 1, revised edition, Dover Publications, New York.

Burrows, A. and J. Hayes (1996). "Pulsar recoil and gravitational radiation due to asymmetrical stellar collapse and explosion," *Phys. Rev. Lett.* **76**, 352. [astro-ph/9511106].

Caldwell, R. R. and B. Allen (1992). "Cosmological constraints on cosmic string gravitational radiation," *Phys. Rev. D* **45**, 3447.

Caldwell, R. R., R. A. Battye and E. P. S. Shellard (1996). "Relic gravitational waves from cosmic strings: Updated constraints and opportunities for detection," *Phys. Rev. D* **54**, 7146. [astro-ph/9607130].

Callan, C. G. Jr. and S. R. Coleman (1977). "The fate of the false vacuum. 2. First quantum corrections," *Phys. Rev. D* **16**, 1762.

Camilo, F. and F. A. Rasio (2005). "Pulsars in globular clusters," in *Binary Radio Pulsars*, ASP Conf. Ser. Vol. 328, F. A. Rasio and I. H. Stairs eds., Astronomical Society of the Pacific, San Francisco. [astro-ph/0501226].

Campanelli, M., C. O. Lousto, P. Marronetti and Y. Zlochower (2006). "Accurate evolutions of orbiting black-hole binaries without excision," *Phys. Rev. Lett.* **96**, 111101. [gr-qc/0511048].

Campanelli, M., C. O. Lousto, and Y. Zlochower (2006). "Spinning-black-hole binaries: The orbital hang up," *Phys. Rev. D* **74**, 041501. [gr-qc/0604012].

Campanelli, M., C. O. Lousto, Y. Zlochower, B. Krishnan and D. Merritt (2007). "Spin flips and precession in black-hole-binary mergers," *Phys. Rev. D* **75**, 064030. [gr-qc/0612076].

Campanelli, M., C. O. Lousto, Y. Zlochower and D. Merritt (2007a). "Merger recoils and spin flips from generic black-hole binaries," [gr-qc/0701164v1].

Campanelli, M., C. O. Lousto, Y. Zlochower and D. Merritt (2007b). "Large merger recoils and spin flips from generic black-hole binaries," *Astrophys. J.* **659** L5. [gr-qc/0701164v3].

Campanelli, M., C. O. Lousto, Y. Zlochower and D. Merritt (2007c). "Maximum gravitational recoil," *Phys. Rev. Lett.* **98**, 231102. [gr-qc/0702133].

Cannon, K., *et al.* (2010). "Singular value decomposition applied to compact binary coalescence gravitational-wave signals," *Phys. Rev. D* **82**, 044025. [arXiv:1005.0012 [gr-qc]].

Cappellaro, E., R. Evans and M. Turatto (1999). "A new determination of supernova rates and a comparison with indicators for galactic star formation," *Astron. Astrophys.* **351**, 459. [astro-ph/9904225].

Cappellaro, E. and M. Turatto (2001). "Supernova types and rates," *Astrophys. Space Sci. Libr.* **264**, 199. [astro-ph/0012455].

Bibliography 757

Caprini, C. and R. Durrer (2006). "Gravitational waves from stochastic relativistic sources: Primordial turbulence and magnetic fields," *Phys. Rev. D* **74**, 063521. [astro-ph/0603476].

Caprini, C., R. Durrer, T. Konstandin and G. Servant (2009). "General properties of the gravitational wave spectrum from phase transitions," *Phys. Rev. D* **79**, 083519. [arXiv:0901.1661 [astro-ph.CO]].

Caprini, C., R. Durrer and G. Servant (2009). "The stochastic gravitational wave background from turbulence and magnetic fields generated by a first-order phase transition," *JCAP* **0912**, 024. [arXiv:0909.0622 [astro-ph.CO]].

Caprini, C., R. Durrer and X. Siemens (2010). "Detection of gravitational waves from the QCD phase transition with pulsar timing arrays," *Phys. Rev. D* **82**, 063511. [arXiv:1007.1218 [astro-ph.CO]].

Caprini, C. *et al.* (2016). "Science with the space-based interferometer eLISA. II: Gravitational waves from cosmological phase transitions," *JCAP* **1604**, 001. [arXiv:1512.06239 [astro-ph.CO]].

Cardoso, V., E. Franzin and P. Pani (2016). "Is the gravitational-wave ringdown a probe of the event horizon?," *Phys. Rev. Lett.* **116**, 171101. Erratum: *Phys. Rev. Lett.* **117** (2016), 089902. [arXiv:1602.07309 [gr-qc]].

Carlton, A. K. *et al.* (2011). "Expansion of the youngest Galactic supernova remnant G1.9+0.3," *Astrophys. J.* **737**, L22. [arXiv:1106.4498 [astro-ph.GA]].

Carroll, B.W. and D.A. Ostlie (2007). *An Introduction to Modern Stellar Astrophysics*, 2nd edition, Addison-Wesley, Reading, MA.

Centrella, J. M., K. C. B. New, L. L. Lowe and J. D. Brown (2001). "Dynamical rotational instability at low T/W," *Astrophys. J.* **550**, L193. [astro-ph/0010574].

Centrella, J., J. G. Baker, B. J. Kelly and J. R. van Meter (2010). "Black-hole binaries, gravitational waves, and numerical relativity," *Rev. Mod. Phys.* **82**, 3069. [arXiv:1010.5260 [gr-qc]].

Cerdá-Durán, P., V. Quilis and J. A. Font (2007). "AMR simulations of the low $T/|W|$ bar-mode instability of neutron stars," *Comput. Phys. Commun.* **177**, 288. [arXiv:0704.0356 [astro-ph]].

Chamberlin, S. J. and X. Siemens (2012). "Stochastic backgrounds in alternative theories of gravity: Overlap reduction functions for pulsar timing arrays," *Phys. Rev. D* **85**, 082001. [arXiv:1111.5661 [astro-ph.HE]].

Chamel, N. and P. Haensel (2008). "Physics of neutron star crusts," *Living Rev. Rel.* **11**, 10. [arXiv:0812.3955 [astro-ph]].

Chandrasekhar, S. (1961). *Hydrodynamic and Hydromagnetic Stability*, Oxford University Press, Oxford. Republished 2005 by Dover Publications, New York.

Chandrasekhar, S. (1969). *Ellipsoidal Figures of Equilibrium*, Yale University Press, New Haven, CT. Republished 1987 by Dover Publications, New York.

Chandrasekhar, S. (1970). "Solutions of two problems in the theory of gravitational radiation," *Phys. Rev. Lett.* **24**, 611.

Chandrasekhar, S. (1975). "On the equations governing the perturbations of the Schwarzschild black hole," *Proc. Roy. Soc. Lond. A* **343**, 289.

Chandrasekhar, S. (1979). "On the equations governing the perturbations of the Reissner–Nordstrom black hole" *Proc. Roy. Soc. Lond. A* **365**, 453.

Chandrasekhar, S. (1983). *The Mathematical Theory of Black Holes*, Oxford University Press, Oxford.

Chandrasekhar, S. (1984). "On algebraically special perturbations of black holes," *Proc. Roy. Soc. Lond. A* **392**, 1.

Chandrasekhar, S. and J. L. Friedman (1972). "On the stability of axisymmetric systems to axisymmetric perturbations in general relativity. II," *Astrophys. J.* **176**, 745.

Chandrasekhar, S. and S. Detweiler (1975). "The quasi-normal modes of the Schwarzschild black hole," *Proc. Roy. Soc. Lond. A* **344**, 441.

Chandrasekhar S. and V. Ferrari (1991a). "On the non-radial oscillations of a star," *Proc. Roy. Soc. Lond. A* **432**, 247.

Chandrasekhar S. and V. Ferrari (1991b). "On the non-radial oscillations of slowly rotating stars induced by the Lense–Thirring effect," *Proc. Roy. Soc. Lond. A* **433**, 423.

Chandrasekhar, S. and V. Ferrari (1992). "On the non-radial oscillations of a star. IV—An application of the theory of Regge poles," *Proc. Roy. Soc. Lond. A* **437**, 133.

Cheng, B., R. I. Epstein, R. A. Guyer and A. C. Young (1995). "Soft gamma ray repeaters and earthquakes," *Nature* **382**, 518.

Chernikov, N.A. and E. A. Tagirov (1968). "Quantum theory of scalar fields in de Sitter space-time," *Ann. Inst. H. Poincaré A* **9**, 109.

Chevallier, M. and D. Polarski (2001). "Accelerating universes with scaling dark matter," *Int. J. Mod. Phys. D* **10**, 213. [gr-qc/0009008].

Ching, E. S. C., P. T. Leung, W. M. Suen and K. Young (1995a). "Late time tail of wave propagation on curved space-time," *Phys. Rev. Lett.* **74**, 2414.

Ching, E. S. C., P. T. Leung, W. M. Suen and K. Young (1995b). "Wave propagation in gravitational systems: Late time behavior," *Phys. Rev. D* **52**, 2118.

Chiotellis, A., K.M. Schure, and J. Vink (2012). "The imprint of a symbiotic binary progenitor on the properties of Kepler's supernova remnant," *Astron. Astrophys.* **537**, A139.

Chomiuk, L. *et al.* (2011). "Pan-STARRS1 discovery of two ultra-luminous supernovae at $z \approx 0.9$," *Astrophys. J.* **743**, 114. [arXiv:1107.3552 [astro-ph.HE]].

Choquet-Bruhat, Y. and J. W. York (1980). "The Cauchy Problem," in *General Relativity and Gravitation*. Vol. 1, A. Held ed., Plenum Press, New York.

Choquet-Bruhat, Y. (2009). *General Relativity and the Einstein Equations*, Oxford University Press, Oxford.

Choquet-Bruhat, Y. (2015). "Beginnings of the Cauchy problem," in *One Hundred Years of General Relativity*, L. Bieri and S.-T. Yau eds., Surveys in Differential Geometry, Vol. 20. International Press, Somerville, MA [arXiv:1410.3490 [gr-qc]].

Chornock, R. *et al.* (2017). "The electromagnetic counterpart of the binary neutron star merger LIGO/VIRGO GW170817. IV. Detection of near-infrared signatures of r-process nucleosynthesis with Gemini-South," *Astrophys. J.* **848**, L19. [arXiv:1710.05454 [astro-ph.HE]].

Choudhury T. R. and T. Padmanabhan (2004). "Quasi normal modes in Schwarzschild-deSitter spacetime: A simple derivation of the level spacing of the frequencies," *Phys. Rev. D* **69** 064033. [gr-qc/0311064].

Christodoulou, D. (1970). "Reversible and irreversible transformations in black hole physics," *Phys. Rev. Lett.* **25**, 1596.

Churazov, E. *et al.* (2014). "First detection of ^{56}Co gamma-ray lines from type Ia supernova (SN2014J) with INTEGRAL," *Nature* **512**, 406. [arXiv:1405.3332 [astro-ph.HE]].

Ciolfi, R. and L. Rezzolla (2012). "Poloidal-field instability in magnetized relativistic stars," *Astrophys. J.* **760**, 1. [arXiv:1206.6604 [astro-ph.SR]].

Clark, D.H. and F.R. Stephenson (1977). *The Historical Supernovae*, Pergamon Press, Oxford.

Clesse, S. and J. García-Bellido (2017). "The clustering of massive primordial black holes as dark matter: Measuring their mass distribution with Advanced LIGO," *Phys. Dark Univ.* **15**, 142. [arXiv:1603.05234 [astro-ph.CO]].

Coleman, S. R. and E. J. Weinberg (1973). "Radiative corrections as the origin of spontaneous symmetry breaking," *Phys. Rev. D* **7**, 1888.

Coleman, S. R. (1977a). "The fate of the false vacuum. 1. Semiclassical theory," *Phys. Rev. D* **15**, 2929. Erratum: *Phys. Rev. D* **16** (1977) 1248.

Coleman S. R. (1977b). "The uses of instantons (1977 Erice Lectures)," in *The Whys of Subnuclear Physics*, A. Zichichi ed., Plenum Press, New York. Reprinted 1985 in *Aspects of Symmetry: Selected Erice Lectures*, Cambridge University Press, Cambridge.

Colladay, D. and V. A. Kostelecky (1998). "Lorentz-violating extension of the Standard Model," *Phys. Rev. D* **58**, 116002. [hep-ph/9809521].

Colleoni, M., L. Barack, A. G. Shah and M. van de Meent (2015). "Self-force as a cosmic censor in the Kerr overspinning problem," *Phys. Rev. D* **92**, 084044. [arXiv:1508.04031 [gr-qc]].

Colpi, M., L. Mayer and F. Governato (1999). "Dynamical friction and the evolution of satellites in virialized halos: The theory of linear response," *Astrophys. J.* **525** 720. [astro-ph/9907088].

Connaughton, V. *et al.* (2016). "Fermi GBM Observations of LIGO gravitational wave event GW150914," *Astrophys. J.* **826**, L6. [arXiv:1602.03920 [astro-ph.HE]].

Cook, G. B. (2000). "Initial data for numerical relativity," *Living Rev. Rel.* **3**, 5.

Cook, G. B., S. L. Shapiro and S. A. Teukolsky (1994). "Rapidly rotating neutron stars in general relativity: Realistic equations of state," *Astrophys. J.* **424** 823.

Cook, G. B. and M. Zalutskiy (2014). "Gravitational perturbations of the Kerr geometry: High-accuracy study," *Phys. Rev. D* **90** 12, 124021 [arXiv:1410.7698 [gr-qc]].

Cooray, A. (2002). "Nonlinear integrated Sachs–Wolfe effect," *Phys. Rev. D* **65**, 083518. [astro-ph/0109162].

Copeland, E. J., R. Easther and D. Wands (1997). "Vacuum fluctuations in axion–dilaton cosmologies," *Phys. Rev. D* **56**, 874. [hep-th/9701082].

Cornish, N. J. and L. Sampson (2016). "Towards robust gravitational wave detection with pulsar timing arrays," *Phys. Rev. D* **93**, 104047. [arXiv:1512.06829 [gr-qc]].

Corsi, A. and B. J. Owen (2011). "Maximum gravitational-wave energy emissible in magnetar flares," *Phys. Rev. D* **83**, 104014. [arXiv:1102.3421 [gr-qc]].

Corvino, G., L. Rezzolla, S. Bernuzzi, R. De Pietri and B. Giacomazzo (2010). "On the shear instability in relativistic neutron stars," *Class. Quant. Grav.* **27**, 114104. [arXiv:1001.5281 [gr-qc]].

Coulter, D. A. *et al.* (2017). "Swope Supernova Survey 2017a (SSS17a), the optical counterpart to a gravitational wave source," *Science*, doi:10.1126/science.aap9811. [arXiv:1710.05452 [astro-ph.HE]].

Cowperthwaite, P. S. *et al.* (2017) . "The electromagnetic counterpart of the binary neutron star merger LIGO/Virgo GW170817. II. UV, optical, and near-infrared light curves and comparison to kilonova models," *Astrophys. J.* **848**, L17. [arXiv:1710.05840 [astro-ph.HE]].

Creminelli, P. and F. Vernizzi (2017). "Dark energy after GW170817 and GRB170817A," *Phys. Rev. Lett.* **119**, 251302. [arXiv:1710.05877 [astro-ph.CO]].

Crittenden, R., J. R. Bond, R. L. Davis, G. Efstathiou and P. J. Steinhardt (1993). "The imprint of gravitational waves on the cosmic microwave background," *Phys. Rev. Lett.* **71**, 324. [astro-ph/9303014].

Crittenden, R., R. L. Davis and P. J. Steinhardt (1993). "Polarization of the microwave background due to primordial gravitational waves," *Astrophys. J.* **417**, L13. [astro-ph/9306027].

Crittenden, R. G. and N. Turok (1996). "Looking for Λ with the Rees–Sciama effect," *Phys. Rev. Lett.* **76**, 575. [astro-ph/9510072].

Crowder, J. and N. J. Cornish (2005). "Beyond LISA: Exploring future gravitational wave missions," *Phys. Rev. D* **72**, 083005.

Cunningham, C. T., R. H. Price and V. Moncrief (1978). "Radiation from collapsing relativistic stars. I — Linearized odd-parity radiation," *Astrophys. J.* **224** 643.

Cusin, G., R. Durrer, P. Guarato and M. Motta (2015). "Gravitational waves in bigravity cosmology," *JCAP* **1505**, 030. [arXiv:1412.5979 [astro-ph.CO]].

Cusin, G., R. Durrer, P. Guarato and M. Motta (2016). "A general mass term for bigravity," *JCAP* **1604**, 051. [arXiv:1512.02131 [astro-ph.CO]].

Cutler, C. (2002). "Gravitational waves from neutron stars with large toroidal B fields," *Phys. Rev. D* **66**, 084025. [gr-qc/0206051].

Cutler, C. and L. Lindblom (1987). "The effect of viscosity on neutron star oscillations," *Astrophys. J.* **314**, 234.

Cutler, C. and L. Lindblom (1992). "Post-Newtonian frequencies for the pulsations of rapidly rotating neutron stars," *Astrophys. J.* **385**, 630.

Cutler, C. and E. E. Flanagan (1994). "Gravitational waves from merging compact binaries: How accurately can one extract the binary's parameters from the inspiral wave form?," *Phys. Rev. D* **49**, 2658. [gr-qc/9402014].

Cutler, C. and D. E. Holz (2009). "Ultra-high precision cosmology from gravitational waves," *Phys. Rev. D* **80**, 104009. [arXiv:0906.3752 [astro-ph.CO]].

Cyburt, R. H., B. D. Fields, K. A. Olive and E. Skillman (2005). "New BBN limits on physics beyond the Standard Model from ^{4}He," *Astropart. Phys.* **23** 313. [astro-ph/0408033].

Dal Canton, T. *et al.* (2014). "Implementing a search for aligned-spin neutron star-black hole systems with advanced ground based gravitational wave detectors," *Phys. Rev. D* **90**, 082004. [arXiv:1405.6731 [gr-qc]].

Dalal, N., D. E. Holz, S. A. Hughes and B. Jain (2006). "Short GRB and binary black hole standard sirens as a probe of dark energy," *Phys. Rev. D***74**, 063006. [astro-ph/0601275].

D'Amico, G. *et al.* (2011) "Massive cosmologies," *Phys. Rev. D* **84**, 124046. [arXiv:1108.5231 [hep-th]].

Damour, T. (1982). "Problème des deux corps et freinage de rayonnement en relativité générale," *C. R. Acad. Sci. Ser. II* **294**, 1355.

Damour, T. (1983). "Gravitational radiation and the motion of compact bodies," in *Gravitational Radiation, Proceedings of the 1982 Les Houches Summer School*, N. Deruelle and T. Piran eds., North-Holland, Amsterdam.

Damour, T. (1984). "The motion of compact bodies and gravitational radiation," in *General Relativity and Gravitation*, B. Bertotti, F. de Felice and A. Pascolini eds., Reidel, Dordrecht.

Damour, T. (2001). *Phys. Rev. D* **64**, 124013. "Coalescence of two spinning black holes: An effective one-body approach," [gr-qc/0103018].

Damour, T. (2010). "Gravitational self force in a Schwarzschild background and the effective one body formalism," *Phys. Rev. D* **81**, 024017. [arXiv:0910.5533 [gr-qc]].

Damour, T. (2016). "Gravitational scattering, post-Minkowskian approximation and effective one-body theory," *Phys. Rev. D* **94**, 104015. [arXiv:1609.00354 [gr-qc]].

Damour, T. and N. Deruelle (1981). 'Lagrangien généralisé du système de deux masses ponctuelles à l'approximation post-post-Newtonienne de la relativité générale, *C. R. Acad. Sci. Ser. II* **293**, 537.

Damour, T. and G. Schäfer (1985). "Lagrangians for N point masses at the second post-Newtonian approximation of general relativity," *Gen. Rel. Grav.* **17**, 879.

Damour, T. and G. Schäfer (1988). "Higher-order relativistic periastron advances and binary pulsars," *Nuovo Cim.* **10**, 123.

Damour, T., M. Soffel and C. Xu (1991). "General relativistic celestial mechanics. 1. Method and definition of reference systems," *Phys. Rev. D* **43**, 3273.

Damour, T., M. Soffel and C. Xu (1992). "General relativistic celestial mechanics. 2. Translational equations of motion," *Phys. Rev. D* **45**, 1017.

Damour, T., B. R. Iyer and B. S. Sathyaprakash (1998). "Improved filters for gravitational waves from inspiralling compact binaries," *Phys. Rev. D* **57**, 885. [gr-qc/9708034].

Damour, T., B. R. Iyer and B. S. Sathyaprakash (2000). "Frequency domain P-approximant filters for time-truncated inspiral gravitational wave signals from compact binaries," *Phys. Rev. D* **62**, 084036. [gr-qc/0001023].

Damour, T., P. Jaranowski and G. Schäfer (2000). "On the determination of the last stable orbit for circular general relativistic binaries at the third post-Newtonian approximation," *Phys. Rev. D* **62**, 084011. [gr-qc/0005034].

Damour, T. and A. Vilenkin (2000). "Gravitational wave bursts from cosmic strings," *Phys. Rev. Lett.* **85**, 3761. [gr-qc/0004075].

Damour, T. and A. Vilenkin (2001). "Gravitational wave bursts from cusps and kinks on cosmic strings," *Phys. Rev. D* **64**, 064008. [gr-qc/0104026].

Damour, T., E. Gourgoulhon and P. Grandclement (2002). "Circular orbits of corotating binary black holes: Comparison between analytical and numerical results," *Phys. Rev. D* **66**, 024007. [gr-qc/0204011].

Damour, T., B. R. Iyer, P. Jaranowski and B. S. Sathyaprakash (2003). "Gravitational waves from black hole binary inspiral and merger: The span of third post-Newtonian effective-one-body templates," *Phys. Rev. D* **67**, 064028. [gr-qc/0211041].

Damour, T. and A. Vilenkin (2005). "Gravitational radiation from cosmic (super)strings: Bursts, stochastic background, and observational windows," *Phys. Rev. D* **71**, 063510. [hep-th/0410222].

Damour, T. and A. Gopakumar (2006). "Gravitational recoil during binary black hole coalescence using the effective one body approach," *Phys. Rev. D* **73**, 124006. [gr-qc/0602117].

Damour, T. and A. Nagar (2007a). "Final spin of a coalescing black-hole binary: An effective-one-body approach," *Phys. Rev. D* **76**, 044003. [arXiv:0704.3550 [gr-qc]].

Damour, T. and A. Nagar (2007b). "Faithful effective-one-body waveforms of small-mass-ratio coalescing black-hole binaries," *Phys. Rev. D* **76**, 064028. [arXiv:0705.2519 [gr-qc]].

Damour, T. and A. Nagar (2008). "Comparing effective-one-body gravitational waveforms to accurate numerical data," *Phys. Rev. D* **77**,
024043. [arXiv:0711.2628 [gr-qc]].

Damour, T. , A. Nagar, E. N. Dorband, D. Pollney and L. Rezzolla
(2008) "Faithful effective-one-body waveforms of equal-mass coalescing black-hole binaries," *Phys. Rev. D* **77**, 084017. [arXiv:0712.3003
[gr-qc]].

Damour, T., A. Nagar, M. Hannam, S. Husa and B. Bruegmann (2008).
"Accurate effective-one-body waveforms of inspiralling and coalescing black-hole binaries," *Phys. Rev. D* **78**, 044039. [arXiv:0803.3162
[gr-qc]].

Damour, T., P. Jaranowski and G. Schäfer (2008a). "Hamiltonian of
two spinning compact bodies with next-to-leading order gravitational
spin–orbit coupling," *Phys. Rev. D* **77**, 064032. [arXiv:0711.1048 [gr-
qc]].

Damour, T., P. Jaranowski and G. Schäfer (2008b). "Effective one
body approach to the dynamics of two spinning black holes with
next-to-leading order spin–orbit coupling," *Phys. Rev. D* **78**, 024009.
[arXiv:0803.0915 [gr-qc]].

Damour, T., B. R. Iyer and A. Nagar (2009). "Improved resummation
of post-Newtonian multipolar waveforms from circularized compact
binaries," *Phys. Rev. D* **79**, 064004. [arXiv:0811.2069 [gr-qc]].

Damour, T. and A. Nagar (2009a). "An improved analytical description of inspiralling and coalescing black-hole binaries," *Phys. Rev. D*
79, 081503. [arXiv:0902.0136 [gr-qc]].

Damour, T. and A. Nagar (2009b). "Relativistic tidal properties of
neutron stars," *Phys. Rev. D* **80**, 084035. [arXiv:0906.0096 [gr-qc]].

Damour, T. and A. Nagar (2010). "Effective one body description
of tidal effects in inspiralling compact binaries," *Phys. Rev. D* **81**,
084016. [arXiv:0911.5041 [gr-qc]].

Damour, T. and A. Nagar (2011). "The effective one body description of the two-body problem," *Fundam. Theor. Phys.* **162**, 211.
[arXiv:0906.1769].

Damour, T., A. Nagar and S. Bernuzzi (2013). "Improved effective-
one-body description of coalescing nonspinning black-hole binaries
and its numerical-relativity completion," *Phys. Rev. D* **87**, 084035.
[arXiv:1212.4357 [gr-qc]].

Damour, T., P. Jaranowski and G. Schäfer (2014). "Nonlocal-in-time
action for the fourth post-Newtonian conservative dynamics of two-
body systems," *Phys. Rev. D* **89**, 064058. [arXiv:1401.4548 [gr-qc]].

Damour, T. and A. Nagar (2014). "New effective-one-body description
of coalescing nonprecessing spinning black-hole binaries," *Phys. Rev.
D* **90**, 044018. [arXiv:1406.6913 [gr-qc]].

Damour, T., P. Jaranowski and G. Schäfer (2015). "Fourth post-
Newtonian effective one-body dynamics," *Phys. Rev. D* **91**, 084024.
[arXiv:1502.07245 [gr-qc]].

Danielsson, U. H. (2002). "On the consistency of de Sitter vacua," *JHEP* **0212**, 025. [hep-th/0210058].

Davis, M. R., R. Ruffini, W. H. Press and R.H. Price (1971). "Gravitational radiation from a particle falling radially into a Schwarzschild black hole," *Phys. Rev. Lett.* **27**, 1466.

Davis, M. R., R. Ruffini and J. Tiomno (1972). "Pulses of gravitational radiation of a particle falling radially into a Schwarzschild black hole," *Phys. Rev. D* **5**, 2932.

Davis, R. L., H. M. Hodges, G. F. Smoot, P. J. Steinhardt and M. S. Turner (1992). "Cosmic microwave background probes models of inflation," *Phys. Rev. Lett.* **69**, 1856 Erratum: *Phys. Rev. Lett.* **70** (1993) 1733. [astro-ph/9207001].

de Andrade, V. C., L. Blanchet and G. Faye (2001). "Third post-Newtonian dynamics of compact binaries: Noetherian conserved quantities and equivalence between the harmonic coordinate and ADM Hamiltonian formalisms," *Class. Quant. Grav.* **18**, 753. [gr-qc/0011063].

de Bernardis, P. *et al.* [Boomerang Collaboration] (2000). "A flat universe from high resolution maps of the cosmic microwave background radiation," *Nature*, **404**, 955. [astro-ph/0004404].

Deffayet, C. and K. Menou (2007). "Probing gravity with spacetime sirens," *Astrophys. J.* **668**, L143. [arXiv:0709.0003 [astro-ph]].

De Felice, A. and S. Tsujikawa (2010). "$f(R)$ theories," *Living Rev. Rel.* **13**, 3. [arXiv:1002.4928 [gr-qc]].

Del Pozzo, W. (2012). "Inference of the cosmological parameters from gravitational waves: Application to second generation interferometers," *Phys. Rev. D* **86**, 043011. [arXiv:1108.1317 [astro-ph.CO]].

Demorest, P., T. Pennucci, S. Ransom, M. Roberts and J. Hessels (2010). "Shapiro delay measurement of a two solar mass neutron star," *Nature* **467**, 1081. [arXiv:1010.5788 [astro-ph.HE]].

Dent, J. B., L. M. Krauss, S. Sabharwal and T. Vachaspati (2013). "Damping of primordial gravitational waves from generalized sources," *Phys. Rev. D* **88**, 084008. [arXiv:1307.7571 [astro-ph.CO]].

de Rham, C. (2014). "Massive gravity," *Living Rev. Rel.* **17**, 7. [arXiv:1401.4173 [hep-th]].

de Rham, C., G. Gabadadze and A. J. Tolley (2011). "Resummation of massive gravity," *Phys. Rev. Lett.* **106**, 231101. [arXiv:1011.1232 [hep-th]].

Deser, S. (2004). *Int. J. Mod. Phys. A* **19S1**, 99. "Some remarks on Dirac's contributions to general relativity," [gr-qc/0301097].

Dessart, L., D. J. Hillier, C. Li and S. Woosley (2012). "On the nature of supernovae Ib and Ic," *Mon. Not. Roy. Astron. Soc.* **424**, 2139. [arXiv:1205.5349 [astro-ph.SR]].

Detweiler, S. L. (1979). "Pulsar timing measurements and the search for gravitational waves," *Astrophys. J.* **234**, 1100.

Detweiler, S. L. and E. Szedenits (1979). "Black holes and gravitational waves. II - Trajectories plunging into a nonrotating hole," *Astrophys. J.* **231**, 211.

Dicke, R. H., P. J. E. Peebles, P. G. Roll and D. T. Wilkinson (1965). "Cosmic black-body radiation," *Astrophys. J.* **142**, 414.

Dicus, D. A. and W. W. Repko (2005). "Comment on damping of tensor modes in cosmology," *Phys. Rev. D* **72** 088302. [astro-ph/0509096].

Diehl, R. *et al.* (2006). "Radioactive ^{26}Al and massive stars in the galaxy," *Nature* **439**, 45 [astro-ph/0601015].

Diehl, R. *et al.* (2015). "SN2014J gamma-rays from the ^{56}Ni decay chain," *Astron. Astrophys.* **574**, A72. [arXiv:1409.5477 [astro-ph.HE]].

Dimmelmeier, H., J. A. Font and E. Müller (2002). "Relativistic simulations of rotational core collapse. 2. Collapse dynamics and gravitational radiation," *Astron. Astrophys.* **393**, 523 [astro-ph/0204289].

Dimmelmeier, H., C. D. Ott, H.-T. Janka, A. Marek and E. Müller (2007). "Generic gravitational wave signals from the collapse of rotating stellar cores," *Phys. Rev. Lett.* **98**, 251101. [astro-ph/0702305].

Dimmelmeier, H., C. D. Ott, A. Marek and H.-T. Janka (2008). "The gravitational wave burst signal from core collapse of rotating stars," *Phys. Rev. D* **78** 064056. [arXiv:0806.4953 [astro-ph]].

Dirac, P. A. M. (1958). "Generalized Hamiltonian dynamics," *Proc. Roy. Soc. Lond. A* **246**, 333.

Dirac, P. A. M. (1959). "Fixation of coordinates in the Hamiltonian theory of gravitation," *Phys. Rev.* **114**, 924.

Dirian, Y. (2017). "Changing the Bayesian prior: Absolute neutrino mass constraints in nonlocal gravity," *Phys. Rev. D* **96**, 083513. [arXiv:1704.04075 [astro-ph.CO]].

Dirian, Y., S. Foffa, N. Khosravi, M. Kunz and M. Maggiore (2014). "Cosmological perturbations and structure formation in nonlocal infrared modifications of general relativity," *JCAP* **1406**, 033. [arXiv:1403.6068 [astro-ph.CO]].

Dirian, Y., S. Foffa, M. Kunz, M. Maggiore and V. Pettorino (2015). "Non-local gravity and comparison with observational datasets," *JCAP* **1504**, 044. [arXiv:1411.7692 [astro-ph.CO]].

Dirian, Y., S. Foffa, M. Kunz, M. Maggiore and V. Pettorino (2016). "Non-local gravity and comparison with observational datasets. II. Updated results and Bayesian model comparison with ΛCDM," *JCAP* **1605**, 068. [arXiv:1602.03558 [astro-ph.CO]].

Di Valentino, E., A. Melchiorri and J. Silk (2016). "Reconciling *Planck* with the local value of H_0 in extended parameter space," *Phys. Lett. B* **761**, 242. [arXiv:1606.00634 [astro-ph.CO]].

Dodelson, S. (2003). *Modern Cosmology*, Academic Press, San Diego.

Doeleman, S.S. *et al.* (2008). "Event-horizon-scale structure in the supermassive black hole candidate at the Galactic centre," *Nature* **455**, 78.

Dolgov, A. D., D. Grasso and A. Nicolis (2002). "Relic backgrounds of gravitational waves from cosmic turbulence," *Phys. Rev. D* **66**, 103505. [astro-ph/0206461].

Doneva, D. D., E. Gaertig, K. D. Kokkotas and C. Krüger (2013). "Gravitational wave asteroseismology of fast rotating neutron stars with realistic equations of state," *Phys. Rev. D* **88**, 044052. [arXiv:1305.7197 [astro-ph.SR]].

Doneva, D. D., K. D. Kokkotas and P. Pnigouras (2015). "Gravitational wave afterglow in binary neutron star mergers," *Phys. Rev. D* **92**, 104040. [arXiv:1510.00673 [gr-qc]].

D'Onofrio, M. and K. Rummukainen (2016). "Standard Model crossover on the lattice," *Phys. Rev. D* **93**, 025003. [arXiv:1508.07161 [hep-ph]].

Doroshkevich, A. G. and I. D. Novikov (1964). "Mean density of radiation in the metagalaxy and certain problems in relativistic cosmology," *Sov. Phys. Dokl.* **9**, 111.

Doroshkevich, A. G., Ya. B. Zel'dovich, and R. A. Sunyaev (1978). "Fluctuations of the microwave background radiation in the adiabatic and entropic theories of galaxy formation," *Soviet Astronomy* **22**, 523.

Dosopoulou, F. and F. Antonini (2017). "Dynamical friction and the evolution of supermassive black hole binaries: The final hundred-parsec problem," *Astrophys. J.* **840**, 31. [arXiv:1611.06573 [astro-ph.GA]].

Dotti, M., A. Sesana and R. Decarli (2012). "Massive black hole binaries: dynamical evolution and observational signatures," *Adv. Astron.* **2012**, 940568. [arXiv:1111.0664 [astro-ph.CO]].

Drasco, S. (2006). "Strategies for observing extreme mass ratio inspirals," *Class. Quant. Grav.* **23**, S769. [gr-qc/0604115].

Drasco, S. and S. A. Hughes (2006). "Gravitational wave snapshots of generic extreme mass ratio inspirals," *Phys. Rev. D* **73**, 024027. Errata: [*Phys. Rev. D* **88** (2013), 109905 and **90** (2014), 109905]. [gr-qc/0509101].

Dreyer, O., B. Krishnan, D. Shoemaker and E. Schnetter (2003). "Introduction to isolated horizons in numerical relativity," *Phys. Rev. D* **67**, 024018. [gr-qc/0206008].

Drout, M. R. *et al.* (2011). "The first systematic study of type Ibc supernova multi-band light curves," *Astrophys. J.* **741**, 97 [arXiv:1011.4959 [astro-ph.CO]].

Drout, M. R. *et al.* (2017). "Light curves of the neutron star merger GW170817/SSS17a: Implications for r-process nucleosynthesis," *Science* . doi:10.1126/science.aaq0049 [arXiv:1710.05443 [astro-ph.HE]].

Dufaux, J. F., A. Bergman, G. Felder, L. Kofman and J. P. Uzan (2007). "Theory and numerics of gravitational waves from preheating after inflation," *Phys. Rev. D* **76**, 123517. [arXiv:0707.0875 [astro-ph]].

Dufaux, J. F., G. Felder, L. Kofman and O. Navros (2009). "Gravity waves from tachyonic preheating after hybrid inflation," *JCAP* **0903**, 001. [arXiv:0812.2917 [astro-ph]].

Dufaux, J. F., D. G. Figueroa and J. García-Bellido (2010). "Gravitational waves from abelian gauge fields and cosmic strings at preheating," *Phys. Rev. D* **82**, 083518. [arXiv:1006.0217 [astro-ph.CO]].

Duncan, R. C. and C. Thompson (1992). "Formation of very strongly magnetized neutron stars — Implications for gamma-ray bursts," *Astrophys. J.* **392**, L9.

Durrer, R. (2008). *The Cosmic Microwave Background*, Cambridge University Press, Cambridge.

Durrer, R. and T. Kahniashvili (1998). "CMB anisotropies caused by gravitational waves: A parameter study," *Helv. Phys. Acta* **71**, 445. [astro-ph/9702226].

Dvali, G., G. Gabadadze and M. Porrati (2000). "4-D gravity on a brane in 5-D Minkowski space," *Phys. Lett. B* **485**, 208. [hep-th/0005016].

Dvali, G. and A. Vilenkin (2004). "Formation and evolution of cosmic D strings," *JCAP* **0403**, 010. [hep-th/0312007].

Eardley, D. M., D. L. Lee, A. P. Lightman, R. V. Wagoner and C. M. Will (1973). "Gravitational-wave observations as a tool for testing relativistic gravity," *Phys. Rev. Lett.* **30**, 884.

East, W. E., V. Paschalidis, F. Pretorius and S. L. Shapiro (2016). "Relativistic simulations of eccentric binary neutron star mergers: One-arm spiral instability and effects of neutron star spin," *Phys. Rev. D* **93**, 024011. [arXiv:1511.01093 [astro-ph.HE]].

Easther, R. and E. A. Lim (2006). "Stochastic gravitational wave production after inflation," *JCAP* **0604**, 010. [astro-ph/0601617].

Easther, R., J. T. Giblin, Jr. and E. A. Lim (2007). "Gravitational wave production at the end of inflation," *Phys. Rev. Lett.* **99**, 221301. [astro-ph/0612294].

Echeverria, F. (1989). "Gravitational wave measurements of the mass and angular momentum of a black hole," *Phys. Rev. D* **40**, 3194.

Eckart, A. (2003). "The mass of the Galactic center black hole," in *The Galactic Black Hole*, H. Falcke and F. W. Hehl eds., IoP Publishing, Bristol.

Eckart, A. and R. Genzel (1996). "Observations of stellar proper motions near the Galactic centre," *Nature* **383**, 415.

Einhorn, M. B. and F. Larsen (2003). "Interacting quantum field theory in de Sitter vacua," *Phys. Rev. D* **67**, 024001. [hep-th/0209159].

Einstein, A. (1916). *Ann. Phys.* **49**, 769.

Eisenstein, D. J. and W. Hu (1998). "Baryonic features in the matter transfer function," *Astrophys. J.* **496**, 605. [astro-ph/9709112].

Enoki, M., K. T. Inoue, M. Nagashima and N. Sugiyama (2004). "Gravitational waves from supermassive black hole coalescence in a hierarchical galaxy formation model," *Astrophys. J.* **615**, 19. [astro-ph/0404389].

Enoki, M. and M. Nagashima (2007). "The effect of orbital eccentricity on gravitational wave background radiation from cosmological binaries," *Prog. Theor. Phys.* **117**, 241. [astro-ph/0609377].

Enqvist, K., J. Ignatius, K. Kajantie and K. Rummukainen (1992). "Nucleation and bubble growth in a first order cosmological electroweak phase transition," *Phys. Rev. D* **45**, 3415.

Enqvist, K., K. Kainulainen and V. Semikoz (1992). "Neutrino annihilation in hot plasma," *Nucl. Phys. B* **374** 392.

Enqvist, K. and M. S. Sloth (2002). "Adiabatic CMB perturbations in pre-big-bang string cosmology," *Nucl. Phys. B* **626**, 395. [hep-ph/0109214].

Epstein, R. (1978). "The generation of gravitational radiation by escaping supernova neutrinos," *Astrophys. J.* **223**, 1037.

Espinosa, J. R., T. Konstandin, J. M. No and G. Servant (2010). "Energy budget of cosmological first-order phase transitions," *JCAP* **1006**, 028. [arXiv:1004.4187 [hep-ph]].

Espinoza, C. M., A. G. Lyne, M. Kramer, R. N. Manchester and V. M. Kaspi (2011). "The braking index of PSR J1734-3333 and the magnetar population," *Astrophys. J. Lett.* **741**, L13. [arXiv:1109.2740 [astro-ph.HE]].

Estabrook, F. B. and H. D. Wahlquist (1975). "Response of Doppler spacecraft tracking to gravitational radiation," *Gen. Rel. Grav.* **6**, 439.

Evans, P. A. *et al.* (2017). "Swift and NuSTAR observations of GW170817: Detection of a blue kilonova," *Science*, doi:10.1126/science.aap9580. [arXiv:1710.05437 [astro-ph.HE]].

Fabbri, R. and M. D. Pollock (1983). "The effect of primordially produced gravitons upon the anisotropy of the cosmological microwave background radiation," *Phys. Lett. B* **125**, 445.

Faber, S. M. *et al.* (1997). "The centers of early-type galaxies with HST. IV. Central parameter relation," *Astron. J.* **114**, 1771.

Falcke, H. and S. B. Markoff (2013). "Toward the event horizon — The supermassive black hole in the Galactic center," *Class. Quant. Grav.* **30**, 244003. [arXiv:1311.1841 [astro-ph.HE]].

Favata, M. (2014). "Systematic parameter errors in inspiraling neutron star binaries," *Phys. Rev. Lett.* **112**, 101101. [arXiv:1310.8288 [gr-qc]].

Favata, M., S. A. Hughes and D. E. Holz (2004). "How black holes get their kicks: Gravitational radiation recoil revisited," *Astrophys. J.* **607**, L5. [astro-ph/0402056].

Faye, G., L. Blanchet and A. Buonanno (2006). "Higher-order spin effects in the dynamics of compact binaries. I. Equations of motion," *Phys. Rev. D* **74**, 104033. [gr-qc/0605139].

Felder, G. N. *et al.* (2001). "Dynamics of symmetry breaking and tachyonic preheating," *Phys. Rev. Lett.* **87**, 011601. [hep-ph/0012142].

Felder, G. N., L. Kofman and A. D. Linde (2001). "Tachyonic instability and dynamics of spontaneous symmetry breaking," *Phys. Rev. D* **64**, 123517. [hep-th/0106179].

Felder, G. N. and L. Kofman (2007). "Nonlinear inflaton fragmentation after preheating," *Phys. Rev. D* **75**, 043518. [hep-ph/0606256].

Ferrand, G. and S. Safi-Harb (2012). *Adv. Space Res.* **49**, 1313. "A census of high-energy observations of Galactic supernova remnants," [arXiv:1202.0245 [astro-ph.HE]].

Ferrarese, L. and D. Merritt (2000). "A fundamental relation between supermassive black holes and their host galaxies," *Astrophys. J.* **539**, L9. [astro-ph/0006053].

Ferrari, V. and B. Mashhoon (1984). "Oscillations of a black hole," *Phys. Rev. Lett.* **52**, 1361.

Ferrari, V., G. Miniutti and J. A. Pons (2003). "Gravitational waves from newly born, hot neutron stars," *Mon. Not. Roy. Astron. Soc.* **342**, 629. [astro-ph/0210581].

Ferrario, L. and D. Wickramasinghe (2006). "Modelling of isolated radio pulsars and magnetars on the fossil field hypothesis," *Mon. Not. Roy. Astron. Soc.* **367**, 1323. [astro-ph/0601258].

Filippenko, A. V. (1997). "Optical spectra of supernovae," *Annu. Rev. Astron. Astrophys.* **35**, 309.

Finn, L. S. (1987). "*g*-modes in zero-temperature neutron stars", *Mon. Not. Roy. Astron. Soc.* **227**, 265.

Finn, L. S. and C. R. Evans (1990). "Determining gravitational radiation from Newtonian self-gravitating systems," *Astrophys. J.* **351**, 588.

Finn, L. S. and D. F. Chernoff (1993). "Observing binary inspiral in gravitational radiation: one interferometer," *Phys. Rev. D* **47**, 2198. [gr-qc/9301003].

Finn, L. S. and Thorne, K. S. (2000). "Gravitational waves from a compact star in a circular, inspiral orbit, in the equatorial plane of a massive, spinning black hole, as observed by LISA." *Phys. Rev. D* **62**, 124021. [gr-qc/0007074].

Finn, L. S. and A. N. Lommen (2010). "Detection, localization and characterization of gravitational wave bursts in a pulsar timing array," *Astrophys. J.* **718**, 1400. [arXiv:1004.3499 [astro-ph.IM]].

Fish, V. L. *et al.* (2011). "1.3 mm wavelength VLBI of Sagittarius A*: Detection of time-variable emission on event horizon scales," *Astrophys. J.* **727**, L36.

Fitchett, M. (1983). "The influence of gravitational wave momentum losses on the centre of mass motion of a Newtonian binary system," *Mon. Not. R. Astron. Soc.* **203**, 1049.

Fitchett, M. and S. Detweiler (1984). "Linear momentum and gravitational waves — Circular orbits around a Schwarzschild black hole," *Mon. Not. R. Astron. Soc.* **211**, 933.

Flanagan, E. E. and S. A. Hughes (2005). "The basics of gravitational wave theory," *New J. Phys.* **7**, 204. [gr-qc/0501041].

Flanagan, E. E. and T. Hinderer (2008). "Constraining neutron star tidal Love numbers with gravitational wave detectors," *Phys. Rev. D* **77**, 021502. [arXiv:0709.1915 [astro-ph]].

Flauger, R., J. C. Hill and D. N. Spergel (2014). "Toward an understanding of foreground emission in the BICEP2 region," *JCAP* **1408**, 039. [arXiv:1405.7351 [astro-ph.CO]].

Foffa, S., M. Maggiore and R. Sturani (1999). "Loop corrections and graceful exit in string cosmology," *Nucl. Phys. B* **552**, 395. [hep-th/9903008].

Font, J. A., H. Dimmelmeier, A. Gupta and N. Stergioulas (2001). "Axisymmetric modes of rotating relativistic stars in the Cowling approximation," *Mon. Not. Roy. Astron. Soc.* **325**, 1463. [astro-ph/0012477].

Foster, R.S. and D. C. Backer (1990). "Constructing a pulsar timing array," *Astrophys. J.* **361**, 300.

Franci, L., R. De Pietri, K. Dionysopoulou and L. Rezzolla (2013). "Dynamical bar-mode instability in rotating and magnetized relativistic stars," *Phys. Rev. D* **88**, 104028. [arXiv:1308.3989 [gr-qc]].

Freese, K., J. A. Frieman and A. V. Olinto (1990). "Natural inflation with pseudo-Nambu-Goldstone bosons," *Phys. Rev. Lett.* **65**, 3233.

Freire, P. C. *et al.* (2001a). "Timing the millisecond pulsars in 47 Tucanae," *Mon. Not. Roy. Astron. Soc.* **326**, 901. [astro-ph/0103372].

Freire, P. C. *et al.* (2001b). "Detection of ionized gas in the globular cluster 47 Tucanae," *Astrophys. J.* **557**, L105. [astro-ph/0107206].

Frieben, J. and L. Rezzolla (2012). "Equilibrium models of relativistic stars with a toroidal magnetic field," *Mon. Not. Roy. Astron. Soc.* **427**, 3406. [arXiv:1207.4035 [gr-qc]].

Friedman, J. L. and B. F. Schutz (1978a). "Lagrangian perturbation theory of nonrelativistic fluids," *Astrophys. J.* **221**, 937.

Friedman, J. L. and B. F. Schutz (1978b). "Secular instability of rotating Newtonian stars," *Astrophys. J.* **222**, 281.

Friedman, J. L. and J. R. Ipser (1992). "Rapidly rotating relativistic stars," *Phil. Trans. Roy. Soc. Lond. A,* **340**, 391.

Friedman J. L. and S. M. Morsink (1998). "Axial instability of rotating relativistic stars," *Astrophys. J.* **502**, 714. [gr-qc/9706073].

Friedman, J. L. and N. Stergioulas (2014). "Instabilities of relativistic stars," *Fundam. Theor. Phys.* **177**, 427.

Friedman, J. L., L. Lindblom and K. H. Lockitch (2016). "Differential rotation of the unstable nonlinear r-modes," *Phys. Rev. D* **93**, 024023. [arXiv:1503.08864 [astro-ph.HE]].

Frolov, V. P. and I. D. Novikov (1998). *Black Hole Physics: Basic Concepts and New Developments.* Kluwer Academic, Dordrecht.

Fröman, N., P. O. Fröman, N. Andersson and A. Hökback (1992). "Black hole normal modes: Phase integral treatment," *Phys. Rev. D* **45**, 2609.

Fromme, L. , S. J. Huber and M. Seniuch (2006). "Baryogenesis in the two-Higgs doublet model," *JHEP* **0611** 038. [hep-ph/0605242].

Fryer, C. L. (1999). "Mass limits for black hole formation," *Astrophys. J.* **522**, 413. [astro-ph/9902315].

Fryer, C. L., S. E. Woosley and D. H. Hartmann (1999). "Formation rates of black hole accretion disk gamma-ray bursts," *Astrophys. J.* **526**, 152. [astro-ph/9904122].

Fryer, C. L. and V. Kalogera, (2001). "Theoretical black hole mass distributions," *Astrophys. J.* **554**, 548. [astro-ph/9911312].

Fryer, C. L., D. E. Holz and S. A. Hughes (2002). "Gravitational wave emission from core collapse of massive stars," *Astrophys. J.* **565**, 430. [astro-ph/0106113].

Fryer, C. L. *et al.* (2012). "Compact remnant mass function: Dependence on the explosion mechanism and metallicity," *Astrophys. J.* **749**, 91. [arXiv:1110.1726 [astro-ph.SR]].

Fuhrmann, K. (2005). "A thick-disc origin for Tycho Brahe's 1572 supernova?," *Mon. Not. Roy. Astron. Soc.* **359**, L35.

Fujita, R. (2012). "Gravitational waves from a particle in circular orbits around a schwarzschild black hole to the 22nd post-Newtonian order," *Prog. Theor. Phys.* **128**, 971. [arXiv:1211.5535 [gr-qc]].

Fujita, R. (2015). "Gravitational waves from a particle in circular orbits around a rotating black hole to the 11th post-Newtonian order," *Prog. Theor. Exp. Phys.* **2015**, 033E01. [arXiv:1412.5689 [gr-qc]].

Fukugita, M. *et al.* (1996). "The Sloan Digital Sky Survey photometric system," *Astron. J.* **111**, 1748.

Gabler, M., P. Cerdá-Durán, N. Stergioulas, J. A. Font and E. Müller (2012). "Magneto-elastic oscillations of neutron stars with dipolar magnetic fields," *Mon. Not. Roy. Astron. Soc.* **421**, 2054. [arXiv:1109.6233 [astro-ph.HE]].

Gabler, M., P. Cerdá-Durán, N. Stergioulas, J. A. Font and E. Müller (2013). "Imprints of superfluidity on magnetoelastic quasiperiodic oscillations of soft gamma-ray repeaters," *Phys. Rev. Lett.* **111**, 211102. [arXiv:1304.3566 [astro-ph.HE]].

Gaertig, E. and K. D. Kokkotas (2009). "Relativistic g-modes in rapidly rotating neutron stars," *Phys. Rev. D* **80**, 064026. [arXiv:0905.0821 [astro-ph.SR]].

Gaertig E. and K. D. Kokkotas (2011). "Gravitational wave asteroseismology with fast rotating neutron stars," *Phys. Rev. D* **83**, 064031. [arXiv:1005.5228 [astro-ph.SR]].

Gair, J. R. (2009). "Probing black holes at low redshift using LISA EMRI observations," *Class. Quant. Grav.* **26**, 094034. [arXiv:0811.0188 [gr-qc]].

Gair, J. R. *et al.* (2004). "Event rate estimates for LISA extreme mass ratio capture sources," *Class. Quant. Grav.* **21**, S1595. [gr-qc/0405137].

Gal-Yam, A. (2012). "Pair-instability explosions: Observational evidence," *IAU Symp.* **279**, 253. [arXiv:1206.2157 [astro-ph.CO]].

García-Bellido, J. and D. G. Figueroa (2007). "A stochastic background of gravitational waves from hybrid preheating," *Phys. Rev. Lett.* **98**, 061302. [astro-ph/0701014].

García-Bellido, J., D. G. Figueroa and A. Sastre (2008). "A gravitational wave background from reheating after hybrid inflation," *Phys. Rev. D* **77**, 043517. [arXiv:0707.0839 [hep-ph]].

Gasperini, M. (2007). *Elements of String Cosmology*, Cambridge University Press, Cambridge.

Gasperini, M. and G. Veneziano (1993a). "Pre-big bang in string cosmology," *Astropart. Phys.* **1**, 317. [hep-th/9211021].

Gasperini, M. and G. Veneziano (1993b). "Inflation, deflation, and frame independence in string cosmology," *Mod. Phys. Lett.* **A8**, 3701. [hep-th/9309023].

Gasperini, M. and G. Veneziano (1994). "Dilaton production in string cosmology," *Phys. Rev. D* **50**, 2519. [gr-qc/9403031].

Gasperini, M., M. Maggiore and G. Veneziano (1997). "Towards a non-singular pre-big bang cosmology," *Nucl. Phys. B* **494**, 315. [hep-th/9611039].

Gasperini, M. and G. Veneziano (2003). "The pre-big bang scenario in string cosmology," *Phys. Rept.* **373**, 1. [hep-th/0207130].

Gebhardt, K. *et al.* (2000). "A relationship between nuclear black hole mass and galaxy velocity dispersion," *Astrophys. J.* **539**, L13. [astro-ph/0006289].

Ghez, A. M., B. L. Klein, M. R. Morris and E. E. Becklin (1998). "High proper-motion stars in the vicinity of Sagittarius A*: Evidence for a supermassive black hole at the center of our Galaxy," *Astrophys. J.* **509**, 678. [astro-ph/9807210].

Ghez, A. M., M. R. Morris, E. E. Becklin, A. Tanner and T. Kremenek (2000). "The accelerations of stars orbiting the Milky Way's central black hole," *Nature* **407**, 349. [astro-ph/0009339].

Ghez, A. M., *et al.* (2008). "Measuring distance and properties of the Milky Way's central supermassive black hole with stellar orbits," *Astrophys. J.* **689**, 1044. [arXiv:0808.2870 [astro-ph]].

Giacomazzo, B., L. Rezzolla and L. Baiotti (2011). "Accurate evolutions of inspiralling and magnetized neutron-stars: Equal-mass binaries," *Phys. Rev. D* **83**, 044014. [arXiv:1009.2468 [gr-qc]].

Gibbons, G. W. and S. W. Hawking (1977). "Action integrals and partition functions in quantum gravity," *Phys. Rev. D* **15**, 2752.

Giblin, J. T. and J. B. Mertens (2014). "Gravitational radiation from first-order phase transitions in the presence of a fluid," *Phys. Rev. D* **90**, 023532. [arXiv:1405.4005 [astro-ph.CO]].

Gillessen, S. *et al.* (2009a). "Monitoring stellar orbits around the massive black hole in the galactic center," *Astrophys. J.* **692**, 1075. [arXiv:0810.4674 [astro-ph]].

Gillessen, S. *et al.* (2009b). "The orbit of the star S2 around Sgr A* from VLT and Keck data", *Astrophys. J.* **707**, L114. [arXiv:0910.3069 [astro-ph.GA]]

Glampedakis, K., L. Samuelsson and N. Andersson (2006). "Elastic or magnetic? A toy model for global magnetar oscillations with implications for QPOs during flares," *Mon. Not. Roy. Astron. Soc.* **371**, L74. [astro-ph/0605461].

Gleiser, R. J., C. O. Nicasio, R. H. Price and J. Pullin (1996). "Colliding black holes: How far can the close approximation go?," *Phys. Rev. Lett.* **77**, 4483. [gr-qc/9609022].

Gleyzes, J., D. Langlois and F. Vernizzi (2015). "A unifying description of dark energy," *Int. J. Mod. Phys. D* **23**, 1443010. [arXiv:1411.3712 [hep-th]].

Gögüs, E. *et al.* (2000). "Statistical properties of SGR 1806-20 bursts," *Astrophys. J.*, **532** L121. [astro-ph/0002181].

Goldstein, A. *et al.* (2017) . "An ordinary short gamma-ray burst with extraordinary implications: Fermi-GBM detection of GRB 170817A," *Astrophys. J.* **848**, L14. [arXiv:1710.05446 [astro-ph.HE]].

Goldstein, H. (1980). *Classical Mechanics*, Addison-Wesley, San Francisco.

González, J. A., U. Sperhake, B. Brügmann, M. Hannam and S. Husa (2007). "Total recoil: The maximum kick from nonspinning black-hole binary inspiral," *Phys. Rev. Lett.* **98**, 091101. [gr-qc/0610154].

González, J. A., M. Hannam, U. Sperhake, B. Brügmann and S. Husa (2007). "Supermassive recoil velocities for binary black-hole mergers with antialigned spins," *Phys. Rev. Lett.* **98**, 231101. [gr-qc/0702052].

González Hernández, J. I. *et al.* (2009). "The chemical abundances of Tycho G in supernova remnant 1572," *Astrophys. J.* **691**, 1. [arXiv:0809.0601 [astro-ph]].

Gorbunov, D. S. and V. A. Rubakov (2011). *Introduction to the Theory of the Early Universe. Cosmological Perturbations and Inflationary Theory*, World Scientific, Singapore.

Gossan, S. E. *et al.* (2016). "Observing gravitational waves from core-collapse supernovae in the advanced detector era," *Phys. Rev. D* **93**, 042002. [arXiv:1511.02836 [astro-ph.HE]].

Gotthelf, E. V., D. J. Helfand and L. Newburgh (2006). "A shell of thermal X-ray emission associated with the young Crab-like remnant 3C58," *Astrophys. J.* **654**, 267. [astro-ph/0609309].

Gottlieb, O., E. Nakar, T. Piran and K. Hotokezaka (2017). "A cocoon shock breakout as the origin of the γ-ray emission in GW170817," [arXiv:1710.05896 [astro-ph.HE]].

Gourgoulhon, E. (2007). "3+1 formalism and bases of numerical relativity," [gr-qc/0703035].

Gourgoulhon, E. (2012). *3+1 Formalism in General Relativity: Bases of Numerical Relativity*. Lecture Notes in Physics, Vol. 846, Springer-Verlag, Berlin.

Gradshteyn, I. S. and Ryzhik, I. M. (1980). *Table of Integrals, Series and Products*. Academic Press, London.

Graham, A. W., C. A. Onken, E. Athanassoula and F. Combes (2011). "An expanded $M_{bh} - \sigma$ diagram, and a new calibration of active galactic nuclei masses," *Mon. Not. Roy. Astron. Soc.* **412**, 2211. [arXiv:1007.3834 [astro-ph.CO]].

Granett, B. R., M. C. Neyrinck and I. Szapudi (2008). "An imprint of super-structures on the microwave background due to the integrated Sachs–Wolfe effect," *Astrophys. J.* **683**, L99. [arXiv:0805.3695 [astro-ph]].

Green, D. A. and F. R. Stephenson (2003). "The historical supernovae," in *Supernovae and Gamma-Ray Bursters*, K. W. Weiler ed., Lecture Notes in Physics, Vol. 598, Springer-Verlag, Berlin, Heidelberg. [astro-ph/0301603].

Green, D. A. (2014). "A catalogue of 294 Galactic supernova remnants," *Bull. Astron. Soc. India* **42**, 47 [arXiv:1409.0637 [astro-ph.HE]].

Greiner, J., J. M. Burgess, V. Savchenko and H.-F. Yu (2016). "On the Fermi-GBM event seen 0.4 sec after GW 150914," *Astrophys. J.* **827**, L38. [arXiv:1606.00314 [astro-ph.HE]].

Griffiths, D. J. (1995). *Introduction to Quantum Mechanics*, Prentice-Hall, Englewood Cliffs, NJ.

Grishchuk, L. P. (1974). "Amplification of gravitational waves in an isotropic universe," *Sov. Phys. JETP* **40** (1975) 409 [*Zh. Eksp. Teor. Fiz.* **67** (1974) 825].

Grishchuk, L. P. and Y. V. Sidorov (1990). "Squeezed quantum states of relic gravitons and primordial density fluctuations," *Phys. Rev. D* **42**, 3413.

Gualandris, A. and D. Merritt (2008). "Ejection of supermassive black holes from galaxy cores," *Astrophys. J.* **678**, 780. [arXiv:0708.0771 [astro-ph]].

Guidorzi, C. *et al.* (2017). "Improved constraints on H_0 from a combined analysis of gravitational-wave and electromagnetic emission from GW170817," *Astrophys. J.* **851**, L36. [arXiv:1710.06426 [astro-ph.CO]].

Gultekin, K. *et al.* (2009). "The $M - \sigma$ and $M - L$ relations in galactic bulges and determinations of their intrinsic scatter," *Astrophys. J.* **698**, 198. [arXiv:0903.4897 [astro-ph.GA]].

Gundlach, C., R. H. Price and J. Pullin (1994a). "Late time behavior of stellar collapse and explosions: 1. Linearized perturbations," *Phys. Rev. D* **49**, 883. [gr-qc/9307009].

Gundlach, C., R. H. Price and J. Pullin (1994b). "Late time behavior of stellar collapse and explosions: 2. Nonlinear evolution," *Phys. Rev. D* **49**, 890. [gr-qc/9307010].

Gundlach, C., J. M. Martín-García, G. Calabrese and I. Hinder (2005). "Constraint damping in the Z4 formulation and harmonic gauge," *Class. Quant. Grav.* **22**, 3767. [gr-qc/0504114].

Guth, A. H. (1981). "The inflationary Universe: A possible solution to the horizon and flatness problems," *Phys. Rev. D* **23**, 347.

Guth, A. H. and S. Y. Pi (1982). "Fluctuations in the new inflationary Universe," *Phys. Rev. Lett.* **49**, 1110.

Guth, A. H. and S. Y. Pi (1985). "The quantum mechanics of the scalar field in the new inflationary Universe," *Phys. Rev. D* **32**, 1899.

Guy, J., P. Astier, S. Nobili, N. Regnault and R. Pain (2005). "SALT: A spectral adaptive light curve template for type Ia supernovae," *Astron. Astrophys.* **443**, 781. [astro-ph/0506583].

Guzzetti, M., C., N. Bartolo, Liguori, M. and S. Matarrese (2016). "Gravitational waves from inflation," *Riv. Nuovo Cim.* **39**, 399. [arXiv:1605.01615 [astro-ph.CO]].

Haehnelt, M. G. and Kauffmann, G. (2002). "Multiple black holes in galactic bulges," *Mon. Not. Roy. Astron. Soc.* **336**, L61. [astro-ph/0208215].

Haggard, D., *et al.* (2017). "A deep *Chandra* X-ray study of neutron star coalescence GW170817," *Astrophys. J.* **848**, L25. [arXiv:1710.05852 [astro-ph.HE]].

Hallinan, G. *et al.* (2017). "A radio counterpart to a neutron star merger," *Science*, doi:10.1126/science.aap9855. arXiv:1710.05435 [astro-ph.HE].

Hanany, S. *et al.* (2000). "MAXIMA-1: A Measurement of the cosmic microwave background anisotropy on angular scales of 10 arcminutes to 5 degrees," *Astrophys. J.* **545**, L5. [astro-ph/0005123].

Hannam, M., S. Husa, U. Sperhake, B. Brügmann and J. A. Gonzalez (2008). "Where post-Newtonian and numerical-relativity waveforms meet," *Phys. Rev. D* **77**, 044020. [arXiv:0706.1305 [gr-qc]].

Hannam, M. *et al.* (2014). "Simple model of complete precessing black-hole-binary gravitational waveforms," *Phys. Rev. Lett.* **113**, 151101. [arXiv:1308.3271 [gr-qc]].

Hanson, D. *et al.* [SPTpol Collaboration] (2013). "Detection of B-mode polarization in the cosmic microwave background with data from the South Pole Telescope," *Phys. Rev. Lett.* **111**, 141301. [arXiv:1307.5830 [astro-ph.CO]].

Harry, G. M., P. Fritschel, D. A. Shaddock, W. Folkner and E. S. Phinney (2006). "Laser interferometry for the Big Bang Observer," *Class. Quant. Grav.* **23**, 4887. Erratum: *Class. Quant. Grav.* **23** (2006) 7361.

Hartle, J. B. (2003). *Gravity. An Introduction to Einstein's General Relativity*, Addison-Wesley, San Francisco.

Hartle, J. B., K. S. Thorne and S. M. Chitre (1972). "Slowly rotating relativistic stars. VI. Stability of the quasiradial modes," *Astrophys. J.* **176**, 177.

Hartle, J. B. and J. L. Friedman (1975). "Slowly rotating relativistic stars. VIII. Frequencies of the quasi-radial modes of an $n = 3/2$ polytrope," *Astrophys. J.* **196**, 653.

Haskell, B., D. I. Jones and N. Andersson (2006). "Mountains on neutron stars: Accreted vs. non-accreted crusts," *Mon. Not. Roy. Astron. Soc.* **373**, 1423. [astro-ph/0609438].

Haskell, B. L., Samuelsson, K. Glampedakis and N. Andersson (2008). "Modelling magnetically deformed neutron stars," *Mon. Not. Roy. Astron. Soc.* **385** 531. Erratum: *Mon. Not. Roy. Astron. Soc.* **394** (2009) 1711. [arXiv:0705.1780 [astro-ph]].

Hassan, S. F. and R. A. Rosen (2012a). "Resolving the ghost problem in non-linear massive gravity," *Phys. Rev. Lett.* **108**, 041101. [arXiv:1106.3344 [hep-th]].

Hassan, S. F. and R. A. Rosen (2012b). "Bimetric gravity from ghost-free massive gravity," *JHEP* **1202**, 126. [arXiv:1109.3515 [hep-th]].

Hawking, S. W. (1982). "The development of irregularities in a single bubble inflationary Universe," *Phys. Lett. B* **115**, 295.

Hawking, S. W. and G. F. R. Ellis (1973). *The Large Scale Structure of Space-Time*, Cambridge University Press, Cambridge.

Hawking, S. W. and G. T. Horowitz (1996). "The gravitational Hamiltonian, action, entropy and surface terms," *Class. Quant. Grav.* **13**, 1487

Heger, A., C. L. Fryer, S. E. Woosley, N. Langer and D. H. Hartmann (2003). "How massive single stars end their life," *Astrophys. J.* **591**, 288. [astro-ph/0212469].

Heger, A., S. E. Woosley and H. C. Spruit (2005). "Pre-supernova evolution of differentially rotating massive stars including magnetic fields," *Astrophys. J.* **626**, 350. [astro-ph/0409422].

Hellings, R. W. and G. S. Downs (1983). "Upper limits on the isotropic gravitational radiation background from pulsar timing analysis," *Astrophys. J.* **265**, L39.

Herrmann, F., I. Hinder, D. Shoemaker, P. Laguna and R. A. Matzner (2007). "Gravitational recoil from spinning binary black hole mergers," *Astrophys. J.* **661**, 430. [gr-qc/0701143].

Hessels, J.W.T. *et al.* (2006). "A radio pulsar spinning at 716 Hz," *Science* **311**, 1901.

Hilbert, S., J. R. Gair and L. J. King (2011). "Reducing distance errors for standard candles and standard sirens with weak-lensing shear and flexion maps," *Mon. Not. Roy. Astron. Soc.* **412**, 1023. [arXiv:1007.2468 [astro-ph.CO]].

Hillebrandt, W. and J. C. Niemeyer (2000). "Type Ia supernova explosion models," *Annu. Rev. Astron. Astrophys.* **38**, 191. [astro-ph/0006305].

Hinderer, T. (2008). "Tidal Love numbers of neutron stars," *Astrophys. J.* **677**, 1216. [arXiv:0711.2420 [astro-ph]].

Hinderer, T. *et al.* (2016). "Effects of neutron-star dynamic tides on gravitational waveforms within the effective-one-body approach," *Phys. Rev. Lett.* **116**, 181101. [arXiv:1602.00599 [gr-qc]].

Hinderer, T., B. D. Lackey, R. N. Lang and J. S. Read (2010). "Tidal deformability of neutron stars with realistic equations of state and their gravitational wave signatures in binary inspiral," *Phys. Rev. D* **81**, 123016. [arXiv:0911.3535 [astro-ph.HE]].

Hindmarsh, M. B. and T. W. B. Kibble (1995). "Cosmic strings," *Rept. Prog. Phys.* **58**, 477. [hep-ph/9411342].

Hindmarsh, M., S. J. Huber, K. Rummukainen and D. J. Weir (2014). "Gravitational waves from the sound of a first order phase transition," *Phys. Rev. Lett.* **112**, 041301. [arXiv:1304.2433 [hep-ph]].

Hindmarsh, M., S. J. Huber, K. Rummukainen and D. J. Weir (2015). "Numerical simulations of acoustically generated gravitational waves at a first order phase transition," Phys. Rev. D **92**, 123009. [arXiv:1504.03291 [astro-ph.CO]].

Hinshaw, G. *et al.* [WMAP Collaboration] (2013). "Nine-year Wilkinson Microwave Anisotropy Probe (WMAP) observations: Cosmological parameter results," *Astrophys. J. Suppl.* **208**, 19. [arXiv:1212.5226 [astro-ph.CO]].

Hjorth, J. *et al.* (2017). "The distance to NGC 4993: The host galaxy of the gravitational-wave event GW170817," *Astrophys. J.* **848**, L31. [arXiv:1710.05856 [astro-ph.GA]].

Hobbs, G., D. R. Lorimer, A. G. Lyne and M. Kramer (2005). "A statistical study of 233 pulsar proper motions," *Mon. Not. Roy. Astron. Soc.* **360**, 974. [astro-ph/0504584].

Hobbs, G. *et al.* (2009). "TEMPO2, a new pulsar timing package. III: Gravitational wave simulation," *Mon. Not. Roy. Astron. Soc.* **394**, 1945. [arXiv:0901.0592 [astro-ph.SR]].

Hobbs, G. *et al.* (2010). "The International Pulsar Timing Array project: Using pulsars as a gravitational wave detector," *Class. Quant. Grav.* **27**, 084013. [arXiv:0911.5206 [astro-ph.SR]].

Hod, S. (1998). "Bohr's correspondence principle and the area spectrum of quantum black holes," *Phys. Rev. Lett.* **81**, 4293. [gr-qc/9812002].

Hofmann, F., E. Barausse and L. Rezzolla (2016). "The final spin from binary black holes in quasi-circular orbits," *Astrophys. J. Lett.* **825**, L19. [arXiv:1605.01938 [gr-qc]].

Hogan, C. J. (1983). "Nucleation of cosmological phase transitions," *Phys. Lett. B* **133**, 172.

Hogan, C. J. (1986). "Gravitational radiation from cosmological phase transitions," *Mon. Not. Roy. Astron. Soc.* **218**, 629.

Hogan C. J. and M. J. Rees (1984). "Gravitational interactions of cosmic strings," *Nature* **311**, 109.

Holz, D. E. and S. A. Hughes (2005). "Using gravitational-wave standard sirens," *Astrophys. J.* **629**, 15. [astro-ph/0504616].

Hopman, C. and T. Alexander (2006). "The effect of mass-segregation on gravitational wave sources near massive black holes," *Astrophys. J.* **645**, L133. [astro-ph/0603324].

Hopman, C., M. Freitag and S. L. Larson (2007). "Gravitational wave bursts from the Galactic massive black hole," *Mon. Not. Roy. Astron. Soc.* **378**, 129. [astro-ph/0612337].

Horiuchi, S., J. F. Beacom and E. Dwek (2009). "The diffuse supernova neutrino background is detectable in Super-Kamiokande," *Phys. Rev. D* **79** 083013. [arXiv:0812.3157 [astro-ph]].

Horowitz, C. J. and K. Kadau (2009). "The breaking strain of neutron star crust and gravitational waves," *Phys. Rev. Lett.* **102**, 191102. [arXiv:0904.1986 [astro-ph.SR]].

Hotokezaka, K. and T. Piran (2015). "Mass ejection from neutron star mergers: Different components and expected radio signals," *Mon. Not. Roy. Astron. Soc.* **450**, 1430. [arXiv:1501.01986 [astro-ph.HE]].

Hotokezaka, K., K. Kyutoku, Y.-i. Sekiguchi, and M. Shibata (2016). "Measurability of the tidal deformability by gravitational waves from coalescing binary neutron stars," *Phys. Rev. D* **93**, 064082. [arXiv:1603.01286 [gr-qc]].

Hotokezaka, K. *et al.* (2016). "Radio counterparts of compact binary mergers detectable in gravitational waves: A simulation for an optimized survey," *Astrophys. J.* **831**, 190. [arXiv:1605.09395 [astro-ph.HE]]

Hu, W. and N. Sugiyama (1995). "Anisotropies in the cosmic microwave background: An analytic approach," *Astrophys. J.* **444**, 489. [astro-ph/9407093].

Hu, W., D. Scott, N. Sugiyama and M. J. White (1995). "The effect of physical assumptions on the calculation of microwave background anisotropies," *Phys. Rev. D* **52**, 5498. [astro-ph/9505043].

Hu, W. and N. Sugiyama (1996). "Small scale cosmological perturbations: An analytic approach," *Astrophys. J.* **471**, 542. [astro-ph/9510117].

Huber, S. J. and T. Konstandin (2008a). "Production of gravitational waves in the nMSSM," *JCAP* **0805**, 017. [arXiv:0709.2091 [hep-ph]].

Huber, S. J. and T. Konstandin (2008b). "Gravitational wave production by collisions: More bubbles," *JCAP* **0809**, 022. [arXiv:0806.1828 [hep-ph]].

Huerta, E. A., S. T. McWilliams, J. R. Gair and S. R. Taylor (2015). "Detection of eccentric supermassive black hole binaries with pulsar timing arrays: Signal-to-noise ratio calculations," *Phys. Rev. D* **92**, 063010. [arXiv:1504.00928 [gr-qc]].

Hughes, S. A., S. Drasco, E. E. Flanagan and J. Franklin (2005). "Gravitational radiation reaction and inspiral waveforms in the adiabatic limit," *Phys. Rev. Lett.* **94**, 221101. [gr-qc/0504015].

Hurley, K. (2010). "Soft gamma repeaters," *Mem. Soc. Astron. Ital.* **81**, 432.

Husa, S., J. A. Gonzalez, M. Hannam, B. Brügmann and U. Sperhake (2008). "Reducing phase error in long numerical binary black hole evolutions with sixth order finite differencing," *Class. Quant. Grav.* **25**, 105006. [arXiv:0706.0740 [gr-qc]].

Husa, S. *et al.* (2016). "Frequency-domain gravitational waves from nonprecessing black-hole binaries. I. New numerical waveforms and anatomy of the signal," *Phys. Rev. D* **93**, 044006. [arXiv:1508.07250 [gr-qc]].

Ignatius, J., K. Kajantie, H. Kurki-Suonio and M. Laine (1994). "The growth of bubbles in cosmological phase transitions," *Phys. Rev. D* **49**, 3854. [astro-ph/9309059].

Iocco, F., G. Mangano, G. Miele, O. Pisanti and P. D. Serpico (2009). "Primordial nucleosynthesis: From precision cosmology to fundamental physics," *Phys. Rept.* **472**, 1. [arXiv:0809.0631 [astro-ph]].

Ioka, K. (2001). "Magnetic deformation of magnetars for the giant flares of the soft gamma-ray repeaters," *Mon. Not. Roy. Astron. Soc.* **327**, 639. [astro-ph/0009327].

Ipser, J. R. and L. Lindblom (1991). "The oscillations of rapidly rotating Newtonian stellar models. II — Dissipative effects," *Astrophys. J.* **373**, 213.

Iyer, S. (1987). "Black hole normal modes: A WKB approach. 2. Schwarzschild black holes," *Phys. Rev. D* **35**, 3632.

Iyer, S. and C. M. Will (1987). "Black hole normal modes: A WKB approach. 1. Foundations and application of a higher order WKB analysis of potential barrier scattering," *Phys. Rev. D* **35**, 3621.

Jaccard, M., M. Maggiore, M. and E. Mitsou (2013). "Bardeen variables and hidden gauge symmetries in linearized massive gravity," *Phys. Rev. D* **87**, 044017. [arXiv:1211.1562 [hep-th]].

Jackson, M. G., N. T. Jones and J. Polchinski (2005). "Collisions of cosmic F- and D-strings," *JHEP* **0510**, 013. [hep-th/0405229].

Jacobson, T. and T. P. Sotiriou (2009). "Over-spinning a black hole with a test body," *Phys. Rev. Lett.* **103**, 141101. Erratum: *Phys. Rev. Lett.* **103** (2009) 209903. [arXiv:0907.4146 [gr-qc]].

Jaffe, A. H. and D. C. Backer (2003). "Gravitational waves probe the coalescence rate of massive black hole binaries," *Astrophys. J.* **583**, 616. [astro-ph/0210148].

Janka, H.-T. (2012). "Explosion mechanisms of core-collapse supernovae," *Annu. Rev. Nucl. Part. Sci.* **62**, 407. [arXiv:1206.2503 [astro-ph.SR]].

Janka, H. T., K. Langanke, A. Marek, G. Martínez-Pinedo and B. Müller (2007). "Theory of core-collapse supernovae," *Phys. Rept.* **442**, 38. [astro-ph/0612072].

Jenet, F. A., A. Lommen, S. L. Larson and L. Wen (2004). "Constraining the properties of the proposed supermassive black hole system in 3c66b: Limits from pulsar timing," *Astrophys. J.* **606**, 799. [astro-ph/0310276].

Jenet, F. A., G. B. Hobbs, K. J. Lee and R. N. Manchester (2005). "Detecting the stochastic gravitational wave background using pulsar timing," *Astrophys. J.* **625**, L123. [astro-ph/0504458].

Jenet, F. A. *et al.* (2006). "Upper bounds on the low-frequency stochastic gravitational wave background from pulsar timing observations: Current limits and future prospects," *Astrophys. J.* **653**, 1571. [astro-ph/0609013].

Jiang, L. *et al.* (2016). "The final SDSS high-redshift quasar sample of 52 quasars at $z > 5.7$," *Astrophys. J.* **833**, 222. [arXiv:1610.05369 [astro-ph.GA]].

Jinno, R. and M. Takimoto (2017). "Gravitational waves from bubble collisions: An analytic derivation," *Phys. Rev. D* **95**, 024009. [arXiv:1605.01403 [astro-ph.CO]].

Jones, D. I. (2002). "Gravitational waves from rotating neutron stars," *Class. Quant. Grav.* **19**, 1255. [gr-qc/0111007].

Jones, N. T., H. Stoica and S. H. H. Tye (2002). "Brane interaction as the origin of inflation," *JHEP* **0207**, 051. [hep-th/0203163].

Jones, N. T., H. Stoica and S. H. H. Tye (2003). "The production, spectrum and evolution of cosmic strings in brane inflation," *Phys. Lett. B* **563**, 6. [hep-th/0303269].

Kajantie, K., M. Laine, K. Rummukainen and M. E. Shaposhnikov (1996a). "Is there a hot electroweak phase transition at m_H larger or equal to m_W?," *Phys. Rev. Lett.* **77**, 2887. [hep-ph/9605288].

Kajantie, K, M. Laine, K. Rummukainen and M. E. Shaposhnikov (1996b). "The electroweak phase transition: A nonperturbative analysis," *Nucl. Phys. B* **466**, 189. [hep-lat/9510020].

Kamionkowski, M., A. Kosowsky and M. Turner (1994). "Gravitational radiation from first order phase transitions," *Phys. Rev. D* **49**, 2837. [astro-ph/9310044].

Kamionkowski, M., A. Kosowsky and A. Stebbins (1997). "Statistics of cosmic microwave background polarization," *Phys. Rev. D* **55**, 7368. [astro-ph/9611125].

Kaplan, D. L. and M. H. van Kerkwiik (2009). "Constraining the spin-down of the nearby isolated neutron star RX J2143.0+0654," *Astrophys. J.* **705**, 798. [arXiv:0901.4133 [astro-ph.HE]].

Kasen, D., N. R. Badnell and J. Barnes (2013). "Opacities and spectra of the r-process ejecta from neutron star mergers," *Astrophys. J.* **774**, 25. [arXiv:1303.5788 [astro-ph.HE]].

Kasen, D., B. Metzger, J. Barnes, E. Quataert and E. Ramirez-Ruiz (2017). "Origin of the heavy elements in binary neutron-star mergers from a gravitational wave event," *Nature* **551**, 80. [arXiv:1710.05463 [astro-ph.HE]].

Kasliwal, M. M. *et al.* (2017). "Illuminating gravitational waves: A concordant picture of photons from a neutron star merger," *Science*, doi:10.1126/science.aap9455. [arXiv:1710.05436 [astro-ph.HE]].

Kaspi, V. M., J. H. Taylor and M. F. Ryba (1994). "High-precision timing of millisecond pulsars. 3: Long-term monitoring of PSRs B1855+09 and B1937+21," *Astrophys. J.* **428**, 713.

Katz, A. and A. Riotto (2016). "Baryogenesis and gravitational waves from runaway bubble collisions," *JCAP* **1611**, 011. [arXiv:1608.00583 [hep-ph]].

Kawamura, S. *et al.* (2011). "The Japanese space gravitational wave antenna: DECIGO," *Class. Quant. Grav.* **28**, 094011.

Khan, F., A. Just and D. Merritt (2011). "Efficient merger of binary supermassive black holes in merging galaxies," *Astrophys. J.* **732**, 89. [arXiv:1103.0272 [astro-ph.CO]].

Khan, S. *et al.* (2016). "Frequency-domain gravitational waves from nonprecessing black-hole binaries. II. A phenomenological model for the advanced detector era," *Phys. Rev. D* **93**, 044007. [arXiv:1508.07253 [gr-qc]].

Khlebnikov, S. Y. and I. I. Tkachev (1997a). "The Universe after inflation: The wide resonance case," *Phys. Lett. B* **390**, 80. [hep-ph/9608458].

Khlebnikov, S. Y. and I. I. Tkachev (1997b). "Relic gravitational waves produced after preheating," *Phys. Rev. D* **56**, 653. [hep-ph/9701423].

Khoury, J., B. A. Ovrut, P. J. Steinhardt and N. Turok (2001). "The ekpyrotic Universe: Colliding branes and the origin of the hot big bang," *Phys. Rev. D* **64**, 123522. [hep-th/0103239].

Khoury, J., B. A. Ovrut, N. Seiberg, P. J. Steinhardt and N. Turok (2002). "From big crunch to big bang," *Phys. Rev. D* **65**, 086007. [hep-th/0108187].

Kibble, T. W. B. (1976). "Topology of cosmic domains and strings," *J. Phys. A* **9**, 1387.

Kidder, L. E. (1995). "Coalescing binary systems of compact objects to (post)$^{5/2}$-Newtonian order. V. Spin effects," *Phys. Rev. D* **52**, 821. [gr-qc/9506022].

Kidder, L. E., C. M. Will and A. G. Wiseman (1993). "Spin effects in the inspiral of coalescing compact binaries," *Phys. Rev. D* **47**, 4183. [gr-qc/9211025].

Kilic, M., C. A. Prieto, W. R. Brown and D. Koester (2007). "The lowest mass white dwarf," *Astrophys. J.* **660**, 1451. [astro-ph/0611498].

Kirkpatrick, J.D. *et al.* (1999). "Dwarfs cooler than M: the definition of spectral type L using discoveries from the 2 Micron All-Sky Survey (2MASS)," *Astrophys. J.* **519** 802.

Kistler, M. D., H. Yuksel, S. Ando, J. F. Beacom and Y. Suzuki (2011). "Core-collapse astrophysics with a five-megaton neutrino detector," *Phys. Rev. D* **83**, 123008. [arXiv:0810.1959 [astro-ph]].

Kistler, M. D., W. C. Haxton and H. Yüksel (2013). "Tomography of massive stars from core collapse to supernova shock breakout," *Astrophys. J.* **778**, 81. [arXiv:1211.6770 [astro-ph.CO]].

Kitzbichler, M. G. and S. D. M. White (2008). "A calibration of the relation between the abundance of close galaxy pairs and the rate of galaxy mergers," *Mon. Not. Roy. Astron. Soc.* **391**, 1489. [arXiv:0804.1965 [astro-ph]].

Klimenko, S., I. Yakushin, A. Mercer and G. Mitselmakher (2008). "Coherent method for detection of gravitational wave bursts," *Class. Quant. Grav.* **25**, 114029. [arXiv:0802.3232 [gr-qc]].

Klimenko, S. *et al.* (2016). "Method for detection and reconstruction of gravitational wave transients with networks of advanced detectors," *Phys. Rev. D* **93**, 042004. [arXiv:1511.05999 [gr-qc]].

Kocsis, B., Z. Frei, Z. Haiman and K. Menou (2006). "Finding the electromagnetic counterparts of cosmological standard sirens," *Astrophys. J.* **637**, 27. [astro-ph/0505394].

Kocsis, B. and A. Sesana (2011). "Gas driven massive black hole binaries: Signatures in the nHz gravitational wave background," *Mon. Not. Roy. Astron. Soc.* **411**, 1467. [arXiv:1002.0584 [astro-ph.CO]].

Kodama, H. and M. Sasaki (1984). "Cosmological perturbation theory," *Prog. Theor. Phys. Supp.* **78**, 1.

Kodama, H. and M. Sasaki (1987). "Evolution of isocurvature perturbations. 2. Radiation dust Universe," *Int. J. Mod. Phys. A* **2**, 491.

Kodama, H. and A. Ishibashi (2003). "A master equation for gravitational perturbations of maximally symmetric black holes in higher dimensions," *Prog. Theor. Phys.* **110**, 701. [hep-th/0305147].

Könnig, F., Y. Akrami, L. Amendola, M. Motta and A. R. Solomon (2014). "Stable and unstable cosmological models in bimetric massive gravity," *Phys. Rev. D* **90**, 124014. [arXiv:1407.4331 [astro-ph.CO]].

Kofman, L., A. D. Linde and A. A. Starobinsky (1994). "Reheating after inflation," *Phys. Rev. Lett.* **73**, 3195. [hep-th/9405187].

Kofman, L., A. D. Linde and A. A. Starobinsky (1997). "Towards the theory of reheating after inflation," *Phys. Rev. D* **56**, 3258. [hep-ph/9704452].

Kojima Y. (1988). "Two families of normal modes in relativistic stars," *Prog. Theor. Phys.* **79**, 665.

Kojima, Y. (1992). "Equations governing the nonradial oscillations of a slowly rotating relativistic star," *Phys. Rev. D* **46**, 4289.

Kojima, Y. (1993). "Normal modes of relativistic stars in slow rotation limit," *Astrophys. J.* **414**, 247.

Kokkotas, K. D. (1991). "Normal modes of the Kerr black hole," *Class. Quant. Grav.* **8**, 2217.

Kokkotas, K. D. (1994). "Axial modes for relativistic stars," *Mon. Not. Roy. Astron. Soc.* **268**, 1015.

Kokkotas K. D. and B. F. Schutz (1986). "Normal modes of a model radiating system," *Gen. Rel. Grav.* **18**, 913.

Kokkotas K. D. and B. F. Schutz (1992). "W-modes: A new family of normal modes of pulsating relativistic stars," *Mon. Not. Roy. Astron. Soc.* **255**, 119.

Kokkotas, K. D. and B. G. Schmidt (1999). "Quasinormal modes of stars and black holes," *Living Rev. Rel.* **2**, 2. [gr-qc/9909058].

Kol, B. and M. Smolkin (2012). "Black hole stereotyping: Induced gravito-static polarization," *JHEP* **1202**, 010. [arXiv:1110.3764 [hep-th]].

Kolb, E. W. and M. S. Turner (1990). *The Early Universe*, Addison-Wesley, Reading, MA.

Komatsu, E. *et al.* [WMAP Collaboration] (2011). "Seven-year Wilkinson Microwave Anisotropy Probe (WMAP) observations: Cosmological interpretation," *Astrophys. J. Suppl.* **192**, 18. [arXiv:1001.4538 [astro-ph.CO]].

Komossa, S., H. Zhou and H. Lu (2008). "A recoiling supermassive black hole in the quasar SDSSJ092712.65+294344.0?," *Astrophys. J.* **678**, L81. [arXiv:0804.4585 [astro-ph]].

Komossa, S. (2012). "Recoiling black holes: Electromagnetic signatures, candidates, and astrophysical implications," *Adv. Astron.* **2012**, 364973. [arXiv:1202.1977 [astro-ph.CO]].

Koppitz, M., *et al.* (2007). "Recoil velocities from equal-mass binary-black-hole mergers," *Phys. Rev. Lett.* **99**, 041102. [gr-qc/0701163].

Kosowsky, A. (1999). "Introduction to microwave background polarization," *New Astron. Rev.* **43**, 157. [astro-ph/9904102].

Kosowsky, A., M. S. Turner and R. Watkins (1992a). "Gravitational waves from first order cosmological phase transitions," *Phys. Rev. Lett.* **69**, 2026.

Kosowsky, A., M. S. Turner and R. Watkins (1992b). "Gravitational radiation from colliding vacuum bubbles," *Phys. Rev. D* **45**, 4514.

Kosowsky, A. and M. S. Turner (1993). "Gravitational radiation from colliding vacuum bubbles: Envelope approximation to many bubble collisions," *Phys. Rev. D* **47**, 4372. [astro-ph/9211004].

Kosowsky, A., A. Mack and T. Kahniashvili (2002). "Gravitational radiation from cosmological turbulence," *Phys. Rev. D* **66**, 024030. [astro-ph/0111483].

Kostelecky, V. A. (2004). "Gravity, Lorentz violation, and the Standard Model," *Phys. Rev. D* **69**, 105009. [hep-th/0312310].

Kotake, K. (2013). "Multiple physical elements to determine the gravitational-wave signatures of core-collapse supernovae," *C. R. Phys.* **14**, 318. [arXiv:1110.5107 [astro-ph.HE]].

Kotake, K. *et al.* (2004). "Gravitational radiation from rotational core collapse: effects of magnetic fields and realistic equation of states," *Phys. Rev. D* **69**, 124004. [astro-ph/0401563].

Kotake, K., K. Sato and K. Takahashi (2006). "Explosion mechanism, neutrino burst, and gravitational wave in core-collapse supernovae," *Rept. Prog. Phys.* **69**, 971. [astro-ph/0509456].

Kouveliotou, C, R. C. Duncan and C. Thompson (2003). "Magnetars," *Scientific American* (Feb. 2003), p. 25.

Kovac, J. *et al.* [DASI experiment] (2002). "Detection of polarization in the cosmic microwave background using DASI," *Nature* **420**, 772. [astro-ph/0209478].

Kramer, M. (2007). "Fundamental physics with the SKA: Strong-field tests of gravity using pulsars and black holes," in *Exploring the Cosmic Frontier: Astrophysical Instruments for the 21st Century*, A. P. Lobanov *et al.* eds., Springer-Verlag, Berlin. [astro-ph/0409020].

Kramer, M. and D. J. Champion (2013). "The European Pulsar Timing Array and the Large European Array for Pulsars," *Class. Quant. Grav.* **30**, 224009.

Krause, O. *et al.* (2005). "Infrared echoes near the supernova remnant Cassiopeia A," *Science* **308**, 1604.

Krause, O. *et al.* (2008a). "The Cassiopeia A supernova was of type IIB," *Science* **320**, 1195.

Krause, O. *et al.* (2008b). "Tycho Brahe's 1572 supernova as a standard type Ia explosion revealed from its light echo spectrum," *Nature* **456**, 617.

Krüger, C. J., W. C. G. Ho and N. Andersson (2015). "Seismology of adolescent neutron stars: Accounting for thermal effects and crust elasticity," *Phys. Rev. D* **92**, 063009. [arXiv:1402.5656 [gr-qc]].

Kulkarni, S. R. (2005). "Modeling supernova-like explosions associated with gamma-ray bursts with short durations," astro-ph/0510256.

Kunz, M. (2012). "The phenomenological approach to modeling the dark energy," *Comptes Rendus Physique* **13**, 539. [arXiv:1204.5482 [astro-ph.CO]].

Kuroyanagi, S., T. Chiba and N. Sugiyama (2009). "Precision calculations of the gravitational wave background spectrum from inflation," *Phys. Rev. D* **79**, 103501. [arXiv:0804.3249 [astro-ph]].

Kuroyanagi, S., K. Nakayama and S. Saito (2011). "Prospects for determination of thermal history after inflation with future gravitational wave detectors," *Phys. Rev. D* **84** 123513 [arXiv:1110.4169 [astro-ph.CO]].

Lacy, J. H., C. H. Townes, T. R. Geballe and D. J. Hollenbach (1980). "Observations of the motion and distribution of the ionized gas in the central parsec of the Galaxy. II". *Astrophys. J.* **241**, 132.

Lagos, M. and P. G. Ferreira (2014). "Cosmological perturbations in massive bigravity," *JCAP* **1412**, 026. [arXiv:1410.0207 [gr-qc]].

Lam, M. T. *et al.* (2017). "The NANOGrav nine-year data set: Excess noise in millisecond pulsar arrival times," *Astrophys. J.* **834**, 35. [arXiv:1610.01731 [astro-ph.HE]].

Landau, L. D. and E. M. Lifshitz (1976). *Course of Theoretical Physics*, Vol. I: *Mechanics*, Pergamon Press, Oxford.

Landau, L. D. and E. M. Lifshitz (1977). *Course of Theoretical Physics*, Vol. III: *Quantum Mechanics*, Pergamon Press, Oxford.

Landau, L. D. and E. M. Lifshitz (1987). *Course of Theoretical Physics*, Vol. VI: *Fluid Mechanics*, Pergamon Press, Oxford.

Lattimer, J. M. and D. N. Schramm (1974). "Black-hole–neutron-star collisions," *Astrophys. J.* **192**, L145.

Lazzati, D., A. Deich, B. J. Morsony and J. C. Workman (2017). "Off-axis emission of short gamma-ray bursts and the detectability of electromagnetic counterparts of gravitational wave detected binary mergers," *Mon. Not. Roy. Astron. Soc.* **471**, 1652. [arXiv:1610.01157 [astro-ph.HE]].

Lazzati, D. *et al.* (2017). "Off-axis prompt X-ray transients from the cocoon of short gamma-ray bursts," *Astrophys. J.* **848**, L6. [arXiv:1709.01468 [astro-ph.HE]].

Leaver, E. W. (1985). "An analytic representation for the quasi-normal modes of Kerr black holes," *Proc. Roy. Soc. Lond. A* **402**, 285.

Leaver, E. W. (1986a). "Solutions to a generalized spheroidal wave equation: Teukolsky's equations in general relativity, and the two center problem in molecular quantum mechanics," *J. Math. Phys.* **27**, 1238.

Leaver, E. W. (1986b). "Spectral decomposition of the perturbation response of the Schwarzschild geometry," *Phys. Rev. D* **34**, 384.

Lee, K. J. *et al.* (2011). "Gravitational wave astronomy of single sources with a pulsar timing array," *Mon. Not. Roy. Astron. Soc.* **414**, 3251. [arXiv:1103.0115 [astro-ph.HE]].

Lehner, L. (2001). "Numerical relativity: A review," *Class. Quant. Grav.* **18**, R25. [gr-qc/0106072].

Lehners, J. L. (2008). "Ekpyrotic and cyclic cosmology," *Phys. Rept.* **465** 223. [arXiv:0806.1245 [astro-ph]].

Leibundgut, B. and N. B. Suntzeff (2003). "Optical light curves of supernovae," in *Supernovae and Gamma-Ray Bursters*, Lecture Notes in Physics, Vol. 598, Springer-Verlag, Berlin. [astro-ph/0304112].

Leins, M., H. P. Nollert and M. H. Soffel (1993). "Nonradial oscillations of neutron stars: A new branch of strongly damped normal modes," *Phys. Rev. D* **48**, 3467.

Lentati, L. *et al.* (2015). "European Pulsar Timing Array limits on an isotropic stochastic gravitational-wave background," *Mon. Not. Roy. Astron. Soc.* **453**, 2576. [arXiv:1504.03692 [astro-ph.CO]].

Lesgourgues, J. (2011). "The Cosmic Linear Anisotropy Solving System (CLASS) I: Overview," [arXiv:1104.2932 [astro-ph.IM]].

Lesgourgues, J., G. Mangano, G. Miele and S. Pastor (2013). *Neutrino Cosmology*, Cambridge University Press, Cambridge.

Le Tiec, A., L. Blanchet and C. M. Will (2010). "Gravitational-wave recoil from the ringdown phase of coalescing black hole binaries," *Class. Quant. Grav.* **27**, 012001. [arXiv:0910.4594 [gr-qc]].

Levin, Y. (2006) "QPOs during magnetar flares are not driven by mechanical normal modes of the crust," *Mon. Not. Roy. Astron. Soc.* **368**, L35. [astro-ph/0601020].

Levin Y. and M. van Hoven (2011). "On the excitation of f-modes and torsional modes by magnetar giant flares," *Mon. Not. Roy. Astron. Soc.* **418**, 659. [arXiv:1103.0880 [astro-ph.HE]].

Lewis, A., A. Challinor and A. Lasenby (2000). "Efficient computation of CMB anisotropies in closed FRW models," *Astrophys. J.* **538**, 473. [astro-ph/9911177].

Lewis, A. and A. Challinor (2006). "Weak gravitational lensing of the CMB," *Phys. Rept.* **429**, 1. [astro-ph/0601594].

Li, T. G. F. *et al.* (2012). "Towards a generic test of the strong field dynamics of general relativity using compact binary coalescence," *Phys. Rev. D* **85**, 082003. [arXiv:1110.0530 [gr-qc]].

Li, L. X. and B. Paczynski (1998). "Transient events from neutron star mergers," *Astrophys. J.* **507**, L59. [astro-ph/9807272].

Libeskind, N.I., S. Cole, C. S. Frenk and J. C. Helly (2006). "The effect of gravitational recoil on black holes forming in a hierarchical universe," *Mon. Not. R. Astron. Soc.* **368**, 1381. [astro-ph/0512073].

Liddle, A. R. and D. H. Lyth (1992). "COBE, gravitational waves, inflation and extended inflation," *Phys. Lett. B* **291**, 391. [astro-ph/9208007].

Liddle, A. R. and D. H. Lyth (2000). *Cosmological Inflation and Large-Scale Structure*, Cambridge University Press, Cambridge.

Liddle, A. R. and D. H. Lyth (2009). *The Primordial Density Perturbation: Cosmology, Inflation and the Origin of Structure*, Cambridge University Press, Cambridge.

[LIGO Scientific Collaboration and VIRGO Collaboration], Aasi, J. *et al.* (2014). "Improved upper limits on the stochastic gravitational-wave background from 2009–2010 LIGO and Virgo data," *Phys. Rev. Lett.* **113**, 231101. [arXiv:1406.4556 [gr-qc]].

[LIGO Scientific Collaboration and Virgo Collaboration] Abbott, B. P. *et al.* (2016a). "Observation of gravitational waves from a binary black hole merger," *Phys. Rev. Lett.* **116**, 061102. doi:10.1103/PhysRevLett.116.061102 [arXiv:1602.03837 [gr-qc]].

[LIGO Scientific Collaboration and Virgo Collaboration] Abbott, B. P. *et al.* (2016b). "GW150914: The Advanced LIGO detectors in the era of first discoveries," *Phys. Rev. Lett.* **116**, 131103. [arXiv:1602.03838 [gr-qc]].

[LIGO Scientific Collaboration and Virgo Collaboration], Abbott, B. P. *et al.* (2016c). "GW150914: First results from the search for binary black hole coalescence with Advanced LIGO," *Phys. Rev. D* **93**, 122003. [arXiv:1602.03839 [gr-qc]].

[LIGO Scientific Collaboration and Virgo Collaboration], Abbott, B. P. *et al.* (2016d). "Properties of the binary black hole merger

GW150914," *Phys. Rev. Lett.* **116**, 241102. [arXiv:1602.03840 [gr-qc]].

[LIGO Scientific Collaboration and Virgo Collaboration], Abbott, B. P. *et al.* (2016e). "Tests of general relativity with GW150914," *Phys. Rev. Lett.* **116**, 221101. [arXiv:1602.03841 [gr-qc]].

[LIGO Scientific Collaboration and Virgo Collaboration], Abbott, B. P. *et al.* (2016f). "The rate of binary black hole mergers inferred from Advanced LIGO observations surrounding GW150914," *Astrophys. J.* **833**, L1. [arXiv:1602.03842 [astro-ph.HE]].

[LIGO Scientific Collaboration and Virgo Collaboration], Abbott, B. P. *et al.* (2016g). "Observing gravitational-wave transient GW150914 with minimal assumptions," *Phys. Rev. D* **93**, 122004. [arXiv:1602.03843 [gr-qc]].

[LIGO Scientific Collaboration and Virgo Collaboration], Abbott, B. P. *et al.* (2016h). "Characterization of transient noise in Advanced LIGO relevant to gravitational wave signal GW150914," *Class. Quant. Grav.* **33**, 134001. [arXiv:1602.03844 [gr-qc]].

[LIGO Scientific Collaboration], Abbott, B. P. *et al.* (2016i). "Calibration of the Advanced LIGO detectors for the discovery of the binary black-hole merger GW150914," *Phys. Rev. D* **95** (2017), 062003. [arXiv:1602.03845 [gr-qc]].

[LIGO Scientific Collaboration and Virgo Collaboration], Abbott, B. P. *et al.* (2016j). "Astrophysical implications of the binary black-hole merger GW150914," *Astrophys. J.* **818**, L22 [arXiv:1602.03846 [astro-ph.HE]].

[LIGO Scientific Collaboration and Virgo Collaboration], Abbott, B. P. *et al.* (2016k). "GW150914: Implications for the stochastic gravitational wave background from binary black holes," *Phys. Rev. Lett.* **116**, 131102. [arXiv:1602.03847 [gr-qc]].

[LIGO Scientific Collaboration and Virgo Collaboration+28 Collaborations], Abbott, B. P. *et al.* (2016l). "Localization and broadband follow-up of the gravitational-wave transient GW150914," *Astrophys. J. Lett.* **826**, L13. [arXiv:1602.08492 [astro-ph.HE]].

[LIGO Scientific Collaboration and Virgo Collaboration], Abbott, B. P. *et al.* (2016m). "GW151226: Observation of gravitational waves from a 22-solar-mass binary black hole coalescence," *Phys. Rev. Lett.* **116**, 241103. [arXiv:1606.04855 [gr-qc]].

[LIGO Scientific Collaboration and Virgo Collaboration], Abbott, B. P. *et al.* (2016n). "Binary black hole mergers in the first Advanced LIGO observing run," *Phys. Rev. X* **6**, 041015. [arXiv:1606.04856 [gr-qc]].

[LIGO Scientific Collaboration and Virgo Collaboration], Abbott, B. P. *et al.* (2016o). "Improved analysis of GW150914 using a fully spin-precessing waveform model," *Phys. Rev. X* **6**, 041014. [arXiv:1606.01210 [gr-qc]].

[LIGO Scientific Collaboration and Virgo Collaboration], Abbott, B. P. *et al.* (2017a) "Upper limits on the stochastic gravitational-wave background from Advanced LIGO's first observing run," *Phys. Rev. Lett.* **118**, 121101. [arXiv:1612.02029 [gr-qc]].

[LIGO Scientific Collaboration and Virgo Collaboration], Abbott, B. P. *et al.* (2017b). "GW170104: Observation of a 50-solar-mass binary black hole coalescence at redshift 0.2," *Phys. Rev. Lett.* **118**, 221101. [arXiv:1706.01812 [gr-qc]].

[LIGO Scientific Collaboration and Virgo Collaboration], Abbott, B. P. *et al.* (2017c). "GW170814: A three-detector observation of gravitational waves from a binary black hole coalescence," *Phys. Rev. Lett.* **119**, 141101. doi:10.1103/PhysRevLett.119.141101. [arXiv:1709.09660 [gr-qc]].

[LIGO Scientific Collaboration and Virgo Collaboration], Abbott, B. P. *et al.* (2017d). "GW170817: Observation of gravitational waves from a binary neutron star inspiral," *Phys. Rev. Lett.* **119**, 161101. [arXiv:1710.05832 [gr-qc]].

[LIGO Scientific Collaboration and Virgo Collaboration], Abbott, B. P. *et al.* (2017e). "GW170608: Observation of a 19-solar-mass binary black hole coalescence," *Astrophys. J.* **851**, L35. [arXiv:1711.05578 [astro-ph.HE]].

[LIGO Scientific Collaboration and Virgo Collaboration and Fermi-GBM and INTEGRAL Collaborations], Abbott, B. P. *et al.* (2017). "Gravitational waves and gamma-rays from a binary neutron star merger: GW170817 and GRB 170817A," *Astrophys. J.* **848**, L13. doi:10.3847/2041-8213/aa920c. [arXiv:1710.05834 [astro-ph.HE]].

[LIGO Scientific Collaboration and Virgo Collaboration and Fermi GBM and INTEGRAL and IceCube and other collaborations], Abbott, B. P. *et al.* (2017). "Multi-messenger observations of a binary neutron star merger," *Astrophys. J.* **848**, L12. [arXiv:1710.05833 [astro-ph.HE]].

[LIGO Scientific Collaboration and Virgo Collaboration and 1M2H and Dark Energy Camera GW-E and DES and DLT40 and Las Cumbres Observatory and VINROUGE and MASTER Collaborations], Abbott, B. P. *et al.* (2017). "A gravitational-wave standard siren measurement of the Hubble constant," *Nature* **551**, 85. [arXiv:1710.05835 [astro-ph.CO]].

Lindblom, L. and S. L. Detweiler (1983). "The quadrupole oscillations of neutron stars," *Astrophys. J. Suppl.* **53**, 73.

Lindblom, L., B. J. Owen and S. M. Morsink (1998). "Gravitational radiation instability in hot young neutron stars," *Phys. Rev. Lett.* **80**, 4843. [gr-qc/9803053].

Linde, A. D. (1977). "On the vacuum instability and the Higgs meson mass," *Phys. Lett. B* **70**, 306.

Linde A. D. (1981). "Fate of the false vacuum at finite temperature: Theory and applications," *Phys. Lett. B* **100**, 37.

Linde, A. D. (1982). "A new inflationary Universe scenario: A possible solution of the horizon, flatness, homogeneity, isotropy and primordial monopole problems," *Phys. Lett. B* **108**, 389.

Linde A. D. (1983a). "Chaotic inflation," *Phys. Lett. B* **129**, 177.

Linde A. D. (1983b). "Decay of the false vacuum at finite temperature," *Nucl. Phys. B* **216**, 421.

Linde, A. D. (1994). "Hybrid inflation," *Phys. Rev. D* **49**, 748. [astro-ph/9307002].

Linder, E. V. (2003). "Exploring the expansion history of the universe," *Phys. Rev. Lett.* **90**, 091301. [astro-ph/0208512].

Lipunov, V. M. *et al.* (2017). "MASTER optical detection of the first LIGO/Virgo neutron star binary merger GW170817," *Astrophys. J.* **850**, L1. [arXiv:1710.05461 [astro-ph.HE]].

Liu, H. and B. Mashhoon (1996). "On the spectrum of oscillations of a Schwarzschild black hole," *Class. Quant. Grav.* **13**, 233.

Liu, K. *et al.* (2016). "Variability, polarimetry, and timing properties of single pulses from PSR J1713+0747 using the Large European Array for Pulsars," *Mon. Not. Roy. Astron. Soc.* **463**, 3239. [arXiv:1609.00188 [astro-ph.HE]].

Lombriser, L. and A. Taylor (2016). "Breaking a dark degeneracy with gravitational waves," *JCAP* **1603**, 031. [arXiv:1509.08458 [astro-ph.CO]].

Lommen, A. N. (2015). "Pulsar timing arrays: The promise of gravitational wave detection," *Rept. Prog. Phys.* **78**, 124901.

Lorimer, D. R. (2008). "Binary and millisecond pulsars," *Living Rev. Rel.* **11**, 8. [arXiv:0811.0762 [astro-ph]].

Lorimer, D. R. (2011). "Radio pulsar populations," in *High-Energy Emission from Pulsars and their Systems*, D. F. Torres and N. Rea eds., Astrophysics and Space Science Proceedings, Springer-Verlag, Berlin. [arXiv:1008.1928 [astro-ph]].

Lorimer, D. R., M. Bailes and P. A. Harrison (1997). "Pulsar statistics — IV. Pulsar velocities," *Mon. Not. R. Astron. Soc.* **289**, 592.

Lorimer, D. and M. Kramer (2005). *Handbook of Pulsar Astronomy*, Cambridge University Press, Cambridge.

Lousto, C. O. and R. H. Price (1997). "Head-on collisions of black holes: The particle limit," *Phys. Rev. D* **55**, 2124. [gr-qc/9609012].

Lovegrove, E. and S. E. Woosley (2013). "Very low energy supernovae from neutrino mass loss," *Astrophys. J.* **769**, 109. [arXiv:1303.5055 [astro-ph.HE]].

Lovelace, G. *et al.* (2010). "Momentum flow in black-hole binaries. II. Numerical simulations of equal-mass, head-on mergers with antiparallel spins," *Phys. Rev. D* **82**, 064031. [arXiv:0907.0869 [gr-qc]].

Lucchin, F., S. Matarrese and S. Mollerach (1992). "The gravitational wave contribution to CMB anisotropies and the amplitude of mass fluctuations from COBE results," *Astrophys. J.* **401**, L49. [hep-ph/9208214].

Lukash, V. N. (1980). "Production of phonons in an isotropic universe," *Sov. Phys. JETP* **52**, 807 [*Zh. Eksp. Teor. Fiz.* **79**, 1601].

Lynch, R. S., P. C. C. Freire, S. M. Ransom and B. A. Jacoby (2012). "The timing of nine globular cluster pulsars," *Astrophys. J.* **745**, 109. [arXiv:1112.2612 [astro-ph.HE]].

Lynch, R., S. Vitale, R. Essick, E. Katsavounidis and F. Robinet (2017). "An information-theoretic approach to the gravitational-wave burst detection problem," *Phys. Rev. D* **95**, 104046. [arXiv:1511.05955 [gr-qc]].

Lyne, A. and F. Graham-Smith (2006). *Pulsar Astronomy*, 3rd edition, Cambridge University Press, Cambridge.

Lyth, D. H. and A. Riotto (1999). "Particle physics models of inflation and the cosmological density perturbation," *Phys. Rept.* **314**, 1. [hep-ph/9807278].

Lyth, D. H. and D. Wands (2002). "Generating the curvature perturbation without an inflaton," *Phys. Lett. B* **524**, 5. [hep-ph/0110002].

Lyth, D. H., C. Ungarelli and D. Wands (2003). "The primordial density perturbation in the curvaton scenario," *Phys. Rev. D* **67**, 023503. [astro-ph/0208055].

Lyutikov, M. (2016). "Fermi GBM signal contemporaneous with GW150914 — An unlikely association," [arXiv:1602.07352 [astro-ph.HE]].

Ma, C.-P. and E. Bertschinger (1995). "Cosmological perturbation theory in the synchronous and conformal Newtonian gauges," *Astrophys. J.* **455**, 7. [astro-ph/9506072].

Maassen van den Brink, A. (2004). "WKB analysis of the Regge–Wheeler equation down in the frequency plane," *J. Math. Phys.* **45**, 327.

MacFadyen, A. I. and S. E. Woosley (1999). "Collapsars: Gamma-ray bursts and explosions in 'failed supernovae'," *Astrophys. J.* **524** 262. [astro-ph/9810274].

MacFadyen, A. I., S. E. Woosley and A. Heger (2001). "Supernovae, jets, and collapsars," *Astrophys. J.* **550**, 410. [astro-ph/9910034].

Madau, P. and E. Quataert (2004). "The effect of gravitational-wave recoil on the demography of massive black holes," *Astrophys. J.* **606**, L17.

Madison, D. R. *et al.* (2016). "Versatile directional searches for gravitational waves with pulsar timing arrays," *Mon. Not. Roy. Astron. Soc.* **455**, 3662. [arXiv:1510.08068 [astro-ph.IM]].

Maeder, A. and G. Meynet (2000). "The evolution of rotating stars," *Annu. Rev. Astron. Astrophys.* **38**, 143. [astro-ph/0004204].

Maggiore, M. (2000). "Gravitational wave experiments and early universe cosmology," *Phys. Rept.* **331**, 283. [gr-qc/9909001].

Maggiore, M. (2005). *A Modern Introduction to Quantum Field Theory*, Oxford University Press, Oxford.

Maggiore, M. (2008). "The physical interpretation of the spectrum of black hole quasinormal modes," *Phys. Rev. Lett.* **100**, 141301. [arXiv:0711.3145 [gr-qc]].

Maggiore, M. (2014). "Phantom dark energy from nonlocal infrared modifications of general relativity," *Phys. Rev. D* **89**, 043008. [arXiv:1307.3898 [hep-th]].

Maggiore, M. (2015). "Dark energy and dimensional transmutation in R^2 gravity," arXiv:1506.06217 [hep-th].

Maggiore, M. (2016). "Perturbative loop corrections and nonlocal gravity," *Phys. Rev. D* **93**, 063008. [arXiv:1603.01515 [hep-th]].

Maggiore, M. (2017). "Nonlocal infrared modifications of gravity. A review," *Fundam. Theor. Phys.* **187**, 221. [arXiv:1606.08784 [hep-th]].

Maggiore, M. and A. Nicolis (2000). "Detection strategies for scalar gravitational waves with interferometers and resonant spheres," *Phys. Rev. D* **62**, 024004. [gr-qc/9907055].

Maggiore, M. and M. Mancarella (2014). "Nonlocal gravity and dark energy," *Phys. Rev. D* **90**, 023005. [arXiv:1402.0448 [hep-th]].

Margutti, R. *et al.* (2017). "The electromagnetic counterpart of the binary neutron star merger LIGO/VIRGO GW170817. V. Rising X-ray emission from an off-axis jet," *Astrophys. J.* **848**, L20. [arXiv:1710.05431 [astro-ph.HE]].

Majumdar, M. and A.-C. Davis (2002). "Cosmological creation of D-branes and anti-D-branes," *JHEP* **0203**, 056. [hep-th/0202148].

Manca, G. M., L. Baiotti, R. De Pietri and L. Rezzolla (2007). "Dynamical non-axisymmetric instabilities in rotating relativistic stars," *Class. Quant. Grav.* **24**, S171. [arXiv:0705.1826 [astro-ph]].

Manchester, R. N., G. B. Hobbs, A. Teoh and M. Hobbs (2005). "The Australia Telescope National Facility pulsar catalogue," *Astron. J.* **129**, 1993. [astro-ph/0412641].

Manchester, R. N. *et al.* (2013). "The Parkes Pulsar Timing Array project," *Publ. Astron. Soc. Austral.* **30**, 17. [arXiv:1210.6130 [astro-ph.IM]].

Mangano, G., G. Miele, S. Pastor and M. Peloso (2002). "A precision calculation of the effective number of cosmological neutrinos," *Phys. Lett. B* **534**, 8. [astro-ph/0111408].

Mangano, G., *et al.* (2005). "Relic neutrino decoupling including flavor oscillations," *Nucl. Phys. B* **729**, 221. [hep-ph/0506164].

Mangilli, A., N. Bartolo, S. Matarrese and A. Riotto (2008). "The impact of cosmic neutrinos on the gravitational-wave background," *Phys. Rev. D* **78**, 083517. [arXiv:0805.3234 [astro-ph]].

Marek, A. and H.-T. Janka (2009). "Delayed neutrino-driven supernova explosions aided by the standing accretion-shock instability," *Astrophys. J.* **694**, 664. [arXiv:0708.3372 [astro-ph]].

Marek, A., H.-T. Janka and E. Müller (2009). 'Equation-of-state dependent features in shock-oscillation modulated neutrino and gravitational-wave signals from supernovae," *Astron. Astrophys.* **496**, 475. [arXiv:0808.4136 [astro-ph]].

Martin, D. *et al.* (2015). "Neutrino-driven winds in the aftermath of a neutron star merger: Nucleosynthesis and electromagnetic transients," *Astrophys. J.* **813**, 2. [arXiv:1506.05048 [astro-ph.SR]].

Martin, J. and C. Ringeval (2010). "First CMB constraints on the inflationary reheating temperature," *Phys. Rev. D* **82** 023511. [arXiv:1004.5525 [astro-ph.CO]].

Martin, J., C. Ringeval and V. Vennin (2014). "Encyclopædia Inflationaris," *Phys. Dark Univ.* **5–6**, 75. [arXiv:1303.3787 [astro-ph.CO]].

Mashhoon, B. (1982). "On the contribution of a stochastic background of gravitational radiation to the timing noise of pulsars," *Mon. Not. Roy. Astron. Soc.* **199**, 659.

McConnell, N. J. *et al.* (2011). "Two ten-billion-solar-mass black holes at the centers of giant elliptical galaxies," *Nature* **480**, 215. [arXiv:1112.1078 [astro-ph.CO]].

McConnell, N. J. and C. P. Ma (2013). "Revisiting the scaling relations of black hole masses and host galaxy properties," *Astrophys. J.* **764**, 184. [arXiv:1211.2816 [astro-ph.CO]].

McDermott, P. N., H. M. van Horn, C. J. Hansen and R. Buland (1985). "The nonradial oscillation spectra of neutron stars," *Astrophys. J.* **297**, L37.

McDermott, P. N., H. M. van Horn and C. J. Hansen (1988). "Nonradial oscillations of neutron stars," *Astrophys. J.* **325**, 725.

McLaughlin, M. A. (2013). "The North American Nanohertz Observatory for Gravitational Waves," *Class. Quant. Grav.* **30**, 224008. [arXiv:1310.0758 [astro-ph.IM]].

McWilliams, S. T., J. P. Ostriker and F. Pretorius (2014). "Gravitational waves and stalled satellites from massive galaxy mergers at $z \leqslant 1$," *Astrophys. J.* **789**, 156. [arXiv:1211.5377 [astro-ph.CO]].

Medved, A. J. M., D. Martin and M. Visser (2004). "Dirty black holes: Quasinormal modes," *Class. Quant. Grav.* **21**, 1393. [gr-qc/0310009].

Melatos, A., J. A. Douglass and T. P. Simula (2015). "Persistent gravitational radiation from glitching pulsars," *Astrophys. J.* **807**, 132.

Melia, F. (2007). *The Galactic Supermassive Black Hole*, Princeton University Press, Princeton, NJ.

Mereghetti, S. (2008). "The strongest cosmic magnets: Soft gamma-ray repeaters and anomalous X-ray pulsars," *Astron. Astrophys. Rev.*, **15**, 225. [arXiv:0804.0250 [astro-ph]].

Mereghetti, S. (2013). "Pulsars and magnetars," *Braz. J. Phys.* **43**, 356. [arXiv:1304.4825].

Mereghetti, S., J. Pons and A. Melatos (2015). "Magnetars: Properties, origin and evolution," *Space Sci. Rev.* **191**, 315. [arXiv:1503.06313 [astro-ph.HE]].

Merritt, D., M. Milosavljević, M. Favata, S. A. Hughes and D. E. Holz (2004). "Consequences of gravitational radiation recoil," *Astrophys. J.* **607**, L7. [astro-ph/0402057].

Messenger, C. and J. Read (2012). "Measuring a cosmological distance–redshift relationship using only gravitational wave observations of binary neutron star coalescences," *Phys. Rev. Lett.* **108**, 091101. [arXiv:1107.5725 [gr-qc]].

Meszaros, P. and M. J. Rees (1993). "Relativistic fireballs and their impact on external matter — Models for cosmological gamma-ray bursts," *Astrophys. J.* **405**, 278.

Metzger, B. D. *et al.* (2010). "Electromagnetic counterparts of compact object mergers powered by the radioactive decay of r-process nuclei," *Mon. Not. Roy. Astron. Soc.* **406**, 2650. [arXiv:1001.5029 [astro-ph.HE]].

Metzger, B. D. (2017). "Kilonovae," *Living Rev. Rel.* **20**, 3. [arXiv:1610.09381 [astro-ph.HE]].

Meylan, G. and D. C. Heggie (1997). "Internal dynamics of globular clusters," *Astron. Astrophys. Rev.* **8**, 1. [astro-ph/9610076].

Mijic, M. B., M. S. Morris and W. M. Suen (1986). "The R^2 cosmology: Inflation without a phase transition," *Phys. Rev. D* **34**, 2934.

Mikkola, S. and Valtonen, M. J. (1992). "Evolution of binaries in the field of light particles and the problem of two black holes," *Mon. Not. R. Astron. Soc.* **259**, 115.

Miller, C. M. (2004). "Probing general relativity with mergers of supermassive and intermediate-mass black holes," *Astrophys. J.* **618**, 426. [astro-ph/0409331].

Mino, Y. (2003). "Perturbative approach to an orbital evolution around a supermassive black hole," *Phys. Rev. D* **67**, 084027. [gr-qc/0302075].

Mino, Y. (2005). "From the self-force problem to the radiation reaction formula," *Class. Quant. Grav.* **22**, S717. [gr-qc/0506002].

Mino, Y., M. Sasaki and T. Tanaka (1997). "Gravitational radiation reaction to a particle motion," *Phys. Rev. D* **55** 3457. [gr-qc/9606018].

Mishra, C. K., K. G. Arun, B. R. Iyer and B. S. Sathyaprakash (2010). "Parametrized tests of post-Newtonian theory using Advanced LIGO and Einstein Telescope," *Phys. Rev. D* **82**, 064010. [arXiv:1005.0304 [gr-qc]].

Misner, C. (1960). "Wormhole Initial Conditions," *Phys. Rev.* **118**, 1110.

Misner, C. W., K. S. Thorne and J. A. Wheeler (1973). *Gravitation*, Freeman, New York.

Mitsou, E. (2011). "Gravitational radiation from radial infall of a particle into a Schwarzschild black hole. A numerical study of the spectra, quasi-normal modes and power-law tails," *Phys. Rev. D* **83**, 044039. [arXiv:1012.2028 [gr-qc]].

Mollerach, S. (1990). "Isocurvature baryon perturbations and inflation," *Phys. Rev. D* **42**, 313.

Mönchmeyer, R., G. Schäfer, E. Müller, and R. E. Kates (1991). "Gravitational waves from the collapse of rotating stellar cores," *Astron. Astrophys.* **246**, 417.

Moncrief, V. (1974). "Gravitational perturbations of spherically symmetric systems. I. The exterior problem," *Ann. Phys.* **88**, 323.

Mooley, K. P. *et al.* (2017). "A mildly relativistic wide-angle outflow in the neutron star merger GW170817," [arXiv:1711.11573 [astro-ph.HE]].

Moore, G. D. and A. E. Nelson (2001). "Lower bound on the propagation speed of gravity from gravitational Cherenkov radiation," *JHEP* **0109**, 023. [hep-ph/0106220].

Moriya, T. J. *et al.* (2014). "Electron-capture supernovae exploding within their progenitor wind," *Astron. Astrophys.* **569**, A57. [arXiv:1407.4563 [astro-ph.HE]].

Mortlock, D. J. *et al.* (2011). "A luminous quasar at a redshift of $z = 7.085$," *Nature* **474**, 616. [arXiv:1106.6088 [astro-ph.CO]].

Mortonson, M. J. and U. Seljak (2014). "A joint analysis of *Planck* and BICEP2 B modes including dust polarization uncertainty," *JCAP* **1410**, 035. [arXiv:1405.5857 [astro-ph.CO]].

Motl, L. (2003). "An analytical computation of asymptotic Schwarzschild quasinormal frequencies," *Adv. Theor. Math. Phys.* **6**, 1135. [gr-qc/0212096].

Motl, L. and A. Neitzke (2003). "Asymptotic black hole quasinormal frequencies," *Adv. Theor. Math. Phys.* **7**, 307. [hep-th/0301173].

Mottola, E. (1985). "Particle creation in de Sitter space," *Phys. Rev. D* **31**, 754.

Mroué, A. H. *et al.* (2013). "Catalog of 174 binary black hole simulations for gravitational wave astronomy," *Phys. Rev. Lett.* **111**, 241104. [arXiv:1304.6077 [gr-qc]].

Mukhanov, V. F. (1985). "Gravitational instability of the universe filled with a scalar field," *JETP Lett.* **41** 493 [*Pisma Zh. Eksp. Teor. Fiz.* **41** (1985) 402].

Mukhanov, V. F. (1989). "Quantum theory of cosmological perturbations in R^2 gravity," *Phys. Lett. B* **218**, 17.

Mukhanov, V. F. (2005). *Physical Foundations of Cosmology*, Cambridge University Press, Cambridge.

Mukhanov, V. F. and G. V. Chibisov (1981). "Quantum fluctuations and a nonsingular Universe," *JETP Lett.* **33** 532 [*Pisma Zh. Eksp. Teor. Fiz.* **33** (1981) 549].

Mukhanov, V. F. and G. V. Chibisov (1982). "The vacuum energy and large scale structure of the Universe," *Sov. Phys. JETP* **56** 258. [*Zh. Eksp. Teor. Fiz.* **83** (1982) 475].

Mukhanov, V. F., H. A. Feldman and R. H. Brandenberger (1992). "Theory of cosmological perturbations," *Phys. Rept.* **215**, 203.

Mukhanov, V. F. and Winitzki, S. (2007). *Quantum Effects in Gravity*. Cambridge University Press, Cambridge.

Müller, E. and H.-T. Janka (1997). "Gravitational radiation from convective instabilities in type II supernova explosions," *Astron. Astrophys.* **317**, 140.

Müller, E., M. Rampp, R. Buras, H.-T. Janka, D. H. Shoemaker (2004). "Toward gravitational wave signals from realistic core-collapse supernova models", *Astrophys. J.* **603**, 221. [astro-ph/0309833].

Müller, B., H.-T. Janka and A. Marek (2013). "A new multidimensional general relativistic neutrino hydrodynamics code of core-collapse supernovae III. Gravitational wave signals from supernova explosion models," *Astrophys. J.* **766**, 43. [arXiv:1210.6984 [astro-ph.SR]].

Muhlberger, C. D. *et al.* (2014). "Magnetic effects on the low-$T/|W|$ instability in differentially rotating neutron stars," *Phys. Rev. D* **90**, 104014. [arXiv:1405.2144 [astro-ph.HE]].

Murphy, J. W., C. D. Ott and A. Burrows (2009). "A model for gravitational wave emission from neutrino-driven core-collapse supernovae," *Astrophys. J.* **707**, 1173. [arXiv:0907.4762 [astro-ph.SR]].

Nadathur, S. and R. Crittenden (2016). "A detection of the integrated Sachs–Wolfe imprint of cosmic superstructures using a matched-filter approach," *Astrophys. J.* **830**, L19. [arXiv:1608.08638 [astro-ph.CO]].

Nagar, A., T. Damour and A. Tartaglia (2007). "Binary black hole merger in the extreme mass ratio limit," *Class. Quant. Grav.* **24** S109. [gr-qc/0612096].

Nakamura, T., K. Oohara and Y. Kojima (1987). "General relativistic collapse to black holes and gravitational waves from black holes," *Prog. Theor. Phys. Suppl.* **90**, 1.

Nakar, E. (2007). "Short-hard gamma-ray bursts," *Phys. Rept.* **442**, 166. [astro-ph/0701748 [astro-ph]].

Nakar, E. and T. Piran (2011). "Radio remnants of compact binary mergers — The electromagnetic signal that will follow the gravitational waves," *Nature* **478**, 82.

Nakar, E. and T. Piran (2017). "The observable signatures of GRB cocoons," *Astrophys. J.* **834**, 28. [arXiv:1610.05362 [astro-ph.HE]].

Nakayama, K., S. Saito, Y. Suwa and J. Yokoyama (2008). "Probing reheating temperature of the Universe with gravitational wave background," *JCAP* **0806** 020. [arXiv:0804.1827 [astro-ph]].

Narayan, R. (2005). "Black holes in astrophysics," *New J. Phys.* **7**, 199. [gr-qc/0506078].

Newman, E. and R. Penrose (1962). "An approach to gravitational radiation by a method of spin coefficients," *J. Math. Phys.* **3**, 566.

Nicholl, M. *et al.* (2017). "The electromagnetic counterpart of the binary neutron star merger LIGO/VIRGO GW170817. III. Optical and UV spectra of a blue kilonova from fast polar ejecta," *Astrophys. J.* **848**, L18. [arXiv:1710.05456 [astro-ph.HE]].

Nishizawa, A. (2017). "Generalized framework for testing gravity with gravitational-wave propagation. I. Formulation," arXiv:1710.04825 [gr-qc].

Nissanke, S., D. E. Holz, S. A. Hughes, N. Dalal and J. L. Sievers (2010). "Exploring short gamma-ray bursts as gravitational-wave standard sirens," *Astrophys. J.* **725**, 496. [arXiv:0904.1017 [astro-ph.CO]].

Nollert, H.-P. (1993). "Quasinormal modes of Schwarzschild black holes: The determination of quasinormal frequencies with very large imaginary parts," *Phys. Rev. D* **47**, 5253.

Nollert, H.-P. (1999). "Quasinormal modes: The characteristic 'sound' of black holes and neutron stars," *Class. Quant. Grav.* **16**, R159.

Nollert, H.-P. and B. G. Schmidt (1992). "Quasinormal modes of Schwarzschild black holes: Defined and calculated via Laplace transformation," *Phys. Rev. D* **45** 2617.

Nollert, H. P. and R. H. Price (1999). "Quantifying excitations of quasinormal mode systems," *J. Math. Phys.* **40**, 980. [gr-qc/9810074].

Obergaulinger, M., M. A. Aloy and E. Müller (2006). "Axisymmetric simulations of magneto-rotational core collapse: Dynamics and gravitational wave signal," *Astron. Astrophys.* **450**, 1107. [astro-ph/0510184].

Oh, S.-H., W. J. G. de Blok, E. Brinks, F. Walter, R. C. Kennicutt, Jr. (2011). *Astron. J.* **141**, 193. "Dark and luminous matter in THINGS dwarf galaxies," [arXiv:1011.0899 [astro-ph.CO]].

Olausen, S. A. and V. M. Kaspi (2014). "The McGill magnetar catalog," *Astrophys. J. Suppl.* **212**, 6. [arXiv:1309.4167 [astro-ph.HE]].

Olmez, S., V. Mandic and X. Siemens (2010). "Gravitational-wave stochastic background from kinks and cusps on cosmic strings," *Phys. Rev. D* **81**, 104028. [arXiv:1004.0890 [astro-ph.CO]].

Onozawa, H. (1997). "A detailed study of quasinormal frequencies of the Kerr black hole," *Phys. Rev. D* **55**, 3593.

Ott, C. D. (2009). "The gravitational wave signature of core-collapse supernovae," *Class. Quant. Grav.* **26** 063001. [arXiv:0809.0695 [astro-ph]].

Ott, C. D., S. Ou, J. E. Tohline and A. Burrows (2005). "One-armed spiral instability in a slowly rotating, post-bounce supernova core," *Astrophys. J.* **625**, L119. [astro-ph/0503187].

Ott, C. D. *et al.* (2007). "3D collapse of rotating stellar iron cores in general relativity with microphysics," *Phys. Rev. Lett.* **98**, 261101. [astro-ph/0609819].

Owen, B. J. *et al.* (1998). "Gravitational waves from hot young rapidly rotating neutron stars," *Phys. Rev. D* **58**, 084020. [gr-qc/9804044].

Paczynski, B. (1998). "Are gamma-ray bursts in star forming regions?," *Astrophys. J.* **494**, L45. [astro-ph/9710086].

Padmanabhan, T. (2001). *Theoretical Astrophysics.* Vol. II: *Stars and Stellar Systems*, Cambridge University Press, Cambridge.

Padmanabhan, T. (2002). *Theoretical Astrophysics.* Vol. III: *Galaxies and Cosmology*, Cambridge University Press, Cambridge.

Padmanabhan, T. (2004). "Quasinormal modes: A simple derivation of the level spacing of the frequencies," *Class. Quant. Grav.* **21**, L1. [gr-qc/0310027].

Pagano, L., L. Salvati and A. Melchiorri (2016). "New constraints on primordial gravitational waves from *Planck* 2015," *Phys. Lett. B* **760**, 823. [arXiv:1508.02393 [astro-ph.CO]].

Pan, Y. *et al.* (2008). "A data-analysis driven comparison of analytic and numerical coalescing binary waveforms: Nonspinning case," *Phys. Rev. D* **77**, 024014. [arXiv:0704.1964 [gr-qc]].

Pan, Y. *et al.* (2014). "Inspiral–merger–ringdown waveforms of spinning, precessing black-hole binaries in the effective-one-body formalism," *Phys. Rev. D* **89**, 084006. [arXiv:1307.6232 [gr-qc]].

Papaloizou, J. and J. E. Pringle (1978). "Non-radial oscillations of rotating stars and their relevance to the short-period oscillations of cataclysmic variables," *Mon. Not. Roy. Astron. Soc.* **182**, 423.

Parker, L. (1969). "Quantized fields and particle creation in expanding universes. 1.," *Phys. Rev.* **183**, 1057.

Passamonti, A., N. Stergioulas and A. Nagar (2007). "Gravitational waves from nonlinear couplings of radial and polar nonradial modes in relativistic stars," *Phys. Rev. D* **75**, 084038. [gr-qc/0702099].

Passamonti, A., B. Haskell, N. Andersson, D. I. Jones and I. Hawke (2009). "Oscillations of rapidly rotating stratified neutron stars," *Mon. Not. Roy. Astron. Soc.* **394**, 730. [arXiv:0807.3457 [astro-ph]].

Passamonti, A. and S. K. Lander (2013). "Stratification, superfluidity and magnetar QPOs," *Mon. Not. Roy. Astron. Soc.* **429**, 767. [arXiv:1210.2969 [astro-ph.SR]].

Passamonti, A. and S. K. Lander (2014) "Quasi-periodic oscillations in superfluid magnetars," *Mon. Not. Roy. Astron. Soc.* **438**, 156. [arXiv:1307.3210 [astro-ph.SR]].

Passamonti A. and N. Andersson (2015). "The intimate relation between the low T/W instability and the corotation point," *Mon. Not. Roy. Astron. Soc.* **446**, 555. [arXiv:1409.0677 [astro-ph.SR]].

Passamonti, A., N. Andersson and W. C. G. Ho (2016). "Buoyancy and g-modes in young superfluid neutron stars," *Mon. Not. Roy. Astron. Soc.* **455**, 1489. [arXiv:1504.07470 [astro-ph.SR]].

Patrignani, C. *et al.* [Particle Data Group Collaboration] (2016). "Review of Particle Physics," *Chin. Phys. C* **40**, 100001.

Pazos-Ávalos, E. and C. O. Lousto (2005). "Numerical integration of the Teukolsky equation in the time domain," *Phys. Rev. D* **72**, 084022. [gr-qc/0409065].

Peacock, J. A. (2007). *Cosmological Physics*, Cambridge University Press, Cambridge.

Peebles, P. J. E. and J. T. Yu (1970). "Primeval adiabatic perturbation in an expanding Universe," *Astrophys. J.* **162**, 815.

Peebles, P. J. E., L. A. Page and R. B. Partridge (2009). *Finding the Big Bang*, Cambridge University Press, Cambridge.

Penrose, R. (1969). "Gravitational collapse: The role of general relativity," *Riv. Nuovo Cim.* **1**, 252.

Penzias, A. A. (1978). 1978 Nobel Lecture in Physics. In *Nobel Lectures, Physics 1971–1980*, S. Lundqvist ed., World Scientific, Singapore, 1992.

Penzias, A. A. and R. W. Wilson (1965). "A measurement of excess antenna temperature at 4080 Mc/s," *Astrophys. J.* **142**, 419.

Peres, A. (1962). "Classical radiation recoil," *Phys. Rev.* **128**, 2471.

Perlmutter, S. *et al.* [Supernova Cosmology Project Collaboration] (1997). "Measurements of the cosmological parameters Ω and Λ from the first 7 supernovae at $z \geqslant 0.35$," *Astrophys. J.* **483**, 565. [astro-ph/9608192].

Perna, R., P. J. Armitage and B. Zhang (2005). "Flares in long and short gamma-ray bursts: A common origin in a hyperaccreting accretion disk," *Astrophys. J.* **636**, L29. [astro-ph/0511506].

Phillips, M. M. (1993). "The absolute magnitudes of type IA supernovae," Astrophys. J. **413**, L105.

Phinney, E. S. (2001). "A practical theorem on gravitational wave backgrounds," [astro-ph/0108028].

Piran, T. (2004). "The physics of gamma-ray bursts," *Rev. Mod. Phys.* **76**, 1143. [astro-ph/0405503].

Piro, A. L. and E. Pfahl (2007). "Fragmentation of collapsar disks and the production of gravitational waves," *Astrophys. J.* **658**, 1173. [astro-ph/0610696].

[Planck Collaboration], Ade, P. A. R. *et al.* (2014a). "*Planck* 2013 results. XV. CMB power spectra and likelihood," *Astron. Astrophys.* **571** A15. [arXiv:1303.5075 [astro-ph.CO]].

[Planck Collaboration], Ade, P. A. R. *et al.* (2014b). "*Planck* 2013 results. XXII. Constraints on inflation," *Astron. Astrophys.* **571**, A22. [arXiv:1303.5082 [astro-ph.CO]].

[Planck Collaboration], Aghanim, N. *et al.* (2014). "*Planck* 2013 results. XXVII. Doppler boosting of the CMB: Eppur si muove," *Astron. Astrophys.* **571**, A27. [arXiv:1303.5087 [astro-ph.CO]].

[Planck Collaboration], Adam, R. *et al.* (2016a). "*Planck* intermediate results. XXX. The angular power spectrum of polarized dust emission at intermediate and high Galactic latitudes," *Astron. Astrophys.* **586** A133. [arXiv:1409.5738 [astro-ph.CO]].

[Planck Collaboration], Adam, R. *et al.* (2016b). "*Planck* 2015 results. I. Overview of products and scientific results," *Astron. Astrophys.* **594** A1. [arXiv:1502.01582 [astro-ph.CO]].

[Planck Collaboration], Ade, P. A. R. *et al.* (2016a). "*Planck* 2015 results. XIII. Cosmological parameters," *Astron. Astrophys.* **594** A13. [arXiv:1502.01589 [astro-ph.CO]].

[Planck Collaboration], Ade, P. A. R. *et al.* (2016b). "*Planck* 2015 results. XIV. Dark energy and modified gravity," *Astron. Astrophys.* **594**, A14. [arXiv:1502.01590 [astro-ph.CO]].

[Planck Collaboration], Ade, P. A. R. *et al.* (2016c). "*Planck* 2015 results. XX. Constraints on inflation," *Astron. Astrophys.* **594** A20. [arXiv:1502.02114 [astro-ph.CO]].

[Planck Collaboration], Aghanim, N. *et al.* (2016). "*Planck* intermediate results. XLVI. Reduction of large-scale systematic effects in HFI polarization maps and estimation of the reionization optical depth," *Astron. Astrophys.* **596**, A107. [arXiv:1605.02985 [astro-ph.CO]].

Pnigouras, P. and K. D. Kokkotas (2016). "Saturation of the f-mode instability in neutron stars: II. Applications and results," *Phys. Rev. D* **94**, 024053. [arXiv:1607.03059 [astro-ph.HE]].

Poelarends, A. J. T., F. Herwig, N. Langer and A. Heger (2008). "The supernova channel of super-AGB stars," *Astrophys. J.* **675**, 614. [arXiv:0705.4643 [astro-ph]].

Poisson, E. (1997). "Gravitational radiation from infall into a black hole: Regularization of the Teukolsky equation," *Phys. Rev. D* **55**, 639. [gr-qc/9606078].

Poisson, E. (2004a). *A Relativist's Toolkit. The Mathematics of Black-Hole Mechanics*, Cambridge University Press, Cambridge.

Poisson, E. (2004b). "The motion of point particles in curved space-time," *Living Rev. Rel.* **7**, 6. [gr-qc/0306052].

Poisson, E. and C. M. Will (1995). "Gravitational waves from inspiraling compact binaries: Parameter estimation using second post-Newtonian wave forms," *Phys. Rev. D* **52**, 848. [gr-qc/9502040].

Poisson, E., A. Pound and I. Vega (2011). "The motion of point particles in curved spacetime," *Living Rev. Rel.* **14**, 7. [arXiv:1102.0529 [gr-qc]].

Poisson, E. and C. M. Will (2014). *Gravity. Newtonian, Post-Newtonian, Relativistic.* Cambridge University Press, Cambridge.

Polarski, D. and A. A. Starobinsky (1996). "Semiclassicality and decoherence of cosmological perturbations," *Class. Quant. Grav.* **13**, 377. [gr-qc/9504030].

Polchinski, J. (1998). *String Theory.* Vol. I: *An Introduction to the Bosonic String*, Cambridge University Press, Cambridge.

Pollney, D. *et al.* (2007). "Recoil velocities from equal-mass binary black-hole mergers: A systematic investigation of spin–orbit aligned configurations," *Phys. Rev. D* **76**, 124002. [arXiv:0707.2559 [gr-qc]].

Polnarev, A. G. (1985). "Polarization and anisotropy induced in the microwave background by cosmological gravitational waves," *Sov. Astron.* **29**, 607.

Popham, R., S. E. Woosley and C. Fryer (1999). "Hyperaccreting black holes and gamma-ray bursts," *Astrophys. J.* **518**, 356. [astro-ph/9807028].

Popov, S. B., J. A. Pons, J. A. Miralles, P. A. Boldin and B. Posselt (2010). "Population synthesis studies of isolated neutron stars with magnetic field decay," *Mon. Not. Roy. Astron. Soc.* **401**, 2675. [arXiv:0910.2190 [astro-ph.HE]].

Press, W. H. (1971). "Long wave trains of gravitational waves from a vibrating black hole," *Astrophys. J.* **170**, L105.

Press, W. H. and S. A. Teukolsky (1973). "Perturbations of a rotating black hole. II. Dynamical stability of the Kerr metric," *Astrophys. J.* **185**, 649.

Preto, M., I. Berentzen, P. Berczik and R. Spurzem (2011). "Fast coalescence of massive black hole binaries from mergers of galactic nuclei: implications for low-frequency gravitational-wave astrophysics," *Astrophys. J.* **732**, L26. [arXiv:1102.4855 [astro-ph.GA]].

Pretorius, F. (2005). "Evolution of binary black hole spacetimes," *Phys. Rev. Lett.* **95**, 121101. [gr-qc/0507014].

Pretorius, F. (2009). "Binary black hole coalescence," in *Physics of Relativistic Objects in Compact Binaries: From Birth to Coalescence*, M. Colpi *et al.* eds., Springer-Verlag, Dordrecht. [arXiv:0710.1338].

Price, R. H. (1972a). "Nonspherical perturbations of relativistic gravitational collapse. I. Scalar and gravitational perturbations," *Phys. Rev. D* **5**, 2419.

Price, R. H. (1972b). "Nonspherical perturbations of relativistic gravitational collapse. II. Integer-spin, zero-rest-mass fields," *Phys. Rev. D* **5**, 2439.

Price, R. H. and K. S. Thorne (1969). "Non-radial pulsation of general-relativistic stellar models. II. Properties of the gravitational waves," *Astrophys. J.* **155**, 163.

Price, R. H. and Pullin (1994). "Colliding black holes: The close limit," *Phys. Rev. Lett.* **72**, 3297.

Price, R. H., S. Nampalliwar and G. Khanna (2016). "Black hole binary inspiral: Analysis of the plunge," *Phys. Rev. D* **93**, 044060. [arXiv:1508.04797 [gr-qc]].

Pritchard, J. R. and M. Kamionkowski (2005). "Cosmic microwave background fluctuations from gravitational waves: An analytic approach," *Ann. Phys.* **318**, 2. [astro-ph/0412581].

Pürrer, M. (2014). "Frequency domain reduced order models for gravitational waves from aligned-spin compact binaries," *Class. Quant. Grav.* **31**, 195010. [arXiv:1402.4146 [gr-qc]].

Pürrer, M. (2016). "Frequency domain reduced order model of aligned-spin effective-one-body waveforms with generic mass-ratios and spins," *Phys. Rev. D* **93**, 064041. [arXiv:1512.02248 [gr-qc]].

Quinlan, G. D. (1996). "The dynamical evolution of massive black hole binaries - I. Hardening in a fixed stellar background," *New Astron.* **1**, 35. [astro-ph/9601092].

Quinn, T. C. and R. M. Wald (1997). "An axiomatic approach to electromagnetic and gravitational radiation reaction of particles in curved space-time," *Phys. Rev. D* **56**, 3381. [gr-qc/9610053].

Quiros, M. (1999), "Finite temperature field theory and phase transitions," Lectures at Summer School in High Energy Physics and Cosmology, ICTP, Trieste, June 29–July 17, 1999. [hep-ph/9901312].

Radice, D., S. Bernuzzi and C. D. Ott (2016). "One-armed spiral instability in neutron star mergers and its detectability in gravitational waves," *Phys. Rev. D* **94**, 064011. [arXiv:1603.05726 [gr-qc]].

Rajagopal, M. and R. W. Romani (1995). "Ultralow frequency gravitational radiation from massive black hole binaries," *Astrophys. J.* **446**, 543. [astro-ph/9412038].

Ramirez-Ruiz, E., A. Celotti and M. J. Rees (2002). "Events in the life of a cocoon surrounding a light, collapsar jet," *Mon. Not. Roy. Astron. Soc.* **337**, 1349. [astro-ph/0205108].

Ravi, V. *et al.* (2012). "Does a 'stochastic' background of gravitational waves exist in the pulsar timing band?," *Astrophys. J.* **761**, 84. [arXiv:1210.3854 [astro-ph.CO]].

Ravi, V., J. S. B. Wyithe, R. M. Shannon and G. Hobbs (2015). "Prospects for gravitational-wave detection and supermassive black hole astrophysics with pulsar timing arrays," *Mon. Not. Roy. Astron. Soc.* **447**, 2772. [arXiv:1406.5297 [astro-ph.CO]].

Read, J. S. *et al.* (2013). "Matter effects on binary neutron star waveforms," *Phys. Rev. D* **88**, 044042. [arXiv:1306.4065 [gr-qc]].

Rebhan, A. K. and D. J. Schwarz (1994). "Kinetic versus thermal field theory approach to cosmological perturbations," *Phys. Rev. D* **50**, 2541. [gr-qc/9403032].

Redmount, I. H. and M. J. Rees (1989). "Gravitational-radiation rocket effects and galactic structure," *Comments Astrophys.*, **14**, 165.

Rees, M. J. and D. W. Sciama (1968). "Large scale density inhomogeneities in the Universe," *Nature* **217**, 511.

Regge, T. and J. A. Wheeler (1957). "Stability of a Schwarzschild singularity," *Phys. Rev.* **108**, 1063.

Regge, T. and C. Teitelboim (1974). "Role of surface integrals in the Hamiltonian formulation of general relativity," *Ann. Phys.* **88**, 286.

Reisenegger, A. and P. Goldreich (1992). "A new class of g-modes in neutron stars," *Astrophys. J.* **395**, 240.

Rest, A. *et al.* (2008). "Scattered-light echoes from the historical Galactic supernovae Cassiopeia A and Tycho (SN 1572)," *Astrophys. J.* **681**, L81. [arXiv:0805.4607 [astro-ph]].

Reynolds, S. P. *et al.* (2007). "A deep *Chandra* observation of Kepler's supernova remnant: A type Ia event with circumstellar interaction," *Astrophys. J.* **668**, L135. [arXiv:0708.3858 [astro-ph]].

Reynolds, S. P. *et al.* (2008). "The youngest Galactic supernova remnant: G1.9+0.3," *Astrophys. J.* **680**, L41. [arXiv:0803.1487 [astro-ph]].

Rezzolla, L., F. K. Lamb and S. L. Shapiro (2000). "R-mode oscillations in rotating magnetic neutron stars," *Astrophys. J.* **531**, L141. [astro-ph/9911188].

Rezzolla, L. *et al.* (2008a). "The final spin from the coalescence of aligned-spin black-hole binaries," *Astrophys. J.* **674**, L29. [arXiv:0710.3345 [gr-qc]].

Rezzolla, L. *et al.* (2008b). "On the final spin from the coalescence of two black holes," *Phys. Rev. D* **78**, 044002. [arXiv:0712.3541 [gr-qc]].

Rezzolla, L. *et al.* (2009). "Modelling the final state from binary black-hole coalescences," *Class. Quant. Grav.* **26**, 094023. [arXiv:0812.2325 [gr-qc]].

Rezzolla, L., L. Baiotti, B. Giacomazzo, D. Link and J. A. Font (2010). "Accurate evolutions of unequal-mass neutron-star binaries: Properties of the torus and short GRB engines," *Class. Quant. Grav.* **27**, 114105. [arXiv:1001.3074 [gr-qc]].

Rezzolla, L., R. P. Macedo and J. L. Jaramillo (2010). "Understanding the 'anti-kick' in the merger of binary black holes," *Phys. Rev. Lett.* **104**, 221101. [arXiv:1003.0873 [gr-qc]].

Rezzolla, L. and K. Takami (2016) "Gravitational-wave signal from binary neutron stars: A systematic analysis of the spectral properties," *Phys. Rev. D* **93**, 124051. [arXiv:1604.00246 [gr-qc]].

Richardson, D. *et al.* (2002). "A comparative study of the absolute-magnitude distributions of supernovae," *Astrophys. J.* **123**, 745. [astro-ph/0112051].

Richardson, D., D. Branch and E. Baron (2006). "Absolute-magnitude distributions and light curves of stripped-envelope supernovae," *Astron. J.* **131**, 2233. [astro-ph/0601136].

Ridley, J. P. and D. R. Lorimer (2010). "Isolated pulsar spin evolution on the $P - \dot{P}$ diagram," *Mon. Not. Roy. Astron. Soc.* **404**, 1081. [arXiv:1001.2483 [astro-ph.GA]].

Riess, A. G., W. H. Press, and R. P. Kirshner (1996). "A precise distance indicator: Type Ia supernova multicolor light curve shapes," *Astrophys. J.* **473**, 88 [astro-ph/9604143].

Riess, A. G. *et al.* (2016). "A 2.4% determination of the local value of the Hubble constant," *Astrophys. J.* **826**, 56. [arXiv:1604.01424 [astro-ph.CO]].

Rodriguez, C. L., C. J. Haster, S. Chatterjee, V. Kalogera and F. A. Rasio (2016). "Dynamical formation of the GW150914 binary black hole," *Astrophys. J.* **824**, L8. [arXiv:1604.04254 [astro-ph.HE]].

Rosado, P. A. (2011). "Gravitational wave background from binary systems," *Phys. Rev. D* **84**, 084004. [arXiv:1106.5795 [gr-qc]].

Rosswog, S. *et al.* (1999). "Mass ejection in neutron star mergers," *Astron. Astrophys.* **341**, 499. [astro-ph/9811367].

Rosswog, S. (2005). "Mergers of neutron star black hole binaries with small mass ratios: Nucleosynthesis, gamma-ray bursts and electromagnetic transients," *Astrophys. J.* **634**, 1202. [astro-ph/0508138].

Rosswog, S. (2015). "The multi-messenger picture of compact binary mergers," *Int. J. Mod. Phys. D* **24**, 1530012. [arXiv:1501.02081 [astro-ph.HE]].

Rosswog, S. *et al.* (2017). "The first direct double neutron star merger detection: implications for cosmic nucleosynthesis," arXiv:1710.05445 [astro-ph.HE].

Ruan, J. J., M. Nynka, D. Haggard, V. Kalogera and P. Evans (2017). "Brightening X-ray emission from GW170817/GRB170817A: Further evidence for an outflow," arXiv:1712.02809 [astro-ph.HE].

Rubakov, V. A., M. V. Sazhin and A. V. Veryaskin (1982). "Graviton creation in the inflationary Universe and the Grand Unification scale," *Phys. Lett. B* **115**, 189.

Ruiz-Lapuente, P. *et al.* (2004). "The binary progenitor of Tycho Brahe's 1572 supernova," *Nature* **431**, 1069. [astro-ph/0410673].

Ryan, F. D. (1995). "Gravitational waves from the inspiral of a compact object into a massive, axisymmetric body with arbitrary multipole moments," *Phys. Rev. D* **52** 5707.

Sá, P. M. and B. Tome (2005). "Nonlinear evolution of r-modes: The role of differential rotation," *Phys. Rev. D* **71**, 044007. [gr-qc/0411072].

Sachs, R. K. (1962). "Gravitational waves in general relativity. VIII. Waves in asymptotically flat space-time," *Proc. Roy. Soc. Lond. A* **270**, 103.

Sachs, R. K. and A. M. Wolfe (1967). "Perturbations of a cosmological model and angular variations of the microwave background," *Astrophys. J.* **147**, 73.

Sago, N., H. Nakano and M. Sasaki (2003). "Gauge problem in the gravitational self-force. I: Harmonic gauge approach in the Schwarzschild background," *Phys. Rev. D* **67**, 104017. [gr-qc/0208060].

Sahni, V. (1990). "The energy density of relic gravity waves from inflation," *Phys. Rev. D* **42**, 453.

Saijo, M., T. W. Baumgarte and S. L. Shapiro (2002). "One-armed spiral instability in differentially rotating stars," *Astrophys. J.* **595**, 352. [astro-ph/0302436].

Sakstein, J. and B. Jain (2017). "Implications of the neutron star merger GW170817 for cosmological scalar–tensor theories," *Phys. Rev. Lett.* **119**, 251303. [arXiv:1710.05893 [astro-ph.CO]].

Salopek, D. S. and J. R. Bond (1990). "Nonlinear evolution of long wavelength metric fluctuations in inflationary models," *Phys. Rev. D* **42**, 3936.

Saltas, I. D., I. Sawicki, L. Amendola and M. Kunz (2014). "Anisotropic stress as a signature of nonstandard propagation of gravitational waves," *Phys. Rev. Lett.* **113**, 191101. [arXiv:1406.7139 [astro-ph.CO]].

Sani, E., A. Marconi, L. K. Hunt and G. Risaliti (2011). "The *Spitzer*/IRAC view of black hole–bulge scaling relations," *Mon. Not. Roy. Astron. Soc.* **413**, 1479. [arXiv:1012.3073 [astro-ph.CO]].

Santamaria, L. *et al.* (2010). "Matching post-Newtonian and numerical relativity waveforms: Systematic errors and a new phenomenological model for non-precessing black hole binaries," *Phys. Rev. D* **82**, 064016. [arXiv:1005.3306 [gr-qc]].

Sarangi, S. and S. H. H. Tye (2002). "Cosmic string production towards the end of brane inflation," *Phys. Lett. B* **536**, 185. [hep-th/0204074].

Sarkar, S. (1996). "Big bang nucleosynthesis and physics beyond the Standard Model," *Rept. Prog. Phys.* **59**, 1493. [hep-ph/9602260].

Sasaki, M. (1986). "Large scale quantum fluctuations in the inflationary Universe," *Prog. Theor. Phys.*, **76** 1036.

Sasaki, M. and T. Nakamura (1982a). "A class of new perturbation equations for the Kerr geometry," *Phys. Lett. A* **89**, 68.

Sasaki, M. and T. Nakamura (1982b). "Gravitational radiation from a Kerr black hole. 1. Formulation and a method for numerical analysis," *Prog. Theor. Phys.* **67** 1788.

Sasaki, M. and H. Tagoshi (2003). "Analytic black hole perturbation approach to gravitational radiation," *Living Rev. Rel.* **6**, 6. [gr-qc/0306120].

Savchenko, V. *et al.* (2016). "INTEGRAL upper limits on gamma-ray emission associated with the gravitational wave event GW150914," *Astrophys. J.* **820**, L36. [arXiv:1602.04180 [astro-ph.HE]].

Savchenko, V. *et al.* (2017). "INTEGRAL detection of the first prompt gamma-ray signal coincident with the gravitational-wave event GW170817," *Astrophys. J.* **848**, L15. [arXiv:1710.05449 [astro-ph.HE]].

Sawyer R. F. (1989). "Bulk viscosity of hot neutron-star matter and the maximum rotation rates of neutron stars," *Phys. Rev. D* **39**, 3804.

Sazhin, M. V. (1978). "Opportunities for detecting ultralong gravitational waves," *Sov. Astron.*, **22** 36.

Schäfer, G. (1985). "The gravitational quadrupole radiation reaction force and the canonical formalism of ADM," *Ann. Phys.* **161**, 81.

Schäfer, G. (1986). "The ADM Hamiltonian at the postlinear approximation," *Gen. Rel. Grav.* **18** 255.

Scheck, L., K. Kifonidis, H.-T. Janka and E. Müller (2006). "Multidimensional supernova simulations with approximative neutrino transport. 1. Neutron star kicks and the anisotropy of neutrino-driven explosions in two spatial dimensions," *Astron. Astrophys.* **457**, 963. [astro-ph/0601302].

Scheel, M. A. *et al.* (2009). "High-accuracy waveforms for binary black hole inspiral, merger, and ringdown," *Phys. Rev. D* **79**, 024003. [arXiv:0810.1767 [gr-qc]].

Scheidegger, S., R. Kaeppeli, S. C. Whitehouse, T. Fischer and M. Liebendoerfer (2010). "The influence of model parameters on the prediction of gravitational wave signals from stellar core collapse," *Astron. Astrophys.* **514**, A51. [arXiv:1001.1570 [astro-ph.HE]].

Schmidt, P., F. Ohme and M. Hannam (2015). "Towards models of gravitational waveforms from generic binaries II: Modelling precession effects with a single effective precession parameter," *Phys. Rev. D* **91**, 024043. [arXiv:1408.1810 [gr-qc]].

Schmidt-May, A. and M. von Strauss (2016). "Recent developments in bimetric theory," *J. Phys. A* **49**, 183001. [arXiv:1512.00021 [hep-th]].

Schnittman, J. D. *et al.* (2008). "Anatomy of the binary black hole recoil: A multipolar analysis," *Phys. Rev. D* **77**, 044031. [arXiv:0707.0301 [gr-qc]].

Schödel, R. *et al.* (2002). "A star in a 15.2-year orbit around the supermassive black hole at the centre of the Milky Way," *Nature* **419**, 694.

Schön, R. and S.-T. Yau (1981). "Proof of the positive mass theorem. 2.," *Commun. Math. Phys.* **79**, 231.

Schumaker, B. L. and K. S. Thorne (1983). "Torsional oscillations of neutron stars," *Mon. Not. Roy. Astron. Soc.* **203**, 457.

Schutz, B. F. (1986). "Determining the Hubble constant from gravitational wave observations," *Nature* **323**, 310.

Schutz, B. F. and C. M. Will (1985). "Black hole normal modes — A semianalytic approach," *Astrophys. J.* **291**, L33.

Schwaller, P. (2015). "Gravitational waves from a dark phase transition," *Phys. Rev. Lett.* **115**, 181101. [arXiv:1504.07263 [hep-ph]].

Schwarz, D. J. and M. Stuke (2009). "Lepton asymmetry and the cosmic QCD transition," *JCAP* **0911**, 025. Erratum: *JCAP* **1010** (2010) E01. [arXiv:0906.3434 [hep-ph]].

Seidel, E. and S. Iyer (1990). "Black hole normal modes: A WKB approach. 4. Kerr black holes," *Phys. Rev. D* **41**, 374.

Seljak, U. (1997). "Measuring polarization in cosmic microwave background," *Astrophys. J.* **482**, 6. [astro-ph/9608131].

Seljak, U. and M. Zaldarriaga (1996). "A line of sight integration approach to cosmic microwave background anisotropies," *Astrophys. J.* **469**, 437. [astro-ph/9603033].

Sellentin, E. and R. Durrer (2015). "Detecting the cosmological neutrino background in the CMB," *Phys. Rev. D* **92**, 063012. [arXiv:1412.6427 [astro-ph.CO]].

Sendra, I. and T. L. Smith (2012). "Improved limits on short-wavelength gravitational waves from the cosmic microwave background," *Phys. Rev. D* **85**, 123002. [arXiv:1203.4232 [astro-ph.CO]].

Sesana, A. (2013). "Systematic investigation of the expected gravitational wave signal from supermassive black hole binaries in the pulsar timing band," *Mon. Not. Roy. Astron. Soc.* **433**, 1. [arXiv:1211.5375 [astro-ph.CO]].

Sesana, A. (2016). "The promise of multi-band gravitational wave astronomy," *Phys. Rev. Lett.* **116**, 231102. [arXiv:1602.06951 [gr-qc]].

Sesana, A., F. Haardt, P. Madau and M. Volonteri (2004). "Low-frequency gravitational radiation from coalescing massive black hole binaries in hierarchical cosmologies," *Astrophys. J.* **611**, 623. [astro-ph/0401543].

Sesana, A., A. Vecchio and C. N. Colacino (2008). "The stochastic gravitational-wave background from massive black hole binary systems: Implications for observations with pulsar timing arrays," *Mon. Not. Roy. Astron. Soc.* **390**, 192. [arXiv:0804.4476 [astro-ph]].

Sesana, A. and A. Vecchio (2010). "Measuring the parameters of massive black hole binary systems with pulsar timing array observations of gravitational waves," *Phys. Rev. D* **81**, 104008. [arXiv:1003.0677 [astro-ph.CO]].

Sesana, A., C. Roedig, M. T. Reynolds and M. Dotti (2012). "Multimessenger astronomy with pulsar timing and X-ray observations of massive black hole binaries," *Mon. Not. Roy. Astron. Soc.* **420**, 860. [arXiv:1107.2927 [astro-ph.CO]].

Sesana, A. and F. M. Khan (2015). "Scattering experiments meet N-body I. A practical recipe for the evolution of massive black hole binaries in stellar environments," *Mon. Not. Roy. Astron. Soc.* **454**, L66. [arXiv:1505.02062 [astro-ph.GA]].

Shannon, R. M. *et al.* (2013). "Gravitational-wave limits from pulsar timing constrain supermassive black hole evolution," *Science* **342**, 334. [arXiv:1310.4569 [astro-ph.CO]].

Shannon, R. M. *et al.* (2015). "Gravitational waves from binary supermassive black holes missing in pulsar observations," *Science* **349**, 1522. [arXiv:1509.07320 [astro-ph.CO]].

Shapiro, C., D. Bacon, M. Hendry and B. Hoyle (2010). "Delensing gravitational wave standard sirens with shear and flexion maps," *Mon. Not. Roy. Astron. Soc.* **404**, 858. [arXiv:0907.3635 [astro-ph.CO]].

Shibata, M. (1999). "Fully general relativistic simulation of coalescing binary neutron stars: Preparatory tests," *Phys. Rev. D* **60**, 104052. [gr-qc/9908027].

Shibata, M. (2005). "Constraining nuclear equations of state using gravitational waves from hypermassive neutron stars," *Phys. Rev. Lett.* **94**, 201101. [gr-qc/0504082].

Shibata, M. (2015). *Numerical Relativity*. World Scientific, Singapore.

Shibata, M. and K. Uryu (2000). "Simulation of merging binary neutron stars in full general relativity: $\Gamma = 2$ case," *Phys. Rev. D* **61**, 064001. [gr-qc/9911058].

Shibata, M. and K. Uryu (2002). "Gravitational waves from the merger of binary neutron stars in a fully general relativistic simulation," *Prog. Theor. Phys.* **107**, 265. [gr-qc/0203037].

Shibata, M., S. Karino and Y. Eriguchi (2002). "Dynamical instability of differentially rotating stars," *Mon. Not. Roy. Astron. Soc.* **334**, L27. [gr-qc/0206002].

Shibata, M., K. Taniguchi and K. Uryu (2003). "Merger of binary neutron stars of unequal mass in full general relativity," *Phys. Rev. D* **68**, 084020. [gr-qc/0310030].

Shibata, M., Y. T. Liu, S. L. Shapiro and B. C. Stephens (2006). "Magnetorotational collapse of massive stellar cores to neutron stars: Simulations in full general relativity," *Phys. Rev. D* **74**, 104026. [astro-ph/0610840].

Shtanov, Y., J. H. Traschen and R. H. Brandenberger (1995). "Universe reheating after inflation," *Phys. Rev. D* **51**, 5438. [hep-ph/9407247].

Silk, J. (1968). "Cosmic black-body radiation and galaxy formation," *Astrophys. J.* **151**, 459.

Sidery, T., A. Passamonti and N. Andersson (2010). "The dynamics of pulsar glitches: Contrasting phenomenology with numerical evolutions," *Mon. Not. Roy. Astron. Soc.* **405**, 1061. [arXiv:0910.3918 [astro-ph.HE]].

Siemens, X., V. Mandic and J. Creighton (2007). "Gravitational wave stochastic background from cosmic (super)strings," *Phys. Rev. Lett.* **98**, 111101. [astro-ph/0610920].

Siemens, X., J. Ellis, F. Jenet and J. D. Romano (2013). "The stochastic background: Scaling laws and time to detection for pulsar timing arrays," *Class. Quant. Grav.* **30**, 224015. [arXiv:1305.3196 [astro-ph.IM]].

Smarr, L. (1979). "Gauge conditions, radiation formulae and the two black hole collision," in *Sources of Gravitational Radiation*, L. Smarr ed., Cambridge University Press, Cambridge.

Smartt, S. J. (2015). "Observational constraints on the progenitors of core-collapse supernovae: The case for missing high mass stars," *Publ. Astron. Soc. Austral.* **32**, e016. [arXiv:1504.02635 [astro-ph.SR]].

Smartt, S. J. *et al.* (2017). "A kilonova as the electromagnetic counterpart to a gravitational-wave source," *Nature* **551**, 75. [arXiv:1710.05841 [astro-ph.HE]].

Smith, N. (2013). "The Crab Nebula and the class of type IIn-P supernovae caused by sub-energetic electron capture explosions," *Mon. Not. Roy. Astron. Soc.* **434**, 102. [arXiv:1304.0689 [astro-ph.HE]].

Smith, N. *et al.* (2007). "The brightest supernova ever recorded, powered by the death of an extremely massive star," *Astrophys. J.* **666**, 1116. [astro-ph/0612617].

Smith, T. L., M. Kamionkowski and A. Cooray (2006). "Direct detection of the inflationary gravitational wave background," *Phys. Rev. D* **73**, 023504. [astro-ph/0506422].

Smith, T. L., E. Pierpaoli and M. Kamionkowski (2006). "A new cosmic microwave background constraint to primordial gravitational waves," *Phys. Rev. Lett.* **97**, 021301. [astro-ph/0603144].

Smoot, G. F. *et al.* (1992). "Structure in the COBE differential microwave radiometer first year maps," *Astrophys. J.* **396**, L1.

Soares-Santos, M. *et al.* [DES and Dark Energy Camera GW-EM Collaborations] (2017). "The electromagnetic counterpart of the binary neutron star merger LIGO/Virgo GW170817. I. Discovery of the optical counterpart using the Dark Energy Camera," *Astrophys. J.* **848**, L16. [arXiv:1710.05459 [astro-ph.HE]].

Sopuerta, C.F., N. Yunes and P. Laguna (2006). "Gravitational recoil from binary black hole mergers: The close-limit approximation," *Phys. Rev. D* **74**, 124010. Errata: [*Phys. Rev. D* **75** (2007) 069903 and **78** (2008) 049901]. [astro-ph/0608600].

Souradeep, T. and V. Sahni (1992). "Density perturbations, gravity waves and the cosmic microwave background," *Mod. Phys. Lett. A* **7**, 3541. [hep-ph/9208217].

Spergel, D. N. *et al.* [WMAP Collaboration] (2003). "First-year Wilkinson Microwave Anisotropy Probe (WMAP) observations: Determination of cosmological parameters," *Astrophys. J. Suppl.* **148**, 175. [astro-ph/0302209].

Spergel, D. N. *et al.* [WMAP Collaboration] (2007). "Three-year Wilkinson Microwave Anisotropy Probe (WMAP) observations: Implications for cosmology," *Astrophys. J. Suppl.* **170**, 377. [astro-ph/0603449].

Sperhake, U. (2009), "Colliding black holes and gravitational waves," in *Physics of Black Holes: A Guided Tour*, E. Papantonopoulos ed., Lecture Notes in Physics, Vol. 769, Springer-Verlag, Berlin.

Stachel, J. (1988). "The Cauchy problem in general relativity — The early years ," in *Studies in the History of General Relativity*, J. Eisenstaedt and A. J. Kox eds., Birkhäuser, Boston.

Stark, R. F. and T. Piran (1985). "Gravitational wave emission from rotating gravitational collapse," *Phys. Rev. Lett.* **55**, 891. Erratum: *Phys. Rev. Lett.* **56** (1986) 97.

Starobinsky, A. A. (1979). "Spectrum of relict gravitational radiation and the early state of the Universe," *JETP Lett.* **30**, 682 [*Pisma Zh. Eksp. Teor. Fiz.* **30** (1979) 719].

Starobinsky, A. A. (1980). "A new type of isotropic cosmological models without singularity," *Phys. Lett. B* **91**, 99.

Starobinsky A. A. (1982). "Dynamics of phase transition in the new inflationary Universe scenario and generation of perturbations," *Phys. Lett. B* **117**, 175.

Starobinsky, A.A. (1985). "Cosmic background anisotropy induced by isotropic flat-spectrum gravitational-wave perturbations," *Sov. Astron. Lett.* **11**, 113.

Starobinsky, A. A. (1986). "Stochastic de Sitter (inflationary) stage in the early Universe," in *Field Theory, Quantum Gravity and Strings*, H. J. de Vega and N. Sánchez eds., Lecture Notes in Physics, Vol. 246. Springer, Berlin, Heidelberg.

Starobinsky, A. A. and J. Yokoyama (1994). "Equilibrium state of a self-interacting scalar field in the de Sitter background," *Phys. Rev. D* **50**, 6357. [astro-ph/9407016].

Stebbins, A. (1996). "Weak lensing on the celestial sphere," [astro-ph/9609149].

Steinhardt, P. J. (1982). "Relativistic detonation waves and bubble growth in false vacuum decay," *Phys. Rev. D* **25**, 2074.

Steinhardt, P. J. and N. Turok (2002). "Cosmic evolution in a cyclic Universe," *Phys. Rev. D* **65**, 126003. [hep-th/0111098].

Steinhoff, J., T. Hinderer, A. Buonanno and A. Taracchini (2016). "Dynamical tides in general relativity: Effective action and effective-one-body Hamiltonian," *Phys. Rev. D* **94**, 104028. [arXiv:1608.01907 [gr-qc]].

Stephenson, F. R. and D. A. Green (2002). *Historical Supernovae and their Remnants*, Oxford University Press, Oxford.

Stergioulas, N. (2003). "Rotating stars in relativity," *Living Rev. Rel.* **6**, 3. [gr-qc/0302034].

Stergioulas, N. and J. L. Friedman (1998). "Nonaxisymmetric neutral modes of rotating relativistic stars," *Astrophys. J.* **492**, 301.

Stergioulas, N., A. Bauswein, K. Zagkouris and H.-T. Janka (2011). "Gravitational waves and nonaxisymmetric oscillation modes in mergers of compact object binaries," *Mon. Not. Roy. Astron. Soc.* **418**, 427. [arXiv:1105.0368 [gr-qc]].

Stewart, E. D. and D. H. Lyth (1993). "A more accurate analytic calculation of the spectrum of cosmological perturbations produced during inflation," *Phys. Lett. B* **302**, 171. [gr-qc/9302019].

Straumann, N. (2004). *General Relativity. With Applications to Astrophysics*. Springer-Verlag, Berlin.

Sukhbold, T., T. Ertl, S. E. Woosley, J. M. Brown and H.-T. Janka (2016). "Core-collapse supernovae from 9 to 120 solar masses based on neutrino-powered explosions," *Astrophys. J.* **821**, 38. [arXiv:1510.04643 [astro-ph.HE]].

Sun, Y. and R. H. Price (1988). "Excitation of quasinormal ringing of a Schwarzschild black hole," *Phys. Rev. D* **38**, 1040.

Szilágyi, B. *et al.* (2015). "Approaching the post-Newtonian regime with numerical relativity: A compact-object binary simulation spanning 350 gravitational-wave cycles," *Phys. Rev. Lett.* **115**, 031102. [arXiv:1502.04953 [gr-qc]].

Takami, K., L. Rezzolla and L. Baiotti (2014). "Constraining the equation of state of neutron stars from binary mergers," *Phys. Rev. Lett.* **113**, 091104. [arXiv:1403.5672 [gr-qc]].

Takami, K., L. Rezzolla and L. Baiotti (2015). "Spectral properties of the post-merger gravitational-wave signal from binary neutron stars," *Phys. Rev. D* **91**, 064001. [arXiv:1412.3240 [gr-qc]].

Takiwaki, T. and K. Kotake (2011). "Gravitational wave signatures of magnetohydrodynamically-driven core-collapse supernova explosions," *Astrophys. J.* **743**, 30. [arXiv:1004.2896 [astro-ph.HE]].

Tammann, G. A., W. Loeffler and A. Schroder (1994). "The Galactic supernova rate," *Astrophys. J. Suppl.* **92**, 487.

Tanvir, N. R. *et al.* (2017). "The emergence of a lanthanide-rich kilonova following the merger of two neutron stars," *Astrophys. J. Lett.* **848**, L27. [arXiv:1710.05455 [astro-ph.HE]].

Taracchini, A. (2014). "Inspiral-merger-ringdown models for spinning black-hole binaries at the interface between analytical and numerical relativity," PhD Thesis, University of Maryland, College Park.

Taracchini, A. *et al.* (2012). "Prototype effective-one-body model for nonprecessing spinning inspiral–merger–ringdown waveforms," *Phys. Rev. D* **86**, 024011. [arXiv:1202.0790 [gr-qc]].

Taracchini, A. *et al.* (2014). "Effective-one-body model for black-hole binaries with generic mass ratios and spins," *Phys. Rev. D* **89**, 061502. [arXiv:1311.2544 [gr-qc]].

Tasson, J. D. (2014). "What do we know about Lorentz invariance?," *Rept. Prog. Phys.* **77**, 062901. [arXiv:1403.7785 [hep-ph]].

Tavani, M. *et al.* (2016). "AGILE observations of the gravitational wave event GW150914," *Astrophys. J.* **825**, L4. [arXiv:1604.00955 [astro-ph.HE]].

Taylor, S. R., J. R. Gair and I. Mandel (2012). "Hubble without the Hubble: Cosmology using advanced gravitational-wave detectors alone," *Phys. Rev. D* **85**, 023535. [arXiv:1108.5161 [gr-qc]].

Taylor, S. R. and J. R. Gair (2012). "Cosmology with the lights off: Standard sirens in the Einstein Telescope era," *Phys. Rev. D* **86**, 023502. [arXiv:1204.6739 [astro-ph.CO]].

Taylor, S. R., *et al.* (2017). "All correlations must die: Assessing the significance of a stochastic gravitational-wave background in pulsar-timing arrays," *Phys. Rev. D* **95**, 042002. arXiv:1606.09180 [astro-ph.IM].

Tegmark, M. *et al.* [SDSS Collaboration] (2004). "The 3-D power spectrum of galaxies from the SDSS," *Astrophys. J.* **606**, 702. [astro-ph/0310725].

Teukolsky, S. A. (1973). "Perturbations of a rotating black hole. I. Fundamental equations for gravitational, electromagnetic, and neutrino-field perturbations," *Astrophys. J.* **185** 635.

Thompson, C. and R. C. Duncan (1995). "The soft gamma repeaters as very strongly magnetized neutron stars — I. Radiative mechanism for outbursts," *Mon. Not. Roy. Astron. Soc.* **275**, 255.

Thompson, C. and R. C. Duncan (1996). 'The soft gamma repeaters as very strongly magnetized neutron stars — II. Quiescent neutrino, X-ray, and Alfven wave emission," *Astrophys. J.* **473**, 322.

Thorne, K. S. (1969a). "Nonradial pulsation of general-relativistic stellar models. III. Analytic and numerical results for neutron stars," *Astrophys. J.* **158**, 1.

Thorne, K. S. (1969b). "Nonradial pulsation of general-relativistic stellar models. IV. The weak field limit," *Astrophys. J.* **158**, 997.

Thorne, K. S. and A. Campolattaro (1967). "Non-radial pulsation of general-relativistic stellar models. I. Analytic analysis for $L \geqslant 2$," *Astrophys. J.* **149**, 591.

Thorsett, S. E. (1992). "Identification of the pulsar PSR1509 − 58 with the 'guest star' of AD 185," *Nature* **356**, 690.

Tichy, W. and P. Marronetti (2008). "The final mass and spin of black hole mergers," *Phys. Rev. D* **78**, 081501. [arXiv:0807.2985 [gr-qc]].

Tominaga, N., S. I. Blinnikov and K. Nomoto (2013). "Supernova explosions of super-asymptotic giant branch stars: Multicolor light curves of electron-capture supernovae," *Astrophys. J.* **771**, L12. [arXiv:1305.6813 [astro-ph.HE]].

Townsend, P. K. (1997). "Black holes," Lecture Notes, DAMTP. [arXiv:gr-qc/9707012].

Traschen, J. H. and R. H. Brandenberger (1990). "Particle production during out-of-equilibrium phase transitions," *Phys. Rev. D* **42**, 2491.

Tremaine, S. *et al.* (2002). "The slope of the black hole mass versus velocity dispersion correlation," *Astrophys. J.* **574**, 740. [astro-ph/0203468].

Tremblay, P.-E., P. Bergeron and A. Gianninas (2011). "An improved spectroscopic analysis of DA white dwarfs from the Sloan Digital Sky Survey data release 4," *Astrophys. J.* **730**, 128. [arXiv:1102.0056 [astro-ph.SR]].

Troja, E. *et al.* (2017). "The X-ray counterpart to the gravitational wave event GW 170817," *Nature* **551**, 71. [arXiv:1710.05433 [astro-ph.HE]].

Turner, M. S. (1978). "Gravitational radiation from supernova neutrino bursts," *Nature* **274**, 565.

Turner, M. S. and F. Wilczek (1990). "Relic gravitational waves and extended inflation," *Phys. Rev. Lett.* **65**, 3080.

Turner, M. S., E. J. Weinberg and L. M. Widrow (1992). "Bubble nucleation in first order inflation and other cosmological phase transitions," *Phys. Rev. D* **46**, 2384.

Turner, M. S., M. J. White and J. E. Lidsey (1993). "Tensor perturbations in inflationary models as a probe of cosmology," *Phys. Rev. D* **48**, 4613. [astro-ph/9306029].

Turok, N. (1984). "Grand unified strings and galaxy formation," *Nucl. Phys. B* **242**, 520.

Ugliano, M., H.-T. Janka, A. Marek and A. Arcones (2012). "Progenitor-explosion connection and remnant birth masses for neutrino-driven supernovae of iron-core progenitors," *Astrophys. J.* **757**, 69. [arXiv:1205.3657 [astro-ph.SR]].

Ushomirsky, G., C. Cutler and L. Bildsten (2000). "Deformations of accreting neutron star crusts and gravitational wave emission," *Mon. Not. Roy. Astron. Soc.* **319**, 902. [astro-ph/0001136].

Usman, S. A. *et al.* (2016). "The PyCBC search for gravitational waves from compact binary coalescence," *Class. Quant. Grav.* **33**, 215004. [arXiv:1508.02357 [gr-qc]].

Vachaspati, T. and A. Vilenkin (1985). "Gravitational radiation from cosmic strings," *Phys. Rev. D* **31**, 3052.

Valenti, S. *et al.* (2017). "The discovery of the electromagnetic counterpart of GW170817: Kilonova AT 2017gfo/DLT17ck," *Astrophys. J.* **848**, L24. [arXiv:1710.05854 [astro-ph.HE]].

van den Bergh, S. and G. A. Tammann (1991). "Galactic and extragalactic supernova rates," *Annu. Rev. Astron. Astrophys.* **29**, 363.

Van Den Broeck, C., M. Trias, B. S. Sathyaprakash and A. M. Sintes (2010). "Weak lensing effects in the measurement of the dark energy equation of state with LISA," *Phys. Rev. D* **81**, 124031. [arXiv:1001.3099 [gr-qc]].

van der Sluys, M. V. *et al.* (2008). "Gravitational-wave astronomy with inspiral signals of spinning compact-object binaries," *Astrophys. J.* **688**, L61. [arXiv:0710.1897 [astro-ph]].

Vandoren, S. and P. van Nieuwenhuizen (2008). "Lectures on instantons," [arXiv:0802.1862 [hep-th]].

van Haasteren, R., Y. Levin, P. McDonald and T. Lu (2009). "On measuring the gravitational-wave background using pulsar timing arrays," *Mon. Not. Roy. Astron. Soc.* **395** 1005. [arXiv:0809.0791 [astro-ph]].

van Haasteren, R. *et al.* (2011). "Placing limits on the stochastic gravitational-wave background using European Pulsar Timing Array data," *Mon. Not. Roy. Astron. Soc.* **414**, 3117. Erratum: *Mon. Not. Roy. Astron. Soc.* **425** (2012), 1597. [arXiv:1103.0576 [astro-ph.CO]].

van Haasteren, R. and Y. Levin (2013). "Understanding and analysing time-correlated stochastic signals in pulsar timing," *Mon. Not. Roy. Astron. Soc.* **428**, 1147. [arXiv:1202.5932 [astro-ph.IM]].

van Kerkwijk, M. H., P. Bergeron and S. R. Kulkarni (1996). "The masses of the millisecond pulsar J1012+5307 and its white-dwarf companion," *Astrophys. J. Lett.* **467**, L89.

Vasiliev, E., F. Antonini and D. Merritt (2015). "Efficient merger of binary supermassive black holes in merging galaxies," *Astrophys. J.* **810**, 49. [arXiv:1505.05480 [astro-ph.GA]].

Vavoulidis, M., K. D. Kokkotas and A. Stavridis (2008). "Crustal oscillations of slowly rotating relativistic stars," *Mon. Not. Roy. Astron. Soc.* **384**, 1711. [arXiv:0712.1263 [gr-qc]].

Veneziano, G. (1991). "Scale factor duality for classical and quantum strings," *Phys. Lett. B* **265**, 287.

Verbiest, J. P. W. *et al.* (2016). "The International Pulsar Timing Array: First data release," *Mon. Not. Roy. Astron. Soc.* **458**, 1267. [arXiv:1602.03640 [astro-ph.IM]].

Vilenkin, A. (1981a). "Cosmological density fluctuations produced by vacuum strings," *Phys. Rev. Lett.* **46**, 1169.

Vilenkin, A. (1981b). "Gravitational radiation from cosmic strings," *Phys. Lett. B* **107**, 47.

Vilenkin, A. (1985). "Cosmic strings and domain walls," *Phys. Rept.* **121**, 263.

Vilenkin, A. and E. P. S. Shellard (2000). *Cosmic Strings and Other Topological Defects*, Cambridge University Press, Cambridge.

Vines, J., E. E. Flanagan and T. Hinderer (2011). "Post-1-Newtonian tidal effects in the gravitational waveform from binary inspirals," *Phys. Rev. D* **83**, 084051. [arXiv:1101.1673 [gr-qc]].

Vink, J. *et al.* (2006). "The X-ray synchrotron emission of RCW 86 and the implications for its age," *Astrophys. J.* **648**, L33. [astro-ph/0607307].

Vishveshwara, C. V. (1970). "Stability of the Schwarzschild metric," *Phys. Rev. D* **1**, 2870.

Vitale, S., R. Lynch, J. Veitch, V. Raymond and R. Sturani (2014). "Measuring the spin of black holes in binary systems using gravitational waves," *Phys. Rev. Lett.* **112**, 251101. [arXiv:1403.0129 [gr-qc]].

Volonteri, M. (2010). "Formation of supermassive black holes, *Astron. Astrophys. Rev.* **18**, 279.

Volonteri, M. and R. Perna (2005). "Dynamical evolution of intermediate mass black holes and their observable signatures in the nearby Universe," *Mon. Not. R. Astron. Soc.* **358**, 913. [astro-ph/0501345].

Volonteri, M. and M. J. Rees (2006). "Quasars at $z = 6$: The survival of the fittest," *Astrophys. J.* **650** 669. [astro-ph/0607093].

Voloshin, M. B., I. Yu. Kobzarev and L. B. Okun (1974). "Bubbles in metastable vacuum," *Yad. Fiz.* 20, 1229 [*Sov. J. Nucl. Phys.* 20, (1975) 644].

Wagoner, R. V. (1979). "Low frequency gravitational radiation from collapsing systems," *Phys. Rev. D* **19**, 2897.

Wald, R. M. (1974). "Gedanken experiments to destroy a black hole," *Ann. Phys.* **82**, 548.

Wald, R. M. (1979). "Note on the stability of the Schwarzschild metric," *J. Math. Phys.* **20**, 1056.

Wald, R. M. (1984). *General Relativity*, University of Chicago Press, Chicago.

Warburton, N., S. Akcay, L. Barack, J. R. Gair and N. Sago (2012). "Evolution of inspiral orbits around a Schwarzschild black hole," *Phys. Rev. D* **85**, 061501. [arXiv:1111.6908 [gr-qc]].

Watanabe, Y. and E. Komatsu (2006). "Improved calculation of the primordial gravitational wave spectrum in the Standard Model," *Phys. Rev. D* **73**, 123515. [astro-ph/0604176].

Watts, A. L., N. Andersson and D. I. Jones (2005). "The nature of low $T/|W|$ dynamical instabilities in differentially rotating stars," *Astrophys. J.* **618**, L37. [astro-ph/0309554].

Watts, A., B. Krishnan, L. Bildsten and B. F. Schutz (2008). "Detecting gravitational wave emission from the known accreting neutron stars," *Mon. Not. Roy. Astron. Soc.* **389**, 839. [arXiv:0803.4097 [astro-ph]].

Weinberg, S. (2004). "Damping of tensor modes in cosmology," *Phys. Rev. D* **69**, 023503. [astro-ph/0306304].

Weinberg S. (2008). *Cosmology.* Oxford University Press, Oxford.

Weir, D. J. (2017). "Gravitational waves from a first order electroweak phase transition: A review," [arXiv:1705.01783 [hep-ph]].

White, M., D. Scott and J. Silk (1994). "Anisotropies in the cosmic microwave background," *Annu. Rev. Astron. Astrophys.* **32**, 319.

Wijers, R. A. M. J. (2005). "On the stellar luminosity of the Universe," [astro-ph/0506218].

Will, C. M. (1998). "Bounding the mass of the graviton using gravitational wave observations of inspiralling compact binaries," *Phys. Rev. D* **57**, 2061.

Will, C. M. (2014). "The confrontation between general relativity and experiment," *Living Rev. Rel.* **17**, 4. [arXiv:1403.7377 [gr-qc]].

Wilson, M.L. and J. Silk (1981). "On the anisotropy of the cosmological background matter and radiation distribution. 1. The radiation anisotropy in a spatially flat universe," *Astrophys. J.* **243**, 14.

Witten, E. (1981). "A simple proof of the positive energy theorem," *Commun. Math. Phys.* **80**, 381.

Witten, E. (1984). "Cosmic separation of phases," *Phys. Rev. D* **30**, 272.

Woods, P. M. *et al.* (1999). "Discovery of a new soft gamma repeater, SGR 1627-41," *Astrophys. J.* **519**, L139. [astro-ph/9903267].

Woods, P. M. and C. Thompson (2006). "Soft gamma repeaters and anomalous X-ray pulsars: Magnetar candidates," in *Compact Stellar X-Ray Sources*, W. Lewin and M. van der Klis eds., Cambridge University Press, Cambridge. [astro-ph/0406133].

Woosley, S. E. (1993). "Gamma-ray bursts from stellar mass accretion disks around black holes," *Astrophys. J.* **405**, 273.

Woosley, S. E. and T. A. Weaver (1995). "The evolution and explosion of massive stars. 2. Explosive hydrodynamics and nucleosynthesis," *Astrophys. J. Suppl.* **101**, 181.

Woosley, S. E., A. Heger and T. A. Weaver (2002). "The evolution and explosion of massive stars," *Rev. Mod. Phys.* **74**, 1015.

Woosley, S. E., S. Blinnikov and A. Heger (2007). "Pulsational pair instability as an explanation for the most luminous supernovae," *Nature* **450**, 390. [arXiv:0710.3314 [astro-ph]].

Woosley, S. E. and A. Heger (2015). "The remarkable deaths of 9–11 solar mass stars," *Astrophys. J.* **810**, 34. [arXiv:1505.06712 [astro-ph.SR]].

Wyithe, J. S. B. and A. Loeb (2003). "Low-frequency gravitational waves from massive black hole binaries: Predictions for LISA and pulsar timing arrays," *Astrophys. J.* **590**, 691. [astro-ph/0211556].

Yamada, S. and H. Sawai (2004). "Numerical study on the rotational collapse of strongly magnetized cores of massive stars," *Astrophys. J.* **608**, 907.

Yamamoto, T., M. Shibata, and K. Taniguchi (2008). "Simulating coalescing compact binaries by a new code SACRA," *Phys. Rev. D* **78**, 064054. [arXiv:0806.4007 [gr-qc]].

York, J. W. (1972). "Role of conformal three geometry in the dynamics of gravitation," *Phys. Rev. Lett.* **28**, 1082.

York, J. W. (1979). "Kinematics and dynamics of general relativity," in *Sources of Gravitational Radiation*, L. Smarr ed., Cambridge University Press, Cambridge.

York, J. W. (1980). "Energy and momentum of the gravitational field," in *Essays in General Relativity: A Festschrift for Abraham Taub*, F. J. Tipler ed., Academic Press, New York.

York, J. W. (1986). "Boundary terms in the action principles of general relativity," *Found. Phys.* **16**, 249.

Young, M. D., R. N. Manchester and S. Johnston (1999). "A radio pulsar with an 8.5-second period that challenges emission models," *Nature* **400**, 848.

Yu, Q. (2002). "Evolution of massive binary black holes," *Mon. Not. Roy. Astron. Soc.* **331**, 935. [astro-ph/0109530].

Yunes, N. and F. Pretorius (2009). "Fundamental theoretical bias in gravitational wave astrophysics and the parameterized post-Einsteinian framework," *Phys. Rev. D* **80**, 122003. [arXiv:0909.3328 [gr-qc]].

Zaldarriaga, M. (2003). "The polarization of the cosmic microwave background," [astro-ph/0305272].

Zaldarriaga, M. and U. Seljak (1997). "An all sky analysis of polarization in the microwave background," *Phys. Rev. D* **55**, 1830. [astro-ph/9609170].

Zel'dovich, Y. B. (1970). "Particle production in cosmology," *JETP Lett.* **12** (1970) 307 [*Pisma Zh. Eksp. Teor. Fiz.* **12** (1970) 443].

Zel'dovich, Y. B. and Starobinsky, A. A. (1971). "Amplification of gravitational waves in an isotropic universe," *Sov. Phys. JETP* **34** (1972) 1159 [*Zh. Eksp. Teor. Fiz.* **61** (1971) 2161].

Zel'dovich, Y. B. (1980). "Cosmological fluctuations produced near a singularity," *Mon. Not. Roy. Astron. Soc.* **192**, 663.

Zerilli, F. J. (1970). "Gravitational field of a particle falling in a Schwarzschild geometry analyzed in tensor harmonics," *Phys. Rev. D* **2**, 2141.

Zhang, Z., E. Berti and V. Cardoso (2013). "Quasinormal ringing of Kerr black holes. II. Excitation by particles falling radially with arbitrary energy," *Phys. Rev. D* **88**, 044018. [arXiv:1305.4306 [gr-qc]].

Zhao, W., C. Van Den Broeck, D. Baskaran and T. G. F. Li (2011). "Determination of dark energy by the Einstein Telescope: Comparing with CMB, BAO and SNIa observations," *Phys. Rev. D* **83**, 023005. [arXiv:1009.0206 [astro-ph.CO]].

Zink, B., P. D. Lasky and K. D. Kokkotas (2012). "Are gravitational waves from giant magnetar flares observable?," *Phys. Rev. D* **85**, 024030. [arXiv:1107.1689 [gr-qc]].

Zinn-Justin, J. (2002). *Quantum Field Theory and Critical Phenomena*, 4th edition, Oxford University Press, Oxford.

Zwerger, T. and E. Müller (1997). "Dynamics and gravitational wave signature of axisymmetric rotational core collapse," *Astron. Astrophys.* **320**, 209.

Index